Norwegian Petroleum Society (NPF), Special Publication No. 3

Basin Modelling: Advances and Applications

Proceedings of the Norwegian Petroleum Society Conference, 13–15 March 1991, Stavanger, Norway

Further titles in this series:

1. R.M. Larsen, H. Brekke, B.T. Larsen and E. Talleraas (Editors)
STRUCTURAL AND TECTONIC MODELLING AND ITS APPLICATION TO PETROLEUM GEOLOGY – Proceedings of Norwegian Petroleum Society Workshop, 18–20 October 1989, Stavanger, Norway

2. T.O. Vorren, E. Bergsager, Ø.A. Dahl-Stamnes, E. Holter, B. Johansen, E. Lie and T.B. Lund (Editors)
ARCTIC GEOLOGY AND PETROLEUM POTENTIAL – Proceedings of the Norwegian Petroleum Society Conference, 15–17 August 1990, Tromsø, Norway

Norwegian Petroleum Society (NPF), Special Publication No. 3

Basin Modelling: Advances and Applications

Proceedings of the Norwegian Petroleum Society Conference, 13–15 March 1991, Stavanger, Norway

Edited by

A.G. Doré et al.

ELSEVIER
Amsterdam – London – New York – Tokyo 1993

ELSEVIER SCIENCE PUBLISHERS B.V.
Sara Burgerhartstraat 25
P.O. Box 211, 1000 AE Amsterdam, The Netherlands

ISBN 0-444-89272-9

This book is printed on acid-free paper.

Printed in The Netherlands

Organizing Committee

ANTHONY G. DORÉ (Chairman)
Conoco Norway Inc., P.O. Box 488, N-4001 Stavanger, Norway

ELISABETH HOLTER (Conference Organizer)
Norsk Petroleums-forening (NPF), P.O. Box 1897, Vika, N-0124 Oslo, Norway

JAN H. AUGUSTSON
Norsk Hydro Research Exploration Office, Storaakeren 11, N-9400 Harstad, Norway

WILLY FJELDSKAAR
Rogaland Research, P.O. Box 2503 Ullandhaug, N-4004 Stavanger, Norway

SIGMUND HANSLIEN
Statoil A/S, P.O. 300, N-4001 Stavanger, Norway

CHRISTIAN HERMANRUD
Statoil A/S, P.O. 300, N-4001 Stavanger, Norway

BENTE NYLAND
Norwegian Petroleum Directorate (NPD), P.O. Box 600, N-4001 Stavanger, Norway

DAVID J. STEWART
A/S Norske Shell, P.O. Box 40, N-4056 Tananger, Norway

ØYVIND SYLTA
IKU Petroleum Research, N-7034 Trondheim, Norway

Editing Committee

ANTHONY G. DORÉ
Conoco Norway Inc., P.O. Box 488, N-4001 Stavanger, Norway

JAN H. AUGUSTSON
Norsk Hydro Research Exploration Office, Storaakeren 11, N-9400 Harstad, Norway

CHRISTIAN HERMANRUD
Statoil A/S, P.O. 300, N-4001 Stavanger, Norway

DAVID J. STEWART
A/S Norske Shell, P.O. Box 40, N-4056 Tananger, Norway

ØYVIND SYLTA
IKU Petroleum Research, N-7034 Trondheim, Norway

Information on Norwegian Petroleum Society Conferences and Publications

This volume is the third in a series of Norwegian Petroleum Society Special Publications. It is recommended that this volume is referred to in one of the following forms, as appropriate:

Doré, A.G. et al. (Editors), 1993. *Basin Modelling: Advances and Applications.* Norwegian Petroleum Society Special Publication No. 3. Elsevier, Amsterdam, 680 pp.

Hermanrud, C., 1993. Basin modelling techniques — an overview. In: A.G. Doré et al. (Editors), *Basin Modelling: Advances and Applications.* Norwegian Petroleum Society Special Publication No. 3. Elsevier, Amsterdam, pp. 1–34.

The conference in Stavanger was the 12th in a series of meetings organized by the Norwegian Petroleum Society (Norsk Petroleumsforening, NPF) comprising:

- Jurassic Northern North Sea Symposium, Stavanger (1975)
- Mesozoic Northern North Sea Symposium, Oslo (1977)
- Norwegian Sea Symposium, Tromsø (1979)
- The Sedimentation of the North Sea Reservoir Rocks, Geilo (1980)
- Norwegian Symposium on Exploration, Bergen (1981)
- North European Margin Symposium, Trondheim (1983)
- Petroleum Geochemistry in Exploration of the Norwegian Shelf, Stavanger (1984)
- Habitat of Hydrocarbons Symposium, Stavanger (1985)
- Correlation in Hydrocarbon Exploration, Bergen (1988)
- Structural and Tectonic Modelling and its Application to Petroleum Geology, Stavanger (1989)
- Arctic Geology and Petroleum Potential, Tromsø (1990)

Information on the first five symposia and their publication can be obtained from: NPF, P.O. Box 1897 – Vika, 0124 Oslo, Norway.

The proceedings of the next four symposia and an "atlas" of hydrocarbon discoveries, containing a full description of 35 of Norway's major oil and gas fields and finds, "Geology of the Norwegian Oil and Gas Fields", have been published by and can be obtained from Graham and Trotman Ltd., Stirling House, 66 Wilton Road, London SW1V 1DE, England.

The proceedings of the final two symposia are published by Elsevier as part of the Norwegian Petroleum Society Special Publication series.

Preface

The Norwegian Petroleum Society's conference "Basin Modelling; Advances and Applications" attracted nearly a hundred papers, 49 of which are published in this volume. It was attended by an international audience of more than 300 delegates.

To my knowledge (and I accept full responsibility if this statement is in error) the conference was the first major international gathering dedicated solely to Basin Modelling. This claim, however, opens the difficult question of how exactly Basin Modelling should be defined. The reader will note a number of varying definitions in this volume. Ideally, the name should be a catch-all for the integrated simulation of basin processes using the computer power that is now routinely available in the industry. In actuality, it is frequently used specifically to mean geochemical modelling, sedimentary modelling, mechanical modelling or (the dream scenario) the integration of all three. The conference organizers opted to emphasize the narrower definition that is most prevalent in the industry; that is, the modelling of burial history, thermal regimes, source rock maturation, hydrocarbon generation, expulsion, secondary migration and entrapment. A number of papers, however, are included to remind us of the importance of other modelling techniques. These include *Levell and Leu*'s overview paper on basin-fill modelling, and *Fjeldskaar et al.*'s discussion of isostatic and tectonic subsidence in the Egersund Basin.

The conference was divided into six sessions, and this volume follows the same general pattern. The first section contains a selection of overview papers. Major reviews by *Hermanrud* and *Lerche* trace the history of basin modelling, the current state of the science, and unresolved problems. Both authors stress the importance of addressing ranges and sensitivities, and the requirement of future modelling packages to model transient states, principally of temperature and pressure. A particularly useful aspect of Hermanrud's paper is his summary of the options available within currently available basin modelling systems. Many of these systems are featured in the current volume, including home-grown Norwegian packages such as SEMI (*Sylta*) and BMT (e.g. *Walderhaug and Fjeldskaar*).

In the second section on thermal modelling a number of papers discuss the usefulness and reliability of organic maturity parameters in reconstructing temperature history (e.g. *Hermansen*, *Throndsen et al.*, *Ritter et al.*). Other papers demonstrate the growing importance of inorganic palaeotemperature indicators such as apatitite fission-tracks (*Naeser*) and fluid inclusions (*Leischner et al.*). The reliability of lithospheric palaeo-heat flow modelling was a pervasive theme of the conference. Papers by *Jensen and Doré* and *Vik and Hermanrud* suggest that the present-day thermal regime of parts of the Norwegian shelf may be anomalous compared to the past, although the approach to this problem is very different between the two papers. Jensen's paper points out that, with most basin modelling packages, a geologically recent "heat flow increase" is required to satisfy both maturity indicators and present-day temperatures. I count ten papers in the current volume, from Norway and elsewhere, that either discuss this phenomenon or feature it as part of their modelling strategy. The consensus view appears to be that the "heat flow increase" may be an artifact of the systems used, perhaps stemming from the vitrinite reflectance–temperature calibration. Whatever the answer, the present writer takes this debate as a welcome indication that geoscientists do not always accept the output of the "black box", and are questioning the algorithms and strategies internal to the standard basin modelling packages.

The third section of the volume is devoted to controls on hydrocarbon generation and expulsion. Kinetic models for kerogen transformation are now widely accepted, and several papers examine the reliability of these models. Both *Ungerer* and *Espitalié* indicate that, using current kinetic models, there is still considerable uncertainty in computing key factors such as "depth to oil window". A need is evident for more optimization of laboratory data against field observations, and to appreciate generation thresholds in many different basin types worldwide. Understanding of expulsion is required to determine the amounts of oil and gas available for secondary migration, and hence is critical to quantitative basin modelling. Most basin modelling systems use simple empirical models that belie the complexity of the process. As Ungerer says in his abstract "... there is still a long and difficult way to go before a comprehensive, satisfactory model is developed."

The fourth section combines porosity, fluid flow and secondary migration. *Waples and Kamata*'s paper on porosity modelling points out that, although a wide spectrum of effects contributes to the reduction of porosity, most integrated systems adopt a very simplistic approach to the problem. A multivariate approach to this critical modelling parameter is suggested in the paper. The potential thermal effect of basinal fluid-flow — a poorly understood subject that is very much on the minds of basin modellers — is examined in papers by *Ludvigsen et al.* and *Demongodin and Vasseur* (see also the papers by *Hermanrud* and *Grigo et al.* elsewhere in the volume, which provide contrasting views on the importance of this phenomenon). On the subject of secondary migration, *Sylta* describes a method for quantitative migration modelling using ray-tracing techniques; and applications of this system to the Norwegian shelf are discussed later in the volume by *Schroeder* and *Skjervøy*. The entrapment of hydrocarbons is addressed in a significant conceptual work by *Sales*, who examines the relationship between vertical closure and seal capacity, and its implications for hydrocarbon accumulation.

While it can perhaps be argued that no major theoretical leaps in basin modelling technology have been made in the last few years, the opposite is surely true for the application of existing techniques. Examples of such case studies occupy the last two sections of this book. In her introduction to the Norwegian case studies, *Nyland* outlines the development of basin modelling as a decision-making tool in concession rounds, and documents the increasing sophistication of techniques used throughout the last decade. However, basin modellers will be chastened (but not surprised) by her comment that large untested structures are still likely to be drilled, even when the prognosis from basin modelling is negative! The Norwegian papers cover most of the explored portions of the Norwegian shelf. *Grigo et al.* provide the only paper in this volume to adopt a "three-dimensional" modelling approach, while *Irwin et al.*'s work on the probabilistic modelling of hydrocarbon charge in the Egersund Basin signals a direction many of us believe basin modelling must eventually take. The Norwegian studies are balanced by a selection of international case histories including some of the world's most important producing basins such as the Western Siberian Basin (*Nemchenko et al.*), the South Caspian Basin (*Narimanov*) and the Gulf of Mexico (*Summa et al.*).

The papers mentioned above serve as examples only, and in no way devalue the contributions made by other authors. On behalf of the Organizing and Editing Committees, I would like to express hearty thanks to all of those who contributed presentations, posters, papers, manuscript reviews and ideas to the conference. On a personal note, I thank my secretary Linda Moen, who handled the vast correspondence associated with this volume.

A.G. Doré
Stavanger, October 1992

Opening and Welcome

It is a great pleasure to wish you all welcome to Stavanger and this International Conference on Basin Modelling, organized by the Norwegian Petroleum Society. More than 300 delegates representing most of the international oil companies, together with scientists from research institutes and consulting companies all over the world, are here today.

Since the beginning of the 1980s, basin modelling has been featured in many international conferences on petroleum geology and geochemistry. However, this is the first time to my knowledge that an international conference has been primarily devoted to advances and applications in basin modelling. It is certainly the first such conference to be held in Norway. Despite the fact that basin modelling is considered to be a very specialized subject, this conference has turned out to be one of the most popular of its kind arranged by the Norwegian Petroleum Society in recent years. This interest was readily apparent when the Conference Committee was offered more than 100 highly qualified papers.

We appreciate very much that some of the pioneers in developing basin modelling are present here today. These and other authors will cover most aspects of basin modelling. There will be presentations covering the physical and chemical processes acting during the burial of sedimentary basins. The last day of the conference is reserved for case studies, both from different parts of the Norwegian Shelf and from selected hydrocarbon provinces worldwide.

Basin modelling is a relatively young discipline in the evaluation of the hydrocarbon potential in an area. The first integrated basin modelling study to be presented was at the end of the 1970s. Since then, a variety of basin modelling tools and techniques have been developed, all with the same goal of trying to quantify the generation and accumulation of hydrocarbons as governed by known physical and chemical processes.

Advanced basin modelling studies as a part of the exploration activity on the Norwegian Shelf have become common during the last five years. The variety of software programs available on the market, and the use of inhouse systems, is demonstrated by the number of regional basin modelling studies performed by the oil companies. Some of these studies are presented to the Norwegian Petroleum Directorate in applications for new exploration licenses, or as special studies under license work commitments.

At the beginning of the 1980s only a few companies performed regional evaluations of hydrocarbon potential with emphasis on source rock potential, level of maturation and migration of hydrocarbons. Today, most of the companies do integrated one-dimensional or two-dimensional studies in their exploration activity, and lately we have also seen the first three-dimensional basin modelling studies.

We have registered with great interest the increased use of basin modelling in exploration activity on the Norwegian Shelf, and its possibilities as a tool to make exploration strategy more cost-effective. The Norwegian Petroleum Directorate has, in recent years, worked with several projects related to basin modelling. We have especially focused on primary and secondary migration problems, and on the causes and effects of the uplift and erosion seen across most of the Norwegian Shelf, in the western part of the Barents Sea, and on Svalbard. We have also allocated resources to develop our own computer basin modelling systems.

Although basin modelling has proved to be a helpful tool in the evaluation of prospectivity in some areas, the inconsistencies between the results from different studies in the same area are also obvious. This observation reminds us of the importance of being aware of the inaccuracies in some models, and of the inherent uncertainty in input data and in calibration parameters. It also clearly indicates the need for further improvement in modelling capability.

These issues and many others will be discussed actively at the conference. I wish you a rewarding exchange of ideas, and hope you will all take home some new concepts to work with. Finally, I wish you all an interesting and pleasant stay in Stavanger, the oil capital of Norway.

Fredrik Hagemann
Director General, Norwegian Petroleum Directorate

List of Contributors

K. AARESKJOLD	*IKU Petroleum Research, N-7034 Trondheim, Norway*
B. ANDRESEN	*Institutt for Energiteknikk, P.O. Box 40, N-2700 Kjeller, Norway*
P. ANDRESEN	*Saga Petroleum A/S, P.O. Box 490, N-1301 Sandvika, Norway*
M.T. ARIENTI	*AGIP S.P.A., P.O. Box 12069–20120 Milan, Italy*
J.H. AUGUSTSON	*Norsk Hydro Research Exploration Office, Storaakeren 11, N-9400 Harstad, Norway*
K. BJØRLYKKE	*Department of Geology, University of Oslo, P.O. Box 1047, 0316 Oslo 3, Norway*
F. BRIGAUD	*Société Nationale Elf Aquitaine (Production), C.S.T.J.F., Avenue Larribau, 64018 Pau Cedex, France*
J. BURRUS	*Institut Français du Pétrole, B.P. 311, 92506 Rueil-Malmaison, France*
S. CAO	*Alberta Research Council, P.O. Box 8330, Postal Station F, Edmonton, Alberta, Canada T6H 5X2* Present address: *Geoscience Department, University of South Carolina, Columbia, SC 29208, U.S.A.*
E.M. CARLSEN	*Statoil A/S, P.O. 300, N-4001 Stavanger, Norway*
S.I. CHEREDEEV	*Arctic Centre, Kola Science Centre, Institute of Economy, Murmansk-23, Russia*
J. COGAN	*1718 35th Street N.W., Washington DC 20007, U.S.A.*
A. COUDRAIN-RIBSTEIN	*Laboratoire de Géologie Appliquée (URA CNRS "Circulations et Transports Hydriques Continentaux"), UPMC, 4 place Jussieu, 75252 Paris Cedex 05, France*
B. DAHL	*Norsk Hydro Research Centre, N-5020 Bergen, Norway*
L. DEMONGODIN	*Centre Géologique et Géophysique (UPR 361, CNRS), USTL, 34095 Montpellier Cedex 5, France*
A.G. DORÉ	*Conoco Norway Inc., P.O. Box 488, N-4001 Stavanger, Norway*
S. DROUET	*Institut Français du Pétrole, B.P. 311, 92506 Rueil-Malmaison, France*
S.J. DÜPPENBECKER	*Institute for Chemistry and Dynamics of the Geosphere (ICG-4) at the Jülich Research Centre (KFA), P.O. Box 1913, D-5170 Jülich, Germany* Present address: *B.P. Research Centre, Chertsey Road, Sunbury-on-Thames, TW16 7LN, Great Britain*
L.T. ERDOĞAN	*ÖMV Aktiengesellschaft Basin Modelling Group, Gerasdorferstraße 151, A-1210 Vienna, Austria*
J. ESPITALIÉ	*Institut Français du Pétrole, B.P. 311, 92506 Rueil-Malmaison, France*
M. FIORANI	*AGIP S.P.A., P.O. Box 12069–20120 Milan, Italy*
W. FJELDSKAAR	*Rogaland Research, P.O. Box 2503 Ullandhaug, N-4004 Stavanger, Norway*
H. FU	*Japan Oil Engineering, Tokyo, Japan*
G.J. GAIL	*Mobil Exploration and Producing Services, Inc., P.O. 650232, Dallas, TX 75265, U.S.A.*

H.H. GANZ	*Koninklijke/Shell Exploratie en Produktie Laboratorium, Rijswijk, The Netherlands; and Institut für Geologie, Technische Universität Berlin, Berlin, Germany*
L.Sh. GIRSHORN	*Imalneftegas Geophysicq, Salekhard, Russian Federation*
P. GOUZE	*Laboratoire de Géologie Appliquée (URA CNRS "Circulations et Transports Hydriques Continentaux"), UPMC, 4 place Jussieu, 75252 Paris Cedex 05, France; and Centre Géologique et Géophysique de Montpellier (UPR CNRS 631), Université Montpellier II, Place E. Bataillon, 34095 Montpellier Cedex 05, France*
K. GRAN	*Saga Petroleum A.S., Kjøbovn 16, P.O. Box 490, N-1301 Sandvika, Norway*
N. GRAVDAL	*Enterprise Oil Norge, P.O. Box 399, N-4001 Stavanger, Norway*
D. GRIGO	*AGIP S.P.A., P.O. Box 12069–20120 Milan, Italy*
C. GUARGENA	*Enterprise Oil Norge, P.O. Box 399, N-4001 Stavanger, Norway*
D. HANEBECK	*Institute of Petroleum and Organic Geochemistry (ICG-4), Forschungszentrum (KFA), Jülich GmbH, P.O. Box 1913, D-5170, Jülich, Germany*
W.J. HARRISON	*Department of Geology and Geological Engineering, Colorado School of Mines, Golden, CO 80401, U.S.A.*
C. HERMANRUD	*Statoil A/S, P.O. 300, N-4001 Stavanger, Norway; and Statoil A/S, 6500, Kristiansund N., Norway* Present address: *Statoil Forskningssenteret, Postuttak, N-7004 Trondheim, Norway*
D. HERMANSEN	*Norsk Hydro Research Centre, Lars Hilles gt. 30, N-5028 Nygårdstangen, Bergen, Norway*
B. HORSFIELD	*Institute for Chemistry and Dynamics of the Geosphere (ICG-4) at the Jülich Research Centre (KFA), P.O. Box 1913, D-5170 Jülich, Germany*
H. IRWIN	*Statoil A/S, P.O. 300, N-4001 Stavanger, Norway*
R.P. JENSEN	*Conoco Norway Inc., P.O. Box 488, N-4001 Stavanger, Norway*
H. KAMATA	*Technology Research Center, Japan National Oil Corporation, 1-2-2 Hamada, Chiba 260, Japan* Present address: *Japan Oil Engineering Company Ltd., 6-2-1, Ginza, Chuo-ku, Tokyo 104, Japan*
K. KODAMA	*Geological Survey of Japan, Tsukuba, Japan*
B.M. KROOSS	*Institute of Petroleum and Organic Geochemistry (ICG-4), Forschungszentrum (KFA), Jülich GmbH, P.O. Box 1913, D-5170, Jülich, Germany*
K.W. LARSON	*Technology Research Center, Japan National Oil Corporation, Chiba, Japan* Present address: *Hydrogeology Program, University of Illnois, Urbana, IL 61801, U.S.A.*
K. LEISCHNER	*Institute for Petroleum and Organic Geochemistry (ICG-4), Research Centre Jülich (KFA), P.O. Box 1913, D-5170 Jülich, Germany* Present address: *RISM, AGIP Spa, Via Bonarelli, 1, 20097 San Donato-Milanese, Italy*
I. LERCHE	*Department of Geological Sciences, University of South Carolina, Columbia, SC 29208, U.S.A.*
W. LEU	*Koninklijke/Shell Exploratie en Produktie Laboratorium, Volmerlaan 6, 2288 GD Rijswijk, The Netherlands*
B. LEVELL	*Koninklijke/Shell Exploratie en Produktie Laboratorium, Volmerlaan 6, 2288 GD Rijswijk, The Netherlands*

D. LEYTHAEUSER	*Institute of Petroleum and Organic Geochemistry (ICG-4), Forschungszentrum (KFA), Jülich GmbH, P.O. Box 1913, D-5170, Jülich, Germany* Present address: *Geologisches Institut, Universität zu Köln. Zülpicher Straße 49, D-5000 Köln 1, Germany*
R. LITTKE	*Institute for Petroleum and Organic Geochemistry (ICG-4), Research Centre Jülich (KFA), P.O. Box 1913, D-5170 Jülich, Germany*
A. LUDVIGSEN	*Det Norske Veritas Sesam A.S.. Veritasveien 1, P.O. Box 300, N-1322 Høvik, Norway*
X. LUO	*Department of Geology, Northwestern University, Xi'an, China; and Centre Géologique et Géophysique, Université de Montpellier II, 34095 Montpellier Cedex, France*
B. MARAGNA	*AGIP S.P.A., P.O. Box 12069–20120 Milan, Italy*
F. MARQUIS	*Institut Français du Pétrole, B.P. 311, 92506 Rueil-Malmaison, France*
M. MARRONE	*IBM ECSEC, V.1e Oceano Pacifico, Rome, Italy*
N. MILLS	*Saga Petroleum A/S, P.O. Box 490, N-1301 Sandvika, Norway*
A. MOSTAFA	*Institut für Geologie, Technische Universität Berlin, Berlin, Germany*
M.B. MYHR	*IKU Petroleum, Research, N-7034 Trondheim, Norway*
N.D. NAESER	*U.S. Geological Survey, MS 424, Federal Center, Denver, CO 80225, U.S.A.*
K. NAKAYAMA	*Japan Petroleum Exploration Co., Ltd. 2-17-22, Akasaka, Minato-ku, Tokyo, 107 Japan*
A.A. NARIMANOV	*Kaspmorneftegas, Republic of Azerbaijan*
N.N. NEMCHENKO	*State Commision for Natural Resources, B. Poljanka 54, Moscow, Russia*
H.H. NIELSEN	*Mobil Exploration Norway Inc., P.O. Box 510, N-4001 Stavanger, Norway*
S.B. NIELSEN	*Department of Earth Sciences, Geophysical Laboratory, University of Aarhus, Finlandsgade 8, DK-8200 Aarhus N, Denmark*
I. NORDVALL	*Statoil A/S, P.O. 300, N-4001 Stavanger, Norway*
B. NYLAND	*Norwegian Petroleum Directorate (NPD), P.O. Box 600, N-4001 Stavanger, Norway*
A. OKUI	*Technology Research Center, Japan National Oil Corporation, 1-2-2 Hamada, Chiba 260, Japan* Present address: *Idemitsu Oil Development Co., Ltd., 5-6-1, Shiba, Minato-ku, Tokyo 108, Japan*
B.K. OSTISTY	*Arctic Centre, Kola Science Centre, Institute of Economy, Murmansk-23, Russia*
E. PALM	*Department of Mathematics, University of Oslo, P.O. Box 1053, N-0316 Oslo 3, Norway*
A. PARISI	*IBM ECSEC, V.1e Oceano Pacifico, Rome, Italy*
R.J. POTTORF	*Exxon Production Research Company, P.O. Box 2189, Houston, TX 77252-2189, U.S.A.*
E. PRESTHOLM	*Rogaland Research, P.O. Box 2503 Ullandhaug, N-4004 Stavanger, Norway*
U. RITTER	*IKU Petroleum Research, N-7034 Trondheim, Norway*
A.S. ROVENSKAYA	*Institute of Geology and Exploitation of Combustible Fuels, Moscow, Russia*
B. RUSSELL	*Marathon Oil Co., P.O. Box 629, Littleton, CO 80160, U.S.A.*
J.K. SALES	*Mobil Exploration Norway Inc., P.O. Box 510, N-4001 Stavanger, Norway*

R.G. SCHAEFER	*Institute for Chemistry and Dynamics of the Geosphere (ICG-4) at the Jülich Research Centre (KFA), P.O. Box 1913, D-5170 Jülich, Germany*
H.J. SCHENK	*Institute for Chemistry and Dynamics of the Geosphere (ICG-4) at the Jülich Research Centre (KFA), P.O. Box 1913, D-5170 Jülich, Germany*
F. SCHMIDT	*ÖMV Aktiengesellschaft Basin Modelling Group, Gerasdorferstraße 151, A-1210 Vienna, Austria*
F. SCHNEIDER	*Institut Français du Pétrole, B.P. 311, 92506 Rueil-Malmaison, France*
L. SCHOU	*IKU Petroleum Research, N-7034 Trondheim, Norway*
F.W. SCHROEDER	*Exxon Production Research Company, P.O. Box 2189, Houston TX 772252-2189, U.S.A.*
T.F. SCHWARZER	*Exxon Company U.S.A., P.O. Box 4778, Houston, TX 77060-2544, U.S.A.*
P. SGUAZZERO	*IBM ECSEC, V.le Oceano Pacifico, Rome, Italy*
A. SKJERVØY	*Conoco Norway Inc., P.O. Box 488, N-4001 Stavanger, Norway*
M. STEPHENSON	*Enterprise Oil Norge, P.O. Box 399, N-4001 Stavanger, Norway*
L.L. SUMMA	*Exxon Production Research Company, P.O. Box 2189, Houston, TX 77252-2189, U.S.A.*
Ø. SYLTA	*IKU Petroleum Research, N-7034 Trondheim, Norway*
J. TANG	*Department of Geological Sciences, University of South Carolina, Columbia, SC 29208, U.S.A.*
N.J. THEIS	*Mobil Exploration Norway Inc., P.O. Box 510, N-4001 Stavanger, Norway*
T. THRONDSEN	*Institutt for energiteknikk, P.O. Box 40, N-2700 Kjeller, Norway*
A.S. UBERG	*IBM Scientific Center, Thormøhlensgate 55, N-5008 Bergen, Norway*
Å. UNANDER	*Institutt for Energiteknikk, P.O. Box 40, N-2700 Kjeller, Norway*
P. UNGERER	*Institut Français du Pétrole, B.P. 311, 92506 Rueil-Malmaison, France*
G. VASSEUR	*Centre Géologique et Géophysique (UPR 361, CNRS), USTL, Université de Montpellier II, 34095 Montpellier Cedex 5, France*
B. VELDE	*Départment Géologie (URA 1316, CNRS), ENS, 24 Rue Lhomond, 75005 Paris, France*
E. VIK	*Statoil A/S, 6500, Kristiansund N., Norway* Present address: *Statoil Forskningssenteret, Postuttak, N-7004 Trondheim, Norway*
J. VOLLSET	*Statoil A/S, P.O. 300, N-4001 Stavanger, Norway*
O. WALDERHAUG	*Rogaland Research, P.O. Box 2503 Ullandhaug, N-4004 Stavanger, Norway*
D.W. WAPLES	*Technology Research Center, Japan National Oil Corporation, 1-2-2 Hamada, Chiba 260, Japan*
D.H. WELTE	*Institute for Petroleum and Organic Geochemistry (ICG-4), Research Centre Jülich (KFA), P.O. Box 1913, D-5170 Jülich, Germany; and Integrated Exploration Systems (IES), Bastionstraße 11–19, D-5170 Jülich, Germany*
S. WOLF	*Institut Français du Pétrole, B.P. 311, 92506 Rueil-Malmaison, France*
M.A. YUKLER	*IBA Inc., 2394 Shady Lake Circle, Carrollton, TX 75006, U.S.A.*
K. ZHAO	*Department of Geological Sciences, University of South Carolina, Columbia, SC 29208, U.S.A.*

Contents

V. Case studies, Norway

VI. Case studies, worldwide

Basin modelling techniques — an overview

Christian Hermanrud

The recognition that hydrocarbon generation and accumulation in sedimentary basins are governed by known physical processes has triggered substantial efforts to quantify these processes and their interactions during the last 15 years. A large variety of models that aim at quantifying hydrocarbon generation and accumulation has been developed. An analysis of the differences between the programmes and the historical development of basin modelling concepts thus seems appropriate.

The early basin models seem to have derived their ideas from reservoir engineering and soil mechanics, and these ideas propagated to later developed programmes. Modelling of the processes (heat and fluid flow, compaction, hydrocarbon generation, expulsion and secondary migration) are the cornerstones of basin modelling. A historical development of modelling of each of these processes can be tracked, and the state of the art documents that several unresolved questions remain. Four different schools of basin modelling developed in the early 1980's (here termed the German, French, South Carolina and Illinois schools). Several properties of published and commercially available basin modelling programmes have been tabulated and analysed to show that the different models of today have inherited their ideas and mathematical formulations from these schools.

The results from basin modelling programmes are supposed to aid management in oil companies to make drilling decisions. Analyses of errors and sensitivity is vital for these purposes, yet little emphasis has been put on these items historically. An examination of the basin modelling literature from the last few years shows that the emphasis on accuracy and sensitivity is growing, although a large potential for improvement still exists.

Introduction

The wording of the term "basin modelling" points to numerical modelling of any physical process which takes place in a sedimentary basin, including basin formation, sediment transport and predictive stratigraphy. A common and somewhat narrower definition, adopted in this paper, reduces the topic to the modelling of heat and fluid flow, compaction, oil and gas generation and hydrocarbon migration. This paper aims at giving an overview of the computer modelling of these processes.

First we will discuss the different mathematical approaches used to model heat and fluid transport in compacting sedimentary basins, and gives a historical record of the mathematical cornerstones in basin modelling programmes. Most models published after 1984 seem to have branched off from four main schools, all developed in the early 1980's. The characteristics of these schools will be examined. Later models have adopted different properties from the main schools and may indicate the relative success of the different approaches. An examination of the choices made in these papers is included together with a discussion of the relative merits of these approaches. A tabulation of several properties of some commercial basin modelling programmes is provided in the Appendix.

Then follows a discussion of published suggestions for modelling the processes of compaction, heat transport, fluid flow, overpressuring, hydrocarbon generation, primary and secondary migration. Three aspects of the modelling of each process are covered: historical development, state of the art, and unresolved questions. The paper does not aim at giving a complete history of the knowledge of each process, but covers this history where it is considered essential to the discussion of the state of the art or the unresolved questions.

The last part of the paper covers the application of basin modelling software and concepts to hydrocarbon exploration. The discussion in this part includes sensitivity analysis, calibration, error analysis and strategies for the design of basin modelling studies.

Mathematical development

Early papers (prior to 1984)

The basic principles of heat and fluid flow were well understood long before the development of computer technology made simulation of these processes in sedimentary basins attractive. The historical development of these principles is beyond the scope of this paper; the reader is referred to Hubbert (1940) for a rigorous treatment of fluid flow in sed-

Basin Modelling: Advances and Applications edited by A.G. Doré et al. NPF Special Publication 3, pp. 1–34, Elsevier, Amsterdam.

iments and to Carslaw and Jaeger (1947, 1959) for mathematical solutions to heat conduction problems. Coupled equations for heat and fluid transport in non-compacting porous media were given by Stallman (1960, 1963).

The thermal balance of the earth was given little systematic attention prior to the monograph of Lee (1965). The monograph of Kappelmeyer and Haenel (1974) serves as a classical reference to geothermics in general, as does Gretener (1981) for the pre-basin modelling applications of geothermics in hydrocarbon exploration. The need for accurate determination of thermal histories for the purpose of calculating hydrocarbon generation was established by Tissot (1969), Lopatin (1971) and Tissot and Espitalié (1975).

Backstripping as a basis for further geohistory analysis was introduced in the mid-1970's (Perrier and Quiblier, 1974; Watts and Ryan, 1976; van Hinte, 1978). This technique allowed different thermal regimes through geological time to be superimposed on the subsidence history of the sediments. Palaeotemperature was modelled as a consequence of crustal stretching and continental breakup by McKenzie (1978) and Royden and Keen (1980). These palaeotemperature models were followed by papers describing the computation of maturation of organic matter for the purpose of aiding hydrocarbon exploration (Royden et al., 1980; Sclater and Christie, 1980). None of these models included fluid flow effects, and the palaeoheat flow from the crustal models was used to perform steady state one-dimensional calculations in the sediments.

Several papers discussed the modelling of separate geological processes in the 1960's: Harbaugh, 1966 (marine sedimentation); Bonham-Carter and Sutherland, 1967 (deltaic sedimentation); Ojakangas, 1967 (three-dimensional compaction and oil trap development). These models appear to have sparked the interest for modelling of sediment deposition, and were later succeeded by models which are today applied extensively within sequence stratigraphy. Sediment infill models have been coupled to hydrocarbon generation and migration (e.g. Nakayama and van Sichlen, 1981). Such coupling has not been included in most later basin models and will not be discussed here.

Ortenblad (1930) appears to have given the first (approximate) solutions for fluid flow in a compacting sedimentary sequence: his equations model a constant deposition rate. Several papers addressed different aspects of fluid flow in compacting sedimentary basins in the period 1958–1973. The first papers which describe an integrated treatment of a transient heat flow equation and a fluid flow equation in compacting sediments were given by Sharp (1976), Sharp and Domenico (1976) and Yukler et al. (1978). These papers and articles from the early 1980's mainly focusing on heat transport in sedimentary basins (Royden et al., 1980; Sclater and Christie, 1980), seem to have formed the basis on which four main "schools" within basin modelling were established in the early 1980's. Subsequent papers on basin modelling describe models which use the general ideas from these schools and can in most cases be closely related to one particular school. Table 1 gives further information on the most significant early papers, the characteristics of the main schools, and properties of a number of late papers within basin modelling. The symbols and abbreviations used in Table 1 are explained in Table 2.

Most of the "early papers" listed in Table 1 have several features in common. They take their general ideas from soil mechanics, based on ideas developed, for example, by Terzaghi (1925) and Biot (1941). These early papers consider porosity reduction as a purely compressive phenomenon, and calculate compaction from knowledge of rock compressibility. They also formulate Darcy's law using hydraulic conductivity (K, dimension L/T, at times called coefficient of permeability) instead of permeability (k, dimension L^2).

The papers of Smith (1971, 1973) were based on reservoir engineering concepts rather than on soil mechanics. They apply an empirical porosity vs. effective stress law, and also formulate Darcy's law using permeability (k) instead of hydraulic conductivity (K). Note that:

$$K = \frac{k\rho g}{\mu} \quad (1)$$

where μ = viscosity, ρ = fluid density and g = the acceleration due to gravity. Fluid flow equations written with k and K as parameters may thus behave identically. The fact that both k, ρ and μ are (different) functions of other parameters such as porosity and temperature implies that neither k or K should be treated as constants in the computation of fluid flow in compacting sedimentary basins. The K vs. k distinction seems to be diagnostic to the background on which the equations are founded, and is therefore tabulated in Table 1.

Most of the early papers in Table 1 describe analytical solutions to the problems being investigated. Authors who applied numerical techniques used the finite difference method, and used a coordinate system with fixed grid size (Eulerian coordinates). De Bremaecker (1983) introduced the application of the finite element method in basin modelling; he also appears to have been (among) the first to treat com-

Three of the thirteen papers listed describe coupled equations of heat and fluid flow, using two- or three-dimensional models. The other ten papers describe programmes which were designed to investigate separate aspects of basin dynamics. A multi-dimensional integration of temperature and fluid flow equations was apparently considered to be unnecessary for the purposes of these studies.

While only one of the pre-1984 papers applied Lagrangian coordinates, about half of the later papers have used such coordinates, including all of the three papers which describe coupled transient equations of heat and fluid flow. The fact that four of the five tabulated papers from the 1990's use Lagrangian coordinates, and that at least one commercial basin modelling programme (BMT, see Table A-1 in the Appendix) is about to change its formulation from Eulerian to Lagrangian coordinates, seems to indicate a growing preference for this method.

While Table 1 shows that various numerical methods dominate over analytical solutions in the investigated papers, no apparent preference of numerical method (finite difference vs. finite element) has emerged as yet. Information on the relative performance of the different methods within basin modelling has so far been limited, and it remains to be seen if the lack of preference for one specific method indicates that the two methods are equally well suited for basin modelling.

The influence from soil mechanics is distinctly less apparent in the "late papers" of Table 1 than was the case for the pre-1980 papers, and most of these late papers adopt the general concepts of fluid flow and porosity reduction from reservoir engineering. Only two of the nine "late" papers which describe fluid flow combine density, viscosity, permeability and the acceleration of gravity to give one parameter (hydraulic conductivity), and only three of twelve "late" papers discuss rock compressibility as the (only) factor controlling the compaction of sediments.

An evaluation of the different approaches

The different options described above lead to programmes with different inherent properties. None of the described approaches are incorrect (in the sense that they necessarily give inaccurate results), but different problems which have to be addressed will emerge depending on the design of the programmes.

A rock sequence is generally treated as a series of different layers in basin modelling. Each of these modelled layers has constant or smoothly varying properties (e.g. porosity decreases smoothly with burial within a layer of constant lithology), while real rock properties may change radically over a lithologic boundary (permeability will in general change by several orders of magnitude over a sand/shale boundary). Such changes in lithological properties favour Lagrangian coordinates. Eulerian (non-compacting) coordinates should only be applied if the vertical spacing of cells is small compared to the thickness of the thinnest layers. Proper simulation of thin layers will thus require a dense grid, implying prolonged computer time.

An Eulerian grid also requires much book keeping (to monitor which layers are actually in the individual cells and calculate the appropriate parameter for that cell). Deming and Chapman (1989) (who did not include fluid flow in their model) demonstrated that the error that was introduced because of the averaging in their Eulerian scheme was about 1°C. This error estimate should not be taken as a general measure of the accuracy of the Eulerian scheme; it is only valid for the layer thicknesses and grid size applied in their investigated model. The results from modelling with an Eulerian scheme will in general converge towards the results from a Lagrangian scheme when the grid is refined.

Some authors (e.g. De Bremaecker, 1983) formulate their equations in an Eulerian frame of reference, and adjust the grid spacing at each time step to preserve uniform lithologies within each cell. This approach has been labelled as Lagrangian in Table 1. An alternative, as suggested by Lerche and Glezen (1984), is to write the equations directly in Lagrangian form through:

$$\xi = \int (1 - \phi)\, \delta z \tag{2}$$

where ξ is the new (Lagrangian) depth axis, z is depth and ϕ is porosity. The latter approach makes the equations less attractive to read and write, but removes the need for regridding during simulations, and thus seems to be best suited for basin modelling.

The new release of the BMT programme (see Table A-1 in the Appendix) is written in Lagrangian coordinates, which denotes a change from previous versions of this programme. The change from Eulerian to Lagrangian coordinates was made because of "the general superiority of Lagrangian coordinates for modelling of a compacting medium, but especially because of the handling of listric faulting" (M. Langfelt, PROCOM, pers. commun., 1992)

Lerche and Glezen (1984) scaled their depth axis from zero to one (fractional coordinates). This technique was recommended by Crank (1984), and has long been known (see Landau, 1950). The fractional coordinates (the coordinates should be scaled from 0 to 1 within each layer) may have their merit. They do, however, make the equations harder to comprehend for non-mathematicians. The fact that none of the "late" papers of Table 1 suggests this approach testifies that stability can be achieved without fractional coordinates. Whether computer time can be saved by moving to such coordinates has not yet been demonstrated.

The different merits of various discrete formulations and solution strategies has been given little attention in the literature, and tabulation of the sparse information on preferred approaches has not been made. The South Carolina school applies a (semi-)explicit method with an alternating direction implicit (ADI) scheme (Mitchell and Griffiths, 1980). Most other papers who describe their solution procedures seem to rely on fully implicit or semi-implicit (e.g. Crank-Nicholson) schemes.

The finite element method (FEM) is generally considered to be superior to the finite difference method (FD) for the simulation of irregular geometries. The FEM may however require more computer time, and also complicates the equations. Comparisons of the FEM and FD versions of the (French school) TEMISPACK model revealed that the FD model produced satisfactory results, although the FEM version was more accurate in cases with highly irregular geometries (Bouvier, 1989; J. Burrus, pers. commun., 1990). The study concluded that the slightly improved accuracy of the FEM version did not justify the extra costs. This conclusion is not necessarily of general validity for all basin modelling programmes, as it depends on the capabilities of the programme to handle complex geometries (such as salt movement and listric faulting) in general. The FD method should certainly work satisfactory for most practical purposes. The (German school) PDI programmes are now available in both FEM and FD versions. A comparison of these versions would further elucidate the relative merits of the two numerical methods in basin modelling.

While the "late" papers clearly have preferred to follow the reservoir engineering (rather than the soil mechanics) approach to compaction and fluid flow simulations, little criticism of the former has been raised. Porosities which are calculated from rock compression only may be as good (or bad) as those calculated from empirical relationships, provided that rock compressibilities are calibrated to match observed porosities (e.g. by the establishment of suitable compressibility vs. porosity curves). A possible explanation for the "late" preference for the reservoir engineering approach may be that the German school (which is based on soil mechanics) has been comparatively restrictive regarding information on the mathematical properties of their programmes. Most papers which have been published from the four main schools within basin modelling have accordingly either described the reservoir engineering approach or have not discussed mathematical formulations in any detail. New workers in a field of science will generally adapt ideas from the literature, and as more recent information has been available on the reservoir engineering approach, this has probably led to the favouring of this approach by new authors.

Since the first three-dimensional (3D) model was introduced in 1981, only one other 3D model has been reported (Grigo et al., this volume). While three dimensions are necessary to perform volumetric calculations, the need for 3D heat and fluid flow equations needs further substantiation. Valid arguments for constructing a fully integrated 3D programme could be that thermal refraction and fluid flow in the third dimension is so significant that 2D calculations of hydrocarbon generation or migration have an unacceptable inaccuracy.

The influence of thermal refraction can be evaluated by running 1D, 2D and 3D models over the same area (with identical rock properties). Such an evaluation has been performed over a salt dome by Statoil (unpublished). In this case the thermal refraction as modelled in 2D increased the temperature above the salt dome by 10°C, and another 3°C was added when 3D modelling was introduced. Thus 3D modelling may give results which differ from 1D or 2D results, and the differences may be larger in other areas than those just mentioned. The extra accuracy comes as a reward for tedious data preparation and prolonged computer time. Whether this time could be used in a more cost-effective way is open to debate. The continuing progress in computer hardware and access to digitized data may make 3D modelling more attractive in the future.

All of the basin modelling programmes described above rely on mass and thermal energy balance equations, and those which handle fluid flow apply Darcy's law. Some authors invoke more advanced concepts like statistical physics (Wong, 1988), effective medium theory (Palciauskas, 1986; Levine and Cuthiell, 1986) and Kriging (Harff et al., 1989). Such methods may well become popular in the future, but have so far not been frequently used in basin modelling programmes.

Modelling compaction

Historical development

The first quantitative approach to sediment compaction was apparently given by Terzaghi (1925), who stated that an increase in the effective stress (stress at grain-to-grain contacts) leads to increased compaction. Bayer (1989) discusses the application of Terzaghi's principle in large-scale systems in some detail. Athy (1930a) established an exponential porosity vs. depth relationship based on 200 samples, and many porosity vs. depth curves have been established since then. Most authors who have presented new porosity vs. depth relationships have suggested that the curves were valid in restricted areas, and have not attempted to show why previously reported curves fail.

The differences in porosity vs. depth for shales as reported by different studies are extremely large. Rieke and Chilingarian (1974, fig. 17) show that values from 5% to 30% have been suggested for shale porosity at 1500 m depth of burial in different parts of the world. They attributed these variations to: (1) the amount of induration in the sediments, (2) geologic age, (3) compositional and textural variations, and (4) tectonic stress history. The different measurement techniques that have been employed also contribute to the variations: log-derived porosities are generally higher than porosities derived from He or Hg injection or weighing/drying of samples.

Magara (1976) and Sclater and Christie (1980) presented methods for compaction assessments in undercompacted (overpressured) shales. Early attempts at modelling compaction in overpressured shales are listed in Table 1.

State of the art

Several new porosity vs. depth (or effective stress) curves have been suggested in the last decade. Again, little reasoning is given as to why the earlier curves were not appropriate as worldwide average curves. Figure 1 shows eight of the "new" porosity vs. depth curves for shales superimposed on the curves collected by Rieke and Chilingarian (1974). The curves of Bethke (1985), Doligez et al. (1986), England et al. (1987), Hermanrud et al. (1990a) and Lerche (1990a) were given as porosity vs. effective stress curves by the authors. They have here been transformed to porosity vs. depth curves by assuming hydrostatic pressure, a grain density of 2.7 g/cm^3 and a constant fluid density of 1.0 g/cm^3. The curve of Lerche

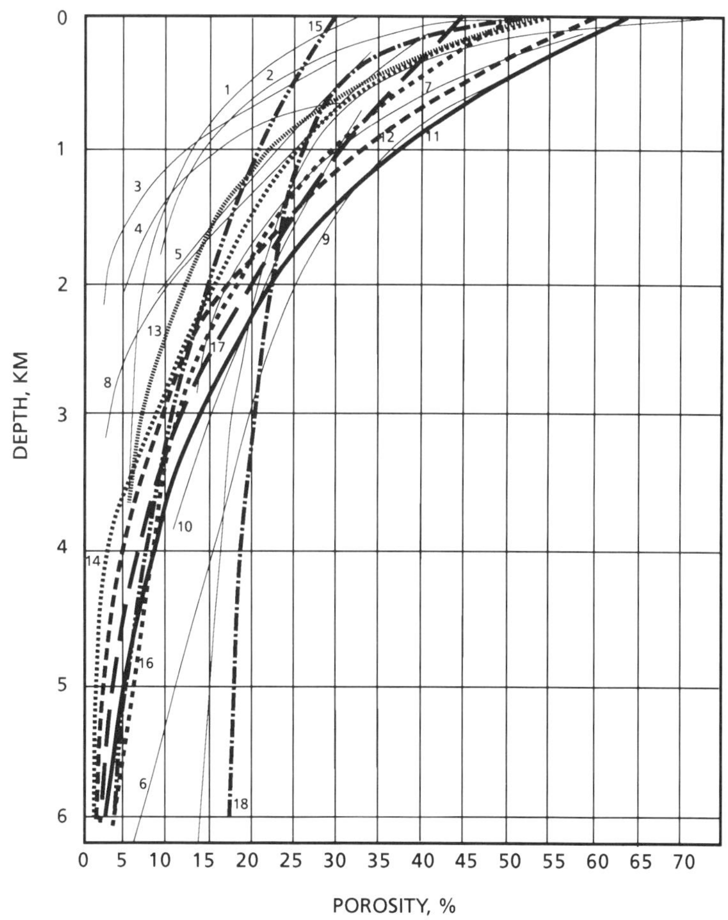

Fig. 1. Suggested porosity vs. depth relationships. The thin curves (and their labelling) are those presented by Rieke and Chilingarian (1974, their fig. 17). *1* = Proshlyakov (1960); *2* = Meade (1966); *3* = Athy (1930a); *4* = Hosoi (1963); *5* = Hedberg (1936); *6* = Dickinson (1953); *7* = Magara (1968); *8* = Weller (1959); *9* = Ham (1966); *10* = Foster and Whalen (1966); *11* = Sclater and Christie (1980); *12* = Bethke (1985); *13* = Doligez et al. (1986); *14* = England et al. (1987); *15* = Deming and Chapman (1989); *16* = Burnham and Braun (1990); *17* = Hermanrud et al. (1990a); *18* = Lerche (1990a).

(1990a) appears to give significantly higher porosities than the other curves. It should be noted that the author of this curve suggests that it is modified to match porosities inferred from well data when applied in real basins (Lerche, 1990a). The shape of the curve is nevertheless markedly different from the other "new" curves of Fig. 1. Additional compaction curves have been given by Baldwin (1971), Magara (1976), Hunt (1979) and Baldwin and Butler (1985), but are not included in Fig. 1 to avoid unnecessary complication of this figure.

At levels shallower than 2 km, the spread in porosity values is approximately ±50% from the mean value at a given depth for the eight "new" curves. With the exception of the Lerche (1990a) curve, the same holds true for depths greater than 2 km. The spread between the compaction curves implies that quite large differences in burial histories will emerge from studies using the various curves. The differences in compaction estimates also give different porosity-dependent rock properties, such as thermal conductivity, permeability and hydrocarbon saturation in mature source rocks.

Unresolved questions

There is a definite need to constrain porosity vs. depth (or effective stress) relationships. This has long been recognized for reservoir rocks. A knowledge of shale porosity is at least equally important in basin modelling, because shales dominate most sequences volumetrically, and because hydrocarbon generation and expulsion take place in shales. The spread in the "new" porosity vs. depth curves of Fig. 1 is not necessarily a measure of the accuracy of porosity estimates, as the basis for most curves is not, or is only vaguely, documented in the quoted papers.

The modelling of shale compaction is uncertain for several reasons. The handling of bound water (and conversion from bound to free water) is barely mentioned in the basin modelling literature (but see Magara, 1975, 1979). Yet the two water forms exhibit different flow properties, densities and thermal conductivities. The presence of bound water is probably also important for wettability (and thereby hydrocarbon adsorption) in shales. The distinction between total and effective (interconnected) porosity is also infrequent in basin modelling, despite its obvious importance for fluid flow.

The measurement of shale porosity constitutes another source of error. Shale porosities inferred from well logs tend to give too high values, especially in smectite-rich clays, because of clay swelling. Porosity measurements by Hg or He injection may give values which are too small if the injected fluid does not penetrate all of the tiny pores. On the other hand, fracturing of the samples (either by the measurement procedures or as a result of transporting the sample from its *in situ* high-pressure environment to the surface) may result in porosity measurements which are too high. Injection has to be performed on well-lithified samples, and such samples may not be representative for the whole sequence (and definitely not for the shallower sections).

The notion of compaction as only a result of compression has not been justified, and diagenetic reactions (functions of mineral assemblage and temperature) may well be significant. Leder and Park (1986) modelled porosity loss from diagenetic considerations alone. Schmoker and Gautier (1989) model porosity as a function of time and temperature, thus taking indirect account of burial depth. Schmoker and Gautier (1988), Ehrenberg (1990) and Oelkers et al. (1992) further discuss the importance of time and temperature for porosity reduction in sandstones.

Only through a more complete understanding of the processes which govern porosity reduction can compaction modelling become more accurate. The present uncertainty in estimates of (shale) porosity is noticeable, and influences markedly the accuracy of modelled water and hydrocarbon flow in sedimentary basins.

Modelling thermal processes

Historical development

The fact that thermal energy is transported from the Earth's interior to the surface has been known for several centuries. Bullard (1965) gives a summary of the early work on terrestrial heat flow (from 1530 to 1965). The monograph of Lee (1965) appears to be the first attempt to focus on geothermics as a separate subject within the earth sciences. The monograph of Kappelmeyer and Haenel (1974) focuses on the application of geothermics to various economically important problems such as geothermal energy extraction, radioactive waste disposal and prediction of temperatures in mines, but does not discuss the importance of subsurface temperature to hydrocarbon exploration or production. Gretener (1981) summarizes the application of geothermics to hydrocarbon exploration prior to the emergence of basin modelling.

An equation which describes heat conduction in a pile of sedimentary layers with different thermal conductivities was given by Bullard (1939). Carslaw and Jaeger (1947) gave an equation which included heat accumulation, generation and diffusion to describe the conservation of thermal energy in a sedimentary basin. Stallman (1960, 1963) introduced a convective term into this equation to account for heat transport by moving ground water. The equation of Stallman (at times rewritten to account for compaction and a moving coordinate system) is presently applied in most basin modelling programmes, including those of the main "schools" of basin modelling. Stallman's (1960) equation with a source/sink term is:

$$c_b \rho_b \frac{\partial T}{\partial t} = \nabla \cdot (K \nabla T) + c_w \rho_w \nabla \cdot (vT) + A \qquad (3)$$

where c is the specific heat, ρ is density, K is thermal conductivity, T is temperature, t is time, v is fluid velocity and A is the source/sink term. The first term of the equation describes heat storage (accumulation), the second term describes heat conduction (diffusion), and the third term describes convective heat transfer (heat transported by moving fluids). The subscript b refers to the bulk sediment property (grains + water), while the subscript w refers to properties of the moving fluid.

The essence of the equations described above is that they predict a non-linear temperature vs. depth relationship which also varies with time. The opposite approach (constant geothermal gradients through time) is nevertheless still widely used, and is also included as an option in several commercially available basin modelling programmes. A constant geothermal gradient through time prevails if only one single, uniform, non-compacting, flat sedimentary sequence with thermal properties independent of pressure and temperature exists in the basin. It is further required that basin dynamics or fluid flow did not influence the thermal state of the basin in the period of time which is investigated, and that the sediments as well as the lithosphere do not contain radioactive isotopes of uranium, thorium, potassium or other minerals. These conditions are never fulfilled in nature.

State of the art

Thermal modelling performed by different individuals using different basin modelling programmes may give significantly different results. Discrepancies may result from different geological interpretations, heat flow equations, coupling of heat flow to other equations, thermal parameters, boundary conditions and calibrations to well data. The latter will be covered in a separate section; the rest of these topics are discussed below.

The differences in geological interpretation include such factors as age of individual horizons, timing and amount of erosion, seismic interpretation and depth conversion. Hermanrud et al. (1990a) described a case where four different seismic interpretation studies differed in depth to the kitchen areas by 300–1400 m. Uncertainties of this magnitude are so severe that they do not warrant sophisticated thermal modelling.

The heat flow equations of the four main schools of basin modelling include terms for heat conduction, heat storage, convective heat transfer and heat generation within the sediments and/or the crust. Several basin modelling programmes that have been developed at later stages neglect a number of these terms, and often only consider steady state heat conduction. The cancellation of (especially) the convective and accumulation terms simplifies the mathematical treatment substantially. Such cancellations have thus been attractive to developers of commercial basin modelling programmes, including commercial versions of the main schools' programmes.

The terms in the heat flow equations do contribute with unequal amounts to the thermal budget. Thermal diffusion (heat conduction) is the dominant heat transporting agent (Kappelmeyer and Haenel, 1974), with the exception of hydrothermal areas (Didyk and Simoneit, 1989). The thermal effects of moving ground water have been proven to be significant (Smith and Chapman, 1983; Person and Garven, 1992), and should accordingly be considered in modelling of onshore areas. The transport of heat by compaction-derived water is generally thought to be of minor importance (Bethke, 1985; Hermanrud, 1986), although Chiaramonte et al. (1987) and Grigo et al. (this volume) state otherwise. The thermal influence of free convection in sediments can safely be ignored in areas where the subsurface temperatures are dominated by burial heating (Ludvigsen et al., this volume).

The contribution of radiogenic heat flow from the sediments is generally small: the equations of Rybach (1986) indicate contributions of a few percent or less. The radiogenic heat production in the crust may however be significant (Chapman, 1986; Jaupart, 1986; Ungerer et al., 1990). The modelled transient thermal effects of rapid subsidence and erosion are sensitive to crustal parameters such as heat capacity and radiogenic heat production, and are comparatively insensitive to the heat capacity of the sediments. Vik and Hermanrud (this volume) model a 10% heat flow reduction in the Haltenbanken area due to the rapid Plio-Pleistocene subsidence. Their results indicate that the transient heat flow should be included in thermal modelling of areas with subsidence rates of about 1 km per million years.

The coupling of heat flow equations to other equations (e.g. through porosity) also influences the modelled thermal history. Models that do not take account of compaction will produce markedly different thermal histories from those that include com-

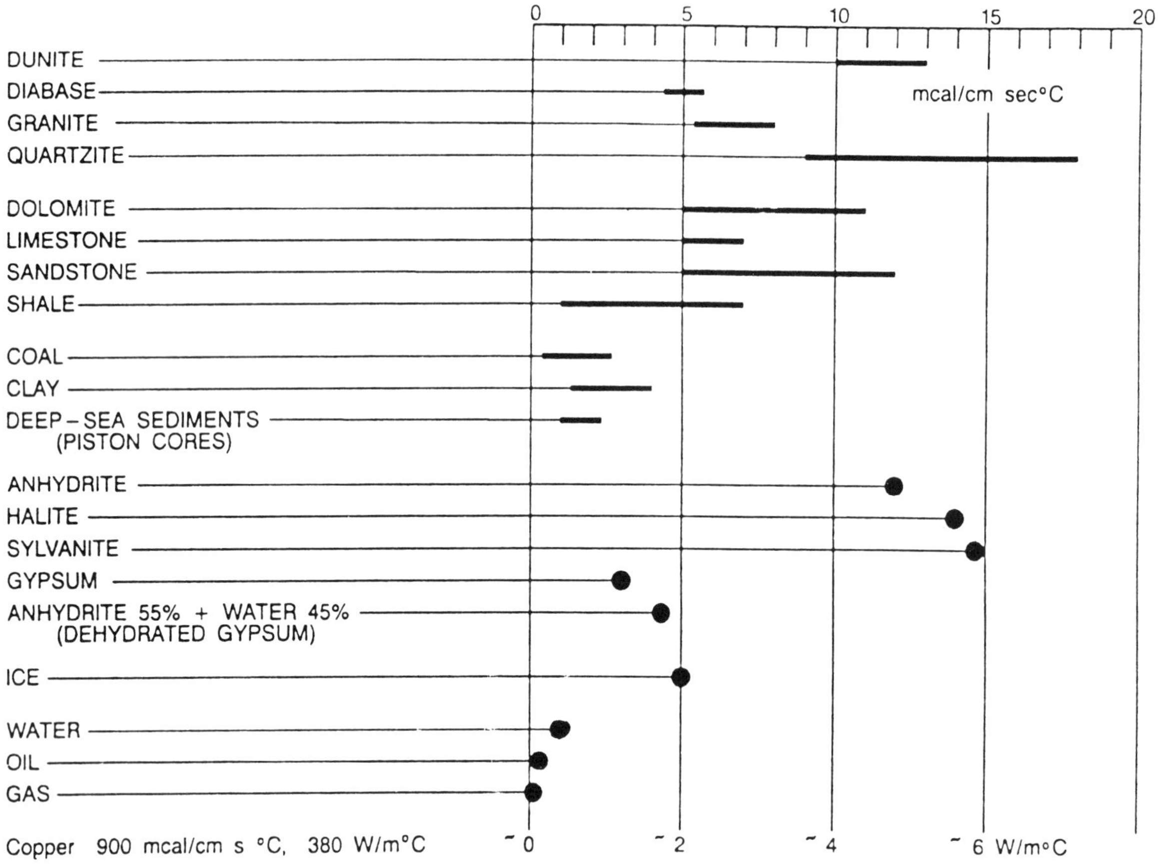

Fig. 2. Thermal conductivities of various rocks and pore fluids at room temperature. From Gretener (1981); courtesy of the American Association of Petroleum Geologists.

paction. Empirical relationships that are calibrated to thermal histories based on non-compacting sediments (such as the TTI calculations of Waples, 1980) should accordingly not be used by models which include compaction (Dykstra, 1987).

While there is a general agreement on which terms could be included in the heat flow equations of basin modelling programmes, there seems to be less control on the parameters that go into these terms. Figure 2 shows how the thermal conductivity differs within different lithologies. The differences in thermal conductivity for a given lithology are a result of differences in mineralogy, texture, grain size and porosity. The influence of porosity has conventionally been modelled by geometric averaging (Assad, 1955; Lewis and Rose, 1970; Sharp and Domenico, 1976), and the fit to measured data was established by Balling et al. (1981). Some basin modelling programmes have standard (end-member) lithologies with predefined thermal conductivities, while other programmes require matrix thermal conductivities as input values.

The effects of temperature and, particularly, pressure on rock thermal conductivities are of secondary importance. The effect of temperature is nevertheless included by most advanced modelling programmes, and its influence on the curvature of the conductivity vs. depth curve for a given lithology warrants such implementation.

The thermal boundary conditions employed also affect the modelled thermal history. The thermal boundary conditions are surface temperature as a function of time, and either a heat flow or a temperature vs. time relationship at the bottom of the domain where temperatures are computed. Most commercially available basin modelling programmes introduce the lower thermal boundary at the base of the sediments. This will result in a modelled adjustment to thermal equilibrium within a period of about 100,000 years. However, Deming and Chapman (1989) and Vejbæk (1990) argue that the transient thermal effects of rapid sediment deposition have typical lifetimes of 50–100 million years, and that the whole lithosphere should be included in the modelling of transient thermal effects.

Palaeoheat flow also varies due to large-scale crustal movements such as stretching or thrusting. Several models have been set up to simulate the thermal effects of such movements. The concepts of McKenzie (1978) are frequently used to give the heat flow history resulting from crustal thinning and subsequent thermal relaxation, although numerous other models which also describe the thermal effects of crustal stretching have been suggested (Kooi, 1991). Models which describe the physical relaxation processes more closely are described by Withjack (1979), Spohn and Schubert (1982), Pedersen and Skogseid (1989), and Nielsen and Balling (1990). Palaeoheat flow in compressional tectonic settings has also been investigated, e.g. by Heasler and Surdam (1985) (subduction zones) and Endignoux and Wolf (1990) (advancing thrust sheet). The thermal effects of short-lived thermal events were discussed by Hermanrud (1986) (rapid pulses of fluid flow) and Lerche (1990b) (thermal effects of igneous intrusions).

Few published basin modelling programmes incorporate analyses of transient thermal effects due to rapid deposition/erosion and large-scale crustal movements (stretching, thrusting) simultaneously. Most basin modelling programmes do accordingly require heat flow history (or asthenospheric temperature plus lithospheric thickness) as input parameters. Deming (1993) gives a comprehensive discussion of subsurface temperature prediction in different geological settings.

Unresolved questions

The accuracy of the input parameters to thermal modelling is, in contrast to the concepts of such modelling, not well established. The accuracy of average thermal conductivities for modelled rock sequences is generally not reported (for an exception, see Chapman et al., 1984). Thermal conductivity estimates for basin modelling are in most cases based on assessments of the relative content of end-member lithologies (sand, shale, ...) based on completion (or other) log information. Such determinations are quite subjective and may result in inaccuracies of the thermal conductivities of 20% or more.

The constantly changing lithologies in a well and the uncertainties concerning matrix conductivities make the idea of finding thermal conductivities from well logs attractive. Numerous different techniques have been applied (Anand et al., 1973; Goss et al., 1975; Houbolt and Wells, 1980; Vacquier et al., 1988; Brigaud et al., 1990; Demongodin et al., 1991; Griffiths et al., 1992). Errors of 10–15% generally seem to be expected by these authors. Such values compare well with error estimates for thermal conductivities derived from measurements and averaged over rock sequences, which are reported as ca. 10% (Sass et al., 1971; Chapman et al., 1981; Andrews-Speed et al., 1984). Leadholm et al. (1985) computed thermal conductivities from seismic interval velocities. Fjeldskaar et al. (1990) included a large thermal conductivity data base in their model, and predicted thermal conductivity (horizontal and vertical) as a function of mineralogy and porosity. An independent study, which compares the different methods of determining thermal conductivities from well logs and seismic sections with each other and with measured data would be most valuable.

Modelled subsurface temperatures are often more influenced by errors in virgin rock temperatures than by errors in thermal conductivity. The sensitivity to thermal conductivity increases as the difference in depth and lateral distance between the area of interest and the calibration point increases. Accurate knowledge of thermal conductivities is thus most important in areas with laterally varying lithologies and few high-quality temperature data.

Modelled thermal histories are highly sensitive to the estimated eroded thickness in areas where the rocks are not presently at their maximum depth of burial. Several methods for determination of eroded thickness have been suggested, including analyses of apatite fission tracks, shale compaction, and maturity indicators. Unfortunately, few studies that aim at quantifying the uncertainty of these methods seem to have been published.

The thermal effects of crustal stretching have been given much attention in the literature. However, these effects have little importance to the calculated hydrocarbon generation when the source rock is at its maximum temperature. The sensitivity of the calculated transformation ratio to palaeoheat flow is, in most such cases, significantly less than the sensitivity to the virgin rock temperature. Only when the palaeoheat flow influences the maximum temperature that the source rocks have experienced is palaeoheat flow a major source of uncertainty. However, in most cases where the source rocks are not presently at their maximum temperature, this is so because of erosion, and the sensitivity to the eroded thickness generally dominates over the sensitivity to the heat flow model. Accurate knowledge of palaeoheat flow may however be crucial in areas where the timing of hydrocarbon generation (e.g. relative to cap rock formation) is important.

In summary, independent assessments of input parameter accuracy have not been sufficient in the thermal modelling of sedimentary basins. Such assessments seem necessary to further improve the accuracy and predictive capability of thermal modelling.

Modelling fluid flow and overpressuring

Historical development

An overview of the development of fluid flow modelling was given earlier in this paper. While the flow of fluids may play a subordinate role in transporting thermal energy, its interaction with other parameters makes fluid flow modelling important to basin evaluation.

Lewis and Rose (1970) noted that fluid overpressuring acts against compaction, resulting in less decrease of porosity in the overpressured formations. As the thermal conductivity of water is lower than that of rock-forming minerals, overpressured rocks will constitute thermal barriers, leading to the blanketing of heat below overpressured rocks. These effects imply that a proper handling of fluid pressure development through time is important for the calculation of thermal history.

The causes of overpressuring have been debated at length. Barker (1972) introduced aquathermal pressuring as a cause of overpressure under some conditions. This concept was later discussed by Daines (1982), Sharp (1983), Bethke (1985), Shi and Wang (1986), and Mudford (1990). It is now generally agreed that aquathermal pressuring plays a subordinate role; Bethke (1985) computed the contribution to the total overpressuring to be less than 1%. The importance of compaction disequilibrium as a main source for overpressuring is thus well established. The pressure effects of hydrocarbon generation (Spencer, 1987; Mudford, 1990), and oil to gas cracking (Barker, 1990) can be quite significant in some cases.

State of the art

All basin modelling programmes which simulate fluid flow seem to apply Darcy's law in the calculations. Shale permeability is conventionally adjusted to match fluid pressures in one or several wells. Permeability is modelled as a function of porosity, in most cases with one free parameter which determines the order of magnitude of the permeability, and one parameter which determines how permeability is modified by porosity changes (Bethke, 1985). Such formulae can be adjusted to match the fluid pressure in a rock at two different depths (for a given lithology), but simulations with these formulae do not guarantee correct permeabilities at other depths (or back in time).

Doligez et al. (1986) applied the Kozeny-Carman relationship between porosity, permeability and specific surface area (S) in all rock types. This formula is well established for clean sands, but has not been verified for low-permeability rocks such as shales. The application to shales requires an estimate of S, which seems to be the only free parameter in the adjustments to fluid pressures. Ungerer et al. (1990) modified the Kozeny-Carman equation for porosities less than 10%, presumably to achieve a more significant permeability decrease for low-porosity rocks.

The relative importance of aquathermal pressuring, hydrocarbon generation and oil to gas cracking can readily be examined in fluid flow models which include water compressibility and hydrocarbon generation and cracking. These topics were included in the models of Lerche and Glezen (1984) and Doligez et al. (1986). The modelled effects of hydrocarbon generation and cracking on fluid overpressuring depend a.o. on solubilities of oil, gas and water in each other and the modelled compressibility of these three (saturated) components. The way hydrocarbon generation and primary migration is modelled also influences the modelled overpressures, as do the formulae which describe the development of porosity and permeability. The modelling of overpressuring and its causes thus requires careful sensitivity analysis before conclusions of overpressuring in real sedimentary basins can be achieved.

Unresolved questions

Neither the porosity vs. effective stress nor the permeability vs. porosity equations which are used in basin modelling have been theoretically verified. Jacquin (1965) investigated the applicability of Darcy's law to shales by flow experiments, and concluded that Darcy's law does not describe the observed fluid flow. The possible fracturing of overpressured shales (Meissner, 1978; Palciauskas and Domenico, 1980; du Rouchet, 1981) is not accounted for in Darcy's law. This obstacle can however be overcome by implementation of separate routines for fracture simulation (Ungerer et al., 1990).

Other processes which take place in shales, but which are generally not considered in today's basin modelling programmes, include osmosis (Magara, 1978), clay mineral transformations (Burst, 1969), and electrokinetic effects and structural changes introduced by the moving fluid (Jacquin, 1965). The actual influence of these processes is poorly known, and modelling of them may be unjustified at the present stage of knowledge. Their existence nevertheless suggests that modelling with Darcy's law alone may not be sufficient to accurately describe fluid flow in shales.

An experimental control of Darcy's law in overpressured shales seems hard to obtain because of the low fluid velocities that operate in such rocks. A thorough check against fluid pressures in sedimentary basins might, however, reveal the appropriateness of this law. Strong support for the validity of Darcy's law would be evident if a given permeability formula could reproduce the fluid pressure in an overpressured shale in both basinal and marginal settings over a range of subsidence histories. The free parameters in the formula should either be identical for all calculations, or they should be derived independently, e.g. from laboratory analyses of the specific surface area of clay particles. The measurements of the rate-limiting specific surface area in an overpressured shale are very hard to obtain, and no study which gives reliable fluid flow parameters independent of measured subsurface pressures has yet been reported.

The failure of matching various fluid pressures in sedimentary basins with only one set of shale parameters does not, on the other hand, invalidate Darcy's law in shales. Different shale formations may have different mechanical and chemical properties, and may require separate treatment. The adjustment to fluid pressures in one- (or even two-)dimensional models may also not be appropriate, as fluid flow in general takes place in three dimensions, and may be significantly influenced by circumstances beyond those included in the models (such as faults and diagenetic changes).

The present status of the handling of overpressure in basin models thus seems to be that Darcy's law is trusted quite blindly, although little evidence has been put forward to validate this law in shales. The interpretation of modelled fluid overpressures should accordingly be made with caution.

Modelling hydrocarbon generation

Historical development

Many theories for oil and gas generation have been put forward. More than 60 years ago, Athy (1930b) noted that "the geological literature is replete with papers concerning the origin, migration and accumulation of oil". The number of papers published on these subjects has accelerated since then. Aspects of the historical development of the different concepts of hydrocarbon generation have been described by Waples (1984), Yukler and Kokesh (1984), Ungerer et al. (1988), Lerche (1990a), Ungerer (1990) and Hunt et al. (1991). A short summary of the main efforts is as follows.

White (1930) has been credited as the first to notice that prolonged exposure time could substitute for high temperatures in petroleum generation. The importance of both time and temperature to hydrocarbon generation was further discussed in the mid-1960's (Habicht, 1964; Philippi, 1965). Karweil (1955) established a chemical-kinetic model for coalification, and Bostick (1973) modified this method to relate thermal history to vitrinite reflectance.

Lopatin (1971) developed a time–temperature index (TTI) to describe coalification in the Ruhr district of Germany, and subsequently applied this method to petroleum source rocks to follow the maturation of kerogen (Hunt et al., 1991). His method, in which a 10°C increase in temperature gives the same contribution to maturity as a doubled exposure time, was made popular in the west by Waples (1980), who related the TTI values to hydrocarbon generation through correlation with vitrinite reflectance and other thermal indicators.

The influence of heating time is pronounced in the Lopatin (1971) method. Hood et al. (1975) developed a method in which only the time when the source rock was within 15°C of its maximum temperature ("effective heating time") was considered. Price (1982, 1983) compiled vitrinite reflectance data for sedimentary basins presently at maximum temperature, and could not find a correlation between increasing R_o and increasing burial time for any temperature interval. He therefore concluded that heating time had no observable effect on hydrocarbon generation, and that vitrinite reflectance could be used as a palaeothermometer.

Several different correlations between time–temperature history and vitrinite reflectance have been suggested since 1975, and modified TTI relationships have also been suggested (Royden et al., 1980; Lerche et al., 1984; Ritter, 1984; Wood, 1988) (Fig. 3). The suggested TTI values for a vitrinite reflectance of 1.0 (peak oil generation according to Waples, 1980) range between 65 and 450. As a doubling of TTI corresponds to a 10°C increase in temperature, the TTI range from 65 to 450 corresponds to a temperature difference of close to 30°C to the estimated temperature of peak oil generation (a doubling of 65 three times gives 520, so a difference in TTI between 65 and 520 corresponds to a temperature difference of 30°C).

Figure 4 shows different suggested combinations of time and temperature that have been proposed as sufficient for the start of oil generation. This figure is based on the assumptions of a constant heating rate and a surface temperature of 20°C. The range of suggested temperatures for a heating period of 10 m.y. is once again close to 30°C.

The assumed correspondence between vitrinite reflectance and oil generation, which is the basis of most of the TTI models, is attractive from an explorationist's point of view. In principle, one would only need to establish a relationship between measured vitrinite reflectance and depth in a basin, and then use this relationship to find the depth to the oil window, through Waples' relationships between R_o and stages of hydrocarbon generation. Several relationships between measured vitrinite reflectance and depth have been suggested for areas in the North Sea (Fig. 5).

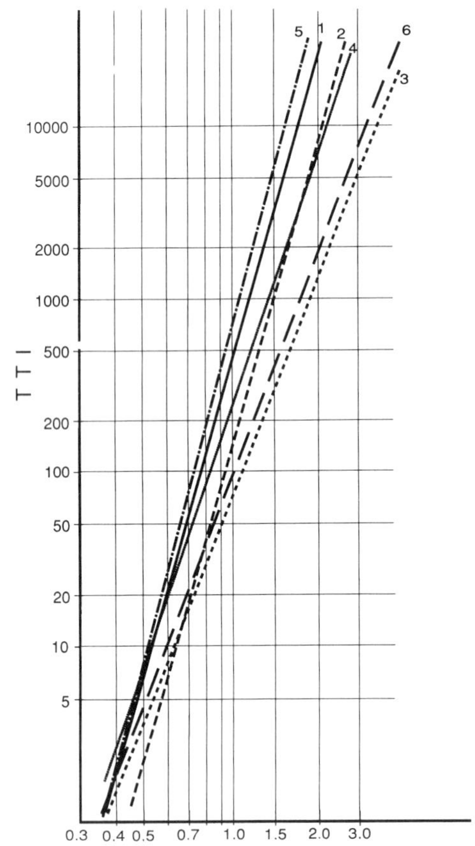

Fig. 3. Suggested relationships between vitrinite reflectance and the time–temperature integral TTI, based on a linearisation of equivalent data. *1* = Hood et al. (1975); *2* = Royden et al. (1980); *3* = Waples (1980); *4* = Goff (1983); *5* = Issler (1984); *6* = Dykstra (1987). Calculated by C. Cornford, IGI Ltd., using Platte River's BasinMod programme; reproduced by permission.

The differences in measured vitrinite reflectance vs. depth which emerge from Fig. 5 are alarming. The depth to R_o = 0.5 differs from 1.5 to 3 km, which is equivalent to a temperature difference of about 50°C in this area. Linking hydrocarbon generation stages to vitrinite reflectance trends vs. depth in sedimentary basins seems to have a limited accuracy potential, given the uncertainty in measured vitrinite reflectance which is reflected by Fig. 5.

The assumed one-to-one correspondence between vitrinite reflectance and hydrocarbon generation which is the basis for the TTI models has not been verified. Yukler and Kokesh (1984) suggested that the correspondence of oil generation to vitrinite reflectance depends on the heating rate. Figure 6 shows how Yalchin and Welte (1988) assumed that peak oil generation takes place in a R_o range of 0.6 to 0.9 for heating rates ranging from 1 to 6°C/m.y. Burnham and Sweeney (1989) and Larter (1989), who developed kinetic models based on chemical considerations, concluded that the mean activation energy for vitrinite maturation near the oil window is close to that for oil generation. The vitrinite reflectance models of Burnham and Sweeney (1989), Larter (1989) and Sweeney and Burnham (1990) were verified by comparing simulated vitrinite reflectance (based on published thermal histories) with measured vitrinite reflectance. The thermal histories which were used for the modelling are generally poorly constrained (Vik and Hermanrud, this volume), and hardly verify the models to a higher accuracy than ±500 m.

Tissot (1969) and Tissot and Espitalié (1975) proposed to simulate the degradation of kerogen to hydrocarbons by parallel, first-order, kinetic reactions. The application of kinetic modelling to the simulation of hydrocarbon generation has since then become widely used, and is summarized by Ungerer (1990, and this volume).

Tissot and Welte (1978) gave six activation energies and frequency factors for three kerogen types. The application of the activation energies of Tissot and Welte (1978) requires that the kerogen is classified as belonging to type I, II or III, based on the O/C and H/C ratios (Tissot et al., 1974). The success of this method depends on the uniqueness of the kerogen classification and on the concept that kerogens of a given type require equal amounts of thermal stress to degrade. The latter condition has been strongly opposed by Lewan (1985, 1989) and Hunt et al. (1991), who investigated the effect of sulphur content on kerogen transformation. Hunt et al. (1991) claim that variations in response to thermal stress within type II kerogen are larger than the differences between the different kerogen types as proposed by Tissot and Welte (1978).

The individual variations within a given kerogen type suggest that activation energies and frequency factors should be determined for each separate source rock. Ungerer (1984) suggested that kinetic parameters for immature source rocks could be determined by non-linear least squares optimization of pyrolysis experiments performed at different heating rates. His technique can be used to find kinetic parameters for an arbitrary number of channelled reactions. This method of determining kinetic parameters has become popular, and is in most cases used to determine 6–18 channelled reactions (each with a specific activation energy).

The optimization technique of Ungerer (1984) generally gives a very good fit between modelled and measured pyrolysis yields. The number of free parameters in the optimization is, however, quite large (one parameter, E_a, for each reaction plus the common frequency factor A_0). A good fit to pyrolysis data comes therefore as no surprise. This fit does not, however, guarantee an accurate simulation of hydrocarbon generation under natural conditions.

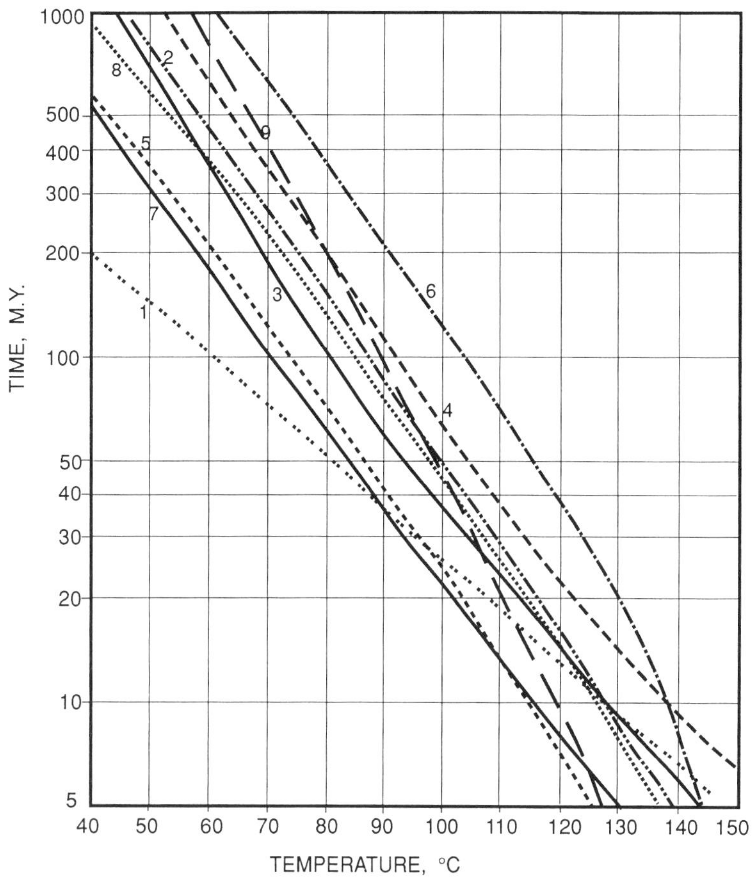

Fig. 4. Time–temperature isolines for the onset of oil generation according to various authors. *1* = Karweil (1955) (R_o = 0.5); *2* = Lopatin (1971) (TTI = 6.17, R_o = 0.5); *3* = Connan (1974); *4* = Hood et al. (1975) (LOM = 9); *5* = Waples (1980) (TTI = 3, R_o = 0.5); *6* = Waples (1980) (TTI = 15, R_o = 0.65); *7* = Yang (1981) (Transformation Ratio = 6%); *8* = Falvey and Deighton (1982) (R_o = 0.5); *9* = He (1985) (R_o = 0.5). From He (1985), used by permission.

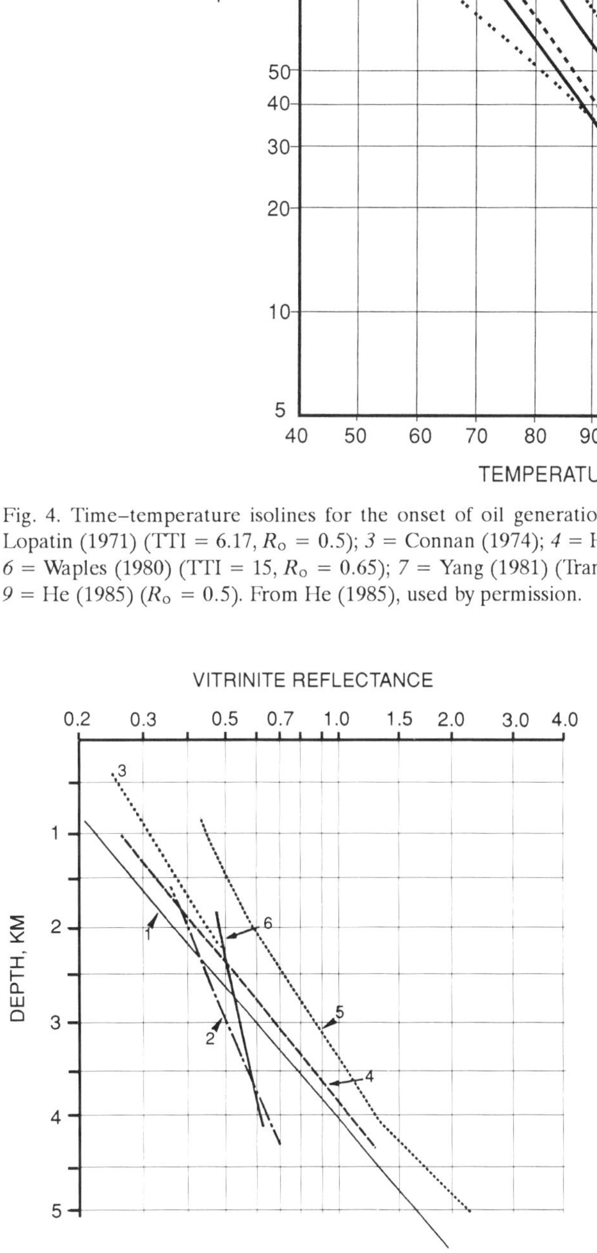

Fig. 5. Measured R_o vs. depth trends from the North Sea as suggested by different authors. *1* = Goff (1983) (60–62°N); *2* = Baird (1986) (56–60°N); *3* = Jensen et al. (1985) (Well U-1, Danish North Sea, 55°30′N); *4* = Leadholm et al. (1985) (56–60°N); *5* = Leadholm et al. (1985) (60–62°N); *6* = Maragna et al. (1985) (Block 33/6, 61°30′N). Compiled by C. Cornford, IGI Ltd., used by permission.

Both experimental shortcomings and extrapolation to geological time may introduce inaccuracies which may not be identified during optimization of kinetic parameters to laboratory analyses.

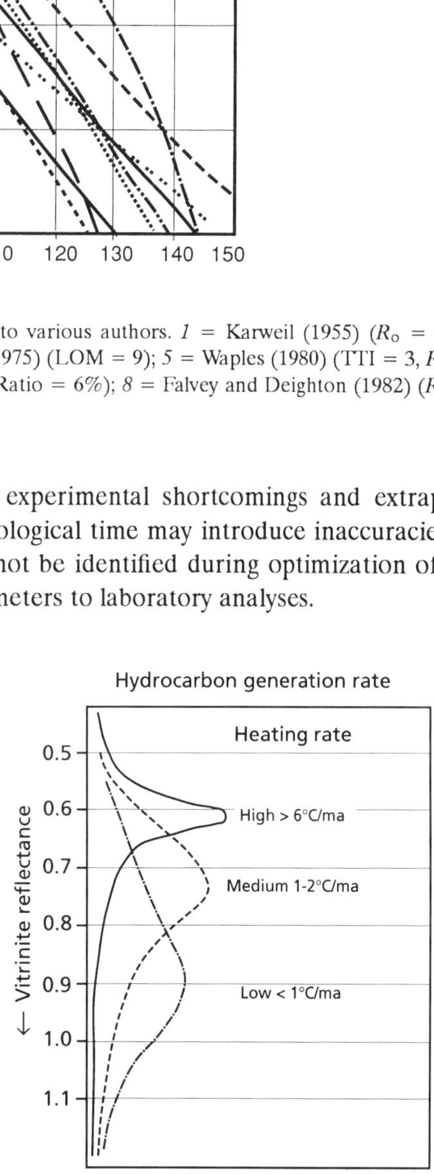

Fig. 6. Suggested shift of the oil generation peak on the maturity (vitrinite reflectance) scale due to different heating rates. Adapted from Yalcin and Welte (1988).

The number of free parameters in the optimization of pyrolysis data can be reduced to three by assuming a distribution of activation energies, and by using the pyrolysis data to determine the mean activation energy, its standard deviation, and the frequency factor. This technique was applied to coal pyrolysis by Pitt (1962). Burnham and Braun (1985) determined a distribution of activation energies for the Green River Shale, and Braun and Burnham (1987) gave a rigorous fitting procedure for the determination of the activation energy distribution from pyrolysis experiments. Since then, a large number of studies described the determination of kinetic parameters of various source rocks with different equipment and under different heating conditions (Burnham et al., 1987; Sweeney et al., 1987; Horsfield et al., 1989; Burnham and Braun, 1990; Klomp and Wright, 1990; Burnham, 1991; Burnham et al., 1991; Reynolds et al., 1991).

Further refinements of the kinetics models include modelling of the compositional characteristics of hydrocarbons (Sweeney et al., 1987; Ungerer et al., 1988; Behar et al., 1991; Forbes et al., 1991). The model of Forbes et al. (1991) describes generation of, and cracking between, the component groups C_1, C_2–C_5, C_6–C_{14}, C_{15+} and coke. This scheme has been adapted by several commercial basin modelling programmes. Figure 7 visualizes the main developments within hydrocarbon generation modelling during recent decades.

There seems to be little doubt about the main mechanisms and the dominant temperature intervals for hydrocarbon generation among basin modellers. The suggestions of juvenile hydrocarbons (Giardini and Melton, 1983), of hydrocarbon generation unrelated to temperature or burial depth (Petzoukha, 1987) or of the biogenic origin of most oils (Wilson, 1990) have not been picked up by the basin modelling community. The predictive success of basin models which do not invoke such sources for hydrocarbon generation argue against the abundant presence of oil of juvenile, stress-induced or biogenic origin.

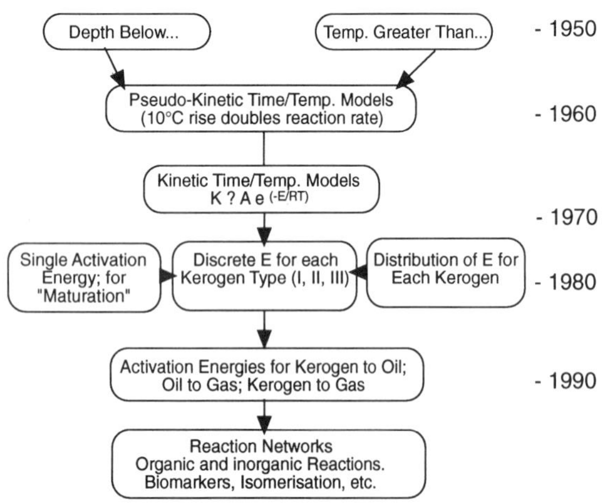

Fig. 7. Historical progress in concepts of maturity modelling. The time scale refers to the development and first publication of each concept — note it often takes from 15 to 25 years for a concept to be accepted and applied by the oil industry. Prepared by J.E. Leonard, Platte River Associates Inc. and C. Cornford, IGI Ltd.; reproduced by permission.

State of the art

While the scientific community generally tends to agree that the kinetic modelling of hydrocarbon generation is superior to extrapolation of vitrinite reflectance trends, most papers which discuss the hydrocarbon potential of sedimentary basins still refer to the oil window, or the zone of peak oil generation, and relate the depth interval in question to vitrinite reflectance. The main reason for the continued use of vitrinite reflectance and TTI modelling may be the general time lag between research and applications, as is seen in most sciences. Another reason may be that the TTI/R_o approach is very easy to use, and most basin modelling programmes were far from being user-friendly in the early and mid-1980's. The TTI/R_o approach will probably lose some of its popularity in the future for at least two reasons. Firstly, Wood (1988) presented a time–temperature index based on Arrhenius' equation, and Hunt et al. (1991) demonstrated how this equation could be solved graphically. This graphic procedure is as easy to use as the TTI method, but is more flexible, and incorporates many of the advantages of the kinetic models. Secondly, basin modelling programmes are becoming easier to use, and have been acquired by most major oil companies.

A number of papers published by researchers from British Petroleum (BP) addresses hydrocarbon generation and migration (Cooles et al., 1986; England et al., 1987; Mackenzie et al., 1987; Mackenzie and Quigley, 1988; Quigley and Mackenzie, 1988; England and Mackenzie, 1989). These papers focus on quantifying the amounts of oil and gas which reach a trap through time. Kerogen is classified as being either labile (oil prone, gas prone at high maturity) or refractory (gas prone). Gaussian distributions of activation energies describe the degradation of the two kerogen types to oil and gas, and also describe cracking from oil to gas. The BP approach, with subsequent calculations of gas/oil ratios, has been followed by several oil companies and also by vendors of commercial basin modelling software.

Most commercially available basin modelling software offers the choice between a few standard sets of kinetic parameters. The calculation of gas generation (either as a product of oil cracking (Tissot and Espitalié, 1975), or as a combined product of cracking and primary generation from kerogen (Sweeney et al., 1987; Espitalié et al., 1988b)) is offered in many commercial basin modelling programmes. Compositional modelling of different hydrocarbon types as demonstrated by Forbes et al. (1991) has been included in the last few years, and may become increasingly popular as advances in kerogen typing proceed.

Unresolved questions

Several different techniques are applied to determine kinetic parameters for hydrocarbon source rocks. Different results may arise from experiments on isolated kerogens and on whole rock samples. Open and closed (aqueous and dry) pyrolysis experiments, too, give different results (Düppenbecker and Horsfield, 1990). Individual components that are released by pyrolysis often have chemical compositions which differ from naturally occurring hydrocarbons (Lewan, 1985, 1989), possibly because the intermediate bitumen stage is not properly accounted for in open pyrolysis systems (Hunt et al., 1991). Pressure may also have a significant effect on hydrocarbon generation under natural and laboratory conditions (Dominé, 1991). Accurate monitoring of the bulk chemical hydrocarbon composition as a continuous function of aqueous pyrolysis has not yet been accomplished.

The mismatch between chemical composition of naturally and artificially matured kerogen indicates that the pyrolysis results may not be totally relevant to reactions in nature, and that therefore the kinetic parameters which are determined from pyrolysis do not apply strictly under geologic time. It can also be questioned if the S_2 yield from pyrolysis of an immature sample gives a valid estimate of the initial hydrocarbon potential, as non-hydrocarbons are also included in the S_2 signal.

The application of numerous discrete activation energies, and the measurement of separate activation energies for each new source rock is only warranted if the inaccuracy of the parameters to be determined is moderate compared to the differences which are obtained by changing from a simple to a more advanced approach. Investigations of the accuracy of kinetic parameters for kerogen degradation conclude that uncertainties of 300–600 m or more may be expected in the calculated depth to a transformation ratio of about 0.5 (Lakshmanan et al., 1991; Nielsen and Dahl, 1991; Espitalié et al., this volume). These uncertainties will hopefully be reduced as research on source rock kinetics proceeds, and as criteria for visual identification of unreliable pyrolysis data are verified and published. The present large uncertainties in kinetic parameter determination still raise serious doubts about the appropriateness of determining source rock kinetics based on pyrolysis experiments for individual source rocks. The application of a multitude of parallel reactions as opposed to a statistical distribution of activation energies (or only three discrete classes, as discussed by Liu and Lerche, 1990) may likewise be questioned.

Verification of modelled transformation ratios cannot easily be achieved. The progressive increase of S_1 and corresponding decrease of S_2 vs. depth for a given source rock would give independent information on the transformation of kerogen under geological conditions. Source rock samples from a large number of wells at different locations in the basins would be required for such a verification, and the thermal history of the area would have to be determined with high accuracy. Several verification studies should be performed before a global verification of the procedures for kinetic parameter determination can be established.

Modelling primary migration

Historical development

The mechanisms responsible for the expulsion of hydrocarbons from a source rock have been, and still are, widely debated (Doligez, 1987; Durand, 1988; Stainforth and Reinders, 1990). The modelling of primary migration is, accordingly, limited by the lack of knowledge of the operating processes. Expulsion modelling has therefore to a large degree been based on empirical relationships. The validity of these relationships should be judgoubts about the appropriateness of determining source rock kinetics based on pyrolysis experiments for individual source rocks. The application of a multitude of parallel reactions as opposed to a statistical distribution of activation energies (or only three discrete classes, as discussed by Liu and Lerche, 1990) may likewise be questioned.

Verification of modelled transformation ratios cannot easily be achieved. The progressive increase of S_1 and corresponding decrease of S_2 vs. depth umber of micelles. Migration of gas solubilized in oil (Nogaret, 1983; Price et al., 1983; Heum et al., 1986) and oil in gaseous solution (Leythaeuser and Poelchau, 1991) may however be important in some cases.

Tissot and Pelet (1971), Meissner (1978) and du Rouchet (1981) suggested that microfracturing of shales was important to primary migration. This view was supported by Mandl and Harkness (1987), who developed a theoretical basis for hydrocarbon migration by hydraulic fracturing. Littke et al. (1988) observed that bitumen-filled fractures occurred in the Posidonia shale at maturities of R_o of 0.68 and above, but not at lower maturity. Talukdar et al. (1988) observed bitumen in microfractures in the La Luna source rock of the Maracaibo Basin, and Leythaeuser et al. (1988) identified compositional gradients away from fractures, and attributed these to preferential depletion of the fractured zones. The role of fracturing as an operating factor in primary migration thus seems well established.

Leythaeuser et al. (1982) suggested that diffusion of hydrocarbons plays an important role in the primary migration of gas, but not of oil. Krooss (1988) performed experimental investigations on the primary migration of light hydrocarbons, and concluded that a combination of first-order disorption and molecular diffusion could account for the experimental results.

McAuliffe (1979) suggested that primary migration takes place as a pressure-driven process through kerogen networks. Stainforth and Reinders (1990) suggested that primary migration is rate-limited by diffusion through organic matter networks. This hypothesis was also discussed by Thomas and Clouse (1990a, b, c). They found that diffusion through organic matter networks was about a hundred times more efficient than diffusion through a water-filled porous rock. The calculated fluxes over geologic time are sufficient to source small and moderate hydrocarbon accumulations, but not large or rapidly formed hydrocarbon accumulations. Pepper (1991) suggested that the distribution of kerogen in the source rocks exerts the prime control on expulsion behaviour by retaining generated petroleum, and demonstrated that geochemical observations match favourably with this expulsion mechanism.

Primary migration has also been postulated to take place as pressure-driven flow governed by Darcy's law for multi-phase flow (Ungerer et al., 1984). As relative permeability for oil is close to zero for low oil saturations, a significant portion of the free pore volume has to be oil-saturated before hydrocarbon expulsion is efficient. The free pore volume may be small because of a low shale porosity at the depths of hydrocarbon expulsion, or significant portions of the pore water may be bound to the mineral surfaces, and therefore not be available for movement out of the source rock (Magara, 1979).

State of the art

Primary migration seems to be a result of several processes, and their relative importance in different geological settings is still elusive. This situation has led to the emergence of empirical methods for the handling of primary migration in shales. Welte and Yukler (1981) suggested the application of (constant value?) expulsion efficiencies of 10–20%, and Welte (1987) and Mackenzie and Quigley (1988) suggested saturation thresholds as fixed fractions of the pore volume. Hermanrud et al. (1990a) suggested that retention by primary migration should be calculated as a fraction of the bulk rock volume. This fraction should be comparable to the S_1 value from pyrolysis, which with all its limitations is supposed to give an estimate of the mass of free hydrocarbons in the source rock.

The commercially available basin modelling programmes that include hydrocarbon expulsion tend to favour the concept of porosity-dependent expulsion efficiencies. The TEMISPACK and Sigma-2 models appear to be the only models listed in Table A-1 which rely on two-phase fluid flow modelling and relative permeability curves for expulsion calculations.

Unresolved questions

The concept of expulsion efficiencies (expelled hydrocarbons/generated hydrocarbons) is not very helpful alone in predictive modelling, as it varies from 0 to at least 88% (Espitalié et al., 1988a) as a function of maturity for a rich source rock. Additional information on how this variation takes place as a function of maturity is thus vital for the simulations. The use of saturation thresholds removes the problem with maturity dependence inherent in the expulsion efficiency concept. The calculated retention of hydrocarbons from saturation thresholds is, on the other hand, dependent on the determination of shale porosity, which may well be in error by several hundred percent.

It should be noted that Doligez et al. (1986) and England et al. (1987), who modelled primary migration as pressure-driven flow of hydrocarbons in a separate phase, predict lower porosities vs. depth than most other authors, as is clear from Fig. 1. These porosity curves are required to avoid modelling of unrealistically large losses of hydrocarbons in the source rocks. The porosity curves of Doligez et al. (1986) and England et al. (1987) are, however, vaguely documented, and their applied relative permeability curves are independent of grain size. Okui and Waples (this volume) suggest that the relative permeability curves should be scaled for grain size, and demonstrate that such scaling gives satisfactory expulsion modelling with porosities which are significantly higher than those of Doligez et al. (1986) and England et al. (1987). The reported success of expulsion modelling with these different relative permeability and porosity relationships has not however documented that hydrocarbons flow through the pore space of the entire source rock. The general scepticism about the use of Darcy's law in shales

(Jacquin, 1965) also applies to modelling of primary migration.

The appropriateness of the concept of pressure-driven primary migration also has an important influence on the calculated directions of primary oil migration. Hydrocarbons will move away (upwards and/or downwards) from a hydraulic head maximum according to this hypothesis. This would, for example, force all generated hydrocarbons to move downwards if the source rock was overlain by a fluid flow barrier. This flow pattern is not a necessary consequence of diffusion of organic matter through kerogen networks. A more complete understanding of primary migration is thus of significant importance to the oil industry.

Modelling secondary migration and leakage

Historical development

The physical principles governing secondary migration (interplays between buoyancy, water movement and capillary pressure) seem to be well established (Hubbert, 1953; Berg, 1975; Schowalter, 1979). The retention of hydrocarbons in carrier beds is nevertheless one of the least constrained parameters in quantitative basin modelling.

Calculation of sealing capacity (the maximum column of oil or gas that a cap rock can hold) requires accurate assessments of water pressure as well as of fracturing criteria and pore throats of the largest interconnected pores through the seal. These quantities can not be given with high accuracy, and only small errors in these quantities will lead to large errors in calculated trapping capacity. Efforts towards calculating leakage to aid hydrocarbon exploration have been made (e.g. du Rouchet, 1984), but seem to have a limited predictive potential.

Direct observations of secondary migration are few, but North (1985) refers to cases where oil has been observed to move with a speed of 40–100 m a month. Indirect evidence of secondary migration in the form of seeps and pockmarks is however common (Hovland and Judd, 1988). Evidence of migration distances or retained volumes during secondary migration has occasionally been given. Lateral migration distances of 100 km or even more were suggested by Dow (1977), Tissot and Welte (1978) and Bethke et al. (1991), and vertical migration distances to commercial hydrocarbon accumulations of 3 km or more have been advocated (Goff, 1983; Clayton et al., 1990).

Several different approaches have been applied to reveal the migration history of hydrocarbon accumulations. The relationship between hydrocarbon filling of reservoir rocks and reservoir diagenesis has been discussed by several authors (e.g. Schwarzkopf, 1990), and deductions about migration history have been made from fluid inclusions (Walderhaug and Fjeldskaar, this volume) and inspection of residual oil below the oil–water contact (Augustsson, 1992).

Few experiments on secondary migration have been reported (Franco, 1986; Dembicki and Anderson, 1989; Catalan et al., 1992). Frette et al. (1992) describe three-dimensional experiments of slow immiscible fluid–fluid displacement in a porous medium, and use a modified invasion-percolation algorithm to describe their results numerically.

Modelling of secondary migration was apparently first attempted by Nakayama and van Sichlen (1981). Further progress in such modelling is discussed below.

State of the art

Secondary hydrocarbon migration must be modelled in a two- or three-dimensional framework. This fact explains why very few basin models include secondary migration. The inclusion of secondary migration in a compacting, two- or three-dimensional, fully integrated basin modelling programme has been demonstrated by the main "schools" of basin modelling (Welte, 1987; Lerche, 1990a; Ungerer et al., 1990; Bethke et al., 1991).

Sylta (1991, and this volume) computed multi-component secondary migration as a function of buoyancy and rate of hydrocarbon supply in three dimensions, without considering water flow in his model. This model is based on the concept of pressure compartments (Buhrig, 1989). Hydrocarbons migrate freely by buoyancy within each such compartment, unaffected by water flow (from compaction-derived water), which is several orders of magnitude less than the hydrocarbon velocity. Movement between different pressure compartments is restricted, but hydrocarbons may move from one pressure compartment to another without leaving behind significant amounts.

A more pragmatic way of modelling retention by secondary migration was suggested by England et al. (1987). They assigned an average saturation to each carrier bed, and report that saturations between 1 and 10%, with a preferred value of 3%, seemed appropriate in most cases. The actual retention during secondary migration is governed by microtraps (both structural and stratigraphic) in the migration route, and is hard to quantify. The retention factor should presumably be modelled as large in an area with lateral stratigraphic variations and extensive faulting, and should be small in uncomplicated areas with a short distance between the source rock and the top of the carrier bed.

Retention during secondary migration is of great importance where the kitchen area is restricted or where migration over a large area is necessary to supply hydrocarbons to the trap. Retention factors of 2 and 5% respectively may in some cases imply that the calculated charge to a structure is changed from more than 100 million Sm^3 to zero. (S. Larter, pers. commun., 1991). The uncertainty of volumetric calculations is thus substantial. Hermanrud et al. (1990a) and Irwin et al. (this volume) give likelihood distributions for the input parameters in their volumetric calculations, and use a Monte Carlo simulation technique to describe the possible range of outcomes. Such stochastic modelling appears to be more appropriate than a purely deterministic approach for volumetric calculations at the present stage of knowledge.

The PVT behaviour of trapped hydrocarbons experiencing different temperature and pressure regimes was modelled by Heum et al. (1986). Such modelling has not so far been implemented in commercial basin modelling software, probably because of the difficulties in obtaining reliable input data. Calibration to reservoir compositions in the area of application is also a prerequisite for successful predictive PVT modelling of subsurface hydrocarbon compositions. The potential of such PVT modelling is best utilized in areas where sufficient amounts of data are available, and may become more widely used in the future (Ungerer, this volume).

Unresolved questions

The conditions for leakage from a reservoir have been given substantial attention, as this represents a major risk in hydrocarbon exploration. Several leakage criteria have been defined (e.g. Smith, 1966; Magara, 1981; Watts, 1987; Sales, this volume). The attempts to predict leakage from a reservoir in a quantitative way have been few. Leythaeuser et al. (1982) computed the diffusion of gas through a cap rock. The determination of diffusion coefficients at reservoir conditions is presently quite inaccurate, resulting in considerable inaccuracy in the simulated diffusion rates.

Ungerer et al. (1987a, b, 1990) modelled the fracturing of cap rocks due to reservoir pressure in the Haltenbanken area. Fracture criteria and subsurface fluid pressures are, however, not very well known prior to drilling. The modelling of fracturing without proper well control may thus not be warranted, but may be instructive where such information is available.

Constraining the hydrocarbon volumes that are retained in the carrier beds appears to be the main challenge for migration modelling in the future. Improved geological and mathematical descriptions of carrier beds seem necessary for a significant improvement in modelling accuracy.

Application of basin modelling to hydrocarbon exploration

The last part of this paper covers (1) the different approaches to application of basin modelling programmes which emerge from the literature, and (2) a discussion of unresolved questions within the application of basin modelling.

Modes of applying basin modelling

Modelling of the development of a sedimentary basin through time requires a general idea of the history of that basin. That history, often termed the conceptual model, serves as a framework for the input parameters which are applied in the modelling of a given area. As modelling results are produced these are to be checked against observations from wells. The degree of mismatch between modelling results and well data (temperatures, porosities, maturity, ...) determines if the conceptual model is considered to be validated or if it should be modified. Only when the conceptual model has been verified will the modelling results be used for predictive purposes in other places in the basin. Figure 8 shows this generalized scheme of progress of a basin modelling study, and Fig. 9 visualizes how a mathematical model can be constructed from a conceptual (geological) model. A detailed discussion of the optimization of the conceptual model was given by Waples et al. (1992a, b). Such discussions are unfortunately rarely included in papers which describe the application of basin modelling techniques.

The fit between observed and modelled parameters is generally the only measure of accuracy which is given from the modelling. The distinction between an acceptable and a non-acceptable fit to well data is generally made on a subjective basis, and accuracy limits of the modelling results are in most cases not given.

The sensitivity of exploration decisions to variations caused by the uncertainty ranges of the various input parameters differs substantially. A systematic search through the various input parameters will reveal which parameters contribute most to the total uncertainty, and which therefore need further attention. Systematic sensitivity analysis of modelling results was demonstrated by Cao and Lerche (1989), Hermanrud et al. (1990a), and Waples et al. (1992a, b).

Basin modelling studies may also be designed as numerical laboratory experiments, to focus on the

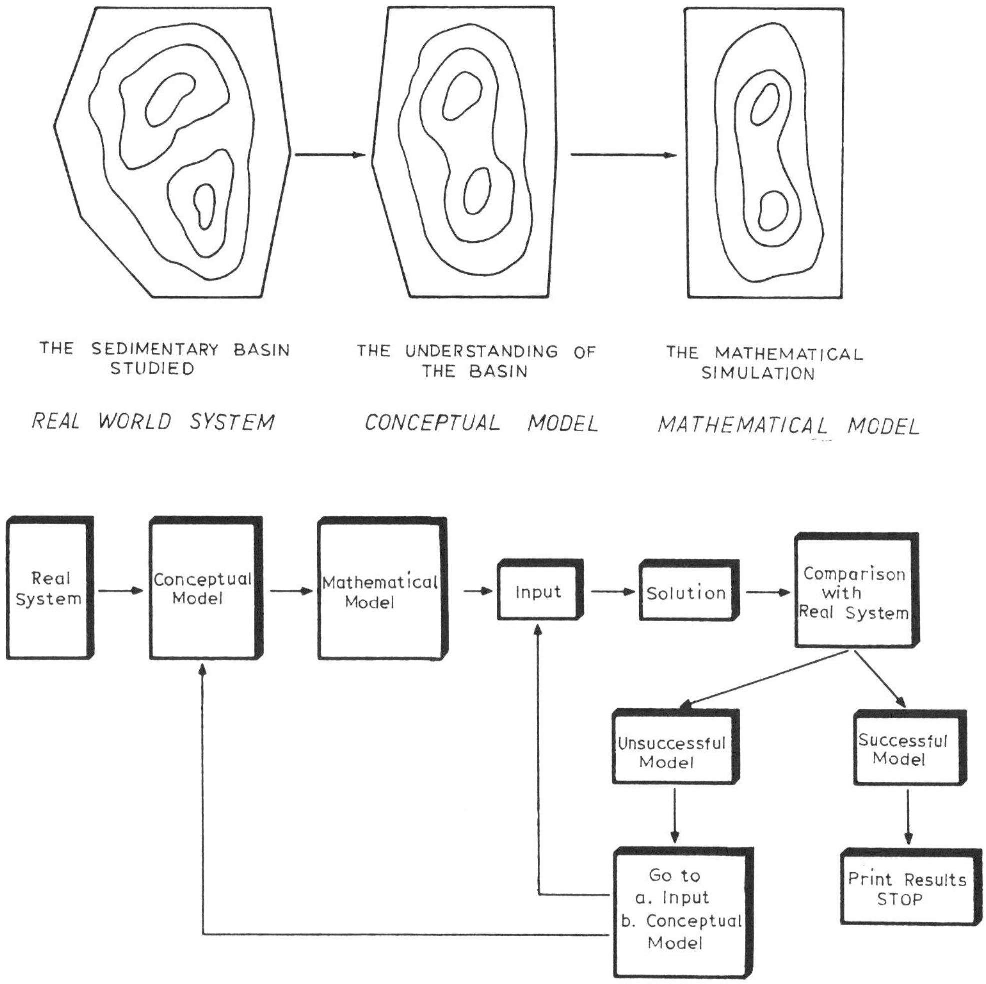

Fig. 8. Flow diagram of a basin modelling project. Adapted from Yukler and Kokesh (1984), and Welte et al. (1983); courtesy of Academic Press and the Geological Society of London.

effect of variations in one single parameter. Such studies can be valuable for hydrocarbon exploration, as demonstrated in the Egersund Sub-basin by Hermanrud (1989). In this case, a well location was to be chosen from two alternatives. The deepest, but smallest prospect (which also had the deepest kitchen) was considered to have the best chance for hydrocarbon charge in this marginally mature basin. One of the licence partners disagreed, and claimed that the 300-m shallower kitchen was hotter because of heat refraction. Two-dimensional basin modelling showed that no reasonable set of thermal parameters could yield sufficient thermal refraction. The well which was sourced from the deeper kitchen was drilled, and made an oil discovery. The optimism caused by this well resulted in drilling of the shallowest structure, which unfortunately turned out to be dry.

The modelling strategy in such studies is quite different from those where prediction of hydrocarbon generation (with maximum accuracy) is sought, as determination of the parameter range for one or a few parameters accounts for the main work load in these cases.

The net mass of solids (rock volume minus water volume) within each formation is chosen such that the lithological boundaries from a modelled well are reproduced by the model. The net solid mass which is required to match these lithological boundaries depends on the modelled porosity, and will have to be adjusted when the modelled porosity is changed. Here, no *a priori* value of net solid mass exists, and adjustments should be made to match well observations without putting any emphasis on the first guess of solid mass.

Calibration of input parameters is frequently made to better reproduce porosity, thermal conductivity, fluid pressure, density, temperature, maturity indicators, and fluid type in nearby structures. Calibration to these parameters should consider the implications to the conceptual model which is introduced by the new parameters (Waples et al., 1992a, b).

The matching of calculated thermal history to maturity indicators illustrates the different options dur-

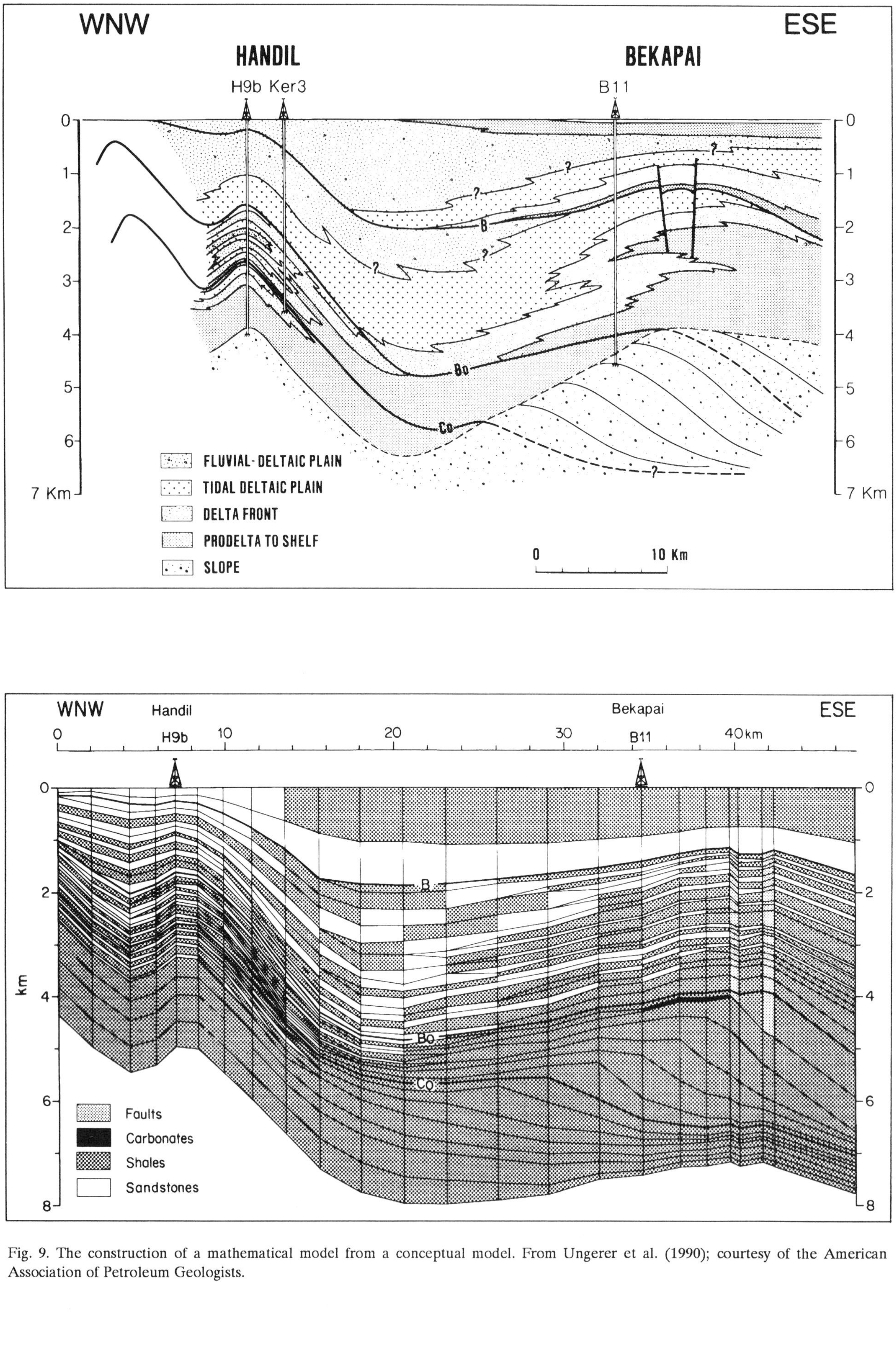

Fig. 9. The construction of a mathematical model from a conceptual model. From Ungerer et al. (1990); courtesy of the American Association of Petroleum Geologists.

ing the calibration process. Ritter et al. (1987) tested several heat flow histories derived from stretching models, but did not force the resulting thermal histories to give the exact same present-day temperature in the sediments. His approach attributes part of the misfit to inaccurate virgin rock temperatures determined from BHT measurements (quantified by Hermanrud and Shen (1989) and Hermanrud et al. (1990b)). Dahl et al. (1987) increased the heat flow by 15% in the last million years to get a correspondence between modelled and observed maturity parameters. Jensen and Doré (this volume) discussed the possible existence of a recent heat flow peak in the Haltenbanken area. The authors do not advocate any physical process which is responsible for such a heat flow peak, but note that a recent heat flow peak is necessary to match calculated vs. observed vitrinite reflectance. Vik and Hermanrud (this volume) model a recent heat flow low as a result of rapid sediment deposition in the same area, and claim that their model cannot be discarded on the basis of vitrinite reflectance data, although the fit between these data and modelled vitrinite reflectance is not perfect. Hermanrud et al. (1990a) did not calibrate their modelled thermal history to maturity indicators at all, because of the (at that time) improper knowledge of the uncertainty of these parameters. Waples (1993a, b) adjusts thermal conductivities within "geologically acceptable limits" to improve the fit between measured and modelled maturity.

This listing of the different choices which have been made in the calibration stage of modelling demonstrates the need for a systematic approach to calibration. At present, the individual modeller chooses one of several options, with significant implications for the final modelling results and, therefore, the assumed hydrocarbon prospectivity. Some suggestions for more systematic approaches have, however, been put forward, as will be briefly discussed below.

The basin modelling group at the University of South Carolina has advocated the application of thermal indicator tomography for some time (He and Lerche, 1989; Lerche, 1990a, b; and references therein). Such tomography methods are designed to find the combined sets of parameters that are most consistent with all well data. The parameters which are optimized include those which enter equations for alteration of maturity indicators. Figure 10 shows how the best parameter value (with error limits) of the linear heat flow vs. time equation is determined. As several of these optimized parameters should be independent of the actual well from which samples are taken, several wells with widely varying geological histories should in the ideal case be inverted

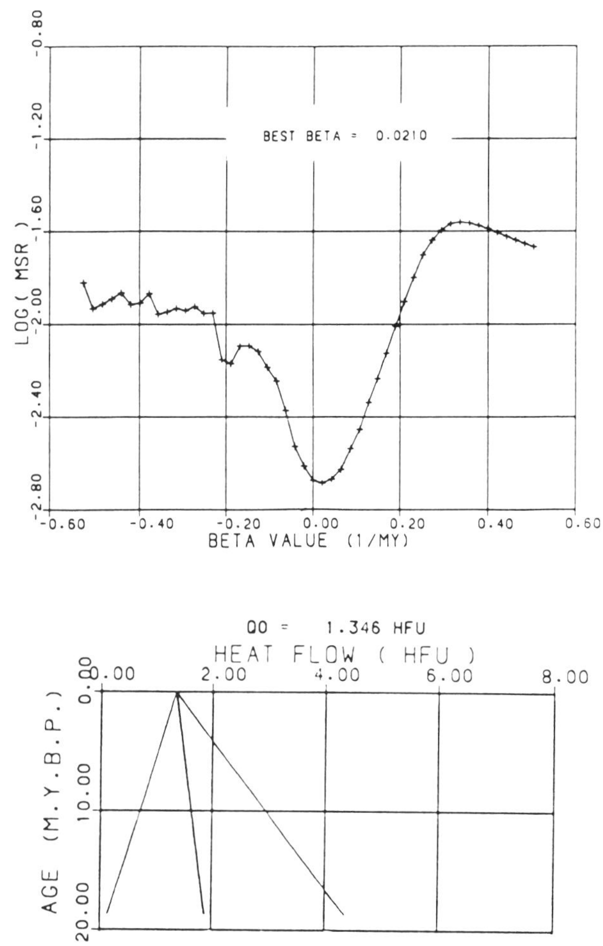

Fig. 10. Non-linear optimization of the beta value in the heat flow equation: Heat flow = heat flow at present × $(1 + \beta t)$, where t is time. The figure shows the mean square residual (MSR) vs. β, and the corresponding palaeoheat flow with error range. From Lerche (1990b); courtesy of Academic Press Inc.

simultaneously to derive the proper parameters. The *a priori* constraints and dependence among the different parameters which can be optimized have not yet been fully explored. At present, the tomography method secures a good fit between observations and modelling results. The application of the optimized parameters to new areas, and their uniqueness, have however not been validated.

An alternative approach, as suggested by Irwin et al. (this volume), requires the determination of an *a priori* confidence interval for a transformation ratio based on external information (maturity or hydrocarbon occurrence). According to this method, a modelled transformation ratio is either accepted or rejected, dependent on whether it falls inside or outside the preset limits. A rejected modelling result implies that some of the input parameters are not correct, and should motivate a reconsideration of these parameters (based on external information). If the outcome is accepted, the final (best estimate) transformation ratio should be derived from weighing the modelled transformation ratio with that extracted from external information.

It is in general not possible to identify *the* erroneous parameter when a modelling outcome is rejected. All input parameters have an uncertainty, and a combination of several parameters which are all within their uncertainty ranges may give an unacceptable result. To overcome this problem, Irwin et al. (this volume) created numerous input files with the help of a random number generator, which picked input parameters from the likelihood distribution of each parameter following the procedures of Hermanrud et al. (1990a). The results from modelling with all input files were tested against the rejection criteria, and only outcomes which were accepted according to these criteria were used in a predictive sense. This method was shown to give volumetric estimates with narrow error ranges close to the discovered hydrocarbon volumes in the Egersund Sub-basin. The main drawback of the procedure of Irwin et al. (this volume) seems to be the large consumption of computer time, which is especially important in two- or three-dimensional studies. This problem may however largely be overcome by linearization of the accuracy estimates (but not of the "best case" model), as suggested by Nielsen (this volume). The application of fuzzy arithmetic to this sort of problem may also become rewarding (Fang and Chen, 1990).

Unresolved questions

The various options for calibration of basin models document the lack of systematic planning of basin modelling studies. This lack of planning, and the general sparsity of sensitivity analysis within basin modelling, has led to a wide variety of different approaches to basin modelling applications. Personal experience with different oil companies on the Norwegian continental shelf and inspection of published modelling studies has shown that some modellers give high priority to modelling a large number of locations in the kitchen area. Others try a multitude of different modelling programmes or options within these programmes (such as kinetic schemes or formulae for maturity indicator conversion). Still others may put emphasis on applying advanced models (e.g. fully integrated 3D models), or on collecting and interpreting large amounts of data.

While all of these approaches have their merits, a systematic discussion of when the different approaches are appropriate is rarely seen. Investigations of the errors which are introduced by simplifying shortcuts are, as a rule, not reported. The determination of such errors is, however, crucial to the planning of a basin modelling project. If the modelling results are more sensitive to inaccurate input parameters (e.g. amount of eroded material), then simplified modelling (e.g. one-dimensional, steady state) may be adequate. Thorough sensitivity analysis seems to be necessary for the planning of a basin modelling study with optimum accuracy in the available time.

The results from basin modelling are supposed to aid management in oil companies to make drilling decisions. The most frequent way of presenting modelling results seems to be by giving numbers for source rock maturity or hydrocarbon generation. Such numbers may, however, not give the manager what he needs the most. The decision to drill or leave a prospect involves calculation of both geological and economic risk, and the allowed outcome ranges from basin modelling are of much larger importance to such evaluations than the single numbers which represent the most likely estimate.

The establishment of accurate error ranges for basin modelling results may not be achieved in the near future. Several (input and calibration) parameters do not have well defined error limits, and these can at best be determined only after thorough examination of large amounts of data. Still more elusive is the assignment of error ranges which aim at covering the uncertainty of understanding a given physical process, or to take account of processes which are presently not known.

The benefit of basin modelling to an oil company ultimately depends on how the results from the modelling are incorporated into the full process of prospect evaluation. This aspect has so far not been covered by the basin modelling literature, and definitely needs more attention in the future.

Conclusions

A substantial development has taken place within basin modelling since the establishment of the main "schools" in the early 1980's. A large number of programmes of different level of sophistication has been presented, the majority of which applies simplifying concepts compared to the programmes of the main "schools". The development of computer hardware and general software have led to substantial improvements in the user-friendliness of basin modelling programmes. More advanced modelling of hydrocarbon generation has been introduced as a result of the progress of science within this field. Progress has also been made in the understanding of basin geothermics, fluid flow, compaction and migration. These advances have not, however, so far resulted in significant modifications of the mathematical simulation of these processes.

Further development of basin modelling may be expected in the future. New items may be introduced as the understanding of subsurface physics proceeds, especially within expulsion modelling. The user front end will probably continue to develop, and direct access to other data (-bases) from basin modelling programmes will probably be a standard rather than an exception.

Continuously improving computer performance will make an integration of basin modelling programmes with other programmes (e.g. sediment deposition simulators) possible. Such integration would, however, make the modelling programmes less easy to operate, and should not be implemented until it has been clearly demonstrated that the integrations aid prospect evaluation. Processes which determine input parameters to basin modelling are, in general, most conveniently simulated in separate programmes. Only when the determination of these parameters is significantly influenced by the calculated development of burial, temperature, fluid flow, hydrocarbon generation or migration will the advantage of determining these parameters internally in the basin modelling programme be clear.

Automatic adjustment to well data such as subsurface temperature, fluid pressure and porosity may be included in future basin modelling programmes. Automatic sensitivity analysis and error analysis may likewise be included. Calibration of the thermal history to maturity indicators is, technically speaking, easy to include in basin modelling programmes. However, it is presently not clear how the best calibrated thermal history can be achieved. This problem will hopefully be solved in the future.

Finally, future basin modelling will probably give output results which can go more directly into routine prospect evaluation than is the case today. Whether the changes will be in the programmes for basin modelling or prospect evaluation (or both) remains to be seen.

Acknowledgements

Discussions with Alan Burnham, Ian Lerche and Eirik Vik have significantly contributed to the development of this paper. Antony M. Spencer and an anonymous referee are thanked for constructive suggestions. Jay Leonard, Jean Burrus, Jean-Claude De Bremaecker, Ian Lerche, Kazuo Nakayama and Tom Pedersen are thanked for giving away unpublished information about their programmes, and Statoil is thanked for permission to publish this paper. Special thanks go to Nina Korbu, Kristine Abelsnes and the staff at Statoil's library, without whose help the reference list would have been substantially shorter.

Discussion

P. Ungerer, IFP: *This morning, two important points have not been mentioned by the various speakers.*

(1) Most applications of thermal models do not include computation of heat transfer within the lithosphere. This means that we neglect the thermal inertia of the lithosphere, and consequently underestimate the time constants of return to the steady state regime, in case of rapid erosion or rapid sedimentation. Besides, an important proportion of the heat flow from the basement is due to radiogenic heat production within the lithosphere.

(2) Maturation models yield the chemical composition of generated petroleum, but the phase state of accumulations depends not only on chemical composition, but also on the pressure and temperature conditions. This must be accounted for by thermodynamic models, if we wish to model the type of accumulation (oil, gas with condensate, dry gas, etc.).

What is the opinion of the speakers about these points?

Author's reply: I fully agree about the first topic you have raised, and am happy to coauthor a paper in this volume which focuses explicitly on the transient thermal effects of rapid subsidence (Vik and Hermanrud, this volume).

I also agree on the second point which was made. Heum et al. (1986) modelled the PVT-effects on phase composition in the Haltenbanken area, and could explain most of the different phase relationships of the studied area by the modelling. Such modelling does, however, require fairly accurate hydrocarbon compositions and reservoir pressures as input to the modelling. PVT modelling can be a valuable exploration tool in a moderately to well explored area, but seems to give less information about reservoir phases in a virgin or poorly explored area.

Appendix — Properties of different basin modelling programmes

Several oil companies have developed their own basin modelling software. Variable amounts of information are publicly available about such programmes (Chia, 1985; Dutta, 1986; Nakayama, 1987; Mudford, 1990; Stainforth and Reinders, 1990; Hermanrud et al., 1991; Grigo et al., this volume).

A number of other basin modelling packages has been developed at public institutions in the U.S. and elsewhere (De Bremaecker, 1983, Rice University; Rosenblat, 1985, Illinois Institute of Technology, Chicago; Keith and Rimstidt, 1985, Virginia Polytechnical Institute; Haugerud, 1986, 1988, U. S. Ge-

TABLE A-1

Properties of several basin modelling programs currently in use

Vendor	Programme name	Coordinate system	Numerical method	Transient eq. in P and/or T.	#Dimensions in transient eq.	Porosity equation	Includes HC generation	Includes HC expulsion	k or K	Sale/Service	Transient T eq. in lithosphere	Thermal conductivities	Interface with other programs	Contact Address
Programmes developed for commercial purposes														
BEICIP-FRANLAB	Matoil (1)	L	D	T	1	Z, Em	+	-	-	Sa/Se	-	Li	+	232 Avenue Napoleon Bonaparte, 92500 Rueill-Malmaison, France
"	Genex (1)	L	D	T	1	Z, Em	+	+	-	Sa/Se	+	Li	+	"
"	Temispack	L	D	P,T	2	Tz, Em	+	+	k	Sa/Se	+	Li	+	"
GAPS Geol. Cons.	Bury	E	D	-	-	Z, Em	+	+	-	Sa/Se	-	Li	+	13 Deodar Rd., London SW15 2NP, UK
IES GmbH	PDI-1D	E	EL	P,T	1	Tz, C	+	+	K	Sa/Se	+	Li, En	-	Bastionstrasse 11-19, D-5170 Juelich, Germany
"	PDI-2D	E	D	P,T	2	Tz, C	+	+	K	Sa/Se	+	Li, En	+	"
"	PDI-3D (2)	E	D	P,T	3	Tz, C	+	+	K	Sa/Se	+	Li, En	+	"
"	PDI-2DM	E	EL	P,T	2	Tz, C	+	+	k	Sa/Se	+	Li, En	+	"
"	PDI/PC (1-D)	E	D	P,T	1	Tz, C	+	+	K	Sa/Se	+	Li, En	+	"
IFE	IFE 1D (3)	L	D	P,T	1	Tz, Em	+	-	k	Sa/Se	+	Li	-	P.O. Box 40, 2007 Kjeller, Norway
IKU	Semi/Baseval	-	Rt	-	-	Z, Em	+	+	-	Se	-	Li	-	7034 Trondheim, Norway
INTERA	Basin	-	Rt	-	-	Z,C	(+)	-	-	Se	-	En	+	Highlands Farm, Greys Rd., Henley-on-Thames, Oxon, RG9 4PS, UK
Japex	BSS [JPX]	E	D	-	-	Tz,Em	+	+	k	Sa/se	-	Li, En	-	2-17-22 Akasaka, Minatoku, Tokyo, Attn. K. Nakayama, Exploration Dept.
"	Geopet [JPX]	L	D	P,T	2	Tz,Em	+	+	k	Se	-	Li, En	-	"
Platte River	Basinmod	L	EL	P,T	1	Z, Em	+	+	k	Sa	+	Li, En	+	2000 West 120th Avenue, Suite 10, Denver, Co 80234, USA
PSTI	Hedera (4)	E	D	T	1	Z, Em	+	+	-	Sa/Se	-	Li	+	Proj. Hederaq, PSTI, Offshore Tech. Park, Exploration Drive, Bridge of Don, Aberdeen, AB23, 8GX, UK
PROCOM	BMT (5)	E	D	-	2	Z, Em	+	-	-	Sa/Se	-	Li, Mi	+	Rogaland Research, P.O.Box 2503 Ullandhaug, 4004 Stavanger, Norway
Yükler, M.A.	YüklerPC	E	D	P,T	1	Tz, C	+	+	K	Sa	+	Li	+	4108 Rainsong, Dallas, TX 75287, USA
"	Yükler1D	E	D	P,T	1	Tz, C	+	+	K	Sa	+	Li	+	"
"	Yükler2D (3)	E	D	P,T	2	Tz, C	+	+	K	Sa	+	Li	+	"
"	Yükler3D (3)	E	D	P,T	3	Tz, C	+	+	K	Sa	+	Li	+	"
Industry - sponsored, University - developed programmes														
Un. of Alabama	BHTH	-	D	T	1	Z,Em	(+)	-	-	Sa	-	Li	-	Prof. P.B.Visscher, Univ. of Al., Dept. of Physics, Box 870324 Tuscaloosa, AL 35487-0324, USA
Un. of Cal., Berkeley	Integrated Basin Mod.	L	EL	P,T	2	Tz,Em	+	+	k	Sa	+	En	-	Prof. C-Y Wang, Univ. of Cal., Dept. Geol. Geophys., Berkeley, CA 94720, USA
Un. of Illinois	Basin 2 (6)	L	D	P,T	2	Tz,Em	+	-	k	Sa	-	Li, En	+	Prof. C. Bethke, Univ. of Ill. at Urbana-Champaign, Dept. Geol., 245 Natural Hist. Building, 1301 West Green Street, Urbana, Il. 61801, USA
Un. of S. Carolina	Geopet 2 (6)	L	D	P,T	2	Tz,Em	+	+	k	Sa	-	Li	-	Prof. I. Lerche, Univ. S. Carolina, Dept. Geol. Sci., Columbia, SC 29201, USA
Cornell/Lamont	Akcess. Basin	L	EL	P,T	2	Tz, C	-	-	k	Sa	-	Li, En	-	Computational Mechanics Corp. (GBRN), 601 Concord Street, Suite 201, TENN 37919, USA
Programmes whose use are restricted by nationality														
JNOC	Sigma-2	L	D	P,T	2	Tz,Em	+	+	k	?	-	Li, En	-	1-2-2 Hamada, Mihama-Ku, Chiba City, 261 Japan

Numbers 1–6 in table denote: (1) = Use a 1D version of TEMISPACK for temperature calculations. Modelling of pressure will be included in 1992. (2) = A finite element version of the 3D model is in preparation. (3) = Product name subject to Trade Mark approval. (4) = 1D fluid flow and 2D with and without fluid flow are also being developed. (5) = New programme version with Lagrangian coordinates and transient temperature calculations (2D, including the lithosphere) to be released in 1992. (6) = This university-based group also has a number of other basin-modelling-related codes for sale, and provides services for consortia members. See Table 2 for further explanation.

ological Survey; Shi and Wang, 1986, University of California at Berkeley; Pedersen, 1987, University of Oslo; Blanchard, 1987, University of Texas at Austin; Deming and Chapman, 1989, University of Utah; Dye et al., 1991, University of Alabama; Özkaya, 1991, Kuwait University). The main properties of these programmes are listed in Table 1.

Several universities have established industrial associations, whereby the members get access to programmes developed at that university (free or for a reduced fee). These programmes are (with the exception of BHTH) very complex. Less advanced (but also very complex) modelling packages are now commercially available for modelling services and/or for sale. Some properties of these programmes are listed in Table A-1. The symbols in this table are explained in Table 2. The information given in Table A-1 has been disclosed by the vendors of the different programmes, and describes the status as of July 1st, 1992.

Table A-1 is not intended as a buyer's checklist, and is included only to give an overview of the general status of the commercial side of basin modelling. Important features of basin modelling programmes also include flexibility (options for maturity, oil and gas generation, plotting, ...), interface with other programmes (e.g. prospect evaluation software, databases), input/output, the ability to model "special effects" (e.g. diagenesis, listric faulting, salt movement, compositional degradation of kerogen and hydrocarbons, palaeoheat flow from basin subsidence, fracturing, thrusting, stress fields, ...) and the sparsity of bugs (which probably exist in all basin modelling programmes). Numerical formulations and stability, computing efficiency and options for automatic calibration to well data (not necessarily including maturity) are also essential for evaluation of a basin modelling programme.

References

Anand, J., Somerton, W.K. and Gomaa, E., 1973. Predicting thermal conductivity of formations from other known properties. Soc. Pet. Eng. J., 13 (1): 107–173.

Andrews-Speed, C., Oxburgh, E.R. and Cooper, B.A., 1984. Temperatures and depth-dependent heat flow in western North Sea. Am. Assoc. Pet. Geol., Bull., 68: 1764–1781.

Angevine, C.L., 1983. Thermal Histories of Sedimentary Basins and Compaction of Sediments. Ph.D. Thesis, Cornell University, Ithaca, N.Y., 133 pp.

Assad, Y., 1955. A Study of Thermal Conductivity of Fluid-Bearing Porous Rocks. Ph.D. Thesis, University of California at Berkeley, Calif., 71 pp.

Athy, L.F., 1930a. Compaction and oil migration. Am. Assoc. Pet. Geol., Bull., 14 (1): 25–35.

Athy, L.F., 1930b. Density. porosity and compaction of sedimentary rocks. Am. Assoc. Pet. Geol., Bull., 14 (1): 1–24.

Augustson, J.H., 1993. A method on classification of oil traps based on heavy oil content in cores with relevance to filling and drainage of Barents Sea oil bearing structures. In: T.O. Vorren, E. Bergsaker, Ø.A. Dahl-Stamnes, E. Lie, B. Johansen and T.B. Lund (Editors), Arctic Geology and Petroleum Potential. Norwegian Petroleum Society (NPF), Special Publication 2. Elsevier, Amsterdam, pp. 691–702.

Baird, R.A., 1986. Maturation and source rock evaluation of Kimmeridge Clay, Norwegian North Sea. Am. Assoc. Pet. Geol., Bull., 70 (1): 1–11.

Baker, E.G., 1959. Origin and migration of oil. Science, 129: 871–874.

Baker, E.G., 1960. A hypothesis concerning the accumulation of sediment hydrocarbons to form crude oil. Am. Assoc. Pet. Geol., Bull., 19: 309–317.

Baker, E.G., 1962. Distribution of hydrocarbons in petroleum. Am. Assoc. Pet. Geol., Bull., 46: 76–84.

Baldwin, B., 1971. Ways of deciphering compacted sediments. J. Sediment. Petrol., 4: 293–301.

Baldwin, B. and Butler, C.O., 1985. Compaction curves. Am. Assoc. Pet. Geol., Bull., 69 (4): 622–626.

Balling, N., Kristiansen, J.I., Breiner, N., Poulsen, K.D. and Rasmussen, R., 1981. Geothermal measurements and subsurface temperature modelling in Denmark. GeoSkrifter, 16.

Barker, C., 1972. Aquathermal pressuring-role of temperature in development of abnormal-pressure zones. Am. Assoc. Pet. Geol., Bull., 56: 2068–2071.

Barker, C., 1990. Calculated volume and pressure changes during the thermal cracking of oil to gas in reservoirs. Am. Assoc. Pet. Geol., Bull., 74 (8): 1254–1261.

Bayer, U., 1989. Sediment compaction in large scale systems. Geol. Rundsch., 78 (1): 155–169.

Behar, F., Ungerer, P., Kressman, S. and Rudkiewicz, J.L., 1991. Thermal evolution of crude oils in sedimentary basins: experimental simulation in a confined system and kinetic modeling. Rev. Inst. Fr. Pét., 46 (2): 151–181.

Berg, R.R., 1975. Capillary pressures in stratigraphic traps. Am. Assoc. Pet. Geol., Bull., 59 (6): 939–956.

Bethke, C.M., 1985. A numerical model of compaction-driven groundwater flow and heat transfer and its application to the paleohydrology of intracratonic sedimentary basins. J. Geophys. Res., 90: 6817–6828.

Bethke, C.M., 1989. Modeling subsurface flow in sedimentary basins. Geol. Rundsch., 78 (1): 129–154.

Bethke, C.M., Harrison, W.J., Upson, C. and Altaner, S.P., 1988. Supercomputer analysis of sedimentary basins. Science, 239: 261–266.

Bethke, C.M., Reed, J.D. and Oltz, D.F., 1991. Long range petroleum migration in the Illinois basin. Am. Assoc. Pet. Geol., Bull., 75 (5): 925–945.

Biot, M.A., 1941. General theory of three-dimensional consolidation. J. Appl. Phys., 12: 155–164.

Blanchard, E., 1987. Fluid Flow in Compacting Sedimentary Basins. Ph.D. Thesis, University of Texas at Austin, 190 pp.

Bonham-Carter, G.F. and Sutherland, A.J., 1967. Diffusion and setting of sediments at river mouths: a computer simulation model. Trans. Gulf Coast Assoc. Geol. Soc., 17: 326–338.

Bostick, N., 1973. Time as a factor in thermal metamorphism of phytoclasts (coaly particles). Congr. Int. Stratigr. Geol. Carbon., C.R., Septième, Krefeld, August 23–28, Maastricht, 2: 183–193.

Bouvier, V., 1989. Modélisation bidimensionnelle des phénomènes de transport dans les bassins sédimentaires, par la méthode des éléments finis. Ph.D. Thesis, Ecole Nationale Supérieure des Mines de Paris, 344 pp.

Braun, R.L. and Burnham, A.K., 1987. Analysis of chemical reaction kinetics using a distribution of activation energies and simpler models. J. Energy Fuels, 1: 153–161.

Bredehoeft, J.D. and Hanshaw, B.B., 1968. On the maintenance of anomalous fluid pressures. 1. Thick sedimentary sequences. Geol. Soc. Am. Bull., 79: 1097–1106.

Brigaud, F., Chapman, D.S. and Le Douaran, S., 1990. Thermal conductivity in sedimentary basins predicted from lithologic data and geophysical well logs. Am. Assoc. Pet. Geol., Bull., 74: 1459–1477.

Buhrig, C., 1989. Geopressured Jurassic reservoirs in the Viking Graben: modelling and geological significance. Mar. Pet. Geol., 6: 31–48.

Bullard, E.C., 1939. Heat flow in South Africa. Proc. R. Soc. London, A, 178: 408–420.

Bullard, E.C., 1965. Historical introduction to terrestrial heat flow. In: W.H.K. Lee (Editor), Terrestrial Heat Flow. Am. Geophys. Union, Geophys. Monogr., 8: 1–6.

Burnham, A.K., 1991. Oil evolution from a self-purging reactor: kinetics and composition at 2°C/min and 2°C/h. J. Energy Fuels, 5: 205–214.

Burnham, A.K. and Braun, R.L., 1985. General kinetic model of oil shale pyrolysis. In Situ, 9 (1): 1–23.

Burnham, A.K. and Braun, R.L., 1990. Development of a detailed model of petroleum formation, destruction and expulsion from lacustrine and marine source rocks. Adv. Org. Geochem. 1989, 16: 27–39.

Burnham, A.K. and Sweeney, J.J., 1989. A chemical kinetic model of vitrinite maturation and reflectance. Geochim. Cosmochim. Acta, 53: 2649–2657.

Burnham, A.K., Braun, R.L. and Gregg, H.R., 1987. Comparison of method for measuring kerogen pyrolysis rates and fitting kinetic parameters. J. Energy Fuels, 1: 452–458.

Burrus, J. and Audebert, F., 1990. Thermal and compaction processes in a rifted basin in the presence of evaporites, the Gulf of Lions case study. Am. Assoc. Pet. Geol., Bull., 74 (9): 1420–1440.

Burst, J.F., 1969. Diagenesis of Gulf Coast clayey sediments and its possible relation to petroleum migration. Am. Assoc. Pet. Geol., Bull., 53: 73–93.

Cao, S. and Lerche, I., 1989. Sensitivity analysis of basin modelling with applications. In: F. Agterberg (Editor), Applications of Statistics in the Earth Sciences. Can. Geol. Soc. Pap., 89-9: 489–504.

Carslaw, H.S. and Jaeger, J.C., 1947. Conduction of Heat in Solids. Oxford University Press, London, 386 pp.

Carslaw, H.S. and Jaeger, J.C., 1959. Conduction of Heat in Solids. Oxford University Press, London, 2nd ed., 510 pp.

Catalan, L., Fu, X., Chatzis, I. and Dullien, F.A.L., 1992. An experimental study of secondary oil migration. Am. Assoc. Pet. Geol., Bull., 76 (5): 638–650.

Chapman, D.S., 1986. Thermal gradients in the continental crust. In: J.B. Dawson, D.A. Carswell, J. Hall and K.H. Wedepohl (Editors), The Nature of the Lower Continental Crust. Geol. Soc. London, Spec. Publ., 24: 63–70.

Chapman, D.S., Clement, M.D. and Mase, C.W., 1981. Thermal regime of the Escalante Desert, Utah, with an analysis of the Newcastle geothermal system. J. Geophys. Res., 6 (11): 735–746.

Chapman, D.S., Keho, T.H., Bauer, M.S. and Picard, M.D., 1984. Heat flow in the Uinta basin determined from bottom hole temperature (BHT) data. Geophysics, 49: 455–466.

Chia, Y.-P., 1985. Numerical simulation of subsurface flow in a sedimentary basin. Am. Assoc. Pet. Geol., Bull., 69 (2): 243–244.

Chiaramonte, M.A., Tamburini, A., Salvatori, T. and Bersani, A., 1987. Geochemical modelling in rapidly subsiding basins. Org. Geochem., 13 (1–3): 181–186.

Clayton, J.L., Spencer, C.W., Koncz, I. and Szalay, A., 1990. Origin and migration of hydrocarbon gases and carbon dioxide, Bèkès Basin, southeastern Hungary. Org. Geochem., 5 (3): 233–247.

Connan, J., 1974. Time–temperature relation in oil genesis. Am. Assoc. Pet. Geol., Bull., 58: 2516–2521.

Cooles, G.P., Mackenzie, A.S. and Quigley, T.M., 1986. Calculation of petroleum masses generated and expelled from source rocks. Advances in Organic Geochemistry 1985. Pergamon, London, pp. 235–245.

Crank, J., 1984. Free and Moving Boundary Problems. Oxford University Press, London, 425 pp.

Dahl, B., Nysæther, G., Speers, G.C. and Yükler, A., 1987. Oseberg area — integrated basin modelling. In: J. Brooks and K. Glennie (Editors), Petroleum Geology of North West Europe. Graham and Trotman, London, pp. 1029–1038.

Daines, S.R., 1982. Aquathermal pressuring and geopressure evaluation. Am. Assoc. Pet. Geol., Bull., 66 (7): 931–939.

De Bremaecker, J.Cl., 1983. Temperature, subsidence, and hydrocarbon maturation in extensional basins: a finite element model. Am. Assoc. Pet. Geol., Bull., 6 (9): 1410–1414.

Dembicki, H., Jr. and Anderson, M.J., 1989. Secondary migration of oil: experiments supporting efficient movement of separate, buoyant oil phase along limited conduits. Am. Assoc. Pet. Geol., Bull., 73 (8): 1018–1021.

Deming, D., 1993. Overburden rock, temperature and heat flow — essential elements of the petroleum system. In: L.B. Magoon and W.D. Dow (Editors), The Petroleum System. From Source to Trap. Am. Assoc. Pet. Geol., (in press).

Deming, D. and Chapman, D.S., 1988. Heat flow in the Utah–Wyoming thrust belt from analysis of bottom-hole temperature data measured in oil and gas wells. J. Geophys. Res., 93 (B11): 13657–13673.

Deming, D. and Chapman, D.S., 1989. Thermal histories and hydrocarbon generation: an example from the Utah–Wyoming thrust belt. Am. Assoc. Pet. Geol., Bull., 73 (12): 1455–1471.

Demongodin, L., Pinoteau, B., Vasseur, G. and Gable, R., 1991. Thermal conductivity and well logs — a case study in the Paris basin. Geophys. J. Int., 105 (3): 675–691.

Dickinson, G., 1953. Reservoir pressures in Gulf Coast Louisiana. Am. Assoc. Pet. Geol., Bull., 37 (2): 410–432.

Didyk, B.M. and Simoneit, B.R.T., 1989. Hydrothermal oil of Guaymas Basin and implications for petroleum formation mechanisms. Nature, 342: 65–68.

Doligez, B.F., 1987. Migration of Hydrocarbons in Sedimentary Basins. Technip, Paris, 681 pp.

Doligez, B.F., Bessis, J., Burrus, J., Ungerer, P. and Chénet, P.Y., 1986. Integrated numerical modelling of sedimentation, heat transfer and fluid migration in a sedimentary basin: the THEMIS model. In: J. Burrus (Editor), Thermal Modeling of Sedimentary Basins. Technip, Paris, pp. 173–195.

Domenico, A. and Palciauskas, V.V., 1973. Theoretical analysis of forced convective heat transfer in regional ground-water flow. Geol. Soc. Am. Bull., 84: 3803–3814.

Dominé, F., 1991. High pressure pyrolysis of n-hexane, 2,4-dimethylpentane and 1-phenylbutane. Is pressure an important geochemical parameter? Org. Geochem., 17 (5): 619–634.

Dow, W.G., 1977. Kerogen studies and geological interpretations. J. Geochem. Explor., 7: 79–99.

Du Rouchet, J., 1981. Stress fields, a key to oil migration. Am. Assoc. Pet. Geol., Bull., 65: 79–99.

Düppenbecker, S. and Horsfield, B., 1990. Compositional information for kinetic modelling and petroleum type prediction. Advances in Organic Geochemistry 1989. Org. Geochem., 16: (1–3) 259–266.

Durand, B., 1988. Understanding of HC migration in sedimentary basins (present state of knowledge). Org. Geochem., 13: 445–459.

Durand, B., Ungerer, P., Chiarelli, A. and Oudin, J.L., 1984. Modélisation de la migration de l'huile. Application à deux exemples de bassins sédimentaires. Proc. 11th World Petroleum Congr., London, 1983, PD1, 1: 1–13.

Dutta, N.C., 1986. Shale compaction, burial diagenesis and geopressures: a dynamic model, solution and some results. In: J. Burrus (Editor), Thermal Modelling in Sedimentary Basins. Technip, Paris, pp. 149–172.

Dye, J., Visscher, B., Fang, J.H., Johnson, S., Davis, A.M.J. and Tipping, R., 1991. Fast PC-based burial/thermal modelling. Am. Assoc. Pet. Geol., Bull., 75 (2): 352.

Dykstra, J., 1987. Compaction correction for burial history curves: application to Lopatin's method for source rock maturation determination. Geobyte, November: 16–23.

Ehrenberg, S.N., 1990. Relationship between diagenesis and reservoir quality in sandstones of the Garn Formation, Haltenbanken, Mid-Norwegian continental shelf. Am. Assoc. Pet. Geol., Bull., 74 (10): 1538–1558.

Endignoux, L. and Wolf, S., 1990. Thermal and kinematic evolution of thrust basins: a 2D numerical model. In: J. Letowzey (Editor), Petroleum and Tectonics in Mobile Belts. Technip, Paris, pp. 181–192.

England, W.A. and Mackenzie, A.S., 1989. Some aspects of the organic geochemistry of petroleum fluids. Geol. Rundsch., 78 (1): 291–303.

England, W.A., Mackenzie, A.S., Mann, D.M. and Quigley, T.M., 1987. The movement and entrapment of petroleum fluids in the subsurface. J. Geol. Soc. London, 144: 327–347.

Espitalié, J., Maxwell, J.R., Chénet, P.Y. and Marquis, F., 1988a. Aspects of hydrocarbon migration in the Mesozoic in the Paris Basin as deduced from an organic geochemical survey. Org. Geochem., 13: 467–481.

Espitalié, J., Ungerer, P., Irwin, H. and Marquis, F., 1988b. Primary cracking of kerogens. Experimenting and modelling C_1, C_2–C_5, C_6–C_{15+} classes of hydrocarbons formed. Org. Geochem., 13: 893–899.

Espitalié, J., Marquis, E. and Drouet, S., 1993. Critical study of modelling parameters. In: A.G. Doré et al. (Editors), Basin Modelling: Advances and Applications. Norwegian Petroleum Society (NPF), Special Publication 3. Elsevier, Amsterdam, pp. 233–242 (this volume).

Falvey, D.A. and Deighton, I., 1982. Recent advances in burial and thermal geohistory analysis. J. Aust. Pet. Explor. Assoc., 22: 65–81.

Fang, J.H. and Chen, H.C., 1990. Uncertainties are better handled by fuzzy arithmetic. Am. Assoc. Pet. Geol., Bull., 74 (8): 1228–1233.

Fjeldskaar, W., Mykkeltveit, J., Johansen, H., Langfeldt, J.M., Christie, O.H.J., Tyvand, P., Skurve, O. and Bjørkum, A., 1990. Interactive 2D basin modelling on workstations. SPE paper 20330, Proc. 5th SPE Petroleum Computer Conf., Denver, Colo., June 25–28, pp. 181–195.

Forbes, P.L., Ungerer, P., Kuhfuss, A.B., Riis, F. and Eggen, S., 1991. Compositional modelling of petroleum generation and expulsion. Trial application to a local mass balance in the Smørbukk Sør field, Haltenbanken area, Norway. Am. Assoc. Pet. Geol., Bull., 75 (5): 873–893.

Foster, J.B. and Whalen, H.E., 1966. Estimation of formation pressures from electrical surveys — offshore Louisiana. J. Pet. Technol., 18 (2): 165–171.

Franco, J.V.B., 1986. Laboratory Simulation of Crude Oil Migration: Possible Implications for Oil Exploration. Ph.D. Thesis, University of Oklahoma at Norman, Okla., 97 pp.

Frette, V., Feder, J., Jøssang, T. and Meakin, P., 1992. Buoyancy-driven fluid migration in porous media. Phys. Rev. Lett., 68 (21): 3164–3167.

Giardini, A.A. and Melton, C.E., 1983. A scientific explanation for the origin and location of petroleum accumulations. J. Pet. Geol., 6 (2): 117–138.

Gibson, R.E., 1958. The progress of consolidation in a clay layer increasing in thickness with time. Geotechnique, 8: 172–182.

Goff, J.C., 1983. Hydrocarbon generation and migration from Jurassic source rocks in the east Shetland basin and Viking graben of the northern North Sea. J. Geol. Soc. London, 140: 445–474.

Goss, R.D., Combs, J. and Timur, A., 1975. Prediction of thermal conductivity in rocks from other physical parameters and from standard geophysical well logs. SPWLA 10th Annu. Logging Symp., pp. 1–21.

Gretener, P.E., 1981. Geothermics. Using temperature in hydrocarbon exploration. Am. Assoc. Pet. Geol., Course Note Ser., 17, 156 pp.

Griffiths, C.M., Brereton, N.R., Beausillon, R. and Castillo, D., 1992. Thermal conductivity prediction from petrophysical data: a Case Study. In: A. Hurst et al. (Editors), Geological Applications of Wireline Logs II.

Grigo, D., Maragna, B., Arienti, M.T., Fiorani, M., Parisi, A., Marrone, M., Sguazzero, P. and Uberg, A.S., 1993. Issues in 3D sedimentary basin modelling and application to Haltenbanken, offshore Norway. In: A.G. Doré et al. (Editors), Basin Modelling: Advances and Applications. Norwegian Petroleum Society (NPF), Special Publication 3. Elsevier, Amsterdam, pp. 455–468 (this volume).

Habicht, J.K.A., 1964. Comment on the history of migration in the Gifhort Trough. Proc. 6th World Petroleum Congr., 1: 480.

Ham, H.H., 1966. New charts help estimate formation pressures. Oil Gas J., 64 (51): 58–63.

Harbaugh, J.W., 1966. Mathematical simulation of marine sedimentation with IBM 7090/7094 computers. Kans., State Geol. Surv., Comput. Contrib., 1, 52 pp.

Harff, J., Schretzenmayr, S., Springer, J., Hoth, P., Eiserbeck, W. and Süssmuth, S., 1989. Mathematisch-numerische Modellierung regionaler bis lokaler Einheiten in sedimentären Becken an einem Beispiel für die Kohlenwasserstofferkundung im Rotliegenden. Z. Geol. Wissensch. Berlin, 17: 747–759.

Haugerud, R.A., 1986. IDT — an interactive, screen-oriented microcomputer program for simulation of 1-dimensional geothermal histories. U.S. Geol. Surv., Open-File Rep., 86-511: 19, and 5-1/4 inch diskette.

Haugerud, R.A., 1988. On numerical modeling of one-dimensional geothermal histories. Comput. Geosci., 15 (5): 825–836.

He, Z., 1985. Hydrocarbon evaluation in Langga Basin, Jizhong, North China. M.S. Thesis, Graduate School of East China Petroleum Institute, Beijing, 233 pp. (in Chinese).

He, Z. and Lerche, I., 1989. Inversion of multiple thermal indicators: quantitative methods of determining paleo heat flux and geological parameters, IV. Case histories using thermal indicator tomography. Math. Geol., 21: 523–542.

Heasler, H. and Surdam, R.C., 1985. Thermal evolution of coastal California with application to hydrocarbon maturation. Am. Assoc. Pet. Geol., Bull., 69 (9): 1386–1400.

Hedberg, H.D., 1936. Gravitational compaction of clays and shales. Am. J. Sci., 5th Ser., 231 (184): 241–287.

Henry, P. and Wang, C.-Y., 1991. Modeling of fluid flow and pore pressure at the toe of Oregon and Barbados accretionary wedges. J. Geophys. Res., 96 (B12): 20109–20130.

Hermanrud, C., 1986. On the importance to the petroleum generation of heating effects from compaction derived water: an example from the northern North Sea. In: J. Burrus (Editor), Thermal Modelling in Sedimentary Basins. Technip, Paris, pp. 247–270.

Hermanrud, C., 1989. Numerical basin modelling. Proc. VISTA Conf. Fundamental Oil and Gas Research, Stavanger, June 1988. Published by Statoil (ISBN 82-991255-2-9).

Hermanrud, C. and Shen, Y., 1989. Virgin rock temperatures from well logs — accuracy analysis from some advanced inversion models. Mar. Pet. Geol., 6: 360–363.

Hermanrud, C., Eggen, S., Jacobsen, T., Carlsen, E.M. and Pallesen, S., 1990a. On the accuracy of modelling hydrocarbon generation and migration: the Egersund Basin oil find, Norway. Org. Geochem., 16 (1–3): 389–399.

Hermanrud, C., Cao, S. and Lerche, I., 1990b. Estimates of virgin rock temperature derived from BHT measurements; bias and error. Geophysics, 55 (7): 924–931.

Hermanrud, C., Eggen, S. and Larsen, R.M., 1991. Investigation of the thermal regime of the Horda platform by basin modelling; implications for the hydrocarbon potential of the Stord basin, northern North Sea. In: A.M. Spencer (Editor), Generation, Accumulation and Production of Europe's Hydrocarbons. Spec. Publ. European Association of Petroleum Geoscientists, 1, Oxford University Press, London, pp. 65–73.

Heum, O.R., Dalland, A. and Meisingset, K.K., 1986. Habitat of hydrocarbons at Haltenbanken (PVT-modelling as a predictive tool in hydrocarbon exploration). In: A.M. Spencer et al. (Editors), Habitat of Hydrocarbons on the Norwegian Continental Shelf. Graham and Trotman, London, pp. 259–274.

Hood, A., Gutjahr, C.C.M. and Heacock, R.L., 1975. Organic metamorphism and the generation of petroleum. Am. Assoc. Pet. Geol., Bull., 59: 986–996.

Horsfield, B., Disko, U. and Leistner, F., 1989. The micro-scale simulation of maturation: outline of a new technique and its potential applications. Geol. Rundsch., 78 (1): 361–174.

Hosoi, H., 1963. First migration of petroleum in Akita and Yamagata Prefectures. Jpn. Assoc. Mineral., Petrol. Econ. Geol. J., 49 (2): 43–55; 49 (3): 101–114.

Houbolt, J.J.U.C. and Wells, P.R.A., 1980. Estimation of heat flow in oil wells based on a relation between heat conductivity and sound velocity. Geol. Mijnbouw, 59: 215–224.

Hovland, M.T. and Judd, A.G., 1988. Seabed Pockmarks and Seepages: Impact on Geology, Biology and the Marine Environment. Graham and Trotman, London, 293 pp.

Hubbert, M.K., 1940. The theory of groundwater motion. J. Geol., 48: 785–944.

Hubbert, M.K., 1953. Entrapment of petroleum under hydrodynamic conditions. Am. Assoc. Pet. Geol., Bull., 37: 1954–2026.

Hunt, J.M., 1979. Petroleum Geochemistry and Geology. Freeman, San Francisco, Calif., 617 pp.

Hunt, J.M., Lewan, M.D. and Hennet, R.J.-C., 1991. Modeling oil generation with time temperature index graphs based on the Arrhenius equation. Am. Assoc. Pet. Geol., Bull., 75 (4): 795–807.

Irwin, H., Hermanrud, C., Carlsen, E.M., Vollset, J. and Nordvall, I., 1993. Basin modelling of hydrocarbon charge in the Egersund Basin, Norwegian North Sea: pre- and post-drilling assessments. In: A.G. Doré et al. (Editors), Basin Modelling: Advances and Applications. Norwegian Petroleum Society (NPF), Special Publication 3. Elsevier, Amsterdam, pp. 539–548 (this volume).

Issler, D.R., 1984. Calculation of organic maturation levels for offshore eastern Canada — implications for general application of Lopatin's method. Can. J. Earth Sci., 21: 477–488.

Jacquin, C., 1965. Interaction entre l'argile et les fluides écoulements à travers les argiles compactes. Rev. Inst. Fr. Pét., 10: 1475–1501.

Jaupart, C., 1986. On the average amount and vertical distribution of radioactivity in the continental crust. In: J. Burrus (Editor), Thermal Modelling in Sedimentary Basins. Technip, Paris, pp. 33–48.

Jensen, P.K., Holm, L. and Thomsen, E., 1985. Modelling burial history, temperature and maturation. In: B.M. Thomas et al. (Editors), Petroleum Geochemistry in Exploration of the Norwegian Shelf. Graham and Trotman, London, pp. 145–152.

Jensen, R.P. and Doré, A.G., 1993. A recent Norwegian shelf heating event — fact or fantasy?. In: A.G. Doré et al. (Editors), Basin Modelling: Advances and Applications. Norwegian Petroleum Society (NPF), Special Publication 3. Elsevier, Amsterdam, pp. 85–106 (this volume).

Kappelmeyer, O. and Haenel, R., 1974. Geothermics with Special Reference to Application. Borntraeger, Berlin, 238 pp.

Karweil, J., 1955. Die Metamorphose der Kohlen vom Standpunkt der Physikalischen Chemie. Z. Dtsch. Geol. Ges., 107: 132–139.

Keith, L.A. and Rimstidt, J.D., 1985. A numerical compaction model of overpressuring in shales. Math. Geol., 17 (2): 115–135.

Klomp, U.C. and Wright, P.A., 1990. A new method for the measurement of kinetic parameters of hydrocarbon generation from source rocks. Advances in Organic Geochemistry 1989. Org. Geochem., 16 (1–3): 49–60.

Kooi, H., 1991. Tectonic Modelling of Extensional Basins. The Role of Lithospheric Flexure, Intraplate Stress and Relative Sea-Level Change. Ph.D. Thesis, The Free University of Amsterdam, 183 pp.

Krooss, B.M., 1988. Experimental investigation of the molecular migration of C_1–C_6 hydrocarbons: kinetics of hydrocarbon release from source rocks. Org. Geochem., 13: 513–523.

Lakshmanan, C.C., Bennet, M.L. and White, N., 1991. Implications of multiplicity in kinetic parameters to petroleum exploration: distributed activation energy models. J. Energy Fuels, 5: 110–117.

Landau, H.G., 1950. Heat conduction in a melting solid. Quart. Appl. Math., 80 (1): 81–94.

Larter, S., 1989. Chemical models of vitrinite reflectance evolution. Geol. Rundsch., 78 (1): 349–359.

Leadholm, R.M., Ho, T.T.Y. and Sahai, S.K., 1985. Heat flow, geothermal gradients and maturation modelling on the Norwegian continental shelf using computer methods. In: B.M. Thomas et al. (Editors), Petroleum Geochemistry in Exploration of the Norwegian Shelf. Graham and Trotman, London, pp. 131–143.

Leder, F. and Park, W.C., 1986. Porosity reduction in sandstone by quarts overgrowth. Am. Assoc. Pet. Geol., Bull., 70 (11): 1713–1728.

Lee, W.H.K., 1965. Terrestrial heat flow. Am. Geophys. Union, Monogr., 8, 276 pp.

Lerche, I., 1990a. Basin Analysis. Quantitative Methods, Vol. 1, Academic Press, San Diego, Calif., 562 pp.

Lerche, I., 1990b. Basin Analysis. Quantitative Methods, Vol. 2. Academic Press, San Diego, Calif., 570 pp.

Lerche, I. and Glezen, W.H., 1984. Deposition, Compaction and Fluid Migration: Time-Dependent Models in One and Two Dimensions. Gulf Research and Development Co., Pittsburgh, Pa., 52 pp.

Lerche, I., Yarzab, R.F. and Kendall, C.C.St.C., 1984. The determination of paleoheat flux from vitrinite reflectance data. Am. Assoc. Pet. Geol., Bull., 68: 1704–1717.

Levine, S. and Cuthiell, D.L., 1986. Relative permeabilities in two-phase flow through porous media: an application of effective medium theory. J. Can. Pet. Technol., Sept.–Oct.: 74–84.

Lewan, M.D., 1985. Evaluation of petroleum generation by hydrous pyrolysis experimentation. Philos. Trans. R. Soc., London, 315: 123–134.

Lewan, M.D., 1989. Hydrous pyrolysis study of oil and tar generation from Monterey shale containing high sulfur kerogen. Abstract from American Chemical Society National Meeting, Division of Geochemistry, Abstr. 94.

Lewis, C.R. and Rose, S.C., 1970. A theory relating high temperatures and overpressures. J. Pet. Technol., January: 11–16.

Leythaeuser, D. and Poelchau, H.S., 1991. Expulsion of petroleum from type III kerogen source rocks in gaseous solution: modelling of solubility fractionation. In: W.A. England and A.J. Fleet (Editors), Petroleum Migration. Geol. Soc. London, Spec. Publ., 59: 33–46.

Leythaeuser, D., Schaefer, R.G. and Yükler, A., 1982. Role of diffusion in primary migration of hydrocarbons. Am. Assoc. Pet. Geol., Bull., 66 (4): 408–429.

Leythaeuser, D., Littke, R., Radke, M. and Schaefer, R.G., 1988. Geochemical effects of petroleum migration and expulsion from Toarcian source rocks in the Hils syncline area, NW-Germany. Org. Geochem., 13: 489–502.

Littke, R., Baker, D.R. and Leythaeuser, D., 1988. Microscopic and sedimentologic evidence for the generation and migration of hydrocarbons in Toarcian source rocks of different maturities. Org. Geochem., 13 (1–3): 549–559.

Liu, J. and Lerche, I., 1990. Inverse methods and kinetic models of hydrocarbon generation, II. Case histories for residual kerogen analysis. Math. Geol., 22: 989–1009.

Lopatin, N.V., 1971. Temperature and geologic time as factors in coalification. Izv. Akad. Nauk SSSR, Ser. Geol., 3: 95–106.

Ludvigsen, A., Gran, K., Palm, E. and Bjørlykke, K., 1993. Effects of convection — currents on heat transfer in sedimentary basins. In: A.G. Doré et al. (Editors), Basin Modelling: Advances and Applications. Norwegian Petroleum Society (NPF), Special Publication 3. Elsevier, Amsterdam, pp. 353–359 (this volume).

Mackenzie, A.S. and Quigley, T.M., 1988. Principles of geochemical prospect appraisal. Am. Assoc. Pet. Geol., Bull., 72: 399–415.

Mackenzie, A.S., Leythaeuser, D., Muller, P., Radke, M. and Schaefer, R.G., 1987. The expulsion of petroleum from Kimmeridge Clay source rocks in the area of the Brae oilfield,

U.K. continental shelf. In: J. Brooks and K.W. Glennie (Editors), Proc. 3rd Conf. Petroleum Geology of NW Europe. Graham and Trotman, London, pp. 865–878.

Magara, K., 1968. Compaction and migration of fluids in Miocene mudstone, Nagaoka Plain, Japan. Am. Assoc. Pet. Geol., Bull., 52 (12): 2466–2501.

Magara, K., 1975. Re-evaluation of montmorillonite dehydration as cause of abnormal pressure and hydrocarbon migration. Am. Assoc. Pet. Geol., Bull., 59 (2): 292–302.

Magara, K., 1976. Thickness of removed sedimentary rocks, paleopore pressure, and paleotemperature, southwest part of Western Canada Basin. Am. Assoc. Pet. Geol., Bull., 60 (4): 554–565.

Magara, K., 1978. Compaction and fluid migration; practical petroleum geology. Developments in Petroleum Science, 9, Elsevier, Amsterdam, 319 pp.

Magara, K., 1979. Structured water and its significance in primary oil migration. Bull. Can. Pet. Geol., 27 (1): 87–93.

Magara, K., 1981. Mechanisms of natural fracturing in a sedimentary basin. Am. Assoc. Pet. Geol., Bull., 6 (1): 123–132.

Mandl, G. and Harkness, R.M., 1987. Hydrocarbon migration by hydraulic fracturing. Deformation of Sediments and Sedimentary rocks, Geol. Soc., Spec. Publ., 29: 39–53.

Maragna, B., Zaro, G. and Pessina, P., 1985. Block 33/6 geochemical evaluation. In: B.M. Thomas et al. (Editors), Petroleum Geochemistry in Exploration of the Norwegian Shelf. Graham and Trotman, London, pp. 197–204.

McAuliffe, C.D., 1979. Oil and gas migration — chemical and physical constraints. Am. Assoc. Pet. Geol., Bull., 65 (5): 761–781.

McKenzie, D., 1978. Some remarks on the development of sedimentary basins. Earth Planet. Sci. Lett., 40: 25–32.

Meade, R.H., 1966. Factors influencing the early stages of compaction of clays and sands — review. J. Sediment. Petrol., 36: 1085–1101

Meinschen, W.G., 1959. Origin of petroleum. Am. Assoc. Pet. Geol., 43: 925–943.

Meinschen, W.G., 1961. Significance of hydrocarbons in sediments and petroleum. Geochim. Cosmochim. Acta, 22: 58–64.

Meissner, F., 1978. Petroleum Geology of the Bakken Formation, Williston Basin, North Dakota and Montana. Williston Basin Symp. Proc., Montana Geological Society, pp. 207–227.

Mitchell, A.R. and Griffiths, D.F., 1980. The Finite Difference Method in Partial Differential Equations. Wiley, Chichester, 272 pp.

Mudford, B.S., 1990. A one-dimensional, two phase model of overpressure generation in the Venture gas field, offshore Nova Scotia. Bull. Can. Pet. Geol., 38 (2): 246–258.

Nakayama, K., 1987. Two-Dimensional Basis Analysis for Petroleum Exploration. Ph.D. dissertation, University of South Carolina, Columbia, S.C., 241 pp.

Nakayama, K. and van Sichlen, D.C., 1981. Simulation model for petroleum exploration. Am. Assoc. Pet. Geol., Bull., 65: 1230–1255.

Nielsen, S.B., 1993. Fast Monte Carlo simulation of hydrocarbon generation. In: A.G. Doré et al. (Editors), Basin Modelling: Advances and Applications. Norwegian Petroleum Society (NPF), Special Publication 3. Elsevier, Amsterdam, pp. 265–276 (this volume).

Nielsen, S.B. and Balling, N., 1990. Modelling subsidence, heat flow, and hydrocarbon generation in extensional basins. First Break 8 (1): 23–31.

Nielsen, S.B. and Dahl, B., 1991. Confidence limits on kinetic models of primary cracking and implications for the modelling of hydrocarbon generation. Mar. Pet. Geol., 8: 483–492.

Nogaret, E., 1983. Solubilité des Hydrocarbures dans les Gas Naturel Comprimé; Application à la Migration du Pétrole dans les Bassins Sédimentaires. Ph.D. Thesis, Ecole des Mines, Paris, 260 pp.

North, F.K., 1985. Petroleum Geology. Allen and Unwin, Winchester, 607 pp.

Oelkers, E.H., Bjørkum, P.A. and Murphy, W.M., 1991. The mechanism of porosity reduction, stylolite development and quartz cementation in North Sea sandstones. Proc. Water–Rock Interaction Seminar No. 7 (WRI-7), 13–18 July, 1991, Park City, Utah.

Ojakangas, D.R., 1967. Mathematical Simulation of Oil Trap Development. Ph.D. Thesis, Stanford University, Calif., 161 pp.

Okui, A. and Waples, D.W., 1993. Relative permeabilities and hydrocarbon expulsion from source rocks. In: A.G. Doré et al. (Editors), Basin Modelling: Advances and Applications. Norwegian Petroleum Society (NPF), Special Publication 3. Elsevier, Amsterdam, pp. 293–301 (this volume).

Ortenblad, A., 1930. Mathematical theory of the process of consolidation of mud deposits. J. Math. Phys., 9: 73–149.

Özkaya, I., 1991. Computer simulation of primary oil migration in Kuwait. J. Pet. Geol., 14 (1): 37–48.

Paaswell, R.E., 1967. Thermal influence on flow from a compressible porous medium. Water Resour. Res., 3: 271–278.

Palciauskas, V.V., 1986. Models for thermal conductivity and permeability in normally compacting basins. In: J. Burrus (Editor), Thermal Modelling of Sedimentary Basins. Technip, Paris, pp. 323–336.

Palciauskas, V.V. and Domenico, P.A., 1980. Microfracture development in compacting sediments: relation to hydrocarbon-maturation kinetics. Am. Assoc. Pet. Geol., Bull., 64: 927–937.

Pedersen, T., 1987. A Computer-Based Approach to Basin Analysis. Cand. Sci. (M.S.) Thesis, University of Oslo, 150 pp.

Pedersen, T. and Skogseid, J., 1989. Vøring plateau volcanic margin: extension, melting, and uplift. Proc. ODP, Sci. Results, 104: 985–991.

Pepper, A.S., 1991. Estimating the petroleum expulsion behaviour of source rocks: a novel quantitative approach. In: W.A. England and A.J. Fleet (Editors), Petroleum Migration. Geol. Soc. London, Spec. Publ., 59: 9–31.

Perrier, R. and Quiblier, J., 1974. Thickness changes in sedimentary layers during compaction history. Am. Assoc. Pet. Geol., Bull., 52: 57–65.

Person, M. and Garven, G., 1992. Hydrologic constraints on petroleum generation within continental rift basins: theory and application to the Rhine Graben. Am. Assoc. Pet. Geol., Bull., 76 (4): 468–488.

Petzoukha, Y., 1987. A new mechano-chemical approach to hydrocarbon generation. Proc. 12th World Petroleum Congr., 2: 251–262.

Philippi, G.T., 1965. On the depth, time and mechanism of petroleum generation. Geochim. Cosmochim. Acta, 29: 1021–1049.

Pitt, G.J., 1962. The kinetics of the evolution of volatile products from coal. Fuel, 41: 267–274.

Price, L.C., 1982. Time as a factor in organic metamorphism and use of vitrinite reflectance as an absolute paleogeothermometer. Am. Assoc. Pet. Geol., Bull., 66: 619–620.

Price, L.C., 1983. Geologic time as a parameter in organic metamorphism and vitrinite reflectance as an absolute paleogeothermometer. J. Pet. Geol., 6: 5–38.

Price, L.C., Wegner, L.M., Ging, T. and Blount, C.W., 1983. Solubility of crude oil in methane as a function of pressure and temperature. Org. Geochem., 4: 201–221.

Proshlyakov, B.K., 1960. Reservoir properties of rocks as a function of their depth and lithology. Geol. Nefti Gaza, 12: 24–29.

Quigley, T.M. and Mackenzie, A.S., 1988. The temperatures of oil and gas formation in the subsurface. Nature, 333: 549–552.

Reynolds, J.G., Crawford, R.W. and Burnham, A.K., 1991. Analysis of oil shale and petroleum source rock pyrolysis by triple quadrupole mass spectrometry: comparisons of gas evolution at a heating rate of 10°C/min. J. Energy Fuels, 5: 507–523.

Rieke, H.H., III and Chilingarian, G.V., 1974. Compaction of Argillaceous Sediments. Developments in Sedimentology, 16, Elsevier, Amsterdam, 424 pp.

Ritter, U., 1984. The influence of time and temperature on vitrinite reflectance. Org. Geochem., 6: 473–480.

Ritter, U., Leslie Leith, L.T., Griffiths, C.M. and Schou, L., 1987. Hydrocarbon generation and thermal evolution in parts of the Egersund basin, northern North Sea. Can. Soc. Pet. Geol., Mem., 12: 75–85.

Rosenblat, S., 1985. A mathematical model of compaction-induced migration. J. Geophys. Res., 90 (B1): 779–789.

Royden, L. and Keen, C.E., 1980. Rifting process and thermal evolution of the continental margin of eastern Canada determined from subsidence curves. Earth Planet. Sci. Lett., 51: 342–361.

Royden, L., Sclater, J.G. and Von Herzen, R., 1980. Continental margin subsidence and heat flow: important parameters in formation of petroleum hydrocarbons. Am. Assoc. Pet. Geol., Bull., 64: 173–187.

Rybach, L., 1986. Amount and significance of radioactive heat sources in sediments. In: J. Burrus (Editor), Thermal Modelling in Sedimentary Basins. Technip, Paris, pp. 311–322.

Sales, J.K., 1993. Closure vs. seal capacity — a fundamental control on the distribution of oil and gas. In: A.G. Doré et al. (Editors), Basin Modelling: Advances and Applications. Norwegian Petroleum Society (NPF), Special Publication 3. Elsevier, Amsterdam, pp. 399–414 (this volume).

Sass, J.H., Lachenbruch, A.H. and Munroe, R.J., 1971. Thermal conductivity of rocks from measurements on fragments and its application to heat flow determinations. J. Geophys. Res., 76: 3391–3401.

Schmoker, J.W. and Gautier, D.L., 1988. Sandstone porosity as a function of thermal maturity. Geology, 16: 1007–1010.

Schmoker, J.W. and Gautier, D.L., 1989. Compaction of basin sediments: modeling based on time–temperature history. J. Geophys. Res., 94 (B6): 7379–7386.

Schowalter, T.T., 1979. Mechanics of secondary hydrocarbon migration and entrapment. Am. Assoc. Pet. Geol., Bull., 63: 723–760.

Schwarzkopf, T.A., 1990. Relationship between petroleum generation, migration and sandstone diagenesis, Middle Jurassic, Gifhorn Trough, N. Germany. Mar. Pet. Geol., 7: 153–170.

Sclater, J.G. and Christie, P.A.F., 1980. Continental stretching: an explanation of the post-mid-Cretaceous subsidence of the Central North Sea basin. J. Geophys. Res., 85: 3711–3739.

Sharp, J.M., Jr., 1974. An Investigation of Energy Transport in Thick Sequences of Compacting Sediments. Ph.D. Thesis, University of Illinois, Urbana, Ill., 151 pp.

Sharp, J.M., Jr., 1976. Momentum and energy balance equations for compacting sediments. Math. Geol., 8 (3): 305–322.

Sharp, J.M., Jr., 1983. Permeability controls on aquathermal pressuring. Am. Assoc. Pet. Geol., Bull., 67 (11): 2057–2061.

Sharp, J.M., Jr. and Domenico, A., 1976. Energy transport in thick sequences of compacting sediment. Geol. Soc. Am. Bull., 87: 390–400.

Shi, Y. and Wang, C.-Y., 1986. Pore pressure generation in sedimentary basins. Overloading versus aquathermal. J. Geophys. Res., 91 (B2): 2153–2162.

Smith, D.A., 1966. Theoretical considerations of sealing and non-sealing faults. Am. Assoc. Pet. Geol., Bull., 50 (2): 363–374.

Smith, J.E., 1971. The dynamics of shale compaction and the evolution of pore fluid pressures. Math. Geol., 3: 239–263.

Smith, J.E., 1973. Shale compaction. Soc. Pet. Eng. J., 13: 12–22.

Smith, L. and Chapman, D.S., 1983. On the thermal effects of groundwater flow, 1. Regional scale systems. J. Geophys. Res., 88 (B1): 593–608.

Spencer, C.W., 1987. Hydrocarbon generation as a mechanism for overpressuring in the Rocky Mountain region. Am. Assoc. Pet. Geol., Bull., 71 (4): 368–388.

Spohn, T. and Schubert, G., 1982. Convective thinning of the lithosphere: a mechanism for the initiation of continental rifting. J. Geophys. Res., 87 (B6): 4669–4681.

Stainforth, J.G. and Reinders, J.E.A., 1990. Primary migration of hydrocarbons by diffusion through organic matter networks, and its effect on oil and gas generation. Advances in Organic Geochemistry 1989. Org. Geochem., 16: 61–74.

Stallman, R.W., 1960. Notes on the use of temperature data for computing groundwater velocity. Société Hydrotechnique de France, Nancy, 6th Assembly on Hydraulics, Rept. 3, pp. 1–7.

Stallman, R.W., 1963. Computation of groundwater velocity from temperature data. In: R. Bentall (Editor), Methods of Collecting and Interpreting Ground-Water Data. U.S. Geol. Surv., Water-Supply Pap., 1544-H: 36–46.

Summa, L.L., Pottorf, R.J., Schwarzer, T.F. and Harrison, W.J., 1993. Palaeohydrology of the Gulf of Mexico Basin: development of compactional overpressure and timing of hydrocarbon migration relative to cementation. In: A.G. Doré et al. (Editors), Basin Modelling: Advances and Applications. Norwegian Petroleum Society (NPF), Special Publication 3. Elsevier, Amsterdam, pp. 641–656 (this volume).

Sweeney, J.J. and Burnham, A.K., 1990. Evaluation of a simple model of vitrinite reflectance based on chemical kinetics. Am. Assoc. Pet. Geol. Bull., 74: 1559–1570.

Sweeney, J.J., Burnham, A.K. and Braun, R.L., 1987. A model of hydrocarbon maturation from type I kerogen: application to Uinta basin, Utah. Am. Assoc. Pet. Geol., Bull., 71: 967–985.

Sylta, Ø., 1991. Modelling of secondary migration and entrapment of a multicomponent hydrocarbon mixture using equation of state and ray-tracing modelling techniques. In: W.A. England and A.J. Fleet (Editors), Petroleum Migration. Geol. Soc. London, Spec. Publ., 59: 111–122.

Sylta, Ø., 1993. New techniques and their applications in the analysis of secondary migration. In: A.G. Doré et al. (Editors), Basin Modelling: Advances and Applications. Norwegian Petroleum Society (NPF), Special Publication 3. Elsevier, Amsterdam, pp. 385–398 (this volume).

Talukdar, S., Gallanango, O., Vallejos, C. and Ruggiero, A., 1988. Observations on the primary migration of oil in the La Luna source rocks of the Maracaibo Basin, Venezuela. Rev. Inst. Fr. Pét., 43 (3): 357–370.

Terzaghi, K., 1925. Erdbaumechanik auf bodenphysikalischer Grundlage. Deuticke, Leipzig, 399 pp.

Thomas, M.M. and Clouse, J.A., 1990a. Primary migration by diffusion through kerogen, I. Model experiments with organic-coated rocks. Geochim. Cosmochim. Acta, 54: 2775–2779.

Thomas, M.M. and Clouse, J.A., 1990b. Primary migration by diffusion through kerogen, II. Hydrocarbon diffusivities in kerogen. Geochim. Cosmochim. Acta, 54: 2781–2792.

Thomas, M.M. and Clouse, J.A., 1990c. Primary migration by diffusion through kerogen, III. Calculation of geologic fluxes.

Geochim. Cosmochim. Acta, 54: 2793–2797.

Tissot, B., 1969. Premières données sur les méchanismes et la cinétique de la formation du pétrole dans les sédiments: simulation d'un schéma réactionnel sur ordinateur. Rev. Inst. Fr. Pét., 24: 470–501.

Tissot, B. and Espitalié, J., 1975. L'évolution thermique de la matière organique des sédiments: applications d'une simulation mathématique. Rev. Inst. Fr. Pét., 30: 713–777.

Tissot, B. and Pelet, R., 1971. Nouvelles données sur les méchanismes et la cinétique de la formation du pétrole dans les sédiments: simulation d'un schéma réactionel sur ordinateur. Rev. Inst. Fr. Pét., 24: 470–501.

Tissot, B. and Welte, D.H., 1978. Petroleum Formation and Occurrence. Springer-Verlag, Berlin, 699 pp.

Tissot, B., Durand, B., Espitalié, J. and Combaz, A., 1974. Influence of nature and diagenesis of organic matter in formation of petroleum. Am. Assoc. Pet. Geol., Bull., 58: 499–506.

Tissot, B., Pelet, R. and Ungerer, P., 1987. Thermal history of sedimentary basins, maturation indices, and kinetics of oil and gas generation. Am. Assoc. Pet. Geol., Bull., 71: 1445–1466.

Ungerer, P., 1984. Models of petroleum formation: how to take into account geology and chemical kinetics. In: B. Durand (Editor), Thermal Phenomena in Sedimentary Basins. Technip, Paris, pp. 235–246.

Ungerer, P., 1990. State of the art of research in kinetic modelling of oil formation and expulsion. Advances in organic geochemistry 1989. Org. Geochem., 16 (1–3): 1–25.

Ungerer, P., 1993. Modelling of petroleum generation and expulsion — an update to recent reviews. In: A.G. Doré et al. (Editors), Basin Modelling: Advances and Applications. Norwegian Petroleum Society (NPF), Special Publication 3. Elsevier, Amsterdam, pp. 219–232 (this volume).

Ungerer, P., Chénet, P.Y., Durand, B., Nogaret, E., Chiarelli, A., Oudin, A. and Perrin, J.F., 1984. Geological and geochemical models in oil exploration: principles and practical examples. In: G. Demaison and R.J. Murris (Editors), Petroleum Geochemistry and Basin Evaluation. Am. Assoc. Pet. Geol., Mem., 35: 53–77.

Ungerer, P., Doligez, B., Chénet, P.Y., Burrus, J., Bessis, F., Lafargue, F., Giroir, G., Heum, O. and Eggen, S., 1987a. A 2-D model of basin scale petroleum migration by two-phase fluid flow. Application to some case studies. In: B. Doligez (Editor), Migration of Hydrocarbons in Sedimentary Basins. Technip, Paris, pp. 415–456.

Ungerer, P., Behar, F., Villalba, M., Heum, O.R. and Audibert, A., 1987b. Kinetic modelling of oil cracking. Advances in Organic Geochemistry 1987, Org. Geochem., 13 (4–6): 857–868.

Ungerer, P., Burrus, J., Doligez, B., Chénet, P.Y. and Bessis, F., 1990. Basin evaluation by integrated two-dimensional modeling of heat transfer, fluid flow, hydrocarbon generation, and migration. Am. Assoc. Pet. Geol. Bull., 74 (3): 309–355.

Vacquier, V., Mathieu, Y., Legendre, E. and Blondin, E., 1988. An experiment on estimating the thermal conductivity of sedimentary rocks from oil well logging. Am. Assoc. Pet. Geol., Bull., 72: 758–764.

Van Hinte, J.E., 1978. Geohistory analysis — applications of micropaleontology in exploration geology. Am. Assoc. Pet. Geol., Bull., 62: 201–222.

Vejbæk, O.V., 1990. Geodynamic modelling of the Danish basin. In: B. Pinet and C. Bois (Editors), The Potential of Deep Seismic Profiling for Hydrocarbon Exploration. Technip, Paris, pp. 63–72.

Vik, E. and Hermanrud, C., 1993. Transient thermal effects of rapid subsidence in the Haltenbanken area. In: A.G. Doré et al. (Editors), Basin Modelling: Advances and Applications. Norwegian Petroleum Society (NPF), Special Publication 3. Elsevier, Amsterdam, pp. 107–117 (this volume).

Walderhaug, O. and Fjeldskaar, W., 1993. History of hydrocarbon emplacement in the Oseberg Field determined by fluid inclusion microthermometry and temperature modelling. In: A.G. Doré et al. (Editors), Basin Modelling: Advances and Applications. Norwegian Petroleum Society (NPF), Special Publication 3. Elsevier, Amsterdam, pp. 485–497 (this volume).

Waples, D.W., 1980. Time and temperature in petroleum formation: application of Lopatin's method to petroleum exploration. Am. Assoc. Pet. Geol., Bull., 64: 916–926.

Waples, D.W., 1984. Thermal models for oil generation. In: J. Brooks and D. Welte (Editors), Advances in Petroleum Geochemistry, Vol. 1. Academic Press, London, pp. 8–67.

Waples, D.W., 1993a. Maturity modeling: thermal indicators, hydrocarbon generation and oil cracking. In: L.B. Magoon and W.D. Dow (Editors), The Petroleum System. From Source to Trap. Am. Assoc. Pet. Geol. (in press).

Waples, D.W., 1993b. Modeling of sedimentary basins and petroleum systems. In: L.B. Magoon and W.D. Dow (Editors), The Petroleum System. From Source to Trap. Am. Assoc. Pet. Geol. (in press).

Waples, D.W., Kamata, H. and Suizu, M., 1992a. The art of maturity modeling, Part 1. Finding a satisfactory geologic model. Am. Assoc. Pet. Geol., Bull., 76 (1): 31–46.

Waples, D.W., Suizu, M. and Kamata, H., 1992b. The art of maturity modeling, Part 2. Alternative models and sensitivity analysis. Am. Assoc. Pet. Geol., Bull., 76 (1): 47–66.

Watts, A. and Ryan, W.B.F., 1976. Flexure of the lithosphere and continental margin basin. Tectonophysics, 36: 25–44.

Watts, N.L., 1987. Theoretical aspects of cap-rock and fault seals for single- and two-phase hydrocarbon columns. Mar. Pet. Geol., 4: 274–307.

Weller, J.M., 1959. Compaction of sediments. Am. Assoc. Pet. Geol., Bull., 43 (2): 273–310.

Welte, D.H., 1987. Migration of hydrocarbons facts and theory. In: B. Doligez (Editor), Migration of Hydrocarbons in Sedimentary Basins. Technip, Paris, pp. 393–413.

Welte, D.H. and Yukler, M.A., 1980. Evolution of sedimentary basins from the standpoint of petroleum origin and accumulation — an approach for a quantitative basin study. Org. Geochem., 2: 1–8.

Welte, D.H. and Yukler, M.A., 1981. Petroleum origin and accumulation in basin evolution — a quantitative model. Am. Assoc. Pet. Geol., Bull., 65 (8): 1387–1396.

Welte, D.H., Yukler, M.A., Radke, M., Leythaeuser, D., Mann, U. and Ritter, U., 1983. Organic geochemistry and basin modelling; important tools in petroleum evaluation. In: J. Brooks (Editor), Petroleum Geochemistry and Exploration of Europe. Geol. Soc. London, Spec. Publ., 12: 237–252.

Welte, D.H., Schaefer, R.G. and Yalcin, M.N., 1988. Gas generation from source rocks: aspects of a quantitative treatment. Chem. Geol., 71: 105–116.

White, D., 1930. Exchange of time for temperature in petroleum generation. Am. Assoc. Pet. Geol., Bull. 14 (2): 1227–1229.

Wilson, H.H., 1990. The case for early generation and accumulation of oil. J. Pet. Geol., 13 (2): 127–156.

Withjack, M., 1979. A convective heat transfer model for lithospheric thinning and crustal uplift. J. Geophys. Res., 84 (B6): 3008–3022.

Wong, P.-Z., 1988. The statistical physics of sedimentary rocks. Phys. Today, Dec.: 24–32.

Wood, D.A., 1988. Relationships between thermal maturity indices calculated using Arrhenius equation and Lopatin method: implications for petroleum exploration. Am. Assoc. Pet. Geol., Bull., 72 (2): 115–134.

Wygrala, B.P., 1988. Integrated computer-aided basin modeling applied to analysis of hydrocarbon generation history in a Northern Italian oil field. Org. Geochem., 13 (1–3): 187–197.

Yalcin, M.N., 1991. Basin modeling and hydrocarbon exploration. J. Sci. Eng., 5: 379–398.

Yalcin, M.N. and Welte, D.H., 1988. The thermal evolution of sedimentary basins and significance for hydrocarbon generation. Bull. Turk. Pet. Geol., 1 (1): 12–26.

Yalcin, M.N., Welte, D.H., Misra, K.N., Mandal, S.K., Balan, K.C., Mehrotra, K.L., Lohar, B.H., Kumar, S.P. and Misra, G.S., 1987. 3-D computer aided basin modelling of Campay Basin, India — a case study of hydrocarbon generation. In: S.P. Kumar, P.D. Wived, V. Banejie and V. Gupta (Editors), Petroleum Geochemistry in the Afro-Asian Region. Balkema, Rotterdam, pp. 417–450.

Yang, W.K., 1981. On quantitative prediction methods of oil and gas occurrences. Oil Gas Geol., 2 (2): 104–113 (in Chinese).

Yukler, M.A., 1976. Analysis of Error in Groundwater Modelling. Ph.D. Thesis, University of Kansas, Lawrence, Kans., 194 pp.

Yukler, M.A. and Kokesh, F., 1984. A review of models used in petroleum resource estimation and organic geochemistry. In: J. Brooks and D. Welte (Editors), Advances in Petroleum Geochemistry, Vol. 1. Academic Press, London, pp. 69–113.

Yukler, M.A., Cornford, C. and Welte, D.H., 1978. One-dimensional model to simulate geologic, hydrodynamic and thermodynamic development of a sedimentary basin. Geol. Rundsch., 67: 960–979.

CHRISTIAN HERMANRUD *Statoil A/S, P.O. Box 300, N-4001 Stavanger, Norway*

Theoretical aspects of problems in basin modelling

I. Lerche

Basin analysis is the study of the interactions of sediment deposition and compaction, thermal changes and fluid flow, tectonic forces and sea-level variations, and hydrocarbon production and accumulation, in evolving sedimentary basins. A quantitative understanding is sought of the dynamical interrelationships of these processes, and the controls exerted by present-day data, to improve on basic geologic knowledge, and to foster oil exploration for future generations.

The framework of sedimentary basin analysis is used to illustrate the impact of quantitative model building versus data control. Three strategies are examined: "forward" models, "inverse" models, and "pseudo-inverse" models. Pragmatic strategies for real case histories are considered using the interwoven aspects of dynamical burial histories, thermal histories, and hydrocarbon generation, migration and accumulation histories. These examples show that the problems of resolution, uniqueness, precision, and sensitivity of any quantitative model are tightly connected to the three factors of model assumptions, model parameters, and quality, quantity, noise and distribution of input data. These problems are not unique to quantitative models within basin analysis, but occur in all branches of science. Using a variety of situations a philosophy and appropriate strategies are discussed for examining the sharpness of information retrieval from limited constructions of basin models and data, in order to illuminate behaviors and dependencies, and to outline aspects of problems which require attention within the discipline of basin analysis.

I. Introduction

The construction is now well advanced of computer codes that simulate fluid flow, sediment compaction, thermal history, hydrocarbon generation, migration and accumulation, in one, two and even three spatial dimensions (Yukler, 1979; Welte and Yukler, 1981; Bethke, 1985; Lerche, 1990a, b).

The variously available codes have technical differences in their methods of calculation (grid sizes, time steps, smoothing and interpolation techniques, and so on) which, of themselves, would make for a fascinating review. However, the emphasis of this review is not so much on the differences of individual computer manipulations, but rather has to do with fundamental difficulties and problems in any endeavor to codify procedures for quantitative basin analysis assessment. The guiding motivation is to provide an appreciation of the strengths and weaknesses, accuracies and uncertainties, of basin analysis models as they relate, directly or indirectly, to potential hydrocarbon accumulation sites in a basin. In addition major fundamental problems yet to be addressed will be considered.

Three input factors control the veracity and usefulness of a quantitative model. These factors are: intrinsic assumptions of the model; implicit or explicit parameter values or functional dependencies in required equations of state; quality, quantity and sampling frequency of data used as controls on the procedure.

These factors may be of a forgivable nature in particular applications, or they may be of paramount importance. Part of the task of basin analysis procedures is to determine under what conditions each of the input factors plays a dominant role and to delimit the ranges of uncertainty of parameter values which are allowed, consistent with observations and inferences from observations.

In this paper, five separate sections illustrate some of the difficulties, problems and consequences that arise as a result of each of the individual input factors being invoked in different manners.

Section II provides an overview of the interlocked relation of burial, thermal and hydrocarbon histories in order to set a basic framework for the later examination of individual facets.

Then, Section III examines the consequences of dynamical factors in impacting the sedimentary evolution component of the system. This discussion is followed, in section IV, by some considerations related to determinations of both present-day, and palaeo, heat flux and temperature. Section V then discusses the impact of hydrocarbon kinetic models, and of primary and secondary migration processes, in influencing the assessments of potential hydrocarbon

Basin Modelling: Advances and Applications edited by A.G. Doré et al. NPF Special Publication 3, pp. 35–65, Elsevier, Amsterdam.

accumulation amounts and sites. In section VI a brief consideration is given of the data controls available on models, both pre and post borehole drilling.

Finally, some fundamental outstanding concerns for future investigations are provided, as well as suggestions for linking the model behaviors to areas impacted by the ability to construct models, such as risk analysis and exploration economics.

II. Overview of basin analysis

Hydrocarbon accumulations in a basin are massively dependent on: (i) the sediment fill and structural evolution with time of a basin including fluid escape; (ii) the thermal history of the sediments; (iii) the organic geochemistry controlling the generation of hydrocarbons from kerogen and the subsequent migration and accumulation in sealed traps of the generated hydrocarbons. Consider each in turn.

Burial history

Overview

Concentrate first on dynamical evolution. At deposition, sediments typically have porosities in the range 50–75%. When drilled at depth, say 5000 m, porosities are typically in the order of 10% or less. Thus an enormous amount of water has escaped from the sediments during their deposition and later evolution. The grain-to-grain contact pressure must then increase in order to support the overlying sediment weight. A problem of fundamental concern, underlying all other aspects of basin analysis, is to understand the dynamical evolution of fluid escape and compaction in a sedimentary basin. In addition, the sediments must be supported on an underlying basement rock, which must respond to the applied load. If the basement material behaves as an incompressible fluid, then isostatic compensation occurs, creating a surficial volume accommodation for further sediments. However, if the basement behaves as a rigid unmoving plate, then no response occurs. Reality must lie between these two extremes, leading to basinal evolution models which encompass tectonic effects due to flexural plate behavior, tectonic rifting, wrenching, compression, and so on.

Dynamical fluid escape depends intricately upon the permeability behavior of the deposited, and evolving, lithologic units. As fluid loss proceeds, porosity decreases, permeabilities become smaller, leading to an ever increasing delay in extracting the residual fluids. The addition of more sedimentary overburden is then compensated for by an increase in the retained fluid pressure over the hydrostatic value, leading to an overpressured situation. If the sedimentary deposition rate is about 30 m/Myr or more, then there is a very strong possibility of significant overpressuring. Since, on average, about 75% of all deposition is low-permeability shale, overpressure is dominant in burial history calculations. Several consequences are immediate. First, compaction of overpressured formations is less than without overpressure, so formations retain thicknesses closer to their depositional values for a longer period of time. In turn this effect changes the structural and stratigraphic shaping of sedimentary units, providing as well a wider (laterally) thoroughfare for fluid migration. Second, the thermal conductivity of an overpressured region is lowered, thereby increasing the vertical temperature gradient (often by as much as about 1°C/100 m from the "normal" value of around 2–2.5°C/100 m) in order to preserve the heat flux. Thus overpressured regions tend to be hotter than non-overpressured regions at the same subsurface depth. Third, as a consequence of the increased temperatures, the solubilities of carbonates and sulphates are lowered, leading to calcite, gypsum, marl and anhydrite precipitation which tend to strengthen formations by diagenetic cementation. Fourth, isolated sand bodies in a surrounding shale matrix will also be overpressured so that a parametric geometry-dependence on overpressuring exists. Fifth, the thermal expansion of hot water in the overpressured formations possibly also provides an excess pressure acting to divert fluid flow, and to pressure the encompassing rocks. If the total fluid pressure exceeds a limiting value, a rock must fracture, thereby releasing fluid. Empirically the total fluid pressure limit appears to be roughly around 0.8–0.85 psi/ft (excess fluid pressure of 0.3–0.35 psi/ft) (1 psi/ft $\equiv$ 0.2 atm/m). Thus the dynamical evolution of sediments is firmly tied to the dynamical fluid flow. The ability to account for this behavior is crucial to any model.

In addition, an increased temperature will also speed hydrocarbon generation from kerogen. These mobile components provide a generated pressure, can flow either in solution with the prevailing waters or as separate phases, and so impact on the dynamical burial history of the sediments.

Apart from fluid effects, major components in burial history evolution are also provided through depositional variation laterally in a basin, by erosional events, overthrusting, rotation of fault blocks, by intrinsic faulting of the sediments as syndepositional growth faults, tectonically induced faulting, or shear failure caused by differential stress on consolidated, lithified formations as the basin evolves. The presence of faults can lead to fluid escape up the faults, if the faults stay open for any length of time, thereby influencing the sediment evolution.

Two other lithological fabrics, salt and carbonates, can also be dominant controls on basinal evolution. Salt normally occurs in an evaporitic basin which is later open to clastic input. The evaporation can leave behind great thicknesses (several km) of salt overcoated with other evaporites. Later sediment deposition on the underlying mobile, constant density, salt causes salt movement laterally and vertically, giving rise to sediment distortion, faulting, and changes in the pattern of later sediment deposition through topographic expression caused by the salt and from dissolution.

Carbonates play a different role. Deposited as the skeletal remains of marine organisms, carbonates basically "grow" by secretion of calcite. Hence carbonate reefs, platforms, pinnacle reefs, and so on represent shallow marine warm water where nutrient supply is abundant and the optical depth to sunlight penetration is high. Large, laterally extensive, carbonate deposition leads to the weight of the deposited carbonates contributing to basinal sag, leaving a volume accommodation for further carbonate deposition. Apart from the continuing impact such a weight has on the evolution of basinal structure, the stress on neighboring formations can often be sufficient to produce major faulting. In addition, the porous carbonates can act as good reservoir sites, can alter further depositional patterns of sedimentation, and can act as aquifers for ground water transport into the deep sub-surface, and for deep basinal waters to reach the surface.

In any attempt to determine the dynamical burial of sediments in a basin we have four major controls at each well, or pseudo-well, location. These controls are (a) age-dated formation lithologies and thicknesses, (b) porosity with depth, (c) fluid pressure with depth, and (d) formation permeability with depth. A model must honor these four controls at each location.

Models and consequences

An inverse procedure: isostatic reconstruction.

Geohistory (Fig. 1a) and burial history plots (Fig. 1b) can be used to predict the evolving thickness of formations (van Hinte, 1978), estimate the amount of material removed at unconformities by erosion (Fig. 1c), and estimate the rate of sediment deposition as a function of time (Fig. 1d).

The data needed to construct burial curves are the geologic ages of the layers, their depths, and the water depth at the time of deposition of each layer. The burial history plot may be obtained by linearly interpolating between the depth today for the first layer and the time at which this layer was at the surface. Ignoring compaction effects, all lower layers move up at the same rate as the first layer. When the first layer was on the surface, depths to all lower layers are calculated. The process is repeated until all layers have been moved to the surface. The geohistory plot is then constructed by adding the depth of water at each time of deposition.

To model compaction history, it is necessary to model the changes in rock porosity with depth and time. One approach is to consider the rate of sediment accumulation (R) with respect to time, as given by van Hinte (1978):

$$R = T_\rho(1 - f)/A(1 - \phi_0) \quad (1)$$

where T_ρ is the thickness at present porosity (ϕ), A is the time duration, and ϕ_0 is porosity at time of deposition.

To model layer thickness as a function of time van Hinte (1978) illustrated a procedure using observed present porosities and back-calculating to arbitrarily chosen initial porosities. In the absence of porosity data from wells, phenomenological models (see below) can be used to relate porosity to depth or load for various lithologies. Actual values of porosity are variable and depend on compositional variations of lithology, overpressure or undercompaction, and diagenesis, but generally decrease with depth. For normal pressures, porosity, ϕ, can be approximated by:

$$\phi = \phi_o \exp(-cz) \quad (2)$$

where ϕ_o is the surface ($z = 0$) porosity. Values of ϕ_o and c from bore-hole porosity measurements can be obtained by fitting via least-squares techniques (Sclater and Christie, 1980). An alternate phenomenological model of porosity has the porosity decrease as a function of the load on a layer which leads to an empirical relation of the form:

$$1/\phi = 1/\phi_o + kz \quad (3)$$

where ϕ_o is the initial uncompacted porosity, z is the depth and k is a lithology-dependent constant.

Well data may be corrected for the effects of compaction as follows. First, the total conserved amount (h_s) of solid material in an interval (z_1 to z_2) is given in the Sclater-Christie model by:

$$h_s = (z_2 - z_1) - (\phi_o/c)[\exp(-cz_1) - \exp(-cz_2)] \quad (4)$$

When the top of a stratal unit is on the surface ($z_1 = 0$), its base (z_2) may be calculated by solving eq. 4. Similarly depths and thicknesses of all lower layers can be obtained sequentially and self-consistently. A similar equation is obtained in the porosity-load

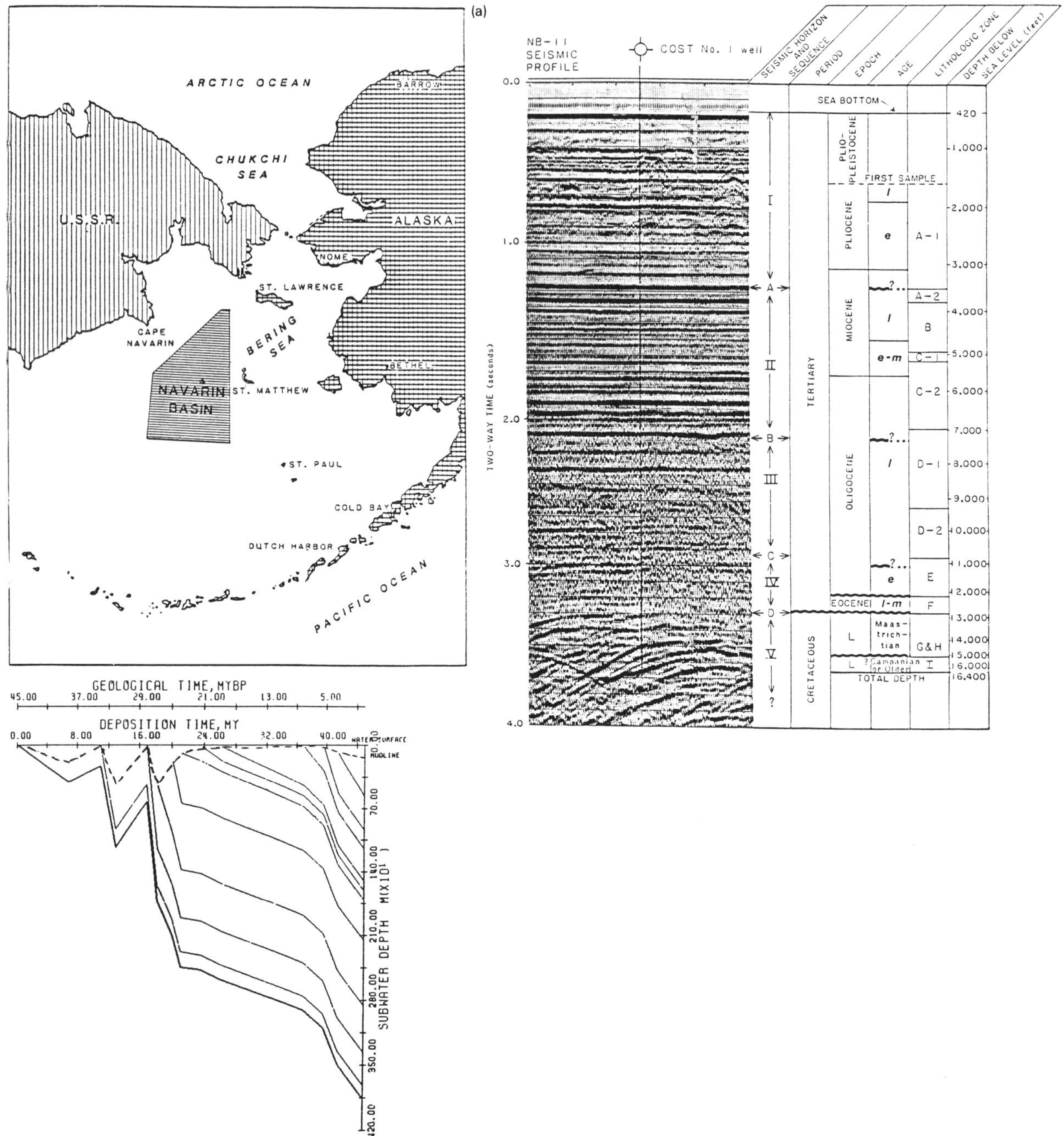

Fig. 1a. Reconstructed geohistory of COST No.1 well (offshore eastern Alaska) allowing for overpressuring with time. Note the rapid subsidence event at about 28 Ma, with relatively slow subsidence thereafter, and two major unconformities prior to the rapid subsidence.

model:

$$z_2 - z_1 = h_s + (1/k) \times \ln[(1/\phi_o + kz_2)/(1/\phi_o + kz_1)] \qquad (5)$$

Given the compaction parameters and drill-hole values of depth, age, and palaeowater depth, a burial history curve (Fig. 1b) is obtained.

Unconformities disrupt any regular porosity versus time or depth model, regardless of the model assumptions or procedures. There are at least three possible ways to model an unconformity: (1) as a depositional hiatus; (2) as a period of deposition followed by erosion in which the thickness of the eroded section and the absolute age marking the beginning of erosion are determined from a knowledge of local geology (see van Hinte, 1978); parameters may be varied to see which model(s) best fit observed data (e.g., present-day vitrinite reflectance and/or behavior of velocity or density logs); (3) as a period of deposition and erosion in which the thickness of the eroded section and the age at the beginning of

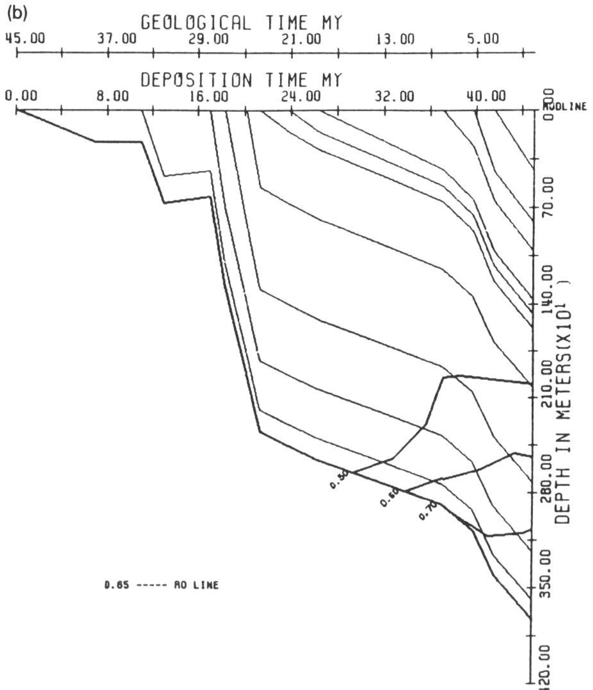

Fig. 1b. Superposition of burial history of the COST No. 1 well and iso-reflectance lines, obtained by using the best palaeoheat-flux variation from thermal inversion. Note that the $R_o = 0.6$ iso-reflectance line (occasionally considered indicative of formations entering the "oil window") varies from 2800 m at about 20 Ma to about 2500 m at the present day, suggesting that hydrocarbon maturity has occurred only for formations deeper than about 2500 m and deposited earlier about 27 Ma.

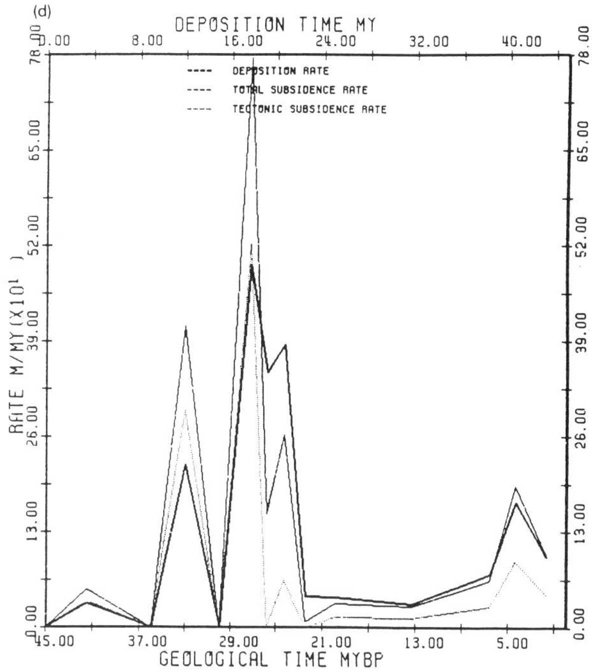

Fig. 1d. The three curves show (i) the rate of subsidence including sediment load (light dashed line); (ii) the rate of basement subsidence with isostatic compensation due to sediment load removed (heavy dashed line); and (iii) the sedimentary deposition rate (light dotted line).

erosion are calculated from sedimentation rates in the immediately overlying and underlying sediment layers.

From the depth and thickness of each layer as a function of time, and from the lithologies and densities of the layers, basement subsidence history is inferred by applying isostasy calculations to replace successive backstripped sediment layers by water. The basement subsidence (Y) in terms of the water depth (W_d) and sediment thickness (S) is given by:

$$Y = S[(\rho_m - \rho_s)/(\rho_m - \rho_w)] + W_d - \Delta_{SL}\,[\rho_m/(\rho_m - \rho_w)] \quad (6)$$

where ρ_m (ρ_w) is the average mantle (water) density, ρ_s is the average sediment density, Δ_{SL} is the change in elevation of mean sea level, and Y is the depth to basement without sediment and water loads, and represents the subsidence caused by tectonic effects. The terms W_d and D_{SL} must be determined by palaeobathymetry. The isostatic reconstruction assumes that the observed present-day porosity variation with depth has been constant through time; effects of fluid overpressuring are not incorporated.

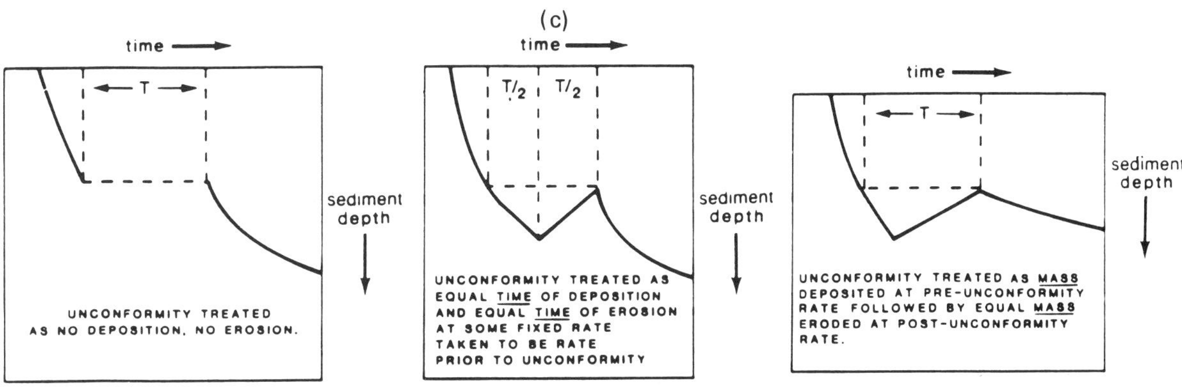

Fig. 1c. Geohistory diagram with unconformities handled in different ways.

A pseudo-inverse procedure: fluid flow/compaction. To incorporate overpressuring, compaction of sediments is taken to be the result of fluid expulsion caused by increasing overburden. Vertical fluid movement through sediment is described by Darcy's equation:

$$\partial(\rho_f e)/\partial t = \partial/\partial\xi[K_z \rho_f \mu^{-1}(1+e)^{-1}\partial P/\partial\xi] \quad (7)$$

where e is the void ratio (related to porosity ϕ by $e(1-\phi)=\phi$), $\rho_f(\mu)$ the density (viscosity) of the fluid, K_z the vertical permeability, and P is the fluid pressure in excess of hydrostatic. The fully compacted depth coordinate, ξ, represents the net thickness of the rock. Darcy's law can then be applied correctly. The physical depth, z, is related to ξ by $d\xi/dz = (1-\phi)$. A multi-dimensional equation for non-vertical flow is also available (Lerche, 1990a).

Excess fluid pressure and frame pressure, P_f, are related by

$$P = \int_{\xi}^{\xi*} g(\rho_s - \rho_f)\,d\xi - (P_f - P_{f*}) \quad (8)$$

where g is gravitational acceleration, ρ_s is the solid density, P_f is the frame pressure (a prescribed function of void ratio for each lithology), and * denotes the appropriate value on the sediment surface $\xi*$ ($P = 0$ on $\xi*$).

All variables are time-dependent and related to the sediment accumulation rate. Solving eq. 7 determines the changes of sediment thickness, porosity, permeability, pressure and fluid flow. Current values constrain the model predictions (Fig. 2 and Table 1).

In the quasi-empirical equations of state for shales:

$$K_z = K_*(e/e_*)^A; \quad P_f = P_{f*}(e/e_*)^{-B}; \quad (9)$$

the two numbers A and B are varied until the observational constraints are met. Empirical evidence from many wells suggests $A = 3 \pm 2$, $B = 3 \pm 3$ (Lerche, 1990a). The procedure is pseudo-inverse because only parameters (not functions) are adjusted until predicted values are in accord with observations as shown in Fig. 2 for the Navarin Basin COST No. 1 well (offshore eastern Alaska). Figure 1b shows the burial history.

Thermal history

Overview

Concentrate next on the determination of sedimentary thermal history, which rests on two fundamental factors: variation of sources of heat flux through, and within, the sediments; and heat transfer (conduction, convection) through the sediments. Significant internal heat sources can be highly radioactive layers (e.g. Kimmeridge clay, North Sea Basin), igneous intrusives, frictional heating by overthrusting, and so on. Such heat sources, dominant locally, appear secondary in magnitude and regional character to basement heat flux (exceptions occur at greater depths in some sedimentary basins, depending on lithology types).

Present-day estimates of sedimentary rock thermal conductivities vary in the rough range 7 ± 3 mcal cm^{-1} s^{-1} C^{-1} with low values for high porosity, unconsolidated rocks, and high values for low porosity, nearly fully compacted sediments (Hunt, 1979). Ignoring lithology effects, an empirical estimate connecting fractional porosity, ϕ, and bulk thermal conductivity, κ, can be obtained from (Hunt, 1979):

$$\kappa = K_R^{1-\phi} K_w^{\phi} \quad (10)$$

where K_w is thermal conductivity of water (~1 mcal cm^{-1} s^{-1} C^{-1}) and K_R is the thermal conductivity of fully consolidated sediments (~10 K_w). Present-day temperature gradients are around 2°C/100 m so that, if conduction dominates, estimates of basement heat flux today are around $Q = 1$ HFU (= 42 mW m^{-2}). A dynamical range of about 0.5–3 HFU is not inappropriate for most sedimentary basins today. The spatial variation of present-day heat flux is estimated from measurements, but palaeoheat flux is also required. Two end-member approaches are possible. (1) From the dynamical evolution of a basin, posit a *model* of basement motion and a *model* relation to heat flux variations, determine any parameters in the proposed basement motion, and apply these directly to the heat flux. Such schemes give rise to a variety of popular models of the stretching-rift sort (McKenzie, 1978), to phase change models, and the like. Most such models end up predicting some sort of high heating in the past ($Q \approx$ 4–10 HFU) with a cool-down to the present day (see later). The weakness is that both a dynamical model/thermal model connection must be made, as well as a dynamical model from which to extract salient parameters. (2) From various thermal indicators measured in wells; if the temperature-dependent physics/chemistry of the indicators along the burial paths is known, then the temporal evolution of temperature can be determined. Lopatin (1971), and later Waples (1980) and Brooks and Welte (1984), have popularized the idea that vitrinite reflectance, $R(z)$, measured with depth z, can be related to a time, t, temperature, $T(t)$, integral (TTI) of the form:

$$\text{TTI} = \int_0^t \exp[(T(t') - T_c)/T_D]\,dt' \quad (11)$$

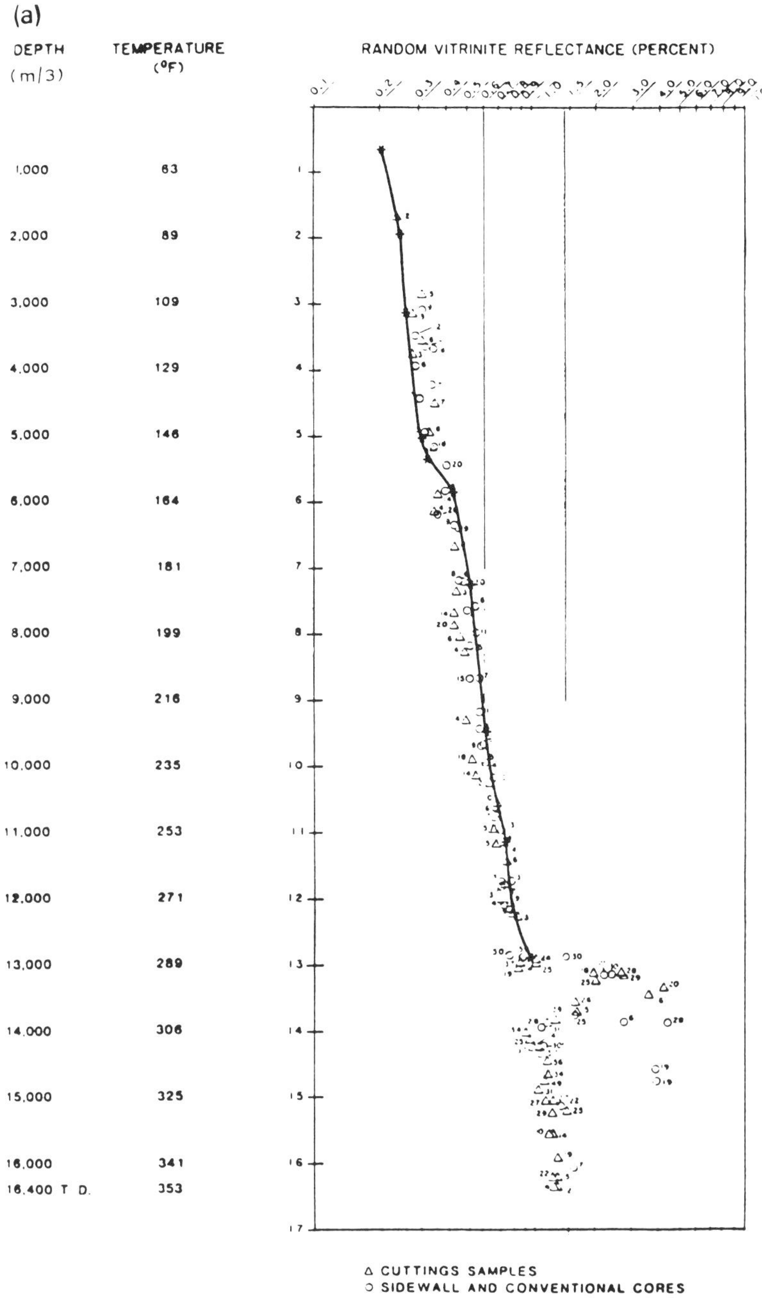

Fig. 2a. Plot of vitrinite reflectance observations with depth (symbols). Superimposed is the solid curve representing the predicted vitrinite reflectance obtained by thermal inversion.

where the integral is taken along the burial path of the sediments from deposition to the present day, and T_c and T_D are two scaling constants. To make TTI a hydrocarbon measure Waples (1980) suggested that T_c is best set at 105°C, and T_D at $10/\ln_e 2$°C. But a connection still needs to be found between $R(z)$ and TTI. If the geothermal gradient is constant for all time, then TTI can be *calculated* for each sedimentary layer and plotted versus the *measured* $R(z)$ values. A quasi-empirical relation is then generated connecting TTI and R. Unfortunately there is no way of knowing whether the geothermal gradient is constant with time, nor is there any way of assessing the choice of parameter values for T_c and T_D. Indeed, a *constant* geothermal gradient with depth today is not possible because of thermal conductivity variations, and palaeogeothermal gradients relevant to a specific layer cannot be identical to present gradients. Waples (1989) has strongly pleaded for kinetic descriptions to replace the Lopatin (1971) TTI. Every thermal indicator (vitrinite, steranes, pollen translucency, hopanes, $^{39}Ar/^{40}Ar$, apatite fission tracks, etc.) has its own particular TTI, peculiarly related to the nature of the individual indicator (Lerche, 1988a, b).

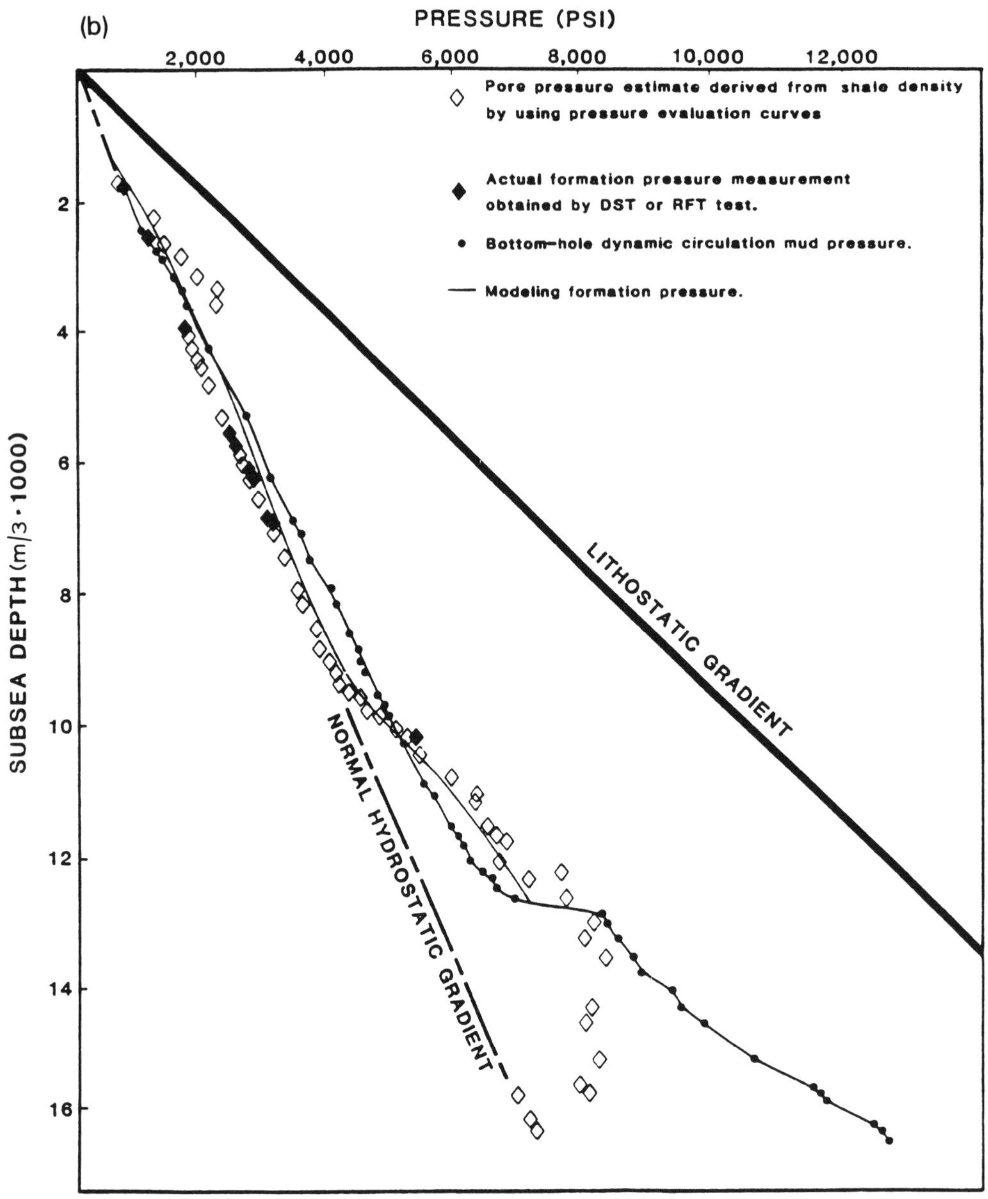

Fig. 2b. Pore pressure measurements with depth. The three different symbols represent different methods of estimating pore pressure. The light solid curve is the predicted variation of pore pressure with depth obtained from running a fluid flow/compaction code.

Inverse schemes, based on multiple thermal indicators, are rapidly replacing the thermal maturity procedure of "calibrating" the Lopatin-Waples TTI to vitrinite reflectance (values of TTI = 15 and TTI = 75 are often taken as marking the onset of entry to the "oil window" and "gas window", respectively; or $R = 0.6$ and $R = 1.1$) (Waples, 1980).

Even with palaeoheat flux known throughout a basin, palaeotemperature determination is still confounded by two major problems.

First, thermal conductivity variations due to lithology and porosity differences act to focus and defocus conductive heat transport. Second, high-permeability domains provide pathways for convective transport of heat, thereby heating and/or cooling regions in amounts considerably different than conduction alone provides. Hydrocarbon generation is affected, as is hydrocarbon transport and further generation of oil to gas after location in a reservoir site, by such convective heat transfer. In addition mineral solubility is changed by high-temperature fluid flow leading to cementation and/or dissolution which, in turn, influence the migration pathways by diagenetic evolution. In turn the transport of heat impacts on fluid expansion and rock failure, as well as on porosity. But the porosity variations relate directly to the thermal conductivity, thereby altering the temperature gradient which then changes the convective heat transfer, and which reflects directly on the burial history. Thus thermal and burial histories are intertwined and must be addressed simultaneously.

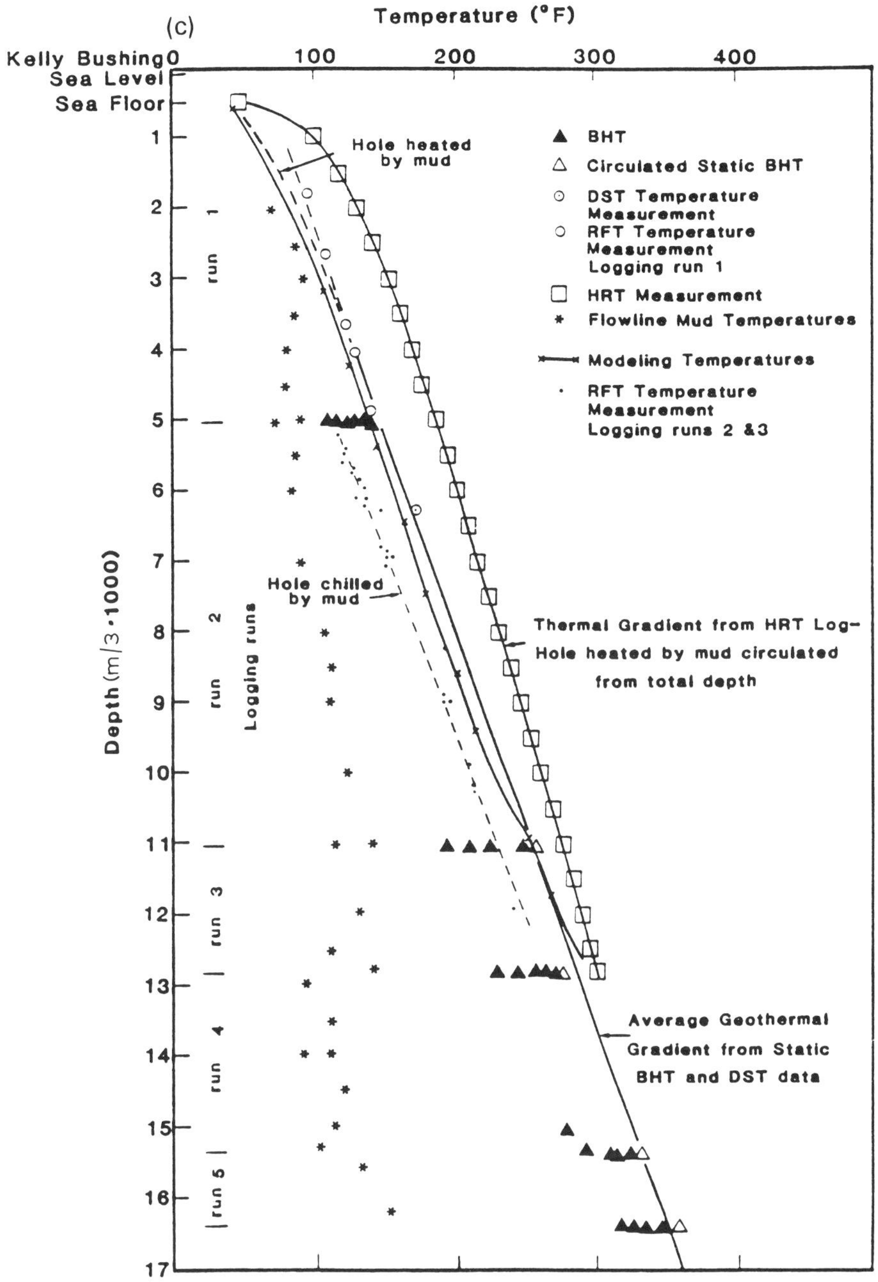

Fig. 2c. Several measured present-day temperature profiles. To obtain a present-day heat flux the curve shown is used.

The only direct constraints on the thermal history models are measurements of multiple thermal indicators and present-day temperature gradient (Figs. 2a, 2c).

Thermal history models

Two end-member models are often used: forward models with undetermined parameters and pseudo-inverse models based on present-day observations.

Forward models of palaeoheat flux. McKenzie (1978) argued that rapid stretching of continental lithosphere produces thinning and passive upwelling of hot asthenosphere, with associated block faulting and subsidence. The lithosphere then thickens by cooling, and the surface slowly subsides. The decreasing rate of subsidence and heat flow depend on the amount of stretching. Sclater and Christie (1980) modified the McKenzie model by adding a heat flow component from the decay of radioactive elements within the crust. An additional thermal blanketing by low thermal conductivity sediments slows the cooling of the lithosphere, perhaps important to studies of older rifted margins, such as the North Sea, where sediment thicknesses occur in the order of 15 km (49,000 ft). Modified stretching models encompass two layers to explain uplift and erosion on the flanks of a rift basin. More recent work has been concerned

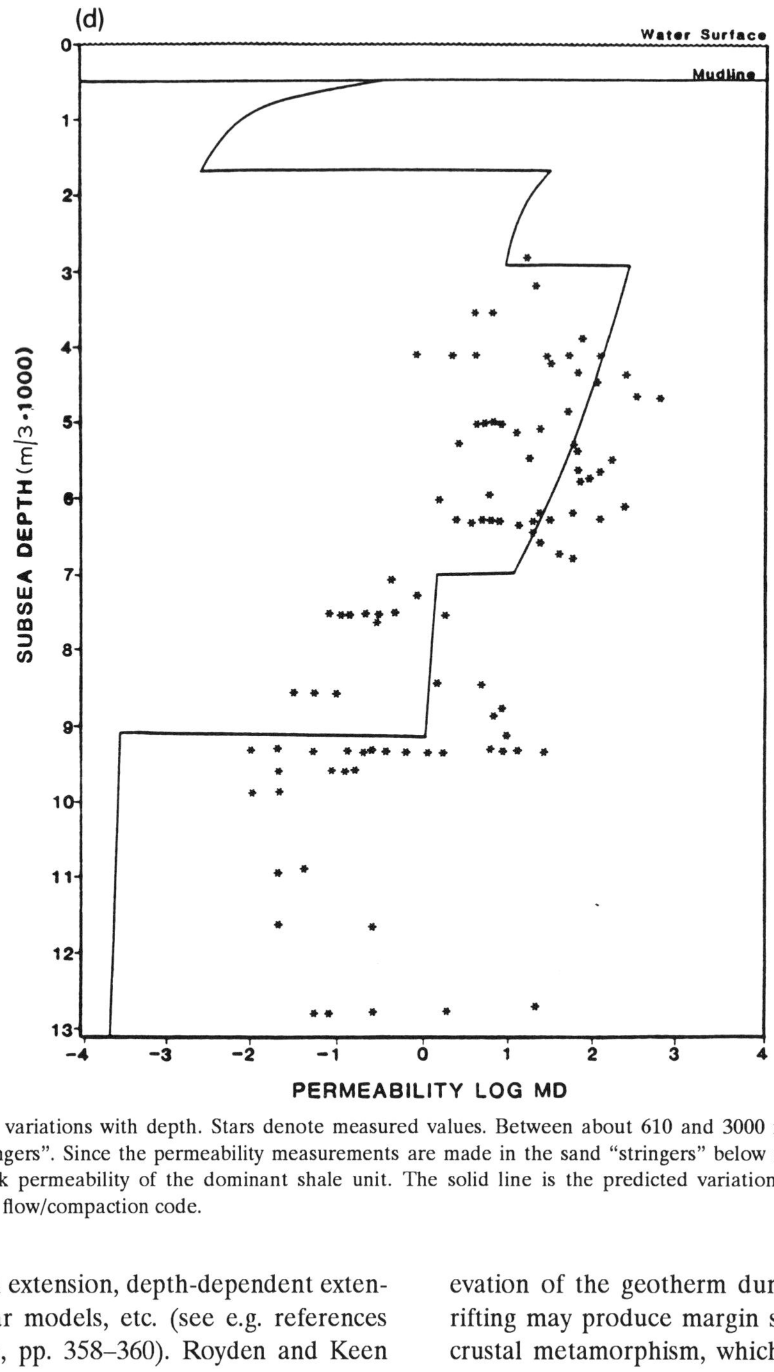

Fig. 2d. Permeability variations with depth. Stars denote measured values. Between about 610 and 3000 m, the dominant lithology is sand with shale "stringers". Since the permeability measurements are made in the sand "stringers" below about 3000 m, they seriously misrepresent the bulk permeability of the dominant shale unit. The solid line is the predicted variation of permeability with depth obtained from a fluid flow/compaction code.

with non-uniform extension, depth-dependent extension, simple shear models, etc. (see e.g. references in Lerche, 1990b, pp. 358–360). Royden and Keen (1980) and Royden et al. (1980) proposed a model of continental lithosphere fracturing and intrusion by dikes from the mantle. Replacement of light crustal rocks by denser ultramafic materials results in initial subsidence and avoids the general problem of uplift.

The subsidence is *inferred* from the burial history. Discordance between predictions and inferred observations is minimized by parameter adjustment, and such parameters are then used directly in the posited heat flux models. The subsidence components of these models are pseudo-inverse, and the heat flux components are forward models.

These theories of basin formation rely on thermal contraction of the crust to produce subsidence. Elevation of the geotherm during continental margin rifting may produce margin subsidence due to deep crustal metamorphism, which occurs predominantly at the greenschist–amphibolite facies boundary. The model requires a period of heating followed by a period of cooling. Subsidence due to deep crustal metamorphism occurs during the later part of the heating period and is followed immediately by subsidence due to thermal contraction during the cooling period. The distinguishing features of this model are that the heat flow *increases* up to the point of breakup and then *decreases*, as opposed to other models in which heat flow always *decreases* after the time of stretching. The heat flow from a metamorphic model is in the order of half that of a crustal extension model, caused by the different model *postulates* that connect heat flux to subsidence (elevation).

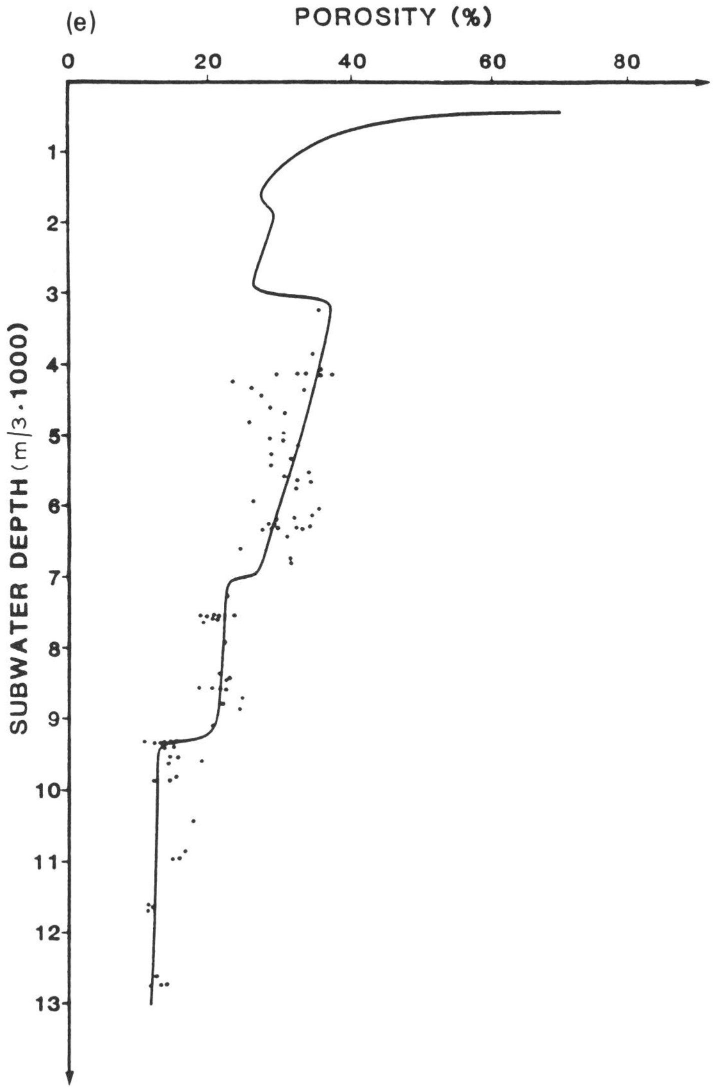

Fig. 2e. Porosity measurements with depth in COST No. 1 (dots) and the predicted porosity variation (solid line) obtained from a fluid flow/compaction code.

Inverse and pseudo-inverse models of palaeoheat flux. Dissatisfaction with forward heat flux models has focused attention on inverse methods applied to measured thermal maturity indicators (Lerche, 1990a).

Thermal indicator inversion schemes for 1D systems are constructed as follows. In a one-dimensional burial history, variations of temperature, $T(z, t)$ occur only with depth z, and with time, t. A thermal indicator, $TI(z)$, is then related to the temperature history of the stratum in which it is measured through a kinetic evolution model. The problem of determining the entire temperature field is replaced with the problem of determining the heat flux, $Q(t)$, at one depth in the basin. The inverse problem takes the thermal indicators backstripped through the layers, and determines the $Q(t)$ consistent with observations of $TI(z)$. Modern basin analysis codes include convective and conductive effects in dealing with real basins to handle the inverse problem appropriately (Lerche, 1988a, b).

The inversion procedures use present-day downhole measurements to determine palaeoheat flux variations subject to several constraints:

(1) The underlying thermal evolution process and temperature dependence controlling maturation of each thermal indicator must be known.

(2) The chemical and physical parameters associated with the evolution process must be provided. (However, inversion procedures normalize out as many of such parameters as possible so that palaeoheat flux determinations are beholden minimally to vagaries of uncertainties in the parameters.)

(3) The burial history of each sedimentary layer is assumed to be known with precision in order that inversion procedures can be used to extract

TABLE 1

Thickness change (m) of each modelled layer with time (Ma)

Layer number:	1 shale	3 shale	5 shale	6 sandy mud	7 sandy	8 sandy	9 sandy	10 sandy	11 sandy	12 gray shale	13 mudstone	Total thickness
Deposition time (Ma)												
7 (1)	230.0											
11 (2)	229.4											
13 (3)	196.8	477.2										
17 (4)	187.5	436.8										
18.3 (5)	181.0	401.1	691.5									
20.2 (6)	171.3	385.0	558.1	800.1								
21.3 (7)	164.4	373.4	530.1	734.1	553.5							
23.8 (8)	161.3	368.3	528.2	729.8	514.6	157.1						
26.4 (9)	159.1	364.8	524.6	726.8	508.7	123.3	154.0					
37.1 (10)	153.3	357.0	515.5	717.5	487.1	118.3	115.8	426.8				
39.7 (11)	151.4	355.9	513.6	712.4	476.0	115.5	113.1	380.5	261.5			
41.6 (12)	148.0	355.1	511.3	705.0	452.9	109.6	107.3	361.8	211.3	440.5		
45.0 (13)	144.6	353.4	508.7	701.8	442.2	107.0	104.8	353.6	206.7	375.0	431.2	3729.0 predicted
Measured thickness	152.4	359.7	502.9	707.1	434.6	104.8	106.7	350.5	207.3	384.1	427.6	3737.7 measured

Note. Layers 2 and 4 are unconformities treated as depositional hiatus.

palaeoheat flux variations with maximum sharpness.

With geological data and their attendant "scruffiness", not too much improvement can be made to the underlying thermal evolution processes. Such functional determinations are determined more properly by controlled laboratory experiments.

But chemical and/or physical parameters of such processes determined in the laboratory, over human lifetimes and under pristine conditions, do not have to be the appropriate values for maturation of thermal indicators over geological time scales and in the uncontrolled environment of an evolving sedimentary basin. So such parameters should be regarded as unknowns and should be determined by inversion procedures.

In a similar vein, geological parameters related to unknown variations in burial history information should also be determinable from inversion procedures.

Thus three basic types of parameters enter the methods used for thermal indicators: (1) parameters such as activation energies, critical temperatures, and rate constants directly associated with the underlying physics and/or chemistry peculiar to each thermal indicator; (2) parameters such as palaeoheat flux variations with time, radioactive heating etc. with time directly associated with changing the thermal maturity of the sedimentary layers; and (3) parameters such as erosional unconformities, palaeo-overpressuring etc., which directly change the burial path of the sedimentary layers with time, and so the temperature, or which directly change the thermal conductivity and so the temperature, even if the palaeoheat flux does not vary with time. They therefore have a role to play in changing the time–temperature integrals and therefore the thermal maturity of the individual thermal indicators with depth at the present day.

Out of this array of parameters, one or two such parameters in a given geological setting are easy to determine by trial-and-error searches. But if more than a few parameters are involved (such as would happen with, say, multiple unconformities where the erosional thicknesses, and/or onset time of erosion and palaeoheat flux variations with time are to be determined), exhaustive trial-and-error searches are not only inefficient but extremely difficult to implement.

The difficulties are clear: how to determine simultaneously a large number (say four or more to fix ideas) of a mixture of chemical, thermal, and geological parameters by inversion of downhole thermal indicator data? And how to know when resolution capability of the data is reached so that it is pointless to search for further parameters?

Fortunately, precisely this type of parameter determination problem has arisen in a variety of other scientific disciplines over the years and a substantial body of information has been culled on appropriate methods of analysis to tackle such problems. In general, the most powerful and accurate techniques are those based on some form of tomographic analysis.

Thermal indicator tomography is a nonlinear inverse problem. Nonlinearity arises because thermal maturity along the paths taken by sediments, and the paths themselves, are affected by parameters to be determined (e.g., palaeoheat flux, insertion times, unconformity erosion thicknesses, etc.). Generally, enough gross information about the palaeostructure of a region is available to approximate roughly the thermal indicators; to this extent, thermal indicator tomography is "almost linear". As usual in nonlinear problems, an initial guess is made for the parameters, the linear inversion is solved for an improved set of parameters, and the process is reiterated. Experience from many disciplines suggests that if the (linear) inversion has been properly regularized (stabilized), and if the intermediate solutions are constrained to be reasonably smooth, this procedure converges satisfactorily in a few iterations (typically 5 to 10, although extreme cases are known when many tens of nonlinear iterations are necessary).

The essence of thermal indicator tomography, then, is to construct a model which approximates past thermal behavior of the subsurface, compute theoretical thermal indicators to each sedimentary layer by forward tracing in time through the model, measure the actual thermal indicators from the appropriate depths, and modify the thermal model through which the theoretical thermal indicators traveled in time to minimize the difference between measured and computed maturations. The model is modified by inverting a version of the least-squares equations. Finally, the entire process is iterated by tracing through the modified model and proceeding as before (Lerche, 1988a, b).

Hydrocarbon generation, migration and accumulation histories

Overview

Concentrate now on the hydrocarbons. General considerations (Hunt, 1979; Brooks and Welte, 1984) suggest that the dominant mode of hydrocarbon production is by thermal degradation of kerogen after sedimentary burial. A "rich" source rock might contain around 1–5% TOC (total organic carbon) capable of producing hydrocarbons. Three major problems need to be addressed: (1) how are hydrocarbons produced from kerogen? (2) how do hydrocarbons migrate out of the (dominantly shale) source beds, and then how do they move thereafter (primary and secondary migration)? (3) what sort of accumulation sites and sealing conditions are available for the hydrocarbons, and do those sites persist through to the present day?

Hydrocarbon production is a peculiar phase change in which solid material (kerogen) is turned into mobile liquids (oils) and gases (mainly methane but with some ethane, H_2S, CO_2, N_2). Oils are virtually incompressible and normally have a specific gravity around 0.8–1.01 — usually less than subsurface brines ($\sim$1.0 to 1.3). Gases are highly compressible, and normally have a specific gravity of around 0.1–0.3 at production. Thus oils and gas are buoyant relative to water (buoyancy pressure $\sim$0.1 psi/ft for oils, $\sim$0.3–0.5 psi/ft for gas). Some oils are completely miscible with water, others are not (Hunt, 1979), while gas tends to be completely miscible until the "bubble point" pressure and temperature are reached when gas exsolves and transports as a separate phase.

The ability to provide a detailed kinetic scheme relating the thermocatalytic breakdown of kerogen to hydrocarbons is difficult — perhaps because such a scheme must handle organic macromolecule chemistry (kerogen) breaking down to methane (molecular weight 16) in the sub-surface when ionic, mineral, and surficial chemistry influence and modify the chemical kinetics. From modified van Krevelen diagrams of H/C and O/C (Hunt, 1979) there are at least three types of kerogen — presumably each having its own kinetic breakdown scheme, and yielding different production efficiencies for oil and gas generation. As the sub-surface temperature increases oil converts to gas (Hunt, 1979).

To by-pass detailed kinetics a "lumped" kinetic scheme was introduced (Tissot, 1969; Tissot and Espitalié, 1975; Tissot and Welte, 1978). Since then, many other "lumped" kinetic frameworks have been suggested. A common theme among them is multichannel reactions, in series or parallel, with specification of many reaction rate constants and activation energies which relate kerogen to oil and gas, and oil to gas. All are temperature-dependent and most involve first-order chemical kinetics. Typical *total* production estimates range from 1 to 500 mg of hydrocarbons per gram of kerogen, while typical estimates of the *onset* of significant oil (gas) production *rates* are round 100°C (150°C).

The question arises as to how the mobile hydrocarbons escape the fine-grained clastic sources beds of generally low permeability (primary migration). Several mechanisms have been proposed. Thermal expansion coefficients of fluid hydrocarbons are larger than rock values. Hence the conversion of kerogen to fluid puts an additional pressure on the encompassing source rock leading to microfracturing and fluid loss through the high-permeability fractures. An alternative suggestion is that the volume occupied by produced hydrocarbons is larger than that of the original kerogen, leading directly to an increase in pressure, again driving fluid motion and microfracturing. A third suggestion is that capillary pressure at sand/shale interfaces has a dominant role to play. Typical pore radii in fine-grained materials are around 10^{-5}–10^{-6} cm, while in sands the size is around 10^{-4} cm. The surface tension difference, T, between oil and water is around 40 dyne/cm so that the capillary pressure, p, at a sand/shale interface is $p \approx 2T(1/r_{\text{fine}} - 1/r_{\text{coarse}}) \approx 10$–100 atm ($\equiv$150–1500 psi). Thus capillary pressure is the largest force at, but decreases rapidly away from, the boundary. Hence oils displace water in coarse-grained beds abutting a fine-grained source, while water displaces oil in the source.

Oil and gas then find themselves in a relatively high-permeability bed and move (secondary migration) under several forces. Any excess fluid pressure provides a hydrodynamic drive. Those fractions of oil and gas which go into solution move with the prevailing fluid. The non-soluble gas/oil fractions have a vertical buoyancy pressure drive acting on them of around 0.1–0.5 psi/ft (1 psi/ft = 0.2 atm/m), causing motion along the path of least resistance, which is not necessarily vertical. Evolution of the structural aspects of bedding geometry now come into play; up-dip migration to structural culminations being favored by oil and gas.

On the other hand Darcy's law notes that the speed of motion is proportional to the permeability as well as to the driving pressure gradient. As the hydrocarbons move, any significant decrease in permeability will effectively bar their further migration. Such decreases can be caused by compaction as the basin develops, by diagenetic cementation, by spatial variations in lithology and, of course, by tectonic forces altering the basin. Thus there are also evolving stratigraphic sites of potential accumulation.

Continued evolution of the basin can lead to aquifer flow attempting to flush hydrocarbons from earlier accumulation sites. Dynamical hydrodynamic transport plays a major role in some hydrocarbon regions. In addition, the sealing conditions around a filled reservoir can change significantly as later deposition and basinal development impact on the stress conditions of the seal, leading to fracturing and faulting and consequent loss of hydrocarbons.

Hydrocarbon models — forward and pseudo-inverse

To determine the amounts of hydrocarbons that have been generated and accumulated and their locations, information needed is: (1) amount of oil and gas generated in each source rock and in every part of the basin; (2) timing of hydrocarbon formation, timing of development of impermeable seals, and timing of folding and faulting, so traps can be identified; (3) amounts of oil and gas that have been expelled from the source rock(s) and how much is still retained; (4) evaluation of the ultimate oil and gas reserves of the sedimentary basin. A quantitative approach provides this information. However, because many of the intermediate steps in petroleum and gas generation from kerogen degradation are unknown, such a scheme has to be phenomenological; parameters have to be determined by least-squares fitting to basin and laboratory data.

The validity of a phenomenological model depends on two factors: (1) the validity of the physics and chemistry involved in the calculation, and (2) the values of the parameters chosen. The Tissot scheme considers three types of initial kerogens which differ in their hydrogen-to-carbon ratio. Type 1 (algal) has a high liquid-generating potential, type 2 (amorphous) has an intermediate liquid-generating potential, and type 3 (woody) has a low petroleum potential. Each type is assumed to have six different bond types that have activation energies in the range 10 to 80 kcal/mole. The amount of organic material (x_i) reacting in the ith reaction (breaking of the ith bond type) for $i = 1$ to 6 is specified for each kerogen type. Calibration for the formation of gas has been made from laboratory experiments on hydrocarbon cracking. A single reaction with an activation energy of 50 kcal/mole was convenient to account for gas generation in the deep parts of sedimentary basins. Tables of the activation energies (E_i), the reaction constants (A_i), and the initial values (x_i) for each kerogen type are given by Tissot and Welte (1978, table V.4.1).

Two problems are apparent. Is the structure of the kinetic model adequate? How are the best kinetic parameters determined? Other kinetic schemes have been proposed (see references in Lerche, 1990b) that also are consistent with limited observational data. At the time of writing,"lumped" kinetic models can be handled in a forward modelling sense or a pseudo-inverse sense. The generation parameters are often determined from measurements made *after* the produced hydrocarbons have moved from their source "kitchen" to reservoirs, confounding the problem even more.

Basement motion

Basement response to loading (and correspondingly, loading response to basement movements!) is one of the major factors that determines the shape and sedimentation fill in sedimentary basins. Early models of lithospheric compensation were based on the principle of isostasy — the idea that crustal loads are in hydrostatic equilibrium with the upper mantle. In effect, the crust is assumed to be supported entirely by buoyant forces in the mantle. Fundamental to this model is the assumption that the lithosphere has zero rigidity. However, the presence of regional free-air gravity anomalies over large-scale features indicates only partial compensation by isostatic forces; some portion of the load is supported by the rigidity of the lithosphere (Turcotte and Schubert, 1982). Two end-members are possible: (1) zero rigidity, loads are entirely compensated by isostatic forces; (2) infinite rigidity, loads are completely supported by lithospheric rigidity.

Proposed flexural models fall into one of three general categories: elastic, viscoelastic and thermally controlled flexure (Keen et al., 1981). Linear elastic models are based on Hooke's law (Turcotte and Schubert, 1982). Deformation occurs instantaneously under an applied stress with complete recovery when the stress is removed (no permanent deformation). Viscoelastic models express strain as a function of both stress and time. Following Maxwell's law, stress is taken to decrease exponentially with time (Turcotte and Schubert, 1982). There is a relaxation of stress with time, resulting in permanent deformation. Thermal flexure models are based on elastic or viscoelastic theory with the additional assumption that the strength of the material decreases with increasing temperature.

All three types of models have been applied to geologic systems with varying degrees of success. Flexural models have been applied to a wide variety of situations: loading due to ice sheets, seamounts, intracratonic basin fill, and thrust sheets in a foreland setting. Foreland basins represent the most clear-cut example of flexure; they are also among the most productive hydrocarbon basins (see references in Lerche, 1990b). The basement is assumed to act as an elastic plate of finite length and constant thickness, anchored at one end in cantilever fashion. Thrust and sediment loading cause the plate to bend. Immediately in front of the thrust front a depression, or foredeep, develops in which sediments are deposited. Cratonward of the foredeep the basement is flexed upward forming a peripheral bulge which may act as a locus for erosion or nondeposition. This configuration is not static because, as the thrust front moves, the peripheral bulge migrates ahead of the front in the direction of thrusting. The shape and position of the bulge depend on the loading history and the properties of the elastic plate.

The development of the peripheral bulge can have a significant effect on the basinal sediment fill. Periods of erosion and/or nondeposition will occur in times and places where the forebulge is exposed above sea level. In cases where the bulge does not rise above sea level there may be appreciable shallowing, an effect which may contribute to the formation of carbonate build-ups. In addition, the forebulge may act as a sediment source for the foredeep depression. If the mechanical properties of the basement are known, the locations of forebulge unconformities and carbonate build-ups can be estimated, even though such locations may be below the resolution of available geological and geophysical data. The properties of an elastic plate in a foreland setting are calculated directly from seismic and well data using quantitative forward and inverse methods (Lerche, 1990b).

Sea level

A sedimentary basin's gross morphology depends upon (1) basement movement, (2) sediment accumulation and compaction, and (3) eustatic sea-level variations. An accurate history of eustasy would aid in basin stratigraphy and help in the prediction of sediment geometry.

Some techniques which attempt to determine the magnitude of eustatic sea-level excursions include the measurement of: (1) the amount of sedimentary onlap onto the continental margins with and without the use of hypsometric curves; (2) the thickness of marine sedimentary cycles and the elevation and distance between indicators of old strandlines; (3) the perturbations on individual thermo-tectonic subsidence curves and stacked crustal subsidence curves; (4) the variations in deep-ocean oxygen isotopes found in sediments; and (5) the size of variables used in graphical and numerical simulations which "invert" the problem of basin fill in terms of tectonic behavior, rates of sediment accumulation, and eustasy. A unique solution for *absolute* eustatic variations is not yet available (Wilgus et al., 1988). Quantitative modelling shows that, at best, only the *sum* of tectonic basement subsidence and sea-level variations can be obtained. Every proposed scheme to measure sea-level excursions assumes some behavior for two of the three underlying processes: tectonic movement of the basement, sedimentary accumulation, and eustatic sea level. Each scheme determines the third process relative to the assumed model behavior of the other two.

An accurate eustatic sea-level variation chart would be useful in frontier basin exploration, particularly where seismic cross-sections and occasional stratigraphic and/or wildcat wells are the only data sources. However, such a sea-level chart cannot yet be made. When *relative* sea-level charts (combining tectonic and eustatic effects) are tied to wells, sedimentary sequences on seismic cross-sections can be projected (Vail et al., 1988, in Wilgus et al., 1988).

"Relative" (tectono/eustatic) sea-level curves generate a "family" of solutions which are the product of a variety of absolute sea-level curves and tectonic models. Each solution has to be assessed in terms of geologic setting and the hypotheses it generates. Recent work (Cloetingh and Wortel, 1988) on the impact of intraplate stress in relation to plate boundary formation makes the difficulty of unravelling the sea-level problem all the more severe.

Summary

The simple recognition that hydrocarbons are produced by thermal degradation of organic material, forces an examination of the interlocked sediment and fluid evolution with the evolving thermal regime in the sub-surface, and with hydrocarbon generation, migration and accumulation.

Quantitative methods in basin analysis examine the inter-relations among the various components, quantify the dynamical behavior so that basinal processes can be modeled, use present-day data to constrain generic processes and their applications to particular basins, and provide assessments of the most likely sites of hydrocarbon accumulations ahead of drilling. Such assessments must also provide a measure of their probability of being right based on the best sensitivity studies possible with the data and models to hand.

In the long range thrust to reach that goal individual problems are addressed, with others being secondary, in order to exhibit as sharply as possible the role of an individual facet under consideration. A gradually more integrated picture of interwoven development is arising as these facets are assimilated into a common body of knowledge.

III. Problems associated with burial history

The ability to fill a basin with sediments rests on four major factors: (i) the rate of supply and compaction of sediments; (ii) the motion of the underlying basement, which determines the structural evolution of the basin as a whole; (iii) the influence of faults and fractures, which partially determine the structural evolution of sedimentary sections and which partially determine the flow of fluids through a sedimentary sequence depending on whether, and for how long, a fault is both active and/or open to flow; (iv) the impact of extraneous factors, such as salt diapirism and salt sheet tectonics, which can be dominant influences throughout the history of particular basins (e.g. the Northern Gulf of Mexico).

Here we examine facets of the above four factors in order to illuminate particular concerns that are often crucial to a consistent approach to describing the evolution of a basin.

Rate of sediment supply and compaction

Neither the rate of sediment supply nor the evolution of compaction in a sedimentary basin are directly measurable. Inferences must be made from present-day observations. Observationally the best information available is on present-day thicknesses, stratigraphic ages, and porosity variations with depth. This information can be used, as shown in Section II, to determine the net amount of residual solid material deposited during a given time interval. But complications and uncertainties arise if the information available is interpreted further in attempts to deduce the physical thicknesses of sedimentary strata at prior times.

First is the problem that there is, most often, limited resolution on geologic ages. Tops and bases of stratal intervals are often the only age information to hand. Thus if erosion occurred *within* an interval then, since only the net present-day solid thickness is available, no knowledge of such an event would be recorded. On this basis, the observed (inferred) rate of solid sediment supply is, most likely, an underestimate. Equally, if post-depositional diagenetic cementation has occurred, then the measured amount of solid material is likely greater than at deposition. (But redistribution/dissolution and/or re-precipitation confound even this interpretation.) On this basis an overestimate of sediment supply would be inferred if diagenetic cementation is not allowed for. A problem which is becoming of some considerable concern is due to the increasing recognition that the estimates of sedimentary ages are themselves uncertain — compounding the problem. For instance, for an Ordovician section in the Cooper Basin the age at total depth (TD) is variously quoted as 400–500 Ma (Kantsler et al., 1983), 350–700 Ma (Battersby, 1976), and 450 ± 50 Ma (Pantano and Lerche, 1990). The youngest to oldest spread in age of TD is, therefore, from 700 to 350 Ma which, taken at face value, would imply a considerable variation (by a factor of two or so) in the rate of deposition of solid material. Clearly, precise and accurate ages are of fundamental importance in basin analysis — and as many such ages as possible.

But even if this major problem is discounted, a second major problem, which limits the ability to reconstruct the stratigraphic and structural variations of sediment evolution with time, is the variation of palaeocompaction. Here the problem, as shown in Section II, is the ability to describe correctly the loss of fluids from the deposited sediments. Pseudo-inverse models, which connect parameters in functional equations of state with present-day observations, would seem to be readily available models which can be made consistent with observations most easily. At each location in a basin, the four dominant types of observable information requiring satisfaction by a model are: (i) observed stratal thicknesses; (ii) observed porosity variations with depth; (iii) excess fluid pressure variations with depth; and (iv) measured permeability variations.

Table 1 shows the results of present-day thickness variations using a 1D pseudo-inverse model applied to the Navarin Basin COST No. 1 well (offshore eastern Alaska), while Figs. 2b–2e show the corresponding predicted versus observed behavior for fluid pressure, downhole temperature, permeability, and porosity, respectively.

Note that consistency with present-day observations is a necessary requirement but is not sufficient because other models (e.g. a 2D compaction/fluid-flow model) could equally provide consistency with the observed values. Clearly other criteria, favoring one class of models over another, are required. We shall return to such selection criteria later.

Basement motion

Apart from the generation of variable structural and stratigraphic behavior in a basin by differential compaction and facies variation across a basin, a major factor influencing structural behavior is the evolution of the basement on which the sediments lie. Turcotte and Schubert (1982) provide a compendium of studies varying from isostatic through elastic flexural plate, thermo-elastic, and elastic-plastic plate behaviors, to viscoelastic plate models. The important point in all such models is that they individually have several parameters which are to be adjusted until minimal discord is obtained between predicted and observed basement positions at the present day. Again, the most flexible models available to date are those of the pseudo-inverse sort. However, even here care must be exercised to be aware of the implicit and explicit assumptions and limitations of a model. For instance, application of an elastic plate model intrinsically presupposes that the data will indeed conform to the model. When the parameters have been optimally adjusted to minimize discord between predictions and observations, then a judgement has to be made as to whether the residual mis-match is within some tolerance level deemed acceptable — and it is this last statement that requires a human judgement call.

For instance, consider the case exhibited in Fig. 3, which reflects the application of a finite-length elastic plate model to the Carpathian foredeep zone of Czechoslovakia (Cogan et al., 1990). The situation depicted in Fig. 3 provides the minimum mis-match between predicted and observed basement deflections, while Fig. 4 shows how the minimum mean square residual mis-match (MSR) varies as each of the parameters is, in turn, varied away from the optimal values. In the case of a finite-length elastic plate, of homogeneous properties along its length (intrinsic assumption), there are seven parameters: rigidity D; bending moment M_0 at the free end of the plate in the absence of sediment load; initial hinge angle, θ_0, of the unloaded plate; initial equivalent end point load force, V_0, on the free end of the plate which mimics non-sedimentary force considerations (Turcotte and Schubert, 1982); the length L of the plate which can be longer than the observed data segment length but presumably cannot be shorter; the vector hinge point position (i.e. both lateral and vertical coordinates of the hinge point have to be determined).

A fairly satisfactory job is done by the pseudo-inverse model in matching the observed loaded basin

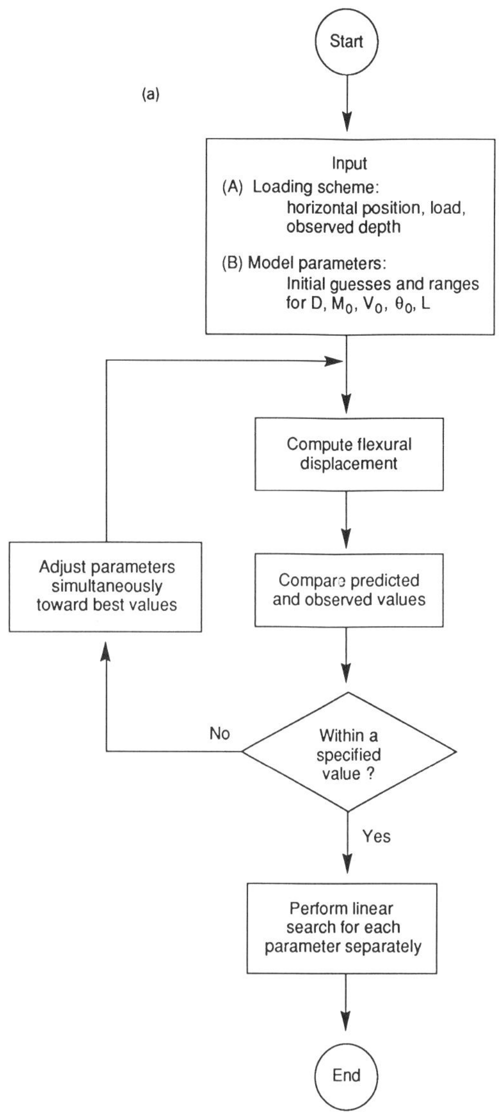

Fig. 3. (a) Flow diagram of the numerical procedure used to determine flexural plate parameters.

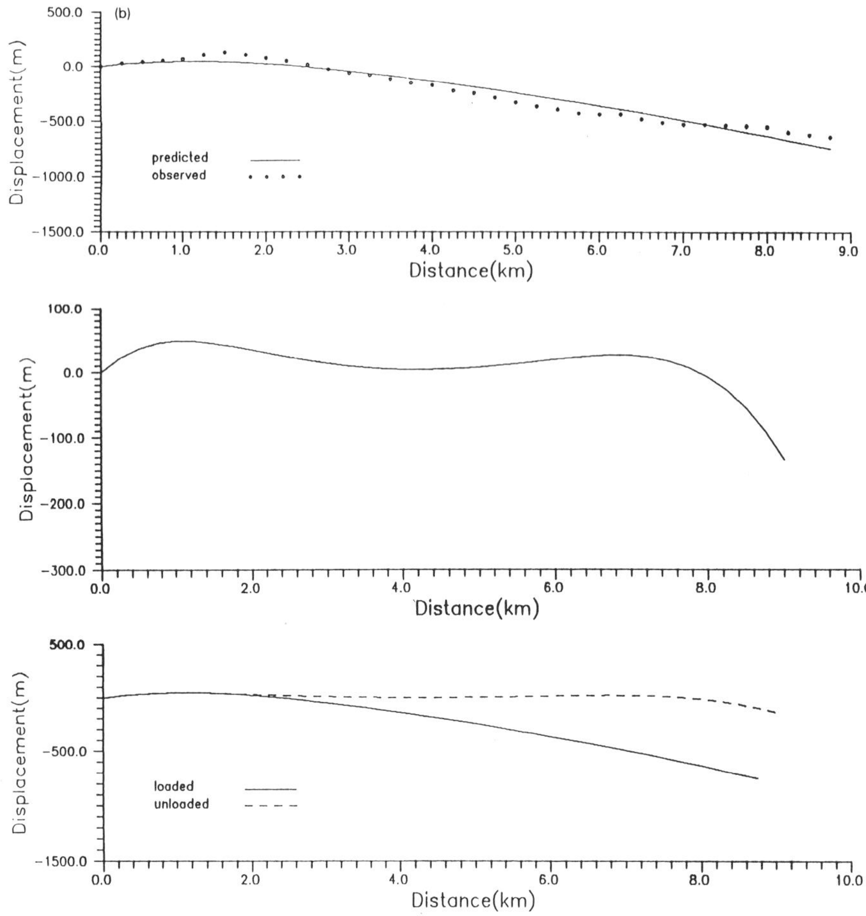

Fig. 3 (continued). (b) predicted versus observed loaded deflections of the basement in the Carpathian foredeep.

shape, as indicated in Fig. 3. Yet the effects of compressive or extensive thrusts, of non-homogeneous plate properties, of plasticity of thermal variations of parameters, or of parameters describing relaxation of effective stress with time, are not included — as is also true of the lack of inclusion of any other intrinsic assumption on the behavior of the basement except that the plate is to be considered as acting like a purely elastic plate. The questions of uniqueness, precision, sensitivity and accuracy of the results to intrinsic model assumptions, to parameters of the model, and to data resolution and finite sampling then arise. A general discussion of such problems is deferred to later in this paper.

Dynamical salt effects

In a large number of sedimentary basins, salt plays a major role in controlling structural, stratigraphic and thermal variations (Lerche and O'Brien, 1987). The reason for this major effect of salt is due to three basic properties of salt:

(1) On a geologic timescale salt has little to no internal structural strength and flows like a fluid under applied stress — such as sediment load.

(2) Salt has a nearly constant density of between about 2 and 2.2 g cm^{-3}, a value intermediate between sediment density at deposition (1.6–1.9 g cm^{-3}) and sediment density at full compaction (2.6–2.8 g cm^{-3}). Thus at some point in the evolution of a basin the sediment density overlying the salt will exceed that of the salt. Thereafter the salt will attempt to rise up through the sediments by buoyancy unless impeded by differential sediment strength (such as a competent formation) or by differential buoyancy decrease (such as from development of an overpressured zone).

Prior to, and post, the neutral buoyancy point being reached the salt will attempt to flow to locations of minimum applied stress if not otherwise impeded. Examples of such behavior from the Gabon area of Africa and from the German–Polish salt basins are provided in Thompson et al. (1989).

(3) Salt has a thermal conductivity 2–3 times higher (at 25°C) than sediments. Thus a salt shape acts to distort heat flow in the salt's vicinity by focusing effects, thereby creating spatially variable domains of enhanced and lowered (relative to regional) temperature which, in turn, impact sediment thermal maturity, sediment draping, and hydrocarbon trapping over and around the mobile salt masses,

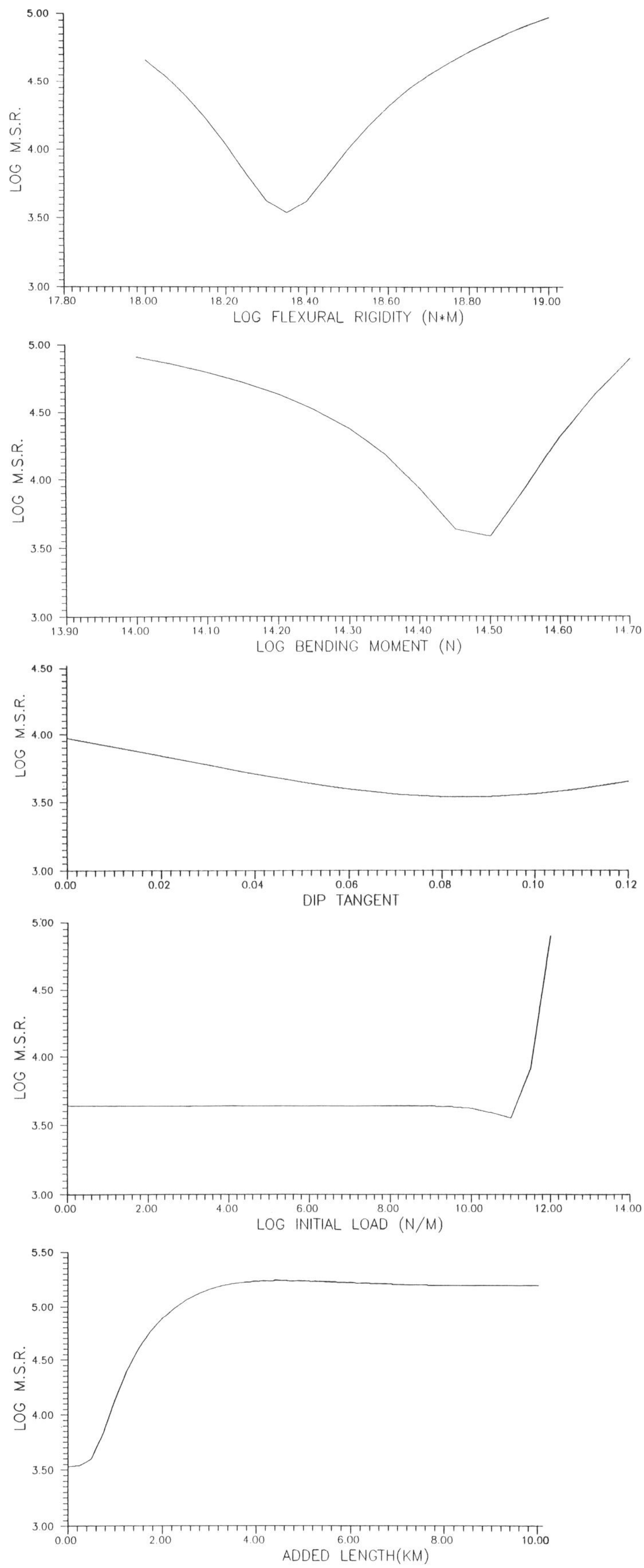

Fig. 4. Sensitivity resolution of the degree of mismatch for each parameter, D, θ_0, M_0, V_0, L as each, in turn, is varied away from the optimal mis-match value.

and which impact thermal fluid convection and salt dissolution (Lerche and O'Brien, 1987).

Thus the effects of salt are felt throughout a sedimentary basin both in terms of regional structural and thermal developments, and in terms of hydrocarbon generation, migration and accumulation. The impact of salt dissolution in producing collapse features in sediments has been ably described by Jenyon (1987).

A problem of some concern is to provide quantitative, self-consistent, dynamical models of salt–sediment interactions which allow for all of the above effects and which are constrained by present-day observations or inferences from observations. It is relatively easy to construct models in which a prescribed salt behavior is investigated for its impact on surrounding sediments (Lerche and O'Brien, 1987), or in which a prescribed sediment behavior forces a corresponding response from a salt mass. But a combined interactive dynamical model which includes the effect of salt on sediments and, equally, the effect of sediments on salt would seem elusive. This problem is of considerable concern if we are not to overlook potential hydrocarbon accumulations.

Faults: open or closed?

A debate of considerable proportions has occasionally been waged over methods for estimating the timing of faults and their motions (active versus non-active) in relation to sedimentary strata, and over the question of whether or not faults act or can act as open conduits for fluid flow, and so hydrocarbon escape, from the basinal system.

From a basin analysis viewpoint, attempts to answer these questions are also of considerable concern. The absolute timing of fault motions throughout the evolution of a basin has impacts on structural blocks (horsts versus grabens), on fault block rotation, on lateral and vertical fluid flow through juxtapositional variations of different lithologies across a fault which are dependent on the throw of the fault, and so on. *Relative* timing of a fault's motion is often inferred by inspecting the locations and strata where faults peter out, and then arguing that the fault activity ceased at the ages of such strata. Activity of a fault is often estimated by comparing similar age stratal displacements on either side of the fault to see if the displacement is constant or varying with age (in the latter case the fault is assigned to the "growth" fault type, evolving syndepositionally). But procedures for estimating, rather than assuming, absolute age dating of fault motions from observations would still seem to be difficult to obtain.

Even putting to one side the question of absolute timing of fault motions, the problem of assessing whether a fault is, or has been, open or shut for geologic time, either along the total length of the fault or along only a part of the fault is, perhaps, of even more pressing concern. Proponents can be found for both points of view, and field evidence (e.g. cement fill versus no fill) suggests that both open and closed faults occur. The problem from basin analysis considerations is not whether a fault is open or closed, but rather whether a fault is sufficiently open that it can have a significant impact on the flow of fluids (and so on both sediment compaction and hydrocarbon loss) throughout the course of a basin's evolution. In addition any dynamic model which ascribes openness or closedness to a fault must have predictable consequences which can be tested by comparison against observations.

Figures 5A and 5B show the effects of running a 2D dynamical basin analysis model, which allows for fluid flow and compaction, in the Pinedale Anticline area of the Green River Basin, Wyoming (Wei and Lerche, 1988). A major fault cuts the sedimentary strata at the position of the Wagon Wheel No. 1 well. Two extreme end-member cases were run: fault open, fault shut. The open fault case, as shown in Fig. 5A, does not permit the presence of any fluid overpressuring today at the Wagon Wheel No. 1 location. By way of contrast the shut fault case permits a major build-up and retention of overpressuring, as shown in Fig. 5B.

Observationally, as seen in Fig. 5C, there is indeed significant overpressuring today at the Wagon Wheel No. 1 site, implying that the closed-fault situation is a better representation of the observational information than the open-fault end-member case. However, the question remains: is it possible that the fault was open at an earlier time, leading to fluid escape, with the shutting of the fault being a phenomenon of fairly recent origin? On the basis of estimates of the recovery time for re-overpressuring (by sealing) of a hydrostatic basin, it would appear that the appropriate answer in the case of the Pinedale Anticline is that the fault has been shut since roughly the time of the Wyoming thrust ($\sim$60 Ma BP) (Wei and Lerche, 1988). However, this argument does not rest directly on observational testing, but rather rests indirectly on the time-scale estimate which is based on the low permeability of shale, i.e. without a major erosional event it is a long, slow, process to bleed overpressure out of shales because of their low permeability.

In any event the twin requirements of providing pragmatic procedures for determining, rather than assuming, both the absolute timing of fault motions and the openness/closedness of faults within a generic framework would seem to be outstanding problems.

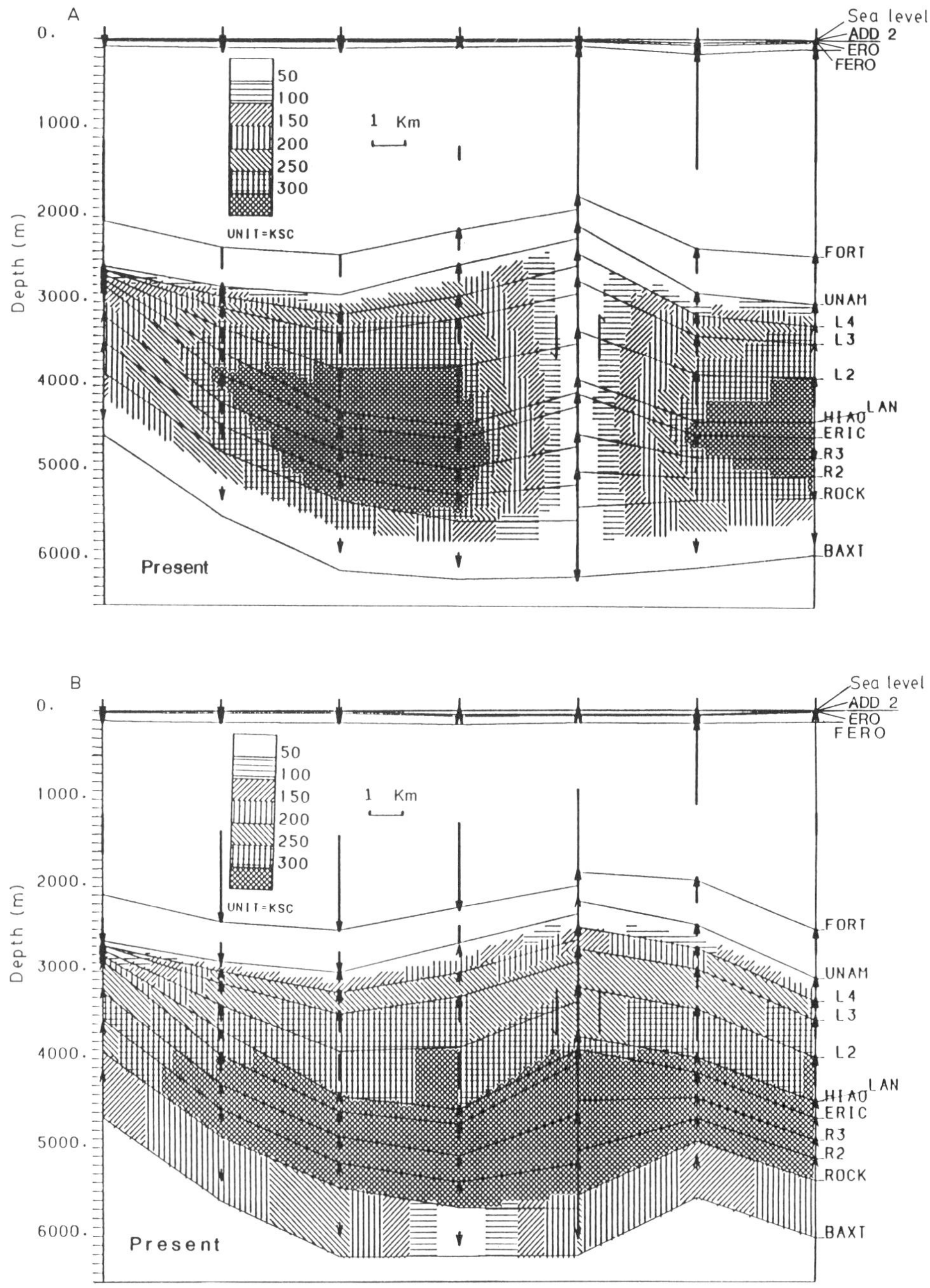

Fig. 5. (A) Basin analysis simulation at present of the Pinedale Anticline, Green River Basin, Wyoming treating the fault, which intersects the Wagon Wheel No. 1 location, as a completely open sand-dike for all time. Note the absence of any predicted overpressure, in opposition to the observations. (B) As for (A) but with the fault considered closed for all time. Note the high degree of overpres suring.

IV. Problems associated with thermal history

While the structural, stratigraphic and fluid-flow aspects of the dynamic evolution of a basin are of over-riding influence in determining the likely migration pathways and potential present-day accumulation sites for hydrocarbons, nevertheless, without the production of hydrocarbons at some time in a basin's history, the geologic evolution, while of scientific interest, is of little economic significance

Current thought (Hunt, 1979; Brooks and Welte, 1984; Waples, 1984) suggests that thermal degradation of kerogen is the dominant mechanism for producing hydrocarbons (oil and/or gas). Given that viewpoint the problem of determining hydrocarbon production hinges on two underpinning facets: (1) Can estimates be provided of palaeotemperature evolution with such estimates constrained by present-day data? (2) Can quantitative models be provided of temperature-dependent kinetic degradation reactions which are based on an observational framework and which can be shown to be extrapolatable to the

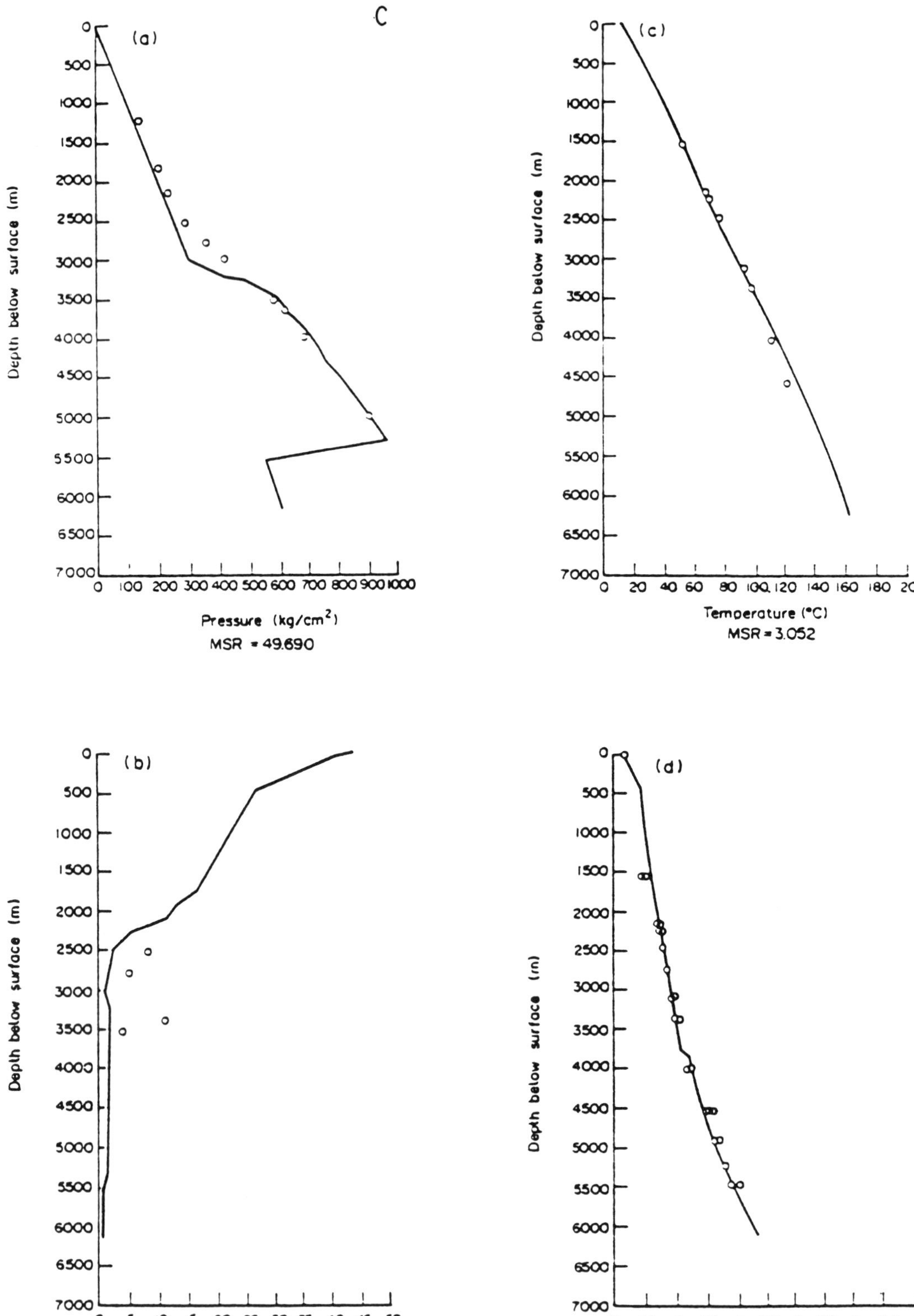

Fig. 5 (continued). (C) Agreement of predicted and observed overpressure (a) at Wagon Wheel No. 1 location, indicating that a closed fault provides a greater degree of consistency with the observed overpressure than does an open fault (KSC = kg cm^{-2}). (b), (c) and (d) show the equivalent variations of predicted and observed porosity, temperature, and vitrinite reflectance, respectively, at the same location.

geologic time frame with a quantifiable degree of resolution? In this section we address the first of these two questions.

Present-day temperature and heat flux

Most models used to construct thermal history rely on present-day thermal data to act as control constraints on either the relative scaling of palaeoheat flux compared to the present-day value or on the temperature evolution relative to present-day values.

One of the major problems that has been put into focus more sharply over the last few years is the difficulty of obtaining accurate formation temperatures. Using several hundred DST measurements from many well bores Hermanrud et al. (1990a, b) have shown that, *at best*, it is unlikely that downhole formation temperatures are measured or inferred to a *statistical* (random) accuracy of better than about ±8 to 10°C, while the *systematic* uncertainty on true formation temperatures is also typically about 10°C. Both sorts of errors then compound the difficulty of being accurate about the scaling of palaeotemperature to present-day values.

In addition, even if these problems are ignored, and present-day formation temperatures taken as both precise and accurate, one of the problems that is then made manifest is the determination of present-day heat flux, Q_0, from the temperature measurements.

In estimating Q_0 several implicit or explicit assumptions are normally made: that thermal conduction dominates; that a steady-state treatment is appropriate; that a one-dimensional spatial model (vertical heat conduction only) is acceptable; that the present-day thermal conductivities of the sediments are well determined. Even within these confining, and likely uncertain, assumptions there are still several procedures for determining Q_0, as has been detailed elsewhere (Lerche, 1990b). Some of these procedures are: the use of only the bottomhole and surface temperatures; use of all the downhole temperature data to construct a constant average thermal gradient and to then use this gradient to assess Q_0; use of temperature measurements only from similar lithologies, so that effects of lithology dependencies of thermal conductivity are minimized; use of a least-squares weighted average procedure to produce a Q_0 which is in least discord with all of the observed temperature data, irrespective of whether the data are from spatial domains that violate the confining assumptions above or are poorly measured.

Assessments of variations that arise in present-day heat flux as a consequence of using one or more of the above procedures indicate that heat flux values are likely overestimated by about 10%, with a most likely smallest statistical uncertainty also of around 10%, at best (Lerche, 1990b).

A problem of some concern is to provide measurement methods and interpolation procedures that will narrow the uncertainties on present-day downhole temperatures and heat flux estimates.

Palaeoheat flux and palaeotemperature

For the purposes of assessing hydrocarbon production in a basin it is not the palaeoheat flux which is of direct concern; rather it is the palaeotemperature. A high palaeoheat flux does not unequivocally imply a high sediment temperature (e.g. the sediments could have been at shallow sub-surface depth at the time of the high heat flux and so would then record only the low temperature of the shallow depth). In addition most modern codes that handle compacting sediments in basins allow for lateral and vertical convective transfer of fluids with the attendant transfer of thermal energy, as well as for focusing and defocusing of heat spatially because of lithology and overpressure variations.

However, if the palaeoheat flux variation with time and spatial position could be determined then, of course, that knowledge could be used in basin analysis codes to predict the evolution of temperature with time for each sediment parcel at each location in the basin. The nub of the problem is to devise a procedure to obtain the required information. As has been remarked already in Section II, two basic and very different procedures are available. Either one assumes a model connecting palaeoheat flux and dynamic basin evolution, and then uses the basinal evolution parameters determined for the particular dynamic model under consideration to connect to the proposed palaeoheat flux variation; or one attempts to use measured present-day thermal indicator information, in an inverse or pseudo-inverse sense (Lerche, 1990a), to obtain an assessment of palaeoheat flux. The difficulty with this second procedure is that each thermal indicator has its own dynamic range of sensitivity, is often sparsely or poorly measured, can be influenced in its development by other than purely thermal factors (e.g. mud additives and sub-surface migration have significant influences on biomarker isomerization and aromatization ratios which are then measured with depth in boreholes), and each thermal indicator has physical and chemical kinetic parameters which often also carry some significant degree of uncertainty.

Even when measurements on multiple thermal indicators are available in the same well, it is often difficult to reduce uncertainty on palaeoheat flux determinations because of the above considerations (Lerche, 1990a).

Difficulties with the first approach are that a double model connection must be made and different models, while consistent with present-day data, lead to quite different estimates of palaeoheat flux.

A problem of some concern is to find a procedure for assessing palaeoheat flux and/or palaeotemperatures which will not only be consistent with downhole measurements of multiple thermal indicators in many wells, but which will also simultaneously provide a capability of separating thermal-basement motion models into classes which are viable from those which are not, based on both dynamical and thermal considerations.

And, in resolving this problem, a pressing concern is to provide an estimate of the resolution with which the palaeoheat flux is determined in order that minimum, average, and maximum ranges can be categorized in an accurate manner.

V. Problems with hydrocarbon kinetics and migration

Most of the more sophisticated basin analysis codes currently in operation have some form of thermally-dependent kinetic degradation scheme which permits the breakdown of solid kerogen to liquid and gaseous mobile hydrocarbons of high buoyancy, which are then capable of migrating away from the source rock(s).

Here two problems are exhibited, the first connected with the form and substance of kinetic schemes employed in kerogen production of hydrocarbons, and the second has to do with hydrocarbon migration problems.

Hydrocarbon kinetics

Two basic approaches have been taken to date in respect of kerogen breakdown to hydrocarbons. The first approach refers to the kerogen only, without being unduly concerned, per se, with the products produced by the kerogen breakdown. This approach is labeled residual kerogen kinetics. It is clear that this form of kinetic scheme is not mass conserving because the analysis considers only the fractional mass of remaining kerogen which gradually disappears as maturity proceeds.

The second keeps track of the products of the kerogen breakdown and of the inter-relation of one product with another (e.g. oil converting to gas) as maturity proceeds. This approach is a pyrolysis product analysis and, if all products are traced, it is a mass conserving kinetic framework. Consider each in turn.

Residual kerogen analysis. The main thrust of such kinetic schemes has changed little since the introduction by Lopatin (1971) of TTI as a measure of kerogen thermal maturity, although technical details have changed significantly in the last few years (Braun and Burnham, 1987; Ungerer et al., 1988).

In essence the procedure is to propose a set of kinetic reactions for a pure end-member component of kerogen (normally done pragmatically by declaring Type I, II or III kerogens to be end-members based on the H/C and O/C ratios in a modified van Krevelen diagram (Tissot, 1969; Tissot and Espitalié, 1975;. Tissot and Welte, 1978)). Each reaction is usually taken to be of the temperature-dependent exponential Arrhenius form so that both an activation energy, E, and a pre-exponential frequency factor, k, are required for each reaction pathway. (Note the intrinsic assumption that the reactions are Arrhenius in form.) Determination of the parameters k and E for each reaction as well as the number, N, of reaction pathways needed is usually accomplished using laboratory measurements of residual kerogen under increasing temperature (Braun and Burnham, 1987), and then attempting some least-squares fit to the laboratory data. The problems with this approach are well known. First, there is the potentially serious difficulty that the number of reaction pathways can be increased systematically and steadily until the assumed kinetic process is bound to be closer and closer to any set of observations. Current models use around 26–30 channels for reactions, suggesting that between 52 and 60 parameters have to be determined, which seems excessive (Liu and Lerche, 1990). Second, even if such a large number of parameters is truly needed to describe residual kerogen analysis, the uncertainty (both statistical and systematic) in the parameters needs to be addressed. Third, the heating rates and temperature regimes of the laboratory analysis are well in excess of those pertaining to geological conditions. How does one prove that high-temperature reactions, done under pristine laboratory conditions, are really the appropriate representations to extrapolate to the slow-heating, low-temperature, non-pristine conditions of geologic basins? And how are such extrapolations to be constrained by data? A detailed sensitivity analysis investigating such considerations would seem to be a problem of some significance to reducing risk in finding hydrocarbons.

Pyrolysis product analysis. Apart from the efforts made to describe the thermal maturation of kerogen, efforts have also been undertaken to consider the products of thermal maturation (and of the further conversion of one product to another) with the ultimate goal of providing a self-consistent kinetic description of hydrocarbon generation (Ungerer et al., 1988). Such kinetic schemes rely on a sequence of parallel or series kinetic reactions, again most often of the first-order Arrhenius form, and the reactions are usually taken to be solely temperature-dependent. Again the problems here are very close, in a formal sense, to those arising in the residual kerogen analysis scheme. For each reaction pathway, both an activation energy and a frequency factor must be specified, and the resolution of each such parameter determined. Again the question arises of extrapolation of kinetic schemes derived from high heating rate laboratory experiments to the low heating rates appropriate to geological conditions. Two further problems arise which are not encountered in the residual analysis kinetics. First, the liquid and gaseous pyrolysis products can migrate in the subsurface so that the thermodynamic mix of materials changes with time, making even more difficult the extrapolation from laboratory conditions; second, the *forward* modelling fits to the laboratory sets of data (Ungerer et al., 1988) systematically undershoot or overshoot (Fig. 6) even when the number of reaction pathways is increased and optimal fit parameter adjustments are used. Recent work (Lerche, 1990c; Liu and Lerche, 1990, 1991) using an *exact* inverse of the general kinetic schemes proposed to date, has shown that the reaction kinetics are intrinsically time-dependent based on application to the laboratory data (Fig. 7). Clearly a greater degree of effort is needed to uncover a kinetic framework that can be extrapolated to geological conditions outside the laboratory arena with some sharp and defined degree of confidence in the veracity of the framework. This problem would seem to be of pressing concern as well.

Hydrocarbon migration and accumulation

As remarked in Section II, the problems of hydrocarbon migration and accumulation are intimately intertwined with the dynamical and thermal evolution of a basin. Recognizing that a detailed discussion of all problems connected with primary and secondary hydrocarbon migration would make for a very long review (Lerche, 1990a, b), here we consider just two problems, one of migration and the other of accumulation, in order to underscore some of the problems that need to be addressed if we are to improve on our ability to understand the flow of hydrocarbons in evolving sub-surface conditions.

Capillary pressure effects. In concert with hydrodynamic forces moving hydrocarbons, Hubbert (1953) has also written extensively on the capability of capillary pressure forcing liquid hydrocarbons out of hydrocarbon-wet, fine-grained source rocks into water-wet, coarse-grained carrier aquifer beds. Later work by Dahlberg (1982) has acted to underpin the essential correctness of Hubbert's pioneering considerations. And yet some points seem to be still underappreciated despite nearly a half century of effort (Lerche and Thomsen, 1992).

Consider a hydrocarbon-bearing fine-grained shale stratum abutted on both sides by an initially water-wet, coarse-grained sandstone.

A capillary pressure of some tens to hundreds of atmospheres exists at the upper and lower interfaces between the shale and the surrounding sands. Accordingly, as Hubbert (1953) has noted, the hydrocarbons are driven by capillary pressure into the sands with water being correspondingly driven into the shale from the sands. Thus *both* the overlying and underlying sands become hydrocarbon-charged. Hydrocarbons in the overlying sand, being buoyant relative to water, will promptly seek to move vertically away from the sand–shale interface; hydrocarbons in the underlying sand, also being buoyant relative to water, will tend to hug the top of the sand and so remain near the underlying shale–sand interface.

But now a major question arises as follows. *If* the sands stay *water-wet* then capillary pressure will continually force hydrocarbons from the shale into both sands. However, *if* the sands become *hydrocarbon-wet* (perhaps as a consequence of their initial charging by hydrocarbons), then the capillary force changes its character completely, as Hubbert (1953) has noted. The initial water-wet conditions imply no capillary impedance to the flow of water across a sand–shale boundary, only an impediment to the flow of hydrocarbons. Equally, hydrocarbon-wet conditions imply no capillary impedance to the flow of hydrocarbons across a sand–shale boundary, only an impediment to the flow of water. Thus, *if* the sands become oil-wet, then not only is there enhancement of the buoyant escape of hydrocarbons from the shale source-rock into the overlying sand, but hydrocarbons previously driven into the underlying sand can now use their buoyancy to transit back across the shale into the overlying sand. Thus the behavior of primary migration of hydrocarbons is particularly, and sensitively, dependent on the water to hydrocarbon wetness of the coarse-grained and fine-grained rocks. Hence a problem of considerable concern is to assess the water-wetness of coarse-grained sand strata as hydrocarbons flow from fine-grained source rocks, because the dynamical evolution of wetness can have a major impact on the ability of sands to act as aquifers for hydrocarbon migration.

Accumulation effects. Suppose that, by whatever mechanism, liquid hydrocarbons, in the form of oils, have reached a structural accumulation site at sometime in a basin's history. Two factors dictate that knowledge of this site is not adequate to assess a low-risk for hydrocarbon exploration today — unless, of course, the oils reached the site in the very recent

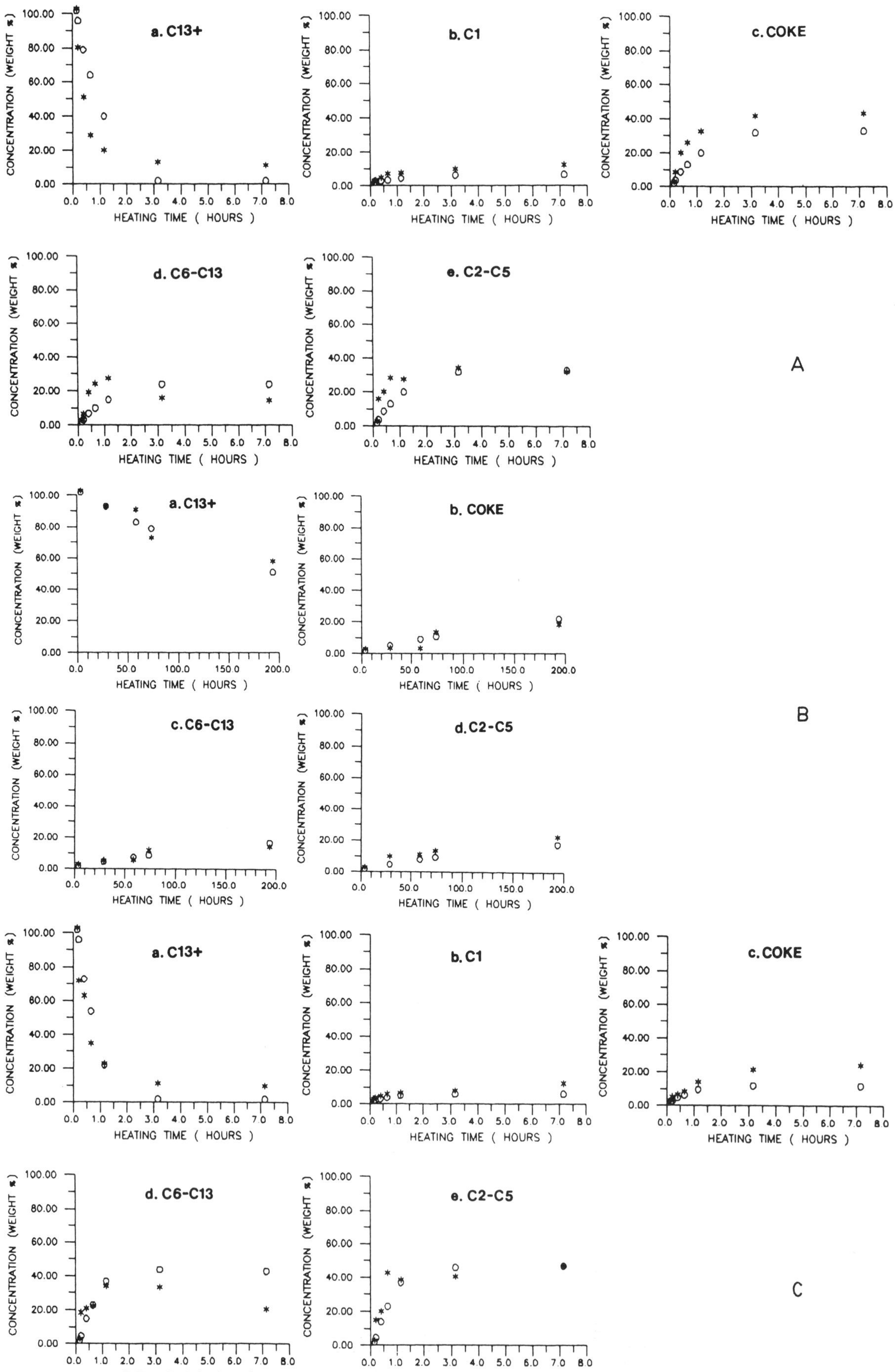

Fig. 6. Forward kinetic model applied to 3 sets of autoclave product pyrolysis data (from Ungerer et al., 1988). Asterisks denote observations, open circles denote predictions.

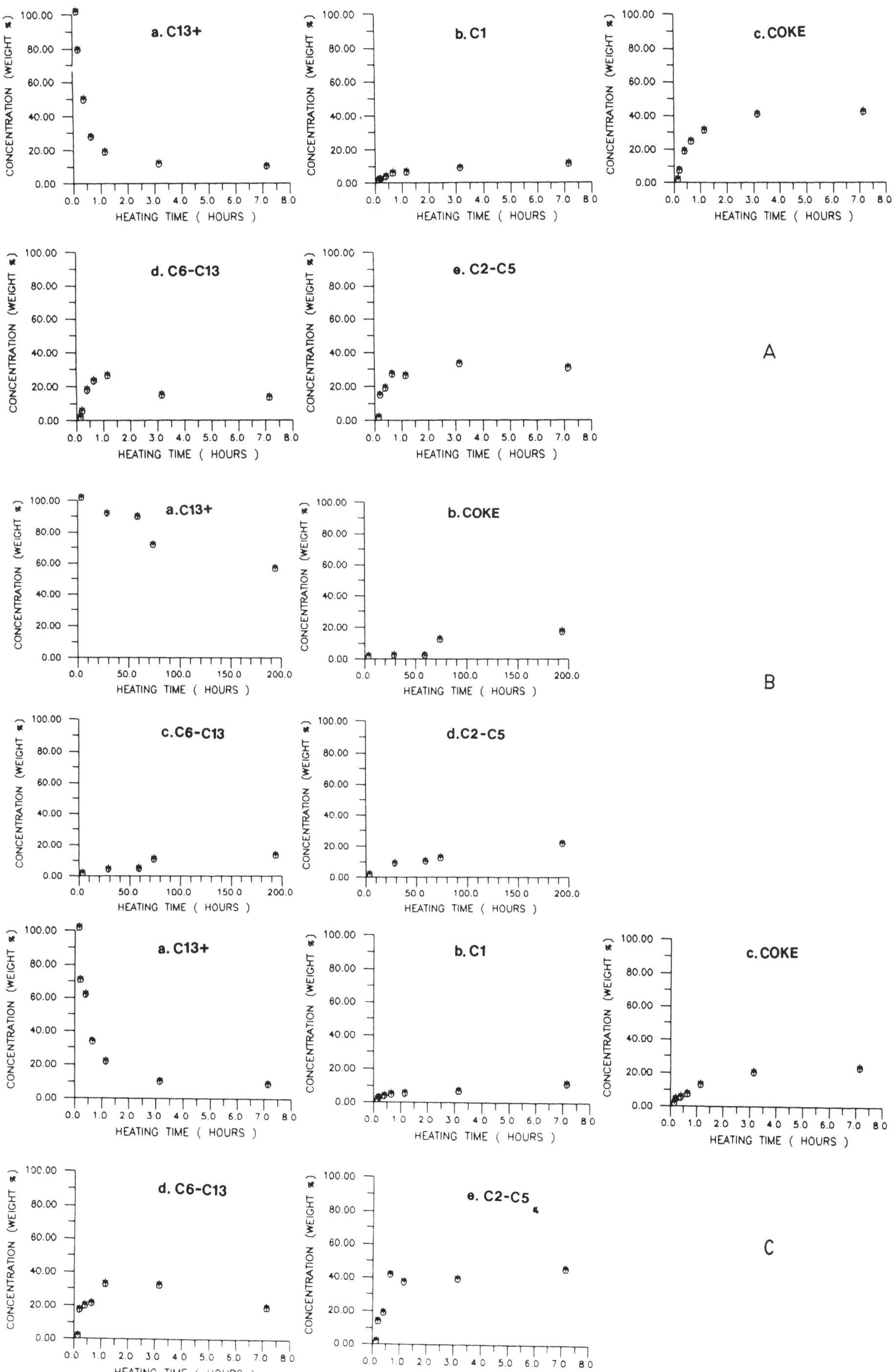

Fig. 7. Kinetic model of Ungerer et al. (1988) applied in an inverse sense to the 3 sets of data of Fig. 6. Note the significant improvement in match to the data (from Liu and Lerche, 1991).

past. First, as the basinal evolution proceeds, the accumulation site may be exhumed, buried deeper, hydrodynamically flushed by meteoric waters, faulted and so on. Thus the retention of hydrocarbons to the present day in a putative accumulation site is intimately intertwined with the continued dynamical evolution of the basin. Second, even if the dynamical evolution of the accumulation site is known after hydrocarbon entrapment, the continued thermo-kinetic evolution of liquid hydrocarbons to gaseous hydrocarbons after emplacement puts its own constraints on the accumulation site. Conversion of oil to gas increases the buoyancy pressure pushing on the overlying seal, and may end up fracturing the trap, thereby releasing hydrocarbons. This effect clearly depends on the total amount of hydrocarbons entrapped, the amount converted to gas, and the evolving strength and permeability of the overlying seal which may be increased or decreased by compaction, stylolitization, diagenetic cementation, dolomitization, etc.

Clearly, a problem of some concern in respect of hydrocarbon accumulations is to provide an integrated picture of hydrocarbon generation and migration in relation to dynamical and thermal evolution of a basin, and in respect of the structural, stratigraphic, thermal, diagenetic, and kinetic considerations particularly and especially relevant to the basin under consideration.

VI. Controls, constraints and uncertainties

The essence of science is model building to account for known data and observations, and to predict new phenomena which are then measured and compared with the model's predictions. When a model ceases to predict observed phenomena, it must be modified, abandoned or replaced such that the perception of science is adjusted to reality. But within this quintessential framework, the philosophies and strategies of model building and model testing are as broad and varied as circumstances and imagination allow.

There is a general understanding that causative agents (sources) are modulated by dynamical processes (filters) to produce end products (outputs). In modelling, the specification of the sources and filters determines the outputs. We mortals have a brief life span compared to the geological time scale, and it is rare when we can observe a system from beginning to end and thus know unequivocally what relations exist among sources, filters and outputs. Instead, usually we are restricted to measuring outputs and inferring filters and sources.

This limitation imposes severe difficulties and constraints on model development. Each model of sources and/or filters depends upon intrinsic assumptions and upon the functional relationships that are assumed to connect one set of physical variables to another. In addition, the functional dependencies among intrinsic and extrinsic variables and parameters are often uncertain, yet they must be provided for or adjusted in models. Determination of these dependencies is one of the major goals of models. The uniqueness, resolution, and sensitivity of a model relative to other competing models then must be assessed by comparing their output behaviors. But effective comparisons of outputs are limited by measurements with finite resolution, noise, sampling, uniqueness and reproducibility.

Pragmatic strategies for assessing model veracity must be tailored both to the proposed sources and filtering behaviors as well as to the quality, quantity and limitations of data available at the time the model is developed. Various observational and theoretical constraints (for example, requiring that positive variables remain positive, or that mass, momentum and energy be conserved) are often effective aids in designing model strategies.

This paper has illustrated by examples the philosophies and strategies that underpin model building in sedimentary basin analysis, although the arguments advanced apply equally to any scientific discipline.

Assumptions and ambiguities

Prediction of the amounts and locations of oil and gas in sedimentary basins is the primary goal of sedimentary basin analysis from an oil-industry viewpoint. To accomplish this goal we require knowledge of three major components of basin evolution: a dynamical burial history of sediments; a thermal history of the basin; and a hydrocarbon generation, migration, and accumulation history. Unfortunately, we can observe only the end result of the interactive processes that contributed to the basin evolution and, even then, only with limited resolution and limited quality and quantity of data. The task is to invert observations made today to obtain the dynamical and thermal evolution of the basin. A better determination can then be made of where and when hydrocarbons were produced in the basin, and where they might be now.

Several factors limit the precision of any inversion scheme. Irrespective of either model behavior or intrinsic assumptions from which a model was derived, measurements of input data are limited in their resolution, accuracy, precision, and uniqueness. Even if measurements of input data were perfect, requirements of finite spatial and temporal sampling (discrete samples) imposed by the models further limit resolution.

If we ignore these limitations of data input quality and sampling, there remain three additional sources of ambiguity that limit precision of inversion models. The first includes all intrinsic assumptions inherent to a class of behaviors. As examples, a one-dimensional model is inherently a poor approximation of the real four-dimensional world and may not behave like the real world; an assumed constant, equilibrium-balanced, conductive heat flux ignores likely changes in heat flow or thermal gradients through time; a time invariant porosity with depth function ignores geographic, stratigraphic and temporal changes in subsurface fluid pressures, and therefore porosity, through time.

The second source of ambiguity is the type of behavior proposed or assumed within an intrinsic assumption class. For example, it might be assumed that the palaeoheat flux of a basin has the particular form of a McKenzie (1978) stretching model with three parameters: time of onset, magnitude "jump" at onset, and thermal decay time scale. Using this assumption, a pseudo-inverse procedure may be employed to determine a value or a family of values for each parameter as precisely as possible. This is termed a pseudo-inverse method because it posits specific forms of the functional behavior. Other models with different functional behaviors and a similar number of parameters might provide equally good fits to the data, and it would not be possible to discriminate among conflicting model behaviors.

Within an intrinsic assumption class, the third source of ambiguity arises from a true inverse procedure in which the functional behavior of a required quantity is determined (rather than assumed as in the pseudo-inverse situation) by direct inverse techniques applied to the data. True inverse techniques also must determine the degree to which a functional behavior can be resolved relative to "noise" in the system.

In addition to these philosophies two other end-member classes are available in forward models. One could build an absolutely determined model and calculate a quantity of interest at the present day that evolved from assumed initial conditions. To assess the validity of the forward model, the calculated quantity is then compared in one of two ways with measured values. First, the calculated quantity may be compared directly with observed data of comparable type (e.g., stratal thickness, lithology or porosity). Alternatively, the calculated quantity may be intrinsically unobservable in which case an extra postulate of the quantity's relation to an observable must be made which can never be proven (e.g., TTI to vitrinite reflectance). In both cases, the forward model may be judged acceptable or not depending on how close the calculated quantity matches the observed data. However, even if the calculated and observed quantities are close and the model is judged sufficient, the condition of necessity is not also necessarily proven — that is, the model may be inaccurate (a different model might be a better one) and its calculated quantity may be only fortuitously similar or identical to observed data.

Sensitivity, noise, resolution and uniqueness

In discussing the forward, pseudo-inverse, and true inverse philosophies for analyzing burial, thermal, and hydrocarbon histories, three underlying considerations were taken for granted, either explicitly or implicitly. First, we assumed that any input data required (e.g., stratigraphic ages, formation thicknesses, porosity variations with depth, thermal indicator measurements, present-day heat flux) were accurate (no error) and continuous with depth (no finite sampling interval problem). Neither assumption is true in general. Second, we assumed that, apart from the parameters or functional forms we were attempting to determine, all other parameters and functional forms were known precisely (e.g., thermal conductivity, permeability scale values, mantle density). This assumption is likely invalid. Third, we assumed that the intrinsic underlying assumptions of a model were accurate enough to describe the real world (e.g., 1D, 2D, 3D, conductive heat flux only, no diagenesis, hydrocarbon kinetic model forms). But this is known to be perhaps the most invalid point of any model. Indeed, even the models connecting heat flux variations to basement subsidence vary amongst themselves quite markedly and all meet constraints of present-day data (Sleep, 1971; Turcotte and Ahern, 1977; McKenzie, 1978; Royden et al., 1980; Falvey and Middleton, 1981).

Each of these problems is associated with uniqueness, resolution, sensitivity, precision and signal versus noise, in forward, pseudo-inverse and true inverse models (Cao, 1987). This area would seem to be one requiring a more detailed investigation and recognition than it has so far received to date.

Perhaps the best that can be hoped for in our attempts to determine hydrocarbon accumulations is to bracket least probable, average, and most probable situations with confidence intervals so that we can at the least quantify precisely the level of our lack of determinism. And even that would be a major step forward.

Conclusion

We have examined the philosophies and practices of forward and inverse models in sedimentary basin analysis. In general forward models suffer from being presented independently of the observed data, but have the advantage of guiding thought about dynamical sedimentary behavior. Forward models often conflict one with another, and occasionally have internal inconsistencies as well. True inverse procedures are often more difficult to obtain. They usually honor some components of data sets, but may be so restrictive in their underpinning assumptions as to omit the dominant physical and geological behaviors of the system. Pseudo-inverse models would seem to represent a compromise between forward models and true inverse models. On the one hand they permit determination of a finite number of parameters for a given functional shape, but on the other hand they also contain the capability of allowing an extended set of functional shapes should the data warrant such an extension.

All models suffer from the vagaries of data noise and sampling, from assumptions concerning parameter values and functional behaviors within a model framework, and from intrinsic model assumptions. Measures of determination which can provide resolution, uniqueness and precision of dynamical evolution in the face of such a lack of resolution of information would seem to be needed rather urgently.

Acknowledgements

The work reported here was supported by the Industrial Associates of the Basin Analysis Group at the University of South Carolina. Support is also greatly appreciated from the Nordic Council of a Nordic Professorship in Petroleum Geology, based in Copenhagen, during July–December 1990, when the bulk of this review was written. The courtesies and support extended to me by Ole Graversen, Head of the Geologisk Centralinstitut, during my stay are also much appreciated.

References

Battersby, D.G., 1976. Cooper Basin gas and oil fields. Aust. Inst. Min. Metal., Monogr. Ser., 7: 321–368.

Bethke, C.M., 1985. A numerical model of compaction-driven groundwater flow and heat transfer and its application to the paleohydrology of intracratonic sedimentary basins. J. Geophys. Res., 90: 6817–6829.

Braun, R.L. and Burnham, A.K., 1987. Analysis of chemical reaction kinetics using a distribution of activation energies and simpler models. Energy Fuels, 1: 153–161.

Brooks, J. and Welte, D., 1984. Advances in Petroleum Geochemistry, Vol. 1. Academic Press, London, 335 pp.

Cao, S., 1987. Sensitivity Analysis of 1-D Dynamical Model for Basin Analysis. Ph.D. dissertation, University of South Carolina, Columbia, S.C.

Cloetingh, S.A.P.L. and Wortel, M.J.R., 1988. On the mechanics of plate boundary formation. In: N.J. Vlaar, G. Nolet, M.J.R. Wortel and S.A.P.L. Cloetingh (Editors), Mathematical Geophysics. D. Reidel, Dordrecht, pp. 363–388.

Cogan, J., Dorman, J., Lerche, I. and Kanes, W., 1990. Flexural plate inversion: application to the Carpathian Foredeep, Czechoslovakia. Mod. Geol. (submitted).

Cross, T.A. (Editor), 1989. Quantitative Dynamic Stratigraphy. Prentice-Hall, Englewood Cliffs, N.J., 625 pp.

Dahlberg, E.C., 1982. Applied Hydrodynamics in Petroleum Exploration. Springer-Verlag, New York, N.Y.

Falvey, D.A. and Middleton, M.F., 1981. Passive continental margins: evidence for a prebreakup deep crustal metamorphic subsidence mechanism. 26th Int. Geol. Congr., Colloque C3.3, Geology of Continental Margins, Suppl. to Vol. 4, pp. 103–114.

Hermanrud, C., Cao, S. and Lerche, I., 1990a. Estimates of virgin rock temperatures derived from BHT measurements. Geophysics, 55: 229–237.

Hermanrud, C., Meisingset, K.K. and Lerche, I., 1990b. Formation temperatures from drill stem tests. J. Pet. Technol. (in press).

Hubbert, M.K., 1953. Entrapment of petroleum under hydrodynamic conditions. Am. Assoc. Pet. Geol., Bull., 37: 1954–2026.

Hunt, J.M., 1979. Petroleum Geochemistry and Geology. W.H. Freeman, San Francisco, Calif., 617 pp.

Jenyon, M.K., 1987. The development by salt diapirs of superficial overhang features, and effects on associated sediments. In: I. Lerche and J.J. O'Brien (Editors), Dynamical Geology of Salt and Related Structures. Academic Press, San Diego, Calif., pp. 679–709.

Kantsler, A.J., Prudence, T.J.C., Cook, A.C. and Zwigulis, M., 1983. Hydrocarbon habitat of the Cooper/Eromanga Basin, Australia. Aust. Pet. Explor. Assoc. J., 23: 75–90.

Keen, C.E., Beaumont, C. and Boutilier, R., 1981. Preliminary results from a thermo mechanical model for the evolution of Atlantic-type continental margins. 26th Int. Geol. Congr., Colloque C3.3, Geology of Continental Margins, Suppl. to Vol. 4, pp. 123–128.

Lerche, I., 1988a. Inversion of multiple thermal indicators: quantitative methods of determining paleoheat flux and geological parameters, I. The theoretical development for paleoheat flux. Math. Geol., 20: 1–36.

Lerche, I., 1988b. Inversion of multiple thermal indicators: quantitative methods of determining paleoheat flux and geological parameters, II. The theoretical development for chemical, physical and geological parameters. Math. Geol., 20: 73–96.

Lerche, I., 1990a. Basin Analysis: Quantitative Methods, Vol. I. Academic Press, San Diego, Calif., 562 pp.

Lerche, I., 1990b. Basin Analysis: Quantitative Methods, Vol. II. Academic Press, San Diego, Calif., 614 pp.

Lerche, I., 1991. Inverse methods and kinetic models of hydrocarbon generation I. Theoretical considerations. Math. Geol., 23: 817–832.

Lerche, I. and O'Brien, J.J. (Editors), 1987. Dynamical Geology of Salt and Related Structures, Academic Press, San Diego, Calif., 832 pp.

Lerche, I. and Thomsen, R.O., 1992. Hydrodynamics of Oil and Gas. Academic Press, San Diego, Calif. (in press).

Liu, J. and Lerche, I., 1990. Inverse methods and kinetic models of hydrocarbon generation. II. Case histories for residual kerogen analysis. Math. Geol., 22: 989–1009.

Liu, J. and Lerche, I., 1991. Inverse methods and kinetic models of hydrocarbon generation, III. Case histories for autoclave product analysis. Math. Geol., 23: 325–347.

Lopatin, N.V., 1971. Temperature and geologic time as factors in coalification. Izv. Akad. Nauk. SSSR, Ser. Geol., 3: 95–106.

McKenzie, D., 1978. Some remarks on the development of sedimentary basins. Earth Planet. Sci. Lett., 40: 25–32.

Pantano, J. and Lerche, I., 1990. Inversion of multiple thermal indicators: quantitative methods of determining paleoheat flux and geological parameters, III. Stratigraphic age determination from inversion of vitrinite reflectance data and sterane isomerization data. Math. Geol., 22: 281–307.

Royden, L. and Keen, C.E., 1980. Rifting process and thermal evolution of the continental margin of eastern Canada determined from subsidence curves. Earth Planet. Sci. Lett., 51: 343–361.

Royden, L., Sclater, J.G. and Von Herzen, R.P., 1980. Continental margin subsidence and heat flow: important parameters in formation of petroleum hydrocarbons. Am. Assoc. Pet. Geol., Bull., 64: 173–187.

Sclater, J.G. and Christie, P.A.F., 1980. Continental stretching: an explanation of the post-mid-Cretaceous subsidence of the central North Sea basin. J. Geophys. Res., 85: 3711–3739.

Sleep, N.H., 1971. Thermal effects of the formation of Atlantic continental margins by continental breakup. Geophys. J. R. Astron. Soc., 24: 325–350.

Thompson, D.T., Lerche, I. and O'Brien, J.J., 1989. Salt basins of Western Europe and Gabon, West Africa: dynamical aspects and hydrocarbon production. In: D. West (Editor), Gulf of Mexico Salt Tectonics, Associated Processes and Exploration Potential. GCS/Soc. Econ. Paleontol. Mineral., Spec. Publ., pp. 122–130.

Tissot, B., 1969. Premières données sur les mécanismes et la cinétique de la formation du pétrole dans les sédiments (First data on the mechanism and kinetics of the formation of petroleum in sediments). Rev. Inst. Fr. Pét., 24: 470–501.

Tissot, B. and Espitalié, J., 1975. L'évolution thermique de la matière organique des sédiments: applications d'une simulation mathématique (Thermal evolution of organic material in sediments: Applications of a mathematical simulation). Rev. Inst. Fr. Pét., 30: 743–777.

Tissot, B. and Welte, D.H., 1978. Petroleum Formation and Occurrence: A New Approach to Oil and Gas Exploration. Springer-Verlag, Berlin, 538 pp.

Turcotte, D.L. and Ahern, J.L., 1977. On the thermal and subsidence history of sedimentary basins. J. Geophys. Res., 82: 3762–3766.

Turcotte, D.L. and Schubert, G., 1982. Geodynamics: Applications of Continuum Physics to Geological Problems. Wiley, New York, N.Y., 450 pp.

Ungerer, P., Behar, F., Villalba, M., Heum, O.R. and Audibert, A., 1988. Kinetic modeling of oil cracking. Org. Geochem., 13: 857–868.

van Hinte, J.E., 1978. Geohistory analysis — application of micropaleontology in exploration geology. Am. Assoc. Pet. Geol., Bull., 62: 201–222.

Waples, D.W., 1980. Time and temperature in petroleum formation: application of Lopatin's method to petroleum exploration. Am. Assoc. Pet. Geol., Bull., 64: 916–926.

Waples, D.W., 1984. Thermal models for oil generation. Adv. Pet. Geochem., 1: 7–68.

Waples, D.W., 1989. Maturity modeling of sedimentary basins: approaches, limitations and prediction of future developments. In: T. A. Cross (Editor), Quantitative Dynamic Stratigraphy. Prentice Hall, Englewood Cliffs, N.J., pp. 181–193.

Wei, Z.P. and Lerche, I., 1988. Quantitative dynamical geology of the Pinedale Anticline, Wyoming, U.S.A.: an application of a two-dimensional simulation model. Appl. Geochem., 3: 423–440.

Welte, D.H. and Yukler, M.A., 1981. Petroleum Origin and Accumulation in basin evolution — a quantitative model. Am. Assoc. Pet. Geol., Bull., 65: 1387–1396.

Wilgus, Ch.K., Hastings, B.S., Kendall, Ch.G.St.C., Posamentier, H., Ross, Ch.A. and Van Wagoner, J. (Editors), 1988. Sea-Level Changes: An Integrated Approach. Soc. Econ. Paleontol. Mineral., Spec. Publ., 42, 412 pp.

Yukler, M.A., 1979. Sensitivity analysis of groundwater flow system and an application to a real case. In: D.F. Merriam (Editor), Geomathematical and Petrophysical Studies in Sedimentology (Computer Geology). Pergamon Press, New York, N.Y.

I. LERCHE *Department of Geological Sciences, University of South Carolina, Columbia, SC 29208, U.S.A.*

Future potential of basin modelling techniques (extended abstract)

M.A. Yukler and B. Dahl

Since the 1970's, modelling techniques have been developed to quantify the geological evolution of sedimentary basins and to determine hydrocarbon occurrences in them. Many different types of models are used for different combinations of geological, geophysical, hydrodynamic and geothermal information. These models can be classified into three groups: physical, symbolic and mathematical.

Physical models are set up to study one variable at a time and, thus, are unsuitable to study complex geological systems. Symbolic models are used for graphical representation of simple inter-relationships among the components of a system as maps, cross-sections and block diagrams. Symbolic models have become particularly important with the development of balanced section techniques (Chamberlin, 1910; Elliot, 1977; Gibbs, 1983; Jamison, 1987).

Mathematical models are divided into two groups: statistical and deterministic. Statistical methods attempt to determine the probability of finding hydrocarbon accumulations of various sizes by making analogies between known and unknown (or partially known) basins (Yukler and Kokesh, 1984).

Deterministic models try to determine the hydrocarbon potential of basins by using well established physico-chemical and chemical principles (Yukler et al., 1978; Ungerer et al., 1984; Nakayama, 1987; Yukler, 1987).

The interest in quantitative basin modelling in the oil industry has increased at an accelerating rate, especially during the last ten years. Consequently, geoscientists with diverse backgrounds and differing scientific goals have been involved in the construction and application of these models. This has, however, led to confusion in the definition of 'basin modelling'. Currently there are at least six different definitions of this process, viz.: (1) basin analysis (qualitative geological modelling); (2) symbolic models (mainly balanced sections); (3) geophysical models (quantification of plate motions, subsidence and heat flow); (4) geochemical models (maturity and hydrocarbon generation evaluation with limited interest in the geological evolution of basins); (5) mathematical models (series of calculations without full integration of the processes in a basin as an integrated system); and (6) fully integrated quantitative geological models (based on system concepts and modelling a basin as a total entity with all the processes operating in it).

Two popular approaches are geochemical modelling and fully integrated quantitative geological modelling. Geochemical modelling does not require an extensive knowledge of geology, hydrodynamics, and mathematics and is generally not used to check the validity of the given geological history. The major shortcoming of this approach is that one can easily model hydrocarbon generation in a basin without understanding fully the processes involved and this may lead to erroneous results.

Fully integrated quantitative geological modelling attempts first to quantify the formation and evolution of the basin using all the available know-how and data from plate tectonics, sedimentology, stratigraphy, palaeogeography, palaeoclimate, hydrodynamics, geothermics, and inorganic and organic geochemistry in an integrated manner. Once the geologic evolution of the basin is optimized by minimizing the differences between the computed and measured sedimentary thicknesses, porosity, pressure, temperature, and maturity indicator values, it tries to quantify generation, migration and accumulation of the hydrocarbons as a function of time and space.

Both of these types of models use organic-maturity indicators such as vitrinite reflectance and biological markers to optimize temperature histories in a given basin. Some of these models have used the empirical vitrinite reflectance versus temperature–time relationship proposed by Lopatin (1971). Other models have attempted to use chemical kinetic approaches for vitrinite reflectance predictions, e.g. Middleton (1982), Wood (1988), Larter (1988), Barker (1989),

Basin Modelling: Advances and Applications edited by A.G. Doré et al. NPF Special Publication 3, pp. 67–70, Elsevier, Amsterdam.

Burnham and Sweeny (1989). The approach by Burnham and Sweeney, using parallel first order kinetics for formation of four species from vitrinite particles, has been used by several authors (Vik and Hermanrud, Hermansen: this volume). However, it is still debatable how good this approach is in describing the vitrinite reflectance evolution under different temperature–time scenarios. Further research and field study is necessary to confirm or improve this chemical kinetic approach.

Molecular marker epimerization predictions are usually calculated by simplistic first-order chemical reaction (e.g. MacKenzie, 1984). According to the authors' experience problems are associated with these parameters due to limited understanding of the impact of the molecular contributions from the various source rock constituents (i.e. bitumen fractions and kerogen). Research to reveal the chemical pathways and mechanisms involved is currently being carried out and will eventually result in improved chemical kinetic models for molecular marker epimerisations.

Hydrocarbon generation has been computed from kinetic equations for the last seventeen years (Tissot and Espitalié, 1975) and studies in various laboratories continue for more precise determination of the kinetic parameters. It is, however, debatable whether the activation energies and frequency factors determined in the laboratory are reliable enough to be used in quantitative basin modelling (Dahl and Yukler, 1989; Nielsen and Dahl, 1991). These techniques employ elevated temperatures to compensate for geologic time. From a "chemical-kinetic" viewpoint, it is likely that other reaction mechanisms with associated reaction orders, activation energies and frequency factors are dominant under geological rather than under laboratory conditions. Studies will certainly continue to determine kinetic models for maturity indicators and hydrocarbon generation and various kinetic models will eventually replace all other simplistic and empirical models.

There are certain problems in all the modelling techniques that are used today. The major problem in the application of statistical models is that each sedimentary basin may have a unique characteristic which is different from the population we think it represents. On the other hand, deterministic models require a perfect understanding of the processes taking place in a basin so that the problem can be solved deterministically. If there are major uncertainties in our knowledge of the processes and/or parameters of the system, the deterministic model applications will be no more than mathematical applications done on computers with impressively coloured graphical outputs. Future quantitative basin models will be

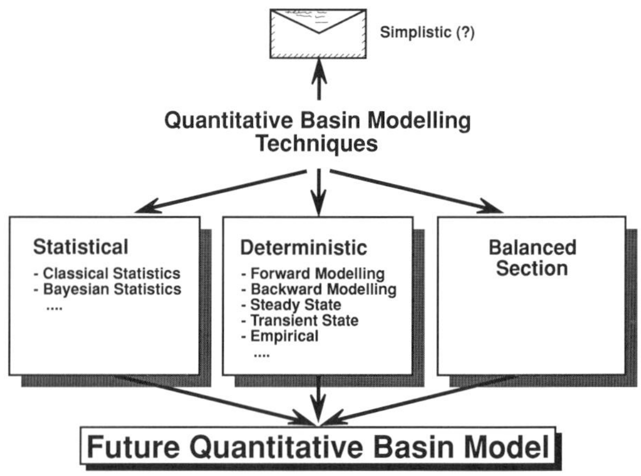

Fig. 1. The future quantitative basin models will attempt to integrate the current "stand alone" models and they will bridge the gap between the deterministic and probabilistic approaches.

constructed as a combination of balanced sections, statistical and deterministic techniques as proposed in Fig. 1. Hermanrud et al. (1990), for example, suggested the application of Monte Carlo simulation in conjunction with basin modelling and prospect appraisal. This suggestion bridges the deterministic and the probabilistic exploration approaches. Such methods will transform the associated geological uncertainties to probability distributions, a technique that is traditionally familiar to explorationists.

Quantitative basin models are currently applied in explored, partially explored, and unexplored basins to determine the hydrocarbon prospects within them. Most of the time modelling is undertaken as a postmortem process because the conceptual model for geologic history and temporal distribution of heat flow has to be optimized based on data from real wells. As basin modelling studies continue, and the input data and modelling results are stored in data bases, conceptual models will be established not only for individual basins, but also for wider regions, continents and eventually on a global basis. Basin models will then be a real exploration tool starting with the location of a wildcat through to development drilling.

Basin modelling software has until now mainly been integrated with graphics software to produce plots and diagrams. Future basin models will be integrated with other computer intensive software such as relational databases, mapping programs, statistical packages, etc. Some suggestions have already been made by Dahl et al. (1988) who proposed the integration of data bases, via a multivariate data analysis program, to improve the quality of input data to basin modelling software.

With the full integration of stress analysis, mechanical compaction of sediments, diagenetic changes, thermal expansion of fluids, generation of hydrocarbons, cracking of oil to gas, and with improvements in the quantitative prediction of porosities, permeabilities, and capillary pressures, basin models will be used more and more in the prediction of subsurface pressures and assessment of drilling prognosis and associated casing and mudweight programs before drilling. This will improve the prospect prediction potential of future models. Such models will also be used in development geology, production geology, geothermal energy assessments, deposition of water-soluble minerals, and environmental problems such as pollution of groundwater systems.

The first improvements in basin modelling will be related to play concepts and better prospect appraisal. One obvious aspect that needs better understanding and quantification is primary migration. Currently primary migration of oil is handled as a pressure-driven mechanism. If the amount of oil generation with time is above a threshold value, the pressure in the oil phase may be higher than in the water phase, resulting in the expulsion of oil until the pressures are equal. Rapid oil generation may also increase formation pressure leading to fracturing which also increases the expulsion efficiency. If the amounts generated are low then the expulsion of oil may take place only by saturation/diffusion processes resulting in low expulsion efficiencies. Once the oil is expelled, secondary migration starts toward potential accumulation locations. Two approaches are used to quantify secondary migration: the first is the computation of secondary migration head as a function of elevation, buoyancy pressure, hydraulic head, capillary pressure and specific weight of oil in three dimensions (Yukler and Welte, 1980). The second is the application of reservoir engineering concepts to model secondary migration (Ungerer et al., 1984). Future migration models will be three-dimensional, multiphase and with matrix fracture flow.

Other important aspects in prospect appraisal are the quantification of sealing mechanisms, capacities of faults and cap rocks, and the charge and accumulation histories. An attempt was made to quantify these processes based on one- and two-dimensional modelling by Dahl and Yukler (1991). In this study they computed the minimum horizontal stresses, pore pressures, and subsidence, pressure and temperature histories for the Oseberg Alpha reservoir–cap rock system of the Oseberg Alpha Field (offshore Norway) to determine the maximum cap-rock retention capacities for oil and gas. Future three-dimensional modelling with multiphase expulsion, multiphase migration, and structural reconstruction (together with techniques to determine potential stratigraphic traps) will enable more accurate predictions to be made of the hydrocarbon potential. Modelling will also improve our understanding of "in-reservoir" processes that result in alteration of oils, oil quality variations and diagenetic features, such as cessation of cementation, creation of secondary porosity etc. Future high-resolution models may even predict reservoir wettability.

As in other fields of science, basin modelling also requires knowledge and experience. It is not only a question of how good our models are, but also how good we are in using them. In this respect, more formal education is needed both at the university level and inside the various user environments than is apparent at the present time.

Acknowledgements

The authors are indebted to G.C. Speers and B.S. Cooper for kindly reading and commenting on this manuscript.

References

Barker, C.E., 1989. Temperature and time in the thermal maturation of sedimentary organic matter. In: N.D. Naeser and T.H. McCulloch (Editors), Thermal History of Sedimentary Basins — Methods and Case Histories. Springer-Verlag, New York, pp. 75–98.

Burnham, A.K. and Sweeney, J.J., 1989. A chemical kinetic model of vitrinite maturation and reflectance. Geochim. Cosmochim. Acta, 153: 2649–2657.

Chamberlin, R.T., 1910. The Appalachian folds of Central Pennsylvania. J. Geol., 18: 225–251.

Dahl, B. and Yukler, M.A., 1989. Calibration of hydrocarbon generation kinetics in the Viking Graben. Poster presentation at the 14th Int. Meet. on Organic Geochemistry, Paris (abstract).

Dahl, B. and Yukler, M.A., 1991. The role of petroleum geochemistry in basin modelling of the Oseberg Area, North Sea. In: R.K. Merrill (Editor), AAPG Treatise of Petroleum Geology Handbook. Source and Migration Processes and Evaluation Techniques. American Association of Petroleum Geologists, Tulsa, Okla., pp. 65–85.

Dahl, B., Kvalheim, O.M., Christie, A.A. and Yukler, M.A., 1988. Multivariate data analysis in quantitative petroleum geology: applications to basin analysis. Abstr., Am. Assoc. Pet. Geol. Res. Conf. Petroleum Potential of Sedimentary Basins, Techniques, Methods and Approaches, Leesburg, Va., April 25–29.

Elliot, D., 1977. The motion of thrust sheets. J. Geophys. Res., 81: 949–963.

Gibbs, A.D., 1983. Balanced cross-section construction from seismic sections in areas of extensional tectonics. J. Struct. Geol., 5: 153–160.

Hermanrud, C., Eggen, S., Jacobsen, T., Carlsen, E.M. and Pallesen, S., 1990. On the accuracy of modelling hydrocarbon generation and migration: the Egersund Basin oil find, Norway. Org. Geochem., 16 (1/3): 389–399.

Hermansen, D., 1993. Optimization of temperature history — aspects of vitrinite reflectance and sterane isomerization. In: A.G. Doré et al. (Editors), Basin Modelling: Advances and Applications. Norwegian Petroleum Society (NPF), Special Publication 3. Elsevier, Amsterdam, pp. 119–126 (this volume).

Jamison, W.R., 1987. Geometric analysis of fold development in overthrust terranes. J. Struct. Geol., 9: 207–219.

Larter, S.R., 1988. Some pragmatic perspectives in source rock geochemistry. Mar. Pet. Geol., 5: 194–204.

Lopatin, N.V., 1971. Temperature and geologic time as factors in coalification. Akad. Nauk SSSR Sibirsk. Otd. Geol. Geofiz., 10: 3–10 (in Russian).

MacKenzie, A.S., 1984. Applications of biological markers in petroleum geochemistry. In: J. Brooks and D.H. Welte (Editors), Advances in Petroleum Geochemistry, Vol. 1. Academic Press, London, pp. 115–214.

Middleton, M.F., 1982. Tectonic history from vitrinite reflectance. Geophys. J. R. Astron. Soc., 68: 121–132

Nakayama, K., 1987. Two-Dimensional Basin Analysis for Petroleum Exploration. Ph.D.Thesis, University of South Carolina, Columbia, S.C.

Nielsen, S.B. and Dahl, B., 1991. Confidence limits on kinetic models of primary cracking and implication for the modelling of hydrocarbon generation. Mar. Pet. Geol., 8: 483–492

Tissot, B. and Espitalié, J., 1975. L'évolution technique de la matière organique des sédiments: application d'une simulation mathématique. Rev. Inst. Fr. Pét., 30: 743–777.

Ungerer, P., Bessis, F., Chénet, P.Y., Durand, B., Chiarelli, A., Oudin, J.L. and Perrin, J.F., 1984. Geological and geochemical models in oil exploration: principles and practical examples. In: G. Demaison and R.J. Murris (Editors), Petroleum Geochemistry and Basin Evaluation. Am. Assoc. Pet. Geol., Mem., 35: 53–77.

Vik, E. and Hermanrud, C., 1993. Transient thermal effects of rapid subsidence in the Haltenbanken area. In: A.G. Doré et al. (Editors), Basin Modelling: Advances and Applications. Norwegian Petroleum Society (NPF), Special Publication 3. Elsevier, Amsterdam, pp. 107–117 (this volume).

Wood, D.W., 1988. Relationships between thermal maturation indexes calculated using Arrenius equations and Lopatin method: implications for petroleum exploration. Am. Assoc. Pet. Geol., Bull., 72 (2): 115–134

Yukler, M.A., 1987. How essential is quantitative basin modelling in petroleum exploration? Proc. 7th Biannual Pet. Congr., Turkey, pp. 392–404.

Yukler, M.A. and Kokesh, F., 1984. A review of models used in petroleum resource estimation and organic geochemistry. In: J. Brooks and D.H. Welte (Editors), Advances in Petroleum Geochemistry. 1: 69–113.

Yukler, M.A. and Welte, D.H., 1980. A three-dimensional deterministic, dynamic model to determine geologic history and hydrocarbon generation, migration and accumulation in a sedimentary basin. In: Fossil Fuels. Technip, Paris, pp. 267–285.

Yukler, M.A., Cornford, C. and Welte, D.H., 1978. One-dimensional model to simulate geologic, hydrodynamic and thermodynamic development of a sedimentary basin. Geol. Rundsch., 67: 960–979.

M.A. YUKLER *IBA Inc., 2394 Shady Lake Circle, Carrollton, TX 75006, U.S.A.*
B. DAHL *Norsk Hydro Research Centre, Lars Hillesgt. 30, N-5000 Bergen, Norway*

Stratigraphic basin modelling — recent advances

Bruce Levell and Werner Leu

Stratigraphic forward-modelling programs offer the explorationist the following advantages:

(1) An opportunity to objectively test geological understanding of a stratigraphic cross-section. This can result in reduction of perceived exploration risk.

(2) The possibility to make logical semi-quantitative stratigraphic predictions of reservoir, seal and source rock away from control and test the results for their sensitivity to assumptions about tectonic evolution, sea-level behaviour etc.

(3) A convenient tool for demonstrating stratigraphic principles.

Shell uses a stratigraphic modelling package consisting of partly empirical and partly theoretical algorithms for the creation and filling of accommodation space in a 2D simulation. New examples of output from the program, from The Netherlands and the Barents Sea, demonstrate both the principles behind the package and its impact on exploration.

Introduction

The explorationist is frequently confronted with the task of predicting the occurrence and distribution of reservoirs, source rocks and seals from very limited data.

Typically stratigraphic predictions are based upon a reconstructed palaeogeographic evolution of an area. However, such reconstructions are often judged as "acceptable" purely on the basis of their similarity to previous ideas or analogies with other settings with which the exploration team is familiar. Stratigraphic simulation programs allow geological interpretations to be checked for internal consistency, in much the same way as palinspastic reconstruction programs in structural geology. Of course, stratigraphic simulations will remain dependent upon certain assumptions about how basins evolve and how sediments are deposited; there is therefore no guarantee that the simulation is either correct or unique. Many see this as a fundamental shortcoming of the simulation approach. However, how correct, unique, or internally consistent are the interpretations that are routinely made on incomplete industry data sets? With a stratigraphic simulation program assumptions are made explicitly, are fully documented and can be changed. Furthermore, stratigraphic simulation programs reconstruct the geometry of time lines. This allows the simulations to be directly compared with real seismic data.

For these reasons, and because of the insight such models can give into interesting theoretical questions concerning the evolution of sedimentary basins, a large number of stratigraphic models has been developed. It is the purpose of this paper to demonstrate some of the ways in which these tools have been used.

The model developed within Shell (Aigner et al., 1989, 1990; Lawrence et al., 1990; Featherstone et al., 1991) is, strictly speaking, a simulation since it does not pretend to have a full theoretical basis in terms of physical processes. The program has been described by Lawrence et al. (1990) and only a brief description will be given here.

In principle the program is very simple: accommodation space is created through sea-level rise (a library of sea-level curves is available), subsidence, and compaction. Subsidence can be specified in a number of ways: as total subsidence (derived from thicknesses in wells), as back-stripped subsidence (using an Airy isostatic model), or as a theoretical subsidence related to flexural loading and a simple rift model or intra-plate stress (e.g. Cloetingh et al., 1987). Subsidence patterns can be interpolated between data points linearly, using Airy isostatic models, or by using flexural models with a user-specified elastic thickness. Compaction is calculated using standard compaction curves.

This accommodation space is filled with sediment up to a base level, which is either sea level or the fluvial graded profile (shape specified by the user). The graded profile is attached to the shoreline. Clastic sediments are supplied according to a user-specified sediment flux, which represents the sediment flux

Basin Modelling: Advances and Applications edited by A.G. Doré et al. NPF Special Publication 3, pp. 71–83, Elsevier, Amsterdam.

at the shoreline. The shoreline must therefore be within the modelled grid at all times. Sediments are distributed seawards from this point according to an exponentially declining sedimentation rate function. Clastic sediments may be derived from the left- or right-hand side or from out of the plane of the modelled section. Grain size is determined by the water depth at the time of deposition in the marine section, and by an alluvial architecture model in the non-marine realm. Erosion down to base level is governed by a user-specified erosion time constant and elevation.

Carbonates are generated by a carbonate growth function which represents the potential production of carbonate sediment with water depth. This potential carbonate sediment may, at a given location, exceed the available accommodation space and in this case it is re-distributed in an analogous manner to clastic sediment, but in this case either landward or seaward from the productive site. Both clastics and carbonates are subject to ponding, and slumping according to slope stability criteria (slope angle, cohesion, angle of internal friction and depth to overpressure). Pelagic sedimentation rates are user-specified.

The approach to the sedimentological questions that arise in generating a simulated stratigraphic cross-section, has been to use "reasonable" empirical rules to formulate algorithms to calculate, for instance, sediment re-distribution. Various of these rules are described by Lawrence et al. (1990). The user is given a high degree of freedom to adjust parameters via a set of menus. The advantage of this approach is that the complexity of real stratigraphy can be approached. The disadvantage is obviously that extensive calibration and data collection are necessary to both formulate and, subsequently, check the various rules.

Applications

Two major roles have been identified for the program: (1) educational; (2) extrapolating away from well control, using an iterative comparison of seismic interpretation and simulation results.

Educational applications

The role of models in leading to new insight into the eustatic school of sequence stratigraphy has been demonstrated by, amongst others, Jervey (1988) and Posamentier et al. (1988). The emphasis within the methodology of sequence stratigraphy on eustatic sea-level variations as the paramount control on the geometry and lithology of sedimentary bodies has, however, been a subject of controversy. We have attempted, by running simulations in a true modelling mode (i.e. disconnecting the empirical rules and algorithms, and using basic physics) to assess, individually, the impact of different controlling parameters on basin fills. In each of the following simulations as many variables as possible have been held constant. Time lines in most simulations are shown for a one million year time step.

One of the most important variables controlling sedimentary evolution is the initial bathymetry of the receiving basin. Through the amplification of subsidence due to isostatic compensation of the sedimentary load, small initial differences in bathymetry have a remarkably long-lived effect on sedimentary geometries. This is particularly true for carbonates, but also holds for clastics. In Fig. 1 a deep and a shallow basin are modelled. The initial shoreline is in both cases situated at the tectonic hinge (100 km). Simple rotational tectonic subsidence about this hinge occurs at the same rate in both models, and lithospheric parameters for the flexural isostatic response, such as elastic thickness, are identical. The sediment supply rate, measured at the shoreline, has been kept constant and is the same for both models. In these simple examples, in order to demonstrate the impact of a single variable as much as possible, no account is taken of sediment compaction or slumping due to slope oversteepening, but sediment deposited above sea level is eroded.

The initial bathymetry mainly influences the mode of progradation and the width/thickness ratio of the final sediment body. In Fig. 1a the system builds out more slowly than in Fig. 1b because additional accommodation space must be filled. The shoreline position eventually stabilizes at a position where the rate of relative sea-level rise equals the rate of sedimentation. This position is different because the amount of loading-induced subsidence is different in the two cases. Furthermore, the amount of flexure is greater in the upper figure where the depocentre stays close to the original shoreline. Thus the initial bathymetry influences not only the sedimentary geometry, but also the basin shape.

The geometries produced by sinusoidal variations of sea level at different frequencies are shown in Fig. 2 and in more detail in Fig. 5. These geometries may be compared with alternative models for the origin of cyclical sedimentation patterns, for example, intra-plate stress (Cloetingh et al., 1987; Kooi and Cloetingh, 1989). In these examples sea level, sediment input and tectonic rotation rate are held constant. The initial bathymetry is shallow and flat (as in Fig. 1), with tectonic rotation about a hinge situated beneath the initial shoreline position at 200 km. There is again neither compaction nor slumping. The

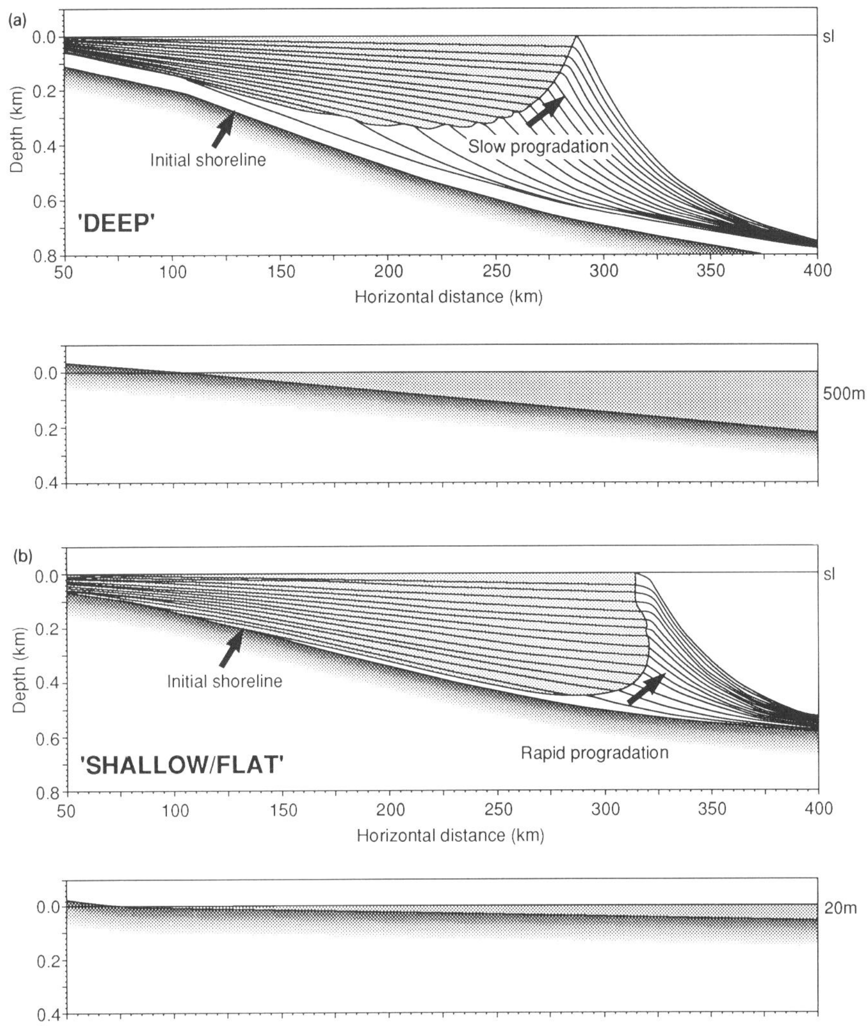

Fig. 1. Two stratigraphic simulations illustrating differences in geometry produced by differences only in the initial bathymetry, which is 500 m deep in (a) and 20 m deep in (b). Other variables have been kept constant. Constant rotational tectonic subsidence about a hinge at 100 km, the initial shoreline position, is amplified by isostatic loading with a constant elastic thickness. Compaction has been ignored. Note differences in the time taken for the shoreline to reach a stable position, where sediment supply equals relative sea-level rise. Note also the change in the shape of the flexure dependent upon the position of the depositional load.

models of in-plane stress (Fig. 3) show the effect of compression and tension. A horizontal compression magnifies the peripheral bulges at the basin flank and leads to regression of the shoreline. A tensional force drowns the basin flank and reduces the subsidence in the basin centre resulting in an overall transgression. Note that the non-marine onlap pattern is a complex function of the in-plane stress and the elastic flexure of the crust as amplified by sediment loading. Figure 3 shows a cycle of compression and tension or vice versa; such cycles produce geometries that resemble those of sinusoidal sea-level change. The question, of course, remains as to whether the frequency of such in-plane stress cycles is similar to that of the proposed third-order global sea-level cycles. Note moreover that the generated sequence boundary (internal onlap surface) is of Vail's type 2 (Vail, 1987), i.e. the shelf edge is not exposed.

A variable that is often ignored in discussing sequence stratigraphy is the sediment supply rate. In Fig. 4 two simulations are shown, one with a sharp increase in sediment supply rate followed by an exponential decay, perhaps simulating an (epi-) orogenic or footwall uplift, and a sinusoidal curve. In both examples elastic flexural subsidence is modelled with constant rotational tectonic subsidence about a hinge at 100 km, the initial shoreline position. The initial bathymetry is shallow and flat. In the upper figure the simulated cycle has its peak progradation some time after the peak sediment input due to an interaction between sediment volume and loading and flexurally induced relative sea-level rise. The sudden decrease in the rate of sediment supply happens too quickly for an equilibrium to be established. In both figures note that no sequence boundary is formed. In other words there is no downward shift of facies or

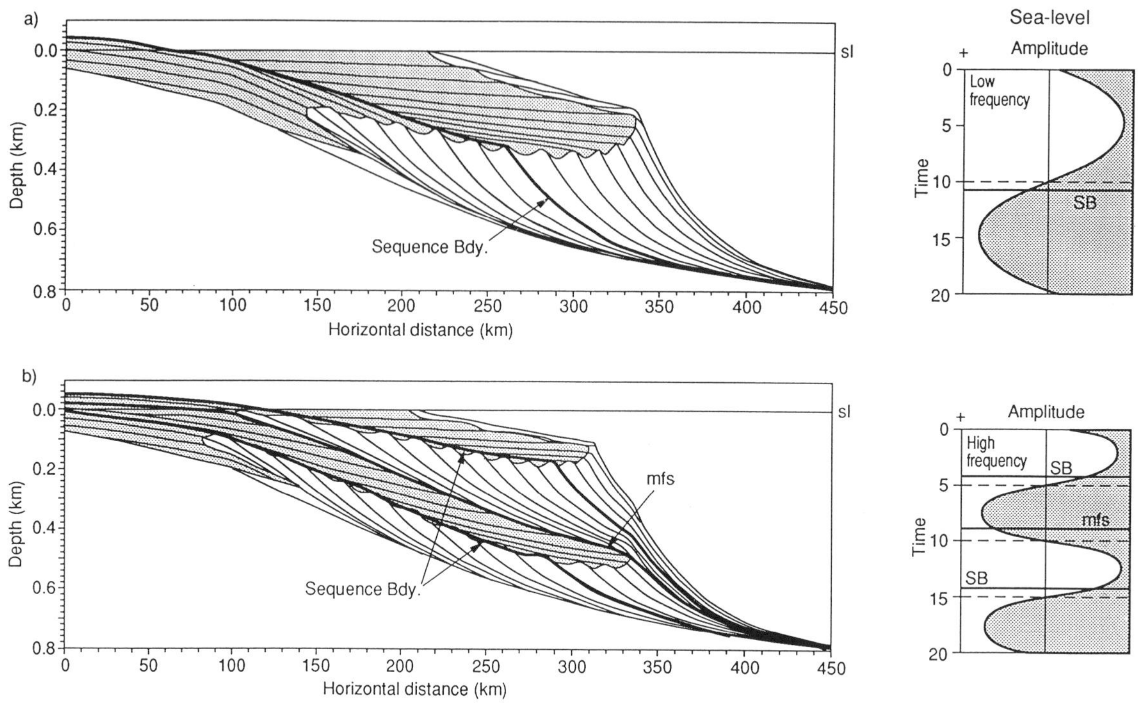

Fig. 2. Stratigraphic simulations with two different rates of sinusoidal eustatic sea-level change. In both cases note the clearly expressed sequence boundaries (*SB*) (onlap surfaces). In the upper simulation the time step is 1 m.y. whereas in the lower the time step is 0.5 m.y.. In the lower figure the maximum flooding surface (*mfs*) occurs at 8.5 Ma, slightly later than the rising limb inflection point at 10 Ma, and as postulated in sequence stratigraphic theory. The transgressive system tract thus spans the interval of 12 to 8.5 Ma, the lowstand system tract (shelf margin system tract in this case) from 14 to 12 Ma, and the highstand system tract from 8.5 to 4 Ma. The delay in formation of the sequence boundary (onlap surface) with respect to the falling limb inflection point in these examples, is an artefact of the sampling of the sea-level curve, and disappears as the sampling rate is increased (cf. Fig. 5).

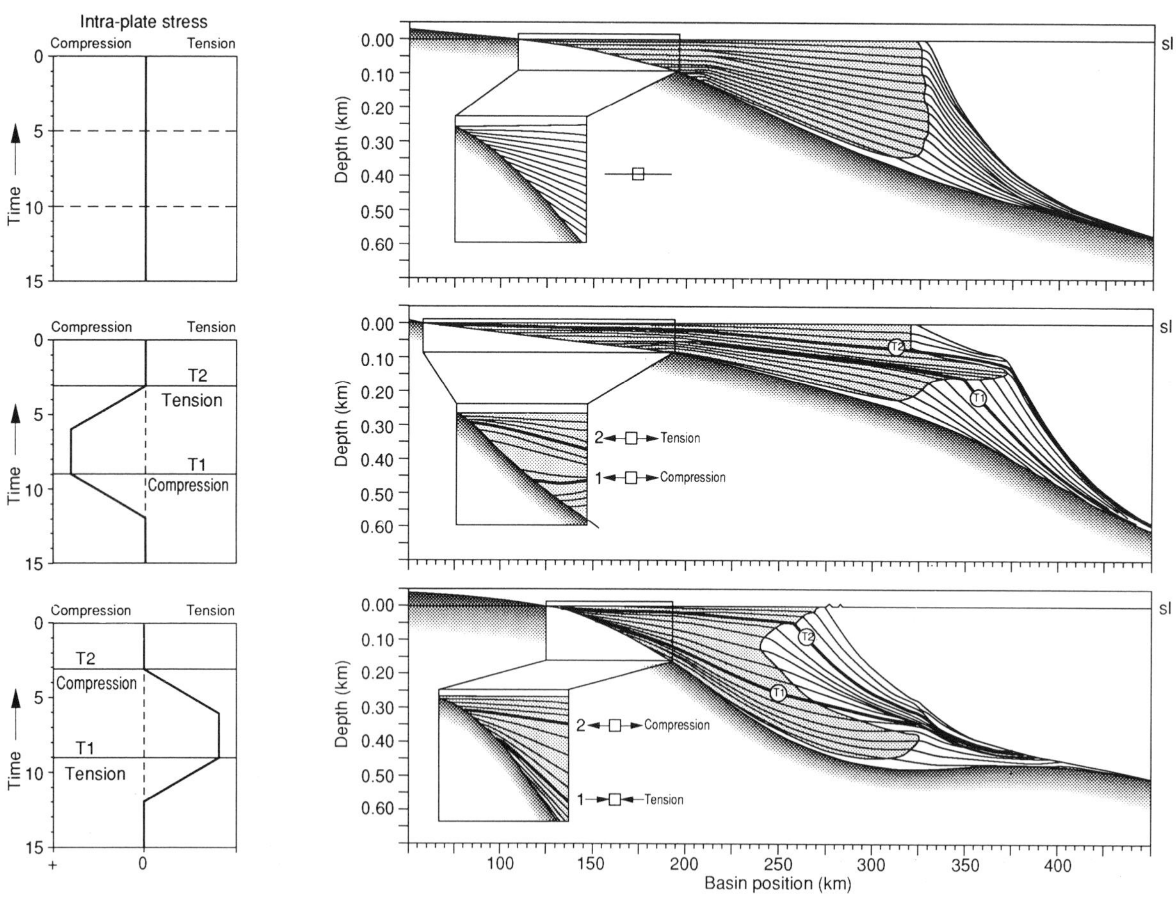

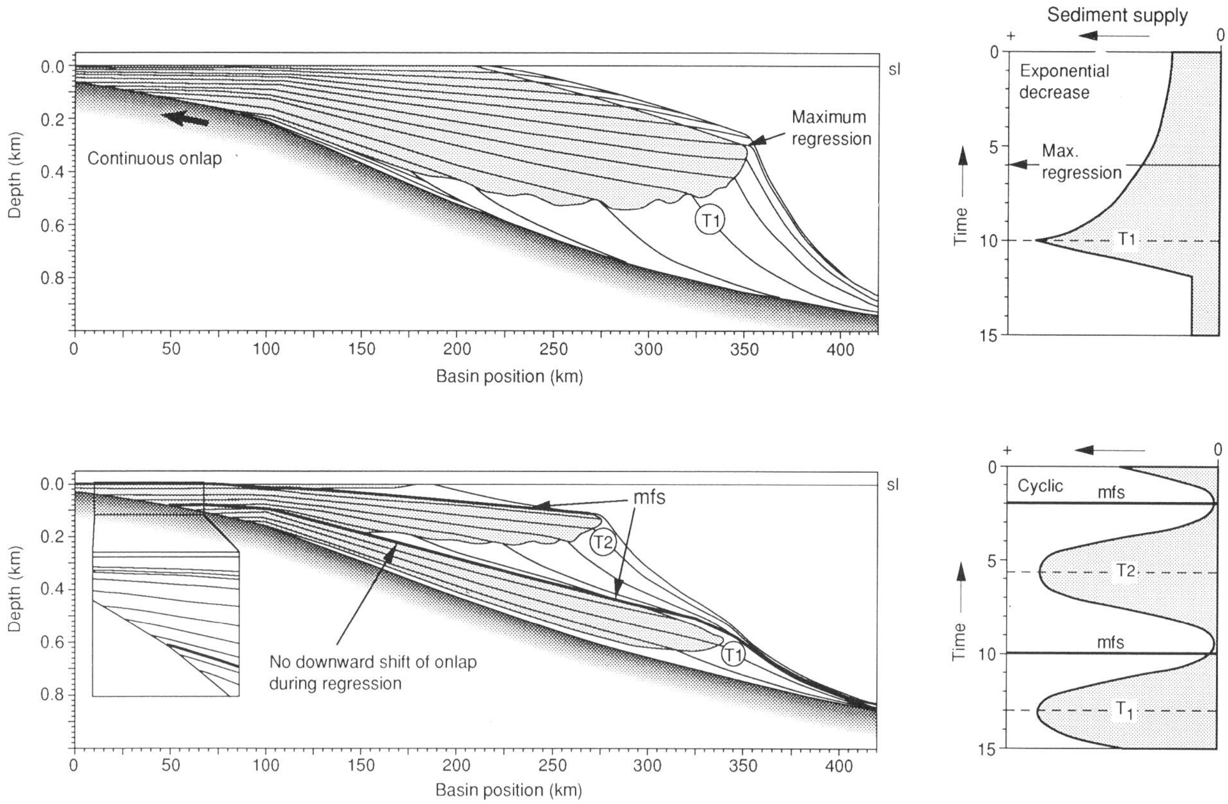

Fig. 4. Stratigraphic simulations illustrating geometries created by changing sediment supply at constant sea level. A sudden increase, perhaps representing footwall uplift, is followed by an exponential decline. Note that the maximum regression occurs later than the peak sediment supply. This is because the peak sediment supply induces subsidence by isostatic adjustment. Due to the fact that with rotational tectonic subsidence about a hinge line at 100 km, the sediment prism experiences greater tectonic subsidence as it progrades. A sinusoidal sediment supply curve produces transgressions and regressions, with well-defined condensed sections at times of zero sediment input. However, no internal onlap surfaces are generated. Nonetheless, cycles caused by such a process might be easy to confuse, on seismic data, with eustatic sequences in areas with high subsidence rates. Unfortunately, little is known quantitatively about the rates and magnitudes of sediment supply changes to real basins.

regional onlap corresponding, for instance, to peak rates of sediment input. This follows from the fact that changes in sedimentation rate cannot reduce the amount of accommodation space; they can only affect the speed with which accommodation space is created via loading of the crust.

In real examples it is often very difficult to analyse which depositional geometry can be attributed to which basin-filling control. The initial conditions may not be known, the data may be incomplete and more than one variable will change in a continuous way so that the system may not reach an equilibrium with the controlling factors. Simulations such as those shown here are a visual way of highlighting end members.

In most simulations it is assumed that only sea-level changes can occur with a frequency and speed that explains the cyclicity and the geometries of sediment packages that were deposited in time intervals of about 5 m.y. or less. Input tectonic subsidence curves show changes of subsidence rate over longer time scales, and sediment supply is normally held constant for longer time periods than this. The "unit cell" of a sea-level-driven sequence, in a case of simple rotational tectonic subsidence with elastic flexure as simulated by the program, is shown in Fig. 5. This simulation closely resembles the diagrams of Vail (1987), even to the extent that the transgressive maximum occurs slightly later than the maximum rate of sea-level rise, due to the flexural res-

Fig. 3 (left). Stratigraphic simulations of the effects of the Cloetingh et al. (1987) intra-plate stress concepts. In the upper figure there is no intra-plate (in-plane) stress. The initial shoreline is at a position of 200 km, which marks a tectonic hinge about which constant rate rotational subsidence occurs. This subsidence is amplified by flexural isostasy with a constant elastic thickness. In the lower figures all parameters are held constant and additional intra-plate stress cycles are applied to the simulation. The enlargements show the onlap patterns. The final flexure is a function of intra-plate stress, rotational "driving" subsidence, sedimentation, and elastic response to the load. Features resembling sequence boundaries and transgressive systems tracts are generated.

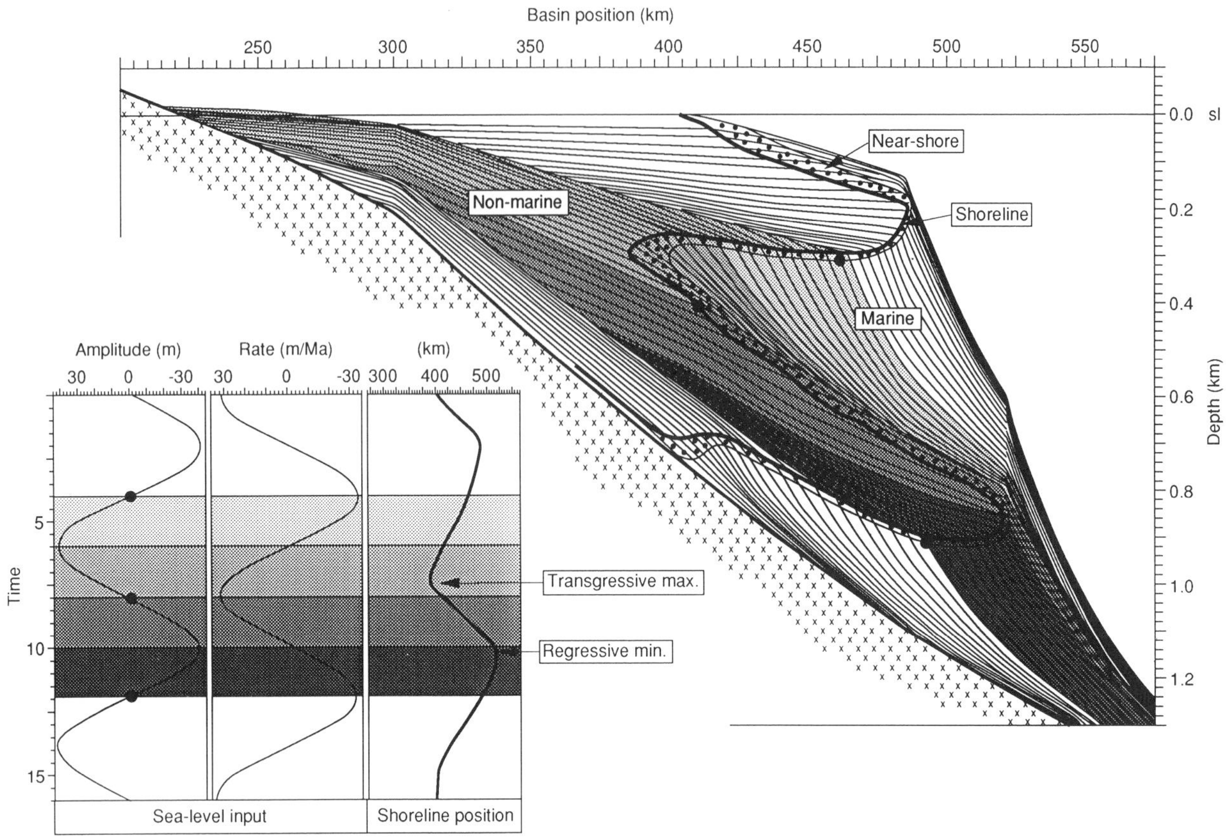

Fig. 5. Simulated cross-section through a clastic basin margin with sinusoidal eustacy, constant sediment input, constant rotational tectonic subsidence, and flexural isostatic compensation. The shadings indicate four successive periods in one sea-level cycle. Note that such a cycle is divided into three system tracts (shelf margin, transgressive and highstand) and that the maximum flooding surface again lags the maximum rate of transgression. Such an idealised sequence architecture forms the building blocks for operational application of the stratigraphic simulation program.

ponse to the sediment load. In this simulation the slumping module was not used and the basin floor fans of the Vail et al. model are therefore not simulated.

More realistic numerical experiments can be undertaken if real datasets are modelled. Figure 6 is a regional NW–SE cross-section of the Oligocene to Lower Miocene section of the Lower Rhine Embayment (SW Netherlands), slightly modified after Zagwijn et al. (1985). This section is based on shallow borehole data. The section was initially (Fig. 6A) simulated using the Haq et al. (1987) sea-level curve, and back-stripped subsidence curves for eight positions along the section. In the first run sediment input was held constant. The simulation resulted in a major Early Miocene transgression at around 16 Ma and, in general, too large a proportion of marine facies in the lowermost Miocene and uppermost Oligocene. A better result was obtained by steering the simulation by specifying a variable sediment supply rate through time (Fig. 6B). This was done by trial and error, but the end result is that high rates of sediment input are associated with high rates of sea-level change, in particular high rates of rise. Little appears to be known about quantitative variations in sediment input, so a third simulation was run, modifying the sea-level curve whilst maintaining a constant rate of sediment input (Fig. 6C). This simulation is less satisfactory than 6B. Chronostratigraphic diagrams generated for part of these simulations illustrate testable differences between the models (Fig. 7). Interestingly, all the simulations associate peat deposition with transgressive intervals.

This exercise raises quantitative questions about sea level and sediment input behaviour. These would clearly have economic implications if the section were from an exploration play: for instance, in terms of the number and pinch-out position of individual sand units.

Operational applications

An attempt to forward simulate the stratigraphy of the Norwegian Barents Sea illustrates how the program can be applied to operational problems. In this case a regional NW–SE seismic line (Fig. 8) was simulated using data from the two wells: 7226/11-1 on the Mid-Barents Sea Platform, and 7321/8-1 in the

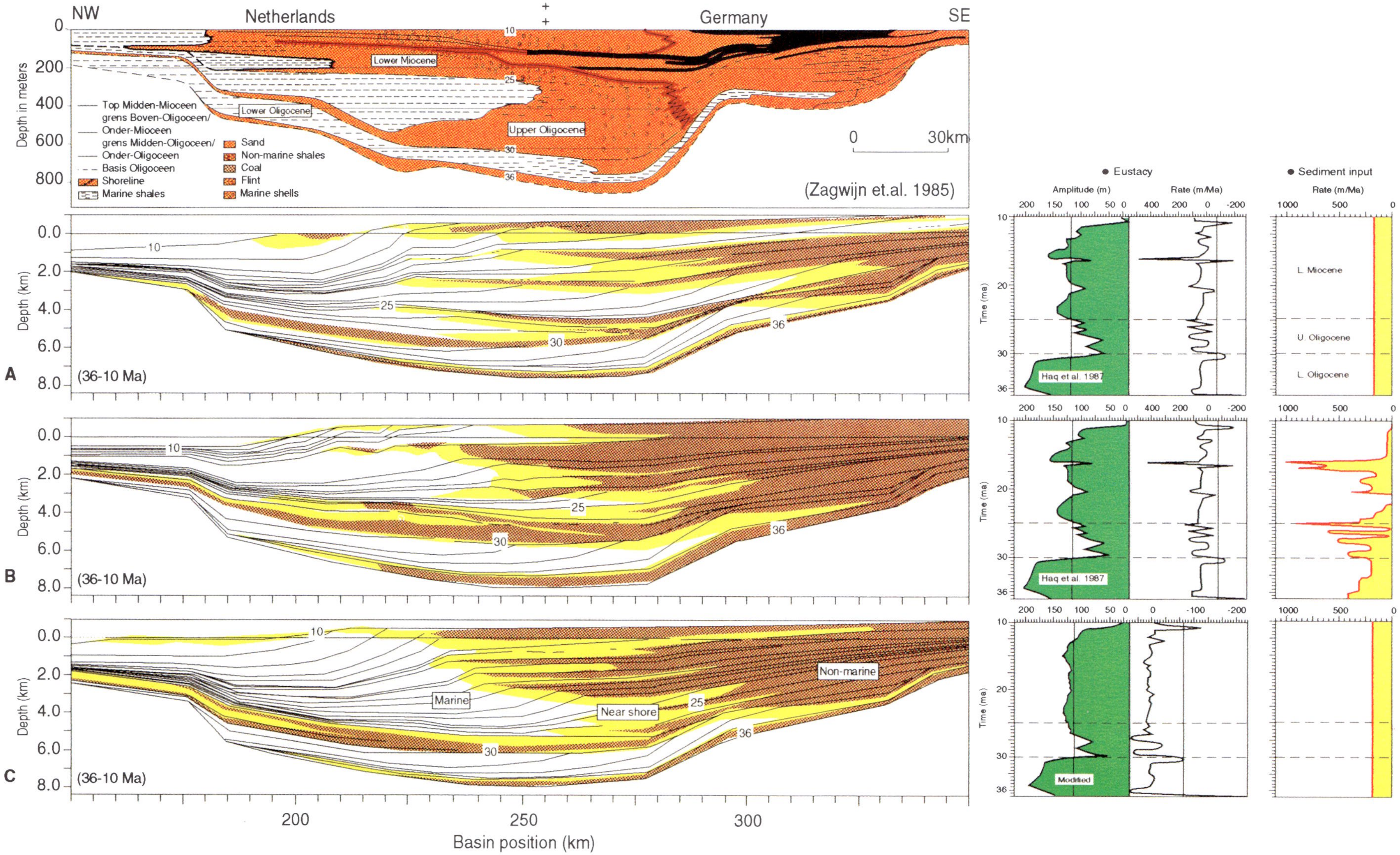

Fig. 6. Cross-section illustrating the Oligocene to Lower Miocene of the Lower Rhine Embayment of the southwestern Netherlands and Germany (slightly modified after Zagwijn et al., 1985). Three simulations are shown for comparison. In (A) the unmodified Haq et al. (1987) sea-level curve is used, and constant sediment input through time. It mimics closely the middle Oligocene regression, but has too high a proportion of marine clays in the upper part of the section. A better simulation is achieved by editing the sediment supply curve, (B), which can be made to fit the data. Alternatively, a moderately good match can be achieved by modifying the amplitudes (and therefore the rates) of the sea-level curve, (C). The acceptability of (B) versus (C) would depend on further investigation into the likelihood of shifts in sediment supply (but see also Fig. 7).

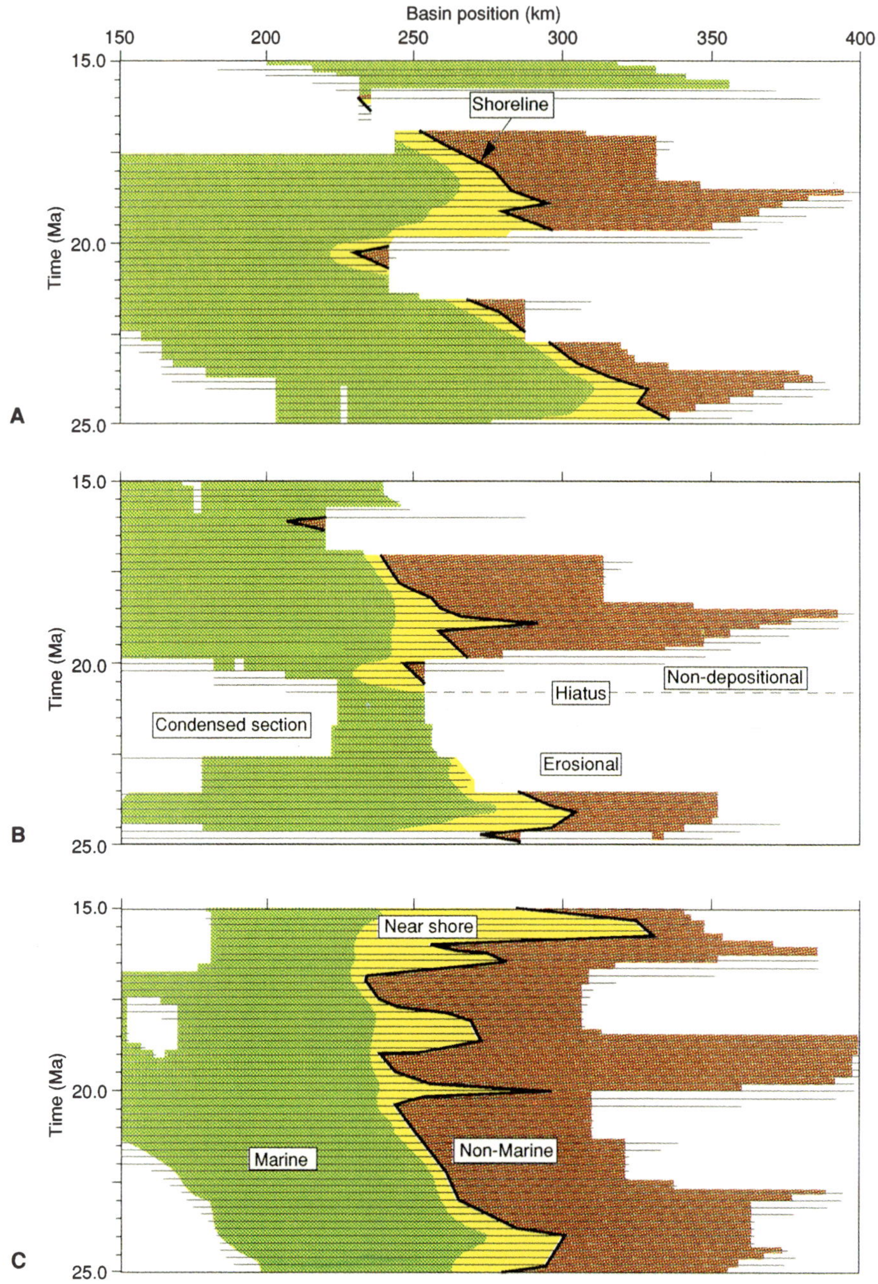

Fig. 7. Chronostratigraphic diagram for the Upper Oligocene–Lower Miocene for the three simulations of Fig. 6. Note that such diagrams provide a detailed basis for deciding between the three simulations of Fig. 6. Reduction in the rates of sea-level change (especially fall) in (C) implies more continuous deposition than in (A) and (B).

Bjørnøya Basin, as well as several shallow coreholes on the NE extension of the Loppa High. Tectonic subsidence was computed by back-stripping nine locations along the interpreted seismic section using a smoothed, long-term sea-level curve. The simulation (Figs. 9 and 10) was run interactively until an acceptable model was achieved. The main additional input variables were the Haq et al. (1987) sea-level curve (together with the geologic time scale attached to the sea-level curve), and the fluctuating rates of sediment input from three sources (Fig. 11).

Sediment input is derived from both ends of the line (Greenland and the Urals) as well as from an out-of-plane clastic source situated to the northeast. The latter was found to be necessary to prevent water depths in the Triassic basin from becoming too deep, in particular in the Lower and Middle Triassic. A step by step display of the simulated transect at particular times (Fig. 10) and the accompanying chronostratigraphic chart (Fig. 12) demonstrate the facies distribution and palaeobathymetry of the basin through time.

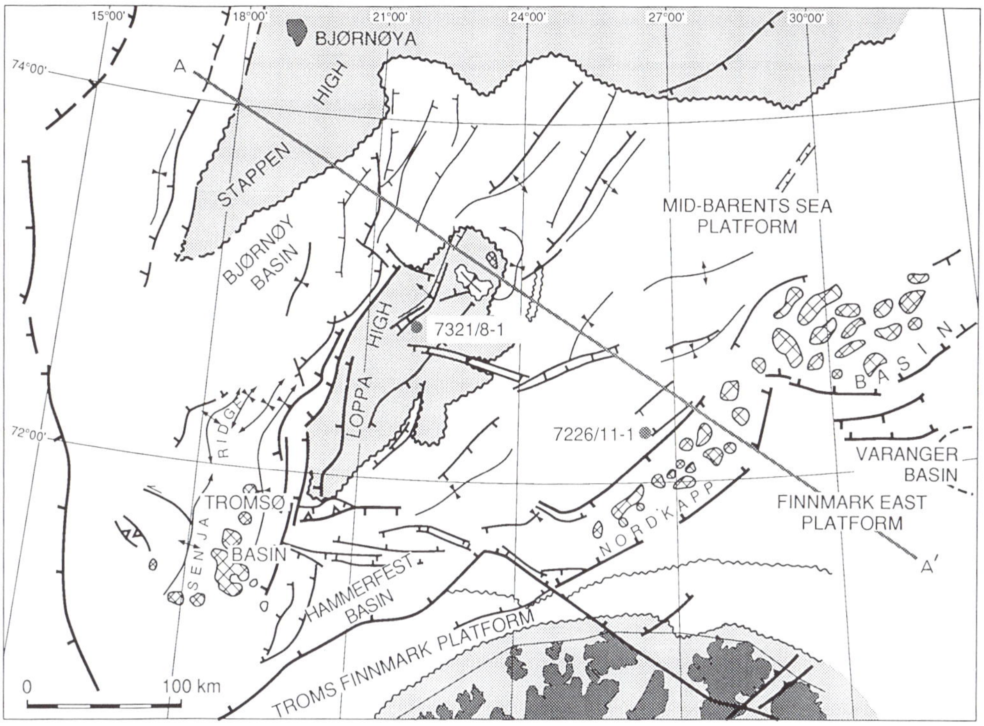

Fig. 8. Location map of the southwestern Barents Sea showing the location of the simulated transect.

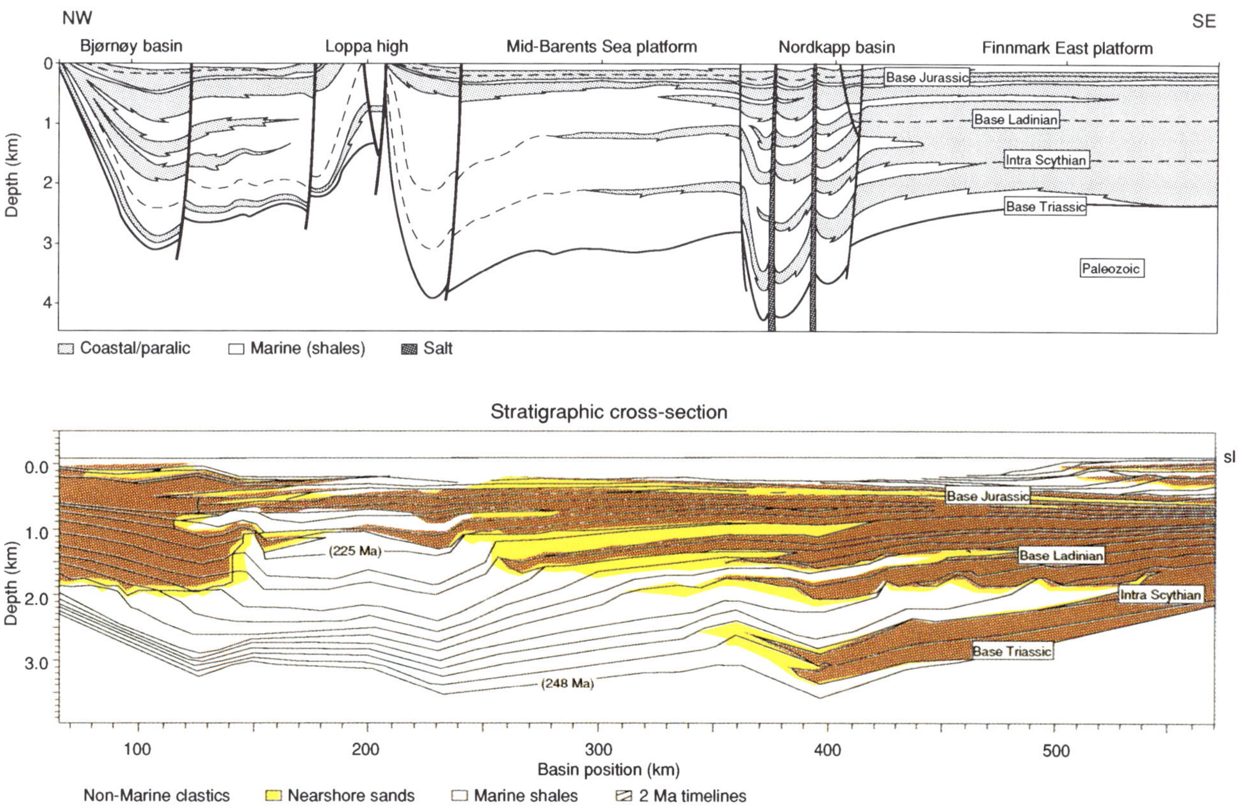

Fig. 9. Stratigraphic cross-section along line A–A' (Fig. 8) with the base Cretaceous as datum. Stratigraphic interpretation is based on seismic, well data and regional geology, but is loosely constrained. The simulation, below, is hung on base Cretaceous sea level.

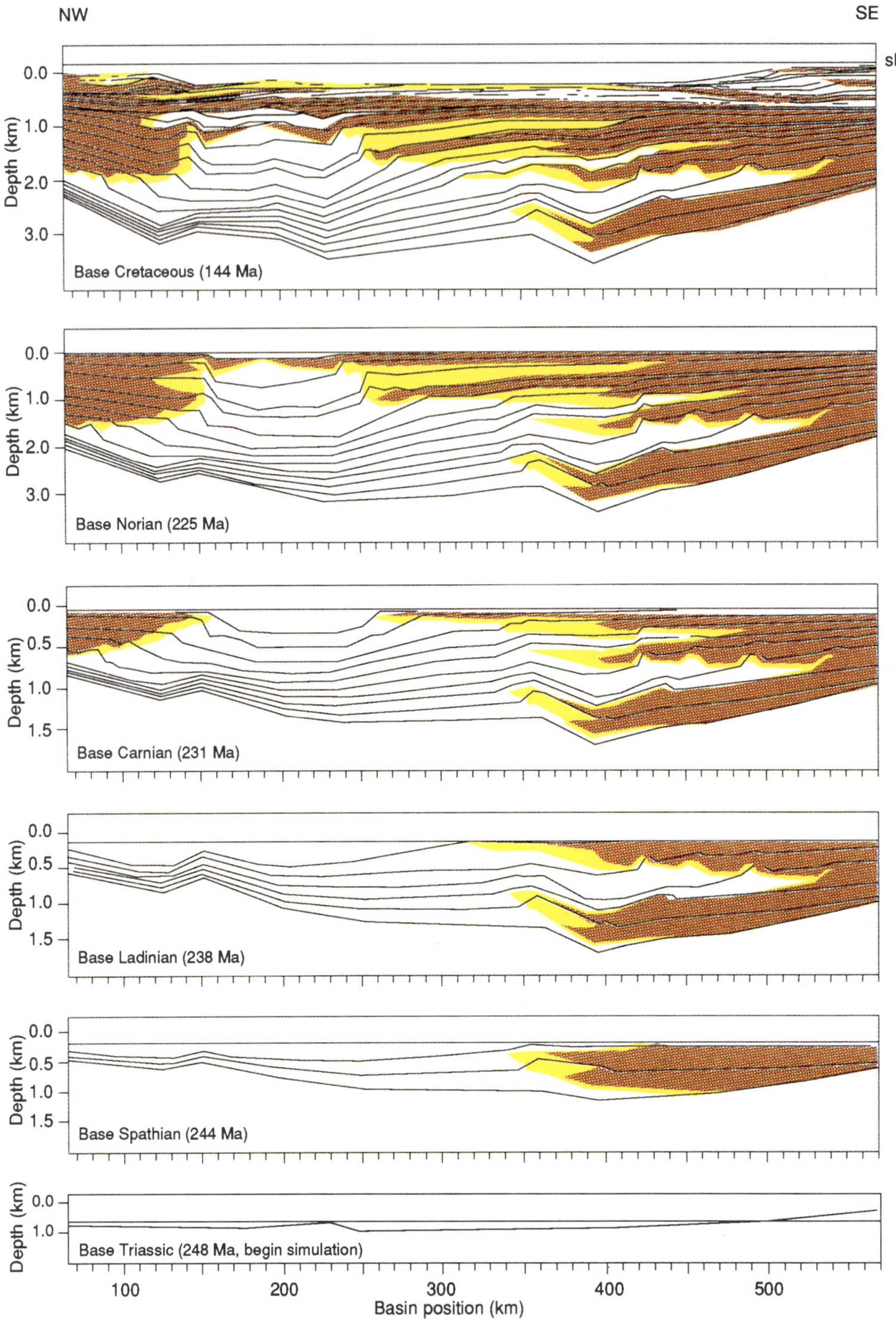

Fig. 10. Step by step reconstructions of the simulation in Fig. 9 for the specified geologic times.

The relative extent of transgressions and regressions is in good agreement with the available data. The thicknesses of individual stages within the Triassic show a westward migration of the basin centre. This is also evident on the seismic data. The modelled intervals in the Bjørnøy and Nordkapp basins are too thin compared with the well data mainly because of the high flexural rigidity of the basement which was found necessary in the simulation to match the observed thicknesses. Similarly, uplift and erosion on the Loppa High are not great enough in the simulation. A substantial increase in clastic sediment input in the Carnian was found to be necessary. A possible explanation for this is simply that the chronostratigraphic interval is too short or the amplitude of the postulated sea-level rise in the early Carnian is too large for the Barents Sea region. There is considerable disagreement about the

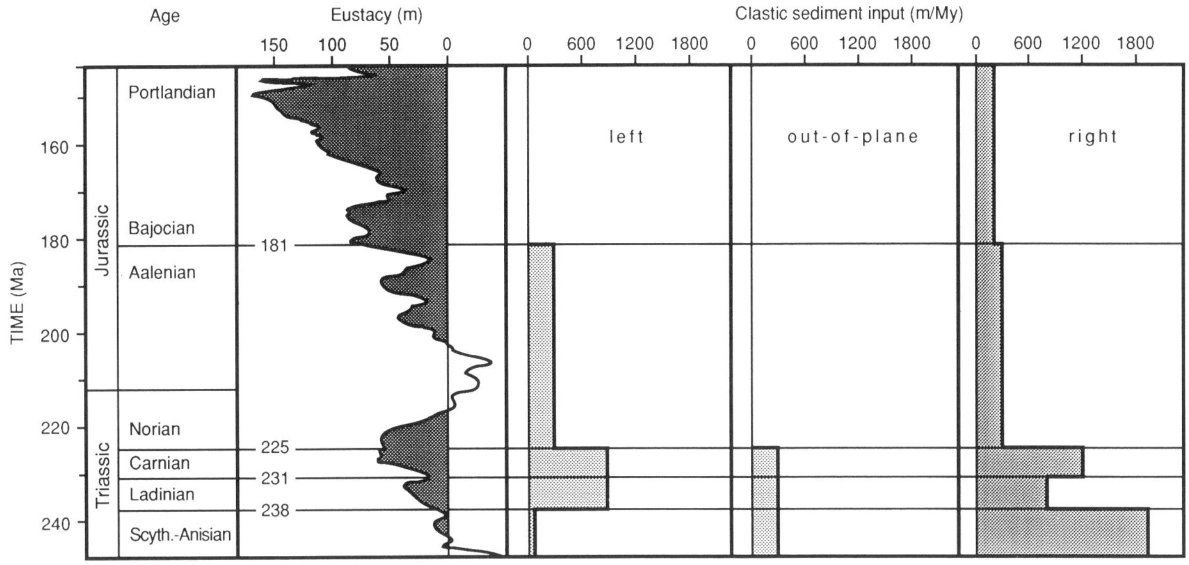

Fig. 11. Input parameters to the simulation of Fig. 9. The eustatic sea-level curve is from Haq et al. (1987), with the time scale of Harland et al. (1982).

duration of Triassic stages (Harland et al., 1989). In the program, as in nature, it is the balance of rates of relative sea-level rise and rates of sediment supply which control progradation versus transgression. Simulation results are thus very sensitive to small changes in stage duration. (In these simulations the Harland et al., 1982 time scale was used.)

In conclusion, the depth/facies and sand/shale relationships (Fig. 13) near the existing wells were reasonably simulated giving some confidence that the simulation was valid beyond the well control. New insight was gained into the timing and magnitude of individual sediment sources. Perhaps most importantly a fully consistent, even if poorly constrained, geologic model was generated for testing.

Discussion

There are numerous major uncertainties involved in stratigraphic modelling at present. These may be summarized as follows.

(1) Uncertainties concerning the relative importance of major controlling parameters operating together: for instance, in-plane stress and flexural subsidence, sea-level change and sediment supply.

(2) Uncertainties concerning feedback between variables: for example, how does sediment supply at the shoreline respond to sea-level change through gradients?

(3) Quantitative uncertainties: What are reasonable lithospheric parameters for elastic thickness, relaxation isotherm? What are reasonable values for sediment input, and how do these vary through time? What is the absolute time scale to be attached to stage boundaries? What is the global sea-level curve? Is there a global sea-level curve (e.g. Fjeldskaar, 1989)?

(4) Uncertainties about the validity of sedimentological rules: for instance, how exactly are carbonates affected by clastic input through water turbidity, and what is the influence of seasonality of sediment input?

(5) To what extent are simulations that "match" data unique?

All of these questions are difficult to answer. Nonetheless, data can be collected to address many of them, and mathematically relatively simple models can be used to convolve the various time series in a single simulation. The future of stratigraphic modelling will very probably lie in inverse models which allow the user to vary numerous free parameters in order to converge on the available data.

Acknowledgements

We would like to acknowledge the authors of the stratigraphic simulation program used in these studies: Mark Doyle, ex KSEPL, Rijswijk, and Dave Lawrence of Shell Dev. Co. in Houston. We are also grateful to Shell Internationale Research Maatschappij, Shell Internationale Petroleum Maatschappij and Norske Shell Exploration and Production A/S for permission to publish this paper. Kees v.d. Zwaan, ex Norske Shell, helped considerably with the Barents Sea simulation. Arthur van Vliet, ex KSEPL, was closely involved with all the simulations presented here.

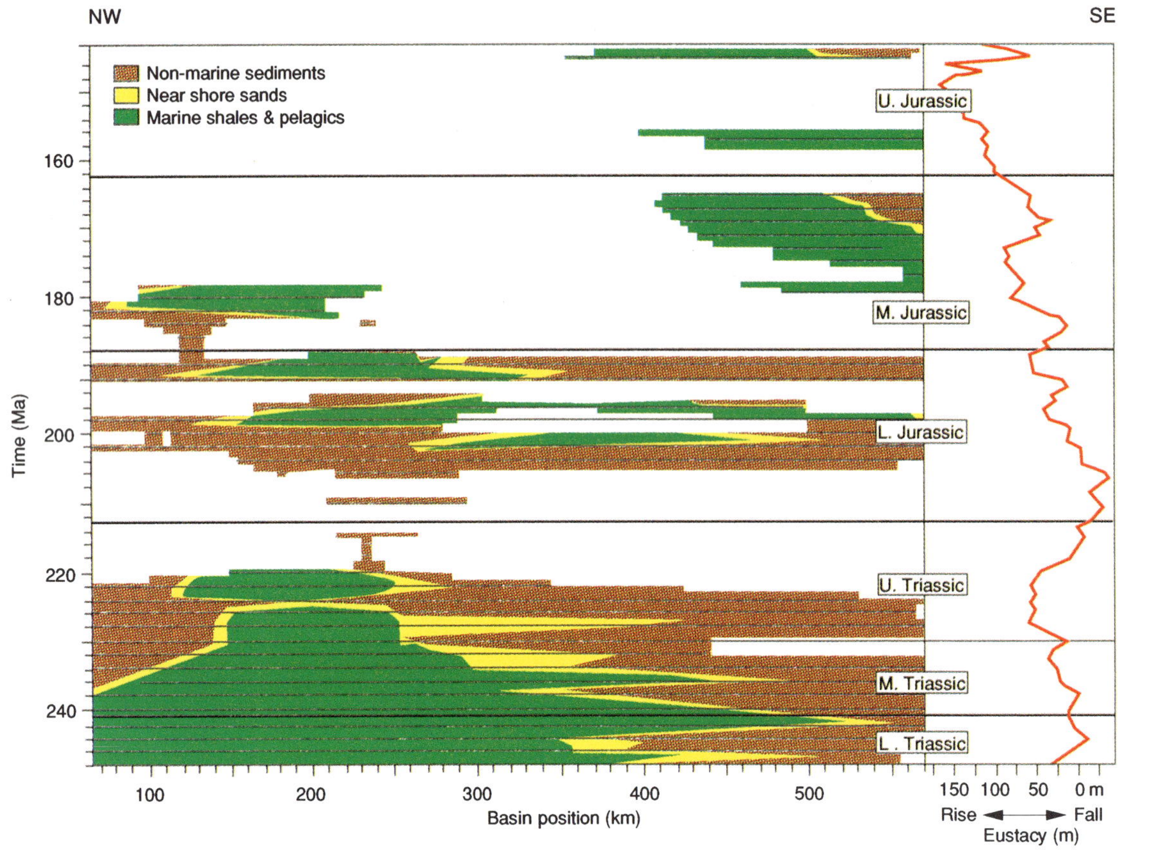

Fig. 12. Chronostratigraphic chart for the simulation of Fig. 9. Note the predictions concerning hiatuses which can be tested with real data as these become available.

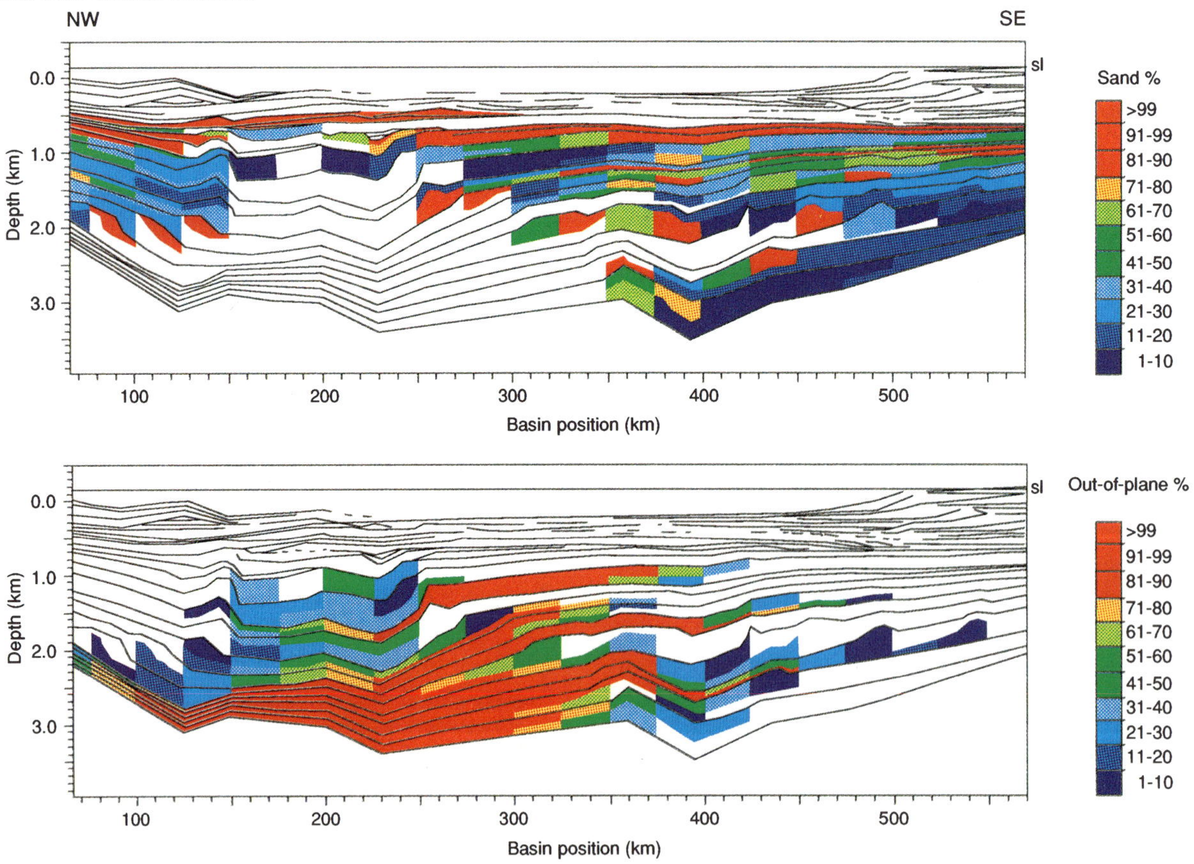

Fig. 13. Lithologic predictions made by the program for the simulation of Fig. 9. The sand percentages are based on depositional processes as well as depositional environment. The lower diagram illustrates the quantitative significance in the simulation of an out-of-plane source from the northeast.

References

Aigner, T., Doyle, M., Lawrence, D., Epting, M. and van Vliet, A., 1989. Quantitative modelling of carbonate platforms: some examples. In: P.D. Crevello, J.L. Wilson, J.F. Sarg and J.F. Read (Editors), Controls on Carbonate Platforms and Basin Development. Soc. Econ. Paleontol. Mineral., Spec. Publ., 44: 27–37.

Aigner, T., Brandenburg, A., van Vliet, A., Doyle, M., Lawrence, D. and Westrich, J., 1990. Stratigraphic modelling of epicontinental basins; two applications. Sediment. Geol., 69: 167–190.

Cloetingh, S., Lambeck, K. and McQueen, H., 1987. Apparent sea level fluctuations and a palaeostress field for the North Sea region. In: J. Brooks and K. Glennie (Editors), Petroleum Geology of North-West Europe, 1. Graham and Trotman, London, pp. 49–57.

Featherstone, P., Aigner, T., Brown, L., King, M. and Leu, W., 1991. Stratigraphic modelling of the Gippsland Basin. Aust. Pet. Explor. Assoc. J., 31 (1): pp. 105–114.

Fjeldskaar, W., 1989. Rapid eustatic changes — never globally uniform. In: J.D. Collinson (Editor), Correlation in Hydrocarbon Exploration. Norwegian Petroleum Society, pp. 13–19.

Haq, B.-U., Hardenbol, J. and Vail, P.R., 1987. Chronology of fluctuating sea-levels since the Triassic. Science, 237: 1156–1166.

Harland, W.B., Cox, A.V., Llewellyn, P.G. and Fancett, K.E., 1982. A Geologic Time Scale. Cambridge Univ. Press, Cambridge, 131 pp.

Harland, W.B., Armstrong, R.L., Cox, A.V., Craig, L.E., Smith, A.G. and Smith, D.G., 1989. Geological Time Scale. Cambridge Univ. Press, Cambridge, 256 pp.

Jervey, M.T., 1988. Quantitative geologic modelling of siliciclastic rock sequences and their seismic expression. In: Ch.K. Wilgus, B.S. Hastings, Ch.G.St.C. Kendall, H. Posamentier, Ch.A. Ross and J. Van Wagoner (Editors), Sea Level Changes — An Integrated Approach. Soc. Econ. Paleontol. Mineral., Spec. Publ., 42: 47–69.

Kooi, H. and Cloetingh, S., 1989. Some consequences of compressional tectonics for extensional models of basin subsidence. Geol. Rundsch., 78 (1): 183–195.

Lawrence, D.T., Doyle, M. and Aigner, T., 1990. Stratigraphic simulation of sedimentary basins: concepts and calibration. Am. Assoc. Pet. Geol. Bull., 74 (3): 273–295.

Posamentier, H.W., Jervey, M.T. and Vail, P.R., 1988. Eustatic controls on clastic deposition, 1. Conceptual framework. In: Ch.K. Wilgus, B.S. Hastings, Ch.G.St.C. Kendall, H. Posamentier, Ch.A. Ross and J. Van Wagoner (Editors), Sea Level Changes — An Integrated Approach. Soc. Econ. Paleontol. Mineral., Spec. Publ., 42: 109–124.

Vail, P.R., 1987. Seismic stratigraphy; interpretation using sequence stratigraphy. In: A.W. Bally (Editor), Atlas of Seismic Stratigraphy. Am. Assoc. Pet. Geol., Stud. Geol., 27: 1–10.

Zagwijn, W.H., Beets, D.J., van den Berg, M., van Montfrans, H.M. and van Rooijen, P., 1985. Atlas van Nederland, 13. Geologie. Staatsuitgeverij, Den Haag, 23 pp.

BRUCE LEVELL *Koninklijke/Shell Exploratie en Produktie Laboratorium, Volmerlaan 6, 2288 GD Rijswijk, The Netherlands*

WERNER LEU *Koninklijke/Shell Exploratie en Produktie Laboratorium, Volmerlaan 6, 2288 GD Rijswijk, The Netherlands*

A recent Norwegian Shelf heating event — fact or fantasy?

Rebecca P. Jensen and Anthony G. Doré

Many published and in-house maturation models using different modelling algorithms on North Sea, Halten Terrace and Barents Shelf well data require an apparent increase in heat flow over the last two million years in order to match both the measured temperature and measured maturity values. Maturity values alone can be matched by modelling large amounts of erosion or higher palaeo-temperatures, but the resulting models do not match present-day temperature values.

Explanations for the apparent heating event fall into two categories — geological and mathematical. Geological causes could include glaciation, rapid subsidence, over-pressuring, fluid flow, and changing pore fluids. In addition, the effect of uncertainties in control data (i.e., temperature and maturity) on the geological and thermal histories must be considered. Mathematically, the modelling algorithms used may not sufficiently model geological processes. Errors within these categories will have significant implications for maturation and hydrocarbon generation modelling.

In this paper we evaluate the apparent heating event, its causes, possible modelling solutions, and significance to modelled oil generation history. We conclude that the Jurassic section, within the over-pressured and transitionally pressured zones on the Halten Terrace, is today anomalously hot relative to the overlying section. There is no compelling evidence, however, to support an increase in heat flow on a lithospheric scale in the Plio-Pleistocene on the Halten Terrace. The elevated Jurassic temperatures on the Halten Terrace are a result of the convergence of recent geological phenomena, including rapid burial, over-pressuring, hydrocarbon generation, and possibly fluid flow. The modelled recent heat flow increase is an artifact of current modelling algorithms which do not adequately handle such complex geological situations; it does not represent an actual change in lithospheric heat flow.

Introduction

The goal of basin modelling is to accurately determine the geological, thermal and generational histories of a basin within the limits of our understanding of geological events, thermal phenomena, hydrocarbon generation mechanisms, and our ability to mathematically model these processes correctly. An apparent recent "heating event" on the Norwegian Shelf is a good example of how these four areas are interrelated.

In order to match present-day temperatures and maturities on the Norwegian Shelf using one-dimensional basin modelling software we have found it necessary to invoke a "heat flow" increase in the last few million years (Fig. 1). The nature of this

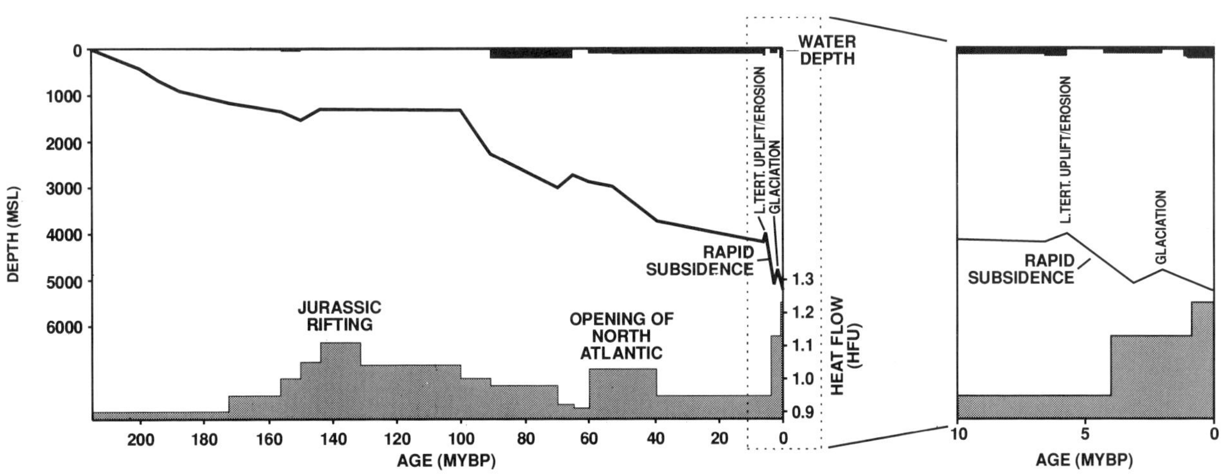

Fig. 1. Modelled subsidence of base Jurassic and heat flow history for the Halten Terrace.

Basin Modelling: Advances and Applications edited by A.G. Doré et al. NPF Special Publication 3, pp. 85–106, Elsevier, Amsterdam.

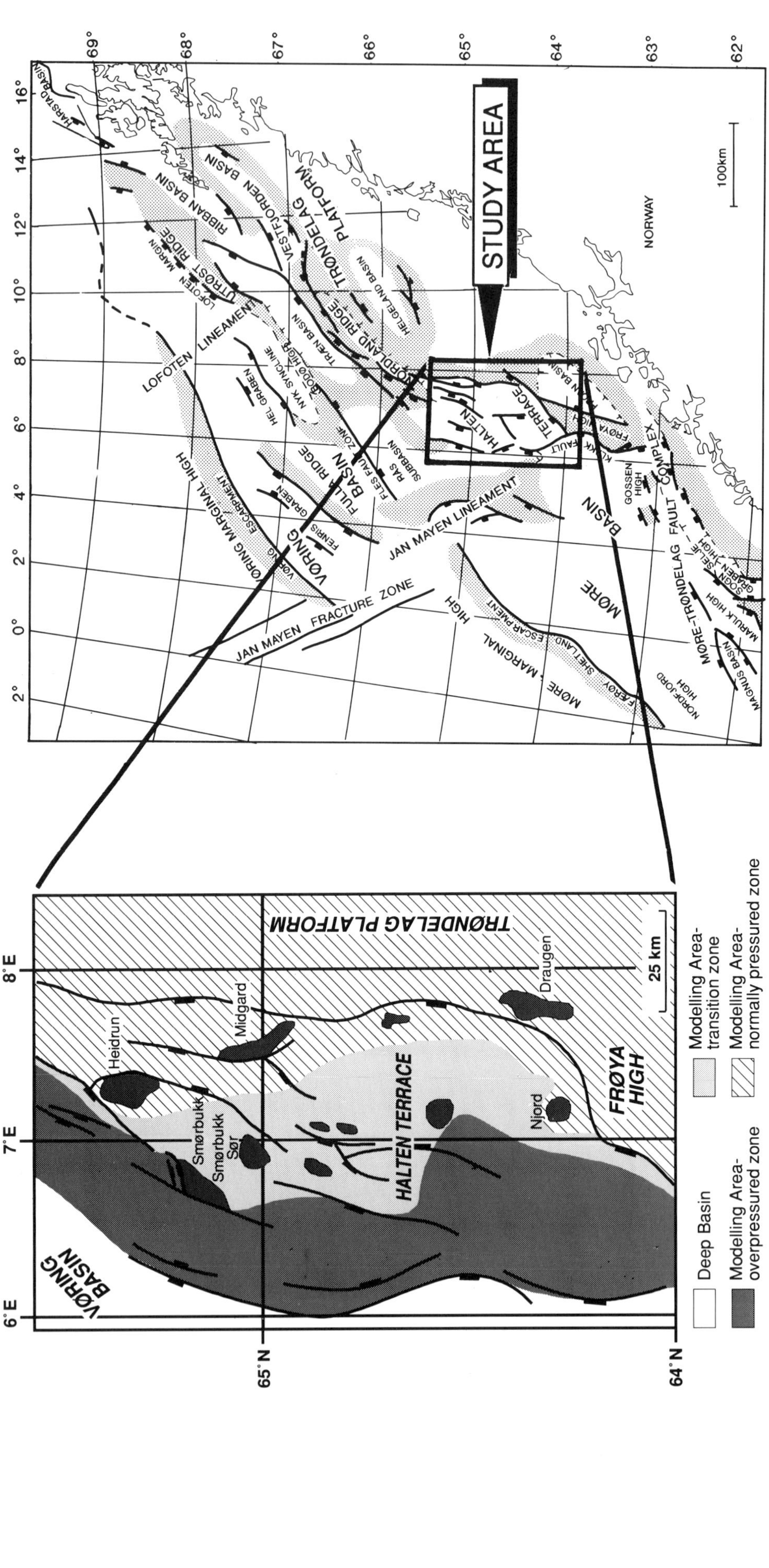

Fig. 2. Location map showing the major structural features of the Halten Terrace area and the primary faults, fields, and modelling area used in this study.

increase has raised questions concerning its possible origins and its validity. In this paper we discuss possible geological and mathematical causes of such an increase in heat flow.

We first noted the apparent increase in heat flow today, relative to the recent geological past, in the North Sea. Present-day temperatures are too hot for the measured maturities. Modelling "confirmed" that the temperatures today were not the key to the past thermal environment. Our work in the Barents Sea also indicated that today's temperature profile could not be projected into the past without overmaturing kerogens. However, in the Barents Sea, unlike the Halten Terrace or the North Sea, sediments reached their highest temperatures prior to the well-documented major Tertiary uplift. In this situation matching of present-day temperatures is necessary for scientific consistency, but not for the purposes of accurately calculating maturities.

On the Halten Terrace in Mid-Norway, our one-dimensional maturation/generation models again indicated that present-day "heat flow" could not be used throughout geological time. A closer examination of the geological history indicated that alternate thermal models were possible, though each would yield significantly different generation histories. This made it clear that a thorough understanding of the geological, thermal, and generational histories was a significant key to understanding hydrocarbon distribution on the Halten Terrace. It is this area (Fig. 2) that will be discussed in detail in this paper.

The apparent recent heat flow increase on the Norwegian Shelf has been mentioned in the literature, commercial reports, license reports, and communications among scientists. Dahl et al. (1987) used a heat flow increase in the Pleistocene in their model of the Oseberg Field area in the North Sea; no mention was made regarding the cause of the heating event. Mo et al. (1989) also used a heat flow increase in the last few million years to model the Midgard Field area in Mid-Norway. This heating event was attributed to the rapid subsidence which occurred at the end of the Tertiary. The cause of the subsidence was not addressed. Papers in this volume (e.g., Dahl and Augustson, and Grigo et al.) and Brigaud et al. (1992) also discuss the apparent anomalous temperatures seen and heat flow increases inferred today on the Norwegian Shelf. Curiously, similar heat flow increases have also been noted in other widely differing basins world-wide; see, for example Chiaramonte et al. (1988) who discuss two African basins, Schmidt and Erdoğan (this volume) who discuss the overthrust region of Austria, and Leischner et al. (this volume) who discusses the lower Saxony Basin. In this paper we attempt to draw together the possible causes of, and solutions to, the "heat flow increase" on the Halten Terrace. A similar evaluation of the geological and thermal processes in other basins would be useful in determining the validity of other reported heat flow increases around the world.

This paper is divided into six sections. In the first section we discuss how heat flow is used in basin modelling, with particular emphasis on palaeo-heat flow modelling. The geological history of the Mid-Norway region is then discussed, with emphasis on the Cenozoic section. The thermal framework for the Mid-Norway area is then developed based on the geological history and our understanding of the palaeo-heat flow history. Causes of the recent anomaly and then the implications for hydrocarbon generation are discussed next. Finally, conclusions are drawn concerning the likely nature of the recent heating event used in modelling.

Thermal modelling — general

The thermal history of a basin is deduced from the items shown in Table 1. Of the items shown, present-day measured temperatures and maturities are readily obtainable within accepted error limits. In addition, the present-day mineralogical components and the porosities of a layer can be determined from well samples and logs. From these it is possible to predict the present-day thermal conductivities of the sediments (Brigaud et al., 1990). The tectonic history, palaeo-climate, and depositional environments can be deduced from a knowledge of the area and detailed evaluation of geophysical and geological data, although variations in interpretation are likely between different scientists. The remaining items — palaeo-porosities and thermal conductivities, heat flow history, and changes in pore filling fluids —

TABLE 1

Parameters used to determine thermal histories

Parameters used to determine thermal histories
Present-day measured temperatures (surface and down-hole)
Present-day measured maturities
Understanding of geologic past tectonic history heat flow history latitude variations depositional environments erosional events
Physical properties of sedimentary horizons, past and present components organic constituents thermal conductivities porosities pore filling components

are more difficult to establish accurately, and are the major area in which basin models will differ (as an example compare this paper with Vik and Hermanrud, this volume).

Palaeo-porosity can be modelled in either a forward or backward direction. The common backwards models (Sclater and Christie, 1980; Falvey and Middleton, 1981) use an exponential function for each lithology type to determine a porosity–depth relationship. One shortcoming of the backwards models is that they inadequately account for over-pressuring as a result of rapid sedimentation. Ignoring over-pressures can significantly alter the calculated thermal histories since undercompaction results in lower thermal conductivities and higher gradients. One-dimensional forward models can also be limited in that they only account for vertical fluid flow. On the Norwegian Shelf, over-pressuring is an important phenomenon which must be fully considered because it may be a major factor in the "thermally anomalous" present-day.

A sediment's thermal conductivity is affected not only by its mineralogical components and porosity, but also by temperature, pressure, and pore filling fluids. In most cases, water is assumed to be the pore filling fluid. However, equation 6 from Brigaud et al. (1990) indicates that changing the pore filling fluid of a sand with 20% porosity from water to oil would result in a total conductivity reduction of 19%; a change in fluid from water to gas would give a 33% reduction. This translates to a 33% increase in an interval gradient when the unit changes from water-filled to gas-filled. The effect of different pore fluids increases with increasing porosity and hydrocarbon saturation, as shown in Table 2.

Thermal histories can be modelled using either geothermal gradients or a combination of thermal conductivities and heat flow values. The latter method allows the thermal parameters of a sedimentary layer to be altered as the layer is compacted and subjected to higher temperatures and pressures. A similar thermal history can be created using interval geothermal gradients that change through time; this method, however, is time-consuming and prone to error as it is a highly iterative process. Thermal conductivities, on the other hand, can be determined from lithologic and/or log parameters (Brigaud et al., 1990) and then modelled mathematically to account for changes in porosity, temperature, etc. In both cases, the present thermal environment is readily determined, whereas the palaeo-thermal environment is unknown and must be constrained during modelling by present-day measurements of thermal indicators.

Today's bottom hole and drill stem temperatures can be used to calculate interval geothermal gradients for the present only. If these data are then combined with present-day thermal conductivities for each sedimentary layer, present-day heat flow values can be calculated. A constant heat flow value for the entire well indicates an equilibrium state for the present. Variations in the calculated heat flow may indicate disequilibrium and should not be dismissed lightly. On the Norwegian Shelf there are indications that the Jurassic section is anomalous, with higher interval geothermal gradients, and, hence, higher, calculated interval heat flow values for this section on certain areas of the shelf (Fig. 3; Carstens and Finstad, 1981; Andrews-Speed et al., 1984; Brigaud et al., 1992). It is this anomalous interval that forces the use of an apparent recent heating event.

Determination of the past thermal environment is more difficult. The temperature history is based on knowledge of the tectonic history, the relative movement of the plates, depositional environments, and the present-day maturities. The tectonic history can be used to predict heat flow changes (see, for example, McKenzie, 1978; Sclater and Christie, 1980; Royden et al., 1980). Plate movements (see, for example, Rowley and Lottes, 1988) and depositional environments are used to predict palaeo-sea floor temperatures (see, for example, Frakes, 1979).

TABLE 2

Change in thermal conductivity with pore filling fluids

Fluid	Porosity (%)	Conductivity[a] (mcal cm^{-1} s^{-1} $°C^{-1}$)	Change (%)	Gradient[b] (°C km^{-1})
Water	20	10.24	–	14.0
Oil	20	8.30	19	17.3
Gas	20	6.84	33	21.0
Water	43	5.81	–	24.7
Oil	43	3.71	33	38.7
Gas	43	2.44	58	58.8

[a] Assumes clean sand with matrix conductivity of 16.75 (mcal cm^{-1} s^{-1} $°C^{-1}$), calculated using equation 6 from Brigaud et al., 1990.
[b] Assumes 1.44 mW m^{-2} heat flow.

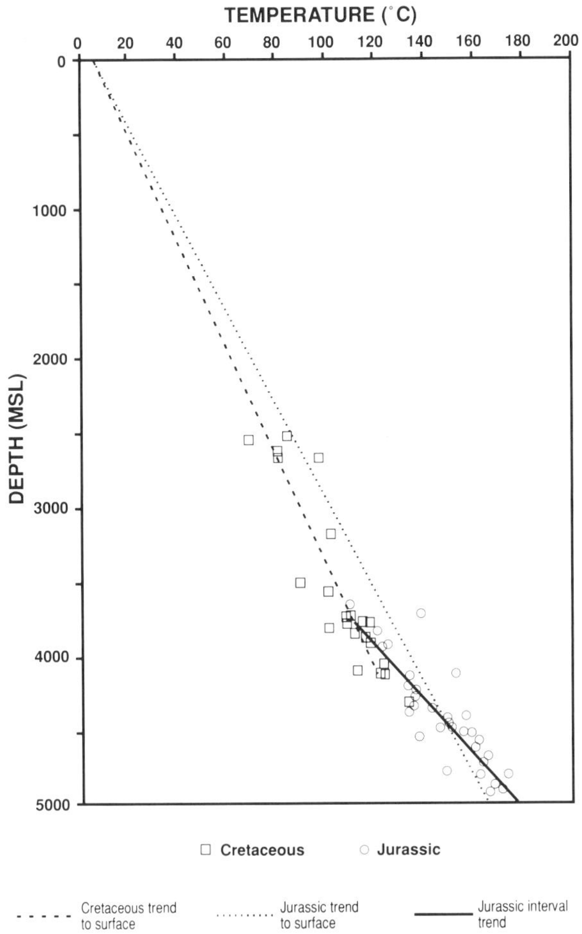

Fig. 3. Corrected bore hole temperature trends on the Halten Terrace for wells in which the top Jurassic is deeper than 3500 m below sea level.

Present-day maturities are used as a check to verify that the thermal history has not, at any time, been either too hot or too cold, so that today's maturities are matched within acceptable error limits of both the data and the modelling capabilities. In general, if the present-day thermal environment can be used for the past, or little is known about the geology of the area, then the thermal model should be kept as simple as possible. On the Norwegian Shelf, as already indicated, the present is not the key to the past, and alternative thermal histories must be investigated.

For this study, corrected present-day bottom hole temperatures and measured maturities were used as controls on the thermal history. It is recognized, however, that multiple modelling scenarios can result in a match to both the present-day measured temperatures and maturities. Further, the algorithms used in this study assume that heat flow is constant with depth, and therefore, by implication, that all heat sources are deep-seated and can be applied to the whole section. This assumption is necessary, given the limitations of current modelling algorithms, but may be incorrect. It is recognized, for example, that radiogenic isotopes, the lateral presence of sedimentary or basement features with contrasting thermal conductivities (Jones and Oxburgh, 1979; Gable, 1984) and convective fluid flow can impose local variations in heat flow.

Within the constraints that have been described, thermal histories were devised by varying the vertical heat flow history based on the interpreted sequence of geologic events. These geological events are described below.

Summary of the geological evolution of the Mid-Norway area

The Mid-Norway shelf is a passive margin that has undergone repeated episodes of extensional deformation since consolidation of the Baltic and Laurentian cratons during the Caledonian orogeny, approximately 400 Ma. Rifting culminated in seafloor spreading in the Early Tertiary, with the result that elements of the complex rift system are now distributed between the East Greenland and Mid-Norway shelves bordering the northeast Atlantic.

A succession of mainly extensional tectonic episodes between the Devonian and Triassic, largely taking place in a continental depositional environment, was followed by the deposition of thick (2 km or more) continental clastic sequences, culminating with the paralic, coaly Åre Formation of a latest Triassic–Early Jurassic age (Fig. 4). This unit is regarded as a significant source rock for

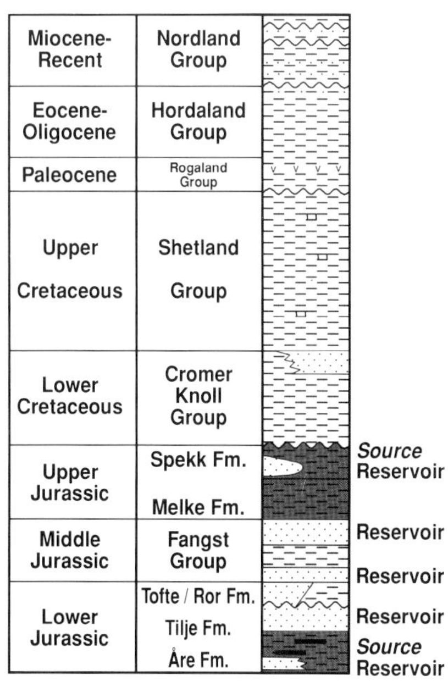

Fig. 4. Simplified Halten Terrace lithostratigraphy.

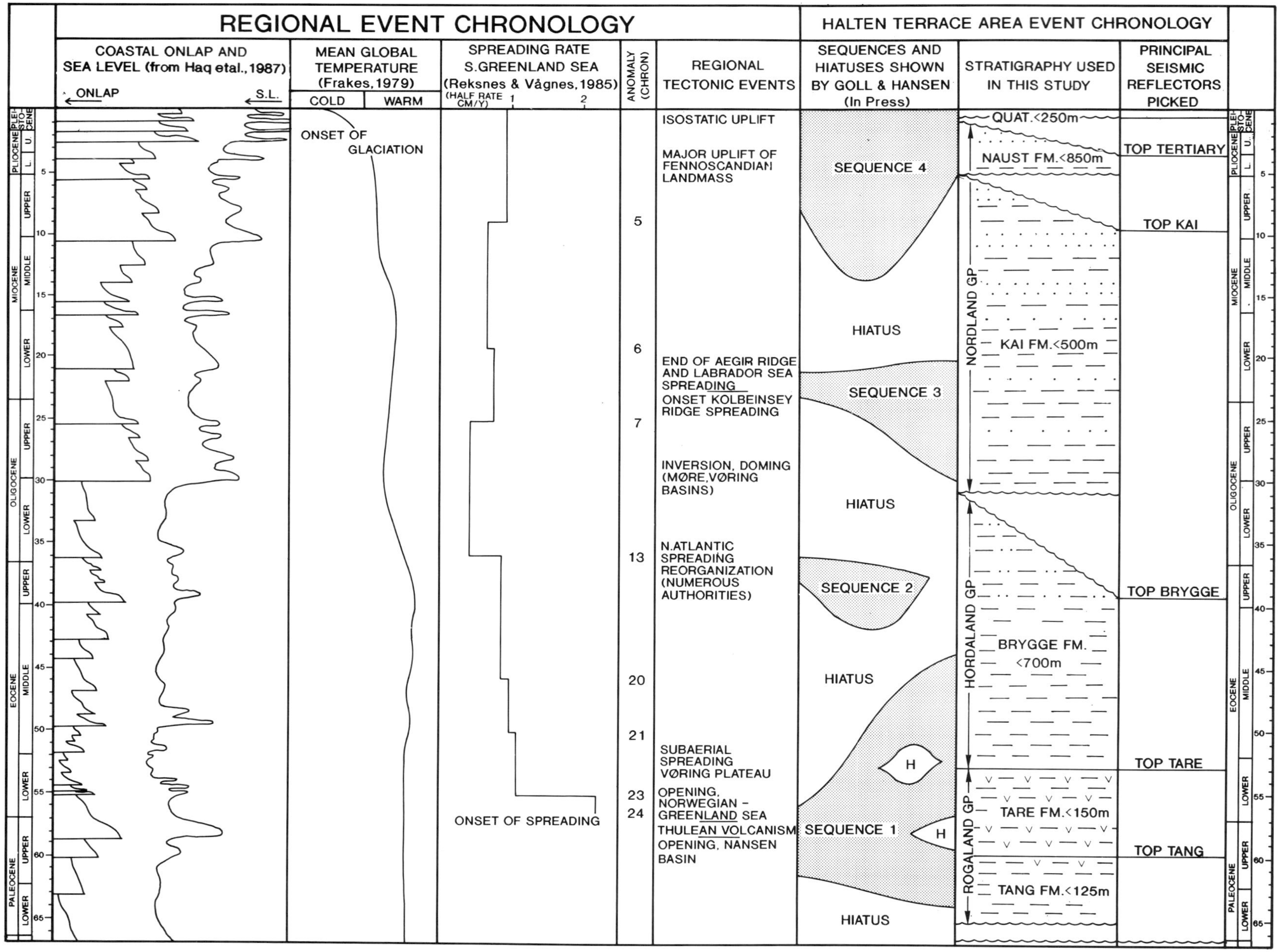

Fig. 5. Summary chart of regional and local (Halten Terrace area) Cenozoic event chronology. Sequences, lithologies and unconformities used in modelling are shown on the right of the chart. Key oceanic, tectonic and climatic events at global/North Atlantic scale are shown on the left.

gas, condensate and some oil (Heum et al., 1986). A continuous overall transgressive regime through the Jurassic resulted in the deposition of marine coarse clastic sequences (Early Jurassic Tilje Formation and Early–Middle Jurassic Fangst Group) which constitute the principal reservoir units of the Halten Terrace petroleum province. The transgression culminated in widespread mudstone deposition in late mid-Jurassic–Late Jurassic times. The shale sequence is capped by the principal oil-prone source rock, the Spekk Formation.

Jurassic deposition occurred against a background of extensional tectonics. Early differential movements, probably including pre-rift updoming and some faulting, can be dated as Early to Middle Jurassic (Doré, 1991). Widespread brittle failure of the crust took place in Late Jurassic–earliest Cretaceous times. This episode created the main structural divisions (Fig. 2) observable on seismic records today. From east to west these include: the Trøndelag Platform, a relatively stable region involving small-scale faulting; the Halten Terrace, a down-faulted region containing the main productive structures of the area; and the deep Vøring Basin, where the Jurassic target horizons are frequently buried to depths of 5000 m or more.

The deep crustal nature of the rifting in Mid-Norway is important in the establishment of a lithospheric heat flow model for the Jurassic and Early Cretaceous. The structural complexity of the province suggests that simple stretching models will inadequately portray the geology. A distinct segmentation of the basin system and polarity reversals of fault sets along the rift trend (in-house studies) suggest the possibility that deep crustal detachment models (e.g. Lister et al., 1986) may apply. A transtensional element to the rifting is also suggested by large-scale plate kinematics (Rowley and Lottes, 1988). However, as will be demonstrated later, the choice of heat flow models has little effect on the maturation history of the Jurassic source rocks until depth of burial exceeds approximately 2000 m towards the end of the Cretaceous for the Åre Formation and later in the Tertiary for the Spekk Formation.

In the Late Cretaceous there was rapid subsidence, and the establishment of a major depocentre, in the Vøring Basin to the west of the study area. More moderate subsidence rates are evident, however, on the Halten Terrace and Trøndelag Platform. The tectonic and sedimentary history of the succeeding Cenozoic Era is of critical importance to the modelling of the area, since the principal source rocks are predicted to have reached maturation and generation thresholds during this time interval.

The complex event chronology of the Cenozoic is summarized in Fig. 5 on both a regional (NE Atlantic) and local (Halten Terrace) scale. Although the essentials of NE Atlantic development are reasonably well understood, the local chronology presents more of a problem. This is exemplified by the comparison between the Cenozoic stratigraphy developed by Goll and Hansen (1992) and the stratigraphy used in our modelling. Major age dating differences between these schemes (Fig. 5) stem principally from inadequate data collection in exploration wells. Such differences have the potential to influence basin models, as discussed later in this report. We emphasize that, if integrated basin modelling is to become a standard procedure on the Norwegian Shelf, the careful logging and sampling of the “top-hole” section in exploration wells is essential. The present section attempts to draw only a broad consensus picture of Cenozoic basin development.

In Paleocene time, thermal uplift of the NE Atlantic rift axis took place at the margin of the Vøring Basin. Crustal stretching and subsequent thermal subsidence appears to have taken place in a comparatively narrow (100–150 km wide) zone in the outer Vøring Basin. The Paleocene marginal events were associated with intense volcanism, probably as a result of continued stretching of an already much thinned portion of continental crust (Skogseid, 1990). In contrast to the outer Vøring Basin, only moderate sediment thicknesses of Paleocene sediments (circa 150 m) accumulated in the Halten Terrace/Trøndelag Platform region. As with the entire Tertiary sequence, argillaceous sediments predominated, with tuffs in the upper part of the Late Paleocene–Early Eocene sequence (Tare Formation; Dalland et al., 1988) representing the onset of volcanic activity in the west.

Crustal separation and sea-floor spreading took place in the earliest Eocene at approximately Anomaly 24 (57.5 Ma) time. The continent–ocean transition is thought to lie just seaward of the Vøring Escarpment (Fig. 2). During the earliest phases of drifting, the western margin of the Vøring Basin remained uplifted and subaerial spreading took place, with continued volcanism. In the basins to the east, including the study area, gentle subsidence took place with accumulation of the predominantly argillaceous Hordaland Group sediments in the Eocene–Early Oligocene interval.

A number of regionally significant events took place in Oligocene–Early Miocene times. Numerous authorities (e.g. Eldholm and Theide, 1980) report a spreading reorganization in the North Atlantic in Early Oligocene (Anomaly 13) times. The reorganization was apparently coincident with a slowing of the spreading rate (Reksnes and Vågnes, 1985). This situation persisted until the Late Oligocene–Early Miocene (Anomaly 7-6) when the spreading rate increased again and further plate reorganization took place, with cessation of activity on the Aegir Ridge and onset of spreading to the west on the Kolbeinsey (Iceland) Ridge. Broad updoming of the sedimentary pile in axial parts of the Vøring Basin (Stuevold, 1989) is likely to have been a concomitant of this latter phase of activity. A marked global fall in sea level in mid-Oligocene times (Fig. 5) may also relate to plate reorganization.

Throughout Cenozoic times, the Fennoscandian landmass contributed sediments to the Mid-Norway basins to the west. The phenomenon of Cenozoic mainland uplift is a subject of intense debate in the Norwegian literature (e.g. Jensen and Riis, 1992) and a detailed discussion is beyond the scope of this paper. However, it is clear that uplift and erosion was episodic in nature. Stuevold (1989) considers that the most important events took place in Late Oligocene–Pliocene times. Principal among these was major uplift and erosion during the Pliocene, with the development of a thick (circa 1000 m) westward-prograding clastic sequence (Nordland Group) in the Halten Terrace area. Parts of the inner Mid-Norway basins (e.g. Trøndelag Platform) were also uplifted, resulting in major lateral thickness changes in the Nordland Group.

Evidence from ODP cores on the Vøring Plateau (Goll and Hansen, 1992; Riis and Fjeldskaar, 1992) suggests that glaciation of the Norwegian mainland and the deposition of glaciomarine sediments commenced during the late uplift phase, about 2.5–3.0 Ma. Glaciation gave rise to intense differential erosion of the mainland which, on melting of the ice caps, resulted in isostatic uplift and elevation of the mainland summit height surface. In-house studies in the Heidrun Field area indicate that grounded ice probably existed on the mid-Norwegian Shelf during the last glaciation. The thin Pleistocene deposits representing the last glaciation lie above a flat-lying regional unconformity that truncates the underlying Neogene sequences.

The events of the Plio-Pleistocene described above are particularly significant to the central theme of this paper: the anomalous present-day thermal regime on the Norwegian Shelf. It can be postulated that these events are interconnected on a regional scale. Major uplift, erosion and sedimentary outbuilding in the Pliocene may be related to an increase in the spreading rate in the adjacent Atlantic, which commenced in the Late Miocene (Fig. 5) and/or to a more widespread Late Neogene episode of spreading reorganization around the Atlantic margins (Cloetingh et al., 1990). Elevation of the Fennoscandian landmass, in combination with the global cooling trend throughout the Cenozoic (Fig. 5) could have triggered glaciation, which in turn resulted in further rapid erosion, redeposition and isostatic uplift.

The value of such speculation can, of course, be questioned. It is nevertheless undeniable that the recent geological past saw a convergence of significant oceanic, tectonic, climatic, erosional and depositional events on the Norwegian Shelf. It is this fact that convinces us that the requirement for significant changes in basin modelling input parameters for the recent past should be examined seriously.

Thermal framework

Halten Terrace thermal modelling methodology

Modelling in this study was performed with three different software packages — two commercially available programs and an in-house program. Porosity was modelled in both a backward sense using the equations of Sclater and Christie (1980) and iteratively in a forward sense using a relationship between present thickness and present-day and initial porosities. Thermal conductivities were calculated automatically by the programs based on lithological composition and changes in porosity and temperature. Heat flow was estimated from tectonic events and adjusted for the present in order to match the present-day measured temperatures.

Maturity calculations were carried out using one of three algorithms. A modified TTI concept was used to calculate an equivalent vitrinite reflectance when calculating porosity in a forwards direction. However, when the Scalter and Christie (1980) decompaction algorithm was used, maturity was predicted using a relationship between vitrinite reflectance evolution and kinetic reactions for the maturation of a Type III kerogen according to the techniques of Burnham and Sweeney (1989) and Leadholm et al. (1985). The calculation of the hydrocarbon generation is based on parallel first-order reactions. The kinetic values used for hydrocarbon generation in this study were determined from Halten Terrace, Spekk and Åre samples using time slice-pyrolyis (in-house data). The values are similar to those published by Braun and Burnham (1987) and Tissot et al. (1987).

The thermal modelling procedure followed in this study involved an initial detailed analysis of all log and DST temperatures for all wells on the Halten Terrace. The bottom hole temperatures were corrected based on drilling depth using the chart shown in Fig. 6. This chart was developed by the AAPG Geothermal Survey of North America Committee

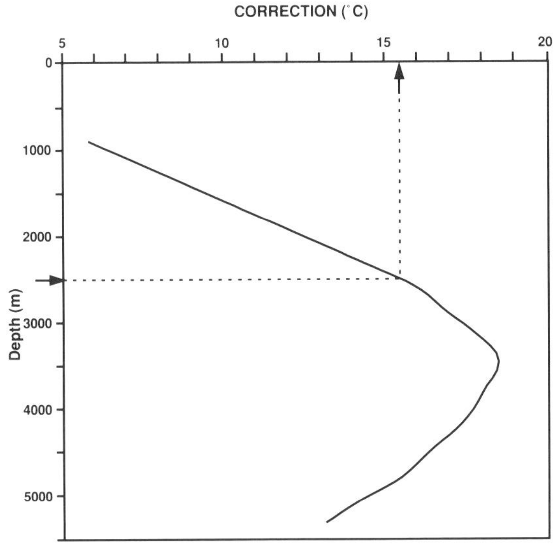

Fig. 6. Bore hole temperature correction chart. Enter the chart at the depth of the temperature measurement, read the corresponding temperature correction, add this value to the borehole temperature. Method should be used only for the first log run at any given depth; circulation time should be less than 10 hours (after Kehle, 1973).

(Kehle, 1973), and is similar to that developed later by Haile et al. (1987) using North Sea data. We have found that the corrections, when added to the first log run (less than 10 hours since circulation) temperatures, result in temperatures which generally match within 3–4°C the DST temperatures. This method can also be used when data limitations prohibit the construction of Horner plots to determine the equilibrium temperature (Dowdle and Cobb, 1975). On the Halten Terrace, and in the northern North Sea, these corrected temperatures are within 5°C of a Horner corrected temperature. In general, the AAPG method assumes that circulation times increase while hole sizes decrease, thus allowing for differences in the cooling rate as a function of these factors. Thus below 3500 m, the hole size is sufficiently small, and circulation sufficiently long, that the correction begins to decrease.

It is recognized that bottom hole temperatures are subject to great variation, and that the corrections are not always accurate (see Hermanrud, 1988, and Carstens and Finstad, 1981, for discussions). However, in this study, the results show consistent trends across the Halten Terrace, lending credence to the accuracy of the corrections applied and the trends observed.

The corrected log temperatures indicate that the present-day Jurassic section is anomalously hot over the majority of the Halten Terrace (Fig. 3). Interval gradients within the Jurassic section are significantly higher than would be expected for a well-compacted, water-filled sand and shale sequence. Interval gradients indicate that a single vertical heat flow value and thermal conductivities appropriate to the lithologies, cannot fit both the present-day Cretaceous and Jurassic measured temperatures, assuming water is the pore filling fluid (Fig. 7). However, all three modelling packages used, require a single heat flow value with depth (assuming that all heat emanates vertically from a single deep-seated source), a constant pore filling fluid, and that temperature is steady-state. Therefore, to meet these requirements, and because the Jurassic interval contains both the source and reservoir, greater confidence was placed in heat flow scenarios which resulted in temperatures which matched present-day Jurassic temperatures, rather than those of the overlying Cretaceous section.

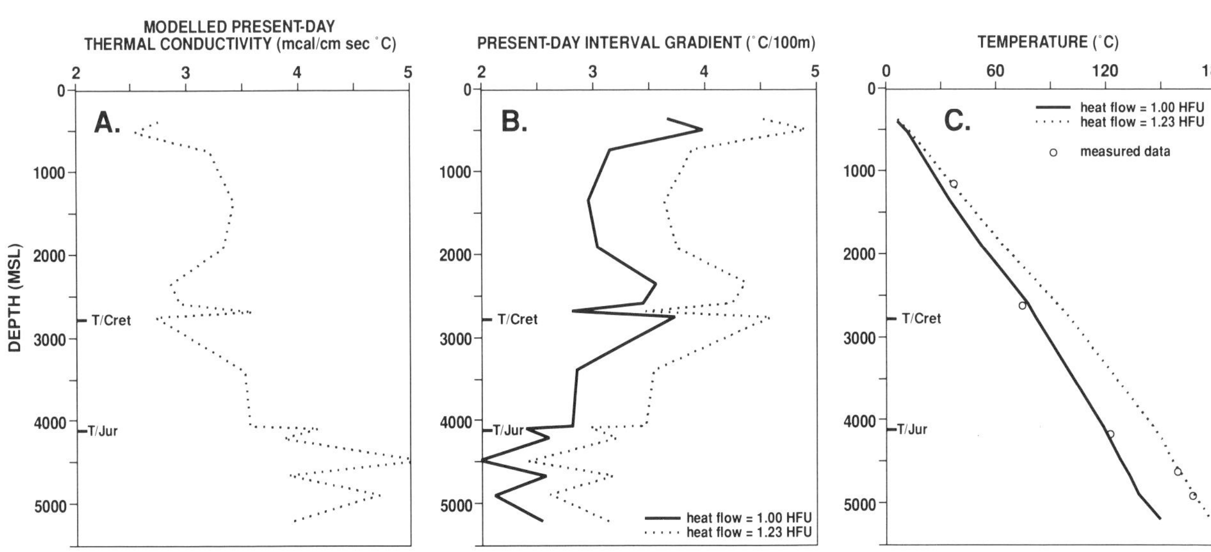

Fig. 7. Relationship between thermal conductivities, interval gradients, and heat flow on Halten Terrace. (A) Present-day modelled thermal conductivities. (B) Modelled interval gradients. (C) Modelled and measured present-day temperatures.

The heat flow values used in the figures in this paper are shown schematically to highlight the values associated with the various tectonic, depositional and erosional events. The modelling programs use a linear interpolation, and not the instantaneous changes depicted, between the input heat flow values (which are most often associated with the onset of a depositional or erosional event). In addition to the variation in heat flow (input) and thermal conductivity (calculated) through time, surface temperature (input) was varied in accordance with changes in sea level and palaeo-latitude.

Modelling scenarios

Figure 8 shows four possible heat flow scenarios for the Halten Terrace, based on significantly different underlying assumptions. Within the limits of the assumptions made, all were designed to match present-day temperatures in either the Jurassic (Models A and B) or the Cretaceous (Models C and D) interval and maturities within both intervals; the results are shown in Figs. 9 and 10.

Model A attempts to match heat flow variation to geological events. Model B assumes that heat flow has not changed throughout time and therefore uses the present heat flow in the past. Model C assumes that temperatures in the Jurassic section today are invalid, and uses a lower present-day heat flow to match the temperatures in the Cretaceous section (Fig. 9). Model C must then use higher palaeo-heat flows in order to ensure that the model reaches appropriate maturity levels today (Fig. 10). Model D assumes that the opening of the North Atlantic was of greater importance than the Jurassic rifting event in terms of thermal effects. Heat flow values before the opening of the North Atlantic in Model D are the

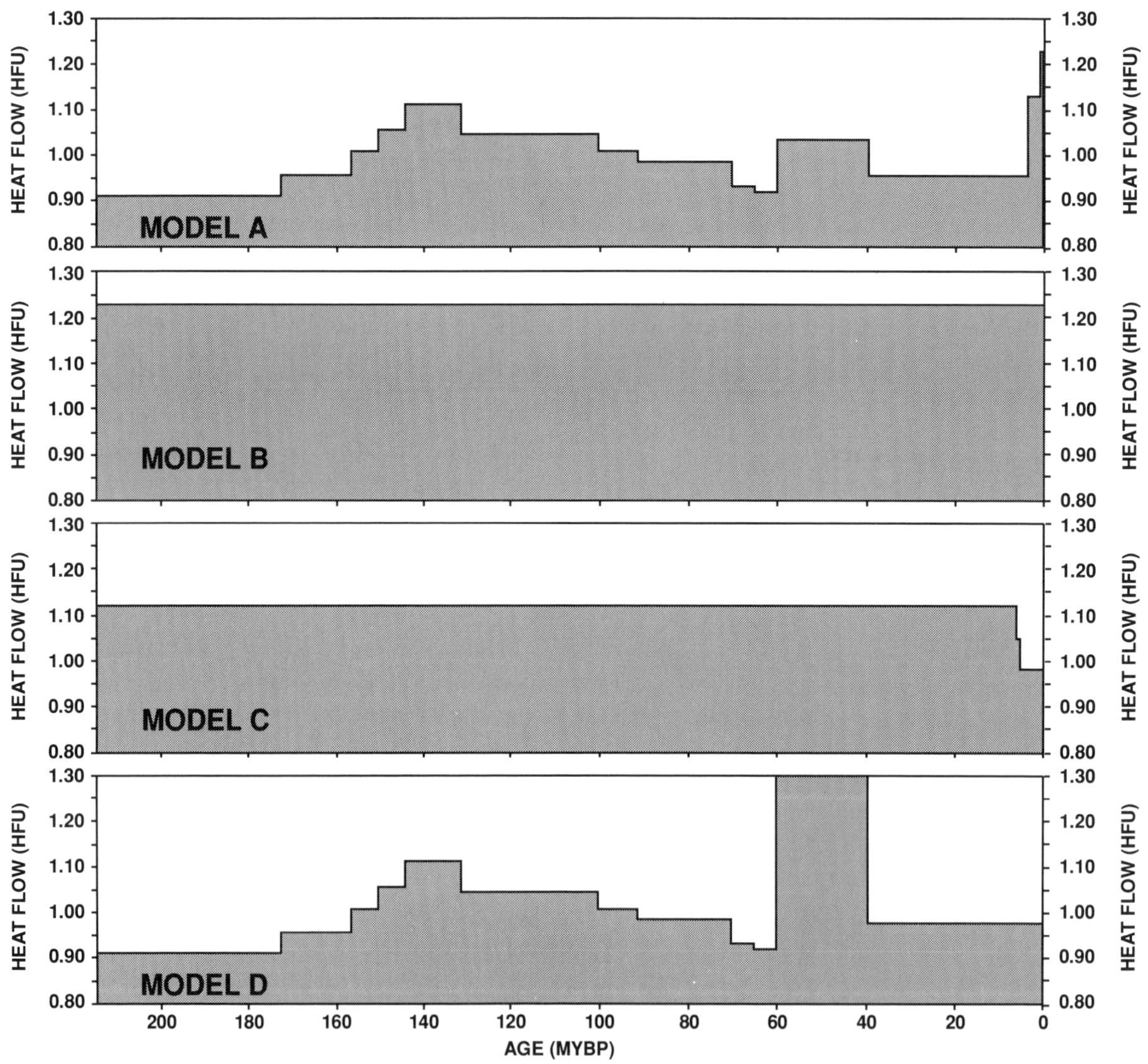

Fig. 8. Heat flow scenarios for the Halten Terrace. Model A represents the best fit to tectonic events. Model B assumes that the present thermal state is the key to the past. Model C uses a decrease in heat flow to reflect the rapid subsidence of the Pliocene (see Fig. 1). Model D assumes that the opening of the North Atlantic was more important than the Jurassic rifting event.

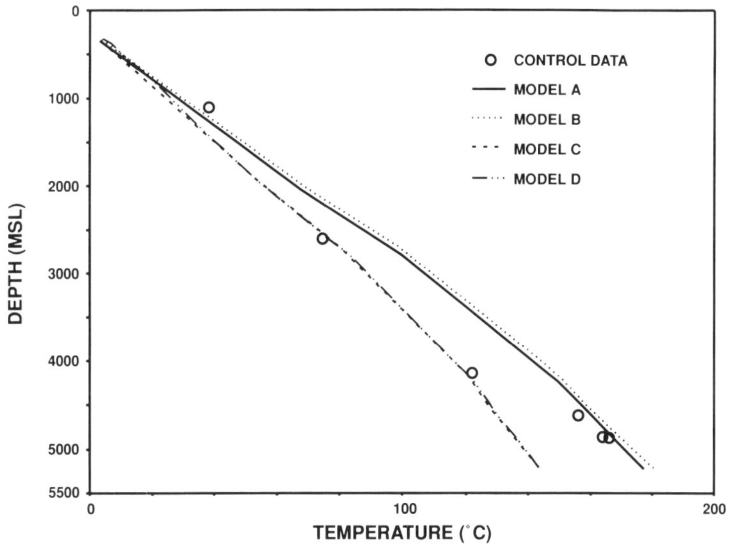

Fig. 9. Present-day measured and modelled temperatures for the heat flow histories shown in Fig. 8; well from the over-pressured zone (see Fig. 2).

same as those used for Model A. However, a present-day heat flow which results in a match to present-day temperatures in the Cretaceous section is then required to model the present-day measured maturities. All models use the same lithologies, porosities and thermal conductivities (although allowances are made for temperature differences); only heat flow is varied in the four scenarios.

A fifth scenario (not shown) with an average heat flow history resulting in temperatures half-way between the present-day Jurassic and Cretaceous temperatures is also feasible. This scenario is not included in this paper, as its results lie between the models depicted. The models shown give the best

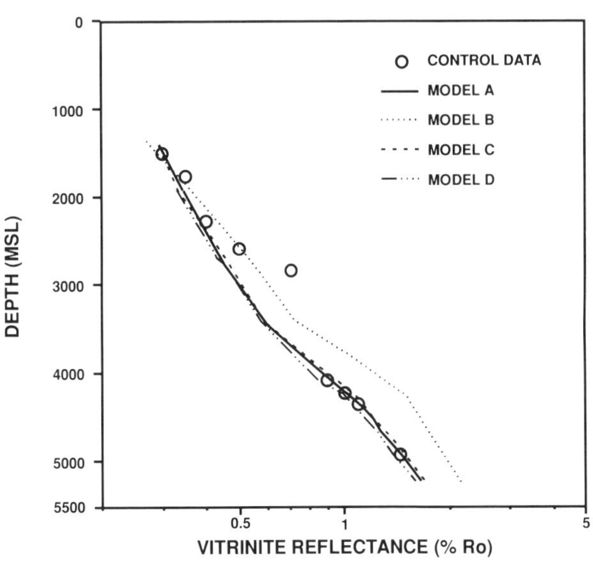

Fig. 10. Present-day measured and modelled maturities for the heat flow histories shown in Fig. 8; same well as in Fig. 9.

indications of the potential "error bar" in the models possible for the area.

The heat flow increase used in Model A for the Late Jurassic/Early Cretaceous rifting event discussed earlier (see also Fig. 1) is based on the stretching theory of McKenzie (1978). Our subsidence studies indicate that a stretching factor (beta) of 1.3 is appropriate for the Halten Terrace. In terms of the model, this results in a maximum heat flow value of 1.14 HFU at the end of rifting. Although the McKenzie (1978) model was used to determine the Jurassic heat flow increase and subsequent thermal decay history, it is recognized that methodologies proposed by other authors (i.e., Sclater and Christie, 1980; Royden et al., 1980) may also be applied in this area.

Model A assumes that heat flow decreased exponentially from the Early Cretaceous to the Early Tertiary. A small increase in heat flow during the Paleocene–Eocene is used to model possible effects from the increased volcanic activity to the west on the Vøring Plateau and the opening of the North Atlantic. From the Eocene through the Miocene, a constant, relatively low heat flow is used. Starting in the Pliocene, the heat flow model increases, with a maximum being reached today. This increase was required for model A in all three computer modelling packages to match the present-day corrected temperatures in the Jurassic section.

Using model A, it became apparent that Jurassic subsidence was such that, even with significantly higher palaeo-heat flow values than shown, the temperatures during and immediately after the rifting phase on the Halten Terrace were insufficient to effect any great increase in the maturities of the source horizons. However, heating during the Tertiary is critical. If the heat flow increase is too early, the modelled maturities are too high; if the increase is too late, the maturities are too low.

Models B and C do not use an increase in palaeoheat flow during the Jurassic/Cretaceous. These models were purposely kept simple, but could easily incorporate the rifting event without any noticeable change in the results of the models.

Model B matches present-day corrected temperatures in the Jurassic interval (Fig. 9) but predicts more mature sediments than measured (Fig. 10). Clearly, the present is not the key to the past on the Halten Terrace.

The rapid subsidence phase of the Pliocene has been modelled with a heat flow low in Model C (Fig. 8). This, however, is not meant to represent a change in the heat emanating from the mantle, but rather a change in surface conditions and the development of transient states as a result of a massive influx of cooler sediments and associated water. This model, in modifying the heat flow history in order to model a transient situation, is able to match present-day maturity values (Fig. 10), but is unable to model present-day temperatures in the Jurassic interval (Fig. 9).

It is difficult to assess the thermal significance of the opening of the North Atlantic on the Halten Terrace. Although the rifting phase prior to opening affected a limited area in the outer Vøring Basin, the volcanism associated with rifting and later sea-floor spreading on the Vøring Plateau was widespread and comparatively long-lived. There is therefore considerable leeway for invoking a more intense or longer lasting heat flow increase for the Early Tertiary, as shown in Model D. Although present-day maturities are matched with Model D (Fig. 9), temperatures are low relative to the Jurassic source and reservoir interval. It should be noted that Model D is a simplification of the potential effect of the opening of the North Atlantic and that further refinements are possible. However, if a significant heating event is invoked in the Paleocene–Eocene, then cooler temperatures must prevail through to the present in order to accurately model the present maturity state.

As mentioned previously, we have chosen to accept the temperatures in the Jurassic section as the most valid, particularly due to the presence of the source and reservoirs within this interval. Therefore, even though we show two alternative scenarios where the temperatures in the Cretaceous section are matched, we do so to show possible variations in the tectonic history, and the extent of "error" in a model as a result of extreme modelling variations. Obviously, alternate models are also possible, but one goal of our basin modelling efforts is to place error limits on the models, and their predictions of maturation, generation, and trap fill.

Causes of the heating anomaly

We have indicated previously, that the present-day Jurassic is anomalously hot relative to the overlying section on the Halten Terrace. We have also shown that in order to model the present-day Jurassic temperatures, we must ignore the present-day Cretaceous temperatures and invoke a heat flow increase in the recent past. We will now discuss the geological factors that could result in an anomalously hot present-day Jurassic, and the aspects of the modelling algorithms which require us to invoke a heat flow increase in order to simulate the present-day thermal environment. In this discussion, we will attempt to separate the geological causes for the real heating of the present-day Jurassic section from the mathematical causes for the modelled recent heat flow increase.

Geological factors

Glaciation, rapid subsidence, over-pressuring, changing pore fluids, and fluid flow could all cause an increase in temperature in the present-day Jurassic section. Uncertainty regarding Tertiary ages and depositional/erosional rates are unlikely to have an effect on the present thermal regime, but could affect the way the thermal history is modelled and thus result in a modelled heat flow increase. Errors in the control data (i.e., temperature and maturity) could also result in the need for an apparent heating event in modelling scenarios.

Glacial effects. Ice loading, glacial scouring and isostatic rebound could affect the present-day temperatures in the Jurassic section. All of these factors would most likely cause transient effects, resulting in a present-day non-steady state system. If this is the case, the mathematical models currently used are inadequate to accurately handle the transient nature of the events.

Grounded ice probably existed on the Mid-Norway Shelf at intervals during the Pliocene. A substantial thickness of ice will have resulted in depression of the crust. In addition, subglacial scouring during glaciation could have eroded substantial quantities of sediment. After glaciation, isostatic compensation for ice loading and sediment removal will have resulted in recent uplift, and a consequent transient temperature state. Hermanrud et al. (1991) in a study of the Troll Field area, show that an uplift of 250 m in the last 200,000 years could explain the abnormally high temperatures seen in this field relative to the surrounding area. Uplift on the Halten Terrace after glaciation has not been modelled in this study. However, it is unlikely that glaciation is the sole cause of the significant recent thermal anomaly that this paper addresses, and certainly cannot give rise to an overall increase in crustal heat flow of the magnitude under discussion (C. Hermanrud, pers. commun., 1991; T. Pedersen, pers. commun., 1991).

Recent subsidence, over-pressuring, changing pore fluids and fluid flow. The Halten Terrace underwent a rapid subsidence phase (>300 m/Ma compacted rate) during the Pliocene. This subsidence is probably related to the recent uplift of the Fennoscandian Shield and rapid sediment influx (see previous sections). Thorne and Watts (1989) indicate that similar recent subsidence rates in the North Sea could imply incipient rifting. If this is the case, a rapid increase in heat flow could be expected. On the other hand, Cloetingh et al. (1990) argue that recent uplift/subsidence events in the northern Atlantic region are indicative of a compressional phase resulting from intraplate stress. In this case, a decrease in heat flow would be expected. Only the former theory could provide the type of temperature anomaly discussed in this paper. However, extensional tectonic activity in the Mid-Norway area at present appears to be confined to small-scale post-glacial readjustment on faults. There is no evidence for the scale of activity that could result in a major lithospheric heat flow increase. We, therefore, argue against a deep-seated crustal event and choose to seek other causes for the modelled heating event and the anomalously hot present-day Jurassic section.

The modelled heat flow increase shown in Model A and the high heat flow used in Model B are caused by the anomalously hot present-day Jurassic section (relative to the overlying Cretaceous section). The wells that show this higher interval gradient (Fig. 3), all fall on the western half of the Halten Terrace where the present-day Cretaceous section thickens substantially and the Jurassic section is usually over-pressured (Fig. 11). Between the normally pressured and the over-pressured Jurassic zones is a transition zone where the Cretaceous thins; here the Jurassic section is only mildly over-pressured and the thermal anomaly is not as apparent as it is further west (Figs. 11C, 11D). The majority of the oil discoveries lies within, and just to the east of, the transition zone (Figs. 2 and 11 A), while the mature Upper Jurassic source rock lies generally to the west of the transition zone. These associations suggest a relationship between over-pressuring, thermal anomaly and hydrocarbon generation.

An isochore for the present-day Pliocene is shown in Fig. 11B. Our modelling indicates that the rapid Pliocene deposition (>300 m/Ma) is the cause of the over-pressuring, within both the present-day Cretaceous and the Jurassic sections. The Early Tertiary and Cretaceous sections are undercompacted as a result of this rapid sedimentation. Therefore, higher than normal geothermal gradients could be expected.

While it is likely that the rapid subsidence and resulting over-pressuring are related to the anomalously high temperatures in the Jurassic, the relationship is not as clear cut as shown in other papers which discuss anomalous North Sea temperatures. Several theories can be proposed, i.e., the effect of over-pressures on the thermal state of sediments, heat transfer as a result of fluid flow (convective or compaction-driven), and thermal conductivity changes as a result of the change in pore filling fluids. Each of these theories has drawbacks which prohibit a complete explanation for the presence/absence of the thermal anomaly in the present-day Jurassic section on the Halten Terrace.

Lewis and Rose (1970) show that for a well with an over-pressured section, interval gradients should increase upon entering the over-pressured section, then return to "normal" on leaving this section (Fig. 12). This relationship on the Halten Terrace is not as clear cut as the theory would indicate (see, for example, Figs. 11C, 11D). In theory, the present-day Paleocene/Cretaceous section should show an increase in interval gradients due to the onset of over-pressure in the Paleocene section. However, this increase is not obvious on the Halten Terrace, perhaps due to a lack of temperature data at the top of the section (casing is generally set above the Paleocene and again at the base of the Cretaceous/top of the Jurassic). There is also no obvious increase in pressure at the base of the Cretaceous to explain the increase in temperatures in the Jurassic. It is possible that the Cretaceous section, being shalier than the underlying Jurassic, acts as an insulator, and that this insulating effect is accentuated when the Jurassic is also over-pressured.

A change in pore filling fluids from water to oil or gas will lower the total thermal conductivity of a section significantly, thus resulting in a higher interval gradient (Table 2). Such a change cannot occur until after hydrocarbons have been generated (for shales) and until they have migrated into carrier beds or reservoirs (for sands). For Model A, generation is predicted to have begun late in the Cretaceous for the Lower Jurassic Åre coals, but very recently for the major source rock, the Upper Jurassic Spekk

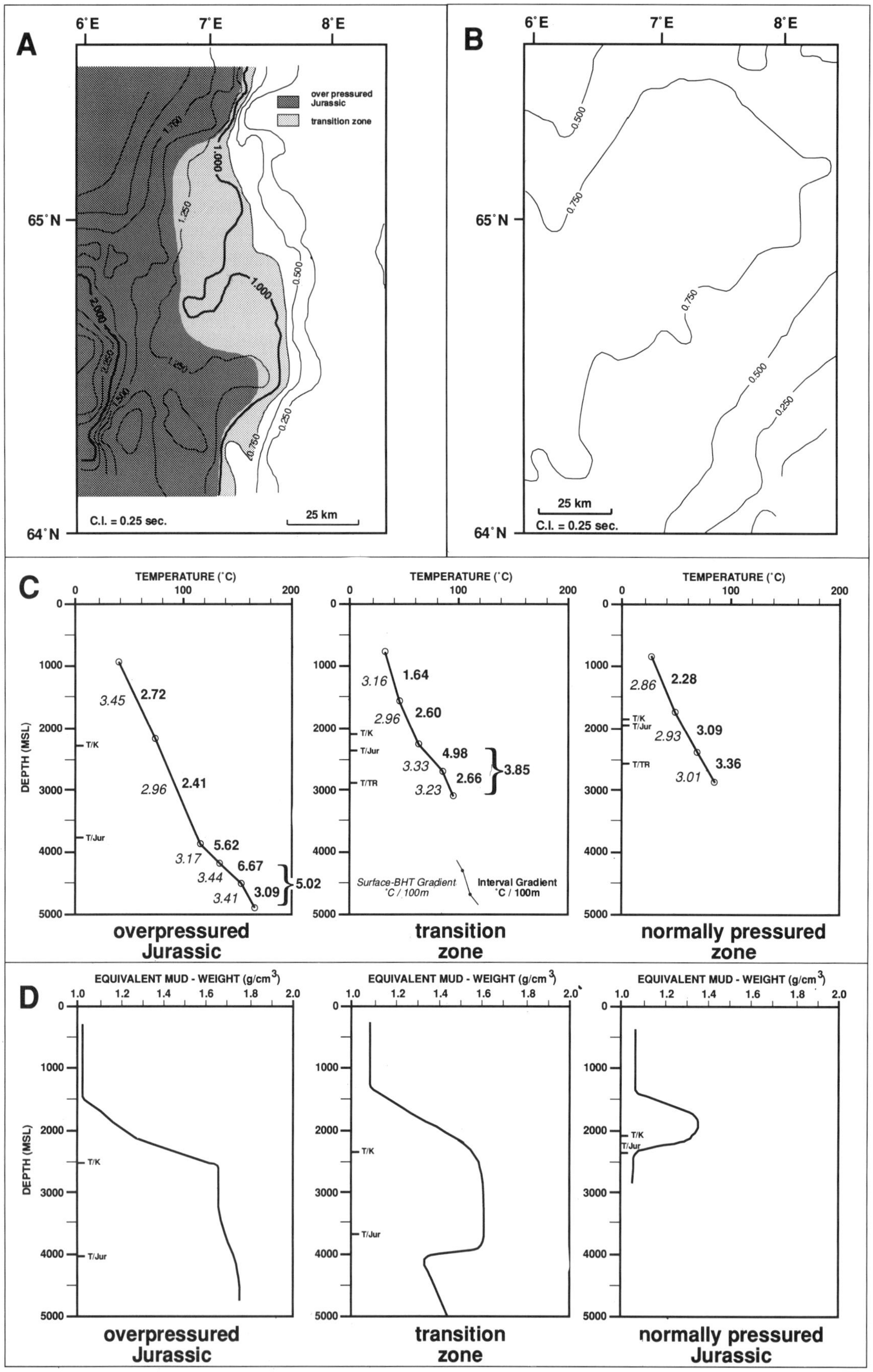

Fig. 11. Relationship between (A) Cretaceous and (B) Pliocene thicknesses, (C) corrected temperature with surface to bottom hole temperature gradients (normal text) and interval gradients (bold text)) vs depth, and (D) mud weight vs depth.

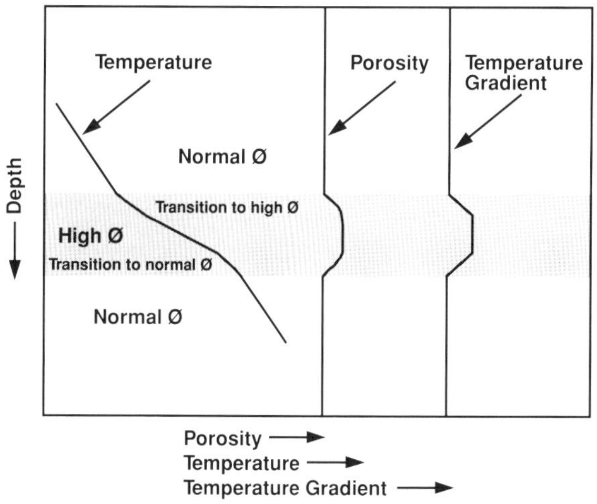

Fig. 12. Effect of over-pressure on temperature, porosity and gradients (after Lewis and Rose, 1970).

Formation (Fig. 13). The other models would predict an earlier onset of generation at both levels. If Model A correctly predicts the generation history (though not necessarily the heat flow history), then the change in pore fluids could partially explain the higher geothermal gradients seen today in the Jurassic section.

If a change in pore filling fluid is the sole explanation for the thermal anomaly, then higher gradients would be expected in all the oil fields. However, the Njord and Heidrun Fields (on the edge of the pressure transition zone) and Draugen Field (in the normally pressured zone), all show mixed results with some wells showing a strong anomaly and others none at all (Fig. 14). On the other hand, many of the dry wells within the over-pressured and transition zones show an anomaly in the present-day Jurassic section. Clearly, a simple change in pore fluids is an insufficient explanation for all of the well data.

It is possible that the temperatures for the shallower fields (within the area of a thin Cretaceous section) are in error. Draugen and Heidrun Fields are both shallow (1500–2400 msl). The potential errors in the temperature measurement or in the applied corrections at these depths are sufficiently large to create or destroy any apparent thermal anomaly in the present-day shallow Jurassic. The consistency of the data in the deeper area continues to argue, however, for the reality of the anomalously high present-day Jurassic temperatures, at least for the area of thick Cretaceous cover.

Fluid flow has been proposed to explain similar thermal anomalies found in the North Sea (Carstens and Finstad, 1981; Andrews-Speed et al., 1984; Brigaud et al., 1992). In the northern North Sea, the high temperatures in the present-day Jurassic appear regardless of pressure regime (Carstens and Finstad, 1981). Fluid flow upwards along faults from the deep Viking Trough to the shallower fault blocks is postulated to result in higher than expected temperatures at shallow depths.

On the Halten Terrace, the area of the higher interval gradients within the Jurassic is roughly related to the area of over-pressure within the Jurassic (the Cretaceous being over-pressured everywhere). This is also the area of active hydrocarbon generation. Oil

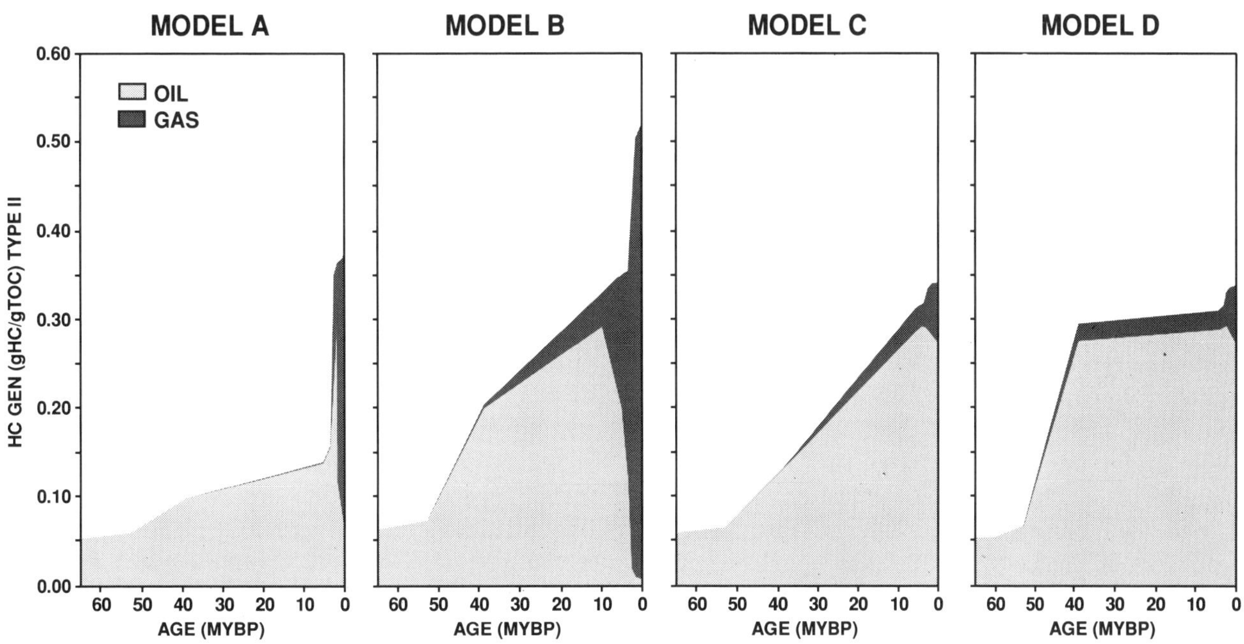

Fig. 13. Modelled hydrocarbon generation histories for the Upper Jurassic Spekk Formation for the heat flow scenarios shown in Fig. 8; same well as in Fig. 9.

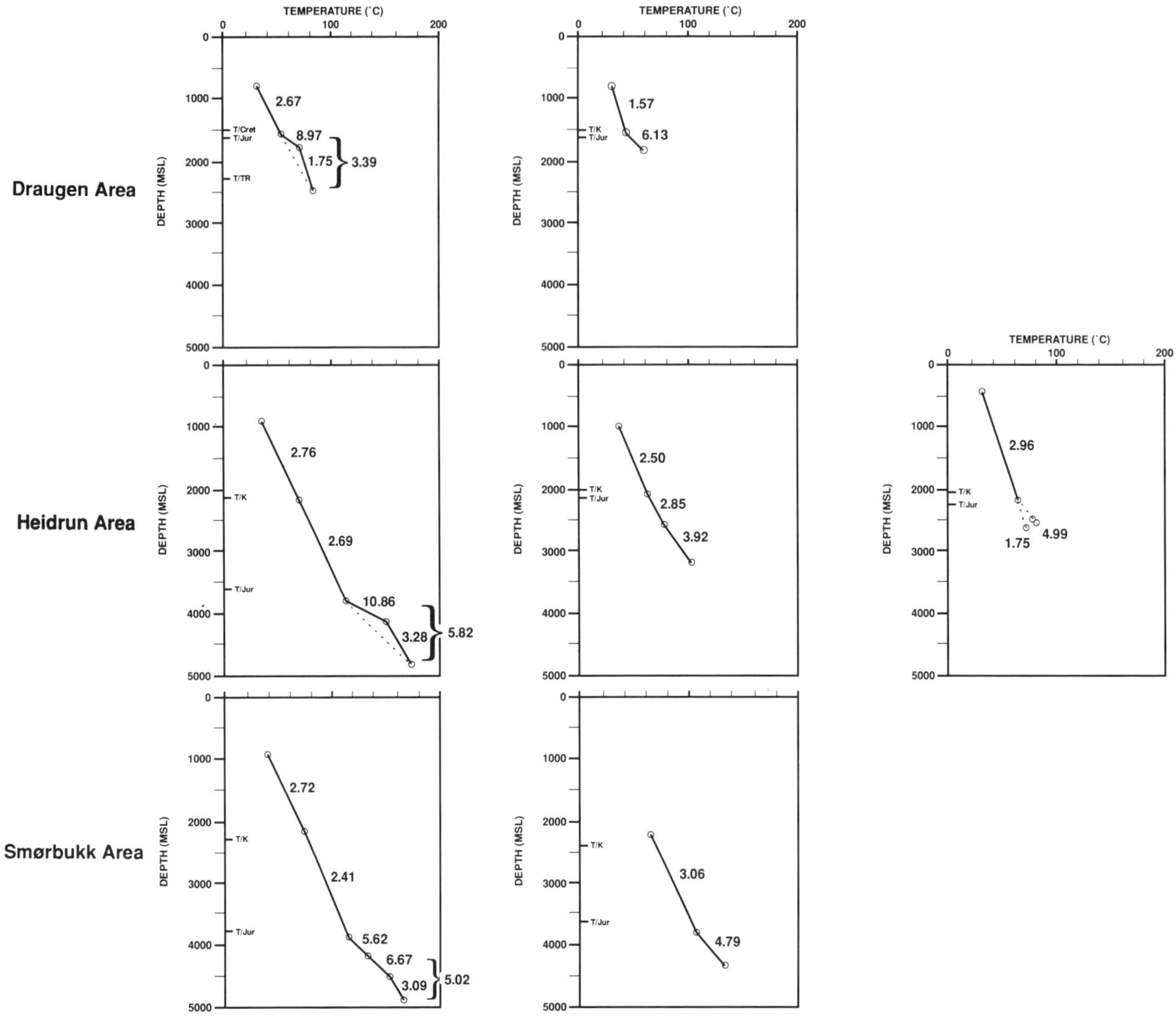

Fig. 14. Comparison of corrected bore hole temperature and interval gradient trends for the Draugen, Heidrun and Smørbukk areas.

and gas, in order to fill the known fields on Halten Terrace, must move from west to east, i.e., from the area of high pressure and active generation to the area of lower pressure and immaturity (see Mann and MacKenzie, 1990). The present pressure regimes would indicate that this is possible but that not all the pressure of the deeper area of the western Halten Terrace has been released.

In the over-pressured Jurassic zone (Fig. 11), generated hydrocarbons may be forced to move upwards out of the highly pressured Jurassic section to the Base Cretaceous Unconformity. This surface may then act as a conduit whereby hydrocarbons can migrate by buoyancy to the lower pressure Jurassic reservoirs in the east. This process would be sufficiently slow that a thermal imprint of the oil's hotter origin is unlikely to be carried along. However, for generation as predicted by Model A, the process must be sufficiently fast to have filled the traps since the onset of hydrocarbon generation within the Upper Jurassic section a few million years ago. If the hydrocarbons are moving both as a result of buoyancy and pressure differential, the absence of reservoired hydrocarbons in the over-pressured zone may be explained.

It is conceivable that fluid flow from the hotter, deeper Vøring Basin to the west of the Halten Terrace is responsible for local heating of the Jurassic section. For such a scenario to be operative, fluid flow to the east would have to be sufficiently impeded by lateral permeability barriers (principally faults) to account for the observed differentiation into pressure domains, with the highest pressures in the west. Alternatively, cooling of such fluids as they move eastwards would account for the change in the thermal regime of the Jurassic interval on the Halten Terrace. The origin of such a fluid flow in this area is highly speculative. It could presumably occur as a result of compaction in the rapidly subsiding Vøring Basin. The calculations of Hermanrud (1986), however, would indicate that there is an insufficient amount of water generated by compaction to transfer the amount of heat in question in this area. Thermal convection, as suggested for the Viking Graben (Brigaud et al., 1992) is a further possibility. It is difficult to see, however, how this mechanism, operating against the numerous vertical permeability barriers in the deep section, could transfer sufficient heat to the Jurassic interval of the western Halten Terrace.

Clearly the mechanism and rate of fluid flow, and the mechanism and timing of secondary migration are critical and interrelated factors that bear upon the filling of the known fields on the eastern half of the Halten Terrace, the development of the over-pressured domain on the western half of the Halten Terrace, and the local heating of the Jurassic rocks. An integrated model for this relationship cannot be derived with one-dimensional basin modelling algorithms. Grigo et al. (this volume), using a three-dimensional model, indicate that fluid flow could cause a thermal anomaly in the Jurassic section, particularly near the Smørbukk area. However, due to the complexity of the highly faulted Jurassic section, a grid smaller than their 8 km would be required to fully model the relationship between over-pressuring, hydrocarbon generation, fluid flow and a thermal anomaly.

Tertiary dating. The Tertiary section on the Halten Terrace, being of little interest to date in terms of hydrocarbons, has been drilled rapidly, with sparse sample collection. This results in coarse stratigraphic subdivisions and a poor understanding of the actual ages of the section. Therefore, limited confidence can be placed in detailed basin models for the Tertiary. This is a matter of concern since we believe that the Tertiary (and Quaternary) represent the principal periods of generation and migration on the Halten Terrace.

Goll and Hansen (1992) have reinterpreted and reanalyzed the Tertiary section in Mid-Norway, and integrated their results with seismic data. This project resulted in significantly different dates than those used in this study (Fig. 5). Using these ages instead of those used throughout this study, results in negligible differences in terms of hydrocarbon generation. It is, therefore, concluded that though the ages are poorly known, a considerable variability in ages prior to burial to a depth greater than 2000–2500 m (the shallowest possible oil generation) is possible without greatly altering the models. The onset of major subsidence in the Pliocene is the dominant influence on maturation and generation. Nevertheless, the potential of inaccurate Tertiary dating for influencing basin models should not be underestimated.

One aspect of the Goll and Hansen study that was not modelled in detail, was the amount of erosion possible as a result of the age redeterminations. In the southeastern portion of the Halten Terrace (near the Draugen Field), significant erosional periods in the Eocene–Pliocene section resulted from a change in basin configuration. If the ages of Goll and Hansen (1992) are used, more erosion than had previously been accounted for would be possible. However, this could then require a cooler past in order to match the present-day temperatures and maturity indicators. Clearly, more work is needed in order to fully understand the Tertiary section, its ages, and its depositional and erosional events.

Measurement limitations. This study is based on the assumption that the corrected bore hole temperatures and measured maturities accurately portray the present and past thermal histories, respectively. However, it is recognized that both these types of measurements are prone to errors which could adversely affect the accuracy of the modelling scenarios presented here.

As mentioned earlier, the temperatures used in this study are derived primarily from bore hole temperatures corrected using the chart shown in Fig. 6. This correction gives temperatures generally within 3–4°C of DST temperatures, and which are usually on the low side in the Halten Terrace wells. Therefore, if the temperatures used are incorrect, they are so on the low side, which would indicate that the modelled heat flow for the present, although high, is potentially too low! However, the temperature data form a consistent and coherent picture in conjunction with the geology, as indicated by the previous discussion on geological factors.

Maturity values, like temperatures, are only available for the present day. They, too, are prone to errors. An example of data taken directly from tables within report summaries for a single well is shown in Fig. 15. Clearly some of the data is incorrect, but which set of data? The data from Company D was the lower of two populations; had the contractor chosen the correct population, and on what basis was the population chosen? The data from Company C was entirely from cuttings data — perhaps a caving problem. The high points on Company A's curve were from reworked samples, but not clearly marked as such in the table. For a single well, these data are relatively consistent. However, it is more common to have only one report (with any one or more of these problems) per well.

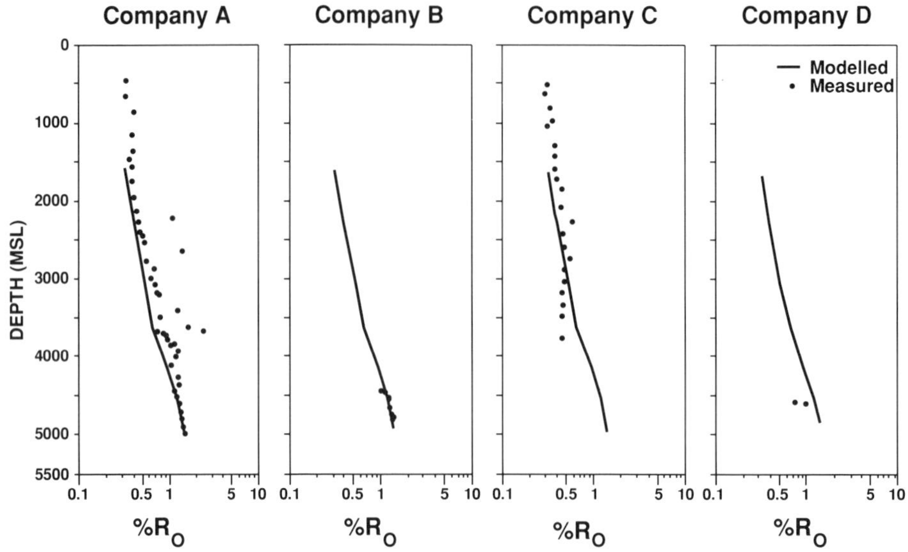

Fig. 15. Measured vitrinite reflectance data for a single well within the pressure transition zone from four different companies.

Maturity data does not appear to be a significant problem on the Halten Terrace. However, it is important to keep the possibility of error in mind and not to accept the data blindly. Service companies are often unaware of age (hence vitrinite can be "found" in the Ordovician). Casing points (the lowest vitrinite population is probably the best beneath these points) and drilling conditions (turbo-drills can overmature a section) should always be borne in mind. Better communication between the oil company and the service company will aid in better vitrinite profiles.

Modelling assumptions/algorithms

Basin models are the result of mathematical algorithms designed to simulate the real world to the best ability of the algorithm and the input data. The programs are, in many instances, oversimplified so that they can handle the majority of the problems encountered. On the Halten Terrace, the lack of integration and interrelationship of such items as porosity reduction and over-pressuring, hydrocarbon generation and thermal conductivity changes, and transient thermal states as a result of rapid changes, has affected the way the area has been modelled.

Over-pressuring, as indicated previously, probably began during Pliocene deposition, concurrent with hydrocarbon generation. The net effect has influenced the thermal regime of the system through undercompaction of the sediments, expulsion of hydrocarbons in the Jurassic section, and retention of water as pressure regimes changed in the Tertiary and Cretaceous sections. The combination of these factors is difficult to model in one dimension and may result in inaccurate models.

Improvements to the modelling algorithms should take into account short-term events such as glaciation or the rapid subsidence of the Pliocene. In the former case, the ability to model an insulating layer of ice, crustal downwarp and isostatic rebound is necessary. In the latter case, the algorithms must accurately relate the generation of over-pressuring, transient temperature states, and, in the case of the Halten Terrace, the change in pore fluids in the deeper section.

As mentioned previously, the change in pore fluids as a result of hydrocarbon generation and migration can significantly alter the thermal conductivities of a sedimentary layer. The computer models used in this study all assumed a water-filled system, and did not allow for a change in the fluid system. Recent testing of modelling algorithms which allow this, indicate that a change in the pore fluid will enable the use of a lower present-day heat flow and match both the Jurassic and Cretaceous temperatures.

Throughout this paper, we have made the assumption that the algorithms used to predict the maturity (reflectance) of the kerogens are correct. The three programs used slightly different algorithms, and yet predicted similar results (well within any error limits of the data themselves). If the maturation algorithms are incorrect, they are at least consistently so. Nevertheless, the possibility of these limitations cannot be fully dismissed, nor can it be addressed further in this paper.

Significance of the heating anomaly to hydrocarbon generation

Regardless of the apparent thermal anomaly today, it is the recent rapid subsidence that has had the most significant role in hydrocarbon generation from the Upper Jurassic source rocks (Fig. 13). Model A

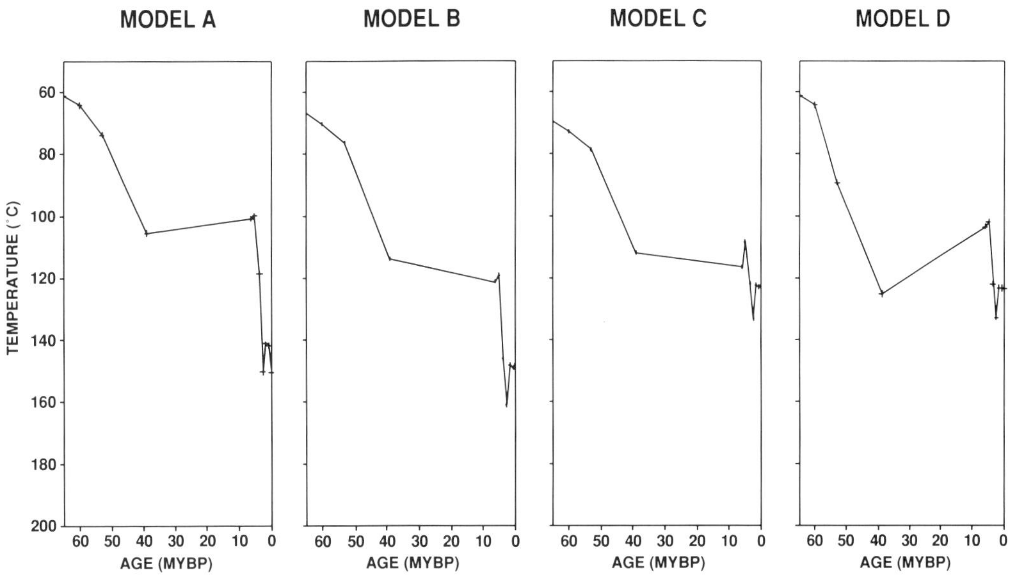

Fig. 16. Upper Jurassic Spekk Formation temperature histories for the heat flow histories shown in Fig. 8; same well as in Fig. 9.

shows a rapid onset and completion of hydrocarbon generation as a result of both the rapid burial and the increase in heat flow. Model B (constant heat flow) shows an earlier onset of oil generation, with an increase in gas generation as a result of the rapid subsidence. Model C shows a slow generation of oil throughout as a result of higher heat flow prior to the rapid subsidence. Even with the decrease in heat flow in conjunction with the rapid subsidence, there is an increase in gas generation as a result of the rapid burial. Model D shows a rapid increase in generation during the assumed heating event associated with the opening of the North Atlantic.

A comparison of the temperature histories (Fig. 16) for the Upper Jurassic section for the four models shows why the generation histories throughout the Tertiary are so different. The temperature history for Model A indicates that temperatures sufficient for active hydrocarbon generation were reached only recently as a result of rapid subsidence and increased heat flow. Models B and C, on the other hand, reached the hydrocarbon generation temperature threshold during the Eocene and then continued to show increases in temperature through to the present as a result of increased burial. Model D reached temperatures sufficient for hydrocarbon generation about 53 Ma but was then exposed to lower temperatures until the recent rapid subsidence of the Pliocene.

The models shown here demonstrate only a few of the scenarios possible for the Halten Terrace, as well as the importance of understanding the apparent "heat flow" increase in the last few million years. Their importance becomes further evident when trying to model the movement of hydrocarbons to their present-day known locations. The known fields all lie to the east of the over-pressured and thermally anomalous Jurassic zone on the Halten Terrace. All hydrocarbon generation, on the other hand, has occurred within the transition and over-pressured zones. It therefore becomes important to accurately model the relationship between deposition, over-pressuring, thermal anomalies, and active hydrocarbon generation and migration. It is necessary to follow several scenarios through a complete basin analysis prior to reaching an understanding concerning the exact relationship between the geological, thermal, and generational phenomena and mathematical modelling artifacts on the Halten Terrace.

Given the one-dimensional modelling capabilities available to us, we believe that Model A is the best model possible in terms of the thermal and generational history of the Jurassic. The heat flow history for Model A is incorrect, but because of other arguments, we believe that the generation history predicted by the model is the most credible. The higher temperatures in the present-day Jurassic are a result of the recent rapid burial of the Pliocene, and are not a long-lived phenomenon. Therefore, it would be unrealistic to apply the present temperatures (heat flow) backwards through time to the Jurassic section. Because these recent higher temperatures are also the main cause of hydrocarbon generation, a change in the pore filling fluid, and lower thermal conductivities within the over-pressured Jurassic section today, temperatures in the Jurassic section are anomalously high relative to the overlying section. Accepting Model A as closest to reality, indicates that many of the reservoired hydrocarbons have only recently been emplaced.

Conclusions

The Jurassic source and reservoir section in the deep portion of the Halten Terrace is presently anomalously hot relative to the overlying section. Placing greater confidence in these temperatures than in the overlying Cretaceous temperatures has resulted in models which require a "heat flow" increase in the recent past in order to match both the Jurassic present-day temperatures and maturities. This "heat flow" anomaly is not unique to the present study, and has featured in numerous published basin models for the Norwegian Shelf as well as elsewhere around the world.

The recent modelled "heat flow" increase on the Halten Terrace is not a result of deep-seated crustal events, but rather is a result of the current modelling algorithms' inability to completely and accurately model the combined effects of lower thermal conductivities as a result of over-pressuring, the change in pore fluids as hydrocarbons are generated in this portion of the basin, and, possibly, the transfer of heat by fluid flow. In short, it is an artificial input required by the current modelling algorithms. Consequently, a heating event which affects the entire section over the last few million years is not geologically valid. The fact that the current modelling software requires such an event, indicates the anomalous nature of the geological events (rapid subsidence and sedimentation, glaciation and isostatic effects) of the late Cenozoic. This artifact also clearly indicates the need for a new generation of basin modelling software designed to better handle the integrated nature of geological, thermal and generational phenomena.

Valid thermal models are necessary in order to model the relationships between hydrocarbon generation, over-pressuring and migration. This is particularly important for the hydrocarbon generation that occurred as a result of the Plio-Pleistocene subsidence. An unwarranted heat flow spike would place the majority of hydrocarbon generation from the Spekk Formation in the last few million years.

Multiple possibilities for input data indicate that accurate measurements of temperature, maturity, and ages are essential if integrated basin models are to be applied seriously in an area such as the Halten Terrace. Attention to data collection must be made a priority, both prior to drilling and at the rig site, in order to aid in later modelling efforts.

Our modelling efforts have resulted in an increased understanding of the interrelated geological and thermal events of the Halten Terrace. As a result, we have been able to discount the "heat flow spike" of the last few million years, while recognizing the thermally anomalous nature of the present-day Jurassic section, and increasing our understanding of the limitations inherent in basin modelling. Similar, thorough investigations of locations world-wide are necessary to ascertain the true nature of recent heat flow increases, and what some might term a "global heating event".

Acknowledgements

The authors wish to thank Conoco for their permission to publish this work, and Sten Lou, Graham Cooper, Odd Senneseth, and James Best for their input to the modelling efforts on the Halten Terrace. We also thank Kate Weissenburger, Christian Hermanrud, Øyvind Sylta, and two anonymous reviewers for their helpful comments.

Discussion

Alfred Kjemperud, Nopec.: *Most of the Norwegian Shelf has probably been glaciated several times during the last two million years. Have you taken any glacial effect into account? If not, is that because you do not think this has any effect?*

Authors' reply: We are unable to model glacial effects with the software currently available to us. We do, however, consider glacial effects to have a possible minor effect on the thermal state of the entire section, as discussed in the text.

James Iliffe, University of South Carolina: *What criteria did you use to determine the maximum values and the shape of the heat flow pulses you indicated for the Upper Jurassic and Paleocene periods?*

Authors' reply: We used a McKenzie (1978) model to determine the stretching factor and the related heat flow history for the Jurassic heat pulse. The Paleocene heat pulses are simplifications of the potential effect the opening of the North Atlantic would have had on the Halten Terrace area. The duration and extent of the heat flow pulse are constrained by the present-day maturities. A short-lived pulse allows for a higher heat flow, while a long-lived pulse requires a lower heat flow. More work in terms of age dating of sediments and structural restorations in the area is required to further constrain the values.

References

Andrews-Speed, C.P., Oxburgh, E.R. and Cooper, B.A., 1984. Temperatures and depth-dependent heat flow in Western North Sea. Am. Assoc. Pet. Geol., Bull., 68: 1764–1781.

Braun, R.L. and Burnham, A.K., 1987. Analysis of chemical reaction kinetics using a distribution of activation energies and simpler models. J. Energy Fuels, 1: 153–161.

Brigaud, F., Chapman, D.S. and Le Douaran, S., 1990. Estimating thermal conductivity in sedimentary basins using lithologic data and geophysical well logs. Am. Assoc. Pet. Geol., Bull., 74: 1459–1477.

Brigaud, F., Vasseur, G. and Caillet, G., 1992. Thermal state in the North Viking Graben (North Sea) determined from oil exploration well data. Geophysics 57 (1): 69–88.

Burnham, A.K. and Sweeney, J.J., 1989. A chemical model of vitrinite maturation and reflectance. Geochim. Cosmochim. Acta, 53: 2649–2657.

Carstens, H. and Finstad, K.G., 1981. Geothermal gradients of the northern North Sea Basin, 59–62°N. In: L.V. Illing and G.D. Hobson (Editors), Petroleum Geology of the Continental Shelf of North-West Europe. Institute of Petroleum, London, pp. 152–161.

Chiaramonte, M.A., Tamburini, A., Salvatori, T. and Bersani, A., 1988. Geochemical modelling in rapidly subsiding basins. In: Advances in Organic Geochemistry, 1987. Org. Geochem., 13: 181–186.

Cloetingh, S., Gradstein, F.M., Kooi, H., Grant, A.C. and Kaminski, M., 1990. Plate reorganization: a cause of rapid Late Neogene subsidence and sedimentation around the North Atlantic. J. Geol. Soc., 147: 495–506.

Dahl, B. and Augustson, J.H., 1993. The influence of Tertiary and Quaternary sedimentation erosion on hydrocarbon generation in Norwegian offshore basins. In: A.G. Doré et al. (Editors), Basin Modelling: Advances and Applications. Norwegian Petroleum Society (NPF), Special Publication 3. Elsevier, Amsterdam, pp. 419–431 (this volume).

Dahl, B., Nysaether, E., Speers, G.C. and Yükler, A., 1987. Oseberg area-integrated basin modelling. In: J. Brooks and K. Glennie (Editors), Petroleum Geology of North-West Europe. Graham and Trotman, London, pp. 1029–1038.

Dalland, A., Worsley, D. and Ofstad, K., 1988. A lithostratigraphic scheme for the Mesozoic and Cenozoic succession, Mid- and Northern-Norway. Norw. Pet. Direct., Bull., 4.

Doré, A.G., 1991. The structural foundation and evolution of Mesozoic seaways between Europe and the Arctic. Palaeogeogr., Palaeoclimatol., Palaeoecol., 87: 441–491.

Dowdle, W.L. and Cobb, W.M., 1975. Static formation temperature from well logs — an empirical method. J. Pet. Tech., Nov. 1975, pp. 1326–1330.

Eldholm, O. and Theide, J., 1980. Cenozoic continental separation between Europe and Greenland. Palaeogeogr., Palaeoclimatol., Palaeoecol., 30: 243–259.

Falvey, D.A. and Middleton, M.F., 1981. Passive continental margins: evidence for a prebreakup deep crustal metamorphic subsidence mechanism. Oceanol. Acta, SP, pp. 103–114.

Frakes, L.A., 1979. Climates Throughout Geologic Time. Elsevier, Amsterdam, 310 pp.

Gable, R., 1984. Temperature disturbances observed in sedimentary basins and their interpretation. In: B. Durand (Editor), Thermal Phenomena in Sedimentary Basins. Technip, Paris, pp. 55–63.

Goll, R.M. and Hansen, J.W., 1992. The Cenozoic sequence stratigraphy of the Halten Terrace and the Outer Vøring Plateau based on seismic and biostratigraphic data. In: L.N. Jensen and F. Riis (Editors), Post-Cretaceous Uplift and Sedimentation along the Western Fennoscandian Shield. Nor. Geol. Tidsskr., Spec. Issue, 72: 295–300.

Grigo, E., Maragna, B., Arienti, M.T., Fiorani, M., Parisi, A., Marrone, M., Sguazerro, P. and Uberg, A.S., 1993. Issues in 3D sedimentary basin modelling and application to Haltenbanken, offshore Norway. In: A.G. Doré et al. (Editors), Basin Modelling: Advances and Applications. Norwegian Petroleum Society (NPF), Special Publication 3. Elsevier, Amsterdam, pp. 455–468 (this volume).

Haile, N.S., Flavell, W.S. and Gorst, M., 1987. Geothermal Database, North Sea. Non-exclusive report published by Roberston Research International Ltd., Llandudno, Gwynnedd.

Haq, B.U., Hardenbol, J. and Vail, P.R., 1987. Chronology of fluctuating sea levels since the Triassic. Science, 235: 1156–1167.

Hermanrud, C., 1988. Determination of Formation Temperature from Downhole Measurements. Ph.D. Dissertation, University of South Carolina, Columbia, S.C., 203 pp.

Hermanrud, C., Eggen, S. and Larsen, R.M., 1991. Investigation of the thermal regime of the Horda platform by basin modelling; implications for the hydrocarbon potential of the Stord basin, northern North Sea. In: A.M. Spencer (Editor), Generation, Accumulation and Production of Europe's Hydrocarbons. Special Publication of EAPG No. 1. Oxford University Press, Oxford, pp. 65–73.

Heum, O.R., Dalland, A. and Meisingset, K.K., 1986. Habitat of hydrocarbons at Halten Terrace (PVT modelling as a predictive tool in hydrocarbon exploration). In: A.M. Spencer et al. (Editors), Habitat of Hydrocarbons on the Norwegian Shelf. Norwegian Petroleum Society, Graham and Trotman, London, pp 259–274.

Jensen, L.N. and Riis, F. (Editors), 1992. Post-Cretaceous Uplift and Sedimentation along the Western Fennoscandian Shield. Nor. Geol. Tidsskr., Spec. Issue, 72: 221–338.

Jones, F.W. and Oxburgh, E.R., 1979. Two-dimensional thermal conductivity anomalies and vertical heat flow variations. In: V. Cermák and L. Rybach (Editors), Terrestrial Heat Flow in Europe. Springer-Verlag, New York, N.Y., pp. 98–106.

Kehle, R.O., 1973. Geothermal Survey of North America 1972 Annual Progress Report. Project supported by the American Association of Petroleum Geologists.

Leadholm, R.H., Ho, T.T.Y. and Sahai, S.K,, 1985. Heat flow, geothermal gradients and maturation modelling on the Norwegian continental shelf using computer methods. In: B.M. Thomas et al. (Editors), Petroleum Geochemistry in Exploration of the Norwegian Shelf. Graham and Trotman, London, pp. 131–143.

Lewis, C.R. and Rose, S.D., 1970. A theory relating high temperatures and over-pressures. J. Pet. Technol., Jan. 1970, pp. 11–16.

Leischner, K., Welte, D.H. and Littke, R., 1993. Fluid inclusions and organic maturity parameters as calibration tools in basin modelling. In: A.G. Doré et al. (Editors), Basin Modelling: Advances and Applications. Norwegian Petroleum Society (NPF), Special Publication 3. Elsevier, Amsterdam, pp. 161–172 (this volume).

Lister, E.S., Etheridge, M.A. and Symonds, P.A., 1986. Detachment faulting and the evolution of passive continental margins. Geology, 14: 246–250.

Mann, D.M. and MacKenzie, A.S., 1990. Prediction of pore fluid pressures in sedimentary basins. Mar. Pet. Geol., 7: 55–65.

McKenzie, D., 1978. Some remarks on the development of sedimentary basins. Earth Planet. Sci. Lett., 40: 319–333.

Mo, E.S., Throndsen, T., Andresen, P., Bäckström, S.A., Forsberg, A., Haug, S. and Tørudbakken, B., 1989. A dynamic deterministic model of hydrocarbon generation in the Midgard Field drainage area Mid-Norway. Geol. Rundsch., 78 (1): 305–317.

Reksnes, P.A. and Vågnes, E., 1985. Evolution of the Greenland Sea and Eurasia Basin. Cand. Sci. Thesis, University of Oslo.

Riis, F. and Fjeldskaar, W., 1992. On the magnitude of the Late Tertiary and Quaternary erosion and its significance for the uplift of Scandinavia and the Barents Sea. In: R.M. Larsen, H. Brekke, B.T. Larsen and E. Talleraas (Editors), Structural and Tectonic Modelling and its Application to Petroleum Geology.

Norwegian Petroleum Society (NPF) Special Publication 1, Elsevier, Amsterdam, pp. 163–186.

Rowley, D.B. and Lottes, A.K., 1988. Plate-kinematic reconstructions of the North Atlantic and Arctic: Late Jurassic to Present. Tectonophysics, 155: 73–120.

Royden, L., Sclater, J.G. and von Herzen, R.P., 1980. Continental margin subsidence and heat flow: important parameters in formation of petroleum hydrocarbons. Am. Assoc. Pet. Geol., Bull., 64: 173–187.

Schmidt, F. and Erdoğan, L.T, 1993. Basin modelling in an overthrust area of Austria. In: A.G. Doré et al. (Editors), Basin Modelling: Advances and Applications. Norwegian Petroleum Society (NPF), Special Publication 3. Elsevier, Amsterdam, pp. 573–581 (this volume).

Sclater, J.G. and Christie, P.A.F., 1980. Continental stretching: an explanation of the post-mid-Cretaceous subsidence of the Central North Sea Basin. J. Geophys. Res., 85, B7: 3711–3739.

Skogseid, J., 1990. Vøring Continental Margin: Magmatic–tectonic Evolution, Sedimentation and Basin Subsidence. Doctor Sci. Thesis, University of Oslo.

Stuevold, L.M., 1989. The Tertiary Fennoscandian Uplift in View of Vertical Movements on the Continental Margin off Mid-Norway. Cand. Sci. Thesis, University of Oslo.

Thorne, J.A. and Watts, A.B., 1989. Quantitative analysis of North Sea subsidence. Am. Assoc. Pet. Geol., Bull., 73: 88–116.

Tissot, B.P., Pelet, R. and Ungerer, P.H., 1987. Thermal history of sedimentary basins, maturation indices, and kinetics of oil and gas generation. Am. Assoc. Pet. Geol., Bull., 71: 1445–1466.

Vik, E. and Hermanrud, C., 1993. Transient thermal effects of rapid subsidence in the Haltenbanken area. In: A.G. Doré et al. (Editors), Basin Modelling: Advances and Applications. Norwegian Petroleum Society (NPF), Special Publication 3. Elsevier, Amsterdam, pp. 107–117 (this volume).

REBECCA P. JENSEN *Conoco Norway, P.O. Box 488, N-4001 Stavanger, Norway*
ANTHONY G. DORÉ *Conoco Norway, P.O. Box 488, N-4001 Stavanger, Norway*

Transient thermal effects of rapid subsidence in the Haltenbanken area

Eirik Vik and Christian Hermanrud

Subsurface temperatures exhibit large lateral variations in five wells in the Haltenbanken area. Most of the temperature variation could be accounted for using a one-dimensional transient thermal model, with a common set of rock properties for end-member lithologies, and with the same boundary conditions for all five wells. These results demonstrate that the model that was used accounts for the major processes which are responsible for the subsurface temperature variations. The simulation results also demonstrate that subsurface temperatures in the Haltenbanken area can be predicted to within 2 to 3°C by the model.

Modelled vitrinite reflectance in well 6407/4-1 was in good agreement with measured data from this well. An inspection of the demonstrated accuracy of the published vitrinite reflectance models and measured vitrinite data showed that the fit between modelled and observed vitrinite reflectance could not alone be taken as supporting evidence for the transient thermal model.

Introduction

Simulation of the thermal history of source rocks is a prerequisite for successful modelling of hydrocarbon generation. Emphasis should thus be put on careful verification of the thermal model used for such work. The study we report here was initiated to aid prospect evaluation in the Haltenbanken area of the Mid Norwegian Shelf (Fig. 1) by providing such a verified thermal model.

The investigated area shows significant lateral variation in subsurface temperature. The present study was designed to test whether these variations could be reproduced using a model with constant temperature at the base of the lithosphere, assuming no lateral flow of fluids or heat, and no lateral variation in lithospheric thickness or radiogenic heat generation. The only variables would then be lithology and geohistory. As we see it, the conditions for successful modelling are: (a) the model must produce calculated maturity indicators in accordance with well data; and (b) the variations in subsurface temperatures should be explicable by a common set of thermal conductivities for the end-member lithologies. In addition, input heat-flow histories should be based on simulations of known physical processes.

Geological description of the Haltenbanken area

A short summary of the geology of the Haltenbanken area is given below. The reader is referred to Bukovics and Ziegler (1985), Heum et al. (1986) and Fagerland (1990) for a more thorough description of the geological history of this area. The investigated area is shown in Fig. 1. It encompasses the northern part of the Haltenbanken area, parts of the Halten Terrace and the slope towards the Trøndelag Platform. A schematic structural cross-section is shown in Fig. 2. This figure also illustrates the main geological events, as described below.

Jurassic and Cretaceous

Deltaic and shallow-marine sandstones of Early and Middle Jurassic age form the main reservoir rocks of the area (the Åre Formation, the Tilje Formation and the Fangst Group). Subsequent deposition of claystones in the Late Jurassic/Early Cretaceous formed the main cap rocks. The Late Jurassic/ Early Cretaceous Spekk Formation was probably deposited in a restricted basin experiencing anoxic conditions, and is one of the two dominant source rocks in the area, the other being coals of the Late Triassic/Early Jurassic Åre Formation.

Extensional movements were initiated in the Late Jurassic. These movements created the main structuring of the area, and formed the block-faulted Jurassic terrain which traps most of the proven oil and gas in the Haltenbanken area. The Late Cretaceous was characterised by thermal subsidence due to the earlier extension. Claystones and minor amounts of sand were deposited.

Basin Modelling: Advances and Applications edited by A.G. Doré et al. NPF Special Publication 3, pp. 107–117, Elsevier, Amsterdam.

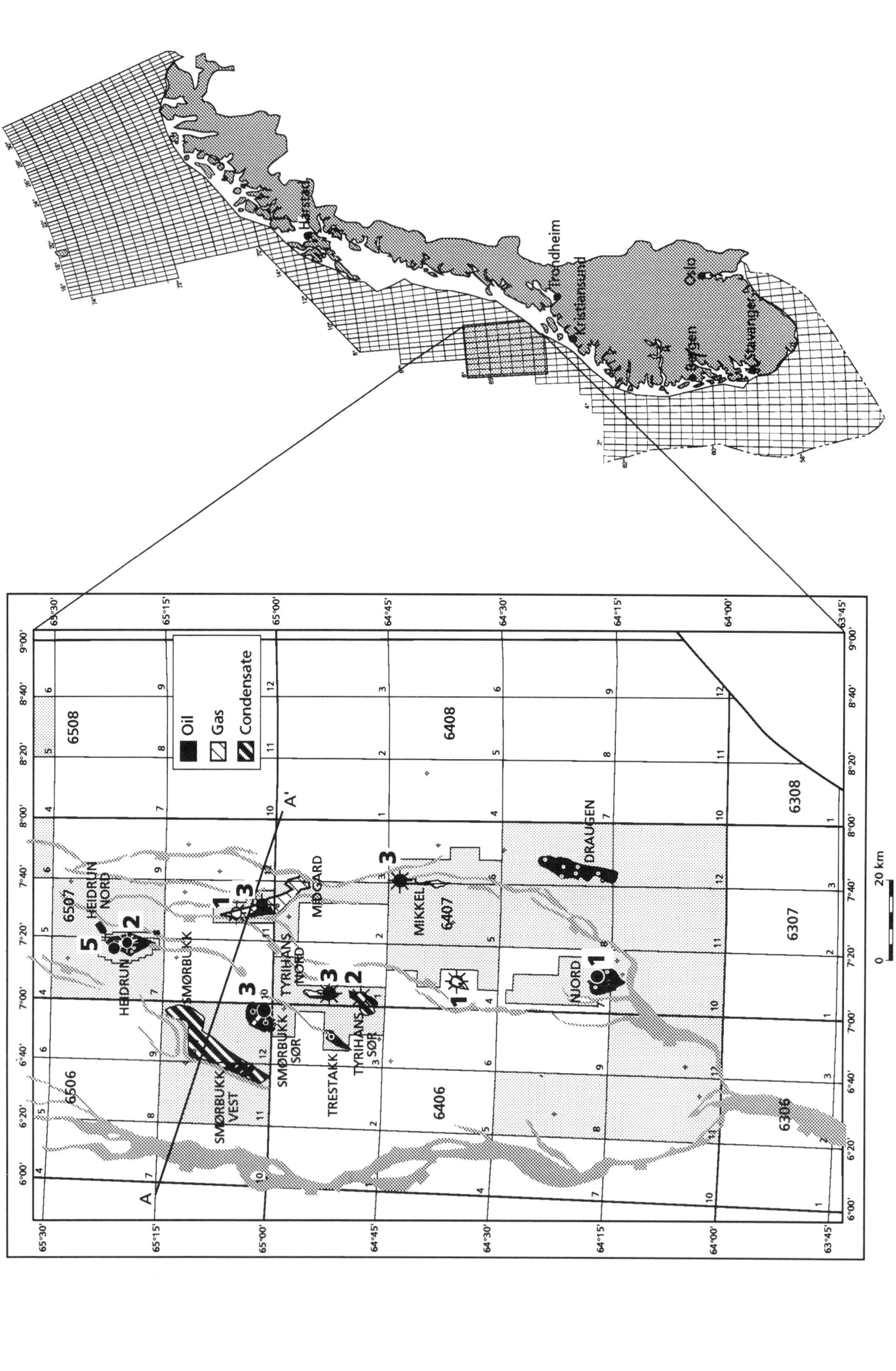

Fig. 1. General map of the Haltenbanken area showing the main structural elements, hydrocarbon discoveries and the wells investigated in this study.

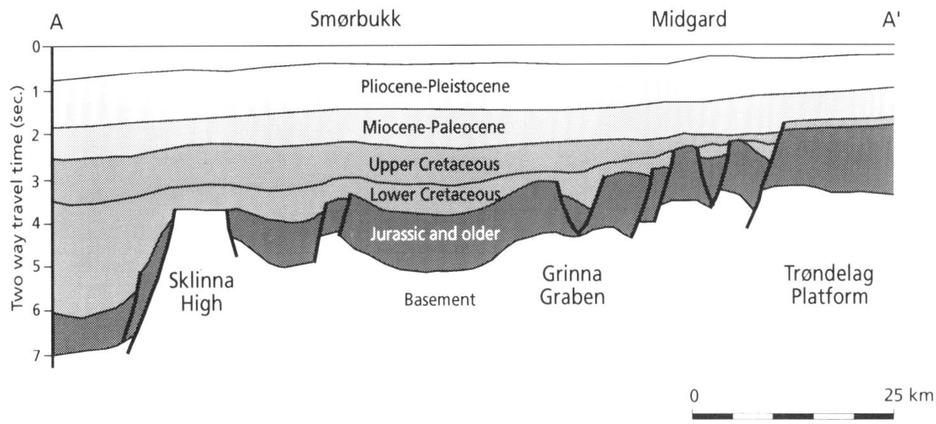

Fig. 2. Structural cross-section over the northern Haltenbanken area. For location see Fig. 1.

Cenozoic

The subsidence and deposition of claystones continued in the Palaeogene. The opening of the Atlantic was accompanied by volcanic activity leading to deposition of tuffaceous sediments in the Late Paleocene. Clay deposition continued throughout most of the Tertiary, interrupted by a regional hiatus of uncertain (and varying) duration in the Miocene.

Very rapid subsidence and sedimentation took place in the Late Pliocene/Early Pleistocene. More than 1 km of sediments, mostly claystones, were deposited during this short period of 1–1.5 m.y. This deposition was probably related to the uplift of the Fennoscandian Shield, which exposed large areas to erosion. The reasons for the uplift are presently a subject of debate (Jensen et al., 1992). This relatively recent episode had a pronounced influence on hydrocarbon generation in the Haltenbanken area, as the subsidence exposed source rocks in the deeper part of the area to temperatures sufficient for hydrocarbon generation. The regional heat flow was also modified by the subsidence, setting up a transient heat flow regime which is still very pronounced, as will be demonstrated later in this paper.

Two- to five-hundred metres of glaciomarine sediments were deposited on top of the Early Pleistocene sediments in the Quaternary.

Temperature database

In the present study only temperatures from drill stem tests were used (DST-temperatures). Temperatures measured during testing have been demonstrated to be more accurate than temperatures recorded during logging (Hermanrud, 1988). DST-temperature measurements have standard deviations of about 1.5°C for single tests under favourable conditions (Hermanrud et al., 1991b); error ranges are further reduced when there are several tests over a restricted depth interval. Even after corrections by the Horner plot method, temperatures derived from well log headings (BHT data) are about 8°C too low, and also have standard deviations of about 8°C (Hermanrud, 1988; Hermanrud et al., 1990). The DST-temperatures used in this study are plotted in Fig. 3. The depth axes in this (and all subsequent) figures are given relative to mean sea level.

Only tests with relatively high flow rates and resulting low drawdown can be expected to yield correct formation temperatures because of the Joule-Thomson effect (Steffensen and Smith, 1973). The Joule-Thomson effect is the change in temperature of a fluid upon expansion, or pressure decrease, in a steady flow process involving no heat transfer or work. This occurs when oil or more notably when gas is flowing into the well bore during testing. Big pressure drops, or drawdown, can give significant temperature perturbations (up to 20°C). The magnitude of the Joule-Thomson effect is also dependent on hydrocarbon composition, temperature and absolute pressure. The heating/cooling resulting from the pressure drops during testing can be computed by equations of state (e.g. Soave, 1972) given information on these parameters. Such calculations were performed for tests with high drawdowns in the investigated wells, suggesting only minor corrections for the oil tests. For some of the gas tests however, the correction was more pronounced, up to 7°C. The corrected temperatures are indicated by dotted lines in Fig. 3.

Several features emerge from an inspection of the DST-temperature data:

(a) The temperature gradient through the reservoirs is relatively constant for all wells, about 2.3 to 2.5°C/100 m.

(b) The temperatures from different tests within a given well show good agreement, confirming that the temperature data are of high quality.

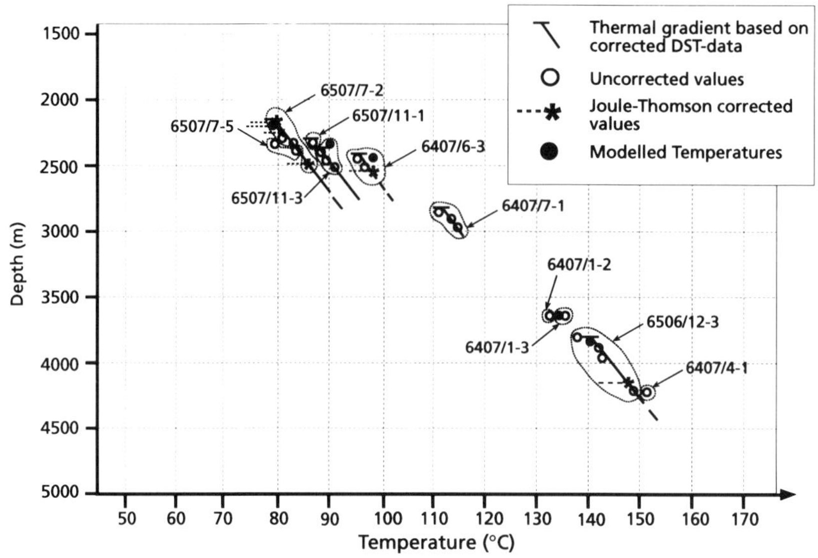

Fig. 3. DST-temperatures in Early and Middle Jurassic reservoirs of the Haltenbanken area.

(c) The (extrapolated) temperatures at the top of the reservoirs in each well scatter around a straight line when plotted vs. depth (Fig. 4). The scatter around this line is 7°C at the most, suggesting that subsurface temperatures in the area have an inaccuracy of about 7°C when based only on temperatures from neighbouring wells.

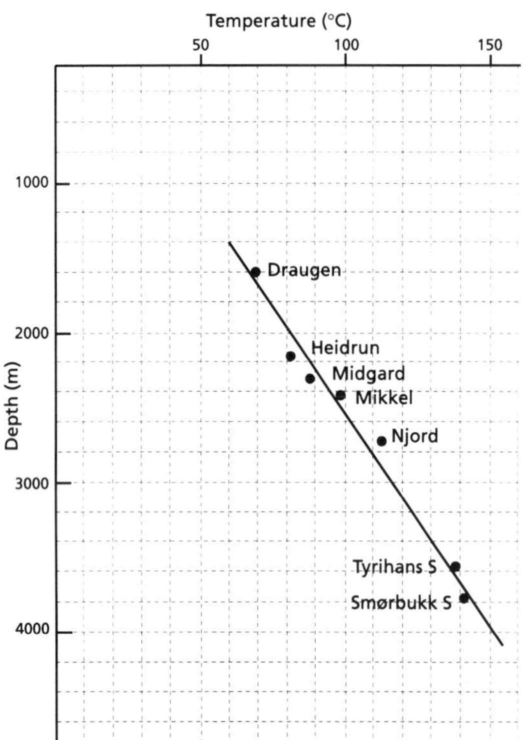

Fig. 4. DST-temperatures extrapolated to top reservoir level at Haltenbanken oil and gas fields. Each datapoint is based on several DSTs and is accurate to within 1°C.

Thermal modelling

Thermal modelling of the wells 6407/1-3, 6407/6-3, 6507/11-1, 6506/12-3 and 6507/7-2 was performed by the one-dimensional version of the basin modelling program described by Hermanrud et al. (1991a).

Input parameters to the calculations are the geohistory of the modelled wells, petrophysical parameters (such as thermal conductivity) and constants in the porosity and permeability functions. Table 1 lists typical end-member values for the most im-

TABLE 1

Input parameters for the thermal modelling

Parameter	Value
C_1, sand, initial porosity (%)	45
C_1, shale, initial porosity (%)	45
C_2, shale (10^7 Pa^{-1})	3.5
C_2, sand (10^7 Pa^{-1})	2.5
Matrix thermal conductivity, shale (W m^{-1} °C^{-1})	1.5
Matrix thermal conductivity, sand (W m^{-1} °C^{-1})	3.5
Lithospheric thickness (km)	110
Temperature at base of lithosphere (°C)	1350
A_0 (W m^{-3})	2.6×10^{-16}
a (km^{-1})	10^{-1}
b (year^{-1})	2.1×10^{-10}

C_1 and C_2 are factors in the porosity vs. effective stress equation:

$$\phi = C_1 \exp(-C_2\sigma_{\text{eff}}) \quad (1)$$

where ϕ is porosity and σ_{eff} is the effective stress. A_0, a and b are parameters in the basement heat generation formula:

$$A = A_0 \exp(-az - bt) \quad (2)$$

where A is heat production, z is depth in km and t is time in years. The values of C_1 and C_2 were determined from inspection of well logs and measurements of He-porosity on cores. The values of A_0, a and b were taken from Morgan (1983) and Turcotte (1980).

portant input parameters. Thermal conductivity was regarded as independent of temperature in the modelling of the five investigated wells. The thermal conductivity of a given formation was thus considered as dependent on lithology and porosity only. The latter was modelled by assuming an exponential relation between porosity and effective stress, while permeability was modelled as a function of porosity.

Parameters in the fluid flow equations were adjusted to ensure agreement between observed and modelled values of porosity, layer thickness and fluid pressure. These adjustments also influence the thermal parameters of the sediments, as the thermal conductivity is a function of porosity.

The lower thermal boundary conditions of the model may be given as a constant heat flow value, as a time-varying palaeo-heat flow, or as a constant temperature at the base of the lithosphere. The last option was applied in this study. Radiogenic heat production was modelled to decrease exponentially with depth in the lithosphere, and the total amount of radiogenic heat production was modified to match the measured well temperatures.

Implicit in this model is the assumption that the heat flow into the base of the model (base of lithosphere) is constant and also that the thickness of the lithosphere is constant through time. These last assumptions are obviously not correct as crustal thinning at the time of stretching will almost certainly have produced high heat flow in the Mesozoic. Sensitivity analysis was carried out to see if these simplifying assumptions affected the calculated vitrinite reflectance and kerogen transformation ratio. Our analysis confirmed that for the present study the consequence of disregarding the elevated Mesozoic heat flow is negligible. This conclusion is not surprising, since the source rocks were deposited at about the same time as the crustal stretching, and were not deeply buried until the thermal effects of this stretching had reduced.

Under steady state conditions, the calculated heat flow is in equilibrium with the temperature at the base of the lithosphere and at the surface, the applied thermal conductivities, and the radioactive heat production of the crust and upper mantle. Deposition (or erosion) of sediments will alter this equilibrium. This is illustrated in a schematic form in Fig. 5. Re-establishment of the thermal equilibrium after sediment deposition will require heating of both the new and the previously deposited sediments, the crust and the upper mantle. As thermal energy is absorbed by the lithosphere during this heating, less heat is available for flow into the sediments, resulting in a reduced heat flow until equilibrium is re-established. A further consequence of this disequilibrium is that heat flow is a function of both time and depth. In the following, heat flow values will always be referred to a specific formation. Unless otherwise specified this will be the top of the Late Jurassic Spekk Formation.

Modelling results

Well 6506/12-3 was used for calibration of the input parameters. The modelled pressures are compared to actual pore pressures (evaluated from drilling parameters) in Fig. 6 to demonstrate the degree of precision of the pressure calibration.

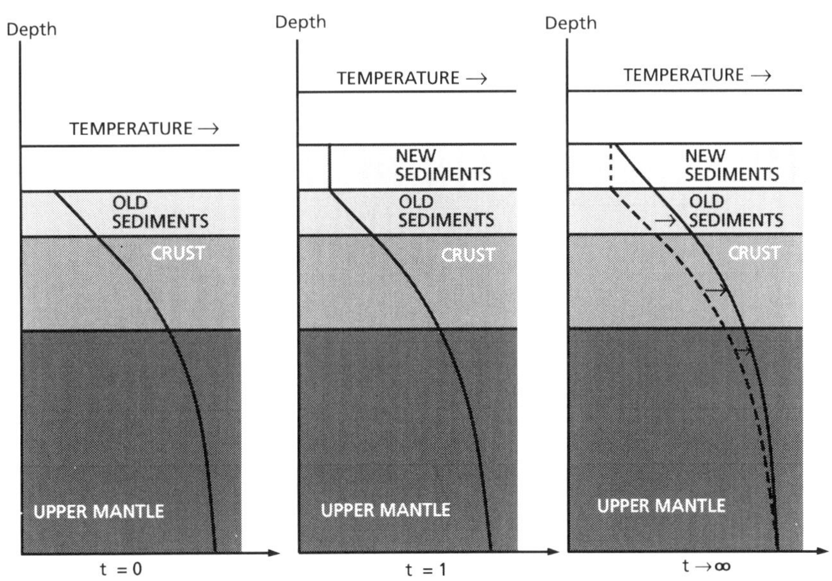

Fig. 5. Simplified model of transient thermal regimes set up by instantaneous deposition of a thick sedimentary layer. Arrows indicate path to new thermal equilibrium.

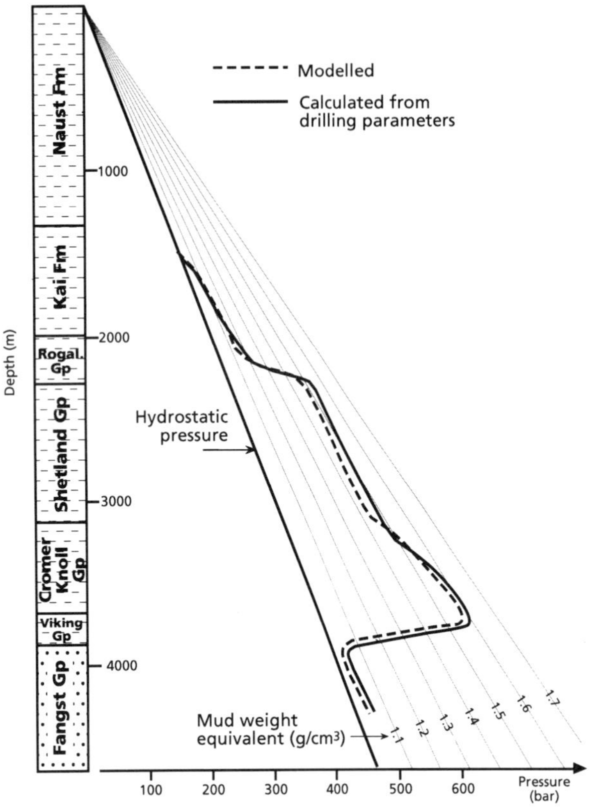

Fig. 6. Modelled and observed pore pressure of well 6506/12-3.

Figure 7 shows the modelled subsidence, temperature and heat flow vs. time for well 6506/12-3 using the procedures described above. The figure visualizes how the progressive burial and heating of the source rock results from the rapid sedimentation during the Pliocene and Pleistocene. Note also that the modelled heat flow in the sediments (top Spekk Formation) was reduced by about 10% because of the rapid burial, and that it is presently in the process of re-approaching thermal equilibrium. The small increase in heat flow at about 3 m.y. BP is a result of reduced seabed temperature at this period.

The sub-recent heat flow depression is accompanied by an equally pronounced temperature increase due to the subsidence. Trial computations indicated that the heat flow depression results in a temperature drop (at this particular location) of about 12°C. Figure 7 also illustrates the importance of timing. An earlier period of rapid deposition than experienced in the Haltenbanken area would have given the system more time to re-equilibrate, while a later depositional event would have made the heat flow anomalies even more pronounced. Mapping and dating of the Plio/Pleistocene units on the Haltenbanken Shelf show that they are highly diachronous, becoming progressively younger westwards, as the sediments build out from the mainland (Goll and Hansen, 1992). The present-day heat-flow response to the rapid sedimentation will thus be dependent on the location relative to the sediment source area.

The four other wells (6407/1-3, 6407/6-3, 6507/11-1 and 6507/7-2) were modelled with the same thermal parameters for end-member lithologies as for the calibration well. The detailed stratigraphy did, however, vary between the four wells. The modelled fluid pressures in these wells turned out to be within 20% of the observed pressures. Such discrepancies could be expected, as one-dimensional modelling will most likely fail to reproduce subsurface fluid pressures accurately, even if rock characteristics are correct. Minor adjustments in the compactional parameters may thus be warranted to match the observed pres-

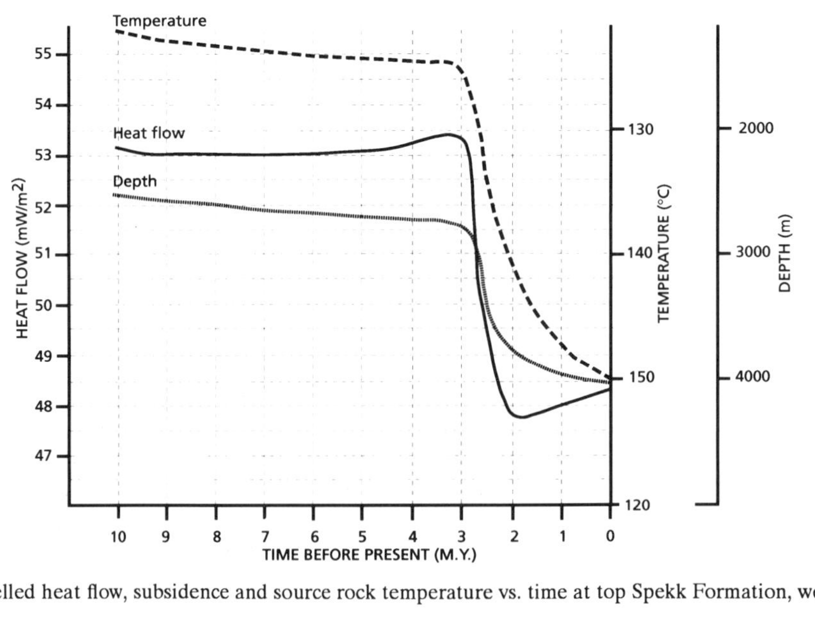

Fig. 7. Modelled heat flow, subsidence and source rock temperature vs. time at top Spekk Formation, well 6506/12-3.

sures. The observed discrepancies between modelled and observed fluid pressures were expected to result in mismatches between predicted and observed subsurface temperatures in the wells, since high fluid pressures result in restricted compaction, higher porosities and thereby low thermal conductivities. The compaction characteristics for the rocks in these wells were therefore modified to match the observed fluid pressures for the purpose of validating the thermal model. Modelling results showed, however, that the modifications only influenced the modelled temperature by 1°C at the most.

The calculated temperatures (after calibration of the modelled pressures to the fluid pressures in the wells) for the four wells are shown in Fig 3. As the figure demonstrates, all temperatures in the wells are now matched within 2°C, while the observed temperature variation between wells at the same depth is about 15°C.

Discussion

Most of the variations in observed subsurface temperature were accounted for by the simulations. The remaining differences may be due to such factors as inaccurate lithological description, uncertainty associated with shale porosities or thermal conductivities, or inaccuracy of the measured temperatures. The reported modelling results still give strong support for the thermal model.

In an attempt to further verify the thermal model, vitrinite reflectance and source rock transformation ratio vs. depth were simulated using the calculated temperature histories as input. As a numerical experiment, a constant heat flow vs. time scenario was modelled as an alternative to the heat flow vs. time scenario which resulted from our simulation of the transient thermal regime.

Simulation of the thermal degradation of kerogen in the Spekk Formation, using the two heat flow scenarios, led to identical predictions of transformation ratios for the Spekk Formation. These results most probably occur because both transient and steady state thermal models are calibrated to match the same present-day temperature, and because hydrocarbon generation kinetics are quite insensitive to temperature differences at shallower depths of burial.

The relinquished well 6407/4-1 was selected for modelling of vitrinite reflectance with the two heat flow scenarios. The measured vitrinite reflectance of this well is given in Fig. 8. The measurements show a significant scatter. Three of the measured values were identified as "good" by the geochemist who carried out the measurements. Two of the three values were derived from coal samples, while the

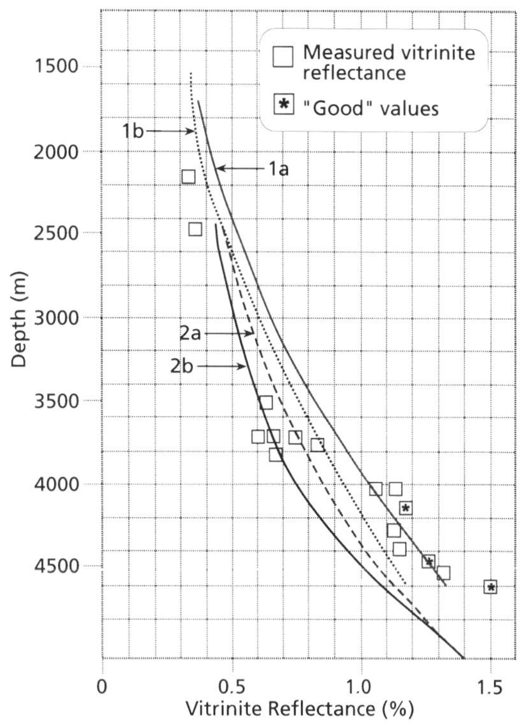

Fig. 8. Modelled and measured vitrinite reflectance of well 6507/4-1. *1* = modelled vitrinite reflectance, *a* = with transient heat flow, *b* = with constant heat flow; *2* = proposed standard R_0 vs. depth curves for the Haltenbanken area, *a* = Odden (1986), *b* = extracted from the data of Khorasani (1989).

last (and deepest) was measured from a cored shale sequence. All other measurements are from sidewall cores of shales. If the "good" measurements of Fig. 8 are to be trusted, then the measurements at about 3.7 km must be too low on average. An alternative explanation might be that the discontinuity of the vitrinite reflectance profile was caused by erosion of Late Cretaceous sediments. This explanation would rely on the eroded thicknesses being in the order of 3–5 km (Katz et al., 1988), and can safely be disregarded in the Haltenbanken area.

The vitrinite reflectance of well 6407/4-1 was modelled with the EasyR_0 programme (Sweeney and Burnham, 1990). The thermal history of this well was modelled as previously described. The virgin rock temperature for this well was less well constrained (estimated to ±3°C) than for the other wells of this study (±1°C), because of thermal disturbances caused by a comparatively large pressure drop during testing.

The modelled vitrinite reflectance profile based on the two heat flow scenarios are superimposed on Fig. 8. The transient heat flow model fits the "good" measurements almost perfectly, while the constant heat flow model underestimates these data with 300–400 m. Whether the modelled vitrinite reflectance curves imply that the transient heat flow model is superior to the constant heat flow model depends on the accuracy of both the measured data and the modelled vitrinite reflectance. If the combined inaccuracy of modelled and measured vitrinite reflectance amounts to 300–400 m or more, then no firm conclusions on the relative merits of the two heat flow scenarios can be made in our case. A brief discussion of the accuracy of measured and modelled vitrinite reflectance is thus required.

The inaccuracy of measured vitrinite reflectance is well documented (Bostick et al., 1978; Tissot et al., 1987; Hermanrud, this volume). Inspection of data from these quoted papers suggests that errors in the order of 500–1000 m may be expected in estimating the depth to a medium-range (say 1.0) vitrinite reflectance. The above error assessments cover both kerogen and coal; a careful selection of telecolenite from coal samples will probably reduce this error significantly (Throndsen, 1990). The significant spread of measured vitrinite reflectance values in well 6407/4-1 is also consistent with our suspicion that the depth to R_o = 1.0 might be shifted by at least 200 m if a different suite of vitrinite particles from the same well was investigated by another technician.

Information from more wells would reduce the uncertainty of the estimated depth to R_o = 1.0. The R_o vs. depth curve given by Odden (1986) and a curve extracted from data given by Khorasani (1989) have been superimposed on Fig. 8 to show how the vitrinite reflectance data measured in well 6407/4-1 compare to average Haltenbanken values. The data from Odden (1986) and Khorasani (1989) were derived from Jurassic coals from the Haltenbanken area. The two superimposed curves are least well constrained in the interval 3.5–4.2 km, as none of the authors had data in this interval. The two standard curves will not necessarily agree well with measurements from the post-Jurassic formations of well 6407/4-1. This is so because the thermal histories of these younger sediments will differ from those of the measured Jurassic coal samples which form the basis of the standard curves.

The Jurassic samples of well 6407/4-1 (the samples below 4 km) exhibit higher vitrinite reflectance than the standard curves. The difference in depth to R_o = 1.0 is about 400 m. Well 6407/4-1 has about 7°C (±3°C) higher reservoir temperature than the average value of Fig. 4 at 4 km depth. One would therefore expect the depth to vitrinite reflectance of 1.0 from well 6407/4-1 to be about 200 m shallower than what is otherwise observed in the area. The R_o values for well 6407/4-1 do accordingly come in about 200 m shallower than one would expect from the standard curves of Fig. 8 and knowledge of the virgin rock temperature of well 6407/4-1. This difference of 200 m is within the resolution of such data (see above). It can therefore not be concluded that the vitrinite reflectances in well 6407/4-1 are anomalous based on the data we present.

It appears from Fig. 8 that the vitrinite vs. depth curve which was modelled with a constant heat flow is about 200 m shallower than the standard curves. This difference is what one would expect from the elevated virgin rock temperatures of well 6407/4-1. The vitrinite reflectance which was modelled with the time-varying heat flow of Fig. 7, seems to overestimate the vitrinite reflectance relative to the standard curves by about 200 m when the differences in subsurface temperature are accounted for.

The constant heat flow model apparently gives a better fit to the standard curves for vitrinite reflectance at Haltenbanken than does the transient heat flow model. The latter, however, gives the best fit to the data from the well which was actually modelled. Whether these observations demonstrate superiority of one of the two heat flow models depends on the accuracy of the measurements which form the basis of the standard curves, the vitrinite reflectance values of well 6407/4-1, and the modelled vitrinite reflectance as performed by the EasyR_o program.

The EasyR_o programme is a simplified version of the VITRIMAT programme which was described by Burnham and Sweeney (1989). This program calculates the evolution of vitrinite based on kinetic modelling of transformation of coal to water, methane, other hydrocarbons and carbon dioxide. The kinetic parameters for these reaction are however only approximately known. The best constrained of these reactions seems to be that from coal to methane. The kinetics for this reaction were derived from pyrolysis experiments. The inversion of pyrolysis data to give kinetic parameters has an inherent uncertainty of about 300 m when extrapolated to burial depths in basins (Espitalié et al., this volume). This estimate does not include assessments of the inaccuracy of extrapolation by 6–10 orders of magnitude in geological time. Such extrapolation is performed when kinetic parameters from laboratory experiments are applied to simulate kerogen degradation under geologic conditions.

The EasyR_o and VITRIMAT models have been verified by application to several wells where vitrinite data were available. These wells cover a variety of heating rates, ranging from 10°C/yr (intrusive metamorphism), through 1°C/1000 yr (geothermal systems) to burial diagenesis with heating rates as low as 0.5°C/m.y. While the good fit to measured vitrinite reflectance from such varying heating rates verifies the model in a qualitative sense, it is not possible to use these results to determine the accuracy of the modelled vitrinite reflectance. Such estimates require independent error estimates of the thermal histories and the vitrinite reflectance data that were used for calibration, and the resulting accuracy estimates should not be extrapolated beyond the data points. An inspection of the accuracy of the thermal histories that were applied in the "verification wells" (with heating rates from normal burial conditions) nevertheless gives information on the demonstrated accuracy of the modelled vitrinite reflectance. The applied thermal histories for the "verification wells" have the following problems.

Cost No. 1 well, Gulf of Mexico

The thermal history for this well, which was also used to verify the model of Larter (1989), was calculated from a surface temperature of 64°F (17.8°C), and a constant geothermal gradient of 1.75°F/100 ft (31.9°C/km) (Waples, 1981). This gradient was determined from BHT measurements. Four BHT measurements from this well (corrected for mud circulation effect, correction procedure not given) are shown by Lerche (1990, p. 218). They show a significant (10–20°C) spread relative to the geothermal gradient given by Waples (1981). The fit between measured and modelled vitrinite reflectance is generally within ±500 m.

The simplified thermal history and the large spread in BHT and vitrinite reflectance data hardly warrant a better estimate of the accuracy of the EasyR_o model than ±500 m (±15°C) based on the Cost No. 1 well only.

Wagon Wheel No. 1, Wyoming

The temperature data of this well are of very high quality (Spencer, 1987), and the modelled vitrinite reflectance appears to be within ±100 m of the majority of the measured data. However, the estimates of post-Eocene erosion in this well vary from negligible (Schuster and Steidmann, 1983) to 900–1500 m (Pollastro and Barker, 1985). Such uncertainties make the well unsuitable for calibration of maturity indicators, as the accuracy of the depth to a given R_o value is at least as uncertain as the spread of the estimates of eroded thickness.

Data from Quigley et al., 1987

These data are from areas with different heating rates (0.5°C/m.y. for the Scotian Shelf, 1°C/m.y. for the North Sea and 30–50°C/m.y. for the Pannonian Basin). The vitrinite vs. depth data show a significant scatter, as would be expected from plotting of a large number of different wells on one figure. The modelled vitrinite reflectance is generally within 700 m (20°C) of the measured data. These data hardly validate the maturation model to a higher accuracy than ±500 m, a.o. because of the simplified and very general thermal histories (constant heating rate through time for large areas), lack of control on virgin rock temperatures, and large spread of vitrinite reflectance data.

Data from Barker, 1989

Vitrinite reflectance data from a large number of wells were compiled by Barker (1989). These were plotted vs. maximum temperature by Sweeney and Burnham (1990). The vitrinite reflectance data scatter substantially. The maximum temperature to R_o = 0.6 varies from 50 to 200°C, which corresponds to a depth range of about 5 km. The modelled vitrinite reflectance appears to satisfy the average of the measured data within ±20°C (±700 m).

The thermal histories which Sweeney and Burnham (1990) used to calculate the vitrinite reflectance of Barker's (1989) data were determined from Barker's functional heating time estimates (the time at which the temperature is within 15°C of the maximum). The maximum temperatures of Barker (1989) were derived from BHT measurements, but most of these are uncorrected. These temperature data are therefore biased towards low values by 5–40°C (Hermanrud, 1988), which corresponds to a depth range of 0.2–1.2 km. The accuracy of the EasyR_o programme cannot be judged to be better than ±700 m based solely on the fit to the data given by Barker (1989).

Data from Dow, 1977

The thermal history used by Sweeney and Burnham (1990) to simulate these data were taken from Hunt (1979), who assumed a 10°C surface temperature and a 25°C/ km geothermal gradient for the Louisiana Gulf Coast. The source of the temperature data, and whether/how they were corrected, was not given by Hunt (1979). This common geothermal gradient for the Cretaceous, Eocene and Miocene data sets is not likely to be of high accuracy, once again hardly better than ±500 m at the depth where R_o = 1.0. The fit between calculated (by EasyR_o) and measured vitrinite data is not at all good (only within 2 km) when the constant geothermal gradient of 25°C is applied. The fit does however improve to ±250 m when the geothermal gradient is varied in time. Sweeney and Burnham (1990) may be right in that the proposed variations in geothermal gradients are "geologically reasonable". However, these proposed changes have not been independently determined, but rather appear to be chosen to achieve a fit between measured and modelled vitrinite reflectance. The fit to the data of Dow (1977) therefore do not give new information on the accuracy of the $EasyR_o$ model.

The $EasyR_o$/VITRIMAT models have also been verified by comparison to previously suggested models, such as those of Hood et al. (1975), Bostick et al. (1978), Waples (1980), Middleton (1982), Wood (1988) and Larter (1989). These models are, however, often determined from theoretical considerations and are generally calibrated to oversimplified thermal histories, based on inaccurate and probably biased temperature data. The agreement between $EasyR_o$/VITRIMAT and the above mentioned models accordingly does not warrant higher accuracy than is suggested from the other studies mentioned above.

In summary, it appears that the $EasyR_o$ model, as well as other vitrinite reflectance models that we are aware of, are calibrated to thermal histories which are either based on simplified concepts (like geothermal gradients) or inferior and often biased temperatures (in many cases uncorrected BHT data), and vitrinite measurements which lack documentation of quality and reproducibility. Although several different data sets have been reproduced to within 100–300 m, the above considerations do suggest that the $EasyR_o$ model has so far only been verified to a higher accuracy than ± 500 m (the $\pm$ range refers to a 95% confidence interval). The widespread use and general acceptance of this method in the oil industry imply that the accuracy of the method is thought to be substantially better than ± 500 m. The arguments put forward in this paper do not disprove such a conclusion; they merely state that high accuracy has not yet been properly demonstrated for any vitrinite reflectance models. They also suggest that any bias of the vitrinite reflectance models may be towards too high modelled values, since most of the thermal histories that were used for verification rely on BHT data. Such data are biased towards too low values even when Horner plot (or most other commonly used) corrections are made (Hermanrud, 1988).

Several authors using steady-state basin modelling programs have introduced recent heat flow spikes in their heat flow history models to derive a better fit between observed and modelled vitrinite reflectance. The validity of introducing such heat flow spikes is discussed in depth by Jensen and Doré (this volume), who conclude that this heat flow spike is probably a modelling artifact in the Haltenbanken area. A vitrinite reflectance model which is biased towards too high modelled values for a given temperature history will need a present-day heat flow spike to fit both measured vitrinite reflectance data and the true formation temperature. The fact that the vitrinite reflectance models currently in use show good fit to the (too cold) thermal histories of the "verification wells" suggests that these models are indeed biased towards too high modelled values of R_o. We thus conclude that the introduction of a recent heat flow pulse at Haltenbanken to improve the fit between measured and modelled vitrinite reflectance is unwarranted.

Conclusions

Subsurface temperatures exhibit significant lateral variations in the Haltenbanken area, giving different geothermal gradients in neighbouring wells. However, thermal modelling of five wells in the area showed that 90% of the differences could be accounted for using the same end-member rock characteristics with the same boundary conditions but allowing for minor adjustments of permeability to match fluid pressures. These results show that the thermal model that was applied does indeed take account of the majority of the processes that are responsible for the varying subsurface temperatures.

Modelling of the transient thermal effects of the rapid subsidence and sedimentation in the Pliocene and Pleistocene showed that these events created a thermal disequilibrium which resulted in a depressed heat flow that is still quite pronounced. The modelled kerogen transformation ratios turned out to be insensitive to these transient thermal effects. The transient thermal model did, however, give a slightly better fit to vitrinite reflectance data than that of the steady state model. The combined uncertainty of modelled and measured vitrinite reflectance do not, however, substantiate any conclusion on the relative merits of the two heat flow scenarios based on comparisons between modelled and measured vitrinite reflectance.

The vitrinite reflectance models we are aware of are verified by fitting modelled vitrinite reflectance based on inferior thermal histories to vitrinite data of unknown quality. There is a definite need to further constrain the accuracy of vitrinite reflectance modelling.

Acknowledgements

We hereby thank Statoil for permission to publish this paper. The work has benefitted substantially from critical reviews by two anonymous referees and from Rebecca Jensen and Tony Doré of Conoco Norway Inc.

References

Barker, C.E., 1989. Temperature and time in thermal maturation of sedimentary organic matter. In: N.D. Naeser and T.M. McCulloh (Editors), Thermal History of Sedimentary Basins

— Methods and Case Histories. Springer, New York, N.Y., pp. 75–98.

Bostick, N.H., Cashman, S.M., McCulloh, T.M. and Waddell, C.T., 1978. Gradients of vitrinite reflectance and present temperature in the Los Angeles and Ventura basins, California. In: D.F. Oltz (Editor), Low Temperature Metamorphism of Kerogen and Clay Minerals. Pacific Section SEPM Special Symposium, pp. 65–96.

Bukovics, C. and Ziegler, P.A., 1985. Tectonic development of the Mid-Norway continental margin. Mar. Pet. Geol., 2: 2–22.

Burnham, A.K. and Sweeney, J.J., 1989. A chemical model of vitrinite maturation and reflectance. Geochim. Cosmochim. Acta, 53: 2646–2657.

Dow, W.G., 1977. Kerogen studies and geological interpretations. J. Geochem. Explor., 1: 79–99.

Espitalié, J., Marquis, F. and Drouet, S., 1993. Critical study of kinetic modelling parameters. In: A.G. Doré et al. (Editors), Basin Modelling: Advances and Applications. Norwegian Petroleum Society (NPF), Special Publication 3. Elsevier, Amsterdam, pp. 233 –242 (this volume).

Fagerland, N., 1990. Mid-Norway shelf-hydrocarbon habitat in relation to tectonic elements. Nor. Geol. Tidsskr., 70: 65–70.

Goll, R.M. and Hansen, J.W., 1992. The Cenozoic sequence stratigraphy of the Halten Terrace and the Outer Vøring Plateau based on seismic and biostratigraphic data. Nor. Geol. Tidskr., 72: 295–299.

Hermanrud, C., 1988. Determination of Formation Temperature from Downhole Measurements. Ph.D. Dissertation, University of Columbia, Columbia, S.C., 203 pp.

Hermanrud, C., 1993. Basin modelling techniques — an overview In: A.G. Doré et al. (Editors), Basin Modelling: Advances and Applications. Norwegian Petroleum Society (NPF), Special Publication 3. Elsevier, Amsterdam, pp. 1–34 (this volume).

Hermanrud, C., Cao, S. and Lerche, I., 1990. Estimates of virgin rock temperature derived from BHT measurements: bias and error. Geophysics, 55: 924–931.

Hermanrud, C., Eggen, S. and Larsen, R.M., 1991a. Investigation of the thermal regime of the Horda platform by basin modelling; implications for the hydrocarbon potential of the Stord basin, northern North Sea. In: A.M. Spencer (Editor), Generation, Accumulation and Production of Europe's Hydrocarbons. Spec. Publ. EAPG 1. Oxford University Press, Oxford, pp. 65–73.

Hermanrud, C., Lerche, I. and Meisingset, K., 1991b. Determination of virgin rock temperatures from drillstem tests. J. Pet. Technol., 43: 1126–1131.

Heum, O.R., Dalland, A. and Meisingset, K.K., 1986. Habitat of hydrocarbons at Haltenbanken (PVT-modelling as a predictive tool in hydrocarbon exploration). In: A.M. Spencer (Editor), Habitat of Hydrocarbons on the Norwegian Continental Shelf. NPS/Graham and Trotman, London, pp. 259–274.

Hood, A., Gutjahr, C.M. and Peacock, R.L., 1975. Organic metamorphism and the generation of petroleum. Am. Assoc. Pet. Geol., Bull., 59: 986–996.

Hunt, J.M., 1979. Petroleum Geochemistry and Geology. W.H. Freeman, San Francisco, Calif., 617 pp.

Jensen, L.N., Riis, F. and Boyd, R. (Editors), 1992. Post-Cretaceous uplift and sedimentation along the western Fennoscandian Shield. Proceedings of the 7th TSGS Conference held in Stavanger, October 1990. Nor. Geol. Tidskr., 72 (3): 221–338.

Jensen, R.P. and Doré, A.G., 1993. A recent Norwegian shelf heating event — fact or fantasy? In: A.G. Doré et al. (Editors), Basin Modelling: Advances and Applications. Norwegian Petroleum Society (NPF), Special Publication 3. Elsevier, Amsterdam, pp. 85–106 (this volume).

Katz, B.J., Pheifer, R.N. and Schunk, D.J., 1988. Interpretation of discontinuous vitrinite reflectance profiles. Am. Assoc. Pet. Geol., Bull., 72: 926–931.

Khorasani, G.K., 1989. Factors controlling source rock potential of the Mesozoic coal-bearing strata from offshore Central Norway: applications to petroleum exploration. Bull. Can. Pet. Geol., 37: 417–427.

Larter, S.R., 1989. Chemical models of vitrinite reflectance evolution. Geol. Rundsch., 78: 349–359.

Lerche, I., 1990. Basin Analysis: Quantitative Methods, 2. Academic Press, San Diego, Calif., 570 pp.

Middleton, M.F., 1982. Tectonic history from vitrinite reflectance. Geophys. J. R. Astron. Soc., 68: 121–132.

Morgan, P., 1983. Constraints on rift thermal processes from heat flow and uplift. Tectonophysics, 94: 277–298.

Odden, W., 1986. A Study of Coal from Haltenbanken. Cand. Sci. (M.S.) Thesis, University of Bergen, Bergen, 222 pp.

Pollastro, R.M. and Barker, C.E., 1985. Geothermometry from clay minerals, vitrinite reflectance, and fluid inclusions — applications to the thermal and burial history of rocks cored from the Wagon Wheel #1 well, Green River Basin, Wyoming. U.S. Geol. Surv., Open-File Rep., 85-0548: 78–94.

Quigley, T.M., Mackenzie, A.S. and Gray, J.R., 1987. Kinetic theory of petroleum generation. In: B. Doligez (Editor), Migration of Hydrocarbons in Sedimentary Basins. Technip, Paris, pp. 649–665.

Schuster, M.W. and Steidmann, J.R., 1983. Origin and development of northern Green River Basin — a stratigraphic and flexural study. Am. Assoc. Pet. Geol., Bull., 67: 1356.

Soave, G., 1972. Equilibrium constants from a modified Redlich-Kwong equation of state. Chem. Eng. Sci., 27: 1197–1203.

Spencer, C.W., 1987. Hydrocarbon generation as a mechanism for overpressuring in Rocky Mountain region. Am. Assoc. Pet. Geol., Bull., 71: 368–388.

Steffensen, R.J. and Smith, R.C., 1973. The importance of Joule-Thomson heating (or cooling) in temperature log interpretation. Soc. Pet. Eng. AIME, Pap. 4636.

Sweeney, J.J. and Burnham, A.K., 1990. Evaluation of a simple model of vitrinite reflectance based on chemical kinetics. Am. Assoc. Pet. Geol., Bull., 74: 1559–1570.

Throndsen, T., 1990. Scatter in vitrinite reflectance data from coal samples in North Sea wells. Nor. Geol. Tidsskr., 70: 223–228.

Tissot, B.P., Pelet, R. and Ungerer, P., 1987. Thermal history of sedimentary basins, maturation indices and kinetics of oil and gas generation. Am. Assoc. Pet. Geol., Bull., 71: 1445–1466.

Turcotte, D.L., 1980. On the thermal evolution of the earth. Earth Planet. Sci. Lett., 48: 53–58.

Waples, D.W., 1980. Time and temperature in petroleum formation: application of Lopatin's method to petroleum exploration. Am. Assoc. Pet. Geol., Bull., 64: 916–926.

Waples, D.W., 1981. Organic Geochemistry for Exploration Geologists. Burgess Publishing, Minneapolis, Minn., 157 pp.

Wood, D.A., 1988. Relationships between thermal maturation indices calculated using Arrhenius equation and Lopatin method: implications for petroleum exploration. Am. Assoc. Pet. Geol., Bull., 72: 115–134.

EIRIK VIK *Statoil A/S, 6500, Kristiansund N., Norway*
Present address: *Statoil Forskningssenteret, Postuttak, N-7004 Trondheim, Norway*

CHRISTIAN HERMANRUD *Statoil A/S, 6500, Kristiansund N., Norway*
Present address: *Statoil Forskningssenteret, Postuttak, N-7004 Trondheim, Norway*

Optimization of temperature history — aspects of vitrinite reflectance and sterane isomerization

Dag Hermansen

The dependence on time and temperature of some of the most common maturity indicators have been examined. The modelling of sterane isomerization and two equations for calculating vitrinite reflection (Modified Lopatin equation and Burnham/Sweeney equation) have been tested at variable heat flow histories and different heating rates.

Sterane isomerization and vitrinite reflection (Modified Lopatin equation) are shown to react very similarly to different temperature histories. Their ranges of activation energies almost coincide at moderate temperatures encountered in sedimentary basins, which make the pair of indicators less valuable for optimization purposes.

The Burnham/Sweeney equation operates with higher activation energies than sterane isomerization, and the differing response to different temperature histories makes this pair of indicators a potentially better optimizing tool. The important requirement for maturity indicators is separation of the activation energy levels.

Due to different kinetics and different calibration sets the two vitrinite reflection equations require different temperature levels to match a given vitrinite reflection distribution. At low and moderate heating rates the Burnham/Sweeney equation demands higher temperatures, usually effected by increasing the heat flow. A better agreement between palaeotemperatures based on the maturity indicator and present-day temperatures is noted, without introducing a recent heat flow peak. It is suggested that the choice of vitrinite reflectance equation may explain the recent heat flow anomalies quoted in the literature.

Introduction

Determination of rock temperatures during deposition and subsequent burial is of prime importance for the prediction of hydrocarbon generation. The temperature history is primarily dependent on surface temperatures, subsurface heat flow and thermal rock properties, through time.

The effect of temperature exposure over time can be measured by thermal indicators, which may be of organic or inorganic origin. A single thermal indicator cannot resolve the time and temperature effects, but the use of multiple indicators may aid the determination of the temperature history.

Vitrinite reflectance and sterane isomerization belong to the organic group of indicators, and are among the most commonly used. Experience with these indicators in Norsk Hydro has disclosed a certain insensitivity to the temperature history, and the aim of this paper is to investigate the effect of time and temperature on them.

Methods

Modelling of vitrinite reflectance employs different concepts. For more than a decade the temperature–time index (TTI) concept has been used (Lopatin, 1971; Waples, 1980). A correlation between TTI and R_o was derived empirically by Waples (1980), and slightly modified by M.A. Yukler (pers. commun., 1985). The latter correlation is used in this study, and the concept is referred to as "Modified Lopatin equation" in this paper.

Larter (1989) and Burnham and Sweeney (1989) have recently published new approaches to vitrinite reflectance modelling. Both models employ the kinetic concept and make use of distributed activation energies. The vitrinite degradation is calculated as parallel first-order differential equations. This paper focuses on the approach of Burnham and Sweeney, and their model is referred to as "Burnham/Sweeney equation". The activation energies range from 150 to 300 kJ/mol (35–70 kcal/mol), and the frequency factor is fixed at 3.15×10^{26} Ma^{-1}.

The sterane isomerization is based on the kinetic approach of MacKenzie and McKenzie (1983), with an activation energy of 91 kJ/mol (21.75 kcal/mol). A single frequency factor of 2.9×10^{11} Ma^{-1} is used.

Hydrocarbon generation is calculated according to the kinetic method of Tissot and Espitalié (1975). Concerning the type II kerogen degradation, the distribution of activation energies is very broad, and dif-

Basin Modelling: Advances and Applications edited by A.G. Doré et al. NPF Special Publication 3, pp. 119–126, Elsevier, Amsterdam.

ferent frequency factors accompany each activation energy level. An important level, probably responsible for most of the oil generation, appears at 209 kJ/mol (50 kcal/mol).

Norsk Hydro's basin model is conceptually based on a commercial model (Yukler and Welte, 1980), but its mathematical part is significantly changed by Norsk Hydro. It belongs to the forward modelling class of deterministic models, and accepts a set of input data for each chronological event. The external heat is provided as heat flow through the basin floor.

Results and discussion

Effect of heat flow variations at a point in a real basin

A relatively deep point in the North Sea was modelled with Norsk Hydro's 1D basin model. Input data for the modelled point are presented in Table 1.

The heat flow history was deliberately manipulated for testing purposes, to examine the sensitivity of the thermal indicators to temperature history. Two scenarios which differ only with respect to the heat flow history were run. A constant heat flow through time was contrasted with a strongly variable heat flow history (Fig. 1). The two scenarios result in very different temperature histories, shown for an arbitrary layer (layer 9) in Fig. 2. There is considerable difference between the subsidence-driven heating rate in the constant heat flow scenario, and the extreme heating rate caused by the heat flow increase in the variable heat flow scenario.

The very different temperature histories did *not* appear in the present-day maturity profiles, i.e. the vitrinite reflectance (Modified Lopatin equation) and sterane isomerization depth trends (Fig. 3). The maturity curves were almost identical, and certainly within the likely standard deviation of real measurements.

The calculation of vitrinite reflectance using the Burnham/Sweeney equation produced clearly distinguishable depth trends as a response to the different temperature histories (Fig. 3).

TABLE 1

Input data for modelling of a North Sea Basin point (heat flow column is given for both the constant and variable heat flow scenarios)

Event No.	Cum. time (Ma)	Duration (Ma)	Lithology[a]	Thickness (m)	Cum. depth (m)	Porosity	Water depth (m)	Surface temperature (°C)	Heat flow (mW/m^2)	
									const.	var.
30	1.0	1.0	14	70.0	190.0	0.50	120.0	7.0	46.0	54.4
29	2.0	1.0	14	90.0	280.0	0.45	100.0	7.0	46.0	54.4
28	3.0	1.0	5	140.0	420.0	0.42	100.0	8.0	46.0	58.6
27	5.0	2.0	5	140.0	560.0	0.39	100.0	9.0	46.0	58.6
26	10.0	5.0	37	50.0	610.0	0.40	50.0	10.0	46.0	58.6
25	15.0	5.0	37	50.0	660.0	0.39	50.0	10.0	46.0	20.9
24	20.0	5.0	37	0.0	660.0	0.32	0.0	11.0	46.0	20.9
23	25.0	5.0	37	50.0	710.0	0.38	100.0	10.0	46.0	20.9
22	30.0	5.0	5	200.0	910.0	0.33	200.0	10.0	46.0	20.9
21	35.0	5.0	14	280.0	1190.0	0.26	400.0	9.0	46.0	20.9
20	40.0	5.0	10	280.0	1470.0	0.20	500.0	9.0	46.0	20.9
19	45.0	5.0	10	300.0	1770.0	0.16	500.0	9.0	46.0	20.9
18	50.0	5.0	10	315.0	2085.0	0.16	400.0	10.0	46.0	20.9
17	55.0	5.0	37	245.0	2330.0	0.24	300.0	10.0	46.0	20.9
16	60.0	5.0	37	130.0	2460.0	0.23	300.0	10.0	46.0	20.9
15	65.0	5.0	3	130.0	2590.0	0.14	400.0	11.0	46.0	20.9
14	70.0	5.0	45	90.0	2680.0	0.15	200.0	12.0	46.0	20.9
13	75.0	5.0	45	90.0	2770.0	0.14	300.0	12.0	46.0	20.9
12	80.0	5.0	10	100.0	2870.0	0.14	300.0	13.0	46.0	20.9
11	85.0	5.0	10	200.0	3070.0	0.14	400.0	13.0	46.0	20.9
10	90.0	5.0	10	320.0	3390.0	0.13	500.0	13.0	46.0	20.9
9	95.0	5.0	10	300.0	3690.0	0.13	400.0	14.0	46.0	20.9
8	100.0	5.0	10	200.0	3890.0	0.12	200.0	15.0	46.0	20.9
7	105.0	5.0	10	200.0	4090.0	0.12	200.0	16.0	46.0	20.9
6	110.0	5.0	10	200.0	4290.0	0.12	200.0	16.0	46.0	20.9
5	115.0	5.0	10	100.0	4390.0	0.12	100.0	17.0	46.0	20.9
4	120.0	5.0	10	100.0	4490.0	0.12	100.0	18.0	46.0	20.9
3	125.0	5.0	3	100.0	4590.0	0.12	50.0	18.0	46.0	20.9
2	130.0	5.0	45	80.0	4670.0	0.12	50.0	18.0	46.0	20.9
1	135.0	5.0	18	10.0	4680.0	0.12	0.0	19.0	46.0	20.9

[a] 3 = sandy shale; 5 = shaly sandstone; 10 = silty shale; 14 = sand and shale; 18 = carbonaceous shale; 37 = silty sandstone; 45 = marl.

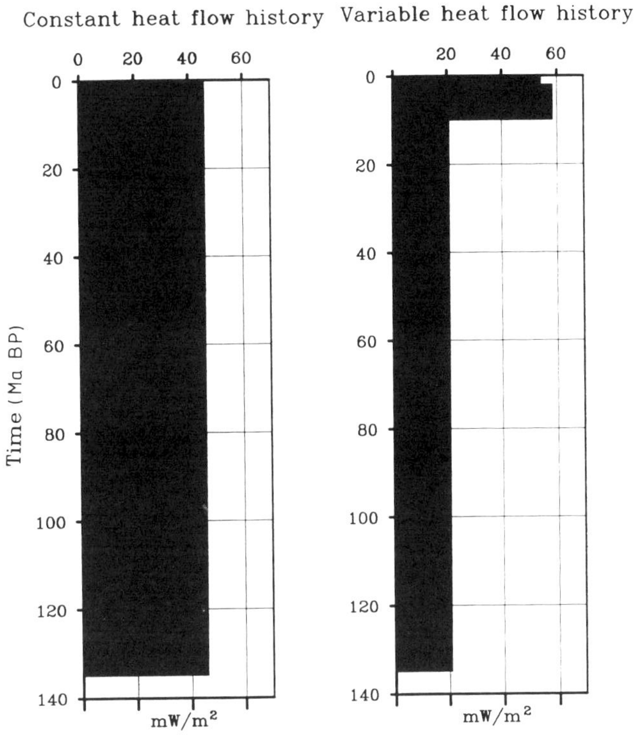

Fig. 1. Distribution of heat flow through time for the constant and variable heat flow scenarios.

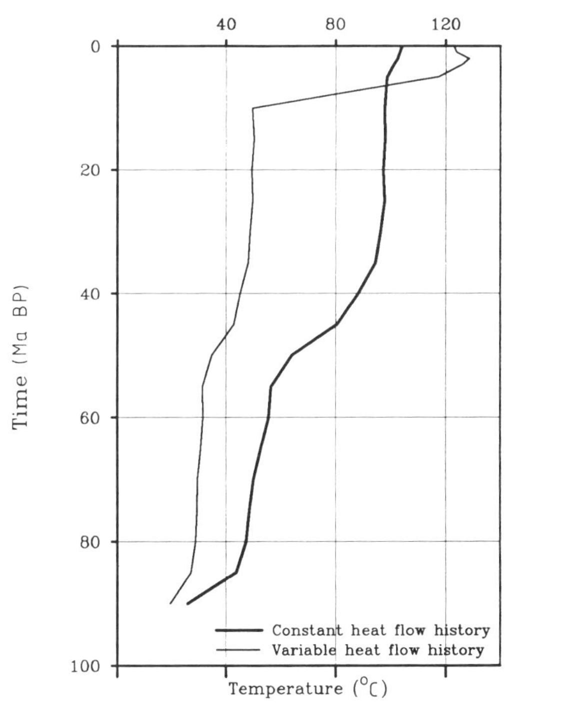

Fig. 2. Temperature history of layer 9 for the constant and variable heat flow scenarios.

Layer 9 was extracted from the depth plot, and its oil generation and maturity development through time is shown in Fig. 4. There is considerable difference between the two scenarios as regards timing of maturity, although both the vitrinite reflectance (Modified Lopatin equation) and sterane isomerization curves nearly merge towards present time. It should be noted that vitrinite reflectance (Burnham/Sweeney equation) is higher, and more oil has been generated, in the variable heat flow scenario due to the generally higher activation energy level of both reaction equations.

This exercise shows that the vitrinite reflectance (Modified Lopatin equation) and sterane isomerization do not adequately resolve the temperature history. Modelling results can be greatly in error unless a firm geological constraint is put on the heat flow history. The main reason for this insensitivity to the temperature history lies in the almost coinciding activation energy levels of the vitrinite reflectance (Modified Lopatin equation) and the current sterane isomerization equation (Fig. 5). The activation energy of the former changes with temperature, while the activation energy of the sterane isomerization reaction remains fixed at 91 kJ/mol (21.75 kcal/mol). The curves intersect at 130°C and stay fairly close within an interval of moderate temperatures (80–160°C), important for the evolution of maturity indicators. Due to the narrow activation energy range, the two maturity indicators are unable to distinguish between even extremely different temperature histories.

The vitrinite reflectance equation of Burnham and Sweeney (1989) employs activation energies far higher than the current sterane isomerization reaction (Fig. 5). It complements the sterane isomerization, thus making the pair of indicators sensitive to temperature history. Given a certain match of calculated to measured sterane isomerization values, the adjustment of calculated and measured vitrinite reflectance values can be accomplished with different temperature histories. The two existent scenarios allow an adjustment within 0.1–0.3% R_o for R_o levels above 0.5% (Fig. 3).

Effect of heating rate variations

Two artificial scenarios have been run to further investigate the effect of heating rate on the modelling of maturity indicators. Constant heating rates of 0.8°C/Ma (0–260°C in 300 m.y.) and 12.5°C/Ma (47–322°C in 22 m.y.) were generated, which embody most heating rates encountered in sedimentary basins. The time development of vitrinite reflectance (Modified Lopatin equation) for one layer is plotted against corresponding sterane isomerization (Fig. 6, stippled lines). The curve separation is poor, indicating that the pair of maturity indicators are not sensitive to different heating rates. To accomplish this, maturity indicators with distinctly different activation energies are needed.

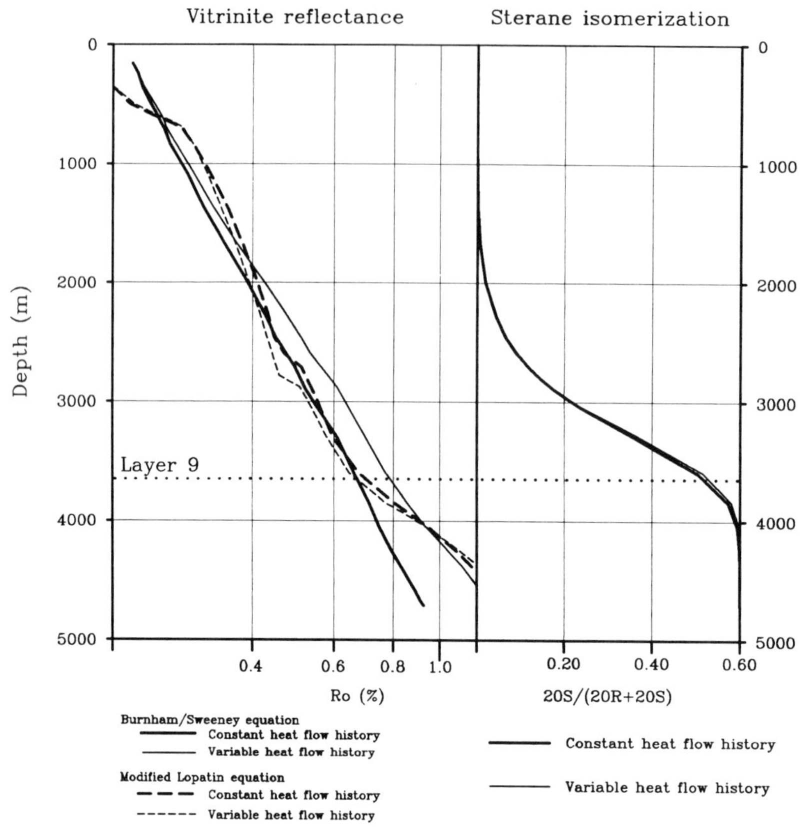

Fig. 3. Vitrinite reflectance (Modified Lopatin equation and Burnham/Sweeney equation) and sterane isomerization depth trends for the constant and variable heat flow scenarios.

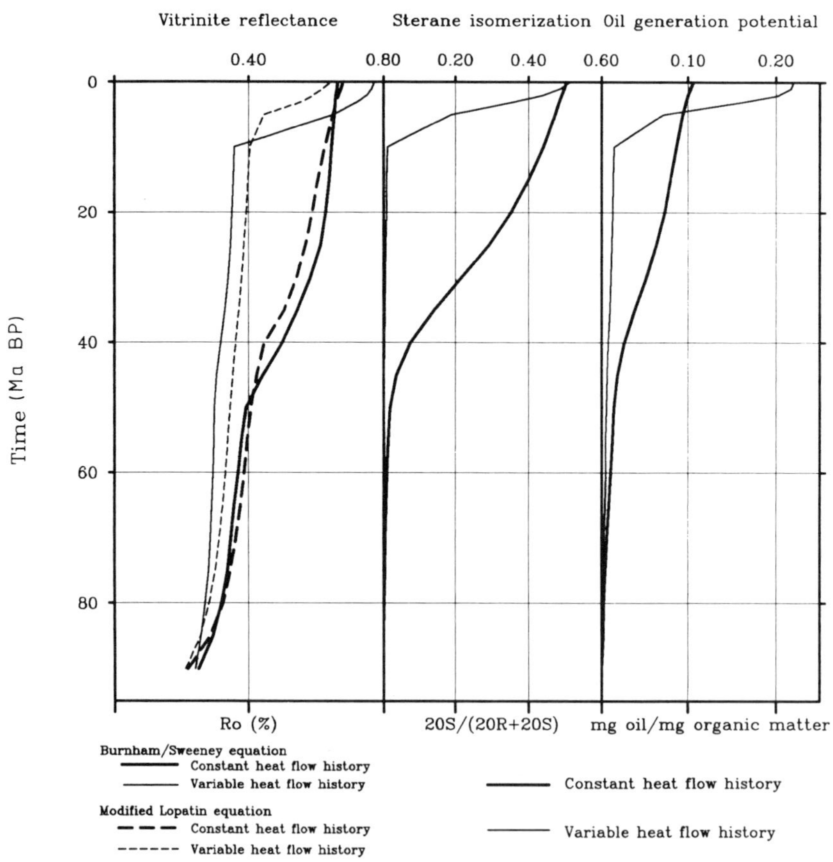

Fig. 4. Time development of vitrinite reflectance (Modified Lopatin equation and Burnham/Sweeney equation), sterane isomerization and oil generation potential, for layer 9, constant and variable heat flow scenarios.

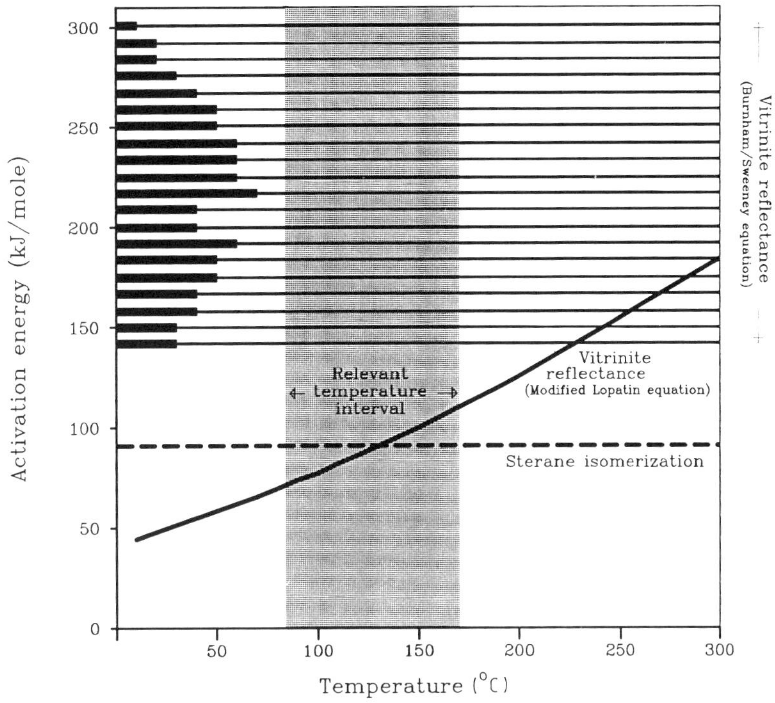

Fig. 5. Activation energy and temperature relations for sterane isomerization and vitrinite reflectance (Modified Lopatin equation and Burnham/Sweeney equation).

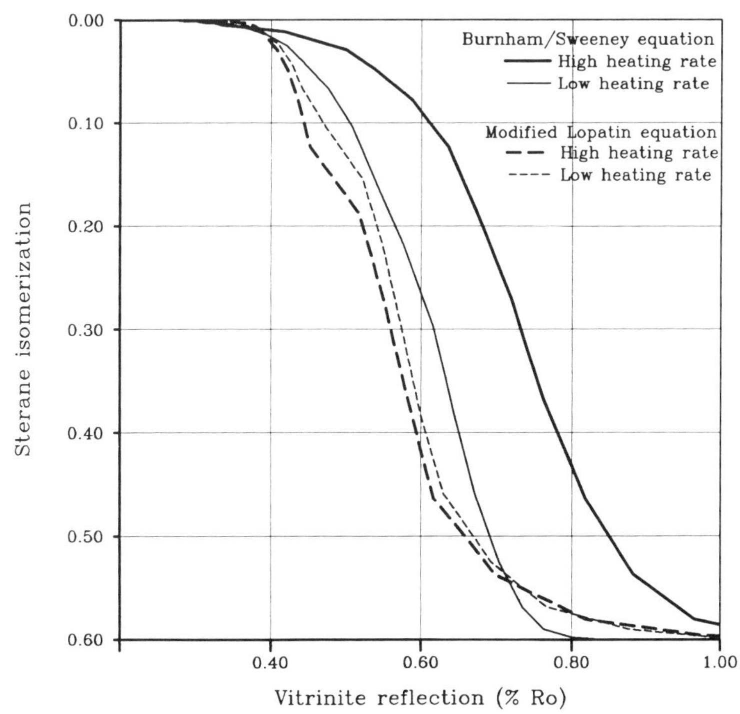

Fig. 6. Sterane isomerization related to vitrinite reflectance (Modified Lopatin equation and Burnham/Sweeney equation) for two different heating rates (12.5° and 0.8°C/Ma).

The R_o-values resulting from calculation with the Burnham/Sweeney equation (same procedure as above) are plotted against sterane isomerization (Fig. 6, solid lines). The wide separation of the curves implies a potential for detecting erroneous temperature histories.

The ability of maturity indicators to resolve different temperature histories depends on the variance in activation energies, disregarding the actual energy level. This was also noted by MacKenzie and McKenzie (1983) in their work on aromatization and isomerization reactions. In this context the sterane isomerization and vitrinite reflection (Burnham/ Sweeney equation) is a suitable pair of maturation indicators. Their relative sensitivity to time and temperature is different, and thus they complement each other.

Implications for the temperature optimization process

The Modified Lopatin equation and Burnham/ Sweeney equation calculate vitrinite reflection differently due to different concepts and calibration of the model equations. The depth–maturity trend for a given set of input data is different, and the result from the two mentioned heating rate scenarios is shown in Fig. 7. The deviation is dependent on the heating rate. As the Burnham/Sweeney equation is more sensitive to temperature, it models higher maturity relative to the Modified Lopatin equation at high heating rates. At high heating rates the Burnham/Sweeney equation calculates the highest reflectance below R_0-values of 1.00%, above which the curves are very similar. At low heating rates the Modified Lopatin equation attains the highest values, and the deviation is pronounced at R_0-values above 0.65%.

The different responses of the two equations is important to bear in mind. If the temperature optimization process is based on either of the equations, the adjustment of calculated and measured vitrinite reflection values will require different palaeotemperature levels. Heating rates on the Norwegian shelf vary in time and space, but average rates tend to lie in the lower part of the heating rate range bracketed by the two artificial scenarios. Consequently, to adjust the calculated R_0-values to measured vitrinite reflectance, the Burnham/Sweeney equation requires higher palaeotemperatures. Depending on the model in use, this can be accomplished by increasing the heat flow or increasing the temperature at the base of the lithosphere.

The heat flow history is an important aspect of the optimization process. The simplest approach is a constant heat flow through time, which may be appropriate in unexplored areas. The heat flow history is normally tied to main geological events. Heat flow pulses may be assigned to major rift phases, following the concept of McKenzie (1978), after which an exponentially decaying heat flow is applied.

Frequently, the temperature history governed by the maturity indicators is not in accordance with the subsurface temperature distribution at present, unless a recent heat flow pulse is applied. This often quoted recent heat flow increase has been explained by various mechanisms: glaciation, overpressure, inadequate modelling algorithms, biased calibration of

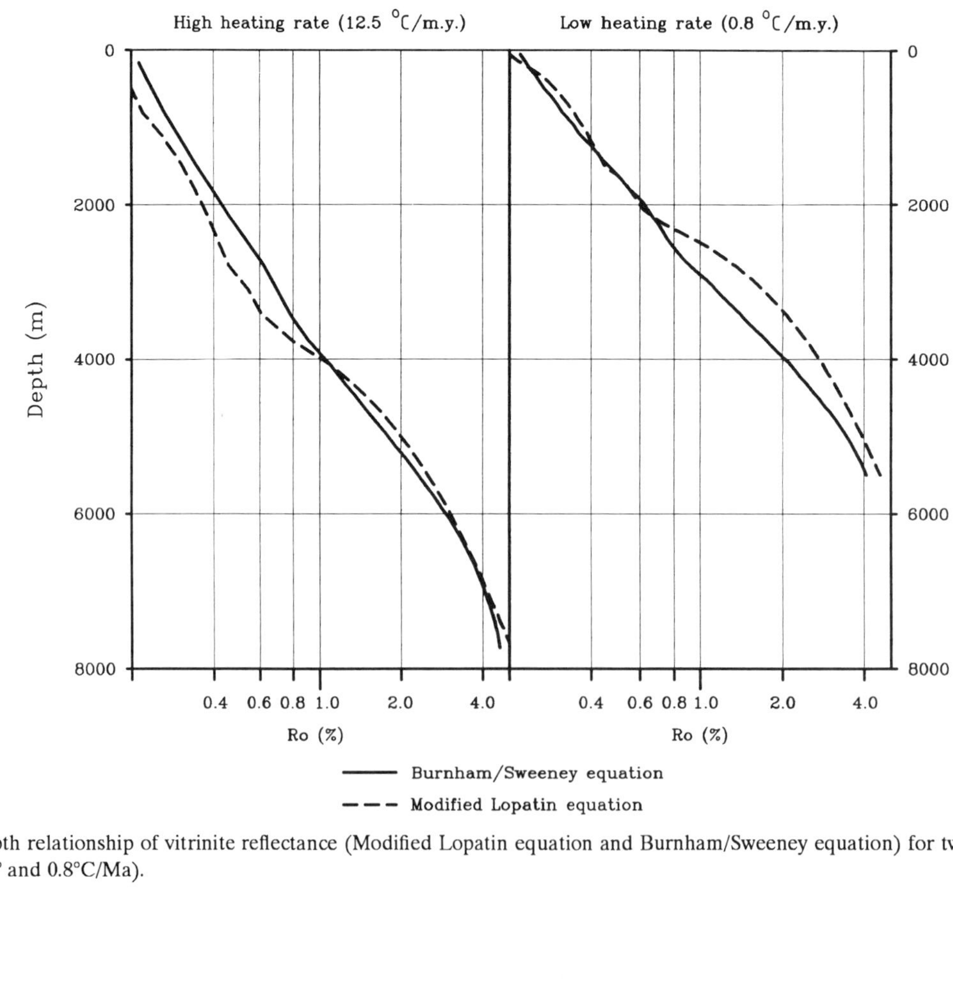

Fig. 7. Depth relationship of vitrinite reflectance (Modified Lopatin equation and Burnham/Sweeney equation) for two different heating rates (12.5° and 0.8°C/Ma).

vitrinite reflectance models, etc. (Dahl et al., 1987; Dahl and Augustson, this volume; Jensen and Doré, this volume; Vik and Hermanrud, this volume). I suggest that part of the heat flow anomaly may be related to the current method of temperature history optimization. To my knowledge several of the above-mentioned authors employ the Modified Lopatin equation for the modelling of vitrinite reflectance. In most cases, the alternative approach of Burnham/Sweeney will require higher palaeotemperatures to adjust the R_o-values, and consequently the match between present-day modelled and measured temperatures will be higher, without introducing a recent heat flow peak. A reasonable match of both vitrinite reflectance and present-day temperatures has been achieved by Vik and Hermanrud (this volume) by using a constant temperature at the base of the lithosphere, equivalent to a constant heat flow at the same point. They applied the EasyR_o programme, which is a simplified version of the Burnham/Sweeney concept. Similar results have been confirmed by Norsk Hydro in some wells on Haltenbanken and in the northern North Sea. Further testing should, however, be made to verify the regional consistency of the Burnham/Sweeney equation.

In a review of the calibration set for the Burnham/Sweeney equation, Vik and Hermanrud (this volume) question the accuracy of the thermal histories of the "verification wells" for the EasyR_o model. They suggest that the model overestimates the R_o-values due to calibration against too low temperatures, as do most other vitrinite reflectance models. Consequently, a recent heat flow spike may be needed to fit both measured vitrinite reflection data and true formation temperatures.

Nevertheless, at low heating rates, the Modified Lopatin equation overestimates the modelled R_o-values to an even higher degree (Fig. 7), often demanding a distinct heat flow spike. The problem may not be entirely solved by applying the Burnham/Sweeney equation, but a considerable improvement is noted.

Vik and Hermanrud (this volume) modelled thermal effects of rapid subsidence during the Pliocene and Pleistocene on Haltenbanken. The results showed a drop in heat flow during the period of transient temperature regime. Confusion may arise if this modelled temporary heat flow pattern is contrasted to the pattern of input heat flow values of other basin models, which relate to steady-state conditions.

Conclusions

The effects of heat flow history and heating rate variations on maturity indicator models have been examined. The indicators considered are sterane isomerization and two equations for calculation of vitrinite reflectance (Modified Lopatin equation and Burnham/Sweeney equation).

The activation energy level of vitrinite reflection (Modified Lopatin equation) increases with temperature, but remains close to the level of sterane isomerization at moderate temperatures (80–170°C). Consequently, this pair of maturity indicators is virtually insensitive to changes in the temperature history. Exposure to very different heating rates (0.8–12.5°C/Ma) confirms their uniform behaviour.

The Burnham/Sweeney equation employs activation energies considerably higher than the sterane isomerization level. The two maturity indicators react differently to various temperature histories, which in this context makes them more suitable for optimization purposes.

The two vitrinite reflectance equations employ different kinetic parameters and also different calibration sets. The resulting depth trends vary according to the heating rate, and the deviation is most pronounced at low heating rates. This affects the optimization of the temperature history, and at relatively low heating rates the Burnham/Sweeney equation requires higher palaeotemperatures to match a given measured R_o-value. Usually, the heat flow is increased to accomplish the desired effect. Normally, the palaeotemperatures derived from the maturity indicator are then in accordance with the present-day temperatures without introducing a recent heat flow peak. It is suggested that the various heat flow anomalies quoted in the literature may, in part, originate from the different vitrinite reflectance equations used.

References

Burnham, A.K. and Sweeney, J.J., 1989. A chemical kinetic model of vitrinite maturation and reflectance. Geochim. Cosmochim. Acta, 53: 2649–2657.

Dahl, B. and Augustson, J.H., 1993. The influence of Tertiary and Quaternary erosion on hydrocarbon generation in Norwegian offshore basins. In: A.G. Doré et al.(Editors), Basin Modelling: Advances and Applications. Norwegian Petroleum Society (NPF), Special Publication 3. Elsevier, Amsterdam, pp. 419–431 (this volume).

Dahl, B., Nysæther, E., Speers, G.C. and Yukler, M.A., 1987. Oseberg area — integrated basin modelling. In: J. Brooks and K. Glennie (Editors), Petroleum Geology of North West Europe. Graham and Trotman, London, pp. 1029–1038.

Jensen, R.P. and Doré, A.G., 1993. A recent Norwegian Shelf heating event — fact or fantasy? In: A.G. Doré et al. (Editors), Basin Modelling: Advances and Applications. Norwegian Petroleum Society (NPF), Special Publication 3. Elsevier, Amsterdam, pp. 85 –106 (this volume).

Larter, S., 1989. Chemical models of vitrinite reflectance evolution. Geol. Rundsch., 78 (1): 349–359.

Lopatin, N.V., 1971. Temperature and geologic time as factors in coalification. Izv. Akad. Nauk SSSR, Ser. Geol., 10: 3–10 (in Russian).

MacKenzie, A.S. and McKenzie, D.P., 1983. Isomerization and aromatization of hydrocarbons in sedimentary basins formed by extension. Geol. Mag., 120: 417–470.

McKenzie, D.P., 1978. Some remarks on the development of sedimentary basins. Earth Planet. Sci. Lett., 40: 319–333.

Tissot, B. and Espitalié, J., 1975. L'évolution thermique de la matière organique des sédiments: applications d'une simulation mathématique. Rev. Inst. Fr. Pét., 30: 743–777.

Vik, E. and Hermanrud, C., 1993. Transient thermal effects of rapid subsidence in the Haltenbanken area. In: A.G. Doré et al. (Editors), Basin Modelling: Advances and Applications. Norwegian Petroleum Society (NPF), Special Publication 3. Elsevier, Amsterdam, pp. 107–117 (this volume).

Waples, D.W., 1980. Time and temperature in petroleum formation: application of Lopatin's method to petroleum exploration. Am. Assoc. Pet. Geol., Bull., 64: 916–926.

Yukler, M.A. and Welte, D.H., 1980. A three-dimensional deterministic dynamic model to determine geologic history and hydrocarbon generation and accumulation in a sedimentary basin. In: Fossil Fuels, 26th Int. Geol. Congr., Paris. Technip, Paris, pp. 267–285.

DAG HERMANSEN *Norsk Hydro Research Centre, Lars Hilles gt. 30, N-5028 Nygårdstangen, Bergen, Norway*

Comparison of different models for vitrinite reflectance evolution using laboratory calibration and modelling of well data

Torbjørn Throndsen, Bjørg Andresen and Åse Unander

The successful use of vitrinite reflectance as a check parameter for palaeotemperature reconstruction in basin modelling is dependent on precise calibration and selection of the correct model. We have tested and calibrated the recently developed models of Larter (1989) and Burnham and Sweeney (1989) by vitrinite reflectance analysis of samples from artificial heating experiments. The resulting laboratory-calibrated versions of the models were applied to wells from the northern North Sea, and calculated and measured vitrinite reflectance were compared. Comparison was also made to the widely used Lopatin (1971) model using the calibration between TTI and vitrinite reflectance of Waples (1984). This showed that the Lopatin (1971) model failed to work when rescaled from basin to laboratory conditions. The new models were more satisfactory in this respect, particularly the model of Burnham and Sweeney (1989) which showed an almost perfect fit with the observed data when extended from laboratory scale to basin scale conditions.

Introduction

Vitrinite reflectance is widely, and often uncritically, used as an organic maturity indicator and control parameter for palaeotemperature reconstruction in basin modelling. Mistakes that are commonly observed are related firstly to inadequate verification of the quality of measured data, and secondly to choice and calibration of model for calculation of vitrinite reflectance from a given time–temperature history. The calculation model is the objective of this study.

A number of models have been proposed to describe the dependence of vitrinite reflectance on time and temperature. The reader is referred to Sweeney and Burnham (1990) for an overview. The purpose of the present study is to test three of the models: namely, the widely used model of Lopatin (1971), and the recently developed models of Larter (1989) and Burnham and Sweeney (1989). The models were tested against vitrinite reflectance data from artificial heating experiments and real wells from the northern North Sea. Thermal histories for the latter were obtained from basin modelling. Furthermore, the models of Larter (1989) and Burnham and Sweeney (1989) were recalibrated against data from the heating experiments.

Experimental

Heating experiments

During the last ten years it has become widely accepted that thermal maturation of kerogen to form oil and gas can be simulated under laboratory conditions using closed vessel hydrous pyrolysis (Winters et al., 1981; Lewan, 1985). We adopted and slightly modified this approach in order to simulate vitrinite reflectance evolution using stepwise heating experiments with temperatures ranging from 150° to 500°C and durations up to 480 h.

Isothermal heating was also tried out in several experiments. However, stepwise or gradual heating starting at fairly low temperatures (150°C) appeared from microscopic examination of vitrinite to give a better approximation of natural processes. This is exemplified in Fig. 1 showing the histogram distribution of vitrinite reflectance values for two samples with approximately equal average reflectance. The values shown in Fig. 1A were obtained from a sample with isothermal heating at 350°C for 72 h. The average vitrinite reflectance after heating was $\%R_m = 1.03$. The distribution of values is relatively broad and erratic. The reason is that during heating the sample attained a spongy appearance, and the vitrinite material became inhomogeneous with

Basin Modelling: Advances and Applications edited by A.G. Doré et al. NPF Special Publication 3, pp. 127–133, Elsevier, Amsterdam.

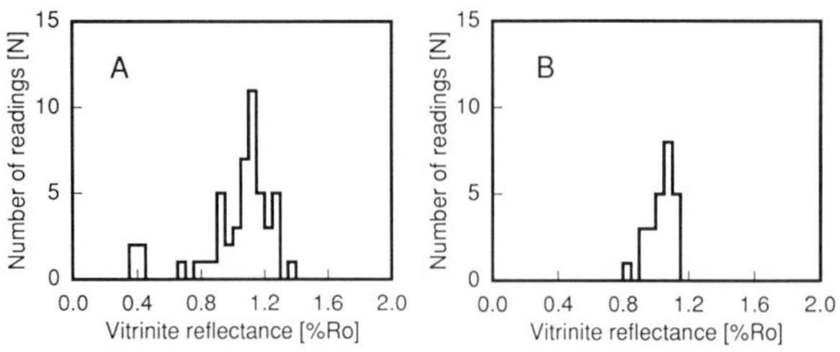

Fig. 1. Vitrinite reflectance histograms of artificially matured coals. (A) Results of isothermal heating at 350°C for 72 h. (B) Results of stepwise heating starting at 150°C and ending at 350°C.

respect to optical properties with low-reflecting central parts and high-reflecting rims. The values in Fig. 1B were obtained from the same type of coal sample, however, with stepwise heating from 150° to 350°C over 288 h. The average reflectance after the heating was $\%R_m = 1.04$. The distribution of values is narrow and as expected from a natural coal sample. During this heating experiment the vitrinite material preserved its natural homogeneous appearance.

The experiments were performed with immature, crushed and homogenized rock samples in sealed glass tubes. The tubes were about 10 cm long and 0.9 cm in diameter. They were heated in the heating chamber of a Carlo Erba Fractovap 2101 AC gas chromatograph. The precision of the heating device was approximately ±1°C. The heating experiments were performed with stepwise and progressive heating starting at 150°C. Two sets of experiments were run, one in steps of 50°C up to 500°C with 72 h at each step, and the other in steps of 25°C up to 365°C with 24 h at each step. The heating and sampling programme is illustrated in Fig. 2.

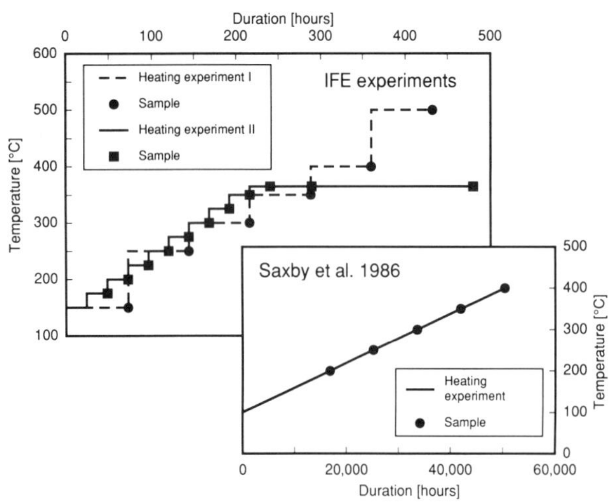

Fig. 2. Laboratory heating experiments showing our stepwise heating and sampling programme (IFE experiments) and the gradual heating and sampling programme of Saxby et al. (1986).

Material

Core samples from exploration wells in the Norwegian sector of the northern North Sea were used in this study.

The majority of the heating experiments was performed with aliquots of a Middle Jurassic humic coal sample. Two Upper Jurassic marine shale samples were used in a few of the experiments. All three were immature at the beginning of the experiment.

The well observations were made exclusively on Jurassic coal samples. All the coal samples were humic and typically banded, with vitrinite as the dominant maceral. Inertinite was more common than exinite.

Vitrinite reflectance techniques

The preparation and analytical techniques were standardized for all samples, both with respect to the heating experiments and the well observations. The samples were embedded directly in epoxy resin to make briquettes. These were subsequently ground flat and polished. The analytical equipment was a Zeiss MPM 03 photometermicroscope equipped with an epiplan-neofluoar 40.0/0.90 oil objective. The sensitive measuring spot was about 2.5 μm in diameter. The measurements were made in oil immersion, through a green filter (546 nm), without a polarizer and using a stationary stage. This procedure is called measurement of random reflectance. Bustin et al. (1985), Davis (1978), Stach et al. (1982), and Ting (1978) provide further information on these techniques. About 25 vitrinite spots were measured in each sample.

Time–temperature histories of the study wells

The well observations of vitrinite reflectance were obtained from narrow coal-bearing intervals in four exploration wells (Table 2). Figure 3 shows recon-

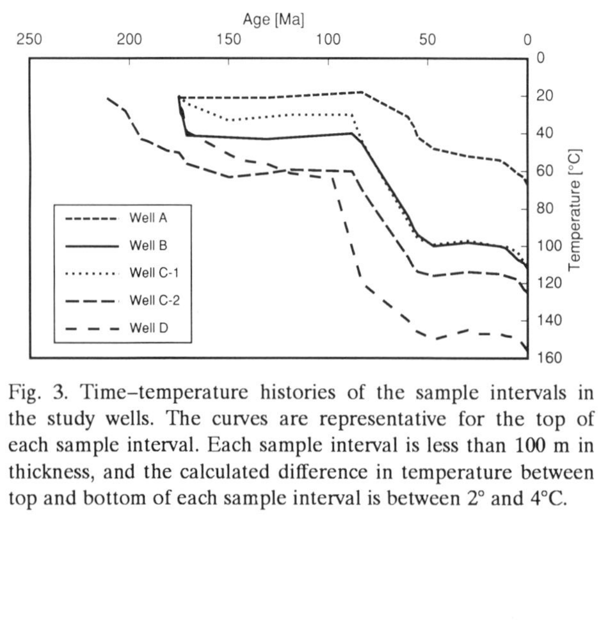

Fig. 3. Time–temperature histories of the sample intervals in the study wells. The curves are representative for the top of each sample interval. Each sample interval is less than 100 m in thickness, and the calculated difference in temperature between top and bottom of each sample interval is between 2° and 4°C.

structed time–temperature histories for the top of each sample interval. The intervals were less than 100 m in thickness, and the temperature difference between the top and bottom of each interval was between 2 and 4°C. The time–temperature histories were reconstructed using a one-dimensional fluid flow basin modelling programme. For simplicity, the basin model was calibrated against present-day corrected bottom-hole temperatures and the heat flow was kept constant versus time.

Results and discussion

The analytical results from the heating experiments and the well samples are listed in Tables 1 and 2, respectively. Also listed in Table 1 are vitrinite reflectance data from a very long-term laboratory experiment by Saxby et al. (1986). This heating and sampling programme is also illustrated in Fig. 2. In this particular experiment a brown coal was heated in sealed glass tubes starting at 100°C with a heating rate of 1°C per week. A tube was removed from the heating vessel at 50°C intervals and analysed.

TABLE 1

Vitrinite reflectance data from heating experiments

Time (h)	Temp. (°C)	Vitrinite reflectance (%R_o)			
		series 1	series 2	series 2	series 4
Heating experiments I					
0	150	0.37	0.29*	0.40*	
144	250	0.57			
216	300	0.63	0.46*	0.62*	
288	350	1.04	0.93*	1.16*	
360	400	1.70	1.92*	1.91*	
432	500	3.22	3.45*	3.59*	
Heating experiments II					
0	150	0.37	0.37	0.37	0.37
48	175	0.48		0.44	
72	200	0.51		0.46	
96	225	0.49	0.45		0.44
120	250	0.52	0.46	0.56	0.56
144	275	0.57	0.48	0.53	0.49
168	300	0.61		0.61	
192	325	0.72	0.73	0.70	0.72
216	350	0.94		0.95	0.88
240	365			1.11	1.06
288	365		1.15	1.53	
456	365				1.20
480	365		1.41		
Saxby et al. (1986) experiments					
0	100	0.28			
16800	200	0.47			
25200	250	0.59			
33600	300	1.00			
42000	350	1.85			
50400	400	2.73			

The results marked * were obtained from experiments with two samples from Upper Jurassic marine shales. The remaining results from experiments I and II were obtained from heating an immature Middle Jurassic humic coal sample. Series 1, 2, 3 and 4 represent parallel heating experiments. Included at the bottom of the table are the results of the very long-term heating experiments of Saxby et al. (1986).

TABLE 2

Vitrinite reflectance data for the well sections

Well section	Thickness of sample interval (m)	Vitrinite reflectance (%R_o)		
		average for sample interval	individual sample values	
A	75	0.41	0.36	0.42
			0.37	0.43
			0.39	0.44
			0.39	0.46
			0.41	0.48
			0.41	
B	87	0.53	0.48	0.53
			0.50	0.53
			0.51	0.53
			0.51	0.56
			0.52	0.56
			0.52	0.58
			0.52	
C-1	40	0.59	0.53	0.59
			0.54	0.60
			0.54	0.64
			0.54	0.64
			0.55	0.68
			0.56	0.71
			0.58	
C-2	1	0.69	0.61	0.68
			0.67	0.78
D	89	0.99	0.90	0.99
			0.94	1.00
			0.94	1.01
			0.94	1.04
			0.97	1.07
			0.97	1.16
			0.98	

Each sample interval is coded for reasons of confidentiality.

The Lopatin (1971) model

Various modifications of the original Lopatin (1971) approach are widely used to calculate vitrinite reflectance evolution, e.g. Ritter (1984) and Waples (1980, 1984). The original model is based on the rather inaccurate chemical assumption that the reaction rate doubles for every 10°C increase in temperature. This means that the activation energy varies with temperature, with very low activation energies for temperatures normally encountered in sedimentary basins, and much higher for the higher temperatures used in laboratory experiments. This makes it impossible to establish a laboratory calibration that also applies to geological conditions unless major modifications are introduced. It is not sufficient to change the doubling factor. The factor must at least be made temperature-dependent. As a

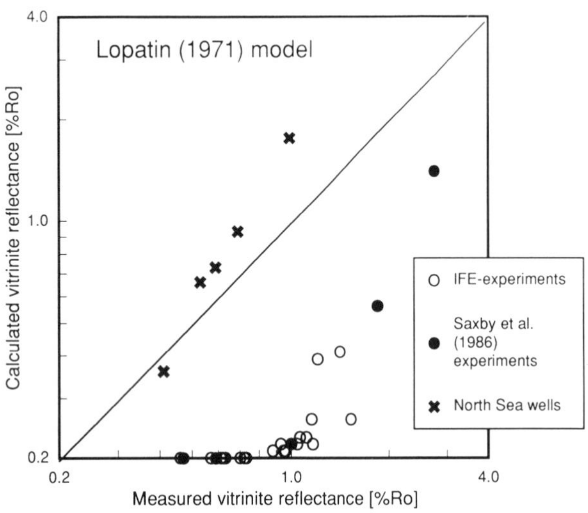

Fig. 4. Comparison between measured and calculated vitrinite reflectance using the Lopatin (1971) model and the calibration of Waples (1980). The line diagonally across the diagram indicates where calculated reflectance is equal to measured values.

consequence the Lopatin (1971) type model can only be calibrated against basin data, and the calibration will not be global. It will apply only for heating rates close to those of the calibration set. For heating rates slower than the calibration set the maturity will be overestimated, and for higher heating rates it will be underestimated.

Figure 4 shows a comparison between measured and calculated vitrinite reflectance using the Lopatin (1971) model and the calibration of Waples (1980) for the temperature–time histories of our experiments, the experiments of Saxby et al. (1986), and the reconstructed temperature–time histories for the sample intervals in the North Sea wells. The solid line diagonally across the diagram indicates the position where calculated vitrinite reflectance is equal to measured values. Data points above the line are overestimated by the model and data points below are underestimated. The results are as predicted above. There is absolutely no fit between calculated and measured vitrinite reflectance for the laboratory experiments, while the correspondence is naturally much better for the basin data.

The calculated values for the heating experiments were highly underestimated due to the very rapid heating. The basin data, on the contrary, were somewhat overestimated. This latter effect may in part be due to wrong assumptions regarding the reconstruction of the temperature histories of the study wells. However, the main effect is due to uncritical use of the calibration set of Waples (1980) which was established without considering sediment compaction. If this calibration is erroneously applied to more advanced models where compaction is accounted for, the resulting vitrinite reflectance will be overestimated as demonstrated in our example. The explanation is that correction for compaction significantly increases the modelled duration of exposure to maximum temperature.

The Larter (1989) model

The model of Larter (1989) for vitrinite reflectance evolution is based on two assumptions. Firstly, vitrinite reflectance is linearly related to the amount of phenol-precursor compounds in vitrinite according to observations in a series of natural vitrinites (Senftle et al., 1986). Secondly, the decrease in phenol-precursors in vitrinite follows chemical kinetics and is formulated as an Arrhenius-type first-order reaction with the rate constant being a function of a single pre-exponential factor, a Gaussian distribution of activation energies and absolute temperature. The model is applicable to vitrinite reflectance values between $\%R_o = 0.45$ and 1.6. Maturities below $\%R_o = 0.45$ are avoided due to interference with the demethoxylation phase of vitrinite precursor diagenesis. Maturities above $\%R_o = 1.6$ cannot be simulated by the model because there are no phenol-precursors left in the vitrinite.

Figure 5 shows a comparison between measured and calculated vitrinite reflectance using the model and calibration of Larter (1989) for our experiments, the experiments of Saxby et al. (1986) and the real wells. The calibration is based on the relationship between vitrinite reflectance and phenol-precursors as observed in natural vitrinites, a pre-exponential factor of $7.48 \times 10^8\ s^{-1}$ and a mean activation energy of 186 kJ/mol with a standard deviation of 13

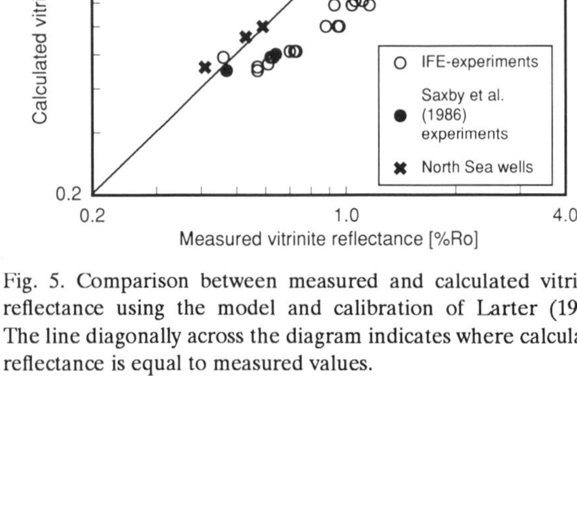

Fig. 5. Comparison between measured and calculated vitrinite reflectance using the model and calibration of Larter (1989). The line diagonally across the diagram indicates where calculated reflectance is equal to measured values.

kJ/mol. The parameters were obtained from calibration against basin data. In performing this calibration Larter (1989) mentions that he found it difficult to locate data sets that have good burial history information. On the other hand, he argued against laboratory calibration, as kerogen-bound oxygen compounds behave quite differently under basin and laboratory heating rates. The model of Larter (1989) gave a much better fit to the observed values than the Lopatin (1971)/Waples (1980) model. However, there are some systematic deviations. The calculated values for the heating experiments were underestimated. This suggests that the mean activation energy is somewhat too low. The correlation is naturally much better for the basin data, although slightly overestimated by the model. This may be due to wrong assumptions in assessing the temperature histories of the study wells.

Figure 6 shows a comparison between measured and calculated vitrinite reflectance where the kinetic parameters were modified in order to fit our experimental data and the experiments of Saxby et al. (1986). The best fit parameters are a pre-exponential factor of $3.17 \cdot 10^{10}$ s^{-1} and a mean activation energy of 190 kJ/mol with a standard deviation of 12.5 kJ/mol. The relationship between vitrinite reflectance and phenol-precursors was kept fixed in accordance with Larter (1989). The resulting correspondence between observed and calculated values for the experimental data was of course quite good; however, the fit for the basin data was slightly poorer than with the original Larter (1989) calibration. The disagreement for the basin data may, as mentioned above, be due to wrong assumptions in assessing the temperature histories, or because the mechanisms for liberation of phenols from vitrinite are different for laboratory and basin conditions. However, it may also be related to limitations in the model. We suspect that it is insufficient to describe vitrinite reflectance evolution in terms of phenol-precursors alone. Vitrinite reflectance is more a bulk parameter and it is necessary to include a series of reactions, some with higher activation energies than inferred by Larter (1989) for the liberation of phenols.

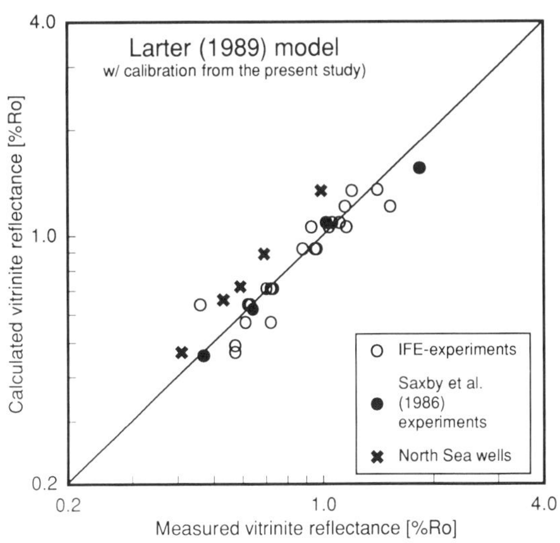

Fig. 6. Comparison between measured and calculated vitrinite reflectance using the Larter (1989) model with calibration against the experimental data of this study. The line diagonally across the diagram indicates where calculated reflectance is equal to measured values.

The model of Burnham and Sweeney (1989)

In the model of Burnham and Sweeney (1989) vitrinite reflectance is related to the elemental (C, H, O) composition of vitrinite. The elemental composition is controlled by separate kinetic models for release of H_2O, CO_2, C_nH_{2n+2} (hydrocarbons) and CH_4. Each kinetic model is formulated as an Arrhenius-type first-order reaction with the rate constant being a function of a single pre-exponential factor, a distribution of activation energies and absolute temperature. In a recent simplification of the model (Sweeney and Burnham, 1990), the four models are joined and presented as one kinetic model.

The relationship between elemental composition and vitrinite reflectance is formulated by two alternative equations. One relates vitrinite reflectance to the carbon content of the vitrinite, and the other relates reflectance to the H/C and O/C ratios. Only the latter will be considered here. It is formulated as follows:

$$\%R_o = 12\exp[-3.3(H/C)] - O/C \qquad (1)$$

Figure 7 shows a comparison between measured and calculated vitrinite reflectance using the model and calibration of Burnham and Sweeney (1989) for our experiments, the experiments of Saxby et al. (1986) and the real well data. The calibration, both with respect to kinetic parameters and to the relationship between elemental composition and vitrinite reflectance, was based on information from a variety of sources (for more information, see Burnham and Sweeney, 1989). The correlation between observed and calculated values is very good, both for the heating experiments and the basin data, and no recalibration of the model seems necessary. The excellent agreement between calculated and measured data suggests that vitrinite reflectance is a bulk parameter and closely related to the overall chemistry of vitrinite.

There is, however, one weak point in the model which is not demonstrated in Fig. 7. It is related to the effect of the O/C ratio as expressed in eq. 1.

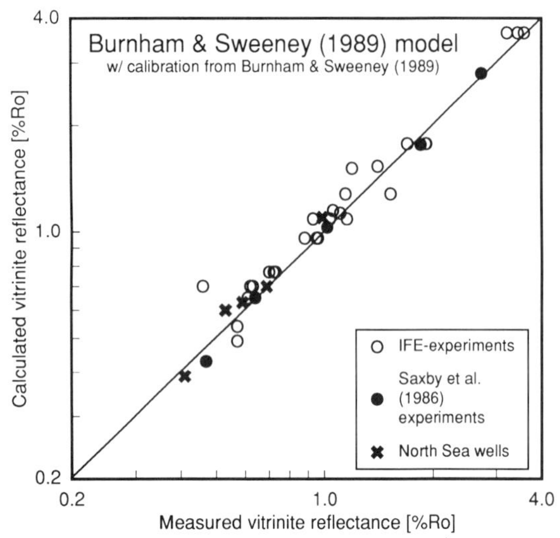

Fig. 7. Comparison between measured and calculated vitrinite reflectance using the model and calibration of Burnham and Sweeney (1989). The line diagonally across the diagram indicates where calculated reflectance is equal to measured values.

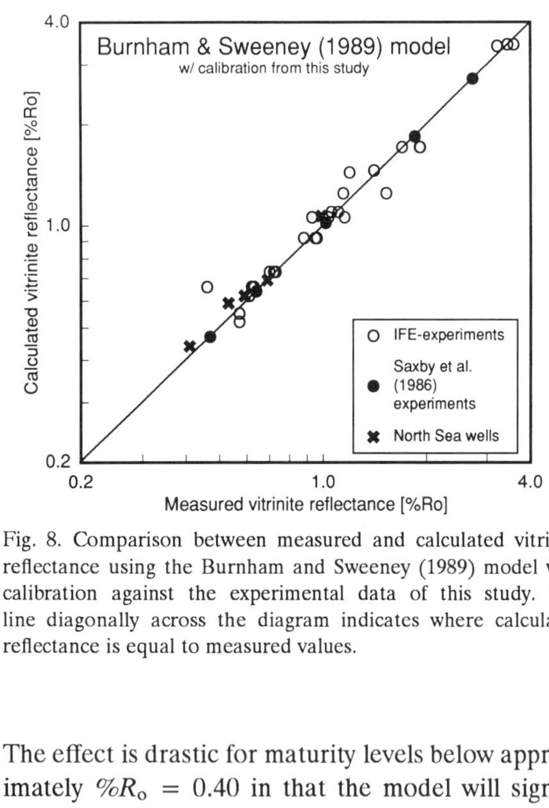

Fig. 8. Comparison between measured and calculated vitrinite reflectance using the Burnham and Sweeney (1989) model with calibration against the experimental data of this study. The line diagonally across the diagram indicates where calculated reflectance is equal to measured values.

The effect is drastic for maturity levels below approximately $\%R_o = 0.40$ in that the model will significantly underestimate calculated vitrinite reflectance. Figure 8 shows a comparison between measured and calculated values where eq. 1 was slightly modified in order to fit both the experimental data and to avoid erratic low calculated maturities for vitrinite reflectance values below $\%R_o = 0.40$. The effect of the suggested modification is not clearly seen in the illustration because no low-maturity samples were included in the data set; however, the correspondence for the higher maturities is similarly good as in the original calibration. The modified equation is as follows:

$$\%R_o = 11\exp[-3.2\,(\mathrm{H/C})] - (\mathrm{O/C})\,[1.3 - 1.5\,(\mathrm{O/C})] \qquad (2)$$

Conclusion

The usefulness of vitrinite reflectance as check parameter for palaeotemperature reconstruction depends on choice of model and precise calibration. Precise calibration can only be performed under controlled time and temperature conditions, which are difficult to ensure outside the laboratory. Laboratory calibration, on the other hand, introduces a scaling problem between experiments performed over a few to some hundred hours, and basin applications where the time scale is in millions of years. Calibration against basin data is an alternative and is necessary for the majority of models for vitrinite reflectance evolution; however, there will always be inherent uncertainties regarding the verification of basin time–temperature histories.

The Lopatin (1971) model can only be calibrated against basin data, and will only yield precise vitrinite reflectance calculations for heating rates close to that of the calibration set.

The Larter (1989) model is based on chemical kinetics. However, the vitrinite reflectance is related to the evolution of a single group of compounds, namely phenol-precursors. This is probably insufficient, and it is questionable to calibrate the model against laboratory experiments for application under basin scale conditions.

In the Burnham and Sweeney (1989) model vitrinite reflectance is related to the bulk chemical composition of the vitrinite in terms of elemental composition, and the relation can be successfully extended from laboratory to basin scale conditions. However, there is a need to refine the model at very low maturities, for vitrinite reflectance values below approximately $\%R_m = 0.40$.

References

Burnham, A.K. and Sweeney, J.J., 1989. A chemical kinetic model of vitrinite maturation and reflectance. Geochim. Cosmochim. Acta, 53: 2649–2657.

Bustin, R.M., Cameron, A.R., Grieve, D.A. and Kalkreuth, W.D., 1985. Coal petrology — its principles, methods and application. Geol. Assoc. Can., Short Course Notes, 3, 230 pp. (2nd ed.).

Davis, A., 1978. The reflectance of coal. In: C. Karr (Editor), Analytical Methods for Coal and Coal Products. Academic Press, New York, N.Y., pp. 27–81.

Larter, S., 1989. Chemical models of vitrinite reflectance evolution. Geol. Rundsch., 78: 349–359.

Lewan, M., 1985. Evaluation of petroleum generation by hydrous pyrolysis experimentation. Philos. Trans. R. Soc., London, Ser. A 315: 123–134.

Lopatin, N.V., 1971. Temperature and geologic time as a factor in coalification. Izv. Akad. Nauk SSSR, Ser. Geol., 3: 95–106.

Ritter, U., 1984. The influence of time and temperature on vitrinite reflectance. Org. Geochem., 6: 473–480.

Saxby, J.D., Bennett, A.J.R., Corcoran, J.F., Lambert, D.E. and Riley, K.W., 1986. Petroleum generation: simulation over six years of hydrocarbon formation from torbanite and brown coal in a subsiding basin. Org. Geochem., 9: 69–81.

Senftle, J.T., Larter, S.R., Bromley, B.W. and Brown, J.H., 1986. Quantitative chemical characterization of vitrinite concentrates using pyrolysis-gas chromatography. Rank variation of pyrolysis products. Org. Geochem., 9: 345–350.

Stach, E., Mackowsky, M.T., Teichmüller, M., Taylor, G.H., Chandra, D. and Teichmüller, R., 1982. Stach's Textbook of Coal Petrology. Borntraeger, Berlin, 3rd ed., 535 pp.

Sweeney, J.J. and Burnham, A.K., 1990. Evaluation of a simple model of vitrinite reflectance based chemical kinetics. Am. Assoc. Pet. Geol., Bull., 74: 1559–1570.

Ting, F.T.C., 1978. Petrographic techniques in coal analysis. In: C. Karr (Editor), Analytical Methods for Coal and Coal Products. Academic Press, New York, N.Y., pp. 3–26.

Waples, D.W., 1980. Time and temperature in petroleum formation: application of Lopatin's method to petroleum exploration. Am. Assoc. Pet. Geol., Bull., 64: 916–926.

Waples, D.W., 1984. Thermal models for oil generation. In: C.J. Brooks and D. Welte (Editors), Advances in Petroleum Geochemistry. Academic Press, London, pp. 7–69.

Winters, J.C., Williams, J.C. and Lewan, M.D., 1981. A laboratory study of petroleum generation by hydrouspyrolysis. In: M. Bjorøy et al. (Editors), Advances in Organic Geochemistry. Wiley Heyden, Chichester, pp. 524–533.

TORBJØRN THRONDSEN *Institutt for Energiteknikk, P.O. Box 40, N-2700 Kjeller, Norway*

BJØRG ANDRESEN *Institutt for Energiteknikk, P.O. Box 40, N-2700 Kjeller, Norway*

ÅSE UNANDER *Institutt for Energiteknikk, P.O. Box 40, N-2700 Kjeller, Norway*

Thermal maturation and burial history of the Multiwell Experiment Site, Piceance Basin, Colorado: application of thermal indicator tomography

K. Zhao and I. Lerche

Using a thermal indicator tomography 1D modelling system, we simulated numerically the geological processes of sedimentary deposition, compaction, fluid flow, and palaeoheat flux. The thermal indicator tomography procedure determines simultaneously the variation of heat flux with time, and the size and timing of a surface unconformity. The results of our modelling were found to be consistent with given geological data and events. Input well data, in the form of present-day formation thicknesses, thermal gradient, and vitrinite reflectance with depth, were used to reconstruct the burial history and thermal history of the Multiwell Experiment Site, Piceance Basin, Colorado. The thermal history results show that the peak heat flux is about 2.7 HFU at 24 Ma (attributed to the volcanic activity in Oligocene–Miocene time). The surface erosion commenced at about 10 Ma and the eroded thickness was approximately 4000 ft (1100 m) which is comparable to the relief of the canyon walls of the Colorado River. In this area hydrocarbon maturation started at about 45 ± 15 Ma and at a depth of about 7500 ± 2500 ft (2290 ± 760 m).

Introduction

In the last few years, a 1D fluid-flow/compaction basin modelling code developed at the University of South Carolina has been applied to different areas, such as the Navarin Basin Cost No. 1 Well, Bering Sea, Alaska (Cao and Lerche, 1989), the Inigok-1 well, National Reserve of Alaska and the Canning Basin of Australia (He and Lerche, 1989), and the NW Danish Central Trough (Thomsen et al., 1990).

Improvements in the modelling procedure based on thermal indicator tomography (a non-linear inversion method, similar to seismic tomography, which inverts thermal indicators to evaluate the parameters involved in burial and thermal histories; Lerche, 1988; He and Lerche, 1989) have allowed the simultaneous determination of several geological, geochemical and geothermal parameters, which makes it relatively easy to reconstruct the geological and thermal histories of sediments in a well. The purposes of this paper are: (1) to apply the 1D model to an area in which there are excellent thermal indicators and a well-defined geological setting; (2) to examine the uncertainty associated with each parameter determined, in particular those parameters connected with palaeoheat flux and erosion events; and (3) to explain or extrapolate the geohistory and thermal history in the area. The MWX Site (Fig. 1), Piceance Creek Basin, has been selected for the study.

The Piceance Creek Basin contains major resources of natural gas (derived from coal formations) in low-permeability (tight) reservoirs. The National Petroleum Council (Spencer and Keighin, 1984) estimated a maximum recoverable gas resource in this area of about 32 TCF (= 10^9 m^3). Major gas reserves occur in fluvial and marine sandstones in the Upper Cretaceous Mesaverde Formation and in fluvial sandstones in the Lower Tertiary Wasatch Formation. Geological and geothermal data are available from USGS Open File Report 84-757 (Spencer and Keighin, 1984) and from vitrinite reflectance well profiles (Law et al., 1989).

For the Multiwell Experiment (MWX) Site, most of the detail geological, geophysical and geochemical data are reported in geological studies of the U.S. Department of Energy Multiwell Experiment, Garfield County, Colorado, by Spencer and Keighin (USGS Open File Report 84-754).

Figure 1 shows the location of the MWX Site in the basin. The wells are designated CER Nos. 1, 2 and 3 MWX and are all located in the northwest quarter of Sec. 34, T.6.S, R.94.W. Garfield County, Colorado. The wells were spudded in the upper part of the Tertiary Wasatch Formation and reached a total depth of 8000 ft (2500 m) at or near the base of the Upper Cretaceous Mesaverde Formation. The three wells deviated only slightly from vertical with only minor lateral drift with depth. Since all

Basin Modelling: Advances and Applications edited by A.G. Doré et al. NPF Special Publication 3, pp. 135–145, Elsevier, Amsterdam.

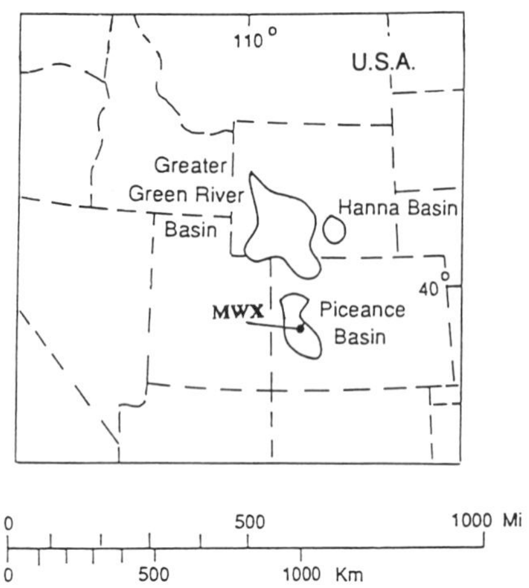

Fig. 1. Index map showing the location of the MWX Site in the Piceance Creek Basin, Colorado, USA (from Law et al., 1989).

three wells were in close proximity and had similar downhole data, they will be treated as a single composite well in the discussion that follows.

Numerous commercial and research geophysical logs were run in all three wells on multiple logging runs. Over 4000 ft (1250 m) of core was cut in the three wells with nearly complete recovery. These cores and their subsequent analyses provide the basic data for a number of research studies, including origin of gases, thermal maturation of source rock,

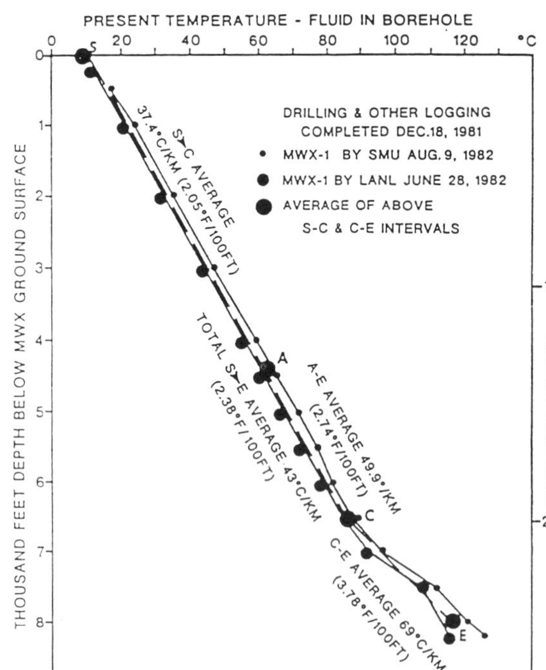

Fig. 2. Present rock temperatures based on temperatures of fluid in the borehole 6 and 8 months after drilling (from Bostick and Freeman, 1984).

well-log interpretation, and vitrinite reflectance from the three wells (Law et al., 1989), while Fig. 2 (taken from Bostick and Freeman, 1984) shows present-day temperature with depth in each borehole.

Geological setting of Piceance Creek Basin, Colorado

The Piceance Creek Basin is a latest Cretaceous through Eocene age sedimentary and structural basin (Johnson and Nuccio, 1986). Prior to the Laramide uplift which broke up the much larger Cretaceous epeiric sea, the basin was not a separate intermontane as it is today; rather it was part of a large, rapidly subsiding, trough that covered much of the central part of the United States and Canada. This trough formed to the east of the actively rising area. The orogenic belt was occupied by an epicontinental seaway for much of the Cretaceous. As much as 5000 ft (1520 m) of marine sediments (Mancos shale and marine portion of the Mesaverde Group) were deposited in the Piceance Creek Basin before the sea retreated to the east in the late Cretaceous. The eastward retreat of the seaway resulted in regressive sequences which include marginal marine, blanket sandstone, coal swamp, and fluvial deposits. A basin-wide unconformity occurs at the top of the Mesaverde, and the upper part of the Mesaverde was deeply weathered. The thickness of Cretaceous rocks removed by the unconformity is unknown, but possibly many hundreds of meters of section were lost.

When sedimentation resumed in the early or middle Paleocene, the unconformity was onlapped by sediments of the Wasatch Formation. The oldest sediments of the Wasatch were coarse clastic, but by the late Paleocene a poorly drained lowland or swamp covered most of the basin.

A permanent lake formed in the north-central part of the basin near the beginning of the Eocene. The lake expanded to cover a larger area (Johnson and Nuccio, 1986) and then formed Lake Uinta in which the widespread Mahogany oil shale zone of the Green River Formation was deposited.

Oligocene magmatism affected the southeastern corner of the Piceance Creek Basin. Shallow intrusions of intermediate composition were emplaced throughout this area from about 34 to 29 Ma. Several of the intrusions apparently vented at the surface forming an extensive sheet of volcanic breccia. Basalts covered much of the central part of the Piceance Creek Basin. About 10 Ma ago major tectonism resumed throughout the Rocky Mountain region. Old Laramide structures were reactivated and some new folds and faults were formed as the entire region was uplifted vertically. Downcutting of the Colorado River Canyon across the middle of the basin began about 9 Ma. The Colorado River system has removed as much as 5000 ft (1525 m) of overburden from the central part of the Piceance Creek Basin.

At the MWX Site (Fig. 2) the preserved formations, from oldest to youngest, are the Mancos shale, Lies Formation, Williams Fork Formation and Wasatch Formation, respectively. In the Mesaverde Formation the sequence is composed predominantly of carbonaceous shale, coal and lenticular sandstone. The lower part, about 900 ft (274 m) of sandy shale, grades up into a more sandy section of overlapping lenticular sandstones interbedded with gray shale and mudstone. The interpreted depositional environment is that of medium- to low-sinuosity fluvial environment. This upper part is about 2500 ft (762 m) thick and is truncated at the top by a regional unconformity.

Overlying the Upper Cretaceous unconformity at the MWX Site is about 3900 ft (1189 m) of late Paleocene through early Eocene age variegated mudstone and lenticular sandstone of the Wasatch Formation. The MWX Site is in the canyon of the Colorado River, and about 4700 ft (1433 m) of Eocene age sediments are still preserved on the nearby canyon walls (Fig. 3, taken from Bostick and Freeman, 1984). Thus a reasonably good estimate of the type and thickness of sediments removed by erosion at the MWX Site can be made. This estimate is necessary in order to construct a burial curve for the MWX Site for use in the thermal maturation studies. Approximately 600 ft (183 m) of early Eocene variegated mudstone and lenticular sandstone of the Wasatch Formation, and 4100 ft (1250 m) of early to late Eocene marlstone, oil shale, and sandstone of the Green River Formation were removed.

Between 10 and 40 Ma volcanic stocks were emplaced about 40 miles southeast of MWX causing some local heating adjacent to the stock. The MWX Site was probably covered by a thin basalt flow which was subsequently cut through by the Colorado River (see Fig. 3). All of these events provide clues to regional heat flux "pulsations" in this period.

Description of the model

The 1D fluid-flow compaction model consists of three parts: a geohistory model, a thermal history model, and a hydrocarbon generation model. The latter was not used in this study because of a lack of available kerogen data.

Geohistory model

In the modelling program, the simulation of sediment compaction is crucial because compaction of sediments either dominates or influences all other aspects of basin evolution. Figure 4 shows the flow diagram of the one-dimensional compaction model. The geohistory model reconstructs the burial history, basement subsidence, vertical fluid flow and the changes of porosity, permeability, pressure and fluid-flow rate with both time and depth. The data required to run the geohistory model are the depth and age of each formation base, the solid mass (derived from input formation thickness and porosity) and lithology of each formation, and the palaeowater depth (see Table 1, the basic input data for Piceance Basin MWX Well).

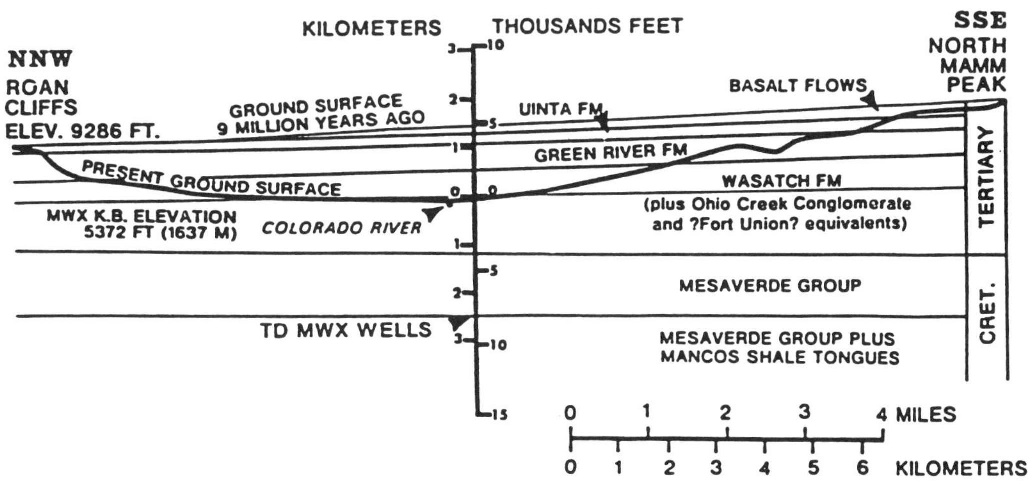

Fig. 3. Schematic stratigraphic cross-section through the MWX boreholes and extending to the valley walls on either side. Formation contacts are extended across the present valley approximately where they were prior to erosion by the Colorado River (from Bostick and Freeman, 1984).

TABLE 1

Basic input data for Piceance Basin MWX Well

Layer No.	Depth at base of formation (m)	Age at base (Ma)	Lithology	Thermal conductivity (cal cm^{-1} s^{-1} °C^{-1})
12	−1220.0	9.8	unconformity	
11	165.0	11.8	shale	3.87×10^{-3}
10	165.0	35.2	hiatus	
9	−1220.0	50.8	unconformity	
8	457.0	55.2	sandy shale	6.5×10^{-3}
7	990.0	57.0	sandy shale	
6	1220.0	60.0	shale	
5	−457.0	63.0	unconformity	
4	−457.0	65.0	unconformity	
3	1463.0	67.0	shale	
2	2377.0	72.7	shale	
1	2560.0	75.0	shale	

Note. Depth is measured from the surface and negative means erosion thickness. Age is based on GSA 1983 Geologic Time Scale. Lithology data from Spencer and Keighin, 1984. Input thermal conductivity from the same lithology is identical.

Thermal history model

Based on the burial history created in the geohistory model, the thermal model, called thermal indicator tomography (a non-linear inversion method), can do the following: (1) construct a model which approximates past thermal behavior of a well; (2) compute theoretical thermal indicators for each sedimentary layer by forward tracing in time through the geohistory model; (3) compare the predicted present-day thermal indicator variations with depth with the actual maturation profile; and then (4) modify the model to minimize the difference between computed and measured maturation.

The model contains three basic types of parameters:

(a) *Heat flow parameters.* The palaeoheat flux function permits the heat flux $Q(t)$, to vary with time t as:

$$Q(t) = Q_o \exp\left[\beta t + a_1 \sin\frac{\pi t}{t_{\max}} + a_2 \sin\frac{2\pi}{t_{\max}} + a_3 \sin\frac{3\pi t}{t_{\max}}\right] \tag{1}$$

where Q_o is the present-day heat flux, β, a_1, a_2, a_3 are parameters to be determined and t_{max} is the age of bottom layer.

The formation temperature at time t and depth $z(t)$ below the sediment–water interface is:

$$T[t, z(t)] = T_s + Q(t) \int_0^{z(t)} \frac{dz}{K(z)} \tag{2}$$

where T_s is the surface temperature and $K(z)$ is the sedimentary thermal conductivity at depth $z(t)$.

Present-day heat flux Q_o is calculated based on downhole measurements of temperature and $K(z)$. The calculation of present-day temperature in eq. 2 should be consistent with the measurements (see later).

(b) *Chemical parameters.* These are intrinsic chemical parameters that describe the kinetic behavior of thermal indicators, e.g., the activation energy for sterane isomerization reactions, rate constants for apatite fission track annealing, etc.

(c) *Geological parameters.* For instance, the amount of sediment eroded at an unconformity, the timing of the unconformity, the stratigraphic age of an undated sedimentary layer as constrained by known ages of layers above and below, etc.

The operating procedure for the two components of the model is divided into three steps. Firstly, a geohistory model is constructed (Fig. 4, from Cao, 1985), and secondly, a thermal model is determined (see Fig. 5). The best thermal history model is obtained by adjusting the three types of parameters above and minimizing discordances between computed and measured thermal indicators (vitrinite reflectance in this paper). The last step uses both the geohistory model and the thermal history model interactively, and reconstructs the best combined model with values or parameters adjusted in step two. This is a particularly necessary step when an erosional event occurs, because the eroded amount and its removal time impact both the burial history of sediments and the thermal history.

The model also provides estimates of the uncertainty in each parameter so that the thermal and geological histories of a well can be better defined (He and Lerche, 1989). Compared with the linear method, tomography (non-linear inversion) generally converges rather rapidly in 5–10 iterations. The model derived after non-linear iterations is a significantly better fit to the data than results of the linearized inversion (Lerche, 1988).

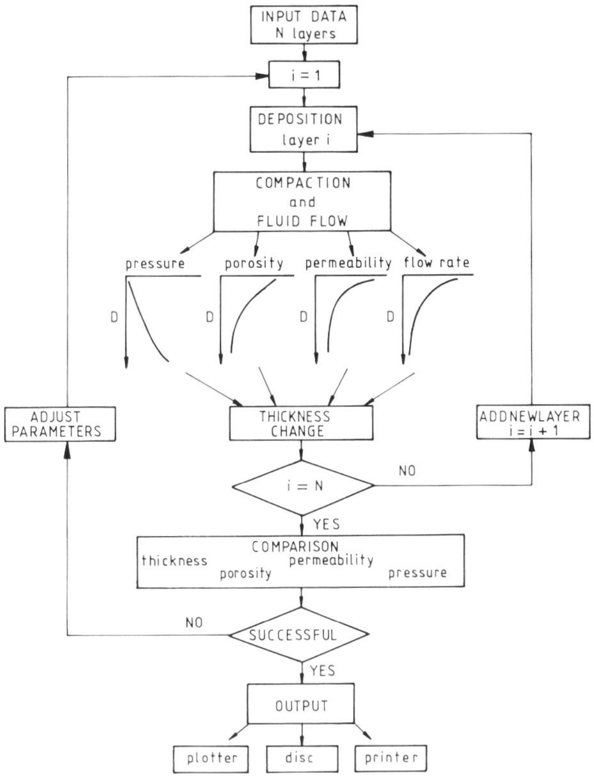

Fig. 4. Flow diagram for 1D compaction model — step 1 (from Cao, 1985).

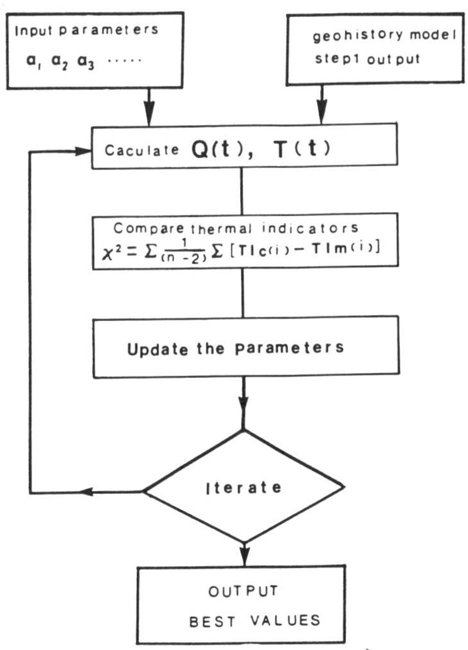

Fig. 5. Flow diagram for 1D thermal indicator tomography — step 2.

Thermal maturation and burial history of the MWX Site in Piceance Basin

Figure 6 shows the burial history in the Piceance Basin at the MWX Site. At 75 Ma the area began to subside. There are three major deposition stages which occurred in the Mesaverde Group (Upper Cretaceous) and Wasatch Formation (Tertiary). Major uplift, starting in the late Miocene, resulted in the present-day surface unconformity.

A long hiatus, of about 25 Ma duration, predated the uplift and a short hiatus (3 Ma) occurred between two major depositional phases. The greatest basement depth is at about 13,000 ft (4500 m) at about 35 Ma. The present-day total depth is 8400 ft (2560 m). Table 2 shows the thicknesses of individual layers comparing input and output data results.

Examining the history of compaction at the MWX Site, we find that the Upper Cretaceous and Tertiary shales and sandy shales were overpressured soon after their deposition and retained overpressure until the present day (Fig. 7). The profiles of pressure with depth at present are shown in Fig. 8.

In general, overpressuring is considered to be the result of compaction caused by the combined effects of three factors: (1) lack of a permeable bed; (2) rapid rate of sedimentation; (3) thick accumulation of sediments (Magara, 1978). In the MWX Site, the burial processes of shale and sandy shale at the early stage of sedimentation could be the reason for overpressure. Alternatively, Law et al. (1989) have emphasized that overpressuring in low-permeability *sandstone* reservoirs in the Rocky Mountain region is due to the accumulation of thermogenic gas. In the thermal indicator tomography procedure, if we consider the thermogenic gas to be completely trapped, predicted fluid pressure will be larger than the presently observed values. For instance, the interpreted pressure for the MWX Site well at about 8000 ft (2430 m) (Spencer, 1987) is 6200 psi (420 atm) which is 1800 psi (120 atm) higher than the calculated fluid pressure (see Fig. 8).

But the general increase in overpressure with depth through the stratigraphic sequences of shales, sandy shales, etc., extending over more than 1000 m, is not due to the thermogenic gas component, which can only serve to balance fluid-overpressure in a sandstone. Rather, there is a general pervasive systematic increase of water overpressuring with depth, in concert with the restricted gas component in sandstone reservoirs.

Because an unconformity represents erosional removal in the well, the trend of measured vitrinite reflectance values with depth shows a discontinuity at the unconformity depth (Armagnac et al., 1989). For the MWX Site the surface measured vitrinite reflectance is 0.6, higher than the general surface depositional value of 0.2, suggesting that erosion happened in the relatively recent past.

In order to fit the vitrinite reflectance data, we

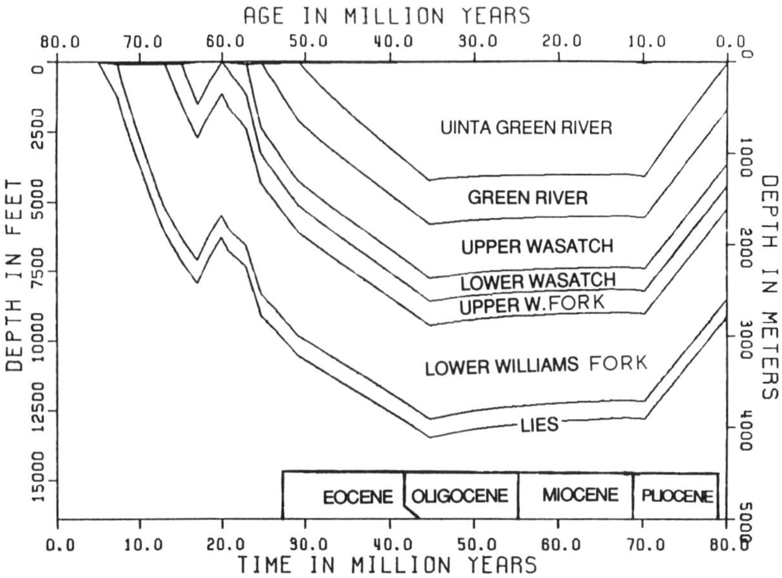

Fig. 6. Geohistory profile at MWX Site, Piceance Creek Basin; horizontal line marks a hiatus; tilted lines represent subsidence or uplift.

TABLE 2

Comparison of input and output thickness and porosity of each layer

Layer	Depth (I) (m)	Depth (O) (m)	Thickness (I) (m)	Thickness (O) (m)	Porosity (I) (fraction)	Porosity (O) (fraction)
Uinte-Green River	61.0	51.4	45.7	45.8	–	0.26
Green River	457.2	448.9	396.2	397.4	–	0.24
Upper Wasatch	990.6	987.5	533.4	538.6	0.23	0.23
Middle Wasatch	1219.1	1216.5	228.6	229.0	–	0.13
Lower Wasatch	1371.5	1368.8	152.4	152.3	–	0.14
Upper Williams Fork	1463.0	1451.3	91.4	82.4	–	0.18
Lower Williams Fork	2316.4	2315.5	853.4	864.3	–	0.16
Cameo Coal Zone	2377.3	2377.1	61.0	61.5	–	0.16
Lies	2560.2	2561.8	182.9	184.7	0.15	0.13

Note. I = input; O = calculated output; – = no data.

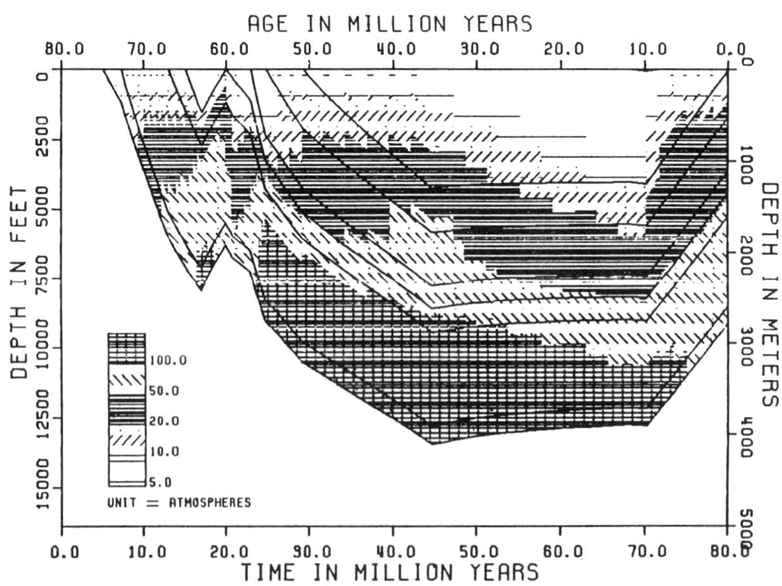

Fig. 7. Excess pressure variation with geohistory.

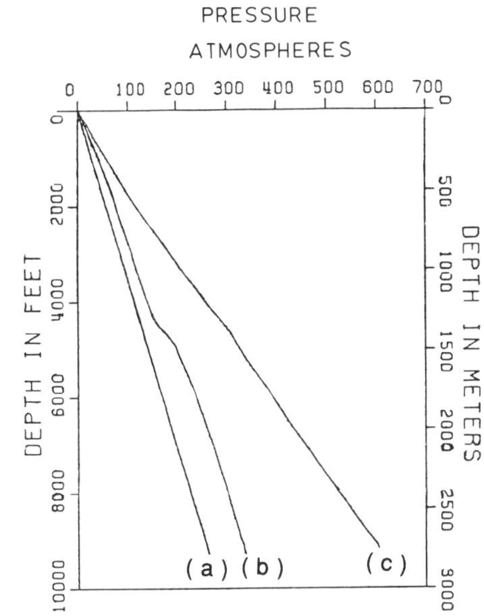

Fig. 8. Three kinds of calculated pressure: (*a*) = hydrostatic pressure; (*b*) = fluid pressure; (*c*) = overburden pressure.

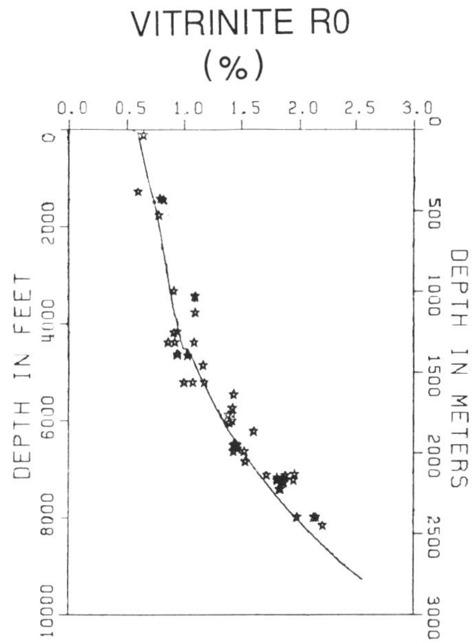

Fig. 9. Vitrinite reflectance profile. The stars represent measured values (Law et al., 1989) while the line shows calculated values.

adjusted both the geohistory model and geothermal model parameters (i.e., the amount of thickness eroded and palaeoheat flux) to obtain the best fit values determined by the mean square residual (MSR) between measured and predicted vitrinite reflectance. Table 3 shows the tomography result for the MWX Site. Predicted present-day vitrinite reflectance is compared with the measured vitrinite reflectance in Fig. 9. The degree of fit is measured by MSR, the mean square mismatch per point between predicted and measured value. The palaeoheat flux pattern, and erosion thickness and erosion timing were determined by the tomography method. Figure 10 shows the heat flux variation with time. The peak heat flux is about 2.74 HFU at about 24 Ma, attributed to the effect of the volcanic activity in Oligocene–Miocene time (Bostick and Freeman, 1984). Shown in Table 3 are the values of eroded thickness and erosion timing, corresponding to the least mismatch, which are basically consistent with the known geologic data and events. Calculated eroded thickness is in the range of 1000 ± 500 m and erosion commenced about 10 Ma.

Concerning the inversion of vitrinite reflectance data, uncertainty of the chemical parameters T_c and T_D, which relate to the calculation of vitrinite reflectance as a thermal indicator (Lerche et al., 1984), causes a variation and uncertainty of heat flux pattern. As shown in Fig. 15 for $T_c < 300$ K and $T_D > 90$ K, MSR has only minor changes. But with the best default values of $T_c = 295$ K and $T_D = 200$ K, the palaeotemperature reached 450°C at the base of the Masaverde Group and the peak heat flow value is 4 HFU at 22 Ma. The range of acceptable T_c and T_D values based on several hundred wells is $T_c = 295 \pm 30$ K and $T_D = 200 \pm 50$ K (Lerche et al., 1984). If one chooses $T_c = 260$ K and $T_D = 160$ K (within the statistical range of uncertainty) the peak heat flow is then only about 2.7 HFU at 24 Ma and the maximum temperature will be 340°C, well below a metamorphic regime, implying that lower T_c and T_D values are more appropriate than the statistical average values for this well. Figures 11, 13, and 14 show the evolution of temperature, Waples' TTI, and vitrinite reflectance, respectively. The present-day temperature is calculated from eq. 2 and Fig. 12 shows the profile of measured (asterisks) and predicted (line) temperature values.

There are three measures to depict the level of maturation:

TABLE 3

Tomography results and error bar of parameters for MWX Site, Colorado

MSR	β (My^{-1})	a_1	T_c (K)	T_D (K)	Eroded thickness (m)	Age erosion begins (Ma)
0.011	$-0.83 \times 10^{-3} \pm 0.1 \times 10^{-3}$	0.55 ± 0.2	260 ± 40	160 ± 30	1250 ± 1000	10.5 ± 2

Note. The error bars were determined by 10% increase of MSR from its minimum value.

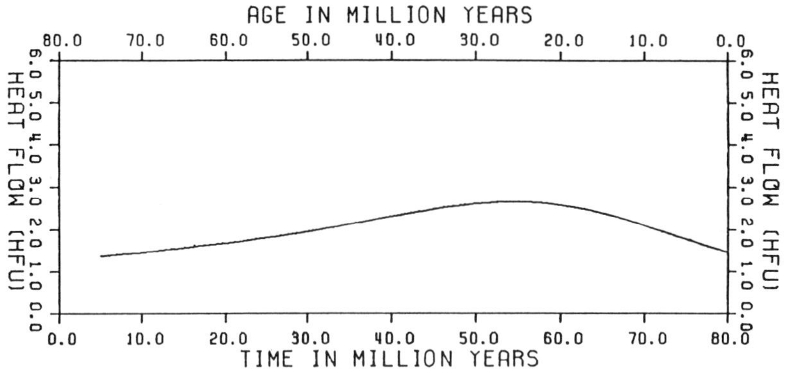

Fig. 10. Heat flow variation with time. Peak heat flux of 2.7 HFU occurred at 24 Ma.

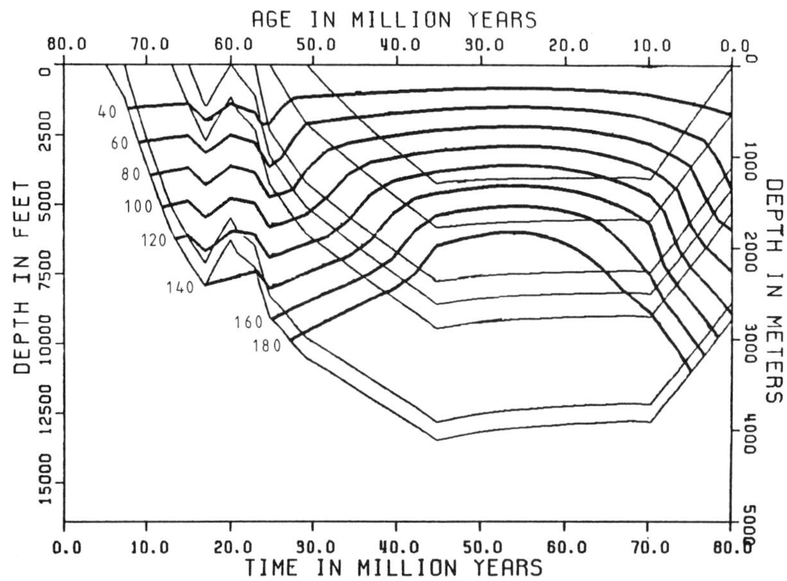

Fig. 11. Temperature variation with geohistory at MWX Site, Piceance Creek Basin, which is a function of heat flux and thermal conductivity.

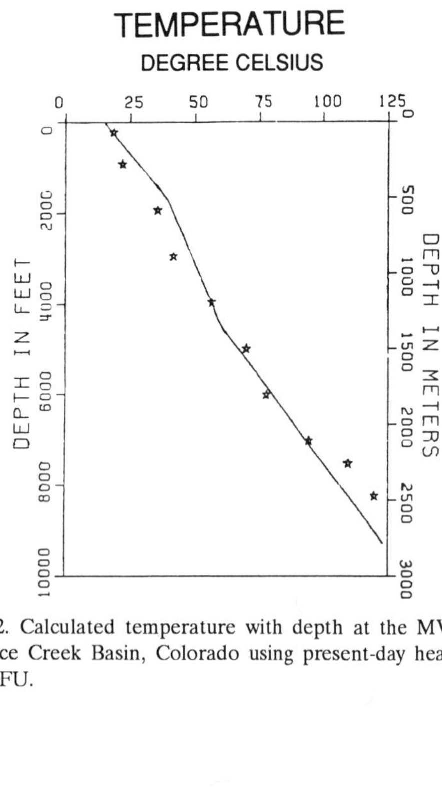

Fig. 12. Calculated temperature with depth at the MWX Site, Piceance Creek Basin, Colorado using present-day heat flux of 1.45 HFU.

(1) The predicted iso-vitrinite reflectance lines, plotted in Fig. 13, show the thermal maturation of each layer. If a vitrinite reflectance value of 0.6 is chosen as defining the onset of the "oil window" (Waples, 1980), the depth and time at which each layer entered the "oil window" indicate that hydrocarbon generation started within the Mesaverde Formation at 45 Ma. The present-day "oil window" range of depths is from the surface to 5250 ft (1600 m), corresponding to a vitrinite reflectance range from 0.6 to 1.2, respectively.

(2) Waples' (1980) TTI is plotted in Fig. 14. If a TTI value of 15 is defined as the top of the "oil window", hydrocarbon generation at the MWX Site started at 58 Ma and the present-day "oil window" reaches the surface, corresponding to TTI values from 15 to 150.

(3) Temperature variation with geohistory is plotted in Fig. 11. If we assume a threshold of 80°C, the onset of generation would occur at about 4000 ft (1220 m) at 72 Ma.

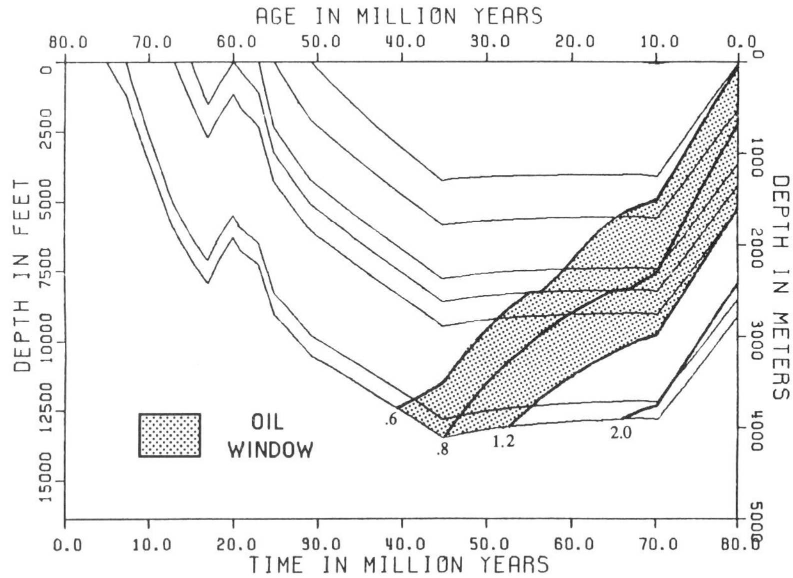

Fig. 13. Predicted vitrinite reflectance values as a function of geologic time and temperature. Iso-vitrinite reflectance lines are labeled.

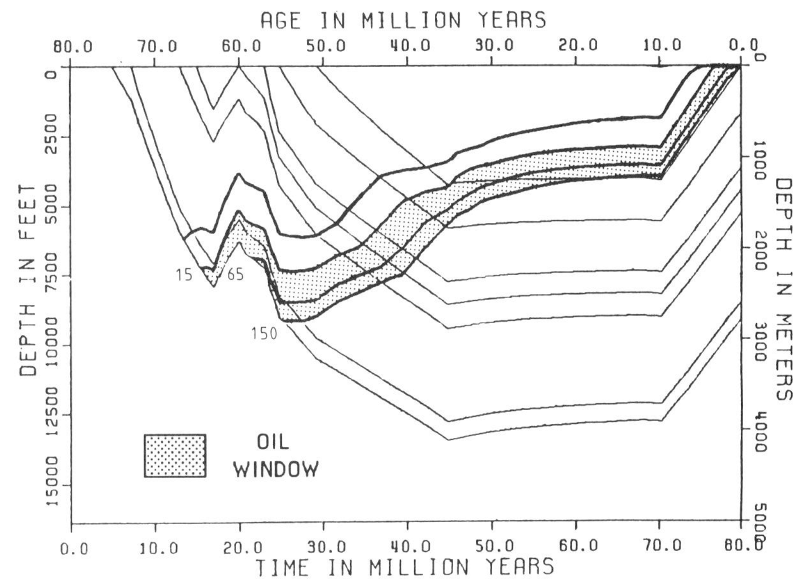

Fig. 14. Waples' TTI variation with geohistory. Iso-TTI lines are labeled.

A naive average of the parameters from these three methods suggests that hydrocarbon generation started at about 45 ± 25 Ma and at a depth of about 7500 ± 2500 ft (2290 ± 760 m). The present-day average depth of the "oil window" is then from the surface to about 5000 ft (1500 m).

Discussion and conclusions

To examine the uncertainties associated with thermal maturation parameters, a system is described whereby one variable is allowed to vary systematically while all other parameters are held fixed at their optimal values. The least squares residual mismatch, MSR, increases when a single parameter deviates from its best value. Results of such an investigation are shown for the MWX Site (see Fig. 15 for the heat flux parameters β, a_1, T_c, and T_D).

All the heat flow parameters appear to be well constrained. For the burial history parameters, the curve of MSR vs. unconformity size (Fig. 15) indicates that the amount of erosion is limited to less than about 2250 m when the time of onset of erosion is varied around its best value (Fig. 15).

The Multiwell Experiment Site, Piceance Basin, Colorado, went through three stages: subsidence, hiatus, and uplift. Both a high sedimentation rate, which resulted in overpressure in the sequence, and high palaeotemperature and rapid uplift are beneficial to hydrocarbon generation, migration, and

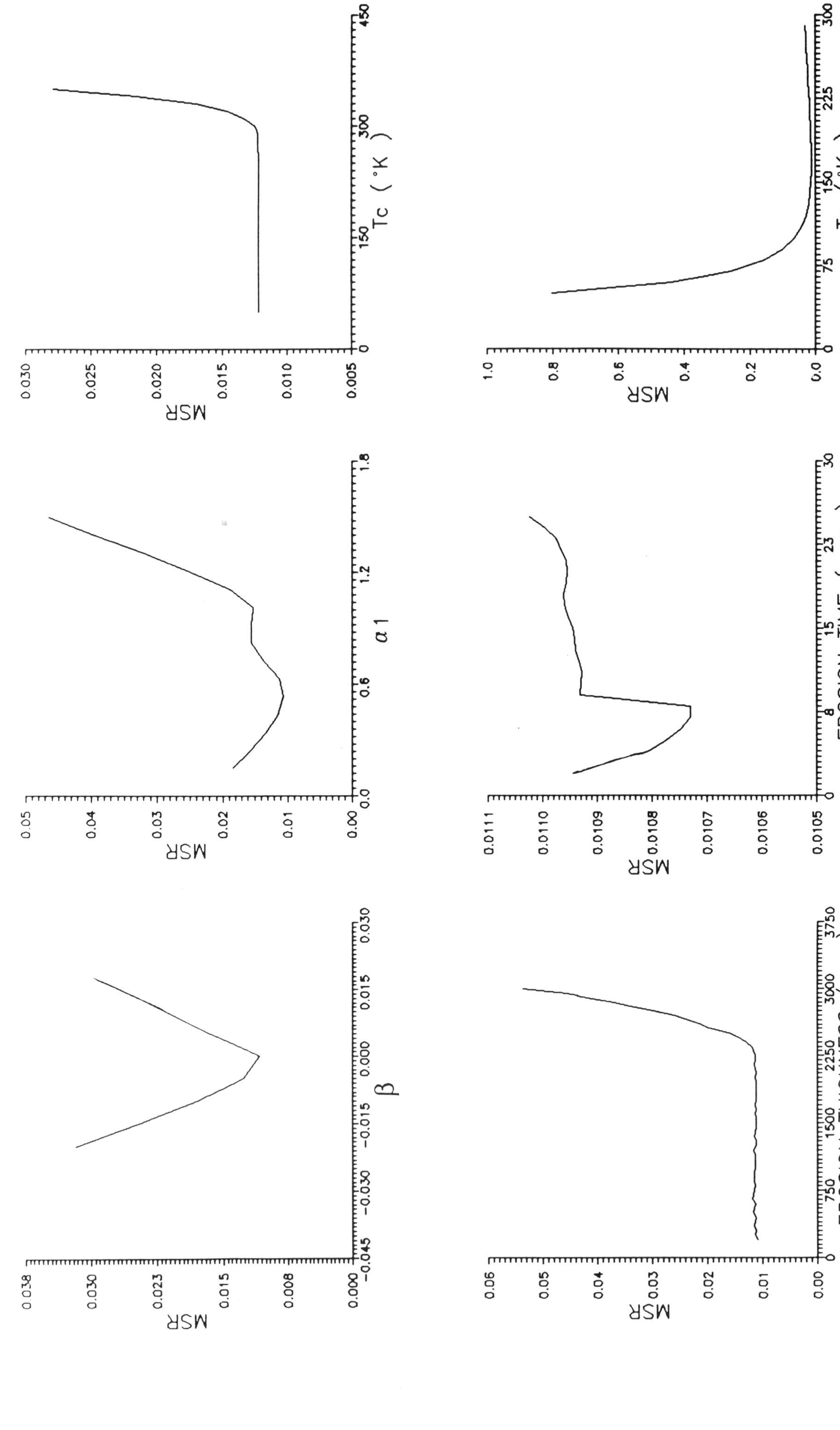

Fig. 15. Sensitivity analysis results of parameters β, a_1, T_c, T_D, erosion thickness and erosion time, which show the MSR variation with each parameter.

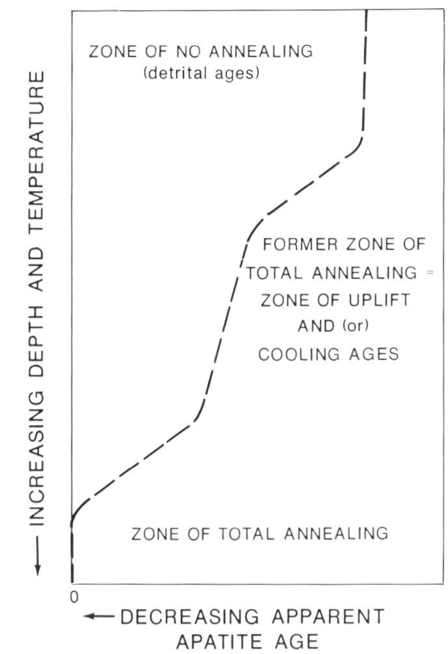

Fig. 2. Expected distribution of apparent apatite fission-track ages in sedimentary rocks after uplift and/or cooling (compare Fig. 1). Modified from Naeser (1979b).

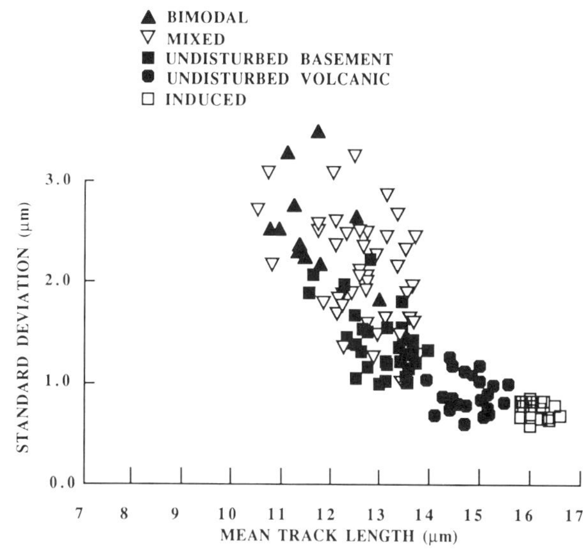

Fig. 4. Relationship between standard deviation and mean track length for the various track-length distributions shown in Fig. 3. Note that bimodal distributions are much broader (have a larger standard deviation) than distributions of undisturbed samples. The mean length and standard deviation of mixed distributions span the range of the bimodal and undisturbed basement samples. Modified from Gleadow et al. (1986b).

"undisturbed volcanic" to "mixed" thermal histories, mean track-length decreases, standard deviation of the track-length distribution increases, and the form of the track-length distribution becomes increasingly complex (Figs. 3 and 4; Gleadow et al., 1986b).

Detrital apatites deposited in a sediment can have any one or a combination of these distributions, depending on the thermal histories of the parent rocks from which they were derived. As detrital apatite is buried and heated to progressively higher temperatures in a basin, tracks that were present when the apatite was deposited in the basin, along with the new tracks that form during burial, are shortened in response to the heating.

In the simplest case of apatites derived from quickly cooled ("undisturbed volcanic") parent rocks and undergoing relatively simple burial, apatites at shallow depths (low temperatures) will have typical "undisturbed volcanic" track-length distributions. With increasing temperature, mean track-length decreases and the spread (standard deviation) of the track-length distribution increases (Fig. 5). The increased spread is caused both by anisotropic annealing of tracks in apatite (tracks parallel to the *c*-axis of apatite shorten more slowly than tracks at high angles to the *c*-axis) and by the fact that, even though existing tracks are being shortened by annealing, new (long) tracks continue to form in the apatite. The spread in track-length distribution is accentuated in apatite suites containing grains with a wide range of annealing susceptibility (most likely related to differences in the Cl-content of the apatite grains; see

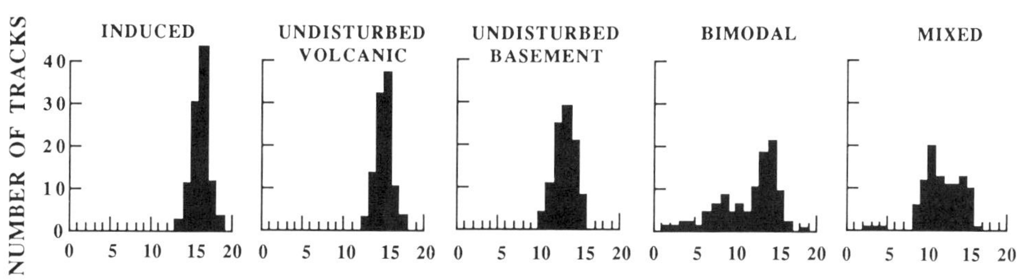

Fig. 3. Comparison of typical "induced", "undisturbed volcanic" (from rocks that have undergone rapid cooling and remained at low temperatures), "undisturbed basement" (from rocks that have undergone slow continuous cooling from deep crustal to ambient surface temperatures), and "mixed" (thermally disturbed) horizontal, confined track-length distributions in apatite. The "bimodal" distribution is a special case of the mixed distribution in which two separate components, reflecting a two-stage thermal history, can be readily identified. All track lengths were measured on polished surfaces essentially parallel to the crystallographic *c*-axis and are not corrected for measurement bias (Laslett et al., 1982). Modified from Gleadow et al. (1986b).

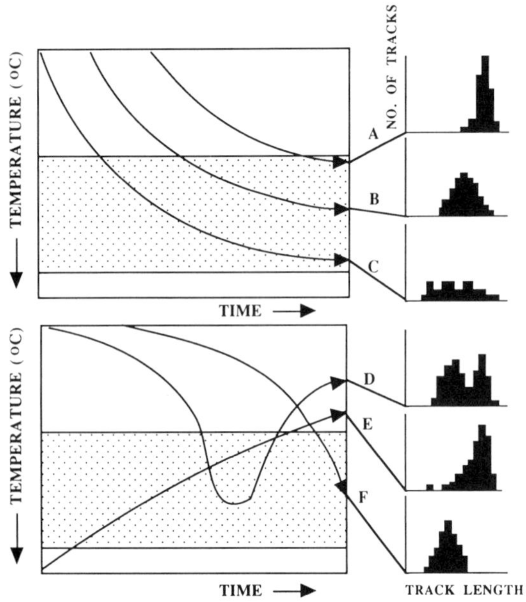

Fig. 5. Temperature–time paths and resulting apatite track-length distributions for rocks of varying thermal history. Upper diagrams (*A–C*) show distributions resulting from simple progressive burial of "undisturbed volcanic" apatite (Fig. 3) to different levels in the zone of partial annealing (shaded). Lower diagram shows distributions for more complicated thermal histories, including (*D*) a past thermal event (bimodal distribution), (*E*) slow cooling, and (*F*) recent heating. Modified from Gleadow et al. (1983).

below) (Green and Durrani, 1977; Gleadow et al., 1986a, b; Green et al., 1986; Donelick et al., 1990; Donelick, 1991).

In the more general case, the actual variation in mean length, standard deviation, and distribution of confined track-lengths with depth in any given basin will reflect the interplay between: (1) the pre-basin (parent-rock) thermal history(s) experienced by the detrital apatite grains; (2) the post-depositional (basin) thermal history; and (3) the range of composition (Cl-content) in the apatite population and its consequent effect on annealing susceptibility.

Variation of mean track-length with standard deviation of the track-length distribution in samples from over 20 basins is plotted in Gleadow et al. (1986b, fig. 8).

Since 1979, apatite fission-track analysis has been used to clarify the thermal history of more than 40 sedimentary basins worldwide, in a wide variety of geologic settings (Fig. 6, Table 1). These basin studies, combined with numerous laboratory annealing studies, have produced extensive information on fission tracks in apatite and their response to the thermal history of sedimentary rocks and have made fission tracks a much more effective tool in basin modelling now than they were 10 years ago. At the same time, these studies have highlighted some limitations of apatite fission-track analysis and questions that remain to be resolved to realize its full potential in basin studies. Problems that are the focus of research include: (1) resolving the kinetics of apatite annealing and the related problem of the relationship between reduction in apatite track length and apparent fission-track age with progressive annealing; (2) dealing with complications introduced by inherited age and thermal history in detrital apatites and their effect on observed patterns of apatite age and track-lengths in basins; and (3) finding practical methods of determining composition, particularly Cl-content, in detrital apatite suites.

Limitations/current research

Annealing kinetics

Two issues involving annealing kinetics are central to the application of apatite fission-track analysis to sedimentary basins: (1) recognition of the role of time in annealing, and (2) establishing the time–temperature correlation for geologically significant annealing conditions.

Although temperature is the dominant controlling factor in annealing, time cannot be ignored in interpreting thermal history. The temperature range for annealing of any given mineral depends on the duration of heating — the longer the heating, the lower the temperature required for annealing. Other factors (such as apatite composition; see below) being equal, decrease in age and track-lengths with progressive annealing (Figs. 1 and 5) will occur at significantly lower temperatures in sedimentary rocks with long heating times (e.g., in the Otway basin) than in rocks with short heating times (found, e.g., in the Los Angeles, Taranaki, and North Sea basins). It is important to remember that *a given degree of annealing is not indicative of a given temperature, only of a temperature–time combination.*

In practice, because an order of magnitude increase in effective heating time decreases annealing temperatures by only about 10° to 15°C, when effective heating times are greater than several tens of millions of years or so, the increase in heating time needed to produce a significant decrease in annealing temperatures becomes so long that fission tracks may be, in effect, an approximate palaeotemperature indicator (e.g., Green et al., 1989b, fig. 6) (assuming factors such as composition and inherited tracks are taken into account). *However*, this is not the case in basins with short effective heating times, which include some of the economically important basins around the world. In such basins, relatively small differences in heating time can result in significant

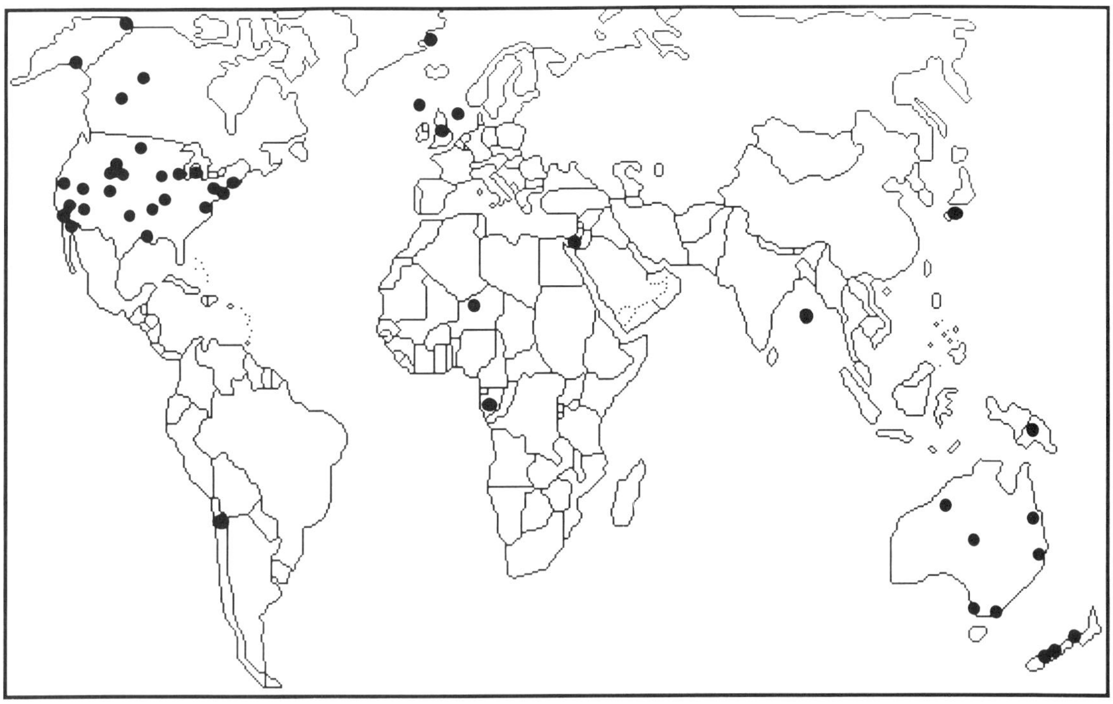

Fig. 6. Location of selected studies (identified in Table 1) using fission-track analysis to evaluate thermal history of sedimentary rocks.

differences in annealing temperatures, as demonstrated both by modelling thermal histories and the resulting track-lengths and fission-track ages and by actual studies in sedimentary basins with different effective heating times (e.g., N.D. Naeser et al., 1989a, 1990b, c; J.-P. Stiberg in Deming et al., 1990; Kamp and Green, 1990). If the role of time in annealing is ignored, we ignore one of the most powerful features of the fission-track method and run a risk of losing valuable information on the thermal history of sedimentary basins.

More research has been devoted to determining the annealing kinetics of apatite than of any other mineral, from laboratory annealing data (e.g., Wagner, 1968, 1969; Naeser and Faul, 1969; Märk et al., 1973; Zimmermann and Gaines, 1978; Burchart et al., 1979; Chaillou and Chambaudet, 1981; Crowley, 1985; Green et al., 1985, 1986, 1989b; Crowley and Cameron, 1987; Laslett et al., 1987; Duddy et al., 1988; Green, 1988; Crowley et al., 1990, 1991; Donelick et al., 1990; Hughes et al., 1990; Donelick, 1991), from observations of annealing behavior of apatite in drill holes in areas where the thermal history of the rocks is reasonably well known (Naeser, 1979a, 1981; Gleadow and Duddy, 1981; Harrison, 1985), and from modelling of physical processes involved in annealing (Carlson, 1990), but uncertainties remain. Drill-hole-derived annealing temperatures for fluorapatite (Fig. 7) indicate that it is effectively totally annealed (yields a zero age) at temperatures that range from about 105°C for relatively long-term heating of ~100 million years duration to 150°C for ~100,000 years heating (Naeser, 1981). This temperature–time correlation agrees reasonably well with the extrapolated laboratory annealing data shown in Fig. 7, although the drill-hole data tend to be displaced towards slightly longer heating times. Harrison (1985) has suggested that the actual effective heating times of rocks in the Eielson and Fenton Hill drill holes (Fig. 7) are shorter than estimated by Naeser (1981). His revised estimates of effective heating time for these two areas make the drill-hole data "virtually coincident" with the extrapolated laboratory data. However, the extrapolation in Fig. 7 is based on only one of a number of annealing models that have been proposed for apatite, based both on additional laboratory annealing studies and on reinterpretation of existing annealing data. Examples in Fig. 8 show that some of these models predict significantly different time–temperature relationships for geologic annealing.

At least part of the difference between apatite annealing models is a direct consequence of differences in experimental procedure used to generate the annealing data (Crowley, 1989). A notable example is the parameter used to characterize annealing. Early models (for example, Fig. 7) were based on studies in which annealing of etchable damage was determined from reduction in track *density*, whereas more recent models have been based on reduction in

TABLE 1

Examples of studies using fission-track analysis to evaluate thermal history of sedimentary rocks

Country/Region	Area	References
AUSTRALIA	Amadeus basin	Duddy et al. (1987)
	Bowen basin	Marshallsea (1986, 1990); Duddy et al. (1987)
	Canning basin	Gleadow and Duddy (1984); Duddy and Gleadow (1985)
	Lennard Shelf Pb–Zn district	Arne (1989); Arne et al. (1989, 1990b)
	Clarence–Moreton basin	Gleadow (1989, 1990)
	Gippsland basin	Gleadow and Duddy (1981); Duddy et al. (1987)
	Otway basin	Gleadow and Duddy (1981); Duddy and Gleadow (1982, 1985); Gleadow et al. (1983); Duddy et al. (1987); Green et al. (1989a); van de Kamp et al. (1990)
CANADA	Pine Point Pb–Zn district, Northwest Territories	Arne (1989, 1991); Arne et al. (1990b)
	Western Canada Sedimentary Basin	Beaumont et al. (1989); Issler et al. (1990)
CHILE/ARGENTINA	North-Chilean Coast Range	Damm and Kelley (1989)
ENGLAND	Cleveland basin, East Midlands, northern England	Green (1986, 1989a, b, 1990); Brown and Green (1991)
GABON	"coastal basin", "interior basin"	Storzer and Selo (1984)
GREENLAND	Jameson Land basin	Hansen (1990)
INDIAN OCEAN	ODP Leg 116, sites 717 and 718	Corrigan and Crowley (1989); Corrigan (1991b)
ISRAEL	Negev area, southern Israel	Feinstein et al. (1986, 1989); Kohn et al. (1990a)
JAPAN	Hohi and Sengan geothermal fields	Tamanyu (1985)
MEXICO	Cerro Prieto geothermal field	Sanford (1981)
NEW ZEALAND	Taranaki basin	Kamp et al. (1988); Kamp and Green (1990); van de Kamp et al. (1990)
	West Coast, South Island: Alpine Fault, Cenozoic basins	Green and White (1985); Kamp et al. (1989, 1992); Seward (1989, 1990); Seward and Nathan (1990)
NIGER	Tim Mersoï basin	Forbes et al. (1989); Meyer et al. (1990)
NORTH ATLANTIC	DSDP Leg 81, site 555	Duddy et al. (1984)
NORTH SEA	offshore Norway[a]	J.-P. Stiberg, in: Deming et al. (1990); Qvale et al. (1990)
PAPUA NEW GUINEA	Papuan fold belt	Hill and Gleadow (1989); Hill (1990)
USA	*Western USA*	
	Alaska, northern	O'Sullivan (1988); O'Sullivan et al. (1989, 1990); McMillen and O'Sullivan (1992)
	Alaska, southern	Kveton (1990)
	Colorado Plateau (Ariz.)	C.W. Naeser et al. (1989b)
	Franciscan Complex (Calif.)	Dumitru (1989, 1991a, b)
	Great Valley sequence (Calif.)	Dumitru (1988, 1991a)
	Huntington Valley (Nev.)	van de Kamp et al. (1990)
	Los Angeles basin (Calif.)	N.D. Naeser et al. (1987a, 1989a, 1990a, b); Geving et al. (1989, 1990a, b, 1991)
	San Joaquin Valley (Calif.)	Briggs et al. (1979, 1981a, b); Naeser (1984); N.D. Naeser et al. (1989b, 1990c)
	Rocky Mountain Region	
	Bighorn basin (Wyo. and Mont.)	Giegengack et al. (1986, 1987, 1990)
	Green River basin (Wyo.)	Naeser (1984, 1986, 1989b); N.D. Naeser et al. (1989b)
	Piceance Creek basin (Colo.)	Kelley and Blackwell (1990)
	Powder River basin (Wyo.)	Naeser (1984, 1992)
	Midcontinent	
	Anadarko basin (Okla.)	Kelley et al. (1991)
	Arkoma basin, Ouachita Mts. (Ark.)	Arne (1989, 1992); Arne et al. (1990a)
	Gulf Coast (La.)	Dokka (1982)
	Illinois basin	Zimmermann (1981, 1986)
	Iowa	Crowley et al. (1986)
	Michigan basin	Crowley (1991)

TABLE 1 (continued)

Country/Region	Area	References
USA	*Midcontinent* (continued)	
	Southeast Missouri lead district and surrounding area	Arne (1989); Arne et al. (1990b)
	Williston basin (N. Dak.)	Crowley et al. (1985); Crowley and Kuhlman (1988)
	Eastern USA	
	Appalachian basin	Lakatos and Miller (1983); Johnsson (1985, 1986); Miller et al. (1986); Duddy et al. (1987); Roden and Miller (1988, 1989b); Roden et al. (1990)
	Hartford basin (Conn.)	Roden and Miller (1989a); Miller et al. (1990)
	Newark basin (N.J.)	Kohn et al. (1988, 1990b); Roden and Miller (1989a); Miller et al. (1990)
	Taylorsville basin (Va.)	Roden and Miller (1989a); Miller et al. (1990)

[a] Specific location not given.

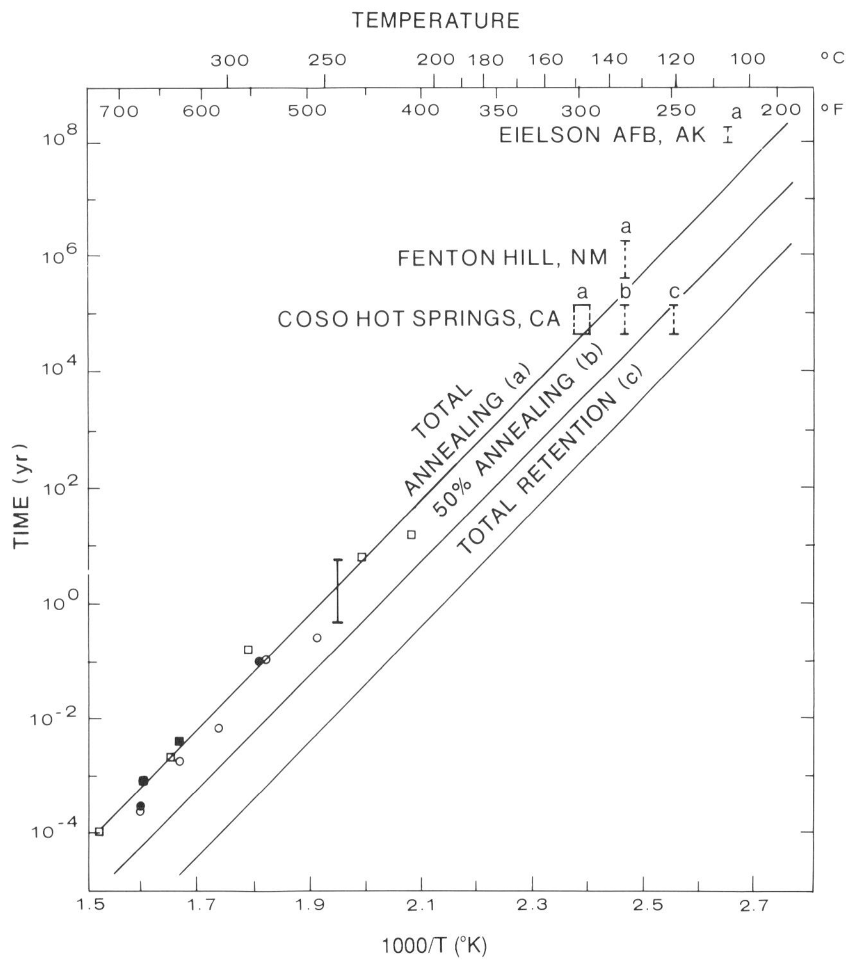

Fig. 7. Temperatures required to anneal fission tracks in apatite over heating durations ranging from short-term laboratory experiments to long-term geologic heating. Temperatures for geologic annealing (dashed vertical bars) were determined by dating apatite in drill holes in three areas where the approximate duration of heating is known (Naeser, 1981). Circles and squares indicate the temperature–time conditions predicted for total annealing using the fading rates determined by the laboratory work of Naeser and Faul (1969) (open circles), Reimer (1972) (solid circles), Märk et al. (1973) (open squares), and Zimmermann and Gaines (1978) (solid squares). Extrapolation of the laboratory data to geologic times and temperatures (solid lines) is based on the rate constant for fission-track annealing determined from a least-squares fit of the laboratory data (Zimmermann and Gaines, 1978). The solid vertical bar is the 95% confidence interval for the least-squares fit. Letters on the solid lines and dashed vertical bars identify data for (*a*) "total annealing", (*b*) 50% annealing, and (*c*) "total retention" of fission tracks. Fluorapatite was used in all of the laboratory studies plotted here and analyses by K.D. Crowley (pers. commun., 1988) indicate that the apatites analyzed in at least two of the drill holes are F-apatite or F-OH-apatite with little or no Cl. Note that Harrison (1985) has suggested revised estimates for the effective heating times at Eielson (~20 to 30 million years) and Fenton Hill (~400,000 years) that bring the drill-hole data into even closer agreement with the extrapolated laboratory data. Figure modified from Sanford (1981, fig. 2).

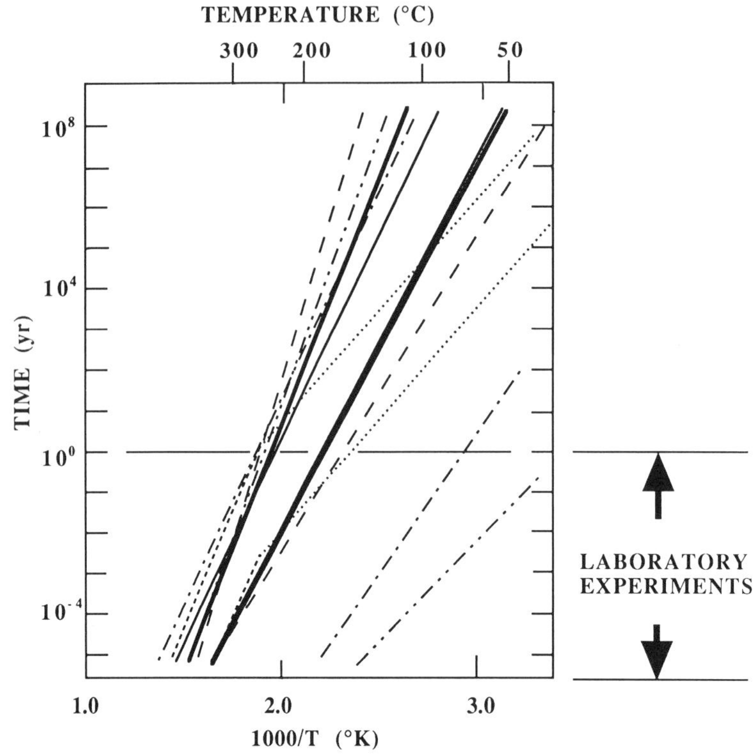

Fig. 8. Comparison of temperature–time correlation for geologic annealing predicted by selected annealing models. Boundaries of the zone of partial annealing (cf. Fig. 7) are shown for five models based on laboratory annealing studies: Naeser and Faul (1969) (dashed lines), Märk et al. (1973) (dotted lines), Zimmermann and Gaines (1978) (light solid lines), Laslett et al. (1987) (dot–dashed lines), and Crowley et al. (1991) (dot–dot–dashed lines). Heavy solid lines (Gleadow and Duddy, 1981) are based on drill-hole annealing data from the Otway and Gippsland basins, Australia, and the laboratory annealing data of Naeser and Faul (1969). The models of Laslett et al. (1987) and Crowley et al. (1991) are based on track lengths; track densities were used in the remaining studies. In the models of Märk et al. (1973), Zimmermann and Gaines (1978), Laslett et al. (1987), and Crowley et al. (1991), lines of total track retention ($r = \rho/\rho_0$ [or l/l_0] = 1) and total annealing ($r = 0$) are approximated by $r = 0.99$ and 0.01, respectively (ρ/ρ_0 [l/l_0] = etchable track density [mean confined track length] expressed in normalized form, where ρ_0 [l_0] = the etchable density [mean confined track length] of unannealed tracks). Note that the beginning of annealing (total retention line) is offset to lower temperatures in the two models based on track lengths, reflecting at least in part the fact that etchable track lengths seem to show the affects of annealing at lower temperatures than track densities (see text). Figure based on Crowley (1989) and K.D. Crowley (written commun., 1991).

confined track *length* (e.g., Green et al., 1985; Laslett et al., 1987; Crowley et al., 1991; Donelick, 1991). A critical question still to be resolved is the relationship between etchable track-length and track density, particularly in the early (low-temperature) stages of annealing. Some workers interpret annealing data as indicating a 1:1 correlation between reduction in length and density during the first 25% or so of track shortening (Green, 1988), whereas other research suggests that there is not a simple 1:1 correlation between length and density and that there may in fact be as much as a 5 to 20% reduction in measured track-length without a significant decrease in track density (C.W. Naeser et al., 1989a). Until the length–density relationship is resolved, the differences between annealing models based on track-lengths versus models based on track densities will not be fully reconciled.

At least nine laboratories worldwide are currently involved in research to more accurately define the kinetics of apatite annealing and improve the practical application of apatite annealing models. As of now, there is *not* a consensus on which of the proposed apatite annealing models best describes annealing for geologically significant time–temperature conditions (see, e.g., Deming et al., 1990; Corrigan, 1991a).

In addition to its importance for assessing various annealing models, the track-length–density relationship is also critical because of the practice of some labs of "correcting" apatite fission-track ages (track densities) of samples based on the extent of track-length reduction. Such corrections have direct bearing on studies in which fission-track ages are used, for example, to constrain the time of uplift and/or time of cooling in sedimentary rocks. Until the length–density issue is resolved, "corrected" apatite fission-track ages should be viewed with caution.

Inherited age and thermal history of detrital apatites

Detrital apatites in sediments are typically derived from parent rocks of widely varying age and thermal history. When detrital minerals enter a depositional basin, they carry with them previously formed fission tracks that reflect the thermal history(s) of the sediment source. Thus, the observed variations in apatite age and track-length distribution with depth in any given sedimentary basin will reflect the age(s) and thermal history(s) of the parent rocks, as well as the thermal history of the depositional basin and the compositional variability of the apatite (see below). These factors, combined with the variation in annealing between basins caused by differences in heating time, must be considered when interpreting apatite fission-track data from sedimentary rocks. Models that assume that all apatites came into a basin with no inherited tracks and that observed variation in age and track-lengths is related only to the thermal history of the basin can produce significant errors in interpreted thermal history in rocks that have not been exposed to temperatures high enough to totally anneal apatite. This is particularly true in basins where only a few isolated samples are available for analysis. The presence of inherited tracks in detrital apatites can be a major limiting factor in the ability to determine a unique thermal history for a sedimentary sequence.

Apatite composition

A factor that may account for the differences between some annealing models, and that also affects the applicability of various models to a given apatite suite, is the role of composition in controlling annealing temperatures. Drill-hole and laboratory annealing data indicate that there is a difference in the annealing susceptibility of individual apatite grains and that this difference is related to the anion composition of apatite, in particular to the Cl-content. For a given heating time, Cl-apatite anneals at temperatures up to ~30°C higher than F-, Sr–F-, and OH-apatite (which, in contrast to Cl-apatite, have very similar annealing characteristics) (Green et al., 1985, 1989a; Crowley and Cameron, 1987; Crowley et al., 1990; Hughes et al., 1990), affecting the observed variation in both apatite age and etchable track-lengths with progressive annealing.

Fortunately, apatite suites are commonly so dominated by F-apatite with insignificant amounts of chlorine (e.g., Berry and Mason, 1959; Deer et al., 1962; Naeser et al., 1987a) that most samples can be reasonably interpreted using F-apatite annealing data. Nevertheless, composition may pose a problem for interpreting thermal history in those basins where the detrital apatites have a significant and variable range in Cl-content.

Apatite composition is normally determined by electron microprobe. The ongoing challenge is to develop an alternative, routine method for determining composition (particularly Cl-content) of individual apatite grains that will eliminate the need for electron microprobe analysis. One method currently being developed uses infrared reflected light microspectroscopy to determine F:Cl composition of individual apatites in the fission-track grain mount (Siddall and Mendelssohn, 1990). Another method under investigation involves measuring track *widths* in the individual apatite grains used for track-length and age determinations. This method is based on the observation that chlorapatites appear to etch more rapidly than other apatite varieties, producing characteristic broad tracks (Crowley et al., 1990). Research is underway to determine if a quantitative correlation can be established between track width and Cl-content (C.W. Naeser, K.D. Crowley and R.A. Donelick, pers. commun., 1991).

Note: zircons in basin analysis

Although apatite has received the most attention and is the most widely used mineral in fission-track studies in sedimentary basins, zircon fission-track analysis also has a role to play in basin modelling. It is true that zircon is generally not as useful as apatite for thermal history studies in sedimentary rocks, for several reasons. Firstly, zircon anneals at higher temperatures than apatite (~200 ± 50°C) (Harrison et al., 1979; Hurford, 1985; Zeitler, 1985), and, therefore, in typical sedimentary rock samples it shows little if any effects of annealing related to burial heating. Secondly, zircon annealing temperatures are not as well known as apatite annealing temperatures, in part because zircon annealing is a more complicated process that is influenced not only by temperature and time but also by the accumulated radiation damage in any given zircon grain (Kasuya and Naeser, 1990). Consequently, if zircon is included in basin thermal history studies at all, which often it is not, it is usually only used to set a general upper limit on maximum palaeotemperatures.

The real contribution of zircon fission-track analysis to basin studies lies in areas other than thermal modelling. For example, the fission-track ages of individual detrital zircon grains can provide information on the provenance of sedimentary rocks and on the uplift and erosion history of the parent rocks (e.g., N.D. Naeser et al., 1987b, 1989b; Cerveny et al., 1988; Hurford and Carter, 1991). Also, fission-track dating of zircon phenocrysts has been widely used to determine the age of volcanic ash beds, which form one of the best marker horizons available for establishing correlation and chronology in sedimentary sequences (reviewed in Naeser and Naeser, 1988).

Conclusions

Uncertainties remain in the application of apatite fission-track analysis to sedimentary basins. Resolving these uncertainties is the subject of current research by laboratories worldwide. Nevertheless, it should be recognized that, in spite of these uncertainties, fission-track analysis has proven to be an extremely useful method for studying sedimentary basins over the last ten years. It has clarified the thermal history in many areas and provided information that would be difficult to obtain otherwise. Uncertainties in the method have really only arisen over the last several years as we have tried to refine the fission-track method to obtain increasingly *quantitative* information on the thermal history of sedimentary basins. While attempting to resolve these problems, we should not lose sight of the valuable role that fission tracks have to play in basin analysis. At the same time, users of fission-track data should be aware of limitations of the method and should be cautious of studies that attempt to overinterpret data beyond the current state of refinement of apatite fission-track analysis.

Acknowledgements

The author wishes to thank William Clendenen, Kevin Crowley, Raymond Donelick, and an anonymous reviewer for helpful comments that improved this manuscript.

References

Arne, D.C., 1989. Thermal Histories of Selected Mississippi Valley-Type Ore Districts from Apatite Fission Track Analysis. Ph.D. Thesis, University of Melbourne, Parkville, Vic., 281 pp.

Arne, D.C., 1991. Regional thermal history of the Pine Point area, Northwest Territories, Canada, from apatite fission-track analysis. Econ. Geol., 86: 428–435.

Arne, D.C., 1992. Evidence from apatite fission track analysis for regional Cretaceous cooling in the Ouachita Mountain fold belt and Arkoma basin of Arkansas. Am. Assoc. Pet. Geol., Bull., 76: 392–402.

Arne, D.C., Green, P.F., Duddy, I.R., Gleadow, A.J.W., Lambert, I.B. and Lovering, J.F., 1989. Regional thermal history of the Lennard Shelf, Canning Basin from apatite fission track analysis: implications for the formation of Pb–Zn ore deposits. Aust. J. Earth Sci., 36: 495–513.

Arne, D.C., Duddy, I.R. and Green, P.F., 1990a. The thermal history of the Arkoma basin and Ouachita Mountains from AFTA (apatite fission track analysis). Am. Assoc. Pet. Geol., Bull., 74: 601 (abstract).

Arne, D.C., Green, P.F. and Duddy, I.R., 1990b. Thermochronologic constraints on the timing of Mississippi Valley-type ore formation from apatite fission track analysis. Nucl. Tracks Radiat. Measurements, 17: 319–323.

Beaty, D.W., Naeser, C.W., Cunningham, C.G. and Landis, G.P., 1988. Genetic model for the Gilman District (Colo.), based on fluid inclusion, stable isotope, geologic, and fission-track time/temperature studies. Geol. Soc. Am., Abstr. Prog., 20 (7): A38.

Beaumont, C., Willett, S.D., Ravenhurst, C. and Donelick, R.A., 1989. Paleotemperature and fluid flow in the Alberta Basin from apatite fission track annealing data. EOS, 70: 1096 (abstract).

Berry, L.G. and Mason, B., 1959. Mineralogy. W.H. Freeman, San Francisco, Calif., 630 pp.

Briggs, N.D., Naeser, C.W. and McCulloh, T.H., 1979. Thermal history of sedimentary basins by fission-track dating. Geol. Soc. Am., Abstr. Prog., 11: 394.

Briggs, N.D., Naeser, C.W. and McCulloh, T.H., 1981a. Thermal history of sedimentary basins by fission-track dating. Nucl. Tracks, 5: 235–237 (abstract).

Briggs, N.D., Naeser, C.W. and McCulloh, T.H., 1981b. Thermal history by fission-track dating, Tejon Oil Field area, California. Am. Assoc. Pet. Geol., Bull., 65: 906 (abstract).

Brown, R.W. and Green, P.F., 1991. Discussion on thermal and tectonic history of the East Midlands shelf (onshore UK) and surrounding regions assessed by apatite fission track analysis (discussion and reply). J. Geol. Soc., London, 148: 785–787.

Burchart, J., Butkiewicz, T., Dakowski, M. and Galazka-Friedman, J., 1979. Fission track retention in minerals as a function of heating time during isothermal experiments: a discussion. Nucl. Tracks, 3: 109–117.

Carlson, W.D., 1990. Mechanisms and kinetics of apatite fission-track annealing. Am. Mineral., 75: 1120–1139.

Cerveny, P.F., Naeser, N.D., Zeitler, P.K., Naeser, C.W. and Johnson, N.M., 1988. History of uplift and relief of the Himalaya during the past 18 million years: evidence from fission-track ages of detrital zircons from sandstones of the Siwalik Group. In: K.L. Kleinspehn and C. Paola (Editors), New Perspectives in Basin Analysis. Springer-Verlag, New York, N.Y., pp. 43–61.

Chaillou, D. and Chambaudet, A., 1981. Isothermal plateau method for apatite fission-track dating. Nucl. Tracks, 5: 181–186.

Corrigan, J.D., 1991a. Inversion of apatite fission track data for thermal history information. J. Geophys. Res., 96: 10347–10360.

Corrigan, J.D., 1991b. Thermal anomalies in the central Indian Ocean: evidence for dewatering of the Bengal fan. J. Geophys. Res., 96: 14263–14275.

Corrigan, J.D. and Crowley, K.D., 1989. Thermal history of sites 717 and 718, ODP Leg 116, central Indian Ocean: constraints from numerical simulations and apatite fission-track data. Geol. Soc. Am., Abstr. Prog., 21 (6): 320.

Crowley, K.D., 1985. Thermal significance of fission-track length distributions. Nucl. Tracks Radiat. Measurements, 10: 311–322.

Crowley, K.D., 1989. Annealing systematics and age interpretation. In: K.D. Crowley, C.W. Naeser and N.D. Naeser, Fission-Track Analysis: Theory and Applications. Geol. Soc. Am. Short Course Notes, pp. 4-1–4-44.

Crowley, K.D., 1991. Thermal history of Michigan Basin and southern Canadian Shield from apatite fission track analysis. J. Geophys. Res., 96: 697–711.

Crowley, K.D. and Cameron, M., 1987. Annealing of etchable fission-track damage in apatite: effects of anion chemistry. Geol. Soc. Am., Abstr. Prog., 19: 631–632.

Crowley, K.D. and Kuhlman, S.L., 1988. Apatite thermochronometry of western Canadian Shield: implications for origin of the Williston Basin. Geophys. Res. Lett., 15: 221–224.

Crowley, K.D., Ahern, J.L. and Naeser, C.W., 1985. Origin and epeirogenic history of the Williston Basin: evidence from fission-track analysis of apatite. Geology, 13: 680–685.

Crowley, K.D., Naeser, C.W. and Babel, C.A., 1986. Tectonic significance of Precambrian apatite fission-track ages from the midcontinent United States. Earth Planet. Sci. Lett., 79: 329–336.

Crowley, K.D., Cameron, M. and McPherson, B.J., 1990. Annealing of etchable fission-track damage in F-, OH-, Cl-, and Sr-apatite, 1. Systematics and preliminary interpretations. Nucl. Tracks Radiat. Measurements, 17: 409–410 (abstract).

Crowley, K.D., Cameron, M. and Schaefer, R.L., 1991. Experimental studies of annealing of etched fission tracks in fluorapatite. Geochim. Cosmochim. Acta, 55: 1449–1465.

Cunningham, C.G. and Barton, P.B., Jr., 1984. Recognition and use of paleothermal anomalies as a new exploration tool. Geol. Soc. Am., Abstr. Prog., 16: 481.

Damm, K.-W. and Kelley, S., 1989. On the thermal history of the central Andean basement — evidence from combined U/Pb, Nd/Sm, Rb/Sr, and fission track data. Geol. Soc. Am., Abstr. Prog., 21 (6): 267.

Deer, W.A., Howie, R.A. and Zussman, J., 1962. Rock-Forming Minerals, Vol. 5. Non-Silicates. John Wiley, New York, N.Y., 371 pp.

Deming, D., Nunn, J.A., Jones, S. and Chapman, D.S., 1990. Some problems in thermal history studies. In: V.F. Nuccio and C.E. Barker (Editors), Applications of Thermal Maturity Studies to Energy Exploration. Rocky Mountain Section of the Society of Economic Paleontologists and Mineralogists, Denver, Colo., pp. 61–80.

Dokka, R.K., 1982. Implications of fission track ages from the Kaplan geothermal-geopressure zone, Vermilion Parish, Louisiana. Gulf Coast Assoc. Geol. Soc. Trans., 32: 455–468.

Donelick, R.A., 1991. Crystallographic orientation dependence of mean etchable fission track length in apatite: an empirical model and experimental observations. Am. Mineral., 76: 83–91.

Donelick, R.A., Roden, M.K., Mooers, J.D., Carpenter, B.S. and Miller, D.S., 1990. Etchable length reduction of induced fission tracks in apatite at room temperature (≈23°C): crystallographic orientation effects and "initial" mean lengths. Nucl. Tracks Radiat. Measurements, 17: 261–265.

Duddy, I.R. and Gleadow, A.J.W., 1982. Thermal history of the Otway Basin, southeastern Australia, from geologic annealing of fission tracks in detrital volcanic apatites. Abstr., Fission-Track Dating Workshop, 5th Int. Conf. Geochronology, Cosmochronology, and Isotope Geology, Nikko National Park, Japan, June 30, pp. 13–16.

Duddy, I.R. and Gleadow, A.J.W., 1985. The application of fission track thermochronology to sedimentary basins: two case studies. Nucl. Tracks Radiat. Measurements, 10: 408 (abstract).

Duddy, I.R., Gleadow, A.J.W. and Keene, J.B., 1984. Fission track dating of apatite and sphene from Paleogene sediments of Deep Sea Drilling Project Leg 81, site 555. In: D.G. Roberts and D. Schnitker (Editors), Initial Reports of the Deep Sea Drilling Project, Vol. 81. U.S. Government Printing Office, Washington, D.C., pp. 725–729.

Duddy, I.R., Gleadow, A.J.W., Green, P.F., Hegarty, K.A., Marshallsea, S.A., Tingate, P.R. and Lovering, J.F., 1987. Quantitative estimates of thermal history and maturation using AFTA (apatite fission track analysis) in extensional and foreland basins — selected case histories. Am. Assoc. Pet. Geol., Bull., 71: 550–551 (abstract).

Duddy, I.R., Green, P.F. and Laslett, G.M., 1988. Thermal annealing of fission tracks in apatite, 3. Variable temperature behaviour. Chem. Geol., 73: 25–38.

Dumitru, T.A., 1988. Subnormal geothermal gradients in the Great Valley forearc basin, California, during Franciscan subduction: a fission track study. Tectonics, 7: 1201–1221.

Dumitru, T.A., 1989. Constraints on uplift in the Franciscan subduction complex from apatite fission track analysis. Tectonics, 8: 197–220.

Dumitru, T.A., 1991a. Effects of subduction parameters on geothermal gradients in forearcs, with an application to Franciscan subduction in California. J. Geophys. Res., 96: 621–641.

Dumitru, T.A., 1991b. Major Quaternary uplift along the northernmost San Andreas fault, Kings Range, northwestern California. Geology, 19: 526–529.

Feinstein, S., Kohn, B.P. and Eyal, M., 1986. Integration of fission-track and vitrinite reflectance measurements, for thermal history analysis of sedimentary basins: an example from southern Israel. Terra Cognita, 6: 119 (abstract).

Feinstein, S., Kohn, B.P. and Eyal, M., 1989. Significance of combined vitrinite reflectance and fission-track studies in evaluating thermal history of sedimentary basins: an example from southern Israel. In: N.D. Naeser and T.H. McCulloh (Editors), Thermal History of Sedimentary Basins — Methods and Case Histories. Springer-Verlag, New York, N.Y., pp. 197–216.

Fitzgerald, P.G. and Gleadow, A.J.W., 1990. New approaches in fission track geochronology as a tectonic tool: examples from the Transantarctic Mountains. Nucl. Tracks Radiat. Measurements, 17: 351–357.

Fleischer, R.L., Price, P.B. and Walker, R.M., 1975. Nuclear Tracks in Solids — Principles and Applications. University of California Press, Berkeley, Calif., 605 pp.

Forbes, P.L., Meyer, A. and Pagel, M., 1989. Modeling of apatite fission tracks formation and annealing: application to thermal evolution in the Guezouman Formation at Akouta, Nigeria. Abstr., 28th Int. Geol. Congr., Washington, D.C., July 9–19, Vol. 1, pp. 499–500.

Geving, R.L., Suchecki, R.K. and Kelley, S.E., 1989. Kinetic controls on the transformation of smectite to illite in the Los Angeles Basin as determined by apatite fission track analysis. Geol. Soc. Am., Abstr. Prog., 21 (6): 349.

Geving, R.L., Kelley, S.E. and Suchecki, R.K., 1990a. Thermal history of the Los Angeles Basin. Geol. Soc. Am., Abstr. Prog., 22 (7): 32.

Geving, R.L., Suchecki, R.K. and Kelley, S.E., 1990b. Kinetic control on clastic diagenesis in the Los Angeles basin as determined by apatite fission track analysis. Am. Assoc. Pet. Geol., Bull., 74: 661 (abstract).

Geving, R.L., Suchecki, R.K., Kelley, S.E. and Lyons, K.T., 1991. Porosity prediction in feldspathic sandstones of the Los Angeles basin using apatite fission track analysis. Am. Assoc. Pet. Geol., Bull., 75: 579 (abstract).

Giegengack, R., Omar, G.I. and Johnson, K.R., 1986. A reconnaissance fission track uplift chronology for the northwest margin of the Bighorn Basin. 1986 Montana Geological Society — Yellowstone Bighorn Research Association Field Conference Guidebook, pp. 179–184.

Giegengack, R., Omar, G.I., Lutz, T.M. and Johnson, K.R., 1987. Thermal history of the Bighorn Basin and flanking highlands, Montana-Wyoming, by fission-track analysis. Geol. Soc. Am., Abstr. Prog., 19: 675.

Giegengack, R., Omar, G.I., Lutz, T.M. and Johnson, K.R., 1990. Tectono-thermal history by fission-track analysis of the Bighorn Basin and adjacent highlands, Montana–Wyoming. Nucl. Tracks Radiat. Measurements, 17: 412–413 (abstract).

Gleadow, A.J.W., 1989. Thermal history of Clarence–Moreton Basin in eastern Australia from apatite fission track analysis. Abstr., 28th Int. Geol. Congr., Washington, D.C., July 9–19, Vol. 1, p. 559.

Gleadow, A.J.W., 1990. Apatite fission track thermochronology and tectonics in the Clarence–Moreton Basin of eastern Australia. Nucl. Tracks Radiat. Measurements, 17: 413–414 (abstract).

Gleadow, A.J.W. and Duddy, I.R., 1981. A natural long-term track annealing experiment for apatite. Nucl. Tracks, 5: 169–174.

Gleadow, A.J.W. and Duddy, I.R., 1982. Fission track lengths in the apatite partial stability zone and the interpretation of mixed ages. Abstr., Fission-Track Dating Workshop, 5th Int. Conf. Geochronology, Cosmochronology, and Isotope Geology, Nikko National Park, Japan, June 30, pp. 19–21.

Gleadow, A.J.W. and Duddy, I.R., 1984. Fission track dating and thermal history analysis of apatites from wells in the north-west Canning Basin. In: P.G. Purcell (Editor), The Canning Basin. Geological Society of Australia and Petroleum Exploration Society of Australia, Perth, W.A., pp. 377–387.

Gleadow, A.J.W., Duddy, I.R. and Lovering, J.F., 1983. Fission track analysis: a new tool for the evaluation of thermal histories and hydrocarbon potential. Aust. Pet. Explor. Assoc. J., 23: 93–102.

Gleadow, A.J.W., Duddy, I.R., Green, P.F. and Hegarty, K.A., 1986a. Fission track lengths in the apatite annealing zone and the interpretation of mixed ages. Earth Planet. Sci. Lett., 78: 245–254.

Gleadow, A.J.W., Duddy, I.R., Green, P.F. and Lovering, J.F., 1986b. Confined fission track lengths in apatite: a diagnostic tool for thermal history analysis. Contrib. Mineral. Petrol., 94: 405–415.

Green, P.F., 1986. On the thermo-tectonic evolution of northern England: evidence from fission track analysis. Geol. Mag., 123: 493–506.

Green, P.F., 1988. The relationship between track shortening and fission track age reduction in apatite: combined influences of inherent instability, annealing anisotropy, length bias and system calibration. Earth Planet. Sci. Lett., 89: 335-352.

Green, P.F., 1989a. Uplift and erosional history of Cleveland Basin and surrounding areas revealed by apatite fission track analysis. Abstr., 28th Int. Geol. Congr., Washington, D.C., July 9–19, Vol. 1, pp. 585–586.

Green, P.F., 1989b. Thermal and tectonic history of the East Midlands shelf (onshore UK) and surrounding regions assessed by apatite fission track analysis. J. Geol. Soc., London, 146: 755–773.

Green, P.F., 1990. Widespread Tertiary uplift of the East Midlands of England revealed by apatite fission track analysis. Nucl. Tracks Radiat. Measurements, 17: 415 (abstract).

Green, P.F. and Durrani, S.A., 1977. Annealing studies of tracks in crystals. Nucl. Tracks, 1: 33–39.

Green, P.F. and White, S.H., 1985. Exhumed “partial” fission-track ages from South Island, New Zealand. Nucl. Tracks Radiat. Measurements, 10: 412 (abstract).

Green, P.F., Duddy, I.R., Gleadow, A.J.W., Tingate, P.R. and Laslett, G.M., 1985. Fission-track annealing in apatite: track length measurements and the form of the Arrhenius plot. Nucl. Tracks Radiat. Measurements, 10: 323–328.

Green, P.F., Duddy, I.R., Gleadow, A.J.W., Tingate, P.R. and Laslett, G.M., 1986. Thermal annealing of fission tracks in apatite, 1. A qualitative description. Chem. Geol., 59: 237–253.

Green, P.F., Duddy, I.R., Gleadow, A.J.W. and Lovering, J.F., 1989a. Apatite fission-track analysis as a paleotemperature indicator for hydrocarbon exploration. In: N.D. Naeser and T.H. McCulloh (Editors), Thermal History of Sedimentary Basins — Methods and Case Histories. Springer-Verlag, New York, N.Y., pp. 181–195.

Green, P.F., Duddy, I.R., Laslett, G.M., Hegarty, K.A., Gleadow, A.J.W. and Lovering, J.F., 1989b. Thermal annealing of fission tracks in apatite, 4. Quantitative modelling techniques and extension to geological timescales. Chem. Geol. (Isot. Geosci. Sect.), 79: 155–182.

Hansen, K., 1990. Thermal history of the Jameson Land basin, East Greenland: a fission track study. Nucl. Tracks Radiat. Measurements, 17: 416 (abstract).

Harrison, T.M., 1985. A reassessment of fission-track annealing behavior in apatite. Nucl. Tracks Radiat. Measurements, 10: 329–333.

Harrison, T.M., Armstrong, R.L., Naeser, C.W. and Harakal, J.E., 1979. Geochronology and thermal history of the Coast Plutonic Complex, near Prince Rupert, British Columbia. Can. J. Earth Sci., 16: 400–410.

Hill, K.C., 1990. Time of uplift and thermal history of the fold-belt in Papua New Guinea, from apatite fission track analysis. Nucl. Tracks Radiat. Measurements, 17: 417 (abstract).

Hill, K.C. and Gleadow, A.J.W., 1989. Uplift and thermal history of the Papuan fold belt, Papua New Guinea: apatite fission track analysis. Aust. J. Earth Sci., 36: 515–539.

Hood, A., Gutjahr, C.C.M. and Heacock, R.L., 1975. Organic metamorphism and the generation of petroleum. Am. Assoc. Pet. Geol., Bull., 59: 986–996.

Hughes, J.M., Cameron, M. and Crowley, K.D., 1990. Annealing of etchable fission-track damage in F-, OH-, Cl-, and Sr-apatite, 2. Correlations with crystal structure. Nucl. Tracks Radiat. Measurements, 17: 417–418 (abstract).

Hurford, A.J., 1985. On the closure temperature for fission tracks in zircon. Nucl. Tracks Radiat. Measurements, 10: 415 (abstract).

Hurford, A.J. and Carter, A., 1991. The role of fission track dating in discrimination of provenance. In: A.C. Morton, S.P. Todd and P.D.W. Haughton (Editors), Developments in Sedimentary Provenance Studies. Geol. Soc. London, Spec. Publ., 57: 67–78.

Issler, D.R., Beaumont, C., Willett, S.D., Donelick, R.A., Mooers, J. and Grist, A., 1990. Preliminary evidence from apatite fission-track data concerning the thermal history of the Peace River Arch region, Western Canada Sedimentary Basin. Bull. Can. Pet. Soc., 38A: 250–269.

Johnsson, M.J., 1985. Late Paleozoic–Middle Mesozoic uplift rate, cooling rate and geothermal gradient for south-central New York State. Nucl. Tracks Radiat. Measurements, 10: 295–301.

Johnsson, M.J., 1986. Distribution of maximum burial temperatures across northern Appalachian Basin and implications for Carboniferous sedimentation patterns. Geology, 14: 384–387.

Kamp, P.J.J. and Green, P.F., 1990. Thermal and tectonic history of selected Taranaki basin (New Zealand) wells assessed by apatite fission track analysis. Am. Assoc. Pet. Geol., Bull., 74: 1401–1419.

Kamp, P.J.J., Hegarty, K.A. and Green, P.F., 1988. Thermo-tectonic history of Taranaki basin (New Zealand) using apatite fission track analysis (AFTA). Am. Assoc. Pet. Geol., Bull., 72: 204 (abstract).

Kamp, P.J.J., Green, P.F. and White, S.H., 1989. Fission track analysis reveals character of collisional tectonics in New Zealand. Tectonics, 8: 169–195.

Kamp, P.J.J., Green, P.F. and Tippett, J.M., 1992. Tectonic architecture of the mountain front-foreland basin transition, South Island, New Zealand, assessed by fission track analysis. Tectonics, 11: 98–113.

Kasuya, M. and Naeser, C.W., 1990. The effect of α-damage on the fission-track annealing in zircon. Nucl. Tracks Radiat. Measurements, 17: 419 (abstract).

Kelley, S.A. and Blackwell, D.D., 1990. Thermal history of the Multi-Well Experiment (MWX) Site, Piceance Creek Basin, northwestern Colorado, derived from fission-track analysis. Nucl. Tracks Radiat. Measurements, 17: 331–337.

Kelley, S.A., Gallardo, J.D., Carter, L.C. and Blackwell, D.D., 1991. Thermal conditions in the Anadarko basin, Oklahoma. Am. Assoc. Pet. Geol., Bull., 75: 607 (abstract).

Kohn, B.P., Wagner, M.E., Lutz, T.M., Organist, G. and Omar, G.I., 1988. Mesozoic rift-related and post-rift uplift of the central Appalachian Piedmont: constraints from fission-track analysis. Geol. Soc. Am., Abstr. Prog., 20 (7): A387.

Kohn, B.P., Feinstein, S. and Eyal, M., 1990a. Cretaceous to present paleothermal gradients, central Negev, Israel: constraints from fission track dating. Nucl. Tracks Radiat. Measurements, 17: 381–388.

Kohn, B.P., Wagner, M.E., Lutz, T.M. and Organist, G., 1990b. Regional thermal overprinting during Mesozoic rifting recorded by sphene and zircon fission track dates, central Appalachian Piedmont, U.S.A. Nucl. Tracks Radiat. Measurements, 17: 420 (abstract).

Kveton, K.J., 1990. Bearing of fission-track dates on the provenance, age, and thermal evolution of the Paleogene Orca Group, S Alaska, U.S.A. Nucl. Tracks Radiat. Measurements, 17: 420–421 (abstract).

Lakatos, S. and Miller, D.S., 1983. Fission-track analysis of apatite and zircon defines a burial depth of 4 to 7 km for lowermost Upper Devonian, Catskill Mountains, New York. Geology, 11: 103–104.

Laslett, G.M., Kendall, W.S., Gleadow, A.J.W. and Duddy, I.R., 1982. Bias in measurement of fission-track length distributions. Nucl. Tracks, 6: 79–85.

Laslett, G.M., Green, P.F., Duddy, I.R. and Gleadow, A.J.W., 1987. Thermal annealing of fission tracks in apatite, 2. A quantitative analysis. Chem. Geol., 65: 1–13.

Märk, E., Pahl, M., Purtscheller, F. and Märk, T.D., 1973. Thermische Ausheilung von Uran-Spaltspuren in Apatiten, Alterskorrekturen und Beiträge zur Geothermochronologie. Tschermaks Mineral. Petrogr. Mitt., 20: 131–154.

Marshallsea, S.J., 1986. Thermal history of the Bowen Basin, Queensland, Australia: an apatite fission track study. Terra Cognita, 6: 118 (abstract).

Marshallsea, S.J., 1990. The thermal history of the Bowen Basin: an apatite fission track study. Nucl. Tracks Radiat. Measurements, 17: 421–422 (abstract).

McMillen, K.J. and O'Sullivan, P.B., 1992. Tectonics and eustatic controls on Paleogene sequence stratigraphy: Beaufort Sea, Alaska and Canada. Am. Assoc. Pet. Geol. Bull. (in press).

Meyer, A.J., Storzer, D. and Pagel, M., 1990. Evidence from fission track analysis of thermal events in a sedimentary basin. Nucl. Tracks Radiat. Measurements, 17: 422–423 (abstract).

Miller, D.S., Duddy, I.R. and Roden, M.K., 1986. Burial and uplift study of northern Appalachian Basin using apatite and zircon fission-track chronology. Terra Cognita, 6: 118 (abstract).

Miller, D.S., Roden, M.K. and Donelick, R.A., 1990. Thermal history of three Mesozoic sedimentary basins, eastern United States. Nucl. Tracks Radiat. Measurements, 17: 423 (abstract).

Naeser, C.W., 1976. Fission track dating. U.S. Geol. Surv. Open-File Rep., 76-190, 65 pp.

Naeser, C.W., 1979a. Fission-track dating and geologic annealing of fission tracks. In: E. Jäger and J.C. Hunziker (Editors), Lectures in Isotope Geology. Springer-Verlag, New York, N.Y., pp. 154–169.

Naeser, C.W., 1979b. Thermal history of sedimentary basins: fission-track dating of subsurface rocks. In: P.A. Scholle and P.R. Schluger (Editors), Aspects of Diagenesis. Soc. Econ. Paleontol. Mineral., Spec. Publ., 26: 109–112.

Naeser, C.W., 1981. The fading of fission tracks in the geologic environment — data from deep drill holes. Nucl. Tracks, 5: 248–250 (abstract).

Naeser, C.W. and Cunningham, C.G., 1984. Age and paleothermal anomaly of the Eagle Mine ore body, Gilman district, Colorado. Geol. Soc. Am., Abstr. Prog., 16: 607.

Naeser, C.W. and Faul, H., 1969. Fission track annealing in apatite and sphene. J. Geophys. Res., 74: 705–710.

Naeser, C.W. and Naeser, N.D., 1988. Fission-track dating of Quaternary events. In: D.J. Easterbrook (Editor), Dating Quaternary Sediments. Geol. Soc. Am., Spec. Pap., 227: 1–11.

Naeser, C.W., Cunningham, C.G., Marvin, R.F. and Obradovich, J.D., 1980. Pliocene intrusive rocks and mineralization near Rico, Colorado. Econ. Geol., 75: 122–127.

Naeser, C.W., Crowley, K.D., McPherson, B.J. and Cameron, M., 1989a. The relationship between fission-track length and fission-track density in apatite. Geol. Soc. Am., Abstr. Prog., 21 (6): 241–242.

Naeser, C.W., Duddy, I.R., Elston, D.P., Dumitru, T.A. and Green, P.F., 1989b. Fission-track dating: ages for Cambrian strata and Laramide and post-middle Eocene cooling events from the Grand Canyon, Arizona. In: D.P. Elston, G.H. Billingsley and R.A. Young (Editors), Geology of the Grand Canyon, Northern Arizona. Field Trip Guidebook T115/315, 28th Int. Geol. Congr., Washington, D.C., July 9–19, pp. 139–144.

Naeser, N.D., 1984. Thermal history determined by fission-track dating for three sedimentary basins in California and Wyoming. Geol. Soc. Am., Abstr. Prog., 16: 607.

Naeser, N.D., 1986. Neogene thermal history of the northern Green River basin, Wyoming — evidence from fission-track dating. In: D.L. Gautier (Editor), Roles of Organic Matter in Sediment Diagenesis. Soc. Econ. Paleontol. Mineral., Spec. Publ., 38: 65–72.

Naeser, N.D., 1989a. Fission-track analysis in petroleum basins. In: L.B. Magoon (Editor), The Petroleum System — Status of Research and Methods, 1990. U.S. Geol. Surv., Bull., 1912: 32–35.

Naeser, N.D., 1989b. Thermal history and provenance of rocks in the Wagon Wheel no. 1 well, Pinedale anticline, northern Green River basin — evidence from fission-track dating. In: B.E. Law and C.W. Spencer (Editors), Geology of Tight Gas Reservoirs in the Pinedale Anticline Area, Wyoming, and at the Multiwell Experiment Site, Colorado. U.S. Geol. Surv., Bull., 1886: E1–E13.

Naeser, N.D., 1992. Miocene cooling in the southwestern Powder River Basin, Wyoming — preliminary evidence from apatite fission-track analysis. U.S. Geol. Surv., Bull., 1917-O, 17 pp.

Naeser, N.D. and McCulloh, T.H. (Editors), 1989. Thermal History of Sedimentary Basins — Methods and Case Histories. Springer-Verlag, New York, N.Y., 319 pp.

Naeser, N.D., Reaves, C.M. and McCulloh, T.H., 1987a. Extreme high temperature burial annealing of apatite fission tracks, Santa Fe Springs oil field, California. Geol. Soc. Am., Abstr. Prog., 19: 784.

Naeser, N.D., Zeitler, P.K., Naeser, C.W. and Cerveny, P.F., 1987b. Provenance studies by fission-track dating of zircon — etching and counting procedures. Nucl. Tracks Radiat. Measurements, 13: 121–126.

Naeser, N.D., McCulloh, T.H., Crowley, K.D. and Reaves, C.M., 1989a. Fission-track analysis as a thermal-history indicator — Los Angeles Basin, California. Am. Assoc. Pet. Geol. Bull., 73: 1168 (abstract).

Naeser, N.D., Naeser, C.W. and McCulloh, T.H., 1989b. The application of fission-track dating to the depositional and thermal history of rocks in sedimentary basins. In: N.D. Naeser and T.H. McCulloh (Editors), Thermal History of

Sedimentary Basins — Methods and Case Histories. Springer-Verlag, New York, N.Y., pp. 157–180.

Naeser, N.D., Crowley, K.D., McCulloh, T.H. and Reaves, C.M., 1990a. High temperature annealing of fission tracks in fluorapatite, Santa Fe Springs oil field, Los Angeles Basin, California. Nucl. Tracks Radiat. Measurements, 17: 424 (abstract).

Naeser, N.D., McCulloh, T.H., Crowley, K.D. and Reaves, C.M., 1990b. Thermal history of the Los Angeles basin — evidence from fission-track analysis. Am. Assoc. Pet., Geol. Bull., 74: 728 (abstract).

Naeser, N.D., Naeser, C.W. and McCulloh, T.H., 1990c. Thermal history of rocks in the southern San Joaquin Valley, California: evidence from fission-track analysis. Am. Assoc. Pet. Geol., Bull., 74: 13–29.

Nuccio, V.F. and Barker, C.E. (Editors), 1990. Applications of Thermal Maturity Studies to Energy Exploration. Rocky Mountain Section of the Society of Economic Paleontologists and Mineralogists, Denver, Colo., 175 pp.

O'Sullivan, P.B., 1988. Apatite Fission Track Study of the Thermal History of Permian to Tertiary Sedimentary Rocks in the Arctic National Wildlife Refuge, Northeastern Alaska. M.S. Thesis, University of Alaska, Fairbanks, Alaska, 184 pp.

O'Sullivan, P.B., Decker, J. and Bergman, S.C., 1989. Apatite fission track thermal history of Permian to Tertiary sedimentary rocks in the Arctic National Wildlife Refuge, northeastern Alaska. Geol. Soc. Am., Abstr. Prog., 21 (6): 126.

O'Sullivan, P.B., Decker, J., Bergman, S.C. and Wallace, W.K., 1990. Constraints by apatite fission track analysis on the cooling times due to uplift of the northeastern Brooks Range, Alaska. Am. Assoc. Pet. Geol. Bull., 74: 733 (abstract).

Qvale, H., Stiberg, J.-P. and Throndsen, T., 1990. Fission track and vitrinite reflectance as thermal indicators in North Sea wells. Nucl. Tracks Radiat. Measurements, 17: 426 (abstract).

Reimer, G.M., 1972. Fission Track Geochronology: Method for Tectonic Interpretation of Apatite Studies with Examples from the Central and Southern Alps. Ph.D. Thesis, University of Pennsylvania, Philadelphia, Pa., 85 pp.

Roden, M.K. and Miller, D.S., 1988. Apatite fission-track thermochronology of the Appalachian Basin. Geol. Soc. Am., Abstr. Prog., 20 (7): A386.

Roden, M.K. and Miller, D.S., 1989a. Thermal history of Hartsford, Newark, and Taylorsville basins using fission-track analysis. Geol. Soc. Am., Abstr. Prog., 21 (6): 266.

Roden, M.K. and Miller, D.S., 1989b. Apatite fission-track thermochronology of the Pennsylvanian Appalachian Basin. Geomorphology, 2: 39–51.

Roden, M.K., Donelick, R.A. and Miller, D.S., 1990. Thermal history of the Appalachian Basin based on apatite fission-track thermochronology. Nucl. Tracks Radiat. Measurements, 17: 426–427 (abstract).

Sanford, S.J., 1981. Dating Thermal Events by Fission Track Annealing, Cerro Prieto Geothermal Field, Baja California, Mexico. M.S. Thesis, University of California, Riverside, Calif., 105 pp.

Seward, D., 1989. Cenozoic basin histories determined by fission-track dating of basement granites, South Island, New Zealand. Chem. Geol. (Isot. Geosci. Sect.), 79: 31–48.

Seward, D., 1990. Cenozoic basin histories determined from basement rocks in New Zealand. Nucl. Tracks Radiat. Measurements, 17: 427–428 (abstract).

Seward, D. and Nathan, S., 1990. Uplift history of South Westland using the fission-track dating technique. N.Z. J. Geol. Geophys., 33: 201–204.

Siddall, R. and Mendelssohn, M.J., 1990. Infra red spectroscopy as a routine method for the semi-quantitative determination of the chemical composition of apatites for fission track analysis. Abstr., 7th Int. Conf. Geochronology, Cosmochronology, and Isotope Geology, Canberra, A.C.T., September 24–29. Geol. Soc. Aust., Abstr., 27: 93.

Storzer, D. and Selo, M., 1984. Toward a new tool in hydrocarbon resource evaluation: the potential of the apatite fission track chrono-thermometer. In: B. Durand (Editor), Thermal Phenomena in Sedimentary Basins. Technip, Paris, pp. 89–110.

Tamanyu, S., 1985. Fission track reset ages recorded at the boring core samples in the geothermal fields. Nucl. Tracks Radiat. Measurements, 10: 435 (abstract).

Van de Kamp, P.C., Kamp, P., Samoun, A. and Green, P., 1990. Petroleum generation modeling with kerogen kinetics and apatite fission-track data. Am. Assoc. Pet. Geol., Bull., 74: 784 (abstract).

Wagner, G.A., 1968. Fission track dating of apatites. Earth Planet. Sci. Lett., 4: 411–415.

Wagner, G.A., 1969. Spuren der spontanen Kernspaltung des ^{238}Urans als Mittel zur Datierung von Apatiten und ein Beitrag zur Geochronologie des Odenwaldes. Neues Jahrb. Mineral., Abh., 110: 252–286.

Waples, D.W., 1980. Time and temperature in petroleum formation: application of Lopatin's method to petroleum exploration. Am. Assoc. Pet. Geol., Bull., 64: 916–926.

Zeitler, P.K., 1985. Closure temperature implications of concordant ^{40}Ar/^{39}Ar potassium feldspar and zircon fission-track ages from high-grade terranes. Nucl. Tracks Radiat. Measurements, 10: 441–442 (abstract).

Zimmermann, R.A., 1981. Where have all the sediments gone?: fission track dating of the Illinois deep hole core. Geol. Soc. Am., Abstr. Prog., 13: 588.

Zimmermann, R.A., 1986. Fission-track dating of samples of the Illinois Basin drill-hole core. U.S. Geol. Surv., Bull., 1622: 99–108.

Zimmermann, R.A. and Gaines, A.M., 1978. A new approach to the study of fission track fading. In: R.E. Zartman (Editor), Short Papers of the 4th International Conference on Geochronology, Cosmochronology, and Isotope Geology. U.S. Geol. Surv., Open-File Rep., 78-701: 467–468.

NANCY D. NAESER *U.S. Geological Survey, MS 424, Federal Center, Denver, CO 80225, U.S.A.*

Fluid inclusions and organic maturity parameters as calibration tools in basin modelling

K. Leischner, D.H. Welte and R. Littke

The calibration of a numerically simulated temperature history is generally performed using vitrinite reflectance (% R_0) and present day temperature data. In order to improve the calibration of temperature histories, investigations on organic maturity parameters and fluid inclusions were carried out on a series of samples from a well in the Lower Saxony Basin. In this well maturity increases from about 0.55% R_0 at 500 m to approximately 2.0% R_0 at 3300 m depth. Vitrinite reflectance values calculated from the methylphenanthrene index (MPI 1) agree with measured vitrinite reflectance data. Fluid inclusions in anhydrite veins and quartz cements yield palaeotemperature values that increase from 140°C at 1860 m depth to approximately 200°C at 3215 m depth. Based on geological observations, these rather elevated palaeotemperatures can be directly related to the intrusion of a magmatic body about 90 million years ago. These palaeotemperature data and the scenario of a magmatic event were used for calibration of the numerically simulated temperature history of the well. Using a kinetic approach, modelled vitrinite reflectance values showed an excellent fit to measured vitrinite reflectance values. If, however, vitrinite reflectance is calculated using the time–temperature index (TTI), the calculated values exceed the measured values by up to 0.4% in terms of R_0-units.

Introduction

Temperature is one of the most critical parameters in hydrocarbon generation. A detailed knowledge of the temperature history in combination with kinetic models for oil and gas generation is, therefore, essential to determine the timing and the rate of hydrocarbon generation in sedimentary basins (Tissot and Welte, 1984; Tissot et al., 1987). Basin modelling techniques, in which all the available geological, geophysical and geochemical data are integrated, can be used to simulate the essentials of basin development (Yukler et al., 1978; Welte and Yükler, 1981; Ungerer et al., 1984; Yalcin et al., 1988; Wygrala, 1989). The calibration of these models is generally achieved using vitrinite reflectance data and present day borehole temperatures. Present-day borehole temperatures allow the calibration of the numerically simulated temperature histories only for the most recent timespan.

Organic maturity parameters such as vitrinite reflectance, however, indelibly record the cumulative effects of thermal influences during the entire evolution of a sedimentary sequence. They evolve according to different, complex, and often unknown reaction kinetics and cannot be related to a specific palaeotemperature. The assessment of maturity is often difficult because the organic maturity parameters are influenced by different organic matter types as well as by the occurrence of allochthonous organic matter such as oil. The problems that arise from these difficulties are discussed in detail by several authors (e.g. Tissot and Welte, 1984; Tissot et al., 1987). Detailed investigations on the quality of the organic matter are necessary and, therefore, the combination of several organic maturity parameters is required to establish an unambiguous maturity trend.

For this reason, independent temperature indicators, derived from geological–petrological studies are essential to further constrain temperature histories. Furthermore, these temperature indicators allow the improved calibration of the numerically simulated temperature histories. Independent palaeotemperatures may be obtained by different approaches. Investigations of fluid inclusions and apatite fission track ages are most promising in this context (Naeser et al., 1989). To a lesser extent, detailed studies of stable isotopes and clay mineral transformations may give information about the palaeotemperatures experienced by sedimentary rocks. A brief overview of the application of these methods is given by Tissot et al. (1987). The use of fluid inclusions as palaeotemperature indicators in sedimentary sequences is discussed in detail by Bodnar and Bethke (1984), McLimans (1987) and Burruss (1987).

To achieve an improved calibration of a numerically simulated temperature history, a well in the Lower Saxony Basin in Germany (well F) was se-

Basin Modelling: Advances and Applications edited by A.G. Doré et al. NPF Special Publication 3, pp. 161–172, Elsevier, Amsterdam.

lected. In this well a wide range in maturity is observed and the burial history is well known. The maturity assessment was performed using several organic geochemical methods such as Rock-Eval pyrolysis and vitrinite reflectance measurements. Furthermore, several molecular maturity parameters were obtained from the aromatic fraction of extracts (Radke, 1987, 1988) and were supplemented by investigations of aromatic and aliphatic hopanoid and steroid molecules (Philp and Lewis, 1987).

The burial and temperature history of well F was numerically simulated utilizing the one-dimensional PDI (Pre-Drilling-Intelligence) software of Integrated Exploration Systems (IES), Jülich, Germany. The goal of our study was to establish palaeotemperatures derived from fluid inclusion measurements for the calibration and reconstruction of the temperature history in well F. The reconstructed temperature history was used to calculate vitrinite reflectance values which were then compared to measured vitrinite reflectance data. Vitrinite reflectance values were calculated following both the EASY%R_0 approach of Sweeney and Burnham (1990) and a modified TTI correlation according to Waples (1980), which is based on the original method of Lopatin (1971). Hence, three different types of vitrinite reflectance values were available; one set measured in the well, and two other calculated independently.

Sampling and analytical methods

A series of fine-grained to coarse-grained detrital rocks, carbonates and evaporites was selected from well F. This well is located in the southwestern part of the Lower Saxony Basin (Fig. 1), which is a major oil and gas province in Germany. Detailed information on the plate tectonic and geological evolution of the Lower Saxony Basin can be found in Kemper (1976), Boigk (1981), Ziegler (1982, 1989) and Betz et al. (1987). Well F is located in a gasfield and was drilled into a horst-like structure. The maturation of the sediments at the southern margin of the Lower Saxony Basin is known to be influenced by deep seated igneous intrusions of late Cenomanian to Turonian age. These intrusions and the corresponding hydrothermal events caused anomalously high vitrinite reflectance values in the sediments, as indicated by the isoreflectance lines (Fig. 1) for the top Carboniferous (Teichmüller et al., 1984).

The complex burial history of well F from the Late Carboniferous (315 million years ago) up to the present is characterized by several phases of subsidence alternating with periods of uplift (Fig. 2). Rapid sedimentation and burial occurred, for example, during the Late Carboniferous. A major uplift and erosion event took place during the Late Cretaceous and is termed the "inversion event" of the Lower Saxony Basin. Maximum burial was attained about 88 million years ago for these sediments, at approximately the same time as when the magmatic bodies were intruded. These intrusions can probably be related to the lateral and vertical movements that occurred during the inversion phase (Betz et al., 1987), which in turn are the result of the orogenic movements in the Alpine region at that time (Ziegler, 1989).

The sequence studied ranges in age from the Late Carboniferous (Westphalian C) to Late Cretaceous. Only the fine-grained rocks (shales and siltstones) were used for the organic geochemical

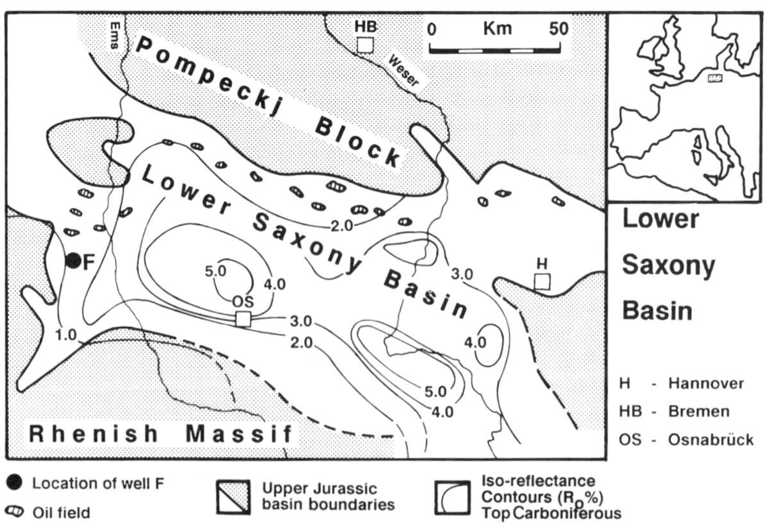

Fig. 1. The Lower Saxony Basin and the sampling location of well F. The intrusive complex of Bramsche and accompanying intrusive bodies are outlined by the isoreflectance lines. See text for details.

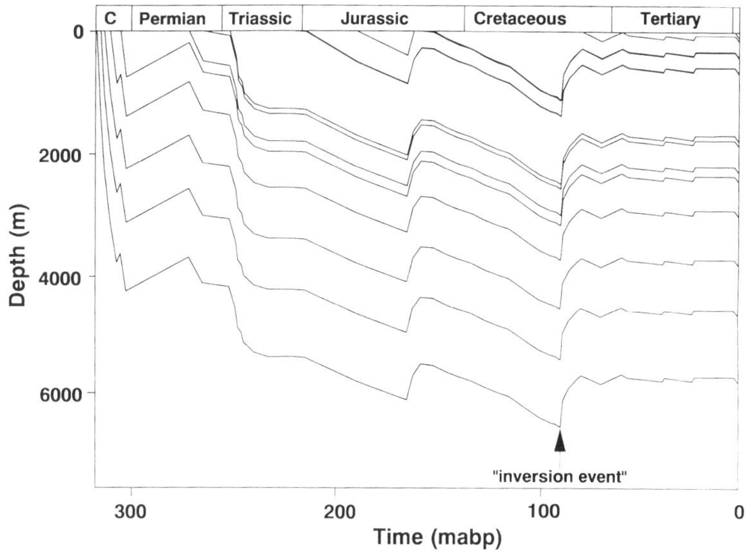

Fig. 2. Simplified burial history diagram of well F. *C*=Upper Carboniferous.

analyses described in Tissot and Welte (1984). Fine- to coarse-grained sandstones were examined using thin sections, SEM–EDX, cathodoluminescence and X-ray diffraction. Fluid inclusion analysis, here considered the most important technique for assessing palaeotemperatures, is discussed later in detail.

Microthermometry for fluid-inclusion analysis

For microthermometry, thin rock slices up to a maximum thickness of 200 μm, were prepared and polished on both sides. Microthermometric analyses were performed on a petrographic microscope (ZEISS Axioplan) equipped with a LINKHAM heating and cooling stage THM 600 (for details see Shepherd et al., 1985). Phase transitions between solid, liquid and vapour phases are observed in the range of −100°C to 200°C and are briefly discussed below. Detailed information concerning the different phase transitions is given by Roedder (1984). The following four phase transitions are observed.

(1) The eutectic melting temperature (T_e) provides information on the dissolved ionic species contained in the fluid inclusion (Roedder, 1984). This phase transition is not discussed in detail in this paper.

(2) The ice melting temperature (T_m) yields information on the approximate concentration of salt components, which are normally expressed as NaCl-equivalents in wt%. If there is evidence for other salt species, it is also possible to express salinities in $CaCl_2$-equivalents in wt%. In some cases the occurrence of gas hydrates (clathrates) and salt hydrates prevents a correct determination of the melting temperatures. Water is trapped inside the clathrates and the fluid is enriched with cations. In this case, the measured melting temperatures are "too low". For this reason, this study uses those fluid inclusion data for which a relatively constant melting temperature is observed and an estimate of salinities is valid.

(3) The dissociation temperatures of clathrates into gas and brine (T_mClath.) may permit the identification of the gas species (e.g. CO_2 or CH_4) contained in the fluid inclusions. The presence of clathrates has a significant influence on the assessment of formation temperatures (Hanor, 1980).

(4) The temperature of homogenization (T_{hom}) between the liquid (l) and the vapour phase (v) is considered as the minimum trapping temperature of the inclusions.

One-dimensional simulation

The 1D/PDI-software package of IES was used for the modelling of well F. The simulation concept for this model is based on previous work by Yukler et al. (1978) and Welte and Yükler (1981). It was later updated by Yalcin et al. (1988). The principles of this software were discussed in detail by Wygrala (1989) and Düppenbecker (1992). Next to the integrated modelling of burial, compaction, and temperature history, this program calculates pore pressures and the amount of hydrocarbons generated and expelled from a source rock.

Results and discussion

Maturity

Vitrinite reflectance values increase from 0.55% R_o at approximately 500 m depth to nearly 2.0% R_o at 3300 m depth (Fig. 3). These values are in good agreement with data of Teichmüller et al. (1984) for samples from the same well. Figure 4 depicts the

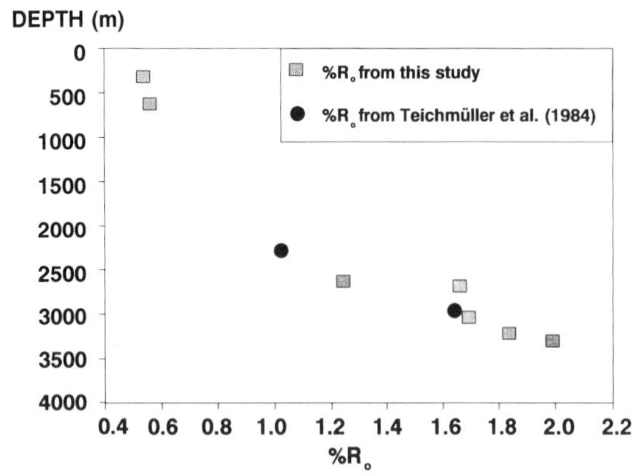

Fig. 3. Vitrinite reflectance values from this study and from Teichmüller et al. (1984).

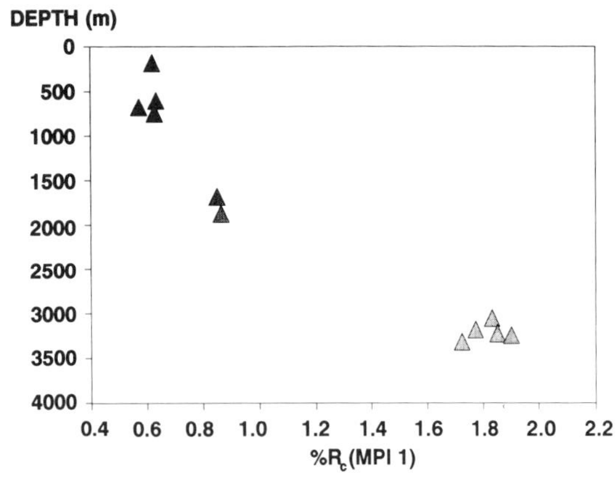

Fig. 4. Vitrinite reflectance values of type III kerogen calculated from the methylphenanthrene index (MPI 1).

calculated vitrinite reflectance values as based on the methylphenanthrene index (MPI 1) after Radke and Welte (1981). The calculated vitrinite reflectance values exhibit the same depth dependence as the measured reflectance data (Fig. 3). Note that for the MPI calculations only samples consisting mainly of type III organic matter and showing no signs of migration of liquid hydrocarbons were used.

Furthermore, Rock-Eval data measured for samples with more than 1% organic carbon in the same depth interval show an increase of T_{max} from 428°C to 525°C, which roughly corresponds to vitrinite reflectance values between 0.5% R_o and 2.0% R_o (Espitalié, 1985). Thus, T_{max} values are a further confirmation of the established maturity trend. This trend was used for calibrating the temperature history of well F.

Fluid inclusions

Fluid inclusions were found in anhydrite veins within the Zechstein and Late Carboniferous and in authigenic and detrital quartz from sediments of Late Carboniferous age.

The maximum diameter of the inclusions in anhydrite veins is 48 μm. Primary or pseudosecondary inclusions (Roedder, 1984) are located along growth zones of anhydrite crystals or show a cluster-like distribution and were entrapped during crystal growth. However, most of the fluid inclusions observed are arranged along healed microfractures that cut across several anhydrite grains. They are called secondary inclusions because they were formed after the growth of the anhydrite crystals. Small inclusions in the anhydrite veins are often regular in shape, whereas large inclusions show irregular shapes (Fig. 5), indicating volume changes subsequent to entrapment.

Primary fluid inclusions within quartz cements show a maximum size of 15 μm. Most of them display an oval to round shape (Fig. 6). These inclusions originated during the formation of quartz cements and are likely to reflect their growth conditions (Roedder, 1984).

Secondary fluid inclusions in sandstone samples show maximum diameters of 10 μm and formed after the growth of the quartz cement. Inclusions of this kind are arranged along healed microfractures that cut across detrital quartz grains and authigenic quartz rims. These inclusions are indicated with arrows in Fig. 6.

The composition of the fluid inclusions varies markedly in different stratigraphic units. Two examples from an anhydrite vein in the Zechstein and a sandstone sample from the Late Carboniferous are discussed below.

Anhydrite vein. Fluid inclusions in an anhydrite vein in the Zechstein at a present depth of 2244 m show low melting temperatures (T_m) between $-11°$ and -14°C. Furthermore, clathrates are observed at the rim of the gas bubble. Clathrate melting temperatures up to +8°C are recorded, indicating that other gas species are regular components in such brine-bearing inclusions. Hanor (1980) and Burley et al. (1989) discuss in detail the presence of gas and its importance for the calculation of palaeotemperatures from T_{hom}-data. According to their results, pressure correction is not necessary in this case.

Homogenization temperatures are in the range of 120–161°C for primary inclusions. This wide scatter of T_{hom} is related to the maximum diameter and presumably to the shape of the primary inclusions. In Fig. 7, the size of the fluid inclusions is plotted versus T_{hom}. The highest homogenization temperature of 161°C is observed for the largest inclusion (48 μm in diameter) and there is a general trend of increasing T_{hom} with increasing size. This relationship leads to the conclusion that the larger inclusions were stretched or decrepitated (re-equilibration

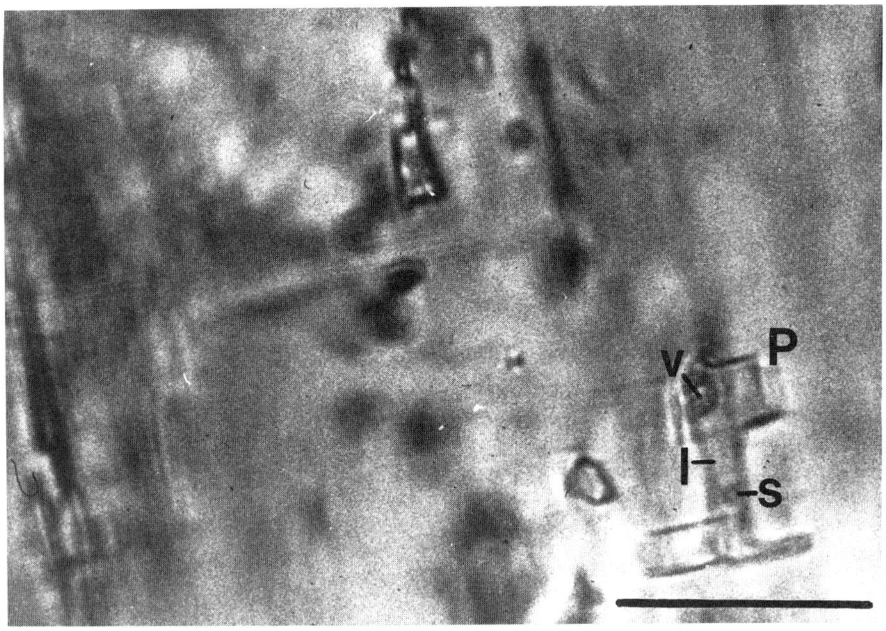

Fig. 5. Photomicrograph of a primary fluid inclusion (*P*) in an anhydrite vein. Note the irregular shape of the large inclusion in the right corner of the figure. The largest diameter of this inclusion is 40 μm. *l* = liquid; *v* = vapour; *s* = solid. Scale bar 50 μm.

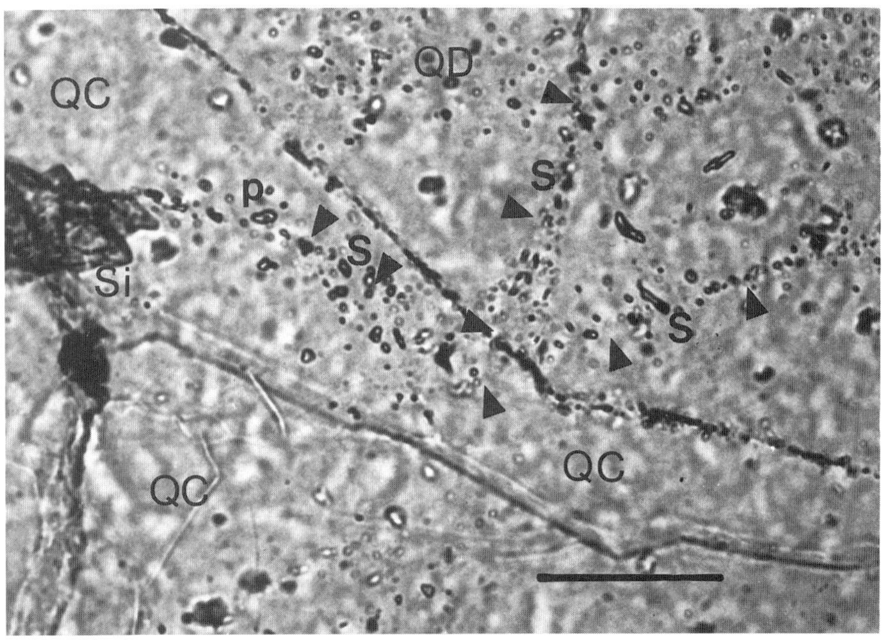

Fig. 6. Photomicrograph of primary and secondary fluid inclusions in a sandstone sample. Primary (*P*) fluid inclusions have a maximum diameter of 15 μm and are located within quartz cement (*QC*). The arrows indicate secondary (*S*) fluid inclusions smaller than 10 μm that are arranged along healed microfractures cutting across detrital grains (*QD*) and quartz cement (*QC*). Siderite (*Si*) is enclosed by quartz cements. Scale bar 50 μm.

to the new *P–T* conditions) after entrapment due to an internal overpressure exceeding the physical strength of the host mineral. Burruss (1987) and Prezbindowski and Tapp (1991) discuss the application of stretched inclusions as maximum palaeothermometers in fluorite and carbonate minerals. These minerals respond very quickly to internal overpressures in fluid inclusions and according to our data the same behaviour is assumed for anhydrite. If this conclusion is correct, homogenization temperatures of large inclusions provide information on maximum palaeotemperatures. The above-mentioned presence of gas confirms that such fluid inclusions would expand when experiencing a later heat pulse.

From these observations it is deduced that the anhydrite veins were formed before the occurrence of

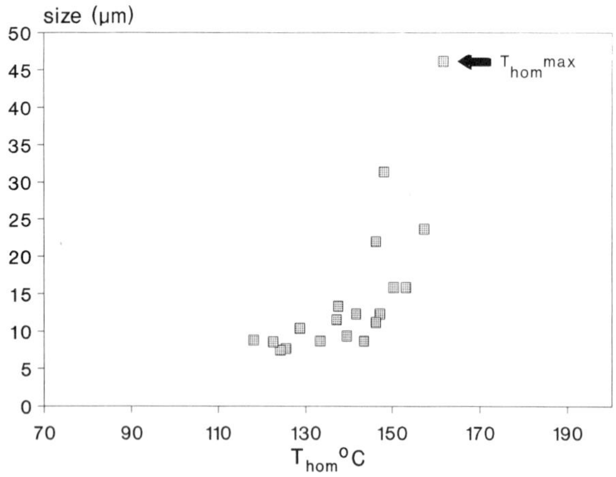

Fig. 7. Homogenization temperatures plotted versus size for fluid inclusions in an anhydrite vein from a depth of 2244 m. A clear relationship between homogenization temperatures and size of fluid inclusions is observed. The arrow indicates the maximum temperature obtained from these data (see text for discussion).

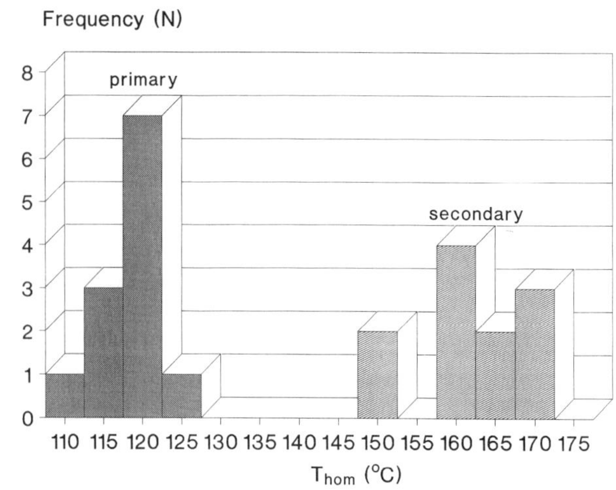

Fig. 8. Histogram of homogenization temperatures for fluid inclusions in a sandstone sample from a depth of 3215 m.

an additional heat pulse during Turonian time. This heat pulse, which was associated with the emplacement of the intrusive body, caused temperatures to rise to 161°C in the largest inclusions. Therefore 160° to 165°C was the maximum palaeotemperature experienced by the sample (present day depth of 2244 m). In Fig. 7 this maximum temperature is marked by an arrow. Two other vein samples from 1860 m and 2500 m revealed maximum palaeotemperatures of 140°C and 175°C, respectively.

Sandstone. Primary fluid inclusions in authigenic quartz from a Late Carboniferous sandstone sample at a present depth of 3215 m were investigated. They contain highly saline (probably Ca^{2+}-dominated) brines with melting temperatures (T_m) of −14°C to −17°C. Even in the largest inclusions, clathrate melting could not be detected. Therefore, gas was absent and a pressure correction was necessary. A maximum pressure of 400 bars was realistic for this sample which corresponds to a burial depth of about 4000 m, provided no significant overpressure existed. The measured homogenization temperatures vary between 112° and 125°C (Fig. 8).

Following the work of Zhang and Frantz (1987), who published PVT-data for Ca^{2+}-solutions, we calculated trapping temperatures of 140° to 150°C. Secondary inclusions in the sandstone samples display melting temperatures of −20 to −25°C. Due to the small size of these inclusions, clathrate melting could not be detected. However, the gas bubble moved around, after the ice melted. This indicates that no clathrate formation occurred and, hence, no gas phase was present. Maximum homogenization temperatures are in the range of 150° to 173°C for secondary inclusions (Fig. 8). As noted above, these inclusions are arranged along microfractures that are often filled with green to bluish luminescing quartz cement, which is possibly related to hydrothermal events as suggested by Leuchs (1985) and R. Neuser (pers. commun., 1990). The green luminescence of quartz cements here may, therefore, indicate that the rock samples studied are also influenced by hydrothermal events. Further evidence for a hydrothermal origin of the fluids is inferred from the high salinity of these inclusions, which seems to be typical for deep basinal brines. Similar compositions were observed by Lüders et al. (1990) in fluid inclusions from fluorite and carbonates from the Harz region, which are also of hydrothermal or deep basinal origin. If the fluids were not directly related to intrusions, it may be that the intrusions heated and mobilized pore fluids within the sedimentary column and caused them to migrate upwards, presumably along faults and microfractures. The influence of hot fluids on the temperature field is described by Burley et al. (1989) and Tilley and Longstaffe (1989).

From these results, we conclude that the fluid inclusions were formed when maximum burial and maximum palaeotemperatures were attained. Under the assumption that no significant overpressure occurred at this time, which is supported by pressure calculations, a pressure correction for a depth of 4000 m and hydrostatic conditions was applied using the dataset of Zhang and Frantz (1987) for Ca^{2+}-dominated fluid inclusions. The resulting trapping temperatures are in the range of 190–200°C for Turonian times, which are approximately the maximum palaeotemperatures for this depth.

Thus, increasing palaeotemperatures between about 140°C and 200°C were obtained for the depth

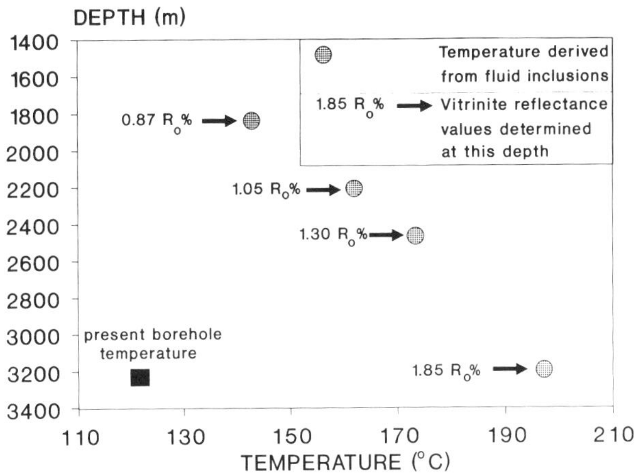

Fig. 9. Maximum temperatures derived from fluid inclusions versus depth for well F. The corresponding vitrinite reflectance values for the examined samples are included in this figure.

interval from 1860 to 3215 m (Fig. 9). Comparing the palaeotemperature data to the corrected present day borehole temperatures reveals a difference of 80°C at a depth of 3215 m. Neglecting heat conductivity, a Turonian palaeogeothermal gradient of approximately 45°C/km results which is significantly higher than the present day geothermal gradient of 34°C/km. Deriving maturity on the basis of vitrinite reflectance, Buntebarth (1985) inferred maximum palaeogeothermal gradients of up to 92°C/km for sediments located in the centre of the Bramscher Intrusive Complex. Our study indicates a significantly lower gradient of 45°C/km for sediments located at the margin of the intrusive body.

Simulation of well F

The simulation of the burial and temperature history of well F was performed using the 1D/PDI-Software. The conceptual model was based on detailed literature and well data, using sediment thicknesses, sediment–water interface temperatures, surface temperatures, heat flux at the base of the sedimentary column, subsidence and uplift phases and taking into account the petrophysical and thermophysical data of the rocks. Based on this data matrix, the burial, compaction, and temperature history can be numerically simulated. Before the calibration of the temperature history was performed, optimization was carried out. Optimization is the procedure of adjusting and balancing the parameters that directly control the compaction of the sedimentary sequence (Wygrala, 1989). During optimization, the simulation run is repeated several times until a reasonable match between the calculated and observed thicknesses is obtained. Thereafter, the calibration is carried out by comparison of the measured and calculated parameters that have a direct influence on thermal history. These parameters are the present day temperature readings in the well and the measured or calculated vitrinite reflectance data.

Calibration without the use of fluid inclusions. Initially, calibration of the simulated temperature history for this well was performed using vitrinite reflectance calculated by a TTI approach modified after Waples (1980). Because the present day borehole temperatures are influenced by the geothermal field of the last few million years, they have minor relevance for the geologic heat flow history and are not considered for calibration. The present heat flow values are in the range of 1.7 HFU (Haenel and Staroste, 1988) for this area and are also related to heat anomalies caused by upward-migrating hot fluids (Schulz, 1989). For this reason, only vitrinite reflectance values were used for the reconstruction of the temperature history.

Using the modified TTI correlation (Waples, 1980), a fit between measured and calculated vitrinite reflectance data is obtained when a constant heat flow of 1.5 HFU is assumed for the entire burial history of well F (Fig. 10). The value of 1.5 HFU is a mean value for continental areas (Lee et al., 1966) and is in accordance with values of 1.4 to 1.45 HFU used by Düppenbecker (1992) for the same area. The calculated palaeotemperature isotherms are shown as dashed lines in Fig. 10. A sample originating from a present depth level of 3215 m would have reached a maximum temperature of 150°C approximately 90 million years ago.

To assume a constant heat flow over the entire geological past, as done in Fig. 10, is not realistic because of indications that hydrothermal events occurred. Furthermore, comparison between the palaeotemperature data derived from TTI calculations and palaeotemperatures derived from fluid inclusions shows a discrepancy of 40° to 50°C for maximum temperatures reached in the Turonian (Fig. 10). Clearly, the TTI calibration and subsequent calculation of palaeotemperatures cannot account for the palaeotemperatures reconstructed from fluid inclusions. Therefore, the temperature history had to be re-calibrated in such a way that palaeotemperature data sets derived from both vitrinite reflectance and fluid inclusions could be taken into account without causing an interpretive conflict.

Calibration using fluid inclusion data. The re-calibrated temperature history incorporating palaeotemperatures derived from fluid inclusions and from vitrinite reflectance is shown in Fig. 11. To achieve a fit between both kinds of "measured" palaeotemperatures and the calculated palaeotem-

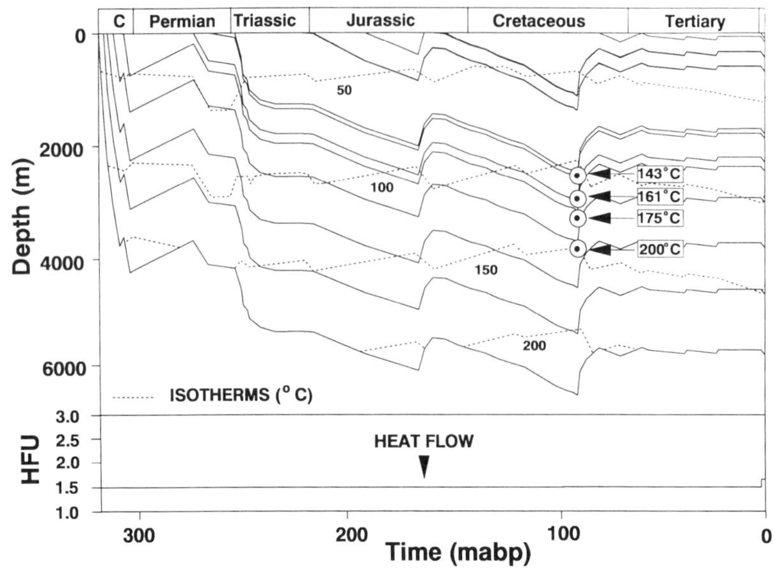

Fig. 10. Simulated burial and temperature history using the TTI correlation for vitrinite reflectance calculation. Note that a constant heat flow of 1.5 HFU is required for the entire burial history to achieve a fit between measured and calculated vitrinite reflectance values. The temperatures indicated by circles and arrows are the temperatures derived from fluid inclusions. They do not agree with the isotherms derived from TTI.

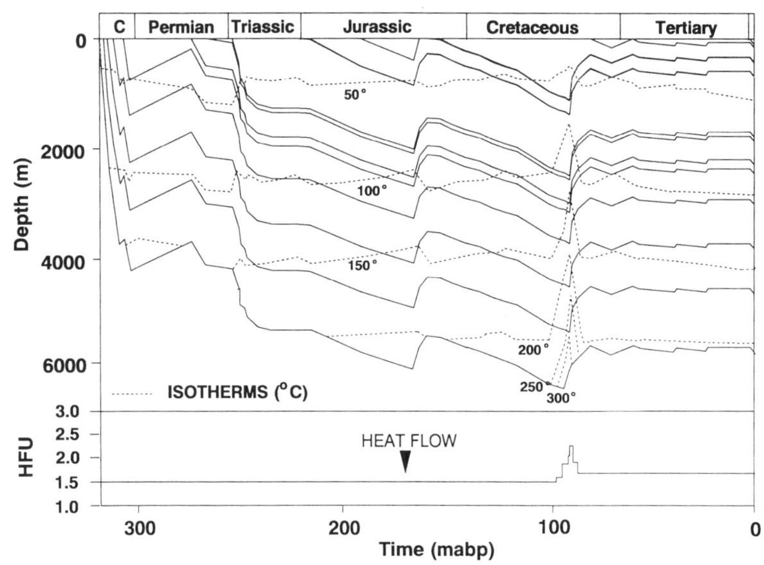

Fig. 11. Re-calibrated temperature history for well F using the palaeotemperatures indicated in Fig. 10. Note that the maximum palaeotemperatures correlate with the commencement of the uplift phase 88 million years ago.

peratures, a heat pulse with a maximum value of 2.25 HFU during the Turonian is required. This heat pulse is related to the intrusion of the magmatic body of Lingen–Bad Bentheim, which is located in the vicinity of the site of well F. This intrusion causes the vitrinite reflectance anomalies in the westernmost part of the Lower Saxony Basin as indicated by the isoreflectance lines in Fig. 1.

This heat flow pulse must have influenced the entire sedimentary sequence beginning with Late Carboniferous sediments. A penetrating quartz cementation in the whole Carboniferous sequence supports this interpretation. In addition, organic geochemical studies on Liassic source rocks further east in the vicinity of the Vlotho magmatic body of the same age revealed that temperature effects were regional rather than local (Rullkötter et al., 1988). The maximum temperatures during this heating event were assumed to last up to one million years. A heat flow value of 2.25 HFU does not appear to be unrealistically high when compared to a heat flow value of 3.5 HFU in the centre of the Bramscher Massif as calculated by Buntebarth (1985) for the Westphalian B strata. This heating event was then followed by a rapid temperature decrease in the order of 25° to 30°C within a period of one million years. These observations are in accordance with results shown by Neugebauer and Walzebuck (1987) on cooling plutons.

This temperature history is confirmed by palaeotemperatures derived from fluid inclusion data. The following explanation shows to what extent the calculations of vitrinite reflectance based on this re-calibrated temperature history match the measured data. For the calculation of vitrinite reflectance, both the EASY%R_o-method (Sweeney and Burnham, 1990) and the TTI-method (modified after Waples, 1980) were used.

EASY%R_o calculation. The EASY%R_o model (Sweeney and Burnham, 1990) is based on the more sophisticated approach of Burnham and Sweeney (1989), called "Vitrimat". In the "Vitrimat" model, a reaction scheme for vitrinite reflectance evolution is used where the loss of water, carbon dioxide, higher hydrocarbons, and methane are considered as four independent parallel reactions. A distribution of activation energies is calculated for each of these reactions. This approach provides the basis for the EASY%R_o model, where a single distribution of activation energies is considered for the four reactions. The calculated activation energies range from 34 to 72 kcal/mol with a maximum at 52 kcal/mol. The composite distribution is applied in the calculations of the vitrinite reflectance, having taken into account the variation in elemental composition.

When applying the EASY%R_o method, a good match is achieved between measured and calculated vitrinite reflectance values (Fig. 12) in the high maturity range above 0.6% R_o. The good fit was achieved with the maximum palaeotemperatures lasting for only a short duration of 200,000 to 400,000 years. This time span is in agreement with Mundry's (1971) calculations for this event but shorter then the 1 million year period assumed above. Combining a shorter time period with our reconstructed palaeotemperatures, calculated vitrinite reflectance values are lower than the measured values. For a longer duration of the maximum temperatures, the calculated vitrinite reflectance values are too high compared to the measured values.

Accordingly, it may be concluded that maximum temperatures were 10°C higher than inferred from fluid inclusion data and lasted for a shorter time period of 200,000 to 400,000 years. The result of 10°C higher palaeotemperatures is in agreement with the estimate of Burruss (1987) that about 10° to 15°C of overheating is needed to cause reequilibration during burial in large fluid inclusions (see above).

TTI calculation. The recalibrated temperature history was also used to calculate vitrinite reflectance values by the TTI approach. Here, no match is observed between calculated and measured vitrinite reflectance values. In addition, only in the range below 0.7% R_o do the TTI-calculated vitrinite reflectance values coincide with the vitrinite reflectance values derived from the EASY%R_o-method. At higher maturities an increasing difference is observed between measured and calculated vitrinite reflectance. The difference reaches a maximum of 0.4% R_o at a depth of 3300 m (Fig. 12).

The difference observed between both methods is not caused by the short time of the heating event. In Fig. 13, % R_o is plotted versus time for the sample at a present depth of 3215 m. The calculated vitrinite reflectance values using the TTI method show a stronger increase between 230 to 90 million years before present. During this period, an average

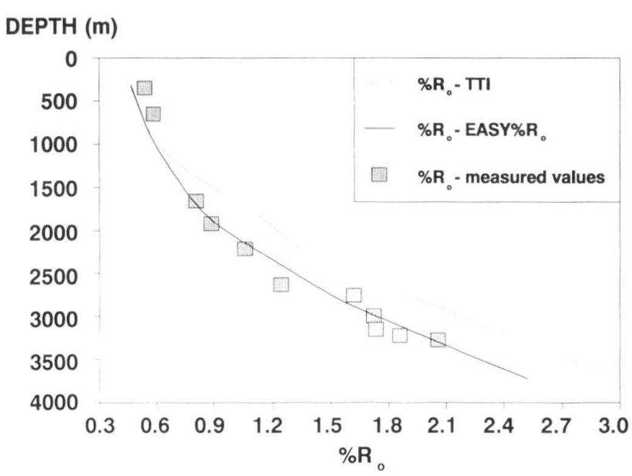

Fig. 12. Comparison of the vitrinite reflectance values calculated by the EASY%R_o approach and TTI applying the temperature history established in this study. See text for details.

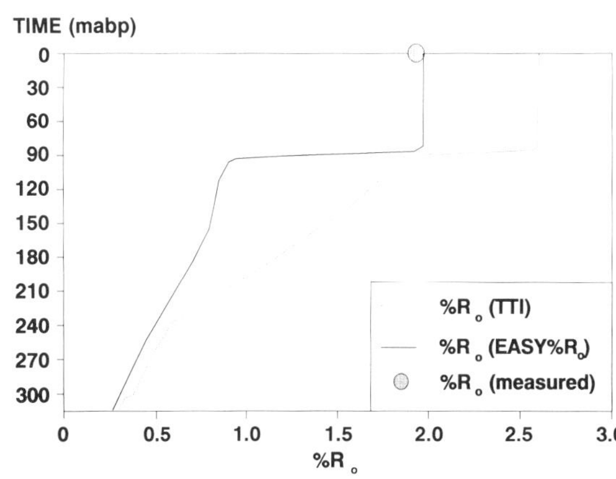

Fig. 13. Comparison of the vitrinite reflectance evolution established by the EASY%R_o and TTI from 310 million years before present to present times for the sample at a depth of 3215 m.

heating rate of less than 0.5°C/Ma is derived. The vitrinite reflectance values increase until a value of 1.95% R_o is reached just before the heat pulse occurs at 88 Ma. A further increase up to 2.4% R_o occurs during the heating event. Although temperatures decrease rapidly thereafter, a slight further increase of vitrinite reflectance to the present day value of 2.42% R_o is found.

In contrast, reflectances calculated by the EASY%R_o method reach 1.1% R_o just before the heating event. From the start to the end of the heat pulse an increase in vitrinite reflectance from 1.1% R_o to 2% R_o results. No further increase is calculated for later times and the 2% value is in agreement with the measured vitrinite reflectance values at 3215 m present depth.

These observations further confirm that the TTI method overestimates the influence of time on the maturation of vitrinite in the case of complex burial and temperature histories, as was discussed earlier by Quigley et al. (1987) and Sweeney and Burnham (1990). The EASY%R_o model, on the other hand, is more temperature-sensitive and thus provides better vitrinite reflectance values.

Conclusions

The most common approach to temperature history simulations of sedimentary basins is to use only vitrinite reflectance data as calibration parameters. The validity of vitrinite reflectance calculations is, however, a subject of long-lasting and ongoing dispute. The importance of time on vitrinite reflectance is assumed to be large in the various methods based on TTI (e.g. Lopatin, 1971; Waples, 1980), but is considered less important in recent kinetic models (Sweeney and Burnham, 1990).

For this reason, in this study we used palaeotemperatures derived from fluid inclusions in addition to vitrinite reflectance as an independent calibration tool for temperature history calculations on well F. From the results obtained for the studied well we conclude, that:

(1) The fluid inclusion technique can be used in order to determine maximum palaeotemperatures when re-equilibrated fluid inclusions are considered. The timing of fluid inclusion entrapment or re-equilibration of fluid inclusions is also critical in order to use these data for temperature history calibrations in basin modelling.

(2) For the burial and heat flow history derived from geologic information and fluid inclusion data, only the kinetic EASY%R_o approach to calculate vitrinite reflectance shows a good fit between measured and calculated vitrinite reflectance data.

(3) Due to the inherent emphasis on time, the calculation of vitrinite reflectance by the TTI method overestimates vitrinite reflectance when combined with the heat flow history derived from fluid inclusions.

(4) In our opinion, the use of at least two independent calibration tools, e.g. the combination of vitrinite reflectance and fluid inclusion data, is needed for the reconstruction of temperature histories. Other palaeogeothermometers, such as apatite fission track parameters, should also be tested complementary to the fluid inclusion technique.

Acknowledgements

The authors would like to thank the German oil companies BEB Erdöl und Erdgas GmbH, Deilmann AG, Preussag AG and Wintershall AG for providing the samples and geological information. The support of this work by colleagues at the Institute of Petroleum and Organic Geochemistry (KFA) at Jülich is gratefully acknowledged. We thank especially D.R. Baker (Rice University, Houston, Texas) and M.N. Yalcin (Research Institute for Basic Sciences Earth Science Department, Gebze, Turkey) for critical reading and comments to this manuscript. We are grateful to B. Stöckhert (Ruhr-Universität Bochum) for his valuable contributions regarding the fluid inclusion analysis. This work was financed by a grant from the Forschungszentrum Jülich (KFA). Thorough reviews of the manuscript by K. Oygard and A.G. Doré are gratefully acknowledged.

References

Betz, D., Führer, F., Greiner, G. and Plein, E., 1987. Evolution of the Lower Saxony Basin. In: P.A. Ziegler (Editor), Compressional Intra-Plate Deformations in the Alpine Foreland. Tectonophysics, 137: 127–170.

Bodnar, R.J. and Bethke, P.M., 1984. Systematics of stretching of fluid inclusions, I. Fluorite and sphalerite at 1 atmosphere confining pressure. Econ. Geol., 79: 141–161.

Boigk, H., 1981. Erdöl und Erdgas in der Bundesrepublik Deutschland. Enke, Stuttgart, 330 pp.

Buntebarth, G., 1985. Das Temperaturgefälle im Dach des Bramscher Massivs aufgrund von Inkohlungsuntersuchungen im Karbon von Ibbenbüren. Fortschr. Geol. Rheinl. Westfalen, 33: 255–264.

Burley, S.D., Mullis, J. and Matter, A., 1989. Timing diagenesis in the Tartan Reservoir (UK North Sea): constraints from combined cathodoluminescence microscopy and fluid inclusion studies. Mar. Pet. Geol., 6: 98–121.

Burnham, A.K. and Sweeney, J.J., 1989. A chemical kinetic model of vitrinite maturation and reflectance. Geochim. Cosmochim. Acta, 53: 2649–2657.

Burruss, R.C., 1987. Diagenetic paleotemperature from aqueous fluid inclusions: re-equilibration of inclusions in carbonate cements by burial heating. Mineral. Mag., 51: 477–481.

Düppenbecker, S.J., 1992. Genese und Expulsion von Kohlenwasserstoffen in zwei Regionen des Niedersächsischen Beckens unter besonderer Berücksichtigung der Aufheizraten. Ber. Forschungszentrum Jülich, 2657, 304 pp.

Espitalié, J., 1985. Use of T_{max} as a maturation index for different types of organic matter: comparison with vitrinite reflectance. In: J. Burrus (Editor), Thermal Modelling in Sedimentary Basins. 1st IFP Exploration Research Conference. Technip, Paris, pp. 475–496.

Haenel, R. and Staroste, E., 1988. Atlas of Geothermal Resources in the European Community, Austria and Switzerland. Publication No. EUR 11026 of the Commission of the European Communities, ESC, EEC, EAEC, Brussels–Luxembourg.

Hanor, J.S., 1980. Dissolved methane in sedimentary brines: potential effect on the PVT properties of fluid inclusion. Econ. Geol., 75: 603–609.

Kemper, E., 1976. Geologischer Führer durch die Grafschaft Bentheim und die angrenzenden Gebiete, mit einem Abriß der emsländischen Unterkreide. 5. erg. Aufl., Nordhorn–Bentheim.

Lee, W.H.K. and Uyeda, S., 1965. Review of heat flow data. In: Terrestrial Heat. Am. Geophys. Union., Geophys. Monogr., 8: 87–190.

Leuchs, W., 1985. Beziehungen zwischen Verquarzung und Dolomitisierung der devonischen Riffkalke von Dornap bei Wuppertal. Neues Jahrb. Geol. Paläontol., Monatsh., pp. 129–152.

Lopatin, N.V., 1971. Temperature and geologic time as factors in coalification. Izv. Akad. Nauk SSSR, Ser. Geol., 3: 95–106 (in Russian).

Lüders, V., Dulski, P., Möller, P. and Hoefs, J., 1990. Zur Genese der hydrothermalen Fluoritmineralisationen des Harzes. In: G. Friedrich (Editor), Metallogenetische Entwicklung im Harz. Report 1, pp. 15–32.

McLimans, R., 1987. The application of fluid-inclusion to migration of oil and diagenesis in petroleum reservoirs. Appl. Geochem., 2: 585–603.

Mundry, E., 1971. Der Temperaturverlauf im Dach des Bramscher Massivs nach der Wärmeleitungstheorie. Fortschr. Geol. Rheinl. Westfalen, 18: 539–546.

Naeser, N.D., Naeser, C.W. and McCulloh, T.H., 1989. The application of fission track dating to the depositional and thermal history of rocks in sedimentary basins. In: N.D. Naeser and T.H. McCulloh (Editors), Thermal History of Sedimentary Basins, Methods and Case Histories. Springer-Verlag, Berlin, pp. 157–180.

Neugebauer, H.J. and Walzebuck, J.P., 1987. A modelling theory for cratonic basins and their hydrocarbon accumulations. Proc. 12th World Petroleum Congress, Houston, Texas, 2: 9–17.

Philp, R.P. and Lewis, C.A., 1987. Organic geochemistry of biomarkers. Annu. Rev. Earth Planet. Sci., 15: 363–395.

Prezbindowski, D.R. and Tapp, J.P., 1991. Dynamics of fluid inclusion alteration in sedimentary rocks: a review and discussion. Org. Geochem., 17: 131–142.

Quigley, T.M., MacKenzie, A.S. and Gray, J.R., 1987. Kinetic theory of petroleum generation. In: B. Doligez (Editor), Migration of Hydrocarbons in Sedimentary Basins. Technip, Paris, pp. 649–665.

Radke, M., 1987. Organic geochemistry of aromatic hydrocarbons. In: J. Brooks and D.H. Welte (Editors), Advances in Petroleum Geochemistry, Vol. 2. Academic Press, London, pp. 141–207.

Radke, M., 1988. Application of aromatic compounds as maturity indicators in source rocks and crude oils. Mar. Pet. Geol., 5: 224–236.

Radke, M. and Welte, D.H., 1981. The methylphenanthrene index (MPI): A maturity parameter based on aromatic hydrocarbons. In: M. Bjørøy et al. (Editors), Advances in Organic Geochemistry 1981. Pergamon Press, Oxford, pp. 504–512.

Roedder, E., 1984. Fluid-inclusions. Rev. Mineral., Mineral. Soc. Am., 12, 644 pp.

Rullkötter, J.D., Leythaeuser, B., Horsfield, B., Littke, R., Mann, U., Müller, P.J., Radke, M., Schaefer, R.G., Schenk, H.-J., Schwochau, K., Witte, E.G. and Welte, D.H., 1988. Organic maturation under the influence of a deep intrusive heat source: a natural experiment for quantitation of hydrocarbon generation and expulsion from a petroleum source rock (Toarcian Shale, northern Germany). Org. Geochem., 13: 847–856.

Schulz, R., 1989. Temperaturverteilung in Nordwestdeutschland. Nachr. Dtsch. Geol. Ges., 41: 72 (abstract).

Shepherd, T.J., Rankin, A.H. and Alderton, D.H.M., 1985. A Practical Guide to Fluid Inclusion Studies. Blackie, London, 239 pp.

Sweeney, J.J. and Burnham, A.K., 1990. Evaluation of a simple model of vitrinite reflectance based on chemical kinetics. Am. Assoc. Pet. Geol., Bull., 74 (10): 1559–1570.

Teichmüller, M., Teichmüller, R. and Bartenstein, H., 1984. Inkohlung und Erdgas — eine neue Inkohlungskarte der Karbon-Oberfläche in Nordwestdeutschland. Fortschr. Geol. Rheinl. Westfalen, 32: 11–34.

Tilley, B.J. and Longstaffe, F.J., 1989. Diagenesis and isotope evolution of porewaters in the Alberta Deep Basin: the Faulher Member and Cadomian Formation. Geochim. Cosmochim. Acta, 53: 2529–2546.

Tissot, B.P. and Welte, D.H., 1984. Petroleum Formation and Occurrence. Springer-Verlag, Berlin, 2nd ed., 699 pp.

Tissot, B.P., Pelet, R. and Ungerer, P.H., 1987. Thermal history of sedimentary basins, maturation indices, and kinetics of oil and gas generation. Am. Assoc. Pet. Geol., Bull., 71 (12): 1445–1466.

Ungerer, P., Bessis, F., Chénet, P. Y., Durand, B., Nogaret, E., Chiarelli, A., Oudin, J. and Perrin, J.F., 1984. Geological and geochemical models in oil exploration: principles and practical examples. In: G. Demaison (Editor), Petroleum Geochemistry and Basin Evaluation. Am. Assoc. Pet. Geol., Mem., 35: 53–77.

Waples, D.W., 1980. Time and temperature in petroleum formation: application of Lopatin's method to petroleum exploration. Am. Assoc. Pet. Geol., Bull., 64 (6): 916–926.

Welte, D.H. and Yükler, A., 1981. Petroleum origin and accumulation in basin evolution. Am. Assoc. Pet. Geol., Bull., 65: 1387–1396.

Wygrala, B.P., 1989. Integrated Study of an Oil Field in the Southern Po Basin, Northern Italy. Ber. Kernforschungsanlage Jülich, 2313, 313 pp.

Yalcin, M.N., Welte, D.H., Misra, K.N., Mandal, S.K., Balan, K.C., Mehrotra, K.L. and Lohar, B.L., 1988. 3-D computer-aided basin modelling of Cambay Basin, India — a case history of hydrocarbon generation. In: R.K. Kumar, P. Dwivedi, P. Banerjee and V. Gupta (Editors), Petroleum Geochemistry and Exploration in the Afro-Asian Region. Balkema, Rotterdam, pp. 417–450.

Yukler, M.A., Cornford, C. and Welte, D.H., 1978. One-dimensional model to simulate geologic, hydrodynamic and thermodynamic development of a sedimentary basin. Geol. Rundsch., 67: 960–979.

Zhang, Yi-gang and Frantz, J.D., 1987. Determination of the

homogenization temperature and densities of super critical fluids in the system $NaCl-KCl-CaCl_2-H_2O$ using synthetic fluid-inclusion. Chem. Geol., 64: 335–350.
Ziegler, P.A., 1982. Geological Atlas of Western and Central Europe. Elsevier, Amsterdam, 130 pp.
Ziegler, P.A., 1989. Geological Evolution of northern Germany — a plate tectonic review. Nachr. Dtsch. Geol. Ges., 41: 81–82 (abstract).

K. LEISCHNER *Institute for Petroleum and Organic Geochemistry (ICG-4), Research Centre Jülich (KFA), P.O. Box 1913, D-5170 Jülich, Germany*
Present address: *RISM, AGIP Spa, Via Bonarelli, 1, 20097 San Donato-Milanese, Italy*
D.H. WELTE *Institute for Petroleum and Organic Geochemistry (ICG-4), Research Centre Jülich (KFA), P.O. Box 1913, D-5170 Jülich, Germany; and Integrated Exploration Systems (IES), Bastionstraße 11–19, D-5170 Jülich, Germany*
R. LITTKE *Institute for Petroleum and Organic Geochemistry (ICG-4), Research Centre Jülich (KFA), P.O. Box 1913, D-5170 Jülich, Germany*

A kinetic interpretation of the smectite-to-illite transformation

G. Vasseur and B. Velde

Many observations have emphasized the progressive loss of smectite content in interstratified illite smectite (I/S) minerals: in deeply buried sediments, the proportion of smectite decreases with depth, terminating at greatest depth with a very illite-rich phase. It is clear that this transformation depends on the time–temperature history of the rocks; this can be deduced from observed series which have been submitted to various diagenetic conditions.

The aim of this study is to take advantage of several series of smectite content measurements in sedimentary sequences which have experienced very different burial rates (from 0.01 to 1 mm per year). After proper reconstruction of the time–temperature history for each case, the data are accounted for by first-order kinetic laws where the thermal activation is of the Arrhenius type. The unknown parameters such as the activation energy(ies) and pre-exponential factor(s) are obtained by minimizing the difference between experimental data and computed values.

A simple transformation reaction from smectite into illite leads to results which greatly vary from one case to another. Analysis of the results obtained at various heating rates shows that a single reaction can explain the beginning of the process but not the whole of it. This analysis suggests that the transformation proceeds in fact through a two-scale serial reaction.

The first step would correspond to the formation and growth of a disordered I/S structure ($R = 0$) at the expense of pure smectite; this reaction would be terminated with pure I/S phase (50% smectite). The second step consists of the disappearance of smectite layers in an ordered I/S phase ($R = 1$). The characteristics (activation energies and pre-exponential factors) of these two reactions are computed using an optimizing procedure and their values are discussed.

Introduction

The transformation of smectite into illite through the intermediate mixed-layer illite/smectite during the burial process has been recognized for a long time (Dunoyer de Segonzac, 1965; Burst, 1969; Perry and Hower, 1970). The understanding of this diagenetic process is crucial, because of the probable relation of this phenomenon with the buildup of excess pressure, hydrocarbon migration and with other phenomena occurring in sedimentary basins (Bruce, 1984).

The exact nature of this transformation is still poorly understood and several possible mechanisms have been proposed. On the basis of natural observations, this transformation appears to be largely controlled by temperature (Perry and Hower, 1972; Velde, 1985). However, the comparison of smectite content profiles obtained in different parts of a sedimentary basin have led to conflicting results (Bruce, 1984; Freed and Peacor, 1989) and the influence of other factors such as the composition of the primary clay and the chemistry of fluids is often proposed.

A few authors have tried to model the transformation smectite-to-illite in terms of a chemical reaction governed by simple kinetics and accounting for the effect of time and temperature (Hower, 1981; Bethke and Altaner, 1986; Dutta, 1986). Most of the data used came from wells in the U.S. Gulf Coast.

A different approach was used by Velde and Espitalié (1989): using series of widely different ages (Early Tertiary to Palaeozoic), they compare the smectite content to the maturity of organic matter in the same samples. They showed that clays in old sedimentary series evolve much more slowly than organics, demonstrating that there is a kinetic factor in the smectite-to-illite transformation. Moreover, the clay reaction is much less temperature-sensitive than organic maturation. Another comparison was attempted between I/S evolution and zeolite facies in Palaeogene sediments (Velde and Iijima, 1988). The temperatures over which illitization reaction proceeds are similar, for time periods of 5–15 Ma, but the zeolites react rapidly and exhibit variation in facies transition as a function of time whereas the clays do not.

These observations show there is a kinetic factor in the change from smectite to illite in diagenesis even though other factors such as the inherited mineralogy may also be of importance. Moreover, this

Basin Modelling: Advances and Applications edited by A.G. Doré et al. NPF Special Publication 3, pp. 173–184, Elsevier, Amsterdam.

kinetic behaviour is different from that governing the reaction rates of other phases present in the same rocks. The purpose of the present paper is to study how simple kinetic assumptions can explain observed I/S series in wells where the time–temperature history may be reconstructed with some confidence. The difference with previous attempts is that the available data cover a large range of time span and temperature history; these data consist of I/S series from eight deep wells with known burial histories.

Experimental data

Samples and XRD method

In this study, suites of samples coming from France (one well called Paris Basin), from Texas (three wells: Carter, Peeler and Mustang Island), from California (one well labelled California) and from Japan (three wells: Masugata, Shimiogarshi and Hamayuchi) have been used. Except for one well in Texas (Mustang Island), the studied wells have been described in Velde and Iijima (1988), Mathieu and Velde (1989) and Velde and Espitalié (1989).

The determination of I/S smectite content makes use of X-ray diffraction (XRD) techniques which are explained in Velde and Vasseur (1992). Two different XRD methods have been used depending on the structure of the material.

For $R = 0$ (disordered) I/S types, the compositional identification was deduced from the maximum of the first-order peak under glycollated conditions. The shape and position of the peak obtained on background-stripped spectra was compared to theoretical X-ray diffractograms of I/S calculated as a function of the percent smectite and of the number of coherently diffracting layers. The basal spacings are assumed to be 16.9 Å for glycollated smectite and 10.0 Å for mica-like layers.

For $R = 1$ (ordered) I/S minerals, the smectite content was estimated using the shape and position of the peak in the air-dried sample assuming that the I/S crystallites have an average of eight coherently scattering layers (see Velde et al., 1986 for details). Repeated measurements of the same sample suggest that the reproducibility errors are on the order of $\pm 5\%$ smectite.

For each of the eight wells, the measured smectite content is plotted as a function of depth in Fig. 1. As described above, the smectite content of the samples was determined by the maximum of the peak representing the aggregate I/S clay mineral assemblage, with sizes less than 2 μm, present in the mudstone samples. The XRD maxima given by the I/S mineral grains are treated as a whole, i.e. no attempt is made to decompose possible composite peaks into individual population components (see Lanson and Champion, 1991). For example, all of the crystallites of both $R = 0$ and $R = 1$ structures present in the same sample which diffract in the same angular region will be assimilated into a single phase. This is particularly important where the $R = 0$ and $R = 1$ phases appear together, as is the case in the 40 to 60% smectite region. The I/S mineral aggregate found before and after this intermediate compositional range is dominated by either the disordered, $R = 0$ smectite-rich or the ordered, $R = 1$ illite-rich mineral which can be considered single-phase assemblages of crystallites in these compositional ranges. The most reliable data then come from the beginning and the end of the smectite-to-illite conversion series.

Thermal history

The burial histories of the eight wells are plotted in Fig. 2 as a function of depth; these histories cover a large range of average burial rate from less than 10 m/Ma to more than 1 km/Ma. In all cases, the correction for compaction of sediments has been neglected: it is probably a second-order effect compared to possible errors in temperature and those of smectite content of the I/S minerals estimations. These data allow a reconstruction of the burial history of each sample, assuming that no erosion took place.

Bottom hole temperatures (BHT) were used to estimate the thermal gradients except in the Paris Basin well where the thermal gradient was deduced from observations at thermal equilibrium (Demongodin et al., 1991). In all cases, the BHT was measured after at least 12 h thermal stabilization time: an error of the order of 5–10°C seems likely but as the depth is great and the temperature is quite high (125°C or higher) the resulting error on the thermal gradient is relatively small. The thermal gradients deduced from these measurements are assumed to be linear, i.e. the effects of thermal conductivity variations are neglected.

In order to estimate the time–temperature history of each sample it is necessary to make some assumptions on possible temporal variations of the thermal gradients. The starting hypothesis used in the following is that the past thermal gradients did not differ markedly from their present values.

Estimation of kinetic parameters

Characterization of the smectite-to-illite transformation

In order to model the change in smectite content of the I/S clay mineral assemblage in terms of reaction kinetics, several assumptions must be made.

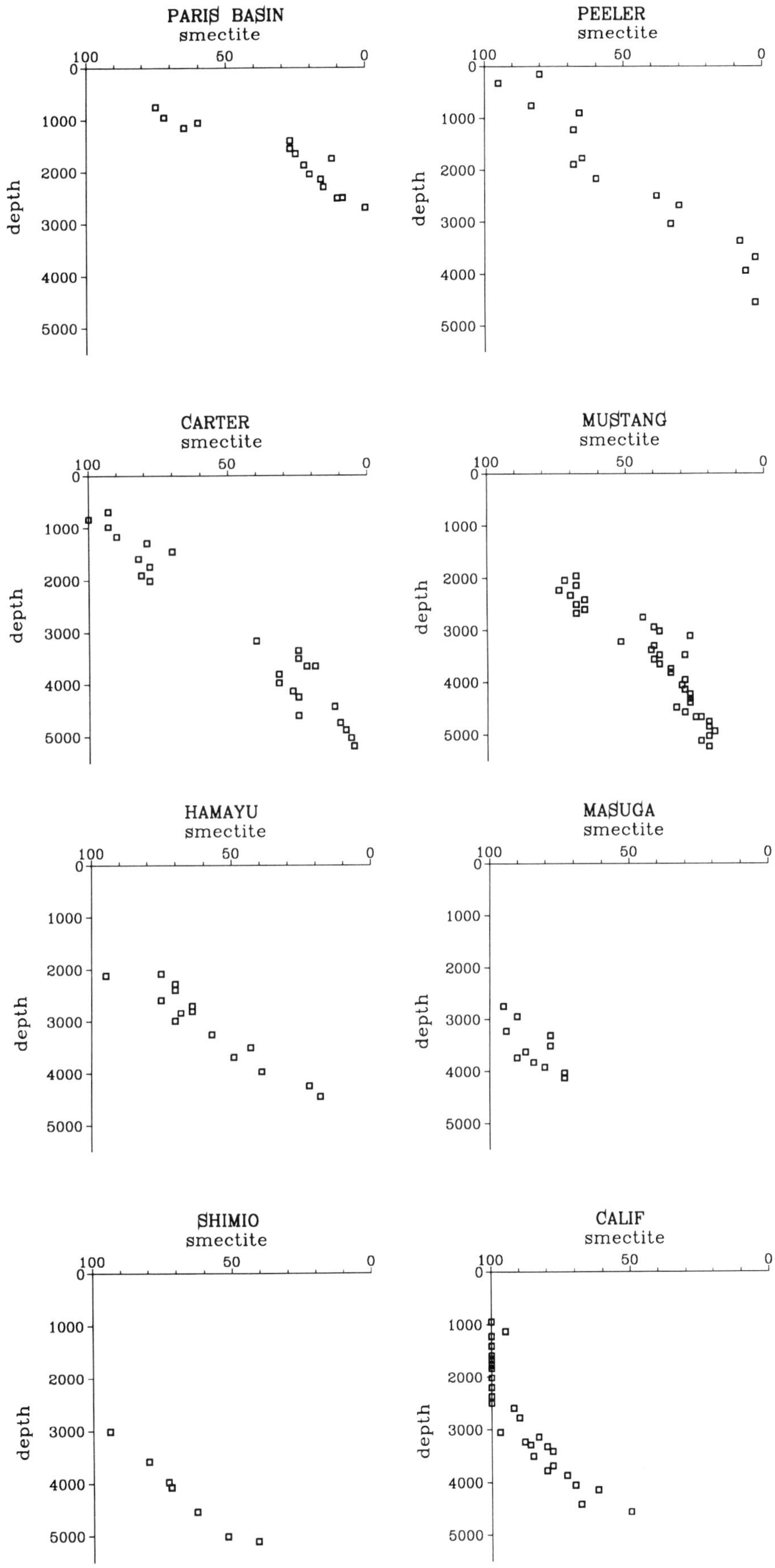

Fig. 1. Results of the XRD determinations of I/S clay mineralogy (smectite content in % as a function of the present depth) for the eight different wells studied. The wells are ordered according to increasing burial rate.

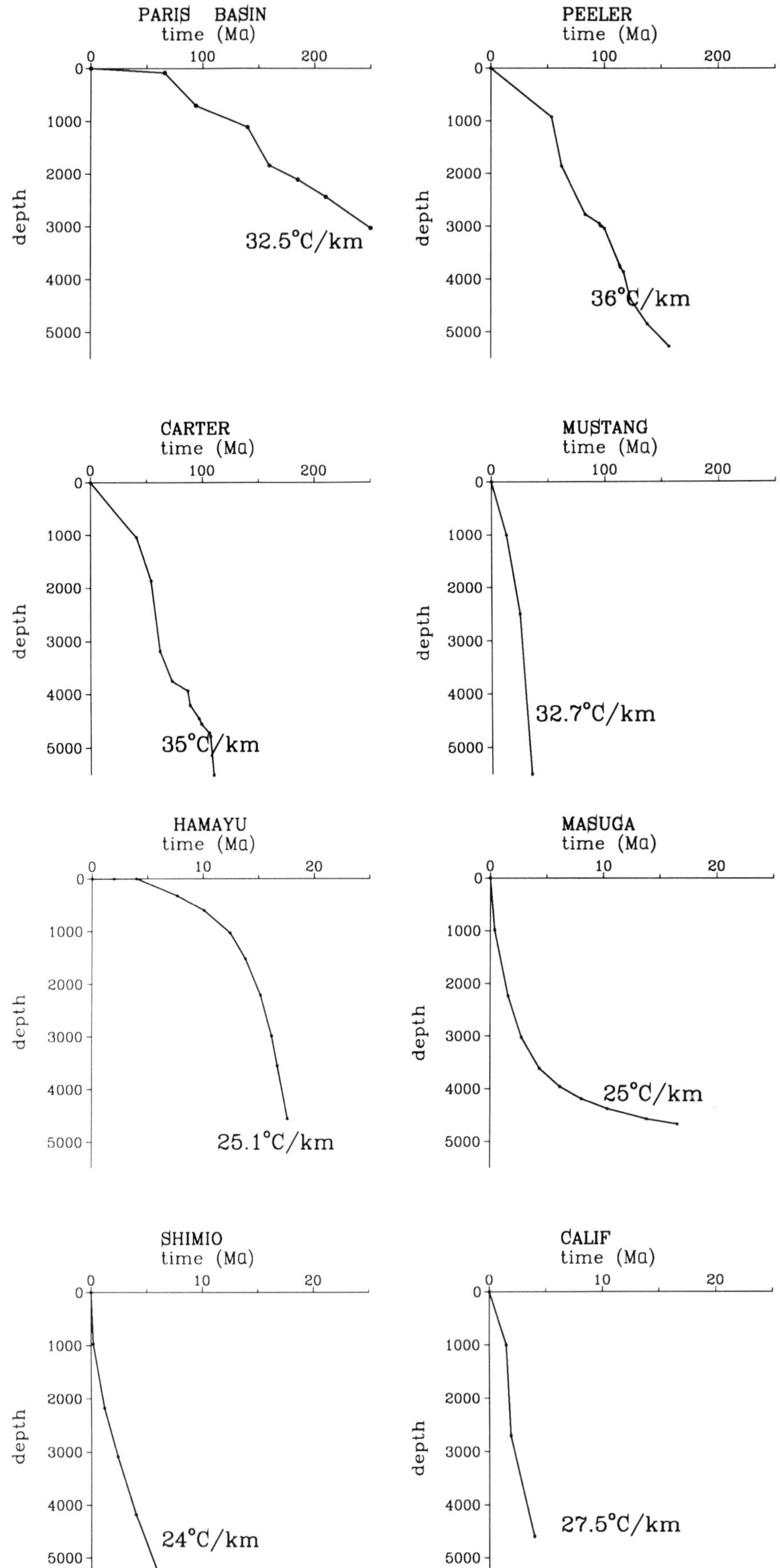

Fig. 2. Plot of age versus depth of sediments in the eight different wells of the present study. Thermal gradients are indicated in each case.

(1) At the beginning of the transformation, the reactant must be well defined: the fundamental assumption is that the reaction starts with 100% smectite in all samples. This assumption is consistent with the data presented in Fig. 1. In fact, observations of Tertiary Gulf Coast and Japanese wells (Hower et al., 1976; Freed, 1982; Velde and Iijima, 1988) show, in the upper portions of the wells, a great variability in smectite content; however, at greater depths, a 100% smectite phase becomes dominant. The upper, heterogeneous, zone represents the detrital, unreacted sediment material; the early change from the variable initial composition to a pure smectite mineral in the wells seems to occur over a range of several hundreds of metres at 1 to 2 km depth in younger wells. At greater depth, the 100% smectite mineral evolves regularly toward an illitic composition; it is this later evolution that is interpreted below, assuming a starting point at 100% smectite.

(2) During the transformation, the reactant products are assumed to be homogeneous. In fact, recent observations have shown (Lanson and Champion, 1992) that in the I/S clay mineral series at least three structurally and morphologically distinct crystalline forms can be distinguished: (a) flake-shaped crystals for smectite-rich forms with an $R = 0$ disordered stacking sequence (100–50% smectite); (b) lath-shaped $R = 1$ ordered structure crystals (50 to near zero % smectite); and (c) hexagonal-shaped, illitic crystals. Moreover, Srodon and Eberl (1984) found that the composition of the illite component is not constant. Therefore, the I/S reaction sequence cannot be considered homogeneous, neither chemically nor structurally. Nevertheless, there is an obvious continuity of the smectite content trends as a function of depth which suggests that one can attempt a modelling of the transformation in terms of reaction kinetics.

(3) The transformation is assumed to be governed by first-order reactions. Such first-order reactions have often been considered to adequately represent complex reaction series where a rate-limiting reaction step predominates in reaction advancement (Lasaga, 1981). In fact, the reaction could be of higher order as suggested by experimental data of Whitney and Northrop (1989) and fitted data for thermal metamorphism of clays (Pytte and Reynolds, 1989). However, there is not enough information to choose a reaction order for the transformation in diagenesis and the first order seems the simplest choice.

Single reaction model

For a first trial, we use the assumption that the complete smectite-to-illite reaction can be modelled as a first-order reaction as already assumed by Dutta (1986) for a single series of Gulf Coast sediments. The basic equations used are:

$$\frac{\mathrm{d}S}{\mathrm{d}t} = -kS \tag{1}$$

$$k(t) = A\exp\left(-\frac{E}{RT(t)}\right) = \exp\left(\log(A) - \frac{E}{RT(t)}\right) \tag{2}$$

where S is the smectite phase reactant, $k(t)$ is the time-dependent reaction constant, t is time (Ma), T temperature in degrees Kelvin, E the energy of activation in kJ/mole, A in Ma^{-1} is the pre-exponential factor (considered constant over the temperature interval studied) and R the universal gas constant.

For a layer deposited at time 0, $S(t)$ (smectite content at time t) is given by:

$$\begin{aligned} S(t) &= S_0 \exp\left[-\int_0^t k(t')\,\mathrm{d}t'\right] \\ &= S_0 \exp\left[-\int_0^t \left(\log(A) - \frac{E}{RT(t')}\right)\mathrm{d}t'\right] \end{aligned} \tag{3}$$

where S_0 is the initial smectite portion in the I/S. In a given well, the temperature history $T(t)$ is evaluated from the burial history and an assumed thermal gradient. It is then possible to compute a theoretical value of S corresponding to any assumed values of parameters A and E.

For each well, the optimal values of the kinetic parameters A and E are deduced from the least-square fit between observed smectite content as a function of depth and calculated values. In order to solve this highly non-linear problem, the fixed point method proposed by Tarantola and Valette (1982) was used to compute iteratively the optimum values for the two parameters A and E. Confidence domains are estimated for the parameters taking into account the experimental data and their errors; the smectite content is assumed to have an error of $\pm 5\%$. For these confidence domains, a linear approximation was used in the vicinity of the optimal value which implies that these domains are elliptic and strongly elongated (shown as polygons in Fig. 3).

The results obtained for the various wells are rather scattered (i.e. E varies from 16 to 49 kJ/mole) and a plot of $\log A$ vs. E reveals that the calculated pairs of values are strongly correlated along a direction which is approximately the direction of the major axis of each domain of uncertainty. This feature is easily explainable by eq. 3: during the diagenetic process, the absolute temperature T does not vary tremendously around an average value $\overline{T}$. Therefore, a small increase ΔE of E compensated by an increase of $\log(A)$ equal to $\Delta E/R\overline{T}$ gives ap-

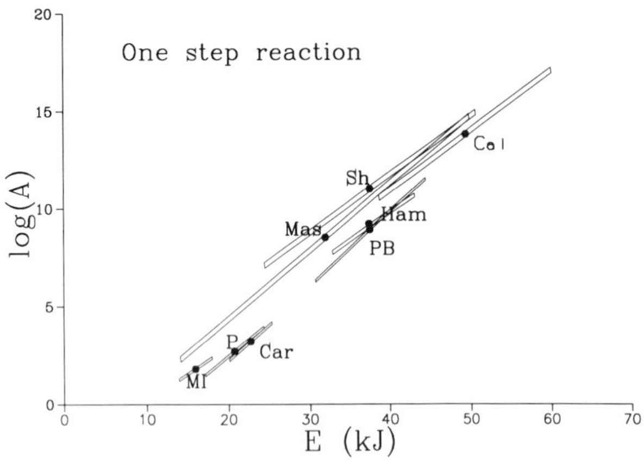

Fig. 3. Confidence domains of the best fit values of the kinetic factors E (activation energy, kJ/mole) and $\log A$ (pre-exponential factor, $\log(\mathrm{Ma}^{-1})$) obtained from the least square fitting for each well using a single-step reaction. Each name refers to one of the wells given in Table 1.

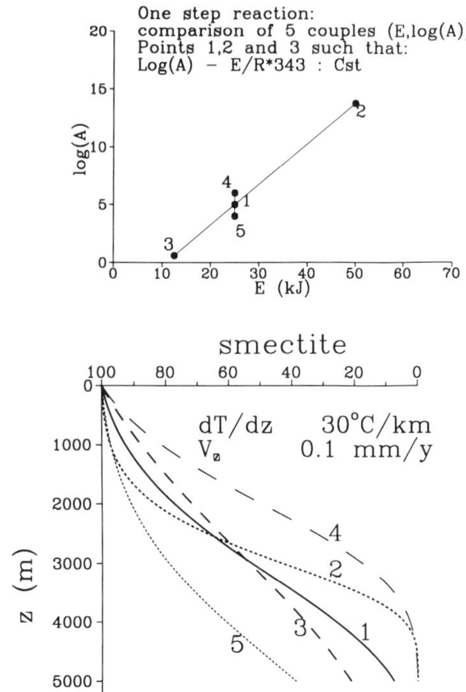

Fig. 4. Comparison of calculated smectite content as a function of depth using the same geological assumptions (burial rate 0.1 mm/y, thermal gradient 30°C/km), but five different kinetic parameters E and $\log A$ shown in the top of the figure. The set of parameters E and $\log(A)$ labelled *1, 2, 3* are such that: $\log(A) - \Delta E/R\overline{T}$ = constant with $\overline{T}$ = 343 K.

proximately the same value of S, at least for small variations.

The respective effects of the parameters E and $\log(A)$ are illustrated in Fig. 4 which presents plots of computed smectite content as a function of depth for different values of E and $\log(A)$, these two parameters being related by $\log(A) - \Delta E/R\overline{T}$ = constant. The various smectite content profiles have a common value around the depth where the temperature is $\overline{T}$ but the slopes of these profiles are different. For experimental data where the slope is poorly defined (this is the case of Masugata as illustrated in Fig. 1), the values of E and $\log(A)$ obtained by the least-square fit have a large uncertainty along the direction where $\log(A) - \Delta E/R\overline{T}$ = constant. Figure 4 also shows that a change of $\log(A)$, E remaining constant, results in large changes in the whole curve.

For these reasons, the various kinetic parameters calculated from the data of individual wells (Table 1 and Fig. 3) lead to predicted values of parameters E and $\log(A)$ which are significantly different. This is illustrated in Fig. 5, where the experimental results (smectite content) are compared to the values computed for four series chosen from the eight wells under study. It is clear from Fig. 5 that it is not possible to fit the whole set of data with a single set of parameters. A single first-order reaction cannot explain the data set and other assumptions are needed.

Dual reaction model

A careful examination of the calculations using a one-reaction model (Fig. 5), shows that the upper part of the depth–smectite content profile (ranging from 100 to 60% smectite in I/S) for the wells agrees

TABLE 1

Results obtained for each well from least square fit of experimental values of smectite content using a one-step reaction model (the symbols refer to Figs. 3 and 7)

Well	Symbol	E (kJ/mole)	$\log(A)$ ($\log(\mathrm{Ma}^{-1})$)	σ (%)
Paris Basin	PB	37.5 ± 7.3	8.9 ± 2.6	8.2
Peeler	P	20.7 ± 3.8	2.7 ± 1.3	8.5
Carter	Car	22.7 ± 2.8	3.2 ± 0.9	7.1
Mustang	MI	15.9 ± 2.1	1.8 ± 0.7	6.6
Hamayuchi	Ham	37.4 ± 5.3	9.2 ± 1.9	6.2
Masugata	Mas	31.9 ± 19.1	8.5 ± 1.8	5.7
Shimiogarshi	Sh	37.5 ± 14.0	11.0 ± 4.7	3.0
California	Cal	49.3 ± 11.1	13.8 ± 3.5	3.8

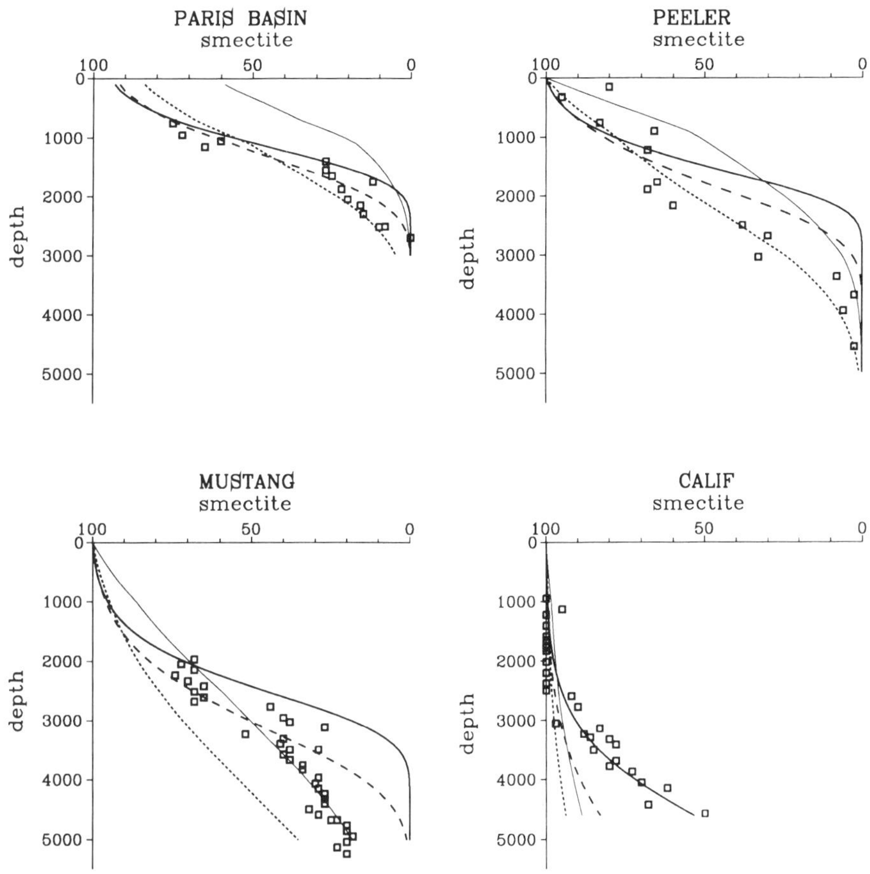

Fig. 5. Plot of the fits of the data points using a kinetic model with a one-step reaction. The thick solid line shows the fit for four wells using parameters fitted for the California well (E = 50 kJ/mole, log(A) = 14); the dashed line shows the predicted curve using data fitted for the Paris Basin well (E = 37.5, log(A) = 9); the dotted line shows predictions using values fitted for the Peeler well (E = 20.7, log(A) = 2.7); the thin solid line shows the predicted curve using data fitted for the Mustang well (E = 16, log(A) = 1.8).

generally with the curve calculated from the data of California, corresponding to E = 50 kJ/mole, log$A$ = 14, even though the fit becomes poorer in the lower part of the profile. This suggests the existence of two kinetic processes. This assumption is also supported by other observations which will be discussed later.

Therefore, the I/S evolution is divided into two reactions which determine the aggregate smectite content of the I/S clays in the sediments. The first reaction involves the $R = 0$ disordered minerals and the second involves the $R = 1$ ordered minerals. The model developed here consists of two serial reactions which are illustrated in Fig. 6.

The clay is assumed to be made of I/S "crystallites" composed of several layers either of smectite or of illite. Initially, all the layers are smectite; it is convenient to call X a subset of two neighbouring smectites as shown in Fig. 6. As diagenesis proceeds, some of the smectite layers may become illite according to one of the two following reactions:

$X \rightarrow Y$: a smectite layer belonging to an X becomes illite. An X disappears but the new illite is associated with its neighbour smectite in order to form a new Y subset composed of one smectite + one illite. Clearly this reaction corresponds to the evolution from the initial $R = 0$ smectite-rich, disordered mineral to an ordered $R = 1$ one. If this reaction was the only one, it would lead, at equilibrium, to a $R = 1$ mineral with 50% smectite as presented in Fig. 6a.

$Y \rightarrow Z$: a smectite layer belonging to an Y becomes illite and forms a subset Z (two illites). This second reaction simulates the progressive illitization of $R = 1$ I/S minerals; it progresses in competition with the former reaction as shown in Fig. 6b.

A somewhat similar description of the transformation has already been proposed by Bethke and Altaner (1986); they used the Monte Carlo simulation in order to simulate the illitization. In the present study, we do not account for the different probabilities of transformation as a function of each neighbour, but we assume that the medium remains statistically homogeneous so that kinetic laws can be applied. Assuming two first-order kinetic laws, one obtains for the concentrations X and Y:

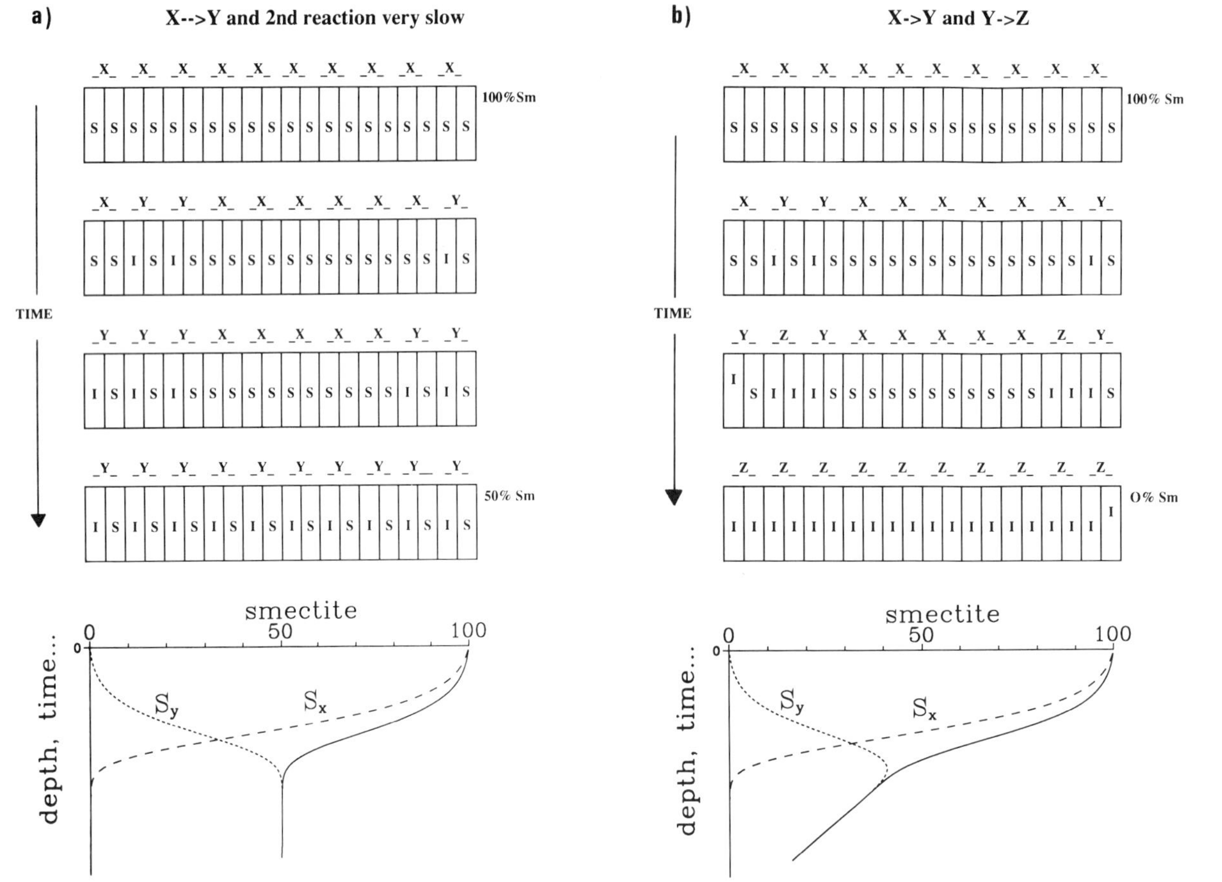

Fig. 6. Sketch of the assumed reactions. The initial crystallite is composed of smectite layers which are submitted to two reactions of illitization. Two assumptions are illustrated in order to explain the respective influence of both reactions. In (a), only the first reaction $X \to Y$ occurs; smectite layers eventually become illite leading to a perfect $R = 1$ structure. As the reaction proceeds (as a function of time, temperature or both), the smectite content evolves from 100% to 50% as shown at the bottom plot. In (b), both reactions $X \to Y$ and $Y \to Z$ occur. As shown in the bottom plot, the $R = 1$ component (S_y) goes through a maximum and finally vanishes. The total smectite content ($S_x + S_y$) decreases from 100% to 0%.

$$\frac{dX}{dt} = -k_1 X \quad \text{with } \log(k_1) = \log(A)_1 - \frac{E_1}{RT} \tag{4}$$

$$\frac{dY}{dt} = k_1 X - k_2 Y$$

$$\text{with } \log(k_2) = \log(A)_2 - \frac{E_2}{RT} \tag{5}$$

where k_1 and k_2 are two kinetic coefficients with two unknowns for each A_1, E_1 and A_2, E_2. It is clear from the definition (Fig. 6) that the smectite content is equal to $X + Y/2$. Starting from pure smectite ($X = X_0$, $Y = Z = 0$), the integration of these kinetic laws can be obtained in closed form:

$$X(t) = X_0 \exp\left(-\int_0^t k_1(t')\,dt'\right) \tag{6}$$

$$Y(t) = X_0 \exp\left(-\int_0^t k_2(t')\,dt'\right) \times \int_0^t k_1(t') \exp\left(\int_0^{t'} [k_2(t'') - k_1(t'')]\,dt''\right) dt' \tag{7}$$

Both X and Y contribute to the total smectite content. This is shown at the bottom of Fig. 6 where the smectite content is plotted versus an evolution parameter (time, temperature or burial depth): when the second reaction is inhibited or very slow (Fig. 6a), the total smectite content tends toward 50% because Y reaches its stable value; when the second reaction is allowed (Fig. 6b), Y goes through a maximum and then decreases so that the total smectite content vanishes; the position of this maximum and the asymptotic trend depend on the k_2 kinetic coefficient.

The problem is now to define the set of four parameters ($\log A_1$, E_1, $\log A_2$, E_2) which explains best the experimental data for the eight wells. In fact, for most individual wells, there are too few data to perform a significant regression on four simultaneous parameters and the solution is very unstable. This was only possible for three wells where the series was quite complete. The resulting parameters are given in Table 2 and illustrated in Fig. 7.

TABLE 2

Results obtained for each well from least square fit of experimental values of smectite content using a two-step reaction model

Well	$\log(A_1)$	E_1	$\log(A_2)$	E_2
Paris Basin	22.0 ± 9.5	70.0 ± 24.7	2.4 ± 2.5	19.5 ± 7.2
Carter	9.2 ± 5.0	37.6 ± 13.7	2.7 ± 2.3	22.6 ± 8.2
Mustang	13.7 ± 6.1	47.7 ± 17.6	2.3 ± 2.2	22.2 ± 7.5

The values of E_1 and $\log(A_1)$ correspond to the first reaction and those of E_2 and $\log(A_2)$ values to the second reaction.

In the three cases under study, the computed uncertainties are important. For the first reaction, the scatter is large but it is clear that this first reaction has coefficients significantly larger than those of the second. The second reaction corresponds to values that appear to be quite close one to another.

The next step is to invert the whole set of data (158 points) simultaneously, i.e. to define a single set of parameters valid for all the data. The result of this inversion appears very robust (it does not depend upon the *a priori* starting point). It leads to the following set of parameters:

$$\log(A_1) = 16.7 \pm 0.6 \log(\mathrm{Ma}^{-1})$$

$$E_1 = 56.2 \pm 1.7 \text{ kJ/mole}$$

$$\log(A_2) = 2.7 \pm 0.6 \log(\mathrm{Ma}^{-1})$$

$$E_2 = 22.2 \pm 2.0 \text{ kJ/mole}$$

The confidence domains on each parameter resulting from the inversion of the whole data set are quite small, which implies that these values are fairly well constrained.

This unique set of parameters accounts quite well for the data of the different wells, as shown in Fig. 8.

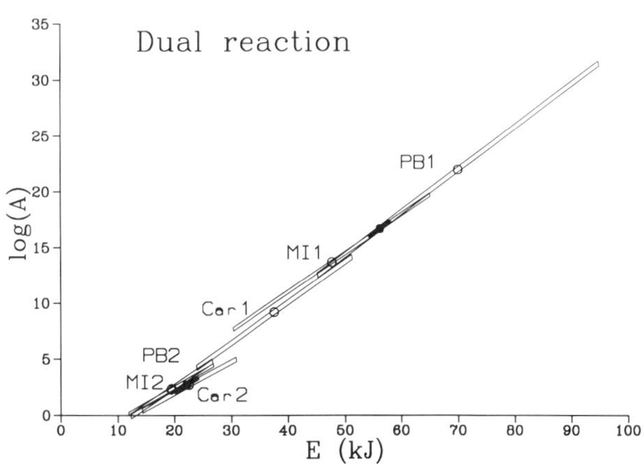

Fig. 7. Kinetic factors calculated with a two-step reaction model. Circles correspond to individual values obtained for both reactions using individual data from Paris Basin (*PB1* and *PB2*, respectively, for the two reactions), Carter (*Car1*, *Car2*) and Mustang Island (*MI1*, *MI2*). The circles corresponding to *PB2* and *MI2* are identical. The polygons represent the confidence area. Dots give the values of respectively E_1 and $\log(A_1)$, E_2 and $\log(A_2)$ which are obtained considering all the data.

The aggregate smectite content curve (solid line) as well as the respective contributions of each of the two reactions to smectite content (proportional to $2X$ and Y, respectively) can be compared to the actual data. In most cases, the agreement between the observed and the predicted values is satisfactory.

In a few wells, a discrepancy between the calculated and observed values for the I/S smectite content–depth profile can be noted: the general trends are similar but the position of the calculated curve is not correct. This is obvious in the Paris Basin, Carter and Shimiogarshi wells. This could indicate significant departure from the proposed model; however, this could also be due to an actual thermal history that was different from the one assumed, for example, because of heat flux variations.

The effects of varying the thermal gradient has been studied on three wells (Fig. 9). In the case of the Paris Basin well, an increase in the gradient of 20% gives a better agreement with the data. In the case of the Texas Carter well, a decrease in the thermal gradient of 20% simulates the data much better. In the Japanese Shimiogarshi well, a 40% increase in the gradient was necessary to fit the data points. However, since these assumptions remain highly speculative, it is not possible to derive firm conclusions.

Discussion and conclusions

The fact that the data can be reconciled with a two-step reaction model, involving two structural types of I/S minerals, is not in itself a demonstration of the reality of this process and has to be discussed in the light of recent studies concerning the transformation of smectite to illite. It seems clear now that the smectite-to-illite reaction or reactions occur(s) through the dissolution of previous minerals grains as suggested by Boles and Franks (1979) and Nadeau et al. (1985). Also, the reaction seems to occur in two major steps, identified by a bulk mineralogical change at near 50% smectite content in the interlayer edifice (Srodon and Eberl, 1984) with a change in crystal morphology (Inoue et al., 1988).

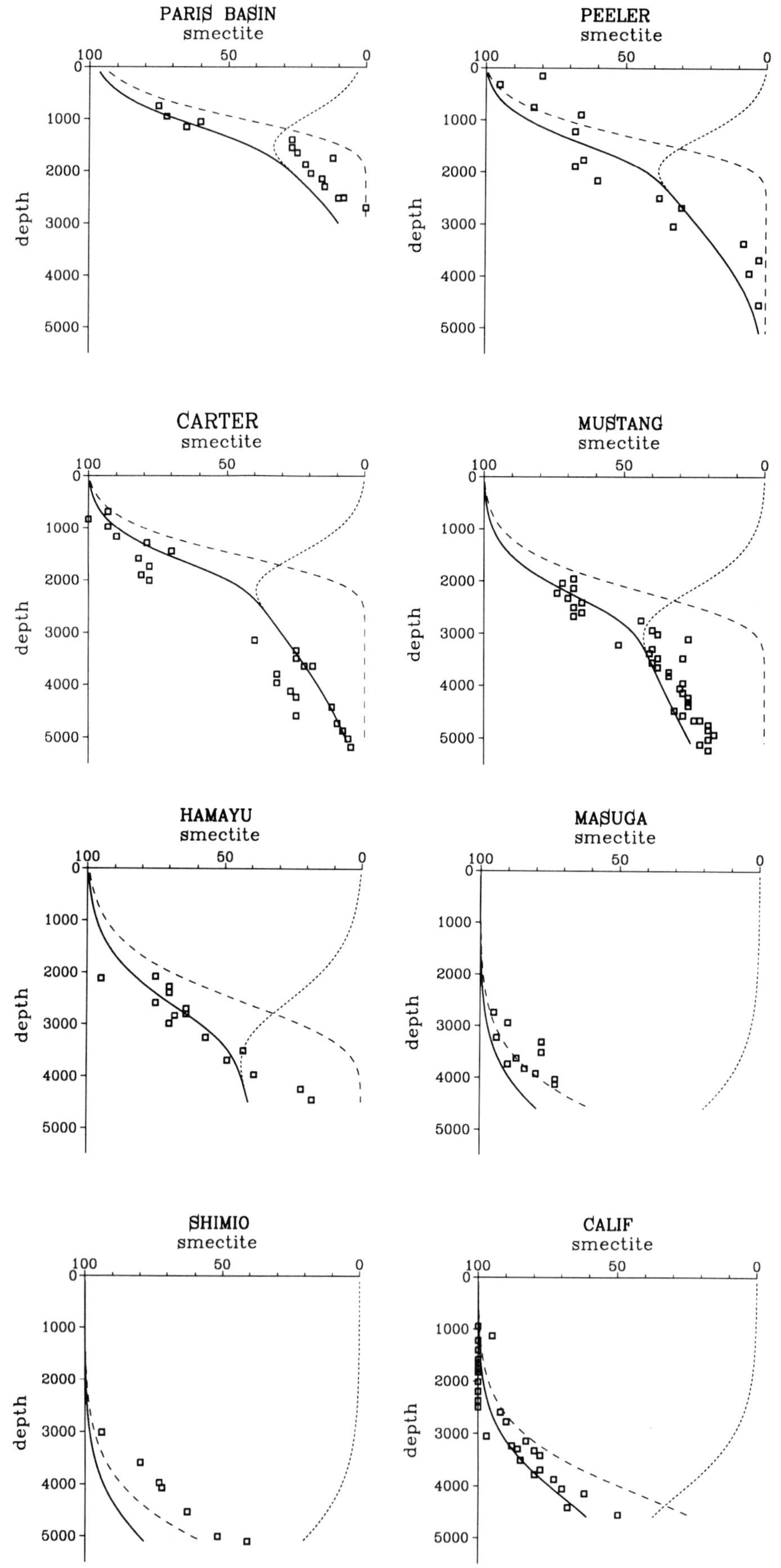

Fig. 8. Comparison of the data (squares) with the calculated values assuming the dual reaction. The solid line is the calculated smectite content using the kinetic values of $E_1 = 56.2$, $\log A_1 = 16.2$ for reaction one, and $E_2 = 22.2$ and $\log A_2 = 2.7$ for reaction two. The contributions of X $(R = 0)$ and Y $(R = 1)$ are shown as dashed line and dotted line.

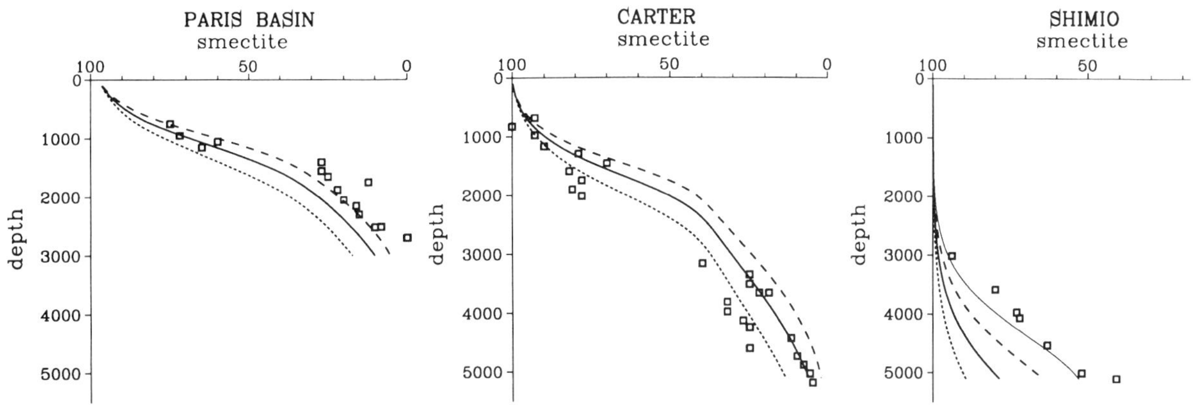

Fig. 9. Adjusted curves of reaction progress for three wells where the calculated curves did not correspond well with the experimental data points of I/S composition. The different curves were obtained by changing the thermal gradient from the measured values (solid lines) by increment of −20% (dotted line), +20% (dashed line) and even +40% for the case of Shimio (thin line).

Whitney and Northrup (1989) found evidence for a two-step reaction and two distinct reaction rates in laboratory experiments using X-ray diffraction observations and measurements of oxygen isotope fractionation. The first mineral series is randomly ordered I/S with compositions ranging from 100% to 50% smectite content; they seem to have large, rather irregular crystal shapes. Nadeau et al. (1985) have concluded that the second part of the reaction series concerns a lath-shaped mineral whose growth shows an increasingly smaller number of expanding layers in the illite/smectite clay aggregate.

Lanson and Champion (1991) have shown that this second part of the reaction series (50% to near 5% smectite) can be described as a crystal growth process. Finally, the last stage of the smectite-to-illite reaction is clearly identified by these authors as a dissolution of the lath-shaped I/S (expandable) minerals which are replaced by hexagonal-shaped illite crystals. Thus, the smectite-to-illite transformation appears as a multistage reaction involving some three distinct crystalline phases: randomly ordered minerals of 100 to 50% smectite content, ordered lath-shaped minerals of 50 to 5% smectite content and illite minerals of hexagonal shape. Change from one crystalline structure to another one seems to be accomplished by dissolution of the old and crystallization of the new phase.

The concordance of mineralogical observations and laboratory experiments substantiates the use of a two-reaction model to simulate the change in smectite content of I/S clays, as proposed in this paper. However, the present work is based on total smectite content data (i.e. the amount of initial smectite and expanding material remaining in the clay crystal aggregates) as determined by X-ray diffraction methods and cannot discriminate the different types of reaction.

Nevertheless, it is interesting to compare the activation energies of 56 and 22 kJ/mole estimated here, with those obtained from laboratory determinations of the smectite-to-illite reaction. The experiments indicate values ranging from 120 to 15 kJ/mole (Eberl and Hower, 1976; Roberson and Lahann, 1981; Howard and Roy, 1985). In all experiments the temperatures exceeded 200°C, well above those experienced by the sedimentary mineral series investigated here. The activation energy values obtained in the present study for the two-part reaction in the smectite-to-illite conversion lie within the experimental estimations; however, the range of these values is so great that it is not possible to use this argument to validate or invalidate the proposed values.

In both steps of the smectite-to-illite transformation, the proposed activation energy, based on observations of the present study, has a relatively small value. For reactions involving disruption of bonds in silicate structures, energies of 80 to 100 kJ/mole are expected (Lasaga, 1981, pp. 33–34). Small values, such as 56 and 22 kJ/mole, are more characteristic of ion exchange or mineral dissolution than transformation (Lasaga, 1981), whereas the higher values obtained in the laboratory (e.g. Eberl and Hower, 1976) would correspond to mineral reaction. The very slow rate of change in the smectite-to-illite reaction of diagenetic rocks seems to indicate that the controlling mechanism is not the same as that in high-temperature environments.

Whatever problem remains in understanding the exact nature of the reaction mechanisms, the model developed here is a useful tool for describing the smectite-to-illite transformation over relatively large ranges of temperature and time histories. Therefore, it can be used to compare different data sets and to make some prediction concerning the effects on clay diagenesis of various possible thermal events.

Acknowledgements

We thank the Texas Bureau of Economic Geology for supplying the cuttings from the Mustang Island well and E. Eslinger, Cities Service Co. for obtaining the cuttings from several other wells. The Institut Français du Pétrole kindly supplied the Paris Basin well cuttings. This work was supported by the Programme Interdisciplinaire de Recherches sur les Sciences pour l'Energie et les Matières Premières of Centre National de la Recherche Scientifique.

References

Bethke, C.M. and Altaner, S.P., 1986. Layer-by-layer mechanisms of smectite illitization and application to a new rate law. Clays Clay Miner., 34: 136–145.

Boles, J.R. and Franks, S.G., 1979. Clay diagenesis in Wilcox sandstones of southwest Texas: implications of smectite diagenesis on sandstone cementation. J. Sediment. Petrol., 49: 55–70.

Bruce, C., 1984. Smectite dehydration — its relation to structural development and hydrocarbon accumulation in Northern Gulf of Mexico Basin. Am. Assoc. Pet. Geol., Bull., 68: 673–683.

Burst, J.F., 1969. Diagenesis of Gulf Coast clayey sediments and its possible relation to petroleum migration. Am. Assoc. Pet. Geol., Bull., 53: 73–93.

Demongodin, L., Pinoteau, B., Vasseur, G. and Gable, R., 1991. Conductivity and well logs: a case study in the Paris Basin. Geophys. J. Int., 105: 675–691.

Dunoyer de Segonzac, G., 1965. Les argiles du Crétacé supérieur dans le bassin de Douala (Cameroun): problèmes de diagénèse. Bull. Serv. Carte Géol. Alsace-Lorraine, 17 (4): 287–310.

Dutta, N.C., 1986. Shale compaction, burial diagenesis and geopressures: a dynamic model solution and some results in thermal modelling in sedimentary basins. In: J. Burrus (Editor), Thermal Modeling in Sedimentary Basins. Technip, Paris, pp. 149–172.

Eberl, D. and Hower, J., 1976. Kinetics of illite formation. Geol. Soc. Am., Bull., 87: 161–172.

Freed, R., 1982. Clay mineralogy and depositional history of the Frio Formation in two geopressured wells, Brazioria County, Texas. Gulf Coast Assoc. Geol. Soc. Trans., 32: 459–463.

Freed, R.L. and Peacor, D.R., 1989. Variability in temperature of illite/smectite reaction in Gulf Coast sediments. Clay Miner., 24: 171–180.

Howard, J.J. and Roy, D.M., 1985. Development of layer charge and kinetics of experimental smectite alteration. Clays Clay Miner., 33: 81–88.

Hower, J., 1981. The influence of mineral diagenetic reaction on the pore water chemistry of shales. Annu. Meet. Geol. Soc. Am., Nov. 2–5, Cincinnati, Ohio.

Hower, J., Eslinger, E.V., Hower, M. and Perry, E., 1976. Mechanism of burial metamorphism of argillaceous sediments, 1. Mineralogical and chemical evidence. Geol. Soc. Am., Bull., 87: 725–737.

Inoue, A., Velde, B., Meunier, A. and Touchard, G., 1988. Mechanism of illite formation during smectite-to-illite conversion in a hydrothermal system. Am. Mineral., 73: 1325–1334.

Lanson, B. and Champion, D., 1991. I/S to illite transformation in diagenesis. Am. J. Sci., 191: 473–506.

Lasaga, A.C., 1981. Rate laws of chemical reactions. In: A.C. Lasaga and R.J. Kirkpatrick (Editors), Kinetics of Geochemical Processes. Mineralogical Society of America, Washington, D.C., pp. 1–67.

Mathieu, Y. and Velde, B., 1989. Identification of thermal anomalies using clay mineral composition. Clays Clay Miner., 24: 591–602.

Nadeau, P.H., Wilson, M.J., McHardy, W.J. and Tait, J.M., 1985. The conversion of smectite to illite during diagenesis: evidence from some illitic clays from bentonites and sandstones. Mineral. Mag., 49: 393–400.

Perry, E.A. and Hower, J., 1970. Burial diagenesis in Gulf Coast pelitic sediments. Clays Clay Miner., 18: 165–177.

Perry, E.A. and Hower, J., 1972. Late stage dehydration in deeply buried pelitic sediments. Am. Assoc. Pet. Geol., Bull., 56: 2013–2021.

Pytte, A.M. and Reynolds, R.C., 1989. The thermal transformation of smectite to illite. In: N.D. Naeser and T.H. McCulloh (Editors), Thermal History of Sedimentary Basins. Springer-Verlag, Berlin, pp. 133–140.

Roberson, H.E. and Lahann, R.W., 1981. Smectite to illite conversion rates: effect of solution chemistry. Clays Clay Miner., 29: 129–135.

Srodon, J. and Eberl, D., 1984. Illite. In: S.W. Bailey (Editor), Micas. Reviews in Mineralogy, 13, Mineralogical Society of America, Washington, D.C., pp. 495–544.

Tarantola, A. and Valette, B., 1982. Generalized non-linear inverse problems solved using least square criterion. Rev. Geophys. Space Phys., 20: 219–232.

Velde, B., 1985. Clay Minerals: A Physico-Chemical Explanation of Their Occurrence. Developments in Sedimentology, Vol. 40. Elsevier, Amsterdam, 427 pp.

Velde, B. and Espitalié, J., 1989. Comparison of kerogen maturation and illite/smectite composition in diagenesis. J. Pet. Geol., 12: 103–110.

Velde, B. and Iijima, A., 1988. Comparison of clay and zeolite mineral occurrences in Neogene age sediments from several deep wells. Clays Clay Miner., 36: 337–342.

Velde, B. and Vasseur, G., 1992. A kinetic model of the smectite-to-illite transformation based on diagenetic mineral series. Am. Mineral., 72: 967–976.

Velde, B., Suzuki, T. and Nicot, E., 1986. Pressure–temperature-composition of illite/smectite mixed layer minerals — Niger delta mudstones and other examples. Clays Clay Miner., 34: 435–441.

Whitney, G. and Northrop, H.R., 1989. Experimental investigation of the smectite to illite reaction: dual reaction mechanisms and oxygen isotope systematics. Am. Mineral., 73: 77–90.

G. VASSEUR *Centre Géologique et Géophysique (UPR 361, CNRS), USTL, Université de Montpellier II, 34095 Montpellier Cedex 5, France*

B. VELDE *Départment Géologie (URA 1316, CNRS), ENS, 24 Rue Lhomond, 75005 Paris, France*

Validation and first application of palaeotemperature models of isomerisation and NMR aromatisation parameters

U. Ritter, M.B. Myhr, K. Aareskjold and L. Schou

Palaeotemperature information at locations in the Viking Graben and the Barents Sea has been obtained from distributed activation energy models of sterane isomerisation at C-20, hopane isomerisation at C-22 and of an Proton-NMR parameter that reflects the percentage of aromatic protons in the aromatic fraction of extracts (FAP). Frequency factor, mean activation energy and distribution factor of hopane- and sterane isomerisation are 1×10^8/s, 37.0 kcal/mol, 2.9%, and 3.5×10^7/s, 37.5 kcal/mol, 2.9%, FAP is characterised by two distributions at 1.3×10^{13}/s, 51.3 kcal/mol, 0.8% and 4.5×10^{10}/s, 43.2 kcal/mol, 12.9%. Application of the indicators to Viking Graben data makes use of the ability of indicator systems to record selectively thermal conditions for different time intervals and with strongly increased sensitivity. The results suggest increasing heat flow during the Tertiary towards the present, although other explanations can be considered equally likely. Integration of geological and isomerisation data at a cross-section near the Svalis Dome, Barents Sea, allows differentiation between the effects of two different heating events which affected the isomerisation profiles. The onset of the reaction resulting from renewed heating depends on the temperatures and extents of reaction reached during the previous event which is in accordance with a model of selective consumption of the reactants according to their reaction temperature. Comparison with accepted hydrocarbon generation models suggests that most of the isomerisation reactions take place prior to the onset of oil generation. In contrast to previous single activation energy models, strongly different mean activation energies in distributed models do not lead to significantly different maturation pathways at different heating rates.

Introduction

In this study we present distributed activation energy models of isomerisation at C-20 in sterane and C-22 in hopane (e.g. Mackenzie et al., 1980, 1982, 1985; Mackenzie and McKenzie, 1983; Beaumont et al., 1985; Sajgo et al., 1988) and a parameter that represents the fraction of aromatic protons (FAP) determined from Proton-NMR spectroscopy and equivalent to the "percentage of aromatic protons", PAP, of Alexander et al. (1978, 1980) and Solli et al. (1985). In addition we use a slightly modified version of the vitrinite reflectance model proposed by Burnham and Sweeney (1989). Unless stated otherwise, all modelled temperatures given in this paper have been calculated at a linear heating rate of 2.5°C/Ma.

To date, the techniques of modelling organic reactions have predominantly been based on TTI- and pseudo-kinetic models and have mainly utilised vitrinite reflectance (e.g. Waples, 1980; Lerche et al., 1984; Ritter, 1984; Armagnac et al., 1989) but also extended to molecular isomerisation and aromatisation reactions (Mackenzie and McKenzie, 1983) or the Pristane Formation Index (Goossens et al., 1988; Burnham, 1989). In the present context, pseudo-kinetic models are defined as having activation energies and frequency factors that allow us to model the reaction under geological, but not under laboratory conditions. It has been shown before (Quigley et al., 1987; Wood, 1988; Ritter, 1988) that such models are rather inflexible and tend to underestimate the reaction at high heating rates while overestimating it at low temperatures and heating rates. Because most thermal histories consist of an alternation of periods of low and of high heating rates a resulting "model" heat flow history could be a completely unpredictable assembly of artifacts, although the overall history may seem reasonably credible.

Background and previous work

Isomerisation at C-20 in sterane and C-22 in hopane hydrocarbons has previously been shown to proceed regularly with increasing temperature (e.g. Mackenzie et al., 1980, 1982, 1985; Mackenzie and McKenzie, 1983; Sajgo et al., 1988). The reactions

Basin Modelling: Advances and Applications edited by A.G. Doré et al. NPF Special Publication 3, pp. 185–199, Elsevier, Amsterdam.

are used as defined in the literature (e.g. Mackenzie et al., 1982; Mackenzie and McKenzie, 1983).

The fraction of aromatic protons (FAP) is determined from Proton-NMR spectroscopy and equivalent to the "percentage of aromatic protons", PAP, of Alexander et al. (1978, 1980) and Solli et al. (1985). Using data from the literature and from our own data base we find regular increases of FAP with depth for all source rocks even though coal and type III kerogen seem to have ratios about 0.1 higher than other source rocks. The initial ratio in very immature samples is around 6% (PAP). Here we use ratios (FAP) normalised by the lowest PAP value of the pre-heating sample and the highest PAP value encountered after pyrolysis (see Appendix).

In first-order reaction kinetics, reaction rate (k) is controlled by the equation:

$$k = A \exp(-E/RT) \qquad (1)$$

where A is the frequency factor, E is the activation energy, R is the universal gas constant, and T is the temperature.

There are three basic reaction mechanisms, schematically displayed in Fig. 1. Although we know that this is not entirely correct, we model hydrocarbon reactions by a distribution of activation energies (Fig. 1) which is strictly only valid for purely parallel reactions of first-order type (Tissot and Welte, 1984; Tissot et al., 1987; Burnham et al., 1988).

To date the isomerisation reaction has been described by most workers in terms of a simple reactant–product relationship (Mackenzie and McKenzie, 1983; Rullkötter and Marzi, 1988; Marzi, 1989; Marzi et al., 1990). Decreasing absolute biomarker concentrations with increasing maturity (e.g. Abbott et al., 1990) and frequently decreasing isomerisation ratios with increasing temperature and duration of pyrolysis (e.g. Lewan et al., 1986; Marzi, 1989; Abbott et al., 1990) suggest a more complex reaction scheme. This may involve alternative reaction mechanisms at different temperatures, or different sensitivity to thermal alteration for compounds bound to a macromolecular network of kerogen or to free biomarkers in extracts. Abbott et al. (1990) presented a scheme of a series of consecutive (serial) first-order reactions:

$$A \rightarrow B \rightarrow C$$

in which A and B are "bound" and "free" biomarkers, and C is the reaction product. They also noted the possibility of parallel reactions by individual processes that take place simultaneously and/or cumulatively to produce the observed overall reactions.

In summary, the mechanisms controlling apparently simple isomerisation reactions seem to involve serial, as well as competing elements. The distribution of the activation energies allows to account for any "width" of the reaction caused by parallel, serial and back reactions and to some extent, also by competing reactions. The crucial point is that all of this can be achieved with "realistic" kinetic constants that maintain the flexibility of the model at all thermal conditions.

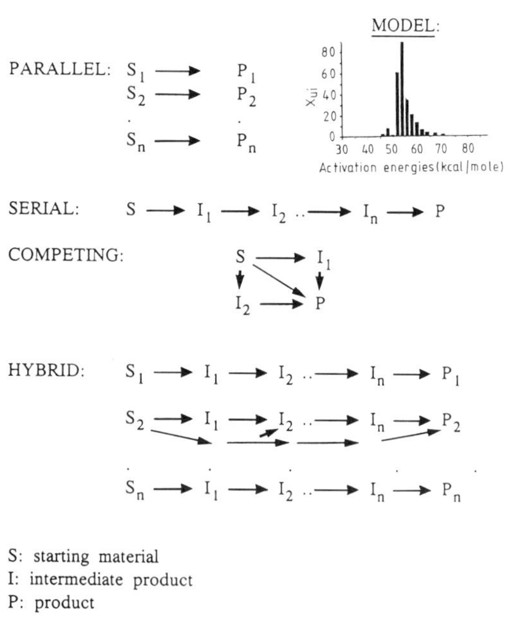

Fig. 1. Possible hydrocarbon reaction mechanisms. Inset: schematic model of distributed activation energies.

Validation of the models

How well do isomerisation and FAP models reflect the past temperature conditions? In order to obtain an idea of the performance of the models we will compare modelled and measured temperatures from suitable wells and with other established thermal indicators.

The kinetic isomerisation models perform well at experimental temperatures (Ritter et al., this volume). The validation against natural data is less straightforward as natural thermal histories are rarely unambiguous. One way of avoiding thermal history ambiguity is to use deep wells in which present temperature is at or close to the maximum and subsidence lasted until the present. In most wells used here maximum isomerisation rate has been reached during Late Tertiary and thermal conditions are therefore likely to be as close as possible to those of the present. Correlation of the modelled and the presently observed temperatures should then provide a measure of the general reliability of the tempera-

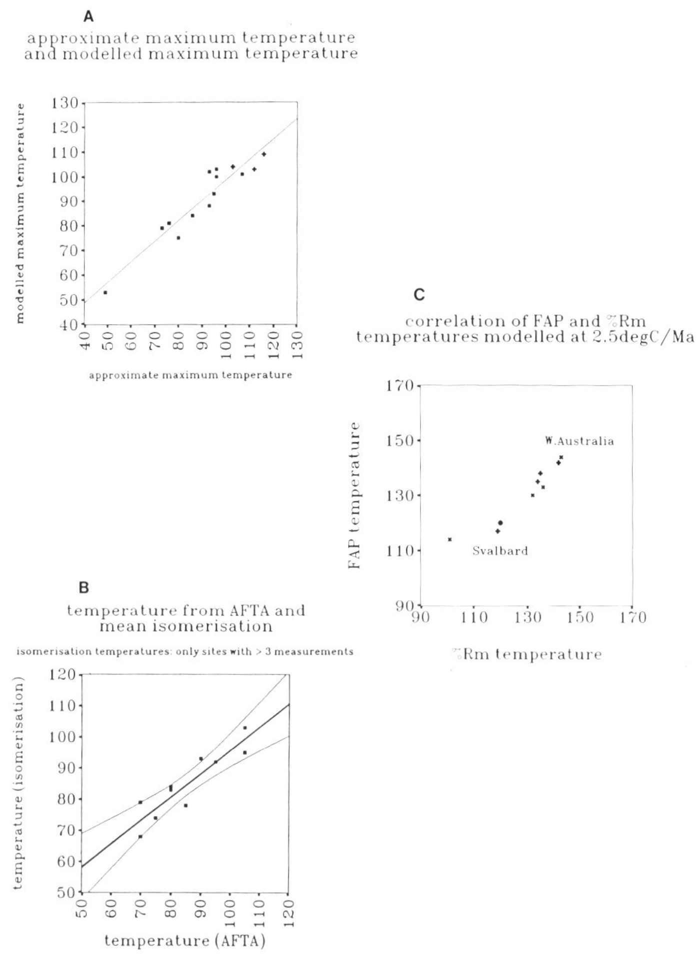

Fig. 2. (A) Modelled temperatures and approximate present temperature in wells where maximum subsidence is reached at present. Isomerisation pairs. (B) Correlation of modelled AFTA temperatures and isomerisation temperatures. To calculate isomerisation temperatures only populations with more than three measurements have been used. Thin line: 66% confidence limit. (C) Correlation of FAP and vitrinite reflectance (%R_m) temperatures.

ture estimates (Fig. 2A). The data and their sources are listed in Table 1. Correlations based on all the data result in a correlation coefficient of 0.84 and a standard error of correlation between 7.5 and 8°C. The error can be considerably decreased if the means of the respective isomerisation couples are used.

The kinetic models appear to underestimate the temperature somewhat below 70°C, possibly due to the asymptotic approach of the isomerisation curve at very low and very high values where small errors in the isomerisation ratio translate into large temperature errors.

To date, only fission track analysis (AFTA) (e.g. Green et al., 1989) and seismic techniques based on interval velocities seem to be sufficiently advanced to satisfy professional standards for estimating palaeotemperature and overburden. Couples of apatite fission track and isomerisation temperatures are available from about 20 wells. Details of the data are confidential and cannot be presented here. Ten of them provide isomerisation temperatures based on populations of 3 or more samples (Fig. 2B). The arithmetic means of these temperatures correlate with AFTA temperatures at a correlation coefficient of 0.91. At temperatures around 100°C AFTA temperatures tend to be about 5°C higher than isomerisation temperatures. At temperatures at or below 70°C they tend to be lower by the same amount.

FAP and %R_m temperatures are compared in Fig. 2C. Both the temperatures from Svalbard and Western Australia have been modelled using a slightly modified version of the vitrinite reflectance model of Burnham and Sweeney (1989) and a linear heating rate of 2.5°C/Ma. There is an excellent agreement along the diagonal between FAP and %R_m temperatures as well as an excellent correlation although further data are necessary to corroborate this result. The data from Western Australia are from Alexander et al. (1978, 1980) and those from Svalbard are from Solli et al. (1985).

TABLE 1

Modelled maximum temperatures from isomerisation ratios (500 m pre-Tertiary subsidence)

Location	Temperature: estimated		modelled*			Source/level	
		mean	mean	20S	22S		
Piper 15/17-P10	71[a] (/76[d])	73	76	78	74	1	Upper Jurassic
Claymore 14/19-4	78[a] (/82[d])	80	90	91	89	1	Upper Jurassic
Beryl 9/13-A1	99[a]/115[b]	107	100	99	102	1	Upper Jurassic
Brent 211/29-1	94[a]/ 92[b]	93	88	89	87	1	Upper Jurassic
Ninian 3/3-1	88[a]/104[b]	96	100	99	101	1	Upper Jurassic
Heather 2/5-8B	81[a]/~90[b]	85	84	89	79	1	Upper Jurassic
Thistle 211/23-4	95[a]/ 96[b]	95	93	96	90	1	Upper Jurassic
Hutton 211/28-1	82[a]/104[b]	93	102	100	105	1	Middle Jurassic
Hutton 211/28-1	85[a]/107[b]	96	103	106	100	1	Lower Jurassic
15/6-5	122[b]/127[d]	124		100		3	Upper Jurassic
15/9-6	114[b]/119[d]	116	109	105	113	3	Upper Jurassic
15/9-12	114[b]/119[d]	116		105		3	Middle Jurassic
15/9-13	103[b]/108[d]	105			91	3	Upper Jurassic
15/9-14	109[b]/115[d]	112	102	108	97	3	Upper Jurassic
15/12-4	101[b]/104[d]	102	104	103	105	3	Middle Jurassic
6/3-1	106[d]/101[e]	103		105		3	Upper Jurassic
6/3-2	109[d]/104[e]	106		105		3	Upper Jurassic
X	98[d]/ 86[e]	92		99		3	Upper Jurassic
Y	98[d]/ 86[e]	92		97		3	Upper Jurassic
9/4-4	78[d]/ 74[e]	76	81	82	80	3	Upper Jurassic
17/11-2	76[d]/ 69[e]	72	79	80	78	3	Upper Jurassic
30/3-4	105[d]/100[b]	102	106	107	106	3	Upper Jurassic
31/4-2	86[d]/ 81[b]	83	86	85	88	3	Upper Jurassic
33/9-14	101[d]/ 96[b]	98	100	95	106	3	Upper Jurassic
33/12-6	98[d]/ 93[b]	95	85	91	80	3	Upper Jurassic
E9818	47[d]/ 42[f]	44	52	54	51	1	Lower Jurassic
E9819	75[d]/ 70[f]	72	75	73	77	1	Lower Jurassic
E9823	83[d]/ 78[f]	80	96	94	99	1	Lower Jurassic
E9833	79[d]/ 74[f]	76	59	54	65	1	Lower Jurassic
Mississauga	90[g]/ 77[h]	83	84	88	80	1	Lower Cretaceous
Wyandot	107[g]/ 90[h]	98	100	102	99	1	Lower Cretaceous
Sable 2.5 km	94[g]/ 79[h]	86		85		1	Lower Cretaceous
Sable 2.8 km	103[g]/ 87[h]	95	100	101	99	1	Lower Cretaceous

* Modelled at 75 Ma duration of subsidence; 20S sterane isomerisation ratio; 22S hopane isomerisation ratio.
[a] From Mackenzie and McKenzie (1983). [b] Present temperature from Carstens and Finstad (1981). [c] IKU, internal data. [d] Maximum temperature. [e] Present temperature, based on regional gradients (Harper, 1971). [f] Present temperature estimated from 30°C/km. [g] Issler (1985), max. gradient 32.7°C/km, 11.8°C surface temperature. [h] Issler (1985), mean gradient 27°C/km, 11.8°C surface temperature.

Palaeoheat flow in the Viking Graben

Thermal indicators from various wells in the northern North Sea obtained during recent years in the course of commercial modelling studies are re-evaluated below.

The thermal history of most North Sea wells is dominated by Tertiary subsidence and thus most thermal indicators will mainly record the conditions during this time. Figure 3 shows that in well 15/9-6 present heat flow overestimates FAP considerably if extrapolated into the past and that an acceptable match is only achieved if palaeoheat flow is reduced considerably. It also indicates that FAP records mainly the thermal conditions of the time between 60 and 30 Ma B.P. Figure 4A shows the maturation history in well 24/9-1 of the *earliest onset* of four thermal indicator reactions. Although it is difficult to delineate the exact time of the thermal record it seems nevertheless clear that the isomerisation reactions tend to record the times younger than 40 Ma whereas FAP mainly records the time between 40 and 50 Ma B.P. and also to a minor extent the preceding time. Vitrinite reflectance, in contrast, shows a more regular increase and is thus likely to record a more integrated history of the whole time range, beginning at a much earlier stage than the other reactions.

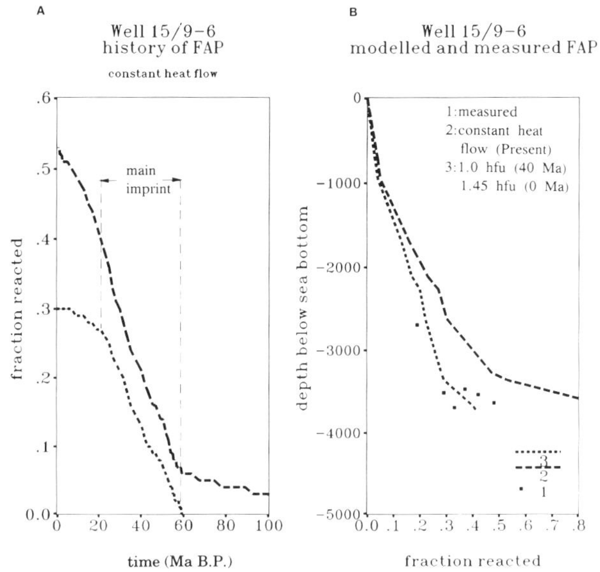

Fig. 3. (A) Maturation history of FAP in well 15/9-6. (B) modelled and measured FAP profiles in well 15/9-6. Heat flow scenario *3* involves a linear increase of heat flow between 40 Ma B.P. and the Present.

Modelled heat flow in Fig. 4B has been obtained from systematic inversion of the respective reactions using algorithm MCHANGE described in the Appendix. This algorithm finds an unambiguous palaeoheat flow if present heat flow is constrained. The time range of the highest reaction rate has been obtained from visual inspection of graphs such as shown in Fig 4A. There is a distinct increase of

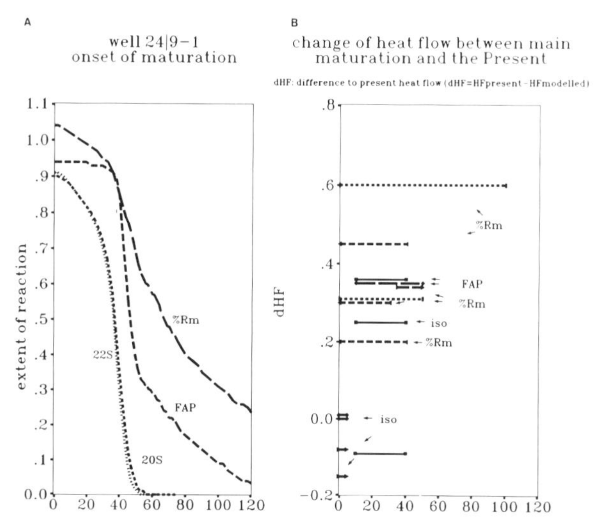

Fig. 4. (A) Periods of significant maturation in well 24/9-1. Example for almost completed reactions. See text for discussion. (B) Correlation of δHF and time of main reaction from wells of the northern North Sea. Time of main reaction (Ma) has been obtained using the period of the highest reaction rate as shown in (A). Modelled heat flow has been obtained from systematic inversion of the respective reactions.

δHF (delta heat flow) with increasing age of the reactions. Most isomerisation reactions reflect Recent and sub-Recent conditions and therefore have the lowest δHF. Vitrinite reflectance on the other hand contains the oldest records of palaeoheat flow and tends to have much higher δHF.

A bias due to different sensitivities of the different reactions cannot be ruled out, but seems less likely since each sub-population seems to display a similar trend. An areal influence also cannot be ruled out but seems less likely in view of the same trend within, for example block 15. The possibility cannot therefore be ruled out that these features in fact represent a trend of regional heating in the North Sea as suggested by Thorne and Watts, 1989, although Jensen and Doré (this volume) doubt the justification of such heating.

Temperature and subsidence history near Svalis Dome, Barents Sea

General

In a cross-section near the Svalis Dome a series of Triassic rocks is exposed that has been tilted during the lateral migration of thick Palaeozoic salt layers during the Jurassic. The section is particularly suited to demonstrate the integration of the low-temperature isomerisation and geological data.

Shallow drilling along digital line D84-144 took place in summer 1986, leading to the acquisition of cores 7323/07-U-01 to 09 (Fig. 5). The stratigraphic sequence reaches from Smithian (Lower Triassic) to early Carnian (Upper Triassic). Structural analysis suggests that the deepest burial was reached during the Jurassic, followed by tilting and uplift during the subsequent salt pillow stage (Løseth et al., 1992; in the following referred-to as "phase I"). Hopane isomerisation ratios at the nearby Mid-Cretaceous core 7324/07-U-01 allow assessment of the effects of renewed subsidence and uplift during the Tertiary, in the following referred to as "phase II". Using the detailed logs of the cores and the available seismic information, a composite depth section was compiled that is representative of phase I subsidence (Table 2). In this context the lateral transposition of the different cores is assumed to be of minor importance.

Isomerisation ratios have been taken from the respective IKU reports. Samples that appeared non-indigenous on the basis of organic-geochemical analysis and samples with possible coelution problems were omitted from further evaluation. The ratios were then tested for consistency, and inconsistent isomerisation couples were omitted.

When have these isomerisation profiles been

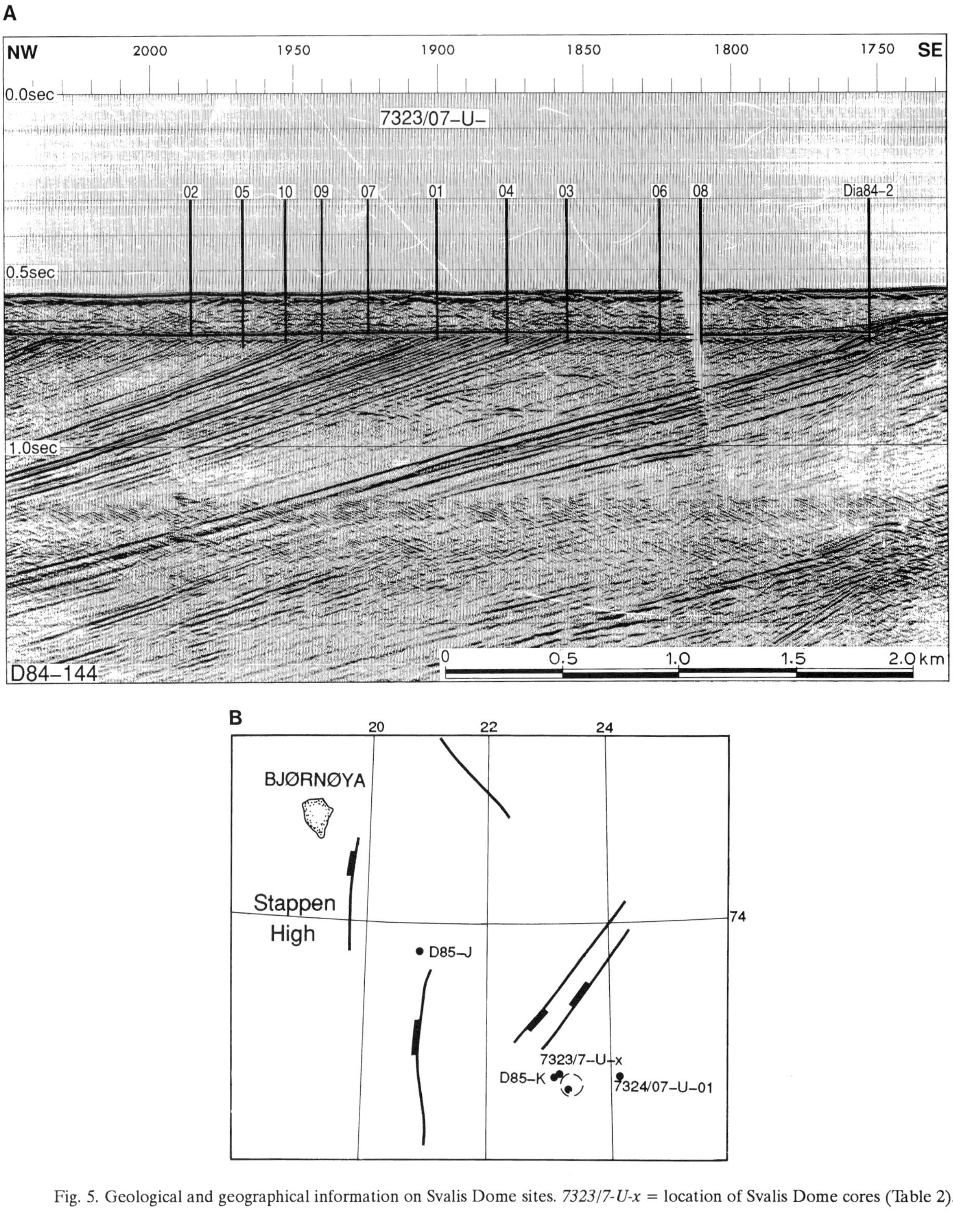

Fig. 5. Geological and geographical information on Svalis Dome sites. *7323/7-U-x* = location of Svalis Dome cores (Table 2).

formed and under what temperature conditions? The isomerisation ratios distinctly decrease with decreasing stratigraphic age (Table 2, Fig. 6). It is evident that such isomerisation gradients could only have formed prior to tilting into the present position. Any significant heating following tilting would have led to equal maturation levels in all cores and would have at least partly obliterated the isomerisation profiles. The apparent geothermal gradients estimated for the isomerisation profiles are 40 and 47°C/km for hopane and sterane, respectively. The geothermal gradient is referred to here as "apparent" as the actual gradient will depend on a variety of factors that will be discussed below. A simple method to determine an apparent interval gradient is given in the Appendix.

Inversion of the isomerisation ratios (Fig. 7) using algorithm HFER supports the estimates from the

TABLE 2

Isomerisation ratios used for modelling at the Svalis Dome cross-section

Site	Depth	20S	22S	Stratigraphic level
7323/07-				
U-08	−956	0.13	0.53	Upper Smithian
U-08	−951	0.17	0.53	
U-06	−836	0.11	0.49	Upper Smithian
U-03	−727	0.06	0.41	Smithian–Spathian
U-03	−718	0.13	0.35	
U-04	−677	0.09	0.40	Spathian–Anisian
U-04	−660	0.05	0.33	
U-01	−599	0.04	0.15	Mid Anisian
U-01	−597	0.05	0.23	
U-01	−593	0.04	0.23	
U-01	−586	0.05	0.18	
U-09	−448	0.10	0.22	Anisian–Ladinian
U-09	−437	0.02	0.24	
D85-K	−400	0.02	0.10	Ladinian
7324/07-U-01		n.a.	0.15	Mid Cretaceous

Depth: composite depth, arbitrary depth scale.
20S: sterane isomerisation ratio. 22S: hopane isomerisation ratio.

isomerisation profiles. It indicates that overburden thicknesses of less than 1650 m to base Smithian will fit the isomerisation profiles relatively well at heat flows higher than 1.5 HFU. At the thermal conductivities used for modelling, an additional eroded overburden of 1650 m and a heat flow of 1.6 HFU during deposition would cause a linear thermal gradient to the surface of 42.3°C/km, which is about the mean of the apparent gradients deduced directly from the isomerisation gradients.

Cumulative effects of different heating episodes

The Svalis Dome cross-section also provides the opportunity to demonstrate the different responses of pseudo- and distributed activation energy models. Unlike pseudo-activation energy or TTI-based models, which accumulate the effects of *all* thermal events regardless of the previous history, distributed models will vary in their response depending on prior events. What are the cumulative effects of repeated heating events on temperature estimates? Figure 8A

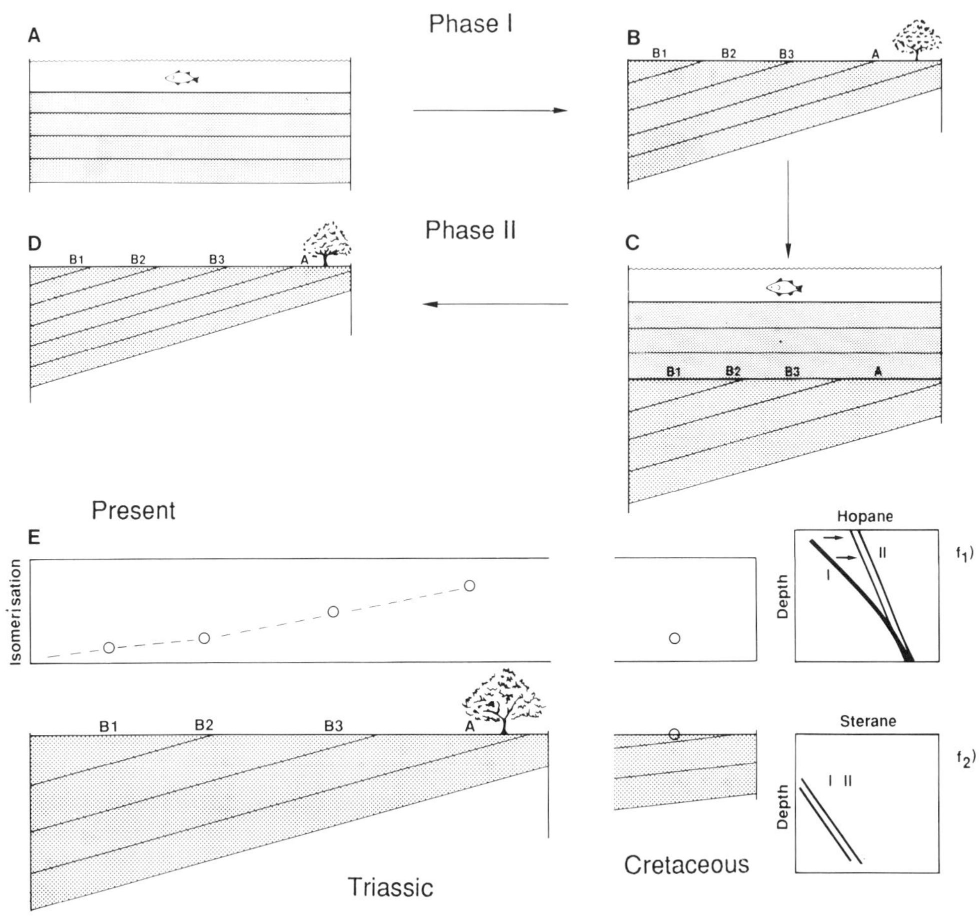

Fig. 6. Subsidence scenario for Svalis Dome cross-section. *B1, B2, B3, A*: representatives of shallow drilling sites; (A) phase I subsidence; (B) phase I uplift and tilting; (C) phase II subsidence; (D) phase II uplift and erosion; (E) schematic present cross-section and Cretaceous site 7324/07-U-01 to constrain phase II thermal conditions; (F) schematic composite isomerisation profiles; arrows indicate selective shift of hopane isomerisation profile during phase II subsidence.

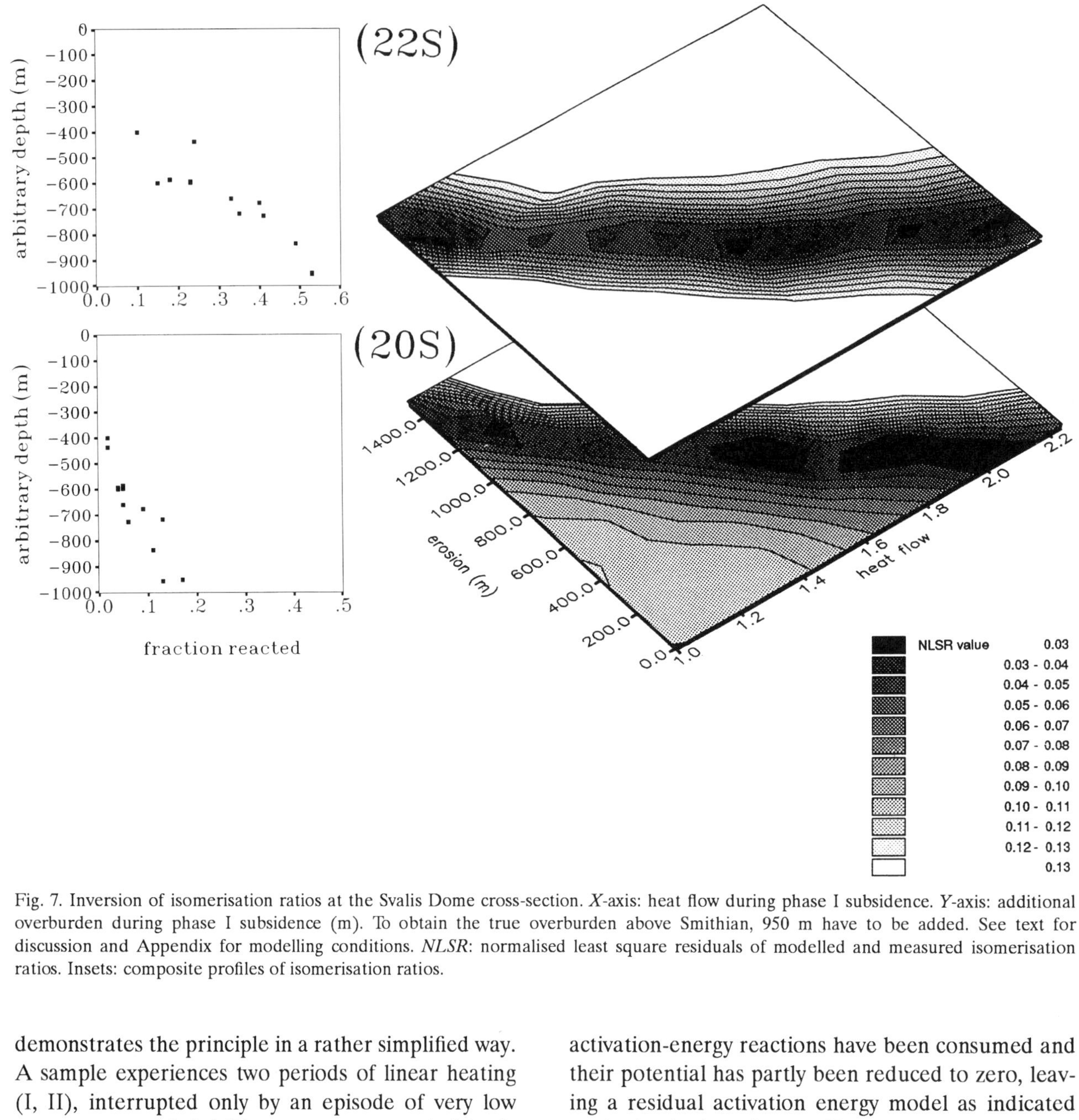

Fig. 7. Inversion of isomerisation ratios at the Svalis Dome cross-section. *X*-axis: heat flow during phase I subsidence. *Y*-axis: additional overburden during phase I subsidence (m). To obtain the true overburden above Smithian, 950 m have to be added. See text for discussion and Appendix for modelling conditions. *NLSR*: normalised least square residuals of modelled and measured isomerisation ratios. Insets: composite profiles of isomerisation ratios.

demonstrates the principle in a rather simplified way. A sample experiences two periods of linear heating (I, II), interrupted only by an episode of very low temperatures. Model (1) represents a pseudo-kinetic model consisting of a single activation energy. Model (2) is represented by a distributed model of activation energies. A and A' are the temperatures at which the respective reactions begin during phase I. B and B' are the temperatures at which the reaction sets in again during phase II. Obviously $A' = B'$ for the pseudo-kinetic model (model 1) while $A < B$ for model 2, i.e. the onset of the reaction in phase II depends on the previous thermal history but its temperature is in any case higher than the initial reaction temperature. This hysteresis-loop-like retardation is easily understood from Fig. 8B which shows the original distribution of activation energies, and the distribution after the reaction has partly been completed. During heating stage I most of the lower-activation-energy reactions have been consumed and their potential has partly been reduced to zero, leaving a residual activation energy model as indicated at the beginning of stage II. Since the onset of the reaction is determined by the lowest available activation energy, the reaction will only proceed further once the critical temperature of the lowest activation energy fraction has been reached. In contrast, the activation energy in a single-activation-energy model (model 1) at the beginning of stage II will necessarily be the same as in stage I and the temperature of the onset of the reaction will therefore be the same, too. Only the rate of further increases will depend on the extent of the reaction already reached.

Phase II heating of model 2 will thus only affect the isomerisation profile if its maximum temperature is higher than B, otherwise it will leave no imprint at all. Model 1 in contrast will respond to all temperatures higher than A' and therefore record the cumulative

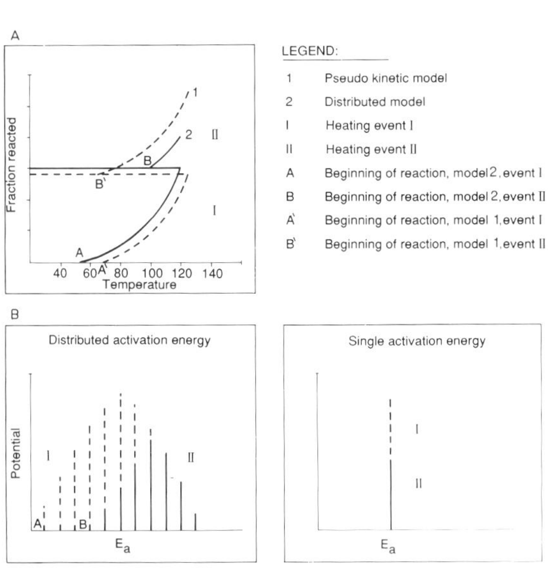

Fig. 8. Schematic view of the effects of repeated linear heating on distributed- and pseudo-activation energy models (hysteresis-loop effect). See text for discussion.

effects of the two events to a greater extent than the distributed models.

Subsidence history of the cross-section

How does phase II subsidence influence the isomerisation profile at the Svalis Dome? The present situation of the cross-section is schematically shown in Fig. 6E and the isomerisation profiles are shown in Fig. 6F. A scenario which could account for the observed isomerisation profiles is shown in Figs. 6A–D. During phase I subsidence, all cores were buried at different depths such that the isomerisation profiles in Fig. 6F could form. After uplift and tilting renewed subsidence led to somewhat lower temperatures (C). Due to previous tilting all cores were now subjected to broadly the same temperatures of about 75°C. As site B1 had been exposed to lower temperature than, for example site A during phase I subsidence, the relaxation temperature is lower for B1 than for A and the hopane isomerisation gradient is thus steepened (Fig. 6) during phase II subsidence.

In order to test this effect quantitatively, subsidences I and II have been modelled in a simplified manner for the uppermost (Ladinian) and the lowermost (Smithian) level of hopane isomerisation, respectively, subsidence II being constrained by Mid-Cretaceous hopane isomerisation in nearby site 7324/07-U-01 (Table 2). In the Ladinian samples this leads to about 5 to 10°C lower isomerisation temperatures during phase I; in the Smithian, phase II subsidence does not affect the hopane ratios at all. The lower part of the profile therefore reflects fully the conditions during phase I whereas the upper part reflects

TABLE 3

Summary of maximum temperatures at the Svalis Dome cross-section and surroundings

Stratigraphic level	Maximum temperature (°C) during subsidence in:	
	phase I	phase II
Svalis Dome cross-section: Smithian	104	75
Svalis Dome cross-section: Anisian–Ladinian	70	75
Core 7324/07-U-01: Cretaceous	n.a.	75
Core 7323/12-U-01, AFTA: Cretaceous		70–80

AFTA temperatures from Løseth et al. (1992). Phase I temperatures refer to the Upper Smithian. All isomerisation temperatures from a linear heating rate of 2.5°C/Ma.

at least partly the conditions during phase II. Modelling of sterane isomerisation in the same manner shows that the whole sterane isomerisation profile has formed during phase I and has not been influenced by phase II heating. This is easily understood from the higher reaction temperature of sterane as compared to hopane isomerisation.

It is therefore likely that the two isomerisation profiles formed during phase I at a geothermal gradient around 60°C/km and reached maximum temperatures around 100°C at the base of the profile (Table 3). Since isomerisation temperatures are reached at relatively shallow depth and in still poorly compacted sediments, such a gradient is by no means particularly high and can easily be reached at a heat flow of 1.5 HFU.

Renewed subsidence following tilting and uplift will have reached about 75°C at the present level of erosion and caused a distinct steepening of the isomerisation gradients, leading to a lower apparent palaeogeothermal gradient.

Discussion

Distributed kinetic models of sterane and hopane isomerisation, and of an aromatisation reaction (FAP), have been developed using time series of hydrous pyrolysis experiments, non-linear regression analysis and calibration against natural data. The optimised models (Appendix, Table A1) accommodate geological as well as experimental conditions. These models are thought to become part of a system of thermal indicators that are highly temperature-sensitive while still covering the important temperature ranges.

The reactions used for thermal history analysis should be represented by models that simulate the actual mechanisms as closely as possible. TTI- or pseudo-activation-energy models are only very rough approximations, the results of which may reflect the shortcomings of the model rather than the thermal history. The parallel-reaction-models used here are also approximations but seem to reflect the actual processes to a much higher degree, although the isomerisation reactions are likely also to involve back- and competing reactions.

Modelling of Viking Graben wells demonstrates that there is potential for indicator systems to record selectively thermal conditions during rather narrow time intervals at high accuracy. Increasing heat flow towards the present is, of course, only one explanation of the results described. Others include the possibility of a systematic bias of the reactions involved, data quality, areal variation of heat flow or an apparent increase of heat flow due to fluid flow or blanketing effects. In particular FAP needs more testing, and the quality of vitrinite reflectance data in the Viking Graben is generally rather poor.

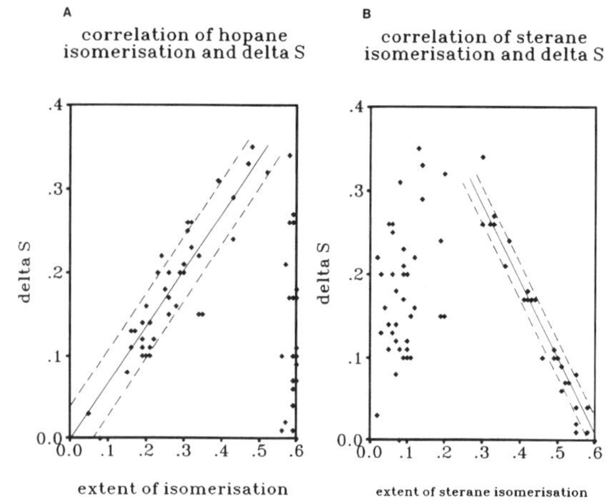

Fig. 9. Relation of isomerisation ratios with δS. To be used to test consistency of isomerisation ratios. Random sample from IKU internal data. Stippled lines: approximate standard error. δS = 22S − 20S. 20S: sterane isomerisation ratio; 22S: hopane isomerisation ratio.

Error sources

The complexity of the reactions discussed so far makes it particularly important to screen prospective samples for aberrant behaviour, i.e. to eliminate those samples the reaction mechanisms of which differ too much from those assumed by the model. The main reasons why isomerisation ratios, and partly also FAP ratios, may not represent the actual thermal conditions are found (1) in the effects of organic facies ("kerogen type"), (2) the reversal or strong scatter of isomerisation ratios once the reaction has reached equilibrium, and (3) the influence of non-indigenous material (migration). Apart from migration most of these effects may be due to interference by back reactions and competing reactions (Fig. 1). Table 4 lists the possible error sources, some of their possible causes and diagnostic features.

Non-indigenous influences can be identified by monitoring the organic-geochemical characteristics (e.g. production index of Rock-Eval pyrolysis, composition of the extracts). Sterane isomerisation ratios which are unusually *low* with respect to hopane ratios (Fig. 9) often also provide an indication, since the hopane reaction usually proceeds at a higher rate than the sterane reaction. This leads to characteristic correlation patterns. Figure 9 has been compiled using random samples of data available from the literature. It allows testing of the consistency of the isomerisation couples.

A particularly important point in assessing isomerisation ratios is the tendency of sterane isomerisation to revert once equilibrium has been reached. This is easily understood in terms of a "back reaction" as outlined in Fig. 1, which under certain conditions may become faster than the forward reaction. Apart from checking against diagrams such as Fig. 9, reversals can easily be recognised once the ge-

TABLE 4

Recognition of isomerisation- and FAP ratios not representative of in-situ thermal conditions (see text for further discussion)

Error due to	Possible explanation	Diagnostic features
Reversal, reaction beyond equilibrium:	Back reaction faster than forward reaction	No positive trend over several 100 m; high scatter without positive trend
Organic facies:	*Isomerisation*: different reaction mechanisms and initial concentrations	Sudden "breaks" of isomerisation trends; unsatisfactory correlation of 20S and 22S
	FAP: higher initial PAP in coals and type III kerogens	High intensities at α-position
Migration:	Migration from more mature rocks	High production index; chemical characteristics of extracts different; apparent maturity abnormally high or low

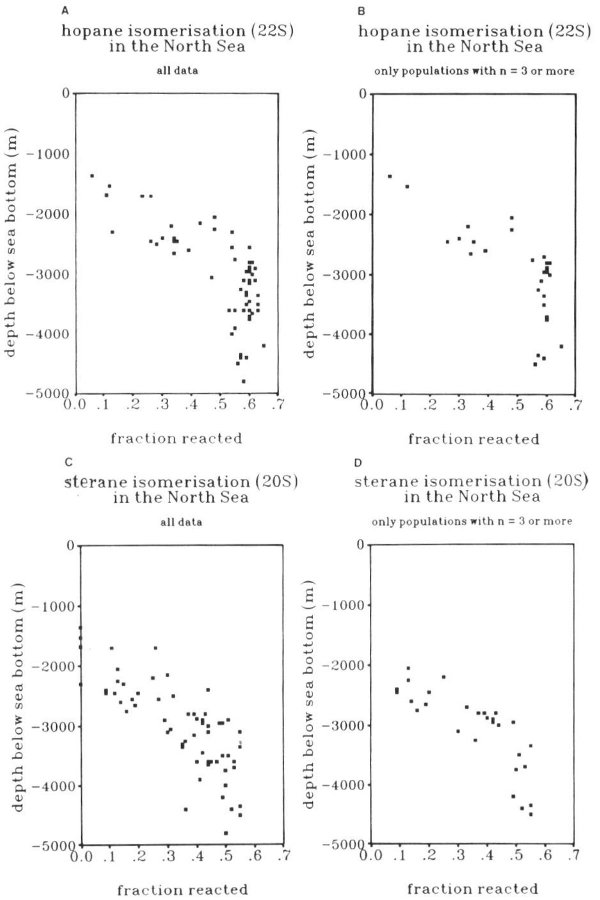

Fig. 10. Variation with depth of sterane and hopane isomerisation ratios in North Sea wells. (A, C) All data populations. (B ,D) Data populations consisting of three and more samples. Scatter in (B) and (D) is considerably reduced.

neral depth of a reaction interval is known. Figure 10 shows these relationships for the Viking Graben. The figures have been compiled using a random sample of available isomerisation data. Hopane isomerisation reaches equilibrium at a depth of 2900 m, sterane isomerisation at about 3300 m below sea bottom.

Unfortunately for thermal analysis purposes, organic-geochemical analyses are preferentially performed in Jurassic source rocks in the North Sea, most of which are deeper than 3000 m and thus contain isomerisation data that lead to a strong underestimate of palaeotemperatures if no reversal has taken place, and totally meaningless results in the case of reversals. In fact more than 80% of the data originally used to compile Fig. 10 were from a depth range not suitable for thermal analysis. Most of the data shown for intermediate and shallow depth have actually been determined specifically for modelling purposes. The figure also suggests that data scatter is considerably stronger for sterane than for hopane ratios (Fig. 10). This is to be expected since hopane isomerisation ratios are obtained from averages of several peaks of the fragmentogram whereas sterane isomerisation ratios are determined from a single pair of peaks.

Evidence for facies influence is contradictory. Moldowan et al. (1986) found a moderate but distinct variation of sterane isomerisation with respect to critical facies parameters, and the data of Curiale and Odermatt (1989) show a decreasing isomerisation ratio with increasing hydrogen index. Strachan et al. (1989) observed distinctly different isomerisation ratios in adjacent coals and shales and explained the opposing signs of the deviations by differential isomerisation due to different kinetic models applicable to coal and shale, respectively. However, these authors use activation energies of 6.2 and 20 kcal/mol for coals and non-coaly source rocks, respectively. Such low activation energies have been shown before to be erroneous if complex reaction mechanisms are expressed by a single pair of activation energy and pre-exponential factor (Braun and Burnham, 1987). It has also been shown before that such low values cannot cope with high heating rates and thus result in erroneous palaeothermal conditions (Quigley et al., 1987; Ritter, 1988). Their results are thus probably pure artifacts (see below).

Our own investigations show that populations of iso-maturity Barents Shelf shallow drilling samples reveal a distinctly increasing sterane isomerisation ratio with increasing hydrogen index (Fig. 11). It also appears that a trend of increasing maturity with increasing depth is superimposed on the trend against the hydrogen index (arrows in Fig. 11). The same samples plotted against TOC could also be

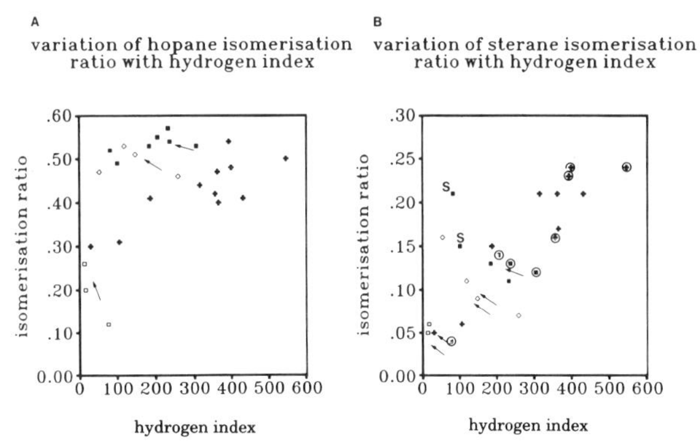

Fig. 11. Correlation of isomerisation ratios with hydrogen index. Samples have been taken from areas of approximate equal maturity. There is a distinct linear increase of sterane isomerisation with hydrogen index (B). (A) Superimposed cross-cutting trend correlates with the respective depth of the samples and is thus possibly related to subtle maturity differences (arrows). Hopane isomerisation shows no corresponding trend (A). A rapid decrease at hydrogen indices smaller than 100 could be due to subtle maturity differences. *S* = siltstone; circles = coal or type 3 kerogen. The different symbols indicate different shallow drilling sites.

interpreted to show a decreasing isomerisation ratio with decreasing TOC, although this correlation is clearly less straightforward than the correlation with hydrogen index. A simple correction factor derived from Fig. 11 so far has not reduced data scatter when independently applied to other populations. This also suggests that the influence of organic matter on isomerisation is more complex and cannot be accounted for by such a simple relationship.

In summary, checking of the thermal indicators for consistency, and elimination of samples suspected of not reflecting *in situ* conditions, will strongly enhance the reliability of the thermal analysis or at least help to explain unusual results. In unknown areas it is essential to ensure that the isomerisation data population has not yet reached equilibrium; tests to eliminate non-indigenous samples should always be carried out.

Isomerisation and hydrocarbon generation

The models presented here also have implications regarding the relative timing of hydrocarbon generation and isomerisation. Recent kinetic models of hydrocarbon generation (e.g. Burnham et al., 1987; Espitalié et al., 1988; Schaefer et al., 1990) suggest that most of the isomerisation reactions take place before oil generation begins (Fig. 12A). Reports of isomerisation ratios used to determine the maturity of oils probably reflect, in most cases, post-equilibrium conditions.

It has been a long-established principle to interpret hydrocarbon reactions in terms of their relative activation energy differences. As demonstrated by Mackenzie and McKenzie (1983) a low activation energy reaction would proceed faster at low heating rates than a high activation energy reaction.

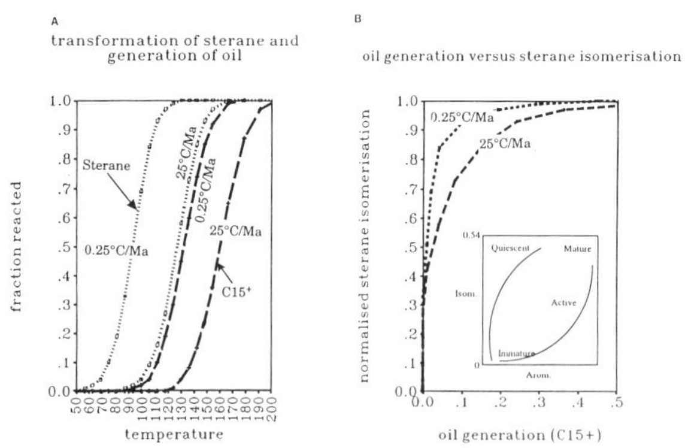

Fig. 12. (A) Sterane isomerisation and hydrocarbon generation at 0.25 and 25°C/Ma linear heating rate. $C15^+$ from Espitalié et al. (1988). Inset: modified from Mackenzie and McKenzie (1983). (B) Sterane isomerisation versus $C15^+$ generation at 0.25 and 25°C/Ma linear heating rate. There is only an insignificant shift at high heating rate towards the reaction with high activation energies.

Figure 12B shows a hydrocarbon reaction ($C15^+$, Upper Jurassic of the North Sea, Espitalié et al., 1988) together with normalised sterane isomerisation. The latter reaction has about 20 kcal/mol higher mean activation energy than the isomerisation reaction; hence it should show the differential maturation rates (as e.g. Fig. 12B, inset) rather well. Instead the heating-rate effect in distributed models is insignificant and hardly goes beyond the possible data scatter. The interval between the two mean activation energies is about the same as the one between isomerisation and aromatisation used by Mackenzie and McKenzie (1983) and there is no significant overlapping of the two distributions. Nevertheless, the activation energies used seem now close enough to obliterate the effects seen in pseudo-activation energies.

A possible explanation would be that there is no or little selective shift with heating rate of the *overall* reaction, although this is certainly the case for each individual reaction. Interpretation of maturity data in terms of differential maturation due to the influence of heating rate should be used with caution, unless it can be demonstrated that the reactions involved permit this effect. If distributed models of hydrocarbon reactions are regarded as more realistic than single-activation-energy models, there is no reason to assume that sterane and hopane isomerisation could reach significantly into the temperature of oil generation.

Conclusions

Distributed models of activation energies for thermal indicator reactions are more realistic than single activation energy or TTI-type models but are probably not the final answer. Even more realistic models could be developed based on specifically designed experiments. Too simplistic models are prone to artifacts. Systems of thermal indicators have many potential advantages but need extremely careful intercalibration if they are not to produce their own set of artifacts. Procedures and rules have to be established to control data quality and to ensure that non-indigenous organic matter and samples affected by different organic facies are recognised.

Examples of modelling strategies have been presented from the North Sea and the Barents Shelf which take account of the different histories of the two environments. North Sea samples generally experienced subsidence up to the Present and therefore, due to the cumulative nature of the indicator reactions used, present conditions tend to obliterate past ones. This can partly be overcome by using indicators that reach their maximum reaction rate at different temperatures and thus tend to record conditions during different time ranges with different intensities. Using an array of multiple thermal indicators, the example presented from the North Sea seems to suggest a gradual increase of heat flow during the Tertiary, although a variety of other factors may have influenced the observed pattern, too.

In contrast, many areas on the Barents Shelf have been uplifted and tilted, and reactions have been "frozen" at least once during their history. Unless reaction temperatures have been reached again at a later stage, the thermal indicators will record the conditions prior to uplift. This may provide relatively good constraints if the time of uplift and cooling can be inferred from other evidence. If reaction temperatures have been reached again, the records from previous episodes may be obliterated or partly overprinted. In the latter case previous thermal conditions may be reconstructed if indicator data not influenced by more than one thermal episode are available. In the former case information on prior events will be lost completely and information from the later event will be partly invalidated.

The isomerisation profiles formed in the Triassic sequence near the Svalis Dome have partly been overprinted by Tertiary re-heating, but the degree of overprinting of the Triassic sequence can be constrained by Tertiary isomerisation temperatures derived from Cretaceous samples. The reconstructed palaeothermal gradients suggest that heat flow during the Triassic was similar or slightly higher than at present.

Acknowledgements

We thank IKU for permission to publish the results of this study and NTNF for partial funding. Two anonymous reviewers provided valuable suggestions. The MR-Centre, Trondheim, performed the NMR analyses.

Appendix

Definition of FAP

$$\mathrm{FAP} = \frac{\left(\frac{\mathrm{PAP}}{100} - R_i\right)}{(R_f - R_i)} \tag{A1}$$

PAP = percentage of aromatic protons in the aromatic fraction (Alexander et al., 1980); R_i = initial fraction, 0.063; R_f = final fraction, 0.28.

Determination of the apparent palaeogeothermal gradient

The apparent gradient may be determined from modelling the temperature at two different depth levels of a thermal indicator profile and dividing the temperature interval by the depth interval:

$$G_a = \frac{(T_b - T_a)}{(Z_b - Z_a)} \tag{A2}$$

G_a = apparent temperature gradient; Z_a, Z_b = depth at levels a and b; T_a, T_b = apparent temperatures deduced from thermal indicators.

For practical purposes any heating rate can be used as long as it is the same for T_a and T_b. Note that the actual temperatures may be very different depending on heating rate but that the difference between level a and b will be the same. The resulting interval gradient is not necessarily comparable to the overall gradient with the intercept temperature of the surface.

Modelling procedure

The proprietary software (BASEVAL), used for reconstructing thermal history from geological data, is a one-dimensional conductive model that employs standard procedures for backstripping, recompaction and calculation of thermal rock conductivities from porosity (Rubey and Hubbert, 1959; Woodside and Messmer, 1961). The model is described in more detail in Ritter et al. (1987) and Ritter (1988). There are several algorithms that modify input files to such an extent that one or a range of optimum thermal histories can be found.

MCHANGE: This algorithm models the effects of magnitude of heat flow variation. The times at which heat flow variation begins and ends are t_1 and t_2, respectively. The heat flows HF_1 prior to t_1, and HF_2 following t_2, do not change with time. HF_1 and HF_2 change at a step width of δHF between HF_{min} and HF_{max}. The total number of thermal histories ($\sum$ Th) is thus:

$$\sum \mathrm{Th} = \sum \mathrm{HF}_1 \times \sum \mathrm{HF}_2 \tag{A3}$$

TABLE A1

Tested and calibrated kinetic models determined from hydrous pyrolysis and optimisation procedures

	A	E_a	σ
Hopane isomerisation (22S):	1.0×10^8	37.0	2.9
Sterane isomerisation (20S):	3.5×10^7	37.5	2.9
FAP:	1.3×10^{13}	51.3	0.8
	4.5×10^{10}	43.2	12.9

A, s^{-1}; E_a, kcal/mol; σ, in % of E_a.

with:

$$\sum HF_1 = \frac{(HF_{1,max} - HF_{1,min})}{\delta HF_1} \qquad (A4)$$

$$\sum HF_2 = \frac{(HF_{2,max} - HF_{2,min})}{\delta HF_2} \qquad (A5)$$

HFER: This algorithm enables us to determine the amount of erosion for any given heat flow that occurred during subsidence of the eroded sediments. If maturity profiles are used, the method may be used to determine a unique heat flow/overburden couple. Subsidence (Z) is defined between Z_1 and Z_2 and varies at a step width of δZ. Heat flow during subsidence (HF) varies between HF_1 and HF_2 at a step width of δHF. The total number of thermal histories ($\sum$ Th) is then:

$$\sum Th = \sum HF \times \sum Z \qquad (A6)$$

with:

$$\sum HF = \frac{(HF_{max} - HF_{min})}{\delta HF} \qquad (A7)$$

$$\sum Z = \frac{(Z_2 - Z_1)}{\delta Z} \qquad (A8)$$

For each history the extent of the reactions is modelled and the root of the normalised least square residuals (NLSR) is calculated.

References

Abbott, G.D., Wang, G.Y., Eglinton, T.I., Home, A.K. and Petch, G.S., 1990. The kinetics of sterane biological marker release and degradation processes during the hydrous pyrolysis of vitrinite kerogen. Geochim. Cosmochim. Acta, 54: 2451–2461.

Alexander, R., Kagi, R.I. and Woodhouse, G.W., 1978. Measurement of thermal maturation of petroleum by proton magnetic resonance spectroscopy. Nature, 276: 598–599.

Alexander, R., Gray, M.D., Kagi, R.I. and Woodhouse, G.W., 1980. Proton magnetic resonance spectroscopy as a technique for measuring the maturity of petroleum. Chem. Geol., 30: 1–14.

Armagnac, C., Bucci, J., Kendall, C.G.S.C. and Lerche, I., 1989. Estimating the thickness of sediment removed at an unconformity using vitrinite reflectance data. In: N.D. Naeser and T.H. McCulloh (Editors), Thermal History of Sedimentary Basins. Springer-Verlag, Berlin, pp. 217–238.

Beaumont, C., Boutilier, R., Mackenzie, A.S. and Rullkötter, J., 1985. Isomerisation and aromatisation of hydrocarbons and palaeothermometry and burial history of Alberta Foreland Basin. Am. Assoc. Pet. Geol., Bull., 69: 546–566.

Braun, R.L. and Burnham, A.K., 1987. Analysis of chemical reaction kinetics using a distribution of activation energies and simple methods. Energy Fuels, 66: 153–161

Braun, R.L. and Burnham, A.K. 1988. Thermal cracking of hydrocarbons. LLNL Internal Rep., UCID 21507.

Burnham, A.K., 1989. On the validity of the Pristane Formation Index. Geochim. Cosmochim. Acta, 53: 1693–1697.

Burnham, A.K. and Sweeney, J.J., 1989. A chemical kinetic model of vitrinite maturation and reflectance. Geochim. Cosmochim. Acta, 53: 2649–2657.

Burnham, A.K., Braun, R.L., Gregg, H.R. and Samoun, H.M., 1987. Comparison of methods for measuring kerogen pyrolysis rates and fitting kinetic parameters. Energy Fuels, 1: 452.

Burnham, A.K., Braun, R.L. and Samoun, A., 1988. Further comparison of methods for measuring kerogen pyrolysis rates and fitting kinetic parameters. In: L. Mattavelli and L. Novelli (Editors), Advances in Organic Geochemistry 1987. Org. Geochem., 13: 830–845.

Carstens, H. and Finstad, K.G., 1981. Geothermal gradients of the northern North Sea basin, 59–62°N. In: L.V. Illing and G.D. Hobson (Editors), Petroleum Geology of the Continental Shelf of North-West Europe. The Institute of Petroleum, London, pp. 152–161.

Curiale, J.A. and Odermatt, J.R., 1989. Short-term biomarker variability in the Monterey formation, Santa Maria Basin. Org. Geochem., 14: 1–13.

Espitalié, J., Ungerer, P., Irwin, H. and Marquis, F., 1988. Primary cracking of kerogens. Experimenting and modelling C_1, C_2 to C_5, C_6–C_{15} and C_{15+} classes of hydrocarbons formed. Org. Geochem., 13: 893–899.

Goossens, H., De Lange, F., De Leeuw, J.W. and Schenck, P.A., 1988. The pristane formation index, a molecular maturity parameter. Confirmation in samples from the Paris Basin. Geochim. Cosmochim. Acta, 52: 2439–2444.

Green, P.F., Duddy, I.R., Gleadow, A.J.W. and Lovering, J.F., 1989. Apatite fission track analysis as a palaeotemperature indicator for hydrocarbon exploration. In: N.D. Naeser and T.H. McCulloh (Editors), Thermal History of Sedimentary Basins. Springer-Verlag, Berlin, pp. 181–195.

Harper, M.L., 1971. Approximate geothermal gradients in the North Sea Basin. Nature, 230: 235–236.

Hermanrud, C., Cao, S. and Lerche, I., 1990. Estimates of virgin rock temperatures from BHT measurements: bias and error. Geophysics, 55: 924–931.

Jensen, R.P. and Doré, A.G., 1993. A recent Norwegian Shelf heating event — fact or fantasy? In: A.G. Doré et al. (Editors), Basin Modelling: Advances and Applications. Norwegian Petroleum Society (NPF), Special Publication 3. Elsevier, Amsterdam, pp. 85–106 (this volume).

Issler, D.R., 1984. Calculation of organic maturation levels for offshore eastern Canada — implications for general application of Lopatin's method. Can. J. Earth Sci., 21: 477–488.

Lerche, I., Yarzab, R.F. and Kendall, C.G.St., 1984. Determination of palaeoheat flux from vitrinite reflectance data. Am. Assoc. Pet. Geol., Bull., 68: 1704–1717.

Lewan, M.D., Boroy, M. and Dolcater, D.L., 1986. Effects of thermal maturation on steroid hydrocarbons as determined by hydrous pyrolysis of Phophoria Retort Shale. Geochim. Cosmochim. Acta, 50: 1977–1987.

Løseth, H., Lippard, J.S., Sættem, J., Fanavoll, S., Fjerdingstad, V., Leith, T.L., Ritter, U., Smelror, M. and Sylta, Ø., 1992. Cenozoic uplift and erosion of the Barents Sea — evidence from the Svalis Dome area. In: In: T.O. Vorren, E. Bergsaker, Ø.A. Dahl-Stamnes, E. Lie, B. Johansen and T.B. Lund (Editors), Arctic Geology and Petroleum Potential. Norwegian Petroleum Society (NPF), Special Publication 2. Elsevier, Amsterdam, pp. 643–664.

Mackenzie, A.S. and McKenzie, D., 1983. Isomerisation and aromatisation of hydrocarbons in sedimentary basins formed by extension. Geol. Mag., 120: 417–470.

Mackenzie, A.S., Patience, R.L., Maxwell, J.R., Vandenbroucke, M. and Durand, B., 1980. Molecular parameters of maturation in the Toarcian shales, Paris Basin, France, I. Changes in the configuration of acyclic isoprenoid alkanes, steranes and triterpanes. Geochim. Cosmochim. Acta, 44: 1709–1721.

Mackenzie, A.S., Lamb, N.A. and Maxwell, J.R., 1982. Steroid hydrocarbons and the thermal history of sediments. Nature, 295: 223–224.

Mackenzie, A.S., Beaumont, C., Boutilier, R. and Rullkötter, J., 1985. The aromatisation and isomerisation of hydrocarbons and the thermal and subsidence history of the Nova Scotia margin. Philos. Trans. R. Soc. London, Ser. A, 315: 203–232.

Marzi, R., 1989. Kinetik und quantitative Analyse der Isomerisierung und Aromatisierung von fossilen Steroidkohlenwasserstoffen im Experiment und in natürlichen Probensequenzen. Dissertation, T.H. Aachen, 1988. Berichte der Kernforschungsanlage Julich Nr. 2264, 169 pp.

Marzi, R., Rullkötter, J. and Perriman, W.S., 1990. Application of the change of sterane isomerisation ratios to the reconstruction of geothermal histories: implications of the results of hydrous pyrolysis experiments. Org. Geochem., 16: 91–102.

Moldowan, J.M. and Fago, F.J., 1986. Structure and significance of novel rearranged monoaromatic steroid hydrocarbon in petroleum. Geochim. Cosmochim. Acta, 50: 343–351.

Quigley, T.M., Mackenzie, A.S. and Gray, J.R., 1987. Kinetic theory of petroleum generation. In: B. Doligez (Editor), Migration of Hydrocarbons in Sedimentary Basins. Technip, Paris, pp. 649–665.

Ritter, U., 1984. The influence of time and temperature on vitrinite reflectance. Org. Geochem., 6: 473–480.

Ritter, U., 1988. Modelling of hydrocarbon generation patterns in the Egersund Basin, North Sea. Org. Geochem., 13: 165–174.

Ritter, U., Leith, T.L., Griffiths, C.M. and Schou, L., 1987. Hydrocarbon generation and thermal evolution in parts of the Egersund Basin (northern North Sea). In: Ch. Beaumont and A.J. Tankard (Editors), Sedimentary Basins and Basin-Forming Mechanisms. Can. Soc. Pet. Geol., Mem., 12: 75–85.

Ritter, U., Myhr, M.B., Aareskjold, K., Ganz, H.H., Mostafa, A. and Schou, L., 1993. Distributed activation energy models from hydrous pyrolysis: towards a second-generation tool of thermal history analysis. In: A.G. Doré et al. (Editors), Basin Modelling: Advances and Applications. Norwegian Petroleum Society (NPF), Special Publication 3. Elsevier, Amsterdam, pp. 201–208 (this volume).

Rubey, W.W. and Hubbert, M.K., 1959. Role of fluid pressure in mechanics of overthrust faulting, II. Overthrust belt in geosynclinal area of western Wyoming in light of fluid pressure hypothesis. Geol. Soc. Am. Bull., 70: 167–206.

Rullkötter, J. and Marzi, R., 1988. Natural and artificial maturation of biological markers in a Toarcian Shale from northern Germany. Org. Geochem., 13: 639–645.

Sajgo, Cs., Horvath, Z.A. and Lefler, J., 1988. An organic maturation study of the Hod-1 borehole (Pannonian Basin). Am. Assoc. Pet. Geol., Mem., 45: 297–309.

Schaefer, R.G., Schenk, H.J., Hardelauf, H. and Harms, R., 1990. Determination of gross kinetic parameters for petroleum formation from Jurassic source rocks of different maturity levels by means of laboratory experiments. Org. Geochem., 16: 115–120.

Solli, H., Schou, L., Krane, J., Skjetne, T. and Leplat, P., 1985. Characterisation of sedimentary organic matter using nuclear magnetic resonance and pyrolysis techniques. In: Petroleum Geochemistry in Exploration of the Norwegian Shelf. Graham and Trotman, London, pp. 309–317.

Strachan, M.G., Alexander, R., Subroto, E.A. and Kagi, R.I., 1989. Constraints upon the use of 24-ethylcholestane diastereomer ratios as indicators of the maturity of petroleum. Org. Geochem., 14: 423–432.

Thomas, B.M., Moller-Pedersen, P., Whitaker, M.F. and Shaw, N.D., 1985. Organic facies and hydrocarbon distributions in the Norwegian North Sea. In: Petroleum Geochemistry in Exploration of the Norwegian Shelf. Norwegian Petroleum Society, Graham and Trotman, London, pp. 3–26.

Thorne, J.A. and Watts, A.B., 1989. Quantitative analysis of North Sea subsidence. Am. Assoc. Pet. Geol., Bull., 73: 88–116.

Tissot, B.P. and Welte, D.H., 1984. Petroleum Formation and Occurrence. Springer-Verlag, Berlin, 699 pp.

Tissot, B.P., Pelet, R. and Ungerer, Ph., 1987. Thermal history of sedimentary basins, maturation indices, and kinetics of oil and gas generation. Am. Assoc. Pet. Geol., Bull., 71: 1445–1466.

Waples, D.W., 1980. Time and temperature in petroleum formation — application of Lopatin's method to petroleum exploration. Am. Assoc. Pet. Geol., Bull., 64: 916–926.

Wood, D.A., 1988. Relationships between thermal maturity indices calculated using Arrhenius equation and Lopatin method: implications for petroleum exploration. Am. Assoc. Pet. Geol., Bull., 72: 115–134.

Woodside, W. and Messmer, J.H., 1961. Thermal conductivity of porous media. J. Appl. Phys., 32: 1688–1699.

U. RITTER *IKU Petroleum Research, N-7034 Trondheim, Norway*
M.B. MYHR *IKU Petroleum Research, N-7034 Trondheim, Norway*
K. AARESKJOLD *IKU Petroleum Research, N-7034 Trondheim, Norway*
L. SCHOU *IKU Petroleum Research, N-7034 Trondheim, Norway*

Distributed activation energy models from hydrous pyrolysis: towards a second-generation tool of thermal history analysis

U. Ritter, M.B. Myhr, K. Aareskjold, H.H. Ganz, A. Mostafa and L. Schou

It is the objective of this ongoing study to search for parameters that change regularly with increasing temperature, to describe their kinetic behaviour in terms of parallel reactions and to test this behaviour against field data. The aim is to develop systems of "stacked" thermal indicators, each confined to narrow but overlapping temperature intervals, which as a whole cover temperature ranges similar to those of conventional indicators (e.g. vitrinite reflectance) but are considerably more temperature-sensitive. We assumed that organic reactions are properly represented by distributed activation energy models. These were derived from non-linear regression analysis of time series of hydrous pyrolysis experiments and optimised against natural data. The models of sterane and hopane isomerisation, and of the ^{1}H-NMR parameter FAP, can be regarded as sufficiently validated to be used in thermal history analysis. The solid-state infrared parameters $W_{\min}$ and C-factor are influenced by kerogen type and require further investigation. The ^{13}C-NMR parameter f_a and the infrared A-factor are strongly dependent on kerogen type and cannot be used for thermal history analysis.

Introduction

To date organic thermal indicators have been represented mainly by TTI-type or pseudoactivation energy models that simulate the chemical processes inadequately, are too inflexible, and tend to produce artifacts (Quigley et al., 1987; Wood, 1988; Ritter, 1988; Ritter et al., this volume). The models can be improved by employing distributions of, rather than single, activation energies. Such improvements rely on specially designed heating experiments, the availability of advanced tools for kinetic analysis, more versatile conceptual models, and indicator systems rather than single indicators.

Some indicators, such as vitrinite reflectance, record thermal events cumulatively over a temperature range, but they are insensitive to subtle temperature variations. Other indicators, such as isomerisation reactions, are highly temperature-sensitive but they proceed only over a rather narrow temperature interval (Fig. 1). The ideal tool would consist of a system of stacked indicators that occupy narrow but overlapping temperature intervals, are independent of kerogen type, and follow the same reaction mechanisms at all temperatures. Such a system would cover a temperature range similar to that of, for example, vitrinite reflectance, but would be about two to four times more temperature-sensitive.

It is the main objective of this ongoing study to search for parameters that are known to change regularly with increasing temperature, to describe their kinetic behaviour in terms of parallel reactions, and to test this behaviour against field data. It is the objective of this paper to provide a short overview of the work presently completed and the underlying concepts and strategies. So far, only the isomerisation parameters have been completely tested and validated. The other parameters represent various intermediate stages of testing and validation. We assumed

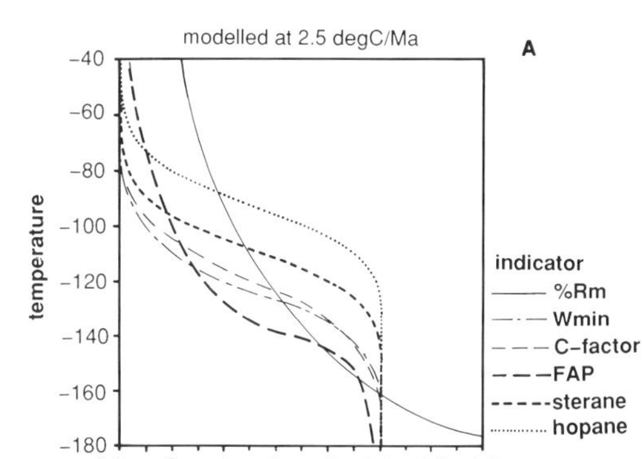

Fig. 1. (A) Temperature range of various reactions for which tested distributed kinetic models are available. The steepest slope corresponds to the highest sensitivity of the reaction. Vitrinite reflectance is given in the original (%R_m) scale.

Basin Modelling: Advances and Applications edited by A.G. Doré et al. NPF Special Publication 3, pp. 201–208, Elsevier, Amsterdam.

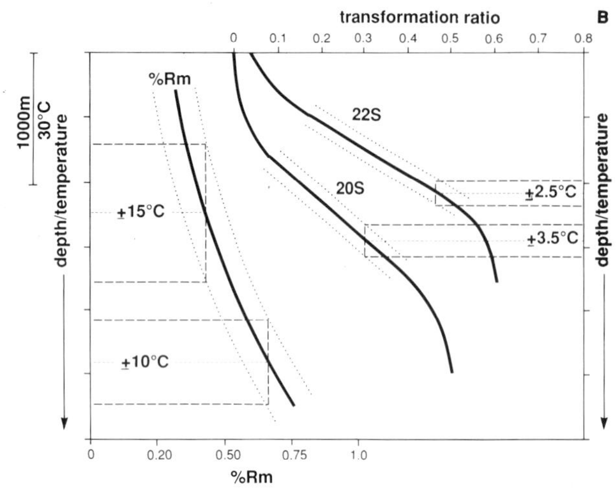

Fig. 1 (continued). (B) Comparison of temperature sensitivity of vitrinite reflectance and isomerisation. The error estimates are based on standard errors of 0.05 for isomerisation and 0.1% R_m.

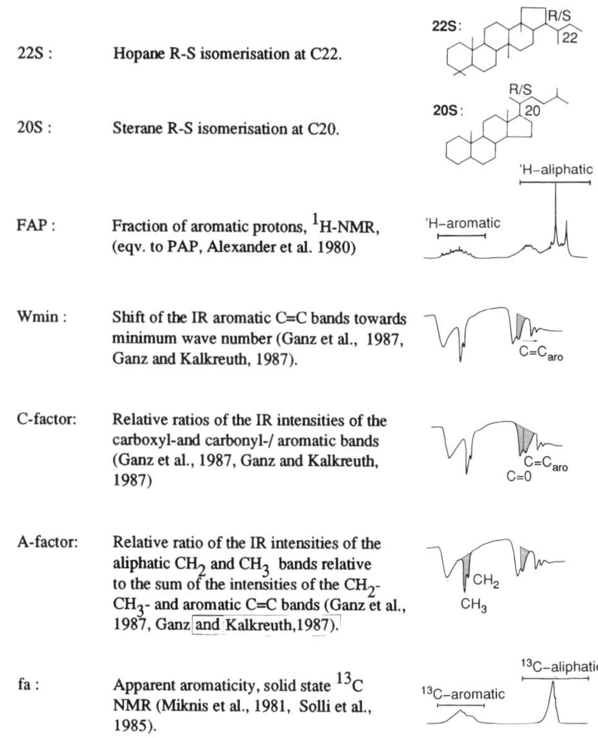

Fig. 2. Parameters investigated in this study. See Mackenzie and McKenzie (1983) for definition of isomerisation reactions.

as a first approximation that organic reactions are made up of parallel sub-reactions and that they are properly represented by distributed activation energy models (Tissot and Welte, 1984; Tissot et al., 1987; Burnham et al., 1988). We chose hydrous pyrolysis for artificial maturation as it is generally regarded as most closely representing natural conditions.

In this paper the terms "parameter" or "indicator" are both used to describe a measurable entity (e.g. vitrinite reflectance, isomerisation ratio). Those parameters that change in a predictable manner when undergoing heating are referred to as "thermal indicators". The term "reaction" is used in the common chemical sense. "Kinetic models" are made up of a set of constants that describe the reaction (e.g. activation energies, frequency factors, distribution factors) and "reaction mechanisms" refer to the manner in which a reaction proceeds (e.g. parallel, serial etc., Braun and Burnham, 1991).

Methodology

The parameters that have been partly or fully investigated are described in Fig. 2. For a detailed discussion of the reactions defining these parameters, we refer to the publications cited in this figure. We employ established technology that is now widely used to derive kinetic models of organic reactions (e.g. Braun and Burnham, 1987, 1991; Tissot et al., 1987; Ungerer et al., 1987; Burnham and Sweeney, 1989; Burnham and Braun, 1990; Sweeney and Burnham, 1990; Sweeney et al., 1990;). Our previous experience with geological application of experimentally derived kinetic models has shown that their accuracy at extrapolation into geological heating rates and temperatures is not sufficient for the requirements of a thermal indicator. We therefore believe that valid kinetic models of thermal indicator reactions can only be obtained if laboratory and field data are properly integrated. For this purpose, we used short and continuous thermal histories in which present-day temperature controls maturation. The limiting factor is the inaccuracy of present-day BHT measurements (e.g. Hermanrud et al., 1990). The flow of information during optimisation is given in Fig. 1. This approach is clearly in contrast to that of other authors (e.g. Alexander et al., 1990; Marzi et al., 1990) who attempted to develop kinetic models of thermal indicators from experiments alone. Although this may lead to acceptable results in many cases, we prefer to have our models "anchored" in natural data as well.

We used three hydrous pyrolysis series with five source rock aliquots each at temperatures of 300, 320 and 350°C which were then investigated by non-linear regression analysis (software KINETICS) to derive models of distributed activation energies. The algorithms used in KINETICS have been described by Braun and Burnham (1987). Figure 4 shows an example for a time-transformation diagram. There is slight scatter in the measured values due to experimental error, but nevertheless these results can be matched by a distributed activation energy model that also may be valid at geological temperatures. Figure 3 shows the experimental design for iso-

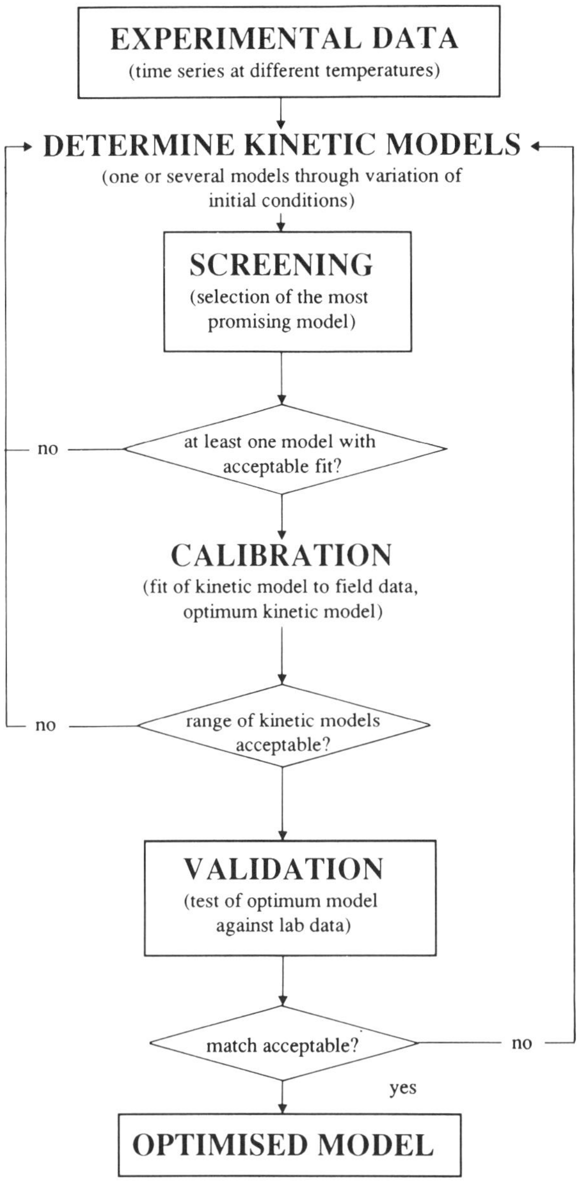

Fig. 3. Optimisation procedures of experimental kinetic models.

merisation reactions. The terms "acceptable fit" and "optimum kinetic model" of this flow chart are used in an informal sense. Modelled and measured data were fitted visually by varying E_a at steps of 0.1 kcal/mol. Particular care was taken to obtain a parallel match of the two curves over the whole range of maturity, because this has an important influence on the distribution factor, i.e. the width of the reaction interval. The optimum kinetic model achieved the closest fit.

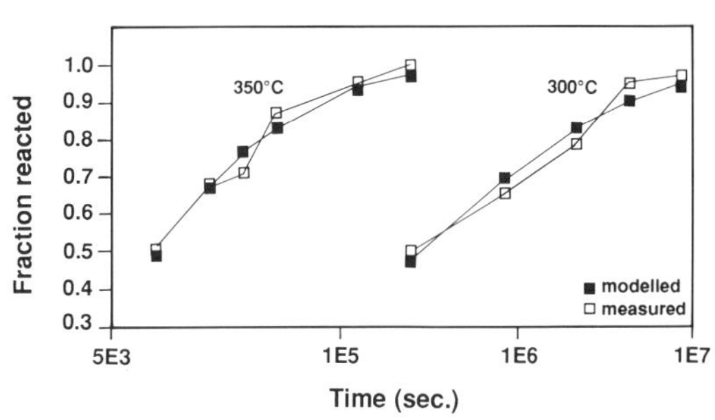

Fig. 4. Example of a time-transformation diagram from hydrous pyrolysis and match of modelled and measured data.

All ratios have been normalised according to the experimental maturities encountered during pyrolysis:

$$F_n = \frac{F_t - F_i}{F_f - F_i}$$

where F_n = normalised fraction reacted; F_t = fraction from experimental heating; F_i = initial fraction before heating; and F_f = highest fraction reacted during pyrolysis.

The kinetic models determined are therefore only valid between these limiting values.

In order to test the stability of the kinetic models the initial parameters used in KINETICS were varied and individual data points were omitted from evaluation (screening). The resulting temperatures under standard conditions (half transformation, T_{50}, at 2.5°C/Ma) were modelled and mean temperature and standard deviation were calculated. The latter is a measure of the stability of the determinations (experimental error in Table 4). The various sets of kinetic parameters (Fig. 5) were checked against a preliminary set of natural temperature data and the sets fitting the natural data best were selected for further refinement.

Carefully selected natural data were used to optimise the kinetic model found during the screening stage (e.g. Table 1). For this purpose, the mean activation energy (E_a) was varied until modelled and measured reaction trends matched best (e.g. Table 2). The models resulting from the calibration stage were again tested against laboratory data. Figure 6 gives an example for hopane isomerisation.

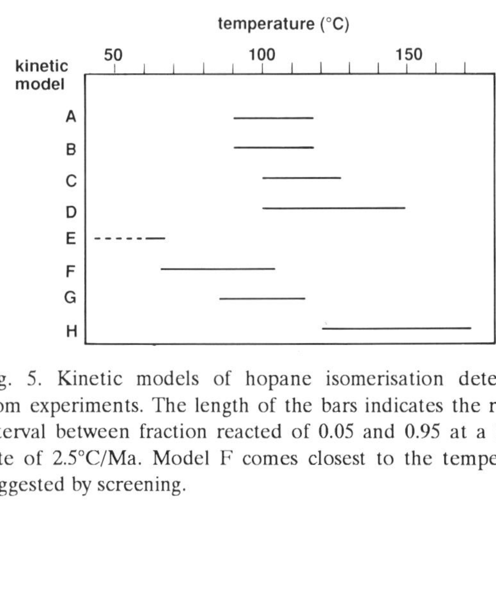

Fig. 5. Kinetic models of hopane isomerisation determined from experiments. The length of the bars indicates the reaction interval between fraction reacted of 0.05 and 0.95 at a heating rate of 2.5°C/Ma. Model F comes closest to the temperatures suggested by screening.

TABLE 1

Data sources used for calibration of isomerisation models

Area	Main subsidence	Temperature	References		
			biomarker	geology	temperature
Handil	10–15 Ma	31°C/km; 34°C/km linear gradient	3	1, 2	3, 4
Linyi	Oligocene, Pliocene	DST,	6	5	confidential report
Hod-1	8 Ma to Present	38°C/km linear gradient	3, 7	3	7
Nova Scotia Margin	150 to 100 Ma	corrected BHT	8	8	8

References: 1 = Durand and Oudin (1979); 2 = Durand et al. (1983); 3 = Mackenzie and McKenzie (1983); 4 = Ungerer et al. (1984); 5 = Tissot and Welte (1984); 6 = Hong et al. (1986); 7 = Sajgo et al. (1988); 8 = Mackenzie et al. (1985).
DST: Drill Stem Test temperature; BHT: bottom hole temperature.

TABLE 2

Optimised isomerisation models derived from calibration against natural thermal histories

Area	A	E_a	σ	T_{05}	T_{50}	T_{95}	
Nova Scotia	3.5×10^7	38.5	1.5	87	114	140	sterane
Margin	1.0×10^8	37.5	1.1	~70	96	118	hopane
Hod-1	3.5×10^7	37.5	1.1	90	104	127	sterane
	1.0×10^8	37.0	1.1	68	90	113	hopane
Linyi	3.5×10^7	38.5	1.1	89	114	130	sterane
	1.0×10^8	36.5	1.1	65	88	108	hopane
Handil (31°C/km)	3.5×10^7	36.0	1.5	65	89	111	sterane
Handil (34°C/km)	3.5×10^7	37.0	1.1	77	98	120	sterane
Optimised models:							
Sterane	3.5×10^7	37.5 ± 1.0	2.9%	80	104	130	
Hopane	1.0×10^8	37.0 ± 0.5	2.9%	68	90	113	
Average T_{50} from calibration:							
Sterane					104 ± 9		
Hopane					91 ± 3.4		
Average T_{50} measured from North Sea data:							
Sterane					104 ± 6		
Hopane					90 ± 8		

A, s^{-1}; E_a, kcal/mol; σ, kcal/mole or % of E_a. Temperatures are given for 5, 50 and 95% of the fractions reacted at a linear heating rate of 2.5°C/Ma. Average North Sea temperature from Ritter et al. (this volume).

Results of individual reactions

Table 3 presents the optimum kinetic model obtained for each parameter. Table 4 compares the average experimental and field errors. The experimental errors have been obtained from the variation of initial conditions during kinetic evaluation of the experiments (e.g. Table 5). The terms "suppressed" and "emphasised" in Table 5 refer to the capability of KINETICS to assign different "weight" to a file. If, for example, the file containing the data for the 300°C experiment is assigned the weight 10 and the other files the weight 1, results will be strongly biased by the 300°C experimental temperatures. The field error is the average deviation of present and modelled temperature.

The experimental isomerisation data (22S) are rather poor and thus the resulting experimental error is high (Table 4 and 6). Due to extensive calibration, however, the field error is rather low. To achieve a best fit during the calibration stage, E_a had to be

TABLE 3

Optimum kinetic models of reactions presented in this paper

Indicator	A	E_a	σ	T_{05}	T_{50}	T_{95}
FAP	1.0×10^{13}	51.3	0.8	53	134	171
	4.5×10^{10}	43.2	12.9			
f_a	5.3×10^{15}	55.7	3.7	110	137	~165
C-factor	4.6×10^{15}	55.6	4.8	101	134	170
W_{min}	3.0×10^{11}	46.3	3.3	95	121	148
A-factor	7.6×10^{21}	75.3	1.8	165	183	>200
Sterane	3.5×10^7	37.5	2.9	80	104	130
Hopane	1.0×10^8	37.0	2.9	68	90	113

σ: distribution factor in % of E_a. T_{05}, T_{50}, T_{95}: temperature at which 5, 50 and 95% of the reaction are completed at a heating rate of 2.5°C/Ma. A in s^{-1}, E_a in kcal/mol.

TABLE 4

Summary of experimental kinetic models of reactions that can be measured in field data and thus allow to estimate the accuracy of the method (for definition of indicators see Fig. 2)

Indicator	Experimental error (°C)	Average field error (°C)	Experimental data quality
22S	23.3	7.5	4
FAP	8.9	8.6	2–3
f_a	<2	31.1	1
C-factor	10.8	9.0	2
W_{min}	2.8	8.5	2
A-factor	3.2	not tested	2

All temperatures in °C. Quality of experimental data: 1 = good; 2 = fair, low scatter; 3 = high scatter; 4 = poor, very high scatter. Average field errors from optimised model where available.

increased by between 0 (Handil, 31°C/km) and 2.5 kcal/mol (Nova Scotia Margin). A and σ remained unchanged. Table 2 shows the respective optimum models, transformation temperatures under standard conditions, and the resulting average models. Subsequently the average isomerisation models resulting from the calibration stage were tested against laboratory data. Figure 6 shows a rather good match of the optimised isomerisation models with our own laboratory data.

The ^{13}C-NMR parameter f_a is experimentally very stable. In fact, regardless of the modifications to the initial conditions of KINETICS the model always converged to the same kinetic constants. The field error is nevertheless unacceptably high, which we suspect is due to strong dependence on kerogen composition. Further investigation is required to investigate whether there are other causes.

W_{min} and C-factor also have relatively low experimental errors in addition to low field errors. However, there are opposing signs of deviation in

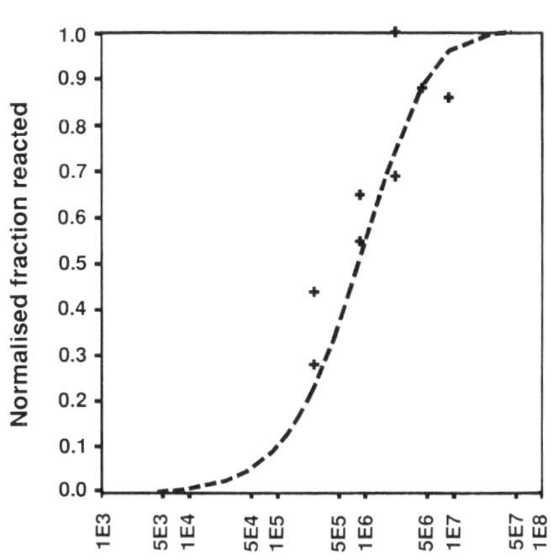

Fig. 6. Match of the hopane isomerisation model with laboratory data at 300°C.

TABLE 5

Experimental conditions of hydrous pyrolysis and of kinetic analysis of hopane isomerisation

Experimental:

Source rock	Time series:	
A	350°C	300°C
B	320°C	300°C
C	320°C	

Time–temperature relationships

Kinetic analysis:

Heat-up time:	60 min, linear
Initial temperature:	20°C

Model version	Weighting conditions
A	all files equal weight
B	320°C suppressed, other files equal weight
C	320°C emphasised (3×), other files equal weight
D	320°C suppressed, sample A emphasised (3×)
E	320°C suppressed, sample B emphasised (3×)
F	300°C emphasised (3×), all other files suppressed
G	320°C emphasised (3×), all other files suppressed
H	350°C emphasised (3×), all others suppressed

Explanation: suppressed, weight = 0.1; equal weight, weight = 1; emphasised, weight = 3.

TABLE 6

Hopane isomerisation ratios determined from hydrous pyrolysis experiments and used for analysis of kinetic constants

Time (days)	350°C	320°C		300°C	
	A	B	C	A	B
0.0	0.12	0.14	0.00	0.12	0.14
0.5	0.46				
1.0	0.52	0.28	0.30		
1.5	0.51				
2.0	0.61				
2.5		0.27	0.40		
3.0				0.32	0.27
5.0			0.57		
10.0				0.44	0.40
25.0				0.56	0.61
50.0				0.55	
100.0				0.54	

A, B, C: sample code. Three immature source rocks from the North Sea.

the two wells we used for testing, which suggests that the influence of kerogen composition or uncertainties of palaeotemperatures are much stronger than is evident from the field errors.

The A-factor shows rather high transformation temperatures, which was expected from previous circumstantial evidence (Ganz and Kalkreuth, 1987). None of the test wells had temperatures high enough to cause significant reaction and we could thus not test the kinetic model against natural thermal histories. Apart, below 160°C the A-factor seems extremely dependent on kerogen type (Ganz and Kalkreuth, 1987; Ganz et al., 1987), thus possibly providing a tool to obtain information on hydrocarbon potential also at higher maturities.

The ^{1}H-NMR parameter FAP shows moderate to low errors. The optimum activation energy model is distinctly bimodal (Table 3) and consists of a low-temperature part with rather wide distribution, and a high-temperature part with narrow distribution, which causes the characteristic shape of the transformation curve (Fig. 1). FAP seems sufficiently tested to be used in thermal history analysis, although more information on possible error sources is required.

Discussion and conclusion

One important aspect that influences the final reliability of the indicator model very strongly is the accuracy of the thermal histories used for calibration. Wireline logs usually attempt to measure temperature during drilling and not formation temperature. Most methods of temperature correction are highly inaccurate (e.g. Hermanrud et al., 1990) and true formation temperatures can be obtained only from Drill Stem Tests or wells shut down long enough to have reached equilibrium. This is normally not the case in commercial exploration wells. Except during precision temperature logging (e.g. Blackwell and Steele, 1989) temperature information is only available for a few points along a profile. At best a linear gradient can be constructed between these points or with the surface, which is inadequate to account for subtle variations due to contrasts of thermal conductivity. Other sources of error include convective heat transfer due to flow of compaction- or meteoric water, or changes of tectonic conditions accompanied by a change of mantle heat flow. The latter can be minimised by choosing wells that have short thermal histories and thus carry a low risk that the tectonothermal regime has changed during the history they represent. The former can only be overcome by careful selection of the wells to be used for calibration.

Unfortunately there are no data sets available that have been prepared specifically for the purpose of calibrating thermal indicators. Most of the wells used here have been selected because of the availability of relatively good geochemical data. Well Hod 1 has been widely used for maturity assessment (Mackenzie and McKenzie, 1983; Sajgo et al., 1988). The temperatures seem to be corrected bottom hole temperatures, although this has not been clearly stated. For well Handil in the Mahakam Delta two slightly different gradients of 31 and 34°C/km are quoted (Mackenzie and McKenzie, 1983; Ungerer et al., 1984). At a depth of 3 km this would translate into an uncertainty of about ±4.5°C. The temperatures at the Nova Scotia Margin seem to be corrected bottom hole temperatures but no information is given about the mode of correction (Mackenzie et al., 1985; Issler, 1984). This is also the history with the largest uncertainty of palaeotemperatures due to its long duration.

Despite these uncertainties the activation energy models required to model these different thermal histories are relatively consistent (Table 2). Sterane isomerisation shows a variability twice as high (±1 kcal/mol) as hopane isomerisation (0.5 kcal/mol). This may also be due to the stronger dependence of sterane isomerisation on kerogen composition (Ritter et al., this volume), which has frequently been observed. For the models investigated here an activation energy interval of 1 kcal/mol roughly corresponds to an interval of the mean (T_{50}) reaction temperature of 9°C. For the hopane isomerisation model, this results in an error of less than ±5°C, which is lower than the expected uncertainty of the measured temperatures. This may suggest that either the natural temperature histories are biased in the same direction, or that the temperature data are better than expected.

Nevertheless, the conclusion is evident that the development of more reliable thermal indicators requires a considerably higher quality of natural data, particularly temperatures. Minimum requirements would have to be Drill Stem Test or equilibrium temperatures, together with good control on cavings. Optimum requirements would include temperatures from precision temperature logging and samples from cores only, in addition to routine well information.

Of the sterane isomerisation models published to date, only those of Mackenzie and McKenzie (1983), Sajgo and Leffler (1986), Marzi (1989) and Marzi et al. (1990) yield realistic temperatures under natural conditions. From these only the sterane model of Marzi (1989), recently improved by additional data (Marzi et al., 1990), is able to predict natural as well as experimental conditions. The mean (T_{50}) reaction temperature derived from our model (104 ± 9°C) and the model of Marzi et al. (1990) (113°C) differ not significantly. Therefore there is no reason to argue which of the models better reflect the mean temperature. There is, however, a significant difference in the reaction intervals described by the two models. This is due to the fact that the model of Marzi et al. (1990) is made up of only a single couple of kinetic constants. It can be shown that the reaction interval in most natural isomerisation profiles is wider than the interval prescribed by the model of Marzi et al.

TABLE 7

Preliminary system of thermal indicator reactions (temperatures in °C at 2.5°C/Ma)

Reaction	Temperature:	
	highest sensitivity	total range
22S	80–110	60– 120
20S	90–120	80– 130
W_{min}, C-factor	110–140	85– 150
FAP	130–150	~50–>180
$\%R_m$	>150	undefined

(1990), that the sterane and hopane isomerisation reactions are more complex than suggested by the simple concept of isomerisation of Mackenzie and McKenzie (1983), which may warrant the assumption of parallel as well as sequential reaction mechanisms (e.g. Abbott et al., 1990). A detailed discussion of these topics as well as details of FAP is beyond the scope of this paper and will be given elsewhere (Ritter et al., 1991, and this volume).

Kinetic models and transformation temperatures of FAP, f_a, C-factor and W_{min} are in the range reported for hydrocarbon generation (Tissot et al., 1987; Espitalié et al., 1988; Burnham et al., 1988), whereas those of isomerisation are distinctly below the temperature of hydrocarbon generation. FAP, sterane and hopane have been sufficiently tested and can be used safely in thermal history analysis if certain precautions are observed (Ritter et al., this volume). f_a is probably too dependent on kerogen composition to serve as a thermal indicator. W_{min} and C-factor are suspected of some facies dependence but have to be tested further.

The indicators obtained up to now cover a temperature range between about 60°C and >180°C at 2.5°C/Ma (Table 7) though the temperature ranges of highest sensitivity do not yet overlap properly if C-factor and W_{min} are omitted. The advantage of a system of stacked reactions is immediately evident when compared with an indicator such as vitrinite reflectance. Figure 1A provides an idea of such a system and Fig. 1B demonstrates the different temperature sensitivities of vitrinite reflectance and isomerisation reactions. Irrespective of the standard errors and temperature gradients used for calculation, the temperature error of the isomerisation reactions is about 10 to 15% of that of $\%R_m$. Similarly high sensitivities of vitrinite reflectance can only be expected at values greater than 1.0% R_m.

Acknowledgements

We thank IKU for permission to publish the results of this study and NTNF for partial funding. Two anonymous reviewers provided valuable suggestions. The MR-Center, Trondheim, performed the NMR analyses, and Bob Goll improved the English of the manuscript.

References

Abbott, G.D., Wang, G.Y., Eglinton, T.I., Home, A.K. and Petch, G.S., 1990. The kinetics of sterane biological marker release and degradation processes during the hydrous pyrolysis of vitrinite kerogen. Geochim. Cosmochim. Acta, 54: 2451–2461.

Alexander, R., Gray, M.D., Kagi, R.I. and Woodhouse, G.W., 1980. Proton magnetic resonance spectroscopy as a technique for measuring the maturity of petroleum. Chem. Geol., 30: 1–14.

Alexander, R., Marzi, R. and Kagi, R.I., 1990. A new method for assessing the thermal history of sediments: a case study from the Exmouth Plateau in Northwestern Australia. Aust. Pet. Explor. J., pp. 364–372.

Blackwell, D.D. and Steele, J.L., 1989. Thermal conductivity of sedimentary rocks: measurement and significance. In: N.D. Naeser and T.H. McCulloh (Editors), Thermal History of Sedimentary Basins. Springer-Verlag, Berlin, pp. 13–36.

Braun, R.L. and Burnham, A.K., 1987. Analysis of chemical reaction kinetics using a distribution of activation energies and simple methods. Energy Fuels, 66: 153–161.

Braun, R.L. and Burnham, A.K., 1992. PMOD: a flexible model of oil and gas generation, cracking and expulsion. Org. Geochem. (in press).

Burnham, A.K. and Braun, R.L., 1990. Development of a detailed model of petroleum formation, destruction and expulsion from lacustrine and marine source rocks. Org. Geochem., 16: 27–39.

Burnham, A.K. and Sweeney, J.J., 1989. A chemical kinetic model of vitrinite maturation and reflectance. Geochim. Cosmochim. Acta, 53: 2649–2657.

Burnham, A.K., Braun, R.L. and Samoun, A., 1988. Further comparison of methods for measuring kerogen pyrolysis rates and fitting kinetic parameters. In: L. Novelli and M. Mattavelli (Editors), Advances in Organic Geochemistry 1987. Org. Geochem., 13: 830–845.

Durand, B. and Oudin, J.L., 1979. Exemple de migration des hydrocarbures dans une série deltaique: le delta de Mahakam, Kalimantan, Indonesie. Proc. 10th World Petroleum Congress, Bucharest, 1979, Vol. 2. Exploration, Supply and Demand. Heyden, London.

Durand, B., Ungerer, Ph., Chiarelli, A. and Oudin, J.L., 1984. Modélisation de la migration de l'huile application à deux exemples de bassins sédimentaires. 11th World Petroleum Congress, London, 1983: Origin, Migration and Accumulation of Hydrocarbons. Wiley and Sons, London, pp. 1–13.

Espitalié, J., Ungerer, P., Irwin, H. and Marquis, F., 1988. Primary cracking of kerogens. Experimenting and modelling C_1, C_2 to C_5, C_6–C_{15} and C_{15+} classes of hydrocarbons formed. Org. Geochem., 13: 893–899.

Ganz, H.H. and Kalkreuth, W., 1987. Application of infrared spectroscopy to the classification of kerogen types and the evaluation of source rock and oil shale potentials. Fuel, 66: 708–711.

Ganz, H.H., Kalkreuth, W., Oener, F., Pearson, M.J. and Small, J.S., 1987. Enhanced geochemical screening analysis — accurate determination of petroleum source rock characteristics and comparison with standard techniques. In: J. Brooks and K. Glennie (Editors), Petroleum Geology of North West

Europe. Graham and Trotman, London, pp. 847–851.
Hermanrud, C., Cao, S. and Lerche, I., 1990. Estimates of virgin rock temperatures from BHT measurements: bias and error. Geophysics, 55: 924–931.
Hong Zhi-hua, Li Hui-xiang, Rullkötter, J. and Mackenzie, A.S., 1986. Geochemical application of sterane and triterpane biological marker compounds in the Linyi Basin. In: D. Leytheauser and J. Rullkötter (Editors), Advances in Organic Geochemistry 1985. Org. Geochem., 10: 433–439.
Issler, D.R., 1984. Calculation of organic maturation levels for offshore eastern Canada — implications for general application of Lopatin's method. Can. J. Earth Sci., 21: 477–488.
Mackenzie, A.S. and McKenzie, D., 1983. Isomerisation and aromatisation of hydrocarbons in sedimentary basins formed by extension. Geol. Mag., 120: 417–470.
Mackenzie, A.S., Beaumont, C., Boutilier, R. and Rullkötter, J., 1985. The aromatisation and isomerisation of hydrocarbons and the thermal and subsidence history of the Nova Scotia Margin. Philos. Trans. R. Soc. London, Ser. A, 315: 203–2321.
Marzi, R., 1989. Kinetik und quantitative Analyse der Isomerisierung und Aromatisierung von fossilen Steroidkohlenwasserstoffen im Experiment und in natürlichen Probensequenzen. Dissertation Technische Hochschule Aachen, 1988, Berichte der Kernforschungsanlage Jülich Nr. 2264, 169 pp.
Marzi, R., Rullkötter, J. and Perriman, W.S., 1990. Application of the change of sterane isomerisation ratios to the reconstruction of geothermal histories: implications of the results of hydrous pyrolysis experiments. Org. Geochem., 16: 91–102.
Miknis, F.P., Sullivan, M., Bartuska, V.J. and Maciel, G.E., 1981. Cross-polarisation magic-angle spinning ^{13}C NMR spectra of coals of varying rank. Org. Geochem., 3: 19–28.
Quigley, T.M., Mackenzie, A.S. and Gray, J.R., 1987. Kinetic theory of petroleum generation. In: B. Doligez (Editor), Migration of Hydrocarbons in Sedimentary Basins. Technip, Paris, pp. 649–665.
Ritter, U., 1988. Modelling of hydrocarbon generation patterns in the Egersund Basin, North Sea. In: L. Mattavelli and L. Novelli (Editors), Advances in Organic Geochemistry 1987. Org. Geochem., 13: 165–174.
Ritter, U., Aareskjold, K. and Schou, L., 1991. Distributed activation energy models of isomerisation reactions from hydrous pyrolysis. Org. Geochem. (submitted).
Ritter, U., Myhr, M.B., Aareskjold, K. and Schou, L., 1993. Validation and first application of paleotemperature models of isomerisation and NMR aromatisation parameters.In: A.G. Doré et al. (Editors), Basin Modelling: Advances and Applications. Norwegian Petroleum Society (NPF), Special Publication 3. Elsevier, Amsterdam, pp. 185–199 (this volume).
Sajgo, Cs. and Lefler, J., 1986. A reaction kinetic approach to the temperature–time history of sedimentary basins. In: G. Buntebarth and L. Stegena (Editors), Lecture Notes in Earth Sciences, Vol. 5. Springer-Verlag, Berlin, pp. 119–151.
Sajgo, Cs., Horvath, Z.A. and Lefler, J., 1988. An organic maturation study of the Hod-1 borehole (Pannonian Basin). Am. Assoc. Pet. Geol., Mem., 45: 297–309.
Solli, H., Schou, L., Krane, J., Skjetne, T. and Leplat, P., 1985. Characterisation of sedimentary organic matter using nuclear magnetic resonance and pyrolysis techniques. In: Petroleum Geochemistry in Exploration of the Norwegian Shelf. Graham and Trotman, London, pp. 309–317.
Sweeney, J.J. and Burnham, A.K., 1990. Evaluation of a simple model of vitrinite reflectance based on chemical kinetics. Am. Assoc. Pet. Geol., Bull., 74: 1559–1570.
Sweeney, J.J., Talukdar, S., Burnham, A.K. and Vallejos, C., 1990. Pyrolysis kinetics applied to prediction of oil generation in the Maracaibo Basin, Venezuela. In: B. Durand and F. Behar (Editors), Advances in Organic Geochemistry 1989. Org. Geochem., 16: 189–196.
Tissot, B.P. and Welte, D.H., 1984. Petroleum Formation and Occurrence. Springer-Verlag, Berlin, 699 pp.
Tissot, B.P., Pelet, R. and Ungerer, Ph., 1987. Thermal history of sedimentary basins, maturation indices, and kinetics of oil and gas generation. Am. Assoc. Pet. Geol., Bull., 71: 1445–1466.
Ungerer, P., Bessis, F., Chénet, P.Y., Durand, B., Nogaret, E., Chiarelli, A., Oudin, J.L. and Perrin, J.F., 1984. Geological and geochemical models in oil exploration; principles and practical examples. Am. Assoc. Pet. Geol., Mem., 34: 53–77.
Ungerer, P., Behar, F., Villalba, M. and Heum, O.D. 1987. Kinetic modelling of oil cracking. Advances in Organic Geochemistry 1987. Org. Geochem., 13: 857–868.
Wood, D.A., 1988. Relationships between thermal maturity indices calculated using Arrhenius equation and Lopatin method: implications for petroleum exploration. Am. Assoc. Pet. Geol., Bull., 72: 115–134.

U. RITTER *IKU Petroleum Research, N-7034 Trondheim, Norway*
M.B. MYHR *IKU Petroleum, Research, N-7034 Trondheim, Norway*
K. AARESKJOLD *IKU Petroleum Research, N-7034 Trondheim, Norway*
H.H. GANZ *Koninklijke/Shell Exploratie en Produktie Laboratorium, Rijswijk, The Netherlands; and Institut für Geologie, Technische Universität Berlin, Berlin, Germany*
A. MOSTAFA *Institut für Geologie, Technische Universität Berlin, Berlin, Germany*
L. SCHOU *IKU Petroleum Research, N-7034 Trondheim, Norway*

Anisotropy of thermal conductivity in clayey formations

L. Demongodin, G. Vasseur and F. Brigaud

Laboratory measurements indicate that the bulk thermal conductivity of sedimentary rocks is controlled by their porosity, their mineralogical content and their temperature. These controlling factors may be accounted for in simple models in order to predict the *in situ* thermal conductivity in sedimentary basins. However, these estimates do not take into account a highly probable anisotropy linked to various structural effects such as preferential crystal orientation (for a given mineral) or small-scale heterogeneities (within a given lithology).

Recent studies, performed on a few wells in the Paris Basin, using well logs and high-resolution temperature logs together, show that the vertical thermal conductivity of shaly formations is significantly lower than predicted by an isotropic model. The magnitude of this phenomenon increases with depth. Using a depth-dependent corrective factor, an effective vertical conductivity can then be computed as a function of depth.

The above feature is interpreted as resulting from anisotropic effects occurring in clays during their burial history: compaction and dewatering, mineralogical transformation (i.e smectite to illite) and structural reorganization. Further arguments are provided by a series of conductivity measurements made on various clayey samples maintained under *in situ* conditions.

Introduction

Because it directly controls the local temperature gradient, thermal conductivity is a key parameter in any thermal study. However, laboratory measurements of thermal conductivity are very scarce and, in most cases, thermal conductivity has to be deduced from well log data which are extensively available in oil exploration boreholes.

Procedures for estimating thermal conductivity from well log data have already been developed. They are generally based on empirical relationships (Anand et al., 1973; Houbolt and Wells, 1980; Vacquier et al., 1988). A different approach, already proposed by Merkel et al. (1976) has been developed by Brigaud et al. (1990). It is based on well log interpretation in terms of mineralogy and porosity; the *in situ* bulk thermal conductivity is then estimated using laboratory calibrated formulae taking into account the mineralogical composition, the porosity and the effect of temperature.

A major drawback of Brigaud et al.'s estimation method is that the predicted conductivity is isotropic and does not take into account the various structural effects. This is particularly serious for clayey formations where structural effects may be important and for which very few significant laboratory measurements are in fact available.

Another approach to the problem of conductivity estimation consists of the use of equilibrium temperature profiles in deep well logs. If vertical conduction dominates, these temperature measurements can be directly interpreted in terms of *in situ* thermal conductivity. From several studies in deep wells (Blackwell and Steele, 1988; Demongodin et al., 1991), it appears that, in clayey formations, the actual *in situ* conductivity value may be significantly smaller than the one predicted by isotropic models.

The present paper aims to critically discuss the estimation of *in situ* thermal conductivity in sedimentary basins with emphasis on the problem of the actual *in situ* conductivity of clayey formations. First the general basis of the method used to predict thermal conductivity as a function of mineralogical composition, porosity and temperature is briefly discussed, based on various published works and in particular on laboratory measurements on isotropic cores by Brigaud and Vasseur (1989) and on cuttings by Brigaud et al. (1990). Then the method for the estimation of *in situ* conductivity from well log data is presented. When applied to a few wells of the Paris Basin (Demongodin et al., 1991), the resulting conductivity estimate can be compared to a detailed profile of the temperature gradient. This comparison permits a further refinement of the conductivity estimation and leads to specific assumptions on the anisotropy of clayey rocks. Finally, provisional results of directional conductivity measurements in "preserved" shaly samples are presented to confirm the inferred anisotropy.

Basin Modelling: Advances and Applications edited by A.G. Doré et al. NPF Special Publication 3, pp. 209–217, Elsevier, Amsterdam.

Thermal conductivity of sedimentary rocks and laboratory measurements

Sedimentary rocks are composed of a mixture of grains of various minerals and of fluid-filled voids. The thermal conductivity of these rocks as a function of mineralogical composition and porosity has been already studied extensively in laboratory (Woodside and Messmer, 1961; Horai, 1971; Brigaud and Vasseur, 1989). A few studies on the influence of temperature are also available (Chapman et al., 1984; Hasan, 1978; Sekiguchi, 1984).

The effect of porosity and mineralogical composition

Experimental studies on composite porous rocks saturated with water have shown that their thermal conductivity can be well accounted for by various composition formulae based on sophisticated analyses (derivation of physical bounds by Hashin and Shtrickman, 1962; use of the effective medium theory by Budiansky, 1970). However, the geometric mean model, where the bulk conductivity is defined by:

$$\lambda = \prod_{i=1}^{m} \lambda_{ci}^{x_i} \tag{1}$$

(λ_{ci} being the conductivity of the individual components, mineral and fluid, and x_i their volumetric fraction) appears to accurately predict the experimental data (Woodside and Messmer, 1961; Horai, 1971)

In order to apply this formula, it is necessary to know the actual values of λ_{ci} (i = 1, ..., m), i.e. the conductivities of the various end terms. Whereas the conductivities are quite well known for non-clay minerals (Horai, 1971), those of clay minerals are poorly defined. Using a set of 143 samples of sedimentary rocks with simultaneous measurements of bulk conductivity, mineralogical composition and porosity, Brigaud and Vasseur (1989) proposed a set of conductivity values for major sedimentary minerals.

These values, valid for room temperature, are given in Table 1. Figure 1 shows a comparison between the experimental conductivity value of the 143 samples used and the values predicted by eq. 1 using the λ_{ci} given in Table 1; the prediction gives an accuracy of about 10%. In this study the samples were isotropic; therefore no structural effect is taken into account in this formula.

Lithology vs. mineralogy

In practical situations the available data set does not give the mineralogy but the lithology. A single lithology may be composed of several minerals with different conductivities (e.g. carbonate = calcite + dolomite + siderite +..., or sandstone = quartz + feldspar +...). It is therefore desirable to calibrate the conductivity of each end term either by mineralogical analyses (i.e. a precise composition of each lithology) or by conductivity measurements on the lithology.

This is illustrated in Fig. 2 from Brigaud et al. (1990) where experimental matrix conductivities of 565 cutting samples of three wells from the Uinta Basin are plotted versus the predicted matrix conductivities (the porosity effects are removed through the use of the geometric model). For the three wells, the samples were described by their lithologies; however, for the first one, mineralogical analyses were also available. In all cases, the prediction agrees to within ±20% with the experimental data; but in the first case, the estimate is significantly improved by the knowledge of the mineralogy and the r.m.s. deviation is only ±10%.

The effect of temperature

Sedimentary rocks may commonly experience a temperature up to 150 or 200°C due to burial. This

TABLE 1

Conductivity of principal rock forming minerals

Mineral	λ ($W m^{-1} K^{-1}$)
Quartz	7.7
Feldspar	2.3
Calcite	3.3
Dolomite	5.3
Anydrite	6.3
Chlorite	3.3
Kaolinite	2.6
Smectite	1.9
Illite	1.8
I/S	1.8

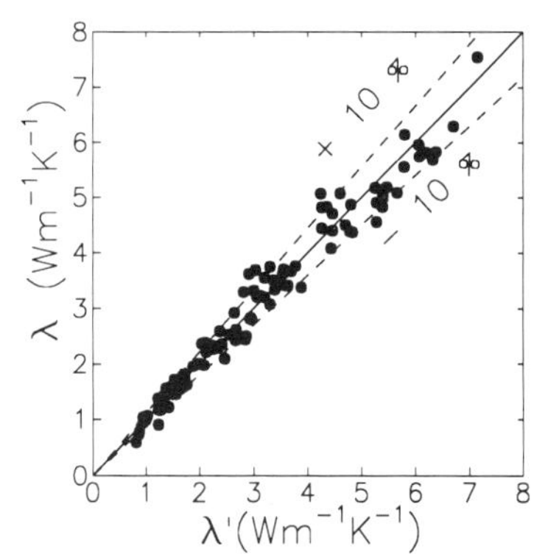

Fig. 1. Comparison of the laboratory-measured conductivities on isotropic rock samples with predicted ones using the geometric model (after Brigaud and Vasseur, 1989).

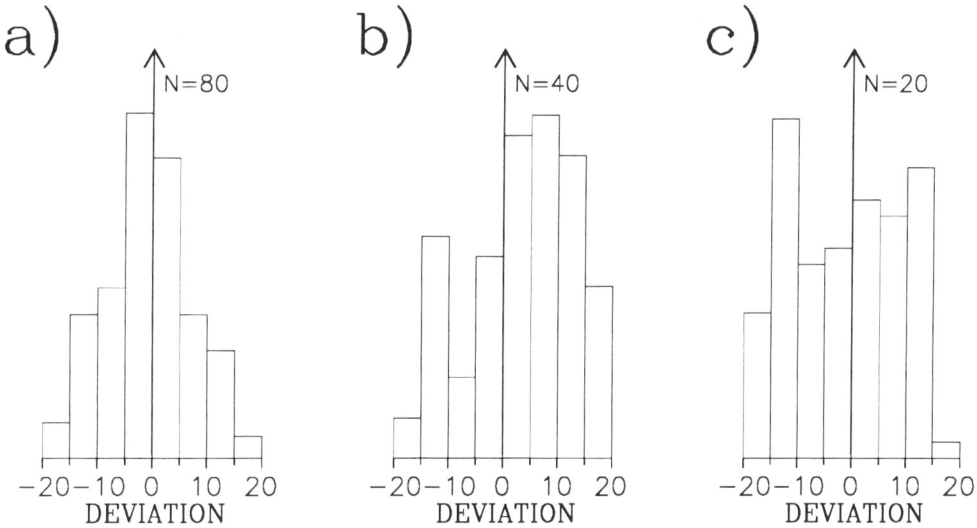

Fig. 2. Comparison of measured matrix conductivities on cuttings with the estimates using the lithological description of cuttings (after Brigaud et al., 1990). The data correspond to three wells of the Uinta Basin. The histograms show the relative deviations $(100 \cdot \lambda_{\text{measur.}} - \lambda_{\text{predict.}})/\lambda_{\text{measur.}}$). N indicates the number of data points used. In the first case (a) the mineralogical composition of various lithological end terms was measured by X-ray diffraction analysis. For the two latter wells (b and c) the lithology was estimated from visual description of the samples.

variation from room temperature induces changes in the conductivity of each component of the rock: the conductivity of minerals generally decreases with increasing temperature whereas that of water increases. However it is reasonable to assume that eq. 1 still holds at temperature T; $\lambda_{ci}(T)$ are then the conductivities of each component at temperature T.

The change in the thermal conductivity of water as a function of temperature is known (Kappelmeyer and Haenel, 1974). Concerning the variation of mineral conductivity with temperature, few experimental data are available. It is clear, however, that this conductivity decreases with temperature. Chapman et al. (1984) and Brigaud et al. (1990) assumed a law involving $1/T$ for all minerals, T being the absolute temperature. From a compilation of available data, Sekiguchi (1984) derives an empirical law which slightly differs for each mineral: the relative temperature effect would be higher for minerals with high conductivity than for those with a low conductivity. The resulting variation is illustrated for major minerals in Fig. 3 and this result is used thereafter.

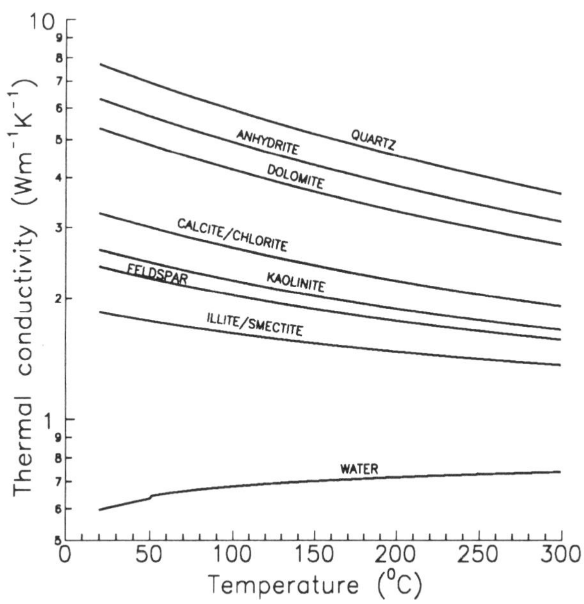

Fig. 3. Variation of conductivity with temperature for the main minerals.

Estimation of *in situ* thermal conductivity from well logs

The method of *in situ* thermal conductivity estimation makes use of the experimental data given previously. The porosity and the volumetric fraction of minerals (i.e. the quantities x_i) can be in principle deduced from well logs; then the use of Table 1 together with the temperature correction (Fig. 3) and eq. 1 provide an estimate of the *in situ* bulk thermal conductivity. Various applications of such estimation have been published (Brigaud et al., 1990; Demongodin et al., 1991). The example shown below is from the Paris Basin where equilibrium temperature logging was available (Demongodin et al., 1991).

Thermal conductivity estimate in a well-log of the Paris Basin

The Paris Basin is an intracratonic basin with sedimentary formations ranging from Triassic to Tertiary where many oil exploration and low-energy geother-

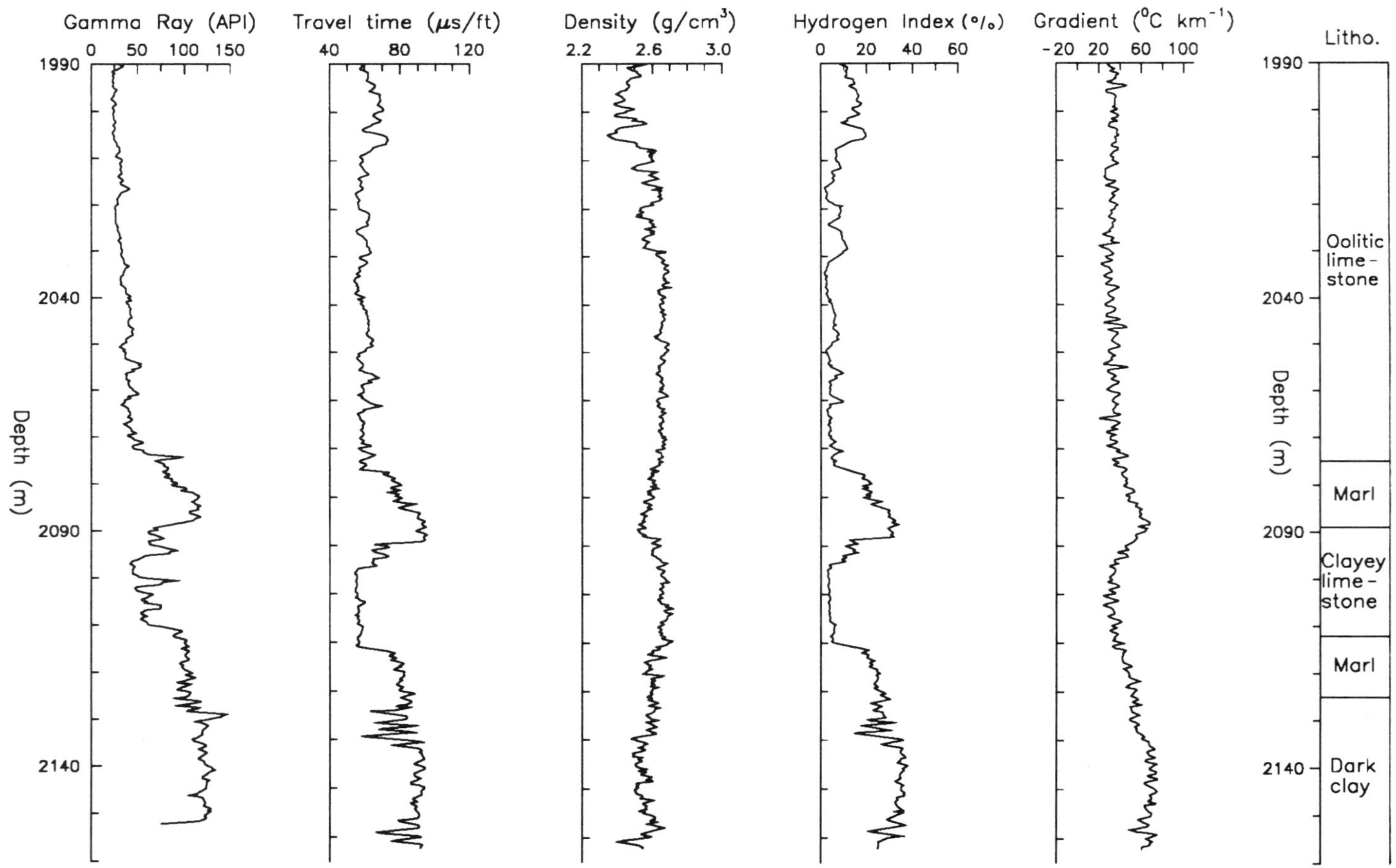

Fig. 4. Well log data available in a well of the Paris Basin with simultaneous temperature logging.

mal wells have been drilled. Re-entry well logging with high-sensitivity temperature sensors has been carried out in some of these wells. A typical example of the available information is illustrated in Fig. 4 for one of these wells (VUS1); it consists of natural gamma-ray, sonic travel time, formation density, neutron porosity and temperature at equilibrium (i.e. a long time after drilling). Moreover, thermal conductivity and mineralogical composition have been measured on a few cutting samples and cores originating from this well and from others nearby .

Using classical well log interpretation, it was possible to estimate, as a function of depth, the mineralogical composition and the porosity (Serra, 1985) and therefore to derive the *in situ* thermal conductivity. The corresponding variations as a function of depth are plotted in Fig. 5. The resulting estimate of thermal conductivity can be compared to the lithological log and to the experimental temperature gradient. It is clear that relatively low thermal conductivities are obtained in places where the lithology is dominated by clays. Such is the case in the two zones around 2080 m and 2120–2150 m depth. These two zones are characterized by relatively high-temperature gradients as expected when the conductivity is low. The other parts of the section are dominated by limestones: their conductivity is relatively high and the thermal gradient is relatively small.

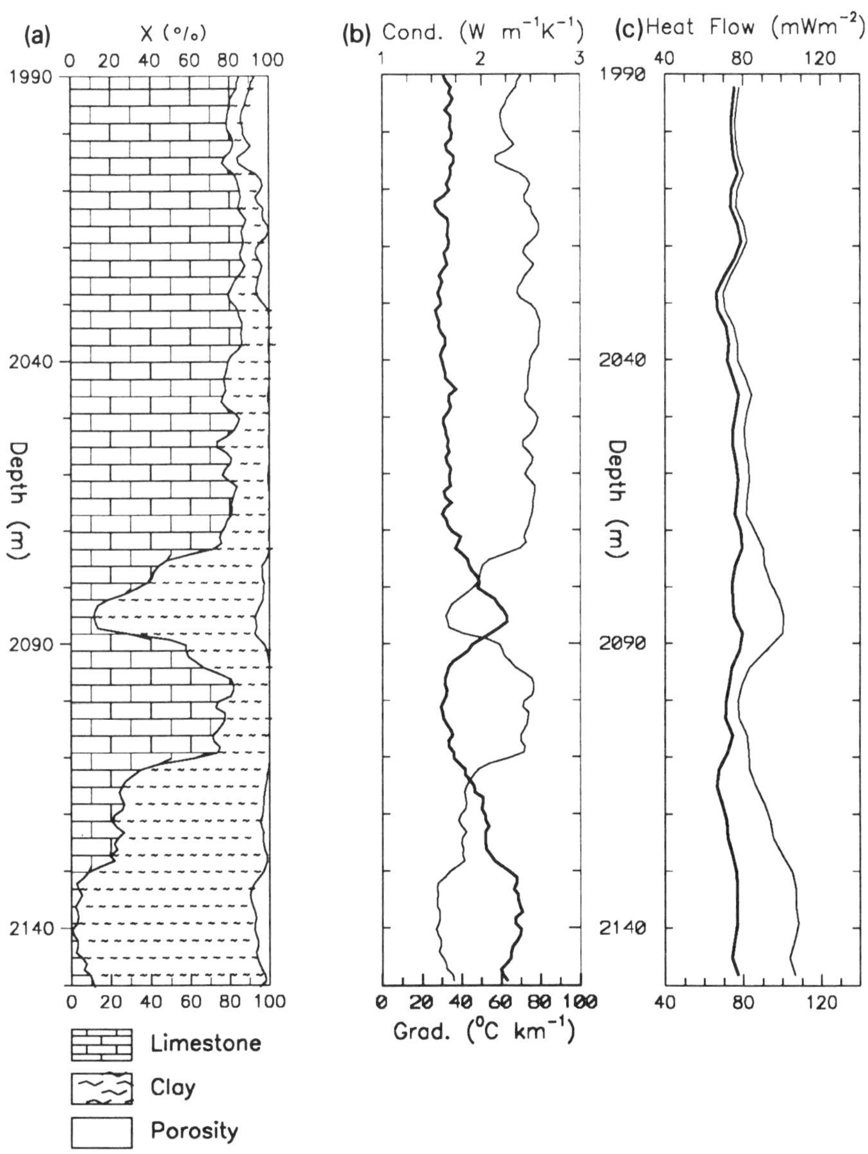

Fig. 5. Estimated conductivity and experimental thermal gradient in a well of the Paris Basin (Demongodin et al., 1991): (a) profile of composition (mineralogy + porosity); (b) comparison of the estimated *in situ* conductivity (thin line) and of the experimental thermal gradient (thick line); (c) comparison between the heat flow obtained from eq. 2 using the uncorrected (thin line) and the corrected (thick line) value of clay conductivity.

If our thermal conductivity estimate is correct, and assuming a pure conductive heat transfer throughout the section, then the apparent heat flow Φ should be constant as a function of depth.

$$\Phi = \lambda \frac{\partial T}{\partial z} \tag{2}$$

In fact, the computed heat flow profile shown in Fig. 5 presents some deviations; however, these deviations are clearly correlated with the abundance of clay. This observation indicates that the conductivity model is not completely satisfactory.

Refining the thermal conductivity model

It is likely that this difficulty is the result of an over-estimation of the thermal conductivity of clayey formations. The arguments for this are provided by laboratory measurements on various limestone samples of the Paris Basin: the value predicted by the model was confirmed for limestones but not for clays.

In order to improve this model, we used the following assumptions (Demongodin et al., 1991): (1) Fourier's law (i.e. eq. 2) holds throughout the section of the well log (for example, that shown in Fig. 5); (2) the bulk *in situ* thermal conductivity is predicted by eq. 1, i.e. the geometric mean model is valid; (3) the conductivity of at least one of the various end terms, λ_{ci}, is well known.

The later statement is essential for the interpretation since the actual value of the terrestrial heat flow is not known with enough accuracy. Then, it is possible to obtain the value of the clay conductivity which explains best the observed temperature gradient, in the least square sense.

The value of λ_{ci} for clay which explains the experimental temperature gradient, is significantly smaller than the one initially assumed: 1.14 W m^{-1} K^{-1} instead of 1.65 W m^{-1} K^{-1}. When using this corrected value, then the calculated heat flow remains much more stable with depth as shown in Fig. 5. Similar observations and interpretations have already been made by Blackwell and Steele (1988) and Jessop and Vigrass (1989).

This suggests the introduction of a corrective multiplicative factor defined as the ratio:

$$\alpha = \lambda_v / \lambda_c \tag{3}$$

where λ_c is the assumed thermal conductivity of clays (Table 1) corrected for temperature effect and λ_v their vertical *in situ* thermal conductivity.

This analysis was applied to three sections from two wells of the Paris Basin and the resulting value of α appears to decrease with increasing depth as

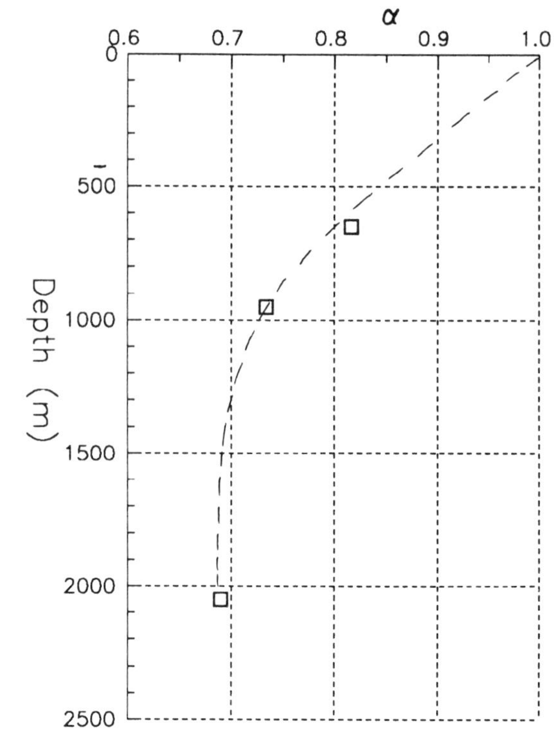

Fig. 6. Corrective factor α to be applied to the conductivity of clays as a function of depth, obtained in the Paris Basin.

illustrated in Fig. 6. This result implies that the thermal conductivity of the clay end term λ_c which contributes to the effective bulk conductivity in eq. 1 is not only a function of temperature but also a function of depth.

This change of clay conductivity as a function of depth can be related to the diagenetic effects, which affect particularly clay minerals and depend on the burial depth, i.e., in the case of the Paris Basin, on the present depth. During their burial history, the clay minerals are submitted to a mechanical rearrangement because of compaction and also to a progressive mineralogical transformation where the superficial smectite is progressively transformed into illite (Velde and Vasseur, 1991). Both transformations result in a progressive reorienting of minerals and therefore in an increasing effect of anisotropy. The coefficient α may be associated with the degree of anisotropy.

Anisotropy of thermal conductivity in clays

Due to their structural organisation most minerals are anisotropic. Their thermal properties are described by a tensor of thermal conductivity defined by its principal conductivities and its eigen-values. For clay minerals, it is reasonable to assume that the normal to the lamellae is a principal direction with an associated conductivity $\lambda_{\perp}$ and that, along the lamellae, the other conductivities are equal to $\lambda_{\parallel}$ (transverse isotropy). This feature is notably observed on minerals of similar structure such as muscovite (Clark, 1966) where $\lambda_{\parallel}=5.10$, $\lambda_{\perp}= 0.84$. It seems likely that for clay minerals $\lambda_{\parallel} > \lambda_{\perp}$.

Along a given direction z making an angle ϕ with the normal to the lamellar plane, the effective conductivity is given by (Carslaw and Jaeger, 1959):

$$\lambda_z = \lambda_{\parallel} \sin^2 \phi + \lambda_{\perp} \cos^2 \phi \tag{4}$$

The bulk thermal conductivity in any direction of a matrix composed of minerals with different orientations is an average of the effective conductivities, along this direction, of the various minerals.

When particles are randomly oriented in the matrix, the bulk conductivity has the same value in any direction and can be approached as the following average:

$$\lambda_{\text{bulk}} = \lambda_{\parallel} \langle \sin^2 \phi \rangle + \lambda_{\perp} \langle \cos^2 \phi \rangle \tag{5}$$

where $\langle\ \rangle$ is the average over all directions. Therefore:

$$\lambda_z = \tfrac{2}{3}\lambda_{\parallel} + \tfrac{1}{3}\lambda_{\perp} \tag{6}$$

When minerals present a preferential orientation, the matrix is anisotropic and the bulk conductivity along a given direction depends on the spatial distribution of the various particles relative to this direction.

Generally, during their deposition, sedimentary minerals have no particular orientation or only a weak one and the matrix may be considered as isotropic. When the particles have an isometric shape, compaction has no effect on their initial orientation and their bulk thermal conductivity does not become anisotropic with burial depth. This has been experimentally evidenced by laboratory measurements on quartz-rich samples (Birch and Clark, 1940; Horai, 1971; Brigaud and Vasseur, 1989). By contrast, clay minerals have irregular shapes and are subject to a progressive mechanical reorientation in the horizontal plane during compactional and diagenetic processes. As a consequence, the bulk conductivity along a vertical direction is progressively controlled by the lower conductivity component $\lambda_{\perp}$ of clay minerals. In the final stage where all the particles would be oriented in the horizontal plane with no other phase, the vertical bulk conductivity would correspond to $\lambda_{\perp}$ and the horizontal bulk conductivity to $\lambda_{\parallel}$. According to this model, the bulk conductivity along a vertical direction of a matrix composed of clay minerals would present an anisotropy that increases with depth.

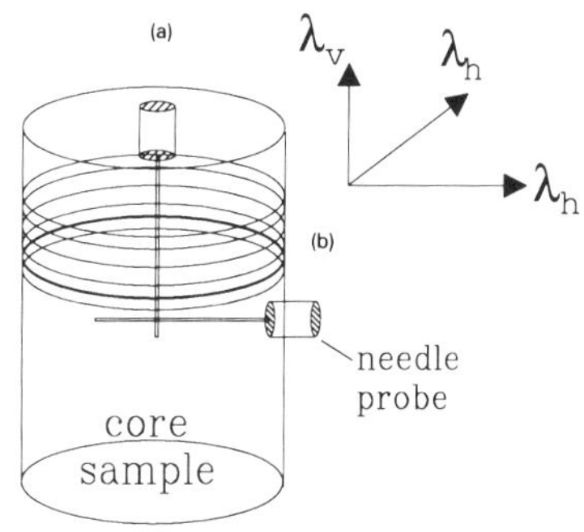

Fig. 7. Thermal conductivity measurements on a core sampled using the needle-probe technique along two perpendicular directions. The bedding is assumed to be perpendicular to the core axis. In position (a), the measure gives the value of λ_h and $\lambda_a = \sqrt{\lambda_h \cdot \lambda_v}$ in position (b).

The principal conductivities of clay minerals ($\lambda_{\perp}$, $\lambda_{\parallel}$) are unknown but the bulk thermal conductivity is known from laboratory measurements on isotropic samples and artificially recompacted pulverized specimens (Brigaud and Vasseur, 1989). The values of mean thermal conductivity of the main rock-forming minerals are reported in Table 1. For anisotropic samples, it is possible to determine a coefficient of bulk thermal conductivity anisotropy, Γ, defined by the ratio between the value of bulk conductivity along the horizontal direction λ_h and the vertical one λ_v, assuming horizontal bedding.

Laboratory measurements have been performed on four clay-rich samples maintained under *in situ* conditions (pressure and fluid saturation). Thermal conductivities have been measured using the needle-probe technique along two perpendicular directions as shown in Fig. 7. In case (a), the measure gives the value of λ_h and in case (b), it corresponds to an average value defined by (Grubbe et al., 1983):

$$\lambda_a = \sqrt{\lambda_v \cdot \lambda_h} \tag{7}$$

The results, reported in Table 2, show a clear increase of the ratio Γ as a function of depth which seems to confirm the model.

Conclusion

Thermal conductivity of sedimentary rocks depends on (1) composition including their mineralogy and their porosity, (2) their temperature, and (3) their structure. The first two effects can be inferred from laboratory measurements whereas the last effect is much more difficult to quantify.

The geometric mean model appears to predict satisfactorily the experimental bulk conductivity of isotropic samples and accounts for the effect of composition. Other models may also be used. However,

TABLE 2

Anisotropy of thermal conductivity (Γ) obtained on 4 core samples

Sample	Depth (m)	λ_a ($W\,m^{-1}\,K^{-1}$)	λ_h ($W\,m^{-1}\,K^{-1}$)	λ_v ($W\,m^{-1}\,K^{-1}$)	$\Gamma = \lambda_h/\lambda_v$
ZDW1	551.6	1.68	1.74	1.62	1.07
ZDW2	553.3	1.51	1.53	1.49	1.02
ZDW3	1204.5	1.13	1.37	0.93	1.47
ZDW4	1505.0	1.54	2.02	1.17	1.73

λ_a and λ_h are the conductivities measured using the needle probe technique; λ_v is deduced from λ_a and λ_h using eq. 7.

the difference in bulk conductivity is generally small; the value of the geometric mean is its simplicity and, in particular its linearity in logarithmic form.

As proposed by Blackwell and Steele (1988), the *in situ* temperature gradient measurements offer a fundamental complementary observation. The simultaneous use of such data and of experimental laws based on laboratory measurements provide a unique way of validating bulk conductivity estimates and of inferring structural effects. The results obtained in the Paris Basin indicate that the vertical matrix conductivity of clays decreases with increasing burial depth. A correcting factor accounting for this effect has been proposed. It is interpreted as an anisotropic effect resulting from structural modifications which affect the arrangement of minerals during compaction and diagenesis. Similar interpretations have already been suggested (Blackwell and Steele, 1988; Jessop and Vigrass, 1989). The data presented here concern the particular geological context of an intracratonic basin with very low subsidence rate (0.01 mm/y). Their applicability to other geological contexts is still to be verified.

This behaviour may be justified as follows. The thermal conductivity of most minerals is anisotropic. However, when the minerals are randomly oriented in a matrix, this anisotropy has no effect on the bulk conductivity. Due to their irregular shape, during the compaction process, clay minerals are subjected to a progressive mechanical rearrangement which introduces a structural anisotropy within the matrix. This preferential orientation is such that the bulk conductivity along a vertical direction is progressively controlled by the lower crystalline conductivity. This behaviour is confirmed by laboratory measurements on a few clay-rich core samples maintained under *in situ* conditions.

Acknowledgements

This study has been supported by the Institut Français du Pétrole, the Bureau de Recherches Géologiques et Minières and the Elf Aquitaine company.

References

Anand, J., Sommerton, W.H. and Gomaa, E., 1973. Predicting thermal conductivities of formations from other known properties. Soc. Pet. Eng. J., 13: 267–273.

Birch, E. and Clark, H., 1940. The thermal conductivity of rocks and its dependence on temperature and composition. Am. J. Sci., 238: 613–635.

Blackwell, D.D. and Steele, J.L., 1988. Thermal conductivity of sedimentary rocks: measurements and significance. In: N.D. Naeser and T.H. McCulloh (Editors), Thermal History of Sedimentary Basins — Methods and Case Histories. Springer-Verlag, Berlin, pp. 13–36.

Brigaud, F. and Vasseur, G., 1989. Mineralogy, porosity and fluid control on thermal conductivity of sedimentary rocks. Geophys. J., 98: 525–542.

Brigaud, F., Chapman, D.S. and LeDouaran S., 1990. Thermal conductivity in sedimentary basins predicted from oil exploration well data. Am. Assoc. Pet. Geol., Bull., 74: 1459–1477.

Budiansky, B., 1970. Thermal and thermoelastic properties of isotropic composites. J. Composite Mater., 4: 286–295.

Chapman, D.S., Keho, T., Bauer, M. and Picard, M.D., 1984. Heat flow in the Uinta basin determined from bottom hole temperature (BHT). Geophysics, 49: 453–446.

Clark, S.P., Jr., 1966. Handbook of Physical Constants. Geol. Soc. Am., Mem., 97 pp.

Carslaw, H.S. and Jaeger, J.C., 1959, Conduction of Heat in Solids. Oxford University Press, New York, N.Y., 2nd ed., 510 pp.

Demongodin, L., Pinoteau, B., Vasseur, G. and Gable, R., 1991. Thermal conductivity and well logs: a case study in the Paris Basin. Geophys. J. Int., 105: 675–691.

Grubbe, K., Haenel, R. and Zoth, G., 1983. Determination of the vertical component of thermal conductivity by lines source methods. Zentralbl. Geol. Paläontol., I (1/2): 49–56.

Hasan, S.E., 1978. Thermophysical properties of rocks. Symp. Rock Mechanics, 1: 210–214.

Hashin, Z. and Shtrickman, S., 1962. A variational approach to the theory of the effective permeability of multiphase materials. J. Appl. Phys., 33: 3125–3131.

Horai, K., 1971. Thermal conductivity of rock forming minerals. J. Geophys. Res., 76: 1278–1308.

Houbolt, J.J. and Wells, P.R.A., 1980. Estimation of heat flow in oil wells based in a relation between heat conductivity and sound velocity. Geol. Mijnbouw, 59: 215–224.

Jessop, A.M. and Vigrass, L.W., 1989. Geothermal measurements in a deep well at Regina, Saskatchewan. J. Volcanol. Geotherm. Res., 37: 151–166.

Kappelmeyer, O. and Haenel, R., 1974. Geothermics with Special Reference to Application. Geoexploration Monograph Ser. 1, 4. Borntraeger, Berlin–Stuttgart, 238 pp.

Merkel, R.H., Maccary, L.M. and Chico, R.S., 1976. Computer

technique applied to formation evaluation. Log Anal., 17: 3–10.

Sekiguchi, K., 1984. A method for determining terrestrial heat flow in oil basinal areas. Tectonophysics, 103: 67–79.

Serra, O., 1985. Diagraphies différées, Bases de l'interprétation, 2. Interprétations des données diagraphiques. Bull. Cent. Rech. Explor. Prod. Mem. Elf Aquitaine, 7, 631 pp.

Vacquier, V., Mathieu, Y., Legendre, E. and Blondin, E., 1988. An experiment on estimating the thermal conductivity of sedimentary rocks from well logging. Am. Assoc. Pet. Geol., Bull., 72: 758–764.

Velde, B. and Vasseur, G., 1991. A kinetic interpretation of the smectite to illite transformation. Am. Mineral. (submitted).

Woodside, W. and Messmer, J.H., 1961. Thermal conductivity of various rocks, I. Unconsolidated sands; II. Consolidated rocks. J. Appl. Phys., 32: 1688–1706.

L. DEMONGODIN *Centre Géologique et Géophysique (UPR 361, CNRS), USTL, 34095 Montpellier Cedex 5, France*

G. VASSEUR *Centre Géologique et Géophysique (UPR 361, CNRS), USTL, 34095 Montpellier Cedex 5, France*

F. BRIGAUD *Société Nationale Elf Aquitaine (Production), C.S.T.J.F., Avenue Larribau, 64018 Pau Cedex, France*

Modelling of petroleum generation and expulsion — an update to recent reviews

P. Ungerer

The purpose of this paper is to complement previous review articles (particularly Ungerer, 1990) about kinetic models of petroleum generation and expulsion.

Although widely accepted now, the classical model involving parallel first-order reactions obeying the Arrhenius law must be considered very empirical, especially when calibrated from open system pyrolysis experiments. Such calibrations yield kerogen-specific parameters, but great care must be taken in such procedures, since the extrapolation of kinetic models to temperatures 200°C lower than pyrolysis is justified empirically, but not theoretically.

Recent research works have improved the ability of geochemical models to quantify the composition of generated products by a combined use of more detailed kinetic schemes (e.g. primary generation of gas from kerogen, several successive secondary reactions, four hydrocarbon fractions or more). Determination of kinetic parameters from closed system pyrolysis is a promising method to improve their accuracy, but is more time-consuming.

Kinetic modelling of maturity parameters such as vitrinite reflectance, sterane isomerization, etc., has also improved in recent years and can be used to constrain basin models by a greater amount of geochemical data. However, the application of vitrinite reflectance models is hampered by the chemical variability of vitrinite.

Modelling of petroleum expulsion together with hydrocarbon generation and secondary reactions is necessary to quantify the amount and composition of petroleum available for secondary migration. In existing zero-dimensional models, the amounts of residual hydrocarbons in the mature source rocks (S1, solvent extract) may be used to calibrate the specific parameters describing expulsion (saturation threshold, relative permeabilities). Although empirical, such models appear to reproduce the composition of accumulations in some documented examples. Noticeable improvements are expected as a result of better consideration of physical and physico-chemical processes (viscosity variations in fluid flow equations, rock mechanics, retention of hydrocarbons on mineral/organic surfaces, etc.). However, there is still a long and difficult way to go before a comprehensive satisfactory model is developed.

Introduction

Basin modelling is now a common exploration tool, and there is no question that it will develop intensively in the forthcoming years. This was not evident ten years ago, when modelling was mostly performed by research teams. The question is no more "Are basin models useful?" but rather "Could basin models be more efficient?". The kinetics of oil formation and primary migration, which constitute an important part of basin models, deserve special attention.

These geochemical aspects have been covered by several recent review papers such as Tissot et al. (1987) and Ungerer (1990) on the kinetics of petroleum generation and expulsion, Bethke (1989) and Ungerer et al. (1990) on basin-scale modelling of fluid flow, England et al. (1987) and Durand (1988) on the physics of petroleum migration. Most of the conclusions of these articles still seem valid today, and they will be summarized here very briefly. This article will mostly be devoted to a prospective view of the most promising lines of research and industrial applications. The following points will provide the framework for the discussion:

Why do we need kinetic models?

What kinetic models (i.e. what kinetic schemes) are the best?

How can we obtain the kinetic parameters of these models, and how can we reduce their uncertainty?

What about vitrinite reflectance modelling?

What models should be used to describe petroleum expulsion from source rocks?

How can we validate basin models by geochemical data?

Do we expect prediction of chemical composition or phase state from basin models?

Why do we need kinetic models?

Why do we need kinetic models, i.e. mathematical models characterizing the influence of temperature

Basin Modelling: Advances and Applications edited by A.G. Doré et al. NPF Special Publication 3, pp. 219–232, Elsevier, Amsterdam.

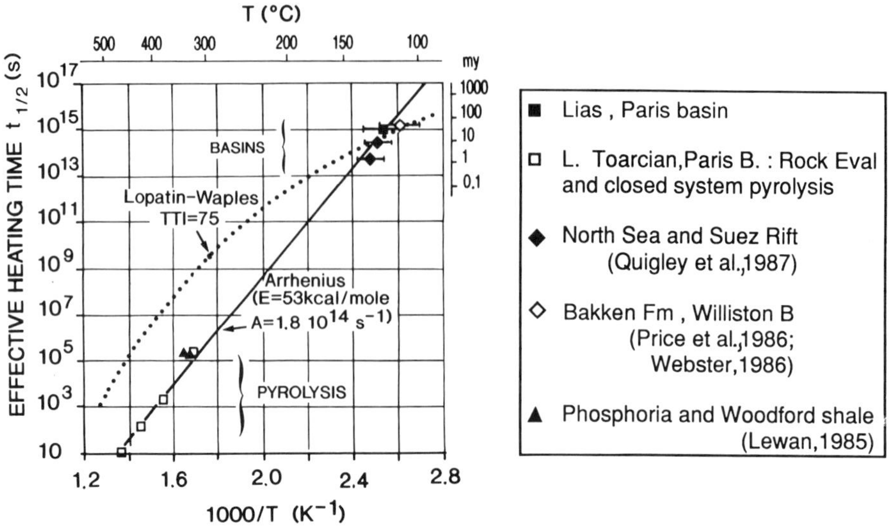

Fig. 1. Arrhenius plot of peak oil generation by type II source rocks in sedimentary basins and in laboratory pyrolysis. For a first-order reaction obeying the Arrhenius law, a straight line with slope E/R represents the time/temperature conditions yielding the same transformation ratio. The heating time plotted here is the time spent at less than 10°C from the maximum temperature reached, because the influence of lower temperatures is comparably minor. Maximum temperatures and heating times reported in this graph represent peak oil generation, assumed to correspond to 50% of initial potential. Error bars around each point indicate the estimated uncertainty on thermal history. Similar plots may be drawn for type III and type I with comparable, yet slightly different, results.

and geologic time? This question is in fact not so easy to answer. As a general feature, kinetically controlled reactions are characterized by a compensation between time and temperature, i.e. a longer heating time compensates for lower temperature to yield the same amount of reaction products. The time/temperature compensation is well illustrated by the comparison of laboratory simulations and natural maturation (Fig. 1) but it is weakly supported by the simple consideration of natural maturation. Indeed, the uncertainty of palaeotemperature reconstruction in natural series is such that the time/temperature compensation may be evidenced only from a qualitative point of view, not from a quantitative one. An indication for this is the low activation energies derived from the observation of natural maturation series in early models (see Ungerer, 1990, for a definition of the various concepts and equations involved in kinetic models: activation energy, reaction order, Arrhenius law, primary vs. secondary reactions, etc.). When considering the activation energy values around 50–55 kcal/mol necessary to describe simultaneously natural maturation and laboratory pyrolysis (Fig. 1), it appears that most of the values lower than 20 kcal/mol in early models were probably due to uncertainties in the palaeotemperature reconstructions of the reference series.

The Lopatin-Waples method (Waples, 1980), which is based on the hypothesis of doubling reaction rates for every 10°C increase, clearly disagrees with laboratory simulations while it is more or less consistent with average basins (Fig. 1). Although it is still used and may give acceptable estimates of maturity in average conditions (i.e. average rate of burial, no erosion), it is a dangerous method in some cases because it overestimates the influence of geologic time compared with temperature (Tissot et al., 1987). For instance, in the case of erosion, it may predict ongoing maturation after the temperature has dropped 20°C, while maturity would be frozen in more recent Arrhenius-based models.

The common opinion, supported by Fig. 1, is that laboratory simulations are necessary to elaborate kinetic models and to determine kinetic parameters, because temperature, time and initial composition are much better controlled than in natural series. However, there is a difficult choice: either we try to base kinetic models on laboratory simulation *and* natural series, or we base them on laboratory experiments only. The latter solution is a tremendous extrapolation, which has very little theoretical support but seems to work empirically (Ungerer and Pelet, 1987). In fact, such an extrapolation involves important uncertainties (see later sections).

Are there important parameters other than time and temperature that should be considered in modelling petroleum generation? Obviously, the *organic type* of the kerogen has an influence on the maximum amount of hydrocarbons and related compounds that can be generated, as it was already taken into account in the precursor model of Tissot and Espitalié (1975). This makes it necessary to account for organic type in kinetic models, but it is not so clear whether the rates of cracking are significantly different from one organic type to another, as discussed below.

A significant influence of *pressure* on the rates of cracking of saturated hydrocarbons has been documented (Fabuss et al., 1964; Dominé, 1989; Dominé et al., 1990). However, most of the influence of pressure seems to occur at moderate pressures, i.e. below 200 bar. Recent work on mixtures has shown that the rates of cracking did not change drastically between 200 and 2000 bar, but heavier reaction products are observed at high pressures (Kressmann, 1991). High pressures are known to cause the absence of unsaturated hydrocarbons in the cracking products, whereas these compounds are important at low pressure.

Few data are available on the kerogens for which Akihisa (1978) showed a moderate influence between 0 and 7 MPa, and on the coals for which Monthioux (1988) did not see significant effects at higher pressures. As a consequence, pressure influence is generally not considered in basin models, except in some secondary reactions of the Lawrence Livermore Laboratory model (Sweeney et al., 1987). This is certainly a line of research in which a lot of investigations are needed to improve the accuracy of existing models.

Similarly, the influence of *water* is rather controversial. Some authors report significant differences when comparing hydrous pyrolysis results with dry pyrolysis (Eglinton et al., 1988; Weres et al., 1988), while others do not (Jones et al., 1987; Monthioux, 1988). Maybe some of these contradictions are due to the difficulty of controlling pyrolysis conditions well enough to quantify the role of a secondary parameter like water: for instance, temperature inside a high-pressure autoclave is generally not uniform, resulting commonly in ±5°C variations which may cause noticeable artifacts. Indeed, small errors in autoclave temperatures frequently cause significant errors in the apparent activation energies.

Finally, *sulphur*-rich kerogens have been sometimes quoted to undergo more rapid maturation than other kerogens of the same type. Such a role of free sulphur has indeed been observed in the processes of thermal cracking which are used to reduce the viscosity of the heavy ends of crude oils (Thomas et al., 1989). Experimental investigations are necessary to quantify this effect, provided it also applies in the case of chemically bound sulphur. In a comparable way, the presence of *oxygen*-containing functions has been shown to accelerate the cracking of carbon–carbon bonds through free radical reactions on a specific ether compound, namely dibutylether (Enguehard et al., 1990). These effects remain to be checked and quantified on kerogens.

The role of clay minerals is probably not as crucial as it was claimed to be but should merit systematic investigations (see discussion in Ungerer, 1990).

As a general feature for all of these secondary parameters like pressure, water, minerals, etc., their role is difficult to evidence because it is often masked by the uncertainties of experimental temperature control.

What kinetic models (i.e. what kinetic schemes) are the best?

The most common model describing petroleum generation from kerogen involves parallel first-order reactions. Such kinetic schemes consider only primary cracking, i.e. they do not consider the secondary reactions which progressively degrade petroleum into gas as the maturation increases. Several versions have been proposed, either based on a discrete distribution or on a continuous distribution. However, both are approximately equivalent, being generally based on the same data, i.e. open system pyrolysis such as the Rock-Eval method. As suggested by the detailed analysis of normal alkane cracking, the activation energies obtained cannot be directly related to those of bond breaking (Ungerer, 1990; Dominé, 1991; Kressmann, 1991). Closed system pyrolysis is increasingly preferred to the open system to elaborate such kinetic models because: (1) the pressure is more representative, (2) it allows for longer heating time and thus lower temperatures, so that less extrapolation is needed to model natural evolution, and (3) pyrolysis conditions are better controlled, particularly the residence time of pyrolysis products in the reactor (see the following section).

Although the discrete distribution is more flexible than the continuous distributions, it suffers from the necessity of determining a great number of parameters and therefore a lack of reproducibility. This is particularly observed if a low spacing of activation energies (1 kcal/mol) is used. As a small spacing of activation energies is needed to model $T_{\max}$ (see later sections), it is thus likely that future models will increasingly consider continuous distributions. However, these should be more flexible than the Gaussian distribution to allow for dissymmetric distributions, as are needed to model the tailing shape of type III pyrolysis curves.

A different type of kinetic model, inspired from polymer chemistry, has been used by Klomp and Wright (1990). In this model, the reaction rate is not computed from several first-order reactions but considered as an empirical function of conversion. The temperature-dependence of the rate constant is however described by the Arrhenius law. Although

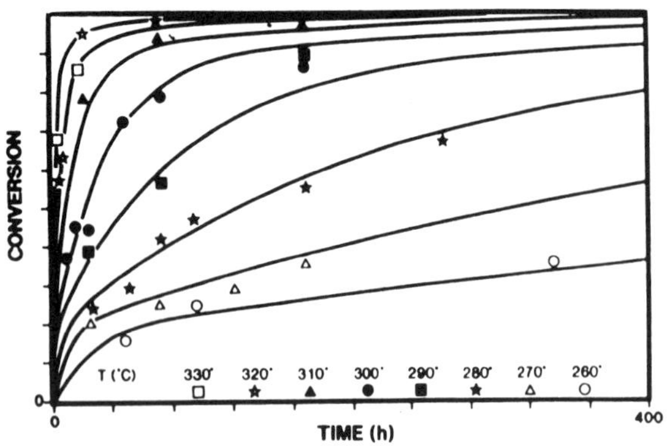

Time-courses of the isothermal degradation of kerogen at different temperatures. The curves correspond to the best fitting of the experimental results.

Example of kinetic model of primary cracking, considering 3 parallel reactions, derived from closed system pyrolysis experiments. $A_i = 10^{16}\ s^{-1}$, standard error estimate 4.3% (Castelli et al., 1990).

Heavy products:	77,5%
Aromatics:	14,4%
Saturates:	3,6%
light products:	4.5%

Relative weight of the single reaction (%)	Activation energy (kcal/mole)
18.4	51.2
69.3	55.4
12.3	57.9

Fig. 2. Closed system pyrolysis data obtained by Castelli et al. (1990) on a type II kerogen and example of kinetic parameter distribution found from these experiments. Conversion is the ratio of observed generated amount to the maximum generated amount of all petroleum fractions.

the model is based on specific temperature programs, the activation energies (46 to 58 kcal/mol) are in reasonable agreement with parallel reaction models based on other open system pyrolysis experiments.

It is also possible to introduce kinetic schemes in which heavy compounds are intermediates in oil and gas generation, as exemplified by Ungerer (1990) and by Braun and Burnham (1989) for type I kerogen.

In fact all these models may be different in their principles but yield similar results, with similar uncertainties, if based on the same data. For this reason, the quantity and quality of reference pyrolysis data is more important than the specific features of kinetic schemes.

The parallel reaction model may be adapted to describe the composition of primary products in terms of chemical fractions (see Ungerer, 1990 for a review). For type III kerogen, higher activation energies are needed to model gas generation (C_1 to C_4) which occurs later than oil generation (C_{5+}). This is not so clear for type II kerogen. For instance, Castelli et al. (1990) did not find any change in the product composition with increasing maturity (Fig. 2).

Secondary cracking is modelled with successive reactions, i.e. kinetic schemes in which the product of a given reaction may be destroyed by another reaction (see Behar et al., 1991 for a review). Now, commercially available models consider either a single secondary reaction as in Tissot and Espitalié (1975) or a greater number (Fig. 3) which allows a better description of the progressive change from oil to dry gas when cracking progresses. However, they have not yet been validated by basin examples (see later section). An alternative method has been used by Düppenbecker and Welte (1991) in which the GOR (gas/oil ratio) is calculated directly as a function of kerogen conversion.

A promising line of research for secondary reactions is to consider fractions by molecular weight and by chemical family (saturates vs. aromatics), as done by Behar et al. (1991). Although the approach is still empirical, it allows modelling of different mixtures with the same set of kinetic parameters, which is not possible when fractions are defined by carbon num-

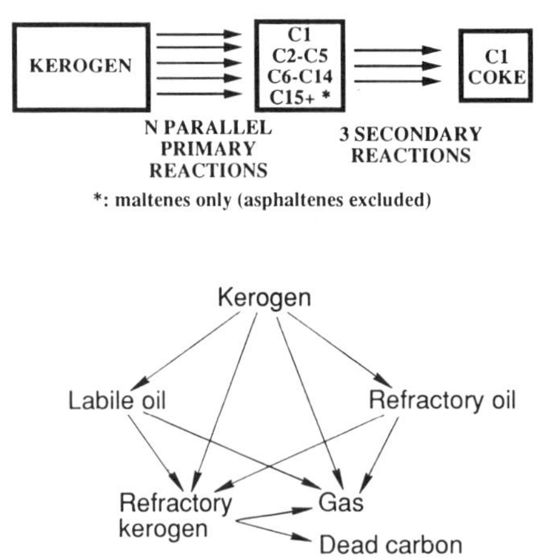

Forbes et al., 1991

Burnham, 1989

Fig. 3. Two examples of simple kinetic schemes. The model used by Forbes et al. (1991), which is now available in commercial 1D basin models, uses four fractions by carbon number, with a different activation energy distribution for each fraction. Three secondary reactions describe the cracking of C_{15+}, C_2–C_5 and C_1. The model of Burnham (1989) considers three petroleum fractions and an intermediate solid fraction, named refractory kerogen. An identical distribution of activation energies is used to model the generation of the various fractions from kerogen (yet with different weights).

ber as in previous models. For instance, the higher coke yield from aromatic than from paraffinic oils may be described by considering fractions by chemical family. However, these models cannot be run on small computers, especially if repeated computations are needed. A standard procedure to elaborate a kerogen-specific simplified model from the detailed model would make for shorter computer time. This would be a significant improvement because there would be no need for specific extensive experiments to derive secondary parameters for a new organic type. The extension of this approach to kerogen is also promising, but more difficult because it requires a good understanding of the relationships between kerogen structure and reactivity. In this respect, network models of kerogen cracking could be useful, as they are for coal pyrolysis (Solomon et al., 1990).

Theoretical models based on free radical mechanisms are useful to understand qualitatively some features such as the low amounts of olefins in natural crude oils, the role of pressure (see a review in Behar et al., 1991) and the role of ether bridges (Enguehard et al., 1990). They will probably not have direct applications in the near future, because they involve heavy computations, still impossible for such complex structures as kerogens. However, they could be used to investigate the validity of extrapolating laboratory-based parameters for model compounds to sedimentary basins conditions.

How can we obtain the kinetic parameters of these models, and how can we reduce their uncertainty?

As discussed in Ungerer (1990), the kinetic parameters of existing models are always obtained empirically by fitting model results with experimental data. From such determinations, it has been found that type II organic matter displays slightly lower activation energies than type I and especially type III organic matter. This results in earlier maturation for type II source rocks, especially if sulphur-rich, than for types I and III. But in fact, the pre-exponential factors[1] found in these determinations are correlated with activation energies (the lower the activation energies, the lower the pre-exponential factor) as seen in Fig. 4. In open system pyrolysis conditions, around 450°C, the models are approximately equivalent for different organic types (i.e. they yield the same reaction rates), but they diverge when extrapolated to basin conditions. Such variations between different samples are probably largely due to the combined uncertainties of the experimental method and the optimization procedure, as confirmed by a theoretical statistical analysis (Nielsen and Dahl, 1991). This is not only valid for open system-based models, but also for closed system-based ones (Ungerer, 1990). This has considerable importance for basin modelling, as these uncertainties may change the computed depth to the oil window by 500 m, if not more (Hermanrud et al., 1990). As a consequence, it is mandatory to validate laboratory-based kinetic parameters by critical applications on well-documented areas, with quantitative control (as discussed in later sections) before applying them without control points.

In order to minimize these uncertainties, it is first necessary to improve the reference data. In the method of Ungerer and Pelet (1987), both low and high heating rate pyrolysis curves must be used. This is easy to understand in Fig. 1, in which the slope is the activation energy divided by the gas constant. A large range of heating rates implies a large range of heating times, which improves the accuracy on the slope of the Arrhenius line. Alternatively, it is possible to use specific temperature programs (Klomp and Wright, 1990) to obtain a good accuracy on kinetic parameters. More generally, a good accuracy on kinetic parameters needs both low-temperature, long heating time experiments and higher-temperature, short heating time experiments. In this respect, open system pyrolysis has specific limitations, because the rates of release are difficult to measure at low temperature. Besides, the volatilization of pyrolysis products plays a role in open system conditions, and the observed rates of release are thus not strictly related to cracking.

Closed system pyrolysis is better controlled and allows maturation to be simulated at a longer time scale and lower temperatures. From such experiments, Castelli et al. (1990) obtained activation energy distributions around 55 kcal/mol (Fig. 2), in reasonable agreement with determinations from open system pyrolysis for the same type (Ungerer and Pelet, 1987).

As mentioned in the preceding section, the systematic differences in the composition of generated products between organic types is rather clear: liquid products generally prevail, gases being however more important from type III than from type II or I kerogens. These features seem better controlled than the differences in activation energies, which may be artifacts as mentioned above. In open system experiments, the continuous monitoring of the composition of pyrolysis products is possible for light hydrocarbons (gases) but approximate for heavier compounds. Closed system experiments allow for a

[1] The pre-exponential factor is the parameter A in the Arrhenius law: $k = A\exp(-E/RT)$, where R is the gas constant (0.001987 kcal/mol), E the activation energy and T is temperature.

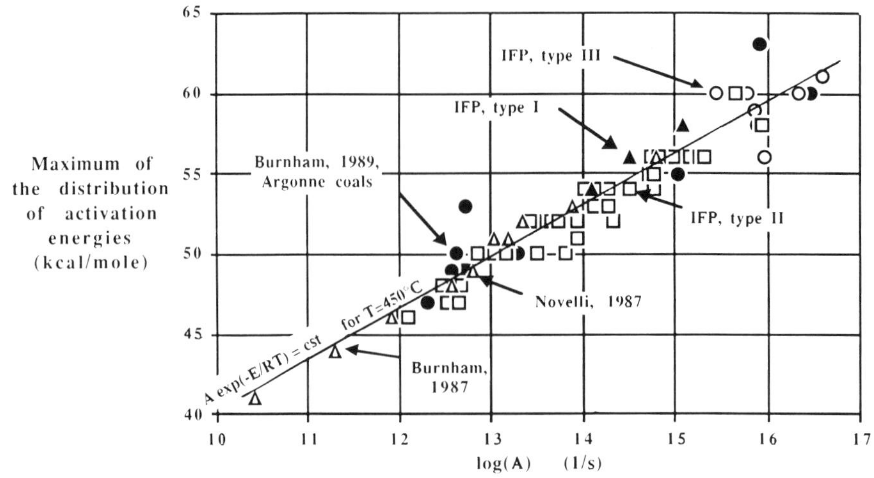

Fig. 4. Compilation of principal activation energy versus pre-exponential factor, obtained from Rock-Eval pyrolysis of various samples. The organization of the various points around the line of equal rate constant at T = 450°C is partly due to the uncertainties of the determination (after Ungerer, 1990).

suitable analysis of maturation products, especially in the C_{15+} fraction, which is necessary for correct modelling of secondary cracking.

For all these reasons, closed system pyrolysis (including hydrous pyrolysis) should be increasingly used to obtain safe kinetic models, at least for a few major organic types, despite the time-consuming character of these methods.

What about vitrinite reflectance modelling?

The method of Lopatin-Waples, still used to model vitrinite reflectance, seems to overestimate the influence of geologic time as discussed above. Now, alternative methods based on type III kinetics are available.

The approach proposed by Tissot and Espitalié (1975) consists of linking vitrinite reflectance with the Transformation Ratio computed from the parallel reaction model. It has since been updated (Fig. 5) to become consistent (1) with laboratory-based distributions of kinetic parameters, and (2) with new data such as those of Quigley et al. (1987). The model of Burnham and Sweeney (1989) is based on a correlation between reflectance and carbon content (Fig. 6), the latter being obtained by a model of H_2O, CO_2, CH_n and CH_4 release. It appears from the correlation used by Burnham and Sweeney (1989) that

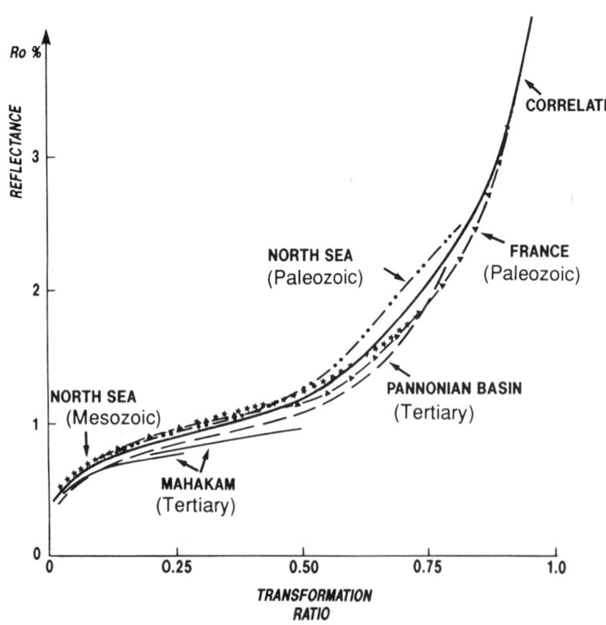

Activation energy (kcal/mol)	Preexponential factor (s^{-1})	Initial potential (mg/g TOC)
46	5.96 10^{14}	2
48	"	3
50	"	4
52	"	5
54	"	9
56	"	28
58	"	19
60	"	12
62	"	8
64	"	6
66	"	5
68	"	4
70	"	3
72	"	3
74	"	2

Fig. 5. Updated correlation between vitrinite reflectance and transformation ratio in IFP's model. The transformation ratio is computed from a parallel reaction model with type III kinetic parameters as shown in the table. Each line represents the average trend for the basin considered. Based on data by Durand et al. (1986), Quigley et al. (1987), among others.

a

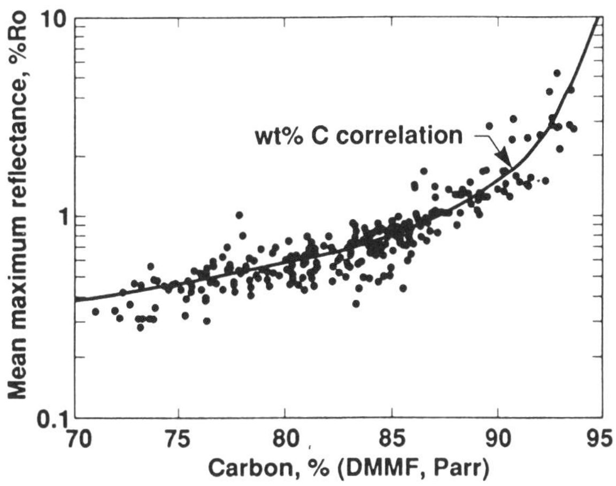

b

Energy kcal/mol	% of species characterized by given E H_2O	CO_2	CH_n	CH_4
38	5			
40	10			
42	15	5		
44	20	15		
46	20	25	5	
48	15	25	10	
50	10	15	20	7
52	5	10	30	12
54		5	20	14
56			10	13
58			5	12
60				11
62				9
64				7
66				5
68				4
70				3
72				2
74				1

A is 2×10^{13} s^{-1} for CH_n and 1×10^{13} s^{-1} for others

Fig. 6. Principles of the model of vitrinite reflectance of Burnham and Sweeney (1989). The carbon content, which is used to determine reflectance from the correlation shown, is modelled by four distributions of parallel reactions, each addressing the release of a specific compound or family.

important discrepancies are sometimes found from this model. The same applies to IFP's model if we consider that the observed trends in Fig. 5 for different basins are only average trends, and dispersion is significant in each basin. Vitrinite reflectance, which should be used in type III series only, cannot be modelled with a great accuracy because of the chemical variability of vitrinites (see the reviews of Durand et al., 1986, and Tissot et al., 1987). The precision is not better than $\pm 0.20\%$ R_0 in the oil window and becomes worse for higher maturity, because anisotropy causes a greater scatter in data.

What models should be used to describe petroleum expulsion from source rocks?

The mechanism of petroleum expulsion from source rocks, also referred to as primary migration, is dominantly a fluid flow in a very low-permeability medium (England et al., 1987; Durand, 1988). Although it is probably minor, diffusion could sometimes be more important than previously considered, namely in the cases when a continuous organic network is present (Stainforth and Reinders, 1990; Thomas and Clouse, 1990).

One way to model expulsion is through basin-scale modelling of compaction and two-phase fluid flow (see reviews in Bethke, 1989, and Ungerer et al., 1990). As a particular feature, Darcy's law is used with relative permeability curves to adapt it to two-phase flow. The advantage of such models is that they provide a consistent reconstruction of the pressure regime, which is important because fluid flow is a function of the pressure gradient in Darcy's law. They describe primary migration, sometimes linked with a pore pressure increase in low-permeability source rocks. Despite this qualitative success, these models are poorly predictive for several reasons: empirical equations for compaction and two-phase fluid flow; poor knowledge of permeability distribution at basin scale; existing models are only two-dimensional (three-dimensional models consider only water flow); secondary reactions are generally neglected or simplified. They are, however, useful for obtaining a synthesis of various geological and geochemical data, and rapid progress is occurring in this field. For instance, a two-dimensional model has been developed recently to include secondary reactions (Faille, 1992). This model also considers composition-related variations of density and viscosity, but it has not yet been applied to real basins.

Alternative models of expulsion use a simplified zero-dimensional treatment, based either on a saturation threshold (Ungerer et al., 1988; see Fig. 7) or on fluid flow equations as in basin-scale models (Braun and Burnham, 1989; Düppenbecker and Welte, 1991). These models account for the density changes of petroleum when its composition changes from oil to gas and for the porosity decrease as a result of compaction (both processes contribute to expulsion) as illustrated by Fig. 8. The kinetic schemes involved in the models of Ungerer et al. (1988) and Burnham and Braun (1990) are more detailed than in most basin-scale models (see Fig. 3 for instance). This allows description of the progressive changes in petroleum composition in source rocks, from oil to wet gas and then dry gas, as maturity increases. The model of Düppenbecker and Welte (1991) stresses the influence of pore size distribution and fracturing.

These simplified zero-dimensional models of expulsion will probably be intensively used in the next

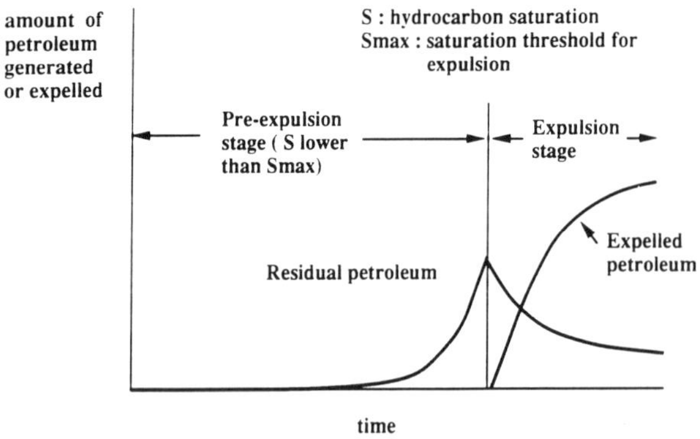

Fig. 7. Principles of the simplified expulsion model used by Forbes et al., 1991. The kinetic scheme includes primary and secondary reactions, similar to Fig. 3. Before the saturation threshold is reached, evolution is the same as in a closed system. Once the expulsion threshold is reached, the content of residual petroleum decreases because of the combined influence of porosity decrease and density decrease. Secondary reactions are applied to the residual petroleum only. No fractionation is considered.

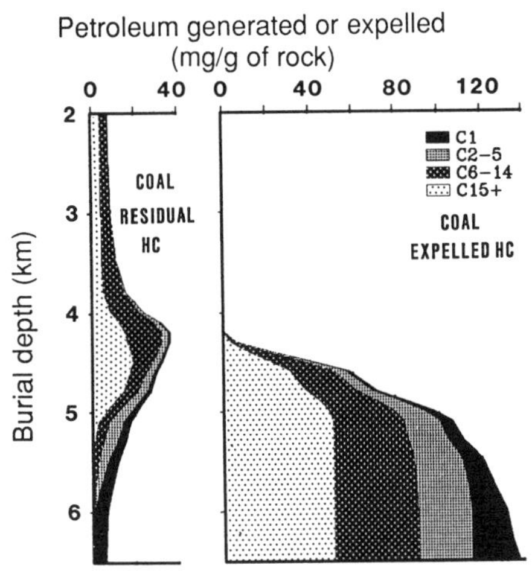

Fig. 8. Results of the simplified model of expulsion of Fig. 7, applied to a typical geological history (constant burial rate 50 m/Ma, 33°C/km geothermal gradient). Petroleum remaining in the coal source rock (right) is progressively cracked to light oil, wet gas, and then dry gas, due to secondary reactions. The amount and composition of expelled petroleum (left) is intermediate between the open system and the closed system (Ungerer et al., 1988).

few years, coupled with one-dimensional models of backstripping and palaeotemperature reconstruction. Indeed, they present several advantages compared with the existing one-dimensional basin models:

(1) Unlike existing models which are based upon the implicit assumption of a closed system, combined models of petroleum generation, destruction and expulsion account for the fact that expelled petroleum generally reaches traps where temperatures are lower and secondary cracking is thus less important. This has important implications when modelling basins or areas where source rocks are overmature at present day. In such zones, existing models overestimate gas generation and underestimate oil generation.

(2) Expulsion models allow the correction of the generated amounts for unexpelled petroleum, and this is important for quantitative studies. Indeed, the observed amount of residual petroleum in source rocks (S1, extract yield) may be used to quantify unexpelled petroleum and then, by subtraction, expelled petroleum (see later sections)

(3) As they compute explicitly the weight of carbon lost in expelled products, expulsion models allow for a safer estimate of the initial organic carbon content of source rocks that are mature at present day.

(4) In cases where no significant changes in composition are likely to occur during secondary migration, the chemical composition of reported accumulations may be used as an independent control of the model. This allows for real "field case studies".

In these models, the physics of expulsion is not known sufficiently well to give predictive values of the saturation threshold or of the relative permeabilities. These must be regarded as empirical parameters, to be calibrated from geochemical data (see later sections). The applicability of the method has been shown by a case study on a Norwegian field (Forbes et al., 1991) as illustrated in Figs. 9 and 13. However, additional case studies are necessary to confirm its general usefulness.

It is possible that some simple concepts in expulsion models will have to be reconsidered. Among these, saturation could be less critical than previously thought, as illustrated by Fig. 10.

Although zero-dimensional models of expulsion will probably bring a significant improvement, it must be borne in mind that the physics of expulsion is still poorly known. Recent progress has occurred in the characterization of the pore system (Mann, 1990), the experimental investigation of hydrocarbon flow through source rocks (Sandvik and Mercer, 1990; Lafargue et al., 1990) and the computation of fluid phase equilibrium of pore fluids during expulsion (Braun and Burnham, 1989). Except in the last case, there may be no practical application of these results in basin models in the near future. Indeed, our knowledge progresses slowly about relative permeabilities, mechanical behaviour, petrophysical properties, and retention on mineral/organic surfaces prevailing in source rocks.

How can we control basin models by geochemical data?

Control of maturation by geochemical data

Apart from vitrinite reflectance, which has been discussed earlier, the main maturity indicators that

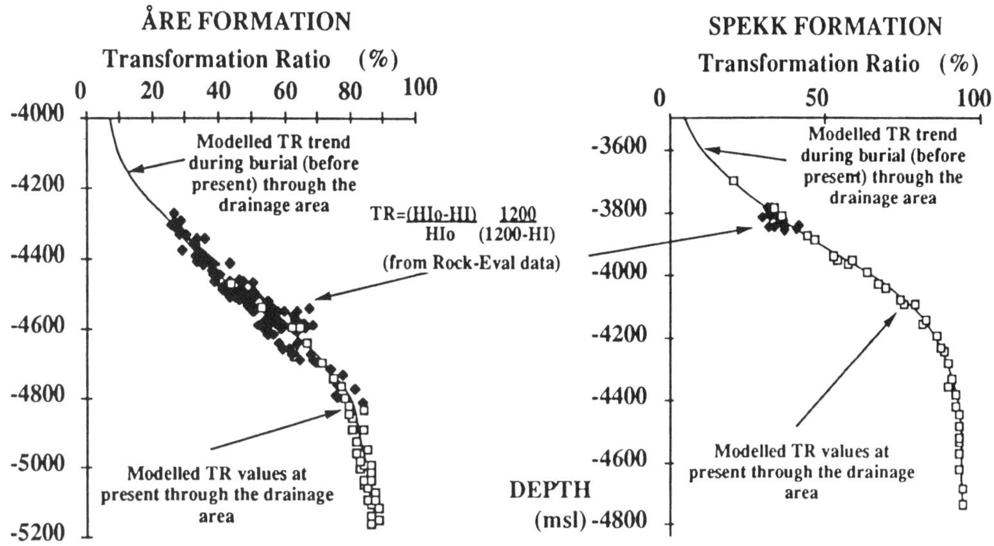

Fig. 9. Calibration of the saturation thresholds from S1 data in Åre coals and Spekk marine shales of the Haltenbanken area, Norway (Forbes et al., 1991). The thresholds are 50% for shales and 80% for coals. Higher values would result in an overestimation of S1 by the model, while lower values would underestimate it.

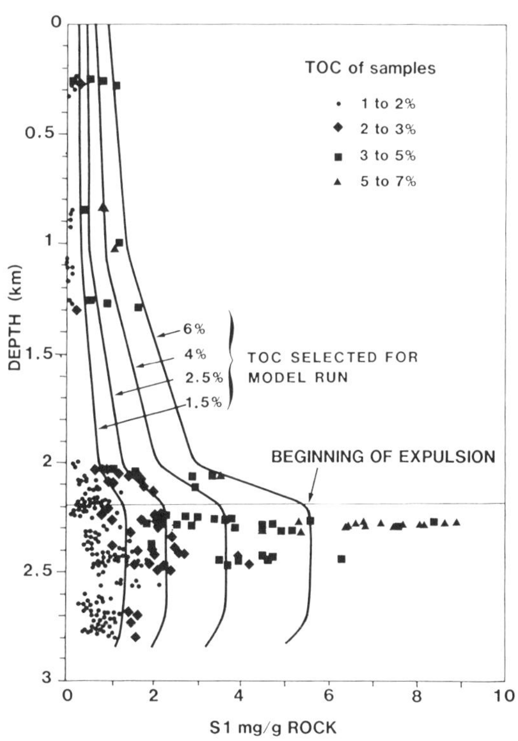

Fig. 10. Comparison of S1 data of 13 exploration wells in the central Paris Basin with a model of generation and expulsion. The samples were analyzed on the rig, a few minutes after issued from drilling, so that the loss of light ends was minimal. The model is based on an empirical but unique threshold on S1/TOC. This model allows reproduction of the higher S1 values in organic-rich samples, together with expulsion below 2200 m. If a saturation threshold is used, it should be different for each class of organic carbon content to match these data (Ungerer, 1990).

can be compared with model results are Rock-Eval data (Hydrogen Index, HI; peak temperature, T_{max}) or the Transformation Ratio (TR) which can be estimated from these data.

The final *Hydrogen Index* (HI, in mg/g) may be computed from:

$$HI = S2_o TR/TOC \quad (1)$$

where $S2_o$ is the initial source rock potential, in mg/g:

$$S2_o = HI_o TOC_o/100$$

TOC is the final organic carbon content (%), which may be different from the initial organic carbon TOC_o if expulsion is considered; and TR is the Transformation Ratio computed by the kinetic model (i.e. final potential/initial potential).

Hydrogen Index data present generally a large scatter but nevertheless allow for a good quantitative control of the model. For other kerogen types than type III, it is clearly a better control than equivalent vitrinite reflectance because it is specific for the organic type considered.

The *Transformation Ratio* may also be used, derived from Rock-Eval data by the equation proposed by Espitalié et al. (1987):

$$TR_{obs} = \frac{(HI_o - HI)\,1200}{HI_o\,(1200 - HI)} \quad (2)$$

where HI_o is the assumed initial Hydrogen Index (mg/g), HI is the measured Hydrogen Index (mg/g), and TR_{obs} is the observed Transformation Ratio.

It is found that the trend of TR_{obs} is often less scattered than the trend of Hydrogen Indices from which it is derived. This makes the comparison easier but may be misleading. Indeed, significant errors may be made while estimating HI_o in the above equation, which is also based on specific hypotheses.

In a general way, it may be stated that a model result should be compared with raw data as far as possible, but not with highly processed data because the data processing may introduce systematic bias.

Rock-Eval T_{max} may be modelled by simulating the shape of the S2 peak of a mature sample (Tissot et al., 1987), so T_{max} data may be compared with model results. This method is particularly valid for type III and type II organic matter, but not for type I because T_{max} does not vary significantly in the latter. It may be noted that T_{max} increases stepwise when the spacing of activation energies is greater than or equal to 2 kcal/mol: a finer spacing must be used to model T_{max} increase more smoothly. Besides, the computed T_{max} values tend to increase somewhat late compared with the real ones. This is not surprising because T_{max} prediction requires an accurate estimation of the rates of generation during pyrolysis, which is a severe requirement for such an empirical model as the parallel reaction model. These problems limit the accuracy of T_{max} as a model control.

Sterane isomerization data may also be used as model control. Related models have been significantly improved recently by considering closed system pyrolysis experiments (Marzi et al., 1990).

Control of expulsion models by geochemical data

As mentioned earlier, S1 data or *solvent extract yields* may be used to control the results of expulsion models. More precisely, these data may be used to constrain empirical parameters such as the saturation threshold (Forbes et al., 1991). If the model predicts excess amounts of residual petroleum in mature source rocks, the saturation threshold must be decreased, and vice versa. However, great care must be taken while using such data, i.e.:

(1) Preliminary control of the maturation model (i.e. primary cracking) must be performed as mentioned above from HI, TR, T_{max} or vitrinite reflectance.

(2) Samples from accumulated source rock intervals (i.e. source rocks containing petroleum migrated from other source rock sections) must be rejected. Such samples may be identified, for instance, by anomalous trends of production indices (Espitalié et al., 1985a, b).

(3) Light ends are often lost by evaporation prior to analysis: for instance, solvent extract does not contain C_1 to C_{14} hydrocarbons. Before comparison is made with data, model results must be corrected for the amounts of such volatile fractions (Fig. 11).

(4) Heavy ends like resins and asphaltenes are not vaporized at 300°C, so that they are not detected

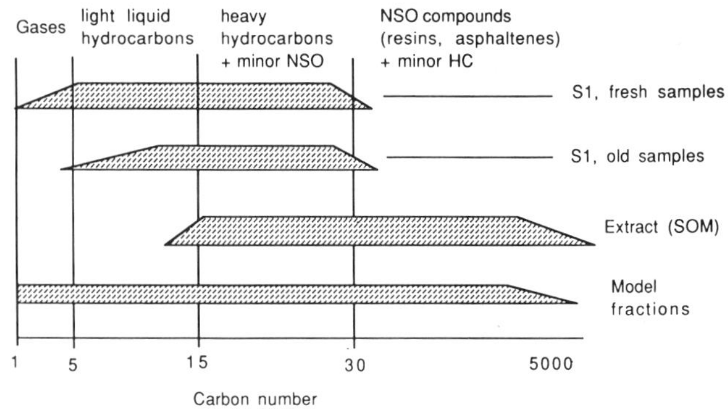

Fig. 11. Range of molecular weight of the compounds present in the Rock-Eval S1 peak and in solvent extract (i.e. bitumen), compared with the range considered in usual kinetic models. The light ends are lost by evaporation before S1 is measured, especially if the samples are analyzed a long time after sampling. The C_1–C_{14} fraction is lost during solvent extraction, unless special procedures are followed. The heavy ends (asphaltenes, resins) are not detected in S1.

in the S1 peak of the Rock-Eval method. Before comparison is made with S1 data, model results must also be corrected for the amounts of heavy fractions (Fig. 11).

The loss of light ends and the absence of heavy ends in S1 may need major corrections. If half of the light oil fraction (C_6–C_{14}) is lost and if heavy compounds are half of the C_{15+} fraction, only 50% of the liquid fraction (C_{6+}) is present in S1. This may be a serious limitation in the use of S1 for model control because the proportion of losses in the light fractions is variable and poorly documented.

Quite often, it is observed that the trends of *Production Indices*, PI = S1/(S1 + S2), are more clearly defined than S1 trends, probably because PI is less influenced than S1 by variations in organic richness. As the information contained in PI data is about the same as in S1 data, they may be efficiently used to control expulsion models (Fig. 12), but suffer the same limitations as related to the loss of light fractions.

Surprisingly, an important control on expulsion models is provided by the organic carbon content (TOC). It has been observed that the final TOC may be significantly different from the initial TOC in mature source rocks that have expelled an important fraction of the initial kerogen weight (Rullkötter et al., 1988). In expulsion models, the final TOC may be explicitly computed and compared with measurements. These data allow the determination of the initial TOC by trial and error. Such a procedure does not need any simplifying assumption; this is a significant improvement compared with mass balance methods (Cooles et al., 1986) which neglect secondary reactions.

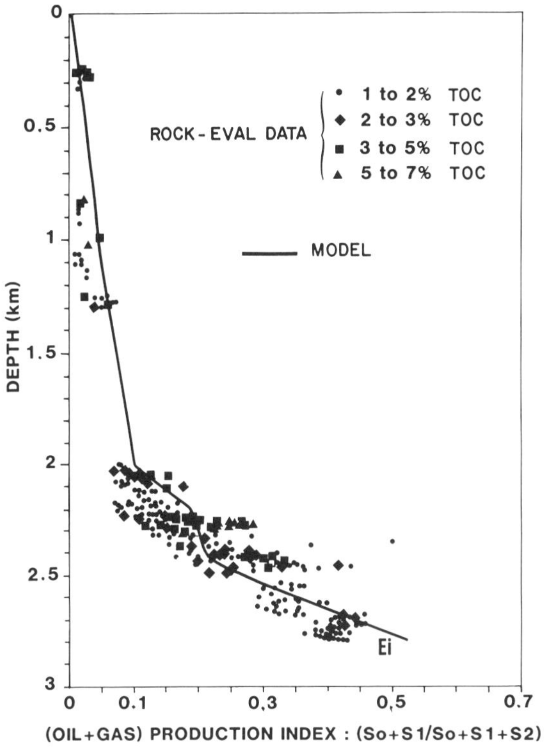

Fig. 12. Control of a simplified expulsion model (Ungerer, 1990) of Lias shales, central Paris Basin, by production indices (PI) determined by the Rock-Eval method (same samples and model as Fig. 10). If a lower expulsion threshold was selected, the modelled production index would be too low, and vice versa.

A final source of data that can be considered is the *size and composition of reported accumulations*. These data are not readily comparable with the amounts and composition of expelled petroleum, because significant changes may be caused by secondary migration (residual amounts in carrier beds, preferential migration of one hydrocarbon phase compared with the other, etc.). If secondary migration is taken into account, these data can be used as a final control of a global model-based migration scheme, as illustrated by Fig. 13.

Do we expect prediction of chemical composition or phase state from basin models?

The phase state of petroleum in the subsurface (oil or gas or both) depends not only on the global chemical composition but also on pressure and temperature. For high pressures (e.g. over 50 MPa in the example of Zhuze et al., 1975), only one supercritical phase is observed apart from water. At lower pressures, two phases (oil and gas) exist. As pressure and temperature decrease, the oil phase contains less gaseous hydrocarbons (C_1 to C_4) and the gas phase contains less liquid hydrocarbons (C_{5+}).

Modelling fluid phase equilibria has great potential importance for the explorationist, because it would provide the link between the chemical composition (computed by existing kinetic models) and the phase state of expected accumulations, depending on the pressures and temperatures encountered on secondary migration paths. However, few investigations have explored the application of thermodynamic models to real basin examples (Nogaret, 1983; Heum et al., 1986; Leythaeuser and Poelchau, 1992). Obviously, this is a field in which further research is needed.

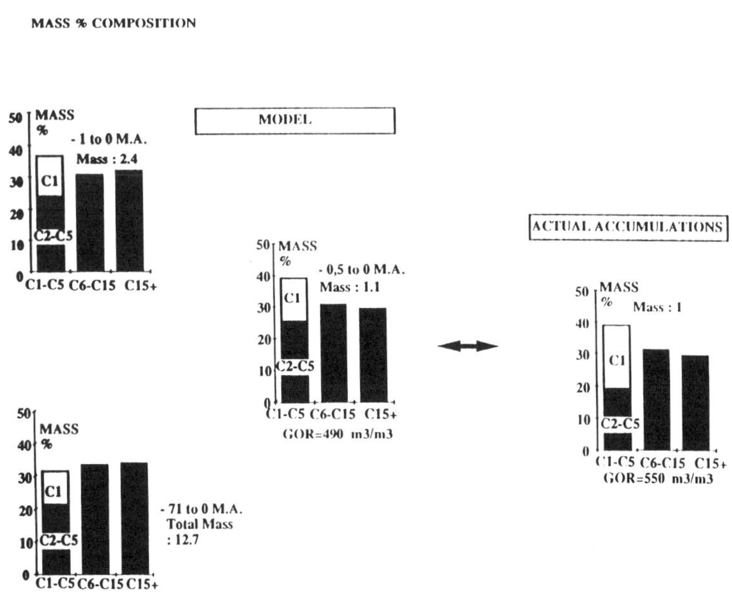

Fig. 13. Comparison between the computed expelled hydrocarbons from the Smørbükk Sør (offshore mid-Norway) drainage area and the actual accumulations (Forbes et al., 1991). This accounts for petroleum left on migration paths. The volume of hydrocarbons expelled is twelve times greater than the capacity of the structure. The composition of the accumulated petroleum is in good agreement with model results for the last 0.5 Ma.

Gas–liquid phase equilibria seem less critical in source rocks, where thermodynamic models indicate that pressure and temperature are probably high enough to prevent individualization of two distinct petroleum phases (Ungerer et al., 1983; Braun and Burnham, 1989). In source rocks, a more critical problem could be the solid–liquid equilibria between kerogen and petroleum. Indeed, the partition of hydrocarbons between the kerogen and the fluid phase and adsorbed phases could be responsible for fractionation effects such as reported by, among others, Leythaeuser et al. (1988) in natural series, or Chakhmachev et al. (1982) in experiments.

Conclusions

(1) There is an obvious need for kinetic models, but some parameters, although presumably of minor importance (e.g. pressure), are still poorly documented.

(2) Empirical kinetic models are continuously improving in their ability to reproduce the detailed chemical composition of generated products.

(3) There is a need for high-quality sets of pyrolysis data (mainly from closed system pyrolysis) to improve the accuracy of kinetic parameters (which is often low) and to investigate more organic types or sub-types.

(4) The low accuracy of existing vitrinite reflectance models seems difficult to improve upon.

(5) Modelling primary migration still presents difficult fundamental problems, but empirical expulsion models have already improved the qualitative and quantitative assessment of the petroleum available for secondary migration.

(6) Numerous geochemical data (S1, solvent extract yields, T_{max}, TOC, composition of reservoir fluids) can be used more intensively for model control.

(7) A lot of research and applications remains to be done to include fluid phase equilibria of oil/gas/water/kerogen systems in basin models.

Acknowledgements

I am particularly indebted to P. Forbes, Jean-Luc Rudkiewicz, O. Lerat and O. Boutry among other colleagues, for their technical contribution to this account. I also thank A.G. Doré and Tanja Barth for their review of the manuscript and for their advice.

References

Akihisa, K., 1978. Etude cinétique des roches mères de pétrole (rapport nr. 4). Formation de produits pétroliers par pyrolyse du kérogène à basse température. J. Jpn. Assoc. Pet. Technol., 44: 26–33.

Behar, F., Ungerer, P., Kressmann, S. and Rudkiewicz, J.L., 1991. Thermal evolution of crude oils in sedimentary basins: experimental simulation in a confined system and kinetic modeling. Rev. Inst. Fr. Pét., 46 (2): 151–182.

Bethke, C., 1989. Modeling subsurface flow in sedimentary basins. Geol. Rundsch., 78 (1) 129–154.

Braun, R.L. and Burnham, A.K., 1989. A mathematical model of oil generation, degradation and expulsion. Lawrence Livermore National Laboratory, Rep. UCRL 101708, September 1989.

Burnham, A.K., 1989. A simple model of petroleum formation and cracking. Lawrence Livermore National Laboratory, Rep. UCID 21665, March 1989.

Burnham, A.K. and Braun, R.L., 1990. Development of a detailed model of petroleum formation, destruction and expulsion from lacustrine source rocks. Org. Geochem., 16 (1–3): 27–40

Burnham, A.K. and Sweeney, J., 1989. A chemical kinetic model of vitrinite maturation and reflectance. Geochim. Cosmochim. Acta, 53: 2649–2657.

Castelli, A., Chiaramonte, M., Beltrame, P., Carniti, P., Delbianco, A. and Stroppa, F., 1990. Thermal degradation of kerogen by hydrous pyrolysis: a kinetic study. In: B. Durand and F. Behar (Editors), Advances in Organic Geochemistry 1989. Org. Geochem., 16: 75–82.

Chakhmachev, V.A., Punanova, S.A., Tverdova, B.A., Zharkov N.I. and Rjabova, G.M., 1982. Etude expérimentale de la modification de la composition de pétrole au cours de son écoulement à travers des milieux poreux. Geokhimija, 7: 1035–1042.

Cooles, G.P., Mackenzie, A.S. and Quigley, T.M., 1986. Calculation of petroleum masses generated and expelled from source rocks. Org. Geochem., 10: 235–245.

Dominé, F., 1989. Kinetics of hexane pyrolysis at very high pressures, 1. Experimental study. Energy Fuels, 3: 89–96.

Dominé, F., 1991. High pressure pyrolysis of *n*-hexane, 2-4-dimethylpentane and 1-phenylbutane. Is pressure an important geochemical parameter? Org. Geochem., 17 (5): 619–634.

Dominé, F., Marquaire, P.M., Muller, C. and Côme, G.M., 1990. Kinetics of hexane pyrolysis at very high pressures, 2. Computer modeling. Energy Fuels, 4: 2–10.

Düppenbecker, S.J. and Welte, D.H., 1991. Petroleum expulsion from source rocks — insight from geology, geochemistry and computerized numerical modelling. Proc. of the 13th World Petroleum Congress, Buenos Aires, October 1991, (3)2, Wiley, London.

Durand, B., 1988. Understanding of HC migration in sedimentary basins (present state of knowledge). Org. Geochem., 13 (4–6): 445–460.

Durand, B., Alpern, J., Pittion, J.L. and Pradier, B., 1986. Reflectance of vitrinite as a control of thermal history of sediments. In: J. Burrus (Editor), Thermal Modeling of Sedimentary Basins. Technip, Paris, pp. 441–474.

Eglinton, T.I., Douglas, A.G. and Rowlands, S.G., 1988. Release of aliphatic, aromatic and sulfur compounds from Kimmeridge kerogen by hydrous pyrolysis: a quantitative study. Org. Geochem., 13 (4–6): 655–663.

England, W.A., Mackenzie, A.S., Mann, D.M. and Quigley, T.M., 1987. The movement of entrapment of petroleum fluids in the subsurface. J. Geol. Soc. London, 144: 327–347.

Enguehard, F., Kressmann, S. and Dominé, F., 1990. Kinetics of dibutylether pyrolysis at high pressure: experimental study and numerical simulation. Org. Geochem., 16 (1–3): 155–160.

Espitalié, F., Deroo, G. and Marquis, F., 1985a. La pyrolyse Rock-Eval et ses applications (1). Rev. Inst. Fr. Pét., 40: 563–579.

Espitalié, F., Deroo, G. and Marquis, F., 1985b. La pyrolyse Rock-Eval et ses applications (2). Rev. Inst. Fr. Pét., 40: 755–784.

Espitalié, F., Deroo, G. and Marquis, F., 1986. La pyrolyse Rock-Eval et ses applications (3). Rev. Inst. Fr. Pét., 41: 73–89.

Espitalié, J., Marquis, F. and Sage, L. 1987. Organic geochemistry of the Paris basin. In: J. Brooks and K. Glennie (Editors), Petroleum Geology of N.W. Europe. Graham and Trotman, London, pp. 71–87.

Fabuss, B.M., Smith, J.O. and Satterfield, C.N., 1964. Thermal cracking of pure saturated hydrocarbons. Adv. Pet. Chem. Refin., 3: 156–201.

Faille, I., 1992. Modélisation bidimensionnelle de la genèse et de la migration des hydrocarbures dans un bassin sédimentaire. Ph.D. Thesis, University of Grenoble.

Forbes, P., Ungerer, P., Kuhfuss, A., Riis, F. and Eggen, S., 1991. Compositional modelling of petroleum generation and expulsion. Application to a local mass balance in the Smørbükk Sør field, Haltenbanken area, Norway. Am. Assoc. Pet. Geol., Bull., 75 (5) 873–893.

Hermanrud, C., Eggen, S., Jacobsen, T., Carlsen, E.M. and Pallesen, S., 1990. On the accuracy of modelling hydrocarbon generation and migration: the Egersund Basin oil find, Norway. Org. Geochem., 16 (1–3): 389–400.

Heum, O.R., Dalland, A. and Meisingset, K.K., 1986. Habitat of hydrocarbons at Haltenbanken (PVT modelling as a predictive tool in hydrocarbon exploration). In: A.M. Spencer (Editor), Habitat of the Hydrocarbons of the Norwegian Shelf. Graham and Trotman, London, pp. 259–274.

Jones, D.M., Douglas, A.G. and Connan, J., 1987. Hydrocarbon distribution in crude oil asphaltene pyrolyzates, 1. Aliphatic compounds. J. Energy Fuels, 1: 468–476.

Klomp, U.C. and Wright, P.A., 1990. A new method for the measurement of kinetic parameters of hydrocarbon generation from source rocks. Org. Geochem., 16 (1–3): 49–60.

Kressmann, S., 1991. Craquage thermique de mélanges d'hydrocarbures à haute pression: étude cinétique expérimentale et modélisation numérique, implications pour la géochimie pétrolière. Thèse de doctorat, Université Pierre et Marie Curie, Paris.

Lafargue, E., Espitalié, J., Jacobsen, T. and Eggen, S., 1990. Experimental simulation of hydrocarbon migration. In: B. Durand and F. Behar (Editors), Advances in Organic Geochemistry 1989. Org. Geochem., 16: 121–131.

Lewan, M.D., 1985. Evaluation of petroleum generation by hydrous pyrolysis experimentation. Philos. Trans. R. Soc. London, Ser. A, 315: 123–134.

Leythaeuser, D. and Poelchau, H.S., 1991. Expulsion of petroleum from type III kerogen source rocks in gaseous solution: modelling of solubility fractionation. In: W.A. England and A.J. Fleet (Editors), Petroleum Migration. Geol. Soc. London, Spec. Publ., 59: 33–46.

Leythaeuser, D., Schaefer, R.G. and Radke, M., 1988. Geochemical effects of primary migration of petroleum in Kimmeridge source rocks from Brae field area, North Sea, 1. Gross composition of C_{15+}-soluble organic matter and molecular composition of C_{15+} saturated hydrocarbons. Geochim. Cosmochim. Acta, 42: 701–713.

Mann, U., 1990. Sedimentological and petrophysical aspects of primary petroleum migration pathways. In: U. Förstner et al. (Editors), Sediments and Environmental Geochemistry — Selected Case Histories. Springer-Verlag, Berlin.

Marzi, R., Rullkötter, J. and Perriman, W.S., 1990. Application of the change of sterane isomer ratios to the reconstruction of geothermal histories: implications of the results of hydrous pyrolysis experiments. Org. Geochem., 16 (1–3): 91–102.

Monthioux, M., 1988. Expected mechanisms in nature and in confined-system pyrolysis. Fuel, 67: 843–847.

Nielsen, S.B. and Dahl, B., 1991. Gaussian confidence limits on kinetic kerogen models. Mar. Pet. Geol. (in press).

Nogaret, E., 1983. Solubilité des hydrocarbures dans le gaz comprimé. Application à la migration du pétrole dans les bassins sédimentaires. Ph.D. Thesis, Ecole des Mines, Paris.

Price, L., Daws, T. and Pawlewicz, M., 1986. Organic metamorphism in the lower Mississipian-upper Devonian Bakken Shale, Part I. Rock-Eval and vitrinite reflectance. J. Pet. Geol., 9 (2): 125–162.

Quigley, T.M., Mackenzie, A.S. and Gray, J.R., 1987. Kinetic theory of petroleum generation. In: B. Doligez (Editor), Migration of Hydrocarbons in Sedimentary Basins. Technip, Paris, pp. 649–666.

Rullkötter, J., Leythaeuser, D., Horsfield, B., Littke, R., Mann, U., Muller, P.J., Radke, M., Schaefer, R.G., Schenk, H.J., Schwochau, K., White, E.G. and Welte, D.H., 1988. Organic matter maturation under the influence of a deep intrusive heat source. A natural experiment of hydrocarbon generation and expulsion from a petroleum source rock (Toarcian Shale, Northern Germany). Org. Geochem., 13 (4–6): 847–856.

Sandvik, E.I. and Mercer, J.N., 1990. Primary migration by bulk hydrocarbon flow. Org. Geochem., 16 (1–3): 83–90.

Solomon, P.R., Hamblen, D.G., Carangelo, R.M., Serio, M.A. and Deshpande, G.V., 1988. General model for coal devolatilization. Energy Fuels, 2: 405–422.

Solomon, P.R., Serio, M.A., Deshpande, G. and Kroo E., 1990. Cross-linking reactions during coal conversion. Energy Fuels, 4: 42–54.

Stainforth, J. and Reinders, J.E.A., 1990. Primary migration of hydrocarbons by diffusion through organic matter networks, and its effect on oil and gas generation. Org. Geochem., 16 (1–3): 61–74.

Sweeney, J.J., Burnham, A.K. and Braun, R.L., 1987. A model of hydrocarbon maturation from type I kerogen: application to Uinta Basin, Utah. Am. Assoc. Pet. Geol., Bull., 71: 967–985.

Thomas, M. and Clouse, J.A., 1990. Primary migration by diffusion through kerogen. Geochim. Cosmochim. Acta, 54: 2775–2797.

Thomas, M., Fixari, B., Le Perchec, P., Princic, Y. and Lena, L., 1989. Visbreaking of Safaniya vacuum residue in the presence of additives. Fuel, 68: 318–322.

Tissot, B.P. and Espitalié, J., 1975. L'évolution de la matière organique des sédiments: application d'une simulation mathématique. Rev. Inst. Fr. Pét., 30: 743–777.

Tissot, B.P., Pelet, R.P. and Ungerer, P., 1987. Thermal history of sedimentary basins, maturation indices, and kinetics of oil and gas generation. Am. Assoc. Pet. Geol., Bull., 71 (12): 1445–1466.

Ungerer, P., 1990. State of the art of research in kinetic modeling of oil formation and expulsion. Org. Geochem., 16 (1–3): 1–25.

Ungerer, P. and Pelet, R., 1987. Extrapolation of the kinetics of oil and gas formation from laboratory experiments to sedimentary basins. Nature, 327 (6117): 52–54.

Ungerer, P., Behar, E. and Discamps, D., 1983. Tentative calculation of the overall volume expansion of organic matter during hydrocarbon genesis from geochemistry data. Implications for primary migration. In: M. Bjørøy (Editor), Advances in Organic Geochemistry 1981. Wiley, New York, N.Y., pp. 129–135.

Ungerer, P., Espitalié, J., Behar, F. and Eggen, S., 1988. Modélisation mathématique des interactions entre craquage et migration lors de la formation du pétrole et du gaz. C.R. Acad. Sci. Paris, Sér. II, pp. 927–934.

Ungerer, P., Burrus, J., Doligez, B., Chénet, P.Y. and Bessis, F., 1990. Basin evaluation by integrated two-dimensional modelling of heat transfer, fluid flow, hydrocarbon generation and migration. Am. Assoc. Pet. Geol., Bull., 74 (3): 309–335.

Waples, D., 1980. Time and temperature in petroleum formation: application of Lopatin's method to oil exploration. Am. Assoc. Pet. Geol., Bull., 64: 916–926.

Webster, R.L., 1986. Petroleum source rocks and stratigraphy of the Bakken Formation in North Dakota. Rocky Mountain Assoc. Pet. Geol. Symp., pp. 57–81.

Weres, O., Newton A.S. and Tsao, L., 1988. Hydrous pyrolysis of alkanes, alkenes, alcohols and ethers. Org. Geochem., 12: 433–444.

Zhuze, T.P., Youchkevitch, G.N. and Tchakhmakhchev, B.A., 1975. L'influence de la température et de la pression sur les compositions des phases dans les systèmes huile–gaz naturel. In: Advances in Organic Geochemistry 1973. Technip, Paris, pp. 464–469.

P. UNGERER *Institut Français du Pétrole, B.P. 311, 92506 Rueil-Malmaison Cedex, France*

Critical study of kinetic modelling parameters

J. Espitalié, F. Marquis and S. Drouet

We know that to use mathematical models of oil and gas generation (MATOIL, for example) in petroleum exploration, it is first necessary to reconstruct burial and geothermal histories of the source rocks studied and to determine their kinetic parameters (activation energies, Arrhenius constant A). All of these parameters can be the source of erroneous results when using models in sedimentary basins for the determination of oil and gas window depths, for the timing of generated hydrocarbons and for the reconstruction of the palaeogeothermal gradients.

Firstly, the determination of kinetic parameters based on a mathematical adjustment of the S2 signals obtained at different heating rates (model OPTKIN 1, for example) will be influenced by several factors which can be gathered together as follows: (a) the geometry of the pyrolyzer furnace, the nature of the carrier gas (He or N_2) and the rate of heating; (b) the selection of the representative sample, its maturity level and its nature (isolated kerogen or whole rock sample); (c) the mathematical adjustment itself which can give, for various S2 curves of the same sample, a different distribution of activation energies.

Secondly, the reconstruction of the burial and thermal histories, generally hard to perform in the case of Palaeozoic units, can also strongly affect the oil and gas window depths and the timing of hydrocarbon generation.

We discuss the effects of these kinetic parameters, and in order to limit the sources of error we propose: (a) to normalize pyrolysis conditions; (b) to adjust the distribution of activation energies of source rocks not only on S2 peaks but also, when possible, on geochemical data (Hydrogen Index values) observed in wells whose burial history is known.

Introduction

Interest in simple mathematical models, which are used to simulate the genesis of hydrocarbons in sedimentary basins, has grown steadily in recent years. One reason for this is that these models can now be employed routinely by the geologist and the geochemist equipped with a simple PC, enabling them rapidly to predict the quantities of hydrocarbons generated by source rocks and the timing of generation. These factors can then be compared with the age of formation of the reservoir structures.

These models obviously require knowledge of the thermal conditions (burial and heat flow) to which the source rocks analyzed were subjected, as well as the kinetic parameters (distribution of activation energies) of their organic matter. As a rule, the geological conditions required can be estimated fairly accurately by the geologist. The same cannot be said of the ancient geothermal conditions, which are more or less unknown. As to the kinetic parameters of the organic matter, they must first be determined from laboratory experiments on artificial maturation (pyrolyses) on the organic matter representative of the source rocks investigated.

In most situations, the geologist using these models employs present geothermal data, and kinetic parameters given by default in the model characteristic of one of the three main types of kerogens known, which are assumed to be those of the source rock analyzed. If geochemical data are available (R_o, T_{max}, Hydrogen Index), the user can even attempt to determine the ancient geothermal gradient by fitting the calculated data to the observational data (Tissot and Espitalié, 1975; Ungerer et al., 1986).

However, the user must keep in mind that the geothermal gradient thus calculated will depend closely on the kinetic parameters selected (since parameters given by default are often quite different from those of the source rock analyzed) and on the mathematical model that uses them. This is commonly a simple model represented by n parallel reactions based on first-order Arrhenius equations (Fig. 1). Even if the kinetic parameters of the source rock are determined from its own organic matter (OM), they would still depend on the accuracy of the artificial maturation used and on the mathematical formalism of optimization.

The methods used to determine these kinetic parameters are normally based on artificial maturation. For these determinations, Tissot and Espitalié (1975) used both natural maturation data (chloroform extract of source rocks at different depths) and arti-

Basin Modelling: Advances and Applications edited by A.G. Doré et al. NPF Special Publication 3, pp. 233–242, Elsevier, Amsterdam.

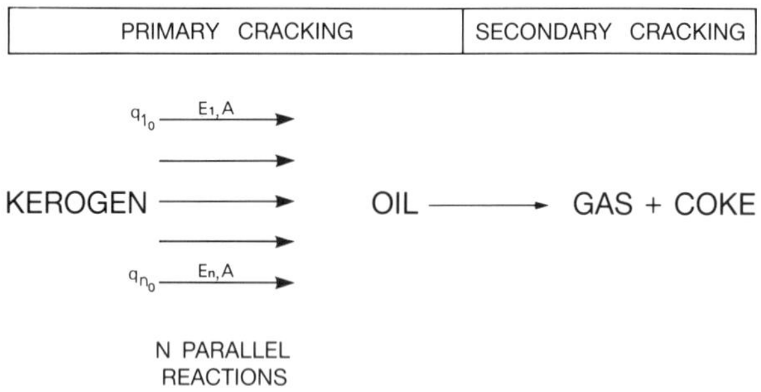

Fig. 1. Kinetic scheme of the mathematical model used in MATOIL.

ficial data (thermogravimetric analysis of kerogens pyrolyzed under nitrogen). More recently, Ungerer (1984), Braun and Burnham (1987), and Burnham et al. (1987, 1988) used artificial maturations in an open system exclusively, with specific detection of hydrocarbons produced during pyrolyses with programming temperature (Rock-Eval S2 peaks). Other types of adjustment have also been proposed using hydrous pyrolysis experiments (Lewan, 1985). It is clear that these kinetic models cannot claim to be "absolute" models of the thermal degradation of kerogens.

The work presented here is based on the results obtained using the OPTKIN-1 model to optimize the kinetic parameters and the MATOIL model to simulate the generation of hydrocarbons. It was devised in order to gain a better understanding of the accuracy and limits of application of these types of mathematical models. Note that similar work was done by Jarvie (1991) using the KINETICS program, and by Nielsen and Dahl (1991).

Experimental

The results presented here were obtained from more than 200 S2 curves resulting from the pyrolysis at different heating rates of four organic matter samples. For each kerogen, all the possible variants of its activation energy distribution were determined using, in OPTKIN-1, several sets of S2 curves, by combination either of the different heating rates, or of the different S2 curves obtained for the same heating rate.

The pyrolyses carried out, the optimization calculations of the kinetic parameters and their applications to concrete cases were achieved using the IFP Petroleum Workstation (Fig. 2). This station consists of an instrument of the Rock-Eval series (RE2 to RE5) connected to a PC equipped with software for acquisition of the parameters and the pyrolysis S1 and S2 signals (ROCKDAT), their interpretation (ROCKINT), the determination of the activation energy distribution of the source rocks (OPTKIN-1), their applications to natural series for the determination of the quantities of oil and gas (MATOIL) and hydrocarbon composition by carbon classes (GENEX).

The samples analyzed were the following rocks and kerogens:

(1) Type II, sample 38977 of the Lias of the Paris Basin (raw rock: TOC = 3.2%, Hydrogen Index (HI) = 450, T_{max} = 427; kerogen: TOC = 49.7%, HI = 400).

(2) Type III, Brent coals from the North Sea (samples 54201 and 54041, with HI values of 220 and 120, and T_{max} of 433 and 441, respectively).

These samples were pyrolyzed at different heating rates:

(1) Under helium between 5° and 50°C/min (interval possible of 1°C/min) with the workstation connected to a Rock-Eval 2 or to an oil show analyzer (OSA).

(2) Under nitrogen and helium between 0.2° and 50°C/min (interval possible of 0.2°C/min) with the Workstation connected to a Rock-Eval 5.

For each pyrolysis, 2000 measurement points corresponding to the S2 signal were memorized, using the ROCKDAT software. 200 values of the 2000 stored were used for kinetic calculations.

The principle used in OPTKIN-1 (Ungerer, 1984) consists of determining the distribution of the activation energies characteristic of the OM analyzed, i.e., the q_n fractions of the organic matter that are degraded with the energy activation values E_n fixed between 40 and 80 kcal/mol (Fig. 3).

The optimization program requires for a given sample two to ten S2 pyrolysis curves, realized at a minimum of two different heating rates. The optimization of the kinetic distribution is terminated when the least-squares difference between observed and predicted data is a minimum for each S2 profile.

Results

The activation energy distributions obtained are generally characteristic of the type of OM (Tissot and Espitalié, 1975). Their shapes (Fig. 4) resemble those of the S2 peaks from which they are derived (Espitalié et al., 1985a, b, 1986; Ungerer and Pelet, 1987; Crumière and Espitalié, 1989). It is worth noting that, for type III, the distribution parameters observed for the high activation energies, above 62 to 69 kcal/mol (Fig. 4), correspond to the fraction of kerogen that only generates gas by primary cracking (Espitalié et al., 1988).

However, the most interesting aspect of distributions determined from artificial maturation per-

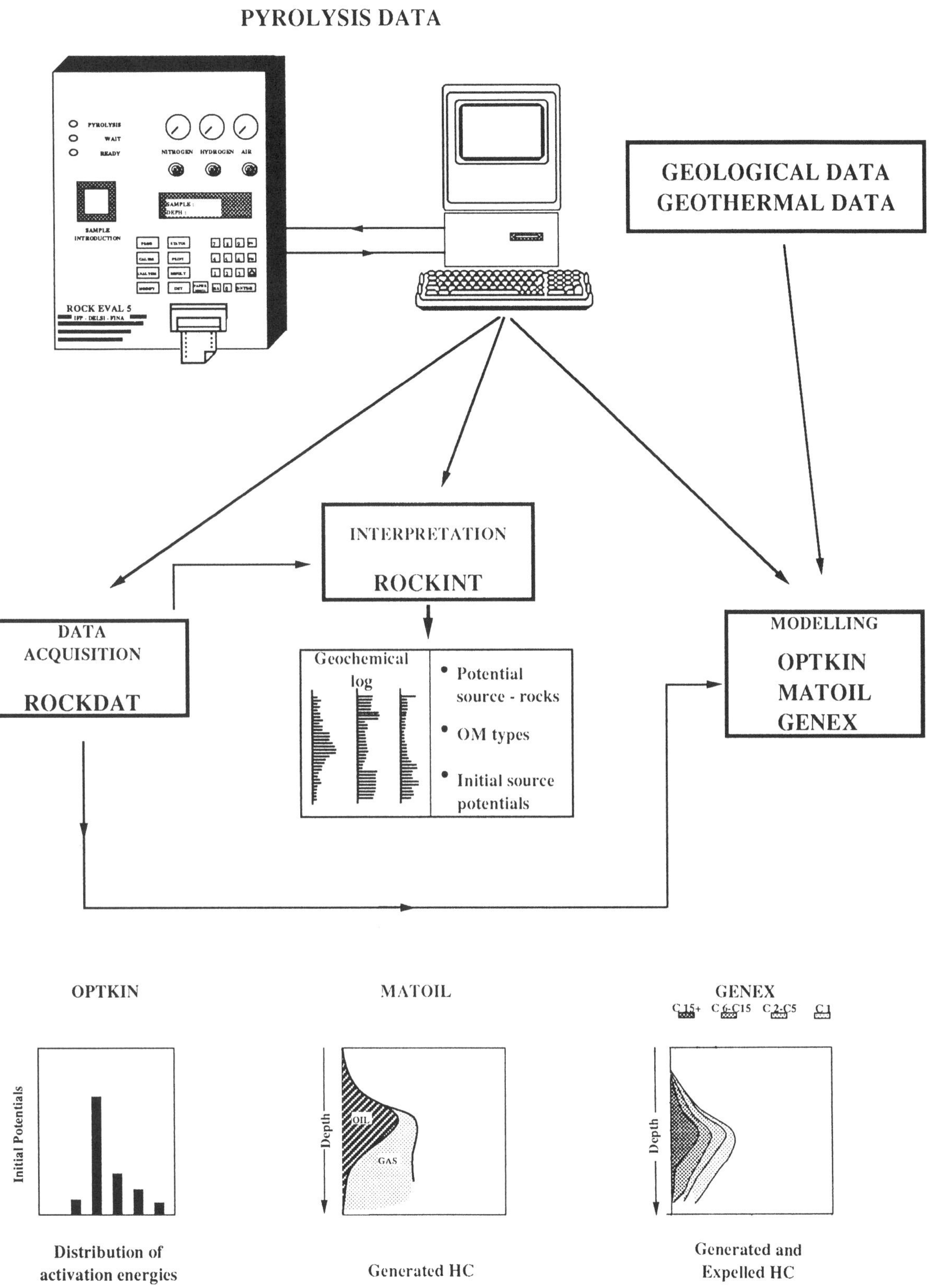

Fig. 2. Flow chart of the Petroleum Workstation.

formed in the laboratory (between 50° and 0.2°C/min) is obviously their usefulness in simulating the natural evolution that occurs at very slow heating rates of about 10^{-11}°C/min. This extrapolation is always difficult to check. However, in the natural series for which geological and geochemical data are available (burial, geothermal gradient, HI, R_o, T_{max}), these distributions are able to simulate natural maturation phenomena.

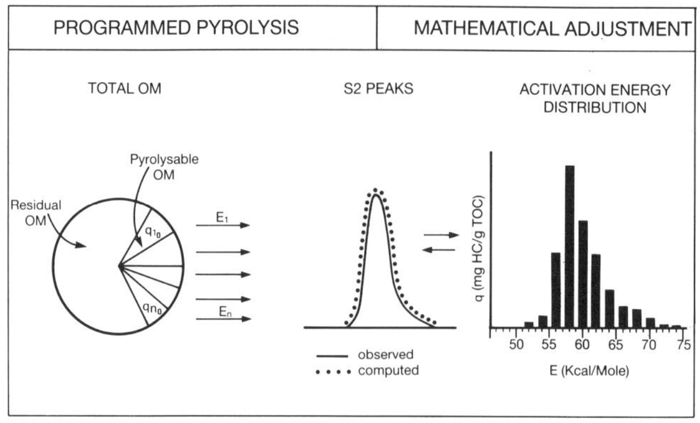

Fig. 3. Principle of the determination of distribution of activation energies by means of the OPTKIN-1 program.

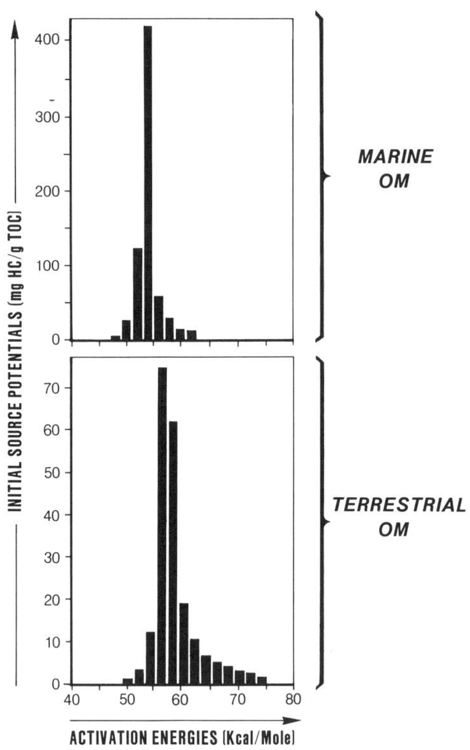

Fig. 4. Examples of distributions of activation energies for kerogen of type II (marine OM) and III (terrestrial OM).

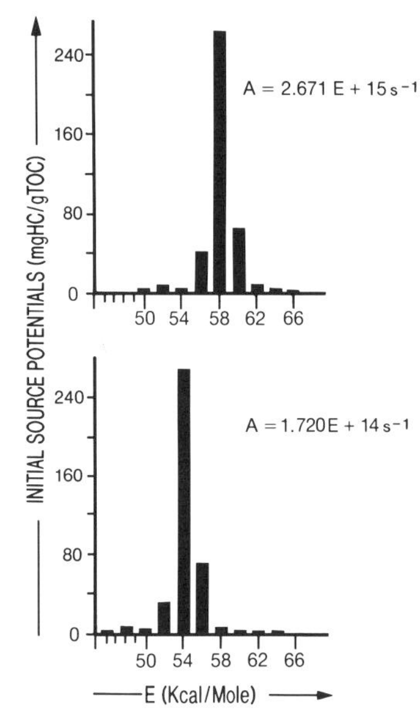

Fig. 5. Observed shift, for a given sample (kerogen type II, sample 38977), of its kinetic distribution when computed with different pyrolysis S2 curves obtained at different heating rates.

Variation of the activation energy distribution for a given sample

This extrapolation to natural series may nevertheless appear dubious owing to the fact that, for the same OM sample and the same pyrolysis oven, the activation energy distribution may vary significantly when calculated from different sets of S2 curves. Figure 5 provides an example of the maximum variations that can be observed for the same sample.

These variations are clearly characterized by a shift in the distribution mode on the scale of activation energies. In the case of Fig. 5, the mode shifts from 58 to 54 kcal/mol without any substantial change in the distribution of the initial potential values.

It can also be observed that the mode shift of the activation energy is offset by a variation of the pre-exponential constant A. When the distribution is centred on the high activation energies, the constant A is the highest (10^{15} s^{-1}), which compensates for the fact that the thermal degradation occurs at higher temperatures when the values of E increase.

These fluctuations in the activation energy distribution may be due to two main causes: (1) the pyrolysis conditions; (2) the mathematical algorithm used to optimize the kinetic parameters.

Pyrolysis conditions

The pyrolysis conditions can undoubtedly induce apparent variations in the kinetic parameters of the source rocks.

The apparatus itself (Fig. 6) may be a source of problems. It is well known, for example, that changes

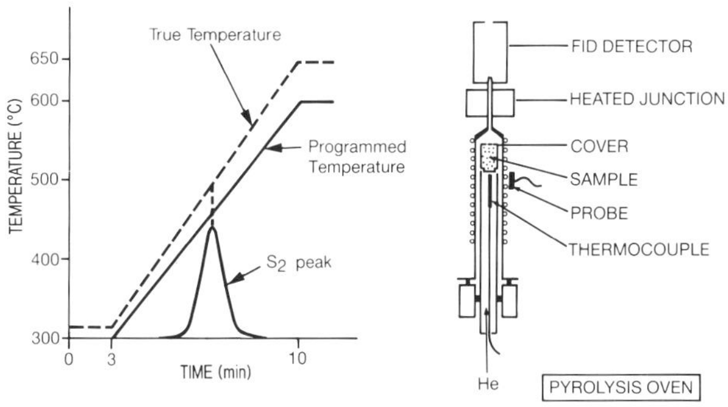

Fig. 6. Effect of the geometry of the pyrolysis oven on the true temperature of the sample.

in porosity of the crucible covers can cause variations in T_{max} of about $\pm 2\%$ (hence overall shifts of the S2 peak). This also applies to the grain size distribution of the sample analyzed, and especially to the amount of rock or kerogen placed in the crucibles (Espitalié et al., 1985a, b, 1986). The junction between the oven and the flame ionisation detector (FID) (Fig. 6) may also be the source of coking. Above all, however, for the Rock-Eval 2, the oven temperature does not correspond to the "true temperature" of the sample. A "true temperature" calibration is therefore necessary and must be performed after each modification involving the oven, the probe or the vector gas flow rates.

The pyrolysis mechanisms themselves are fairly poorly understood. Their results may vary according to the heating rate, as shown below, with secondary cracking occurring at low pyrolysis rates. At the low heating rates, the quantity of oil produced by pyrolysis also decreases, and this is often (probably wrongly) attributed to faulty linearity of the FID.

Determination of true pyrolysis temperatures

The determination of the kinetic parameters obviously requires a knowledge of the "true pyrolysis temperature" of the samples analyzed. In the Rock-Eval 2 and the OSA (Fig. 6), the position of the thermocouple located under the crucible does not allow a good measurement of the true temperature of the pyrolyzed sample. Moreover, owing to a temperature gradient within the oven (Espitalié et al., 1985a, b, 1986; Burnham et al., 1988), this true temperature is about +30°C higher than the temperature controlled by the programmer (programmed temperature, Fig. 6) through the probe located on the outside wall of the oven.

The *true temperature history* (Fig. 6) of the sample pyrolyzed can be determined by calibrating the instrument against standard rocks whose true temperature values of the top of their S2 peaks are known. The ROCKDAT software then determines the coefficients a and b of the relation between these true temperatures and those given by the programmer (programmed temperatures) such that:

$$\text{true temperature} = a\,(\text{programmed temperature}) + b$$

In the new Rock-Eval 5 instruments, the programmer probe is located in the oven wall at the level of the sample. The difference between the true temperature of the sample and that of the oven (programmed temperature) is hence reduced. The microprocessor also helps to compensate for the differences liable to appear during heating. In these conditions, coefficient a of the above law is practically equal to 1 and b to 0.

Effects of an erroneous measurement of the true pyrolysis temperature

In OPTKIN-1 software, by varying the initial and final true temperatures of the S2 curves with a value comprised between -5 and $+5$°C but without changing the heating rates of these curves, it is possible to analyze the effects of an erroneous measurement of the true pyrolysis temperature on the shift in the kinetic parameters. The results for the type II kerogen analyzed show that the parameters of the distribution of activation energies (value of the mode of the activation energy and constant A) are relatively unaffected by the variations in temperature of the S2 peak, even though these are significant, in the range of ± 2°C (Fig. 7). The results are comparable if one merely modifies the true temperatures of one pyrolysis curve out of two in the sets of curves used for optimization. These results show good agreement with those of Burnham et al. (1988), who, investigating the effect of heating rates, found a shift of 4 kcal/mol in the activation energies for a change of 3°C in the initial or final temperature.

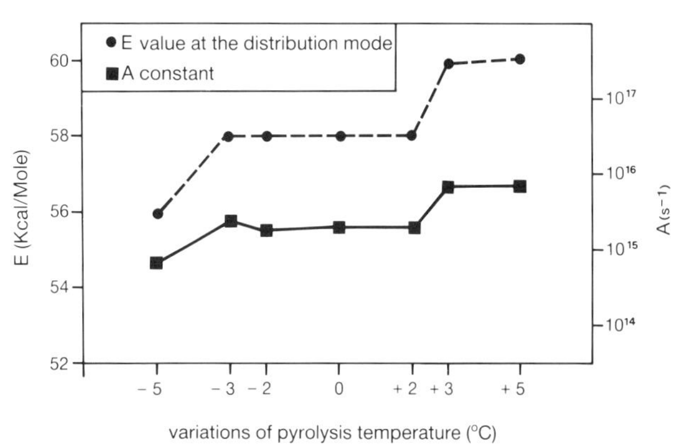

Fig. 7. Effects of erroneous measurements of the pyrolysis true temperature on the shift of the activation energy distribution of an OM sample (type II kerogen, sample 38977 pyrolyzed at 50°, 30°, 20°, 5° and 1°C/min, under nitrogen with a Rock-Eval 5).

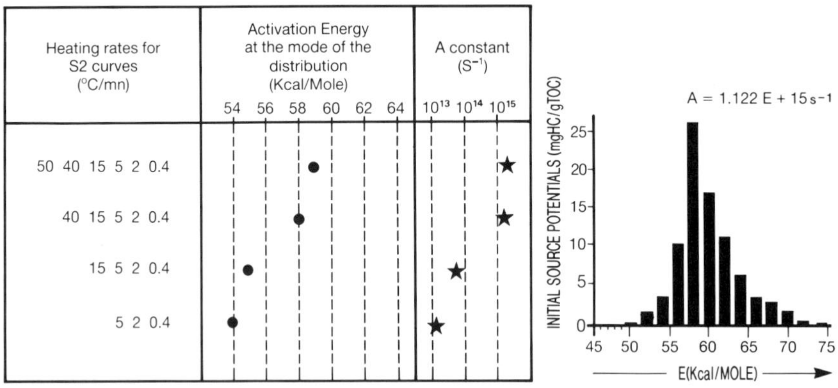

Fig. 8. Effect of the pyrolysis heating rates on the shift of the distribution of activation energies (type III kerogen, sample 54041 pyrolyzed under helium with a Rock-Eval 5).

A preliminary conclusion is that, for the same equipment, using good-quality crucibles (covers of similar porosity) and comparable sample weights between 10 and 30 mg, the fluctuations in the activation energy distribution for a given sample are probably not due to variations in the pyrolysis temperature. Moreover, only pyrolyses at large time intervals (several weeks) could cause such "shifts" in the distribution if the true pyrolysis temperature is modified (e.g., if the calibration is not checked after a displacement of the programmer probe at the surface of the oven).

Effects of heating rates and gas vectors

The heating rates have much more drastic effects on the fluctuations of activation energy distributions. Figure 8 provides an example of these fluctuations for a type III kerogen pyrolyzed under nitrogen. On the whole, the distribution mode shifts towards the low activation energies (from 59 to 54 kcal/mol) if increasingly slow heating rates are used (0.4°, 2° and 5°C/min). As usual, the constant A correlatively decreases, going from 10^{15} to 10^{13} s^{-1}.

One possible explanation for these observations is that, at low heating rates, the products obtained by pyrolysis do not result exclusively from "primary cracking" of the kerogens but also from "secondary cracking" of the heavy products formed, which can no longer be volatilized as in the case of a rapid rise in temperature. They accordingly reside longer in the sample, in which they are recracked. This assumption appears to be substantiated by compiling the balance of pyrolyzed carbon and residual carbon (Fig. 9). In these tests performed under nitrogen on a kerogen, it is observed that, at heating rates less than 1°C/min, coking processes are increased (decrease in pyrolizable organic carbon calculated from S2, increase in residual organic carbon in the sample after pyrolysis). Raw rock tests are more complex due to interactions between organic carbon with

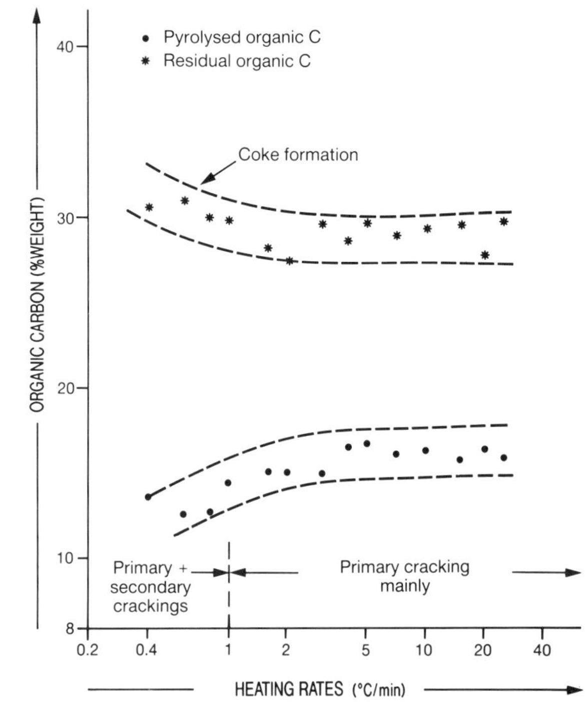

Fig. 9. Effects of heating rates on the primary and secondary cracking of a kerogen during pyrolysis (type II kerogen, sample 38977 pyrolyzed under nitrogen with a Rock-Eval 5).

CO_2 which occur at the low rates, and above 500°C due to the decomposition of the carbonate minerals. Pyrolysis under helium appears to aggravate this process.

The shift of the kinetic distribution in connection with this coking process at low heating rates then could be due to a change in the shape of the S2 peak (narrowest S2 peak) owing to the retention of the heavy hydrocarbon compounds generated at the beginning of the pyrolysis. At higher heating rates these compounds will be more or less volatilized creating a broader S2 peak. Comparative pyrolyses carried out on kerogens under helium and nitrogen as gas vectors have shown, at 30° and 40°C/min, important differences in the shape of S2 peaks

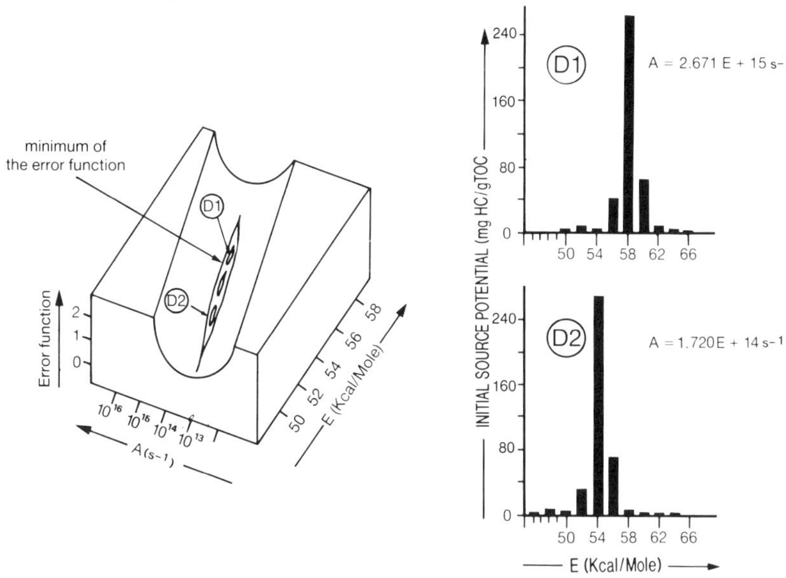

Fig. 10. Effect of the mathematical algorithm of optimization on the shift of the Activation Energy distribution (type II kerogen, sample 38977). The D1 distribution is optimized from S2 peaks pyrolyzed at 40°, 30°, 20°, 10°, 5° and 1°C/min; the D2 distribution from S2 peaks at 40°, 30°, 25°, 20°, 15°, 10°, 5°, 2° and 1°C/min. Pyrolyses are done in a Rock-Eval 5 under helium gas flow.

of certain kerogens. With nitrogen, the volatilization of heavy hydrocarbon compounds always seems more predominant than under helium, effectively creating broader S2 peaks, and giving in that case kinetic distributions shifted toward higher activation energies.

Effect of the optimization mathematical algorithm

For a given OM, it was nevertheless found that the fluctuations in its activation energy distribution persist even if calculated from S2 curves obtained for heating rates between 1° and 40°C/min (Fig. 10).

Mathematical analysis of the error function for different values of E and A (A. Printzhofer, pers. commun., 1990) shows that the geometric locus of this error function is a valley-shaped surface. Several possible minima exist (D1, D2, etc., Fig. 10), each corresponding to a different distribution.

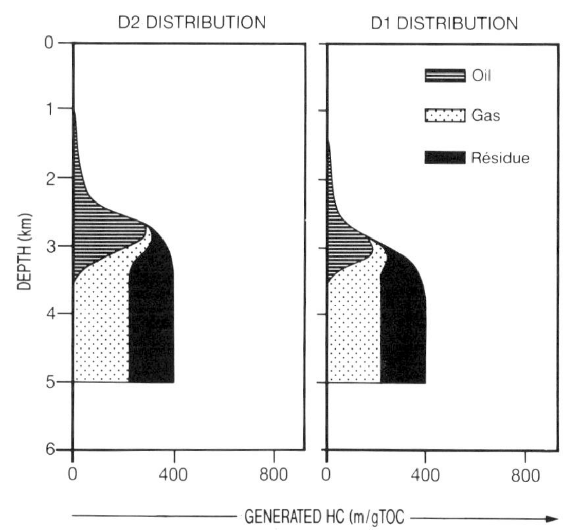

Fig. 11. Comparison of the hydrocarbons generated with depth computed by MATOIL with the kinetic distributions D1 and D2 previously determined (see Fig. 10).

Consequences for simulation in sedimentary basins

When these distributions are used to simulate the formation of hydrocarbons in a sedimentary basin, we observe that small differences at laboratory conditions yield large differences at basin conditions.

The simulations shown here (Figs. 11–13) were performed using MATOIL for a burial rate of 20 m/Ma, a heat flow of 75 mW/m^2 (geothermal gradient of about 30°C/km) and the two extreme distributions D1 and D2 of the foregoing case. Note that a discrepancy of about 300 m exists in the depth of hydrocarbon generation (oil zone, gas zone) calculated using these two distributions (Fig. 12). Moreover, if only one of these distributions is used, the geothermal flux must be increased or decreased by 10% to allow a good coincidence between the geochemical parameters (Transformation Ratio, $T_{\max}$) calculated with the other distribution.

Although the "absolute" kinetic distribution of the OM analyzed is unknown, it can nevertheless be stated that in simulating results for a sedimentary basin the error is a maximum of ±300 m for the determination of the depth of the oil window and ±10% for that of the heat flows or the geothermal gradients. Nielsen and Dahl (1991) also report a

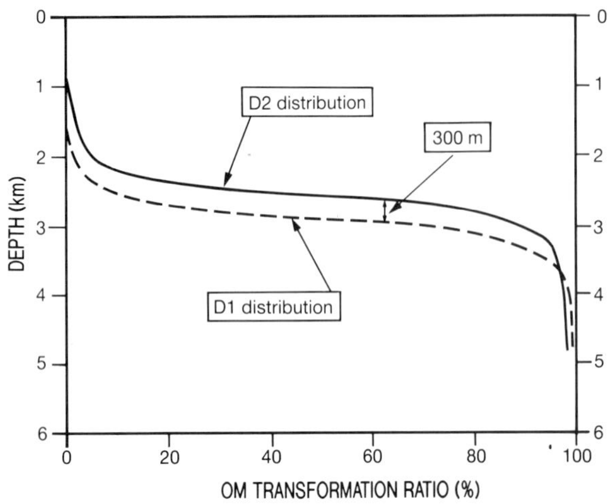

Fig. 12. Comparison of the Transformation Ratios computed by MATOIL with the D1 and D2 kinetic distributions.

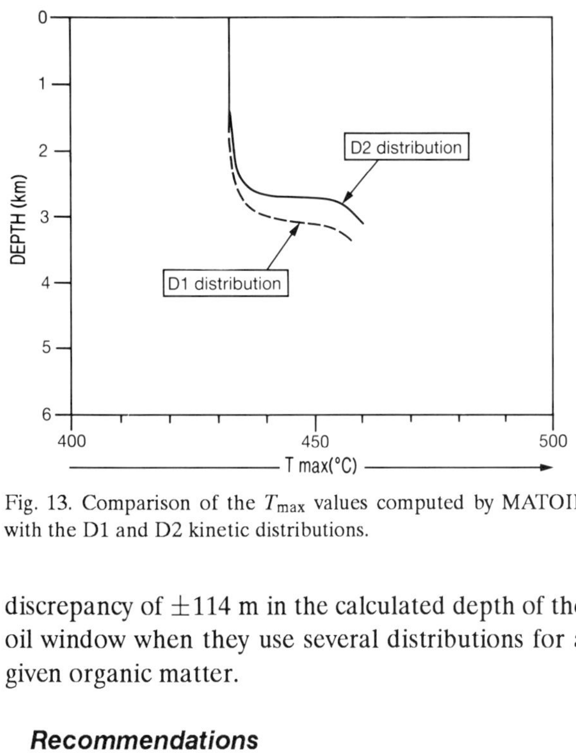

Fig. 13. Comparison of the T_{max} values computed by MATOIL with the D1 and D2 kinetic distributions.

discrepancy of ±114 m in the calculated depth of the oil window when they use several distributions for a given organic matter.

Recommendations

The foregoing results, and also validations by MATOIL of the kinetic distributions of Liassic organic matter of types II and III using well known sequences from the Paris Basin and the North Sea, allow the following recommendations to be made concerning the determination of kinetic parameters and their applications.

(1) The representative samples of the source rocks investigated must be selected in accordance with the following criteria. (a) The Hydrogen Index should correspond to the average HI value of the series. (b) The degree of maturity should, if possible, be similar to the beginning of the oil window (T_{max} between 420° and 435°C, R_o between 0.5 and 0.7%). This is especially important for type III organic matter. Here

6 HEATING RATES (°C / min)

30 15 10 8 5 2

Initial temperature : 300°C
Isothermal time at 300°C : 15min
Final temperature : 650°C

Gas flow : Helium

Fig. 14. Recommended pyrolysis conditions with a Rock-Eval 5 for optimization of kinetic distributions.

the high content of oxygenated compounds (OH, COO, COOH groups) could, in the most immature stages, disturb the pyrolysis (irregular S2 peaks, very low HI).

(2) The sample can be a kerogen or a raw rock if its TOC is at least greater than 1.5%. Prior extraction by an organic solvent is recommended.

(3) The pyrolysis conditions must be standardized. For the Rock-Eval 5, it is suggested that the conditions shown in Fig. 14 are used. The weights used are about 10 mg for a kerogen or a coal, and 30 mg for a rock. With a Rock-Eval 2 or an OSA, the heating rates can be the same, but are limited to 5°C/min for the lower heating rates. The vector gas is helium. In all the instruments, excessively high pyrolysis rates (50°C/min) should be avoided because temperature control is often difficult and because of the risk of saturation in the detector (Jarvie, 1991).

(4) It is preferable, if possible, to optimize the kinetic parameters using geological data (Tissot et al., 1987). Figure 15 provides such an example of fitting for the Lias of the Paris Basin. It is necessary to know the burial history. Optimization is carried out from the S2 curves of an immature sample representative of the series, and decreasing values of the Hydrogen Index. The geothermal gradient can also be varied until a minimum error function is obtained. It is clear that the kinetic distribution thus obtained must then be used with the geothermal gradient which allowed the best kinetic fit for simulating oil and gas generation in the investigated part of the studied basin.

(5) Concerning the simulation results (MATOIL), it must be recalled that it is not possible to obtain the absolute values of the heat flows or geothermal gradients. At the scale of a basin, however, it is possible to determine the relative values of the palaeogradients by fitting computed with observed geochemical data (vitrinite reflectance or Transformation Ratio, T_{max}, for example). This obviously entails the use of a single kinetic distribution, the one corresponding to the OM investigated: fitting to R_o or T_{max} for the adjustment of the palaeogradient in one case and then computing the Transformation Ratio in the other with two different distributions must be avoided.

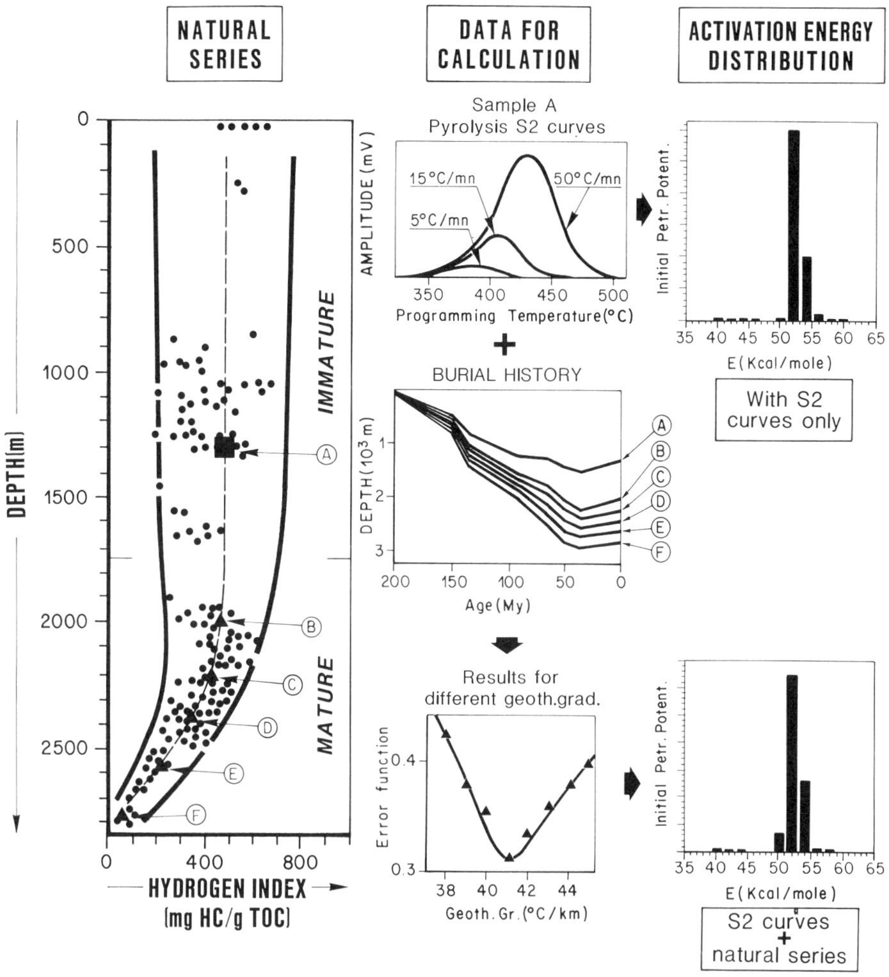

Fig. 15. Example of a geological/geochemical determination of a kinetic distribution (Lias of Paris Basin, type II kerogen).

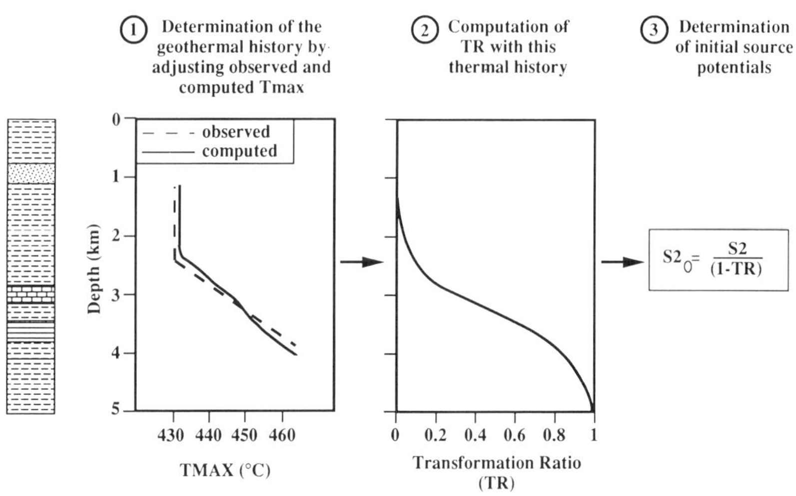

Fig. 16. Example of computation by MATOIL with a given kinetic distribution of the Transformation Ratio of the OM of a series after adjustment on observed T_{max} values.

For a given sedimentary sequence, in which the values of T_{max} or R_o are available as a function of depth, it is possible, after fitting to one of these maturation criteria, to determine, by means of the optimized kinetic distribution, the Transformation Ratio of the OM (Fig. 16). This is an important geochemical parameter for calculating the initial petroleum potentials. But the same activation energy distribution must be used for the fitting to R_o or T_{max}, and for the calculation of the Transformation Ratio.

Conclusions

The use of OPTKIN-1 software to determine the kinetic distributions of different kerogens from more than 200 pyrolysis curves and their use in MATOIL demonstrated the following.

(1) For a given OM sample, fluctuations in its kinetic distribution are generally observed if the optimization is performed from different combinations of its S2 pyrolysis curves obtained at different heating rates. These variations are chiefly due to the optimization conditions (several possible minima for the error function). They appear to be aggravated by certain pyrolysis conditions (coking at low heating rates). To minimize these fluctuations, it is therefore proposed to standardize the pyrolysis conditions (between 2° and 40°C/min).

(2) In sedimentary basins, the use of kinetic distributions determined from S2 curves alone leads to a discrepancy of about ±300 m in the hydrocarbon generation depth and ±10% in the estimation of the geothermal gradient (the real value of which cannot be determined).

(3) It is always preferable to adjust the results to the observed geological and geochemical data (R_o, T_{max}, HI) to determine a "kinetic distribution/geothermal gradient" combination or to calculate another geochemical parameter (e.g., Transformation Ratio). The same kinetic distribution must be used in adjusting the results to the observed data, and in calculating parameters such as the Transformation Ratio.

Acknowledgements

The authors wish to thank Alain Prinzhofer and Jean-Luc Rudkiewicz for very helpful discussions.

References

Braun, R.L. and Burnham, A.K., 1987. Analysis of chemical reaction kinetics using a distribution of activation energies and simpler models. Energy Fuels, 1: 153–161.

Burnham, A.K., Braun, R.L., Gregg, H.R. and Samoun, A.M., 1987. Comparison of methods for measuring kerogen pyrolysis rates and fitting kinetic parameters. Energy Fuels, 1: 452–458.

Burnham, A.K., Braun, R.L., Gregg, H.R. and Samoun, A.M., 1988. Further comparison of methods for measuring kerogen pyrolysis rates and fitting kinetic parameters. In: L. Mattavelli and L. Novelli (Editors), Advances in Organic Geochemistry 1987, Part II. Org. Geochem., 13: 839–845.

Crumière, J.P. and Espitalié, J., 1989. Characterization of organic depositional types based on the variation in the shape of the S2 peak of Rock-Eval pyrolysis. Results compared with data of optical surveys. C.R. Acad. Sci. Paris, Sér. II, 309: 1413–1417.

Espitalié, F., Deroo, G. and Marquis, F., 1985a. La pyrolyse Rock-Eval et ses applications (1). Rev. Inst. Fr. Pét., 40 (5/6): 563–579.

Espitalié, F., Deroo, G. and Marquis, F., 1985b. La pyrolyse Rock-Eval et ses applications (2). Rev. Inst. Fr. Pét., 40 (5/6): 755–784.

Espitalié, F., Deroo, G. and Marquis, F., 1986. La pyrolyse Rock-Eval et ses applications (3). Rev. Inst. Fr. Pét., 41 (1): 73–89.

Espitalié, J., Ungerer, P., Irwing, I. and Marquis F., 1988. Primary cracking of kerogens. Experimenting and modeling C1, C_2–C_5, C_6–C_{15} and C_{15+} classes of hydrocarbons formed. In: L. Mattavelli and L. Novelli (Editors), Advances in Organic Geochemistry. Org. Geochem., 13: 893–899.

Jarvie, D.M., 1991. Factors affecting Rock-Eval derived kinetic parameters. Chem. Geol., 93: 79–99.

Lewan, M.D., 1985. Evaluation of petroleum generation by hydrous pyrolysis experimentation. Philos. Trans. R. Soc. London, Ser. A, 315: 123–134.

Nielsen, S.B. and Dahl, B., 1991. Confidence limits on kinetic models of primary cracking and implications for the modelling of hydrocarbon generation. Mar. Pet. Geol., 8 (4): 483–492.

Tissot, B.P. and Espitalié, J., 1975. L'évolution de la matière organique des sédiments: application d'une simulation mathématique. Rev. Inst. Fr. Pét., 30: 743–777.

Tissot, B.P., Pelet, R. and Ungerer, P., 1987. Thermal history of sedimentary basins, maturation indices, and kinetics of oil and gas generation. Am. Assoc. Pet. Geol., Bull., 71 (12): 1445–1466.

Ungerer, P., 1984. Models of petroleum formation: how to take into account geology and chemical kinetics. In: B. Durand (Editor), Thermal Phenomena in Sedimentary Basins. Technip, Paris, pp. 235–246.

Ungerer, P. and Pelet, R., 1987. Extrapolation of oil and gas formation from laboratory experiments to sedimentary basins. Nature, 327 (6117): 52–54.

Ungerer, P., Espitalié, J., Marquis, F. and Durand, B., 1986. Use of kinetic models or organic matter evolution for the reconstruction of paleotemperatures. Application to the Gironville Well (France). In: J. Burrus (Editor), Thermal Modelling in Sedimentary Basins. Proc. 1st IFP Res. Conf., Carcans, June 1985. Technip, Paris, pp. 531–546.

J. ESPITALIÉ *Institut Français du Pétrole, B.P. 311, 92506 Rueil-Malmaison Cedex, France*
F. MARQUIS *Institut Français du Pétrole, B.P. 311, 92506 Rueil-Malmaison Cedex, France*
S. DROUET *Institut Français du Pétrole, B.P. 311, 92506 Rueil-Malmaison Cedex, France*

Kerogen typing concepts designed for the quantitative geochemical evaluation of petroleum potential

B. Horsfield, S.J. Düppenbecker, H.J. Schenk and R.G. Schaefer

Kerogen typing provides the central framework to which geochemical data are related and upon which routine decision-making in geochemical modelling is based. Ideally, the needs of the explorationist govern which criteria should be selected for practical kerogen typing. Because petroleum yields, gas versus oil potential and the timing of petroleum formation are of ultimate concern to the explorationist, a tripartite kerogen typing scheme (genetic potential, kerogen quality and maturation characteristics) is demanded. An extension of concepts to petroleum fractions is also desirable. The present article discusses maturation characteristics, as deduced from closed versus open system pyrolysis, and the extension of kerogen quality criteria to rock bitumens and petroleum. It is shown that the classification of kerogen according to open system pyrolysis kinetic parameters for bulk petroleum generation appears meaningful despite large compositional disparities between such pyrolysates and natural petroleums. By extending kerogen quality criteria to other macromolecular fractions, it is also revealed that both the detailed type of kerogen in the source rock and the gross compositional characteristics of the petroleum it expelled may be inferred from the analysis of degraded petroleums.

Introduction

Basic quantitative information on the size, composition (e.g. gas/oil ratio (GOR)) and timing of petroleum charges are required on a frequent and routine basis by the explorationist. To help obtain this information, all available drill cuttings, cores and oils are analysed in order to ascertain the source rock properties and maturity level of the stratigraphic section and to identify the active source(s). Undrilled areas are then evaluated by extrapolation and interpolation of the analytical results using organic facies considerations (Jones, 1987) and kinetics, mass balance and phase/volume modelling techniques (Goff, 1984; Larter, 1985, 1990; Heum et al., 1986; Ungerer and Pelet, 1987; Mackenzie and Quigley, 1988; Rullkötter et al., 1988; Sweeney and Burnham, 1990; Hermanrud et al., 1990). Because all of these considerations rely heavily upon a knowledge of source rock composition, kerogen typing is of fundamental importance in modern petroleum geochemistry.

"Kerogen typing" simply consists of classifying the bulk indigenous fraction of sedimentary organic matter according to useful criteria. It can be performed in various ways, depending on the application. Microscopy used to feature strongly in this regard, but since the mid-seventies petroleum geochemists have almost exclusively relied upon the elementally based (H/C versus O/C) classification scheme which defines kerogen types I, II and III (Tissot et al., 1974). The development of the Rock-Eval ensured that these kerogen types could be identified quickly and cheaply using whole rock samples (Espitalié et al., 1977).

The effectiveness of any kerogen typing scheme can be judged according to how it is directly used in problem solving. In this case, we refer to the demands of basin modelling. The elemental scheme for kerogen classification discriminates solely according to genetic potential (total pyrolysate yield per g TOC). This is well-suited to predicting the masses of petroleum expelled from source rock kitchens because the classification criterion and mass balance input data are the same, i.e. hydrogen indices. But the explorationist has no means for predicting petroleum compositions (e.g. GOR) and maturation characteristics using only elemental classification because the types of petroleum generated by kerogens and the maturation levels needed to bring about generation may be substantially variable within the classical kerogen types. Choosing both kerogen quality parameters and kinetic parameters is therefore problematical where phase/volume and kinetic modelling must be performed without direct access to calibration samples. For this reason, a classification of organic matter based on not only genetic potential but also maturation characteristics and kerogen quality is highly desirable for geochemical modelling (Hors-

Basin Modelling: Advances and Applications edited by A.G. Doré et al. NPF Special Publication 3, pp. 243–249, Elsevier, Amsterdam.

field, 1989). Both these additional attributes can be determined by molecular typing using pyrolysis methods. Furthermore, because molecular similarities occur between genetically related asphaltenes, heterocompounds and kerogens (Rubinstein et al., 1979; Béhar and Pelet, 1984; Solli and Leplat, 1986), molecular kerogen typing criteria can be extended to the macromolecular fractions of crude oils and rock bitumen. In this way reservoir and source facies can be classified using a single frame of reference for applications in basin modelling.

The present article addresses two aspects of kerogen typing. These are: (1) the degree to which laboratory pyrolysis conditions may control the extrapolation of maturation characteristics such as kinetic parameters to geological systems, and (2) the extension of kerogen quality criteria to macromolecular petroleum fractions. A collation and discussion of results for the Posidonia Shale (Lias epsilon) of northern Germany is used as a common thread to link these themes. On a related topic, the reader is referred to Larter (1990) for a discussion on how the molecular characterisation of kerogen can aid mass balance and phase/volume modelling.

Materials and methods

All Posidonia Shale (Toarcian) samples were obtained from the Lower Saxony Basin of northern Germany (Fig. 1). Their maturity range encompasses

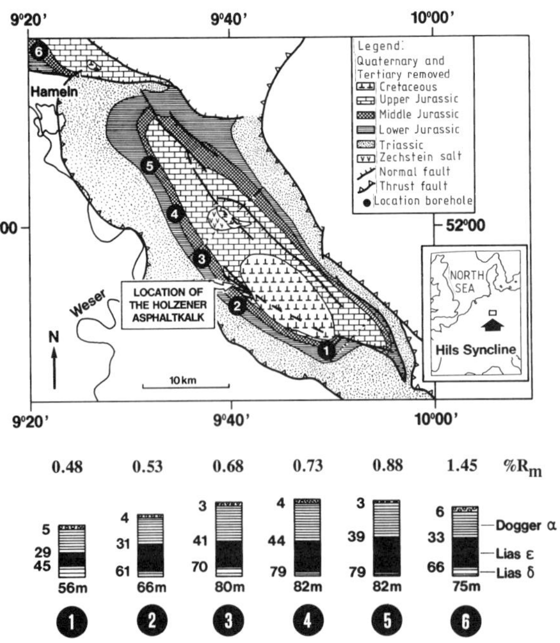

Fig 1. Location map illustrating near-surface stratigraphy and maturity of the Posidonia Shale (Lias epsilon, Lower Jurassic) at selected sites in the Hils Syncline, northern Germany. The location of the Holzener Asphaltkalk is also indicated.

R_m = 0.48–1.45%. Detailed background information can be found in Rullkötter et al. (1988) and Düppenbecker et al. (1991).

Several pyrolysis methods are referred to. These are Rock-Eval pyrolysis, pyrolysis–gas chromatography (py–gc) and microscale sealed vessel (MSSV) pyrolysis. The Rock-Eval technique was used to rapidly measure the yield of total pyrolysis products generated from whole rock samples under programmed (25°C/min) temperature conditions up to 550°C (Espitalié et al., 1977). A modification of the approach, using three heating rates per sample (0.1°, 0.7° and 5.0°C/min) was used to determine kinetic parameters under open system conditions (Schaefer et al., 1990). The analytical configuration for our programmed temperature py–gc method is given in Horsfield (1989): prominent aliphatic and aromatic components in the C_1–C_{35} range were discriminated and quantified as described in the same publication. The closed system MSSV pyrolysis method was used for simulating maturation (Horsfield et al., 1989). Isothermal heating, at temperatures in the range 280–370°C and for durations of 1–10 days, was used for kerogen quality and related assessments. Programmed heating at 0.1°, 0.7° and 5.0°C/min followed by gc analysis and differentiation of cumulative yield curves were used for closed system kinetic studies (Schenk and Horsfield, 1993).

Activation energy distributions from laboratory pyrolysis

Chemical reactions constituting petroleum generation are popularly considered to obey first-order reaction kinetics and to proceed faster with increasing temperature. Because this chemical system is exceedingly complex, a set of parallel reactions, each with its own set of laboratory-determined kinetic parameters (distributed or Gaussian), is commonly used for simplicity (for review see Ungerer, this volume). The approach works on the assumption that these reactions are the same as those involved in natural petroleum formation. Regarding this point, the most common way of gathering kinetic data for so-called bulk petroleum generation is by open system Rock-Eval style pyrolysis at different heating rates (e.g. Ungerer and Pelet, 1987). A means of classifying kerogens according to kinetic parameters is also based on open system pyrolysis (Delvaux et al., 1990). However, the products of open system pyrolysis include olefinic constituents and are not, in this respect, oil-like. The assumption that the laboratory pyrolysis reactions are the same as those leading to petroleum formation in nature therefore must be questioned. The overriding reason for choosing

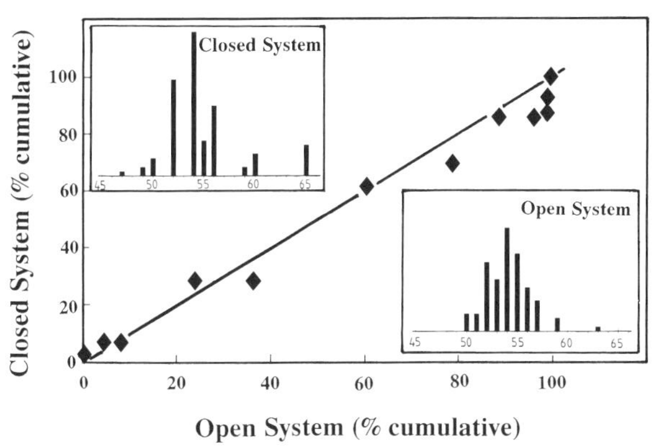

Fig 2. Activation energy distributions for open and closed system pyrolysis of Posidonia Shale kerogen. The distributions themselves are shown as insets. A quantitative comparison of the distributions is shown as cumulative proportions of petroleum potential with increasing activation energy. The line illustrates a 1 : 1 correspondence and is not a calculated best fit line.

open system pyrolysis is, after all, that of convenience (Ungerer, 1990). Closed system pyrolysis is less convenient to perform because cumulative yields are measured, but the products are compositionally more similar to petroleums found in nature because olefin yields are low (Lewan, 1985; Monthioux et al., 1986; Horsfield et al., 1989).

While kinetic parameters from certain open system and hydrous pyrolysis configurations have been shown to be different, the cause has been linked to poorly understood transportation phenomena rather than to the open or closed nature of the reaction system (Burnham et al., 1988 and references therein). However, using directly comparable on-line data, Schenk and Horsfield (1993) showed that the frequency factors for total petroleum generation from immature Posidonia Shale were the same for the open and closed system experiments (1.1×10^{16} min^{-1}) and that the activation energies distributions were similar (Fig. 2). Using these kinetic parameters and an appropriate thermal history from Düppenbecker et al. (1991), it can be seen (Fig. 3) that the timing of petroleum generation in the Emsland area of the Lower Saxony Basin is unaffected by the open or closed nature of the method used for kinetic parameter definition. Bulk petroleum generation, as equated with bulk pyrolysate generation, is therefore simulated equally using the kinetic parameters from either technique. The uniformly high polarity of laboratory pyrolysates, as compared to natural petroleums, irrespective of closed versus open, hydrous or anhydrous conditions is consistent with the reactions proceeding via similar pathways (see Urov, 1980; Castelli et al., 1990; Larter and Horsfield, 1993). For the conditions of these experiments, it appears that the presence or absence of olefinic compounds is only of superficial importance in determining whether a pyrolysate is oil-like or not.

The major outcome, in the context of the present discussion, is that closed system pyrolysis conditions are not prerequisite for determining the kinetic constants for bulk petroleum generation (as equatable to detected unchromatographed pyrolysate) from marine source rocks. Therefore, either method may be used to infer the true petroleum-generating characteristics of source rocks.

The extension of kerogen quality criteria to petroleum

To illustrate this application, the kerogen in the Posidonia Shale was classified according to genetic potential and kerogen quality. Kerogen quality pa-

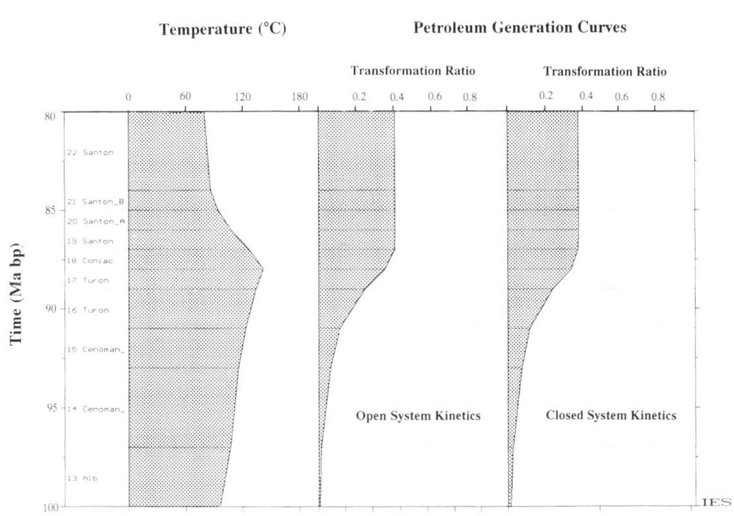

Fig 3. Time plot (100–80 Ma) showing petroleum generation and thermal history for part of the Emsland area, Lower Saxony Basin (after Düppenbecker et al., 1991). Results for open and closed system pyrolysis kinetics are identified accordingly.

rameters were then extended to the bitumen components. Finally, these findings were applied to an exhumed and altered petroleum deposit so as to determine the origin and nature of the original unaltered accumulation. The py–gc results of Muscio et al. (1991) for Posidonia Shale kerogen, asphaltenes and heterocompounds and the isothermal MSSV pyrolysis results of Düppenbecker and Horsfield (1990) on Posidonia Shale kerogen, and the MSSV results of Horsfield et al. (1991) for Holzener Asphaltkalk asphaltenes, provided the basis for this discussion.

Step 1: kerogen typing considerations

The Posidonia Shale of the Lower Saxony Basin was deposited in an epicontinental sea whose water column was stratified and anoxic at the bottom. In the immature state, about 80% of its kerogen consists of alginite; less than 10% is of terrigenous origin. It has a type II composition, with atomic H/C and O/C ratios of around 1.35 and 0.16, respectively. Its genetic potential is very high (immature hydrogen index = 663±43 mg HC/g TOC). Mass balance modelling with normalisation to either mineral matrix or "dead carbon", using the approaches of Cooles et al. (1986) and Rullkötter et al. (1988), revealed that the kerogen in the study area has generated and expelled substantial amounts of petroleum (ca. 60–95% efficiency over R_m = 0.48–1.45%) as a result of an Upper Cretaceous heat pulse (Düppenbecker et al., 1991). The generated petroleum is of the paraffinic-naphthenic type (Tissot and Welte, 1978, p. 374).

Kerogen quality assessment was performed using py–gc and MSSV pyrolysis. In common with the Woodford Shale (Oklahoma, USA), Chattanooga Shale (Tennessee, USA), Kimmeridge Clay (Dorset, U.K.), the lower Toarcian Shales (Paris Basin, France) and many other organic-rich rocks deposited under anoxic marine conditions, the Posidonia Shale kerogen's chain length distribution of *n*-alkenes and *n*-alkanes falls within the Low Wax Paraffinic-Naphthenic-Aromatic Oil-Generating category (P-N-A Oil Low Wax) of Horsfield (1989), as shown in Fig. 4. Despite a relative enrichment of methyl groups in the residual aliphatic fraction of Posidonia Shale kerogen during maturation (Witte et al., 1988), the pyrolysis data points fall in this same category throughout catagenesis and into metagenesis (R_m = 0.48–1.45%), an observation confirmed by artificial maturation studies (Düppenbecker and Horsfield, 1990; Horsfield and Düppenbecker, 1991). This means that original kerogen types can be extrapolated even when hydrogen index is very low and that the original oil-generating potential of highly overmature sediments may be recognised. Importantly, the criterion is also one which can be extended

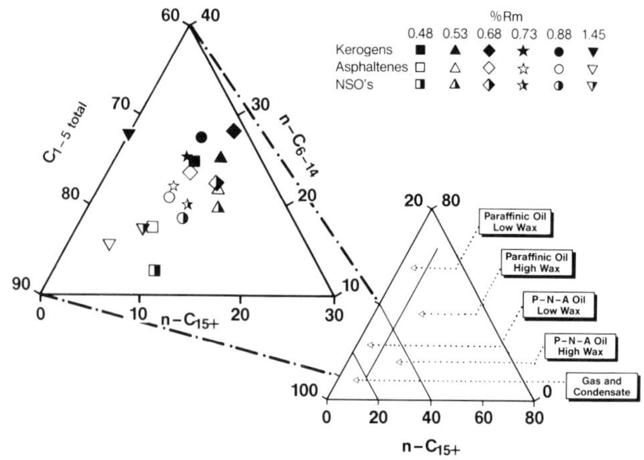

Fig 4. Pyrolysis–gas chromatography of macromolecular organic matter from the Posidonia Shale: relative abundances of C_1–C_5 total hydrocarbons, C_6–C_{14} *n*-alkanes and C_{15+} *n*-alkanes (after Muscio et al., 1991).

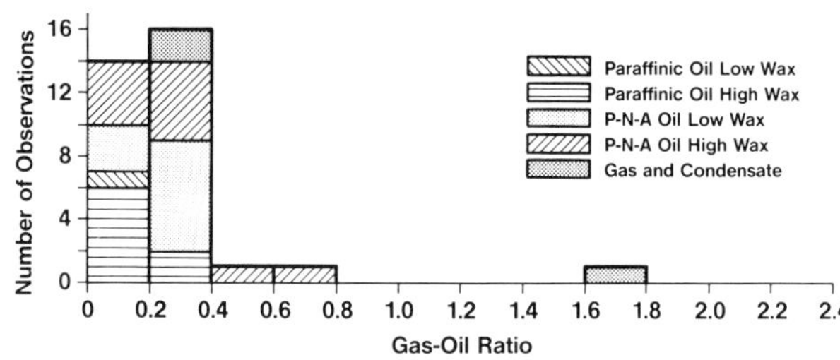

Fig 5. Pyrolysate gas/oil ratio as a function of kerogen quality (categories of Horsfield, 1989).

to rock bitumens and degraded petroleums, as discussed below.

In common with analytical pyrolysates in general (Fig. 5), the ratio of gaseous (C_1–C_5) and heavier (C_{6+}) eluted components (expressed as a gas/oil ratio) in Posidonia Shale kerogen pyrolysates is low and in the same range as that of undersaturated marine-derived petroleums (England and Mackenzie, 1989). Analysis of the aforementioned suite of naturally matured Posidonia Shale samples using py–gc and artificial maturation of the least mature members using MSSV pyrolysis revealed that the gas/oil ratio increases only slightly during catagenesis (ca. 0.15–0.30). The simulation data, expressed as a function of Transformation Ratio, have been used to calculate petroleum densities and establish phase behaviour for the modelling of primary and secondary migration in the Lower Saxony Basin (Düppenbecker et al., 1991). The data for the natural series, shown in Fig. 6, has been extended to rock bitumens and petroleums, as discussed below.

Step 2: extension to rock bitumen

Pyrolysis data for the kerogen, asphaltene and heterocomponent fractions of the Posidonia Shale

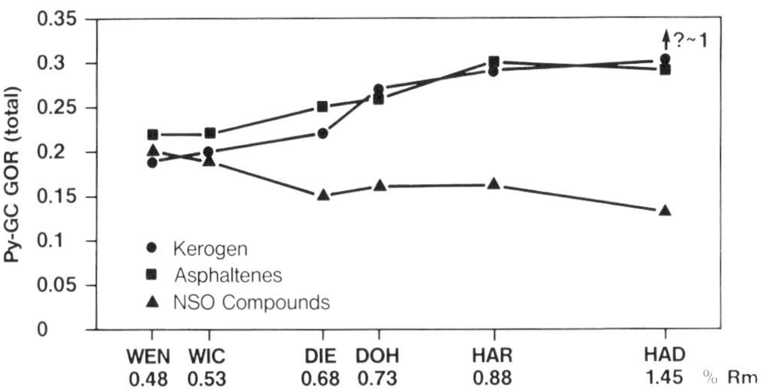

Fig 6. Influence of maturity on macromolecular organic matter composition; py–gc gas/oil ratios for Posidonia Shale.

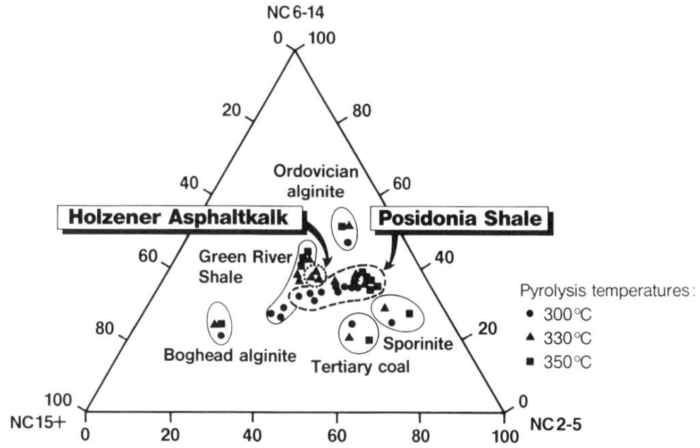

Fig 7. Chain length distribution of normal alkanes in MSSV pyrolysates of immature Posidonia Shale kerogen and Holzener Asphaltkalk asphaltenes.

is shown in Figs. 4 and 6. The pyrolysate chain length distribution for the asphaltenes and heterocompounds are very similar to each other and, as noted for the kerogen fraction, fall in a restricted part of the ternary plot despite the large maturity span (Fig. 4). While the points actually fall in a crescent-shaped evolution path, the data points covering petroleum generation (R_m = 0.53–0.88%) fall in an extremely restricted range. This range is within the Low Wax P-N-A Oil category, directly adjacent to the tight cluster of kerogen data points.

It has been shown by other workers that, despite apparent differences in average molecular weight, the asphaltenes in crude oils show a strong compositional resemblance to those in their source rocks and thence to the parent kerogen (Pelet et al., 1986; Bjorøy et al., 1988; Jones et al., 1988). From this, we deduce that the pyrolysate chain length distribution of asphaltenes in Posidonia Shale-derived crude oils will be the same as that of the asphaltenes in the source rock bitumen at the time of expulsion. Because the data scatter for the oil-generating maturity range is extremely narrow and closely adjacent to the equivalent kerogen data points (see Fig. 4), we believe that organic matter quality can be determined either from kerogens or from genetically related macromolecular organic matter in asphaltene and heterocomponent fractions of rock bitumens and crude oils. Furthermore, the essentially identical pyrolysate gas/oil ratios for the kerogen and asphaltene fractions (Fig. 6) for maturities up to 0.88% R_m suggest that petroleum gas/oil ratio may be determined from both the bitumen and kerogen components of source rocks and, by inference, from asphaltenes in reservoir lithologies.

Step 3: application

The Holzener Asphaltkalk of northern Germany consists of a series of Kimmeridgian–Portlandian bitumen-impregnated limestones which probably represent an exhumed and highly altered multi-pay oil accumulation (Herrmann, 1971). The source of this deposit and the relative mass of unaltered petroleum originally present could be inferred from pyrolysis data using Holzener Asphaltkalk asphaltenes, Posidonia Shale asphaltenes and Posidonia Shale kerogens and the concepts described above.

Carbon isotope data and regional geochemical considerations point towards the original petroleum as having been derived from the Posidonia Shale source rock. MSSV pyrolysis at 330°C/72 h and 350°C/72 h revealed that Posidonia Shale kerogens and asphaltenes plot close together on a ternary diagram. Simulation experiments confirm that the data scatter is very small in the case of mature (>20% conversion) samples and that data points are close to but distinct from those for Holzener Asphaltkalk asphaltenes (Fig. 7). The kerogen-based calibration points in Fig. 7 indicate that an active algal lacustrine source rock may have co-sourced the Holzener Asphaltkalk petroleum. This possibility is currently under examination (Heckers, 1993).

The original form of the accumulation was significantly different to that of today because most of the original reservoired oil was lost, preferentially from pores >25 nm, during post-Cretaceous uplift. Assuming that the original petroleum had a GOR close to 0.2 kg kg^{-1}, based on py–gc data for Holzener Asphaltkalk asphaltenes (Heckers, 1993) and Posidonia Shale kerogens and asphaltenes, and using equating buoyancy and capillary pressure, an original oil column of 100 m was estimated (Horsfield et al., 1991). Petroleum density, water density and interfacial tension values were calculated using a maximum temperature of 100°C and a hydrostatic pressure of up to 15 MPa.

The amount of petroleum lost during alteration was calculated assuming that the gross C_{15+} composition of Posidonia Shale-derived crude oils (Lower

Saxony Basin) and the Holzener Asphaltkalk were representative of the unaltered starting material and altered end-product, respectively. It was also assumed that the total amount of C_{1+} original petroleum could be inferred from the ratio of light to heavy products (C_{15-}/C_{15+}) from pyrolysis. The mass balance calculation revealed that a 50–90% mass loss of petroleum could be attributed to alteration phenomena. GOR values of 0.2 kg kg^{-1} and lower were considered. Figure 5 indicates that this GOR range is valid, even allowing for an algal lacustrine co-source.

These preliminary findings point to macromolecular organic matter in petroleum accumulations, seeps and stains as being a source of data from which the basin modeller might determine, on the one hand, kerogen type/source facies and, on the other, gross compositional characteristics of the original unaltered petroleum expelled from the source rock.

Conclusions

Because it is important to predict the amount and type of petroleum entering traps, and the timing of this event, it is now clear that routine kerogen typing must be based on three practical criteria: namely, genetic potential, kerogen quality and maturation characteristics. The conclusions reached in this paper regarding application of this concept are:

(1) Bulk petroleum generation kinetics determined under closed and open systems are closely similar. Therefore, the classification of kerogen based on open system kinetic parameters is as meaningful an indication of bulk petroleum-generating characteristics as is that based on closed system pyrolysis.

(2) Kerogen quality determinations can be extended to the macromolecular fractions of rock bitumens and petroleums. This means that the modeller might determine (a) kerogen type and source facies, and (b) gross compositional characteristics of original unaltered petroleum charges, based on the analysis of degraded petroleums.

Acknowledgements

The technical support provided by the staff of ICG-4 is gratefully acknowledged, with special thanks to Mr. H. Hardelauf, Mr. F. Leistner and Ms. A. Richter. We thank Prof. D.H. Welte for his support and express our gratitude to Dasag Deutsche Naturasphalt GmbH, BEB Erdgas und Erdöl GmbH, Mobil Erdgas–Erdöl GmbH, Deutsche Texaco AG, C. Deilmann AG, Deutsche Schachtbau- und Tiefbohrgesellschaft mbH and Wintershall AG who provided samples for, and/or who sponsored, the research described here related to the Posidonia Shale.

References

Béhar, F. and Pelet, R., 1984. Characterization of asphaltenes by pyrolysis and chromatography. J. Anal. Appl. Pyrol., 7: 121–135.

Bjorøy, M., Williams, J.A., Dolcater, D.L. and Winters, J.A., 1988. Variation in hydrocarbon distribution in artificially matured oils. Org. Geochem., 13: 901–913.

Burnham, A.K., Braun, L. and Samoun, A.M., 1988. Further comparison of methods for measuring kerogen pyrolysis rates and fitting kinetic parameters. Org. Geochem., 13: 839–845.

Castelli, A., Chiaramonte, M.A., Beltrame, P.L., Carniti, P., Del Bianco, A. and Stroppa, F., 1990. Thermal degradation of kerogen by hydrous pyrolysis. A kinetic study. Org. Geochem., 16: 75–82.

Cooles, G.P., Mackenzie, A.S. and Quigley, T.M., 1986. Calculation of petroleum masses generated and expelled from source rocks. Org. Geochem., 10: 235–245.

Delvaux, D., Martin, H., Leplat, P. and Paulet, J., 1990. Geochemical characterization of sedimentary organic matter by means of pyrolysis kinetic parameters. Org. Geochem., 16: 175–187.

Düppenbecker, S. and Horsfield, B., 1990. Compositional information for kinetic modelling and petroleum type prediction. Org. Geochem., 16: 259–266.

Düppenbecker, S.J., Dohmen, L. and Welte, D.H., 1991. Numerical modelling of petroleum expulsion in two areas of the Lower Saxony Basin, Northern Germany. In: W.A. England and A. Fleet (Editors), Petroleum Migration. Geol. Soc. London, Spec. Publ., 59: 47–64.

England, W.A. and Mackenzie, A.S., 1989. Some aspects of the organic geochemistry of petroleum fluids. Geol. Rundsch., 78: 291–303.

Espitalié, J., Laporte, J.L., Madec, M., Marquis, F., Leplat, P., Paulet, J. and Boutefeu, A., 1977. Méthode rapide de caractérisation des roches mères, de leur potentiel pétrolier et de leur degré d'évolution. Rev. Inst. Fr. Pét., 32: 23–42.

Goff, J.C., 1984. Hydrocarbon generation and migration from Jurassic source rocks in the East Shetland Basin and Viking Graben of the northern North Sea. In: G. Demaison and R.J. Murris (Editors), Petroleum Geochemistry and Basin Evaluation. Am. Assoc. Pet. Geol., Mem., 35: 273–302.

Heckers, J., 1993. Organisch-geochemische und karbonatpetrographische Untersuchungen zur Entstehung des Asphaltkalkvorkommens von Holzen/Ith. Ph.D Thesis, RWTH Aachen.

Hermanrud, C., Eggen, S., Jacobsen, T., Carlsen, E.M. and Pallesen, S., 1990. On the accuracy of modelling hydrocarbon generation and migration: the Egersund Basin oil find, Norway. Org. Geochem., 16: 389–399.

Herrmann, A., 1971. Die Asphaltkalk-Lagerstätte bei Holzen/Ith auf der Südwestflanke der Hils-Mulde. Beih. Geol. Jahrb., 95, 125 pp.

Heum, O.R., Dalland, A. and Meisingset, K.K., 1986. Habitat of hydrocarbons at Haltenbanken (PVT-modelling as a predictive tool in hydrocarbon exploration). In: A.M. Spencer et al. (Editors), Habitat of Hydrocarbons on the Norwegian Continental Shelf. Graham and Trotman, London, pp. 259–274.

Horsfield, B. 1989. Practical criteria for classifying kerogens: some observations from pyrolysis–gas chromatography. Geochim. Cosmochim. Acta, 53: 891–901.

Horsfield, B. and Düppenbecker, S.J., 1991. The decomposition of Posidonia Shale and Green River Shale kerogens using Microscale Sealed Vessel (MSSV) pyrolysis. J. Anal. Appl. Pyrol., 20: 107–123.

Horsfield, B., Disko, U. and Leistner, F., 1989. The microscale simulation of maturation: outline of a new technique and its potential applications. Geol. Rundsch., 78: 361–374.

Horsfield, B., Heckers, J., Leythaeuser, D., Littke, R. and Mann, U., 1991. A study of the Holzener Asphaltkalk, Northern Germany: observations regarding the distribution, composition and origin of organic matter in an exhumed petroleum reservoir. Mar. Pet. Geol., 8: 198–211.

Jones, D.M., Douglas, A.G. and Connan, J., 1988. Hydrous pyrolysis of asphaltenes and polar fractions of biodegraded oils. Org. Geochem., 13: 981–993.

Jones, R.W., 1987. Organic Facies. In: J. Brooks and D.H. Welte (Editors), Advances in Petroleum Geochemistry, Vol. 2. Academic Press, London, pp. 1–89.

Larter, S.R., 1985. Integrated kerogen typing and the quantitative evaluation of petroleum source rocks. In: B.M. Thomas et al. (Editors), Petroleum Geochemistry in Exploration of the Norwegian Shelf. Graham and Trotman, London, pp. 269–286.

Larter, S.R., 1990. Molecular characterisation of kerogen — applications to primary and secondary migration studies and to maturation modelling. Rev. Palaeobot. Palynol., 65: 379–391.

Larter, S.R. and Horsfield, B., 1993. Determination of structural components of kerogen using analytical pyrolysis methods. In: M. Engel and S. Macko (Editors), Organic Geochemistry. Plenum, New York, N.Y. (in press).

Lewan, M.D., 1985. Evaluation of petroleum generation by hydrous pyrolysis experimentation. Philos. Trans. R. Soc. London, Ser. A 315: 123–134.

Mackenzie, A.S. and Quigley, T.M., 1988. Principles of geochemical prospect appraisal. Am. Assoc. Pet. Geol., Bull., 72: 399–415.

Monthioux, M., Landais, P. and Durand, B., 1986. Comparison between extracts from natural and artificial maturation series of Mahakam delta coals. Org. Geochem., 10: 299–311.

Muscio, G.P.A., Horsfield, B. and Welte, D.H., 1991. Compositional changes in the macromolecular organic matter (kerogens, asphaltenes and resins) of a naturally matured source rock sequence from northern Germany as revealed by pyrolysis methods. In: D. Manning (Editor), Organic Geochemistry Advances and Applications in Energy and the Natural Environment. Manchester University Press, Manchester, pp. 447–449.

Pelet, R., Behar, F. and Monin, J.C., 1986. Resins and asphaltenes in the generation and migration of petroleum. Org. Geochem., 10: 481–498.

Rubinstein, I., Spyckerelle, C. and Strauz, O.P., 1979. Pyrolysis of asphaltenes: a source of geochemical information. Geochim. Cosmochim. Acta, 43: 1–6.

Rullkötter, J., Leythaeuser, D., Horsfield, B., Littke, R., Mann, U., Müller, P.J., Radke, M., Schaefer, R.G., Schenk, H.J., Schwochau, K., Witte, E.G. and Welte, D.H., 1988. Organic matter maturation under the influence of a deep intrusive heat source: a natural experiment for quantitation of hydrocarbon generation and expulsion from a petroleum source rock. Org. Geochem., 13: 847–856.

Schaefer, R.G., Schenk, H.J., Hardelauf, H. and Harms, R., 1990. Determination of gross kinetic parameters for petroleum formation from Jurassic source rocks of different maturity levels by means of laboratory experiments. Org. Geochem., 16: 115–120.

Schenk, H.J. and Horsfield, B., 1993. Kinetics of petroleum generation from open versus closed system pyrolysis experiments. Geochim. Cosmochim. Acta (in press).

Solli, H. and Leplat, P., 1986. Pyrolysis–gas chromatography of asphaltenes and kerogens from source rocks and coals — a comparative structural study. Org. Geochem., 10: 313–330.

Sweeney, J. and Burnham, A.K., 1990. Evaluation of a simple model of vitrinite reflectance based on chemical kinetics. Am. Assoc. Pet. Geol., Bull., 74: 1559–1570.

Tissot, B. and Welte, D.H., 1978. Petroleum Formation and Occurrence. Springer-Verlag, Berlin, 538 pp.

Tissot, B., Durand, B., Espitalié, J. and Combaz, A., 1974. Influence of the nature and diagenesis of organic matter in the formation of petroleum. Am. Assoc. Pet. Geol., Bull., 58: 499–506.

Ungerer, P., 1990. State of the art of research in kinetic modelling of oil formation and expulsion. Org. Geochem., 16: 1–25.

Ungerer, P., 1993. Modelling of petroleum generation and expulsion — an update to recent reviews. In: A.G. Doré et al. (Editors), Basin Modelling: Advances and Applications. Norwegian Petroleum Society (NPF), Special Publication 3. Elsevier, Amsterdam, pp. 219–232 (this volume).

Ungerer, P. and Pelet, R., 1987. Extrapolation of the kinetics of oil and gas formation from laboratory experiments to sedimentary basins. Nature, 327: 52–54.

Urov, K.E., 1980. Thermal decomposition of kerogens mechanism and analytical application. J. Anal. Appl. Pyrol., 1: 323–338.

Witte, E.G., Schenk, H.J., Müller, P.J. and Schwochau, K., 1988. Structural modifications of kerogen during natural evolution as derived from ^{13}C CP/MAS NMR, IR spectroscopy and Rock-Eval pyrolysis of Toarcian shales. Org. Geochem., 13: 1039–1044.

B. HORSFIELD — *Institute for Chemistry and Dynamics of the Geosphere (ICG-4) at the Jülich Research Centre (KFA), P.O. Box 1913, D-5170 Jülich, Germany*

S.J. DÜPPENBECKER — *Institute for Chemistry and Dynamics of the Geosphere (ICG-4) at the Jülich Research Centre (KFA), P.O. Box 1913, D-5170 Jülich, Germany*
Present address: *B.P. Research Centre, Chertsey Road, Sunbury-on-Thames, TW16 7LN, Great Britain*

H.J. SCHENK — *Institute for Chemistry and Dynamics of the Geosphere (ICG-4) at the Jülich Research Centre (KFA), P.O. Box 1913, D-5170 Jülich, Germany*

R.G. SCHAEFER — *Institute for Chemistry and Dynamics of the Geosphere (ICG-4) at the Jülich Research Centre (KFA), P.O. Box 1913, D-5170 Jülich, Germany*

An inverse method for determining kinetic parameters for kerogens from routine pyrolysis data, and application to the Peace River arch area, Alberta, Canada

Kazuo Nakayama and Song Cao

Because mathematical models for basin analysis are widely used in exploration, the quality of input parameters, especially those for hydrocarbon generation models, has become important. Results obtained from generation models often depend critically upon the quality of input kinetic parameters.

A method is proposed here to estimate the kinetic parameters for hydrocarbon generation from kerogens using Hydrogen Index (HI) values supplied by routine Rock-Eval analyses. The HI represents the residual potential of the source rock for hydrocarbon generation from kerogen. By trial-and-error comparison of residual potentials calculated using our kinetic parameters with observed residual potentials (HI), we can find suitable kinetic parameters that minimize the MSR (mean square residual) between calculated and observed residual potentials.

We define the term "geochemical facies" as a group of rocks with similar kinetic parameters for hydrocarbon generation. Given the kinetic parameters for the immature kerogen of a geochemical facies and its subsequent temperature history, the amount of hydrocarbons that has been generated can be calculated easily.

In this way, kinetic parameters can be determined from routine, inexpensive Rock-Eval analyses of a significant number of samples from a geochemically homogeneous facies. Multiwell studies are required when a single well does not contain samples from a single geochemical facies covering a significant range of organic maturity.

Application of this method to the Cretaceous and Mississippian source formations in the Peace River arch area, Alberta, Canada, leads us to propose the existence of five areally distinct geochemical facies. We have also derived kinetic parameters for an average kerogen representative of each geochemical facies and have estimated the generation curves for each kerogen in order to determine the hydrocarbon potential and the timing of generation. Sedimentary controls on each geochemical facies are proposed.

Introduction

Use of mathematical models of hydrocarbon generation has become very common in petroleum exploration. Therefore, it has also become increasingly important to know well the geochemical parameters on which the models are based. Such parameters are usually estimated by pyrolysis of immature rock samples in the laboratory. There are three potential problems with this experimental method.

(1) Are the conditions of laboratory pyrolysis valid when applied to natural geologic conditions? In other words, can pyrolysis at high temperature over short time be used to represent a geologic history (low temperature over long time)?

(2) Is the kerogen in an immature rock sample the same as that in the effective source rock which has generated hydrocarbons? The depositional environment of the immature rock in a shallow part of a basin may be different from the mature rock deposited at the basin centre. This facies difference is likely to result in a difference in quality of the source-rock kerogen.

(3) Such analyses are relatively expensive.

On the other hand, much routine Rock-Eval data for various source rocks at various levels of maturation are often available. In this paper we derive a method for estimating the kinetic parameters of kerogen using routinely collected Rock-Eval data instead of through separate, specialized, analyses. Through the application of this method to source rocks from western Canada, we use this new method to delineate the areal distribution of "geochemical facies"; that is, rocks that contain a single type of kerogen with the same kinetic parameters. This paper also speculates on sedimentological controls of these geochemical parameters in western Canada.

Hydrocarbon generation model

Our mathematical model assumes that hydrocarbon generation consists of a set of n parallel chemical

Basin Modelling: Advances and Applications edited by A.G. Doré et al. NPF Special Publication 3, pp. 251–263, Elsevier, Amsterdam.

reactions, each of which follows the kinetics of kerogen degradation (Tissot and Welte, 1984). For the jth reaction, first-order chemical kinetics are assumed:

$$\frac{dC_{k_j}}{dt} = -k_j C k_j \tag{1}$$

where C_k = remaining concentration of kerogen capable of undergoing this reaction, k = reaction rate constant, and t = time.

In turn, k can be expressed by the Arrhenius equation:

$$k_j = A_j \, e^{-E_j/RT} \tag{2}$$

where A = frequency factor, E = activation energy, R = gas constant, and T = absolute temperature. If the initial concentration of kerogen is X_0, integrating eq. 1 gives:

$$C k_j = X_{0_j} \exp[-k_j t] \tag{3}$$

Equation 4 gives the amount of hydrocarbons C_{h_j} generated by the jth reaction over a time period t:

$$\begin{aligned} C_{h_j} &= X_{0_j} - \int_0^t C_{k_j} \, dt \\ &= X_{0_j} \left(1 - \int_0^t \exp[-k_j \cdot t] \, dt\right) \end{aligned} \tag{4}$$

Integrating eq. 2 into eq. 4, we obtain:

$$C_{h_j} = X_{0_j} \left\{1 - \int_0^t \exp\left[-A_j \cdot e^{-E_j/RT} t\right] dt\right\} \tag{5}$$

These equations form the basic concept of the so-called "kinetic" models for hydrocarbon generation. When temperature T is specified at any time (from knowledge of the burial and thermal history), the unknown parameters in eq. 5 are original genetic potential X_0, activation energy E and frequency factor A.

The total amount of hydrocarbons generated (C_H) is given by summing up C_{h_j} for all n reaction series (Fig. 1, eq. 6).

$$\begin{aligned} C_H &= \sum_{j=1}^{n} C_{h_j} \\ &= \sum_{j=1}^{n} \left\{X_{0_j} \left[1 - \int_0^t \exp\left(-A_j \, e^{-E_j/RT} t\right) dt\right]\right\} \end{aligned} \tag{6}$$

Because we have three variables for each reaction set, for the n parallel reaction sets assumed for kerogen degradation, we need to specify $3n$ unknown values.

Although Tissot and Welte (1978, p. 504) proposed the values of these parameters for three types of kerogen, Akihisa (1977) found different lower activation energies for the kerogen from the Japanese oil/gas fields. It was suggested that the different values for geochemical parameters should be used depending on the geohistory of the modelled area. The values may vary from basin to basin according to their geological background (Tissot and Welte, 1984).

Method of estimating kinetic parameters for kerogen

The Hydrogen Index (HI) value from Rock-Eval pyrolysis analysis represents the present-day residual hydrocarbon-generation potential of a rock sample (Fig. 1). Therefore, if we know the initial hydrocarbon potential of this rock ($\sum X_{0_j}$), the difference between the initial and residual potentials should be the amount of hydrocarbons that has been generated from this rock sample in the past ($C_{H_{ob}}$).

$$C_{H_{ob}} = \sum_{j=1}^{n} X_{0_j} - \mathrm{HI} \tag{7}$$

In practice, for a given type of kerogen, if HI is plotted against a maturity indicator (for example, TTI or vitrinite reflectance), HI values generally decrease as maturity increases. By extrapolating this trend into the immature zone a value for initial hydrocarbon potential can be estimated.

The error function tau (τ) is defined here as the mean square residual (MSR) between the observed value ($C_{H_{ob}}$) and the value (C_H) calculated from eq. 6:

$$\begin{aligned} \tau &= \sum_{i=1}^{m} \left[\frac{(C_{H_{ob\,i}} - C_{H_i})^2}{m}\right] \\ &= \sum_{i=1}^{m} \frac{\left\{\mathrm{HI}_i - \left[\sum_{j=1}^{n} X_{0_j} \int_0^t \exp\left(-A_j \, e^{-E_j/RT} t\right) dt\right]\right\}^2}{m} \end{aligned} \tag{8}$$

where m is the number of measured values. This equation can be interpreted as follows. If we assume values for X_{0_j}, A_j, and E_j for all j sets of reactions, τ is a function of the m measured HI values, provided that a temperature history of the area can be derived properly from other geological data. If the assumptions about the geochemical parameters are correct, the error function τ theoretically becomes zero. Therefore, by changing values of X_{0_j}, A_j, and E_j by trial and error, we can find the best set of param-

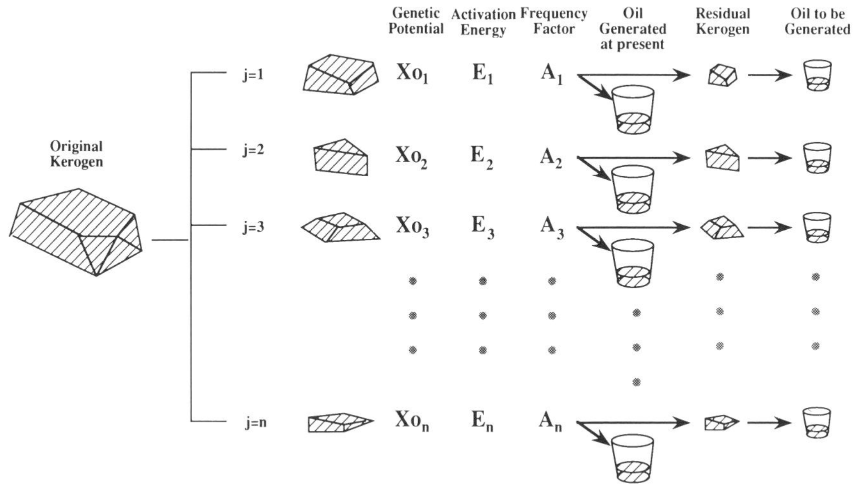

Fig. 1. Schematic diagram explaining parallel reaction series in kerogen. Provided that temperature history and geochemical parameters are given, the model simulates the generation process in the past and calculates the amount of oil remaining to be generated (the equivalent of the Hydrogen Index of Rock-Eval analysis).

eters (that is, the one that gives the minimum in the mean square residual (MSR) between observed and calculated HI values).

It is unrealistic, however, to expect to determine $3n$ independent parameters by trial and error when the number n of parallel reactions is large ($n > 6$ in all existing generation models; e.g., Ungerer, 1984; Ungerer et al., 1986; Burnham and Braun, 1988), even if a systematic search by computer is used. In this paper we decrease the number of parameters by taking into account the following constraints from fundamental geochemistry.

(1) A distribution of activation energies (E_j) and initial hydrocarbon potentials (X_{0_j}) can be treated as following a normal Gaussian distribution (Ungerer and Pelet, 1987; Burnham and Braun, 1988; D. Waples, pers. commun., 1990). Once a normal distribution is assumed, the distribution can be defined by specifying only three parameters regardless of the number of reactions. These three parameters are average activation energy (E_{ave}), standard deviation (σ), and total genetic hydrocarbon potential ($\sum X_{0_j}$).

(2) Because the activation energy (E) relates to the strength of reactive bonds in kerogen, a range of activation energies should exist under natural conditions. It is reported from recent experiments that the range of activation energies is limited to 40–70 kcal/mol (Burnham and Braun, 1988), 45–60 kcal/mol (Ungerer, 1990) or 45–56 kcal/mol (Klomp and Wright, 1990).

(3) It is possible to treat the frequency factor A as a constant. For example, Ungerer and Pelet (1987) found that the frequency factor lies between 3.16×10^{27} and 3.16×10^{31} $(\mathrm{m.y.})^{-1}$.

By applying the above constraints, the total hydrocarbon-generation potential ($\sum X_{0_j}$), average activation energy (E_{ave}), and standard deviation (σ) can be optimized from HI values of Rock-Eval data. Therefore, in the method described below the amounts of hydrocarbons generated are estimated from the data given by naturally matured samples instead of by pyrolysis of immature samples. It should be noted that this method assumes that the burial and thermal histories can be obtained accurately from regional geology, measured temperatures, and thermal indicators (e.g. Waples et al., 1992).

Kinetic parameters and their geological nature

From the above discussion, it can be expected that kerogens of the same origin will follow the same generation path during maturation (Tissot and Welte, 1984). Therefore, if we plot HI against a suitable maturity indicator for all the data obtained from a single area, kerogens will fall into distinct groups. HI observations at various maturation levels are necessary to distinguish between groups of kerogens. Kinetic parameters of each group of kerogens can then be determined using the method described in the previous section.

Such a group of kerogens can be defined as a "geochemical facies". Each geochemical facies is described by the geochemical parameters used in defining it, and by other observed data (including geochemical parameters such as TOC). The areal and stratigraphic extent of each geochemical facies can then be delineated in the study area. Total organic carbon (TOC) represents the quantity of kerogen, whereas a set of kinetic parameters for kerogen is an indicator of the quality of kerogen and its capacity to produce hydrocarbons.

Factors that affect the geochemical character of

kerogen include both depositional environment and early diagenetic processes. If organic matter is deposited under oxic conditions, some part would have been degraded before burial. Differences in the average activation energy (E_{ave}) and total hydrocarbon-generative potential ($\sum X_{0_j}$) may be due to differences in degree of oxidation, as well as to differences in type of input organic matter. Large standard deviations (σ) suggest that the facies consists of a mixture of different original kerogens, because pure kerogens have narrow activation-energy distributions.

The term "geochemical facies" is somewhat different from "organic facies" (Demaison and Holck, 1983) or "organofacies" (Burwood et al., 1990). A geochemical facies is a group of kerogens that have the same hydrocarbon-generation behaviour during maturation.

Application to Peace River arch area, Alberta, Canada

It should be possible to deduce information on geological environments including both depositional and early diagenetic situations if we can differentiate "geochemical facies" by analyzing the data from a large area. For this purpose the data from eight wells in the Peace River arch area were selected (Fig. 2). The data used are S2 and TOC values from Rock-Eval analysis, geothermal gradient, formation tops and the corresponding absolute ages.

The regional geology of this area is summarized as follows. A major unconformity of pre-Cretaceous age divided the geologic sequence into two large units: a Palaeozoic unit and a Cretaceous/younger unit. The sediments in both units thin eastward toward the Canadian Shield and thicken westward into the Rocky Mountains, where the formations are abruptly cut by the thrust zone. The isopachous map of two typical formations of Mississippian and Early Cretaceous age suggests that subsidence was more active westward towards the edge of the Rocky Mountains and that the formations dipped westward through most of geologic time. As a consequence, the deepest part of the basin was always located in the west, and shallow shelf environments were dominant in the east. In this paper, the areal distributions of different geochemical facies have been delineated for samples from these two typical source formations.

For the first step, the TTI value for each sample was calculated as an indicator of maturity, assuming the burial history from the formation tops and ages. No correction was made for compaction, and a constant value for geothermal gradient through all the geologic time was assumed based on the present-

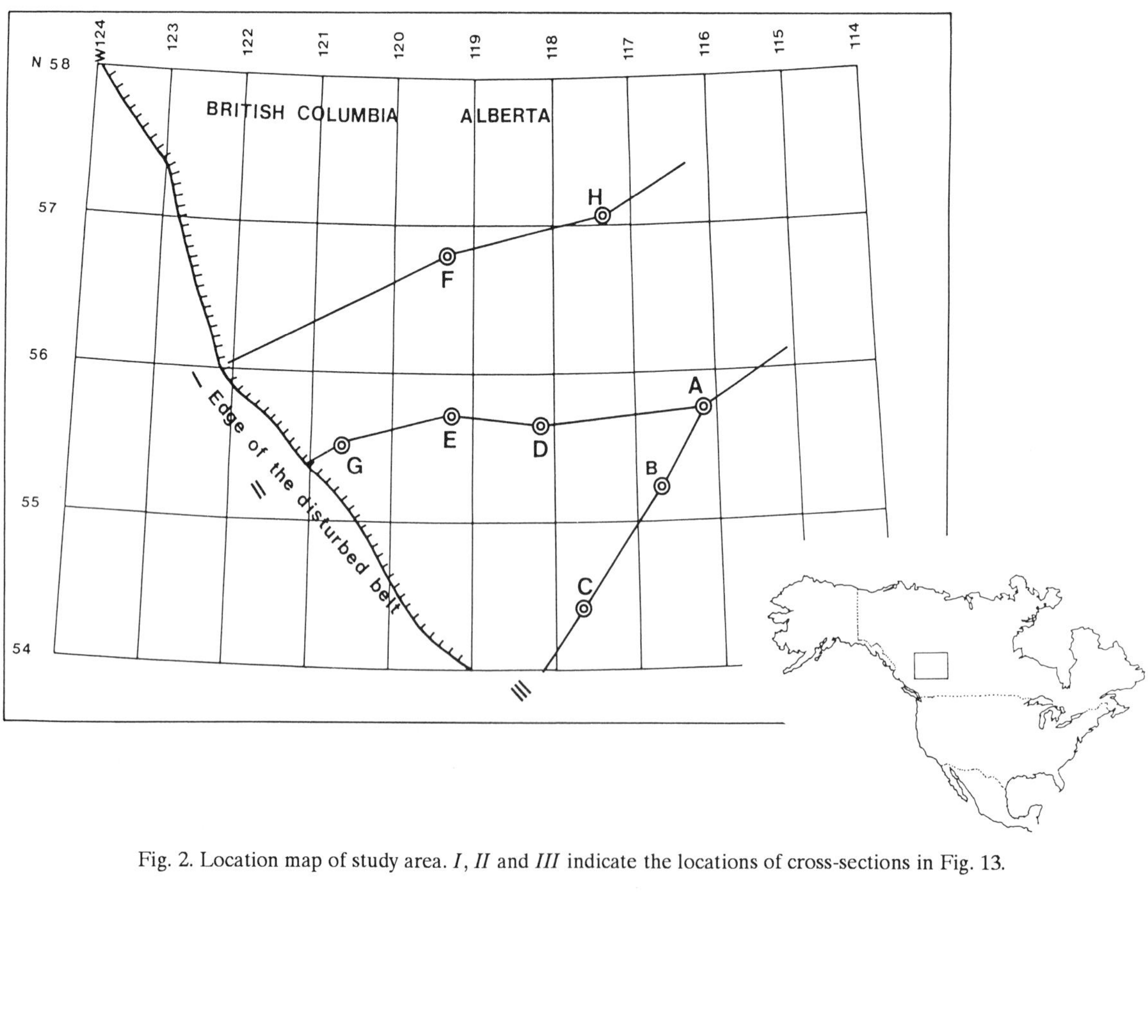

Fig. 2. Location map of study area. *I*, *II* and *III* indicate the locations of cross-sections in Fig. 13.

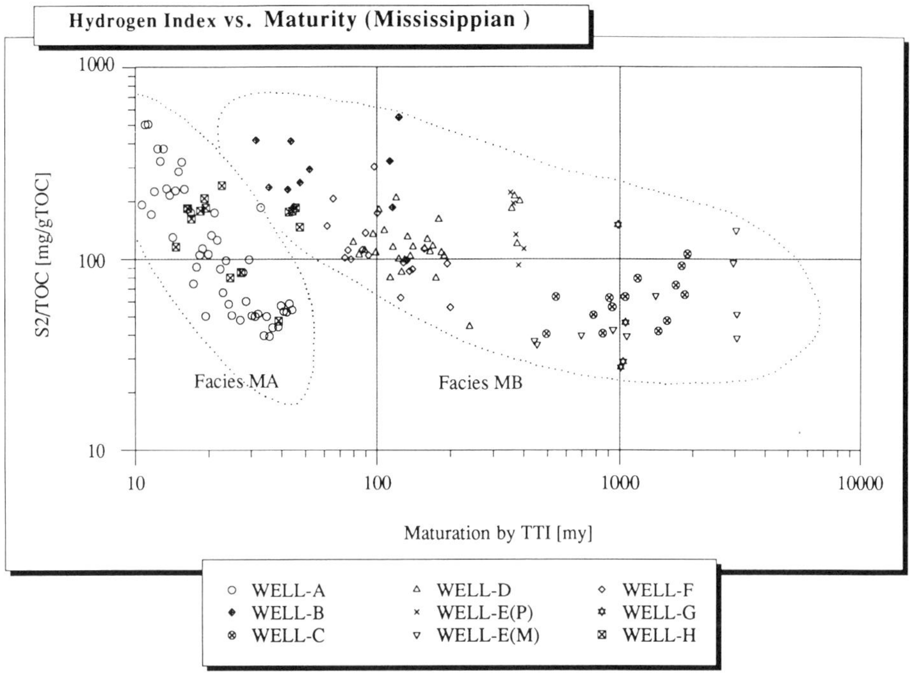

Fig. 3. Plot of Hydrogen Index vs. maturation for Mississippian samples. Geochemical Facies MA and MB are proposed.

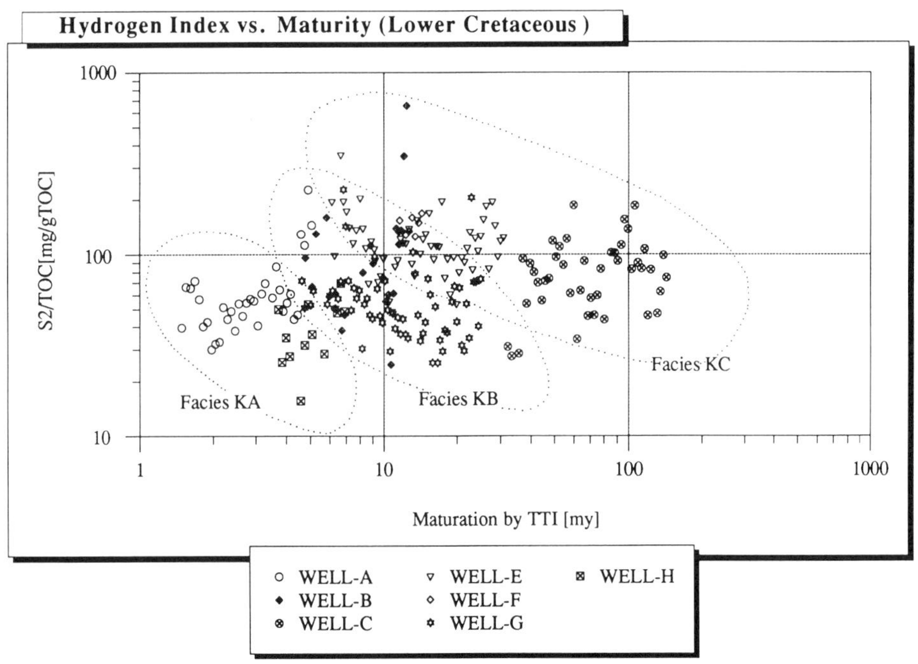

Fig. 4. Plot of Hydrogen Index vs. maturation for Lower Cretaceous samples. Geochemical Facies KA, KB and KC are proposed.

day gradient. Erosion of 1500 m as the last stage of the geohistory was assumed at all locations. The assumption of a constant geothermal gradient and erosion are probably not fully justified. With these limitations, it must be noted that the results are only effective as a first approximation. However, the distinctions we made among the various geochemical facies are, we believe, reliable.

For the next step, HI values were plotted against TTI values. We omitted data where TOC values were lower than 0.5% because these samples were not functioning as important source rocks. As explained in the previous section, the HI values are expected to decrease as maturation increases. With this trend in mind, the plotted values can be differentiated into two groups for the Mississippian and three groups for the Lower Cretaceous (Figs. 3 and 4). Within each group the samples contain kerogen of similar

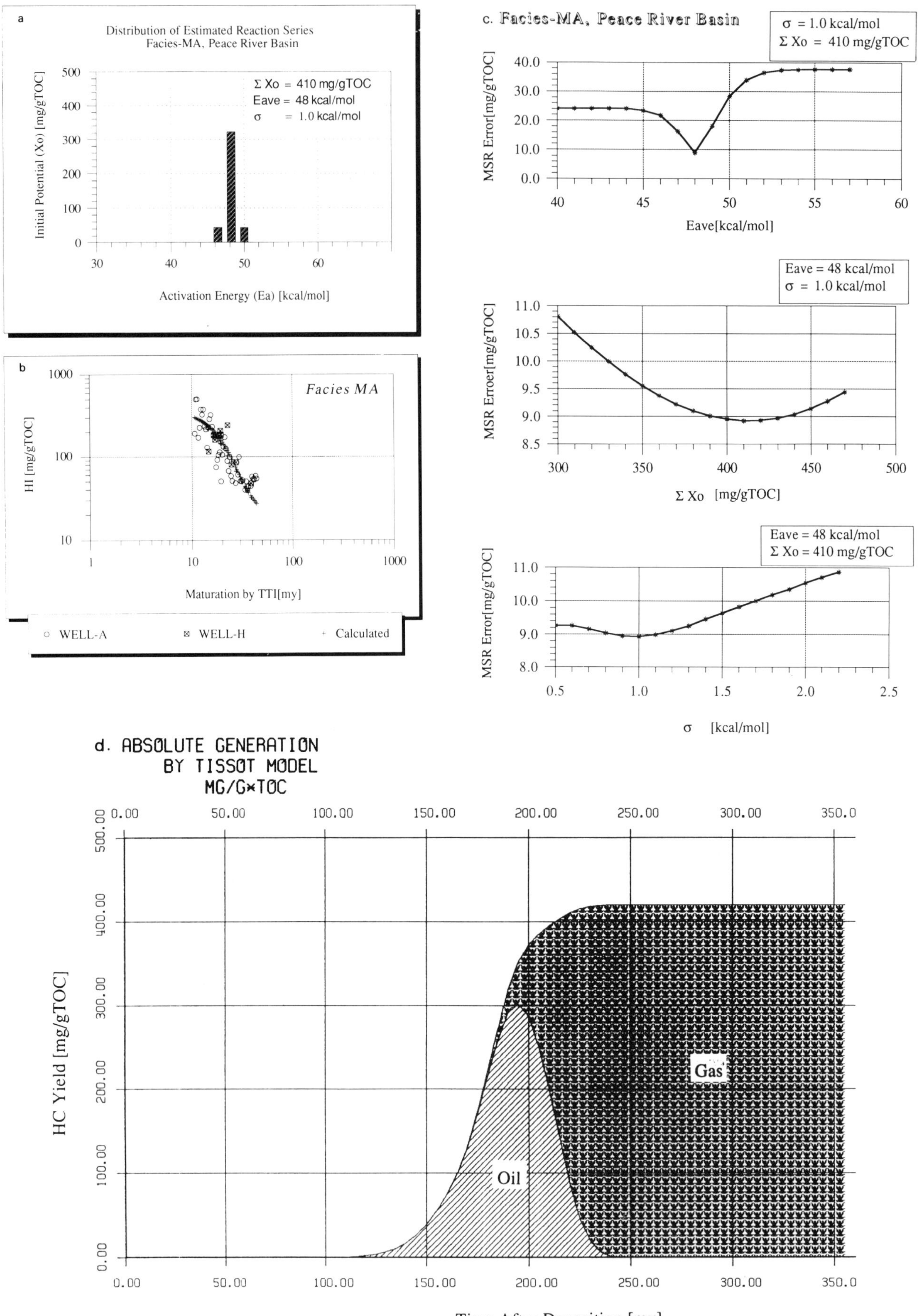

Fig. 5. Geochemical character of Facies MA. (a) Distribution of activation energies. (b) Plot of Hydrogen Index against maturity (observed and calculated). (c) Graph showing mean square residual for the parameters (E_{ave}, ΣX_0, σ). (d) Estimated generation curve (heating rate = 0.5/m.y.).

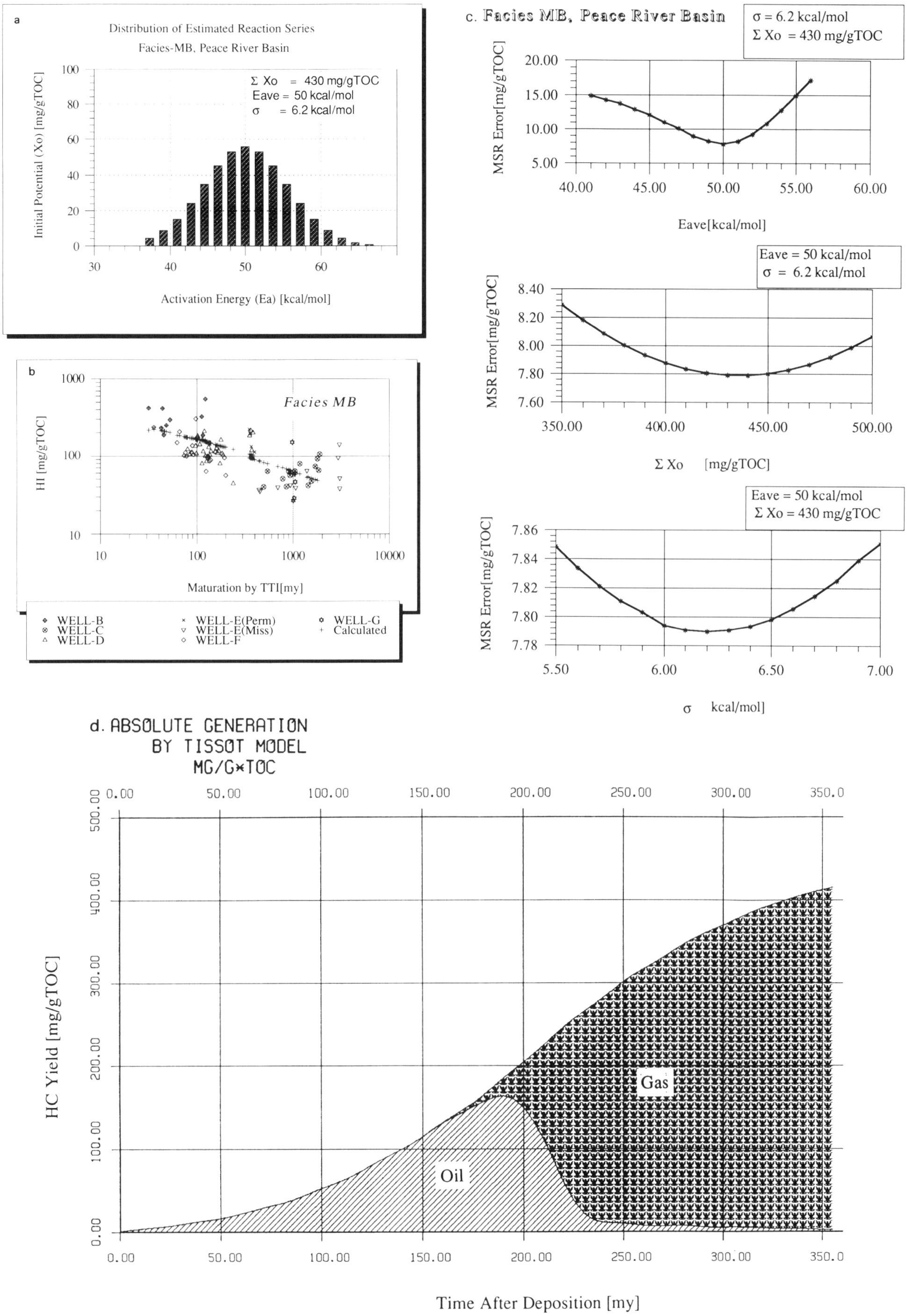

Fig. 6. Geochemical character of Facies MB. (a) Distribution of activation energies. (b) Plot of Hydrogen Index against maturity (observed and calculated). (c) Graph showing mean square residual for the parameters (E_{ave}, ΣX_0, σ). (d) Estimated generation curve (heating rate = 0.5/m.y.).

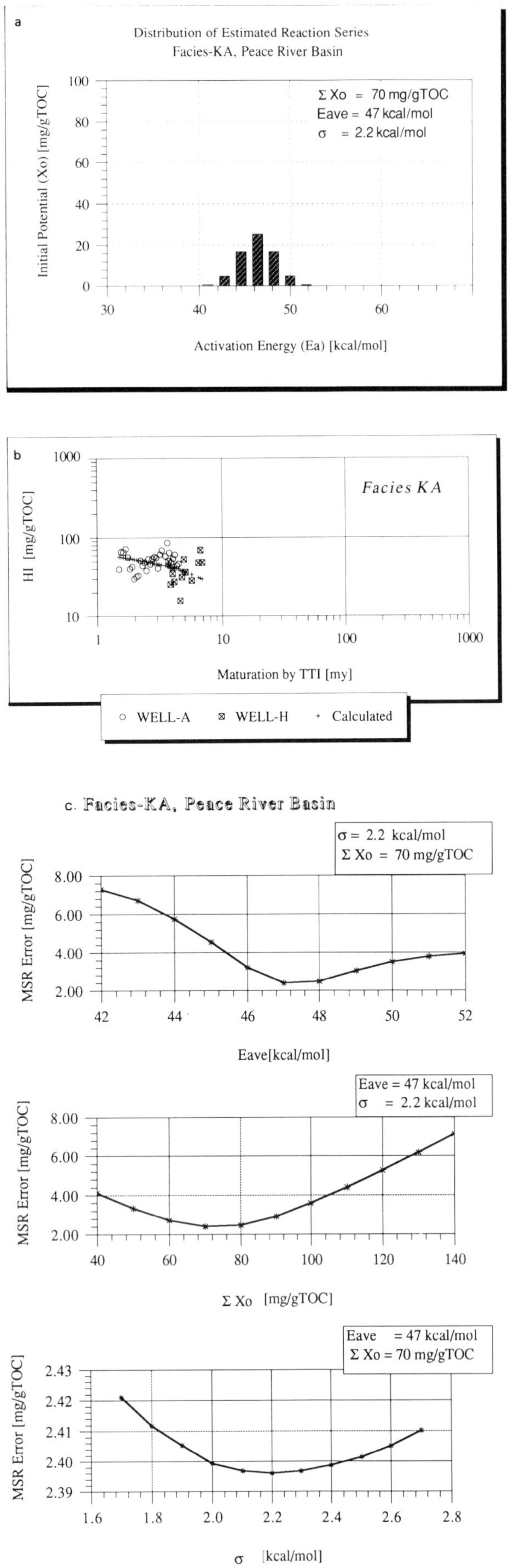

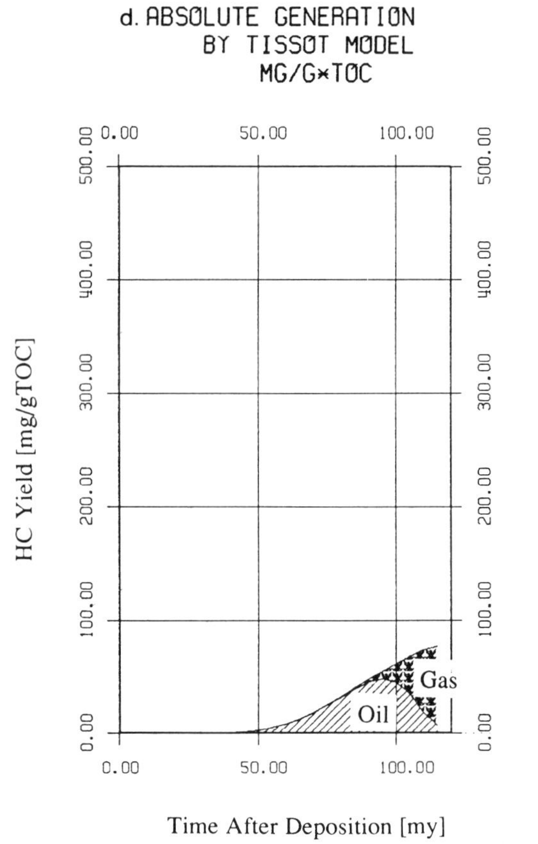

Fig. 7. Geochemical character of Facies KA. (a) Distribution of activation energies. (b) Plot of Hydrogen Index against maturity (observed and calculated). (c) Graph showing mean square residual for the parameters (E_{ave}, ΣX_0, σ). (d) Estimated generation curve (heating rate = 1.0/m.y.).

geochemical character at their immature stage. Each group can be referred to as a "geochemical facies", as defined in the previous section.

The next step is to determine the kinetic parameters for the kerogen in each group by using the method previously described. The characteristics of the kerogens are shown in Figs. 5–9, including the resultant distribution of activation energies (a), graphs showing the fit between observed and calculated HI values (b and c), and hydrocarbon-generation curves (d). The total genetic potential ($\sum X_{0_j}$) indicates that Facies KA is not as good a source rock as Facies KB and KC. Facies KB may be a mixture of Facies KA and KC because of its intermediate parameter values. Although the total genetic potentials are the same for Facies MA and MB, the generation curves are significantly different because of differences in the distribution of activation energies. The standard deviation σ, which controls the distribution of activation energies, is an indicator of the homogeneity of the kerogen. Facies MA is therefore interpreted as consisting of kerogen of a single origin, whereas Facies MB is less homogeneous.

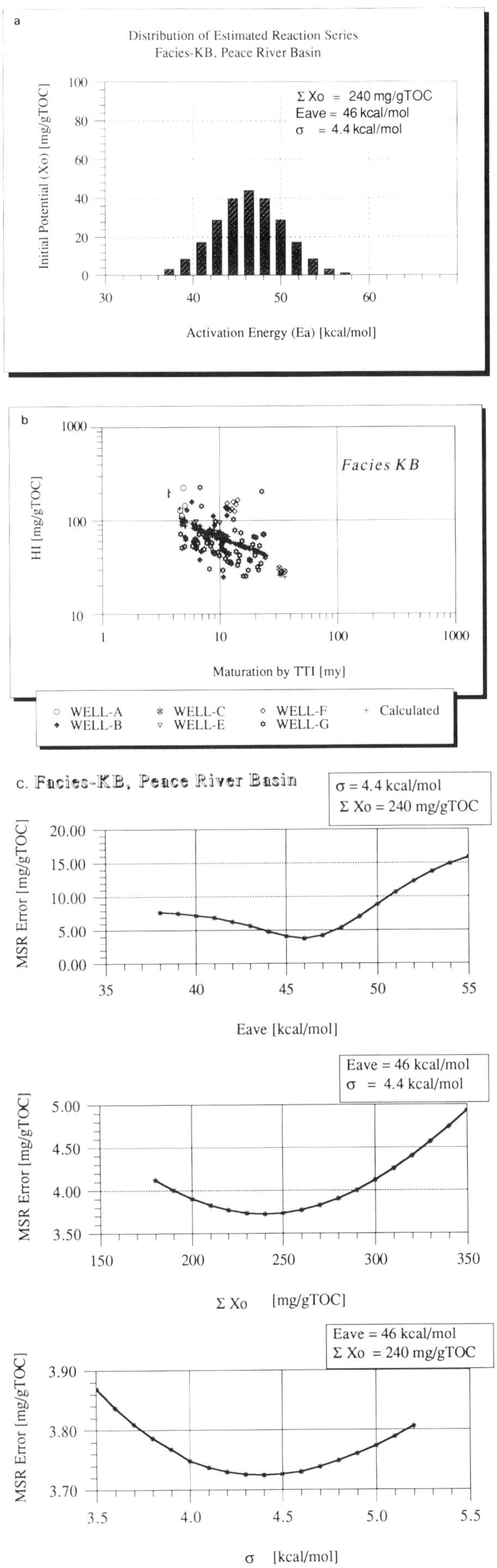

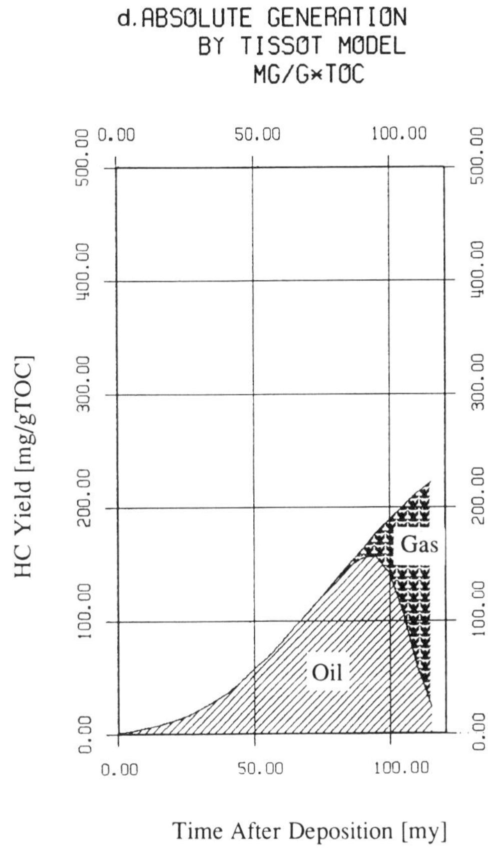

Fig. 8. Geochemical character of Facies KB. (a) Distribution of activation energies. (b) Plot of Hydrogen Index against maturity (observed and calculated). (c) Graph showing mean square residual for the parameters (E_{ave}, ΣX_0, σ). (d) Estimated generation curve (heating rate = 1.0/m.y.).

The areal distribution of each geochemical facies is shown in Figs. 10–13. The kerogens with high genetic potential are mainly concentrated in the thicker parts of each formation, whereas those with low genetic potential correlate with the thinner parts. We infer from this observation that the character of kerogen is mainly controlled by its depositional environment. In the Early Cretaceous, oxic conditions were prevalent on the eastern edge of the study area where less sediment was deposited. The areal distribution of Facies KB implies that some mixing, probably by turbidite flow, has occurred in the southern part of the area. The estimated flow was from east to west, and the sediment was then spread in a north–south direction by basinal currents.

Thus "geochemical facies" can be a new indicator for interpreting sedimentologic environments within a basin, and can provide information which is completely different from that obtained from conventional geological measurements on physical properties of rocks.

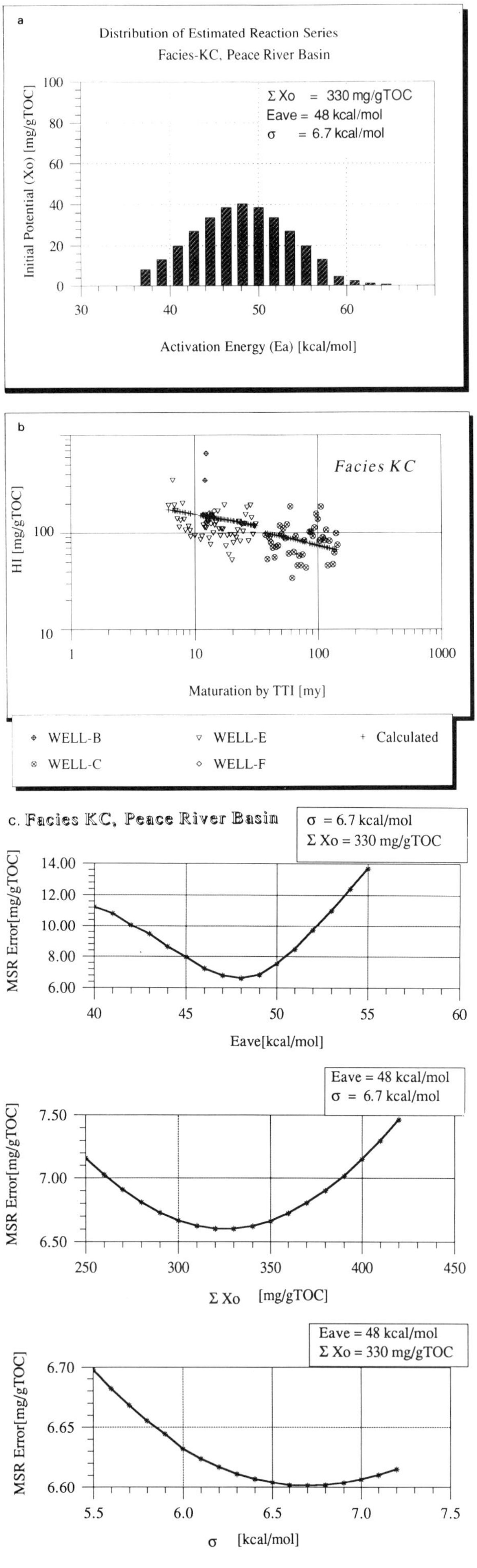

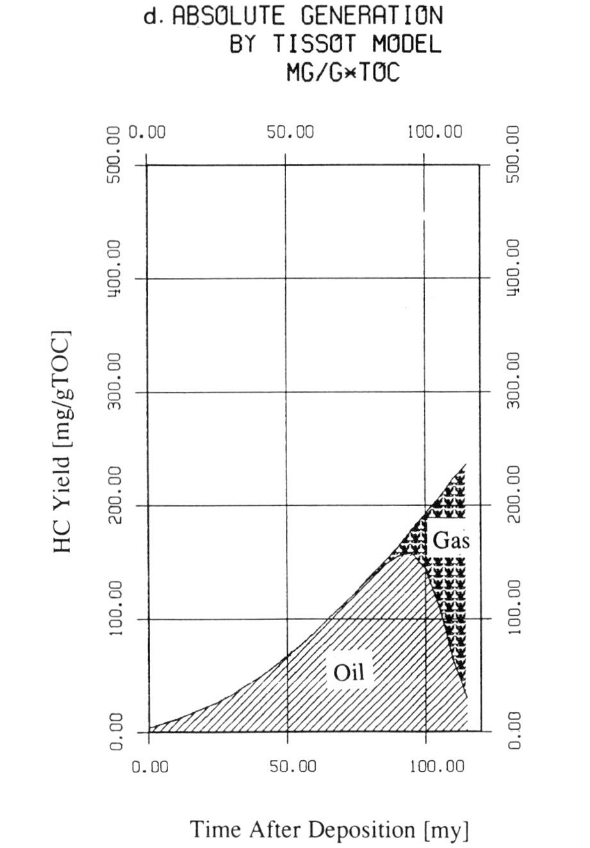

Fig. 9. Geochemical character of Facies KC. (a) Distribution of activation energies. (b) Plot of Hydrogen Index against maturity (observed and calculated). (c) Graph showing mean square residual for the parameters (E_{ave}, ΣX_0, σ). (d) Estimated generation curve (heating rate = 1.0/m.y.).

Conclusions

The kinetic parameters of kerogen can be derived from routine pyrolysis (Rock-Eval) data of naturally matured source rocks. This method employs a least-squares approximation to estimate mean values for such parameters. The advantages of this method are that: (1) no special treatment is necessary to collect the basic data; and (2) the same samples are used that have actually generated hydrocarbons in a basin. That is, data are obtained from naturally matured rocks instead of by experimentally heating immature source rocks in the laboratory over short time periods at high temperatures.

A source rock facies which contains kerogens with the same kinetic parameters is called a "geochemical facies". A geochemical facies depends on depositional environment and early diagenesis. "Geochemical facies" can be regarded as new indicators which are based on somewhat different properties from conventional sedimentary facies or organic facies. A "geochemical facies" indicates not only the chemical characteristics of source rocks (knowledge of which

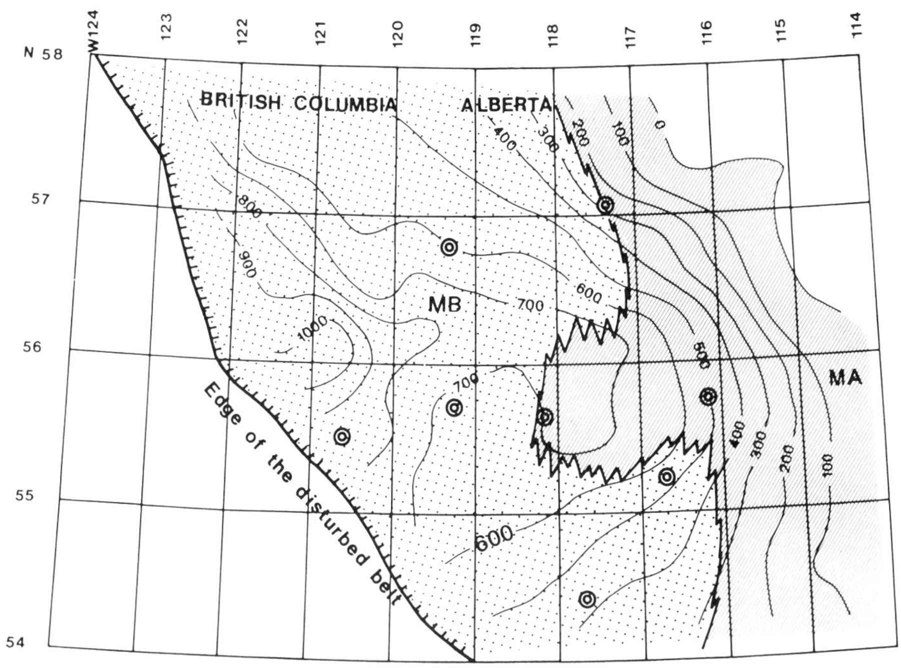

Fig. 10. Areal distribution of each facies (Mississippian). Distribution of Facies MA corresponds to the eastern parts of the basin, whereas Facies MB occupies the western parts of the basin. Thin sediments in the eastern parts are considered to have been deposited in a shallow environment. Contours show the thickness of the formation in meters.

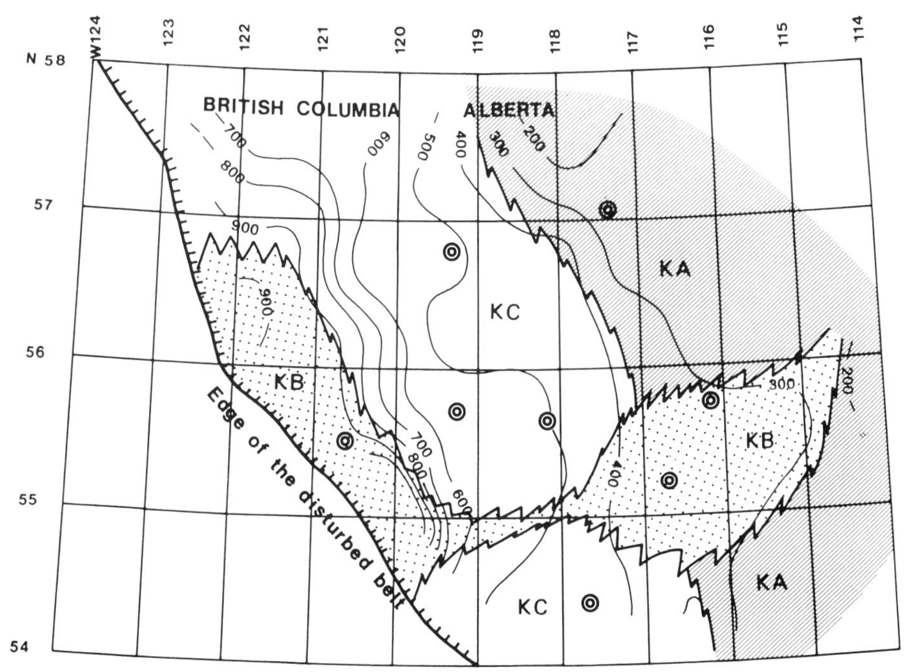

Fig. 11. Areal distribution of each facies (lower part of Lower Cretaceous). Distribution of Facies KA is limited to the eastern parts, and Facies KC is spread over the western parts. Facies KB, considered to be a mixture of KA and KC, occurs on the southern slope and in the deep part of the basin. Facies KB seems to be related regionally to distribution by turbidity currents. (For further explanation see Fig. 10.)

can increase the accuracy of estimating hydrocarbon generation potential), but also of sedimentary environments (which may be useful for reconstructing a basin history). When this method was applied to the Peace River arch area of Canada, source rocks in the Cretaceous and Mississippian strata could be divided into areally consistent groups.

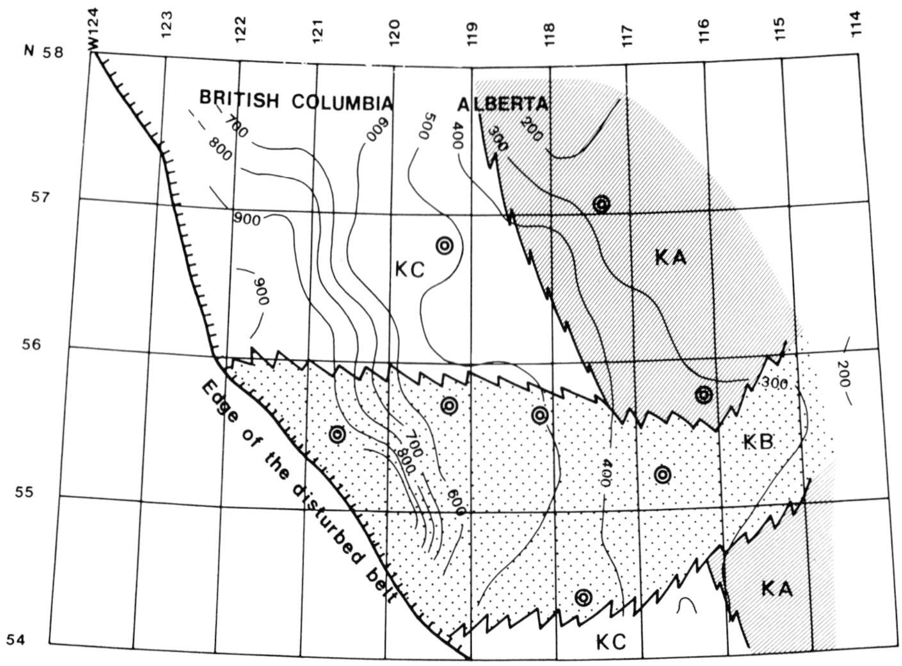

Fig. 12. Areal distribution of each facies (upper part of Lower Cretaceous). The same pattern of distribution for Facies KA, KB and KC is observed as for the lower part of the Lower Cretaceous. The area for Facies KB increases, possibly as a result of turbidite activity. (For further explanation see Fig. 10.)

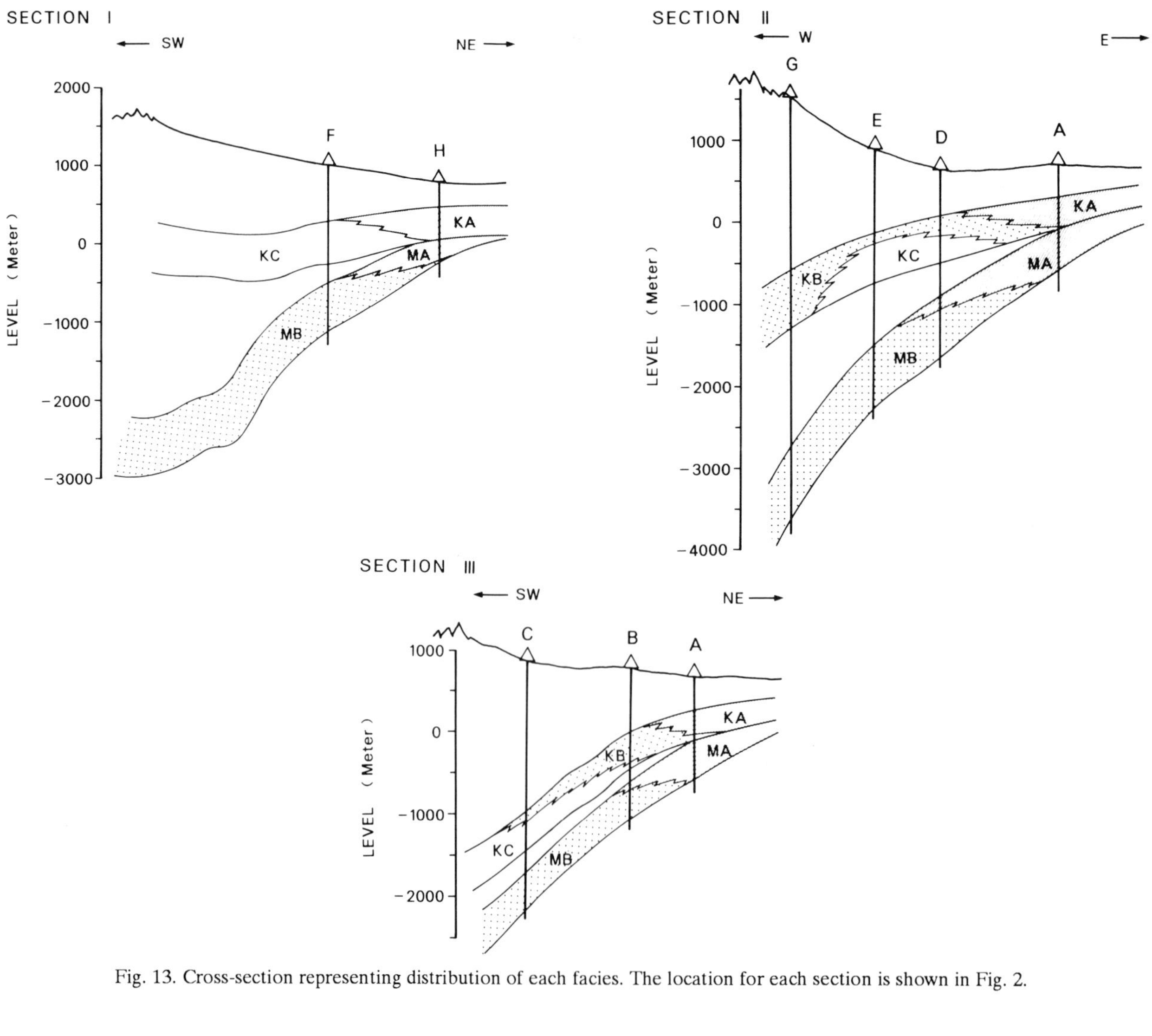

Fig. 13. Cross-section representing distribution of each facies. The location for each section is shown in Fig. 2.

Acknowledgements

We thank Japan Petroleum Exploration Co., Ltd. and the Alberta Research Council for permission to publish this article. We also thank D. Waples of Japan National Oil Corporation for discussions and useful suggestions for the chemical constraints on kerogen kinetics. He reviewed this manuscript and corrected the English. Our gratitude is extended to A.G. Doré of Conoco Norway, editor for this volume, and anonymous reviewers for their critical comments on the first version.

References

Akihisa, K., 1977. Kinetic study of petroleum source rocks (2nd Report): experimental study of the thermal decomposition of kerogen accompanied with the generation of O–N–S compounds. J. Jpn. Assoc. Pet. Technol., 42: 363–369.

Burnham, A.K. and Braun, R.L., 1988. A detailed kinetic and thermodynamic model of petroleum formation, destruction, and expulsion. Lawrence Livermore National Laboratory Report UCRL-99776.

Burwood, R., Cornet, P.J., Jacobs, L. and Paulet, J., 1990. Organofacies variation control on hydrocarbon generation: a Lower Congo Coastal Basin (Angola) case history. Org. Geochem., 16: 325–338.

Demaison, G.J. and Holck, A.J.J., 1983. Predictive source bed stratigraphy: an application of the organic facies concept to basin evaluation. 11th World Petroleum Congress.

Klomp, U.C. and Wright, P.A., 1990. A new method for the measurement of kinetic parameters of hydrocarbon generation from source rocks. Org. Geochem., 16: 49–60.

Tissot, B.P. and Welte, D.H., 1978. Petroleum Formation and Occurrence. Springer-Verlag, Berlin, 538 pp.

Tissot, B.P. and Welte, D.H., 1984. Petroleum Formation and Occurrence. Springer-Verlag, Berlin, 2nd ed., 699 pp.

Ungerer, P., 1984. Models of petroleum formation: how to take into account geology and chemical kinetics. In: G. Demaison (Editor), Thermal Phenomena in Sedimentary Basins. Technip, Paris, pp. 235–246.

Ungerer, P., 1990. State of the art of research in kinetic modeling of oil formation and expulsion. Org. Geochem., 16: 1–25.

Ungerer, P. and Pelet, R., 1987. Extrapolation of the kinetics of oil and gas formation from laboratory experiments to sedimentary basins. Nature, 327: 52–54.

Ungerer, P., Espitalié, J., Marquis, F. and Durand, B., 1986. Use of kinetic models of organic matter evolution for the reconstruction of paleotemperatures. Application to the Gironville well (France). In: J. Burrus (Editor), Thermal Modeling in Sedimentary Basins. Edition Technip, Paris, pp. 531–546.

Waples, D.W., Kamata, H. and Suizu, M., 1992. The art of maturity modeling, Part 1: Finding a satisfactory geologic model. Am. Assoc. Pet. Geol., Bull., 76: 31–46.

KAZOU NAKAYAMA *Japan Petroleum Exploration Co., Ltd. 2-17-22, Akasaka, Minato-ku, Tokyo, 107 Japan*

SONG CAO *Alberta Research Council, P.O. Box 8330, Postal Station F, Edmonton, Alberta, Canada T6H 5X2*

Present address: *Geoscience Department, University of South Carolina, Columbia, SC 29208, U.S.A.*

Fast Monte Carlo simulation of hydrocarbon generation

Søren B. Nielsen

Kinetic parameters, heat flow history, thermal conductivity structure, subsidence history, and the surface temperature history are important input parameters in hydrocarbon modelling. The parameter values are not precisely known and can be considered as random variables characterized by probability density distributions. The induced variability in modelling results can be obtained by Monte Carlo simulation.

This paper discusses the structure of the uncertainties in: (1) the average kinetic parameters of kerogen degradation; and (2) the source rock temperature history. It is shown how a family of possible kinetic models of kerogen degradation and a family of possible source rock temperature histories for use in Monte Carlo simulation of hydrocarbon generation can be derived from a calibrated kerogen model and a calibrated temperature history.

The uncertainty in kinetic kerogen models can be parameterized by a joint Gaussian distribution of average activation energy and logarithmic Arrhenius factor. Uncertainties in the source rock temperature history are parameterized by uncertainties in the surface temperature and heat flow histories, the thermal conductivity structure, and the source rock burial depth. By means of the concept of stochastic processes, a statistical model is developed which makes it computationally efficient to obtain random realizations of possible source rock temperature histories. The time taken to generate a temperature history realization from the statistical model is in the order of 10^2 times faster than the time taken to produce one by the author's 1D thermal modelling program. Much more time can be saved in 2D and 3D modelling.

The methodology is demonstrated by Monte Carlo simulation of the generation history of the Upper Jurassic Farsund Formation in the Danish region of the Central Trough.

Introduction

The reliability of basin modelling results depends on the reliability of the input parameter values. Indeed, the major effort in modelling studies is concerned with the extraction of sensible input parameter values from the vast amount of geological, geophysical, and geochemical information available. Nevertheless, input parameter values inevitably have errors, which must be considered when use is made of the modelling results. Due to the number of variable parameters and the complexity of basin models, it is often obscure how the interplay of different errors affects the results. In order to reveal this effect it is possible to run several forward models with different reasonable combinations of input parameter values. This process is known as Monte Carlo simulation and was applied to investigate the accuracy of basin modelling results by Hermanrud et al. (1990).

The basic requirements of the Monte Carlo method are a powerful computer, a consistent physical and chemical model relating input parameters to results, and knowledge of the nature of variability in the input parameters. This variability is the ultimate cause of variability in the results, and if it is not correctly represented, it is likely that the induced statistical characteristics of the results will be wrong. It is therefore necessary to have well-founded statistical representations of the important modelling parameters.

One of the fundamental tasks in basin modelling is the modelling of hydrocarbon generation. The reliability of migration and charge modelling depends on the reliability of the modelled generation history, which therefore must be assessed. Monte Carlo simulation of the generation history requires: (1) a family of possible source rock kinetic models; and (2) a family of possible source rock temperature histories.

The possible source rock temperature histories in principle should be derived by Monte Carlo simulation of the evolutionary history of the sedimentary basin considered. This, however, easily becomes very time-consuming due to the forward computation time of basin models, especially in 2D or 3D modelling. For the Monte Carlo procedure to become a practical tool for the assessment of generation history, it is therefore paramount that approximate techniques are developed which reduce the computational effort spent in the thermal reconstruction part of the process.

Basin Modelling: Advances and Applications edited by A.G. Doré et al. NPF Special Publication 3, pp. 265–276, Elsevier, Amsterdam.

It is the purpose of this paper to illustrate how Monte Carlo simulation of hydrocarbon generation can be made computationally efficient. This is achieved by utilizing a recent development regarding the representation of uncertainties in the kinetic model of primary cracking (Nielsen and Dahl, 1991; Nielsen and Barth, 1991) and a new method for obtaining a family of possible source rock temperature histories once the "best guess" temperature history has been obtained by calibration of a thermal basin model.

The procedure is illustrated by a modelling example from the Danish section of the Central Trough.

Uncertainty in the primary cracking model

The kinetic parameters (activation energy E_a and Arrhenius factor A) of primary cracking are usually obtained by empirical calibration of a number of parallel first-order reactions against laboratory pyrolysis data (Juntgen and Klein, 1975; Tissot and Espitalié, 1975; Ungerer and Pelet, 1987; Tissot et al., 1987; Burnham et al., 1987). The justification of the approach is due to the demonstration (Ungerer and Pelet, 1987; Quigley and Mackenzie, 1988) that first-order kinetic models can simultaneously satisfy data obtained at laboratory and sedimentary basin heating rates, and that calibrations based solely on laboratory heating rates yield similar kinetic parameters.

However, since laboratory pyrolysis data and basin geochemical information have small errors, there is a range (a family) of kerogen kinetic models which are equally valid: each member of the family is consistent with the calibration data within the error bounds of the calibration data. To be able to randomly draw members from this family is one of the prerequisites for Monte Carlo simulation of hydrocarbon generation.

Nielsen and Dahl (1991) argued that least-squares variance analysis applied to the method of heating rate variation analysis (Braun and Burnham, 1987) yields a useful characterization of the family of equally valid kinetic models. The characterization is in terms of uncertainties in the average kinetic parameters.

The method of heating rate variation states that for a single first-order kinetic reaction, the activation energy E_a, logarithmic Arrhenius factor $\log_e(A)$, heating rate α, and pyrolysis T_{max} are related according to:

$$\log_e\left(\frac{\alpha}{(T_{max})^2}\right) = -\frac{E_a}{R}\frac{1}{T_{max}} + \log_e(A) - \log_e\left(\frac{E_a}{R}\right) \quad (1)$$

where R is the ideal gas constant. Equation 1 shows that pyrolysis for at least two different heating rates and observation of the associated T_{max} values in principle allows for drawing a straight line, the slope of which defines the activation energy and the intercept the logarithmic Arrhenius factor (Fig. 1A). When eq. 1 is applied to associated values of α and T_{max} for general kerogens, which are usually modelled by several independent first-order kinetic reactions, E_a and $\log_e(A)$ can be identified with the average kinetic parameters of the general kerogen (Braun and Burnham, 1987; Nielsen and Dahl, 1991). Since T_{max} and the shift of T_{max} with heating rate constitute the prime data which determine the value of the average kinetic parameters, it is also the uncertainty in T_{max} (Espitalié, 1986) which is responsible for the uncertainty in the average kinetic parameters.

The generalized least-squares scheme and problem formulation applied by Nielsen and Dahl (1991)

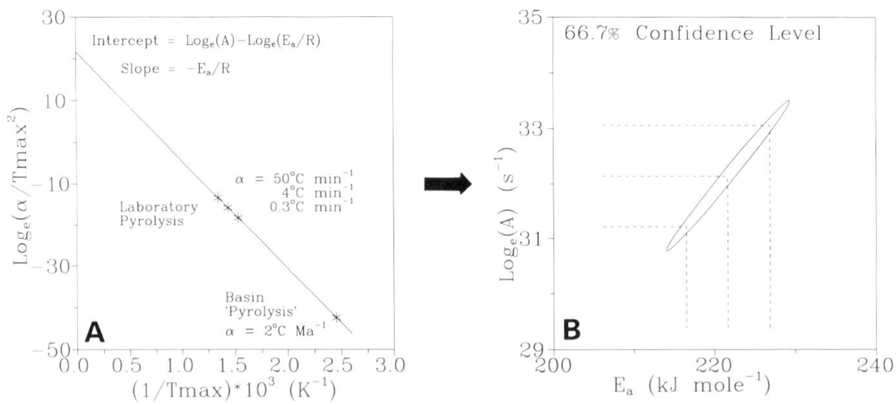

Fig. 1. (A) Heating rate analysis allows the average kinetic parameters of a kerogen to be determined from the slope and intercept of a straight line through associated points $[1/T_{max}, \log_e(\alpha^2/T_{max})]$. (B) Confidence ellipsoid for the average kinetic parameters based on least-squares variance analysis of the uncertainties in T_{max}.

quantifies how experimental uncertainty in T_{max} propagates into uncertainty in the average kinetic parameters. By their method, a set of T_{max} values with associated uncertainties are used to derive the standard deviations σ_e, σ_{lna}, and the correlation coefficient ρ in the joint Gaussian distribution (eq. A5) of average E_a and $\log_e(A)$. The confidence ellipsoid of Fig. 1B is a level curve of this probability density function. It defines the region in the average $E_a - \log_e(A)$ plane which, with probability 2/3, contains the true average kinetic parameters. Due to non-linearities in the heating rate variation problem the ellipsoidal confidence region is an approximation. However, numerical calculations of the true non-linear probability density function show that the Gaussian approximation generally is satisfactory with errors of less than 5% in σ_e and σ_{lna}.

It is important that basin geochemical information is included in the kinetic calibration (Ungerer, 1990; Espitalié et al., this volume). In the present formulation this information can be included as the value of T_{max} at one or more geological heating rates (Fig. 1A). T_{max} at a geological heating rate in some cases can be obtained as the temperature at the depth where the transformation ratio has the value 0.5. Some necessary requirements for this to be correct are that the source rock is vertically homogeneous and that the present source rock temperature is at its maximum value ever. However, it is better to use the values of the transformation ratios at different depths and the associated source rock temperature histories directly in the variance analysis, rather than lumping this detailed information together in one geological T_{max} value.

Figure 1 could illustrate the following situation. A kinetic kerogen model with N (e.g. $N = 8$) discrete activation energies and a global Arrhenius factor has been derived from laboratory pyrolysis data with heating rates of 0.3, 4, and 50°C min^{-1}. This is the "best guess" kinetic model on which the least-squares variance analysis is performed. When subjected to a geological heating rate of 2°C Ma^{-1} this kinetic kerogen model produces a T_{max} of 134°C, which is believed to be consistent within ±5°C with basin geochemical information (e.g. the measured transformation ratio depth-profile). The laboratory pyrolysis measurements have small errors which are assessed to be in the range ±2°C in the reproducibility of T_{max} and ±2°C in the absolute temperature calibration of the pyrolysis equipment (Espitalié, 1986). A variance analysis along the lines of Nielsen and Dahl (1991) yields $\sigma_e = 5.1864$ kJ mol^{-1}, $\sigma_{lna} = 0.9235$ s^{-1}, and $\rho = 0.9916$, which allows for drawing the confidence ellipsoid of Fig. 1B. If the basin information suggests a more trustworthy geological T_{max} than the one produced by the laboratory calibrated kerogen, this better estimate is used in the variance analysis instead. The analysis would then shift the average kinetic parameters and place the confidence ellipsoid elsewhere in the average E_a–$\log_e(A)$ plane. The size of the confidence ellipsoid depends on the number of T_{max} values, their uncertainties and the span of heating rates available.

Given values of σ_e, σ_{lna}, and ρ it is now possible from the "best guess" kinetic model to generate other kinetic models which could equally well have been the result if the kerogen calibration was repeated by the same procedure with another data set based on the same source rock.

To randomly draw a member from the family of equally valid kinetic models for use in one Monte Carlo run proceed as follows. Generate two uncorrelated Gaussian variates x and y (Appendix C) with $\langle x\rangle = \langle y\rangle = \langle x \cdot y\rangle = 0$ and $\langle x^2\rangle = \langle y^2\rangle = 1$ ($\langle\cdot\rangle$ means expectation). Calculate the perturbations DE_a and $D \log_e(A)$ to the average activation energy and logarithmic Arrhenius factor according to:

$$
\begin{aligned}
DE_a &= \sigma_e \cdot x \\
D\log_e(A) &= R \cdot \sigma_{lna} \cdot x + \sigma_{lna} \cdot y\sqrt{1-\rho^2}
\end{aligned} \tag{2}
$$

It is easy to see that by the construction of eq. 2 $\langle DE_a\rangle = 0$, $\langle (DE_a)^2\rangle = \sigma_e^2$, $\langle D\log_e(A)\rangle = 0$, $\langle (D\log_e(A))^2\rangle = \sigma_{lna}^2$, and $\mathrm{Cov}(DE_a, D\log_e(A)) = \langle (DE_a - \langle DE_a\rangle)\cdot(D\log_e(A) - \langle D\log_e(A)\rangle)\rangle = \sigma_e \cdot \sigma_{lna} \cdot \rho$. Let E_{ak}, $k = 1, \ldots, N$, be the N activation energies of the "best guess" kinetic kerogen model and let $\log_e(A)$ be the logarithmic Arrhenius factor. The kinetic parameters $\hat{E}_{ak}$, $k = 1, \ldots, N$, and $\hat{\log}_e(A)$ of the perturbed kinetic model kerogen are now derived from the calibrated model by:

$$
\begin{aligned}
\hat{E}_{ak} &= E_{ak} + DE_a \quad \text{for } k = 1, \ldots, N \\
\hat{\log}_e(A) &= \log_e(A) + D\log_e(A)
\end{aligned} \tag{3}
$$

Equation (3) simply represents a constant shift of the activation energies and a correlated shift of the global Arrhenius factor. The procedure yields a family of kinetic kerogen models, the individual members of which exhibit a variability similar to the experimentally derived variability observed by Espitalié et al. (this volume). By analysing different combinations of data sets for the same source rock they find that the location of the kinetic model on the activation energy axis is uncertain and correlated to the value of the Arrhenius factor. This was also observed by Ungerer (1990). In Espitalié et al. (this volume) the relative weights of individual kinetic reactions remain relatively unaffected. This lends support to the usefulness of the least-squares error analysis of Nielsen and Dahl (1991), which does not involve the relative weights of individual reactions, in characterizing the family of possible kinetic kerogen models for use in Monte Carlo simulation.

The confidence ellipsoid of Fig. 1B indicates the variability in the family of equally valid kinetic models: two thirds of the random draws generated by eqs. 2 and 3 will have average values of E_a and $\log_e(A)$ within the confidence ellipsoid. One third will fall outside, mainly in the elongated direction. The size of the confidence ellipsoid depends strongly on the amount and quality of the calibration data. It is likely that the new recommendations of Espitalié et al. (this volume) regarding pyrolysis procedure (ten different heating rates) will reduce significantly the ambiguity in kinetic models of kerogen degradation, although the use of basin geochemical information is still necessary.

Uncertainty in the source rock temperature history

In this section a new statistical model of uncertainty in the source rock temperature history is developed in order to reduce the computer time consumed by the many forward calculations of thermal evolution which are necessary in Monte Carlo simulations. The model is developed with reference to a case study, which is discussed first. For location refer to Fig. 2.

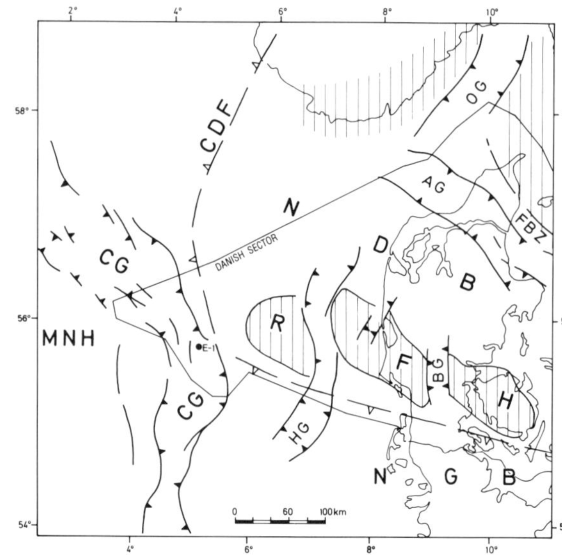

Fig. 2. Location of the well E-1 (lat. 55°43′52.35″N, long. 4°51′4.20″E) in the Danish Central Trough. *AG* = Aalborg Graben; *BG* = Brande Graben; *CDF* = Caledonian Deformation Front; *CG* = Central Graben; *FBZ* = Fennoscandian Border Zone; *HG* = Horn Graben; *MNH* = Mid North Sea High; *NDB* = Norwegian–Danish Basin; *NGB* = North German Basin; *OG* = Oslo Graben; *RFH* = Ringkøbing-Fyn High (after Nielsen and Balling, 1990).

Figure 3 demonstrates the calibration of the thermal evolution at the well E-1 from the Danish region of the Central Trough by means of borehole temperatures and vitrinite reflectances. The computer program for thermal reconstruction was developed by the author. It is based on the principles of thermal conduction (Carslaw and Jaeger, 1959) and makes use of the finite element method (e.g. Becker et al., 1981) to solve the transient one-dimensional heat equation. The porosities of the sediments follow the exponential law with empirically determined lithology-dependent parameters (e.g. Sclater and Christie, 1980). Formation thermal conductivities are calculated by the geometrical formula from the lithology-dependent matrix conductivity and the conductivity of water. Thermal conductivities are taken from our internal data base of rock conductivities measured by the needle probe method (Kristiansen, 1982). Conductivities and heat capacities are temperature-dependent.

The computer program uses least-squares inversion to shift the *a priori* given heat flow history by a constant value in order to reproduce present borehole temperatures. In the present example the *a priori* heat flow history is a constant. This is justified by the principle of favouring the least complicated heat flow history that is consistent with the data. Furthermore, model studies (Nielsen and Balling, 1990) show that rift-induced increases in heat flow (McKenzie, 1978) are strongly suppressed by rapid sedimentation. When more and accurate borehole temperatures are available the computer program can refine the temperature fit by adjusting the thermal conductivity structure within specified soft bonds. This, however, is very rarely possible. The borehole temperatures are equilibrium temperature estimates obtained by the cylindrical source method of Nielsen et al. (1990). This method yields optimized formation temperature estimates with confidence limits. The surface temperature is assumed to have been constant at 15°C prior to 100 Ma after which it decreased linearly to a value of 0°C at present. Due to the Pleistocene glaciations, present borehole temperatures below 1.5 to 2 km depth are likely to be in equilibrium with ice-age surface temperatures rather than the present-day average of around 6–8°C (Balling et al., 1981).

Vitrinite reflectance is modelled by the kinetic vitrinite model of Burnham and Sweeney (1989), which in the experience of the author gives good results. It is apparent that the reflectance data (courtesy of Mærsk Oil and Gas A/S) are reproduced satisfactorily. For this region of the Central Trough it is usual that calibration of thermal models to present-day

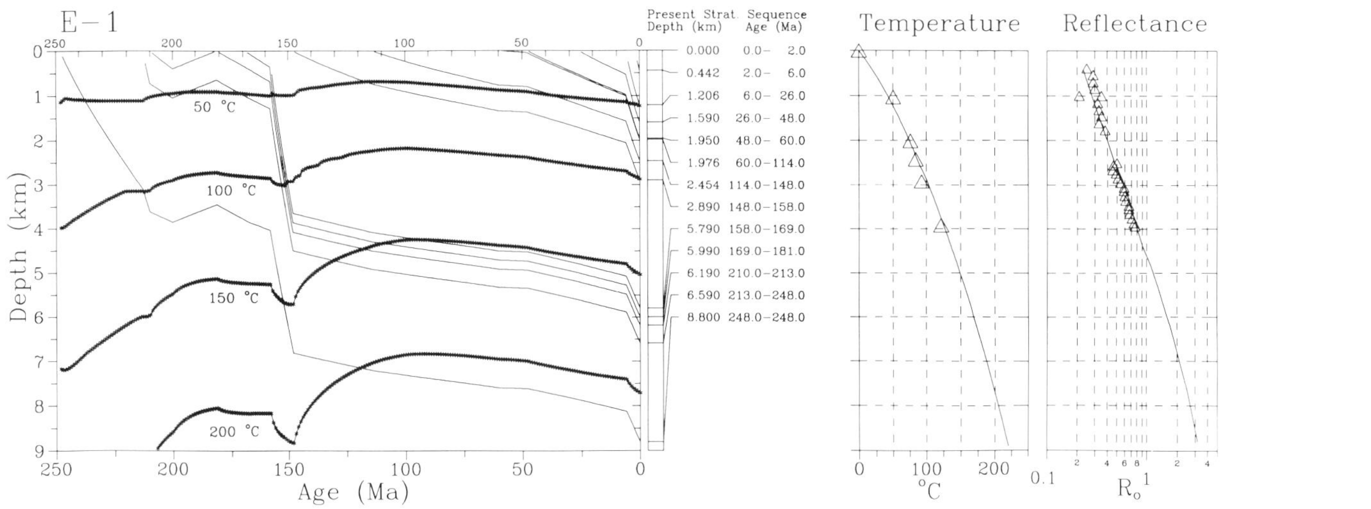

Fig. 3. Thermal reconstruction for the well E-1.

temperatures is a sufficient condition for reproducing the vitrinite reflectance data. This indicates that the vitrinite samples experienced their maximum temperature at present.

The subsidence history of Fig. 3 is characterized by rift-induced rapid subsidence in Upper Jurassic time with deposition of large thicknesses of the Farsund Formation which is time equivalent to the upper part of the Kimmeridge Clay Formation (Michelsen et al., 1987). This formation is believed to be one of the more prolific source rocks in this area. In the Danish region the upper Farsund Formation also interfingers with the informally defined Hot Unit of late Volgian–early Ryazanian age (Michelsen et al., 1987).

Figure 4 shows the temperature histories of the base, the middle and the top of the Farsund Formation extracted from the thermal reconstruction of Fig. 3. Since the thermal reconstruction is based on the best of our knowledge and is consistent with the data available, we shall refer to the extracted temperature histories of Fig. 3 as the "expected" or "best guess" source rock temperature histories. The term

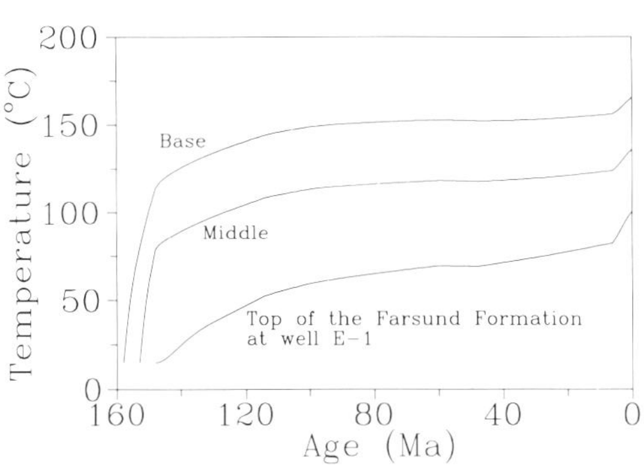

Fig. 4. Expected source rock temperatures for base, middle, and top of the Farsund Formation (Kimmeridgian to early Ryazanian).

"expected" indicates that the temperature histories are the ones we trust the most.

The question we now pose is what kind of variability is realistic in these expected temperature histories when the thermal modelling parameters are allowed to have uncertainties?

Let $\langle T(t)\rangle$ denote the expected source rock temperature history. The objective is to construct a family of temperature histories which has $\langle T(t)\rangle$ as its average and whose members have a probability of departure from $\langle T(t)\rangle$ which is controlled by the theory of thermal conduction and uncertainties in the thermal modelling parameters. We proceed as follows.

Apart from very extreme situations of ground water flow, the *sensitivity* of the subsurface temperature T to errors in the basic thermal parameters is well characterized by:

$$T = T_s + q \cdot r \cdot z \tag{4}$$

where T_s is the surface temperature, q is the surface heat flow density, and r is the average thermal resistivity (reciprocal thermal conductivity) of the rock column above depth z. It is emphasized that eq. 4 is not the computational basis for the thermal reconstruction of Fig. 3. It merely serves to quantify the sensitivity of subsurface temperatures to errors in the basic thermal parameters.

By considering T_s, q, r, and z as independent stochastic processes (e.g. Cramér and Leadbetter, 1967) which are functions of geological time, it can be shown (Appendix B) that the following family of temperature histories satisfies the above stated objective:

$$T(t) = \langle T(t)\rangle + \sigma_s\zeta(t) + \sigma_b[\langle T(t)\rangle - \langle T_s(t)\rangle]\,\eta(t) \tag{5}$$

The right-hand side of eq. 5 is the "best guess" source rock temperature history $\langle T(t)\rangle$ with superimposed random functions representing errors in the surface temperature history $(\sigma_s\zeta(t))$ and errors in the burial component of the temperature history $(\sigma_b(\langle T(t)\rangle - \langle T_s(t)\rangle)\,\eta(t))$.

σ_s is the standard deviation of errors in the expected surface temperature history $\langle T_s(t)\rangle$ used in thermal reconstruction. $\zeta(t)$ is a zero mean unit variance stochastic process with a correlation structure which models the time fluctuations of errors in the surface temperature history.

$\sigma_b(\langle T(t)\rangle - \langle T_s(t)\rangle)$ is the standard deviation of errors in the burial component of the temperature history. $(\langle T(t)\rangle - \langle T_s(t)\rangle)$ is the expected burial temperature history. σ_b is the relative standard deviation which depends on the relative errors in the heat flow density, thermal resistivity structure, and the depth of burial (Appendix B). $\eta(t)$ is a zero mean unit variance stochastic process with a correlation structure which models the time fluctuation of errors in the burial component of the temperature. The correlation structure depends on the time fluctuation of errors in the heat flow history, the thermal resistivity structure, and the depth of burial. This is elaborated in Appendix B.

The point is that it is much faster and more handy to generate possible source rock temperature histories by eq. 5 and the procedure of Appendix C than to feed the same family of surface temperature and heat flow history boundary conditions, and thermal resistivity models, into a numerical basin model and perform numerous forward simulations.

It is emphasized that the family of possible source rock temperature histories are derived as perturbations to the "best guess" source rock temperature history $\langle T(t)\rangle$, which is derived by means of numerical basin modelling and hence contains all the transient thermal effects associated with processes such as surface temperature and heat flow changes, sedimentation, erosion, and water flow as well as the effect of radiogenic heat production in sediments and crystalline crust. Only the amplitudes of the possible departures from the expected temperature history are controlled by the simpler principle of steady-state thermal conduction.

Figure 5 shows some realizations from the three families of possible source rock temperature histories associated with the top, the middle, and the base of the Farsund Formation. The realizations were obtained from eq. 5 by the procedure of Appendix C with the parameter values of Table 1. The stochastic processes $\zeta(t)$ and $\eta(t)$ have been realized by spectral colouring of Gaussian pseudo white noise. The correlation function $\rho(t_1,t_2) = \exp\{-[(t_1 - t_2)/\tau]^2\}$ of the basic process T_s, q, r, and z was chosen in order to yield sufficiently smooth realizations; it is not likely that any of the processes exhibit abrupt changes. The correlation time scale τ controls the frequency content of the stochastic process with correlation function $\rho(t_1,t_2)$. A small value of τ yields a rapidly fluctuating process while a large value yields slower fluctuations.

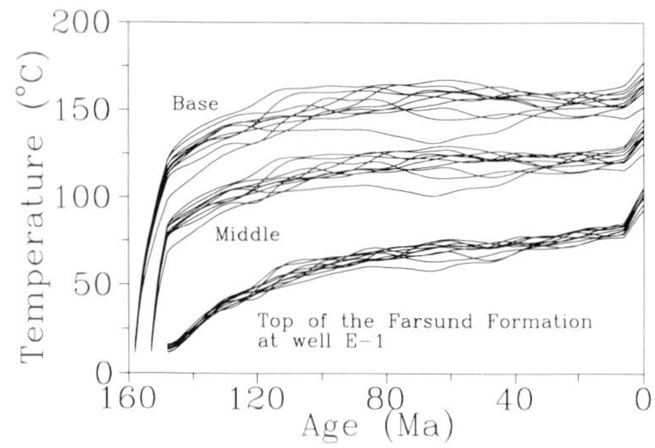

Fig. 5. Ten realizations of possible source rock temperature histories for base, middle and top of the Farsund Formation.

The correlation scale of surface temperature errors is τ = 10 Ma. This means that temperature fluctuations shorter than about 5 Ma are not considered. They of course occur in nature, but they are somewhat damped at the depth of hydrocarbon generation. The amplitude of surface temperature errors is controlled by the standard deviation of σ_s = 2°C. This means that for about two thirds of the burial

TABLE 1

Parameters for Monte Carlo simulation

The family of temperature histories [a]

Parameter type		Uncertainty relative or absolute	Correlation scale
T_s:	surface temperature	2°C	10 Ma
q:	heat flow	3%	50 Ma
r:	thermal resistivity	3%	80 Ma
z:	burial depth	3%	80 Ma

The family of kinetic models

σ_e, $\sigma_{\ln a}$, and ρ are the parameters of the joint Gaussian distribution of average E_a and $\log_e(A)$.
Standard deviation of the average activation energy σ_e = 5.1864 kJ mole^{-1}.
Standard deviation of the logarithmic Arrhenius factor $\sigma_{\ln a}$ = 0.9235 s^{-1}.
Coefficient of correlation ρ = 0.9916.

[a] The correlation function of the stochastic processes T_s, q, r, and z is $\rho(t_1, t_2) = \exp\{[(t_1 - t_2)/\tau]^2\}$. Only the correlation time scale τ changes between processes.

history the expected surface temperature $\langle T_s(t)\rangle$ is believed to be less than ± 2°C in error. For about one third of the burial history the error in $\langle T_s(t)\rangle$ is larger than ± 2°C.

The correlation scale of heat flow errors is $\tau = 50$ Ma, which is in the range of the time constant of the lithosphere (McKenzie, 1978). This models the concept that changes in basal heat flow are due to long time scale lithospheric processes.

The correlation scales of thermal resistivity and burial depth errors have somewhat arbitrarily been given the value $\tau = 80$ Ma. In standard Monte Carlo simulations this correlation scale would be even longer: a change in compaction parameters or matrix thermal conductivities yields a temperature perturbation which is correlated along the whole evolution history.

It is apparent from Fig. 5 how the amplitude of the temperature fluctuations increases with increasing depth (i.e. temperature). This reflects the fact that the uncertainty in temperature extrapolation increases with depth. Clearly, a small heat flow or average thermal resistivity perturbation gives an effect which increases with depth.

The temperature families of Fig. 5 could have been conditioned to reproduce present-day borehole temperatures. However, present borehole temperatures are uncertain (Nielsen et al., 1990) and generally boreholes are representative only of a relatively limited neighbourhood. Often generation modelling is performed in deep and undrilled kitchen areas without direct temperature control. If it is desirable to construct temperature families whose members all reproduce kinetically based maturity indicators such as vitrinite reflectance or isomerization and aromatization ratios, it is sufficient to condition the family members to reproduce the geological maximum temperature (present or past) within a few degrees centigrade.

The variability in hydrocarbon generation modelling

The variability in the modelling of hydrocarbon generation for the Farsund Formation in the region of the E-1 well is now investigated by randomly selecting kerogen models according to the procedure of eqs. 2 and 3 and combining them with randomly selected temperature histories such as those of Fig. 5 (refer to Appendix C for procedure).

Figure 6 (top, middle, base) shows the outcome of this procedure in terms of transformation ratio and generation rate histories. It is apparent that while the generation histories of top and base are quite conclusive (showing, respectively, insignificant and total transformation of the potential) the generation history for the middle of the formation is resolved much more poorly. The reason is that while the expected temperature histories of base and top are sufficiently extreme to ensure almost total (base) and insignificant (top) generation in spite of perturbations to the temperature and to the kinetic parameters, the expected temperature history for the middle level slowly passes through the prime temperature interval of kerogen degradation where temperature fluctuations in the order of 5°C and activation energy fluctuations in the order of 5 kJ mol^{-1} together with correlated fluctuations in $\log_e(A)$ are crucial.

Given the selected uncertainties, it can be concluded that the base of the Farsund Formation at this location generated hydrocarbons for a period of about 30 Ma following deposition, i.e. until Aptian/Albian time. The middle of the formation most probably generated hydrocarbons from 120 Ma to 60 Ma, with increased generation activity at present. Only one generation history out of ten suggests significant generation during the last 40 Ma and is hence considered to be relatively unlikely. The top of the Farsund Formation is immature, possibly with incipient hydrocarbon generation at present. This is consistent with the coring of immature to early mature Hot Unit in the E-1 well (Bjarne Skovbro, Mærsk Oil and Gas A/S, Copenhagen, Denmark, pers. commun., 1991). Doubling the uncertainties of Table 1 would not alter this picture significantly.

The computational efficiency of the present approach makes it an easy task to generate histograms of generation rate or transformation ratio at selected points in time. When properly normalized the histograms approximate the probability density distributions of generation rate or transformation ratio induced by uncertainty in the fundamental modelling parameters. This information can thus form the basis for Monte Carlo simulation of expulsion, migration and entrapment.

Conclusions

Integrated numerical basin models are useful tools in hydrocarbon exploration. Due to the inevitable errors in the input parameters it is not possible to forecast, for example, the precise amount of oil generated. What can be obtained is the statistical distribution of modelling results induced by the statistical distributions of the input parameters and by the nature of the basic physical and chemical laws. However, the resulting statistical distributions will not be realistic if the statistical distributions describing the variability in the input parameters are not.

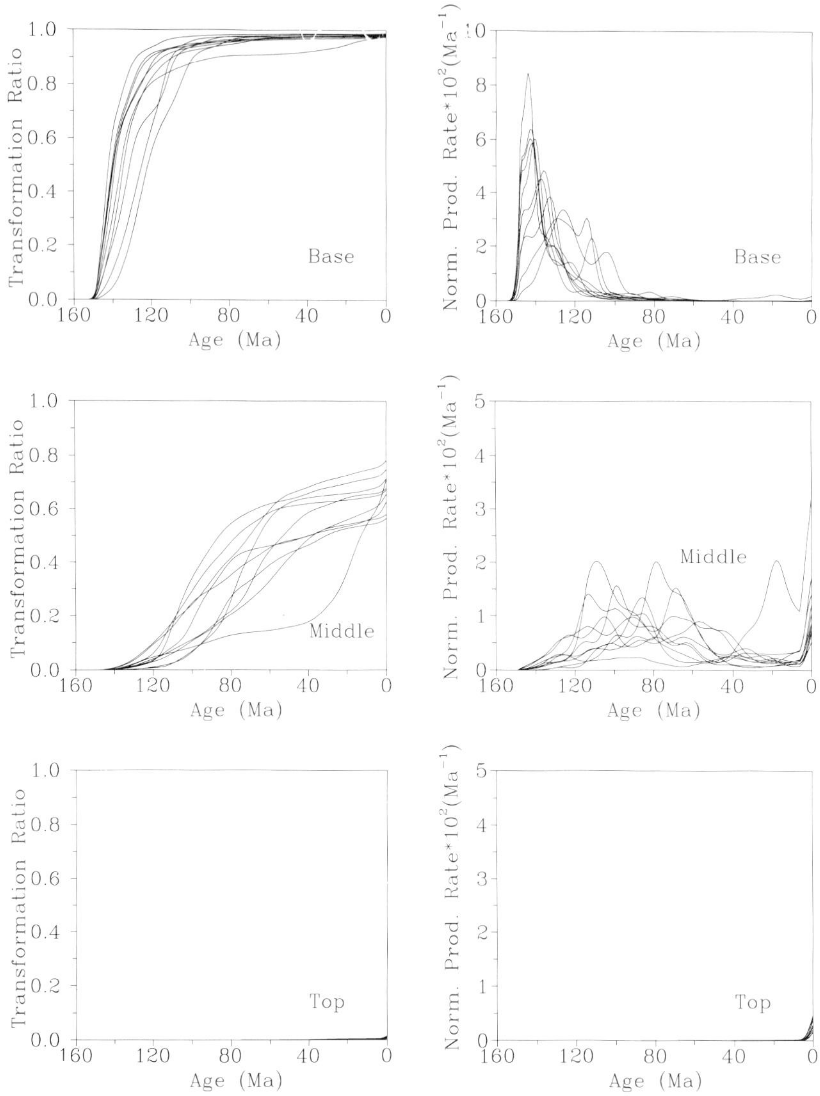

Fig. 6. Transformation ratio and generation rate histories for base, middle, and top of the Farsund Formation.

This paper has discussed the ambiguity in the kinetic parameters of primary cracking and the source rock temperature history. It has been shown that statistical distributions of possible kinetic models and source rock temperature histories for use in fast Monte Carlo simulation can be derived from a calibrated kerogen model and a calibrated temperature history.

The errors in the average kinetic parameters approximately follow a joint normal probability density function which can be derived from least-squares variance analysis of the heating rate variation procedure. The associated confidence regions in the average E_a–$\log_e(A)$ plane are ellipsoidal in shape due to a high degree of correlation ($\rho > 0.99$) between the errors in average E_a and $\log_e(A)$. This description of kinetic uncertainty is consistent with the results of Espitalié et al. (this volume). They find that the ambiguity in kinetic models is characterized by the correlated uncertainty in the average kinetic parameters while the relative weights of the individual kinetic reactions are less affected. The Gaussian description makes it a simple task to generate kinetic kerogen models for use in Monte Carlo simulation of hydrocarbon generation.

The derivation of the distribution of possible source rock temperature histories makes use of the fact that the heat flow, the depth of burial, the average thermal resistivity of the rock column above the point considered, and the surface temperature generally are the main factors controlling the subsurface thermal regime. This principle, combined with stochastic theory, was used to model how uncertainties in these factors can cause the temperature to depart from the expected source rock temperature. The parameters which control the amplitude of departures from the "best guess" temperature history comprise relative uncertainties in heat flow, burial depth, average thermal resistivity, and absolute uncertainty in the surface temperature. The influence of other factors such as uncertainty in porosity and matrix thermal resistivity can be studied by specifying how they on average affect the uncertainty in the average thermal resistivity, which is part of the present model.

The frequency content of the variability in individual processes (T_s, q, r, and z) is controlled by the correlation function and the correlation time scale. A family of possible correlation functions is given by $\rho(|t_1 - t_2|) = \exp[-(|t_1 - t_2|/\tau)^p]$ for $p > 0$. In the present work $p = 2$ was used, which gave smooth temperature history realizations. For example, $p = 1$ would yield more rugged temperature histories. Although the production rate histories would also become more rugged, the general conclusion regarding timing of hydrocarbon generation would not be affected because kerogen degradation is more sensitive to the amplitudes of temperature fluctuations than to how they occur.

The theory was applied to Monte Carlo simulation of the generation of hydrocarbons in an Upper Jurassic source rock in the Danish section of the Central Trough. It was apparent that the generation histories associated with the extreme temperature histories of top and base of the Farsund Formation were not severely influenced by the imposed variability in modelling parameters. However, the generation history associated with the middle of the Farsund Formation was not well resolved.

The time spent in the thermal reconstruction part of the process was reduced by a factor of about 10^2 as compared to the author's 1D thermal reconstruction computer program. In 2D and 3D modelling the source rock can be divided into a number of blocks with associated "best guess" temperature histories. It would then be relevant to introduce a spatial correlation in the temperature history realizations of individual blocks. This, however, is beyond the scope of this paper. The speed of computation and the ease in simulating temperature history makes this a useful method for fast Monte Carlo simulation of hydrocarbon generation.

Acknowledgements

Thank are due to The Nordic Ministry Council for financing participation in the NPF meeting in Stavanger, upon which this volume is based.

Appendix A — Basic statistics utilized

In this appendix only the very basic definitions are outlined. Refer to Rao (1965) and Cramér and Leadbetter (1967) for an introduction to basic probability theory and stochastic processes.

Random variables

A random variable (or stochastic variable) is useful for quantitative modelling of experiments which have an outcome with an unpredictable component. An obvious example is the measurement of pyrolysis $T_{\max}$. Even though the measurement procedure is precisely reproduced, repeated measurements of $T_{\max}$ on samples from the same source rock yield slightly different results each time.

Let X denote a real-valued random variable which simulates the outcome of an experiment. The probabilistic behaviour of X is specified by means of a cumulative distribution function (c.d.f) $F(x)$ defined as:

$$F(x) = P\{X \leq x\}$$

$F(x)$ is the probability of the event between the brackets (the event that an experiment yields a value of X which is less than the real number x). The derivative $f(x) = \partial F(x)/\partial x$, when it exists, is denoted the probability density function (p.d.f.) of X. The mean μ and variance σ^2 of the random variable X can be defined from the p.d.f. as:

$$\begin{aligned} \mu &= \langle X \rangle &&= \int_{-\infty}^{\infty} t f(t)\,\mathrm{d}t \\ \sigma^2 &= \langle (X-\mu)^2 \rangle &&= \int_{-\infty}^{\infty} (t-\mu)^2 f(t)\,\mathrm{d}t \end{aligned} \tag{A1}$$

$\langle \cdot \rangle$ means the "expectation". Let X and Y be two random variables with mean values μ_x and μ_y and variances σ_x^2 and σ_y^2. The covariance Cov (X, Y) between X and Y is defined according to:

$$\operatorname{Cov}(X, Y) = \langle (X - \mu_x)(Y - \mu_y) \rangle \tag{A2}$$

The coefficient of correlation ρ between X and Y is given by:

$$\rho = \frac{\operatorname{Cov}(X, Y)}{\sigma_x \sigma_y} \tag{A3}$$

The Gaussian case

If X is a Gaussian (or "normal") random variable, the p.d.f. $f(x)$ is given by:

$$f(x) = \frac{1}{\sigma\sqrt{2\pi}} \exp\left[\frac{(x-\mu)^2}{2\sigma^2}\right] \tag{A4}$$

The Gaussian probability model is characterized fully by the mean μ and the variance σ^2. If X and Y are two jointly distributed Gaussian variables they follow the bivariate Gaussian p.d.f. $f(x, y)$ given by:

$$f(x,y) = \frac{1}{2\pi\sigma_x\sigma_y\sqrt{1-\rho^2}} \times \exp\left\{-\frac{1}{2(1-\rho^2)}\left[\left(\frac{x-\mu_x}{\sigma_x}\right)^2 - 2\rho\left(\frac{x-\mu_x}{\sigma_x}\right)\left(\frac{y-\mu_y}{\sigma_y}\right)+\left(\frac{y-\mu_y}{\sigma_y}\right)^2\right]\right\} \quad \text{(A5)}$$

Stochastic processes

A stochastic process $X(t)$ is a family of random variables which is indexed by a discrete or continuous variable t (for our purposes t represents time and is continuous). For each t, $X(t)$ is a random variable for which the above statements and definitions hold. That is, the probabilistic behaviour of $X(t)$ at time t is characterized by a c.d.f. and the mean and variance are defined as in eq. A1. However, mean $\mu(t)$ and variance $\sigma^2(t)$ are now generally functions of time t. The definition of the covariance function $r(t_1,t_2)$ follows eq. A2:

$$r(t_1,t_2) = \mathrm{Cov}\,[X(t_1), X(t_2)] \quad \text{(A6)}$$

$r(t_1,t_2)$ is symmetrical in t_1 and t_2. $X(t)$ is a stationary process when $\mu(t)$ is independent of time and $r(t_1,t_2)$ depends only on the time lag $|t_1 - t_2|$. The correlation function $\rho(t_1,t_2)$ is the covariance function normalized to value 1 for $t_1 = t_2$ (compare eq. A3):

$$\rho(t_1,t_2) = \frac{\mathrm{Cov}\,[X(t_1),X(t_2)]}{\sigma_x(t_1)\sigma_x(t_2)} \quad \text{(A7)}$$

The Gaussian case

A stochastic process $X(t)$ is a Gaussian stochastic process if, for arbitrary values t_1 and t_2 of time, the two random variables $X(t_1)$ and $X(t_2)$ follow a joint Gaussian p.d.f. (eq. A5). The correlation coefficient ρ between $X(t_1)$ and $X(t_2)$ in the joint Gaussian distribution follows from the correlation function of eq. A7.

Appendix B — Statistical model of uncertainty in the temperature history

The covariance structure of errors in the source rock temperature history can be obtained by considering the relationship:

$$T = T_s + q \cdot r \cdot z \quad \text{(B1)}$$

Consider T_s, q, r, and z to be uncorrelated stochastic processes which are functions of geological time. Values of the expectations $\langle T_s\rangle$, $\langle q\rangle$, $\langle r\rangle$, and $\langle z\rangle$ were used in thermal reconstruction to find the "best guess" source rock temperature history. Let t_1 and t_2 be two times during the burial history of the source rock and let for ease of notation any of the quantities of eq. B1 taken at time t_1 or t_2 be indexed by numbers 1 or 2. By eq. A2 the covariance between the source rock temperatures T_1 and T_2 is given by:

$$\begin{aligned}\mathrm{Cov}\,(T_1,T_2) &= \langle (T_1 - \langle T_1\rangle)(T_2 - \langle T_2\rangle)\rangle \\ &= \langle T_1\,T_2\rangle - \langle T_1\rangle\langle T_2\rangle\end{aligned} \quad \text{(B2)}$$

Introducing eq. B1 into eq. B2 and using the case where the stochastic processes T_s, q, r, and z are uncorrelated (meaning that e.g. $\langle q_1 r_1 z_1\rangle = \langle q_1\rangle\langle r_1\rangle\langle z_1\rangle$), eq. B2 becomes:

$$\begin{aligned}\mathrm{Cov}\,(T_1,T_2) = &\\ &\langle T_{1s}T_{2s}\rangle - \langle T_{1s}\rangle\langle T_{2s}\rangle + \langle q_1 q_2\rangle\langle r_1 r_2\rangle\langle z_1 z_2\rangle \\ &- \langle q_1\rangle\langle r_1\rangle\langle z_1\rangle\langle q_2\rangle\langle r_2\rangle\langle z_2\rangle\end{aligned} \quad \text{(B3)}$$

Equation B3 can be restated in terms of correlation functions $\rho(t_1,t_2)$ of the respective process by noting that for a general stochastic process $X(t)$ (from eq. A7):

$$\langle X_1 X_2\rangle - \langle X_1\rangle\langle X_2\rangle = \sigma_{1x}\sigma_{2x}\cdot\rho_x(t_1,t_2) \quad \text{(B4)}$$

where σ_{1x} and σ_{2x} are the standard deviations of the process $X(t)$ at times t_1 and t_2

By use of eq. B4, eq. B3 becomes:

$$\begin{aligned}\mathrm{Cov}\,(T_1,T_2) &= \sigma_{1s}\sigma_{2s}\cdot\rho_s(t_1,t_2) \\ &+ [\sigma_{1q}\sigma_{2q}\cdot\rho_q(t_1,t_2) + \langle q_1\rangle\langle q_2\rangle] \\ &\times [\sigma_{1r}\sigma_{2r}\cdot\rho_r(t_1,t_2) + \langle r_1\rangle\langle r_2\rangle] \\ &\times [\sigma_{1z}\sigma_{2z}\cdot\rho_z(t_1,t_2) + \langle z_1\rangle\langle z_2\rangle] \\ &- \langle q_1\rangle\langle r_1\rangle\langle z_1\rangle\langle q_2\rangle\langle r_2\rangle\langle z_2\rangle\end{aligned} \quad \text{(B5)}$$

where the indices identify time and process. Using that $\langle q_1\rangle\langle r_1\rangle\langle z_1\rangle = \langle T_1\rangle - \langle T_{1s}\rangle$ and $\langle q_2\rangle\langle r_2\rangle\langle z_2\rangle = \langle T_2\rangle - \langle T_{2s}\rangle$ (eq. B1) and assuming that the relative uncertainty in each of the processes q, r, and z and the absolute uncertainty in T_s are constant in time, eq. B5 reduces to:

$$\begin{aligned}\mathrm{Cov}\,[T(t_1), T(t_2)] = &\\ &\sigma_s^2\rho_s(t_1,t_2) + \sigma_b^2\,[\langle T(t_1)\rangle - \langle T_s(t_1)\rangle] \\ &\times [\langle T(t_2)\rangle - \langle T_s(t_2)\rangle]\,\rho_b(t_1,t_2)\end{aligned} \quad \text{(B6)}$$

The parameters of eq. B6 have the meaning:

σ_s = standard deviation of errors in the surface temperature history (°C)

$\rho_s(t_1, t_2)$ = correlation function of errors in the surface temperature history

σ_b = relative standard deviation of errors in the burial component of the temperature history

$$= \left\{ \left[\left(\frac{\sigma_q}{\langle q \rangle} \right)^2 + 1 \right] \cdot \left[\left(\frac{\sigma_r}{\langle r \rangle} \right)^2 + 1 \right] \times \left[\left(\frac{\sigma_z}{\langle z \rangle} \right)^2 + 1 \right] - 1 \right\}^{1/2}$$

$\rho_b(t_1, t_2)$ = correlation function of errors in the burial component of the temperature history

$$= \frac{1}{\sigma_b^2} \left\{ \left[\left(\frac{\sigma_q}{\langle q \rangle} \right)^2 \rho_q(t_1, t_2) + 1 \right] \times \left[\left(\frac{\sigma_r}{\langle r \rangle} \right)^2 \rho_r(t_1, t_2) + 1 \right] \times \left[\left(\frac{\sigma_z}{\langle z \rangle} \right)^2 \rho_z(t_1, t_2) + 1 \right] - 1 \right\}$$

$\sigma_q/\langle q \rangle$ = relative standard deviation of errors in the heat flow density

$\rho_q(t_1, t_2)$ = correlation function of errors in the heat flow density

$\sigma_r/\langle r \rangle$ = relative standard deviation of errors in the thermal resistivity

$\rho_r(t_1, t_2)$ = correlation function of errors in the thermal resistivity

$\sigma_z/\langle z \rangle$ = relative standard deviation of errors in the burial depth

$\rho_z(t_1, t_2)$ = correlation function of errors in the burial depth.

A family af temperature histories that has the covariance structure of eq. B6 is now given by:

$$T(t) = \langle T(t) \rangle + \sigma_s \cdot \zeta(t) + \sigma_b \cdot [\langle T(t) \rangle - \langle T_s(t) \rangle] \cdot \eta(t) \tag{B7}$$

where $\zeta(t)$ and $\eta(t)$ are uncorrelated zero mean unit variance stochastic processes with correlation functions given by $\rho_s(t_1, t_2)$ and $\rho_b(t_1, t_2)$, respectively. The easiest way to see that this is true is to form its covariance function of the process of eq. B7 by eq. A2.

Appendix C — Sampling a stationary Gaussian stochastic process

In order to perform Monte Carlo simulation of hydrocarbon generation by the present procedure, we must be able to make random draws of source rock temperature histories from the family of possible source rock temperature histories. This family is characterized by a sum of two stochastic processes with certain covariance functions (eq. B7). We hence need to be able to obtain samples of the stationary processes $\zeta(t)$ and $\eta(t)$ with correlation functions $\rho_s(t_1, t_2) = \rho_s(|t_1 - t_2|)$ and $\rho_b(t_1, t_2) = \rho_b(|t_1 - t_2|)$.

Let $X(t)$ be a stationary zero mean Gaussian stochastic process with correlation function $\rho(|t_1 - t_2|)$. How do we construct a time series which is a sample of the stochastic process $X(t)$?

A general technique is to modify the frequency content of zero mean Gaussian pseudo white noise by filtering. There are two problems: (1) how to generate a sequence of Gaussian pseudo white noise, and (2) how to filter the sequence.

(1) Most programming languages are equipped with a random number generator, which returns random numbers uniformly distributed between 0 and 1. Let u_k be such a number. The random variable X defined by:

$$X = \sqrt{\frac{12}{n}} \cdot \left(\sum_{k=1}^{n} u_k - \frac{n}{2} \right) \tag{C1}$$

is approximately distributed as a normalized zero mean Gaussian random variable for, say, $n = 12$. That is, a sum of computer-generated uniformly distributed random numbers shifted and scaled according to eq. C1 simulates a standardized (zero mean and unit variance) Gaussian variable. A sequence of such variables approximates a standardized Gaussian pseudo white noise sequence.

(2) The Fourier transform of the correlation function of a stochastic process yields the power spectrum of the process. The power spectrum of a white noise sequence ideally is constant, i.e. all wave lengths in the sequence have the same power (hence "white noise"). Multiplication (= filtering) of the Fourier transform of the white noise sequence by the square root of the power spectrum of the correlation function of interest yields a sequence which has the desired correlation function. Proceed as follows:

(A) Sample the correlation function $\rho(|t_1 - t_2|)$ equidistantly as a function of lag $|t_1 - t_2|$. Sample densely enough to avoid aliasing.

(B) Perform a discrete Fourier transform of the sampled correlation function. This should yield a real-valued function (zero imaginary part). If it does not, then the correlation function was probably fed erroneously to the discrete Fourier transform algorithm.

(C) Perform a discrete Fourier transform of the white noise sequence and multiply by the the square root of the filter of step B (normalized to value 1 at frequency 0 Hz).

(D) Perform the inverse discrete Fourier transform of the filtered white noise sequence. The resulting sequence has zero mean and unit variance and the desired correlation function.

References

Balling, N., Kristiansen, J.I., Poulsen, K.D. and Saxov, S., 1981. Geothermal measurements and subsurface temperature modelling in Denmark. Geoskrifter, 16, 172 pp.

Becker, E.B., Carey, G.F. and Oden, J.T., 1981. Finite Elements. An Introduction, Vol I. Prentice-Hall, Englewood Cliffs, N.J., 258 pp.

Braun, R.L. and Burnham, A.K., 1987. Analysis of chemical reaction kinetics using a distribution of activation energies and simpler models. Energy Fuels, 1: 153.

Burnham, A.K. and Sweeney, J.J., 1989. A chemical kinetic model of vitrinite reflectance and maturation. Geochim. Cosmochim. Acta, 53: 2649–2657.

Burnham, A.K., Braun, R.L., Gregg, H.R. and Samoun, A.M., 1987. Comparison of methods for measuring kerogen pyrolysis rates and fitting kerogen kinetic parameters. Energy Fuels, 1: 452–458.

Carslaw, H.S. and Jaeger, J.C., 1959. Conduction of Heat in Solids. Oxford University Press, London, 2nd ed., 510 pp.

Cramér, H. and Leadbetter, M.R., 1967. Stationary and Related Stochastic Processes. Wiley, New York, N.Y., 348 pp.

Espitalié, J., 1986. Use of T_{max} as a maturation index for different types of organic matter. Comparison with vitrinite reflectance. In: J. Burrus (Editor), Thermal Modeling in Sedimentary Basins. Technip, Paris, pp. 475–496.

Espitalié, J., Marquis, F. and Drouet, S., 1993. Critical study of kinetic modelling parameters. In: A.G. Doré et al. (Editors), Basin Modelling: Advances and Applications. Norwegian Petroleum Society (NPF), Special Publication 3. Elsevier, Amsterdam, pp. 233–242 (this volume).

Hermanrud, C., Eggen, S., Jacobsen, T., Carlsen, E.M. and Pallesen, S., 1990. On the accuracy of modelling hydrocarbon generation and migration. Egersund Basin oil find, Norway. In: B. Durand and F. Behar (Editors), Organic Geochemistry in Petroleum Exploration. Org. Geochem., 16: 389–399.

Juntgen, H. and Klein, J., 1975. Entstehung von Erdgas aus kohligen Sedimenten. Erdöl und Kohle, 2: 65–73.

Kristiansen, J.I., 1982. The Transient Cylindrical Probe Method for Determination of Thermal Parameters of Earth Materials. Ph.D. Thesis, Department of Earth Sciences, University of Aarhus, Geoskrifter, 18, 154 pp.

McKenzie, D., 1978. Some remarks on the development of sedimentary basins. Earth Planet. Sci. Lett., 55: 87–98.

Michelsen, O., Frandsen, N., Holm, L., Jensen, T.F., Møller, J.J., and Vejbæk, O.V., 1987. Jurassic–Lower Cretaceous of the Danish Central Trough — depositional environments, tectonism, and reservoirs. Geol. Surv. Denmark, Ser. A, 16, 45 pp.

Nielsen, S.B. and Balling, N., 1990. Modelling subsidence, heat flow, and hydrocarbon generation in extensional basins. First Break, 8: 23–31.

Nielsen, S.B. and Barth, T., 1991. An application of least-squares inverse analysis in kinetic interpretations of hydrous pyrolysis experiments. Math. Geol., 23: 565–582.

Nielsen, S.B. and Dahl, B., 1991. Confidence limits on kinetic models of primary cracking and implications for the modelling of hydrocarbon generation. Mar. Pet. Geol., 8: 483–492.

Nielsen, S.B., Balling, N. and Christiansen, H.S., 1990. Formation temperatures determined from stochastic inversion of borehole observations. Geophys. J. Int., 101: 581–590.

Quigley, T.M. and Mackenzie, A.S., 1988. The temperature of oil and gas formation in the sub-surface. Nature, 333: 549–552.

Rao, C.R., 1965. Linear Statistical Inference and its Applications. Wiley, New York, N.Y., 625 pp.

Sclater, J.G. and Christie, P.A.F., 1980. Continental stretching: an explanation of the post-mid-Cretaceous subsidence of the central North Sea. J. Geophys. Res., 85: 3711–3739.

Tissot, B.P. and Espitalié, J., 1975. L'évolution de la matière organique des sédiments: applications d'une simulation mathématique. Rev. Inst. Fr. Pét., 30: 743–777.

Tissot, B.P., Pelet, R. and Ungerer, Ph., 1987. Thermal history of sedimentary basins, maturation indices, and kinetics of oil and gas generation. Am. Assoc. Pet. Geol., Bull., 71: 1445–1466.

Ungerer, P., 1990. State of the art of research in kinetic modelling of oil formation and expulsion. In: B. Durand and F. Behar (Editors), Organic Geochemistry in Petroleum Exploration. Org. Geochem., 16: 1–25.

Ungerer, P. and Pelet, R., 1987. Extrapolation of kinetics of oil and gas formation from laboratory experiments to sedimentary basins. Nature, 327: 5254.

SØREN B. NIELSEN *Department of Earth Sciences, Geophysical Laboratory, University of Aarhus, Finlandsgade 8, DK-8200 Aarhus N, Denmark*

Experimental measurement of the molecular migration of light hydrocarbons in source rocks at elevated temperatures

B.M. Krooss, D. Hanebeck and D. Leythaeuser

An experimental setup has been constructed to investigate petroleum migration processes in source, carrier and reservoir rocks at elevated pressure and temperature conditions (max. 20 MPa and 200°C). This new experimental technique can be applied for the investigation of various problems related to the molecular transport of light hydrocarbons in the subsurface including primary migration and diffusive loss from petroleum and gas accumulations and their impact on reservoir stability. In an initial study the mobility of C_5–C_7 hydrocarbons in source rocks of different type and maturity was measured at a temperature of 150°C. The experimental data indicate an enhanced molecular mobility of aromatic hydrocarbons (benzene and toluene) as compared to the saturated species in both a Torbanite and a mature type II source rock. These results are in accordance with light hydrocarbon concentration patterns observed in source rock/carrier bed systems during previous studies.

Introduction

Numerical modelling of hydrocarbon migration in sedimentary basins requires, besides a thorough understanding of the relevant transport processes, input parameters which characterize, to the highest possible degree, the envisaged transport phenomena under subsurface temperature and pressure conditions. Experimental simulations under well constrained laboratory conditions can help to provide an improved understanding of the transport processes involved and yield input parameters needed for numerical simulations, although, admittedly, they duplicate the processes in geological systems and on a geological time scale only to a limited extent. In the context of a comprehensive study of the geochemical aspects of petroleum migration, a set of experimental tools is being developed which can be used in a flexible way to investigate migration processes under various boundary conditions. A first application of this experimental equipment was directed towards the investigation of molecular transport processes in sedimentary rocks, with special focus on the primary migration and expulsion of light hydrocarbons from source rocks.

Relevance of molecular transport in petroleum migration

The potential importance of diffusive transport of hydrocarbons in the subsurface was recognized at a relatively early stage in the history of petroleum geology and organic geochemistry. First systematic studies in this field were performed in the early fifties by Antonov (1954), who measured the diffusion coefficients of C_1–C_8 hydrocarbons in clastic rocks of different lithologies. These experiments were later extended by Antonov (1968, 1970a) and Stklyanin and Litvinova (1971). The primary interest of these early experimental studies was focused on the assessment of caprock leakage of natural gas reservoirs as input to gas surveys and surface geochemical prospecting for petroleum. A first attempt to incorporate the diffusive transport of hydrocarbons into a comprehensive mathematical model of natural gas migration was reported by Stklyanin et al. (1968).

Molecular diffusion in primary migration

The rate of expulsion from source rocks into adjacent carrier beds is an essential boundary condition for secondary petroleum migration. Presently, the numerical simulation of petroleum migration in the framework of basin models has to rely largely on assumptions and estimates regarding the expulsion rate and efficiency. In order to reduce this potential source of error an improved understanding of the mechanisms of primary petroleum migration aiming at a qualitative and, ultimately, quantitative description of this process is highly desirable. The influence of primary migration and expulsion on the chemical composition of petroleum is another important aspect that deserves particular attention. The role of molecular migration, i.e. diffu-

Basin Modelling: Advances and Applications edited by A.G. Doré et al. NPF Special Publication 3, pp. 277–291, Elsevier, Amsterdam.

sion, in primary petroleum migration and expulsion has long been — and still is — a matter of controversial discussion. Improved experimental data could contribute to a settling of this dispute.

Leythaeuser et al. (1980) used observed light hydrocarbon concentration profiles in near-surface source rock type sediments (Campanian/Maastrichtian shale outcrop in W. Greenland) and an estimated exposure time of 6000 years available for diffusion from the outcrop studied to calculate effective diffusion coefficients for C_2–C_7 hydrocarbons. These diffusion coefficients, which ranged between 6.4×10^{-7} and 4.2×10^{-8} $cm^2\ s^{-1}$, were in a later study extrapolated to assess the significance of diffusive transport in the primary migration of light hydrocarbons from source rocks (Leythaeuser et al., 1982). The results of this study indicated that the quantities of gas escaping from those volumes of mature source rocks, which are commonly encountered in hydrocarbon field drainage areas, by molecular diffusion alone and within geologically short periods of time (<1 Ma) are comparable to the gas reserves present in certain gas fields. However, this assessment of the gas expulsion efficiency by molecular diffusion was based on a number of too simplified assumptions. For example, the temperature-dependence of the diffusion rate was ignored and likewise the replenishment of diffusive losses by continuing generation of hydrocarbons during the expulsion process. Still it could be shown that commercially relevant quantities of hydrocarbon gases, in particular methane, can migrate from source rocks by diffusion through the water-saturated pore space. In contrast, due to their low diffusivity, a molecular transport mechanism was not considered relevant for the expulsion of petroleum range hydrocarbons.

This point of view is shared by many workers and, according to the presently most widely accepted theory, primary migration and expulsion of petroleum occurs as a separate organic fluid phase through the pore system of the source rock and/or through microfractures.

The combined geochemical effects of hydrocarbon generation and expulsion were studied by Mackenzie et al. (1987) for Upper Jurassic Kimmeridge Clay Formation source rocks in the Brae field area on the UK continental shelf. These authors obtain an estimated expulsion flux of 5×10^{-15} m^3 (petroleum) $m^{-2}\ s^{-1}$ from a 10 m Kimmeridge Clay source rock interval. With an assumed surface petroleum density of 600 kg m^{-3} this corresponds to a mass flux of petroleum of 3 ng $m^{-2}\ s^{-1}$. The calculation was based on the assumption of continuous, steady-state expulsion over a period of 5 Ma. Using the effective diffusion coefficients of Leythaeuser et al. (1982), Mackenzie et al. (1987) conclude that the molecular flow rate of molecules lighter than C_9 exceeds the linear velocity of an assumed pressure-driven expulsion flux of the bulk petroleum phase from the 10 m source rock interval considered and is capable of eliminating concentration gradients at this length scale within a period of 5 Ma.

Stainforth and Reinders (1990) attribute an even higher relevance to the molecular transport of petroleum hydrocarbons during petroleum expulsion by suggesting that thermally activated diffusion through organic matter networks is generally the rate-limiting transport mechanism during primary migration. A numerical model of generation and expulsion (GENEX) based on this concept produced results that were found to be in agreement with both results from laboratory experiments and observations in natural systems by Leythaeuser et al. (1988).

The validation of the various migration models involving molecular diffusion requires more reliable data on the mobility of individual hydrocarbon compounds within the source rock as a function of temperature, pressure and the content as well as the type of organic matter. Another important parameter is the migration resistance of the source-rock/carrier bed boundary. This resistance, which was implicitly taken to be zero in the model of Leythaeuser et al. (1982), is accounted for by the Biot number in the model of Stainforth and Reinders (1990).

Caprock leakage by molecular diffusion

The second important aspect of molecular diffusion in petroleum and gas migration is the assessment and quantification of hydrocarbon losses through the caprocks of petroleum reservoirs. Apparently this topic was first addressed from an experimental and a theoretical point of view by Antonov (1968, 1970b). Smith et al. (1971), using a one-dimensional model for diffusive losses from two representative gas reservoirs, compute residence times for C_1–C_4 hydrocarbons in the range of 74 to 253 Ma depending on the geometry of the reservoirs. Their diffusion coefficients were derived from the aqueous diffusivities of the corresponding hydrocarbons by applying a porosity and tortuosity correction.

Leythaeuser et al. (1982) estimated the rate of caprock leakage by diffusion for the Harlingen gas field in Holland. As this estimate was based on an erroneous assumption for the boundary conditions at the reservoir/caprock interface a recalculation was performed (Krooss et al., 1992) yielding a time of 70 Ma required for the diffusive loss of half of the present-day methane in place.

An approach to constrain the time of filling of a gas-condensate reservoir based on observed light hydrocarbon concentration profiles in the overlying caprock was presented by Krooss et al. (1988).

In general, the results of studies aimed at the assessment of caprock leakage indicate that diffusive losses may be of economic importance under certain conditions, for instance in reservoirs with large closure areas and small pay zone thickness. For an estimation of the gas losses from a particular reservoir specific characteristics such as structure, geometry, temperature, pressure, caprock lithology etc. must be taken into consideration. An attempt to quantify gas losses by diffusion is reasonable only for those situations where the caprock integrity has not been affected over extended periods of time. Compressible Darcy flow of a free gas phase through disturbed caprocks results in an almost instantaneous loss of the reservoir content on a geologic time scale.

Experimental data on the molecular migration of hydrocarbons in sedimentary rocks

The experimental data base on the diffusion behaviour of hydrocarbons in sedimentary rocks under realistic geological pressure and temperature conditions is very limited. In particular the earlier experimental works contain no information on the temperatures at which the diffusion measurements were carried out. A systematic investigation of the temperature-dependence of diffusion coefficients in water-saturated sedimentary rocks was performed by Krooss (1985, 1988). These experiments were performed with C_1–C_5 hydrocarbons and covered only the temperature range between 30 and 70°C. As a first approach, an extrapolation of the measured diffusion coefficients to higher temperatures is possible. However, an experimental study at elevated temperatures is necessary in order to assess the diffusive behaviour of hydrocarbon gases and higher homologues under realistic conditions. The influence of pressure on the molecular transport of hydrocarbons in sedimentary rocks must take into account two factors. The *pore pressure* determines the phase composition (gas, liquid) of the pore fluids and thus the mobility of the diffusing species, which is substantially higher in a gas phase than in a liquid phase. The *lithostatic pressure* or *effective stress* affects the structure and connectivity of the pore system and the formation or closure of microfractures. In the present experiments only the pore pressure was controlled, and maintained above the vapour pressure of water and the light hydrocarbon mixtures used. Meanwhile a new setup has been constructed which permits the performance of experiments under lithostatic stress.

Experimental

Based on previous experience with the determination of molecular migration of light hydrocarbons in sedimentary rocks (Krooss and Schaefer, 1987; Krooss, 1988; Krooss and Leythaeuser, 1988) a new experimental apparatus was constructed to investigate the mobility of light hydrocarbons in sedimentary rocks. While earlier diffusion experiments were limited to a maximum temperature of 70°C and ambient pressure, it was attempted to extend the temperature range of the new setup to 200°C in order to be able to establish thermal conditions comparable to those in actively generating and expelling source rocks. Furthermore, the assessment of diffusive gas loss through the caprock of deep and hot reservoirs which constitutes an economically relevant problem can be addressed with this new experimental equipment.

One principal goal in the construction of the experimental apparatus was the capability of frequent and reproducible on-line sampling and analysis of light hydrocarbons. For this purpose a new sampling method based on the "purge and trap" technique was designed and linked to the experimental system. With this method, on-line analysis of the purgeable hydrocarbons, i.e. low molecular-weight hydrocarbons up to a carbon number of 11, can be performed.

The flow cell (Fig. 1) which constitutes the central part of the experimental setup, was constructed at the Institute for Applied Technology (ZAT) of KFA Jülich. The cell accommodates cylindrical rock samples of 30 mm diameter and a maximum length of 30 mm. The usual thickness of the cylindrical rock slices ranges around 10 mm. The samples are placed between two cylindrical sample holders and the cylinder walls of both the sample and the sample holders are covered with a two-layered sleeve consisting of a Teflon shrink-tube and a thin-walled aluminium tube to which a confining pressure is applied. Tests with metal disks showed that this technique provides a diffusion-tight seal for C_{5+} hydrocarbons up to temperatures of more than 200°C. For hydrocarbon gases ($<C_5$) and high-temperature ($>$200°C) applications a graphite-based sealing material has been successfully tested.

The sample holders, which are equipped with two 1/16″ stainless steel tubes each for introduction and removal of fluids, form two chambers separated by the rock sample. In a modified version of the sample holders the rock sample is confined between two stainless steel frits which serve as an axial support. This arrangement prevents mechanical damage to the sample which may be caused by excessive confining pressure.

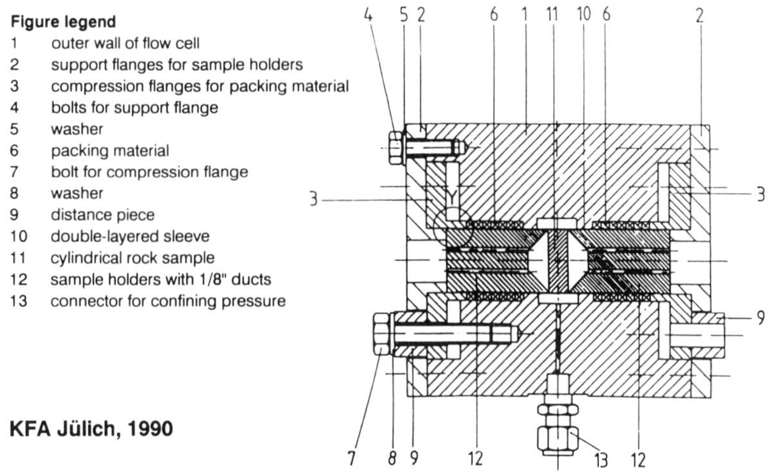

Fig. 1. Flow cell used for primary migration experiments at elevated temperatures (max. 200°C) in the present study.

In order to assure equal pressures at both sides of the rock samples — a prerequisite to avoid interference of pressure-driven volume flow (Darcy flow) with the diffusive transport to be measured — the two cell halves are connected by a communicating tube system. If desired, a pressure gradient across the sample can be readily established to determine the permeability of the rock samples. Permeabilities down to the nanodarcy range were measured in test experiments. These permeability measurements can be combined with a simultaneous analysis of the hydrocarbons within one of the cell halves.

The experimental setup can be used in a "release" and a "permeation" mode. In the release mode, fluid phase from both chambers of the flow cell can be sampled and analyzed at predetermined time intervals. Thus, the amounts of hydrocarbons released from the sample at a given temperature can be monitored as a function of time. In the permeation mode, a chemical potential gradient is established across the rock sample by introducing a fluid phase containing elevated concentrations of selected hydrocarbon compounds into one chamber and monitoring the changes in concentration at the opposite side (sampling side).

In order to maintain liquid saturation within the experimental system (rock sample, cell chambers, tubing) and prevent the formation of a vapour phase at elevated temperatures, the system pressure has to be kept above the vapour pressure of the fluids throughout the experiment. Up to an experimental temperature of 150°C a fluid pressure around 3 MPa (30 bar) was generally used.

Figure 2 shows the flow scheme of the experimental setup in the permeation mode. It can be subdivided into a high-pressure system (flow cell, pump, valve V1), a sampling section (sampling valves

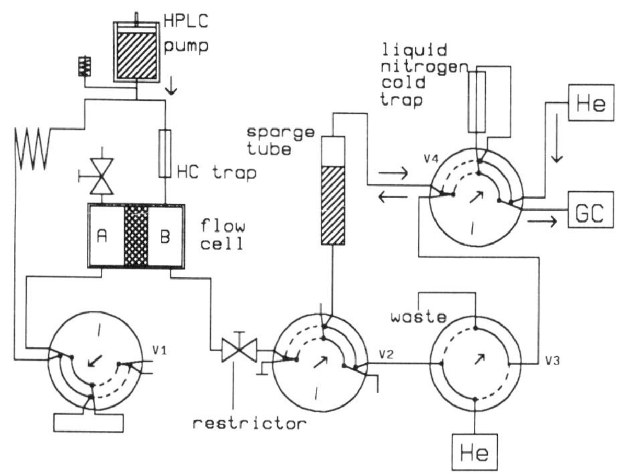

Fig. 2. Flow-scheme of the experimental setup in the "permeation" mode.

V2, V3, V4, sparge tube, liquid nitrogen cold trap) and the analytical system (GC).

The flow cell is pressurized by means of a commercial pump as used in liquid chromatography (HPLC) with a maximum pressure of 50 MPa (500 bar) and a maximum flow rate of 10 ml/min. The pump is equipped with an adjustable relief valve to control the system pressure. Through the 6-port switching valve (V1) with a sample loop a certain volume of liquid hydrocarbon mixture, the "source phase", can be introduced into the source chamber (A) of the flow cell. Special precautions were taken to prevent backward diffusion of the hydrocarbons through the communicating tubing system into the sampling chamber (B). Thus, a 3-m stainless steel tube (1/8″ i.d.) was placed between the switching valve (V1) and the T connector joining the pump with the two cell halves in order to increase the diffusion distance along this bypass. In addition, a hydrocarbon trap containing an adsorbent was installed in front of the sampling chamber (B) to remove hydrocarbons from the water current entering it. The sampling chamber (B) is connected to the sampling valve (V2) via a flow restrictor which prevents a pressure drop in the flow cell during the sampling step. At fixed time intervals a certain amount of aqueous phase (approx. 1 ml) is withdrawn from the sampling chamber and transferred to the sparge tube by opening the sampling valve V2. In a subsequent "purge and trap" step a helium current is passed through the water in the sparge tube for a period of 3–5 min. The purgeable hydrocarbons are stripped from the water and carried to a liquid-nitrogen cold trap where they are retained and concentrated at a temperature of 77.4 K (−195.8°C). The cold trap, which is located within the thermostatic oven of a gas chromatograph (GC), is then heated rapidly to approximately room temperature by means of pressurized air and subsequently to 180°C by a programmed temperature increase of the oven. The hydrocarbons mobilized from the cold trap are analyzed on a 3-m packed stainless steel column (10% DC200 on Chromosorb A-NAW 60/80 mesh). Tests conducted with water spiked with different hydrocarbons showed that *n*-alkanes up to C_{11} can be analyzed with this method. The whole experimental setup is controlled by a personal computer.

Results

Various migration experiments have been performed with the new experimental setup in order to investigate different aspects of the molecular transport of low molecular-weight hydrocarbons in source rocks. During the initial stage of the experimental programme special attention was paid to the release of autochthonous light hydrocarbons from source rocks and the determination of the molecular mobility of individual hydrocarbon species. These experiments bear immediate relevance for the assessment of expulsion rates of light hydrocarbons and the efficiency of primary migration, but they can also be expected to elucidate the importance of diffusive losses through caprocks, which in many locations have source rock characteristics.

Due to inherent difficulties in handling two-phase (gas–liquid) hydrocarbon systems under the experimental conditions envisaged it was decided to perform separate sets of experiments for hydrocarbon gases and liquids. With respect to the almost complete absence of relevant experimental data, the first set of experiments was performed with liquid hydrocarbons (C_5–C_{12} hydrocarbon mixtures). The molecular transport of C_1–C_4 hydrocarbons at elevated temperatures will be addressed in a later study.

The samples used in the experiments reported here, a Torbanite and a mature type II source rock, were kindly provided by the Koninklijke/Shell Exploration and Production Laboratories in Rijswijk (NL).

Release experiments

A first set of experiments was carried out in the "release" mode to test the performance of the system and improve its reliability for long-time measurements. As demonstrated by Krooss (1988), the release rate of autochthonous light hydrocarbons from source rocks can be used to obtain information on their molecular mobility and the transport mechanisms involved. The experimental arrangement can be considered as an analogue of a source rock sequence interbedded between two active aquifers (carrier beds) in which the light hydrocarbons released from the source rock are carried away. It corresponds, in essence, to the model system discussed by Leythaeuser et al. (1982, fig. 5). The temperature range of the release experiments extended from 100°C to 180°C. Sampling intervals varied between 1 and 12 h.

Contrary to previous release experiments with hydrocarbon gases (Krooss, 1988) it was impossible under the present experimental conditions to determine a release rate in terms of the quantity of light hydrocarbons released from the source rock samples per unit time. The hydrocarbon concentrations registered in the water sampled from the flow cell at a given temperature were essentially constant and independent of the duration of the sampling interval, i.e. of the flow rate of water within the adjacent "aquifers". This result is demonstrated in Fig. 3a which shows the GC traces of the light hydrocarbons released from a Torbanite, a rich Permian source rock (60% TOC), at 150°C after sampling intervals of 1 h and 3 h, respectively. The two chromatograms were recorded with the same attenuation and are practically identical.

The two main peaks in these chromatograms were identified as cyclohexane and methylcyclohexane. The set of chromatograms recorded at 100°C system temperature (Fig. 3b) shows the same result. However, all peaks are substantially smaller here than at 150°C.

It is concluded from these release experiments that a partition equilibrium of the hydrocarbons between the source rock and the adjacent aqueous phase is established very rapidly as compared to the minimum residence time of the aqueous phase in the flow cell (approx. 1 h). The measured hydrocarbon concentrations in the aqueous phase are equilibrium

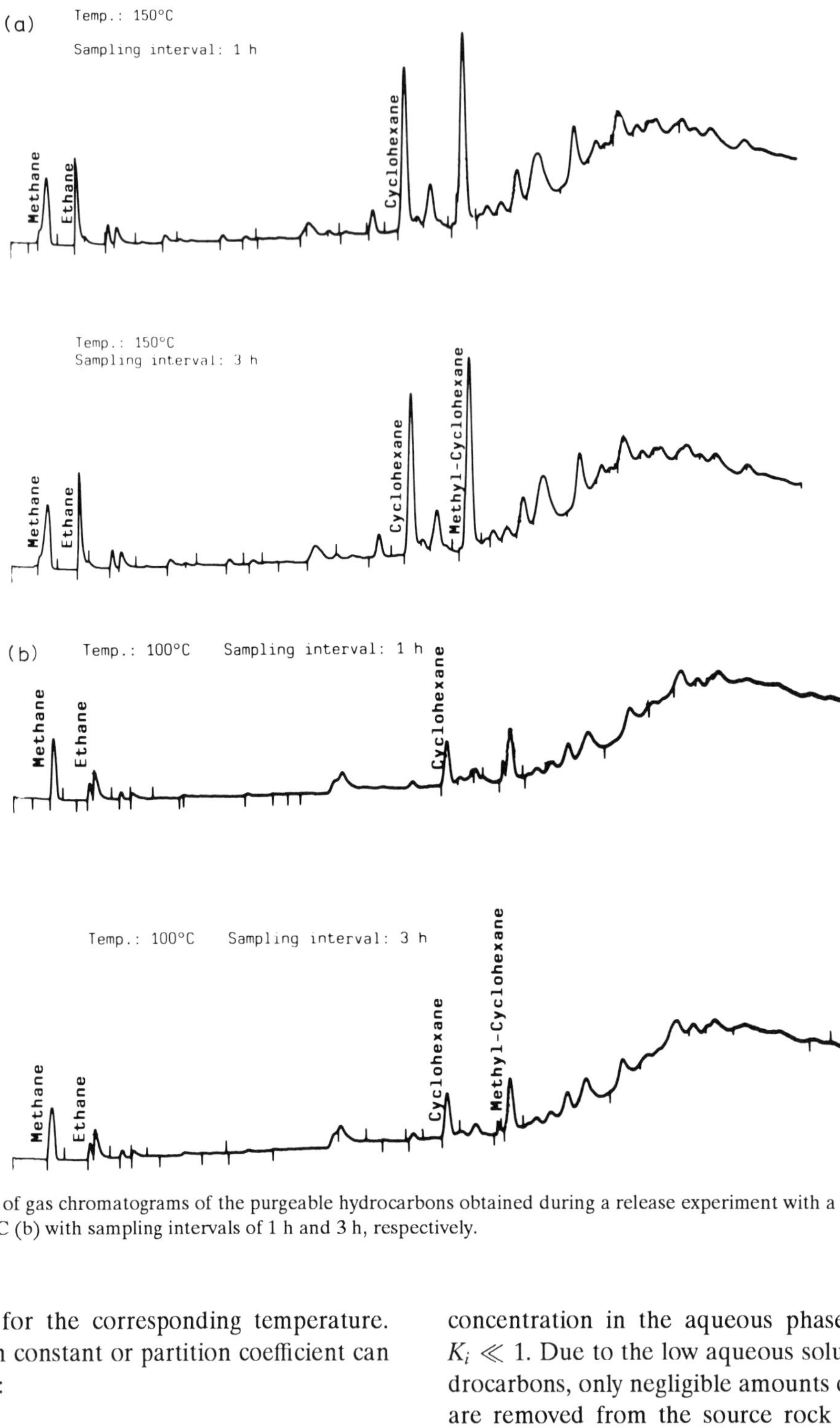

Fig. 3. Comparison of gas chromatograms of the purgeable hydrocarbons obtained during a release experiment with a torbanite sample at 150°C (a) and 100°C (b) with sampling intervals of 1 h and 3 h, respectively.

concentrations for the corresponding temperature. The equilibrium constant or partition coefficient can be expressed as:

$$K_i = \frac{C_i\,(\text{water})}{C_i\,(\text{SR})} \tag{1}$$

where C_i(water) and C_i(SR) denote the concentration of the hydrocarbon i in the aqueous phase and the source rock, respectively. As evidenced by the chromatograms in Fig. 3, this equilibrium constant increases significantly with temperature. The fact that even a large number of repetitive equilibration steps did not affect the measured hydrocarbon concentration in the aqueous phase indicates that $K_i \ll 1$. Due to the low aqueous solubility of the hydrocarbons, only negligible amounts of hydrocarbons are removed from the source rock sample in each equilibration step as compared to the amounts of hydrocarbons in the source rock and C_i(SR) remains practically constant. In summary, the following conclusions were derived from the results of the release experiments:

(1) With the experimental arrangement for the "release mode" only a partitioning equilibrium for light hydrocarbons between the source rock sample and the aqueous phase could be measured. Due to the very rapid establishment of this equilibrium

no information can be obtained with this method on the dynamics of light hydrocarbon release from the source rock or a possible onset of hydrocarbon generation at the experimental temperatures.

(2) At the experimental temperatures and flow rates considered here the water used as a sampling liquid is not capable of removing the amounts of light hydrocarbons which can be potentially released from the source rock sample. The maximum release rate cannot be determined because the transport capacity of the aqueous sampling phase is too small at the sampling rates used in these experiments.

(3) Due to the very low aqueous solubility of the hydrocarbons the water within the "carrier bed" inhibits the transfer of hydrocarbons from the source rock and acts as a practically "impermeable" boundary for the hydrocarbons within the sample. The small, but readily measurable, hydrocarbon concentrations in the aqueous phase are related to the corresponding concentrations within the source rock by the partition coefficients and therefore can be used as a probe to measure the concentration at the source rock surface without significantly affecting the hydrocarbon concentration within the source rock.

The second conclusion relates primarily to the experimental system with the artificially created source rock surface and an average flow rate of water past this surface in the order of magnitude of 1 ml per 10 cm^2 surface area per hour (10^{-3} m/h). In terms of the model of Stainforth and Reinders (1990) this suggests a very small mass Biot number, *Bi*, corresponding to a high mass transfer resistance of the source rock/carrier bed boundary. It remains to be examined to what extent these results can be applied to natural source rock/carrier bed interfaces. In a water-saturated carrier bed the velocity of the hydrodynamic flow, the partition coefficient, and, to some extent, the rate of diffusion in the formation water would affect the migration resistance and determine the release of light hydrocarbons from the source rock.

Conclusion 3 has important implications for the evaluation procedure of the permeation experiments as will be outlined below.

Permeation experiments

As the release experiments did not permit the determination of molecular mobilities of light hydrocarbons in source rocks, a different approach had to be chosen to achieve this goal. Permeation experiments as used, for instance, by Krooss (1985) represent another common method to measure diffusion coefficients in solids and porous media. To this end a chemical potential gradient, corresponding to a first approximation to a concentration gradient, has to be established across the rock sample and maintained throughout the duration of the experiment. First attempts to use hydrocarbon saturated aqueous phases as source phases in the diffusion experiments failed because, due to the low aqueous solubility of the hydrocarbons, no sufficiently large concentration gradients could be established. In order to obtain signals that significantly exceed the background of the autochthonous hydrocarbon concentrations in the source rocks it was decided to use liquid hydrocarbon mixtures as source phases in these experiments. In essence, this approach can be envisaged as representing the case of a small local accumulation of light hydrocarbons generated within the source rock and diffusing away from their place of origin.

No data on the order of magnitude of diffusion coefficients for light hydrocarbons in sedimentary rocks at elevated temperatures (>100°C) were available prior to this study. Therefore, several test runs had to be performed to optimize the experimental conditions (sample thickness, duration of sampling intervals and entire experiment) for the permeation experiments. The results of two experiments will be discussed here in more detail in order to outline the progress made with this experimental approach and the problems still associated with it.

Experimental boundary conditions and mathematical model

Based on the experimental boundary conditions and the results from the release experiments a new method had to be established for the evaluation of the experimental data. The diffusion model and the corresponding initial and boundary conditions are shown schematically in Fig. 4.

The rock sample is treated as a plane sheet with thickness l. One surface of this sheet ($x = l$) is

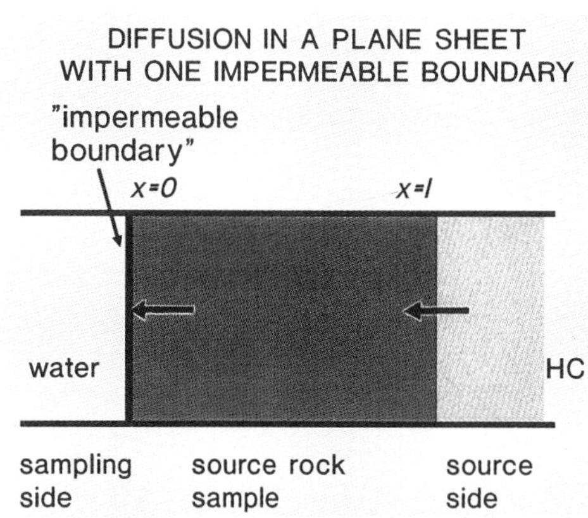

Fig. 4. Scheme of the boundary conditions of the experimental system.

in contact with a hydrocarbon mixture representing the "source phase". The composition of this liquid hydrocarbon mixture determines the concentration of the diffusing substance, C_1, just within the source rock sample at the position $x = l$. C_1 is assumed to be constant during the whole experiment:

$$C(x, t) = C_1 = \text{constant} \quad x = l \quad t \geq 0 \tag{2}$$

The opposite plane of the rock sample at the position $x = 0$ is in contact with the aqueous sampling phase. As stated above, due to their low aqueous solubility, only a negligible fraction of the hydrocarbons within the source rock sample enters the adjacent aqueous phase. Therefore, as a first approximation, the surface of the rock sample at $x = 0$ can be assumed impermeable for the hydrocarbons within the source rock. The diffusion problem corresponds to a "diffusion into a plane sheet with one impermeable boundary" and is essentially an inversion of the problem treated by Leythaeuser et al. (1982) (C_1–C_9 diffusion from source rocks). The aqueous phase serves as a probe to monitor the hydrocarbon concentration just within the rock sample at the position $x = 0$. The analytical expression for the concentration of diffusant in the rock sample under these conditions as a function of time and distance is given by Crank (1975, equation 4.17):

$$\frac{C - C_0}{C_1 - C_0} = 1 - \left(\frac{4}{\pi}\right) \sum_{n=0}^{\infty} \left(\frac{-1^n}{2n+1}\right) \times \exp\left[\frac{-D(2n+1)^2\pi^2 t}{4l^2}\right] \cos\left[\frac{(2n+1)\pi x}{2l}\right] \tag{3}$$

Here D is the diffusion coefficient and C_0 denotes an initial uniform concentration of diffusing species within the plane sheet which can be expressed by the initial condition:

$$C(x, t) = C_0 = \text{constant} \quad 0 \leq x \leq l \quad t < 0 \tag{4}$$

For samples containing no or only negligible amounts of the diffusing species, C_0 is taken as zero; otherwise it is determined in the evaluation of the experimental data together with D and C_1.

Figure 5 shows the normalized concentration profiles at different stages of the diffusion process for the system envisaged here, calculated with eq. 3.

The concentration of diffusing species at the "impermeable" boundary ($x = 0$) is initially C_0. When substance diffuses into the plane sheet from the source side ($x = l$) it gradually increases and approaches C_1. The concentration of diffusing species at $x = l$ as a function of time up to a value of $Dt/l^2 = 1.2$ is plotted in Fig. 6. Here C_0 was assumed to be zero.

Selected results

One of the first permeation experiments was performed with a Torbanite sample, in order to investigate the molecular transport of hydrocarbons within the organic matter (kerogen) which constitutes the bulk of this material. The thickness of the sample was 6.9 mm and the experimental temperature 150°C. The composition of the liquid hydrocarbon phase, which consisted of seven different hydrocarbons with n-nonane as the main component, is listed in Table 1.

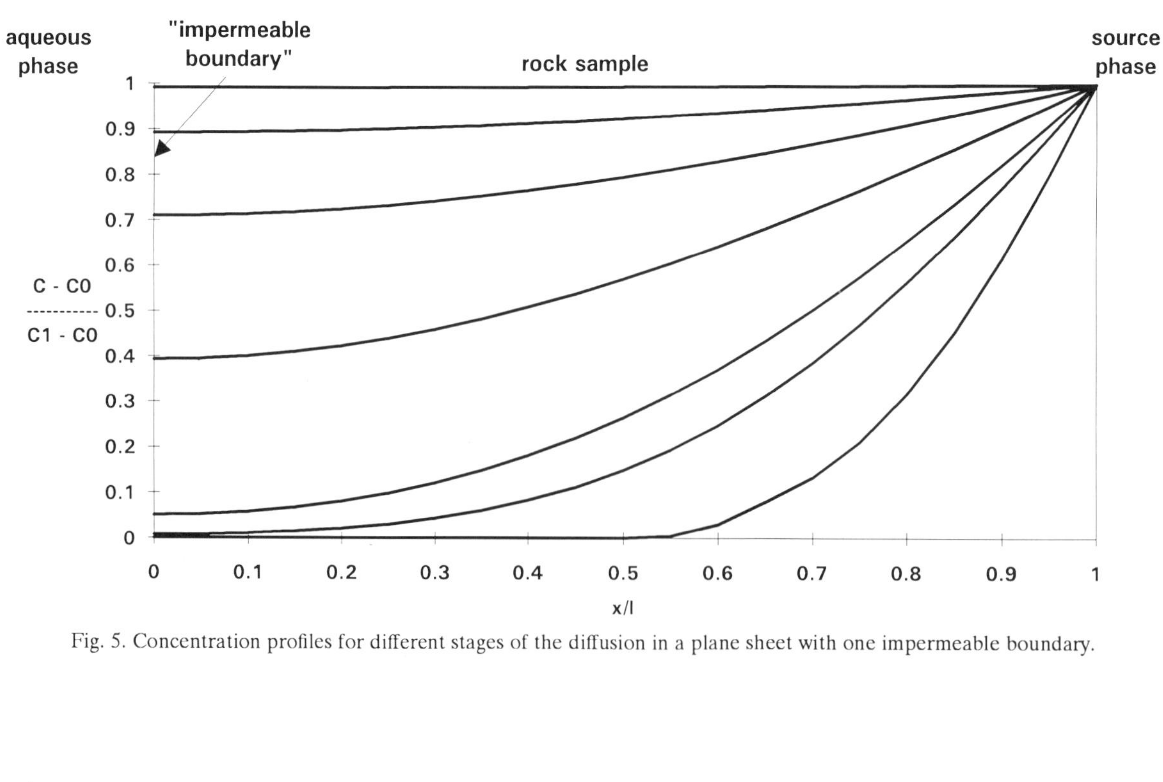

Fig. 5. Concentration profiles for different stages of the diffusion in a plane sheet with one impermeable boundary.

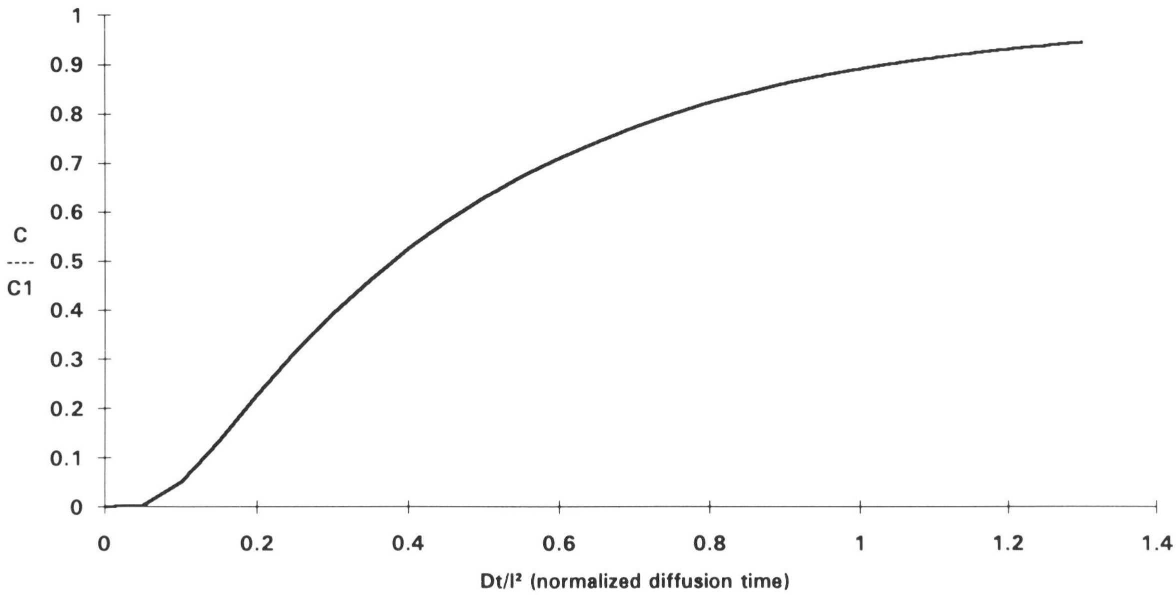

Fig. 6. Evolution of the concentration at the impermeable boundary ($x = 0$) as a function of time for the diffusion in a plane sheet with one impermeable boundary.

TABLE 1

Composition of the hydrocarbon source phase used in migration experiment I (torbanite sample)

Hydrocarbon	Mass (%)	Volume (%)
n-Pentane	0.41	0.47
n-Hexane	4.01	4.34
n-Heptane	4.04	4.21
n-Octane	4.01	4.07
n-Nonane	86.72	86.24
Benzene	0.42	0.34
Toluene	0.40	0.23

During this migration experiment the breakthrough of five hydrocarbons was registered. Figure 7 shows the experimental curves and the corresponding fitted diffusion functions according to eq. 2 for *n*-pentane, *n*-hexane, *n*-heptane, benzene and toluene. Source phase hydrocarbons first appeared at the sampling side of the cell about 20 h after the start of the experiment.

None of the experimental curves reached a plateau value within the duration of the experiment (70 h). The fitted diffusion functions show a relatively good agreement with the experimental values in most cases. The diffusion coefficients obtained from this evaluation are listed in Table 2.

The values lie in the range of 2×10^{-11} m^2 s^{-1} to 7×10^{-11} m^2 s^{-1} and show a steady decrease for the saturated hydrocarbon homologues from *n*-pentane through *n*-heptane. The highest diffusion

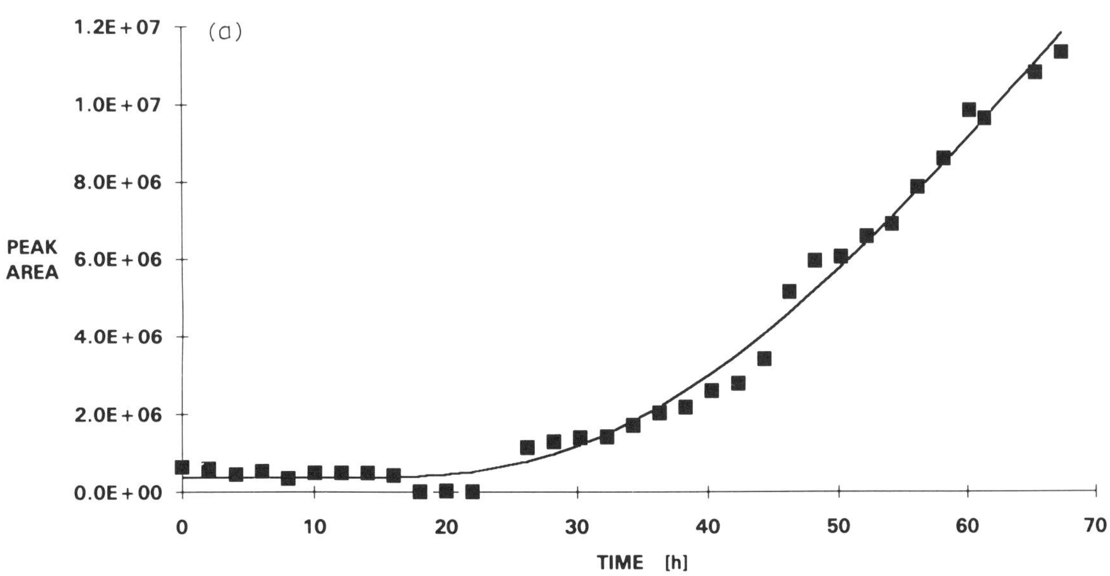

Fig. 7. Experimental data and fitted diffusion function for (a) *n*-pentane.

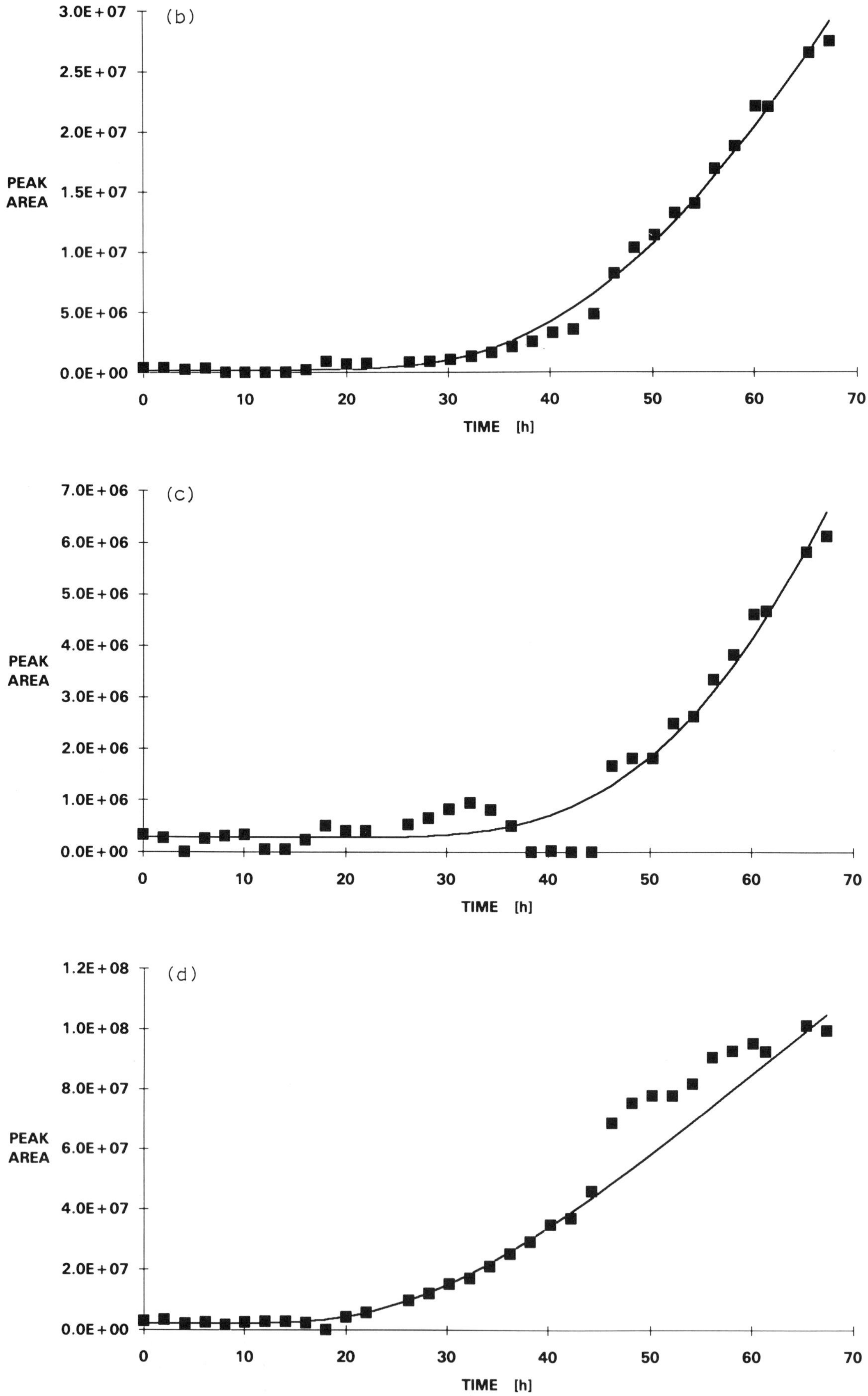

Fig. 7 (continued). Experimental data and fitted diffusion functions for (b) *n*-hexane, (c) *n*-heptane, and (d) benzene.

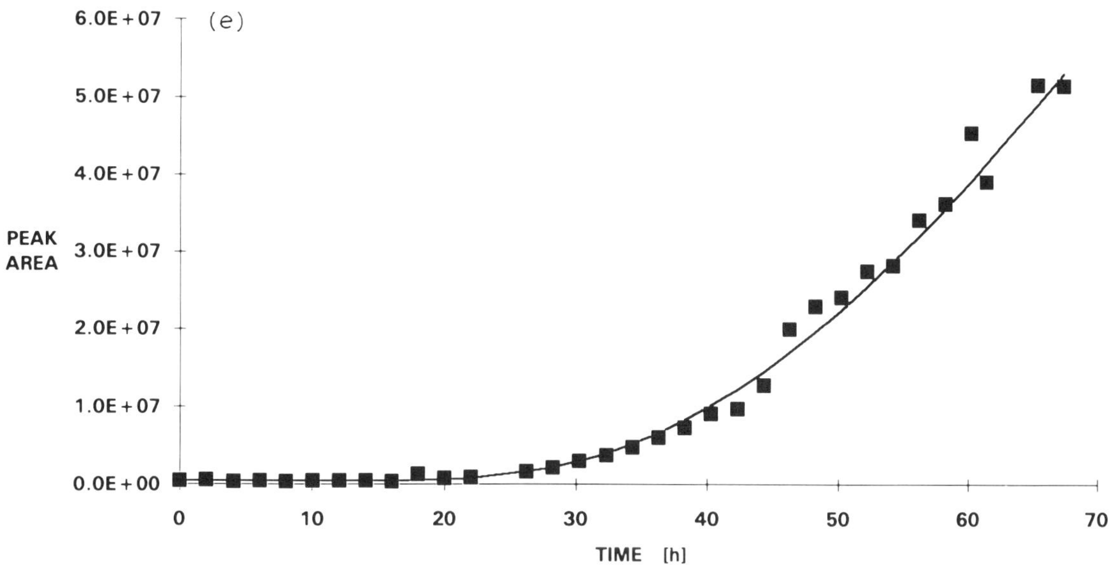

Fig. 7 (continued). Experimental data and fitted diffusion function for (e) toluene of permeation experiment I; artificially matured torbanite, 150°C.

TABLE 2

Diffusion coefficients for light hydrocarbons in a torbanite sample at 150°C

Hydrocarbon	Diffusion coefficients ($m^2\ s^{-1}$)
n-Pentane	4.93×10^{-11}
n-Hexane	3.55×10^{-11}
n-Heptane	2.51×10^{-11}
Benzene	6.38×10^{-11}
Toluene	4.16×10^{-11}

coefficient was obtained for benzene whereas the value for toluene is intermediate between *n*-pentane and *n*-hexane.

Another diffusion experiment was performed with a mature source rock sample bearing type II kerogen. The liquid source phase used in this experiment consisted of eight different hydrocarbons with *n*-pentane as the major component. The composition of the source phase is given in Table 3. The experimental temperature was 150°C.

The most interesting finding in this experiment was that of the eight source-phase components only the most water-soluble hydrocarbons, namely *n*-pentane, benzene and toluene penetrated the rock sample over a time period of 230 h. No significant concentration increase of any of the other source compounds was registered at the sampling side of the flow cell during this experiments. The breakthrough-curves for the three components including the fitted diffusion functions are shown in Fig. 8.

As in the first experiment, the transport process was interpreted as "diffusion in a plane sheet with one impermeable boundary". However, this time the

TABLE 3

Composition of the hydrocarbon source phase used in migration experiment II (mature type II source rock)

Hydrocarbon	Mass (%)	Volume (%)
n-Pentane	78.08	80.94
n-Hexane	3.11	3.06
n-Heptane	3.10	2.94
n-Octane	3.22	2.97
n-Nonane	3.15	2.85
Cyclohexane	3.15	2.62
Benzene	3.09	2.29
Toluene	3.11	2.33

evaluation was performed with the function proposed by Crank (1975) as a solution for small times for diffusion in a plane sheet:

$$\frac{C - C_0}{C_1 - C_0} = 2 \sum_{n=0}^{\infty} (-1)^n \operatorname{erfc}\left[\frac{(2n+1)\,l}{2\sqrt{Dt}}\right] \tag{5}$$

Again, C_0 denotes initial concentration of the diffusing substance in the sample and C_1 is its final concentration after infinite time. A nonlinear least-squares fitting procedure was used to adjust the three parameters D/l^2, C_0, and C_1 and obtain a best fit of the experimental data. The diffusion coefficients resulting from this evaluation range between 2×10^{-11} and $7 \times 10^{-11}\ m^2\ s^{-1}$ (Table 4).

The best fits were obtained for the two hydrocarbons *n*-pentane and benzene. Due to a relatively large scatter of the experimental data for toluene the diffusion coefficient of $6.09 \times 10^{-11}\ m^2\ s^{-1}$ is considered less reliable.

In order to investigate the observed preferential breakthrough of the most water-soluble hydrocar-

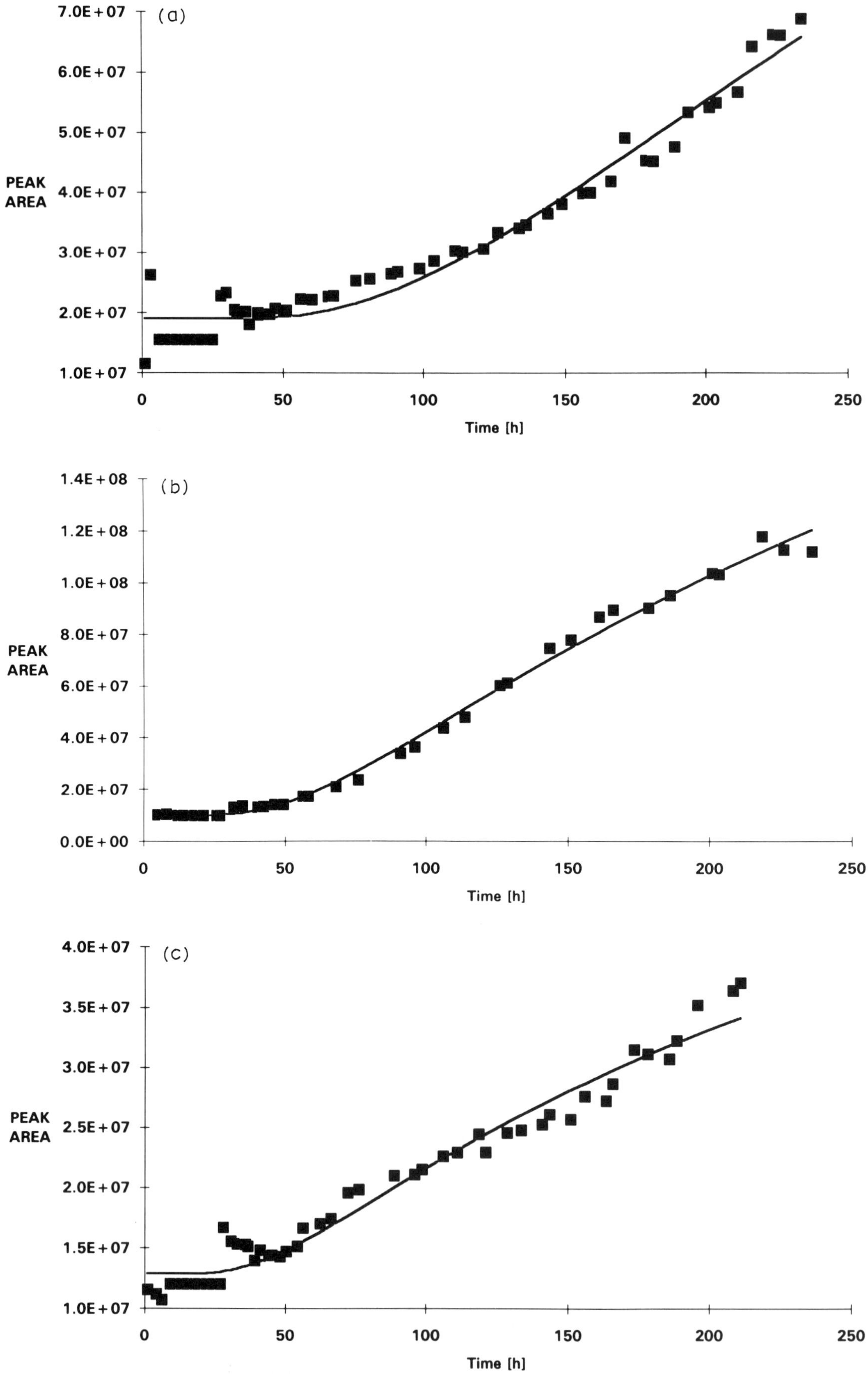

Fig. 8. Experimental data and fitted diffusion functions for *n*-pentane (a), benzene (b), and toluene (c) of permeation experiment II; type II source rock, 150°C.

TABLE 4

Coefficients for light hydrocarbons in a mature type II source rock sample at 150°C

Hydrocarbon	Diffusion coefficients ($m^2 s^{-1}$)
n-Pentane	2.71×10^{-11}
Benzene	4.62×10^{-11}
Toluene	6.09×10^{-11}

bons in more detail, the rock sample was sectioned into three slices after the experiment. Material from these three slices (source side, centre, sampling side) was analysed by headspace gas chromatography (Schaefer et al., 1984). In addition, the original sample material was analysed in the same way. Methylcyclohexane, which represented a major component in the source rock material but was not present in the hydrocarbon source mixture, was used as an internal standard to compare the resulting chromatograms. Figure 9a shows the concentrations for benzene and toluene in the three sample sections normalized to their content in the original sample. The benzene concentration at the source side of the sample after the end of the experiment was found to be approximately 60 times higher than in the original sample.

At the sampling side of the rock slice it has still a value corresponding to about 18 times the initial concentration. Obviously, the benzene diffusion

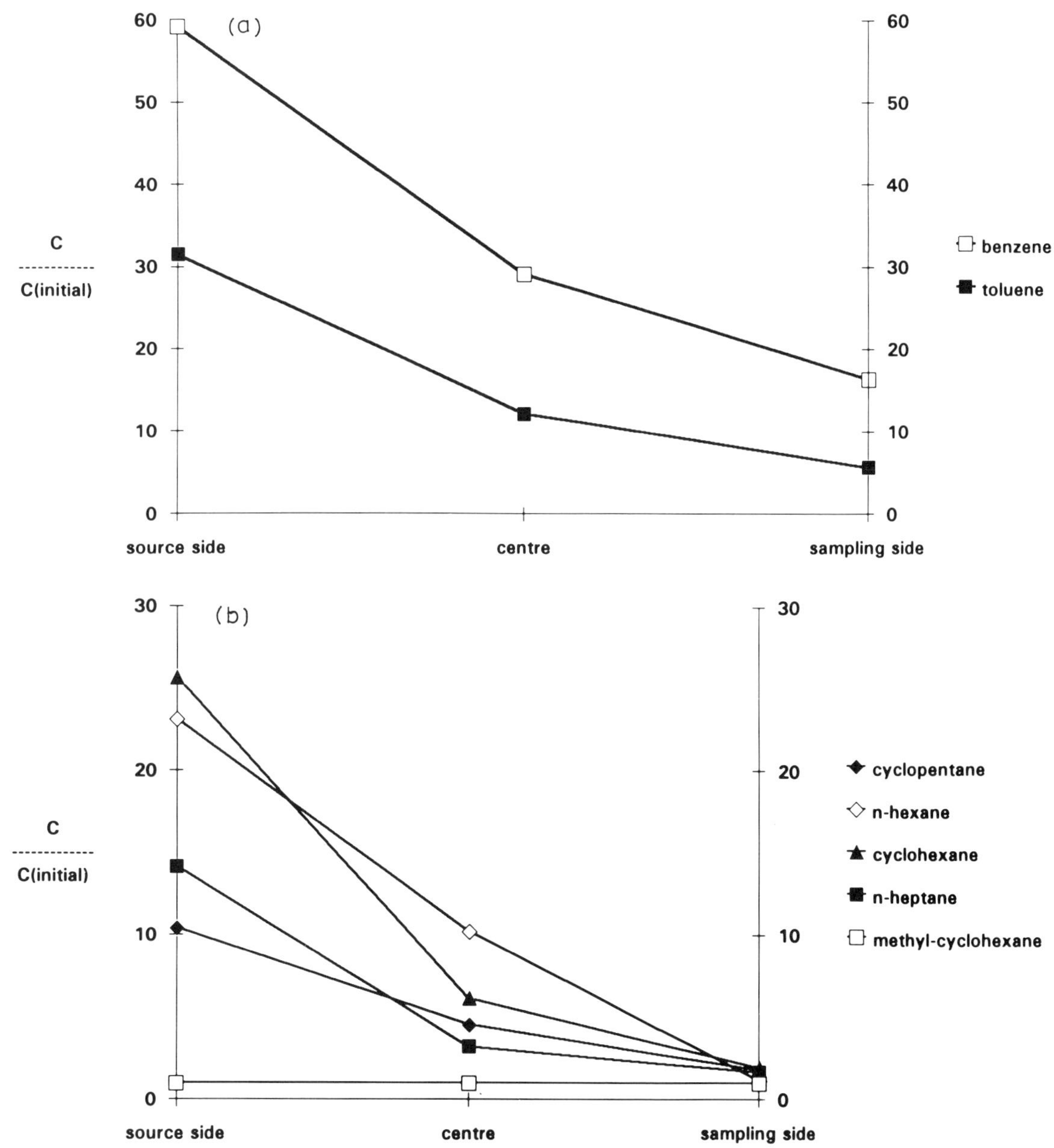

Fig. 9. Concentration profiles obtained from post-experiment headspace GC analysis of the sections of the rock sample used in permeation experiment II. Concentrations are normalized to contents of the corresponding hydrocarbons in the original sample.

front penetrated into the sample to a significant extent and caused a substantial increase in benzene concentration throughout the sample. The same is true, though to a lesser extent, for toluene. Here the source side concentration amounts to slightly more than 30 times the initial concentration and drops to a value of about 5 times the initial concentration at the sampling side.

The results for the nonaromatic compounds, shown in Fig. 9b are much different. Both *n*-hexane and cyclohexane are enriched at the source side of the sample by a factor of approximately 25 with respect to their initial concentration. The corresponding value for *n*-heptane is approximately 15. Cyclopentane, which was present as a major impurity in the pentane used for preparing the hydrocarbon source phase, has a concentration at the source side about 10 times higher than its content in the original sample. The evaluation could not be performed for *n*-pentane due to problems with the integration of the chromatographic peak areas for this compound. Also shown in the diagram is the normalized content of the reference compound, methylcyclohexane.

Contrary to the aromatic compounds (Fig. 9a) the naphthenic and aliphatic hydrocarbons have penetrated into the rock sample to a much lesser extent. Their concentrations at the sampling side are practically unaffected and correspond, within the limits of experimental error, to the initial concentrations. Only *n*-hexane has diffused to some extent towards the centre of the sample as evidenced by a ten-fold increase with respect to the original concentration in the central section.

The headspace GC data confirm the results of the permeation experiment in a qualitative sense and demonstrate very clearly the preferential transport of the more water-soluble hydrocarbons in this particular case.

The results obtained here furnish an impressive experimental support of previous findings by Leythaeuser et al. (1983) who analysed the concentration profiles of light hydrocarbons across a source rock/carrier bed boundary. The abnormally low "depletion ratios" for benzene and toluene were interpreted by these authors as an indication for very high transportation rates of these two aromatic compounds as compared to the aliphatic and naphthenic hydrocarbons of similar molecular weight. They suspected that this behaviour could not be explained by diffusive transport alone. In the experiments performed here, the occurrence of an additional transport process (e.g. pressure-driven volume flow) was ruled out by careful control of the boundary conditions, and with a very high probability it is the higher molecular mobility of the aromatic hydrocarbons within source rocks that accounts for the observed "anomalies". It is suggested that the elevated aqueous solubility of the aromatic compounds plays a substantial role in their transport behaviour. This would mean that water-saturated pathways occur across source rocks which are predominantly accessible to the more water-soluble compounds and restrict the mobility of nonpolar hydrocarbons. It can be expected that future experiments of this kind will reveal more details on the type and transport capacity of different migration pathways within source rocks.

Conclusion

The experimental method presented here was developed in the context of a research programme on "processes in source, carrier and reservoir rocks". Among the multitude of phenomena that still require detailed investigation, the present study was focused on molecular transport of light hydrocarbons in source rock types and organic matter-rich materials. For the first time, to our knowledge, molecular transport parameters for light hydrocarbons (C_5–C_7) in source rocks at temperatures in excess of 100°C have been measured. Experimental temperatures reached with this device are comparable to those prevailing in the subsurface during the principal stage of petroleum generation and expulsion. The first experiments conducted with this apparatus have provided useful information on the mobility of different light hydrocarbons in source rocks and have revealed interesting differences in the transport behaviour of aliphatic and aromatic hydrocarbons. These results are in good agreement with observations previously made in natural systems (e.g. Leythaeuser et al., 1983; Krooss et al., 1988). Furthermore, valuable technical experience was gained for the conduction of future migration experiments under conditions even more closely resembling geological situations. Meanwhile, an enhanced setup with the capability of simulating migration processes under subsurface pressure conditions has been constructed.

The coupling of migration experiments with automatic sampling and subsequent analysis that has been achieved here for light hydrocarbons and experimental temperatures up to 200°C, is a promising approach for various reasons. Repetitive sampling and analysis under constant experimental conditions reduces the risk of analytical errors and improves the reproducibility. Analytical problems become immediately evident and can be traced readily. Another important advantage is that the continuous observation of the experiment under controlled temperature and pressure conditions provides immediate access to the dynamics of the migration process.

Acknowledgements

This project was financed by the German Ministry of Research and Technology (BMFT) in the context of the German–Norwegian Research Cooperation on Basin Modeling and Reservoir Studies (BMFT Grant No. ET 6906A4). Financial support and sample material was kindly provided by the Shell Research B.V. in Rijswijk (NL). The work benefited greatly from stimulating discussions with Drs. J. Stainforth and J.E.A. Reinders, of Koninklijke/ Shell and comments by an unnamed reviewer. The authors wish to acknowledge the support received by Dr. R.G. Schaefer, J. Hoeltkemeier and H.J. Sittardt (KFA-Jülich) in the analysis and identification of light hydrocarbons.

References

Antonov, P.L., 1954. On the diffusion permeability of some claystones. Tr. NIIGGR, Gostoptekhizdat, Moscow, 2: 39–55 (in Russian).

Antonov, P.L., 1968. Some results of the investigation of molecular migration of hydrocarbons in rocks. Tr. VNIIYaGG, 4: 132–154 (in Russian).

Antonov, P.L., 1970a. Results of the investigation of the diffusion permeability of sedimentary rocks for hydrocarbon gases. Tr. VNIIYaGG, 8: 51–79 (in Russian).

Antonov, P.L., 1970b. On the gas saturation of rocks above reservoirs of finite thickness. Depletion of reservoirs during the process of upward diffusion of gases. Tr. VNIIYaGG, 8: 38–50.

Crank, J., 1975. The Mathematics of Diffusion. Clarendon Press, Oxford, 2nd ed., 414 pp.

Krooss, B.M., 1985. Experimentelle Untersuchung der Diffusion niedrigmolekularer Kohlenwasserstoffe in wassergesättigten Sedimentgesteinen. Ph.D. Thesis, RWTH Aachen, 215 pp.

Krooss, B.M., 1988. Experimental investigation of the molecular migration of C_1–C_6 hydrocarbons: kinetics of hydrocarbon release from source rocks. Org. Geochem., 13: 513–523.

Krooss, B.M. and Leythaeuser, D., 1988. Experimental measurements of the diffusion parameters of light hydrocarbons in water-saturated sedimentary rocks, II. Results and geochemical significance. Org. Geochem., 12: 91–108.

Krooss, B.M. and Schaefer, R.G., 1987. Experimental measurements of the diffusion parameters of light hydrocarbons in water-saturated sedimentary rocks, I. A new experimental procedure. Org. Geochem., 11: 193–199.

Krooss, B.M., Leythaeuser, D. and Schaefer, R.G., 1988. Light hydrocarbon diffusion in a caprock. Chem. Geol., 71: 65–76.

Krooss, B.M., Leythaeuser, D. and Schaefer, R.G., 1992. The quantification of diffusive hydrocarbon losses through cap rocks of natural gas reservoirs — a reevaluation. Am. Assoc. Pet. Geol., Bull., 76: 403–406.

Leythaeuser, D., Schaefer, R.G. and Yükler, A., 1980. Diffusion of light hydrocarbons through near-surface rocks. Nature, 284: 522–525.

Leythaeuser, D., Schaefer, R.G. and Yükler, A., 1982. Role of diffusion in primary migration of hydrocarbons. Am. Assoc. Pet. Geol., Bull., 66: 408–429.

Leythaeuser, D., Schaefer, R.G. and Pooch, H., 1983. Diffusion of light hydrocarbons in subsurface sedimentary rocks. Am. Assoc. Pet. Geol., Bull., 67: 889–895.

Leythaeuser, D., Schaefer, R.G. and Radke, M., 1988. Geochemical effects of primary migration of petroleum in Kimmeridge source rocks from Brae field area, North Sea, I. Gross composition of C_{15+} soluble organic matter and molecular composition of C_{15+} saturated hydrocarbons. Geochim. Cosmochim. Acta, 52: 701–713.

Mackenzie, A.S., Price, I., Leythaeuser, D., Müller, P., Radke, M. and Schaefer, R.G., 1987. The expulsion of petroleum from Kimmeridge clay source-rocks in the area of the Brae Oilfield, UK continental shelf. In: J. Brooks and K. Glennie (Editors), Petroleum Geology of North West Europe. Graham and Trotman, London, pp. 865–877.

Schaefer, R.G., Pooch, H. and Leythaeuser, D., 1984. Automatic headspace gas chromatography of C_1–C_7 hydrocarbons in sedimentary rocks. Application in petroleum geochemistry. Chromatographia, 19: 377–381.

Smith, J.E., Erdmann, J.G. and Morris, D.A., 1971. Migration, accumilation and retention of petroleum in the earth. Proc. 8th World Petroleum Congress, 2: 13–26.

Stainforth, J.G. and Reinders, J.E.A., 1990. Primary migration of hydrocarbons by diffusion through organic matter networks, and its effects on oil and gas generation. In: B. Durand and F. Behar (Editors), Advances in Organic Geochemistry 1989. Org. Geochem., 16: 61–74.

Stklyanin, Yu. I. and Litvinova, V.N., 1971. On the parameters of methane diffusion through a water saturated core. Geol. Nefti Gaza, 15: 19–22 (in Russian).

Stklyanin, Yu. I., Breeva, T.F., Vasilieeva, L.B., Domnina, N.A., Kotsonis, A.N., Kondratov, L.C., Antoshina, I.N., Zavyalova, L.M., Krishchenko, V.I. and Shlepova, A.I., 1968. Conditions, parameters and pathways of ascending natural gas migration from oil and gas source beds. Several theoretical migration models. Tr. VNIIYaGG, 4: 58–131 (in Russian).

B.M. KROOSS *Institute of Petroleum and Organic Geochemistry (ICG-4), Forschungszentrum (KFA), Jülich GmbH, P.O. Box 1913, D-5170, Jülich, Germany*

D. HANEBECK *Institute of Petroleum and Organic Geochemistry (ICG-4), Forschungszentrum (KFA), Jülich GmbH, P.O. Box 1913, D-5170, Jülich, Germany*

D. LEYTHAEUSER *Institute of Petroleum and Organic Geochemistry (ICG-4), Forschungszentrum (KFA), Jülich GmbH, P.O. Box 1913, D-5170, Jülich, Germany*

Present address: *Geologisches Institut, Universität zu Köln. Zülpicher Straße 49, D-5000 Köln 1, Germany*

Relative permeabilities and hydrocarbon expulsion from source rocks

Akihiko Okui and Douglas W. Waples

In creating numerical basin models it is often easier to write equations for physical processes than it is to provide input data for those equations. A particular problem in describing hydrocarbon expulsion from source rocks has been to specify relative-permeability curves for source rocks. It is highly doubtful that measured relative-permeability data from reservoir rocks with absolute permeabilities above 5 mD can be used to describe relative permeabilities of fine-grained, low-permeability rocks.

We have studied the problem of relative permeability in the expulsion of oil and gas from source rocks using computer simulations for a variety of kerogen and oil types and source-rock richnesses. Simulations based on a simple model in which relative permeability in source rocks is assumed to be similar to that in reservoir rocks, and in which the only driving force for fluid expulsion is the hydraulic potential, do not permit oil expulsion except at oil saturations that are much too high.

One solution is to add capillary pressure (P_c) and buoyancy (P_b) forces to the hydraulic potential. If carrier beds remain water wet, both P_c and P_b will promote hydrocarbon flow compared to water. Using relative-permeability curves like those for reservoir rocks, we calculate the sum of $P_c + P_b$ must be about 30 times the hydraulic potential for light oils and 350 times for heavy oils. However, we cannot justify the large discrepancy between ($P_c + P_b$) values for different oil types.

Our preferred solution is to develop a new relative-permeability curve for source rocks. Extrapolation of measured relative-permeability data for clean sandstones provides a source-rock relative-permeability curve that is much different from those for reservoir rocks. Results of simulations using the new curve agree substantially with measured data and concepts about expulsion.

Introduction

Numerical basin models that predict hydrocarbon generation, expulsion, and migration (e.g., Willett and Chapman, 1987; Ungerer et al., 1990) have become popular in both research and exploration branches of many oil companies. Describing fluid flow has been a particularly difficult part of constructing such models, and one of the most problematic parts of describing fluid flow is quantifying hydrocarbon expulsion. In recent years most workers have agreed that the concept of relative permeability to different fluid phases should be a foundation of expulsion models (e.g., Nakayama and Lerche, 1987; Ungerer et al., 1990). However, there remain many unanswered questions regarding relative permeability, including: (1) How many phases must one consider? (2) What do relative-permeability curves for each of these phases look like? (3) At what hydrocarbon saturation does hydrocarbon flow start? (4) At what hydrocarbon saturation does most hydrocarbon expulsion occur? (5) How much can we rely on measured relative-permeability data for reservoir rocks to construct a model of relative permeability in fine-grained source rocks?

In an effort to answer these questions, we have carried out computer simulations of hydrocarbon generation and expulsion. We also developed ways to extrapolate empirical data on relative permeabilities of coarse-grained, high-permeability reservoir rocks to fine-grained, low-permeability source rocks. These simulations and data analyses are described and reported in this paper.

Computer model

In order to study the sensitivity of the expulsion system to changes in various parameters (shapes and characteristics of relative-permeability curves for hydrocarbons and water, hydrocarbon viscosity, gas/oil ratio, absolute amounts of hydrocarbon generated in the source rock, and kerogen type), we developed a computer program to calculate relative permeabilities, saturations of hydrocarbons in the source rock, and volumes of each fluid expelled as a function of time for a source rock undergoing compaction, heating, and hydrocarbon generation. The main features of the model are given below.

(1) Burial and heating are constrained to occur at constant rates. Our expulsion program is not linked to a complex burial-history program.

Basin Modelling: Advances and Applications edited by A.G. Doré et al. NPF Special Publication 3, pp. 293–301, Elsevier, Amsterdam.

(2) Initial porosity is assumed to be 60%. Porosity ϕ_i at the end of the ith time step is calculated using an exponential equation (Athy, 1930):

$$\phi_i = 0.6 \exp(-bZ_i) \quad (1)$$

where Z_i is the depth (m) at the end of the ith time step and b is a constant. We use a variety of b values to simulate different compaction rates, but most of our simulations were done using $b = 0.0006$ to simulate typical porosity reduction in shales (e.g., Baldwin and Butler, 1985).

(3) Pressures are assumed to be normal at all times. Therefore, the total amount of fluid expelled in any time step is calculated from the volume of porosity reduction that occurred (Waples, 1991), modified by the changes in fluid volume due to hydrocarbon generation and cracking and by density changes due to changes in temperature and pressure. We therefore never actually calculate fluid pressures in the source rock.

(4) TOC values are allowed to range from 1% to 10%, but most simulations were run with 3% TOC. These values were chosen to represent a reasonable range and most typical value for effective source rocks. 5% of the initial TOC is assumed to be present as bitumen, which we consider to be equivalent to oil.

(5) Pure Type II, Type III, and Type II-S (high-sulfur Monterey- type) kerogens are used. We do not consider mixtures.

(6) Three different oils ("Heavy Californian", "Californian", and "Wyoming" as characterized in Ungerer et al., 1990) are used in our calculations. Selected data are presented in Table 1.

(7) Kerogen conversion to oil and gas is in accordance with published kinetic parameters (Burnham, 1989) for pure Type II and Type III kerogens, and from Tissot et al. (1987) for Monterey kerogen. For Type II and Type III kerogens, gas can be formed either directly from kerogen or by cracking of oil, but Type II-S kerogen generates only oil.

TABLE 1

Selected data for water and oils used in simulations

Fluid	Viscosity (cP, 100°C)	Gravity[a] (°API)	Density[a] (g/cm^3)	D[b]
Water	0.28	4.6	1.04	–
"Heavy Californian" oil	270	15	0.966	350
"Californian" oil	50	30	0.876	100
"Wyoming" oil	2.5	40	0.825	30

[a] At surface conditions.
[b] Using relative-permeability curve for reservoir rocks. See text for explanation of D.

(8) 50% of any oil generated from Type II or Type III kerogens that is cracked, is converted to gas and 50% to residue (Burnham, 1989). However, cracking of Monterey oil yields 32% gas and 68% residue (Behar et al., 1988). Residue is combined with exhausted kerogen in our calculations.

(9) Density of kerogen is assumed to be 1.2 g/cm^3. Oil densities are calculated from assumed (by us) gravities of 15° API (0.966 g/cm^3) for "Heavy Californian crude", 30° API (0.876 g/cm^3) for "Californian crude", and 40° API (0.825 g/cm^3) for "Wyoming crude" at all temperatures and pressures. Density of water is assumed to be 1.04 g/cm^3 at the surface and to decrease by 0.0267 g/cm^3 per km. Density of gas, ρ_g, in g/cm^3 at a depth of Z_i m and a temperature of T_i (°K) in the ith time step is given by:

$$\rho_g = 0.001 \left(\frac{T_s}{T_i}\right) \left(\frac{Z_i + 10}{Z_s + 10}\right) \quad (2)$$

where T_s is the surface temperature (°K) and $Z_s = 0$. Equation (2) was derived using the ideal gas law assuming a surface density of 0.001 g/cm^3 and a surface pressure of 1 atm (equal to about 10 m of water). Although more sophisticated expressions for gas density could be used, considering the various uncertainties about the amount and nature of free gas in the system (e.g., about ratios of oil and gas generated from kerogen, and about gas composition), the ideal gas expression may not lead to major errors.

(10) Viscosities (in cP) of water (μ_w) and oils (μ_o) in the ith time step are calculated using the Andrade equation (Latil, 1980). For example, eq. 3 calculates μ_w. To calculate oil viscosities the constants in eq. 3 are changed to fit the data of Ungerer et al. (1990) for each oil type.

$$\mu_w = 0.00357 \exp\left(\frac{1627}{T_i}\right) \quad (3)$$

Viscosity of gas (μ_g) is calculated using:

$$\mu_g = 0.01 + 0.000015 Z_i - 0.00022 (T_i - T_s) \quad (4)$$

Although some basin models (e.g., Ungerer et al., 1990) consider gas viscosity to be constant, we developed eq. 4 empirically as a simple way to approximately consider the effects of temperature and pressure on viscosity, based on data presented by Slider (1976). More elaborate methods are available, but once again, considering the uncertainties about gas composition and the relative unimportance of this correction, do not seem to be warranted.

(11) All gas generated is assumed to exist in the gas phase; that is, no solution of gas in oil is considered.

This assumption is almost certainly incorrect, but once again, because the boundary between oil and gas molecules is both gradational and changing, and because there are many uncertainties about ratios of gas to oil produced during generation, it probably introduces minimal additional error.

(12) Fluid saturations are calculated based on volumes of each fluid present in the pore space. Volume of fluid expelled in each time step is calculated as the difference between the pore volume at the end of the time step and the volume of fluid present at the start of the time step, plus the volumes of oil and gas generated during the time step. Normal pressure (that is, normal compaction) is assumed at all times. We intend to examine overpressuring in a separate study.

(13) Source rocks are assumed to be water wet. This assumption may be locally false within some or all source rocks, but overall, considering that most source rocks contain mainly mineral matter, that organic matter is often found in local concentrations, and that the bound water in clays is likely to be very difficult to remove, we consider this assumption to be reasonable.

(14) The relative permeability for water is calculated from its saturation using the drainage curve in Fig. 1. We define the curved lines in Fig. 1 mathematically as parabolas. If only oil and water are present, the relative permeability to oil is also calculated from the curve in Fig. 1. Gas can be the only nonwetting phase present only after all oil has been cracked.

If both oil and gas are present as separate phases, the relative permeabilities to those two phases are calculated using Fig. 2 (derived from Longeron, 1987). In such a case the sum of the relative permeabilities to oil and gas ($k_{ro} + k_{rg}$) is less than or equal to the relative permeability to oil in a water–oil system (Fig. 1). Longeron's experimental system involved displacement of oil from reservoir rocks by gas drive, a process which seems to us to be similar to the evolution occurring in source rocks, where oil generation largely precedes gas generation. Longeron's model assumes that there is an irreducible oil saturation as well as an irreducible water saturation.

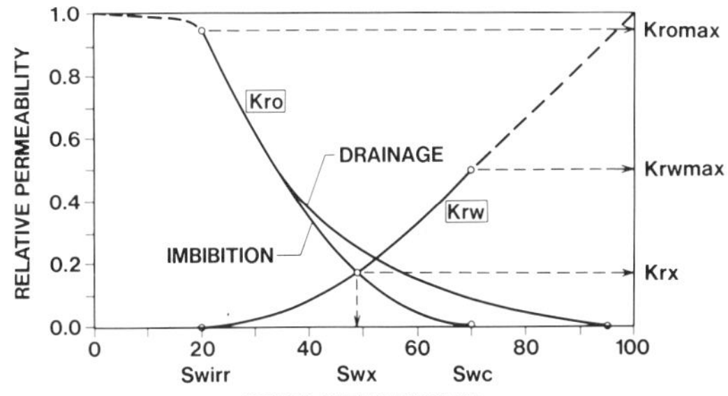

Fig. 1. Typical relative-permeability curves for oil and water in a reservoir rock. (Imbibition curves are much more commonly measured during core analysis.) Points referred to in this paper are labelled.

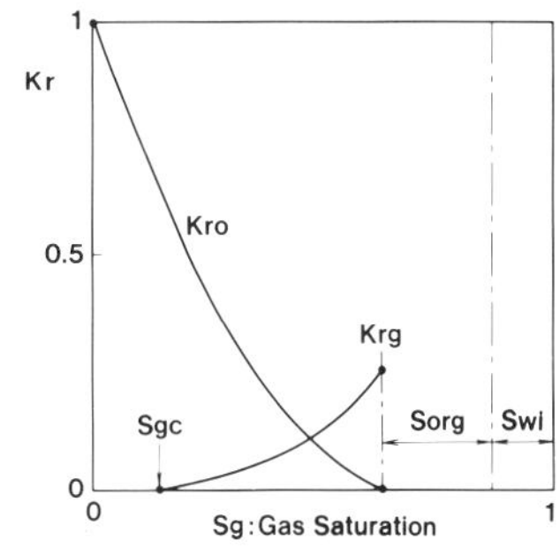

Fig. 2. Relative-permeability curves for oil and gas in the presence of irreducible water saturation (S_{wi}). S_{org} refers to irreducible oil saturation. From Longeron (1987); reprinted with permission of Editions Technip.

Initial simulations

Hydrodynamic driving force only

Our initial simulations were simplistic in two respects. Firstly, we used relative-permeability curves taken directly from measurements on reservoir rocks (Fig. 1), with no effort to correct for differences between reservoir rocks and source rocks. Secondly, we considered the difference in fluid potential created by accumulation of sediment and by hydrocarbon generation to be the only driving force for fluid expulsion. In these simulations we ignored possible contributions from buoyancy and capillary pressure differences between source rock and adjacent strata. The fluid potential in these simulations therefore affects all phases equally.

Numerous simulations were run using a variety of TOC values, kerogen types, and oil types, as well as varying slightly the characteristics of the relative-permeability curve. Results of all simulations led to the conclusion that our initial simplistic model did not adequately replicate the characteristics of hydrocarbon expulsion. Figure 3, which represents generation of medium-gravity "Californian" oil from Type II kerogen from a source rock with 3% TOC, shows a typical result. The main problems are: (1) rapid expulsion begins far too late in the hydrocarbon-generation sequence; (2) rapid expulsion does not occur until source-rock porosity is reduced to a lower level (4–5%) than we think appropriate for most source rocks (although there must be considerable natural variation, we envision 8–10% as more probable); and (3) much of the oil generated is cracked to gas in the source rock because of delayed expulsion. Our highly oil-prone source rock thus becomes a combined oil–gas source.

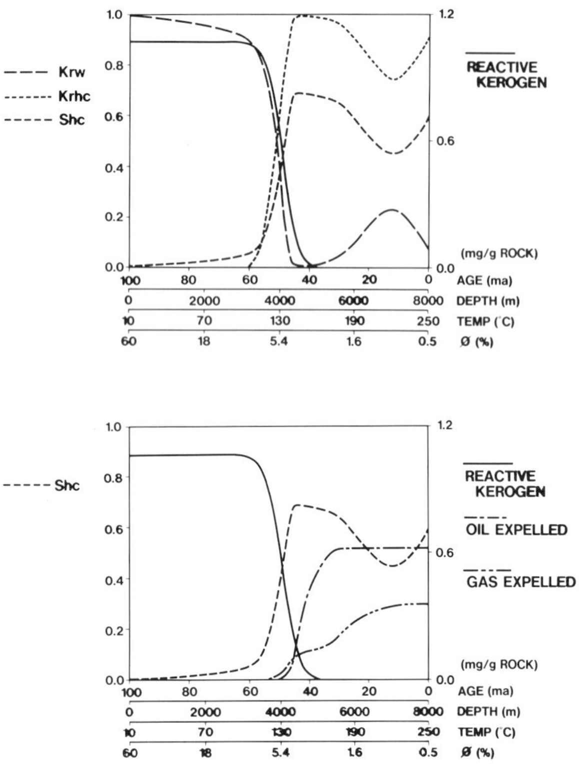

Fig. 3. Relative-permeability curves for oil and water, hydrocarbon saturation, and concentration of reactive kerogen (top) and oil expelled and gas expelled (bottom) for medium-gravity "Californian" oil generated from Type II kerogen, with TOC = 3%. Burial is at a constant rate of 80 m/Ma; geothermal gradient is 3.0°C/100 m with a surface temperature of 10°C. Relative-permeability curves for oil and water were obtained from Fig. 1 (reservoir-rock model). Expulsion is driven by hydrodynamic forces only.

The principal cause of these problems in our initial model is that the viscosities of oils are much higher than that of water. For example, medium-gravity oil is nearly 20 times as viscous as water at the temperatures of hydrocarbon generation (Table 1). According to Darcy's law, therefore, water should have an intrinsic mobility about 20 times as high as that of medium- gravity oil. In order for medium-gravity oil and water to move simultaneously in comparable amounts, the oil saturation would have to be about 20 times as great as the water saturation; in other words, oil saturation would have to be in the neighborhood of 95%.

Capillary, buoyancy, and hydrodynamic driving forces

In order to improve our model, we must find a way to promote hydrocarbon expulsion over water expulsion. Inclusion of capillary pressure (P_c) and buoyancy pressure (P_b) to supplement hydrodynamic pressure is one possible solution, since as long as the carrier bed remains water wet and its pores are larger than those of the source rock, both buoyancy and capillary pressure will favour hydrocarbon movement over water movement (Lerche, this volume). Oil will therefore be sucked out of the source rock by capillary forces. Buoyancy forces, on the other hand, assume that hydrocarbons are less dense than water and that expulsion of hydrocarbons is dominantly upward.

Because we do not calculate hydrodynamic pressure (P_h) in our model, we cannot simply calculate P_c and P_b and add them to P_h. However, we can consider the combined effects of ($P_c + P_b$) quite precisely in an indirect way. Since P_h is the same for water, oil, and gas, we can include ($P_c + P_b$) for oil and gas by multiplying the hydrodynamic pressure by some constant D. For example, if we multiply P_h by 2, the total driving force for oil ($P_h + P_{co} + P_{bo}$) is twice that for water. By simple arithmetics, therefore, when we multiply by 2, ($P_c + P_b$) = P_h. On the other hand, if D = 100, the sum of the capillary + buoyant pressures is 99 times the hydrodynamic pressure, and thus supplies virtually all the driving force for oil expulsion.

Figure 4 shows the results of another simulation using the medium-gravity "Californian" oil from a Type II kerogen with 3% TOC. Burial and thermal histories are identical to those in Fig. 3. We adjusted the value of D to 100 in order to bring the hydrocar-

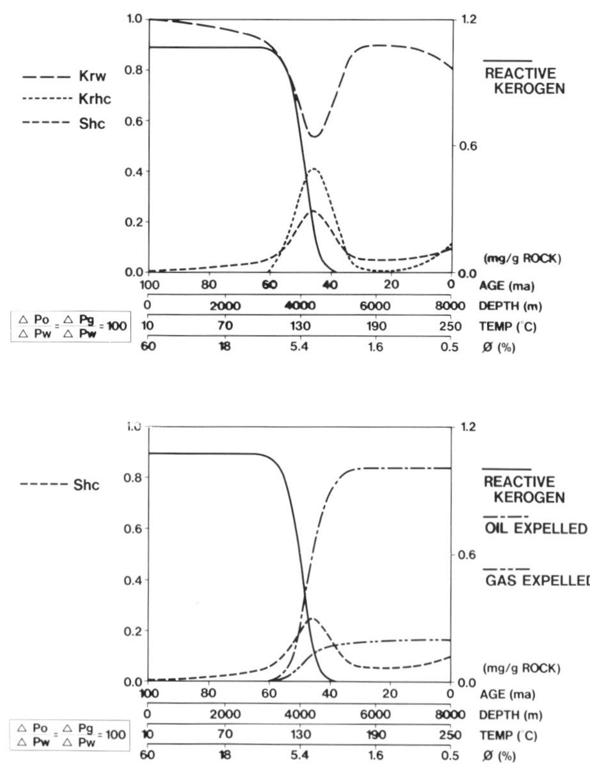

Fig. 4. Results of the same simulation as shown in Fig. 3, except that capillary pressure + buoyant pressure provide 99% of the total driving force for oil expulsion (D = 100).

bon saturation required for effective expulsion down into our target range of <30%. Timing of oil expulsion has also improved slightly, with rapid expulsion occurring at about peak oil generation at a (still low) porosity of 6%. Furthermore, much less cracking of oil to gas occurs because oil expulsion is earlier.

Other oil types required different values of D to achieve our target maximum oil saturations of less than 30%. Results in Table 1 show that for our high-viscosity "Heavy Californian crude" D must be 350, and even for our lightest "Wyoming" oil D must be 30. Thus even where oil viscosity is extremely low (probably lower than for most oils actually being expelled from source rocks), the combined effects of capillary and buoyant forces must be ten times the hydrodynamic force, and must contribute more than 97% of the total driving force for fluid expulsion.

Since by adjusting D we are only modelling the combined effects of capillary plus buoyant pressures, it is logical to inquire which of the two is in general more dominant. The data in Table 1 indicate that the greatest contribution from capillary + buoyant pressures (that is, the highest D value) is required for the oil with the highest viscosity and density (the "Heavy Californian crude"). However, since this oil has the greatest density, it also has the lowest buoyant advantage compared to formation water. Conversely, the least-viscous and least-dense oil ("Wyoming crude"), which has the greatest buoyant advantage, requires the least contribution (lowest D value) from ($P_c + P_b$). We therefore conclude that buoyant pressure is unlikely to be the dominant driving force. Moreover, since buoyant force depends on height of hydrocarbon column as well as on density contrast (e.g., Schowalter, 1979), if buoyant pressures were to dominate, we would come to the unpalatable conclusion that thick source-rock beds should expel more efficiently than thin ones.

On the other hand, if capillary pressure is the dominant force controlling hydrocarbon expulsion, then the main control on expulsion lies outside the source rock. The size of source-rock pores is likely to be fairly uniform compared to possible variation in pore sizes in the beds adjacent to source rocks. The magnitude of the capillary pressure will therefore depend mainly on the size of reservoir-rock pores and the nature of the source-rock/reservoir-rock interface. These phenomena would be difficult to model due to lack of detailed information in most cases, and in any case are not intrinsic properties of the source rock itself. Therefore, if capillary pressure is dominant, it will be almost impossible to model hydrocarbon expulsion. Fortunately, as we shall see later, in our new model capillary pressure does not necessarily play a dominant role.

This second series of simulations has shown that hydrocarbon expulsion may be achieved at reasonable hydrocarbon saturations and with more or less appropriate timing if we provide a large driving force that favours hydrocarbon expulsion over water expulsion. However, in these simulations capillary forces must be many times as large as hydrodynamic forces, and the magnitude of these capillary forces must vary greatly depending on oil viscosity.

According to Lerche (1990), the only property of an oil that affects capillary pressure is interfacial tension. Muskat (1949) has noted that interfacial tensions of oils do indeed vary slightly and unsystematically as a function of API gravity (and presumably viscosity). From Schowalter (1979) we conclude that differences in interfacial tension among various oil types are too small to account for the large differences in D values in Table 1. We therefore conclude that simply including the effects of capillary and buoyant forces cannot provide a fully satisfactory hydrocarbon-expulsion model.

Relative-permeability curves

Another possible source of the discrepancy between the results of our simulations and our conceptual model of expulsion is our assumption that the relative-permeability curve adapted from reservoir rocks (Fig. 1) is valid for source rocks. Nakayama (1987, and pers. commun., 1991) has suggested that relative-permeability curves for source rocks might look quite different from those for reservoir rocks, as a consequence of much greater surface area/volume ratios for shale particles compared to sand grains, and because of differences in pore sizes and shapes between shales and sandstones. Morgan and Gordon (1970) have provided data to support these ideas. However, since relative-permeability curves cannot be measured for rocks with absolute permeabilities as low as those of source rocks, some alternate method must be found to create such a curve.

Relative-permeability data for reservoir rocks typically show a great deal of scatter. For example, although a general trend in the data set shown in Fig. 5 is evident, it would be difficult to quantify or extrapolate this trend with confidence.

However, more detailed analysis of the data of Morgan and Gordon (1970) suggests that relative-permeability curves for sandstones change in predictable ways as a function of grain size and of permeability (Fig. 6). We used these observations as the basis for our extrapolation of measured relative-permeability data for reservoir rocks to fine-grained source rocks.

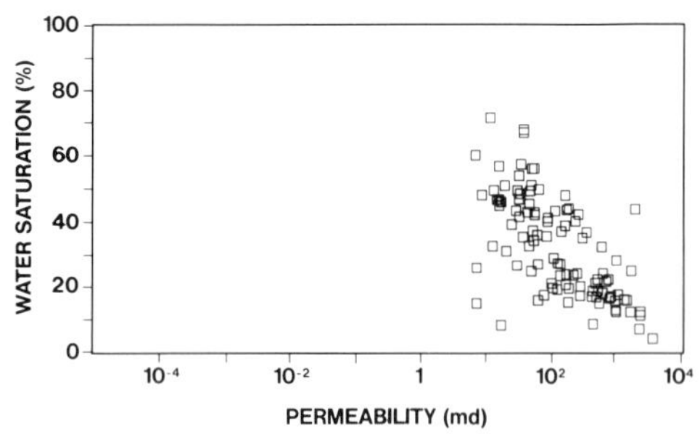

Fig. 5. Plot of S_{wirr} versus log absolute permeability (air) for sandstones in our data base. Sandstones are from four different basins and differ considerably in clay content.

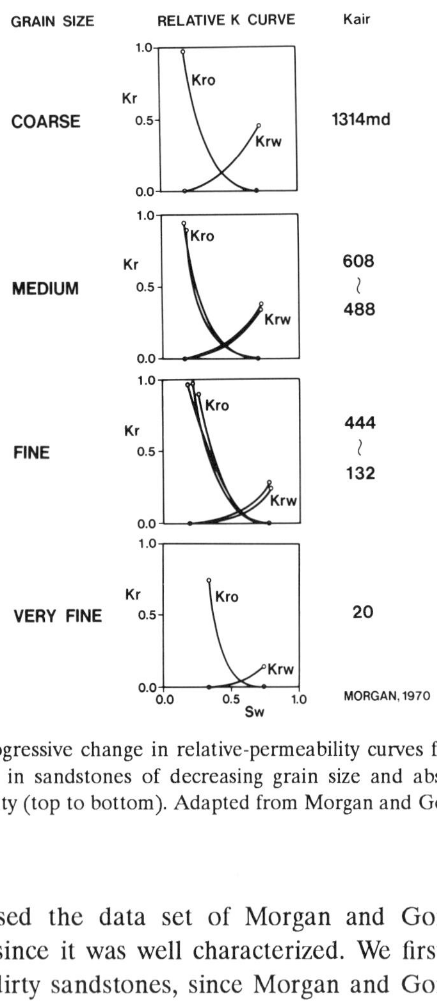

Fig. 6. Progressive change in relative-permeability curves for oil and water in sandstones of decreasing grain size and absolute permeability (top to bottom). Adapted from Morgan and Gordon (1970).

We used the data set of Morgan and Gordon (1970), since it was well characterized. We first excluded dirty sandstones, since Morgan and Gordon (1970) had confirmed that the clays admixed in sandstones would have significant effects on relative permeabilities. We believe that admixing fine clays into sandstones makes pore geometry heterogeneous. Such complexity is unlikely in source rocks, which have fairly uniform pores. The data for clean sandstones in Fig. 7 provide a much clearer trend of the change in S_{wirr} as a function of absolute permeability (and of grain size).

We repeated this analysis for two other points in the relative-permeability curves of Morgan and Gordon (1970): the value of S_w where the k_{ro} and k_{rw} curves cross (S_{wx} in this paper), and the critical water saturation, S_{wc} (the water saturation at which oil flow stops in the imbibition process during core analysis) (Fig. 1). We then extrapolated the trends of these three sets of relative-permeability data as functions of absolute permeability to the low permeabilities (10^{-4} mD) expected in source rocks (Sandvik and Mercer, 1990), as shown in Fig. 7. We constrained our extrapolations to prevent the lines from crossing.

We then estimated the relative permeability to oil (k_{ro}) at S_{wirr} and S_{wx}, and the relative permeability to water (k_{rw}) at S_{wc} for a source rock with 10^{-4} mD permeability using an exponential equation of the form:

$$k_r = a \exp[b \log(k_{air})] \quad (5)$$

where a and b are constants and k_{air} is the absolute permeability to air (Fig. 8). We assume that oil flow during expulsion starts at a water saturation just below 100%. These extrapolations and assump-

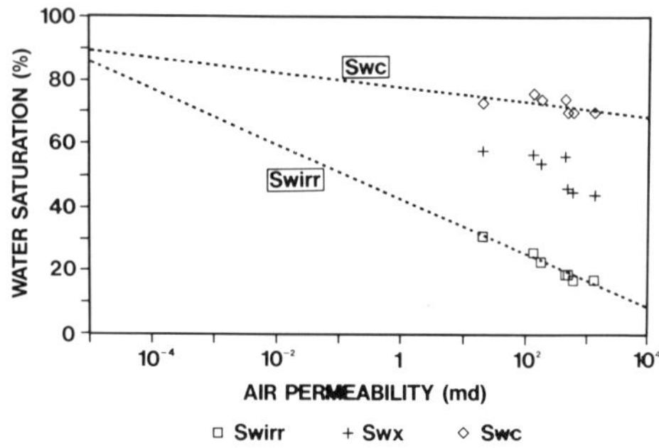

Fig. 7. Water saturations at three different points in the k_{ro} curve for clean sandstones (only) from the data of Morgan and Gordon (1970). Squares represent S_{wirr}; crosses represent the point at which the k_{ro} and k_{rw} curves cross; and diamonds represent S_{wc}, the water saturation at which oil flow stops in the imbibition process during core analysis. Dotted lines show extrapolation to lower absolute permeabilities.

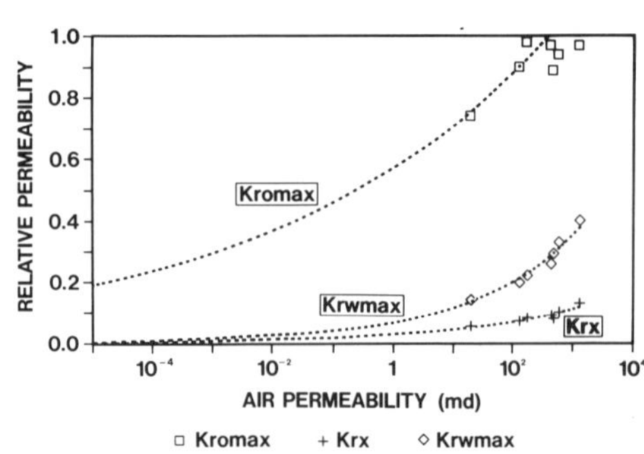

Fig. 8. Exponential extrapolation of measured values of k_{ro} for clean sandstones to lower absolute permeabilities. Squares represent k_{ro} at S_{wirr}; crosses represent k_{ro} at the point at which the k_{ro} and k_{rw} curves cross; and diamonds represent k_{rw} at S_{wc}, the water saturation at which oil flow stops in the imbibition process during core analysis. Refer to Fig. 1.

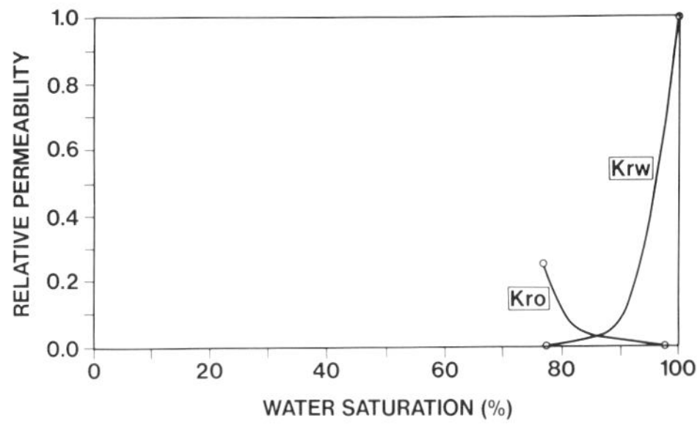

Fig. 9. New relative-permeability curves for oil and water in a "typical" source-rock shale with absolute permeability = 10^{-4} mD.

tions provided values at three critical points in the relative-permeability curve for a source rock, and thus enabled us to construct the curves for both k_{ro} and k_{rw} for a typical source rock (Fig. 9). Although we chose a parabolic interpolation between S_{wx} and S_{wirr}, we also tested a linear interpolation and found that our results were almost insensitive to the interpolation algorithm.

The relative-permeability curve in Fig. 9 looks quite different from that in Fig. 1. Hydrocarbons first become mobile at a water saturation near 100%, in agreement with the ideas of Nakayama (1987). The maximum relative permeability to oil (and cessation of water flow) is also achieved at a much lower oil saturation (23% S_o = 77% S_w). All of these changes promote hydrocarbon movement in source rocks compared to reservoir rocks.

However, the maximum relative permeability to oil (0.24) is considerably lower than in a reservoir rock (>0.90), as a consequence of the greater role of bound water in fine-grained rocks with smaller pores.

Final simulations

We then ran a final series of simulations using our new relative-permeability curves for our source rock. The simulation for our medium-gravity "Californian" oil generated from a Type II kerogen in a source rock with 3% TOC is shown in Fig. 10. Burial and thermal histories are identical to those in previous illustrations. We used a value of D = 1 for all three oil types, and with a single value of D were able to achieve maximum hydrocarbon saturations in the target range of 10–30%. Conclusions were not materially changed by using values of D up to 2, indicating that: (1) the contribution of ($P_c + P_b$) to the total driving force need not overwhelm the contribution of hydrodynamic pressure; and (2) the relatively small differences in interfacial tensions expected among oils of different gravities will not greatly affect the total driving force that oil droplets feel within the source rock.

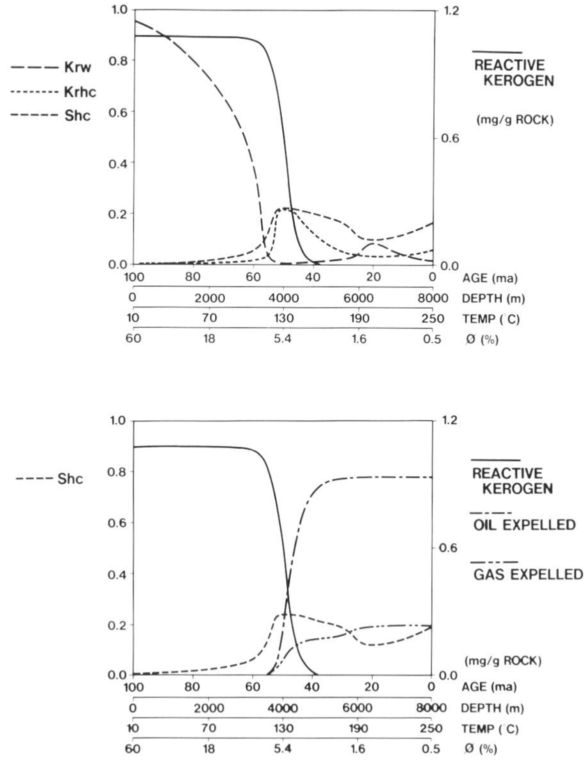

Fig. 10. Results of the same simulation as shown in Fig. 3, except that relative-permeability curves for oil and water are taken from Fig. 9. Contribution from capillary and buoyant forces to total driving force for expulsion is minor (D = 1).

Figure 10 shows that use of our new relative-permeability curve also allows expulsion to occur near peak oil generation when porosity is about 7%. Moreover, cracking is minor, and overall oil-expulsion efficiency is very high (>90% in some simulations), in general agreement with current ideas for source rocks of this quality (e.g., Cooles et al., 1986; Mackenzie and Quigley, 1988). In most of the simulations we have tested, the results of using our new relative-permeability curve with a constant or nearly constant value of D for all oil types are quite similar to the results obtained using the original relative-permeability curve and allowing D to vary greatly from one oil type to another (compare Figs. 4 and 10). However, we believe that our new relative-permeability curve used with constant values of D provides more reasonable physical conditions.

Our simulations also show that when oil expulsion is rapid, very little water is flowing with the oil. For example, in the simulation in Fig. 10, during rapid oil expulsion between 54 and 45 Ma, about 95% of the fluid expelled was oil, and only 5% was water. The low water flow is a direct result of the low relative permeability to water when water saturations approach S_{wirr}.

Oil expulsion occurs earlier (and therefore at higher porosities) when TOC values are higher. Fig-

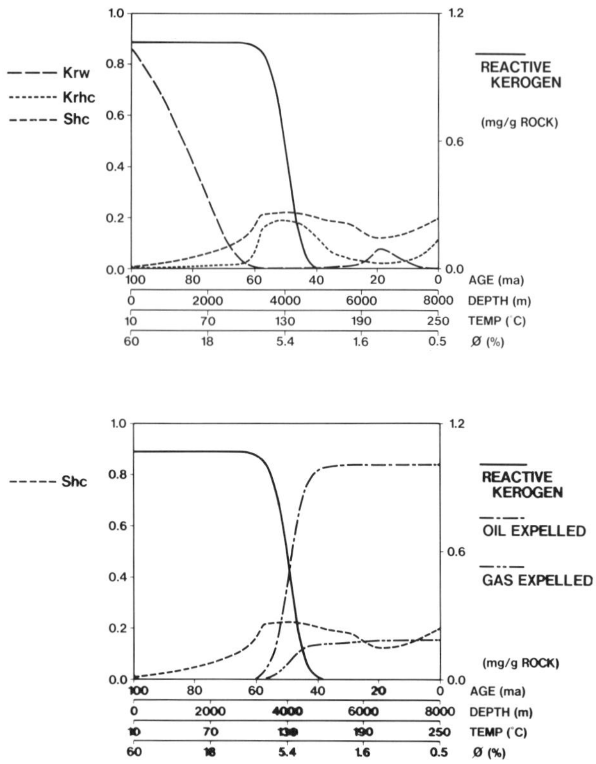

Fig. 11. Results of the same simulation as shown in Fig. 10, except that TOC = 10%.

ure 11 shows generation of medium-gravity "Californian" oil from Type II kerogen with 10% TOC. (Burial and thermal histories are the same as for other figures.) Rapid expulsion begins 60 Ma, when porosity is about 9% (compared to 54 Ma with 7% porosity when TOC = 3%). Simulations of generation of "Heavy Californian" oil from Monterey-type (II-S) kerogen (not shown) predict rapid expulsion beginning about 67 Ma under the same burial and thermal conditions. Because of the earlier generation from Type II-S kerogen, the porosity at the start of rapid expulsion is about 13%. Our model is therefore capable of making reasonable predictions about the source-rock porosities when expulsion begins.

Conclusions

We have shown through computer simulations that relative-permeability curves imported from studies of reservoir rocks are not good analogs for relative-permeability curves in source rocks. We have extrapolated relative-permeability data for clean sandstones of varying grain size but with fairly homogeneous pores, to produce a relative-permeability curve that we believe to be more appropriate for source rocks. Testing of our new relative-permeability curve gives results that agree rather well with our data and concepts on hydrocarbon expulsion.

Our studies have led us to examine the importance of capillary and buoyant pressures in promoting hydrocarbon expulsion. Using our new relative-permeability curve the effects of buoyancy and capillary pressures need not necessary dominate over hydrodynamic pressure, whereas use of the reservoir-rock-based relative-permeability curves required a dominance of capillary-pressure effects in order to overcome the effects of high viscosities of oils compared to water.

Our simulations indicate that most oil expulsion occurs when oil saturations are near their maximum possible values. Therefore, the shape of the relative-permeability curve is of little importance. However, the value of S_{wirr} and k_{ro} at S_{wirr} are very important. During times when oil expulsion is rapid, the fluid flow consists almost entirely of oil, and thus is very close to a single-phase flow.

Further work should be done to refine the concepts developed in this paper. It would be particularly useful to have more information about the porosities at which expulsion actually occurs, as an additional constraint on the model. We intend to examine more empirical data on relative permeabilities and would like to test these concepts in a more sophisticated 2D model. We anticipate extending this work to examine expulsion during and after overpressuring events, and to develop a mathematical relationship between absolute permeability and relative-permeability curves. Since permeability is often calculated from porosity, we would then have a link between source-rock porosity and relative permeability that could be easily integrated into maturity models and basin models to provide more reliable estimates of hydrocarbon expulsion.

Acknowledgements

We thank Hiromi Kamata for technical assistance, Kazuo Nakayama for extremely helpful discussions, and JNOC for support and permission to present and publish this work.

References

Athy, L.F., 1930. Density, porosity, and compaction of sedimentary rocks. Am. Assoc. Pet. Geol., Bull., 14: 1–24.

Baldwin, B. and Butler, C.O., 1985. Compaction curves. Am. Assoc. Pet. Geol., Bull., 69: 622–626.

Behar, F., Ungerer, P., Audibert, A., and Villalba, M., 1988. Experimental study and kinetic modelling of crude oil pyrolysis in relation with enhanced oil recovery processes. In: F. Meyer and J. Wiggins (Editors), 4th UNITAR/UNDP Conf. Heavy Crude and Tar Sands. Alberta Oil Sands Technology and Research Authority, pp. 747–759.

Burnham, A.K., 1989. A simple kinetic model of petroleum formation and cracking. Lawrence Livermore National Laboratory Rep., UCID-21655, 11 pp.

Cooles, G.P., Mackenzie, A.S. and Quigley, T.M., 1986. Calculation of masses of petroleum generated and expelled from source rocks. In: D. Leythaeuser and J. Rullkötter (Editors), Advances in Organic Geochemistry 1985. Pergamon Press, Oxford, pp. 235–246.

Latil, M., 1980. Enhanced Oil Recovery. Gulf Publishing Co., Houston, Texas, 236 pp.

Lerche, I., 1990. Basin Analysis: Quantitative Methods, Vol. 1. Academic Press, San Diego, Calif., 562 pp.

Lerche, I., 1993. Theoretical aspects of problems in basin modelling. In: A.G. Doré et al. (Editors), Basin Modelling: Advances and Applications. Norwegian Petroleum Society (NPF), Special Publication 3. Elsevier, Amsterdam, pp. 35–65 (this volume).

Longeron, D., 1987. Study of the influence of thermodynamic parameters on two-phase flow in porous media. In: B. Doligez (Editor), Migration of Hydrocarbons in Sedimentary Basins. Technip, Paris, pp. 257–280.

Mackenzie, A.S. and Quigley, T.M., 1988. Principles of geochemical prospect appraisal. Am. Assoc. Pet. Geol., Bull., 72: 399–415.

Morgan, J.T. and Gordon, D.T., 1970. Influence of pore geometry on water–oil relative permeability. J. Pet. Technol., 22: 1199–1208.

Muskat, M., 1949. Physical Principles of Oil Production. McGraw Hill, New York, N.Y., 922 pp.

Nakayama, K., 1987. Hydrocarbon-expulsion model and its application to Niigata area, Japan. Am. Assoc. Pet. Geol., Bull., 71: 810–821.

Nakayama, K. and Lerche, I., 1987. Two-dimensional basin analysis. In: B. Doligez (Editor), Migration of Hydrocarbons in Sedimentary Basins. Technip, Paris, pp. 597–611.

Sandvik, E.I. and Mercer, J.N., 1990. Primary migration by bulk hydrocarbon flow. In: B. Durand and F. Behar (Editors), Advances in Organic Geochemistry 1989. Pergamon, Oxford, pp. 83–89.

Schowalter, T.T., 1979. Mechanisms of secondary hydrocarbon migration and entrapment. Am. Assoc. Pet. Geol., Bull., 63: 723–760.

Slider, H.C., 1976. Practical Petroleum Reservoir Engineering Methods. Petroleum Publishing Company, Tulsa, Okla., 559 pp.

Tissot, B.P., Pelet, R. and Ungerer, P., 1987. Thermal history of sedimentary basins, maturation indices, and kinetics of oil and gas generation. Am. Assoc. Pet. Geol., Bull., 71: 1445–1466.

Ungerer, P., Burrus, J., Doligez, B., Chénet, P.Y. and Bessis, F., 1990. Basin evaluation by integrated two-dimensional modeling of heat transfer, fluid flow, hydrocarbon generation, and migration. Am. Assoc. Pet. Geol., Bull., 74: 309–335.

Waples, D.W., 1991. Innovative methods for quantifying poorly known parameters necessary for numerical basin modeling. J. Jpn. Assoc. Pet. Technol., 56: 96–107.

Willett, S.D. and Chapman, D.S., 1987. Temperatures, fluid flow and the thermal history of the Uinta Basin. In: B. Doligez (Editor), Migration of Hydrocarbons in Sedimentary Basins. Technip, Paris, pp. 533–551.

AKIHIKO OKUI *Technology Research Center, Japan National Oil Corporation, 1-2-2 Hamada, Chiba 260, Japan*
Present address: *Idemitsu Oil Development Company Ltd., 5-6-1, Shiba, Minato-ku, Tokyo 108, Japan*

DOUGLAS W. WAPLES *Technology Research Center, Japan National Oil Corporation, 1-2-2 Hamada, Chiba 260, Japan*

Modelling porosity reduction as a series of chemical and physical processes

Douglas W. Waples and Hiromi Kamata

In basin modelling, porosity reduction is normally taken to be a function only of burial depth (or net stress) for a given rock type. Such equations are designed to provide reasonable descriptions of the main features of normal porosity reduction over a wide range of depths. However, this purely empirical approach lumps together the effects of several different porosity-reducing processes, some of which may depend on additional factors such as time or temperature.

We have developed an empirical set of equations grounded as much as possible in theory to take into account the porosity-reducing effects of seven different processes in limestone, chalk, shale or siltstone, and sandstone: *Repacking* of grains from the original inefficient packing mode, *Crushing* of microfossils, *Ductile Flow* and deformation of grains, *Pressure Solution*, *Welding* following Pressure Solution, *Clay Diagenesis* (including formation of authigenic clays), and *Cementation* by minerals precipitated from moving fluids. We use six of these processes in limestones, chalks, and sandstones, and three in shales.

Since our goal is to provide a more universally applicable and convenient method of predicting porosity reduction for basin modelling, ease of use of our model is important. To this end we have determined preliminary values for the various constants in these equations for each lithology. Although our calibration data were limited in number, they ranged in age from Pleistocene to Carboniferous, in depth from 200 to nearly 5000 m, in temperature up to 160°C, and in measured porosity from 2% to 45%. The quality of fit between measured and calculated porosities using the default values for all constants is better than fits obtained using conventional porosity-reduction equations.

Our porosity-reduction equation can easily be incorporated into existing basin models. The main difficulty is that our model does not assume conservation of solid volume. Changes in solid volume are generally less than 5%, however.

Application of our model to North Sea chalks shows that when the effects of overpressuring are taken into consideration, porosities can be predicted within a few percent. Overpressuring in other lithologies has not yet been examined. Results for chalks suggest that the porosities of the North Sea chalks are not actually abnormally high when their entire burial histories are considered, and that the effect of overpressuring on porosity retention varies from negligible to considerable.

Our method should provide more accurate estimates of true porosities (and hence permeabilities) in a wide variety of basins than the current depth-dependent functions, which require local calibration. However, it only attempts to model the regular and general progression of porosity reduction as an aid in basin modelling, and is not designed to predict anomalous events (including porosity enhancement) or to account for natural data scatter within a single lithologic unit. Since reservoir characteristics are often strongly influenced by anomalous or local events, we do not recommend the use of our model at its current stage of development to predict production characteristics of specific reservoirs.

Statement of the problem

For basin modelling it is important to be able to specify the porosity of each rock layer through time with a reasonable degree of accuracy. Porosity strongly affects thermal conductivity of rocks because of the much greater resistivity of pore fluids compared to rock minerals. Inaccurate porosity estimates can therefore strongly affect estimates of rock temperatures, and hence of organic maturation and hydrocarbon generation and expulsion. Moreover, matrix permeabilities are often calculated as an exponential or power function of porosity. Errors in porosity estimates will therefore translate into large errors in matrix permeabilities, which in turn may strongly affect predicted flow rates and pathways for hydrocarbons and water, development and maintenance of abnormal pressures, and convective heat transfer.

There are several approaches to predicting porosity reduction during a rock's burial history. The approach used in basin modelling is to select for each lithology a generalized equation for porosity (ϕ) or porosity reduction ($\Delta\phi$) as a function of burial depth and initial porosity (e.g., Athy, 1930; Roll, 1974; Sclater and Christie, 1980; Falvey and Middleton, 1981; Baldwin and Butler, 1985). Such equations usually are of the form :

$$\phi = \phi_0 \, e^{-bZ} \tag{1}$$

Basin Modelling: Advances and Applications edited by A.G. Doré et al. NPF Special Publication 3, pp. 303–320, Elsevier, Amsterdam.

where Z is depth and b is a constant, although power equations (e.g., Baldwin and Butler, 1985) and harmonic equations (e.g., Falvey and Middleton, 1981) have also been used.

An alternative but similar approach that has not yet been applied to basin modelling is to express porosity as a function of thermal maturity. Maxwell and Verrall (1954) were probably the first to propose a relationship between porosity and time/temperature history. Subsequent investigations of this relationship include Van de Kamp (1976), Lyons (1978, 1979), and Siever (1983). Schmoker (1984), Schmoker and Gautier (1988, 1989), and Schmoker and Hester (1990) have expressed carbonate, sandstone, and shale porosity as a power function of thermal maturity (M):

$$\phi = AM^{B} \tag{2}$$

where A and B are constants, and thermal maturity is represented by either TTI or vitrinite reflectance (R_0). Bloch (1991), in a variation on this idea, has suggested that porosity of sandstones can be expressed adequately as a linear function of R_0. In these approaches porosity reduction is a direct function of only time and temperature; burial depth plays no role except as it influences temperature. Schmoker and Gautier (1989) have cautiously suggested that local calibration and even calibration by lithology may not be necessary, but Bloch (1991) disputes this idea.

More-complex equations that include terms based on petrographic data (e.g., Scherer, 1987) have also been suggested. In addition, empirical models calibrated locally with learning data sets have been used to predict porosities of sandstone reservoirs (Bloch et al., 1990; Bloch, 1991), but this approach is probably not practical for basin modelling.

In nature, porosity reduction is the result of the interplay of numerous processes, which are commonly divided, somewhat imperfectly, into “mechanical compaction” and “chemical compaction”. The depth-dependent equations for porosity reduction implicitly assume that mechanical compaction processes are dominant over the entire burial history of the sediment (that is, compaction occurs as a response to overburden stress only). Schmoker and Gautier (1989), in contrast, have pointed out that porosity reduction occurs more or less continuously during the entire history of a rock.

In contrast, the TTI/R_0-dependent equations emphasize a dominance by chemical compaction (overburden stress plays no role, but time and temperature do, as they do in kinetically controlled processes such as chemical reactions). In view of the documented dependence of porosity reduction on both mechanical and chemical processes (e.g., Houseknecht, 1987), and of the differing importance of mechanical and chemical processes in rocks with different burial and thermal histories, it would be desirable to be able to express porosity reduction as a function of all three variables: time, temperature, and depth.

Existing methods of predicting porosity reduction in basin models suffer from other difficulties as well. The various types of depth-dependent curves all require local calibration, presumably at least in part because they ignore the effects of time and temperature. Until the calibration factor is worked out for a particular area, prediction of porosity reduction is rather uncertain. Furthermore, none of the models considers porosity enhancement through diagenetic processes, since none allows porosity to increase with time. None of the models attempts to deal with natural data scatter due to deviation of individual beds from the normal average history being modelled. Even where overall porosity reduction is modelled adequately, the timing of porosity reduction may be incorrectly estimated, leading to an inaccurate assessment of permeability, thermal conductivity, and other properties through time.

Finally, since none of the existing approaches is truly process-oriented (and indeed they all explicitly ignore whole families of processes), they do not really attempt to explain mechanistically what is occurring during porosity reduction. They therefore do not form the basis for further scientific investigation of porosity reduction.

A process-oriented approach to describing porosity reduction

As our first goal and as an initial step in attempting to develop a new method of predicting porosity reduction, we describe porosity reduction as the sum of seven processes which are capable of operating simultaneously or sequentially as a sediment is buried and heated. Three of these processes (Repacking, Crushing, and Ductile Flow) are mechanical and do not involve chemical reactions. Three of the processes (Welding, Clay Diagenesis, and Cementation) are chemical in nature: they involve chemical reactions, including transformations of clays and other minerals, and precipitation of calcium carbonate and quartz. The seventh process, Pressure Solution, is commonly considered to be part of mechanical compaction because it is promoted by effective stress (e.g., Houseknecht, 1987), but could also be considered as chemical because it involves chemical reactions involved in mineral dissolution. Each of these processes and the equations defining them are discussed in more detail below and in the Appendix.

We have not coupled these seven processes mathematically in our model, but rather calculate the effects of each process sequentially in each time step in our simulations. However, we do allow significant feedback among the processes. Details of these feedback relationships will be discussed below for the individual processes.

Our initial study has been limited to four lithologies: limestone, chalk, shale/siltstone, and sandstone. For limestone we take shelf muds as our model and consider six of the seven processes (Repacking, Crushing, Ductile Flow, Pressure Solution, Welding, and Cementation); for chalks we consider the same six processes as for limestone; for shale, three processes (Repacking, Clay Diagenesis, and Cementation); and for sandstone, six processes (Repacking, Ductile Flow, Pressure Solution, Welding, Clay Diagenesis, and Cementation). Our sandstone model considers compositions ranging from pure quartz to lithic sandstones, but is not yet designed to describe muddy sandstones.

Our descriptions of each process are grounded in theory as much as possible. However, even where the theory is reasonably well developed (Pressure Solution, for example), many variables are poorly quantified. Furthermore, we have neglected several factors (e.g., grain size and shape, sorting, detailed petrographic composition) that undoubtedly exert important controls on both initial porosity and porosity reduction, because such detailed information is not generally available for basin modelling. It is therefore important to recognize that our model is highly empirical.

As the second goal of the study, we determined default values for each of the constants in the equations describing the seven processes by trial and error by comparing measured porosities with predicted porosities for our suite of calibration data. These data covered a range of burial and thermal histories for each lithology.

It is important to recognize that at this stage of development of our model we are attempting to model only those processes which contribute to regular porosity reduction. Our efforts may be viewed as a small step along one possible road to "integrated diagenetic modelling" as described by Surdam and Crossey (1987). The present version of the model does not include any capability for describing porosity enhancement, although possible modifications to describe porosity enhancement are discussed briefly at the end of this paper. Our model also does not consider any effects of organic matter, such as formation of organic acids (e.g., Surdam et al., 1989) or the retardation of cementation in the presence of hydrocarbons (Nagtegaal, 1980). Nor does it attempt to describe local or anomalous events that create abnormally high or low porosities compared to surrounding rocks (that is, we do not attempt to remove natural data scatter). In fact, since reservoir quality often depends on positive or negative factors that are local (i.e., vary on a smaller scale than the scale being modelled) or otherwise abnormal, we agree fully with the philosophy of Harrison (1990), and thus recommend that this porosity-reduction model should not be used to make predictions about the properties of specific reservoir rocks. It is intended primarily to provide better and more universal estimates of general porosity trends as an aid in basin modelling.

Definitions of compaction processes

Introduction

Simple descriptions of each of the compaction processes are included below. Extended mathematical development for each process is found in the Appendix.

Repacking (limestone, chalk, sandstone, and shale)

Repacking is a mechanical process during which grains move into a more stable packing arrangement from an initial loose and inefficient one. In carbonate shelf muds, Repacking ceases at burial depths of about 100 m, after causing a decrease in porosity from 70% to 40% (Moore, 1989). In chalks, in contrast, initial porosities are assumed to be 80%, and Repacking is only allowed to continue until porosity reaches 70% at 50 m (Moore, 1989).

Repacking in sandstones is considered to reduce an initial porosity of 41% to 34% at 2400 m and to the stable-packing porosity of 26% (Houseknecht, 1987) at 2900 m. Our model for Repacking of shales conceptually follows the ideas of Burst (1969) and Roll (1974), starting with an initial total porosity of 77%. Shale porosity is assumed to be 40% at 200 m, and Repacking ceases when porosity reaches 30% at about 750 m.

However, if other processes are simultaneously causing porosity reduction, the amount of reduction due to Repacking will be diminished. This effect is particularly important where early Cementation is occurring, where it reflects the formation of stable grain linkages (see Appendix).

Crushing (limestone and chalk only)

Crushing of weak and hollow grains, particularly microfossils, is particularly important in chalks, where it may lead to an overall decrease in porosity of up to about 10% (Moore, 1989). We therefore allow Crushing to reduce porosity from 70% to 60% in chalks between 50 m and 200 m burial depth. Data for shelf muds, however, suggest that in those sediments Crushing is limited to only a few percent (Shinn and Robbin, 1983; C.H. Moore, pers. commun., 1990).

Our model does not consider Crushing for shales (where individual clay particles cannot be crushed and where the "crushing" of the house of cards structure has already been included in Repacking) or for sandstones, where crushing of individual grains seems to be insignificant (Taylor, 1950). In sandstones, grain deformation and crushing of minor components such as micas are included in Ductile Flow.

Ductile Flow (limestone, chalk, and sandstone only)

Ductile Flow is a response of ductile grains to high shear stress (Taylor, 1950), and is essentially an alternative to Pressure Solution for insoluble materials in sandstones. We also consider Ductile Flow for the deformation of soft material (organic matter, faecal pellets, etc.) in limestones. However, we do not consider Ductile Flow for shales or chalks, where grain sizes and strengths are relatively similar.

Rittenhouse (1971b) has suggested that some threshold amount of ductile grains must be present before porosity reduction by Ductile Flow can occur. He proposed a value of about 5% for ideal grains, but concluded that for real grains some porosity reduction would probably occur at all concentrations. We have compromised by using a threshold value of 2% ductile grains for the initiation of Ductile Flow. We have used the model of Mitra and Beard (1980) in calculating the amount of porosity reduction associated with a given amount of Ductile Flow.

Pressure Solution (limestone, chalk, and sandstone only)

We use the term "Pressure Solution" to refer to the dissolution of mineral matter and ensuing collapse of the mineral structure. Our definition of Pressure Solution therefore does not include reprecipitation of any dissolved minerals; this reprecipitation is represented by what we call "Welding" (see below).

We have adopted the view of Weyl (1959), de Boer (1977), and Spiers et al. (1990), who considered the rate of Pressure Solution, at least for some lithologies, to be controlled by the rate of diffusion of dissolved species away from grain contacts. Diffusion is slow and rate-controlling because it must occur through thin layers of bound water adjacent to the grain contact before it can reach the more-open pore space. Concentrations of dissolved species near the grain contact depend on net effective stress and temperature, and those in the pore space depend on temperature and fluid pressure. Stress in our model in turn depends in a simple way on fluid pressure, and on depth and grain surface areas, which increase as Pressure Solution progresses (Weyl, 1959). In a future modification we intend to vary effective stress as a continuous and more-complex function of pore-fluid pressure, and thus should be able to account directly for decrease in stress at grain–grain contacts during overpressuring.

In our model, porosity reduction due to Pressure Solution depends not only on the rate of solution and diffusion, but also on the rate of collapse of the pore structure as rock is removed. Pressure Solution as we define it (dissolution of mineral matter plus ensuing collapse of the eroded structure) will always lead to a decrease in total pore volume, but because it also leads to a decrease in rock volume, the porosity (that is, the fraction of total volume occupied by pores) will decrease only if the volume of dissolving grains is small compared to the pore volume destroyed by structure collapse.

There does not seem to be complete agreement about the depths required for initiation of Pressure Solution. However, some workers suggest a minimum depth of 600–1000 m for Pressure Solution of limestones (Moore, 1989). Our model for limestones produces results in general conformity with those ideas. Houseknecht (1987) has observed slight amounts of Pressure Solution in sandstones with porosities of about 30%; that is slightly prior to attainment of the stable packing geometry. We therefore do not allow Pressure Solution to occur in sandstones until porosity is less than 32%. However, Pressure Solution becomes important quite early in chalks, where we allow it to begin at 300 m, once Repacking and Crushing are both finished.

Our current model for Pressure Solution does not allow development of enhanced secondary porosity, since we assume that Pressure Solution always leads to collapse of the grain structure. In a future version, however, we would like to allow porosity enhancement through preferential dissolution of less-stable framework grains without ensuing structural collapse (e.g., Giles and de Boer, 1990; Bloch, 1991).

Our decision to treat Pressure Solution as a diffusion-controlled process is also controversial. It has been suggested that for some lithologies the rate of Pressure Solution is actually controlled by the rate of dissolution rather than the rate of diffusion (e.g., Raj, 1982). In a future version we will evaluate whether Pressure Solution could be better described in this way.

Welding (limestone, chalk, and sandstone only)

Our model distinguishes between cements formed (1) by local reprecipitation of minerals dissolved from nearby grains suffering Pressure Solution (e.g., Rittenhouse, 1971a; Bathurst, 1975; Mitra and Beard, 1980; Scholle and Halley, 1985), and (2) by precipitation of minerals brought in from outside the rock in question. We call porosity reduction by the former type of cement "Welding", after Hatch et al. (1938). (The second type of cement, deposited by what we call "Cementation", is discussed below.) Our concept of Welding is closely related to that of "redistributional secondary porosity" (Giles and de Boer, 1990). We ignore Welding for shales, which we assume contain little soluble material that could later be precipitated. Any welding-like effects in shales would involve more-complex chemical reactions, and would therefore be included in Clay Diagenesis.

We assume after Rittenhouse (1971a) that the amount of mineral matter dissolved by Pressure Solution and diffusing to the open pore space in any time step places an upper limit on the amount of Welding that can occur in that time step. We also assume that since the bulk pore fluids are probably saturated with the material dissolved during Pressure Solution, the amount precipitated during Welding depends inversely upon the volume of pore fluid. We expect that most of material dissolved during Pressure Solution will become Welded, since the limited volume of pore fluid and its saturation or near-saturation with respect to the dissolved mineral will severely limit its ability to expel the minerals dissolved during Pressure Solution (see Mitra and Beard, 1980). Furthermore, any mineral material that is expelled will be lost from the rock, and thus will reduce the total mineral volume. In order to maintain reasonable thicknesses of rocks, the Pressure Solution–Welding system must be rather conservative (see Giles and de Boer, 1990), except perhaps where extensive stylolitization has occurred.

In a practical sense it may be difficult to distinguish cement that has been formed by Welding from cement formed by Cementation. However, we feel that the two sources of cement are distinct enough to warrant consideration as separate processes in our model. In most of our simulations the amount of porosity reduction predicted due to Welding greatly exceeds that due to Cementation.

Clay Diagenesis (shale and sandstone only)

What we term "Clay Diagenesis" is actually a collection of many different reactions that will generally involve formation or transformation of a variety of clay minerals. Future development of this model might focus on subdivision of this general category into several types of reactions. For example, although Clay Diagenesis was included in our model for both shales and sandstones, the processes in the two rock types could easily be considered as two distinct processes. In shales we define Clay Diagenesis as the transformation of existing clay minerals to new clay minerals, essentially *in situ*. In sandstones, in contrast, we consider primarily the formation of new clay minerals from feldspars, lithic fragments, etc. These processes may lead to considerable formation of overgrowths (e.g., Wilson and Pittman, 1977). Clay Diagenesis in shales is probably more akin to Repacking, in that much of the effect may be due to the release of bound water as clay structures collapse. In sandstones, in contrast, Clay Diagenesis may be more like Welding. We describe Clay Diagenesis in both shales and sandstones as a series of kinetically controlled first-order chemical processes.

Cementation (limestone, chalk, shale, and sandstone)

As noted earlier, the minerals precipitated during Cementation are all assumed to come from outside the system, transported by water flow through the rock. Our model at present only considers calcite and silica, and we allow both kinds of cement to form in all four lithologies. Cementation thus contrasts with Welding, in which the precipitating minerals originate within the rock system, and are therefore always of the same composition as the material dissolved by Pressure Solution.

Most workers who model chemical diagenesis consider that Cementation occurs under equilibrium or near-equilibrium conditions. Detailed reaction-path models for chemical diagenesis (see Harrison, 1990, for a review of various such models) calculate equilibrium constants for all possible reactions and then precipitate or dissolve minerals to attain full equilibrium for the system. This approach is far too complex for basin modelling at the present time, however. Our model instead assumes that the amount of Cementation depends on the volume of water flowing through the rock, reasoning that the larger the volume, the more times equilibrium is perturbed and reestablished, and the more Cementation occurs. From Darcy's law, the amount of Cementation is therefore directly proportional to the rock's permeability and fluid potential, and inversely proportional to water viscosity. Cementation also depends upon how saturated the pore fluid is with respect to the cementing mineral.

In shales and sandstones we do not permit Cementation to begin until burial depths exceed 200 m. Above this level we assume that pore fluids are mainly meteoric water that is not saturated in calcium carbonate or silica. In limestones and chalks, in contrast, we permit Cementation starting at the surface, since early Cementation in limestones is probably rather common and important (Lasemi et al., 1990).

In most of our simulations the amount of porosity reduction due to Cementation is much less than that due to Welding. However, when rocks remain at relatively shallower burial depths for a long time, their high permeabilities may allow a considerable influx of fluids, and hence may lead to considerable porosity loss due to Cementation.

Termination of porosity reduction

We terminate mechanical compaction (Repacking, Crushing, and Ductile Flow) immediately whenever the sedimentation rate becomes zero or negative (that is, during hiatuses and erosional events). Hiatuses and erosional events do not affect Pressure Solution, Welding, or Clay Diagenesis, except as they influence subsurface pressures and temperatures, and do not directly affect Cementation at all.

We also terminate all types of porosity reduction whenever significant overpressuring develops (2.5–10 atm above hydrostatic pressure as calculated using the Yukler 1D™ model). Our model is simplistic in assuming that no fluid flow occurs while overpressuring exists. In a future modification of this model we intend to allow Pressure Solution to proceed normally, albeit at a lower rate proportional to the lower grain–grain stress, and to permit some of the dissolved mineral material to be expelled. However, Cementation, which requires that fluids from outside sources enter the rock, would be unlikely while overpressures exist.

Finally, we terminate porosity reduction when porosity reaches 1%. At such low porosities it is doubtful whether sufficient permeability exists to move water out of the rock to accommodate further porosity reduction. Furthermore, the various empirical equations used in this model (e.g., porosity–permeability relationships) are not necessarily even approximately valid at such low porosities.

Data and simulations

All computer simulations were run on an IBM 5551 personal computer. Burial, thermal, and pressures histories were provided by the Yukler 1D™ software program. An interface connected to the Yukler 1D™ program allowed us to calculate the porosity of a single selected layer in any given simulation using our porosity-reduction equations. Calculation intervals were usually 1 m.y. (but smaller for young sediments), and depth intervals were between 10 and 50 m. Choosing smaller calculation intervals did not materially affect the results.

Porosity calculations using the method employed in the Yukler 1D™ program were done using that software's default parameters. Calculations using the equations of Falvey and Middleton (1981) and Sclater and Christie (1980) were done using the BasinMod™ software program (version 2.55) from Platte River Associates, Inc., employing default compaction constants supplied by that program.

Burial and thermal histories corresponding to the porosity data of Schmoker (1984) for limestones and chalks were kindly supplied by Jim Schmoker. Burial and thermal histories for data from individual wells were calibrated with measured vitrinite-reflectance data.

Porosity data were interpolated as average values from the data of Schmoker (1984) for the Smackover (Gulf Coast), Madison (Williston Basin), and South Florida limestones, which fit our lithologic criteria for our limestone model, and for the Niobrara Formation (Denver Basin) as a chalk. Porosity data for individual wells were obtained both from logs and from direct core measurement. Measured porosity data ranged from Quaternary to Early Carboniferous in age, from 250 m to 5000 m in present-day depth, and up to 160°C in present-day temperature.

Results and discussion

The various constants in Table 1 were determined by empirical calibration between measured porosities and porosities calculated using the method described above. Some cautions are in order in using the constants in Table 1, however. Firstly, these values are preliminary and should be verified by further comparison between measured and calculated porosities. Secondly, the precise values of the constants will depend on the units used, and on the equations used for viscosity as a function of temperature and permeability as a function of porosity. The choice of porosity–permeability relationships will be of particular importance. Thirdly, these constants are for four end-member lithologies only: limestones derived from shelf muds, chalks, quartz sandstones, and shales/siltstones. Applications to other lithologies, such as limestone oolites, lithic sandstones, shaly sands, etc., should be done with caution, and

TABLE 1

Empirically determined constants used in porosity-reduction equations

Constant	Process	LS	CK	SH	SS	Notes
C_P	Repacking	−0.003	−0.002	–	−0.0000304	A, C
ϕ_o	Repacking	0.70	0.80	0.77	0.41	B, C
C'_P	Repacking id.	–	–	–	−0.000154	A, D
ϕ'_o	Repacking	–	–	–	0.337	B, D
C_R	Repacking	–	–	0.00473	–	A, C
B	Repacking	–	–	6.778	–	E, F
C'_R	Repacking	–	–	8409	–	F, G
C_{CR}	Crushing			–	–	A
X_o	Crushing	0.05	0.58	–	–	B
C_D	Ductile Flow	0.005	0.005	–	0.0005	A
D_o	Ductile Flow	0.20	0.05	1.00	0.05	B, R
V_c	Pressure Solution	37	37	–	–	H
V_s	Pressure Solution	–	–	–	22.7	B, I
F_o	Pressure Solution	0.20	0.30	–	0.10	B
ϕ_{PS}	Pressure Solution	0.40	0.60	–	0.32	B
ΔV_o	Pressure Solution	−58.54	-58.54	–	−9.088	J
ΔK^o	Pressure Solution	−0.015	−0.015	–	0.0	K
$\Delta H^o/R$	Pressure Solution	−657.8	−657.8	–	1242	L, M
$\Delta S^o/R$	Pressure Solution	−10.53	−10.53	–	0.168	E, M
C_{PS}	Pressure Solution	0.036	0.17	–	0.007	F
C_{WC}	Welding	0.005	0.15	–	–	N
C_{WS}	Welding	–	–	–	0.09	N
C_{CL}	Clay Diagenesis	–	–	2×10^5	6×10^{13}	O
E_0	Clay Diagenesis	–	–	10.	25	P
J	Clay Diagenesis	–	–	0.8	1	E
C_{EC}	Cementation	8×10^{-9}	2×10^{-10}	0.01	5×10^{-10}	Q
C_{ES}	Cementation	7×10^{-9}	2×10^{-10}	0.002	1×10^{-9}	Q
"Carb"	Cementation	100	100	10	5	E
"Qtz"	Cementation	0	0	10	5	E

Notes. A: in m^{-1}; B: unitless fraction; C: used from surface; D: used from 2400 m; E: unitless; F: used from 100 m; G: in m; H: for calcite; I: for silica (quartz); J: in cm^3/mol (Millero, 1982) for calcite and Walther and Helgeson (1977, cited in Nordstrom and Munoz, 1985) for quartz; K: in cm^3/mol bar (Owen and Brinkley, 1941, cited in Millero, 1982), neglected for quartz; L: in degrees; M: Nordstrom and Munoz (1985); N: t and Δt in m.y, μ in cP, k in darcies, G is a unitless fraction; O: in s^{-1}; P: in kcal/mol; Q: k in darcies, Δt in m.y., "Carb" and "Qtz" are unitless, μ in cP; R: also used for Clay Diagenesis.

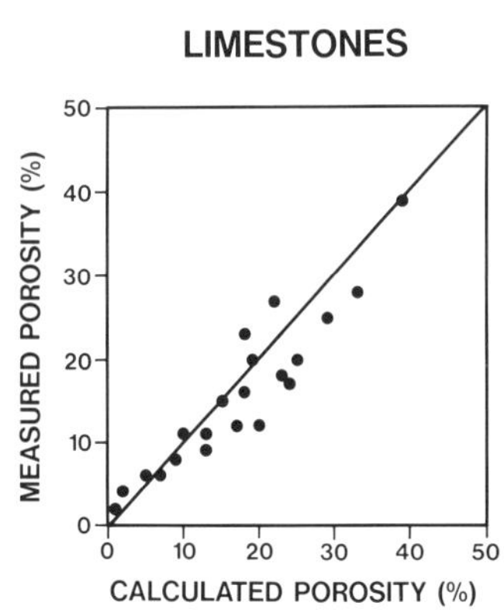

Fig. 1. Measured limestone porosities versus porosities calculated using our seven-process model. The straight line corresponds to perfect agreement.

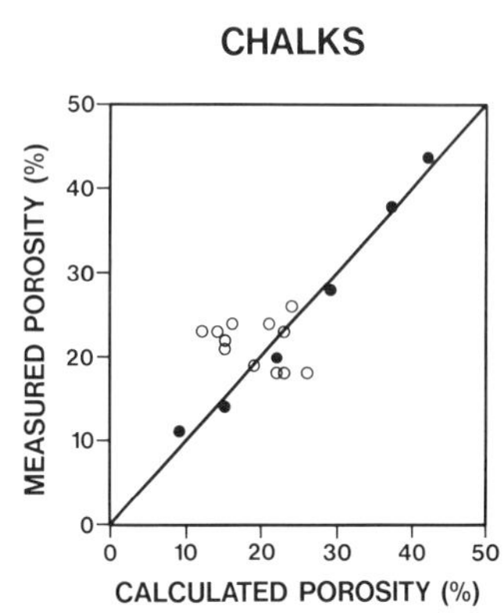

Fig. 2. Measured chalk porosities versus porosities calculated using our seven-process model. The straight line corresponds to perfect agreement. Dots represent normally pressured chalks; circles represent North Sea chalks which are overpressured today.

individual calibrations for new lithologies should be developed wherever possible.

Results of our calibration for the four lithologies are shown in Figs. 1–4, in which our calculated porosities are plotted against measured values. The line in each figure represents perfect agreement between measured and calculated porosities. Although some data scatter exists, the correlations for each lithology are encouraging. Agreement between measured and calculated values within a few percent at

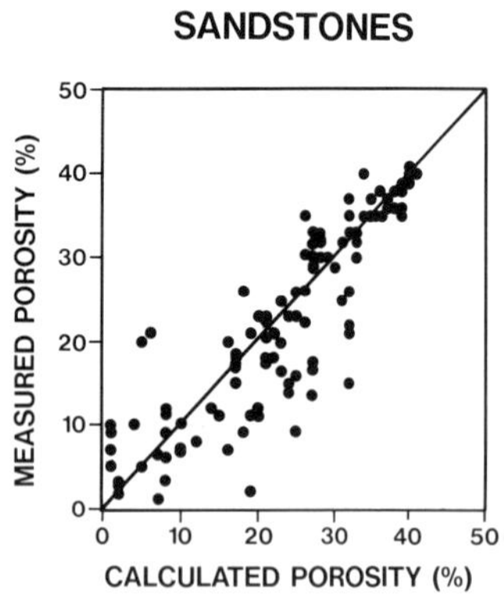

Fig. 3. Measured sandstone porosities versus porosities calculated using our seven-process model. The straight line corresponds to perfect agreement.

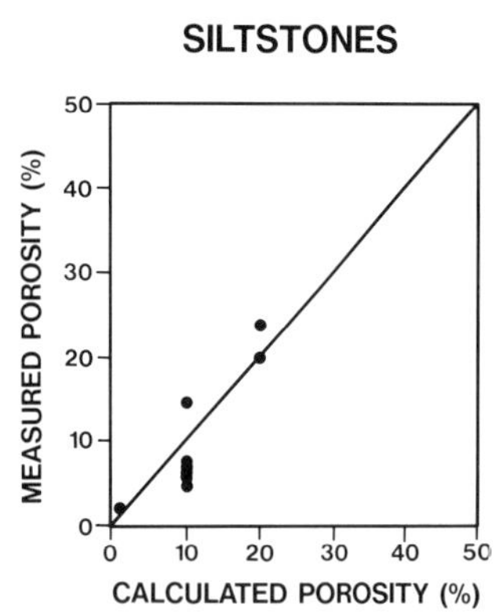

Fig. 4. Measured siltstone porosities versus porosities calculated using our seven-process model. The straight line corresponds to perfect agreement.

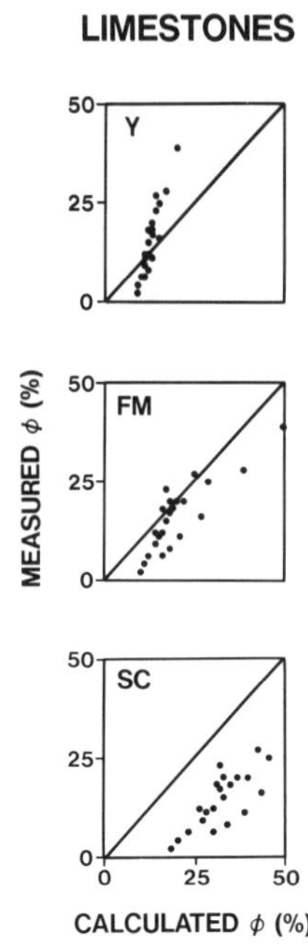

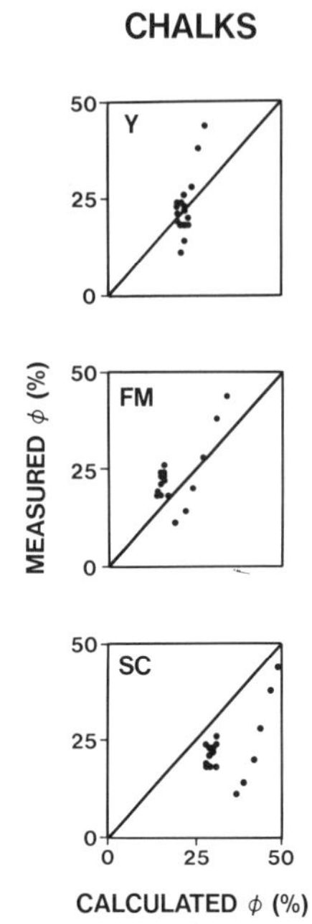

Fig. 5 (left). Measured limestone porosities versus porosities calculated using three different methods. The straight lines correspond to perfect agreement. Top: method included in Yukler 1D™ software program. Middle: equation of Falvey and Middleton (1981) using default constant from BasinMod™ software program. Bottom: equation of Sclater and Christie (1980) using default constant from the BasinMod™ software program.

Fig. 6 (right). Measured chalk porosities versus porosities calculated using the Yukler (top), Falvey-Middleton (middle) and Sclater-Christie methods. The straight lines correspond to perfect agreement. See Fig. 5 and text for description of the three methods.

porosities ranging from over 40% to less than 5% were achieved in each case. (The range for shales and siltstones is much narrower due to lack of data.)

For comparison, porosities calculated using the default compaction parameters provided by the software programs for three other methods are shown for each of the four lithologies in Figs. 5–8. The Yukler 1D™ program calculates porosity reduction based on net effective stress, whereas the Sclater-Christie (1980) and Falvey-Middleton (1981) methods consider only depth. All three methods give less satisfactory fits to the measured data than does our method.

It is important to recognize that in Figs. 1–8 we are comparing the results a user would have obtained for our calibration data set using default values for all compaction parameters for all four methods. Since in practice users will choose the default values most of the time (often due to lack of detailed geologic information), Figs. 1–8 therefore compare the quality of results that most modellers would have obtained with these data sets. However, a knowledgable modeller equipped with good information about actual porosities, actual lithologies, or both could change the default compaction parameters in any of the four methods to provide a locally calibrated system that will give better results. We hope that our model will remove the critical need for local calibration evident in Figs. 5–8.

Porosity reduction in our model is not simply a depth-dependent phenomenon. Figure 9 shows the effect of temperature by comparing calculated porosities for each lithology for the same burial history (constant subsidence rate for 80 m.y. to a present-day depth of 2000 m). Limestones and chalks both lose porosity more slowly when the geothermal gradient is higher, as a consequence of the solubility behaviour of calcite. The difference is greater for chalks, which rely more strongly on Pressure Solution to reduce porosity.

Sandstones and shales, in contrast, lose porosity more rapidly at higher temperatures. The trend in sandstones is due to the solubility behaviour of quartz, whereas that of shales is due to more-rapid Clay Diagenesis at higher temperatures.

Figure 10, in which each lithology was buried to

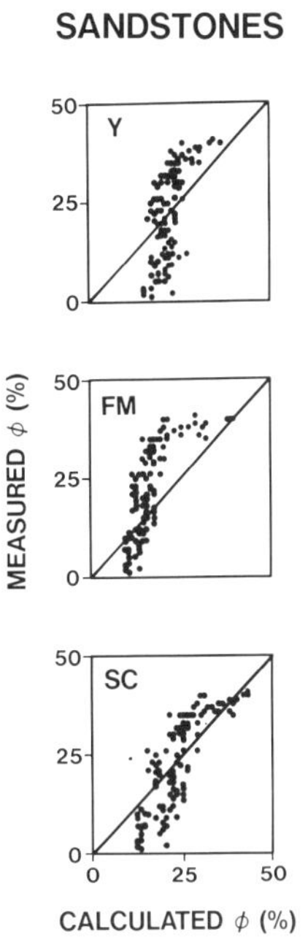

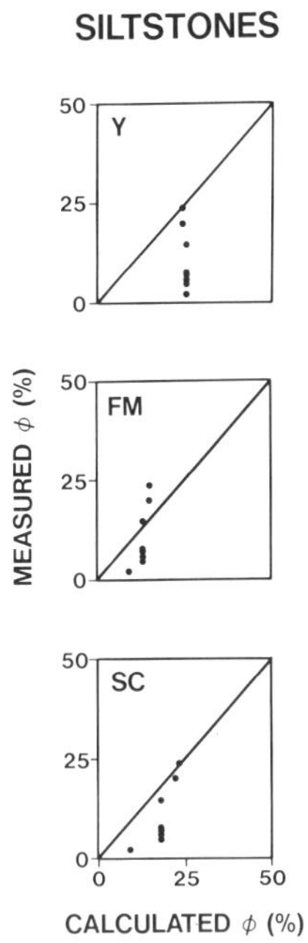

Fig. 7 (left). Measured sandstone porosities versus porosities calculated using the Yukler (top), Falvey-Middleton (middle) and Sclater-Christie methods. The straight lines correspond to perfect agreement. See Fig. 5 and text for description of the three methods.

Fig. 8 (right). Measured siltstone porosities versus porosities calculated using the Yukler (top), Falvey-Middleton (middle) and Sclater-Christie methods. The straight lines correspond to perfect agreement. See Fig. 5 and text for description of the three methods.

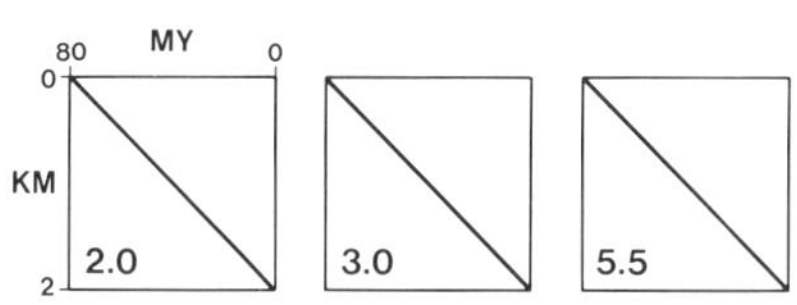

	2.0	3.0	5.5
LS	24	25	26
CK	12	18	26
SS	33	33	23
SH	19	14	4

Fig. 9. Porosities (%) calculated for limestone (*LS*), chalk (*CK*), sandstone (*SS*), and shale (*SH*) using the seven-process model for three different geothermal gradients (2.0, 3.0, and 5.5°C/100 m). In each case the rock was assumed to be 80 Ma old and to have been buried at a constant rate to a present-day depth of 2000 m.

2000 m, shows the effect of time on porosity reduction. Rock ages vary from 10 to 400 Ma, and burial rates are inversely proportional to ages. Porosities of all four lithologies are predicted to decrease as age increases. The effect is sandstones is smallest, followed in order by shale, limestone, and chalk.

Figure 11 compares rocks of the same age and present-day depth and temperature, but with dif-

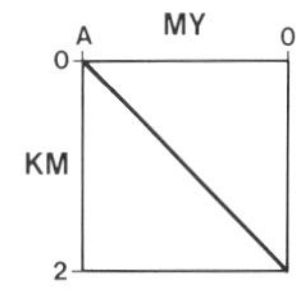

A (MY)	LS	CK	SS	SH
10	30	55	35	26
40	27	33	35	21
80	24	12	33	19
160	18	1	25	17
400	1	1	20	11

Fig. 10. Porosities (%) calculated for limestone (*LS*), chalk (*CK*), sandstone (*SS*), and shale (*SH*) using the seven-process model for rocks at 2000 m today. Rock ages vary from 10 to 400 Ma, and burial rates, which are constant for each simulation, vary inversely with age.

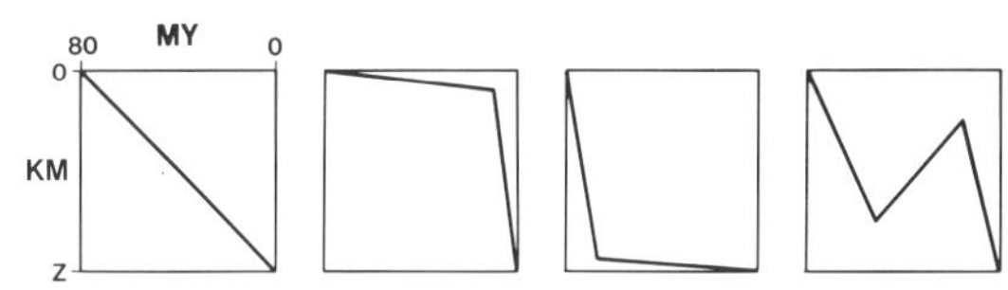

	Z (KM)				
LS	3	20	26	11	20
CK	2	13	37	1	11
SS	4	19	25	8	19
SH	4	8	23	3	9

Fig. 11. Porosities (%) calculated for limestone (*LS*), chalk (*CK*), sandstone (*SS*), and shale (*SH*) using the seven-process model for four different burial histories. In each case the present-day depth is the same for all examples of each lithology. In each case the geothermal gradient is 2°C/100 m. Burial histories are shown for each case. Maximum depths vary among the lithologies to emphasize differences within each lithology.

ferent burial (and hence temperature and pressure/stress) histories. Chalk once again is most sensitive to burial conditions, with predicted porosities at 2000 m ranging from 1% to 37%. The other lithologies show less-dramatic but significant differences.

The timing of porosity reduction by each individual process is shown in Fig. 12 for one burial history for each lithology. Repacking and Cementation are always the earliest processes, followed, where applicable, by Ductile Flow and Crushing. Clay Diagenesis, Pressure Solution, and Welding begin last. Pressure Solution and Welding account for most porosity reduction after the initial dewatering stages in limestones, chalks, and sandstones. As in the model of Mitra and Beard (1980), the role of

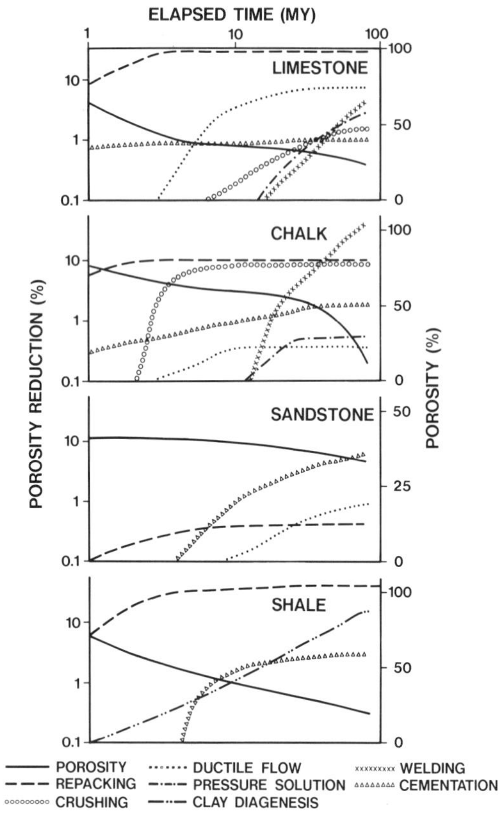

Fig. 12. Porosity reduction by each of seven processes for limestone, chalk, sandstone, and shale. All rocks were buried to 2000 m at a constant rate for 80 m.y., with a geothermal gradient of 2°C/100 m. Porosity reduction and time are plotted on log scales for better resolution. Also shown is total porosity, plotted on a linear scale at the right.

Pressure Solution is more important than Welding in the early stages of Pressure Solution, but the relative importances are reversed as Pressure Solution proceeds. Differences in the relative importances of these two processes in the two models may represent a weakness in our current model, but may also be due in part to the use of perfectly spherical grains in the model of Mitra and Beard (1980).

The relative importances of the different porosity-reduction processes can vary considerably within a single lithology as a function of each rock's burial and thermal history. Figure 13 contrasts porosity reduction for two sandstones at the same 2000 m burial depth today. One sandstone was buried very slowly and suffered considerable early Cementation (e.g., Sullivan and McBride, 1991) which retarded Repacking, whereas the other was buried rapidly without much Cementation. Final porosity of the slowly buried sandstone is considerably higher in spite of (or because of) its early Cementation, due to less Pressure Solution and Welding.

Figure 14 contrasts porosity reduction in two limestones with different burial histories but the same present-day depth. Once again, the slowly buried

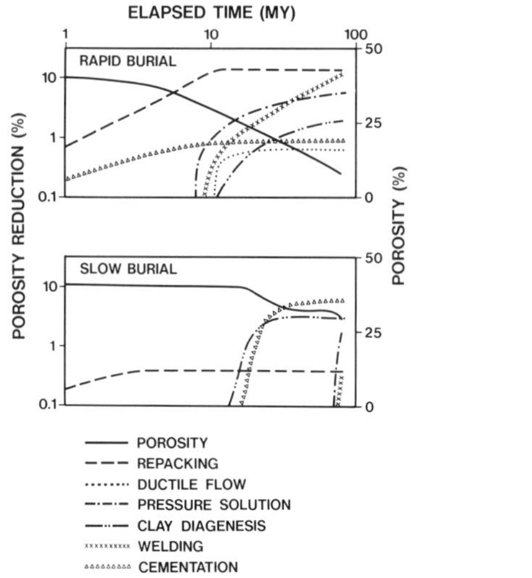

Fig. 13. Porosity reduction by each of six processes for sandstones buried at the fastest (top) and slowest rates shown in Fig. 11. Porosity reduction and time are plotted on log scales for better resolution. Also shown is total porosity, plotted on a linear scale at the right.

rock suffers much more Cementation, which in turn affects the progress of other processes both during early Cementation and later in the rock's history (e.g., Scholle et al., 1991).

Unlike existing maturity models and basin models, our model does not assume constant mineral volume during compaction. In general, however, changes in mineral volume due to Cementation and to Pressure Solution without reprecipitation as Welding is fairly small (generally less than 5%: Fig. 15), and therefore need not cause severe problems for such models.

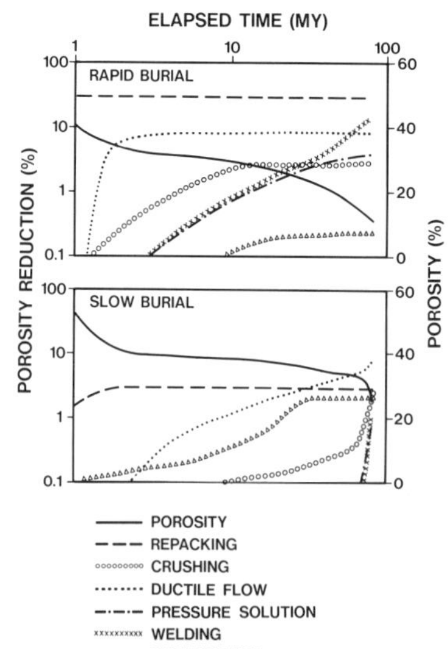

Fig. 14. Porosity reduction by each of six processes for limestones with the same burial histories as in Fig. 13. Porosity reduction and time are plotted on log scales for better resolution. Also shown is total porosity, plotted on a linear scale at the right.

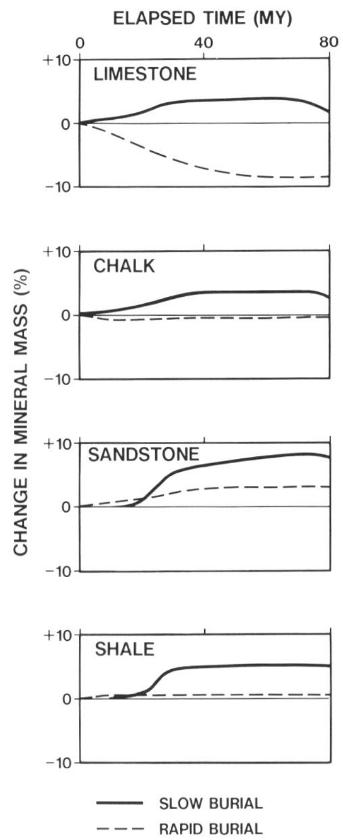

Fig. 15. Change in mineral mass (expressed as percentage of original mineral mass) as a function of time for two cases each of limestone, chalk, sandstone, and shale. The same fast and slow burial histories used in Figs. 13 and 14 were used.

Figure 15 also shows the trends of mineral volume in the four lithologies through time for two burial histories (rapid and slow). Since shales in our model contain no soluble material, they will increase in mineral mass in proportion to the amount of Cementation. In both examples in Fig. 15, sandstone also increases in mineral mass due to Cementation, but in other cases it could decrease. Chalks and limestones both lose mineral mass due to Pressure Solution when burial rates are high, but gain mineral matter due to a dominance of Cementation over Pressure Solution at low burial rates.

Figure 2 includes data from a number of overpressured North Sea chalks, whose porosities have been noted to be abnormally high for their burial depth when compared with other chalks world wide (Scholle, 1977). The results in Fig. 2 were obtained by stopping porosity reduction when overpressuring occurred. Although there is some scatter in the correlation between measured and calculated porosities, the chalks nevertheless were modelled with approximately the same degree of accuracy as limestones and sandstones (compare Figs. 1 and 3). These results suggest that our model can be applied to overpressured as well as normally pressured rocks. We have not yet had an opportunity to test its ability to predict porosity reduction in overpressured sandstones, shales, or limestones, however.

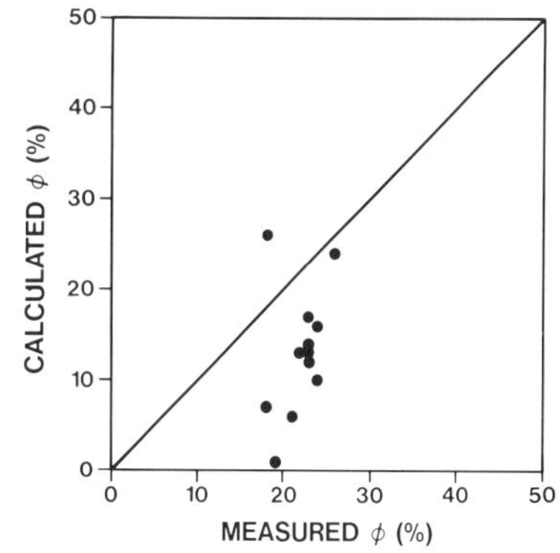

Fig. 16. Measured porosities compared with porosities calculated using the seven-process model for overpressured North Sea chalks assuming that pressures remained normal during the entire burial history. Line represents perfect agreement. Compare with Fig. 2.

We also calculated the porosities the North Sea chalks would have had today had they not developed overpressuring. Figure 16 shows the correspondence between measured porosities and porosities calculated assuming no overpressuring, and Fig. 17 shows the difference $\Delta\phi$ between porosities calculated with and without overpressuring plotted against depth. In some cases, in the absence of overpressuring the porosity would have been very low ($\Delta\phi \approx 20\%$), but in others it would have been essentially normal ($\Delta\phi \approx 0\%$). The importance of overpressuring in maintaining porosity in these chalks thus appears to be quite variable.

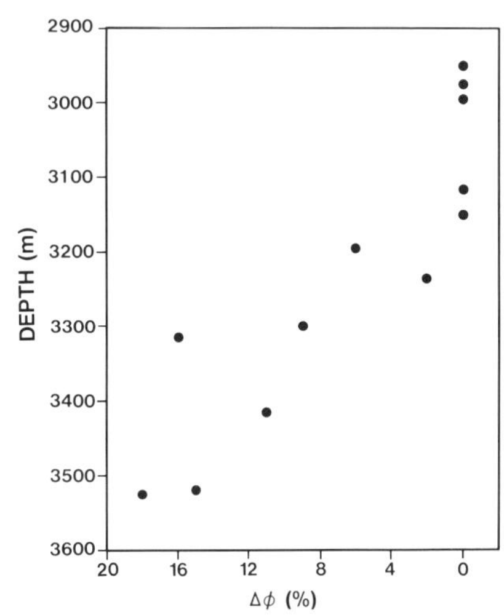

Fig. 17. Present-day (maximum) burial depth versus the difference ($\Delta\phi$) between porosities of overpressured North Sea chalks calculated considering overpressure and ignoring overpressure. Shallow samples show little or no difference (that is, little effect of overpressure on porosity), whereas deep samples show large differences in calculated porosities.

The "high" porosities of at least some of the North Sea chalks are therefore not due to overpressuring. These conclusions contrast with those of Scholle (1977), who had implicitly assumed that chalk diagenesis is fundamentally a depth-dependent process, and that there is a "normal" chalk porosity for a particular depth. However, as Figs. 9–11 show, chalks with different burial and thermal histories can have greatly different porosities at the same depth, even when pressures are normal.

In retrospect, this conclusion is not at all surprising, since most of the porosity reduction in chalks is due to Welding induced by Pressure Solution (Fig. 12). The rate of Pressure Solution, which ultimately controls Welding, is controlled by the rate of diffusion of dissolved material rather than by stress, and thus must be dependent on time and temperature (eq. A18). Although some of the deepest North Sea chalks in our study (3500 m) owe most of their porosity to overpressuring, those near 3000 m would still have porosities of about 20% today even in the absence of overpressuring. The reason that these 3000 m chalks have porosities far above the values near 0% predicted by Scholle (1977) is that they have been at their present burial depths for only a few million years. There simply has not yet been enough time to reduce the porosities to near-zero levels.

Conclusions

Our porosity-reduction model shows promise of being able to estimate porosities of compacting limestones, chalks, shales, and sandstones rather accurately without local calibration. For the data set we examined it is more successful than three other common depth-dependent porosity-reduction functions. Its main application is intended to provide better predictions of porosity for numerical basin models. Since it considers only regular and predictable average porosity-reduction trends, it is not appropriate for application to evaluation of reservoir characteristics, which will often be strongly influenced by processes not included in our model, or by abnormal occurrences of the processes we seek to model. For reservoir evaluation, models based on local learning sets are much more appropriate (e.g., Bloch et al., 1990; Bloch, 1991). However, once we are able to model grain dissolution without structural collapse, we may be better prepared to predict reservoir character.

Our model will also help us understand and study the various processes which contribute to porosity reduction. Furthermore, it can be useful in understanding the porosity-reduction history of individual strata or formations. Application to overpressured North Sea chalks suggests that the retarding influence of overpressuring on porosity reduction in those rocks is rather variable, and that some overpressured chalks do not have abnormally high porosities when the effects of time and temperature on porosity are considered.

The calibration of our model reported here should be considered preliminary, and further testing should improve its reliability. Several possible modifications to the present conceptual model should be investigated, and its general application to overpressured strata should be verified.

Acknowledgements

We thank Jim Schmoker of the USGS for providing porosity data and burial histories used in calibrating our model and for valuable suggestions about the manuscript, Clyde Moore for helpful discussions of porosity reduction in carbonates, Chris Spiers for bringing to our attention alternative opinions on the kinetics of Pressure Solution, and JNOC for support and permission to publish this work.

Appendix — Mathematical descriptions of porosity-reduction processes

Repacking

The choice of a mathematical description of rapid porosity reduction due to Repacking at shallow burial depths is of surprisingly large importance, particularly for sandstones, because an accurate description of early porosity reduction is critical for properly evaluating the importance of early Cementation. For limestones, chalks, and sandstones we have used straight-line relationships (eq. A1) between burial depth (Z) and porosity (ϕ) derived empirically from our data to describe Repacking.

$$\phi = C_R Z + \phi_o \tag{A1}$$

where ϕ_o is the initial porosity and C_R is a constant. For limestones and chalks we use a single equation, but for sandstones we use two. The first equation controls the observed slow Repacking down to 2400 m; the second describes the rapid decrease in the next 500 m. The constants in these equations are given in Table 1.

For shales, at depths less than 100 m we use the standard Athy-type equation (A2). Below 100 m we use eq. A3, from Baldwin and Butler (1985).

$$\phi = \phi_o \exp(-C_R Z) \tag{A2}$$

$$Z = C'_R(1 - \phi)^B \tag{A3}$$

Values of the constants in these equations are also given in Table 1.

In our model, if porosity is reduced by other mechanisms, the amount of porosity reduction due to Repacking will be reduced. Therefore, if the porosity of the sediment is below that predicted by eqs. A1–A3 at any given depth, we do not allow any Repacking to occur during that time step. Repacking will not commence until the porosity is again above the value appropriate to that depth calculated from eqs. A1–A3.

Three processes may compete with Repacking for early porosity reduction. Cementation will tend to bind grains together and thus prevent Repacking. In contrast, we view other stress-sensitive porosity-reduction processes (Ductile Flow and Crushing) as alternatives to Repacking in the effort to release excess stress. We therefore require that these processes operate in parallel with Repacking, rather than in series.

Crushing

Lacking any detailed physical description of the Crushing process, we model porosity reduction during Crushing using the delta form of the derivative of eq. A2:

$$\Delta\phi_r = -C_{CR}(\phi - X)\Delta Z \qquad (A4)$$

Values for C_{CR} are given in Table 1.

In eq. A4 true porosity is reduced by an amount X that is proportional to the amount of (Cementation + Welding) that has occurred, because precipitation of mineral matter via those two processes will strengthen the weak microfossils and thus slow or prevent crushing (Moore, 1989). We calculate X as:

$$X = X_0 - 5\sum\Delta\phi_W - 0.5\sum\Delta\phi_{CE} \qquad (A5)$$

where $\sum\Delta\phi_W$ and $\sum\Delta\phi_{CE}$ represent the total porosity reduction (negative values) due to Welding and Cementation, respectively, in all previous time steps. We consider Welding to be more effective in preventing Crushing than an equivalent amount of Cementation because Welding will be more pervasive in the rock matrix than Cementation (which probably occurs mainly within the most-permeable layers). Further discussion of this point is given in the sections on Welding and Cementation. Our use of 5% porosity as the value for X_0 is equivalent to saying that all Crushing would cease when porosity reaches 5% even in the absence of Cementation or Welding.

Ductile Flow

The change in porosity volume during Ductile Flow is given by:

$$\Delta V_{\phi D} = -C_D(D_f - 0.02)\Delta Z \qquad (A6)$$

where C_D is a constant (Table 1) and D_f is the fraction of ductile grains that have not yet suffered Ductile Flow or Clay Diagenesis. D_f is given by:

$$D_f = D_o + \sum\Delta\phi_{DF} + \sum\Delta\phi_{CD} \qquad (A7)$$

where $\sum\Delta\phi_{DF}$ and $\sum\Delta\phi_{CD}$ are the total amounts of porosity reduction (negative numbers) due to Ductile Flow and Clay Diagenesis, respectively. Default values for D_o are given in Table 1. If transformation of a grain by either Ductile Flow or Clay Diagenesis results in porosity reduction, that grain cannot be used again to reduce porosity in the other process. The term 0.02 in eq. A8 is after Rittenhouse (1971b), as explained in the text.

After Mitra and Beard (1980), we calculate the new porosity after Ductile Flow to be:

$$\phi = 1 - \frac{V_{rock}}{V_{rock} + V_o + \Delta V_{\phi D}} \qquad (A8)$$

where V_o is the volume of porosity prior to Ductile Flow.

Jim Schmoker (pers. commun., 1991) has suggested that Ductile Flow should be a function of time and temperature as well as depth. Ductility would be a function of temperature, and flow rates might be controlled by time. These ideas could be explored in a future modification.

Pressure Solution

The rate of Pressure Solution R_{PSm} for a mineral m is given by:

$$R_{PSm} = D\frac{d[I]}{dX} \qquad (A9)$$

where $d[I]/dX$ is the concentration gradient of diffusing ions from the grain contact to the bulk pore solution and D is the diffusion coefficient. We determine the concentration gradient by assuming that the bulk pore fluid is saturated with dissolved mineral (concentration = a_m), and that near the grain contact the concentration is a_{Pm}.

According to de Boer (1977), the solubility a_{Pm} of a mineral m suffering stress effects is given by:

$$a_{Pm} = a_m + \Delta a_m \qquad (A10)$$

where a_m is the solubility at hydrostatic pressure and Δa_m is the increased solubility due to the greater stress in a particular direction (in our case vertical lithostatic pressure). Furthermore,

$$a_{Pm} = a_m \exp(V_m P_e / FRT) \qquad (A11)$$

where V_m is the molar volume of the mineral, P_e is the net effective stress, the difference between lithostatic and hydrostatic pressures, R is the gas constant, and T the absolute temperature. The term F is the fraction of cross-sectional area covered by grain–grain contacts, and provides a way to correct the average stress P_e for the fact that grain–grain contacts are not continuous and do not cover the entire area of any horizontal cross section. Thus:

$$\Delta a_m = a_m \exp(V_m P_e / FRT) - a_m \tag{A12}$$

The concentration gradient between the grain contact and the open pore space is given by:

$$\frac{d[I]}{dX} = \frac{\Delta a_m}{\Delta X} \tag{A13}$$

The diffusion distance ΔX in eq. A13 is taken to be the radius of the grain at the point of contact. It is assumed to increase as Pressure Solution proceeds, because as grain faces become broader the distance from the point of Pressure Solution to the bulk pore fluid increases (Weyl, 1959). Since F is a measure of the surface area of the grain, we increase ΔX proportionally to $\sqrt{F}$. Thus:

$$\frac{d[I]}{dX} = \frac{c' \Delta a_m}{\sqrt{F}} \tag{A14}$$

Grain–grain contact areas increase during the process of Pressure Solution, leading to decreases in stress on the grain contacts (Lowry, 1956). We have not attempted to calculate precisely the change in surface area of grain contacts as Pressure Solution proceeds. In principle such a calculation might be carried out with difficulty for a homogeneous rock of identical, regularly shaped grains (e.g., Weyl, 1959), but it would be impossible for any real rock (see Mitra and Beard, 1980). We therefore estimate the contact surface area as:

$$F = (1 - F_o) P_{PS}^{2/3} + F_o \tag{A15}$$

where F_o is the initial fraction of surface area covered by grain-to-grain contacts (0.1–0.3, depending on lithology: see Table 1), and P_{PS} is the relative amount of Pressure Solution that has occurred. We calculate P_{PS} as:

$$P_{PS} = 1.0 - \frac{\phi}{\phi_{PS}} \tag{A16}$$

where ϕ is the porosity and ϕ_{PS} was the porosity where Pressure Solution began. We use P_{PS} to the 2/3 power in eq. A14 to convert a volume (P_{PS}) to an area (F). P_{PS} is a reasonable index of the amount of Pressure Solution because in limestones, chalks, and sandstones most of the porosity reduction after the beginning of Pressure Solution is due directly or indirectly to Pressure Solution. In addition, any Cementation or Welding that has occurred will further increase the surface area at the grain contacts, and thus decrease the stress (Sibley and Blatt, 1976). We have ignored this last effect in our present model, but it could be included in a future modification.

The diffusion coefficient D is a function of temperature:

$$D = cT \tag{A17}$$

Combining eqs. A9–A17, we obtain:

$$R_{PSm} = \frac{a_m C_m T[\exp(V_m P_e / FRT) - 1]}{\sqrt{F}} \tag{A18}$$

where C_m is a constant for each mineral. We assume the molar volume is constant, and that $P_e = 0.57Z$ (P_e in psi, depth in feet) $= 0.1275Z$ (P_e in atm, depth in m) $= 12.92Z$ (P_e in kPa, depth in m). The solubility a_c of calcium carbonate in the bulk pore fluid is given by:

$$a_c = K_c^{1/2} \tag{A19}$$

and that of silica, a_s, by:

$$a_s = K_q \tag{A20}$$

The equilibrium constant K_m (K_c for calcite or K_q for silica) is a function of fluid pressure P and temperature T (Millero, 1982):

$$\log K_m = \log K_0 \frac{-\Delta V^\circ P_L}{RT} + \frac{\Delta K^\circ P_L^2}{2RT} \tag{A21}$$

where R is the gas constant, P_L is the fluid pressure, and ΔV° and ΔK° are constants (Table 1). We neglect K° for silica, since that term makes only a minor contribution for calcite. We assume a grain density of 2.7 g/cm^3 in calculating stress.

K_0 is the equilibrium constant as a function of temperature at standard pressure. Thus:

$$\log K_0 = \frac{-\Delta H^\circ}{RT} + \frac{\Delta S^\circ}{R} \tag{A22}$$

The volume of mineral matter mobilized during Pressure Solution, ΔV_m, is calculated as:

$$\Delta V_m = -C_{PS} R_{PSm} \Delta t \tag{A23}$$

where C_{PS} is a constant that includes C_m (Table 1), Δt is the length of the time step in the calculation, and ΔR_{PSm} is the rate of Pressure Solution. Since Pressure Solution itself involves removal only of mineral matter, it does not automatically lead to a decrease in porosity. However, removal of mineral matter leads to collapse of the rock structure, which in turn leads to loss of porosity if the pre-collapse porosity of the particular collapsed structure was greater than average for the whole rock. The collapse of such a structure will lead to a rapid loss of porosity in the initial stages of dissolution, but the rate will gradually slow as the remaining grains occupy a greater proportion of the total rock volume.

Mitra and Beard (1980) have calculated this geometric effect precisely for sandstones with idealized grain shapes. To simulate this geometric effect in our model, however, we simply include the factor F again. Thus the change in total sediment volume (mineral plus pore fluid), ΔV_{totPS}, during Pressure Solution and the collapse that follows is given by:

$$\Delta V_{totPS} = \frac{\Delta V_m}{F} \tag{A24}$$

The change in volume of pore space during Pressure Solution is given by the difference between V_{totPS} and V_m:

$$V_{\phi PS} = V_{totPS} - V_m \tag{A25}$$

Finally, the change in pore volume must be converted to change in porosity by considering the change in rock volume due to Pressure Solution. Change in rock volume is given by the difference between the change in total volume and the change in pore volume.

Welding

Since the material precipitated during Welding must have been previously dissolved by Pressure Solution, the amount of Welding in any time step is limited by the volume of mineral (ΔV_m, from eq. A22) dissolved in that time step (Δt). We also consider that as the pore volume (V_ϕ) decreases, the rate of Welding will increase, since saturation will be easier to achieve with a given amount of dissolved material. We therefore describe the volume of porosity lost by Welding as:

$$\Delta V_{\phi W} = \frac{C_W \Delta V_m \Delta t}{V_\phi} \tag{A26}$$

where C_W is a constant (Table 1). This formulation is consistent with the observation (e.g., Bjørlykke, 1984; Bloch, 1991), that leaching (conceptually the opposite of Welding in our model) is most prevalent where primary porosities are highest.

As with Pressure Solution, the change in pore volume during Welding must be converted to the change in porosity. Although pore volume decreases during Welding, mineral volume increases by an equal and opposite amount. Total rock volume therefore remains constant because precipitated minerals simply replace pore water.

Clay Diagenesis

The change in pore volume due to the kinetically controlled reactions included in Clay Diagenesis, $\Delta V_{\phi CL}$, is given by:

$$\Delta V_{\phi CL} = -C_{CL} C_{rx} \exp(E/RT) \Delta t \tag{A27}$$

where C_{CL} is a constant (Table 1), R is the gas constant, E is the apparent activation energy (Table 1), C_{rx} is the concentration of reactive species, and Δt is the length of the time step in the calculation.

Since Clay Diagenesis in shales probably includes numerous reactions with a wide range of true activation energies, the apparent activation energy E that we use changes as a function of the amount of Clay Diagenesis that has occurred:

$$E = E_0(1 - J \sum \Delta\phi_{CL}) \tag{A28}$$

where E_0 is the activation energy of the earliest reaction, J is a constant, and $\sum \Delta\phi_{CL}$ is a negative number (the sum of all previous porosity reduction, expressed as a fraction, due to Clay Diagenesis).

The empirically determined value for E_0 that we use for shales is low (10,000 cal/mol) compared to measured values for such reactions as the conversion of smectite to illite (e.g., 19,300 cal/mol: Dutta, 1987). However, as Jüntgen and Klein (1975) and Waples (1984) have shown, when a group of parallel reactions with different activation energies is treated as a single reaction, the apparent activation energy will be far lower than that for any of the real reactions. Thus our choice of a low apparent activation energy in eqs. A26 and A27 is not in conflict with chemical principles.

The concentration of reactive species, C_{rx}, is calculated differently for shales and sandstones. It is assumed for shales that the initial concentration is 100% and that the concentration of reactive clays does not decrease during early Clay Diagenesis, since clays are converted to other clays that are capable of undergoing further reaction. Thus in shales at porosities above 16% (the end of the "second dehydration stage" of Burst, 1969), C_{rx} is assumed to have the constant value of 0.15. As porosity ϕ drops below 16%, the clays are assumed to be converted to unreactive clays. Thus under these conditions:

$$C_{\mathrm{rx}} = \phi - 0.01 \quad \text{(A29)}$$

The term 0.01 is included in order to stop Clay Diagenesis when porosity is reduced to 1%.

In sandstones we specify the original fraction of lithic fragments and other reactive material in the sandstone. From the fraction of reactive material (D_{f} in eqs. A6 and A7) we can easily calculate the original volume of reactive material. As Clay Diagenesis proceeds, D_{f} is decreased by the amount of Clay Diagenesis ($\sum \Delta V_{\mathrm{CL}}$) plus Ductile Flow ($\sum \Delta V_{\mathrm{D}}$) that has occurred. Thus:

$$\text{Reactant} = D_{\mathrm{f}} + \sum \Delta V_{\mathrm{CL}} + \sum \Delta V_{\mathrm{D}} \quad \text{(A30)}$$

where $\sum \Delta V_{\mathrm{CL}}$ and $\sum \Delta V_{\mathrm{D}}$ are negative numbers. The apparent activation energy E_0 for sandstones (25,000 cal/mol) is higher than for shales, and is more in line with what we expect for true activation energies for such processes (e.g., Dutta, 1987). Because it represents something closer to a true activation energy, it increases very little as porosity destruction by conversion of reactive material occurs: the value of J in eq. A28 is 1 (Table 1).

Cementation

We assume that the rate of Cementation depends on the rate of fluid flow through the rock. Using Darcy's law, we see that the rate of Cementation will depend directly on permeability and fluid potential, and inversely on viscosity. Fluid potential could be considered automatically if our porosity-reduction function were incorporated in a 2D or 3D basin model (e.g., Ungerer et al., 1990). In this study, however, fluid potential was assumed to be constant and the same for all rocks that were not overpressured. Thus the change in volume of pore space due to Cementation, $\Delta V_{\phi\mathrm{CE}}$, is given by:

$$\Delta V_{\phi\mathrm{CE}} = \frac{\mu \Delta V_{\phi\mathrm{PS}}(C_{\mathrm{Cm}}) G \Delta t}{k} \quad \text{(A31)}$$

where μ is the viscosity of water calculated using some suitable relationship between viscosity and temperature, C_{Cm} equals C_{Cc} or C_{Cs} for calcite or silica (Table 1), G is the ratio of volume of material dissolved by Pressure Solution in this time step to the pore volume, Δt is the length of the time step in the calculation, and k is permeability.

We used the Andrade equation (Latil, 1980) for water viscosity as a function of temperature. Permeabilities (in darcies) were calculated as a function of porosity (fractional) for each lithology using equations obtained from Scholle (1977) for chalks, Dutta (1987) for shale and a modified form of Mann and Mackenzie's (1990) equation for sandstone:

Limestone: $$k = 5 \times 10^{8\phi - 6} \quad \text{(A32)}$$

Chalk: $$k = 10^{6.25\phi - 4.6875}) \quad \text{(A33)}$$

Shale: $$k = 3.4 \times 10^{-6} [\phi/(1-\phi)]^5 \quad \text{(A34)}$$

Sandstone: $$k = 10^4 \phi^7 \quad \text{(A35)}$$

Thus:

$$\Delta V_{\phi\mathrm{CE}} = \frac{-k\Delta t \left(\frac{C_{\mathrm{CEc}}\text{“Carb”}}{a_{\mathrm{c}}} + \frac{C_{\mathrm{CEq}}\text{“Qtz”}}{a_{\mathrm{s}}} \right)}{\mu} \quad \text{(A36)}$$

where Δt is the length of the calculation time step, C_{CEc} and C_{CEq} are constants for carbonate and quartz, respectively (Table 1), a_{c} and a_{s} are the solubilities of calcite and silica as discussed for Welding, and "Carb" and "Qtz" represent the availability of carbonate and silica, respectively, in the flowing formation water.

The terms "Carb" and "Qtz" represent the most difficult input values for the user. Table 1 shows the values actually used in this study, but we envision that they might be varied between 0 and 100 in extreme cases. If "Carb" and "Qtz" are small, little Cementation will occur, whereas a large value for either "Carb" or "Qtz" will result in greater Cementation and more rapid porosity reduction. We would use a high value of "Carb" when waters are flowing through carbonate facies or have previously flowed through carbonate rocks. Choice of a value for carbonate could be made based upon data on carbonate contents or on detailed petrographic analysis of actual rocks (e.g., Heling, 1970; Houseknecht, 1987).

Since Cementation by quartz is usually less common and less pervasive than Cementation by carbonate, we would generally use a low value for "Qtz". However, when there is independent petrographic evidence of quartz cement in excess of that which could be accounted for by Pressure Solution + Welding, we would use a higher value.

If the Cementation model is incorporated into a 2D or 3D basin model, it might be possible to have the program choose a value for "Carb" or "Qtz" automatically. In such a scheme the migration history of water volumes would have to be checked to see whether the moving waters have ever been exposed to carbonate rocks and thus saturated with calcium and carbonate species.

Prediction of "Qtz" may also prove feasible based on analogous logic. Quartz Cementation, however, may depend critically on vertical migration of hydrothermal fluids through faults and fractures. Since fluid flow through fractures is ignored by most basin models (but see Larson et al., this volume), our ability to predict silica saturations is probably poor.

References

Athy, L.F., 1930. Density, porosity, and compaction of sedimentary rocks. Am. Assoc. Pet. Geol., Bull., 14: 1–24.

Baldwin, B. and Butler, C.O., 1985. Compaction curves. Am. Assoc. Pet. Geol., Bull., 69: 622–626.

Bathurst, R.G.C., 1975. Carbonate Sediments and their Diagenesis, Developments in Sedimentology, Vol. 12. Elsevier, Amsterdam, 2nd ed., 658 pp.

Bjørlykke, K., 1984. Formation of secondary porosity: how important is it? In: D.A. McDonald and R.C. Surdam (Editors), Clastic Diagenesis. Am. Assoc. Pet. Geol., Mem., 37: 277–286.

Bloch, S., 1991. Empirical prediction of porosity and permeability in sandstones. Am. Assoc. Pet. Geol., Bull., 75: 1145–1160.

Bloch, S., McGowen, J.H., Duncan, J.R. and Brizzolara, D.W., 1990. Porosity prediction, prior to drilling, in sandstones of the Kekiktuk Formation (Mississippian), North Slope of Alaska. Am. Assoc. Pet. Geol., Bull., 74: 1371–1385.

Burst, J.F., 1969. Diagenesis of Gulf Coast clayey sediments and its possible relation to petroleum migration. Am. Assoc. Pet. Geol., Bull., 53: 73–93.

de Boer, R.B., 1977. On the thermodynamics of pressure solution — interaction between chemical and mechanical forces. Geochim. Cosmochim. Acta, 41: 241–256.

Dutta, N.C., 1987. Fluid flow in low permeable porous media. In: B. Doligez (Editor), Migration of Hydrocarbons in Sedimentary Basins. Technip, Paris, pp. 567–588.

Falvey, D.A. and Middleton, M.F., 1981. Passive continental margins: evidence for a prebreakup deep crustal metamorphic subsidence mechanism. Oceanol. Acta, Spec. Issue, pp. 103–114.

Giles, M.R. and de Boer, R.B., 1990. Origin and significance of redistributional secondary porosity. Mar. Pet. Geol., 7: 378–397.

Harrison, W.J., 1990. Modeling fluid/rock interactions in sedimentary basins. In: T.A. Cross (Editor), Quantitative Dynamic Stratigraphy. Prentice-Hall, Englewood Cliffs, N.J., pp. 195–231.

Hatch, F.H., Rastall, R.H. and M. Black, M., 1938. The Petrology of the Sedimentary Rocks. Thomas Murby and Company, London, 3rd ed.

Heling, D., 1970. Micro-fabrics of shales and their rearrangement by compaction. Sedimentology, 15: 247–260.

Houseknecht, D.W., 1987. Assessing the relative importance of compaction processes and cementation to reduction of porosity in sandstones. Am. Assoc. Pet. Geol., Bull., 71: 633–642.

Jüntgen, H. and Klein, J., 1975. Formation of natural gas from coaly sediments. Erdöl Kohle-Ergas-Petrochem., 28: 65–73 (in German).

Larson, K.W., Waples, D.W., Fu, H. and Kodama, K., 1993. Predicting tectonic fractures and fluid flow through fractures in basin modelling. In: A.G. Doré et al. (Editors), Basin Modelling: Advances and Applications. Norwegian Petroleum Society (NPF), Special Publication 3. Elsevier, Amsterdam, pp. 373–383 (this volume).

Lasemi, Z., Sandberg, P.A. and Boardman, M.R., 1990. New microtextural criterion for differentiation of compaction and early cementation in fine-grained limestones. Geology, 18: 370–373.

Latil, M., 1980. Enhanced Oil Recovery. Gulf Publishing Co., Houston, Texas, 236 pp.

Lowry, W.D., 1956. Factors in loss of porosity by quartzose sandstones of Virginia. Am. Assoc. Pet. Geol., Bull., 40: 489–500.

Lyons, D.J., 1978. Sandstone reservoirs: petrography, porosity–permeability relationship, and burial diagenesis — application of general relationships to reservoir quality prediction in the Cretaceous of Central Hokkaido, Japan. Japan National Oil Corporation, Technology Research Center, Rep. 8, 69 pp.

Lyons, D.J., 1979. Organic metamorphism and sandstone porosity prediction from acoustic data. Japan National Oil Corporation, Technology Research Center, Rep. 9, pp. 1–51.

Mann, D.M. and Mackenzie, A.S., 1990. Prediction of pore fluid pressures in sedimentary basins. Mar. Pet. Geol., 7: 55–65.

Maxwell, J.C. and Verrall, P., 1954. Low porosity may limit oil in deep sands. World Oil, 138 (5): 106–113; 138 (6): 102–104.

Millero, F.J., 1982. The effect of pressure on the solubility of minerals in water and seawater. Geochim. Cosmochim. Acta, 46: 11–22.

Mitra, S. and Beard, W.C., 1980. Theoretical models of porosity reduction by pressure solution for well-sorted sandstones. J. Sediment. Petrol., 50: 1347–1360.

Moore, C.H., 1989. Carbonate Diagenesis and Porosity. Elsevier, Amsterdam, 338 pp.

Nagtegaal, P.J.C., 1980. Diagenetic models for predicting clastic reservoir quality. Rev. Inst. Invest. Geol., 34: 5–19.

Nordstrom, D.K. and Munoz, J.L., 1985. Geochemical Thermodynamics. Benjamin, Menlo Park, Calif., 477 pp.

Owen, B.B. and Brinkley, S.R., 1941. Calculation of the effect of pressure upon ionic equilibria in pure water and in salt solutions. Chem. Rev., 29: 461–474.

Raj, R., 1982. Creep in polycrystalline aggregates by matter transport through a liquid phase. J. Geophys. Res., 87: 4731–4739.

Rittenhouse, G., 1971a. Pore space reduction by solution and cementation. Am. Assoc. Pet. Geol., Bull., 55: 80–91.

Rittenhouse, G., 1971b. Mechanical compaction of sands containing different percentages of ductile grains: a theoretical approach. Am. Assoc. Pet. Geol., Bull., 55: 92–96.

Roll, A., 1974. Long-term reduction of the thickness of sedimentary rocks and its effects — an overview. Geol. Jahrb., A, 14: 2–76 (in German with abstracts in English, French, and Russian).

Scherer, M., 1987. Parameters influencing porosity in sandstones: a model for sandstone porosity prediction. Am. Assoc. Pet. Geol., Bull., 71: 485–491.

Schmoker, J.W., 1984. Empirical relation between carbonate porosity and thermal maturity: an approach to regional porosity prediction. Am. Assoc. Pet. Geol., Bull., 68: 1697–1703.

Schmoker, J.W. and Gautier, D.L., 1988. Sandstone porosity as a function of thermal maturity. Geology, 16: 1007–1010.

Schmoker, J.W. and Gautier, D.L., 1989. Compaction of basin sediments: modeling based on time–temperature history. J. Geophys. Res., 94: 7379–7386.

Schmoker, J.W. and Hester, T.C., 1990. Regional trends of sandstone porosity versus vitrinite reflectance — a preliminary framework. In: V.F. Nuccio and C.E. Barker (Editors), Applications of Thermal Maturity Studies to Energy Exploration. RMS-SEPM, Denver, Colo., pp. 53–60.

Scholle, P.A., 1977. Chalk diagenesis and its relation to petroleum exploration: oil from chalks, a modern miracle? Am. Assoc. Pet. Geol., Bull., 61: 982–1009.

Scholle, P.A. and Halley, R.B., 1985. Burial diagenesis: out of sight, out of mind. In: N. Schneidermann and P.M. Harris

(Editors), Carbonate Cements. Soc. Econ. Paleontol. Mineral., Spec. Publ., 36: 309–334.

Scholle, P.A., Stemmerik, L. and Ulmer, D.S., 1991. Diagenetic history and hydrocarbon potential of Upper Permian carbonate buildups, Wegener Halvo area, Jameson Land Basin, East Greenland. Am. Assoc. Pet. Geol., Bull., 75: 701–725.

Sclater, J.G. and Christie, P.A.F., 1980. Continental stretching: explanation of the post-mid-Cretaceous subsidence of the central North Sea basin. J. Geophys. Res., 85: 3711–3739.

Shinn, E.A. and Robbin, D.M., 1983. Mechanical and chemical compaction in fine-grained shallow-water limestones. J. Sediment. Petrol., 53: 595–618.

Sibley, D.F. and Blatt, H., 1976. Intergranular pressure solution and cementation of the Tuscarora Orthoquartzite. J. Sediment. Petrol., 46: 881–896.

Siever, R., 1983. Burial history and diagenetic reaction kinetics. Am. Assoc. Pet. Geol., Bull., 67: 684–691.

Spiers, C.J., Schutjens, P.M.T.M., Brzesowsky, R.H., Peach, C.J., Liezenberg, J.L. and Zwart, H.J., 1990. Experimental determination of constitutive parameters governing creep of rocksalt by pressure solution. In: R.J. Knipe and E.H. Rutter (Editors), Deformation Mechanisms, Rheology and Tectonics. Geol. Soc. London, Spec. Publ., 54: 215–227.

Sullivan, K.B. and McBride, E.F., 1991. Diagenesis of sandstones at shale contacts and diagenetic heterogeneity, Frio Formation, Texas. Am. Assoc. Pet. Geol., Bull., 75: 121–138.

Surdam, R.C. and Crossey, L.J., 1987. Integrated diagenetic modeling: a process-oriented approach for clastic systems. Annu. Rev. Earth Planet. Sci., 15: 141–170.

Surdam, R.C., Crossey, L.J., Hagen, E.S. and Heasler, H.P., 1989. Organic–inorganic interactions and sandstone diagenesis. Am. Assoc. Pet. Geol., Bull., 73: 1–23.

Taylor, J.M., 1950. Pore-space reduction in sandstones. Am. Assoc. Pet. Geol., Bull., 34: 701–716.

Ungerer, P., Burrus, J., Doligez, B., Chenet, P.Y. and Bessis, F., 1990. Basin evaluation by integrated two-dimensional modeling of heat transfer, fluid flow, hydrocarbon generation, and migration. Am. Assoc. Pet. Geol., Bull., 74: 309–335.

Van de Kamp, P.C., 1976. Inorganic and organic metamorphism in siliciclastic rocks. Am. Assoc. Pet. Geol., Bull., 60: 729.

Walther, J.V. and Helgeson, H.C., 1977. Calculation of the thermodynamic properties of aqueous silica and the solubility of quartz and its polymorphs at high pressures and temperatures. Am. J. Sci., 277: 1315–1351.

Waples, D.W., 1984. Thermal models for oil generation. In: J. Brooks and D. Welte (Editors), Advances in Petroleum Geochemistry, Vol. 1. Academic Press, London, pp. 7–67.

Weyl, P.K., 1959. Pressure solution and the force of crystallization — a phenomenological theory. J. Geophys. Res., 64: 2001–2025.

Wilson, M.D. and Pittman, E.D., 1977. Authigenic clays in sandstones: recognition and influence on reservoir properties and paleoenvironmental analysis. J. Sediment. Petrol., 47: 3–31.

DOUGLAS W. WAPLES *Technology Research Center, Japan National Oil Corporation, 1-2-2 Hamada, Chiba 260, Japan*

HIROMI KAMATA *Technology Research Center, Japan National Oil Corporation, 1-2-2 Hamada, Chiba 260, Japan*
Present address: *Japan Oil Engineering Company Ltd., 6-2-1, Ginza, Chuo-ku, Tokyo 104, Japan*

Compaction coefficients of argillaceous sediments: their implications, significance and determination

X. Luo, F. Brigaud and G. Vasseur

Fine-grained sediment compaction, which is usually expressed by a relationship between burial depth and porosity (e.g. Athy's equation: $\phi = \phi_0 \, e^{-cz}$), exhibits high spatial variability. Even in a given sedimentary basin, for the same stratum in different districts or for different strata in the same district, the compaction coefficients in Athy's equation may change drastically.

The compaction coefficients are closely related to sediment fabric, which is largely affected, during the depositing process and during the early compaction stage, by the composition of sediments, depositional environments, sedimentation rate, properties of interlayer cations and electrolytes, existence of organic materials, and so on.

Such a variability has a major impact on the occurrence of geopressure. A series of computations performed with different compaction coefficients demonstrates that these coefficients affect very much the resulting pressure field. Therefore, even for the same lithology, the use of a single set of compaction coefficients for different parts and for different strata of a sedimentary basin may cause serious errors in the estimation of regional geopressure distribution, in the prediction of ground-water flow, and then finally, in the prediction of petroleum accumulation. To avoid these errors, compaction coefficients must be determined by using *in situ* data.

As it is difficult to measure directly the actual porosity in fine-grained sediments, indirect estimation methods based on well logging may be applied. A formula is proposed for computing porosity from sonic logging data which is calibrated by other data such as corrected density and neutron logging data.

Introduction

For argillaceous sediments, the principal diagenetic process is mechanical compaction, which results in a decrease of porosity with increasing overloading (Magara, 1978). Argillaceous sediments are often subject to overpressure, or high pore pressure, i.e. fluid pressure higher than the expected hydrostatic value. Since Dickinson's work on the American Gulf Coast, many mechanisms have been proposed to explain the occurrence of overpressure. However, most workers believe that compaction is the basic mechanism of this phenomenon (Dickinson, 1953; Magara, 1978; Chapman, 1980; Sharp, 1983; Shi and Wang, 1986; Bethke, 1986; Chen and Luo, 1988). Therefore, in most basin models, compaction is treated as the major factor of geopressure generation (Ungerer et al., 1984; Bethke, 1985).

The compaction is usually expressed by the relationship between porosity and depth (Athy, 1930; Falvey and Middleton, 1981; Ungerer et al., 1984; Buryakovskiy et al., 1991). The porosity–depth law proposed by Athy (1930) is mostly applied:

$$\phi = \phi_0 \, e^{-cz} \tag{1}$$

ϕ and ϕ_0 are porosity at burial depth z and at the depositing surface, respectively; c (in m^{-1}) is a constant.

In fact, this empirical relationship cannot represent the compaction phenomenon in areas where high pore pressure exists. Based on arguments coming from soil mechanics (Terzaghi, 1925; Hubbert and Rubey, 1959), the compaction of sediments is more appropriately described as a function of the effective stress by the following "compaction law":

$$\phi = \phi_0 \, e^{-b\sigma} \tag{2}$$

where σ is effective stress and b is another form of the compaction coefficient. This expression, frequently used in basin models is, in fact, a generalization of eq. 1. Assuming that the total density of the sediments may be treated as a constant, b and c are related as follows: $b = c/[(\rho_b - \rho)g]$ (Dickinson, 1953; Sclater and Christie, 1980), where ρ_b is the average bulk density of the rocks, and ρ the pore fluid density, g being the acceleration due to gravity.

Since a long time, it has been noted that the compaction coefficients c and ϕ_0 greatly vary in different basins (Foster and Whalen, 1966; Magara,

Basin Modelling: Advances and Applications edited by A.G. Doré et al. NPF Special Publication 3, pp. 321–332, Elsevier, Amsterdam.

1978; Bethke, 1986); nevertheless, in a given basin, they are usually assumed constant.

The compaction coefficients c and ϕ_0 in eq. 1 are controlled by the depositional environment of the sediments (Magara, 1978). Chen and Luo (1988) studied the compaction of fine-grained sediments in Chinese continental basins and pointed out that lithological characteristics have a large influence on the value of the compaction coefficients. Even in the same sedimentary basin, these two coefficients of the same fine-grained sediments seem to vary laterally. However, this problem does not seem to have been addressed.

These observations lead to the two main objectives of the present paper, which are: (i) to discuss the relationships between the compaction law and the sedimentary environment; and (ii) to assess quantitatively the effect of the compaction coefficient variation on the evolution of geopressure.

For shaly formations, bulk porosity laboratory measurements are seldom since they are difficult to measure. In such formations, geophysical well logs are particularly useful for predicting continuously the bulk porosity (Brigaud et al., 1992). Therefore, the final purpose of this paper is to present a method for estimating *in situ* porosity and compaction coefficients from sonic logs.

The geological significance of compaction coefficients

From a mechanical point of view, the reduction of sediment porosity during the burial process is the result of the interaction between the effective stress brought by overloading to the grains, and the resistance capacity of the grains. When the effective stress increases to such an extent that the grains cannot support it, the grains adjust their contacts through sliding, bending, fracturing and breaking, thereby augmenting their resistance; the grains get closer one to another, and compaction occurs. This compaction process is obviously governed by the sediment fabric (grain physical properties, mutual arrangement and contact characteristics) which should be reflected in the compaction law.

Fabric of argillaceous sediments

According to Rieke and Chilingarian (1974), the clay-microstructure arrangements may have three fundamental forms: *random structures*, *oriented structures*, *and flocculated structures*. In the former two cases, the clay grains are primary clay mineral particles and in the last case, the sediments are made of flocculated clay grains, each of which is a domain composed of primary clay particles. The flocculated structures are in fact a kind of random structure; within the aggregated grains the clay particles are oriented whereas for the bulk sediment, the grains are unoriented (Engelhardt and Gaida, 1963; Meade, 1964; Moon and Hurst, 1984).

When sediments possess an oriented structure, the lamellar clay minerals particles tend to arrange in the direction perpendicular to the compaction stress (Fig. 1a), and the contacts between particles tend to be of "face-to-face" type (Meade, 1964). The adjustment of the grains occurs principally through sliding and, eventually, through bending of particles (Figs. 1c and 1e). As a consequence, the resistance capability of the sediments is relatively low and, under small effective stress, the sediments become strongly compacted: the compaction coefficient c is then relatively large.

When the sediments possess a random structure, the film-like clay particles arrange in various directions and they tend to make contact in a "face-to-edge" way (Fig. 1b). The adjustment between grains under increasing effective stress occurs by sliding, bending and breaking of the clay particles (Figs. 1d and 1f). This behaviour greatly increases the resistance capability of the grains to support the effective stress, and the compaction process certainly occurs more slowly. Then the compaction coefficient c is smaller than in the previous case.

When the sediments possess a flocculated structure, the aggregated grains have a greater resistance capability than primary clay mineral particles (Engelhardt and Gaida, 1963; Moon and Hurst, 1984) and, under the same effective stress, these sediments retain more porosity than in the two other cases (Engelhardt and Gaida, 1963; Rieke and Chilingarian, 1974). When the effective stress continues to increase, the aggregated grains may be dispersed into primary clay particles (Engelhardt and Gaida, 1963). In this situation, the clay particles released from the flocculated grains tend to arrange in random structures. Moon and Hurst (1984) emphasized that the aggregates are not easily destroyed: they rather merge into one another under the effect of overloading.

Influence of environmental factors and definition of compaction stages

Sediment fabric is formed mainly during the deposition process and is greatly changed during the whole burial diagenesis process by many chemical and physical factors (Meade, 1964; Müller, 1967; Singer and Müller, 1983; Moon and Hurst, 1984).

During the deposition of sediment, sedimentation conditions that can affect the initial fabric include: composition of sediments, depositional envi-

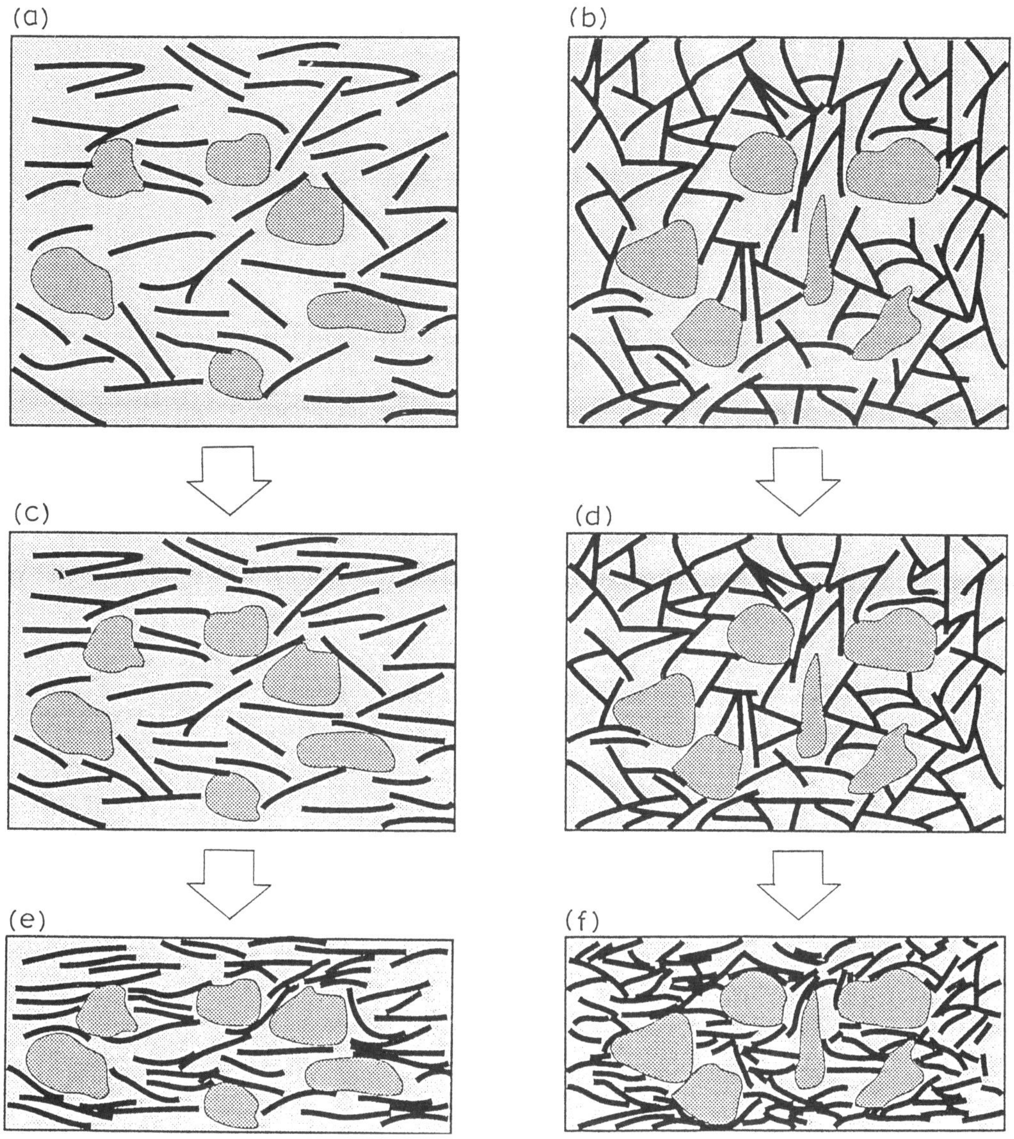

Fig. 1. The compaction behaviours of the particles in argillaceous sediments with different arrangements: (a), (c) and (e) show preferentially orientated arrangement; (b), (d) and (f) show random arrangement. Note that, in the latter situation, the clay particles may bend and break, but in the former situation these deformations seldom occur.

ronments (pH, Eh, salinity, etc.), depositing velocity, properties of exchangeable cations and electrolytes and existence of organic materials (Meade, 1964; Müller, 1967; Rieke and Chilingarian, 1974; Moon and Hurst, 1984). Generally, finer-size clayey compositions, fresher water deposition and slower depositing velocity tend to increase the orientation (Meade, 1964; Müller, 1967; Rieke and Chilingarian, 1974; Magara, 1978).

When fine-grained sediments settle on the water bottom, their porosity may reach 90%. But within a few metres, much of the initial water is lost and the porosity decreases to about 60–65% (Von Engelhardt, 1977; Magara, 1978). This early quick dehydration stage is in fact the continuation of the deposition process.

It is generally considered that, when their porosity reaches 60%, the fine-grained sediments' normal compaction stage begins. In this stage, the pore pressure is always hydrostatic and the sediment grains are gradually compressed together, mainly by the effective stress. Therefore, the corresponding decrease of porosity with burial depth is regular and may be expressed by Athy's porosity–depth relationship (eq. 1) in which the compaction coefficients c and ϕ_0 represent the sediment compaction characteristics.

In fact, in many areas, this statement is not true: in the upper part of this stage, Athy's law is usually not describing the actual situation and more complex equations should then be used to express the porosity–depth relationship (Dzevanshir et al., 1986; Buryakovskiy et al., 1991).

A number of laboratory experiments show that, besides effective stress, various environmental parameters can significantly influence the porosity-reducing process in this stage (Meade, 1964; Müller, 1967; Rieke and Chilingarian, 1974; Von Engelhardt, 1977; Moon and Hurst, 1984; Tessier, 1991). These parameters include: initial fabric, composition of sediments, properties of interlayer cations and electrolytes, pH values, Eh values, existence of organic materials, and so on. However these factors can only be efficient in a limited depth interval: when the effective stress becomes greater, their influence gradually decreases.

Meade (1964) suggested that, at the beginning of the early compaction stage, the space between the particles is large enough to generate a diffuse electric double layer composed of clay particles, exchangeable ions and water (Gouy-Chapman theory). The electric field generated in this double layer can effectively prevent the clay particles from getting closer to one another, and this kind of effect is easily influenced by various environmental conditions. As a result of compaction, the distance between the particles becomes smaller and smaller and the electric double layer gradually loses its repulsive effect. Meade (1964) proposed that this repulsive process ends when the overburden pressure reaches 50 bars, i.e. for a burial depth of about 250 m. However, it is clear that, in sediments with different fabrics, the limiting depth may be different.

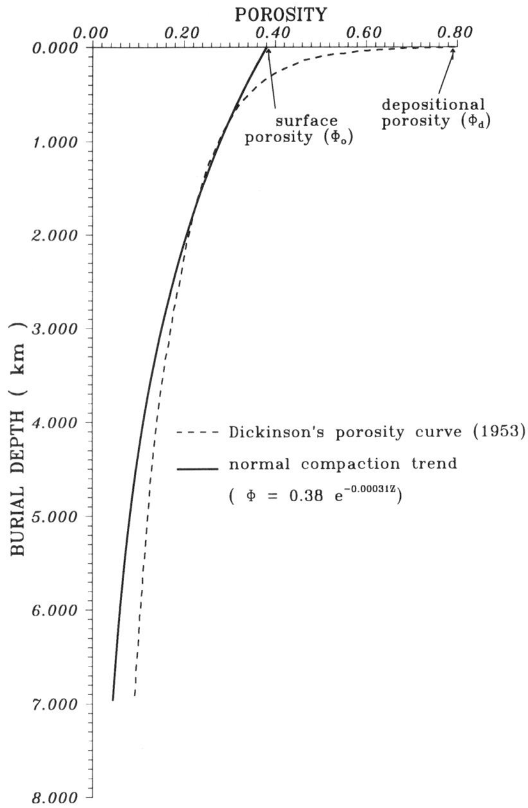

Fig. 2. The normal compaction trend and the actual porosity decrease with depth. Dickinson's curve is taken as an example. The compaction coefficients are determined within the normal compaction zone (depth range from 550 m to 2500 m). Note the difference between ϕ_0 and ϕ_d.

It seems likely that, below this depth, the effects of chemical parameters greatly decrease and even become too small to be considered so that mechanical effects become predominant. If such is the case, it is only in this lower part of the normal compaction stage that the porosity–depth relationship expressed by Athy's law would be better applied.

When applying Athy's law (eqs. 1 or 2) to this lower compaction stage, the compaction coefficient ϕ_0 (obtained by extrapolating $\phi(z)$ to the ground surface) should not be confused with the actual "depositional porosity" ϕ_d. As shown in Fig. 2, ϕ_0 usually falls into the range 0.25–0.50 (Bethke, 1986), much smaller than the actual depositional porosity ϕ_d which varies from 0.90 to 0.60 within a few metres burial (Von Engelhardt, 1977; Magara, 1978).

At greater depth, when abnormal overpressure appears in sediments, porosity no longer decreases following the previous porosity–depth relationship (eq. 1). In this new compaction stage, pore fluid cannot be easily squeezed out; part of the overburden load that was supported by grains, is now supported by pore fluid (Magara, 1978), i.e. the compaction process is greatly influenced by overpressure. However, it is generally admitted that the fabric type established during the depositional and early compaction stages remains intact and that the porosity–effective-stress relationship (eq. 2) can then be used to represent the compaction process (Magara, 1978; Sclater and Christie, 1980; Ungerer et al., 1984; Shi and Wang, 1986).

In summary, Athy's law in the form of eq. 1 or eq. 2 seems to be valid only below a certain depth ranging from a few tens to a few hundreds of metres. Both compaction coefficients c and ϕ_0 can be estimated from experimental $\phi(z)$. However, these coefficients should be obtained within the stable later compaction stage, where change in porosity is mainly a function of effective stress. The compaction characteristics in this stage not only reflect the sediment fabrics, but also determine the occurrence of overpressure in abnormal compaction stages. The fabrics in this stage are determined during the depositional stage, the early dehydration stage and the early compaction stage. Since during these stages environmental factors are very active, variation of the two compaction coefficients from area to area is inevitable.

The effects of compaction coefficients on geopressure

Hydrodynamic model

The hydrodynamics of compacting sedimentary basins have been the subject of many theoretical works (Sharp, 1983; Bethke, 1986; Shi and Wang, 1986). In the present work, the basic hydrodynamic equation (Appendix) used to compute quantitatively the evolution of geopressure is:

$$\left(\phi\beta + \frac{\alpha_\phi}{1-\phi}\right)\frac{\mathrm{d}P}{\mathrm{d}t} = \frac{1}{\rho}\nabla\left[\frac{\rho k}{\mu}\nabla(P - \rho g)\right] + \frac{\alpha_\phi}{1-\phi}\frac{\mathrm{d}S}{\mathrm{d}t} + \alpha\phi\frac{\mathrm{d}T}{\mathrm{d}t} + Q_w \qquad (3)$$

ϕ is the porosity; β is the isobaric coefficient of thermal expansion of pore fluid (Pa^{-1}); α_ϕ is the pore compressibility (Pa^{-1}) which is derived from Athy's law (eq. 2); P is the pore pressure (Pa); ρ is the pore fluid density ($kg\ m^{-3}$); k (D) is the intrinsic permeability of sediments assumed to vary as ϕ^5 (Jacquin and Poulet, 1973; Rieke and Chilingarian, 1974); μ is the dynamic viscosity of pore fluid ($kg\,m^{-1}\,s^{-1}$); g is the acceleration due to gravity; z is the depth; S is the total overloading (Pa); α is the isobaric coefficient of thermal expansion of pore fluid ($°C^{-1}$) and T is the temperature (°C); Q_w is the internal source rate of fluid volume input, which includes transformation of smectite to illite (Burst, 1969) and maturation of organic materials (Hedberg, 1974). In eq. 3, $\mathrm{d}P/\mathrm{d}t$ represents the pressure derivative following the solid movement ($\partial P/\partial t + V\nabla P$, V being the solid velocity).

The compaction coefficients play a fundamental role in the various terms of eq. 3: the porosity influences each term of the equation, as both the pore compressibility α_ϕ (Appendix) and the permeability are functions of the porosity (Jacquin and Poulet, 1973; Rieke and Chilingarian, 1974; Sharp, 1976; Walder and Nur, 1984; Ungerer et al., 1984; Shi and Wang 1986; Luo and Vasseur, 1992).

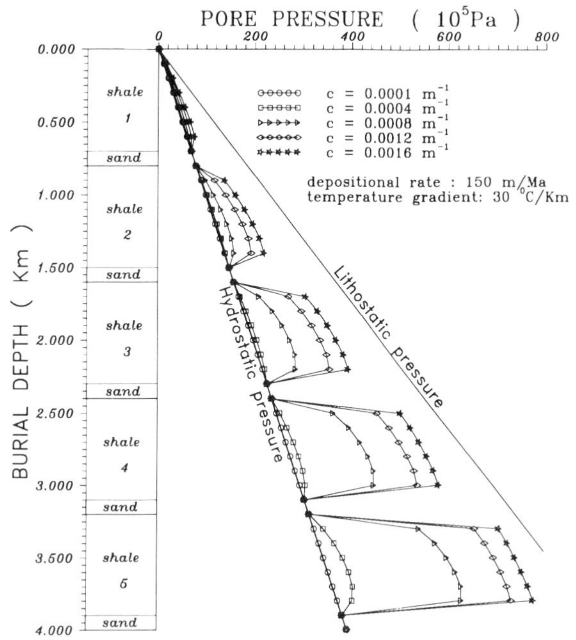

Fig. 3. Pressure profiles illustrating the influence of the compaction coefficient c on the pore pressure.

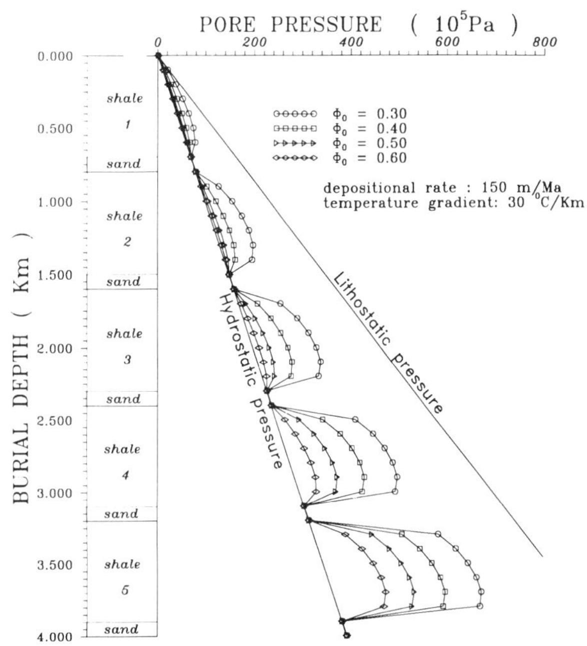

Fig. 4. Pressure profiles illustrating the influence of the compaction coefficient ϕ_0 on the pore pressure.

Geopressure occurrence on a 1D profile

Besides compaction, the occurrence of overpressure is also influenced by other geological conditions, such as: thickness of the argillaceous beds, deposition rate, temperature gradient and effects of chemical diagenesis (Magara, 1978; Bethke, 1986; Luo and Vasseur, 1992). In order to test the effect of the compaction coefficients c and ϕ_0 on overpressure occurrence, a conceptual 1D sediment profile of 4000 m thickness was used in which five fine-grained sediment beds, 700 m thick, are separated by sand beds 100 m thick (Figs. 3 and 4). The sandy beds were assumed to be connected with the ground surface, so that the pressures in these beds were always maintained at hydrostatic values. Moreover, if we assume that the various argillaceous beds have the same lithological properties, then the present distribution of geopressure in these beds also reflects the history of geopressure evolution of the deepest bed during the course of its burial.

According to Bethke (1986), the compaction coefficient c of clay sediments all over the world falls into a range from 0.00014 m^{-1} to 0.0016 m^{-1}. Values of the coefficient c of 0.0001, 0.0004, 0.0008, 0.0012, 0.0016 m^{-1}, respectively, were imposed in our calculations. The following values of the other related parameters were assumed: deposition rate, 150 m Ma^{-1}; temperature gradient, 30°C km^{-1}; surface porosity, 0.50.

Figure 3 illustrates the distribution of the geopressure as a function of depth; for a given c value, the overpressure increases with burial depth because the permeability decreases with decreasing porosity. The greater the value of c, the greater the excess pressure: when $c = 0.0001$ m^{-1}, the pressure value does not deviate from the hydrostatic gradient line; when $c = 0.0004$ m^{-1}, the pressure is slightly higher than hydrostatic only in the deepest bed; when $c = 0.0008$ m^{-1}, the excess pressure is much larger, and when $c = 0.0016$ m^{-1}, even in the most shallow bed, abnormal pressure appears.

From Fig. 3, one can also follow the evolution of geopressure of a given bed as a function of its burial depth. When $c = 0.0001$ m^{-1}, excess pressure never appeared. When $c = 0.0004$ m^{-1}, there has been only very little overpressure within shales 4 and 5; when $c = 0.0008$ m^{-1}, the overpressure evidently appears within shale 2 at 1600 m and continuously increases with burial depth; if c becomes greater than 0.0012 m^{-1}, the overpressure appears even in shale 1.

The effect of the other compaction coefficient, the surface porosity ϕ_0, on the occurrence of overpressure was also studied. In the calculations, this coefficient takes values of 0.30, 0.40, 0.50 and 0.60, respectively, with the c value of 0.0006 m^{-1}. The values of the other parameters are the same as in previous calculations. Figure 4 shows that the surface porosity ϕ_0 greatly influences the distribution and evolution of overpressure.

In fact, the compaction coefficients for argillaceous sediments may have different values in different beds. An example of such a situation is illustrated in Fig. 5. The values of compaction coefficient c in successive shaly beds are taken as 0.0005, 0.0005, 0.0010, 0.0008 and 0.0003 m^{-1}, respectively, the other parameters being the same as in Fig. 3. Figure 5 shows that the distribution of overpressure is completely different from that in Fig. 3; the overpressure of shale 3 is larger than that of shale 4 and much larger than that of shale 5, because the value of c for shale 3 is more important.

Geopressure occurrence on an 2D profile

In actual sedimentary basins, the lithological properties change not only vertically in different sedimentary beds, but also horizontally in the same bed. Therefore, lithological properties and in particular compaction coefficients may vary laterally (Chen and Luo, 1988; Faas, 1984). These lateral variations of compaction coefficients are expected to disturb the distribution of geopressure.

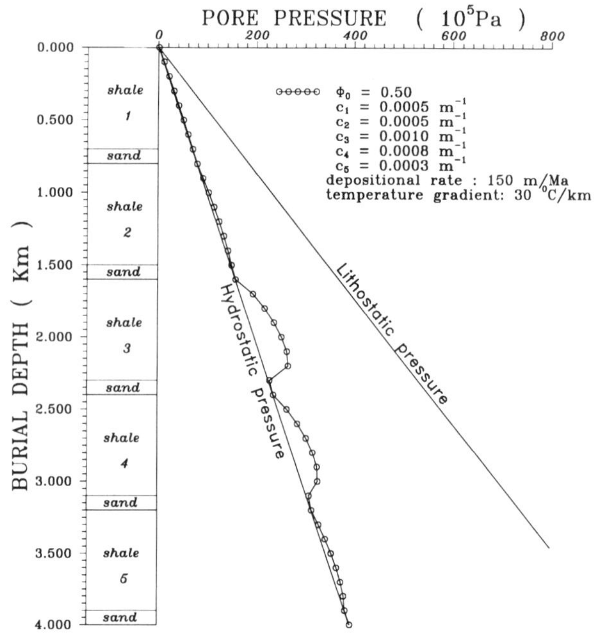

Fig. 5. Pressure profile with different compaction coefficient c in each shale bed.

This effect is illustrated using a synthetic cross-section in which the left side is a gentle slope, the boundary on the right being impermeable; at the centre of the section, there is a gentle dome. The deepest burial depth in this section is 4000 m and the vertical lithology distribution is similar to that of the 1D profiles (Figs. 3– 5). Within the ranges of the slope and the dome, the burial depths are less than 4000 m, but the number of beds does not change; only the thickness of each bed is reduced in proportion.

In the calculations, the compaction coefficient c is assumed to change horizontally whereas the other parameters are treated as constant: deposition rate, 100 m Ma^{-1}; temperature gradient, 30°C km^{-1}; surface porosity, 0.50, respectively.

First the value of c is taken as 0.0006 m^{-1} in all argillaceous sediments, and Fig. 6 illustrates the overpressure distribution obtained. In Fig. 6a, the distribution of the excess pressure (defined as the difference between the fluid pressure and the hydrostatic pressure) in the same argillaceous bed reflects the horizontal topography change. On the left of the profile, the excess pressure reduces gradually toward the top of the slope; within the dome, the excess pressure decreases toward the top.

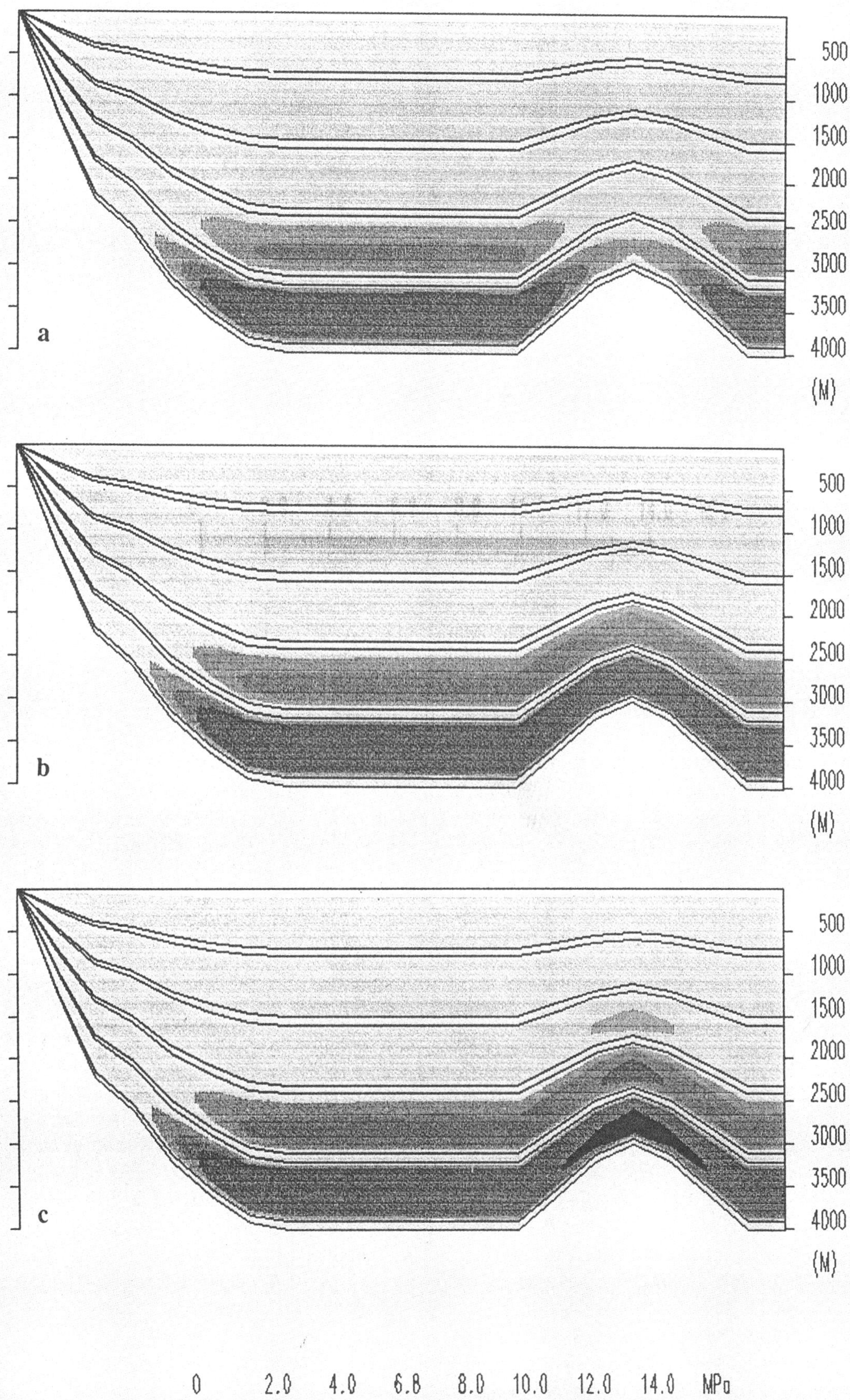

Fig. 6. Cross-sections of overpressure profiles illustrating the lateral influence of the compaction coefficient c on the pore pressure. (a) Value of c is constant laterally. (b) Value of c at the top of the dome is higher and decreases gradually toward both feet of the dome. (c) The value of c at the top of the dome is further increased compared to case (b).

Then the value of the compaction coefficient c in the argillaceous sediments located on the dome is changed: at the top of the dome, the value is assumed to be 0.0009 m^{-1}, and this value decreases gradually to 0.0006 m^{-1} toward both flanks of the dome. In this case, illustrated in Fig. 6b, although the value of the compaction coefficient c over the dome suffers little change, the distribution of the excess pressure over the dome is completely different from that of the previous case (Fig. 6a). In the deeper argillaceous beds, the excess pressure keeps the same values throughout the cross-section.

If the value of compaction coefficient c at the top of the dome is further increased to 0.0010 m^{-1}, the result within the dome is completely the opposite of the situation in Fig. 6a: the excess pressure at the top of the dome becomes greater than on the flanks (Fig. 6c). This means that, horizontally, the pore fluid flows in the opposite direction.

The comparison of Figs. 6a, 6b and 6c points out the large influence of the distribution of the compaction coefficient c on the distribution of abnormal pressure. This emphasizes the importance of knowing precisely the variation of the compaction coefficients. In practice, such variation of the compaction coefficient values in a sedimentary basin is common, especially in continental basins (Chen and Luo, 1988).

In the previous calculations, although only the effect of the compaction coefficient c on the evolution of the overpressure is investigated, the effects of the other compaction coefficient ϕ_0 may be reckoned by comparing these results with those of Figs. 3 and 4.

Estimating porosity from geophysical well logs

In practice, it is very difficult to obtain systematical porosity data from laboratory measurements (Magara, 1978). For obtaining *in situ* values of the compaction coefficients, indirect methods should be used.

For a long time, geophysical logs have been widely and successfully used in compaction studies (Hottman and Johnson, 1965; Ham, 1966; Fertl, 1976; Magara, 1978; Serra, 1985; Chen and Luo, 1988). In these works, the trend of each log vs. depth is used to define a normal characteristic of compaction; the appearance of an abnormal zone is determined empirically as a deviation from this trend. However, in order to estimate numerically the evolution of geopressure by solving the hydrodynamic equation (eq. 3), the porosity–depth relationship must be obtained.

In petroleum exploration basins, accurate porosity estimates, over the complete porosity range (0 to 70%), are usually provided by the Formation Density Compensated, Compensated Neutron logs and Borehole Compensated Sonic log. Therefore, *in situ* porosity can be deduced from the bulk density (ρ) read on the density log, from the hydrogen index (HI) read on the neutron log, and the travel time (Δt) read on the sonic log, using linear relationships (Wyllie et al., 1956, 1958; Serra, 1979; Raymer et al., 1980):

$$\phi = \frac{\rho - \rho_m}{\rho_f - \rho_m} \tag{4}$$

$$\phi = \phi_{HI} - HI_m V_m \tag{5}$$

$$\phi = \frac{\Delta t - \Delta t_m}{\Delta t_f - \Delta t_m} \tag{6}$$

where ρ_f and Δt_f are the fluid density and travel time; ρ_m, HI_m and Δt_m are, respectively, the density, hydrogen index and travel time of the formation matrix. Quantities ρ_m, HI_m and Δt_m are approximated assuming the following mixing laws:

$$\rho_m = \sum \rho_i V_i \tag{7}$$

$$HI_m = \sum HI_i V_i \tag{8}$$

$$\Delta t_m = \sum \Delta t_i V_i \tag{9}$$

where ρ_i, HI_i and Δt_i are the density, hydrogen index and travel time of the ith lithologic end-term of the formation (for instance, the three major lithological end-terms will be sandstone, carbonate and shale); V_i is the matrix volume content.

Usually, only the sonic transit-time log is recorded throughout the entire well, whereas density and neutron logs are run only in potential reservoirs. For this reason, transit-time log data have been most widely used in compaction research and in porosity calculations (Raymer et al., 1980; Serra, 1985; Brigaud et al., 1992). However, the accuracy of this log is only considered acceptable for low porosity (less than 25–30%; Serra, 1979). For high porosities (greater than 25–30%), the sonic porosity tends to overestimate the "true" formation porosity, and the calculated porosity results cannot be used unless they are calibrated (Magara, 1978; Raymer et al., 1980; Luo et al., 1987; Brigaud et al., 1992).

An empirical correction factor may be applied to eq. 4 to give the corrected sonic porosity $\phi_{sonic\,(corr)}$ (Serra, 1979; Brigaud et al., 1992):

$$\phi_{sonic\,(corr)} = \phi_{sonic} \frac{1}{C_p} \tag{10}$$

The correction factor C_p may be estimated by comparing the "raw" sonic porosity with a relevant porosity estimate (i.e. density or neutron logs):

$$C_p = \frac{\phi_{\text{sonic}}}{\phi_{\text{density}}} \quad \text{or} \quad \frac{\phi_{\text{sonic}}}{\phi_{\text{neutron}}} \tag{11}$$

Our data set includes twelve wells in the Norwegian Viking Graben (North Sea). It consists of density, neutron and sonic well log curves, lithological descriptions of the formations, and some mineralogical characterizations of these lithologies. The wells range in depth from sea bottom down to 4.6 km. The key factor in the data set is that it includes thick, uncompacted sedimentary intervals: (a) where the porosity covers a wide range (up to 70%); and

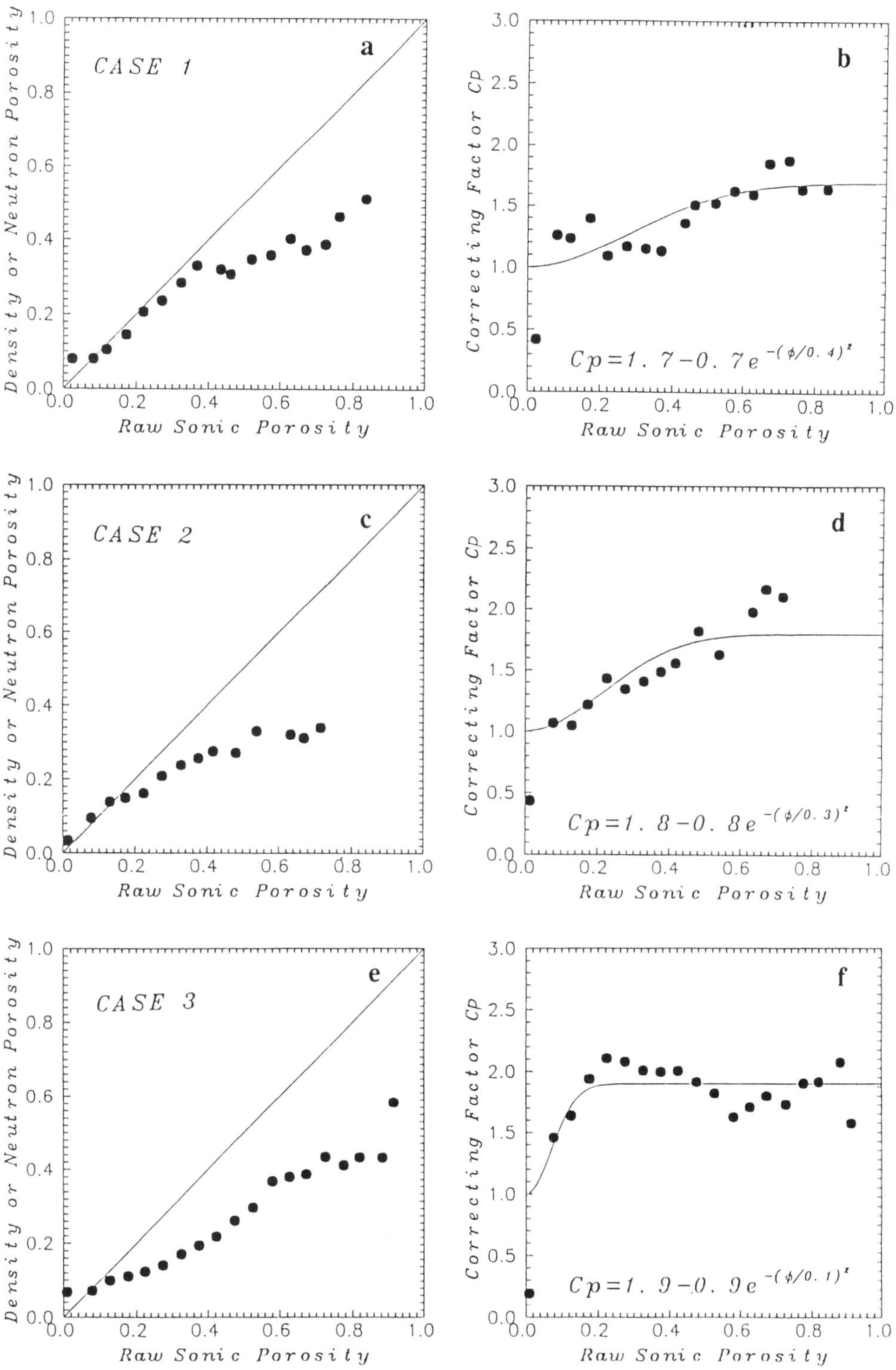

Fig. 7. (a), (c) and (e) Plots of the raw sonic porosity (abscissa) vs. the porosity deduced from compensated density or compensated neutron (ordinate). (b), (d) and (f) The correction coefficient C_p applicable to raw sonic porosity values. Dots correspond to data averaged over each 5% sonic porosity interval and curves to fitted data. *Case 1* = clay content $< 20\%$; *Case 2* = clay content $> 20\%$ and $< 50\%$; *Case 3* = clay content $> 50\%$.

(b) where density, neutron and sonic logs are all available, permitting calibrations for the C_p factor (Brigaud et al., 1992).

Three subsets were distinguished with regard to their shale content: (1) formations with a shale content lower or equal to 20%; (2) formations with a shale content between 20% and 50%; (3) formations with a shale content higher than 50%. Samples containing organic matter or volcanic tuffs were not included in the treatment.

Porosities were computed from eqs. 4–6; from there the correction factor C_p was deduced using eq. 8. Computations were averaged over each 5% porosity interval. Density or neutron porosities were plotted versus the "raw" sonic porosity for each subset in Figs. 7a, 7c and 7e; the corresponding C_p factors were plotted as a function of the "raw" sonic porosity in Figs. 7b, 7d and 7f. Experimental estimates of C_p were adjusted with an empirical fitting function that gives C_p as a function of the "raw" sonic porosity. The functions are displayed in Fig. 7. Results indicate that the bias (overestimate of porosity from the sonic) is not only a function of the porosity but also a function of the shale content: departure of the "raw" sonic porosity from other derived porosities (Fig. 7) occurs for lower porosity values and with a greater bias as the shale content increases. Thus the correction factors C_p become higher for greater shale contents.

Conclusion

From the above comparisons and analyses, the following conclusions can be derived:

(1) For estimating empirically values of the compaction coefficients c and ϕ_0 needed in pressure calculations, one should use the original porosity–depth data observed within the lower normal compaction stage.

(2) The porosity–depth relationship in this lower compaction stage is determined by the sediment fabric. The latter is established in the depositional process and is greatly influenced by various environmental factors during the early dehydration stage and the early compaction stage. All these factors vary very much in actual geological situations. Therefore, the compaction coefficients must certainly be different from place to place.

(3) Variations of compaction parameters greatly influence the occurrence of overpressure: vertically, a larger value of compaction coefficient c and a smaller value of compaction coefficient ϕ_0 result in an earlier abnormal pressure occurrence and in a larger rate of pressure increase with depth within the abnormal zone; horizontally, lateral compaction coefficient variations can significantly change the distribution of geopressure and therefore the underground lateral fluid flow.

(4) The proposed formula for predicting the "true" bulk porosity from a single sonic log, over the complete porosity range, may offer a powerful tool in geopressure model calculation. As the basic principles of the sonic log and seismic measurements are similar, the corrected sonic transit-time–porosity relationship may be applied to subsurface pressure prediction by using the seismic data. This can be very significant in petroleum exploration and drilling safety.

Acknowledgements

The authors thank Elf Petroleum Norge A/S for providing them with geophysical well logs, lithologic descriptions and X-ray diffraction analyses. They also thank J. Montésinos, J. Chery and B. Célérier for their help on image processing.

Appendix

For a porous medium, if the density of solid grains remains constant, and if some second-order effects such as tectonic stress are neglected (Shi and Wang, 1986), the conservation of the mass of fluid and solid phases with respect to fixed space coordinates may be expressed mathematically by:

$$\frac{\partial \rho\phi}{\partial t} + \nabla\,(\rho\phi \boldsymbol{u}_\mathrm{f}) = \rho Q_\mathrm{f} \tag{A1}$$

$$\frac{\partial \rho_\mathrm{s}(1-\phi)}{\partial t} + \nabla\,[\rho_\mathrm{s}(1-\phi)\,\boldsymbol{u}_\mathrm{s}] = \rho_\mathrm{s} Q_\mathrm{s} \tag{A2}$$

where $\boldsymbol{u}_\mathrm{f}$ and $\boldsymbol{u}_\mathrm{s}$ are velocities of fluid and solid grains with respect to fixed coordinates ($\mathrm{m\,s^{-1}}$), respectively; ρ and ρ_s are densities of fluid and solid grains ($\mathrm{kg\,m^{-3}}$), respectively; ϕ is the porosity; Q and Q_s are specific discharges of fluid and solid ($\mathrm{s^{-1}}$).

Eq. A1 may be written as:

$$\frac{\partial \phi\rho}{\partial t} + \nabla\,[\rho(\phi\boldsymbol{u}_\mathrm{f} - \phi\boldsymbol{u}_\mathrm{s})] + \nabla\,(\rho\phi\boldsymbol{u}_\mathrm{f}) = \rho Q_\mathrm{f} \tag{A3}$$

where $\phi\boldsymbol{u}_\mathrm{f} - \phi\boldsymbol{u}_\mathrm{s}$ is volume flow of fluid relative to solid matrix. According to Darcy's law, we have:

$$(\phi\boldsymbol{u}_\mathrm{f} - \phi\boldsymbol{u}_\mathrm{s}) = -\frac{k}{\mu}\,\overrightarrow{\mathrm{grad}}\,(P - \rho\boldsymbol{g}) \tag{A4}$$

where k is the intrinsic permeability ($\mathrm{m^2}$), μ is the fluid dynamic viscosity ($\mathrm{kg\,m^{-1}\,s^{-1}}$) and $\boldsymbol{g}$ is the acceleration due to gravity.

If we assume that ρ_s is constant (that is, the solid grains are incompressible), and that $Q_\mathrm{s} = 0$ (during the process of diagenesis, there is no solid material obtained or lost), eq. A2 becomes:

$$-\frac{\partial \phi}{\partial t} - \boldsymbol{u}_{\mathrm{s}} \overrightarrow{\text{grad}}\, \phi + (1-\phi) \operatorname{div} \boldsymbol{u}_{\mathrm{s}} = 0 \qquad \text{(A5)}$$

Applying the relationship $\mathrm{d}()/\mathrm{d}t = \partial()/\partial t + \boldsymbol{u}_{\mathrm{s}} \overrightarrow{\text{grad}}\, ()$, eq. A5 may be written as:

$$\operatorname{div} \boldsymbol{u}_{\mathrm{s}} = \frac{1}{1-\phi} \frac{\mathrm{d}\phi}{\mathrm{d}t} \qquad \text{(A6)}$$

Substituting eq. A6 into eq. A3 and arranging the result, we get:

$$\frac{1}{1-\phi}\frac{\mathrm{d}\phi}{\mathrm{d}t} + \frac{\phi}{\rho}\frac{\mathrm{d}\rho}{\mathrm{d}t} + \operatorname{div}[\rho(\phi \boldsymbol{u}_{\mathrm{f}} - \phi \boldsymbol{u}_{\mathrm{s}})] = Q_{\mathrm{f}} \qquad \text{(A7)}$$

For a compressible fluid the density is a function of pressure and temperature (Domenico and Palciauskas, 1979):

$$\frac{1}{\rho}\frac{\mathrm{d}\rho}{\mathrm{d}t} = \beta \frac{\mathrm{d}P}{\mathrm{d}t} - \alpha \frac{\mathrm{d}T}{\mathrm{d}t} \qquad \text{(A8)}$$

where P is the pore pressure, T the temperature, α the isobaric coefficient of thermal expansion for pore fluid ($°\mathrm{C}^{-1}$) and β its isobaric coefficient of compressibility ($\mathrm{m\,s^2\,kg^{-1}}$).

From eq. 1, assuming that sediment density changes very little (Dickinson, 1953), the porosity term in eq. A7 may be further expressed as a function of effective stress (Sclater and Christie, 1980):

$$\frac{\mathrm{d}\phi}{\mathrm{d}t} = \frac{\mathrm{d}(\phi_0\, \mathrm{e}^{-b\sigma})}{\mathrm{d}t} \qquad \text{(A9)}$$

where ϕ_0 is the surface porosity and $b = [c/(\rho_{\mathrm{b}} - \rho)]g$ (Pa^{-1}) the compaction coefficient. ρ_{b} is the average density of sediments and σ is the effective stress. Differentiating the right side, and substituting into eq. 2, we obtain:

$$\frac{\mathrm{d}\phi}{\mathrm{d}t} = -b\phi \frac{\mathrm{d}\sigma}{\mathrm{d}t} \qquad \text{(A10)}$$

According to $S = P + \sigma$ (Hubbert and Rubey, 1959), eq. A10 becomes:

$$\frac{\mathrm{d}\phi}{\mathrm{d}t} = \alpha_\phi \frac{\mathrm{d}S}{\mathrm{d}t} - \alpha_\phi \frac{\mathrm{d}P}{\mathrm{d}t} \qquad \text{(A11)}$$

where α_ϕ is the porosity compressibility; it is used here to replace $b\phi$ in eq. A10.

Substituting eqs. A3, A8, A11 into A7 and arranging it, we get our final hydraulic equation:

$$\left(\phi\beta + \frac{\alpha_\phi}{1-\phi}\right)\frac{\mathrm{d}P}{\mathrm{d}t} = \frac{1}{\rho}\nabla\left[\frac{k\rho}{\mu}\nabla(P - \rho \boldsymbol{g})\right] + \frac{\alpha_\phi}{1-\phi}\frac{\mathrm{d}S}{\mathrm{d}t} + \alpha_\phi \frac{\mathrm{d}T}{\mathrm{d}t} + Q_{\mathrm{f}} \qquad \text{(A12)}$$

In this equation, the first term on the right side is the "diffusion term"; it represents the capacity of fluid expulsion of sediments. The second term in the brackets on the left side and the second term on the right side represent the effect of compaction, and the third term on the right side represents the aquathermal effect. The last term represents the variation of fluid volume which tends to increase the fluid volume in pores (by transformation of clay minerals and by maturation of organic materials). We also assume that, in fine-grained sediments, there is no groundwater flow caused by topographic relief, i.e. during the process of diagenesis, water outside fine-grained beds cannot enter pores to replace pore fluid.

References

Athy, L.F., 1930. Density, porosity, and compaction of sedimentary rocks. Am. Assoc. Pet. Geol., Bull., 14: 1–21.

Bethke, C.M., 1985. A numerical model of compaction-driven groundwater flow and heat transfer and its application to the paleohydrology of intracratonic sedimentary basins. J. Geophys. Res., 90: 6817–6828.

Bethke, C.M., 1986. Inverse hydrologic analysis of the distribution and origin of Gulf Coast-type geopressured zones. J. Geophys. Res., 91: 6535–6545.

Bredley, J.S., 1973. Abnormal formation pressure. Am. Assoc. Pet. Geol., Bull., 59: 957–973.

Brigaud, F., Vasseur, G. and Caillet, G., 1992. Thermal state in the north Viking Graben (North Sea) determined from oil exploration well data. Geophysics, 57: 69–88.

Burst, J.F., 1969. Diagenisis of Gulf Coast sediments and its possible relation to petroleum migration. Am. Assoc. Pet. Geol., Bull., 53: 73–93.

Buryakovskiy, L.A., Dzevanshir, R.D. and Chilingarian, G.V., 1991. Mathematical simulation of sediment compaction. J. Pet. Sci. Eng., 5: 151–161.

Chapman, R.E., 1980. Mechanical versus thermal cause of abnormal high pore pressure in shales. Am. Assoc. Pet. Geol., Bull., 64: 2179–2183.

Chapman, R.E., 1982. Mechanical versus thermal cause of abnormal high pore pressure in shales: reply. Am. Assoc. Pet. Geol., Bull., 66: 101–102.

Chen, H.L. and Luo, X.R., 1988. The quantitative calculation of abnormal fluid pressure in argillaceous and arenaceous rocks and its geological application. Geol. Rev., 34: 54–63 (in Chinese).

Daines, S., 1982. Aquathermal pressuring and geopressure evaluation. Am. Assoc. Pet. Geol., Bull., 66: 931–939.

Dickey, P.A., 1975. Abnormal formation pressure: discussion. Am. Assoc. Pet. Geol., Bull., 60: 1124–1128.

Dickinson, G., 1953. Geological aspects of abnormal reservoir pressure in Gulf Coast Louisiana. Am. Assoc. Pet. Geol., Bull., 37: 410–432.

Domenico, P.A. and Palciauskas, V.V., 1979. Thermal expansion of fluids and fracture initiation in compacting sediments. Geol. Soc. Am., Bull., 90: 953–979.

Dzevanshir, R.D., Buryakovskiy, L.A. and Chilingarian, G.V., 1986. Simple quantitative evaluation of porosity of argillaceous sediments at various depths of burial. Sediment. Geol., 46: 169–175.

Engelhardt, W.V. and Gaida, K.H., 1963. Concentration changes of pore solutions during the compaction of clay sediments. J.

Sediment. Petrol., 33: 919–930.

Faas, R.W., 1984. Plasticity and compaction characteristics of the Quaternary sediments penetrated on the Guatemalan transect, DSDP Leg 67. In: D.A.V. Stow and D.J.W. Piper (Editors), Fine-Grained Sediments; Deepwater Processes and Facies. Geol. Soc. London, Spec. Publ., 15: 563–577.

Falvey, D.A. and Middleton, M.F., 1981. Passive continental margins: evidence for a pre-breakup deep crustal metamorphic subsidence mechanism. I.G.C. Colloq. C3, Oceanol. Acta, 4 (Suppl.): 103–114.

Fertl, W.H., 1976. Abnormal Formation Pressures — Implications to Exploration, Drilling, and Production of Oil and Gas Resources. Elsevier, Amsterdam, 382 pp.

Foster, J.B., and Whalen, H.E., 1966. Estimation of formation pressures from electrical surveys. J. Pet. Technol, 18: 165–171.

Gretener, P.E., 1981. Pore pressure: fundamentals, general ramifications and implications for structural geology. Am. Assoc. Pet. Geol., Cont. Educ. Course Note Ser., 4, 131 pp. (2nd rev. ed.).

Ham, H.H., 1966. New charts help estimate formation pressures. Oil Gas J., 64: 58–63.

Hedberg, H.D., 1974. Relation of methane generation to undercompacted shales, shale diapirs, and mud volcanoes. Am. Assoc. Pet. Geol., Bull., 58: 661–673.

Hottman, C.E. and Johnson, R.K., 1965. Estimation of formation pressures from log-derived shale properties. J. Pet. Technol., 16: 717–722.

Hubbert, M.K. and Rubey, W.W., 1959. Mechanics of fluid filled porous solids and its application to overthrust faulting, 1. Role of fluid pressure in mechanics of overthrust faulting. Geol. Soc. Am., Bull., 70: 115–166.

Jacquin, C. and Poulet, M., 1973. Essai de restitution des conditions hydrodynamiques régnant dans un bassin sédimentaire au cours de son évolution. Rev. Inst. Fr. Pét., 28: 269–297.

Lambe, T.W. and Whitman, R.V., 1969. Soil Mechanics. Wiley, New York, N.Y., 553 pp.

Luo, X.R., 1987. Subsurface fluid pressure. Nat. Mag., 10: 203–207 (in Chinese).

Luo, X. and Vasseur, G., 1992. Contributions of compaction and aquathermal pressuring to geopressure and the influence of environmental conditions. Am. Assoc. Pet. Geol., Bull., 70: 1550–1559.

Luo, X.R., Chen, H.L. and Wang, F.Q., 1987. The reconstruction of the burial development of sediment formations. 2nd National Petroleum Migration Conference, Xi'an (in Chinese).

Magara, K., 1978. Compaction and Fluid Migration — Practical Petroleum Geology. Elsevier, Amsterdam, 319 pp.

Meade, R.H., 1964. Removal of water and rearrangement of particles during the compaction of the clayey sediments. U.S. Geol. Surv., Prof. Pap., 450-B: B1–B23.

Meade, R.H., 1966. Factors influencing the early stages of the compaction of clays and sand review. J. Sediment. Petrol., 36: 1085–1101.

Mercer, J.W., Pinder, G.F. and Donalson, I.G., 1975. A Galerkin-finite Element Analysis of the hydrothermal system at Wairakei, New Zealand. J. Geophys. Res., 80: 2608–2621.

Moon, C.F. and Hurst, C.W., 1984. Fabrics of muds and shales: an overview. In: D.A.V. Stow and D.J.W. Piper (Editors), Fine-Grained Sediments; Deepwater Processes and Facies. Geol. Soc. London, Spec. Publ., 15: 579–593.

Müller, G., 1967. Diagenesis in argillaceous sediments. In: G. Larson and G.V. Chilingar (Editors), Diagenesis in Sediments. Developments in Sedimentology, Vol. 8. Elsevier, Amsterdam, pp. 127–177.

Plumley, W.J., 1980. Abnormal high fluid pressure: survey of some basic principles. Am. Assoc. Pet. Geol., Bull., 64: 414–430.

Raymer, L.L., Hunt, E.R. and Gardner, J.S., 1980. An improved sonic transit time-to-porosity transformation. Trans. SPWLA, Annual Logging Symposium, Pap. GG, pp. 1–13.

Rieke, H.H., III and Chilingarian, G.V., 1974. Compaction of Argillaceous Sediments. Developments in Sedimentology Vol. 16. Elsevier, Amsterdam, 424 pp.

Sclater, T.G. and Christie, A.F., 1980. Continental stretching: an explanation of the post-mid-Cretaceous subsidence of the Central North Sea Basin. J. Geophys. Res., 85: 3711–3739.

Serra, O., 1979. Diagraphies différées, bases de l'interprétation, 1. Acquisition des données diagraphiques. Bull. Cent. Rech. Explor. Prod. Mem., 1, Elf Aquitaine, 328 pp.

Serra, O., 1985. Diagraphies différées, bases de l'interprétation, 2. Interprétation des données diagraphiques. Bull. Cent. Rech. Explor. Prod. Mem., 7, Elf Aquitaine, 631 pp.

Sharp, J.M., Jr., 1976. Momentum and energy balance equations for compacting sediments. Math. Geol., 8: 305–322.

Shi, Y.L. and Wang, C.Y., 1986. Pore pressure generation in sedimentary basins: overloading versus aquathermal. J. Geophys. Res., 91: 2153–2162.

Singer, A. and Müller, G., 1983. Diagenesis in argillaceous sediments. In: G. Larsen and G.V. Chilingarian (Editors), Diagenesis in Sediments and Sedimentary Rocks, Vol. 2. Elsevier, Amsterdam, pp. 115–212.

Smith, J.E., 1973. The dynamics of shale compaction and evolution of pore-fluid pressures. Math. Geol., 3: 239–263.

Terzaghi, K., 1925. Principles in soil mechanics, III. Determination of the permeability of clay. Eng. News Rec., 95: 832–836.

Tessier, D., 1991. Behaviour and microstructure of clay minerals. In: M. de Boodt, M. Hayes and A. Herbillon (Editors), Soil Colloids and their Associations in Aggregates. Plenum, New York, N.Y., pp. 387–415.

Ungerer, P., Bessis, F., Chenet, P.Y., Durand, B., Nogaret, E., Chiarelli, A., Oudin, J.L. and Perrin, J.K., 1984. Geological and geochemical models in oil exploration: principles and practical examples. In: G. Demaison (Editor), Petroleum Geochemistry and Basin Evaluation. Am. Assoc. Pet. Geol., Mem., 35: 53–57.

Von Engelhardt, W., 1977. The origin of sediments and sedimentary rocks, III. In: W. von Engelhardt, H. Füchtbauer and G. Müller (Editors), Sedimentary Petrology. Wiley, New York, N.Y., 359 pp.

Walder, J. and Nur, A., 1984. Porosity reduction and crustal pore pressure development. J. Geophys. Res., 89: 11539–11548.

Wyllie, M.R.J., Gregory, A.R. and Gardner, G.H.F., 1956. Elastic wave velocities in heterogeneous and porous media. Geophysics, 21: 41–47.

Wyllie, M.R.J., Gregory, A.R. and Gardner, G.H.F., 1958. An experimental investigation of factors affecting elastic wave velocities in porous media. Geophysics, 23: 459–493.

Yau, Y.C., Peacor, D.R., Essene, E.J., Lee, J.H., Kuo, L.C. and Cosca, M.A., 1987. Hydrothermal treatment of smectite, illite, and basalt to 460°C: comparison of natural with hydrothermally formed clay minerals. Clays Clay Miner., 35: 241–250.

X. LUO — *Department of Geology, Northwestern University, Xi'an, China; and Centre Géologique et Géophysique, Université de Montpellier II, 34095 Montpellier Cedex, France*

F. BRIGAUD — *Elf Aquitaine Production, C.S.T.J.F., Avenue Larribau, 64018 Pau Cedex, France*

G. VASSEUR — *Centre Géologique et Géophysique, Université de Montpellier II, 34095 Montpellier Cedex, France*

Modelling overpressures by effective-stress/porosity relationships in low-permeability rocks: empirical artifice or physical reality?

F. Schneider, J. Burrus and S. Wolf

Since publication of the consolidation theory (Terzaghi, 1923), the consolidation of soils has been described by the effect of increasing frame pressure or effective stress. Later (Smith, 1971), the theory of the consolidation of soils was extended to the compaction of sedimentary rocks.

In the TEMISPACK model, a 2D control volume model, overpressuring is considered as the net effect between a driving term (increased stress due to loading), and a relaxation term (escape of overpressured fluids through the permeability network), and is calculated by coupling the equations of conservation for both the solid matrix and the water with the equation of flow (Darcy), and a mechanical equation (stress–strain relationship). A sensitivity analysis carried out shows that the most sensitive parameters are the sedimentation rate, the effective-stress/porosity relationships, and the specific surface area used in the Koseny-Carman formula which relates permeability to porosity. We explain how these parameters are calibrated in real case studies.

The stress–strain relationship is given for each lithofacies by an empirical effective-stress/porosity curve. This approach is physically valid under certain hypotheses but may be criticized because of the choice of the empirical function. This choice is partly constrained by the data available, which can be used to calibrate the function. Up to now, the effective stress has been computed according to the concept put forward by Terzaghi. Some considerations show that the effective stress must be computed as it is in Biot's theory, with the difference that Biot's coefficient cannot be constant during compaction. In conclusion, we argue that the effective-stress/porosity relation has a physical meaning when mass transfers (e.g. diagenesis) can be neglected. The definition of effective stress should be improved by the introduction of Biot's coefficient, which may at least be a function of porosity. Further work is needed to determine whether this approach remains valid when the solid matrix is modified by diagenesis.

Introduction

Geologic modelling can be used to reconstruct the geometry of basins through geologic time, by taking into account the effects of sedimentation, erosion, and compaction together with the history of temperature, overpressure and fluid flow (e.g. Welte and Yukler, 1981; Nakayama and Lerche, 1987; Bethke, 1989; Ungerer et al., 1990). Explorationists wish to address overpressure reconstruction in order to estimate the risks of drilling or the migration pathways governed by overpressure gradients.

Many such models do not address the mechanical aspect of compaction, which is physically and numerically more complex (e.g. Perrier and Quiblier, 1974; Watts and Ryan, 1976; Beaumont et al., 1982; De Bremaecker, 1983; Hutchinson, 1985; Lucazeau and LeDouaran, 1985; Novelli et al., 1987; Wygrala, 1987; Yalcin and Welte, 1987). In such models, palaeogeometries are reconstructed by using porosity/depth curves. This approach is valid only when sediments have been normally pressured during geologic time, since overpressuring and undercompaction are usually linked (Magara, 1978).

In a different class of models, geopressures are computed assuming that they are mainly governed by the mechanical compaction disequilibrium. Following pioneer work by Terzaghi (1923), several authors (e.g. Dickinson, 1953; Smith, 1971; Magara, 1978; Bredehoef et al., 1988) have attempted to describe compaction processes by a series of coupled equations: an effective-stress equation (relating effective stress to depth or to porosity), and a flow equation (Darcy's equation). The coupling is due to the fact that the deformation of the rock and its permeability are both very closely related to the porosity. The nonlinear nature of the relationships requires special numerical treatment. This approach requires the calibration of parameters such as those that define the effective-stress/porosity and the permeability/porosity relationships. Such calibrations are, in general, not directly possible in the laboratory. Indeed, rock mechanics specialists are accustomed to carrying out experiments on rocks with much higher

Basin Modelling: Advances and Applications edited by A.G. Doré et al. NPF Special Publication 3, pp. 333–341, Elsevier, Amsterdam.

permeabilities and higher deformation rates than geological conditions require. Therefore, geologists need an empirical calibration of these parameters against well data (e.g. overpressure, density). This empirical calibration procedure ignores the conceptual framework usually adopted in rock mechanics (e.g. elasticity, viscoplasticity).

The effects of mineral diagenesis and solid mass transport are not considered in this paper. Such an assumption is certainly not valid for carbonates (De Charpal and Purser, 1984). Furthermore, the thermal expansion of water (Burst, 1969; Barker, 1972; Bradley, 1975; Magara, 1978; Plumley, 1980; Sharp, 1983) and water release by clay mineral transformation (Power, 1967; Burst, 1969; Schmidt, 1973; Magara, 1975; Dutta, 1986; Hutcheon, 1986; Colten-Bradley, 1987) are not taken into account. Hydrocarbon generation can sometimes be associated with overpressure generation (Meissner, 1980; Momper, 1980; Law and Dickinson, 1985; Meloche, 1985; Spencer, 1987; Barker, 1988).

In this paper, we review the principles of the "mechanical" approach based on the effective-stress/porosity relationship, and discuss its empirical calibration. The numerical model used in the discussion is IFP's TEMISPACK basin model, described elsewhere (Doligez et al., 1986; Burrus and Audebert, 1990; Ungerer et al., 1990). In the first section, we briefly present equations and principles and we summarize some results from sensitivity analysis and real case studies. In the second section, we discuss whether the stress based compaction theory is empirical only in nature or can be tied to more general principles of rock mechanics.

Physical concepts, mathematical formulation, sensitivity analysis, and case studies

Concepts and equations

In the mechanical approach, porosity and related geopressures are computed taking the following equations into account:

Conservation laws:

Water:

$$\frac{\partial}{\partial t}(\phi\rho_w) + \text{div}\,(\phi\rho_w \boldsymbol{U}_w) = Q_w \qquad (1)$$

where ϕ is the porous medium porosity, ρ_w is the water density assumed to be constant $\boldsymbol{U}_w$ is the water velocity, and Q_w is the water source term.

Matrix:

$$\frac{\partial}{\partial t}[(1-\phi)\rho_s] + \text{div}\,[(1-\phi)\rho_s \boldsymbol{U}_s] = Q_s \qquad (2)$$

where ρ_s is the solid density assumed to be constant, $\boldsymbol{U}_s$ is the solid velocity, and Q_s is the solid source term.

Darcy's law:

$$\boldsymbol{W} = \phi(\boldsymbol{U}_w - \boldsymbol{U}_s) = -\frac{[K_{ij}]}{\mu_w}\,[\,\overrightarrow{\text{grad}}\,(P) - \rho_w \boldsymbol{g}] \qquad (3)$$

where $\boldsymbol{W}$ is the Darcy velocity, $[K_{ij}]$ is the permeability tensor, μ_w is water viscosity assumed to be a function of temperature and given by Bingham's formula (Latil, 1975), P is the water pressure, and $\boldsymbol{g}$ is gravity acceleration.

Permeability: To account for anisotropy, permeability is given as a 2D tensor:

$$[K_{ij}] = \begin{pmatrix} \Theta_s & 0 \\ 0 & \Theta_a \end{pmatrix} \cdot K(\phi) \qquad (4)$$

where Θ_s is the anisotropy coefficient of permeability parallel to the sedimentary layer and Θ_a is the anisotropy coefficient of permeability perpendicular to the sedimentary layer.

Permeability (K) is related to porosity and specific surface area (S_0) by the Koseny-Carman equation.

$$K(\phi) = \frac{0.2\,\phi^3}{S_0^2\,(1-\phi)^2} \qquad (5)$$

Mechanical aspects: The maximum principal stress is assumed to be vertical and equal to the overburden weight Sg. Following Terzaghi and Peck (1948) and Hubbert and Willis (1957), the effective stress σ^e is taken equal to:

$$\sigma^e = Sg - P \qquad (6)$$

According to Smith (1971), porosity, for each lithology, is related to effective stress through an empirical equation.

$$\phi = \phi(\sigma^e) \qquad (7)$$

Numerical aspects of model solving

The previous equations are not solved in the spatial coordinates system (Eulerian) but in the material coordinates system (Lagrangian). The equations of conservation are first integrated on the volume of each cell. Then the integrated equations are discretized using a finite difference method. The linear system is solved for the variation of porosity. The nonlinearities are processed by adjusting the time step.

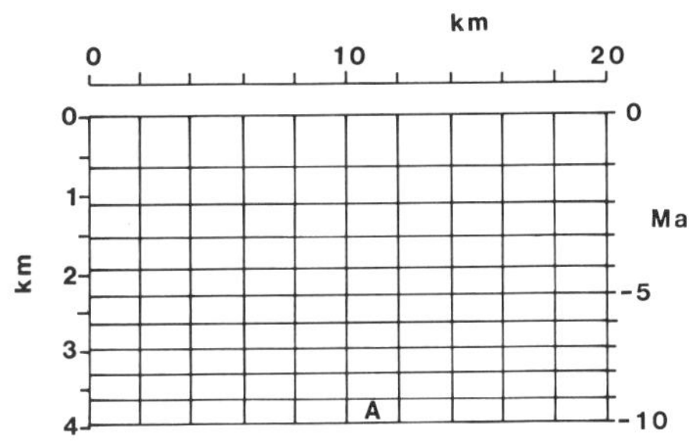

Fig. 1. Mesh used for the tests described in this paper for the sensitivity analysis. The grid is composed of 10 sedimentary layers subdivided into 10 columns. Each cell has the same solid thickness (300 m). The deposition time is generally the same for the 10 layers.

Sensitivity analysis of model response

A sensitivity analysis is carried out with the grid defined in Fig. 1, which represents a 10 km thick basin, with variable sedimentation rates and permeability. Results are shown in Figs. 2 and 3.

Figure 2 shows the overpressure profiles computed with three different porosity/effective-stress curves (lithology is assumed to be shale). It can be seen that the computed overpressures are influenced by the porosity/effective-stress given in the input. It is therefore important to ensure as good a calibration as possible of the porosity/effective-stress curves.

Besides effective-stress/porosity relationships, Fig. 3 illustrates the paramount effect of permeability (through the specific surface area) and of the sedimentation rate. The overpressures generated at the bottom of the sedimentary pile shown in Fig. 1 are plotted as functions of the log of the specific surface area and the log of the sedimentation rate. This figure shows that significant overpressuring (>5 MPa) develops if the sedimentation rate is above the following lithology-dependent threshold: (a) for shale-type permeabilities (10^{-9} to 10^{-7} D) there is no minimum, (b) for intermediate permeabilities (10^{-6} to 10^{-4} D) the minimum sedimentation rate is several 100 m/Ma, (c) for higher permeabilities, the minimum sedimentation rate would be reached only for unrealistic geologic sedimentation rates.

Figures 3B and 3C show the effect of changing the sedimentation rate from 100 m/Ma to 1000 m/Ma (keeping shale-type permeability constant: $S_0 = 10^8$ m^2/m^3). The net effect on overpressures at the bottom of the 10 km thick basin is an increase of overpressure from 5 to 15 MPa.

Calibration in case studies

For each case study, it is of great importance for the sensitive parameters to be effectively calibrated. The validity of a simulation is a function of the quantity and quality of the physical data used to constrain the inputs in the model. We suggest, based on experience gained from the case studies carried out at IFP during the last five years (Burrus and Audebert, 1990; Forbes et al., 1990; Burrus et al., 1991; Ungerer et al., 1990), that the agreement we found between observed and calculated overpressures is not the result of a mathematical artifice (too many free parameters), since independent checks on some

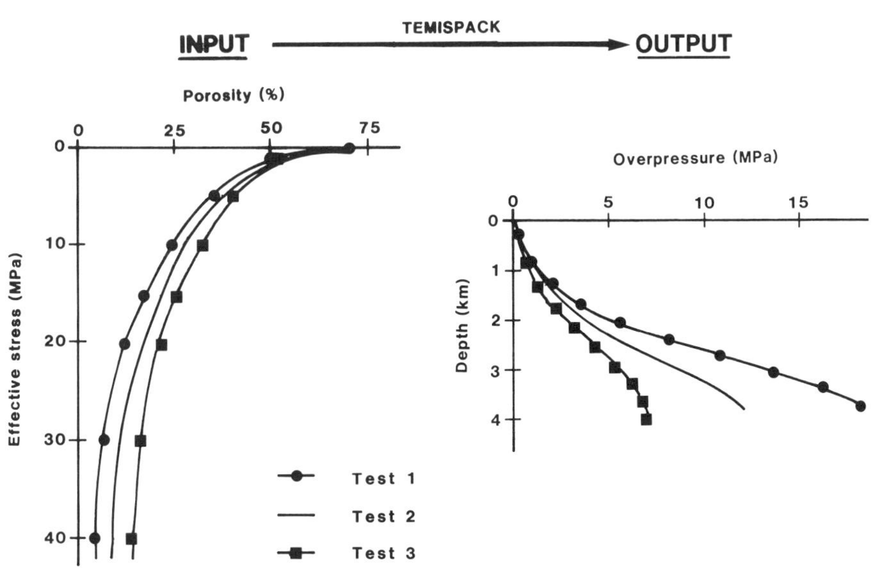

Fig. 2. Influence of effective-stress/porosity relationship variation on overpressure generation. These tests were carried out with the same sedimentation rate and with a single lithofacies with a high specific surface area (10^8 m^2/m^3). The test performed with the porosity/effective-stress represented by dots gives overpressures higher than the others. This is due to a greater decrease of porosity during compaction, which causes a greater decrease of permeability according to the Koseny-Carman formula. The same tests performed with a lower specific surface area (10^5 m^2/m^3) reveal no overpressure generation; water pressure is normal but final geometries are different.

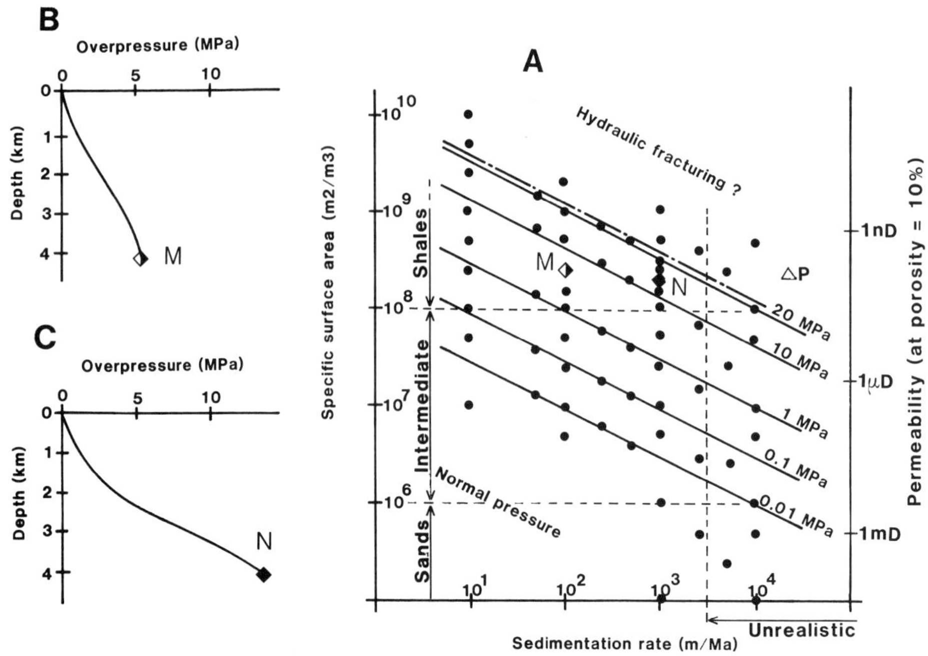

Fig. 3. Sensitivity analysis of the influence of the sedimentation rate and specific surface area on overpressure generation. On the left side, results are shown from two different runs. On the right side, results from several runs are synthesized; the dots correspond to overpressure values computed by the model in cell A (Fig. 1) at the present time.

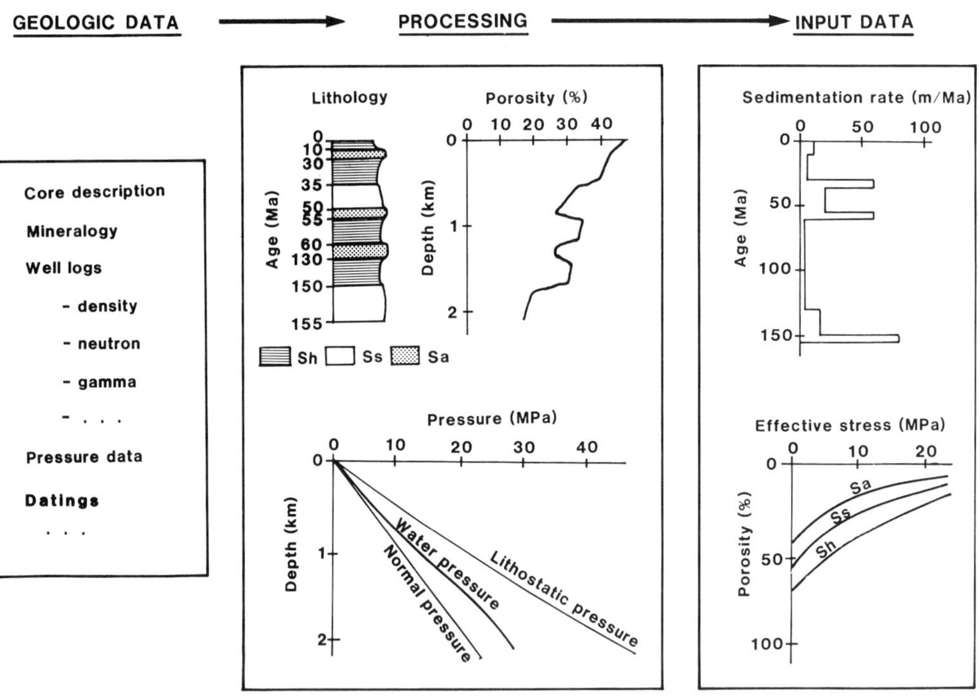

Fig. 4. Preprocessing of geologic data to transform them into input data for the model. The two main inputs here are the sedimentation rate and the porosity/effective-stress relationships.

inputs are available (e.g. permeabilities of both sands and shales, well logs, datings).

Sedimentation rate. Generally the sedimentation rate can be estimated accurately enough. It is derived from porosity/depth curves and from datings (see Fig. 4).

Porosity/effective-stress relationship. When there are enough data, the relationship between effective stress and porosity is calibrated, for each lithology, using in-situ measurements (density, porosity/depth curve and pressure/depth curve). Effective stress is computed using eq. 6 in which the overburden weight is derived from the porosity/depth curve and from rock densities (see Fig. 4).

Generally, pressure data are not available, hence porosity/depth relationships from normally pressured areas are used with density data. Water pressure is directly deduced from depth.

Specific surface area. Permeability is related to porosity and specific surface area by the Koseny-Carman formula. This concept has been widely used for sands and sandstones (Scheidegger, 1960; Guegen et al., 1987; Bourbié and Zinsner, 1985). Its application to shale can be criticized because the specific surface area has no real physical meaning. Nevertheless we have extended this concept to shale, and the specific surface area of shales is calibrated, as fitting parameters, to account for physical observations (observed overpressures). When it is possible, lab measurements have been used to constrain permeability.

Furthermore, it is well known that permeability and its anisotropy for a heterogeneous porous medium depend on the size of the volume considered. Thus, specific surface and anisotropy coefficients are, on a sedimentary basin scale, used as fitting parameters not only for impermeable lithofacies but also for permeable ones.

Example of case study with no calibration. In the Gulf of Lions study (Burrus and Audebert, 1990), the undercompaction and overpressuring of shaly sediments containing evaporites were simulated. The total thickness of the basin considered is 9 km, deposited with an average sedimentation rate of 500/1000 m/Ma. In this study, permeabilities and porosity/effective-stress laws were not measured or adjusted. All parameters (permeabilities, porosity/effective-stress) were derived from observations in the North Sea. Nevertheless, the profile of overpressure, recorded by mud density during drilling, appears to be in good agreement with the predicted theoretical overpressures, which reach 25 MPa (Fig. 5). Similarly, in the deep basin, the calculated undercompaction of shales located below the evaporites gives porosities of 15–20%. This amount seems to be consistent with the anomaly of the seismic velocity reported below the evaporites.

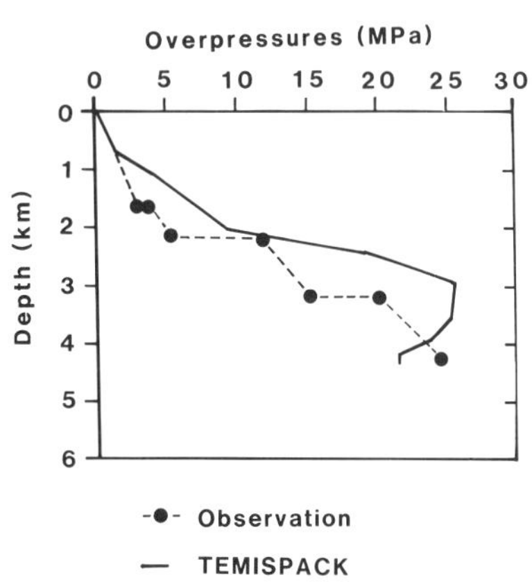

Fig. 5. Comparison between overpressures observed from mud density (dots) and overpressures calculated at the present time in the Gulf of Lions. The compaction parameters (porosity/effective-stress relationships, specific surface area and anisotropies of permeability) were not calibrated for this study but were taken from previous case studies. From Burrus and Audebert (1990).

Discussion

The sensitivity analysis highlights the importance of the sedimentation rate, porosity/effective-stress relationships, and specific surface area data input in controlling overpressures. The case studies illustrate how these inputs can be obtained: (a) the sedimentation rate can be derived from porosity/depth curves and datings, (b) the porosity/effective-stress relationships are ideally calibrated with in-situ measurements or taken from previous studies (i.e. closely studied zones: North Sea, Mahakam delta) and then extrapolated elsewhere, (c) specific surface area and permeability anisotropy need to be calibrated locally, against observed overpressures.

We discuss below whether the concept introduced to simulate compaction (porosity/effective-stress law) is acceptable from the point of view of a specialist in rock mechanics.

Equivalence between the porosity/effective-stress relationship and a rheological law

Athy (1930) was one of the first to consider that compaction of a sediment could be represented by a straightforward porosity/depth relationship. Subsequent experience has shown that similar lithofacies could have different porosities at the same depth. When the sediment is overpressured, porosity is higher than when the porous medium is normally pressured. Thus, the porosity/depth curve may not represent the compaction pathway of the sediment.

In order to account for the effect of pore pressure in the compaction of porous media, the first idea was to correct the porosity/depth relationship by taking into account the pressure–depth relationships.

To correct the effect of pressure it must be assumed that there is a relationship between porosity and pressure at a given reference depth, or a porosity/effective-stress relationship must be defined. Another way is to perform the corrections along "isoeffective stress" curves.

Nevertheless, the corrected curve is valid for use in modelling only if the sedimentary column is composed of a single lithology because the total stress, at a given depth, depends on the densities of the overlying sedimentary layers.

The first approach suggested, accounting for both the effect of pressure and the fact that there are generally more than one lithofacies in each column, was the concept of porosity/effective-stress relationship. When Smith (1971) made the hypothesis that for each lithology there is a relationship between porosity and effective stress, in fact he defined a global stress–strain relation.

If we assume that the solid volume (V_s) included in a volume (V) of the porous medium considered remains constant during compaction, the volumetric strain (ε) can be written:

$$\varepsilon = \frac{dV}{V} = \frac{d\phi}{1-\phi} \tag{8}$$

because:

$$V = \frac{V_s}{1-\phi} \text{ and } V_s = \text{constant} \tag{9}$$

Introducing eq. 7 in eq. 8 and assuming that $\phi(\sigma)^e$ is differentiable, we obtain:

$$\varepsilon = \frac{\phi'(\sigma^e)}{1-\phi(\sigma^e)} d\sigma^e \tag{10}$$

Hence, the porosity/effective-stress relationship can be considered as a nonlinear stress–strain relationship.

This rheology is valid in numerous case studies, but this representation cannot properly take into account the mechanical effect of erosion. Indeed, during erosion, the total stress decreases but the sediment does not regain its original porosity. In order to model this phenomenon, a different porosity/effective-stress relationship has to be defined for any erosional period. Porosity rebound is kept negligible during erosion. Practical experience shows that the porosity/effective-stress needs then to be defined to be numerically stable, rather than physically justified.

Definition of the porosity/effective-stress relationship

There is no theory to support any choice (such as elasticity, visco-elasticity) for the porosity/effective-stress function. Thus we use an empirical function that must have the following properties.

At deposition depth, the effective stress equals 0 and porosity equals the initial porosity ϕ_0:

$$\phi(0) = \phi_0 \tag{11}$$

Porosity decreases with the increasing effective-stress:

$$\phi'(\sigma^e) = \frac{d\phi}{d\sigma^e} < 0 \tag{12}$$

The limit of porosity when the effective-stress increases is 0.

One of the simplest mathematical functions (but not the only one), which satisfies the previous conditions, is:

$$\phi = \phi_0 \exp(-\sigma^e/\sigma_0^e) \tag{13}$$

where σ_0^e is a constant that characterizes the porous medium.

Calibration of the porosity/effective-stress relationship

Up to now, we have never used laboratory experiments to calibrate the porosity/effective-stress relationship because the laboratory time scale is too dissimilar to the geological time scale. Thus, the calibrations performed through experiments may not be usable for case studies.

The porosity/effective-stress relationship is calibrated with present-time log data as has been explained before. Doing this, we make the hypothesis that the mechanical behaviour of the sediments did not change during geologic time.

This calibration is quite easy for normally pressured sediments. We present in Fig. 6 and Table 1 some calibrations carried out for hydrostatic sands and shales from the North Sea and the Mahakam delta. However, if the same lithofacies is both normally pressured in some parts of the studied section and abnormally pressured in other parts, the porosity/effective-stress diagram seems not to be straightforward (see Fig. 7), suggesting the need for a refined effective-stress definition.

Definition of the effective stress

Terzaghi (1923) introduced the concept of effective stress to take into account the effect of pore pressure on the mechanical behaviour of porous media. For saturated sands he defined the effective stress as it is in eq. 6. Some years later, Biot (1941) proposed, in his general theory of three-dimensional consolidation, in the general framework of elasticity, to define the effective stress as follows:

TABLE 1

Some values of ϕ_0 and σ_0^e calibrated from a set of core data

Lithofacies	ϕ_0	σ_0^e (MPa)
Mahakam sands	0.50	37
Mahakam shales	0.45	26
North Sea sands (Brent)	0.43	27
North Sea shales (Jurassic)	0.58	18

Some values of ϕ_0 and σ_0^e calibrated from a set of core data.

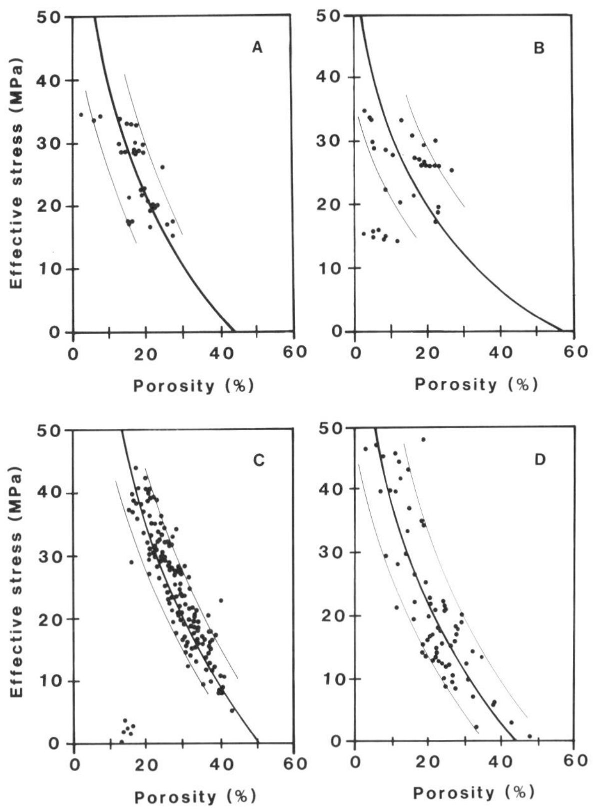

Fig. 6. Porosity/effective-stress relationships calibrated with present-time data from well logs: (A) sands from the North Sea (Brent); (B) shales from the North Sea (Jurassic); (C) sands from the Mahakam delta; (D) shale from the Mahakam delta.

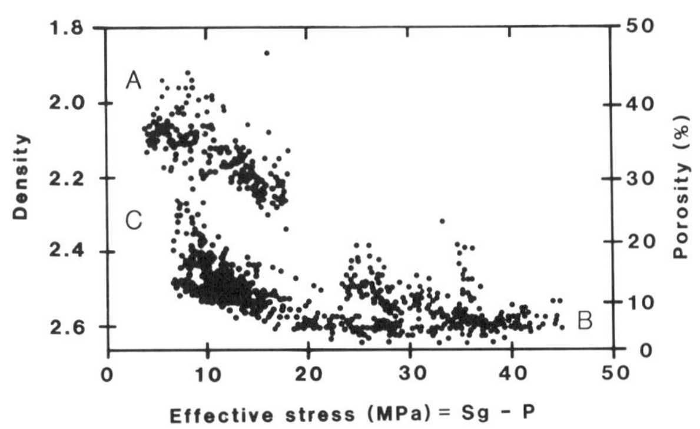

Fig. 7. Density (porosity) versus effective-stress measurements in shales from a real case study (Mahakam delta, from Burrus et al., 1992). Water pressure is normal for data for trend *A–B*. Data corresponding to trend *C–B* are measured in overpressured shales; trend *A–B* and trend *C–D* should be superposed according to Smith's theory.

$$\sigma^e = Sg - \alpha P \tag{14}$$

where α is Biot's coefficient.

If we consider that porosity characterizes the deformation state of the porous medium, α should be at least a function of porosity. Furthermore, independently of the physical property considered, $\alpha(\phi)$ should have the following properties.

If we assume that the effective stress is equal to 0 at the top of the sedimentary pile, we have:

$$\alpha(\phi_0) = 1 \tag{15}$$

This assumption is proved, by theory and experiments in soil mechanics, to be valid.

Research on phenomenological laws and laboratory calibrations of Biot's coefficient for different lithologies (Zimmerman et al., 1986; Laurent et al., 1990; Bouteca et al., 1991) show that α seems to decrease when porosity decreases.

Calibration of Biot's coefficient

In the example described in Fig. 7 (taken from Burrus et al., 1992), porosity data are plotted versus effective stress for the same deltaic shale. The effective stress is computed according to Terzaghi as the difference between the overburden weight and water pressure. Two trends are defined in this figure. The first one (*A–B*) is composed of data for which water pressure is normal hydrostatic; the second trend (*C–B*) represents data from overpressured shales. Geological and geodynamical considerations in this area allow us to accept the hypothesis that these shales were always under increasing loading. Thus, the gap between the two trends should not exist in Smith's theory (1971). Changing the definition of the effective stress by the introduction of the α coefficient, the two curves can however be translated in the diagram. In this example, for porosities between 10% and 35% the two curves overlap when $\alpha = 0.67$ (Fig. 8), suggesting the need to refine the definition of effective stress as proposed by Biot. The previous mean value of α, and eq. 15 may be used to calibrate an empirical or a phenomenological α versus porosity relationship.

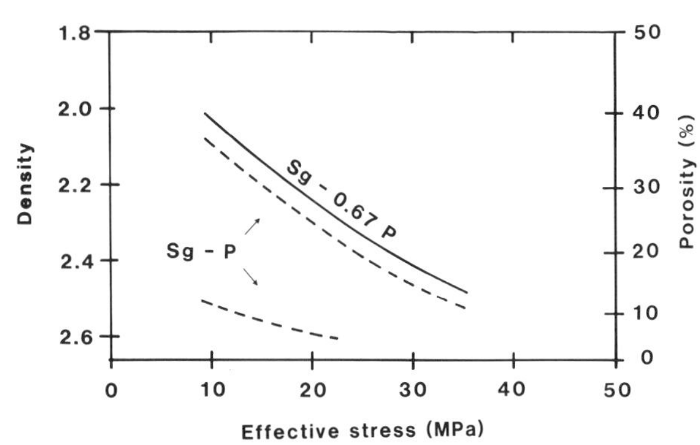

Fig. 8. From data shown Fig. 7, the introduction of Biot's coefficient $\alpha = 0.67$ in the computation of the effective stress leads to only one curve instead of two. The two dotted curves represent a fit in the initial data with $\alpha = 1$ (Smith's theory). These two curves overlap when $\alpha = 0.67$ (Biot's theory).

Conclusions

(1) The porosity/effective-stress relationship allows adequate modelling of overpressure generation in low-permeability rocks when mass transfer can be neglected. This relation seems to be quite general for a given lithofacies. Thus it may be used to provide a first-order estimate of overpressures (without any particular calibration) as shown in the Gulf of Lions example.

(2) This relationship is, with the hypothesis that the solid volume remains constant during time, a true nonlinear stress–strain relationship. For the moment, this relationship is represented by an empirical function that has some physical constraints on it.

(3) A better understanding of the phenomenology of compaction has to be found. A necessary step seems to be the definition of effective stress as it is in Biot's theory. Biot's coefficient may be calibrated in the laboratory or, as is done here, with in-situ measurements. Biot's coefficient can be constrained with some theoretical considerations. This coefficient seems to be, for a given lithofacies, a function of porosity.

(4) The description of rock compaction by concepts taken from soil mechanics is limited to deformation with conservation of the solid volume. Many workers claim that chemical processes (preferential dissolution at grain contact) play a significant role during compaction (e.g. Dewers and Ortoleva, 1990).

Acknowledgements

We thank M. Bouteca and J.P. Sarda (IFP) for useful discussions about the mechanics of rock deformation. We thank the Total Oil Company who provided us with the data shown in Fig. 7. We thank A.G. Doré (Conoco) and an anonymous referee for their great contribution in the writing of this paper.

References

Athy, L.F., 1930. Density, porosity and compaction of sedimentary rocks. Am. Assoc. Pet. Geol., Bull., 14: 1–24.

Barker, C., 1972. Aquathermal pressuring role of temperature on the development of abnormal pressure zones. Am. Assoc. Pet. Geol., Bull., 56: 2068–2071.

Barker, C., 1988. Generation of anomalous internal pressures in source rocks and its role in driving petroleum migration. Rev. Inst. Fr. Pét., 43: 349–355.

Beaumont, C., Keen, C.E. and Boutilier, R., 1982. On the evolution of rifted continental margins; comparison of models and observation for the Nova Scotia margin. Geophys. J. R. Astron. Soc., 70: 667–715.

Bethke, C.M., 1989. Modelling subsurface flow in sedimentary basins. Geol. Rundsch., 78 (1): 129–154.

Biot, M.A., 1941. General theory of three-dimensional consolidation. J. Appl. Phys., 12: 155–164.

Bourbié, T. and Zinsner, B., 1985. Hydraulic and acoustic properties as a function of porosity in Fontainebleau sandstone. J. Geophys. Res., 90 (B13): 11524–11532.

Bouteca, M.J., Sarda, J.P. and Laurent, J., 1991. Rock mechanics contribution to the determination of fluid flow properties. Second European Core Analysis Symposium (Eurocas II), London, 20–22 May, 1991.

Bradley, J., 1975. Abnormal formation pressure. Am. Assoc. Pet. Geol., Bull., 59: 957–973.

Bredehoef, J.B., Djevanshir, R.D. and Belitz, K.R., 1988. Lateral fluid flow in a compacting sand–shale sequence: South Caspian Basin. Am. Assoc. Pet. Geol., Bull., 72 (4): 416–424.

Burrus, J. and Audebert, F., 1990. Thermal and compaction processes in a young rifted basin containing evaporites: Gulf of Lions, France. Am. Assoc. Pet. Geol., Bull., 74 (9): 1420–1440.

Burrus, J., Kuhfuss, A., Doligez, B. and Ungerer, P., 1991. Are numerical models useful in reconstructing the migration of hydrocarbons? A discussion based on the Northern Viking Graben. In: W.A. England and A.J. Fleet (Editors), Petroleum Migration. Geol. Soc. London, Spec. Publ., pp. 89–109.

Burrus, J., Brosse, E., Choppin de Janvry, G. and Grosjean, Y., 1992. Regional basin modelling in the Mahakam delta. Am. Assoc. Pet. Geol., Bull. (submitted).

Burst, J.F., 1969. Diagenesis of Gulf Coast clay sediments and its possible relation to petroleum migration. Am. Assoc. Pet. Geol., Bull., 53: 73–93.

Colten-Bradley, V.A., 1987. Role of pressure in smectite dehydration: effects on geopressure and smectite-to-illite transformation. Am. Assoc. Pet. Geol., Bull., 71: 1414–1427.

De Bremaecker, J.C., 1983. Temperature, subsidence and HC maturation in extensional basins: a finite element model. Am. Assoc. Pet. Geol., Bull., 67: 1410–1414.

De Charpal, O. and Purser, P., 1984. L'influence des contraintes tectoniques et lithostatiques sur la compaction et la diagénèse des calcaires. Inst. Fr. Pét., Internal Rep. 31732.

Dewers, T. and Ortoleva, P., 1990. Pressure solution during diagenesis: a unified model for intergranular pressure solution, stylolites, and differential compaction/cementation. In: I. Meshri and P. Ortoleva (Editors), The Prediction of Reservoir Quality through Geochemical Modelling. Am. Assoc. Pet. Geol., Mem., 49: 147–160.

Dickinson, G., 1953. Geological aspects of abnormal reservoir pressures in Gulf Coast, Louisiana. Am. Assoc. Pet. Geol., Bull., 37: 410–432.

Doligez, B., Bessis, F., Burrus, J., Ungerer, P. and Chénet, P.Y., 1986. Integrated numerical modelling of the sedimentation heat transfer and fluid migration in a sedimentary basin: the Themis model. In: J. Burrus (Editor), Thermal Modelling of Sedimentary Basins. Technip, Paris, pp. 173–195.

Dutta, N., 1986. Shale compaction, burial diagenesis and geopressures: a dynamic model, solution and some results. In: J. Burrus (Editor), Thermal Modelling in Sedimentary Basins. Technip, Paris, pp. 149–171.

Forbes, P., Ungerer, P. and Mudford, B.S., 1990. A 2D model of overpressure development and gas accumulation in the Venture Field (Canada). Am. Assoc. Pet. Geol. (submitted).

Guegen, Y., Darot, M. and Reuschle, T., 1987. Permeability evolution under stress. In: B. Doligez (Editor) Migration of Hydrocarbon in Sedimentary Basins. Technip, Paris, pp. 281–296.

Hubbert, K. and Willis, D.G., 1957. Mechanics of hydraulic fracturing. Soc. Pet. Eng. AIME, Trans., 210: 153–163.

Hutcheon, I., 1986. The relationship of diagenetic mineral reactions to an overpressured zone on the Scotian Shelf. In: Basins of Eastern Canada and Worldwide Analogues. Programme with Abstracts, Atlantic Geoscience Society, August 1986, Halifax, N.S., p. 56.

Hutchinson, I., 1985. The effects of sedimentation and compaction on oceanic heat flow. Geophys. J. R. Astron. Soc., 82: 439–459.

Latil, M., 1975. Récupération assistée. Technip, Paris, 273 pp.

Laurent, J., Bouteca, M. and Sarda, J.P., 1990. Pore pressure influence in the poroelastic behaviour of rocks: experimental studies and results. Europec 90, Soc. Pet. Eng., Pap.. SPE 20922, pp. 385–391.

Law, B. and Dickinson, W., 1985. Conceptual model for origin of abnormally pressured gas and accumulations in low-permeability reservoirs. Am. Assoc. Pet. Geol., Bull., 69: 1295–1304.

Lucazeau, F. and LeDouaran, S., 1985. The blanketing effect of sediments in basins formed by extension. Earth Planet. Sci. Lett., 74: 92–102.

Magara, K., 1975. Reevaluation of montmorillonite dehydration as a cause of abnormal pressure and hydrocarbon migration. Am. Assoc. Pet. Geol., Bull., 59: 292–302.

Magara, K., 1978. Compaction and Fluid Migration — Practical Petroleum Geology. Developments in Petroleum Science, Vol. 9. Elsevier, Amsterdam, 319 pp.

Marsily, G. de, 1981. Hydrogéologie quantitative. Masson, Paris, 215 pp.

Meissner, F., 1978. Petroleum geology of the Bakken formation, Williston basin, North Dakota and Montana. In: G. Demaison (Editor), Petroleum and Basin Evolution. Am. Assoc. Pet. Geol., Mem., 35: 159–180.

Meissner, F., 1980. Examples of abnormal fluid pressure produced by hydrocarbon generation. Am. Assoc. Pet. Geol., Bull., 64: 749 (abstract).

Meloche, J., 1985. Diagenetic evolution of overpressured sandstones, Scotian Shelf. Paper presented at Geological Association of Canada and Mineralogical Association of Canada Joint Annual Meeting, Fredericton, N.B., May 15–17.

Momper, J.A., 1980. Generation of abnormal pressures through organic matter transformations. Am. Assoc. Pet. Geol., Bull., 64: 753 (abstract).

Nakayama, K. and Lerche, I., 1987. Basin analysis by model simulation: effects of geological parameters on 1D and 2D fluid flow systems with application to an oil field. Gulf Coast Assoc. Geol. Soc., Publ., 37: 175–184.

Novelli, L., Chiaramonte, M., Mattavelli, L., Pizzi, G., Sartori, L. and Scotti, P., 1987. Oil habitat in the NW Po Basin. In: B. Doligez (Editor), Migration of HC in Sedimentary Basins. Technip, Paris, pp. 27–57.

Perrier, R. and Quiblier, J., 1974. Thickness changes in sedimentary layers during compaction history. Am. Assoc. Pet. Geol., Bull., 58: 507–520.

Plumley, W., 1980. Abnormally high fluid pressure: survey of some basic principles. Am. Assoc. Pet. Geol., Bull., 64: 414–430.

Power, M.C., 1967. Fluid release mechanisms in compacting marine mudrocks and their importance in oil exploration. Am. Assoc. Pet. Geol., Bull., 51: 1240–1254.

Scheidegger, A.E., 1960. The Physics of Flow through Porous Media. Toronto University Press, Toronto, Ont., 420 pp.

Schmidt, G., 1973. Interstitial water composition and geochemistry of deep Gulf Coast shales and sandstones. Am. Assoc. Pet. Geol., Bull., 57: 321–337.

Sharp, J., 1983. Permeability controls on aquathermal pressuring. Am. Assoc. Pet. Geol., Bull., 67: 2057–2061.

Smith, J.E., 1971. The dynamic of shale compaction and evolution of fluid pressure. Math. Geol., 3 (3): 239–269.

Spencer, C., 1987. Hydrocarbon generation as a mechanism for overpressuring in Rocky Mountain region. Am. Assoc. Pet. Geol., Bull., 71: 368–388.

Terzaghi, K., 1923. Die Berechnung der Durchlässigkeitsziffer des Tones aus dem Verlauf der hydrodynamischen Spannungserscheinungen. Sitzungsgber. Akad. Wiss. Wien, Math. Naturwiss. Kl., IIa, 132 (3/4): 125–138.

Terzaghi, K. and Peck, R.B., 1948. Soil Mechanics in Engineering Practice. Wiley, New York, N.Y., 566 pp.

Ungerer, P., Burrus, J., Doligez, B., Chénet, P.Y. and Bessis, F., 1990. Basin evaluation by integrated 2D modelling of heat transfer, fluid flow, HC generation and migration. Am. Assoc. Pet. Geol., Bull., 74 (3): 309–335

Watts, A. and Ryan, W.B.F., 1976. Flexure of the lithosphere and continental margin basins. Tectonophysics, 36: 25–44.

Welte, D.H. and Yukler, M.A., 1981. Petroleum origin and accumulation in basin evolution, a quantitative model. Am. Assoc. Pet. Geol., Bull., 65: 1387–1396.

Wygrala, B.P., 1987. Integrated computer-aided basin modelling applied to analysis of hydrocarbon history in a Northern Italian oil field. Adv. Org. Chem., 13 (1/3): 187–197.

Yalcin, M.N. and Welte, D.H., 1987. 3D computer aided modelling of Cambay basin, India. A case history of hydrocarbon generation. In: S.P. Kumar (Editor), Petroleum Geochemistry in the Afro-Asian Region. Balkema, Rotterdam, pp. 417–450.

Zimmerman, R.W., Somerton, W.H. and King, M.S., 1986. Compressibility of porous rocks. J. Geophys. Res., 91 (B12): 12765–12777.

F. SCHNEIDER *Institut Français du Pétrole, B.P. 311, 92506 Rueil-Malmaison, France*
J. BURRUS *Institut Français du Pétrole, B.P. 311, 92506 Rueil-Malmaison, France*
S. WOLF *Institut Français du Pétrole, B.P. 311, 92506 Rueil-Malmaison, France*

Quantitative approach to geochemical processes in the Dogger aquifer (Paris Basin): from water analyses to the calculation of porosity evolution

Philippe Gouze and Anne Coudrain-Ribstein

Description of geochemical processes as a result of fluid flow is possible if *in situ* thermodynamical equilibrium relations between phases are determined. In the central part of the Paris Basin's Dogger aquifer, twenty borehole fluid samples are available and provide Al, Ca, Cl, F, HCO_3, K, Mg, Na, Si, SO_4, HS, B content and pH for temperatures ranging from 48° to 81°C. The relations between fluid composition, solid media and gaseous phase were investigated with the help of the speciation model CHIMERE.

Results emphasize that, whatever the chlorinity and the temperature of the sampled solution, thermodynamical equilibrium is achieved for eight minerals commonly found in low-enthalpy geothermal environments (kaolinite, fluorite, chalcedony, K-feldspar, anhydrite, calcite, Na-feldspar and disordered dolomite).

Therefore, the rate of fluid–mineral mass transfers is controlled by transport phenomena which are weak compared to fluid–mineral kinetic rate. Computed partial pressure of CO_2 ranges between $10^{-0.75}$ and $10^{-0.95}$ and satisfies an empirical relation with the temperature fitted to $\log P_{CO_2} = -1.23 + 0.006T$. Several fluid–clay thermodynamical equilibria have been investigated but none are fully satisfactory in order to explain the $P_{CO_2}-T$ trend throughout the whole studied zone.

Nevertheless, P_{CO_2}–temperature relationship, equilibrium with the eight minerals, chlorinity content and neutrality condition sufficiently constrain the model to simulate transfer of matter resulting from regional transport along a synthetic cross-section of the Dogger aquifer, where recent work confirms the present existence of regional fluid flow. Due to the geometry of the basin, fluid flow crosses isotherms and isochlores, exchanging mass with the rock. Computation allows the determination of a zone with little transferred quantities and zones characterised by important amounts of dissolution and precipitation.

If fluid flow remains at steady state during 0.1 million years, rock porosity increases by a maximum of 1.5% in the eastern zone whereas less than 1.0% of the total volume is cemented by authigenic minerals in the zone where fluid is cooling while uprising.

Introduction

When fluid flows through sedimentary layers, mass transfer between fluid and solid phases may become important. Thus, the description of the spatial and temporal evolution of dissolution and precipitation is necessary to evaluate the local redistribution of porosity.

Sedimentary aquifers are strongly heterogeneous in terms of mineralogy and hydrodynamic parameters because of the depositional, burial and diagenetic history. In such media, the notion of Representative Elementary Volume (see, de Marsily, 1986) is useful to describe the main hydraulic system, but the sedimentary description of the medium requires detailed laboratory analyses of a large number of cores. Fluid samples are often the first analyses available to study present diagenesis. When flowing, the fluid is in contact with the mineral surfaces, and the ionic content of the solution is representative of the rock mineral assemblage. In this case, a finite number of local equilibria between minerals (or gases) and fluid is sufficient to characterise the present interaction between the phases. The local fluid–mineral relationships can be computed from composition given by analyses. If a set of equilibrium relations controls the fluid composition over a large zone, the coupling of transport and reaction models makes it easy to predict the amount of mass transferred between phases as a function of time and space.

In this paper, the redistribution of porosity is studied along a cross-section of the aquifer part of the Middle Jurassic (Dogger) of the Paris Basin (France) where accurate fluid analyses are available. This aquifer, exploited for energy resources, is composed of limestone and dolomite. It covers an area of 300 km^2 and ranges in depth from 0 to 2000 m. Isotopic studies (Criaud et al., 1989) and noble gas analyses (Marty et al., 1988) suggest that the fluid is a mixture of meteoric and connate waters. Hydraulic tests

Basin Modelling: Advances and Applications edited by A.G. Doré et al. NPF Special Publication 3, pp. 343–351, Elsevier, Amsterdam.

(Aubertin et al., 1987) and hydraulic mathematical modelling (Wei et al., 1990) establish the existence of a regional flow. Meteoric water recharge zones are located in the eastern and southeastern extremities of the basin and saline water appears to be injected from the underlying Trias through fractured zones. Discharge is by leakage from lower to upper aquifers and also occurs at the western extremity of the basin. Furthermore, computation of the flowing balance of the aquifer (Wei, 1990) permits evaluation of the mean hydraulic parameters along the cross-section.

Sedimentary studies describe some evidence of dissolution and precipitation in the most permeable facies; the oolitic sandstones (Rojas et al., 1989) and chemical study explains the composition of the Dogger fluid by local equilibrium with minerals commonly encountered in sedimentary environments (Michard and Bastide, 1988). Because of the interdependence of all reactions, an exhaustive determination of the geochemical system (i.e. all the species in the media and the controlling equilibrium) is necessary. Thus, before considering the effects of fluid flow, the first part of this paper presents the determination of the geochemical processes which control the water composition of the Dogger aquifer.

Geochemical processes

The geochemical code CHIMERE (Coudrain-Ribstein and Jamet, 1989) computes speciation of a solution at any temperature in the range of validity of equilibrium constants and provides saturation indices of the solution with respect to mineral phases. It may also be used to determine the theoretical composition of a solution and the dissolved or precipitated quantities related to a set of constraints (temperature, equilibrium with respect to minerals and/or gaseous phases).

Water samples come from twenty geothermal boreholes located in the centre of the basin (Fig. 1). Sampling was realised at the wellhead and corrected for CO_2(g) degassing during sampling. Corrections (Bastide, 1985) are based on measurements of gas released and constancy of alkalinity and allow accurate estimate of pH and carbonate contents.

Analyses (published in Michard and Bastide, 1988) provide Al, Ca, Cl, F, HCO_3, K, Mg, Na, Si, SO_4,

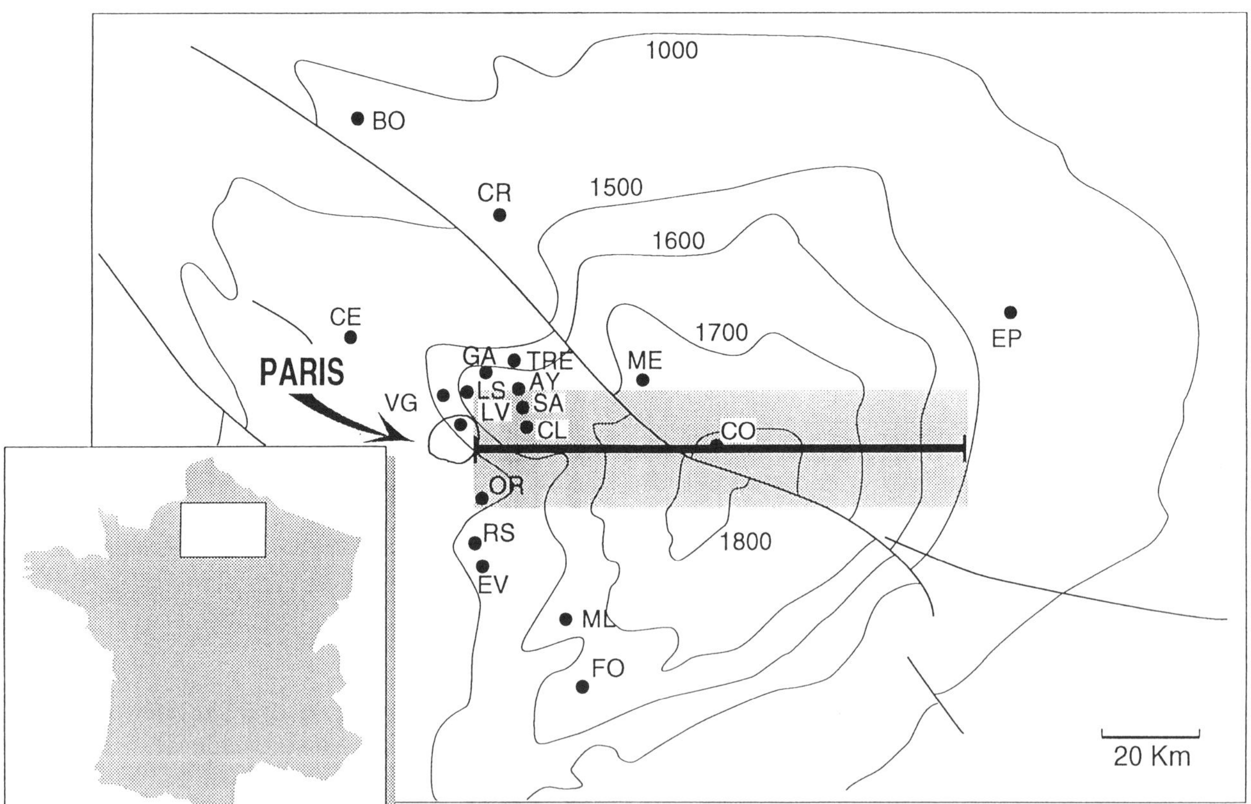

Fig. 1. Locations of geothermal drillholes and isobaths at the top of the Dogger layer in the Paris Basin (from Housse and Maget, 1976). *AY* = Aulnay; *CL* = Clichy; *CO* = Coulommiers; *EV* = Evry; *FO* = Fontainebleau; *GA* = Garges; *LS* = La Courneuve; *LV* = La Villette; *OR* = Orly; *RS* = Ris-Orangis; *SA* = Sevran; *TRE* = Tremblay-les-Gonesse; *VG* = Villeneuve la Garenne. The grey zone is the band in which the hydrodynamic parameters are averaged and projected on the section.

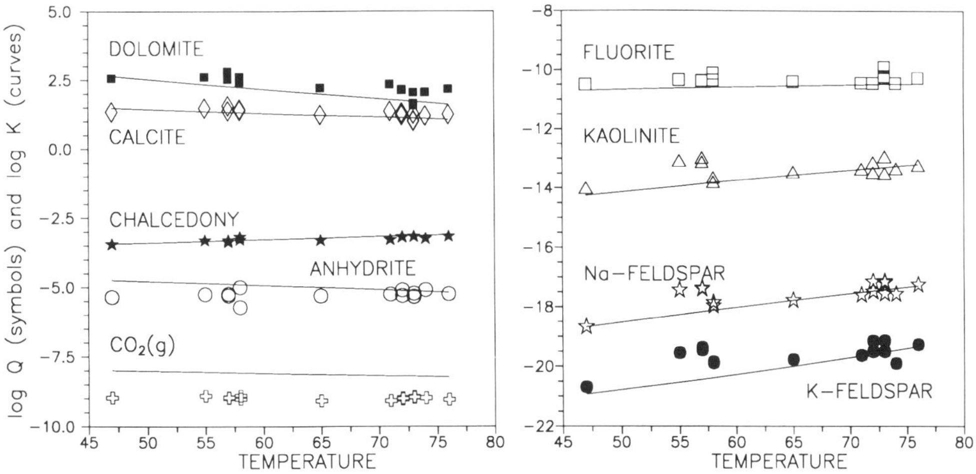

Fig. 2. Ionic activity product (Q) computed from Dogger solution analyses and solubility product (K) computed from thermodynamic data. Mineral formulae and thermodynamic data are shown in Table 1.

HS, B content and pH for large ranges of chlorinity (0.07 to 0.6 mol/l) and temperature (48° to 80°C). Initial speciations including HS and B establish that the presence of those elements has no influence on the speciation of major elements; therefore, these elements are neglected in this study. Measured contents of trace elements such as Li, Sr, Br and I are not taken into account because of their very low concentration.

Mineral saturation indices are computed from each borehole sample. Results emphasize that fluids are close to equilibrium with eight minerals (Fig. 2) which partially control the fluid composition. These results are fully consistent with the conclusions of Michard and Bastide (1988). For these minerals, kinetic rates of heterogeneous reaction are large compared to the fluid renewal rates at fluid–mineral contacts. Then, the rate of exchange between these minerals and the fluid is controlled by transport phenomena including advection, mechanical dispersion and molecular diffusion.

The eight minerals with which the fluid is in equilibrium comprise two carbonates (calcite and dolomite), an oxide (chalcedony), a sulphate (anhydrite), fluorite and three alumino-silicates (albite, kaolinite and adularia). A similar mineral assemblage is suggested by Nagy et al. (1990) to control sandstone diagenetic evolution of some sedimentary formations. According to the classical stability field diagrams for the system Al–K–Si–Na–H_2O, all the samples are located in the kaolinite field but close to the curve where kaolinite and K-feldspar are in metastable equilibrium. Very little quantitative data on the three alumino-silicates are available from cores (Rojas et al., 1989), but altered feldspars and neogene feldspars are observed. For comparison, petrographic analyses on core material from the Dogger reservoir of Gifhorn Trough (North Germany) show the presence of quartz, calcite, dolomite and kaolinite associated with K-feldspar and plagioclase (Schwarzkopf, 1990).

To describe successfully the thermodynamical processes that control the content of the eleven solute elements (Al, Ca, Cl, F, HCO_3, K, Mg, Na, Si, SO_4, H^+) it is necessary to have eleven constraints (Coudrain-Ribstein, 1988). The eleven elements can be understood as eleven components of the whole set of chemical species under consideration (Morel, 1983). Fluid equilibrium with respect to eight minerals is already determined. The electroneutrality condition and the total chloride content which is not controlled by heterogeneous reactions provide two other constraints (Table 2).

The theoretical partial pressure of CO_2 (P_{CO_2}) may give indications of the operating chemical processes (see Giggenbach, 1980, 1981; Hutcheon, 1990). P_{CO_2} is calculated by the following relation:

$$P_{CO_2} = \frac{[HCO_3^-][H^+]}{K_{CO_2}} \tag{1}$$

where K_{CO_2} the equilibrium constant (Table 1).

P_{CO_2} computed from Dogger data appears to vary slightly with temperature (Fig. 3) around a value of $10^{-0.85}$ atm and may be fitted by the following relation:

$$\log_{10}(P_{CO_2}) = -1.23 + 0.006\,T \tag{2}$$

where T is temperature in °C (48° < T < 81°).

The fact that the system is closed with respect to one of the components could be the eleventh constraint necessary to investigate the origin of this trend. For this purpose, one simulates the evolution of P_{CO_2} with the help of CHIMERE. From

TABLE 1

Chemical formulae and equilibrium constants of formation of minerals and CO_2 from the set: [H^+, Cl^-, $Al(OH)_4^-$, F^-, H_4SiO_4, K^+, Na^+, SO_4^{2-}, Ca^{2+}, Mg^{2+}, HCO_3^-]

Mineral	Formulae	$\log_{10}(K)$		
		25°C	40°C	80°C
Adularia [a]	$KAlSi_3O_8$	−21.70	−21.22	−19.08
Albite [a]	$NaAlSi_3O_8$	−19.98	−19.07	−17.13
Anhydrite [a]	$CaSO_4$	−4.46	−4.64	−5.25
Chalcedony [a]	SiO_2	−3.71	−3.51	−3.07
Calcite [a]	$CaCO_3$	1.81	1.59	1.01
Chlorite [c]	$Mg_5Al_2Si_3O_{10}(OH)_8$	18.35	16.37	11.76
Dioxide gas [b]	$CO_2(g)$	−7.81	−7.93	−8.26
Dis. dolomite [a]	$CaMg(CO_3)_2$	3.47	2.90	1.47
Fluorite [a]	CaF_2	−10.97	−10.77	−10.46
Dis. kaolinite [c]	$Al_2Si_2O_5(OH)_4$	−37.4	−35.70	−32.09
Laumontite [a]	$CaAl_2Si_4O_{12},4H_2O$	−30.68	−29.48	−27.00

[a] From Michard (1983).
[b] From Fritz (1981).
[c] Adjusted from data of Fritz (1981) and Michard (1983).

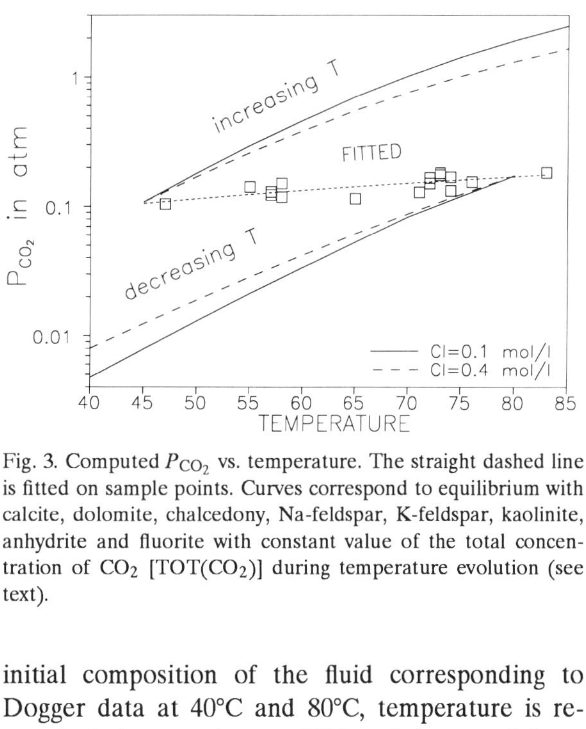

Fig. 3. Computed P_{CO_2} vs. temperature. The straight dashed line is fitted on sample points. Curves correspond to equilibrium with calcite, dolomite, chalcedony, Na-feldspar, K-feldspar, kaolinite, anhydrite and fluorite with constant value of the total concentration of CO_2 [TOT(CO_2)] during temperature evolution (see text).

TABLE 2

Solute elements and constraints = $-1.23 + 0.006\,T$

H^+	$\Sigma+ = \Sigma-$
Cl^-	$\nabla[\overline{\overline{D}} \cdot \nabla C) - UC] = 0$
$Al(OH)_4^-$	dis. kaolinite
F^-	fluorite
H_4SiO_4	chalcedony
K^+	K-feldspar
Na^+	Na-feldspar
SO_4^{2-}	anhydrite
Ca^{2+}	calcite
Mg^{2+}	dis. dolomite
HCO_3^-	X

Eleven constraints are needed (including eight fluid–mineral equilibria) to control the eleven solute element concentrations. $\Sigma-/+$ denotes the sum of the electric charges, $\overline{\overline{D}}$ the dispersion tensor, U the Darcy's velocity and C the total chloride concentration. In the coupled model X is the empirical relation $\log_{10}(P_{CO_2})$

initial composition of the fluid corresponding to Dogger data at 40°C and 80°C, temperature is respectively increased up to 80°C and decreased down to 40°C maintaining equilibrium with the eight minerals previously determined. In these simulations, the eleventh constraint is the concentration constancy of the eleventh component. A convenient choice for this component is CO_2 ($X = CO_2$ in Table 2), but this choice has no influence on the results (Coudrain-Ribstein and Morel, 1987). With increasing temperature from 40°C to 80°C, computed P_{CO_2} increases up to 1 atm. When temperature decreases from 80°C to 40°C, the P_{CO_2} reaches 10^{-2} atm. Both of these T–P_{CO_2} trends (Fig. 3) are not consistent with data. In conclusion, the eleventh constraint must allow dissolution and precipitation of another phase (the system is open with respect to the eleventh controlling component).

Similar relationships between P_{CO_2} and T (eq. 6) are quoted by Smith and Ehrenberg (1989) for the U.S. Gulf Coast ($\log P_{CO_2} = -1.50 + 0.018\,T$) and Norwegian continental shelf ($\log P_{CO_2} = -1.45 + 0.019\,T$). The authors suggest that these relationships are the result of equilibrium between the fluid and the association of feldspar(s), carbonate(s) and clay minerals. Following this idea, some fluid–minerals could be tested in order to determine the eleventh constraint.

Laumontite ($CaAl_2Si_4O_{12}.4H_2O$) is a clay which is frequently invoked in low-temperature and hydrothermal environments to control the composition of solutions (Crossey et al., 1984) when P_{CO_2} is low.

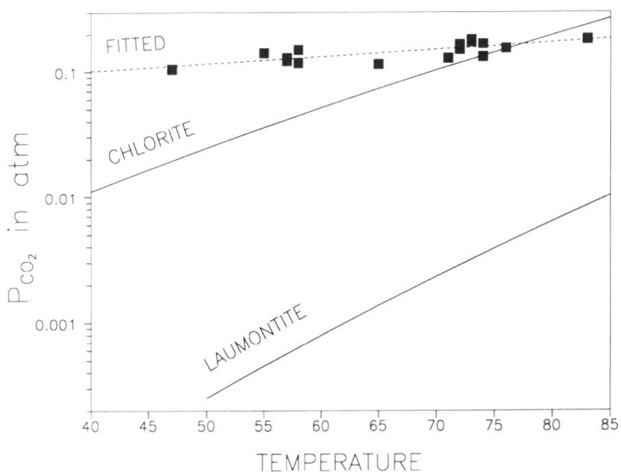

Fig. 4. Computed P_{CO_2} vs. temperature. Dashed line is fitted to the P_{CO_2} computed from Dogger samples (squares). Curves are computed assuming equilibrium with calcite, dolomite, chalcedony, Na-feldspar, K-feldspar, kaolinite, anhydrite, fluorite plus chlorite or laumontite.

Computation of the speciations, assuming that the fluid is in equilibrium with the previous eight minerals plus laumontite, shows strong variation of P_{CO_2} with T in a range of 10^{-4} to 10^{-2}, which is much lower than the P_{CO_2} calculated with Dogger samples (Fig. 4).

Chlorite ($Mg_5Al_2Si_3O_{10}(OH)_8$) is cited as an important mineral in diagenetic environments by Millot (1970) and is proposed, in association with carbonates and kaolinite, to control some values of P_{CO_2} in Iceland volcanic fluids (Arnorsson et al., 1983; Hutcheon, 1990). Results of simulations where the fluid is in equilibrium with the eight minerals plus chlorite show that Dogger samples are almost at equilibrium with this mineral for the highest temperatures, but induce a stronger evolution of computed P_{CO_2} vs. T than the one calculated for Dogger fluids (Fig. 4). All the theoretical ionic contents of a fluid at equilibrium with the eight minerals plus chlorite are in good agreement with Dogger data except for HCO_3^- (Fig. 5).

One concludes from these tests that neither fluid–chlorite nor fluid–laumontite equilibrium constitutes a fully satisfactory constraint of the system. The assumptions of P_{CO_2} buffered by organic matter or gas reactions cannot be tested with the thermodynamical model and are not discussed in this paper. However, whatever the external processes controlling P_{CO_2}, the P_{CO_2}–T relationship is characteristic of the studied zone and the system is fully determined by adding the constraint $\log(P_{CO_2}) = -1.23 + 0.006\,T$ to the eight fluid mineral equilibrium constraints, electroneutrality condition and the total concentration of Cl^- determined by the transport processes (Table 2).

Reaction–transport coupled model

The STELE model (Coudrain-Ribstein, 1988) allows the solution of equations controlling transport of solute species involved in thermodynamic processes, including heterogeneous reactions. This model is a combination of the geochemical static model CHIMERE and the transport finite element model METIS (Goblet, 1981) based on a continuum approach to describe, in two dimensions, the transfer processes through porous media.

In the simulation presented below, the hydraulic regime is computed assuming steady state. Within a fixed volume, the continuity equation is:

$$\nabla \cdot (U) + q = 0 \qquad (3)$$

where q is the source term and U Darcy's velocity:

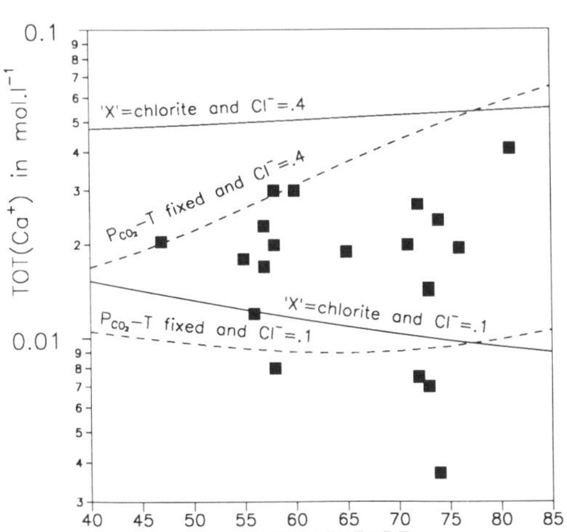

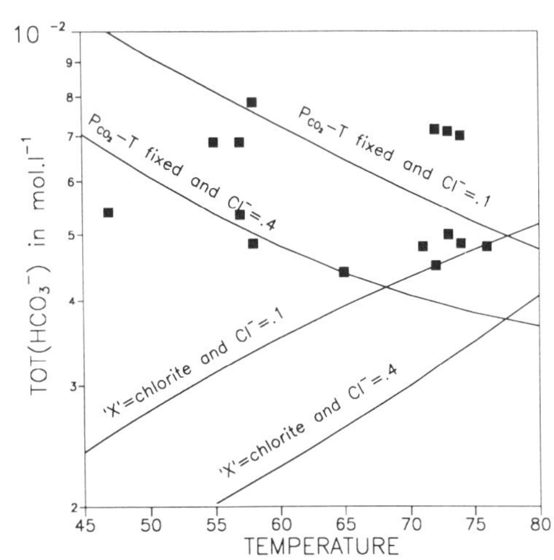

Fig. 5. Total concentrations of calcium and carbonate vs. temperature. Curves are total concentrations computed for equilibrium with calcite, dolomite, chalcedony, Na-feldspar, K-feldspar, kaolinite, anhydrite and fluorite plus the P_{CO_2}–T relationship (eq. 2) or plus equilibrium with chlorite. Curves are given for Cl^- = 0.4 and 0.1 mol/l. Squares correspond to the Dogger solution analyses.

$U = -K \cdot \nabla h$, where K is the hydraulic conductivity and h the hydraulic head.

Assuming that each aqueous species has the same mobility (access to the same porosity and same dispersion coefficient), the global mass balance (advection–dispersion reaction) is given, for a weakly dispersive medium, by the equation (de Marsily, 1986; Bear and Bachmat, 1990):

$$\omega \frac{\partial}{\partial t}[\mathrm{TOT}(C_i)] = \nabla \cdot \left\{ \overline{\overline{D}} \cdot \nabla[\mathrm{TOT}(C_i)] - \boldsymbol{U}\,\mathrm{TOT}(C_i) \right\} + \omega \phi_i \quad (4)$$

where $\overline{\overline{D}}$ denotes the dispersion tensor ($\overline{\overline{D}} = \omega d + \overline{\overline{\alpha}}|\boldsymbol{U}|$) including the molecular diffusion d and the intrinsic dispersion tensor $\overline{\overline{\alpha}}$. In eq. 4, ω is the porosity and ϕ_i is the geochemical sink/source term which accounts for the mass transfers between the solution and the solid or gaseous phases.

With the assumption of local equilibrium, kinetics of reactions are not taken into account. Along the section, solute species concentrations are fully controlled by equilibrium relations and transport–reaction processes are described by the steady-state formulation of eq. 4. As a consequence, ϕ_i values are explicitly computed at each node by solving eq. 4 where the $\mathrm{TOT}(C_i)$ are the at-equilibrium $\mathrm{TOT}(C_i)$.

The synthetic cross-section

Model setting

Modelling is limited to a 100 km long cross-section (Fig. 1). Temperature and depth along the section are average values (from House and Maget, 1976 and Menjoz et al., 1989) of a 10 km wide band centred on the section.

Water-productive thickness (layers formed of oolite limestone) is about 10% of the Dogger total thickness (300 m). Figure 6 shows the structural geometry, which is discretised vertically into 6 elements and into 1000 elements along the horizontal direction.

Average hydraulic conductivity is calculated from Dogger thickness and transmissivity data. Taking into account knowledge on the Dogger vertical structure (Menjoz et al., 1989), 90% of the flow rate is provided by the 10 m central bed of the model structure. Hydraulic conductivity of surrounding strata is set to 10^{-6} m/s (porosity set to 5%) and for the central strata is set to 10^{-5} m/s (porosity set to 10%). The steady-state hydraulic head distribution is computed with recharge and discharge boundary conditions in order to fit the average linear Darcy's velocity computed for the regional model of Wei (1990). Longitudinal Darcy's velocity is presented in Fig. 6.

Although the major part of the data is centred on the western zone of the section except Coulommiers and Meaux (centre of the section) and

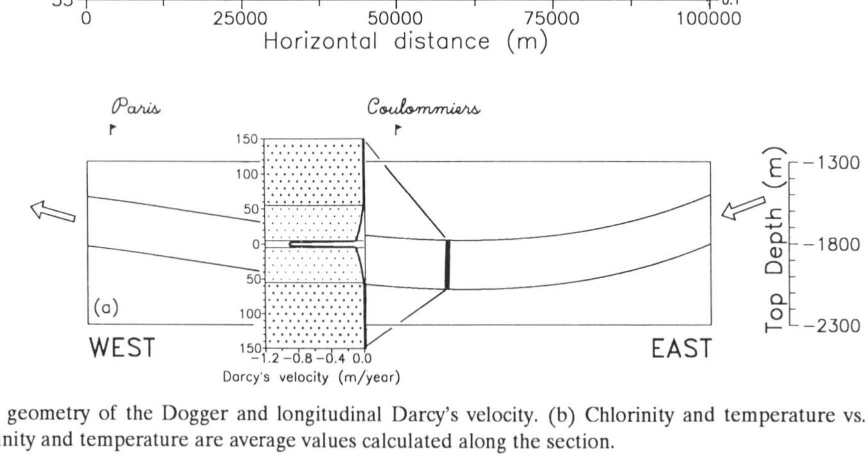

Fig. 6. (a) Aquifer geometry of the Dogger and longitudinal Darcy's velocity. (b) Chlorinity and temperature vs. horizontal distance. Permeability, chlorinity and temperature are average values calculated along the section.

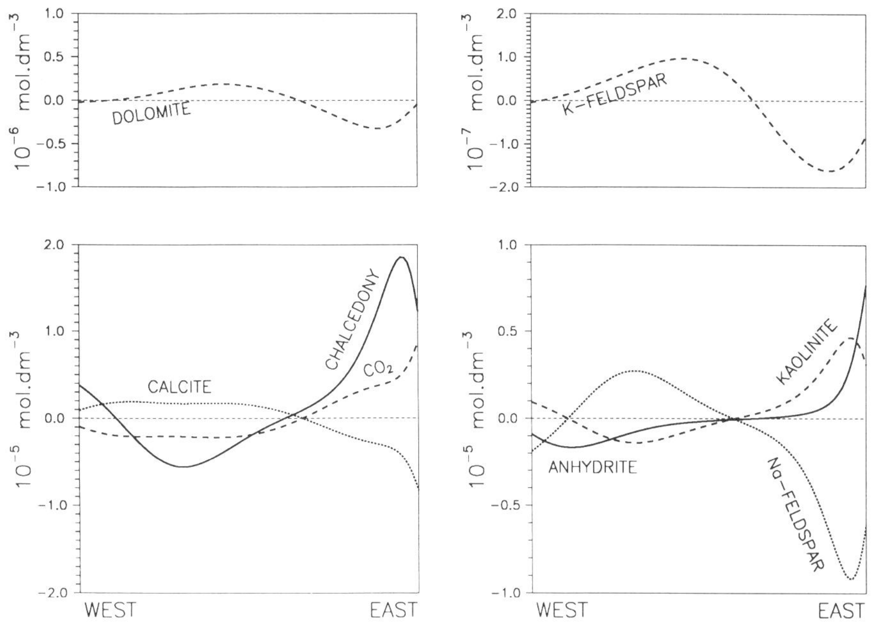

Fig. 7. Computed quantities (in mol/dm^3) of mineral dissolved (<0) or precipitated — outgassed for CO_2 — (>0), during 1 year, at the centre axis of the aquifer where the linear Darcy's velocity is maximum.

Epernay (northeast of the section), the same fluid–mineral equilibria (kaolinite, fluorite, chalcedony, K-feldspar, anhydrite, calcite, Na-feldspar and disordered dolomite) and the same P_{CO_2}–T relationship (eq. 2) are used along the whole section. Following the conclusion of Wei (1990), the transport of Cl^- is close to steady state. In order to obtain a synthetic section, the Cl^- concentrations reported along the section are the orthogonal projections of values averaged on a 20 km band (grey zone in Fig. 1) from available maps (Maget, 1983). Dispersivity tensor and diffusion coefficients are fixed at classical values for such a configuration. These are, respectively, $\alpha_L = 100$ m, $\alpha_T = 10$ m and $d = 10^{-9}$ m^2 s^{-1}.

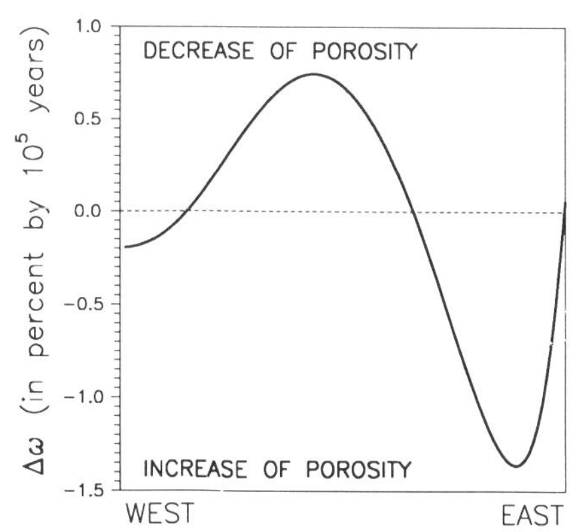

Fig. 8. Porosity evolution (in percent of the total volume) after 0.1 m.y. vs. horizontal distance.

Results of the simulation.

Fluid–solid quantities transferred into the central bed are presented in Fig. 7. Quantities of fluorite are negligible. Dolomite and K-feldspar transfers are, respectively, 1 and 2 orders of magnitude weaker than transfers of the five other minerals. Important mass transfers are localized in sloping areas where current crosses isotherms and isochlores with maximum angle.

Along the eastern part of the section, where the fluid flows while heating, the main feature is the dissolution of feldspars associated with authigenic kaolinite and silica precipitation. This mineral transformation induces the formation of a large amount of secondary porosity (Fig. 8). In the same zone, carbonates are dissolved as a result of water heating. The mobilized calcium ions contribute to the precipitation of anhydrite and to the possibility of CO_2 degassing.

The upstream zone is characterised by dissolution of anhydrite and precipitation of carbonates. Chalcedony, kaolinite and anhydrite dissolution induce precipitation of feldspars and carbonates. The maximum cementation occurs 25 km east of Paris. Closer to Paris, no redistribution of magnesium and potassium occurs and alumino-silicate transfers are in inverse order. In this zone, porosity enhancement is associated with dissolution of anhydrite and Na-feldspar even though formation of authigenic calcite and kaolinite occurs.

Assuming a constant rate of flow and a constant temperature field, the evolution of porosity due to mineral redistribution is computed for a geologically significant time of 0.1 million years (Fig. 8). The model predicts that a 1.5% increase in porosity is reached at the eastern side of the section whereas about 1.0% of the total volume is cemented between 40 and 50 km east of Paris.

Conclusions

A general approach is proposed to calculate the effect on porosity of flow through the Dogger aquifer. Fluid composition is a function of temperature, pressure, mineralogy and chlorinity (Cl^- concentration is controlled by transport processes) along the flow path. The composition of the fluid is locally a function of precipitation and dissolution. The quantitative study of present diagenesis throughout the aquifer is based on chemical analyses of the fluid and may be carried out even where sedimentologic studies are not available.

Using twenty water analyses, a set of mineral–fluid equilibria was used as evidence for the central part of the aquifer. The number of independent constraints that are necessary to control the fluid composition is equal to the number of simple ions for which the concentration is measured *in situ*. In the case of the Dogger, the composition in terms of major solute elements (Al, Ca, F, HCO_3, K, Mg, Na, Si and SO_4) as well as the proton activity may be driven by eight mineral–solution equilibria, a $P_{CO_2}-T$ relation and the electric neutrality of the solution. The Cl^- content is controlled by the steady-state transport phenomena.

Because of the increasing mineral solubility with temperature, none of the investigated clay minerals are able to control the $P_{CO_2}-T$ trend by thermodynamical equilibrium.

Alumino-silicate (in particular kaolinite and Na-feldspar) mass transfers ensure a strong control on porosity evolution. This result emphasizes the importance of accurate measurement of aluminium content.

Calculation of the spatial evolution of porosity over 0.1 m.y. shows that porosity enhancement can locally reach 1.5% of the total volume. To extrapolate the present approach over several million years, it would be necessary, at least, to take into account the evolution of the hydraulic parameters due to matter redistribution.

Acknowledgements

This study was made possible through the financial contribution by the "Action de Recherche Coordonnée — Programme Interdisciplinaire de Recherches sur les Sciences pour l'Energie et les Matières premières" (CNRS AFME) and the Research Group "Transferts dans les Bassins Sédimentaires" (CNRS). Part of the computation was performed with the support of the "Centre de Compétences en Calcul Numérique Intensif" at the electronic computation centre CNUSC, Montpellier, France.

References

Arnorsson, S., Gunnlaugsson, E. and Svavarsson, H., 1983. The chemistry of geothermal waters in Iceland, II. Mineral equilibria and independent variables controlling water compositions. Geochim. Cosmochim. Acta, 47: 547–566.

Aubertin, G., Cordier, E., Doillon, F., Fabris, H., Gable, R., Gaillard, B., Ledoux, E., de Marsily, G. and Menjoz, A., 1987. Détermination expérimentale de la vitesse naturelle d'écoulement de la nappe géothermique du Dogger en région Parisienne. Bull. Soc. Géol. Fr., III (5): 991–1000.

Bastide, J.-P., 1985. Etude géochimique de la nappe du Dogger du bassin Parisien. Thèse Doct., Univ. Paris 7, 152 pp.

Bear, J. and Bachmat, Y., 1990. Introduction to Modeling of Transport Phenomena in Porous Media. Kluwer, Dordrecht, 553 pp.

Coudrain-Ribstein, A., 1988. Transport d'éléments et réactions géochimiques dans les aquifères. Thèse Doct. es Sci., Ecole Nat. Sup. des Mines de Paris, Univ. Paris 6, 381 pp.

Coudrain-Ribstein, A. and Jamet, P., 1989. Choix des composantes et spéciation d'une solution. Bilingual note, C.R. Acad. Sci. Paris, Sér. II, 309: 293–244.

Coudrain-Ribstein, A. and Morel, F.M.M., 1987. Modélisation du transfert d'éléments majeurs réactifs dans un champ de température variable: méthodologie et exemples simples. Bull. Soc. Géol. Fr., 8-III (5): 1009–1017.

Criaud, A., Fouillac, C. and Brach, M., 1989. Etude Géochimique. Rep. BRGM-AFME, IRG SGN 89, Orléans, pp. 41–100.

Crossey, L.J., Frost, B.R. and Surdam, R.C., 1984. Secondary porosity in laumontite-bearing sandstones. Am. Assoc. Pet. Geol., Mem., 37: 225–237.

De Marsily, G., 1986. Quantitative Hydrology. Academic Press, New York, N.Y., 440 pp.

Fritz, B., 1981. Etude thermodynamique et modélisation des réactions hydrothermales et diagénétiques. Sci. Géol. Mem., 65, 197 pp.

Giggenbach, W.F., 1980. Geothermal gas equilibria. Geochim. Cosmochim. Acta, 44: 2021–2032.

Giggenbach, W.F., 1981. Geothermal mineral equilibria. Geochim. Cosmochim. Acta, 45: 393–410.

Goblet, P., 1981. Modélisation des transferts de masse et d'énergie en aquifère. Thèse Doct. Ing., Ecole Nat. Sup. des Mines de Paris, Univ. Paris 7., 199 pp.

Housse, B. and Maget, P., 1976. Le potentiel géothermique du bassin Parisien. Rep. BRGM-SNEAP, 74-70990, 125 pp.

Hutcheon, I., 1990. Clay–carbonate reactions in the Venture area, Scotian Shelf, Scotia, Canada. Geochem. Soc., Spec. Publ., 2: 199–212.

Maget, P., 1983. Potentiel thermique basse température en France. Rep. BRGM, 83 SNG 375 SPG., 319 pp.

Marty, B., Criaud, A. and Fouillac, C., 1988. Low enthalpy geothermal fluids from the Paris sedimentary basin, 1. Characteristics and origin of gases. Geothermics, 17 (4): 619–633.

Menjoz, A., Martin, J.C. and Lambert, M., 1989. Modélisation du réservoir. Rep. BRGM-AFME, IRG SGN 89, Orléans, pp. 105–205.

Michard, G., 1983. Recueil des données thermodynamiques concernant les équilibres eaux-minéraux dans les réservoirs hydrothermaux. Rep. CCE, EUR 8590 FR., 155 pp.

Michard, G. and Bastide, J.-P., 1988. Etude géochimique de la nappe du Dogger du bassin Parisien. J. Volcanol. Geotherm. Res., 35: 151–163; Am. Assoc. Pet. Geol., Mem., 49: 85–101.

Millot, G., 1970. Geology of Clays. Springer-Verlag, New York, N.Y., 429 pp.

Morel, F.M.M., 1983. Principles of Aquatic Chemistry. Wiley, New York, N.Y., 446 pp.

Nagy, K.L., Steefel, C.I., Blum, A.E. and Lasaga, A.C., 1990. Dissolution and precipitation kinetics of kaolinite: initial results at 80°C with application to porosity evolution in a sandstone. Am. Assoc. Pet. Geol., Mem., 49: 85–101.

Rojas, J., Giot, D. and Le Nindre, Y., 1989. Etude Sédimentologique. Rep. BRGM-AFME, IRG SGN 89, Orléans, pp. 1–39.

Schwarzkopf, T.A., 1990. Relationship between petroleum generation, migration and sandstone diagenesis, Middle Jurassic, Gifhorn Trough, N Germany. Mar. Pet. Geol., 7: 153–170.

Smith, J.T. and Ehrenberg, S.N., 1989. Correlation of carbon dioxide abundance with temperature in clastic hydrocarbon reservoirs: relationship to inorganic chemical equilibrium. Mar. Pet. Geol., 6: 129–135.

Wei, H.F., 1990. Modélisation tridimentionnelle du transfert d'eau, de chaleur et de masse dans l'aquifère géothermique du Dogger dans le bassin de Paris. Thèse Doct., Ecole Nat. Sup. des Mines de Paris. Univ. Paris 6, 228 pp.

Wei, H.F., Ledoux, E. and de Marsily, C., 1990. Regional modelling of groundwater flow and salt and environmental tracer transport in deep aquifers in the Paris basin. J. Hydrol., 120: 341–358.

PHILIPPE GOUZE — *Laboratoire de Géologie Appliquée, URA CNRS "Circulations et Transports Hydriques Continentaux", UPMC, 4 place Jussieu, 75252 Paris Cedex 05, France; and Centre Géologique et Géophysique de Montpellier, UPR CNRS 631, Université Montpellier II, Place E. Bataillon, 34095 Monrpellier Cedex 05, France*

ANNE COUDRAIN-RIBSTEIN — *Laboratoire de Géologie Appliquée, URA CNRS "Circulations et Transports Hydriques Continentaux", UPMC, 4 place Jussieu, 75252 Paris Cedex 05, France*

Effects of convection-currents on heat transfer in sedimentary basins

Arild Ludvigsen, Kjetil Gran, Enok Palm and Knut Bjørlykke

The possibility for significant effects of thermal convection on heat transfer is investigated, using mathematical models that take into account the interaction effects between the convective flow and the temperature distribution. The conclusion is that efficient heat transfer is not likely to occur.

Observations of salinity gradients in the southern Norwegian North Sea and Haltenbanken are additional evidence against any important effects of large-scale convection. Heat transport in the basins studied and probably in other, similar, basins is essentially controlled by conduction, and advective transport plays a minor role.

Introduction

Heat can be transported in sedimentary basins by thermal conduction (diffusion) through the solid phase (minerals) and through the pore water. In addition, pore water flow may transfer heat and thereby affect the temperature distribution in a basin. We may distinguish between three main types of pore water flow in sedimentary basins. (1) Meteoric water flow into sedimentary basins driven by a hydrostatic head defined by the elevation of the ground water table above the sea level. (2) Pore water flow driven by gradual compaction and porosity loss in sediments with increasing burial. (3) Pore water flow driven by differences in the densities of pore water, i.e., thermal convection and thermohaline convection.

Meteoric flow into sedimentary basins is characterized by much higher flow rates than compactional pore water flow and has in some basins been shown to depress geothermal gradients significantly (Smith and Chapman, 1983).

It can be shown that pore water flow driven by compaction normally does not exceed the rate of subsidence, so that there is no net upwards flow relative to the sediment–water interface (Caritat, 1989). Therefore, only locally, where there is strongly upwards-focused flow along special permeable conduits, may compactional water flow produce significant temperature anomalies.

We will here examine two types of thermal convection: Rayleigh convection and non-Rayleigh convection. Rayleigh convection can be defined as a type of vertical fluid circulation driven by vertical differences in the density as a function of temperature; in other words, the type of convection with which most scientists are familiar. It is well known in fluid dynamics that a finite horizontal temperature gradient always generates a motion. This motion is here called non-Rayleigh convection, and in contrast to Rayleigh convection starts at the slightest horizontal variation of the temperature. For the configuration shown in Fig. 1 this motion is represented as two-dimensional.

The effects of these types of convection on heat transfer and temperature distribution in sedimentary basins are investigated quantitatively, based on the mathematical model described below. The effect of fluid flow on salinity distribution is also discussed, qualitatively, and salinity–depth relationships for the North Sea and Haltenbanken basins are presented.

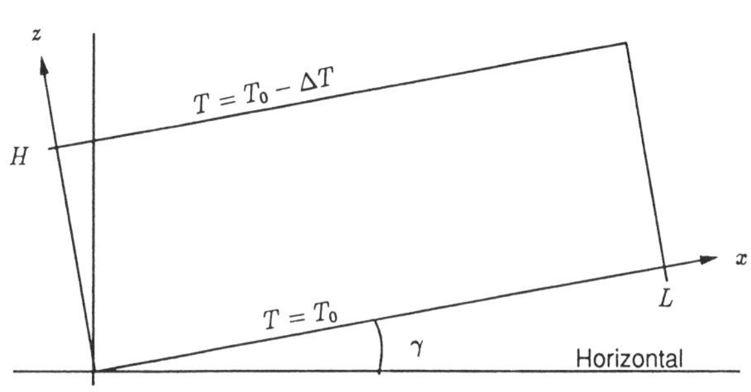

Fig. 1. The porous layer. The thickness and length of the layer is H and L, respectively. A coordinate system with axes parallel with the layer boundaries is introduced. The layer makes an angle γ with the horizontal.

Basin Modelling: Advances and Applications edited by A.G. Doré et al. NPF Special Publication 3, pp. 353–359, Elsevier, Amsterdam.

Modelling

We consider a permeable slab of thickness H and length L, saturated with water and making an angle γ with the horizontal (Fig. 1). The isotherms are sloped. We assume that the temperatures are kept constant at the boundaries ($z = 0$ and $z = H$), being T_0 at the lower boundary and $T_0 - \Delta T$ at the upper boundary. The side walls ($z = 0$ and $z = H$) are then assumed to be perfect conductors and the end walls ($x = 0$ and $x = L$) to be insulating. These boundary conditions are also appropriate for a long, thin layer if the thermal conductivity of the layer is much less than that of the surroundings (Ludvigsen et al., 1991). Furthermore, we assume that the motion is relatively slow so that the isotherms are approximately parallel with the side boundaries. For a long, thin layer, the temperature gradient is then in the z-direction and is equal to $\Delta T/H$.

The fluid flow is described by the Rayleigh-Darcy equations. Applying the Boussinesque approximation, the process is governed by the equations:

$$\nabla \cdot \boldsymbol{v} = 0 \tag{1}$$

$$-\nabla p + \rho g - \frac{\mu}{K}\boldsymbol{v} = 0 \tag{2}$$

$$\boldsymbol{v} \cdot \nabla T = \kappa \nabla^2 T \tag{3}$$

$$\rho = \rho_0 [1 - \alpha (T - T_0)] \tag{4}$$

where $\boldsymbol{v}$ is specific discharge, p is pressure, T is temperature, ρ is mass density of the fluid, K is permeability, κ is thermal diffusivity, μ is dynamic viscosity coefficient and α is cubic expansion coefficient for the fluid. ρ_0 and T_0 are reference values for the mass density of the fluid and the temperature, respectively. The specific discharge, $\boldsymbol{v}$ is defined as the volume flux per unit area of the porous medium normal to the flow direction. The thermal diffusivity for a fluid-filled porous medium is defined in the usual way as the thermal conductivity for the fluid-filled porous medium, divided by the density and specific heat capacity of the fluid.

Equations 1–3 are the equations of continuity, Darcy's law and the energy equation, respectively, while 4 is the equation of state. In the latter equation, the effect of pressure is neglected and only a linear dependence on temperature is considered.

The temperature distribution satisfying eq. 3 is written:

$$T = -\frac{\Delta T}{H} z + T_0 + \vartheta \tag{5}$$

where ϑ is small and is due to the fluid motion. The vertical component of the temperature gradient attributable to a purely diffusive temperature distribution is $\beta = (\Delta T/H) \cos \gamma$.

For horizontal layers ($\gamma = 0$) a Rayleigh number Ra is defined as:

$$Ra = \frac{\rho_0 \, \alpha g \, \Delta T \, KH}{\kappa \mu} \tag{6}$$

For the sloping model shown in Fig. 1, we define the Rayleigh number as $R = Ra \cos \gamma$, or equivalently:

$$R = \frac{\rho_0 \, \alpha g \beta \, KH^2}{\kappa \mu} \tag{7}$$

If the temperature difference across the layer ΔT, or more precisely, the Rayleigh number R, exceeds a certain critical value, Rayleigh convection in the form of two-dimensional rolls grows up, super-positioned on the unicellular non-Rayleigh flow, with the axes of the rolls parallel with the x-axis of the layer in Fig. 1 (Weber, 1975). Thus, the convection is now three-dimensional. However, the rolls constitute a two-dimensional cell-pattern in the plane normal to the x-direction.

As values of parameters in eq. 7, characteristic for sedimentary basins, $\alpha = 7.1 \times 10^{-4}$ K^{-1}, $g = 9.8$ m s^{-2}, $\beta = 3 \times 10^{-2}$ km^{-1}, $\kappa = 3.5 \times 10^{-7}$ m^2 s^{-1} and $\nu = \mu/\rho_0 = 3 \times 10^{-7}$ m^2 s^{-1} may be assumed. The Rayleigh number then depends only on the permeability and the height of the layer.

Rayleigh convection

Bories and Combarnous (1973) found that the critical Rayleigh number R_c for a long, thin sloping layer was given by:

$$R_c = 4\pi^2 \tag{8}$$

This is the same as the result of Lapwood (1948) for horizontal layers, except that $g \cos \gamma$ has to be substituted for g. The critical value of KH^2 for Rayleigh convection then occurs as:

$$KH^2 = 2.0 \times 10^{-8} \text{ m}^4 \tag{9}$$

If the permeability is 10 D, a thickness of about 50 m is then required for Rayleigh convection to occur. Primarily because interbedded low-permeability layers (shale) are rather common, the conditions for such highly permeable layers are rarely met in sedimentary basins.

On the other hand, if layers with thicknesses a hundred times greater than the above exist, with a permeability of only 1/10,000 of the value mentioned above, the conditions for Rayleigh convection would obviously also be satisfied. Such values are more in line with the configurations discussed by Swift et al. (1988) and by Rabinowicz et al. (1985). In high configurations the temperature difference over the layer is large and the temperature dependence of thermal diffusivity, expansion coefficient and viscosity will affect the critical Rayleigh number. Also, the permeability may vary with the height of the layer due to, for example, compaction. Assuming that the parameters in the Rayleigh number are representative averages for the configuration, the temperature dependence, or spatial dependence, of the parameters do not have important effects on the value of the critical Rayleigh number.

As pointed out by Weber (1975), the results for Rayleigh convection in a horizontal layer may be used directly for longitudinal rolls in a sloped layer, except that $g \cos\gamma$ has to be substituted for g. From Palm et al. (1972), the temperature distribution due to convective motion, ϑ, for our model is then:

$$\vartheta = \frac{2\Delta T}{\pi}\sqrt{\frac{R-R_c}{R}}\cos\frac{\pi y}{H}\sin\frac{\pi z}{H} \tag{10}$$

where y is the coordinate along the direction normal to the xz-plane. (The y-axis is pointing normally into the plane in Fig. 1.) The maximum contribution of the temperature due to fluid motion is then:

$$\vartheta_{\max} = \frac{2}{\pi}\sqrt{\frac{R-R_c}{R}}\Delta T \tag{11}$$

and is localized at the middle of the lateral, adjacent "walls" of the convection cells. Differentiation of eq. 10 with respect to z, gives the contribution to the temperature gradient due to Rayleigh convection. The maximum of this contribution to the temperature gradient across the layer is:

$$\left(\frac{\partial\vartheta}{\partial z}\right)_{\max} = 2\sqrt{\frac{R-R_c}{R}}\frac{\Delta T}{H} \tag{12}$$

and is seen to be localized in the corners of the cells. Assuming $K = 10$ D, and $H = 47$ m, corresponding to the critical Rayleigh number being exceeded by 11%, $\vartheta_{\max}$ is 20% of ΔT and $(\partial\vartheta/\partial z)_{\max}$ is 64% of the static temperature gradient, $\Delta T/H$.

Let Q denote the heat transport in the z-direction across the layer averaged over the xy-plane. As a dimensionless measure for the heat transport we introduce the Nusselt number, Nu, defined by:

$$Nu = \frac{Q}{Q_c} \tag{13}$$

where Q_c is the heat transfer due to the purely conductive part of the temperature distribution.

From Palm et al., 1972 we have (for $R > R_c$):

$$Nu = 1 + 2\frac{R-R_c}{R} \tag{14}$$

If, for example, R exceeds R_c by 10%, it is seen that the convective contribution to the Nusselt number is 18%.

For a purely conductive state (no flow), which would be the case for $\gamma = 0$ and $R < R_c$, the thermal conductivity, k, of the porous layer, can be defined by:

$$k = \frac{Q_c}{\Delta T/H} \tag{15}$$

In the case of fluid motion, we may define an effective conductivity, k_{eff} as:

$$k_{\text{eff}} = \frac{Q}{\Delta T/H} \tag{16}$$

The ratio between k_{eff} and k is then equal to Q/Q_c which is, by definition, equal to Nu. Thus, the Nusselt number expresses both the ratio between heat transports and the ratio between effective conductivities in the actual and purely conductive case. Calculations of the Nusselt number is frequently found in literature on thermal convection. The Nusselt number, being an average quantity, is a good representation of the heat flux for Rayleigh convection in long, thin layers, when the heat flux from a large number of cells is considered.

The largest local heat fluxes are, however, antisymmetrically distributed over the cell and, due to cancellation, do not affect the Nusselt number. Locally, the heat flux q in z-direction out of the layer is given by:

$$q = k\frac{\Delta T}{H}\left(1 + 2\sqrt{\frac{R-R_c}{R}}\right) \tag{17}$$

Thus, the convective contribution to q has the same amplitude as the convective contribution to the temperature gradient and has extreme values at the ends of the cells, where the vertical component of the fluid motion is the largest. For large convection cells the local heat flux q will be a more important quantity than the Nusselt number, since the observed heat flux is then averaged over a scale much smaller than the extent of the convection cell.

Non-Rayleigh convection

If the Rayleigh number is less than R_c, Rayleigh convection will not occur. When the isotherms are non-horizontal ($\gamma \neq 0$), a kind of forced convection will however exist, which we here call non-Rayleigh convection.

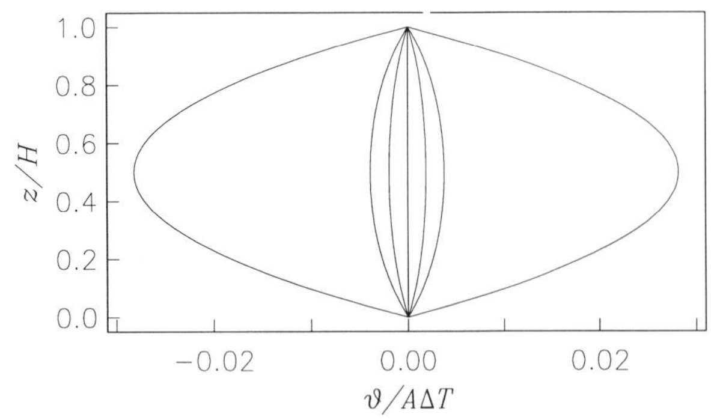

Fig. 2. The non-linear distortion of the temperature profile scaled with $A\Delta T$. From left to right, the profiles are shown for x/L = 0, 0.25, 0.375, 0.5, 0.625, 0.75 and 1. b = 0.1.

This kind of convection is discussed in some detail in Ludvigsen et al., 1991, where a perturbation procedure is used to derive an analytical expression for ϑ. The expression for ϑ for non-Rayleigh convection is quite lengthy and is not given here. ϑ is found to be proportional to $A \equiv R \tan\gamma$, in contrast to $\sqrt{R - R_0/R}$ obtained for Rayleigh convection (see formula 10).

In Fig. 2, $\vartheta/A\Delta T$ is shown as a function of z/H for several values of x/L. The maximal contribution, $\vartheta_{\max}$ to the temperature field due to the fluid motion is here seen to be localized at the end walls and is in our calculations found to be:

$$\vartheta_{\max} = 0.028A\,\Delta T \tag{18}$$

for b = 0.1. The distortion effect depends primarily on permeability (K), height of the permeable layer (H) and the slope of the isotherms (γ). For K = 10 D, H = 25 m and γ = 5°, we would have:

$$\vartheta_{\max} = (0.03)\,\Delta T \tag{19}$$

The calculations give the maximum distortion of the temperature gradient, $\Delta T/H$ due to the fluid motion as:

$$\left(\frac{\partial\vartheta}{\partial z}\right)_{\max} = 0.117A\,\frac{\Delta T}{H} \tag{20}$$

which for the values of KH^2 and γ given above yields:

$$\left(\frac{\partial\vartheta}{\partial z}\right)_{\max} = 0.13\frac{\Delta T}{H} \tag{21}$$

In other words, the distortion of the temperature gradient due to non-Rayleigh convection is found to be 13% for these values of the physical parameters.

If the height of the layer was 40 m instead of 25 m, the maximal deviations in temperature and temperature gradient would be 8% (of ΔT) and 33% (of $\Delta T/H$), respectively. The values of K, H, and γ are believed to be smaller than the values used in the examples above and the deviations in temperature and temperature gradients are probably less than 5% and 10%, respectively.

In Figs. 3 and 4 we have shown the complete temperature profile, here setting A = 25 (a quite exaggerated value) in order to see the effect clearly. The Nusselt number is a function of A and the aspect ratio b. For b = 1, we have:

$$Nu = 1 + 8.5 \times 10^{-4}A^2 \tag{22}$$

For b = 0.1 we have:

$$Nu = 1 + 2.6 \times 10^{-4}A^2 \tag{23}$$

For K = 10 D, H = 40 m and γ = 5° this gives:

$$Nu = 1.007 \tag{24}$$

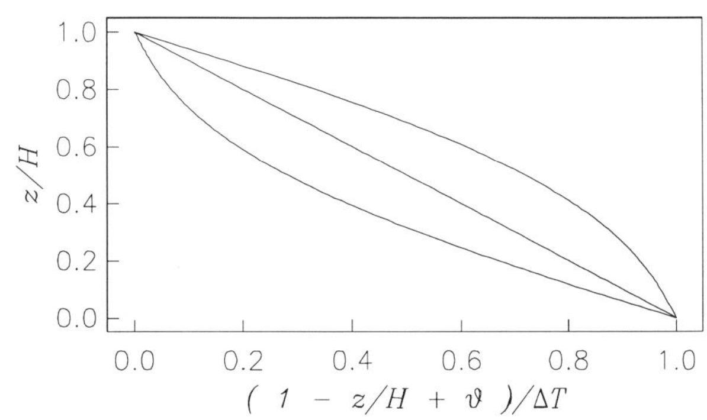

Fig. 3. The temperature profile, scaled with ΔT. A is, for demonstration purposes, set equal to 25. b = 1. The temperature profiles are plotted for (from left to right) x/L = 1/4, 1/2 and 3/4.

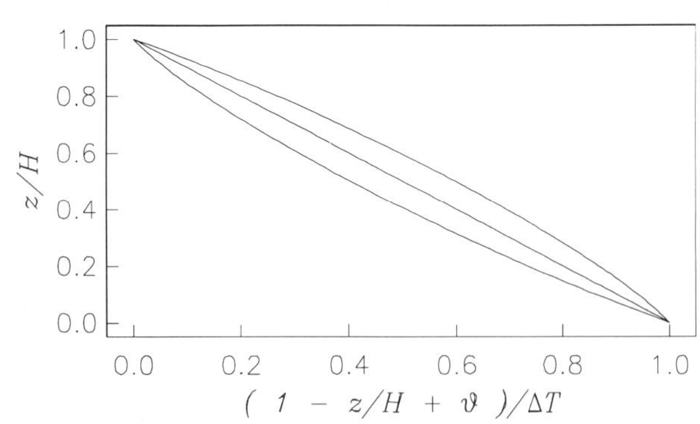

Fig. 4. The temperature profile, scaled with ΔT. A is, for demonstration purposes, set equal to 25. b = 0.1. The temperature profiles are plotted for (from left to right) x/L = 1/4, 1/2 and 3/4. For this (small) value of the aspect ratio, the contribution from ϑ is confined (because the angle between the isotherms and the layer are here set equal to 0) to regions near the end walls and is less noticeable at $x = 1/4L$ and $x = 3/4L$ (compare with the case b = 1 above).

and

$$Nu = 1.002 \tag{25}$$

for aspect ratios of 1 and 0.1, respectively. The contribution to the Nusselt number due to fluid flow is definitely small and, when the isotherms are parallel with the layer, becomes very small for small aspect ratios. This is because non-Rayleigh convection, having a unicellular form (Bories and Combarnous, 1973), only gives significant heat transporting vertical flow in narrow regions near the end walls, whereas the conductive heat transport goes on in the whole extent of the layer.

For analyzing the effects of fluid flow on temperature distribution near the end walls ($x = 0, L$) the assumption that the isotherms are parallel with the layer is not important qualitatively. As in Rayleigh convection, convective contributions proportional to A then cancel each other in the Nusselt number. The local heat flux, q, has a convective contribution, anti-symmetric about $x = L/2$, going from $-0.117Ak\,\Delta T/H$ at the end wall where the fluid goes down, to $0.117Ak\,\Delta T/H$ at the end wall where the fluid goes up.

However, for long, thin layers, the significant effects of fluid flow are located far from the end walls. The temperature distortions are then of the same order of magnitude as given above, except that they are proportional to $\sin(\gamma - \gamma_I)$, where γ_I is the slope angle of the isotherms (Ludvigsen). When $\gamma \neq \gamma_I$, the temperature distortions are then distributed evenly throughout the layer and are not restricted to end-wall regions. The local heat flux far from the end walls is constant and is expected to have a convective contribution of the order $A^2 \sin(\gamma - \gamma_I)$.

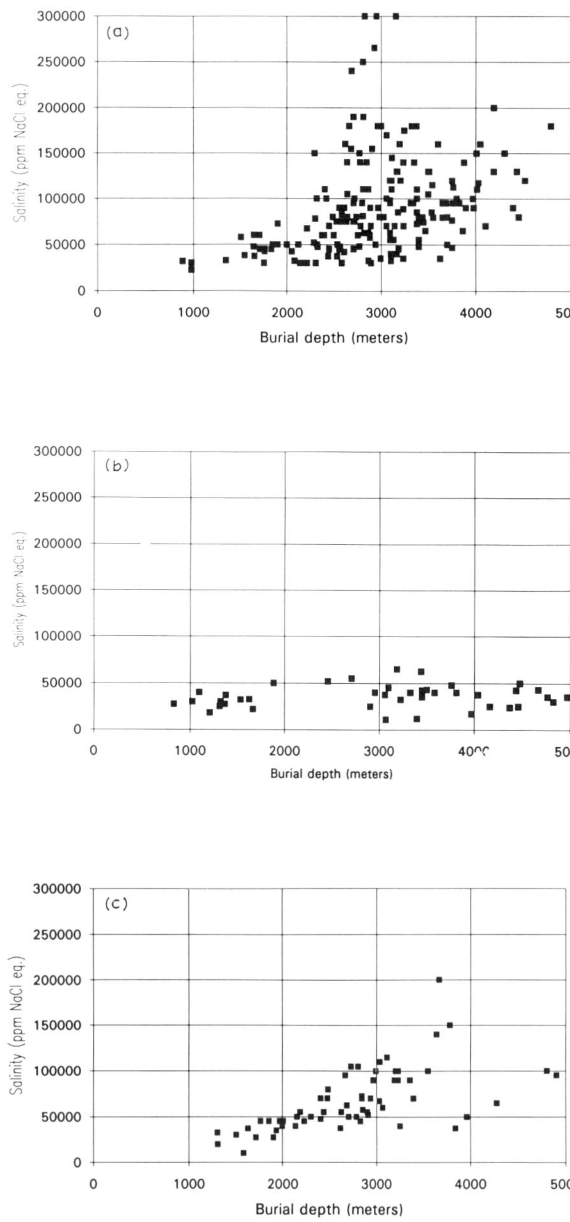

Fig. 5. Salinity versus depth of burial. (a) Southern Norwegian North Sea. (b) Northern Norwegian North Sea. (c) Haltenbanken.

Salinity–depth relationships on the Norwegian shelf; relevance to heat transfer mechanisms

Like heat, salt also may be transported both by diffusion and advection. Convection driven by both salinity and temperature gradients is known as thermohaline (or double diffusive) convection. If thermohaline convection has been going on in a certain region, the salt concentration will tend to be homogenized in this region. Signs of advective transport of salt will indicate if efficient advective heat transfer mechanisms have been present.

Measurement of pore water salinity as a function of burial depth may be important in constraining water–rock interactions and mineral diagenesis. Formation water analyses from the North Sea and Haltenbanken reservoir rocks (Egeberg and Aagaard, 1989) show large compositional variations. Unfortunately, such analyses are generally restricted to single samples from specific reservoir levels over a limited depth interval.

Although less precise than actual formation water analysis, determination of salinity from well log data is very useful for regional studies. Examination of more than 90 wells in the Norwegian North Sea and Haltenbanken basins reveals distinct variation in pore water evolution (Fig. 5). The southern Norwegian North Sea (56–59°N) and Haltenbanken areas are underlain by Permian and Triassic evaporite sediments, respectively. This is clearly reflected in the salinity patterns (Figs. 5a and 5c), with salt concentration gradients increasing with depth, sometimes approaching halite saturation near the top of the

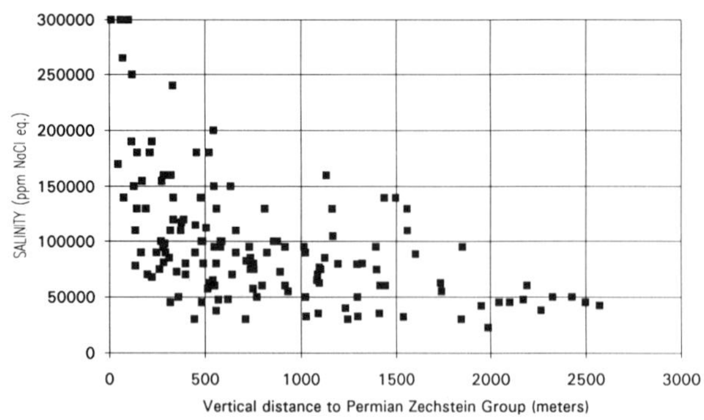

Fig. 6. Formation water salinities versus vertical distance to the top of the Permian Zechstein Group.

evaporites. In the northern North Sea (59–62°N), the formation water evolution is distinctly different (Fig. 5b), with salinity remaining virtually unchanged with increasing burial depth. The log-derived salinities show excellent correlation with reported salinity values from formation water analyses, giving added confidence to these results.

In addition to reflecting the lithological differences between these basins, the concentration gradients limit the various fluid flow models that can be applied in these sediments. Although convection systems have been suggested for transport of solids in North Sea reservoir rocks (Haszeldine et al., 1984; Rabinowicz et al., 1985), mathematical models (Bjørlykke et al., 1988) show that the requirements needed for such flow systems are very unlikely to be satisfied. The concentration gradients observed in the southern Norwegian North Sea and Haltenbanken are additional evidence against large-scale convection, as this would lead to a homogenization of the pore water composition. Increasing salinities with depth cause a density increase, which will partly or totally compensate for the decreased density caused by thermal expansion. A plot of southern North Sea salinity values as a function of vertical distance to the Permian Zechstein Group (Fig. 6) shows that the most rapid change in concentration takes place within the first few hundred metres above the evaporites. Lack of highly saline shallowly buried sequences (<2500 m) (Fig. 5c), indicates that advective flow due to compaction also may not contribute significantly to fluid flow. Instead, with the observed salinity distribution in the North Sea and Haltenbanken basins, it is suggested that diffusion is the main mechanism for transport of solids in these sediments.

Summary

We have discussed the possibility for, and importance of, heat transfer by different types of pore water flow in sedimentary basins. The heat transfer properties of Rayleigh and non-Rayleigh convection are discussed, using mathematical models that take into account the interaction effects between the convective flow and the temperature distribution. We have given quantified results and interpreted them for sedimentary basin conditions.

The conclusions for Rayleigh convection are that it is not very likely to occur in basins like the Norwegian North Sea, where the isotherms slope rather gently. If Rayleigh convection should occur, however, it would, even for small super-critical Rayleigh numbers, be a significant heat transfer mechanism inside the convecting unit. In terms of overall upwards heat flow it would not, however, be very important. The heat flux on a large scale would still be determined by the regional heat flux.

Non-Rayleigh convection will always occur in permeable, fluid-saturated rocks with sloping isotherms, but it is demonstrated to be a less efficient heat transfer mechanism.

Observations of vertical salinity gradients in the southern Norwegian North Sea and Haltenbanken are additional evidence against large-scale convection. Lack of shallowly buried sequences with high salinity, suggests that advective fluid flow due to compaction has also not contributed significantly to re-distribution processes in sedimentary basins. Heat transport in the basins studied (Norwegian North Sea and Haltenbanken) and probably other, similar, basins is essentially controlled by conduction, and advective transport plays a minor role.

Acknowledgements

This research has been supported by VISTA, a research cooperation between the Norwegian Academy of Science and Letters and Den Norske Stats Oljeselskap (Statoil).

References

Bjørlykke, K., Mo, A. and Palm, E., 1988. Modelling of thermal convection in sedimentary basins and its relevance to diagenetic reactions. Mar. Pet. Geol., 5: 338–351.

Bories, S.A. and Combarnous, M.A., 1973. Natural convection in a sloping porous layer. J. Fluid Mech., 57: 63–79.

Caritat, P. de, 1989. Note on the maximum upward migration of pore water in response to sediment compaction. Sediment. Geol., 65: 371–377.

Davis, S.H., Rosenblat, S., Wood, J.R. and Hewett, T.A., 1985. Convective fluid flow and diagenetic patterns in domed sheets. Am. J. Sci., 285: 707–777.

Egeberg, P.K. and Aagaard, P., 1989. Origin and evolution of formation waters form oil fields on the Norwegian shelf. Appl. Geochem., 4: 131–142.

Haszeldine, R.S., Samson, I.M. and Cornford, C., 1984. Quartz diagenesis and convective fluid movement; Beatrice oil field, U.K. North Sea. In: D.J. Morgan and P.F. Rawson (Editors),

Patterns of Mineral Diagenesis on the Northwest European Continental Shelf and Their Relation to Facies and Hydrocarbon Accumulations. Clay Miner., 19: 391–402.
Hermanrud, C., 1986. On the importance to the petroleum generation of heating effects from compaction derived water; an example from the northern North Sea. In: J. Burns (Editor), Thermal Modeling of Sedimentary Basins. Technip, Paris, pp. 247–269.
Lapwood, E.R., 1948. Convection of a fluid in a porous medium. Proc. Camb. Philos. Soc., 44: 508–521.
Ludvigsen, A., Palm, E. and McKibbin, R., 1991. Convective mass transport in porous, sloping layers. Institute of Mathematics, University of Oslo, Preprint Series, 4: 1–31.
Palm, E., 1990. Rayleigh convection, mass transport, and change in porosity in layers of sandstone. J. Geophys. Res., 95 (B6): 8675–8679.
Palm, E., Weber, J.E. and Kvernvold, O., 1972. On steady convection in a porous medium. J. Fluid Mech., 54: 153–161.
Rabinowicz, M., Dandurand, J.-L., Jacubowski, M., Schott, J. and Cassan, J.-P., 1985. Convection in a North Sea reservoir: inferences on diagenesis and hydrocarbon migration. Earth Planet. Sci. Lett., 74: 387–404.
Smith, L. and Chapman, D.S., 1983. On thermal effects of ground water flow, 1. Regional scale systems. J. Geophys. Res., 88: 593–608.
Swift, M., Stagg, H.M.J. and Falvey, D.A., 1988. Heat flow regime and implications for oil maturation and migration in the offshore Northern Carnarvon Basin. In: P.G. Purcell and R.R. Purcell (Editors), The North West Shelf, Australia. Proc. Symp. Petroleum Exploration Society, Perth, W.A., pp. 539–551.
Trevisan, O.V. and Bejan, A., 1990. Combined heat and mass transfer by Natural convection in a porous medium. Adv. Heat Transfer, 20: 315–352.
Weber, J.E., 1975. Thermal convection in a tilted porous layer. Int. J. Heat Mass Transfer, 18: 474–475.

ARILD LUDVIGSEN *Det Norske Veritas Sesam A.S.. Veritasveien 1, P.O. Box 300, N-1322 Høvik, Norway*
KJETIL GRAN *Saga Petroleum A.S., Kjøbovn 16, P.O. Box 490, N-1301 Sandvika, Norway*
ENOK PALM *Department of Mathematics, University of Oslo, P.O. Box 1053, 0316 Oslo 3, Norway*
KNUT BJØRLYKKE *Department of Geology, University of Oslo, P.O. Box 1047, 0316 Oslo 3, Norway*

Thermal disturbances arising from fluid circulation through thin and inclined aquifers

L. Demongodin and G. Vasseur

Convective heat transport may significantly influence the thermal state of a sedimentary basin and is frequently suggested as an explanation for particularities of the present temperature field or observed anomalies in palaeotemperature indicators. Therefore, the convection of heat by fluids should be accounted for in integrated basin models. However, there is a lack of comprehensive rules which would indicate, on the basis of a few parameters, the actual influence of convection in heat transfer. Moreover, the actual groundwater flow is probably much more variable in time and space than that predicted by hydraulic models; because of the low thermal diffusivity of rocks, this could lead to a wrong evaluation of the associated thermal effect.

The aim of this paper is to describe the steady and transient temperature disturbance related to the circulation of fluids trough a thin, inclined and permeable zone which could simulate either a thin aquifer or a fault zone. Such a structure would focus the groundwater flow system and would transfer heat by convection across the surrounding rocks where conduction dominates.

A 2D model is presented in which the flow is characterised by its flow rate, the geometry of the fluid conduit and its thickness. Assuming that horizontal conduction can be neglected, a semi-analytical solution is obtained where the temperature disturbance is described as a function of the Péclet number, of the slope ϕ of the conduit and of the duration of the flow. The validity of the assumption of no horizontal conduction is discussed by comparing with a numerical solution using the finite element method where both horizontal and vertical conduction are considered. The result shows that the agreement depends on the Péclet number α and the slope ϕ. It is better for high α and low ϕ. For a value of slope not exceeding 30°, the analytical solution gives quite acceptable results. Finally, simple rules for estimating the flow required for a given temperature anomaly are proposed.

Introduction

The study of the temperature distribution in sedimentary basins is of great interest, in particular because of its potential application to petroleum exploration. It is therefore highly desirable to assess correctly the role of the different factors governing the transfer of heat in sedimentary basins: heat flow of crustal origin, contrast in the thermal conductivity of sedimentary rocks, convection by active ground water movement and so on.

Recently, much effort has been devoted to the analysis of the present temperature field of sedimentary basins and of its interpretation in terms of thermal processes (Majorowicz and Jessop, 1981; Chapman et al., 1984; Majorowicz et al., 1985; Speece et al., 1985; Willett and Chapman, 1987; Blackwell and Steele, 1988; Brigaud et al., 1992). A common result of such studies, as well as from related theoretical work, is the conclusion that underground water circulation may disturb significantly the temperature field in a basin. Water movement can be the result of various processes. Hydraulic head variation related to topography generates underground flows of meteoric water which can be considered as steady. Another important process is the expulsion of fluid during the burial of sedimentary rocks (Bethke, 1986). The exact nature of this flow is still the object of speculation; for some authors the flow driven by compaction may be episodic (Bonham, 1980; Cathles and Smith, 1983).

Many theoretical and numerical studies have been carried out on the temperature disturbance caused by topography-driven fluid flow (Torrance et al., 1980; Smith and Chapman, 1983; Garven, 1985) or compaction-driven flow (Cathles and Smith, 1983; Bethke, 1986; Hermanrud, 1986). Other works on the problem of injection of fluid in geothermal fields (Bodvarsson and Tsang, 1982) provide interesting approaches. However, analytical and numerical models are generally complicated so that the domain of application is often restricted to simple cases such as horizontal aquifers or steady state. There is a lack of comprehensive and simple rules that would indicate, on the basis of a few parameters, the actual influence of convection in heat transfer.

The purpose of the present paper is to compute the transient and steady-state thermal disturbance

Basin Modelling: Advances and Applications edited by A.G. Doré et al. NPF Special Publication 3, pp. 361–371, Elsevier, Amsterdam.

related to fluid flow, of a given flow rate, within an uneven conduit with sub-horizontal geometry. The case of a horizontal conduit carrying fluids injected along faults has already been investigated by Bodvarsson and Tsang (1982), Ziagos and Blackwell (1986) and Cermak (1989). The geometry considered in the present paper is more complex and includes sloping fluid conduits; this geometry has already been studied by Cathles (1987) but only for the case of a steady-state thermal regime. In the present work, we develop a simple comprehensive model which allows us to define the transient temperature disturbance related to a fluid flow as a function of a small number of parameters. We first define the model and discuss the assumptions made. Then the case of sloping aquifers is considered. Both steady-state and transient effects are discussed through particular solutions showing the fundamental role of the flow rate and geometry, and in particular of the dip angle. The model is also carried out on more realistic geometries commonly encountered in sedimentary basins such as anticlines and synclines. Finally, some practical rules are proposed in order to estimate the actual effect of fluid circulation on the thermal state of a sedimentary basin.

Physical model

The physical model is presented in Fig. 1. It consists of a homogeneous, impermeable, semi-infinite medium crosscut by a thin and permeable fluid conduit. In the matrix, the temperature field is assumed to be entirely governed by conduction. Water circulation is strictly localized in a thin conduit of thickness a and of uneven geometry, cylindrical along direction y and characterized by its depth $h(x)$.

In addition to the assumptions presented above, the behaviour of the system is assumed to satisfy the following conditions.

(1) The thickness a of the conduit is negligible but its specific heat $(\rho c)_c$ is considered.

(2) The physical parameters of the matrix — heat capacity $(\rho c)_r$, thermal conductivity λ — as well as those of the conduit and of the fluid — $(\rho c)_c$ and $(\rho c)_f$ — are constant.

(3) The flow rate Q along the x–z plane is constant in space and time (for $t > 0$).

(4) The thermal resistance between the conduit and the matrix is negligible.

(5) The boundary conditions are imposed at $z = 0$ (ground surface) and $z = +\infty$ (z being positive downwards):

$$T(x,0,t) = 0 \text{ and } \partial T/\partial z|_{+\infty} = \gamma \tag{1}$$

where γ is the temperature gradient.

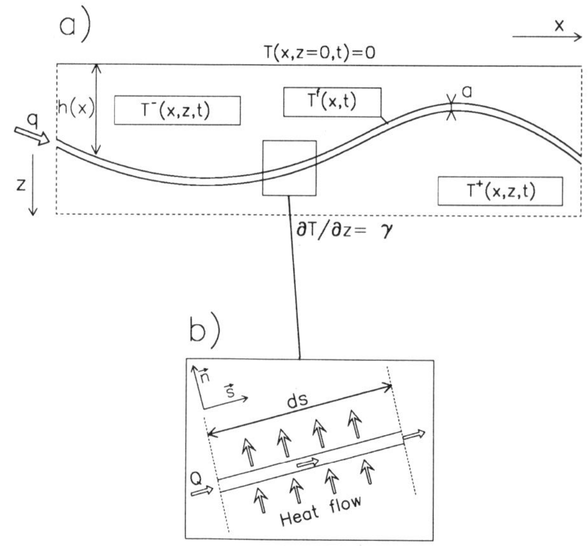

Fig. 1. (a) Simplified sketch of the general model used. The 2D (x–z) cross-section presents a conduit with a geometry characterized by its topography $z = h(x)$, its thickness a and the fluid flow rate Q. (b) Zoom of an element of the conduit showing the energy balance of the various components of the heat flow. The conductive components are represented by arrows perpendicular to the conduit and the convective ones by parallel arrows. For an element of length ds, the rate of heat content can be written as:

$$a(\rho c)_c \, ds \frac{\partial T^f}{\partial t} = ds\lambda \left(\frac{\partial T^+}{\partial n} - \frac{\partial T^-}{\partial n} \right) - ds \, Q(\rho c)_f \frac{\partial T^f}{\partial s}$$

Equation 5 is then easily deduced.

(6) The initial condition corresponds to the conductive equilibrium:

$$T(x,z,t = 0) = \gamma z \tag{2}$$

(7) Above and below the conduit the temperature field (respectively T^- and T^+) satisfies the 2D conduction equation:

$$(\rho c)_r \frac{\partial T^\pm}{\partial t} = \lambda \left(\frac{\partial^2 T^\pm}{\partial x^2} + \frac{\partial^2 T^\pm}{\partial z^2} \right) \tag{3}$$

In eq. 3, the conductivity is isotropic. A more general assumption which will be referred to later, is that the horizontal conductivity λ_x may differ from the vertical one λ_z.

(8) The temperature field has to satisfy a continuity condition at $z = h(x)$:

$$\begin{aligned} T^f(x,t) &= T^+ \quad (x,z = h(x),t) \\ &= T^- \quad (x,z = h(x),t) \end{aligned} \tag{4}$$

where T^f is the fluid temperature.

(9) Along the conduit — $z = h(x)$ — the normal derivative of $T^\pm$ is discontinuous. The condition of discontinuity is obtained by performing an energy

balance in a control volume of the fluid conduit (cf. Fig. 1b):

$$\left.\frac{\partial T}{\partial n}\right|_{-}^{+} = -\alpha\left(\frac{\partial T^{\mathrm{f}}}{\partial s} + \frac{\epsilon}{U}\frac{\partial T^{\mathrm{f}}}{\partial t}\right) \tag{5}$$

where s is the curvilinear coordinate along the conduit in the direction of the fluid flow, n its orientated normal, $\partial T/\partial n|_{-}^{+}$ designates the discontinuity of $\partial T/\partial n$ between the caprock and the bedrock, U the fluid velocity ($U = Q/a$), ϵ the ratio of heat capacity:

$$\epsilon = \frac{(\rho c)_{\mathrm{c}}}{(\rho c)_{\mathrm{f}}} \tag{6}$$

and α the dimensionless Péclet number defined as:

$$\alpha = Q\,\frac{(\rho c)_{\mathrm{f}}}{\lambda} \tag{7}$$

Additional assumption of negligible horizontal conduction

Simultaneous solution of the equations is not analytically possible and must be obtained by a numerical technique. However, a very useful hypothesis is that horizontal conduction may be neglected; indeed, this is justified in most geological situations where vertical gradients are much larger than the horizontal ones. Mathematically, this assumption is introduced by assuming an anisotropic thermal conductivity with $\lambda_x = 0$. In this case, a quasi-analytical solution can be found using the Laplace transform. The inversion from the Laplace domain to real space is possible using a numerical inverter. The Stehfest's algorithm (1970a, b) was found to be very efficient and to give sufficient accuracy, better than 1%.

In order to check on the validity of this simplifying hypothesis of negligible horizontal conduction, a full finite-element numerical approach has been developed, where both horizontal and vertical conduction are allowed. The numerical solutions accounting for both vertical and horizontal conduction are labelled: $\lambda_x = \lambda_z$.

Upward flow along a dipping conduit

In this example, sketched in Fig. 2, the sloping fluid conduit starting at $x = 0$, $z = h_0$ is characterized by its positive angle ϕ with the horizontal. The fluid flow is directed from left to right (toward $x > 0$) and we assume that, at $x = 0$, the temperature of the incoming fluid does not differ from the conductive temperature (i.e. $\Delta T_0 = T_0 - \gamma h_0 = 0$). The temperature disturbance associated with this flow is a result of the vertical component of the fluid movement and of the vertical temperature gradient.

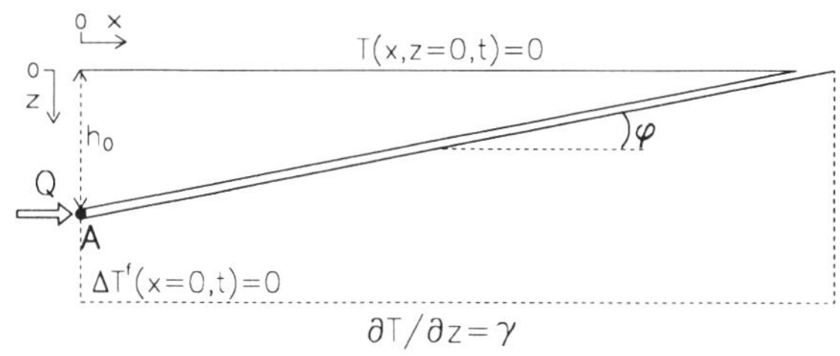

Fig. 2. Cross-section presenting the model of a sloping conduit characterized by the angle ϕ with the horizontal. The entrance of the conduit corresponds to point A: $x = 0$, $z = h_0$. The boundary conditions are imposed at $z = 0$ ($T(x, 0, t) = 0$), at $z = +\infty$ ($\partial T/\partial z = \gamma$) and at the entrance of the conduit ($\Delta T_0 = 0$).

Steady-state solution when using the assumption $\lambda_x = 0$

The disturbed temperature field is the sum of the conductive equilibrium and the disturbance generated by the fluid circulation through the conduit. Because horizontal conduction is neglected, the disturbance is limited to the rising part of the conduit and, in the bedrock, it is independent of depth.

In the conduit, the temperature disturbance $\Delta T^{\mathrm{f}}(x, t = \infty)$ can be conveniently expressed as a function of the depth $h(x)$ of the conduit, for the case ΔT_0 (Cathles, 1987):

$$\Delta T^{\mathrm{f}}(x,t) = \frac{\alpha\gamma\,\tan\phi}{1-\alpha\,\tan\phi}\left[h(x) - h_0\left(\frac{h_0}{h(x)}\right)^{-1/\alpha\,\tan\phi}\right] \tag{8}$$

In Fig. 3, the temperature disturbance in the conduit ΔT^{f}, normalized to γh_0, is plotted as a function of the normalized horizontal distance $x\tan\phi/h_0$ and normalized depth $h(x)/h_0$, for various values of α tan ϕ. The disturbance ΔT^{f} is null for $h = h_0$ (at $x = 0$,

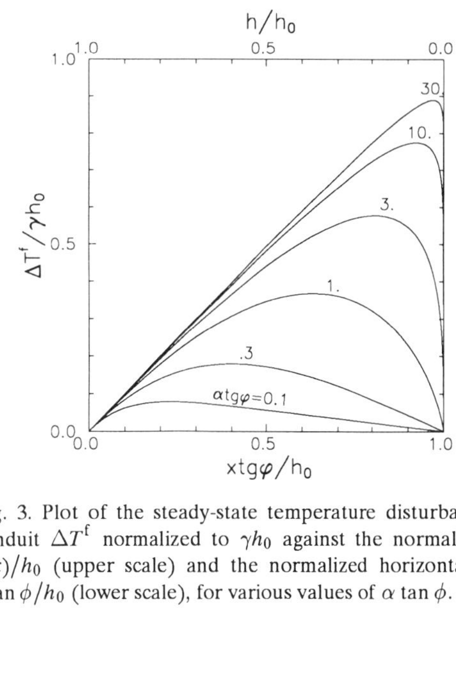

Fig. 3. Plot of the steady-state temperature disturbance in the conduit ΔT^{f} normalized to γh_0 against the normalized depth $h(x)/h_0$ (upper scale) and the normalized horizontal distance $x\tan\phi/h_0$ (lower scale), for various values of α tan ϕ.

origin of the fluid) and for $h = 0$ (discharge). Its amplitude increases with increasing value of $\alpha \tan\phi$. It can also be noted that strong disturbances are asymmetric with maximum values close to the discharge area. The important point here is that the only relevant parameter is $\alpha \tan\phi$ and that the disturbance becomes significant only for values of $\alpha \tan\phi$ larger than 0.1.

Comparison with the steady-state numerical solution

In order to illustrate the effects of horizontal conduction, comparisons are made between the temperature field disturbances obtained using the analytical model ($\lambda_x = 0$) and the numerical ones where both vertical and horizontal conduction are allowed ($\lambda_x = \lambda_z$). An example is presented in Fig. 4 for the same value of $\alpha \tan\phi = 1$ and two values of slope: ϕ=15° and ϕ=30°. For $\lambda_x = 0$, the disturbance is limited to the rising part of the conduit and, in the bedrock, it is independent of depth. By contrast, when the horizontal conduction is allowed (Figs. 4b and 4c), the disturbance propagates in the region $x < 0$ and, below the conduit, the isolines are not vertical but close at some distance. Moreover, the temperature in the host rock and the conduit is systematically lower when horizontal conduction is considered. The effect of the horizontal conduction increases with increasing values of ϕ. This is also illustrated on vertical profiles (Fig. 4d) obtained at the half way point of the fluid conduit. These profiles are characterized by a discontinuity of their derivative at the depth $h(x)$ of the conduit (eq. 5). When horizontal conduction is neglected, the temperature disturbance varies linearly with z from 0 to $T^f(x,\infty)$ for $0 < z < h(x)$ and becomes constant below. When the horizontal conduction is considered, $\lambda_x = \lambda_z$, the amplitude of the disturbance in both the host rock and the conduit is reduced and, below the conduit, it depends on depth.

The effects of the horizontal conduction may be also described by a systematic parametric study of

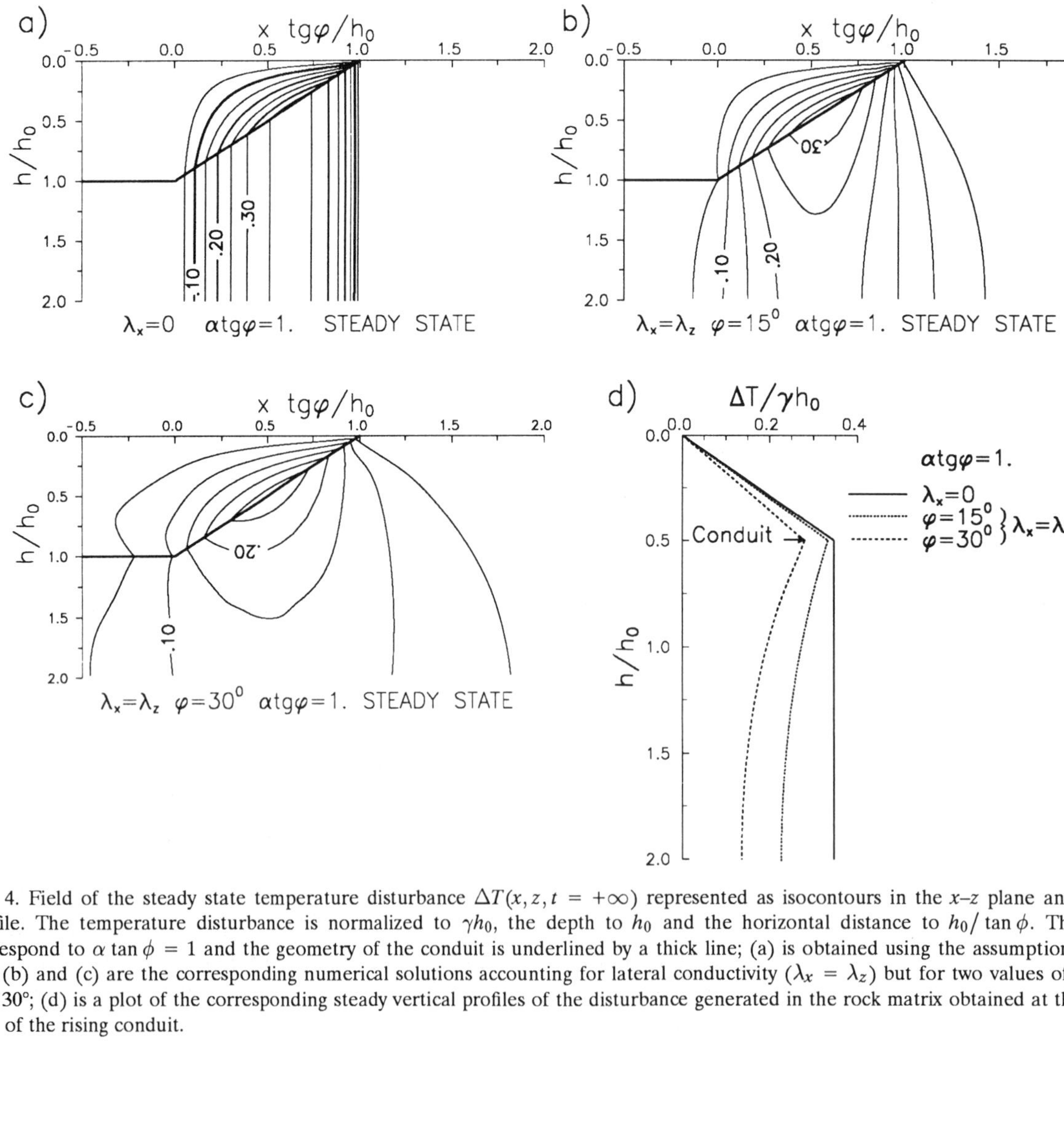

Fig. 4. Field of the steady state temperature disturbance $\Delta T(x, z, t = +\infty)$ represented as isocontours in the x–z plane and vertical profile. The temperature disturbance is normalized to γh_0, the depth to h_0 and the horizontal distance to $h_0/\tan\phi$. The figures correspond to $\alpha \tan\phi = 1$ and the geometry of the conduit is underlined by a thick line; (a) is obtained using the assumption $\lambda_x = 0$, and (b) and (c) are the corresponding numerical solutions accounting for lateral conductivity ($\lambda_x = \lambda_z$) but for two values of ϕ = 15° and 30°; (d) is a plot of the corresponding steady vertical profiles of the disturbance generated in the rock matrix obtained at the middle part of the rising conduit.

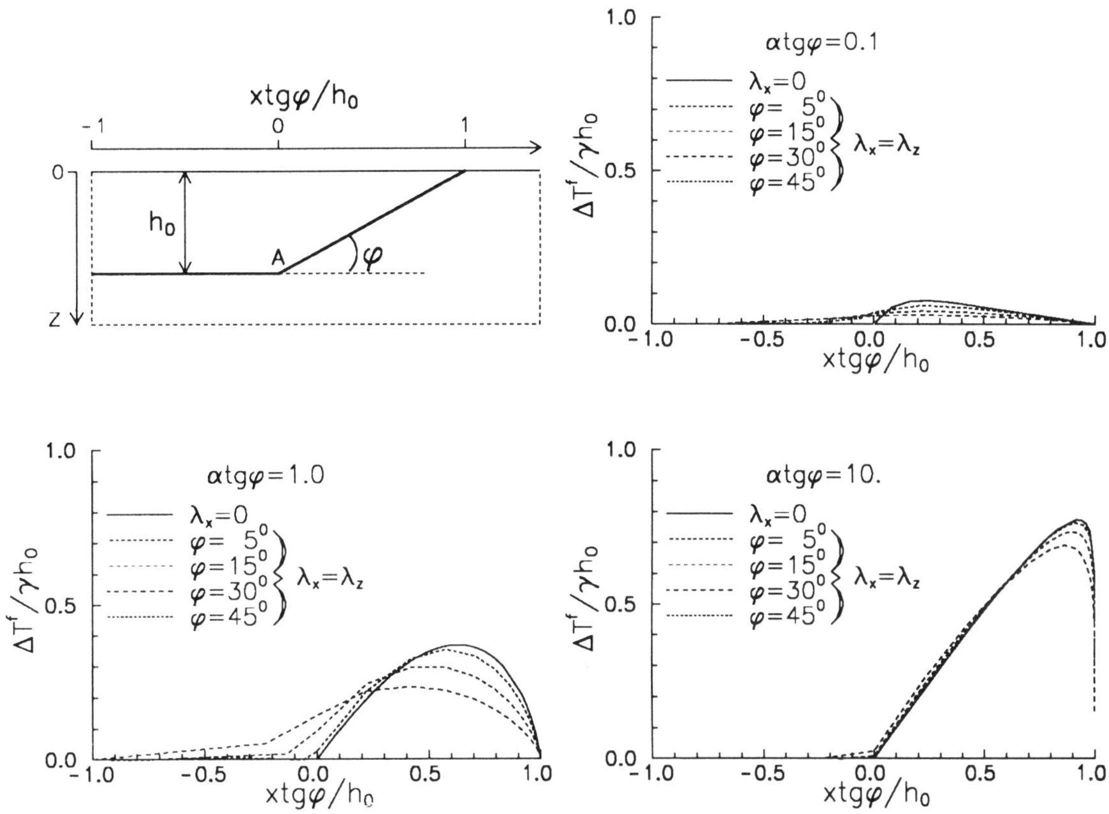

Fig. 5. Plot of the steady-state temperature disturbance ΔT^{f} in the conduit, normalized to γh_0, as a function of the horizontal distance x normalized to $h_0/\tan\phi$. The geometry is summarized on the upper left. The various plots correspond to three values of $\alpha\tan\phi$: 0.1, 1, and 10. On each figure, the disturbance is plotted for different values of ϕ (5°, 15°, 30° and 45°) as well as the quasi-analytical solution (continuous line).

the temperature disturbance along the conduit as a function of the various parameters. Figure 5 shows that, when the horizontal conduction is considered, the disturbance begins in the horizontal part of the conduit (feed back effect) and that its amplitude is lower than that predicted by the $\lambda_x = 0$ model. The difference between the two solutions increases with increasing angle ϕ and decreases with increasing Péclet number. However, for small values of ϕ (5–15°) the solutions are very similar whatever the Péclet number.

The same exercises are performed for $\phi > 45°$. In this case, the approximation of negligible horizontal conduction breaks down. The different solutions obtained for various values of ϕ are presented in Fig. 6. In this case, the chosen parameter is not $\alpha\tan\phi$ but α. In the sloping conduit, the disturbance depends merely on α and the dependence on ϕ tends to vanish.

Transient solution when using the assumption $\lambda_x = 0$

Assuming $\lambda_x = 0$, the transient evolution of temperature following the onset of the flow can be computed numerically. In this case, the Laplace transform of ΔT^{f} depends not only on $\alpha\tan\phi$ but also on the thickness a of the conduit. Figure 7 presents the reduced temperature disturbance $\Delta T^{\mathrm{f}}(x,t)/\gamma h_0$ as a function of the reduced distance $x\tan\phi/h_0$ or of the height in the conduit $h(x)/h_0$, for various times. The time t is normalized to the characteristic diffusion time t_{c} given by $t_{\mathrm{c}} = h_0^2/\kappa$. Three values of $\alpha\tan\phi$ have been tested (0.3, 1, and 3) as well as two values of the ratio a/h_0 (0.01 and 0.1). In each case, the reduced time t' takes the values 0.01, 0.1, 1 and ∞ (steady state).

The major influence of $\alpha\tan\phi$ on the amplitude of the disturbance can be noted, as for the steady-state case. The influence of the thickness parameter a can be appreciated only for very small values of t'.

An interesting behaviour for small values of t' is that ΔT^{f} tends towards a constant value as a function of the distance; in this particular thermal regime, the temperature disturbance remains almost constant along the conduit (except at its beginning and at its end). Analytical developments indicate that, when x is larger that $\alpha\sqrt{\kappa t}$, the thermal disturbance in the conduit ΔT^{f} does not depend on x. For such values of x, it is possible to demonstrate that the thermal disturbance (both in the conduit and the rock matrix)

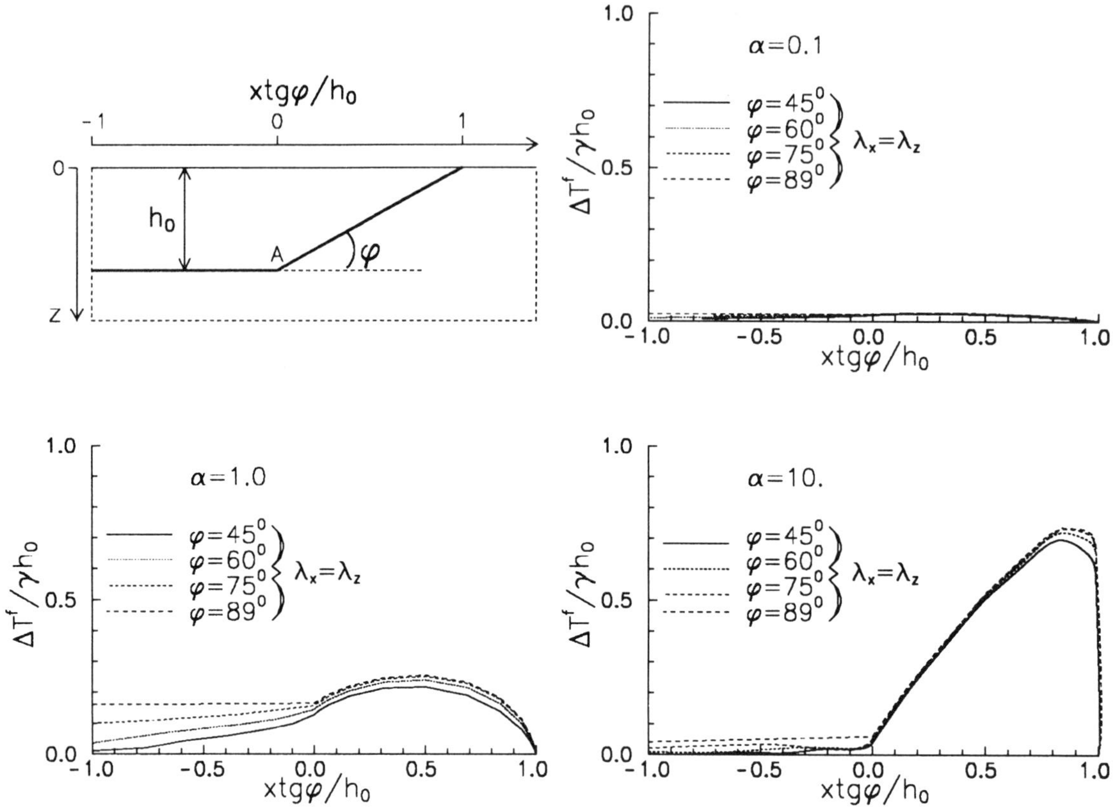

Fig. 6. Plot of the steady-state temperature disturbance ΔT^{f} normalized to γh_0 as a function of the horizontal distance x normalized to $h_0/\tan\phi$, for large values of ϕ. The geometry is summarized on the upper left. The disturbance is plotted for three values of α (0.1, 1, and 10) and different slopes ϕ (45°, 60°, 75° and 90°).

is similar to the perturbation generated by a 1D heat source at $z = h(x)$ (Carslaw and Jaeger, 1959, p. 263). Therefore, at early time, the transient influence of a sloping conduit is similar to the effect of a continuous plane source with a strength defined by:

$$q = \lambda\gamma\alpha\tan\phi \tag{9}$$

It is interesting to note that this strength is precisely equal to the product of the background heat flow ($\lambda\gamma$) multiplied by the Péclet number ($\alpha\tan\phi$); this emphasizes the importance of this latter parameter.

Comparison with the transient numerical solution: $\lambda_x = \lambda_z$

As for the case of steady state, the assumption of negligible horizontal conduction can be discussed through the numerical solution of the heat transfer equation. Vertical profiles, plotted in Fig. 8, are obtained at the half way of the conduit. They show the development with time of the perturbation in the rock matrix. For $t < h(x)^2/\kappa$, the characteristic length of the disturbance is approximately given by $\Delta z = \sqrt{\kappa t}$. When horizontal conduction is allowed, the disturbance is systematically lower and its wavelength is reduced.

A systematic comparison has been performed for a wide range of these parameters. As already noticed for the steady-state solution, the agreement between the numerical solution and the analytical one is quite good for $\phi < 30°$ and becomes worse beyond.

More-complex models

In this section we consider more realistic situations of fluid flow along particular structures that may be encountered in sedimentary basins, such as anticlines and synclines. Due to the complexity of the possible geometries, only a few examples are given. The geometries considered here are made of a succession of straight segments. The semi-analytic method was adapted to a conduit geometry made of successive segments characterized by the same flow rate (same α but different dips ϕ).

Sloping connection between two horizontal levels

Fractures and faults are discontinuities which may be highly permeable pathways. The presence of such structures in sedimentary basins could by-pass the pressure regime between various aquifers and lead to significant flow.

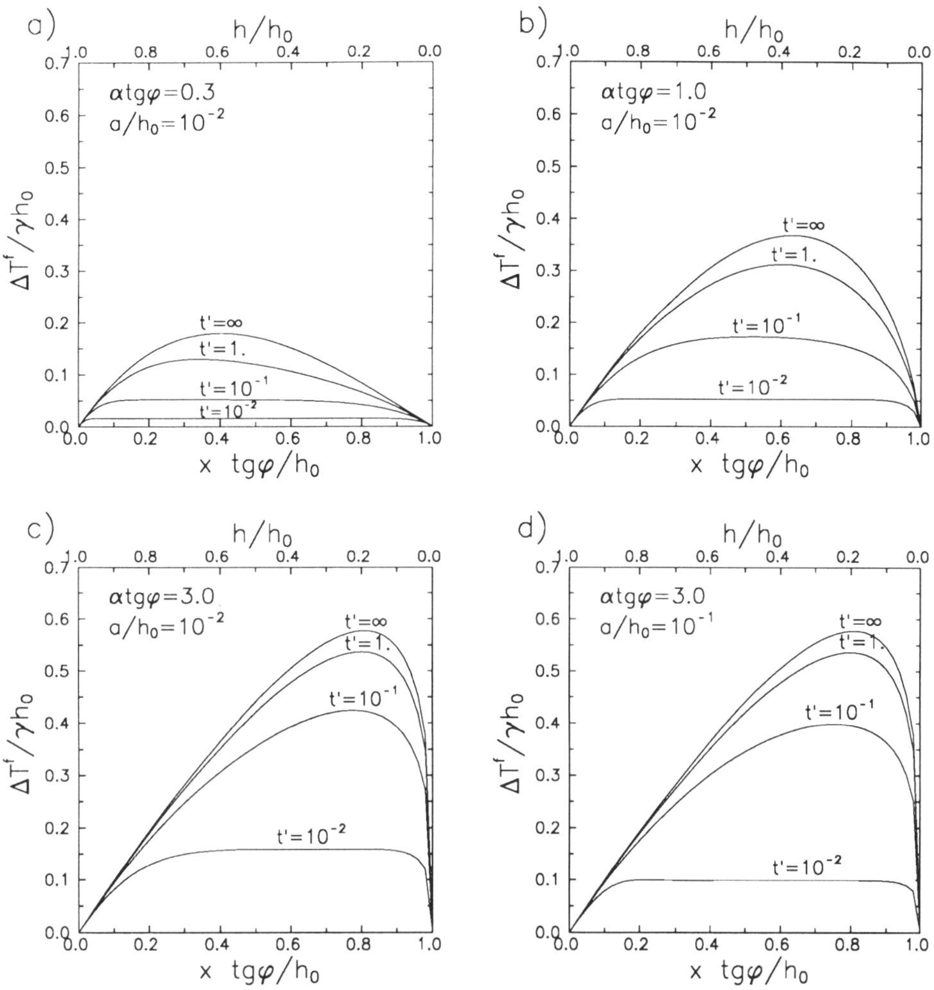

Fig. 7. Normalized temperature disturbance in the conduit $\Delta T^{\mathrm{f}}(x,t)/\gamma h_0$ plotted as a function of the reduced horizontal distance $x \tan\phi/h_0$ for various reduced times $t' = t/t_c$. The upper scale corresponds to the reduced depth of the conduit $h(x)/h_0$. Three values of $\alpha \tan\phi$ have been tested (0.3, 1, and 3) as well as two values of the ratio a/h_0 (0.01 and 0.1: (c) and (d)). In each case, the reduced time t' takes the values 0.01, 0.1, 1 and ∞ (steady state).

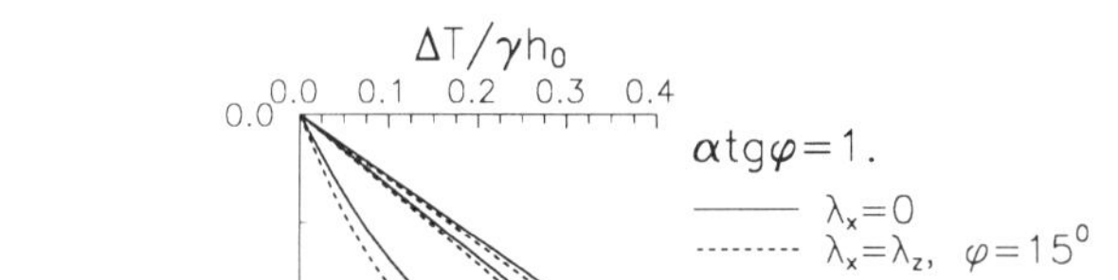

Fig. 8. Plot of the vertical profiles of the transient disturbance generated in the rock matrix obtained at the middle part of the rising conduit for $\alpha \tan\phi = 1$ and three values of reduced time t' (0.1, 1 and ∞). Numerical solutions ($\lambda_x = \lambda_z$) correspond to $\phi = 15°$.

The assumed model is a combination of three connected straight segments: a horizontal semi-infinite line at depth h_0 ($-\infty < x < 0$, $z = h_0$), a sloping segment AB characterized by its angle ϕ with the horizontal, and again a horizontal semi-infinite line at depth h_1 ($[h_0 - h_1]/\tan\phi < x < +\infty$, $z = h_1$). The fluid flow is directed from left to right (toward $x > 0$) and, for $x \to -\infty$ is in conductive equilibrium. For $h_1 < h_0$ (Fig. 9a), the flow has an upward component and, due to the vertical temperature gradient, its effect is that of warming the rock matrix; the reverse applies for $h_1 > h_0$ (Fig. 9b).

Figure 9a corresponds to the case of a rising conduit with $h_1 = h_0/2$ and Fig. 9b to that of a descending one with $h_1 = 2h_0$. The steady-state temperature disturbance along the conduit is normalized to the maximum temperature difference resulting from the depth difference between A and B ($\gamma|(h_0 - h_1)|$). The horizontal coordinate x' is reduced to $h_0 - h_1)/\tan\phi$ so that the sloping part of

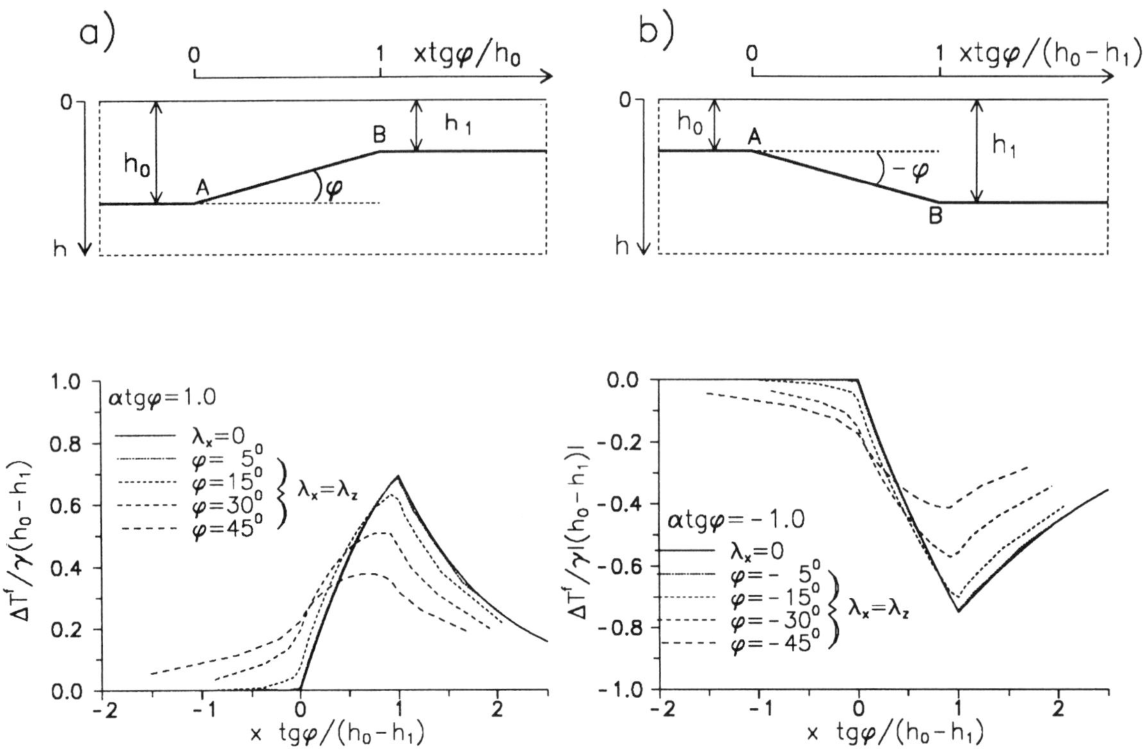

Fig. 9. The upper figures of (a) and (b) define respectively the geometry of a rising and descending conduit connecting two horizontal ones at depths h_0 and h_1 with $h_1 = h_0/2$ and $h_1 = 2h_0$. The corresponding lower plots display the steady state temperature disturbance ΔT^{f} in the conduit, normalized to $\gamma|(h_0 - h_1)|$, as a function of the horizontal distance x normalized to $(h_0 - h_1)/\tan\phi$ for $\alpha \tan\phi = 1$ and different values of ϕ (5°, 15°, 30° and 45°) as well as the quasi-analytical solution (continuous line).

the conduit corresponds to $0 < x' < 1$.

When horizontal conduction is neglected, the value of $\Delta T^{\mathrm{f}}/\gamma|(h_0 - h_1)|$ remains strictly equal to 0 for $x' < 0$ in the initial horizontal part of the conduit; for $0 < x' < 1$, i.e. in the sloping part of the conduit, the temperature disturbance increases in amplitude then it decreases exponentially in the third horizontal part of the conduit. For a given value of $\alpha \tan\phi$ these results do not depend on the actual value of ϕ.

When horizontal conduction is allowed, the resulting curve has the same general behaviour; the main difference is that the temperature disturbance begins for $x < 0$ and that its amplitude is smaller than that obtained for $\lambda_x = 0$. These differences, due to lateral conduction, are relatively smaller for high values of the flow rate characterized by $\alpha \tan\phi$.

Anticlinal and synclinal structures

The anticline case shown in Fig. 10a consists of a symmetric anticline structure starting from depth h_0 and reaching the depth $3h_0/4$ for $x = 0$ (dip angle $|\phi| = 14°$). Only the quasi-analytic solution (approximation $\lambda_x = 0$) is presented as plots of the temperature disturbance in the conduit $\Delta T^{\mathrm{f}}(x,t)/\gamma h_0$ as a function of the reduced horizontal distance x/h_0. The total temperature $T^{\mathrm{f}}(x,t)/\gamma h_0$ in the conduit is also presented. The value of α is fixed at 1.0 and different reduced times t' are considered.

The thermal disturbance develops in two stages, first positive for the upward flow, then negative. For a short reduced time t', this effect corresponds to the regime where ΔT^{f} is independent of x as already noted for the sloping conduit. However, the total temperature in the conduit remains lower than γh_0 and, even at thermal equilibrium, the negative effect prevails for large x values: the flow retains a "memory" of the cooling undergone in the anticline.

The same exercise may be repeated with the same parameters for a syncline. This is illustrated in Fig. 10b. Similar phenomena are observed with a general warming downstream of the structure. It is interesting to note that, as a function of time, the temperature at point C (downstream) reaches a maximum for $t' = t/t_c \sim 1$.

Practical rules and implications

The previous results show that the thermal effect of a fluid circulation can be described by a very small number of parameters: the thickness and the geometry of the conduit for the fluid, the Péclet number which is proportional to the flow rate, and the time since flow initiation. Using dimensionless parameters

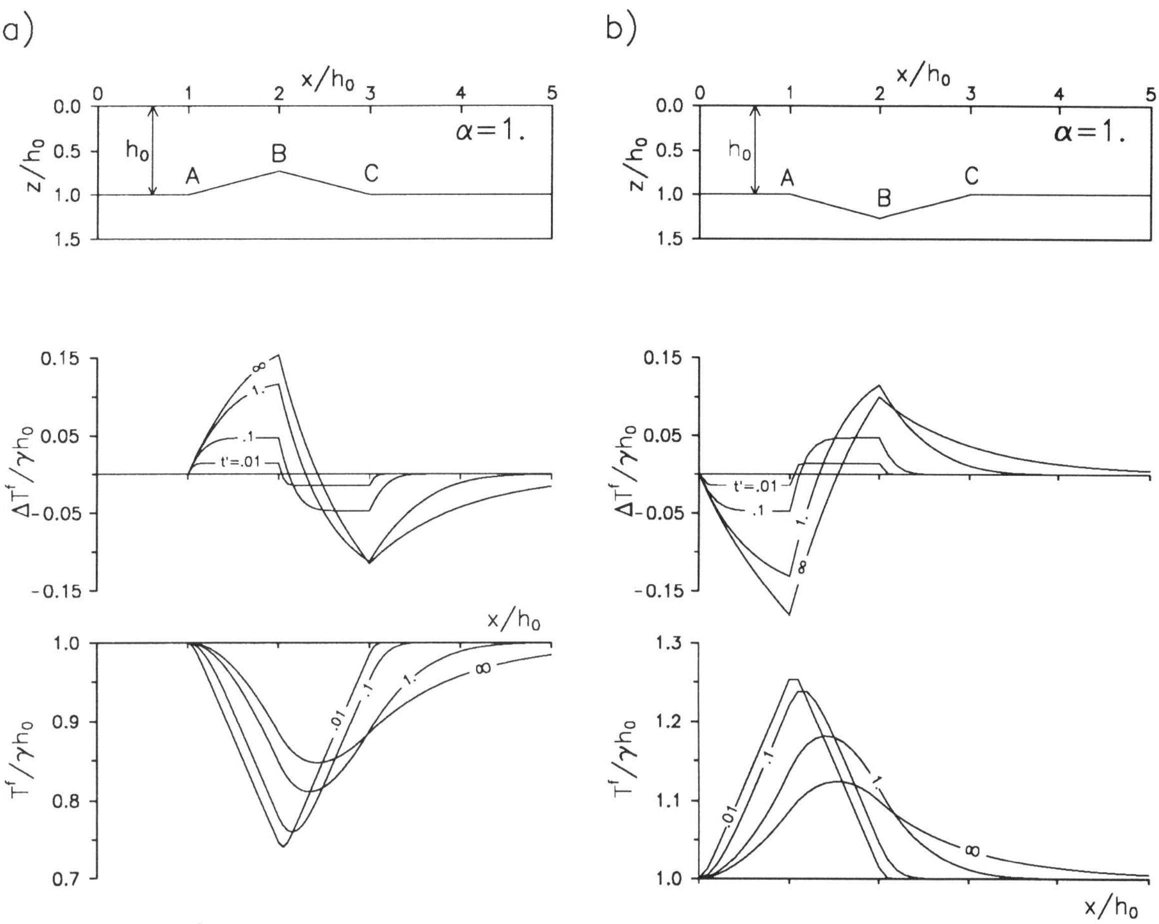

Fig. 10. Transient and steady-state disturbance associated with a fluid flow in a symmetrical anticline (a) and syncline (b). For each case, the upper figure illustrates the corresponding geometry, the middle one gives the variation of the temperature disturbance in the conduit, ΔT^{f} normalized to γh_0, as a function of x/h_0, and the lower one represents the full temperature in the conduit $T^{\mathrm{f}}/\gamma h_0$. Four values of t/t_c ($t_c = \sqrt{h_0^2/\kappa}$ are considered: 0.01, 0.1, 1 and ∞ for the same value of $\alpha \tan\phi = 1$.

it is then possible to propose practical rules which allow prediction of the behaviour of the thermal disturbance in both the conduit and the host rock:

(1) For small values of a/h (10^{-2} or less), the thickness of the conduit has no substantial effect on the temperature field disturbance, except at very early times.

(2) For low values of time, the characteristic thickness of the disturbed zone in the host rock is $\sqrt{\kappa t}$. For a simple geometry such as that of a sloping fluid conduit, the disturbance generated along a vertical profile is similar to that produced by a 1D heat source at depth h, with intensity given by $q_0 \alpha \tan\phi$, where q_0 is the undisturbed heat flow; in particular this disturbance remains constant along the conduit.

(3) For gentle slopes ($|\phi| < 30°$), the thermal disturbance is only controlled by $\alpha \tan\phi$ (and time for the transient case). Small values of $\alpha \tan\phi$ result in negligible thermal disturbances. The development of a 10% thermal disturbance (compared to the undisturbed temperature) requires a value of $\alpha \tan\phi$ of the order of 0.1.

(4) For steep slopes ($|\phi| > 30°$), the effects of the horizontal conduction become important and the quasi analytical model is no longer valid. Numerical solutions show that, for $|\phi| > 45°$, the disturbance along the conduit mainly depends on α.

Large-scale fluid movements in sedimentary basins are frequently suggested from various hydraulic (Bethke, 1989), geothermal (Kilty and Chapman, 1981; Smith and Chapman, 1983) or geological observations (Garven, 1985). These circulations would correspond to stable flow along quasi-horizontal or slightly folded aquifers.

A direct application of the above rules indicates that $\alpha \tan\phi$ must be larger than 0.1 to effect significant redistribution of heat by convection. Since the layer slopes in sedimentary basins are generally small (less than 6°), this would require that α is larger than 1. Using average values of thermal parameters such as:

$$\epsilon' = \frac{(\rho c)_{\mathrm{f}}}{(\rho c)_{\mathrm{r}}} = 2 \text{ and } \kappa = 10^{-6}\ \mathrm{m^2\ s^{-1}}$$

the flow rate corresponding to $\alpha = 1$ is $Q = 15$ m^3 y^{-1}. This flow rate has to be compared to flow rates occurring in actual basins.

In fact, even the present pattern of the deep fluid velocity in sedimentary basins is poorly known and is generally derived from basin-scale hydrodynamic models. An example of such study is that on the deep aquifer layer (Dogger) of the Paris Basin, recently published by Wei et al. (1990). The hydrodynamic modelling led the authors to propose a flow rate in the order of 3 m^3 y^{-1}; since the dip angle of this layer is very small (less than 4°), such a flow rate has a negligible influence on the thermal regime of the basin. However, this conclusion must be modulated by two points:

(1) Aquifers such as the Dogger of the Paris Basin are complex structural systems with spatially varying permeability and fluid velocity. Moreover, geochemical arguments lead one to suspect very rapidly fluctuating fluid flow (Criaud et al., 1988). Therefore, the flow rate may locally reach a sufficient value to disturb the conductive equilibrium. Occurrence of such transient fluid flows seems to be confirmed by continuous thermal logging of a few wells in the basin (Demongodin et al., 1991).

(2) Fractures and faults can also be regarded as potential driving fluid conduits in sedimentary basins. Considering that such structures may have a dip angle much larger than the sedimentary layers, they can be more efficient for carrying heat by convection. However, there is a serious lack of experimental information on their actual role in hydrodynamic processes and particularly on their hydraulic properties.

Conclusions

A quasi-analytical model is developed for computing both transient and steady thermal effects of a 2D geometry fluid flow through thin and inclined conduits. A quasi-analytical solution of heat transport equations is obtained using the approximation that the horizontal conduction is negligible. The validity of this simplifying assumption is systematically verified by comparing the solutions obtained with those of a pure numerical method. The results show that, whatever the flow rate of the circulation, the model is quite acceptable when the slope ϕ of the conduit does not exceed 30° and can be thus applied in most geological contexts.

The prime value of the model is that it describes the thermal disturbance using only four parameters (the geometry and the thickness of the conduit for the fluid, the Péclet number related to the flow rate, and, for the transient case, the time since flow); it also provides some practical rules that allow either to compute the temperature disturbance induced by a given fluid flow or to estimate the required flow rate for explaining observed temperature anomalies. Finally, this work emphasizes the particular value of thermal investigations in sedimentary basins as an indicator of fluid movement occurrences in permeable zones.

Acknowledgements

This work was supported by the Action de Recherche Coordonnée: "Programme Interdisciplinaire de Recherches sur les Sciences pour l'Energie et les Matières premières" (PIRSEM), CNRS AFME; and the Research Group: "Transfert dans les Bassins" (TRABAS).

References

Bethke, C.M., 1986. Hydrologic constraints on the genesis of the upper Mississippi Valley Mineral District from Illinois Basin brines. Econ. Geol., 81: 233–249.

Bethke, C.M., 1989. Modeling subsurface flow in sedimentary basins. Geol. Rundsch., 78: 129–154.

Blackwell, D.D. and Steele, J.L., 1988. Thermal conductivity of sedimentary rocks: measurements and significance. In: N.D. Naeser and T.H. McCulloh (Editors), Thermal History of Sedimentary Basins — Methods and Case Histories. Springer-Verlag, Berlin, pp. 13–36.

Bodvarsson, G.S. and Tsang, C.F., 1982. Injection and thermal breakthrough in fractured reservoirs. J. Geophys. Res., 87: 1031–1048.

Bonham, L.C., 1980. Migration of hydrocarbons in compacting basins. Am. Assoc. Pet. Geol., Bull., 64: 549–567.

Brigaud, F., Vasseur, G. and Caillet, J., 1992. Thermal state in the Viking graben (North Sea) determined from oil exploration well data. Geophysics, 57: 69–88.

Carslaw, H.S. and Jaeger, J.C., 1959. Conduction of Heat in Solid. Oxford University Press, New York, N.Y., 2nd ed., 510 pp.

Cathles, L.M., 1987. A simple analytical method for calculating temperature perturbations in a basin caused by the flow of water through thin, shallow-dipping aquifers. Appl. Geochem., 2: 649–655.

Cathles, L.M. and Smith, A.T., 1983. Thermal constraints on the formation of Mississippi Valley type lead–zinc deposits and their implications for episodic basin dewatering and deposit genesis. Econ. Geol., 78: 983–1002.

Cermak, V., 1989. Heat flow in a sedimentary basin in Czechoslovakia: evaluation of data with special attention to hydrogeology. In: Beck, A.E., Garven, G. and Stegena, L. (Editors), Hydrogeological Regimes and their Subsurface Thermal Effects. IUGG, Geophys. Monogr., 47, 2: 75–80.

Chapman, D.S., Keho, T., Bauer, M. and Picard, M.D., 1984. Heat flow in the Uinta basin determined from bottom hole temperature (BHT) data. Geophysics, 49: 453–466.

Criaud, A., Fouassier, P., Fouillac, C. and Brach, M., 1988. Natural flow and vertical heterogeneities in a sedimentary geothermal reservoir (Paris Basin, France): geochemical investigation. Proc., 13th Workshop Geothermal Reservoir Engineering, Stanford Univ., pp. 133–139.

Demongodin, L., Pinoteau, B., Vasseur, G. and Gable, R., 1991. Thermal conductivity and well logs: a case study in the Paris Basin. Geophys. J. Int., 105: 675–691.

Garven, G., 1985. The role of regional fluid flow in the genesis of the Pine Point deposit, Western Canada sedimentary Basin. Econ. Geol., 80: 307–324.

Hermanrud, C., 1986. On the importance to the petroleum generation of heating effects from compaction-derived water: an example from the northern North Sea. In: J. Burrus (Editor), Thermal Modeling in Sedimentary Basins. Technip, Paris, pp. 247–270.

Kilty, K. and Chapman, D.S., 1981. Convective heat transfer in selected geologic situations. Groundwater, 18: 386–394.

Majorowicz, J.A. and Jessop, A.M., 1981. Regional heat flow patterns in the Western Canadian Sedimentary Basin. Tectonophysics, 74: 209–238.

Majorowicz, J.A., Jones, F.M., Lam, H.L. and Jessop, A.M., 1985. Regional variations of heat flow differences with depth in Alberta, Canada. Geophys. J. R. Astron. Soc., 81: 479–487.

Smith, L. and Chapman, D.S., 1983. On the thermal effects of groundwater flow, 1. Regional scale systems. J. Geophys. Res., 88: 593–608.

Speece, M.A., Bowen, T.D., Folcik, J.L. and Pollack, H.N., 1985. Analysis of temperatures in sedimentary basins: the Michigan Basin. Geophysics, 5: 1318–1334.

Stehfest, H., 1970a. Algorithm 368. Numerical inversion of Laplace transforms. Comm. Assoc. Comp. Mach., 13: 47–49.

Stehfest, H., 1970b. Remark on Algorithm 368. Numerical inversion of Laplace transforms. Comm. Assoc. Comp. Mach., 13: 624.

Torrance, G., Chan, V.W. and Turcotte, D.L., 1980. A model of hydrothermal convection in an aquifer. J. Geophys. Res., 85: 593–608.

Wei, H.F., Ledoux, E. and de Marsily, G., 1990. Regional modelling of groundwater flow and salt and environmental tracer transport in deep aquifers in the Paris Basin. J. Hydrol., 120: 341–358.

Willett, S.D. and Chapman, D.S., 1987. Analysis of temperatures and thermal processes in the Uinta basin. In: C. Beaumont and A.J. Tankard (Editors), Sedimentary Basins and Basin-Forming Mechanisms. Can. Soc. Pet. Geol. Mem., 12: 447–461.

Ziagos, J.P. and Blackwell, D.D., 1986. A model for the transient temperature effects of horizontal fluid flow in geothermal systems. J. Volcanol. Geotherm. Res., 27: 371–397.

L. DEMONGODIN *Centre Géologique et Géophysique (UPR 361, CNRS), USTL, 34095 Montpellier Cedex 5, France*

G. VASSEUR *Centre Géologique et Géophysique (UPR 361, CNRS), USTL, 34095 Montpellier Cedex 5, France*

Predicting tectonic fractures and fluid flow through fractures in basin modelling

Kurt W. Larson, Douglas W. Waples, Han Fu and Kisaburo Kodama

The main value of quantitative 2D and 3D basin models for petroleum exploration is their ability to predict expulsion and migration of hydrocarbons on a basin scale. In calculating fluid flow, however, existing basin models ignore the contributions of fracture permeability. Fracture permeability may in fact be much larger than matrix permeability, particularly for fine-grained rocks. Wherever significant fracturing occurs, ignoring fracture permeability may cause important errors in prediction of groundwater flow rates and directions, cap-rock permeabilities, migration pathways for hydrocarbons, and thermal history.

We have developed a model to predict the formation of tectonic fractures during basin simulations. We evaluate the strain between horizontally adjacent nodes and estimate the intensity of fracturing that will occur as a function of rock ductility and applied stress. Our model is currently only capable of considering stresses induced by vertical block faulting.

Our model calculates fracture-system permeability in either of two ways: (1) from strain directly using an empirical calibration, or (2) from consideration of fracture aperture, density, and orientation. The relative merits of these two approaches still need to be evaluated more fully. Total permeability is then calculated as the sum of matrix permeability and effective fracture-network permeability.

Fractures are allowed to close after termination of the event which created them. Currently, our very simple model for fracture closure assumes an exponential decrease in fracture aperture with time (theoretical method), or in fracture permeability with time (empirical method).

Our model is a highly simplified version of an exceedingly complex reality. Future development of this model could focus on making fracturing dependent on additional factors — such as porosity, temperature, and effective stress — not considered in our model. Future models should also improve the simulation of fracture-closing by distinguishing between tectonic and diagenetic closure. Fracturing caused by concentric folding should also be considered.

Predictions of our model regarding strain have been successfully compared with those of a model based entirely on tectonic stresses. We are encouraged about the potential for inclusion of fracture permeability in basin models in the future.

Application of our model confirms that fluid flow can be strongly affected by fracture permeability. In areas where fracture permeability is important, decisions about hydrocarbon exploration made on the basis of models that ignore fracture permeability may be erroneous.

Introduction

Statement of the problem

Basin fluid-flow models have previously proven their ability in many regions to predict present-day fluid-flow conditions from reconstructions of geologic history (e.g., Bethke, 1985, 1986a, b; Garven, 1985, 1989; Bethke and Marshak, 1990; Ungerer et al., 1990). Most such models calculate fluid flow on the basis of matrix permeability alone. In regions where rocks have been fractured, however, fracture permeability may exceed matrix permeability by several orders of magnitude. Ignoring fracture permeability may therefore lead to erroneous conclusions regarding flow directions and durations, thermal history, diagenetic alteration, and reservoir and cap-rock quality. Some models permit fluid flow through fault systems (e.g., Ungerer et al., 1990), but to our knowledge none attempts to deal with fracture permeability that is not associated with major faults. Furthermore, models which do allow flow through fault zones do not attempt to predict the occurrence of such faults, but rather require the user to recognize their presence independently and insert them into the model.

The present study had two goals: to develop a method to predict the occurrence of fracturing based on fundamental principles of tectonics and petrophysics, and to predict the permeabilities of such fractures. To meet the needs of those wishing to carry out fluid-flow modelling as an aid to exploration, our method had to be simple, scientifically acceptable, and capable of integration into existing basin models without adversely affecting the complexity or expense of those models.

Basin Modelling: Advances and Applications edited by A.G. Doré et al. NPF Special Publication 3, pp. 373–383, Elsevier, Amsterdam.

Possible solutions

There are several ways in which the effects of fracture permeability can be included in basin models. One approach would be to develop a model that solves for the effects of fracturing on fluid flow through the use of independent variables for fracture and fluid flow coupled through governing equations. This approach, while perhaps the most valid scientifically, presents several difficult technical problems (derivation of coupled governing equations, determination of properties to use in the model, and computation time), and would require a very long development time.

A second approach would be to combine through the use of an interface an existing model that predicts fracturing (e.g., Kodama et al., 1985) with an existing basin model that considers only matrix permeability. A data grid of fracture characteristics or fracture permeability through space and time would be created by the fracture model. The interface would then interpolate the data grid into the fluid-flow model before the fluid-flow model is executed. This approach, while technically feasible, would be cumbersome to use.

A third approach is to begin with an existing fluid-flow model and attempt to estimate tectonic fracturing effects through the use of information that has already been calculated (or can be calculated with only minor modification) by the fluid-flow model. However, this approach is technically feasible and scientifically valid only if the occurrence of fractures in time and space can be calculated reasonably accurately from simple computations that fit easily into the structure of the fluid-flow model. We chose this third method for our study, and used Basin2 (Bethke, 1985) as the fluid-flow model for this research.

Strain

Definition

Strain is induced in a rock when the rock is stretched or compressed relative to some initial state. For our purposes, strain is defined as the change in dip-direction width of a nodal block, divided by the reference dip-direction width (Fig. 1). Calculation of the reference dip-direction width is discussed later.

In Basin2 the spacing in the x-direction is constant, so strain is effectively a function of variation in subsidence rates between horizontally adjacent nodal blocks. Strain within a single nodal block due to vertical compression, compaction, dilation, or other processes is not considered. Strain as defined here is strain due to vertical simple shear and not pure shear.

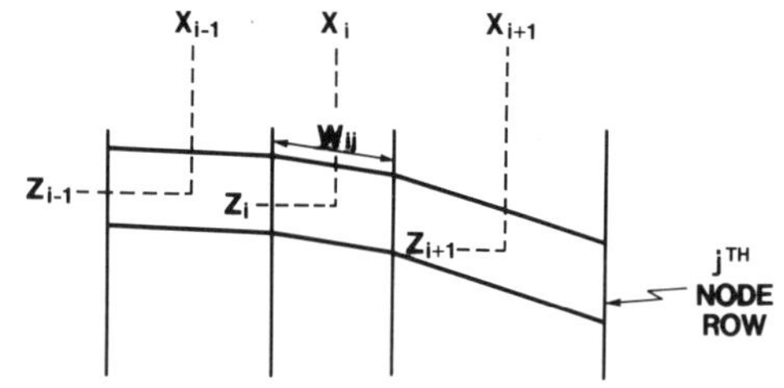

$$W_{ij} = \frac{\sqrt{(X_{i-1}-X_i)^2+(Z_{i-1}-Z_i)^2} + \sqrt{(X_i-X_{i+1})^2+(Z_i-Z_{i+1})^2}}{2}$$

$$\text{STRAIN}_{ij} = \frac{W_{ij}-W_{refij}}{W_{refij}}$$

WHERE W_{refij} IS W_{ij} AT THE LAST STRAIN REVERSAL

Fig. 1. Dip widths (W) and strain in our fracture model. The nodal block of interest is designated ij; that is, the ith block in the jth row. X and Z refer to the x and z coordinates of the centres of the nodal blocks. The subscript ref refers to the reference nodal block.

Basin-scale effects

Kodama et al. (1985) have studied the distribution of strain due to essentially vertical basement deformation in fractured clastic and volcaniclastic rocks of the Niigata Basin, northwestern Japan. Their approach, called the Virtual Basement Displacement (VBD) method, uses governing equations based on rock mechanics and a finite-element numerical solution. It allows nodal points to move in both the x (horizontal) and z (vertical) directions. The VBD method moves the nodes of the virtual basement (which represents a surface of convenience that need not coincide with real basement) through deformation vectors and solves for strain in all elements of the grid. VBD vectors are constrained in magnitude by known surface conditions such as sediment thickness and topography.

Because the governing equations for the VBD method in its current state of development are not time-dependent (that is, rates of change are not considered), this model cannot be directly linked to transient fluid-flow models, in which time rates of change are important. Moreover, the current version of the VBD method does not consider several important geological phenomena that may affect rock strength and certainly affect fluid flow (e.g., fluid pressures and sediment compaction).

The VBD method helps us understand the relationship between basement deformation and regions of highly strained or fractured rock. In their study of the Niigata Basin Kodama et al. (1985) found that strained rocks occurred predominantly in vertical columns above those regions where basement

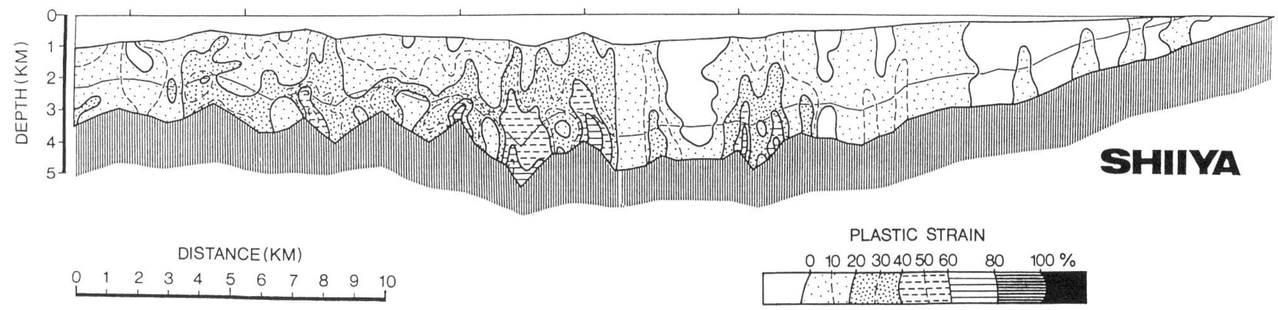

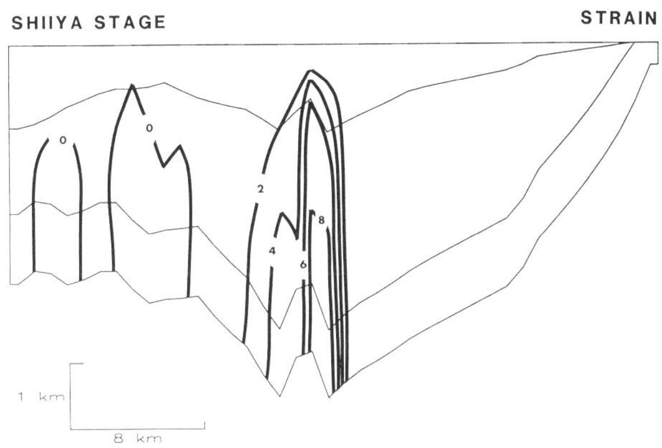

Fig. 2. Comparison of strain calculated during the Shiiya stage (late Late Miocene) in the Niigata Basin, Japan, using (top) the VBD method of Kodama et al. (1985) and (bottom) the model described in this paper. Although strain is shown in percent in both cases, actual strain numbers are not directly comparable due to different modes of calculation related to fundamental dissimilarities between the two models. However, patterns and trends should be similar.

subsidence rates varied greatly across the domain (Figs. 2–4). Strain tended to reach a maximum in the deepest strata and decrease upward. Surface rocks several km above intensely strained regions were still strained a large amount, whereas in regions of the basin where subsidence rates were relatively uniform, surface rocks showed little strain (e.g., Fig. 3).

Method of basin deformation

The deformation model used in Basin2 (and most other basin models: e.g., Ungerer et al., 1990) differs fundamentally from the VBD method, primarily because most basin models are not designed to consider basement deformation as a cause of other geologic phenomena. Basin2 instead implicitly considers basement deformation to be the result of sediment deposition, sediment compaction, topographic changes of the sediment surface, and sea-level changes. In simulations using Basin2, basement is moved automatically through time and space to accommodate the processes listed above; the user has no direct control over basement deformation. Moreover, nodal points in Basin2 are fixed in the x dimension and are free only in the z dimension.

In spite of this lack of direct control over basement deformation, Basin2 can to some degree simulate vertical basement deformation by proper manipulation of sedimentation rates, topography, and water-depth changes across the domain surface. Complex, small-scale basement features cannot be modelled, however. The important consideration for our purposes is that Basin2 can mimic basin deformation of the type studied in the Niigata Basin by Kodama et al. (1985). Because strained rocks occur predominantly in vertical columns above areas of basement deformation, Basin2 can predict strain by comparing the subsidence rates of adjacent columns of nodal blocks.

Comparison of strain predicted by Basin2 and the VBD method

In order to compare strain calculated using Basin2 with that calculated using the VBD method, we duplicated in Basin2 the stratigraphic thicknesses and sediment-surface morphology as shown in the Niigata Basin simulation of Kodama et al. (1985). We did not fit the rock properties (e.g., matrix permeability, compressibility, etc.) of our model to real data for the Niigata Basin, so our flow model may not be correct. We nevertheless believe that our modelling and conclusions are qualitatively valid.

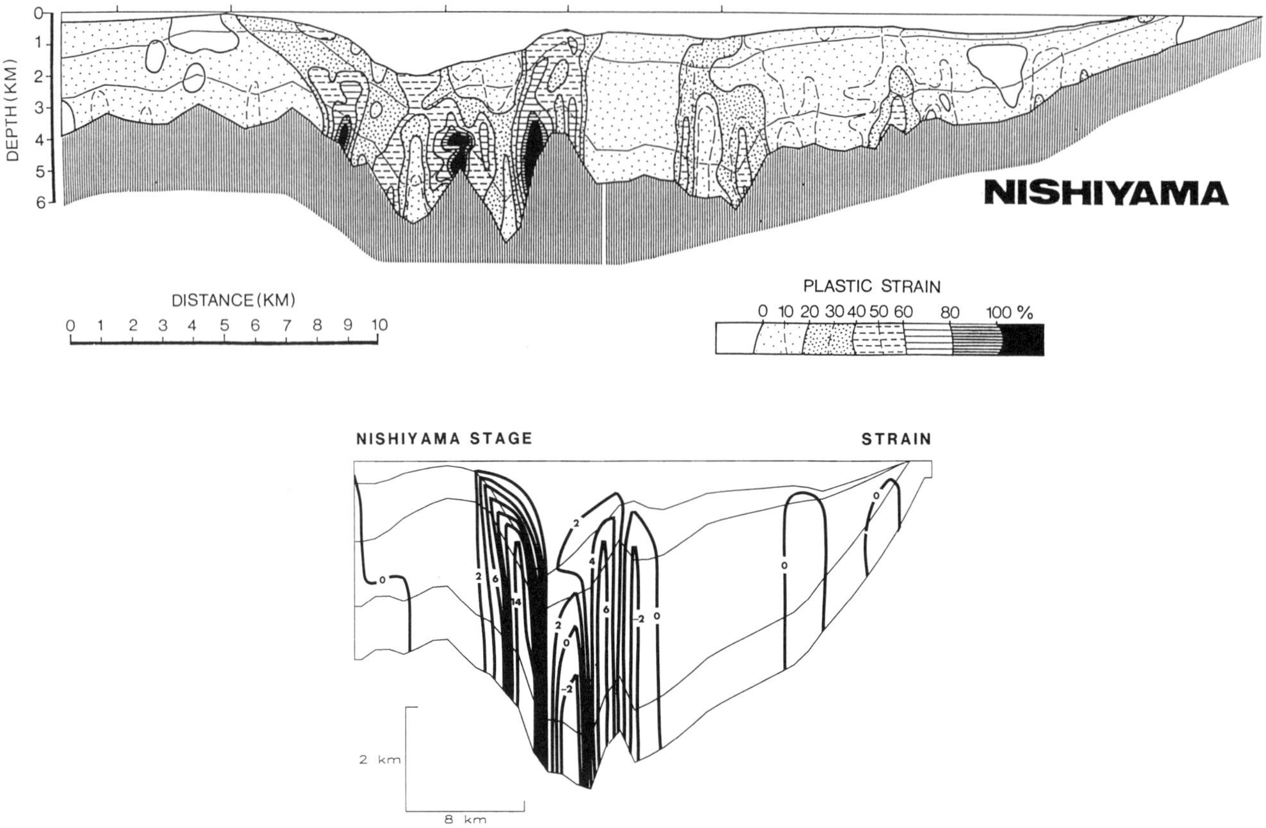

Fig. 3. Comparison of strain calculated during the Nishiyama stage (Pliocene) in the Niigata Basin, Japan, using (top) the VBD method of Kodama et al. (1985) and (bottom) the model described in this paper. See comments in Fig. 2.

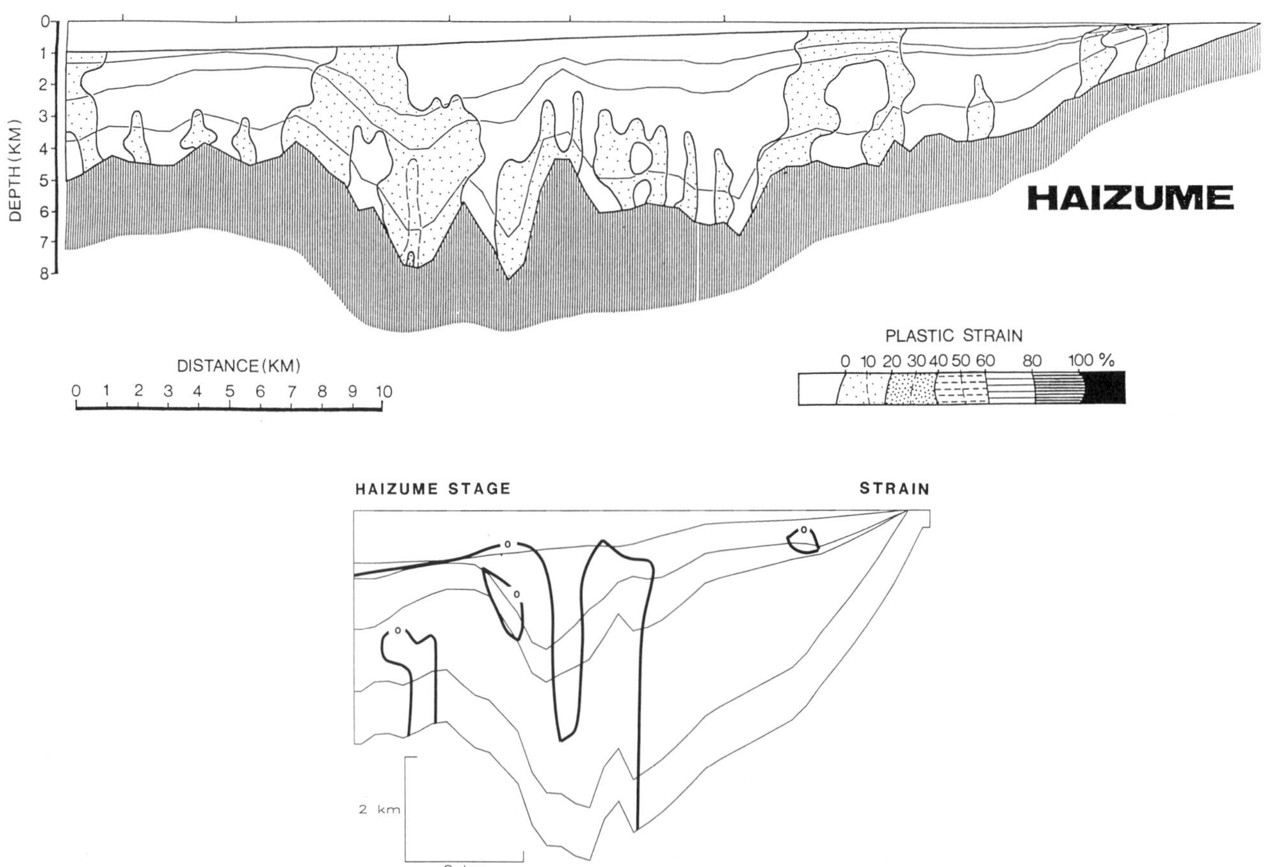

Fig. 4. Comparison of strain calculated during the Haizume stage (Early Pleistocene) in the Niigata Basin, Japan, using (top) the VBD method of Kodama et al. (1985) and (bottom) the model described in this paper. See comments in Fig. 2.

Comparison of the results from the two simulations (Figs. 2–4) shows that Basin2 can predict strain in approximately the same regions as the more rigorous VBD method. (The formulae used to calculate strain in the two models are quite different and the actual numbers plotted in Figs. 2–4 cannot be compared directly.) However, because Basin2 cannot recreate local details in basement morphology that the VBD method handles easily, some regions of strain predicted by the VBD method cannot be duplicated exactly by Basin2.

Behaviour of strained rocks

Rocks accommodate strain through several physical processes. At low total strain (<1–2%), rock is elastic and no permanent deformation remains after the strain-inducing stress is removed. No fracturing occurs during elastic strain. At higher total strain, the rock will deform permanently. Strain may be accommodated by plastic deformation, during which the shape and spatial relationship of grains change without fracture. Alternatively, one of several types of fracturing may occur.

The response of a given rock to strain in excess of its elastic limit is highly variable, but in most cases rock will deform plastically for the first few percent of strain, and then will fracture. With increasing temperature and effective confining pressure, rocks accommodate increasingly more strain through the plastic deformation process. The amount of strain, including elastic deformation, that a rock can accommodate prior to faulting is termed "ductility". The ductility of several rock types as a function of depth (corrected for temperature and fluid pore pressure) has been compiled in Fig. 5 (Handin et al., 1963). Exponential curves expressing ductility as a function of effective stress (Table 1) have been fit to the data in Fig. 5 using eq. 1:

$$\text{Ductility} = k_1 \times \exp(k_2 \times \text{effective stress}) \qquad (1)$$

where k_1 is the ductility under surface conditions (effective stress = 0).

TABLE 1

Constants used in eq. 1 to calculate ductile limits for various rock types at different subsurface depths, assuming normal temperature gradients (from Handin et al., 1963)

Rock type	k_1	k_2
Sandstone	1.25	9.43×10^{-4}
Shale	1.25	1.60×10^{-3}
Limestone	1.25	2.17×10^{-3}
Dolomite	1.25	5.36×10^{-4}

Ductility is expressed as a percent of original dimension; effective stress in atmospheres.

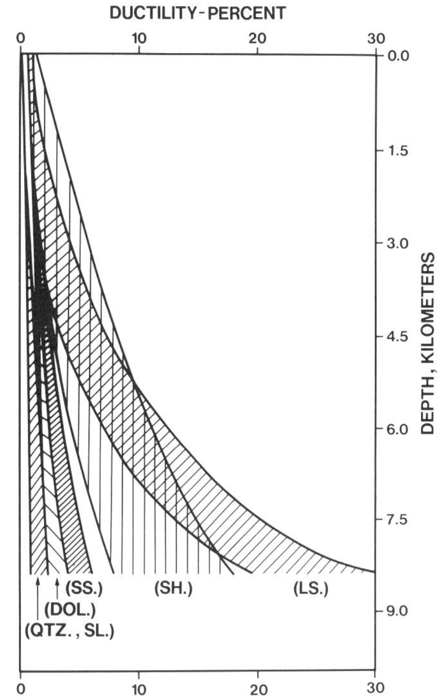

Fig. 5. General trends of ductility of several rock types as a function of burial depth. Adapted from Handin et al. (1963).

The curves in Fig. 5 and constants in Table 1 describe ductility in rocks with negative (compressive) strain. In many cases, however, we will need the corresponding ductility curve for rocks in extension. The relationship between compressional deformation along one axis and the corresponding extensional deformation along the axis normal to the compressional axis can be calculated through the use of Poisson's ratio. For a natural body of rock whose horizontal breadth greatly exceeds its vertical thickness, the horizontal extension corresponding to a vertical compression C will be less than C. Rocks in horizontal extension will therefore fracture sooner than the ductility curves in Fig. 5 indicate.

At the present time we are using the ductility curves in Fig. 5 and constants in Table 1 for rocks in both compressional and extensional strain. We will thus slightly underestimate the number of fractures that occur in rocks under extensional strain. The formula used to calculate ductility should in the future be modified to consider Poisson's ratio.

Total strain, ductility, and fracturing

During a basin simulation, Basin2 will calculate the strain in each nodal block. (The reference frame for the strain will be changed periodically for certain types of simulations, as discussed later under "Strain history and fracture events".) In addition, ductility for each rock type will be calculated for every node at each time step. We assume that any strain less than the ductility will not cause fracturing, and that any strain greater than ductility will.

Permeability

Theoretical vs. empirical methods

We can use one of two methods to calculate the effect of strain on the fracture permeability. The first approach, which we call the theoretical method, calculates the number of fractures that will occur as a function of: (1) fracture orientation; (2) fracture aperture; (3) horizontal displacement across the plane of fracture; and (4) the difference between total strain and ductility. We can specify functions that relate aperture, horizontal displacement, and orientation to depth, effective stress, temperature, etc., in order to simulate the transition from one type of fracturing to another (for example, from single shear to conjugate network) as the rock is buried. The theoretical method then calculates fracture permeability as a function of fracture aperture, orientation, and number of fractures. The second approach, termed here the empirical method, calculates fracture permeability directly as a function of the difference between total strain and ductility for each rock type.

Theoretical method

Introduction. The theoretical method calculates fracture permeability in a series of steps, beginning with computation of strain and ductility for each nodal block. If strain is greater than ductility, fractures form according to the strain and ductility history of the rock (see "Strain history and fracture events" below).

As noted above, we calculate the number of fractures in each rock type in each nodal block from considerations of fracture orientation, horizontal displacement on fracture surfaces, and fracture aperture. At present these critical variables are handled as defaults for each rock type, but the fracture model should eventually specify these variables as functions of porosity, effective stress, etc. (e.g., Hattori et al., 1984).

If the strain conditions in a nodal block change from compressive to extensional, or vice versa, new fractures with different properties are allowed to form. Conversely, old fractures begin to close due to changes in stress regime and to precipitation of mineral cements. We have included a very simplistic model for fracture closure that will be discussed later.

Fracture aperture, horizontal displacement, and orientation. If strain is greater than ductility,

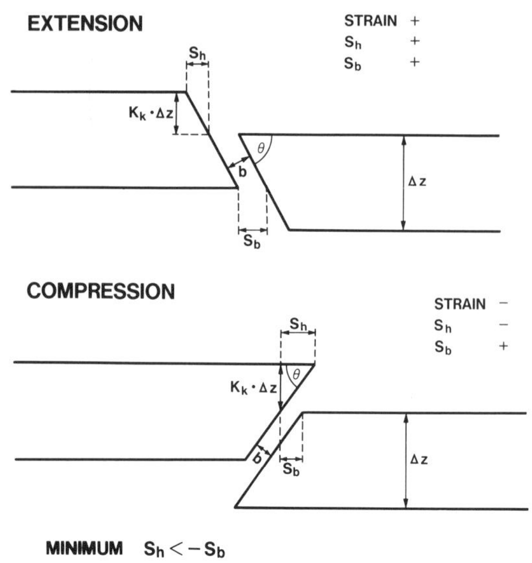

Fig. 6. Fracture parameters for extensional faulting (top) and compressional faulting (bottom). See text for discussion.

we must accommodate the excess strain S_f (the difference between total strain and ductility) through fracturing. The amount of horizontal distance X_f in a nodal block that must be accommodated by fractures is equal to S_f times the reference dip distance (W_{ref}) of the nodal block (Fig. 6).

Each fracture will have an aperture, vertical and horizontal displacement, and orientation associated with it. We define b as the aperture measured perpendicular to the fracture planes, S_b as the apparent aperture in the direction of the horizontal strain axis, and θ as the angle of the fracture to a horizontal plane.

We define S_h to be the displacement of the fault in the direction of the horizontal strain axis (here horizontal strain axis is better defined as the dip-direction strain axis). The sum of S_h plus S_b for all fractures in a nodal block must equal X_f for the nodal block. The magnitude of X_f will thus control the number of fractures that must form in the rock type being considered if we specify the aperture of those fractures in a horizontal direction and S_h of each fracture.

The total number of fractures N created in the nodal block by this strain can now be expressed as:

$$N = S_f \frac{W_{ref}}{S_b + S_h} \tag{2}$$

where

$$S_b = \frac{b}{\sin \theta} \tag{3}$$

and W_{ref} is the reference dip-direction width of the nodal block.

S_h is calculated in our present model as a function of nodal block thickness (ΔZ), the fraction L_k of each rock type in the nodal block, fracture orientation (θ), and a user-supplied constant K_k for each rock type. K_k is positive for extension and negative for compression. The formula for a nodal block containing n rock types is:

$$S_h = \frac{\Delta Z}{\tan \theta} \sum_{k=1}^{n} (L_k K_k) \qquad (4)$$

The conceptual basis for eq. 4 is that the horizontal displacement on a fracture depends on the thickness of the bed being fractured. At present we make the simplifying assumption that all the rock of a given rock type within each nodal block lies in a single bed. Thus a nodal block which contains three rock types is considered to consist of three beds. Thicker beds are observed empirically in the field to contain fewer fractures than thinner beds of the same lithology. A thicker bed of a given lithology, represented by a greater proportion of that lithology in the nodal block, will therefore have fewer fractures than a thinner bed of the same lithology under the same conditions.

Our simplifying assumption that all the rock of a single lithology within a given nodal block is in a single bed is obviously weak. A better approach that we contemplate adopting in the future is to include in the model some information about the thickness of the beds for each rock type in the node.

The use of trigonometric functions in the denominators of eqs. 3 and 4 precludes the use of horizontal or vertical fractures in our basin model. This limitation is probably not serious, however, because except on a very small scale, fractures are not uniformly horizontal or vertical. In real rocks, therefore, fluid flow in both the horizontal and vertical directions is affected by natural fracturing of any orientation.

Our current assumption of uniform fracture behaviour for a given rock type of a given thickness is obviously wrong. We hope that in the future we will be able to calculate fracture parameters for each rock type as functions of rock porosity, effective stress, temperature, or other variables available within the basin model.

It is apparent from inspection of the equations describing fracturing that there is a minimum vertical displacement S_h required for each compressional fracture. Because the aperture b is positive, S_b is always positive (eq. 3). During compression the excess strain S_f is negative. Because the number of fractures N must also be positive, the denominator in eq. 2 must always be negative during compression, and S_h must therefore be less than $-S_b$. The requirement of some vertical and horizontal displacement associated with each compressional fracture arises from the fact that apertures of compressional fractures represent extensional features in a compressional environment. We currently set S_h for compression equal to $-S_b$ and then add the horizontal component of vertical displacement calculated using eq. 4. K_k for extension may be zero, but K_k for compression must be less than zero (see eq. 4).

Strain history and fracture events. In the course of a modelling simulation, a nodal block may experience several reversals in strain direction due to changes in basement configuration. We observe in nature that a single rock may be fractured in many different ways and at different times due to various events in its structural and loading history. We therefore break the strain history of our rocks into a number of "fracture events" and "nonfracture events" (Fig. 7). Each reversal in strain direction (points marked R in Fig. 7) defines the end of one event (either fracture or nonfracture) and the beginning of a new nonfracture event. Our theoretical method assumes that any fractures formed during previous fracture events are inactive in all subsequent events, and that fractures which form during a new fracture event will have different characteristics than those formed earlier.

The theoretical method terminates a fracture or nonfracture event (that is, creates a strain reversal) whenever the absolute value of the total strain is

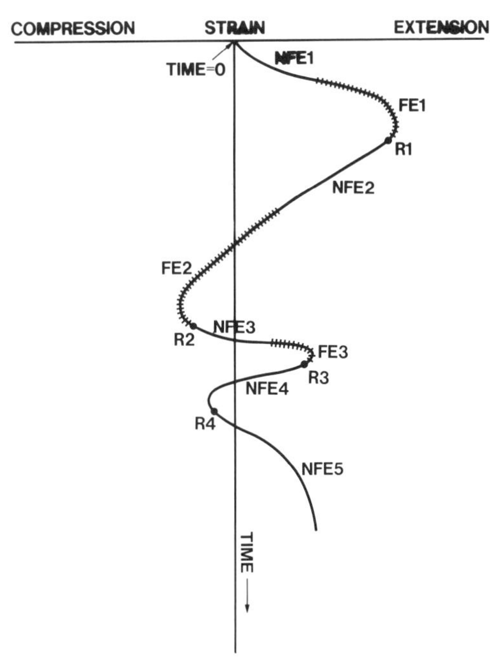

Fig. 7. Fracture (*FE*) and nonfracture events (*NFE*) and reversals (points marked *R*) superimposed on a hypothetical strain history. See text for discussion.

less than 99% of the absolute value of the maximum total strain attained during the event. We chose a value of 1% relaxation to define reversals because: (1) at that point the elastic portion of the total strain during the event will have largely been accounted for and new plastic deformation can begin; and (2) occasionally the iterative process in a basin model will create artificial reversals of very small magnitude in strain direction. Setting a reversal point at 1% relaxation prevents the creation of numerous phantom reversals.

Whenever a reversal occurs, both the theoretical and empirical fracture models reset the reference dip-direction width used in the calculation of strain to the current dip-direction width of the nodal block. The strain and maximum strain variables are reset to zero. If and when the strain accumulated during the new nonfracture event following the reversal exceeds ductility, a new fracture event begins.

Total number of fractures during a fracture event. Consider a simulation in which total strain first exceeds ductility during the nth time step. The rock will then fracture according to the logic and equations outlined above. During time step $(n + 1)$, however, both total strain and ductility will probably change. If the ductility increases faster than the strain (that is, if S_f decreases), the rock may not fracture again, although since strain is increasing, we are still within the same fracture event. (If the strain increases faster than the ductility, in contrast, the rock will probably continue to fracture.) We calculate the total number of fractures for a fracture event as the number of fractures that result from the maximum S_f which the nodal block experienced during the fracture event. The total number of fractures is not simply the sum of fractures calculated independently for each time step within the fracture event, since once fractures have developed, much of the future strain will be accommodated by widening these fractures rather than by creating new ones.

Quantification of fracture permeability. The intrinsic permeability k_r of a single fracture is the permeability of the fracture in the direction of fluid flow. From van Golf-Racht (1982), for horizontal flow (x-direction) it is given by:

$$k_{rx} = \tfrac{1}{12}(\cos\theta)^2 b^2 \tag{5}$$

and for vertical flow (z-direction) it is:

$$k_{rz} = \tfrac{1}{12}(\sin\theta)^2 b^2 \tag{6}$$

The Basin2 flow model calculates fluid flux across the faces of a nodal block. The calculated intrinsic fracture permeability k_r, however, only applies to that area of the nodal block actually occupied by the fracture aperture. We therefore calculate k_{ef}, the effective fracture permeability due to this one fracture, in order to average the permeability effects of fracture over the area of the nodal block used in the fluid-flow calculation. To conserve the volume of fluid transported, we apply the following formulae:

$$V_f = \frac{A_f\, k_{rf}\, \dfrac{dH}{dl}}{\mu} \tag{7}$$

$$V_n = \frac{A_n\, k_{ef}\, \dfrac{dH}{dl}}{\mu} \tag{8}$$

where V_f is the volume of fluid passing through the fracture, V_n is the volume of fracture-conducted fluid passing through the node face of interest, dH/dl is the hydraulic gradient across the node, μ is the viscosity of water, k_{rf} is the intrinsic permeability of the one fracture in a direction perpendicular to the node face, A_f is the area of the fracture in the node face considered, and A_n is the area of the node face. Since by fluid conservation $V_n = V_f$, and since dH/dl and μ are constants, setting eq. 7 equal to eq. 8 and rearranging gives:

$$k_{ef} = \frac{A_f}{A_n} k_{rf} \tag{9}$$

where A_f/A_n is equivalent to fracture porosity.

If we have N different equivalent fractures in a fracture set, then the effective permeability k_{efN} of the entire fracture set is:

$$k_{efN} = N\, k_{ef} \tag{10}$$

In our current model we assume that each fracture event j will create a different set of fractures. We calculate the total effective network permeability k_{efNJ} of J sets of fractures in a given rock type as the sum of the effective permeabilities of each fracture set. Thus:

$$k_{efNJ} = \sum_{j=1}^{J} N_j\, k_{efj} \tag{11}$$

Replacing k_{efj} using eqs. 5, 6, and 9, we have:

$$k_{efNJ} = \tfrac{1}{12}\sum_{j=1}^{J} N_j \frac{A_{fj}}{A_{nj}} b_j^{\,2}\, [\mathrm{fn}_j(\theta_j)]^2 \tag{12}$$

Here N_j represents the number of fractures from the jth fracture event with common orientation θ_j and aperture b_j; A_{fj} $(= A_{fi})$ is the area of the node surface covered by these fractures; and $\text{fn}(\theta_j)$ is either $\sin\theta_j$ or $\cos\theta_j$, depending on the orientation of the node face in question (see eqs. 5 and 6).

The total permeability of a rock unit that we use to describe fluid flow in our fracture-permeability basin model is given as the matrix permeability plus $k_{\text{ef}\,NJ}$, the sum of the effective network permeabilities for all fracture events. Fracture porosity is not considered in our fluid-flow equation. In most cases ignoring porosity should not cause problems, but if matrix porosity is very low there may be some errors, especially in cases where transient fluid-flow phenomena are considered, because changes in porosity affect the storage characteristics of rocks.

Nodal blocks in our simulations are often composed of mixtures of several different rock types. In order to consider fracturing in nodal blocks containing more than one rock type, we calculate the effective permeability for each rock type separately. We then calculate an average fracture permeability for the nodal block by geometrically averaging the permeabilities contributed by each rock type.

Reduction of fracture aperture. We know from field observations that the apertures of fractures open to fluid flow are reduced with time due to the formation of minerals in the fracture voids. Processes such as calcite or quartz precipitation or clay formation will increase the roughness and irregularity of a fracture void and also decrease the aperture. Closure may also occur as a result of changes in tectonic stress. The permeability of fractures will thus decrease with time. We are currently simulating reduction of fracture apertures using the following equation:

$$S_{\text{b}} = S_{\text{b}\,0} \exp -k_3 t \tag{13}$$

where t is the time since closure began, k_3 is a constant related to the rate of fracture closure, $S_{\text{b}\,0}$ is the aperture at the time when closure began, and S_{b} is the aperture at time t. In our present scheme the starting time for closure of fractures formed during the jth fracture event is the end of the jth fracture event.

Our justification for using eq. 13 is not strong. We propose this exponential function mainly as a way to introduce the concept of fracture closure into our model. However, an exponential decay in the rate of aperture reduction may be a reasonable first guess, since the rate at which any diagenetic minerals form in the fracture voids will probably be proportional to permeability. However, in the future it might be desirable to distinguish between tectonic closure and diagenetic closure. We would assume that tectonic closure begins at each strain reversal (points R in Fig. 7), whereas diagenetic closure begins as soon as the fractures begin to form.

Empirical method

We are at present evaluating several ways of relating effective fracture permeability k to S_{f}, the strain in excess of ductility. These functions include:

$$\log k_{\text{ef}} = k_4 S_{\text{f}} + k_5 \tag{14}$$

$$\log k_{\text{ef}} = \frac{k_4}{S_{\text{f}}} + k_5 \tag{15}$$

where k_4 and k_5 are constants for each rock type. These equations calculate only the effective fracture permeability, and do not compute fracture porosity, fracture density, or other values that could be obtained using the theoretical approach discussed previously. For our fluid-flow model, however, effective fracture permeability is the only parameter we really need, and any approach to the problem is valid if it can accurately determine the effective fracture permeability. Unfortunately, we know of few references that can help us determine plausible relationships between strain and permeability.

Results and discussion

Theoretical method

Figure 8 shows fracture permeability in both the horizontal and vertical directions during the Shiiya stage in the Niigata Basin, calculated using our theoretical approach. Permeabilities of more than 1000 darcies in both directions are calculated in the centre of the basin, where strain is greatest. Small areas of high horizontal fracture permeability are also predicted in deep strata near both margins of the cross-section. Fracture permeabilities are highest in the deep strata, and decrease upward.

Figure 9 shows for comparison the horizontal and vertical matrix permeabilities during the Shiiya stage. Permeability trends are highly depth-dependent, and matrix permeabilities are many orders of magnitude lower than the fracture permeabilities in the areas of intense strain in Fig. 8. The patterns of matrix and fracture permeability thus contrast sharply in both magnitude and distribution.

The fluid-flow pattern calculated considering both matrix and fracture permeability (Fig. 10, left) is quite different from that calculated using matrix permeability alone (Fig. 10, right), as a consequence of the extremely high fracture permeability in the central part of the basin. The strong water flow from left to right in an unfractured system (Fig. 10, right) that could lead to hydrocarbon accumulations

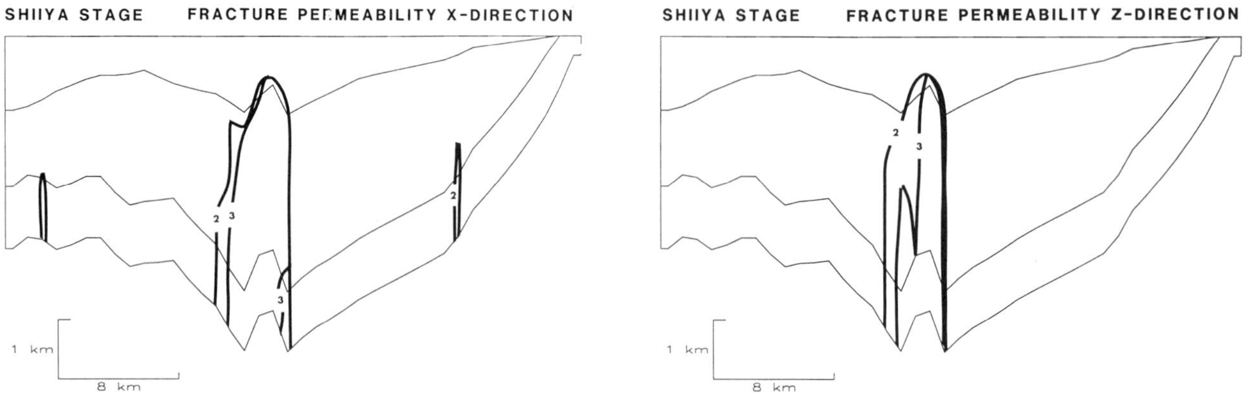

Fig. 8. Log horizontal fracture permeability (left) and log vertical fracture permeability (right) predicted for the Niigata Basin during the Shiiya stage using the theoretical method (cf. Fig. 9). Contour interval = 1.

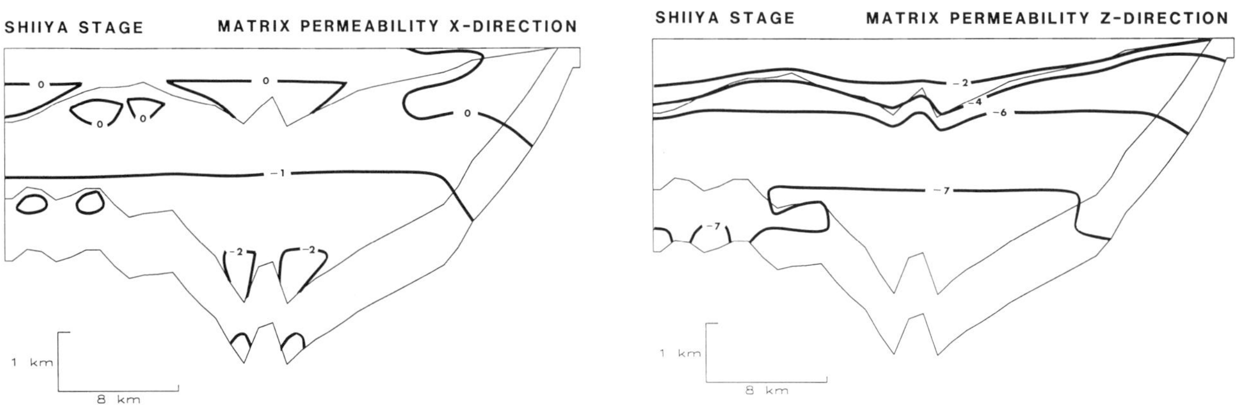

Fig. 9. Log horizontal matrix permeability (left) and log vertical matrix permeability (right) predicted by Basin2 for the Niigata Basin during the Shiiya stage (cf. Fig. 8). Contour interval = 1 (left) and variable (right).

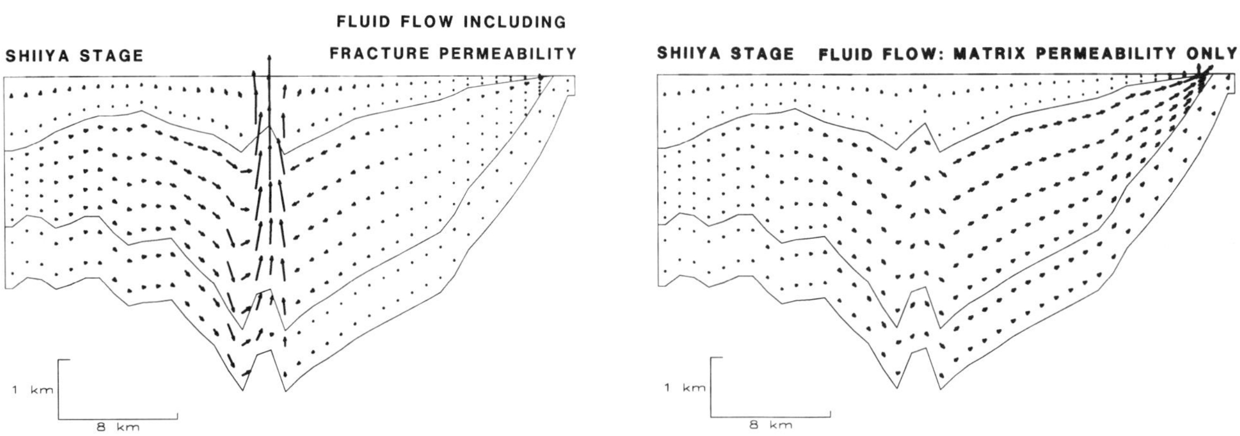

Fig. 10. Left: Fluid-flow patterns predicted by our model for the Niigata Basin during the Shiiya stage. Fluid flow is controlled by vertical fracture permeability in the centre of the basin (Fig. 8 bottom). Right: Fluid flow through matrix permeability predicted by Basin2. Lengths of arrows are directly proportional to fluid flux.

in shallow strata at the right margin of the cross-section are interrupted. Instead, fluid flows from the right-hand side down into the basin centre, and from there to the surface. It is evident that one would employ different hydrocarbon-exploration strategies in the Niigata Basin using the fracture-permeability model and the matrix-permeability model.

Empirical method

In simulations that we have run, the results of the empirical model and the theoretical model could be made quite similar through appropriate choice of constants in both models (compare Fig. 11 with Fig. 8). We therefore believe that the empirical method will prove to be useful.

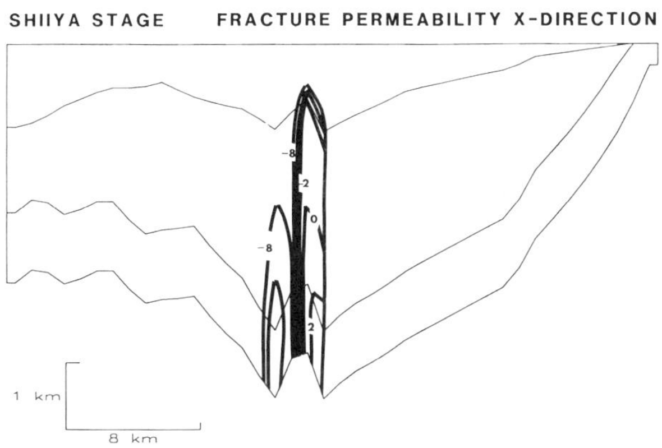

Fig. 11. Log horizontal fracture permeability predicted for the Niigata Basin during the Shiiya stage using the empirical method (cf. Fig. 8, top) and eq. 14: $\log k_{\mathrm{ef}\,x} = 2S_{\mathrm{f}} - 12$. Nearly identical results (not shown) were obtained in calculating vertical fracture permeability using $\log k_{\mathrm{ef}\,z} = 1.6S_{\mathrm{f}} - 12$. Contour interval = 2.

Conclusions

We believe that permeability associated with tectonic fracturing should be included in basin models. We have developed a method to predict the occurrence and intensity of tectonic strain using information already available in basin models. This method is based on differential subsidence of adjacent nodal blocks. Comparisons of the predictions of our model are qualitatively similar to those of another quite different model designed specifically to predict strain.

We have also developed two ways to convert our calculated strain values into fracture permeabilities that can be used to regulate fluid flow. One method is based on the geometry, number, and apertures of fractures; the other is a purely empirical relationship between strain and permeability. The two methods can be calibrated to give similar results.

Preliminary application of our models indicates that where tectonic fracturing occurs, fluid-flow patterns (including hydrocarbon migration) may be completely different than in the absence of fracturing. Use of a model which ignores fracture permeability may thus lead to erroneous exploration decisions.

Our current model only considers fracturing caused by vertical faulting. We are currently working on a model that can also handle strain and fracturing caused by concentric folding.

Acknowledgements

We thank Hiroshi Ohkuma, Mamoru Hara, and Masaki Hattori for insightful discussions about the nature of fracturing and its effects on permeability, and JNOC for support and permission to publish this work.

References

Bethke, C.M., 1985. A numerical model of compaction-driven groundwater flow and heat transfer and its application to the palaeohydrology of intracratonic sedimentary basins. J. Geophys. Res., 90: 6817–6828.

Bethke, C.M., 1986a. Hydrogeologic constraints on the genesis of the Upper Mississippi Valley mineral district from Illinois Basin brines. Econ. Geol., 81: 233–249.

Bethke, C.M., 1986b. Inverse hydrologic analysis of the distribution and origin of Gulf Coast-type geopressured zones. J. Geophys. Res., 91: 6535–6545.

Bethke, C.M. and Marshak, S., 1990. Brine migrations across North America — the plate tectonics of groundwater. Annu. Rev. Earth Planet. Sci., 18: 287–315.

Garven, G., 1985. The role of regional fluid flow in the genesis of the Pine Point deposit, Western Canada sedimentary basin. Econ. Geol., 80: 307–324.

Garven, G., 1989. A hydrogeologic model for the formation of the giant oil sands deposits of the Western Canada sedimentary basin. Am. J. Sci., 289: 105–166.

Handin, J., Hager, R.V., Jr., Friedman, M. and Feather, J.N., 1963. Experimental deformation of sedimentary rocks under confining pressure: pore pressure tests. Am. Assoc. Pet. Geol., Bull., 47: 1–50.

Hattori, M., Hoshino, K. and Inami, K., 1984. Relation of fracturing of the volcanic rocks to mechanical properties under high pressure. J. Jpn. Assoc. Pet. Technol., 49: 20–28 (in Japanese with English abstract).

Kodama, K., Long, X.-M. and Suzuki, Y., 1985. Structural analysis of deep-seated volcanic rock reservoirs by tectonic simulation. U.N. ESCAP CCOP Tech. Bull., 17: 61–79.

Ungerer, P., Burrus, J., Doligez, B., Chenet, P.Y. and Bessis, F., 1990. Basin evaluation by integrated two-dimensional modeling of heat transfer, fluid flow, hydrocarbon generation, and migration. Am. Assoc. Pet. Geol., Bull., 74: 309–335.

Van Golf-Racht, T.D., 1982. Fundamentals of Fractured Reservoir Engineering. Developments in Petroleum Science, Vol. 12. Elsevier, Amsterdam, 710 pp.

KURT W. LARSON *Technology Research Center, Japan National Oil Corporation, Chiba, Japan*
Present address: *Hydrogeology Program, University of Illnois, Urbana, IL 61801, U.S.A.*
DOUGLAS W. WAPLES *Japan National Oil Corporation, 1-2-2 Hamada, Chiba 260, Japan*
HAN FU *Japan Oil Engineering, Tokyo, Japan*
KISABURO KODAMA *Geological Survey of Japan, Tsukuba, Japan*

New techniques and their applications in the analysis of secondary migration

Øyvind Sylta

Computer techniques can be used to model secondary migration of oil and gas. In this paper, migration along decompacted maps of a single carrier bed is modelled. Hydrocarbon flow is simulated by means of a ray-tracing technique which makes it possible to show where the hydrocarbons in a trap have migrated from. The ray-tracing procedure can therefore be utilized in the development of a secondary migration toolbox for the accurate and efficient assessment of secondary migration and entrapment.

Palaeohydrocarbon flow path, drainage area outline, drainage system, alternative spill path, and oil and gas flow rate maps provide important information for exploration decisions. The maps must be produced for many time steps because trap restructuring frequently takes place and trap capacity may increase or decrease in an unpredictable manner. Structuring may increase the trap capacity whereas compaction decreases the capacity.

Three-dimensional secondary migration modelling allows for quantitative predictions of the amounts of oil and gas in prospects not yet drilled. In 1987 we performed a basin modelling study in which we tried to predict the hydrocarbon content in the Norwegian 12th Round Block 35/9. Secondary migration modelling used the results from tectonic modelling, source rock facies modelling, thermal calibration and hydrocarbon generation modelling. Uncertainties in the geologic model were investigated by varying important modelling input parameters. The migration modelling results were calibrated with published reserves and we concluded that the most likely scenario yielded both oil and gas in the 35/9 structure. Later drilling by Norsk Hydro confirmed an oil and gas discovery.

Introduction

Most techniques used in the appraisal of secondary migration are still manual. These include the construction of drainage area maps, risk assessment of leakage from traps, fault leakage, etc. The evaluation of the potential for oil and gas within a particular prospect or block is often based upon a hydrocarbon generation modelling study, and occasionally two-dimensional fluid flow modelling along a cross-section. However, due to a lack of tools for three-dimensional hydrocarbon migration modelling, the last step in the assessment, the estimation of amount and type of hydrocarbons within undrilled prospects has to rely on methods other than deterministic modelling, i.e. statistical procedures, etc. The well known Monte Carlo simulation technique has been used by some oil companies to integrate the various results involved in making the prospectivity assessment (White and Gehman, 1979; Irwin et al., this volume).

This paper first discusses some new computerized techniques where the main benefit is that the results can be obtained at lower costs and quicker than by manual procedures. An important benefit is also that the procedures are used on palaeodepth maps, and therefore a better understanding of the secondary migration history within a basin can be obtained.

The second part of this paper shows how secondary migration modelling techniques can be used to predict type and amount of hydrocarbons within an undrilled license. The uncertainty which exists in the input parameters, i.e. source rock model, cap rock sealing efficiencies etc., are reflected in the results because several simulations with different input are modelled. Norwegian Block 35/9, which had not yet been drilled when this work was done, but which has now been drilled by Norsk Hydro, serves as an example for this discussion.

Computerized analysis techniques

The paper is based upon secondary migration modelling methodology previously presented by this author (Sylta, 1987, 1989). Buoyancy-driven, two-phase secondary migration of oil and gas is modelled along a single carrier bed in three dimensions by means of a ray-tracing technique. Trap volumetrics, gas density changes, gas dissolved in oil, gas leakage, oil losses during migration etc., are computed during subsidence.

Basin Modelling: Advances and Applications edited by A.G. Doré et al. NPF Special Publication 3, pp. 385–398, Elsevier, Amsterdam.

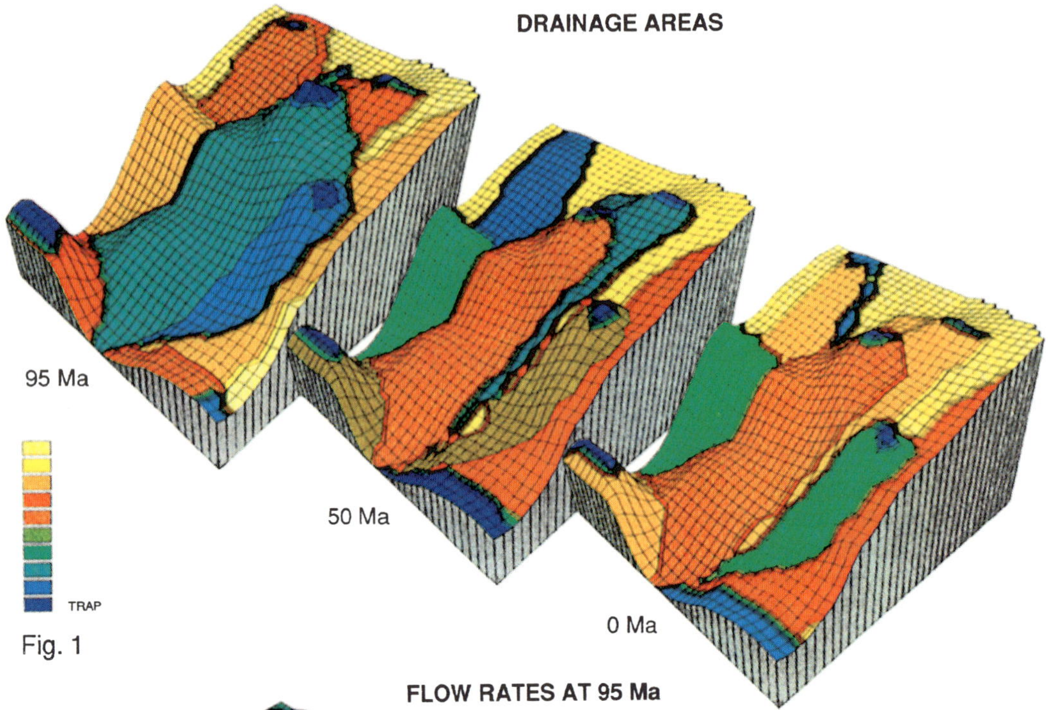

Fig. 1

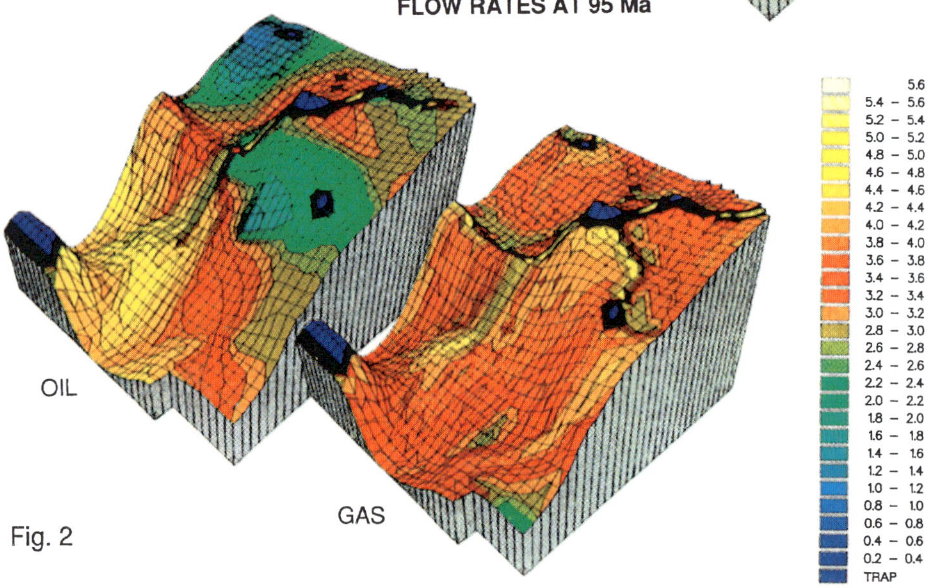

Fig. 2

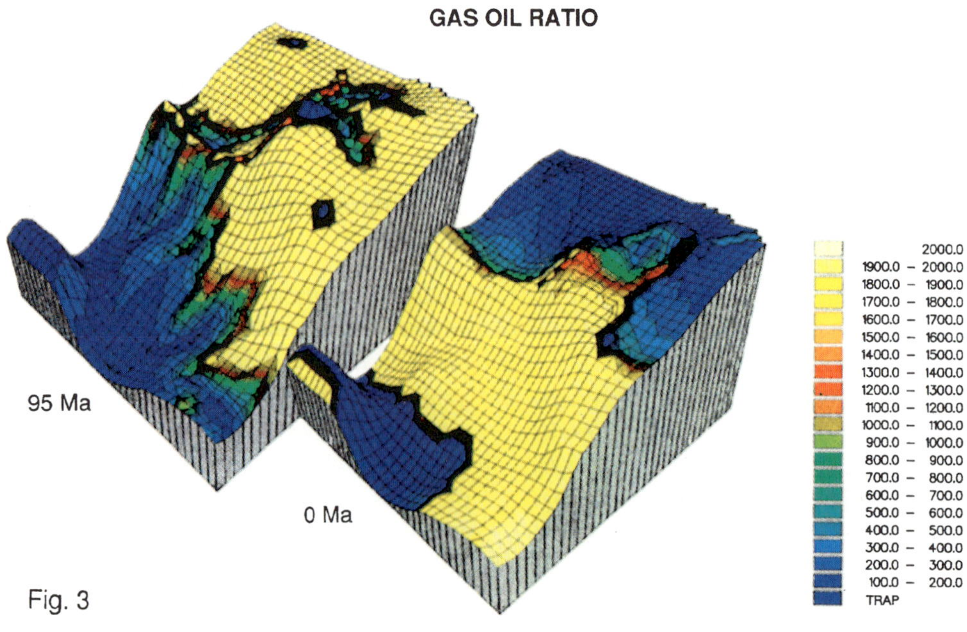

Fig. 3

Overpressure is abundant in many petroleum provinces. In the North Sea, overpressure tends to be distributed within zones of almost equal amounts of overpressure and lateral pressure communication within individual, rotated, fault blocks is inferred (Buhrig, 1989). The pressure compartments are bounded by pressure transition zones where the amount of overpressure changes rapidly (Powley, 1990). Pressure transition zones may be faults which inhibit lateral flow, or low permeability layers which reduce the vertical flow. Because of minor changes in overpressure, hydrocarbon flow will often be buoyancy-driven within each compartment, whereas migration between each compartment will be hydrodynamically driven.

The morphology of oil–water, gas–oil and gas–water contacts are influenced by the presence of hydrodynamic forces. Thus, tilted hydrocarbon contacts are evidence that hydrodynamism cannot be neglected in secondary migration simulations. However, flat contacts may be taken as indicative of (close to) static conditions today.

Palaeopressure conditions need not be the same as present-day observations suggest, and hydrodynamic conditions may have been prevalent even though present-day observations are interpreted to represent a close-to-static system. Unfortunately, we are not yet able to measure or observe palaeopressures. The application of the methodology presented in this paper should, however, be restricted to sedimentary basins where water flow does not significantly alter buoyancy-driven hydrocarbon flow.

One important property of the ray-tracing technique as used here is that the whole migration pathway, i.e. from the point of entrance into the carrier bed to the top of the trap, is available for further analysis. Information on where oil and gas migrate from is utilized in some very useful map types for secondary migration assessment: the palaeoflow direction maps, hydrocarbon flow rate maps, drainage area outline maps, and drainage area system maps.

Palaeoflow direction maps are simple ray-tracing maps where buoyancy-driven migration is modelled along backstripped carrier bed maps for any particular point in time. Changes in the oil and gas flow pattern within a basin can then be analyzed. Many commercial mapping packages provide ortho-contouring options which serve the same purpose if applied to decompacted maps. The palaeodrainage area outline maps (Fig. 1) are constructed from the palaeoflow maps and enable the assessment of catchment areas of individual prospects. Changes through time may be observed easily. We frequently model drainage areas in North Sea basins to be smaller during the early phases of migration than at later stages. Initially, many small traps are present and they are quite often filled with oil or gas. During the main phase of subsidence many of the small traps disappear due to structural reorientation and increase in dip or because they combine into larger traps due to more oil and gas migration.

Hydrocarbon flow rate maps can be used to obtain an understanding of the migration pattern through time. Hermans et al. (1992) presented a video of oil flow through time in the Haltenbanken area which was based upon the methods of Lehner et al. (1988). The same information can be presented as maps and three types of information may be useful in the evaluation of an area:oil flow rate, gas flow rate (Fig. 2) and gas/oil ratio (GOR) (Fig. 3). We use the term "GOR maps" for maps which are obtained by computing the relationship between gas and oil flow rates within each grid node. High GOR values indicate a higher probability of finding gas-prone structures whereas low values indicate that oil may be more likely. When the oil and gas migrate along the same pathways, liquid hydrocarbons may be preserved in traps when the GOR maps show low values, because the gas remains in solution at the particular pressure and temperature. Pressure and temperature vary with spatial location and time. However, oil and gas may also migrate along different pathways, in particular if the carrier bed contains many permeability barriers, flow restrictions or is structurally complex.

The ray-tracing procedure adds the volume of oil and gas passing each grid node when it passes the area of that particular grid node (Fig. 4A). The cumulative sum of all the rays passing the node becomes the resulting cumulative oil flow. The size of the grid node area influences the cumulative amounts passing. Thus, if the grid node length is doubled, i.e. the area of a grid node becomes four times larger, the amount of oil passing each grid node doubles. Thus, the cumulative oil flow map has to be divided by the length of the grid nodes as well as the time step used in order to arrive at the oil flow rate map. Oil flow rate maps are conveniently

Fig. 1 (Top). Palaeodrainage area map at 95, 50 and 0 Ma.

Fig. 2 (Middle). Flow rate maps: oil and gas using $\log_{10}$ scale (units: $m^3\ km^{-2}\ Ma^{-1}$ and $1000\ Sm^3\ km^{-2}\ Ma^{-1}$).

Fig. 3 (Bottom). Flow rate maps: gas/oil ratio at 95 and 0 Ma (units: $Sm^3\ km^{-2}\ Ma^{-1}/m^3\ km^{-2}\ Ma^{-1}$).

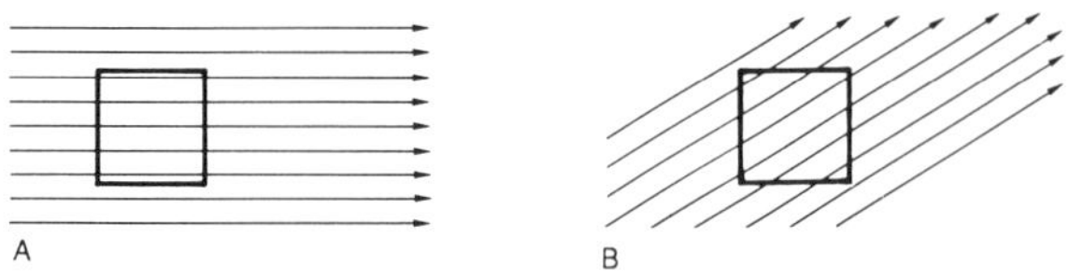

Fig. 4. Flow rate maps: the effect of flow path directions.

presented in units of m^3 km^{-2} Ma^{-1}, whereas gas flow rate maps use Sm^3 km^{-2} Ma^{-1}.

The ray path direction through grid nodes influences the amount of oil and gas recorded. Thus, if paths are straight east–west or north–south, i.e. normal to the grid node boundaries, the number of rays passing a grid node is less than when the rays pass at an angle. These effects are shown in Fig. 4. When the rays pass at a 45° angle (Fig. 4B), the computed amounts passing through the grid node are larger than for 0° angles (Fig. 4A). This is a grid orientation problem, and if the flow within each grid node is aligned (as in Fig. 4B), the problem can be corrected easily. However, most of the time the ray paths are not parallel, and the problem can therefore not be corrected. The interpreter has to recognise how the flow rate maps are constructed and take into account the uncertainties thus introduced.

The amounts of hydrocarbons migrating through an area vary significantly. Often, the use of constant contour intervals may hide information from the interpreter because of a lack of resolution in areas with low flow rates (Fig. 5). Therefore, a logarithmic scale is preferred. The use of $\log_{10}$ eases the conversion from map values to the actual values and a proper selection of contour intervals (0.5, 0.2, etc.) provides enough resolution for areas with high flow rates and low flow rates.

Within a "drainage area system" all traps spill updip and into the same final trap. Drainage area systems are, however, not stationary through time. Most of the time they grow, because more and more traps are filled to spill point and start contributing HC's to updip traps. Structural reorientation, especially changes in dip during subsidence, may change the spill paths significantly, in particular in flat areas, but also in high dip areas. Thus, traps may belong to different drainage area systems through time. Changes in system configuration do not happen often, but significant changes in hydrocarbon types and amounts may result when such does occur. Recent changes are important because secondary migration assessment is often performed on present-day structural depth maps. Therefore, recent changes in the migration pattern may disguise the migration flow pattern through most of the history of the basin.

Computerized techniques allow for a systematic investigation of uncertainties in the structural mapping. Minor errors in the mapping within a drainage area may be unfortunate because a potential closure is overlooked, but the errors in the overall hydrocarbon migration pattern is often minimal. The migration systems of sedimentary basins are often very sensitive to the position of the spill points of the traps. Errors in the structural mapping which cause the spill point to be modelled inaccurately, may result in oil and gas spillage into the wrong drainage area and thus predictions of the type and amounts of hydrocarbons become wrong. The uncertainty in the spill point mapping can be evaluated by means of maps showing the "depth to alternative spill point" (DASP) of each trap. The colour-coded DASP map (Fig. 6) shows how much deeper the modelled spill points for each drainage area have to be for the traps to spill into alternative drainage areas. These maps may suggest where the structural depth maps require special attention and extra quality control.

The study of the filling history of fields and prospects is facilitated by the use of secondary migration modelling. The volume of trapped oil and gas (free and in solution) is computed for each time step modelled, and can be used to reconstruct trap fill histories. Trap fill data may conveniently be shown as posted data for each time step (Schroeder and Sylta, this volume) or as cross-sections along major migration paths. The latter display type may include palaeo-OWC and GOC and may thus facilitate the conceptual understanding of the migration history. However, cross-sections of this type are subject to the same limitations that two-dimensional modelling techniques suffer from: major migration pathways often change through time. Cross-sections should therefore be used with caution. Flow rate maps provide another type of information to the analysis of trapped hydrocarbons. The migration pattern into and out of drilled traps (Fig. 7) can be utilized together with production data on oil and gas densities, GOR etc. History matching using conventional three-dimensional reservoir simulation techniques may utilize the flow rate input from secondary migration modelling as boundary conditions and thereby help in reservoir management.

The filling history of some of the North Sea traps is modelled to be more complex than one might expect. The filling history of many traps may be described as a series of transitions from water-filled, via oil-filled, two-phase reservoirs (gas and oil) and finally a gas-only reservoir. This behaviour results from the continuous increase in temperature that dominates

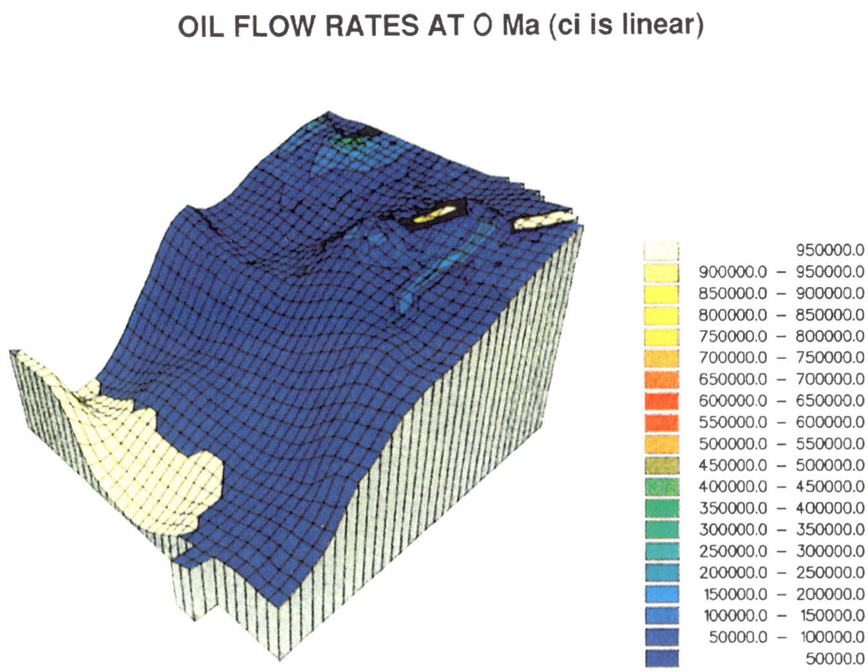

Fig. 5. Flow rate map: linear contour intervals provide low resolution (compare with Fig. 2).

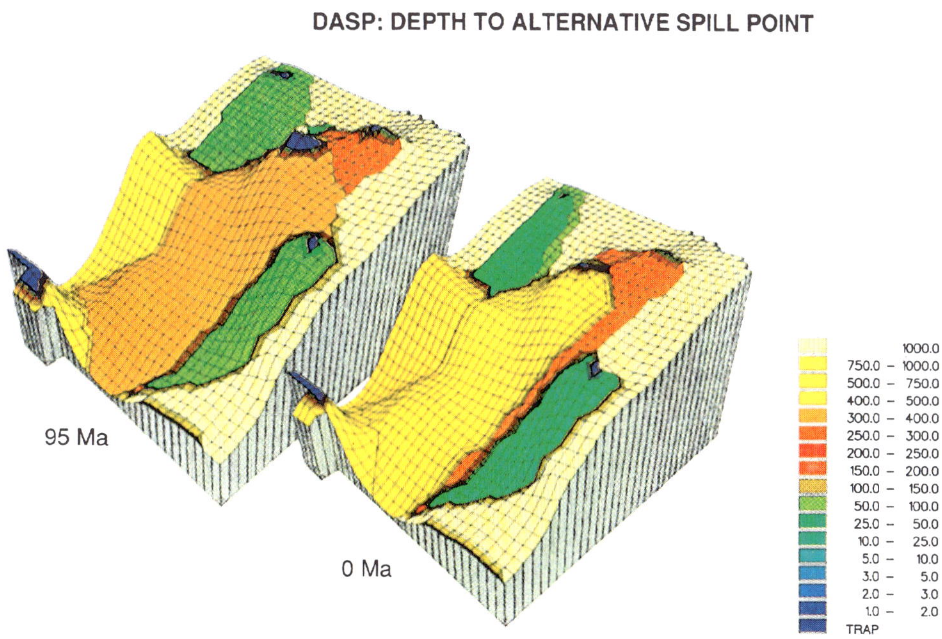

Fig. 6. Depth to alternative spill points (DASP) at 95 and 0 Ma.

both the hydrocarbon generation in source rocks and the traps in most subsiding basins, including the North Sea. The decrease in porosity during subsidence further enhances this trend. However, some modifications to this behaviour have been observed in modelling studies. Structural reorientation of fault blocks may occasionally increase the available pore space and thereby make room available for either oil or gas. During subsidence, the source rock within a drainage area may pass from producing mainly gas and back to producing mainly oil if it covers a significant depth range. A statistical analysis of the drainage area versus depth will sometimes show a higher percentage of the area at shallow depths (Fig. 8) and when the area of highest abundance passes through the oil window (A in Fig. 8), the increase in the amount of oil flowing through the associated trap dissolves some or all of the free gas. The trap becomes oil-filled until zone A reaches the gas window. In our modelling studies, we find that

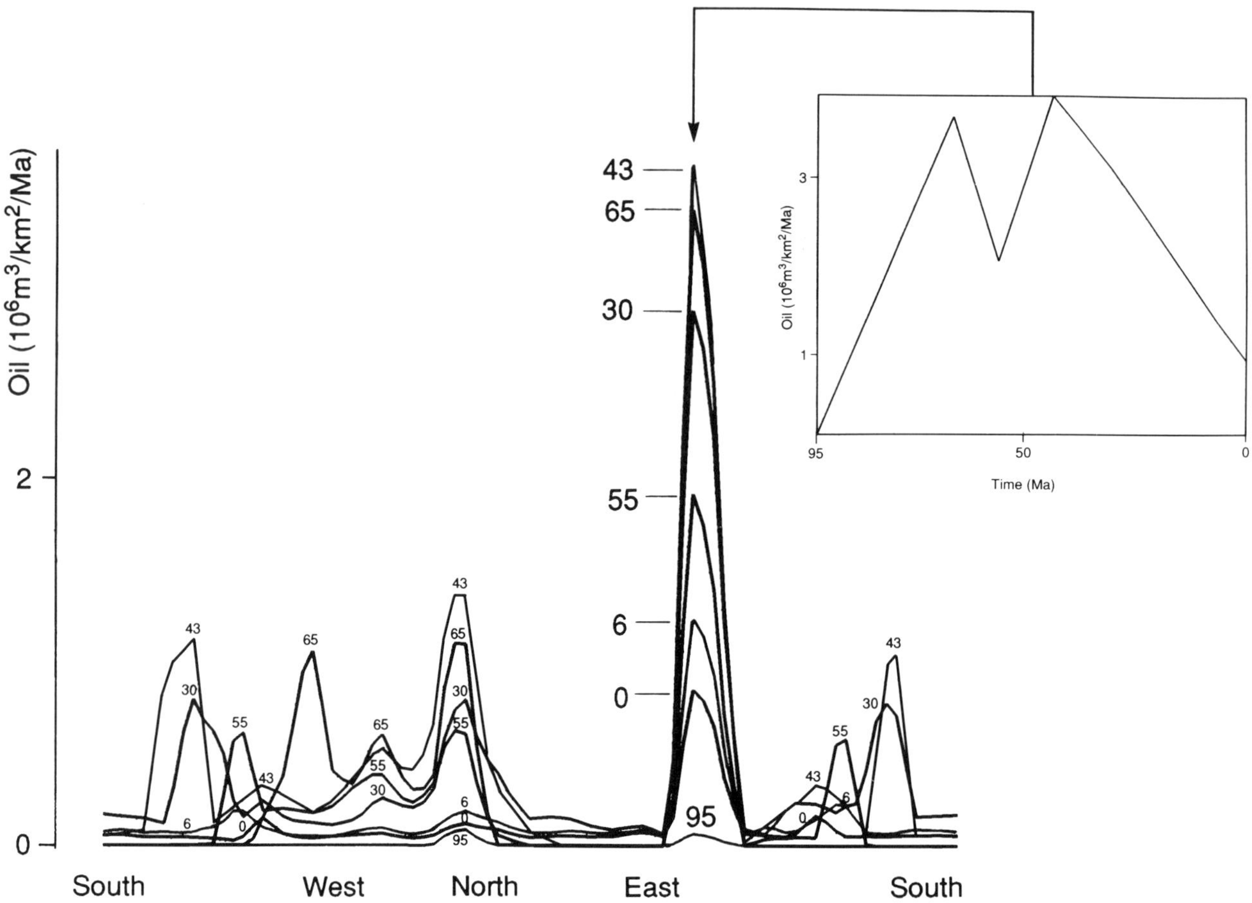

Fig. 7. Oil flow into and out of trap. Curves show flow rates in 6 modelled time periods (65 means from 95 to 65 Ma). The temporal change for the major spill path out of the trap is plotted up to the right.

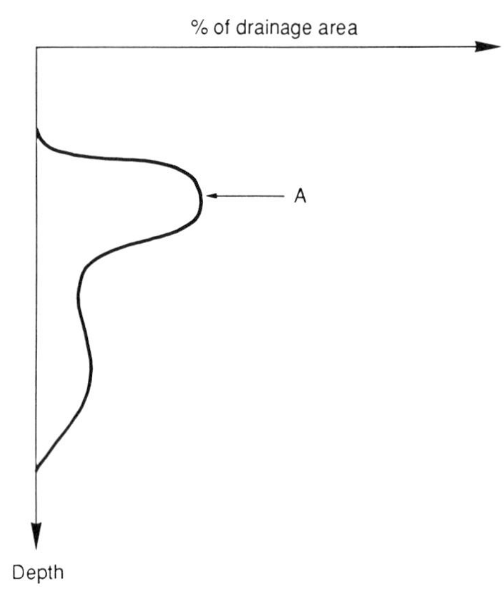

Fig. 8. Drainage area (%) versus depth for a single trap. A definite maximum at one depth (*A*) is observed in most drainage areas.

this effect quite often produces two-phase reservoirs, and that the oil is modelled to be contained within a thin oil leg, even in areas which are dominated by gas generation and migration.

Quantified predictions: Block 35/9

Quantitative estimates of amount and type of hydrocarbons in undrilled traps can be made by means of three-dimensional deterministic modelling techniques. Predictions become more reliable if other fields in the area can be used to calibrate the migration model. The following is a discussion of the predictions of amounts of oil and gas in the Norwegian Block 35/9 (Fig. 9) prior to drilling. We aim to present the technique and concepts rather than to present the 35/9 case study in full detail.

Quadrant 35 is situated from 61° to 62°N and 3° to 4°E, on the east flank of the North Viking Graben. The main structural elements within the quadrant are the Sogn Graben, the Sogn Spur and the northwestern edge of the Horda Platform. The area lies adjacent to one of the most productive regions of the North Sea and discoveries have already been made immediately north and south of Block 35/9, i.e. at Agat and Troll, respectively.

Attention is focused on the migration of oil and gas from Jurassic source rocks into traps in Jurassic reservoir rocks. Secondary migration is modelled

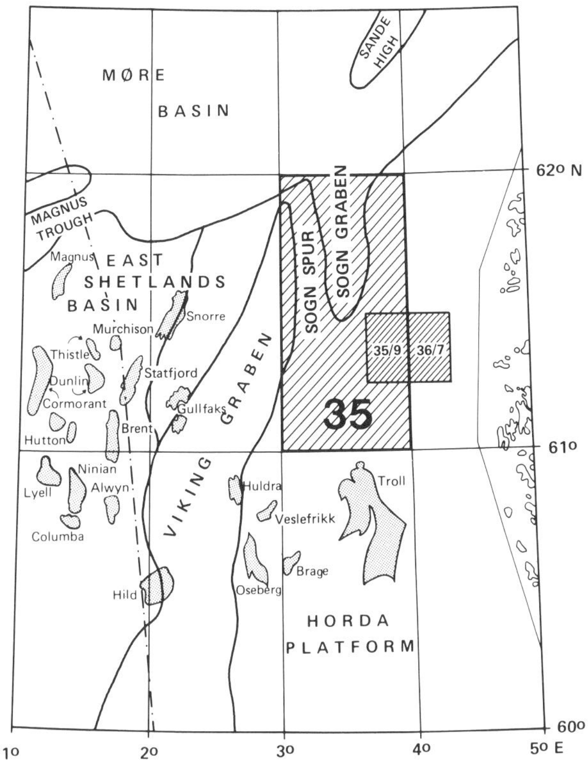

Fig. 9. Location of Block 35/9.

along the Middle Jurassic Brent reservoirs. Results of the hydrocarbon migration simulations are compared to the finds in 35/8-1, 35/8-2, the Troll Field and the Agat Field. Thus, a calibration of the expulsion efficiencies and oil losses during migration is achieved and a prognosis for the hydrocarbon content of traps in 35/9 and 36/7 is made.

Input to secondary migration modelling

Tectonic subsidence curves have been compared to theoretical curves derived from a model of finite rate, uniform pure shear extension after the method of Jarvis and McKenzie (1980) which predicts basement subsidence and heat flow of a water-filled basin. Extension factors (β values) were thus derived and a β value map was constructed. Since each β value corresponds to a unique heat flow history, this map was then used to determine the heat flow history of any point in the area. These heat flow histories were used as input for the hydrocarbon generation modelling.

The source rock model consists of seven and three source rock segments in the Viking and Dunlin Groups, respectively. Each segment is characterized by a constant hydrogen index. The TOC content and thickness of a segment vary and the segment boundaries are time-transgressive.

Two scenarios distinguished by two different sets of kinetic parameters for gas generation were modelled. Gas generation in scenario B sets in about 15° to 20°C earlier than gas generation from scenario A.

Expulsion efficiency of oil differs from that of gas. The oil which is left behind in the source rock due to incomplete expulsion will be transformed into gas upon further temperature increase during burial. The expulsion efficiencies for gas and oil are assessed as a part of the calibration to field data performed later.

The dips of the Brent carrier beds have undergone a change during Tertiary times. The palaeostructure maps of the Brent carrier beds are constructed at 60 Ma by means of a backstripping method. For times later than 60 Ma, however, water depth variations within the area cannot be neglected. Sediments of Tertiary age are mainly composed of north–west-prograding sequences showing significant, systematic variations in thickness. Palaeostructure maps for the Top Brent at later times than 60 Ma were therefore generated by adding a percentage of the difference between depth of the reservoir at 60 Ma and present-day over the whole area. The percentages were determined from the recorded thicknesses in well 35/8-1. This procedure simulates a steady increase in westerly tilt throughout Tertiary times, and eliminates the need to map, in detail, the palaeowater depths within the study area.

The net rock thickness of the reservoir is modelled to be constant throughout the area. Porosities and gross thicknesses change during subsidence according to the decompaction technique.

The net–gross ratio (% sandstone in the carrier bed) is set to 50% based on calculations from the 35/8-1 well. The pore space available for trapping of hydrocarbons is further reduced by the presence of irreducible water. In the calculations, an irreducible water saturation of 15% is used. This compares with values from the Statfjord Field where oil saturation is reported to be 84% (Buza and Unneberg, 1987).

Permeability is often the most important factor controlling hydrocarbon losses during secondary migration (Sylta, 1989). Permeability may be sensitive to porosity variations. In Fig. 10 a plot of published permeability versus porosity data is shown. From this plot, it seems valid to use a formula of the type:

$$\log_{10} k = 20\,\Phi - b$$

where Φ = porosity (0.01%), k = permeability (mD) and b is a constant.

The constant b is a function of the quality of the reservoir. In very permeable reservoir rocks, $b = 2$, whereas b increases to 4 in the less permeable "CS" sand (Fig. 10) resulting in a permeability difference of 2 orders of magnitude. The quality of the Brent reservoir decreases towards the more distal northern

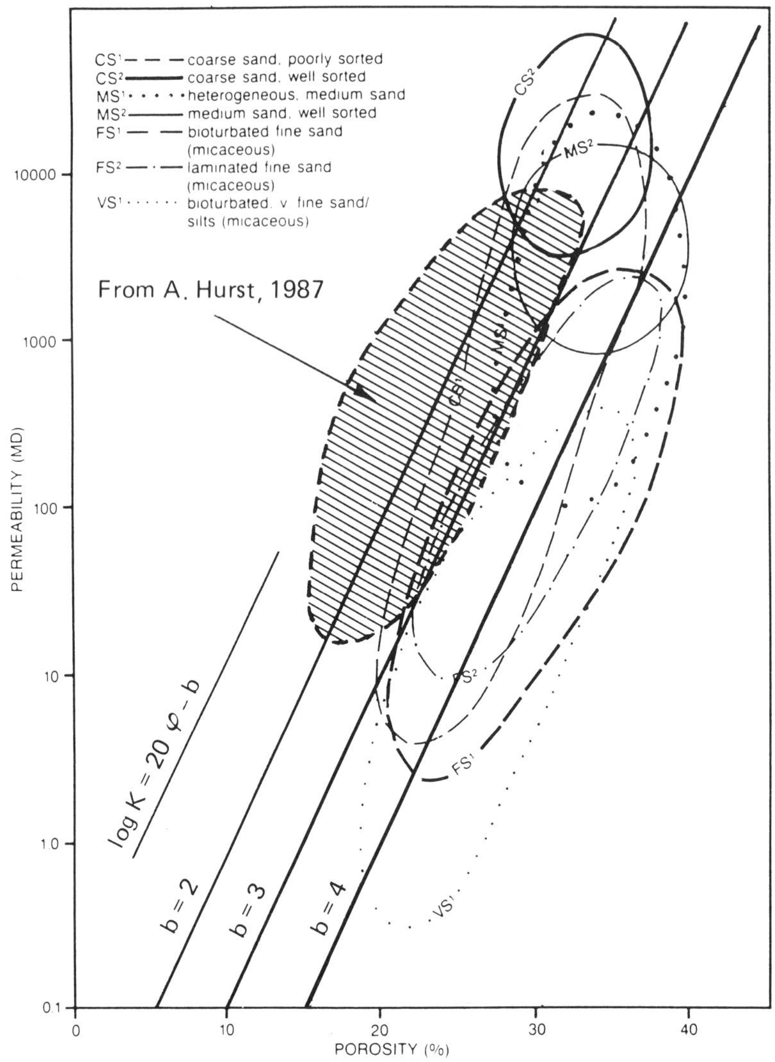

Fig. 10. Relationship between porosity, permeability and lithofacies, after Osborne and Evans (1987). Log k lines used in this study.

parts of the Brent deltaic system. Migration is considered to be concentrated in the upper layers of the Brent sequence, i.e. in the Tarbert Formation where it is present. Thus, a range of possible permeability values may be correct and must be accounted for in the simulations.

Oil and gas migrate in a water-wet system and, although hydrocarbon migration is considered to take place in a separate zone along the top of the carrier bed, some water will also be present in this zone. England et al. (1987) suggested that oil migration can take place in a system where the water saturation is up to approximately 50%. When hydrocarbons migrate at high water saturations, the permeabilities of the system with respect to oil and gas are reduced (Amyx et al., 1960). It is assumed that the oil saturation during migration will be slightly higher than that at which oil starts to migrate and therefore an oil saturation during migration of 60% is used. At 40% water saturation, the relative permeability of oil is reduced to 30% (Fig. 11) which is used in the oil loss calculations (Sylta, 1989).

The density of oil is modelled to be constant. Gas density changes are important when spill of oil and gas from traps is calculated. In the simulations, gas density is calculated from critical temperature (= 207 K) and pressure (= 45 atm). These parameters may not fully honour the effects of the condensates, but the resulting errors are not thought to be greater than other inaccuracies inherent in the modelling. However, the non-linearity of the system may make the answer sensitive in some traps, and further research is therefore required.

The density contrast between gas and water is usually five to ten times greater than that between oil and water at reservoir conditions, and the viscosity of the oil is approximately ten times that of gas. Therefore, the buoyancy force controlling gas flow is frequently 50–100 times that of oil, and as a result gas migrates faster than oil. The losses of gas during

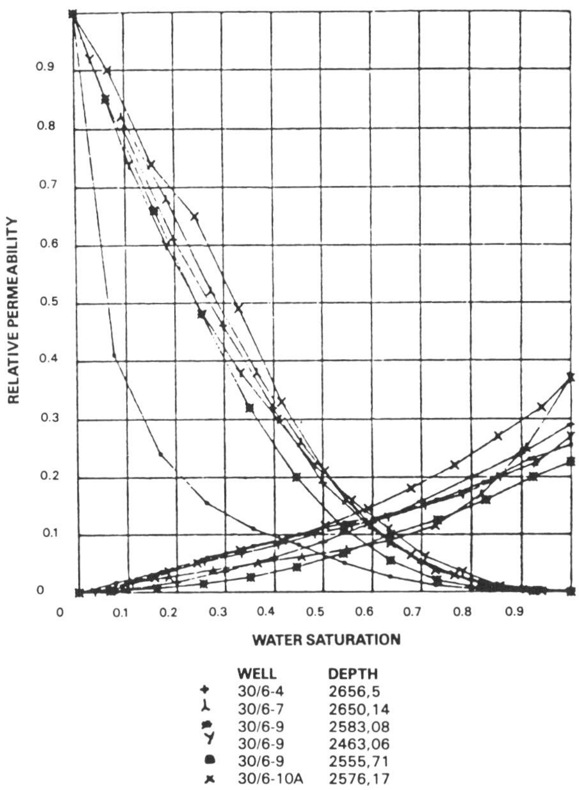

Fig. 11. Water–oil relative permeability curves after normalization (Oseberg data), after Torvund and Nipen (1987).

secondary migration due to saturation of the pathways are therefore significantly lower than for oil (see Sylta, 1989), and were excluded from the calculations. In the deeper parts of sedimentary basins gas losses may have to be included in the calculations.

For quantitative calculations of gas migration and entrapment, loss of gas through the cap rock is important. A simple model for the loss of gas due to leakage through the trap seal is used (see also Appendix):

$$\frac{dV}{dt} = -fl\,\Delta\rho HV$$

where dV/dt = rate of volume loss through seal, V = gas volume in trap, H = maximum gas column height in trap, $\Delta\rho$ = density contrast between gas and water and fl = gas loss factor.

Calibration data

Model results can be calibrated against reservoir data from the Troll area, the condensate/gas finds in wells 35/8-1 and 35/8-2 and the Agat Field.

The Troll Field is a giant oil and gas field which can be subdivided into three provinces based on the hydrocarbon content. In the Troll West oil province (Fig. 12), oil with a small gas cap is present (Osborne and Evans, 1987). Within Troll East, the oil column varies from 11 m in the north to 0 m in the central and southern regions. The Troll West gas province lies between the Troll West oil province and the Troll East province and contains significant amounts of both oil and gas. Some statistics for the Troll Field are listed in Table 1.

The maximum gas columns of the field are 207 m in Troll West and 247 m in Troll East. The present-day regional aquifer map of Fig. 12 shows that the Troll Field can be sourced from the south, west and north. The lack of gas in the western oil province suggests that this area was sourced from the more oil-prone area to the west of the field. The Troll West gas province and the Troll East province were probably sourced mainly from the north and south. The thinning of the oil column towards the south in Troll East is taken as an indication that the oil in this area has been drained from the north. We estimated the rest of the petroleum in the Troll Field to be 50% sourced from the study area.

The results of these considerations are summarized in Table 1 and yield the following calibration data for the Troll Field.

Oil: $\simeq 600 \times 10^6$ m^3 (uncertainty $\pm 200 \times 10^6$ m^3).
Gas: $\simeq 900 \times 10^9$ Sm^3 (uncertainty $\pm 400 \times 10^9$ Sm^3)

The assessed uncertainties are large, i.e. ±30% for the oil and ±45% for the gas, because we have not modelled how much oil and gas migrate into Troll from outside the study area. Nevertheless, the Troll Field offers important constraints on the migration modelling, as will be shown later.

The gas/condensate discoveries of wells 35/8-1 and 35/8-2 and the Agat Field are significantly smaller than the Troll Field. These fields may only constrain simulation results in that no oil should be present today and gas should accumulate.

Simulation procedure

The migration history was modelled in seven time steps (for drainage). The duration of each time step varies between 37 and 7 Ma and the structure of the basin (depth to reservoir, porosity, flowline directions, etc.) is considered static within each time step. The migration history was simulated for Cretaceous and Tertiary times.

Migration simulation using the parameters described was performed in a step-by-step procedure. Each new step involved new and complicating factors and thus an increase in the use of computer resources. The simple model required the use of 8 min CPU (μvax-II) whereas the most complicated simulations took 2 h of CPU to run. Each case was simulated many times with changes in input parameters in order to optimize the results with respect to observed field data.

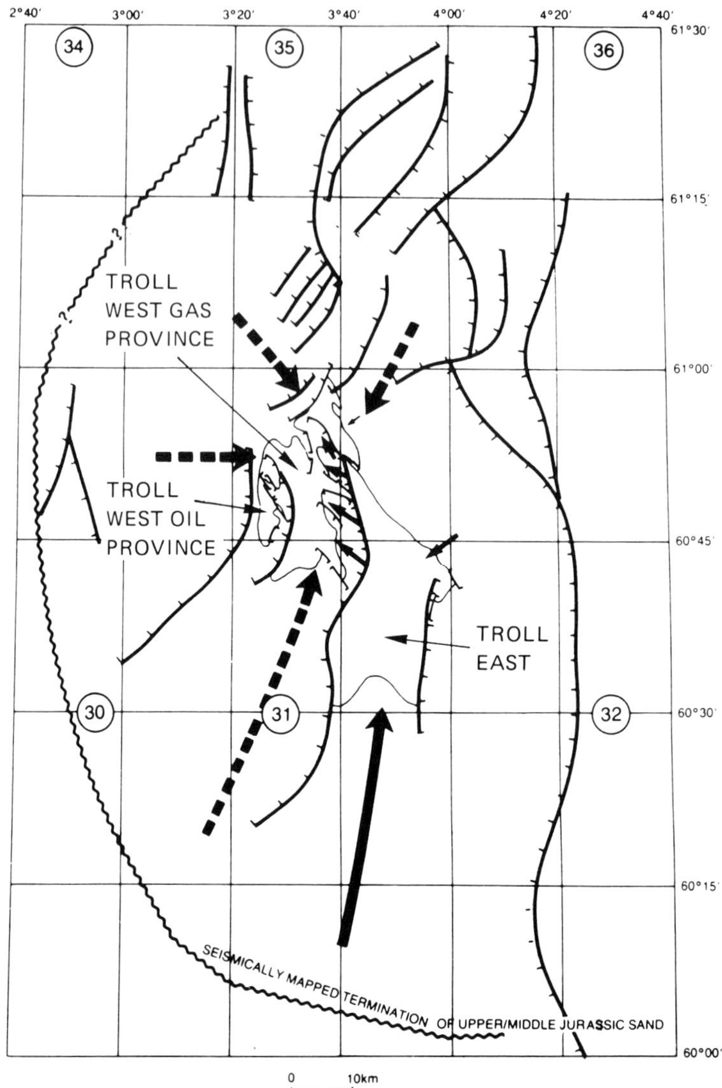

Fig. 12. Troll Field: provinces and regional aquifer, modified from Osborne and Evans (1987). Troll receives hydrocarbons from the south, west and north.

TABLE 1

Oil and gas volumes in the Troll Field (modified from Osborne and Evans, 1987)

	Oil (10^6 m^3)			Gas (10^9 Sm^3)		
	in place	from study area		in place	from study area	
		%	vol		%	vol
Troll West oil province	165	0	0	[a]		
Troll West gas province	462	50	231	632	50	316
Troll East	343	100	343	1156	50	578
Total Troll	970		600 [b]	1788		900 [b]

[a] Included in the Troll West gas province.
[b] Rounded off values (see text).

One of the most important conclusions a modelling study can give is a quantification of the sensitivity of the results. Unfortunately the model, as described here, did not allow a sufficient number of scenarios to be modelled for a thorough study of the sensitivity. This was an economic constraint, and not a scientific one. With the new generation of powerful workstations, it will indeed be possible to routinely include extensive sensitivity testing in studies of this type. Changes in input parameters were therefore determined from geological considerations, and the objective was always to match observed hydrocarbon field volumes. The chosen approach was the only method capable of quantifying some of the uncertainties in the results at the time of the study.

Results

Almost all scenarios that were modelled resulted in gas only accumulating in 35/8-1, 35/8-2 and the Agat Fields. Therefore, we concluded that these fields were of limited value to constrain the modelling results. Fortunately, the content of the Troll Field is very sensitive to changes in the modelling input parameters. Enough oil and gas migrate into Block 35/9 to completely fill the structure and the block would be gas-prone if all gas is retained within the block.

In order to assess the oil potential of Block 35/9, a crossplot between the amount of oil in 35/9 (O-35/9) and the amount of gas migrating towards Troll (G-Troll) is very useful and enables the direct estimation of the amount of hydrocarbons in 35/9 (Fig. 13). Each scenario modelled results in a line in Fig. 13 which shows the relationship between O-35/9 and G-Troll. We previously assessed the most likely value for G-Troll (900 $\pm$ 400 10^9 Sm^3) and can therefore use Fig. 13 to constrain input parameters and predict the hydrocarbon content in 35/9.

The two gas generation scenarios modelled define an area in Fig. 13: a minimum O-35/9 line for scenario B and a maximum O-35/9 line for scenario A. The distance between the lines increases towards larger G-Troll values and therefore the predictions of

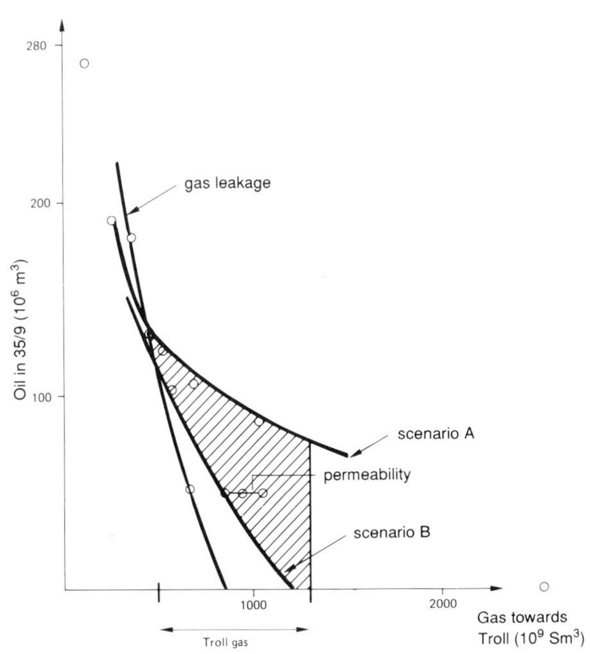

Fig. 13. Trapped oil in 35/9 versus gas migrating towards the Troll Field. Simulated cases (circles) mostly fall within the shaded area (see text).

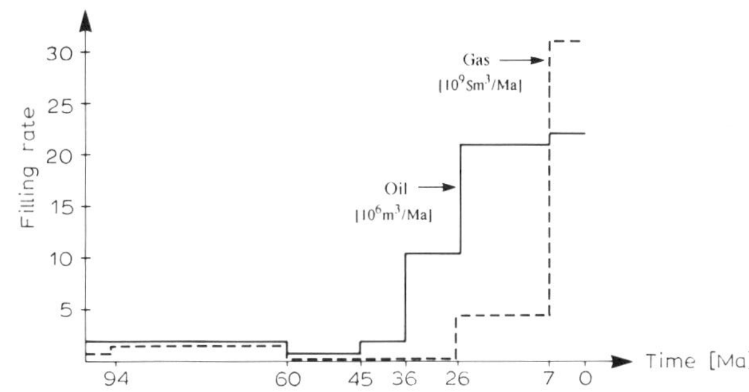

Fig. 14. Rate of oil and gas migration from the study area towards the Troll Field versus time.

oil and gas in 35/9 are more accurate if less of the gas in Troll is sourced from the study area. For very large G-Troll values, the possibility for gas only in 35/9 increases.

The secondary migration modelling results show that gas accumulates earlier in 35/9 than in the Troll Field. Gas fills the 35/9 as early as 110 Ma, whereas major gas accumulation in the Troll Field takes place in the last 26 Ma, with most of the gas accumulating in the last 7 Ma (see Fig. 14). For this reason, the amount of gas in 35/9 decreases more than the amount of gas in Troll when the gas leakage increases. The effects of varying the gas loss factor are shown in Fig. 13 for one particular gas generation scenario. The uncertainty in the predictions of amount of oil in 35/9 is high if we cannot constrain the gas loss factor, and no oil is preserved in 35/9 when the cap rock seal is sufficiently tight.

45% gas expulsion and the gas generation scenario B provide a good match to the gas calibration data for Troll (Table 1) and are therefore used for calibrating the loss of oil during secondary migration with respect to the oil migrating into the Troll Field. Values of b = 2, 2.5, 3 and 4 were used in the permeability equation. The b = 4 value results in very low permeabilities and hence simulates the maximum oil losses during secondary migration. A 20% reduction in the amount of gas migrating into Troll is observed when b is changed from 2 to 3, whereas no further reduction is observed when b is increased to 4. The amount of oil changes from 1520 $\times$ 10^6 m^3 (b = 2) to a low value of 80 $\times$ 10^6 m^3 for b = 4. The change in gas volumes observed stems from the fact that, as less oil reaches the traps due to increased losses, less oil is transformed into gas and hence less gas spills into Troll. It is interesting to note that these changes do not occur in 35/9. Present-day trapped oil in 35/9 remains constant for all simulated oil losses because of an abundance in migrating oil from the 35/9 kitchen area.

Based upon the simulation results shown in Fig. 13 it is now possible to estimate the most probable petroleum potential of 35/9. The hatched area in Fig. 13 shows the minimum (gas generation scenario B) and maximum (gas generation scenario A) oil volume trapped in 35/9 as a function of volume of gas migrating towards the Troll Field. For the most likely Troll gas volume of 900×10^9 Sm^3, the amount of trapped oil in 35/9 is between 50 and 95×10^6 m^3. The two gas generation scenarios represent the two extreme cases of minimum and maximum gas generation. Hence, a most likely oil volume in 35/9 of 70×10^6 m^3 and standard deviation of 20×10^6 m^3 was chosen. The remaining capacity is filled with gas.

The secondary migration modelling suggested that the potential for oil in Block 35/9 and the neighbouring blocks had not been fully investigated with previous exploration wells in the area. 35/9 was the first awarded block with a significant oil potential in this part of the North Sea.

Conclusions

Secondary migration modelling offers new tools in the analysis of secondary migration and prospect assessment. The ability to analyse the migration history and not be constrained to present-day structural depth maps increases the quality of the analysis. Potential high risk areas can be pinpointed at an early stage and additional mapping efforts may then reduce the uncertainties.

The flow pattern of gas and oil through time can be studied in detail and areas with focused flow and therefore higher potential may be found. The fill history of traps can be quantified and input to reservoir simulation studies in producing fields can be obtained.

Quantitative estimates in undrilled prospects are possible and the accuracy of the predictions is significantly increased if results can be calibrated with other discoveries in the area. The amount of gas and oil in Block 35/9 was predicted and uncertainties assessed by means of secondary migration modelling. The successful prediction of the hydrocarbon content in Block 35/9 shows that the methodology is useful, but more case histories are needed before one can determine how accurate in general the predicted results are.

Acknowledgements

NOPEC A/S provided the structural two-way-time maps. Noelle Odling did the tectonic modelling. Wenche Krokstad constructed the source rock model and Ulrich Ritter made the hydrocarbon generation modelling. Good results in secondary migration modelling rely on correct input data and the predictions made were entirely dependent on the quality of their work.

Appendix — A simple relationship for gas leakage from traps

Leakage of oil and gas from traps is controlled by hydrodynamic pressures, capillary pressures in the cap rocks, absolute and relative permeabilities of the cap rock, rock mechanics properties, i.e. fracturing etc., and *in situ* stress conditions. It has not been possible, yet, to account for and determine all the parameters which determine these relationships, and as a consequence, cap rock leakage is one of the least studied processes in the field of basin modelling.

Capillary pressures are determined at laboratory conditions in experiments of short duration relative to geologic times. We can not exclude the possibility that molecular diffusion processes reduce capillary pressures during a time span of several million years. Therefore the simple model presented here, which only aims to model first-order effects of gas leakage, does not account for capillary pressure in the cap rocks. The model does not incorporate any of the other relationships described above, except for the concept of permeability, because they cannot be quantified properly yet. The model as described tries to model the leakage once it has started and does not concern itself with how leakage is initiated.

During gas and oil entrapment, a pressure differential between the hydrocarbons in the reservoir and the water in the cap rock will be present due to the density differences (Berg, 1975; Watts, 1987). At the interface between the cap rock and the reservoir, the difference in pressure can be expressed as (Fig. A1):

$$\Delta p = \Delta \rho g h \tag{A1}$$

where Δp = pressure difference, $\Delta \rho$ = density

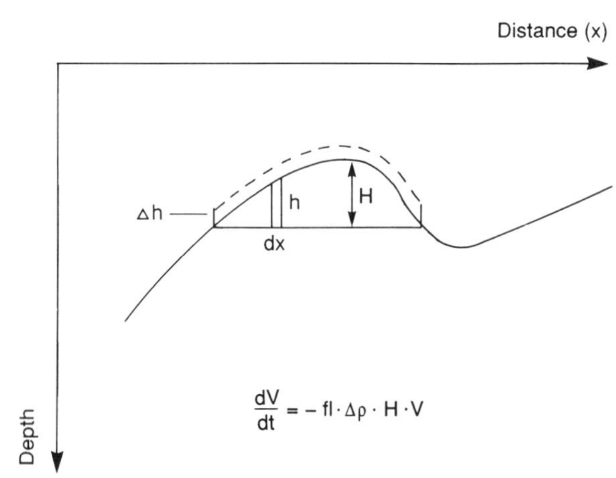

Fig. A1. Model and parameter definition for gas leakage through cap rock (see text).

contrast, h = gas column height, g = constant of gravity.

This results in a driving mechanism for gas (and oil) leakage through the cap rock. If Darcy's law applies, then:

$$v = \frac{k}{\mu}\frac{\Delta p}{\Delta h} \tag{A2}$$

where v = gas velocity, k = permeability, μ = viscosity and Δh is a transitional pressure zone above the trap.

The resulting amount of leakage through the cap rock is (Fig. A1):

$$\begin{aligned} dV &= -\int_t\int\int_A v \,dt\,dx\,dy \\ &= -\int_t\int\int_A \frac{k}{\mu}\frac{\Delta\rho g h}{\Delta h}\,dt\,dx\,dy \\ &= -\int_t \frac{k}{\mu}\frac{\Delta\rho g}{\Delta h}\left(\int\int_A h\,dx\,dy\right) dt \\ &= -\int_t \frac{k}{\mu}\frac{\Delta\rho g V}{\Delta h}\,dt \end{aligned} \tag{A3}$$

where V = volume of trapped gas (Rm^3), t is time, A is closure area and dV = volume of leaking gas. Here, we assume that Δh, $\Delta\rho$, k and μ remain constant or vary very slowly within each trap.

When the gas column increases, fractures in the cap rock may strongly influence the permeability of the cap rock. The amount of leakage is therefore dependent on the pressure difference Δp, i.e. the gas column height. Figure A2 shows some possible functional relationships between permeability and gas column height. On a microscopic scale, fractures tend to open and close instantly (curve *a*). On a larger scale, however, the cumulative effects may not be so sudden, and the bulk permeability may not change so quickly when the gas column increases. Once fractures have been generated, the permeability may stay high even if the column height decreases and a permeability evolution history as shown in curve *b* (Fig. A2) results. However, we are not yet able to constrain the required parameters for such a complex model of cap rock permeability behaviour, and therefore the simple relationship. $k = k' \times H$ (curve *c* in Fig. A2) is proposed. H is the maximum gas column and k' is a constant. The loss equation is thus reformulated to:

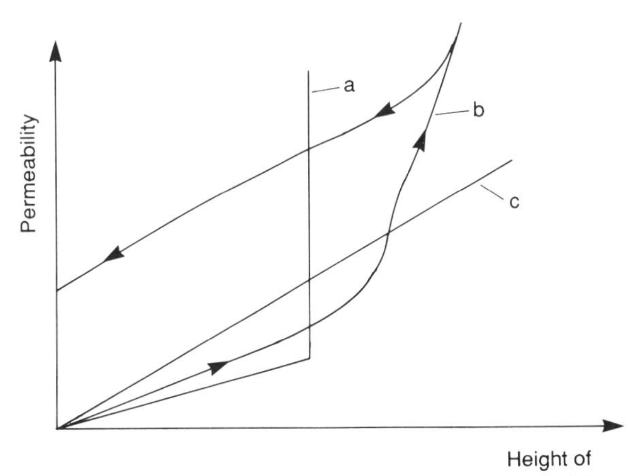

Fig. A2. Permeability versus gas column height. Three possible functional relationships are: instantly opening and closing fractures (*a*); non-closing fractures (*b*); simple relationship used (*c*).

TABLE A1

Gas loss factors versus "halving times" ($\Delta\rho$ = 0.75 g cm^{-3}, h = 200 m)

Halving time, t_h (Ma)	Gas loss factor, fl (10^{-6} m^2 kg^{-1} Ma^{-1})
20	0.231
40	0.1155
80	0.05775

$$\frac{dV}{dt} = -fl\,\Delta\rho H V \tag{A4}$$

where the fl parameter includes the effects of cap rock permeability, gas viscosity, and Δh. These parameters are poorly constrained in cap rocks.

The gas loss factor fl is used to describe traps with regard to their "ability" to leak gas. The density contrast between gas and water is higher than between oil and water, and gas leakage is therefore more important to model. Diffusion mechanisms are also important for gas leakage and similar types of equations result.

The concept of continuous gas loss through the seal has not been much studied quantitatively, and the author is not aware of published data which constrain the loss factor (fl). At any particular time, the expression resembles an exponential decay function and the concept of "halving time" can be used (i.e. how much time is needed to reduce the gas volume of a trap to 50%). Therefore, the loss factor can be derived from considerations of expected halving times for gas leakage from traps. Halving times (t_h) within the range 20 Ma to 80 Ma for a trap with a gas cap height of 200 m and a density contrast of 0.75 g cm^{-3} are shown in Table A1.

References

Amyx, J.W., Bass, D.M. and Whiting, R.L., 1960. Petroleum Reservoir Engineering. McGraw-Hill, New York, N.Y., 610 pp.

Berg, R.R., 1975. Capillary pressure in stratigraphic traps. Am. Assoc. Pet. Geol., Bull., 59 (6): 939–956.

Buhrig, C., 1989. Geopressured Jurassic reservoirs in the Viking Graben: modelling and geological significance. Mar. Pet. Geol., 6: 31–48.

Buza, J.W. and Unneberg, A., 1987. Geological and reservoir engineering aspects of the Statfjord Field. In: J. Kleppe, E.W. Berg, A.T. Buller, O. Hjelmeland and O. Tarsæther (Editors), North Sea Oil and Gas Reservoirs. Norwegian Institute of Technology, Graham and Trotman, London, pp. 23–38.

England, W.A., Mackenzie, A.S., Mann, D.M. and Quigley, T.M., 1987. The movement and entrapment of petroleum fluids in the subsurface. J. Geol. Soc. London, 144: 327–347.

Hermans, L., Van Kuyk, A.D., Lehner, F.K. and Featherstone, P.S., 1992. Modelling secondary hydrocarbon migration in Haltenbanken, Norway. In: R.M. Larsen, H. Brekke, B.T. Larsen and E. Talleraas (Editors), Structural and Tectonic Modelling and its Application to Petroleum Geology. Norwegian Petroleum Society (NPF), Special Publication 1. Elsevier, Amsterdam, pp. 305–323.

Irwin, H., Hermanrud, C., Carlsen, E.M., Vollset, E. and Nordvall, I., 1993. Basin modelling of hydrocarbon charge in the Egersund Basin, Norwegian North Sea: pre- and post-drilling assessments. In: A.G. Doré et al. (Editors), Basin Modelling: Advances and Applications. Norwegian Petroleum Society (NPF), Special Publication 3. Elsevier, Amsterdam, pp. 539–548 (this volume).

Jarvis, G.T. and McKenzie, D.P., 1980. Sedimentary basin formation with finite extension rates. Earth Planet. Sci. Lett., 48: 42–52.

Lehner, F.K., Marsal, D., Hermans, L. and Van Kuyk, A., 1988. A model of secondary hydrocarbon migration as a buoyancy-driven separate phase flow. Rev. Inst. Fr. Pét., 43 (2): 155–164.

Osborne, P. and Evans, S., 1987. The Troll field: reservoir geology and field development planning. In: J. Kleppe, E.W. Berg, A.T. Buller, O. Hjelmeland and O. Tarsæther (Editors), North Sea Oil and Gas Reservoirs. Norwegian Institute of Technology, Graham and Trotman, London, pp. 75–88.

Powley, D., 1990. Hydraulic regimes in the crust. Presented at SEG Research Workshop on Permeability, Fluvial Pressure and Pressure Seals in the Crust, August 5–8, 1990, Denver, Colo.

Schroeder, F.W. and Sylta, Ø., 1993. Modelling the hydrocarbon system of the North Viking Graben: a case study. In: A.G. Doré et al. (Editors), Basin Modelling: Advances and Applications. Norwegian Petroleum Society (NPF), Special Publication 3. Elsevier, Amsterdam, pp. 469–484 (this volume).

Sclater, J.G. and Christie, P.A.F., 1980. Continental stretching: an explanation of the post-mid-Cretaceous subsidence of the Central North Sea Basin. J. Geophys. Res., 85: 3711–3739.

Sylta, Ø., 1987. SEMI — A Program for the Modelling of Buoyancy-Driven, Secondary Migration of Oil and Gas by Means of a Ray-Tracing Technique. Continental Shelf and Petroleum Technology Research Institute (IKU) Rep. 25.2403.00/01/87, 51 pp. (open).

Sylta, Ø., 1989. Modelling of secondary migration and entrapment of a multicomponent hydrocarbon mixture using equation-of-state modelling techniques. Petroleum Migration Meet., Burlington House, London, September, 15 pp.

Torvund, T. and Nipen, O., 1987. The Oseberg Field: potential for increased oil recovery by gas injection in a high permeability environment. In: J. Kleppe, E.W. Berg, A.T. Buller, O. Hjelmeland and O. Tarsæther (Editors), North Sea Oil and Gas Reservoirs. Norwegian Institute of Technology, Graham and Trotman, London, pp. 61–74.

Watts, N.L., 1987. Theoretical aspects of cap-rock and fault seals for single- and two-phase hydrocarbon columns. Mar. Pet. Geol., 4: 274–307.

White, D.A. and Gehman, H.M., 1979. Methods of estimating oil and gas resources. Am. Assoc. Pet. Geol., Bull., 63 (12): 2183–2192.

ØYVIND SYLTA *IKU Petroleum Research, N-7034 Trondheim, Norway*

Closure vs. seal capacity — a fundamental control on the distribution of oil and gas

John K. Sales

A classification of traps based on sealing capacity vs. closure may help explain the type and proportion of hydrocarbons and their spilling and leaking from traps. If at least one trap volume of both oil and gas enters the trap it will contain gas (Class 1), oil and gas (Class 2), or predominantly oil (Class 3), depending on sealing capacity, closure, and buoyancies. Thus, traps function as valves which store and distribute hydrocarbons predictably. Five types of seal, based on nature of the failure path, and including seals that fail by hydraulic fracturing, are believed to honor this trap classification. Because gas overlies oil, they all fail similarly — lower relative closure allows gas to spill oil, higher relative closure allows oil to force gas to leak, limiting the accumulation of gas.

Class 1 Traps have excess sealing capacity and can spill both oil and gas updip. (These are "Gussow-type" traps.) *Class 2 Traps* spill oil, leak gas and have a two-phase fill because closure exceeds the capacity gas column but not the capacity two-phase column. At capacity, oil spills updip, gas leaks through the seal, and the gas/oil contact stabilizes in mid-trap. *Class 3 Traps* have closure greater than the oil-dominated capacity column, leak both oil and gas and spill neither. Both fluid contacts stabilize within the closure.

During uplift, Class 1 Traps may be gas-flushed, Class 2 Traps might gain additional capacity for oil, and Class 3 Traps might modestly lose or gain capacity for oil. While the concepts should be generally applicable, local examples are given from the Norwegian North Sea. Troll-East is probably a Class 1 Trap, Oseberg a Class 2 Trap, and Gullfaks and Ekofisk are Class 3 Traps.

Preliminary analysis shows that Norwegian traps are skewed toward single-phase fills independent of depth and distance from Upper Jurassic gas. Gas traps tend to be filled-to-spill, oil traps do not, thus supporting the proposed classification. There is, however, much overlap and scatter between gas field and oil field column heights, suggesting much variation in seal capacity. Tentatively, it is postulated that over-ample source, maturation and migration potential have driven most Norwegian traps through their Gussow-evolution to equilibrium by trapping principles, with a sub-equal distribution of Class 1 Traps flushed with gas, and Class 3 traps venting gas while retaining oil. Predictably, there are fewer Class 2 Traps with mixed fills.

Introduction

The amount and type of hydrocarbon in a trap are influenced by many factors including generation, expulsion, migration, and trap performance. This paper emphasizes and attempts to systematize trap performance: the way traps spill, leak and retain hydrocarbons. For more thorough discussions of trapping theory, see Berg (1975), Schowalter (1979, 1991), Ungerer et al. (1990), and especially Watts (1987). An excellent source reference for papers on trapping is Foster and Beaumont, 1988.

The relation between sealing capacity, buoyancy and closure dominates in controlling trap performance and influences column heights, proportions of oil and gas, and spilling and leaking of both to shallower traps. Conversely, shallower traps can give important clues about deeper traps.

"Gussow-type" traps (Gussow, 1954) are Class 1 Traps as defined here. They have excess sealing capacity relative to closure, spill both oil and gas, and leak neither. Class 2 and Class 3 Traps have excess closure relative to sealing capacity. Class 2 "balanced" Traps leak gas but spill oil. Class 3 Traps leak both oil and gas and spill neither. Thus, traps are, in a sense, valves which store and distribute oil and gas in a logical and predictable way.

Column heights, fluid contacts, fluid densities, sealing capacity, and closure are interdependent. This often allows any one of them to be estimated if the other factors are known or assumed. The relationships are most clearly displayed and easily manipulated on pressure–depth diagrams. The extended caption of Fig. 3 explains a versatile and simple method of customizing such diagrams to any trap and its related fluid parameters.

Concepts are introduced in Figs. 1–3, and summarized in Table 1. Figure 1, based on Watts (1987) and Mobil in-house data, qualitatively summarizes the sealing capacities of gas, oil, and two-phase columns for membrane seals. As will be discussed in detail under seal types in relation to the classification, it is believed that hydraulic seals (Watts, 1987), that fail by pressure-induced microfracturing, also are compati-

Basin Modelling: Advances and Applications edited by A.G. Doré et al. NPF Special Publication 3, pp. 399–414, Elsevier, Amsterdam.

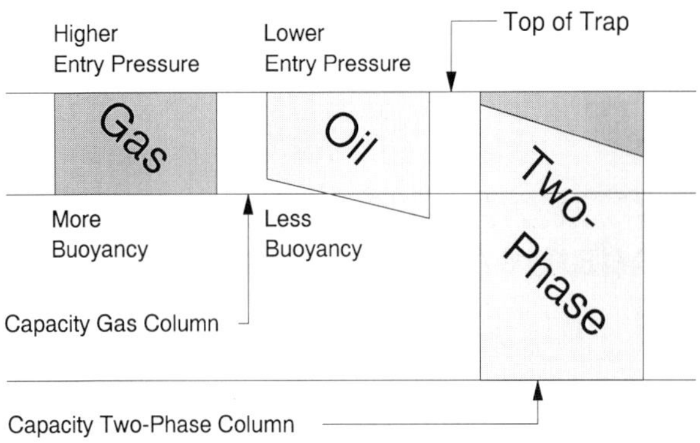

Fig. 1. Diagrammatic column heights for membrane seals. The capacity oil column may be shorter or longer than the capacity gas column. The capacity two-phase column, taking advantage of the higher entry pressure to gas, and oil's lesser buoyancy, is assumed to be longer than either capacity single-phase column. Based on Watts, 1987 and Mobil in-house data.

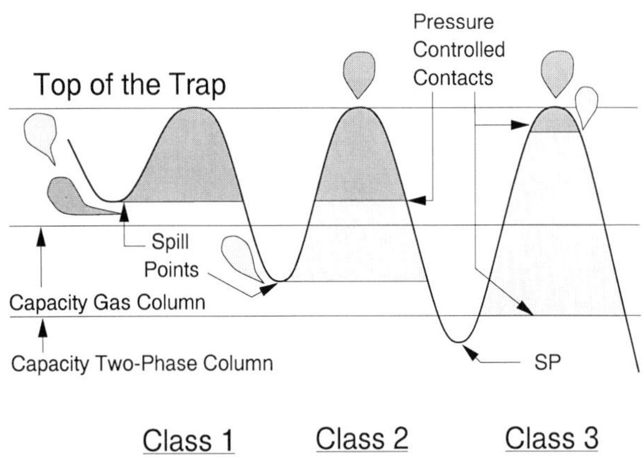

Fig. 2. Diagrammatic cross-section showing the three classes of traps. The depth of the trap's spillpoint in relation to the two thresholds — the capacity gas column and the capacity two-phase (maximum) column — determine the class of the trap. Bubbles indicate the type of fluid that is spilled or leaked from the trap.

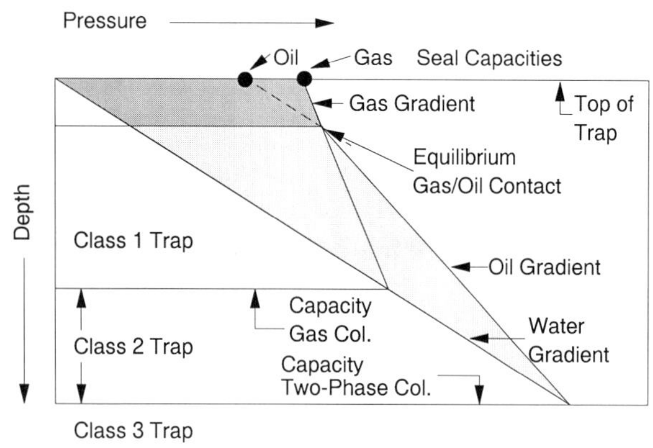

Fig. 3. Pressure–depth diagram characterizing the three classes of trap. Note that the pressure/depth diagram has the same three fundamental horizontal lines as Figs. 1 and 2: the top of the trap, capacity gas column, and capacity two-phase column. The diagonal of the graph is arbitrarily taken as the hydrostatic gradient, with the sealing capacity and density scales adjusted to it. The sealing capacity scale can be calibrated by multiplying a convenient column height by the hydrostatic gradient. As a rule of thumb, a 700 m column of nearly fresh water exerts about 1000 psi pressure. The fluid density is scaled along the bottom of the diagram. Technically, a density of 1 should only be used for pure water. Adjust slightly for the more normal salt water. The oil and gas gradients are constructed from this scale. Oil density is usually given relative to water, but gas density is often given relative to air, and an adjustment has to be made to determine its density relative to water at reservoir conditions. The dashed line projected downward from the oil sealing capacity to the gas gradient parallel to the hydrostatic gradient is a construction line allowing oil's sealing capacity to be transferred onto the actual column. Conversely, if the trap is at its sealing capacity for oil — if it is dominantly oil, leaking, and not filled-to-spill — this same construction line can be used to estimate oil's sealing capacity.

TABLE 1

Characteristics of the trap classes

	Class 1	Class 2	Class 3
Dominant Fluid	Gas	Both	Oil
Spills	Both	Oil	Neither
Leaks	Neither	Gas	Both
At Seal Cap.	No	Yes - Gas	Yes - Both
Fill-to-Spill	Yes	Yes	No
Controls GOC	Charge Vol.	Pressure	Pressure
Controls OWC	Spill Point	Spill Point	Pressure
Trap Updip	Gas	Oil	Water Wet
Uplift	Gas Flushes	Improves	Variable

ble with the basic classification, although the proportional column heights are not as displayed in Fig. 1.

The ease with which a fluid can penetrate a capillary is proportional to the wettability of that fluid compared with the fluid that saturates the capillary. Wettability in turn is proportional to the "interfacial angle", usually visualized as the angle at which the side of a bubble intersects the surface it is resting on. As two familiar extreme cases, mercury has a very high surface tension, resulting in an extremely high interfacial angle that forces the bubble to be nearly a sphere. It is difficult to force mercury through a capillary saturated with air. At the other extreme, water forms no interfacial angle at all when dropped onto a surface that is already water-wet, rather it simply flows into and becomes a part of that surface. Thus, water passes freely through a capillary that is already water-wet.

Because gas has a greater interfacial angle than oil and is therefore less wettable, it takes more pressure to overcome capillary resistance to forcing the gas through the water-wet pores in the seal. Thus, gas has a higher entry pressure than oil under most conditions (Schowalter, 1979). Counter to this, gas stresses the seal more per unit of column height because it has a lower density than oil and is therefore more buoyant under shallower reservoir conditions. This may reverse at greater depth, with the cross-over being in the 1.5–3 km depth range. Thus, the height of the maximum oil column relative to the maximum gas column varies with fluid properties and with depth, sometimes being greater and sometimes less (Watts, 1987). This is reflected in the slanted bottom of column two in Fig. 1.

Watts (1987) presents two models based on analysis of oils and gases from several divergent areas. For membrane seals, however, the highest possible column is always a two-phase column (cf. Watts, 1987, fig. 8). Such a two-phase column takes advantage of the gas's greater entry pressure at the base of the seal and the oil's lesser buoyancy-induced stress throughout the remainder of the column. This maximum column height that sealing capacity allows (capacity two-phase column) is attained in Class 3 Traps.

In Class 3 Traps, oil occupies most of the column height, though there may be considerable additional gas in solution. Because neither fluid can spill, both accumulate until all sealing capacity is utilized for each fluid. At that time both oil and gas begin to leak at the rate of input of each fluid into the trap. The two important column heights, the capacity gas column and the capacity two-phase column, are taken as the thresholds for the trap classes (Figs. 1–3). Class is determined by depth of the spillpoint relative to these thresholds.

The concepts do not violate "differential entrapment" (Gussow, 1954; Schowalter, 1979): the principle that gas can displace oil out of a trap. They merely stress that if too much oil is preserved in a province exposed to excessive gas, differential entrapment, as defined by Gussow, cannot be the only principle. Alternatively, traps might leak gas to make room for oil (Class 2 and Class 3 Traps). Gussow's portrayal covered Class 1 Traps, but not Class 2 and Class 3 Traps as defined here. Gussow's terminology is here broadened to include leakage of gas and oil, the combined case in which gas leaks but oil spills, and stabilization of fluid contacts within the closure.

Assumptions

The following simplifying assumptions are adhered to. Except when discussing a specific quantitative case, the term "sealing capacity" is left abstract, with no number assigned to it. For brevity, and by analogy to a chain of lakes, multiple traps along a migration route are called a "migration chain" or simply a "chain".

(1) Gas, oil, and water are treated as simple fluids of differing density.

(2) Hydrocarbons are assumed to migrate as separate phases, controlled by buoyancy (England et al., 1987; Tissot, 1987; Durand, 1988; Ungerer et al., 1990).

(3) The mode of seal failure is assumed to be the more common capillary breakthrough of membrane seals (Berg, 1975). More problematically, hydraulic fracturing (Watts, 1987) should also be compatible with the classification scheme suggested here. (See discussion under "Seal types related to trap classes".)

(4) For membrane seals, the maximum hydrocarbon column is assumed to be a two-phase column that takes advantage of the gas's greater capillary entry pressure and the oil's lesser buoyancy-induced stress (Watts, 1987). For hydraulic seals the column-height relationships change. Hydraulic fracture failure is initiated at a set pressure for any given trap regardless of which fluid is in contact with the seal. Thus, it takes a longer oil column than gas column to build up to this failure pressure, with a mixed column having intermediate length. Nevertheless, the fundamental principle driving the classification does not change. Because gas overlies oil it is forced against the most highly stressed part of the seal by the oil. Thus, oil can differentially force gas through the seal if closure is great enough relative to sealing capacity, regardless of the type of seal.

(5) It is assumed that a minimum of one trap volume of both oil and gas have been available to the trap. If any less has been available, the trap's fill is still evolving by Gussow-type principles, and the trap has not yet come to equilibrium by the trapping principles suggested here.

(6) Inactive traps, no longer receiving hydrocarbons, are assumed to have had their fill character determined during previous active charging.

(7) Leaking and spilling may cease when charging ceases, with minimal shortening of oil and gas columns from what they were during maximum charging. The alternative that seals become hydrocarbon-wet, drain, and do not repair themselves until the column is dissipated seems unlikely. There are too many high hydrocarbon columns in areas requiring long-term retention for that alternative to be the general rule.

(8) Volumetrics of charging phases are assumed to determine the fill in Class 1 Traps, and in a migration chain of Class 1 Traps — if they have been exposed to enough gas, they will have been flushed by gas.

(9) In Class 2 and Class 3 Traps it is assumed that fluid contacts and rates of spilling and leaking continually equilibrate. This seems to be substantiated by rapid rates of migration relative to geological processes (Dembicki and Anderson, 1989; Schowalter, 1991).

(10) Except for Class 1 Traps, fluid contacts not controlled by a spillpoint are assumed to be in equilibrium, with balanced input and output from the trap. This is because only one position of the

contact allows column buoyancy to match sealing capacity.

(11) Variations in rate of fluid input are assumed to be subordinate to the relationship between sealing capacity and closure in determining fluid contact levels. Variations in rate of input to traps at sealing capacity (Class 2 and Class 3 Traps) will be compensated by adjustments in rates of leakage and spillage. Thus, while a temporary surge might disturb this equilibrium, the shift of fluid contacts should be minimal and short-lived because of the downward increase in perimeter of most traps' seals. If, for example, a trap is forced to leak more gas because of a surge of gas into the trap, the surface area of the seal exposed to gas leakage can be doubled with only a nominal lowering of the gas/oil contact because of this strong downward increase in the perimeter of the seal.

(12) In the discussion of uplift it is assumed that sealing capacity remains constant. Although this is probably not true, it is necessary to focus the discussion.

Classes of trap

The principle is widely accepted that gas can displace oil out of a trap (Gussow, 1954; Schowalter, 1979). Less familiar is the principle that, if the closure is large enough relative to sealing capacity, oil can force gas out of the trap by crestal leakage. If closure exceeds all possible column heights, once at capacity, both oil and gas must leak through the seal at the rate they are input into the trap. Depending on seal capacities, properties of the two hydrocarbons, and depth of the trap, this may occur with oil occupying 2/3 to 3/4 of the column height in shallow traps (J.R. Dixon, pers. commun., 1991). In traps of normal depth, however, the equilibrium gas cap will either be very small or all gas will be forced into solution in the oil.

Between the extremes of gas displacing oil by spillage and both oil and gas being displaced by crestal leakage, there should be a range of closure relative to sealing capacity in which the gas/oil contact equilibrates in mid-trap, with gas leaking through the seal while oil spills at the spillpoint. There are only three cases. These are depicted in a diagrammatic cross-section (Fig. 2) and in a generalized pressure–depth diagram (Fig. 3).

There should be three types of fluid contacts, controlled by spillpoint, charge volumetrics, and pressure. Pressure-controlled fluid contacts, constrained by the combination of closure, sealing capacity, and buoyancy, stabilize within the trap and do not equate to the trap's spillpoint. They should retain their positions even while fluids are entering and exiting the trap. The gas/oil contact in Class 2 Traps and both fluid contacts in Class 3 Traps are pressure-controlled.

Class 1 Traps

Class 1 Traps are Gussow-type traps with excess sealing capacity. Their seal could support a higher gas column, the most buoyant and seal-stressing column, if the trap had more closure. They spill both oil and gas. Because of excess sealing capacity, Class 1 Traps leak neither hydrocarbon through the seal to higher stratigraphic levels. Gas accumulates to the spillpoint, and oil is spilled by a process often described as "gas flushing" (Gussow, 1954). After the trap fills with gas, oil flows under the gas and spills without accumulating — oil "underspills" the trap.

All gas-filled traps, filled-to-spill, by definition, are Class 1 Traps. Class 2 and Class 3 Traps, by definition, do not have enough sealing capacity to allow them to be filled with gas to the spillpoint. In fact, Class 3 Traps do not have enough sealing capacity to allow even the longest (capacity) two-phase column to fill the trap. Some partly filled or oil-filled traps, and some traps with a two-phase fill might have enough sealing capacity to be a Class 1 Trap, but they may have been exposed to insufficient gas to flush them — they are still in the process of Gussow-evolution.

A trap containing only gas, but not filled to its spillpoint is problematical because there is room for oil. The charge may have contained no oil (in general an unlikely case unless the source beds are completely gas-prone); a previously full gas charge may have been compressed by subsidence; some gas may have diffused through the seal; or oil may have been cracked to gas in the trap. Since spillpoints are hard to define, a likely alternative is simply that the spillpoint was misjudged and/or undetectable and that the trap is actually a Class 1 Trap filled-to-spill.

Class 2 Traps

With intermediate closure relative to sealing capacity, Class 2 Traps leak gas but spill oil. Because the spillpoint is deeper than the capacity gas column, gas leaks before the trap can be filled with gas. There is not enough vertical distance between the gas/oil contact and the spillpoint, however, to cause the oil column to leak — the trap is not at sealing capacity for oil. Leaking of gas and spilling of oil keep pace with the rate of charge into the trap.

In such a trap, the proportion of oil to gas cannot change without disrupting the equilibrium of the trap. Because the trap is filled-to-spill the length of the hydrocarbon column does not change, but the column's buoyancy is proportional to the amount of gas. Only one gas/oil contact allows buoyancy of the

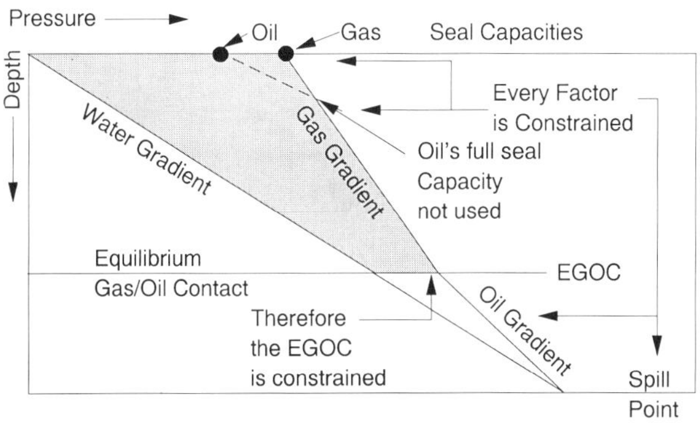

Fig. 4. Pressure–depth diagram showing the reason for the stability of the equilibrium gas/oil contact (EGOC) in Class 2 Traps. Because the oil gradient is constrained by spillpoint and the gas gradient by sealing capacity, only one GOC is allowable. Thus, the EGOC has to stabilize in mid-trap.

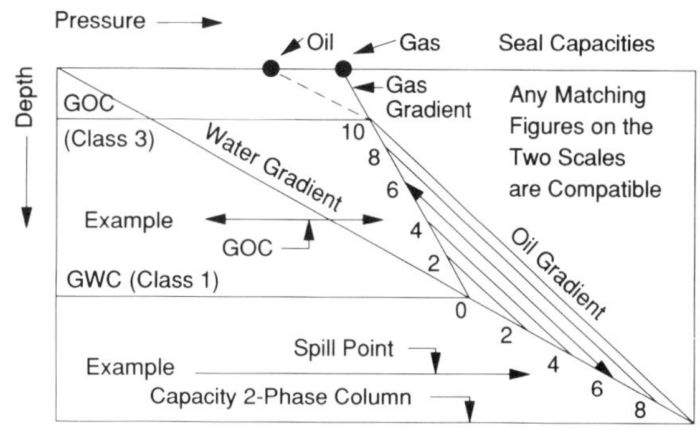

Fig. 5. Pressure–depth graph showing the EGOC that correlates with any specific closure in relation to the capacity gas column and the capacity two-phase column. It is laid out so that corresponding numbers on the two scales are compatible closures and EGOCs.

hydrocarbon column to match sealing capacity. This is the only gas/oil contact that is stable. This level is termed the "equilibrium gas/oil contact" (EGOC).

The reasoning for the stability of the equilibrium gas/oil contact, even in the face of hydrocarbons entering and exiting the trap, shows clearly on a pressure–depth diagram (Fig. 4). Every factor is constrained and no degrees of freedom exist. The gas column is constrained by sealing capacity and gas gradient. The oil column is constrained by the spillpoint and oil gradient. Thus, the two gradient lines can intersect at only one level, the EGOC.

Class 2 Traps can be thought of as "balanced" or transitional traps that span the range of closure between Class 1 and Class 3 Traps. As closure approaches Class 1, their fill-ratio approaches 100% gas. As closure approaches Class 3, the ratio of oil to gas increases until oil dominates, either with a small gas cap, or with gas only allowable in solution in the oil. At that threshold both oil and gas start to leak through the seal at the rate at which they enter the trap. At this closure value, leakage becomes dominant and spilling updip in the carrier bed ceases.

A pressure–depth graph allows determination of the equilibrium gas/oil contact appropriate for any specific spillpoint (Fig. 5). The scales are for reference only and have no absolute numerical value. The graph can be used to estimate the EGOC and the percentage of oil and gas in the trap, if sealing capacity and closure are known or assumed. Conversely, it can be used to estimate sealing capacity if closure and the position of the equilibrium gas/oil contact are known, and indications of crestal leakage of gas suggest that the trap's seal is at capacity. If there are no indications of leakage and it is assumed that it is a Class 1 Trap that is still in Gussow-evolution, the graph still gives a viable minimum for sealing capacity. The graph is arranged so that an appropriate equilibrium gas/oil contact to match any spillpoint depth can be determined by matching numbers on the equilibrium gas/oil contact scale and on the closure scale.

Class 3 Traps

Class 3 Traps, the other extreme from Class 1 Traps, have weak seals relative to closure and leak both oil and gas. Because they spill neither hydrocarbon-type, leakage of both has to equate with input into the trap. Neither the oil/water nor the gas/oil contact is controlled by spillpoint and there is only one combination of column heights that allows oil's buoyancy to match oil's sealing capacity and the buoyancy of the two-phase column to match gas's sealing capacity — both fluid contacts are pressure-controlled. This equilibrium may allow a large oil column, but only a modest gas cap, or no gas cap at all (Figs. 1–3). Because most traps narrow upward, a trap that has even 2/3 oil by column height will normally be 80–95% oil by volume, although additional gas may be in solution in the oil. Thus, Class 3 Traps can be characterized as predominantly oil traps.

There should be almost no lower limit to the size of a Class 3 Trap. A weak spot along a migration route might have two meters of closure but only one meter of sealing capacity; it would therefore function as a Class 3 Trap. Traps updip from such a weak spot along a migration route might either be dry or would be charged only with fluids entering the chain updip. Alternatively, if that weak spot functioned like a Class 2 Trap, it might leak gas vertically, but spill oil to traps updip.

Examples of classes of traps

While the concepts presented in this paper should be generally applicable, examples are presented from the northern North Sea (Fig 6).

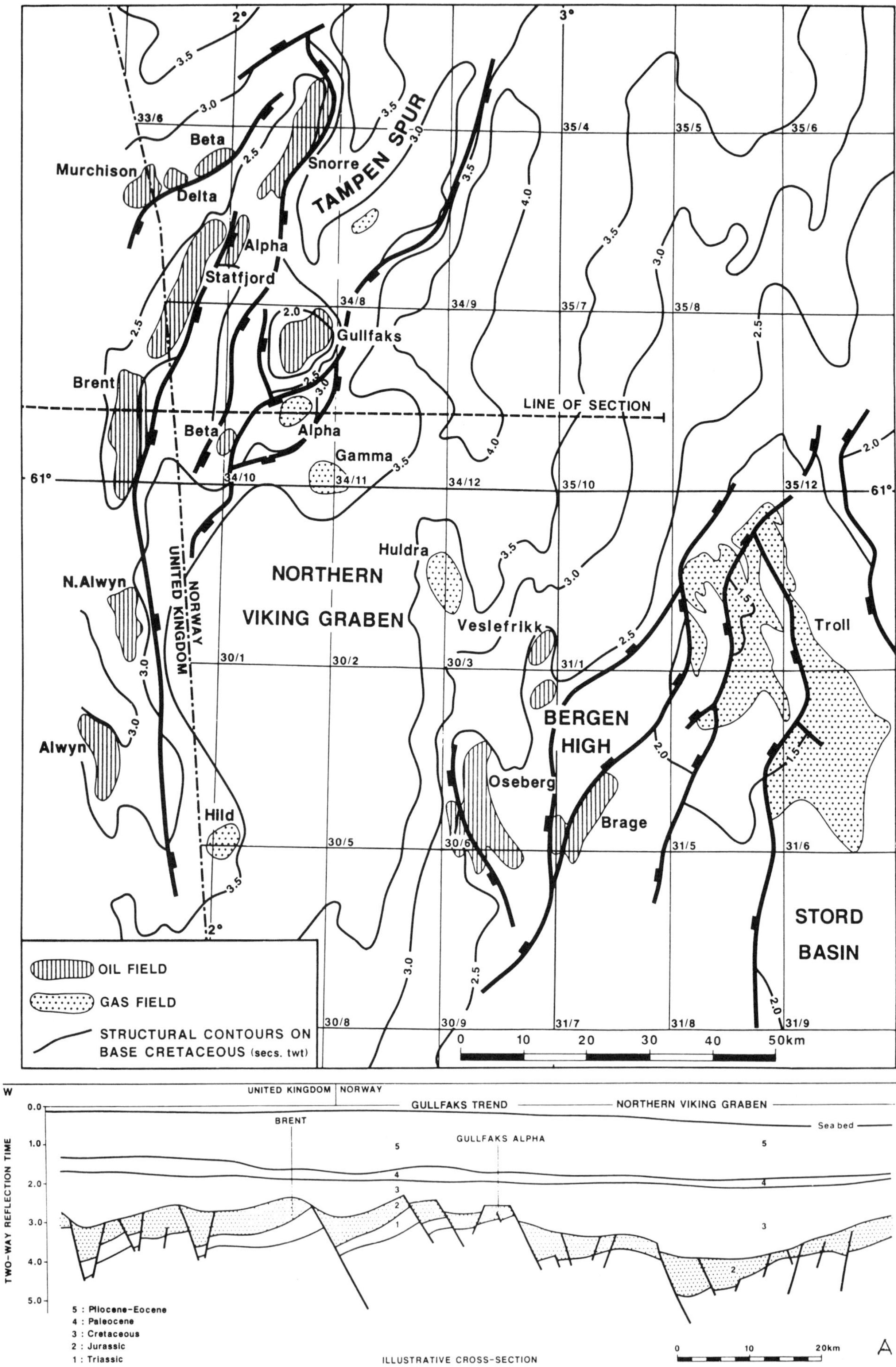

Fig. 6. Index maps of northern Viking Graben (A) and Central Graben (B) in the Norwegian North Sea showing fields discussed. Fields are displayed in terms of their major fluids with no gas caps or oil rims shown. Generation, driven by rapid Neogene sedimentation, is at

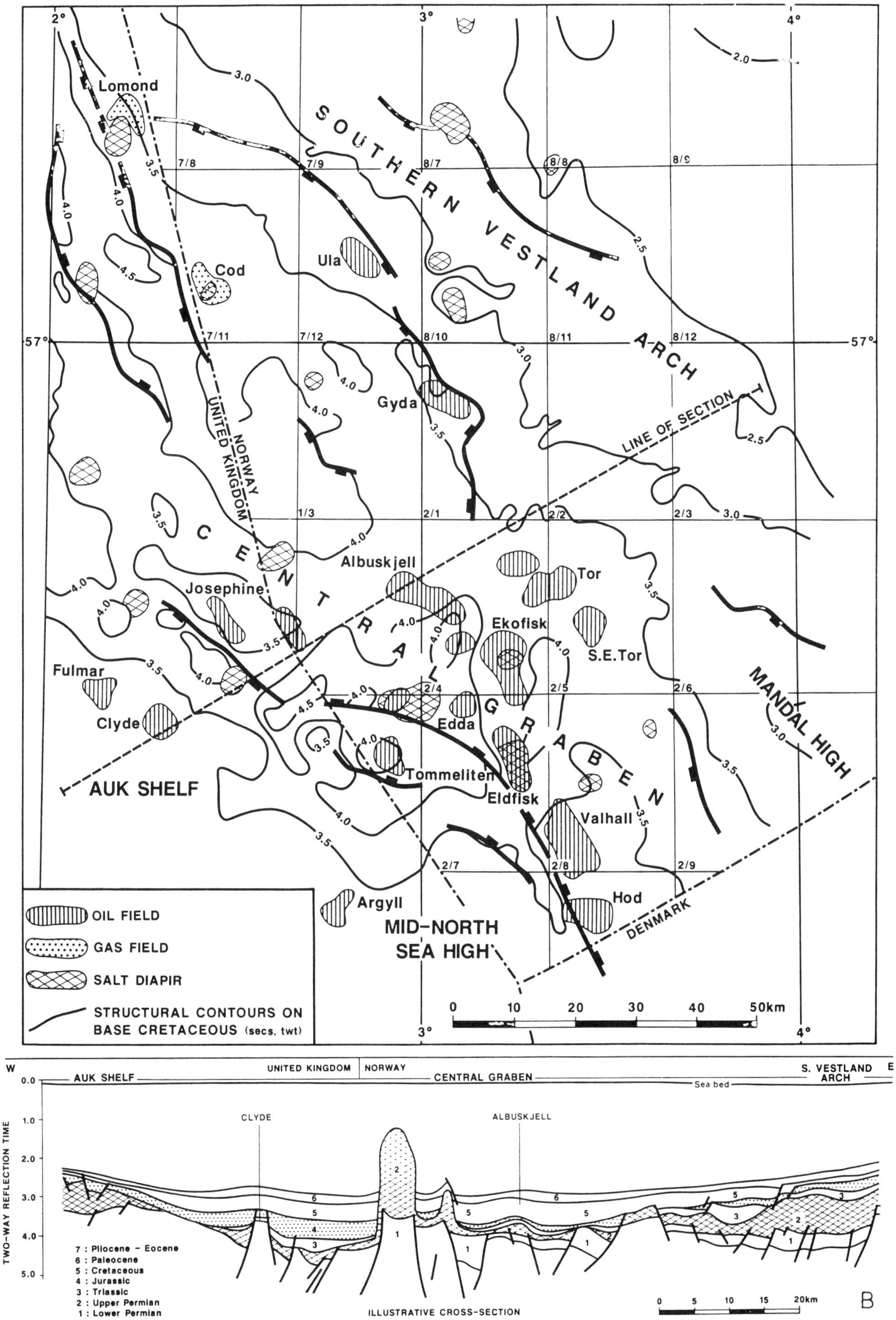

maximum and continuing. Base of the oil window for the main Kimmeridge-equivalent Upper Jurassic source beds varies between the 3.5 km and 4 km structure contour on this Base Cretaceous map horizon, depending on the model used. In (A) note distance of Troll Field from Troll Kitchen in axis of Viking Graben to the northwest. In (B) the Central Graben is dominated by the chalk-reservoired oil fields in proximity to gas generation and with gas chimneys. From Campbell and Ormaasen (1987).

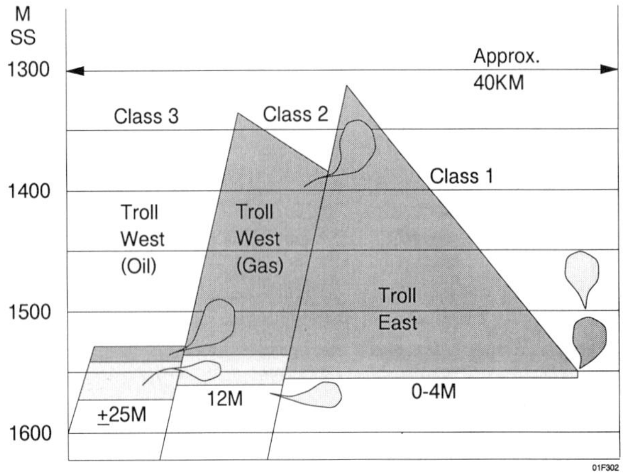

Fig. 7. Diagrammatic cross-section of trap–spill relationships in the Troll Field (Fig. 6A). Note extreme vertical exaggeration. Based on data in Gray (1987). See text for discussion.

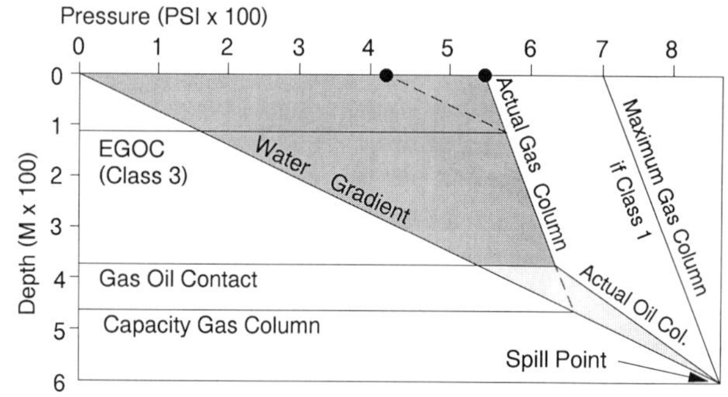

Fig. 8. Pressure–depth diagram showing generalized seal–spill relationships for the Oseberg Field (Fig. 6A). Based on data from Nipen (1987). See text for discussion.

Class 1 — Troll-East

The Troll Field (Gray, 1987) consists of three composite blocks having different fluid contacts (Figs. 6A and 7). Each more westerly block has a slightly deeper hydrocarbon/water contact. Troll-East has 230 m of gas and 0–4 m of oil and is the largest Class 1 Trap in Norway. Troll's central composite block has nearly the same gas column and 12 m of oil. Troll's western composite block has a 22–28 m oil column and a "small gas cap" (Gray, 1987).

The 0–4 m oil column of Troll-East may indicate that it is a borderline Class 2 accumulation. Alternatively, Troll-East may be a Class 1 Trap, and simply have that much oil in transit, underspilling the giant trap. If the latter were true, it would have intriguing possibilities for quantifying present generation and migration along one of the most prolific migration routes in Norwegian waters. If this were the case, a detailed isopach of the oil leg might show thinning toward the spillpoint. This is because a slope should be required at the base of the oil to allow the oil to flow across the horizontal under-surface of the gas.

The western two composite blocks or provinces have the right proportions of gas/oil, and appropriate contact levels to be Class 2 and Class 3 Traps, respectively (Fig. 7). Rather than leaking vertically to higher stratigraphic levels, they probably leak across the fault plane into adjacent blocks, as depicted in Fig. 7.

Class 2 — Oseberg

The Oseberg Field (Fig. 8) is a possible Class 2 Trap. It has a 375 m gas and 602 m total column and is apparently filled to its spillpoint (Nipen, 1987). It has one of the most highly stressed seals in the area (Spencer and Larsen, 1990). Evidence of leakage includes apparent gas blooms, hydrocarbon shows in Cretaceous limestone stringers, and areas of poor data quality on seismic sections (Nipen, 1987).

If Oseberg were a Class 1 Trap, still in its Gussow-evolution, and capable of becoming filled with gas, it would be necessary to project the gas gradient directly from the hydrostatic gradient at the spill-point depth to the upper pressure scale. This would extrapolate to a seal stress about 1/4 greater than that caused by the actual two-phase column. An oil column that stressed the seal the same amount as that hypothetical 600 m gas column would be found by projecting the oil gradient from this higher sealing capacity to the hydrostatic gradient. It would be 800–1000 m. Since Oseberg probably has the best performing seal in the area, it is unreasonable to expect it to be able to perform 1/4 better to accommodate these higher stresses and column heights, especially in combination with the indications of leakage. As will be seen in the later statistical section, Oseberg stands by itself as an exceptional trap.

The more logical alternative is that Oseberg's gas/oil contact was, prior to production, in equilibrium in mid-trap. The seal was at capacity and leaking at the rate of charging. Oseberg's oil column is not proportionally long enough for a Class 3 Trap. In summary, it fulfils the criteria for a Class 2 "balanced" Trap.

Class 3 — Gullfaks and Ekofisk

The Gullfaks Field (Fig. 6A), as well as the Ekofisk Field, and similar oil-dominated chalk fields surrounding Ekofisk (Fig. 6B), are here interpreted as Class 3 Traps.

It has been suggested (Eriksen et al., 1987) that large bounding faults may preclude significant gas from entering Gullfaks. However, the trap has great exposure to gas from the Troll, Oseberg, and possibly Møre Basin kitchens. It is not filled-to-spill, and pore pressures in the reservoir are in equilibrium with the Cretaceous shale seal over the field; the seal is at capacity (Karlsson, 1986). Pock marks and acoustic disturbances over the field seen on sparker profiles (Hovland and Judd, 1988) suggest that the seal is leaking. At Gullfaks, relief wells have been drilled under the platforms to vent gas. Indications of gas occur in the shallower strata (Miles, 1990, fig. 13). Some, but probably not the majority, of this gas may be biogenic. Gullfaks has a small gas cap. Thus, the main criteria which define a Class 3 Trap are satisfied: (1) crestal gas leakage, (2) not filled to the spillpoint, and (3) a column dominated by oil.

In the Central Graben the usual concepts of vertical closure, spillpoint, and a single oil/water contact are difficult to apply to the Cretaceous chalk-reservoired fields (D'Heur, 1984). The fractured chalk reservoir at Ekofisk, for example, is a zone of sub-vertical fractures caused by salt doming (Van den Bark and Thomas, 1981; Watts, 1983; Pekot and Gersib, 1987). These fields contain oil sourced from Upper Jurassic shales. The oil probably migrated initially through Middle or Upper Jurassic carrier beds.

This migration conduit is sedimentologically discontinuous and highly faulted. High overpressures provide additional evidence for this confinement. Vertical fracture-assisted migration dominates once the hydrocarbons leave the sand and by this route they are believed to gain access to the Cretaceous chalk traps. The above characteristics limit the need for long-range lateral migration in the main Jurassic carrier bed. Jurassic oil accumulations are close to Jurassic oil kitchens and Jurassic gas accumulations are close to Jurassic gas kitchens (Cayley, 1987).

Cretaceous oil accumulations in the Central Graben, however, are often located over and in close proximity to areas of Upper Jurassic gas generation (Fig. 6B), suggesting that gas has access to these oil-filled traps. This includes most of the chalk-reservoired oil fields around Ekofisk. Even more definitively, many of these traps have highly developed seismic chimneys caused by gas streaming from their crests.

The chalk-reservoired traps like Ekofisk depend to some extent on fracturing for their effective porosity and permeability. This fracturing is generated by, and proportional to, the flexing produced by subjacent salt swells. Fracturing decreases off-structure and there is little effective permeability in the unstressed chalk away from the areas of bending over the structure. Thus, these chalk reservoirs are sub-vertical zones of fracturing that should bear little relationship to, and not be limited by, structural closure on the top of the chalk. This is emphasized by most of these fields' hydrocarbon columns being greater than their structural closure.

The chalk-reservoired fields of the Central Graben probably have effective closure which approximates the thickness of the chalk (Ekofisk 600 m), plus the height of the structural closure (Ekofisk 270 m). With a hydrocarbon column height of about 400 m, this gives a possible closure of 400 m in excess of sealing capacity, supporting Ekofisk's classification as a Class 3 Trap that is not filled-to-spill. With such a reservoir geometry, both interconnection of the labyrinth of fractures and specific sealing capacity right over sub-areas of interconnected fractures should be variable. These factors, along with variations in capillary characteristics of the fractures, give logical reasons for the variability of column heights and oil/water contacts.

The concepts presented here explain Cayley's (1987) statement that " ... in common with several chalk fields vertical migration of gas has created a seismic 'chimney' (while) ... in most cases the less mobile oil has remained in the reservoir." Most of these fields would be filled-to-spill with gas, rather than oil, if they were not Class 3 Traps.

Seal types related to trap classes

Both the trapping processes and the trap classification proposed here are driven by the principle that gas lies above oil in the crest of the trap where the seal is most highly stressed. When sealing capacity is exceeded gas is preferentially leaked because of its position. In Class 3 Traps oil may also finally leak. Before it does, however, it will have driven most of the gas out of the trap. After the trap comes to equilibrium and both oil and gas are entering and exiting the trap at a steady state, the oil will continue to prevent gas accumulating in the trap because it underlies the gas and utilizes all sealing capacity.

Thus both major classes of seals — (1) membrane seals that fail when capillary entry pressure is exceeded and (2) those seals that fail when hydraulic pressure opens microfractures through the seal (Watts, 1987) — should honor the principles and classification suggested here.

One of the findings of the statistical aspects of the paper, however, is that there is significant variation in sealing capacity from trap to trap. This may be because seals can fail on capillaries that have several origins (Fig. 9). (1) They may be flaws inherited from sedimentary processes. (2) They may be a weakness produced by load flattening or flowage in the plastic field. (3) They may be a path along a hairline fault created in the shear field. (4) They may utilize a joint or open fracture in rocks deformed in the brittle field. (5) Fluid failure may follow microfractures opened by the increasing fluid pressure itself.

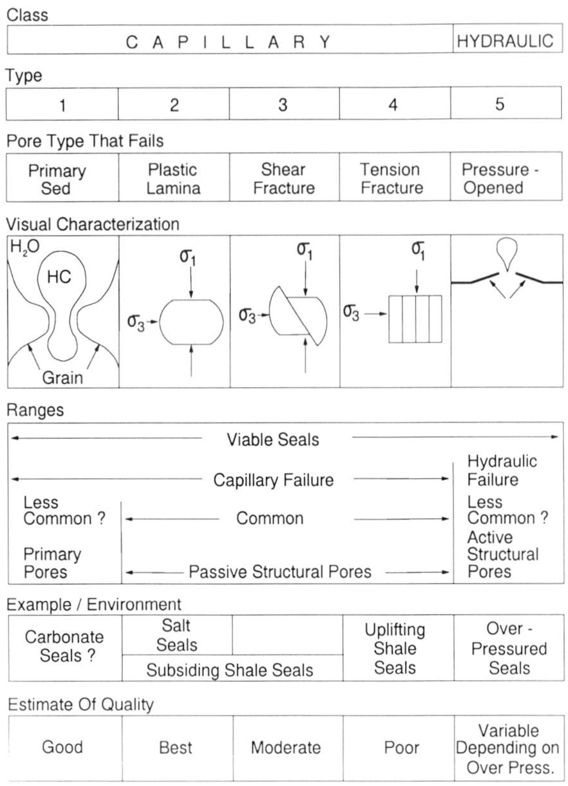

Fig. 9. Proposed working classification of seals based on the nature, cause, and timing of the failing capillary. Based on Griggs and Handin (1960), Berg (1975), Schowalter (1979) and Watts (1987). See text for discussion.

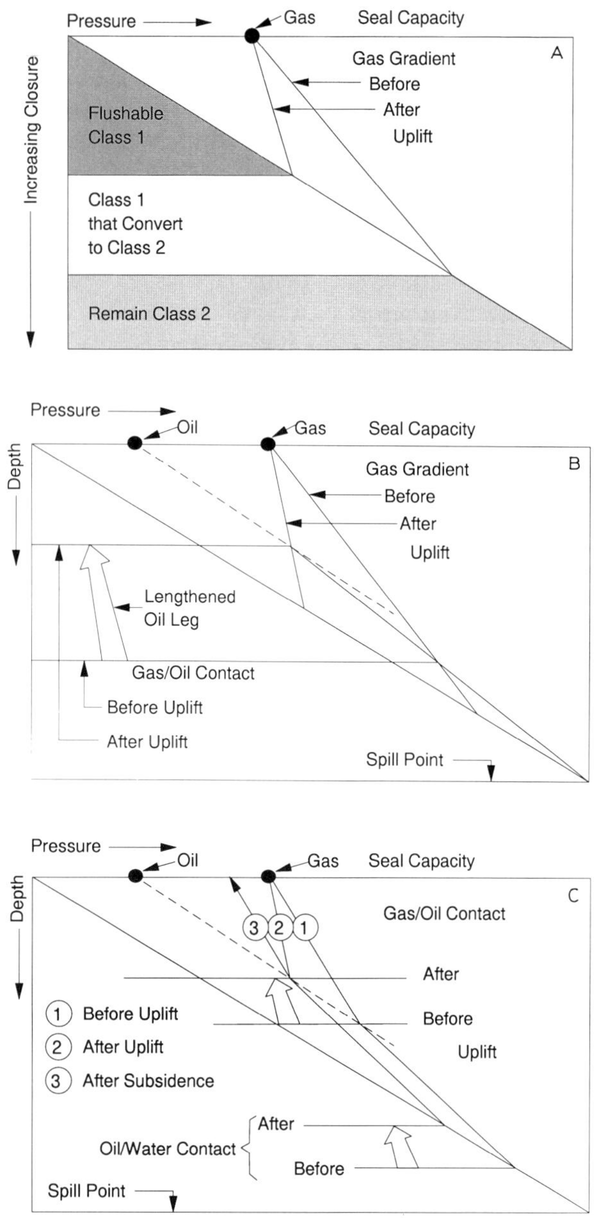

Fig. 10. Pressure–depth diagrams showing the effect of uplift on traps. (A) Class 1 Trap's excess sealing capacity allows gas flushing. If marginal, however, the trap may convert to Class 2 during uplift. (B) Class 2 Traps gain room for oil as their gas becomes less dense and less height of the gas cap is required to match sealing capacity. (C) Class 3 may lose or gain capacity for oil. See text for discussion.

In the first four cases the potential flaw that is finally utilized exists before the seal failure episode, and is simply passively utilized as the failing capillary during the episode. In the fifth case (failure by true hydraulic fracturing) the flaw that provides the failure path either did not exist prior to seal failure, or was one of the other types of flaws that was significantly enlarged by hydraulic pressure that caused the seal to fail.

Performance of trap classes during uplift and erosion

The three trap classes should be affected differently by uplift, with gas being the critical variable (Fig. 10). During uplift and erosion, as pressure drops gas expands and becomes more buoyant, and the same column height stresses the seal more. This steepens the gas gradient on pressure diagrams. Primary generation might be arrested as kitchens cool, but gas expansion downdip may produce a secondary supply of both oil and gas. Additionally, the trap might be uplifted but, because of continuing structural development or tilting, the kitchen might continue to subside and generate.

There should be two cases for traps that were Class 1 prior to uplift. (1) Relatively stronger Class 1 Traps, with their excess sealing capacity relative to closure, may withstand the increased stress of the more-buoyant post-uplift gas and become gas-flushed. (2) Relatively weaker Class 1 Traps might convert to Class 2 and not be flushed completely by gas if the added stress exerted by the post-uplift gas exceeds sealing capacity. These "convertible" Class 1 Traps should start to leak gas before all of their oil is flushed (Fig. 10A).

In a chain of Class 1 Traps each higher trap would

have its pressure reduced proportionally more by the same amount of uplift. This might cause a "flushing catastrophe", forcing all oil and much gas out of the migration chain. Should these traps later subside, they might be oil-stained but only have a small gas cap. This mechanism has been suggested for the Barents Shelf (Skagen, 1993).

Class 2 Traps might improve during uplift. They were at sealing capacity for gas and leaking gas prior to uplift. They were not at sealing capacity for oil prior to uplift, and were spilling rather than leaking oil. Assuming that sealing capacity does not change during uplift, the gas cap has to decrease as the gas gradient steepens. The oil column can increase because Class 2 Traps have excess sealing capacity for oil (Fig. 10B). Thus, Class 2 Traps may gain capacity for oil during uplift.

Class 3 Traps might modestly improve or degrade during uplift. They are at capacity for both fluids prior to uplift. Two cases depend on the presence or absence of a free gas cap. (1) If conditions allowed a free gas cap prior to uplift, the gas cap should decrease during uplift, because the less-dense post-uplift gas has to match the same sealing capacity. Being at capacity, the oil leg cannot extend as it could in Class 2 Traps. Rather, it will merely follow the gas to higher areas of the trap that have less volume, decreasing the amount of oil that the trap is capable of holding, modestly degrading the trap. (2) If conditions allowed no free gas cap prior to uplift, the oil leg would not change position during uplift. The decreased pressure should remove some of the gas from solution, however, allowing relatively more oil in the trap. This less gas-rich oil might be more dense, stress the seal less per unit of column height, and allow the oil column to lengthen slightly, improving the trap.

Performance of seals during uplift

Several scenarios increase the chance for seal failure during uplift: (1) passage of a seal from plastic through shear and into tension-failure strength fields, progressively increasing susceptibility to failure; (2) true hydraulic failure precipitated by a decrease in overburden pressure, assuming uniform uplift; (3) tilting can generate additional overpressure in deep parts of a pressure compartment at the same time as the overburden is being removed from above marginal areas of the compartment, giving double susceptibility to hydraulic fracturing; (4) additional stressing can be caused by associated deformation.

Continually subsiding shale seals should remain near the plastic-shear strength interface because the

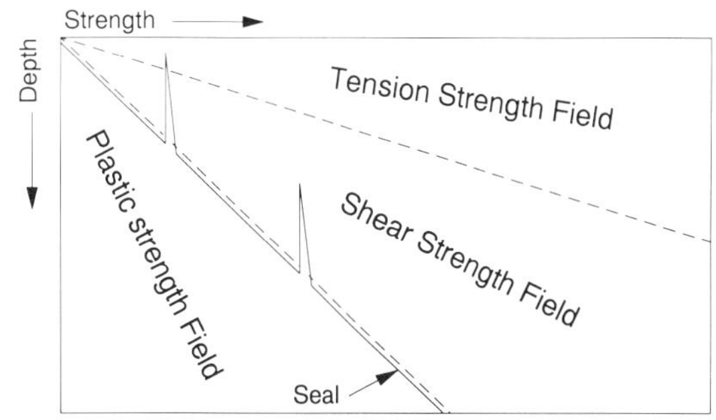

Fig. 11. Cartoon cross-plot of pressure vs. depth relationships that control shale's behavior as a seal. Shales subjected to continual burial should remain near the plastic-shear strength field interface. Upon uplift they enter the shear and then the tension strength fields, being progressively degraded as seals. Because the breadth of the strength-fields decreases upward, less uplift can cause failure in shallow traps.

shale's strength, a function of compaction and dewatering, should stay matched to the overburden stress. When uplifted they should retain a strength inherited from maximum burial. They should progress through the shear and finally the tension-failure strength fields, being progressively degraded as they become more brittle.

Such a seal might remain viable in the shear field, especially in areas like the Viking Graben where active generation decreases the need for long-term retention and allows some leakage. When the seal enters the tension-failure strength field, it should become susceptible to catastrophic damage. Thus, the shear-failure strength field may represent the strength "cushion" available to the seal during uplift. Shallow traps can stand less uplift, both because proportionally more overburden is removed and because boundaries of the strength fields converge upward (Fig. 11).

Glaciation, like tectonic uplift, removes overburden and confining pressure. Due to more rapid onset, erosion and melting, glaciation may stress seals more than tectonic uplift, with less time for pressures to equilibrate. Because topography is highest, most weathered, and retains the youngest and most easily eroded strata prior to glaciation, the first glaciation might be expected to strip off the most overburden. While ice remains, its weight partially substitutes for the weight of rock removed. Thus, the greatest stress on seals might occur during the first deglaciation. This phenomenon may have occurred in parts of the Barents Shelf.

Statistical analysis of traps

Preliminary statistical analysis seems to confirm that, for most traps in Norwegian waters, trapping processes significantly influence the distribution of

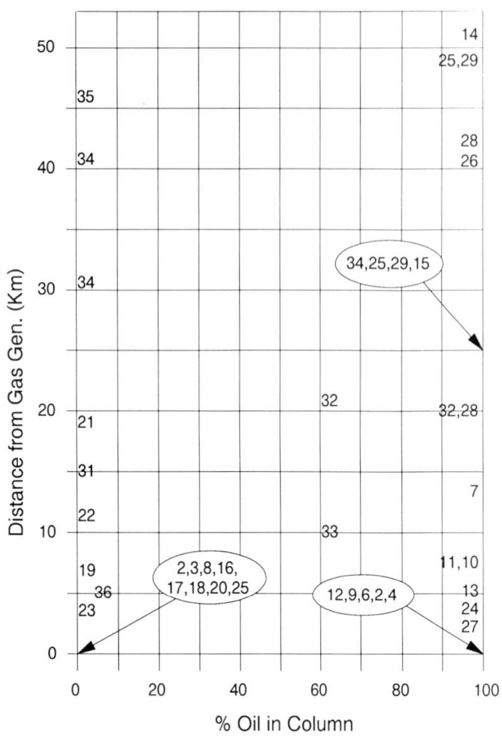

Fig. 12. Cross-plot of distance (near edge of gas kitchen to the near edge of trap) vs. fill character. Actual migration distances for gas generated from the Upper Jurassic could have been considerably greater, but migration distances for gas generated from deeper horizons could have been less. Note lack of relationship between fill character and distance from gas generation. Numbers are keyed to the chapters dealing with fields in Spencer et al., 1987.

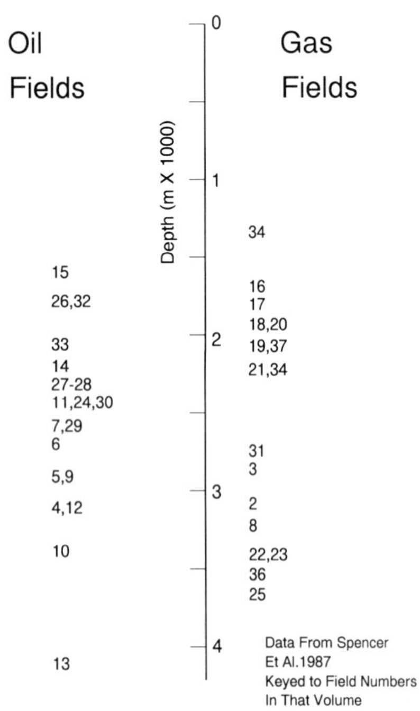

Fig. 13. Graph of trap depth vs. fill character. Note that the shallowest trap (Troll East) is gas and the deepest trap (7/11-5 Discovery) is oil, and that there is no apparent relationship between trap depth and fill character.

fluids. Based on this admittedly crude first pass, the following observations seem correct.

(1) There is little correlation between fill character and distance from generating kitchens (Fig. 12), or depth of the trap (Fig. 13).

(2) Trap fills are skewed toward gas or oil dominance, with fewer traps having sub-equal amounts of both fluids — the inverse of a bell-curve distribution.

(3) There are a sub-equal number of gas-dominated and oil-dominated traps, while there are only a few traps with sub-equal fills of both fluids.

(4) Gas traps tend to be filled-to-spill, oil traps not filled-to-spill.

(5) There is significant overlap and scatter in column heights between oil-dominated and gas-dominated fields (Fig. 14).

The following tentative conclusions might explain the observations.

(1) Most of Norway's traps and migration chains, having been exposed to over-ample supplies of both oil and gas, have gone through their Gussow-

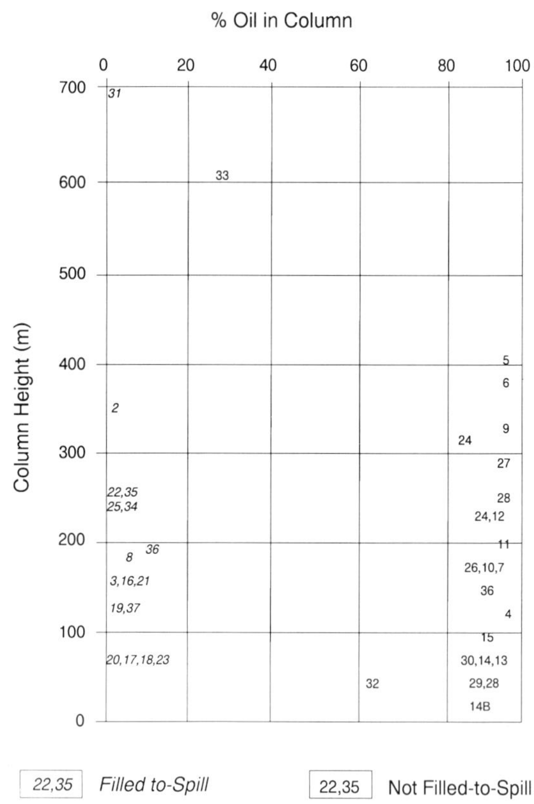

Fig. 14. Cross-plot of hydrocarbon column heights vs. percent oil in the column by column height for Norwegian Fields. Numbers are keyed to the field chapter numbers in Spencer et al. (1987). Italicized numbers denote traps that appear to be filled-to-spill, upright numbers denote those not believed to be filled-to-spill. In most cases this judgement has been gathered from the text and not from the simpler but less reliable "lowest closing contour" figure. Note skewing of the trap population toward all-gas and all-oil traps, and the strong overlap of column heights in these two populations. See text for discussion.

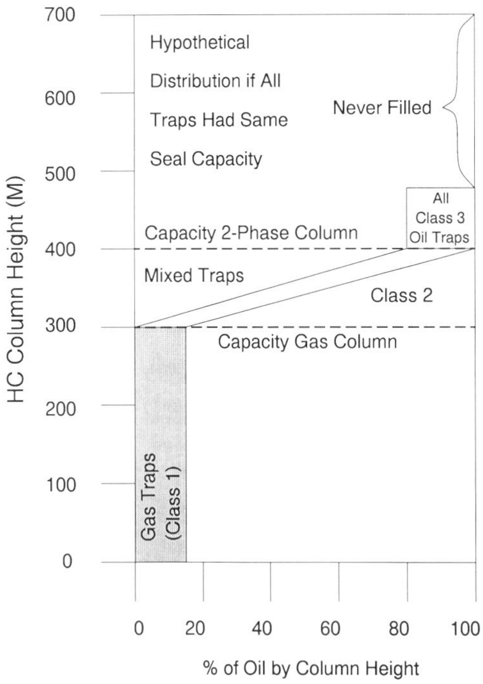

Fig. 15. Idealized pressure–depth plot of relations if all traps had the same sealing capacity and ample access to both fluids.

evolution and have equilibrated by trapping principles suggested here. They have been gas-flushed to the extent that they can be gas-flushed. Traps that retain oil do so because they, or traps downdip, vent gas (Class 2 and 3 Traps).

If all traps were Class 1, with fill determined by stage of Gussow-evolution, and if fluids were limited, fill would correlate with distance from kitchens and trap depth. If similar traps had been exposed to excess fluids, all traps would have been gas-flushed. With just the right amount of oil and gas, the right stage of Gussow-evolution could fortuitously produce the skewed distribution seen (about half gas traps, half oil traps, and with a few two-phase traps), but then gas should be concentrated downdip from oil.

The reasoning above is substantiated by volumetric estimates, as exemplified in the following quote: "These data from the deep mature kitchen areas suggest that charge vastly exceeds the available volume in adjacent traps" (Thomas et al., 1985, p. 15).

(2) The significant scatter and overlap of column heights between the oil-dominated and the gas-dominated sides of the distribution seem most likely to be explained by real variations in sealing capacity between traps. If this scatter had involved only the gas side of the distribution (that tends to be filled-to-spill) it might be explained by a natural distribution of closure heights. Equal scatter occurs, however, on the oil-dominated side of the distribution, where it is less likely to be related to closure, since this set of traps tends not to be filled to the spillpoint. Tentatively, the variations in (a) potential failure paths (see section on seals), (b) primary sedimentation, (c) degree of structural breakup, and (d) pressure regimes make it reasonable to expect this apparently real variation in sealing capacity. Had the principles worked perfectly and had there been no variation of seal strength, a simpler idealized distribution might have been expected (Fig. 15).

Discussion

Trapping processes should be most important in areas where traps have been exposed to excess gas. Class 2 and Class 3 Traps should not be gas-flushed even when exposed to excessive gas. Instead they leak the gas, making room in the trap for the oil. If a trap, migration chain, or region has been exposed to marginal amounts of each fluid, generation and migration should be the more important variables, and the stage of Gussow-evolution the most important consideration.

Since, as stated earlier, most volumetric estimates for the northern North Sea indicate that hydrocarbon charge vastly exceeds available trap volume, it should not be surprising that most traps appear to have gone through their Gussow-evolution and come to equilibrium by closure versus sealing capacity principles. The results of the statistical analysis seem to substantiate this. If anything, the fit seems too perfect — part of the distribution must be the result of migration routing and the stage to which the trap and migration chain have evolved by Gussow-type reasoning.

Many oil fields show gas leakage in the form of: (1) stratigraphically higher accumulations of gas; (2) gas chimneys; (3) gas blooms; (4) gas-induced velocity pull-downs; (5) seismic data degradation; (6) seafloor "pock marks" and gas cones; (7) positive readings on "sniffers" and drop cores (Hovland and Judd, 1988); (8) well blowouts and other drilling problems; and (9) gas accumulation under platforms. All of these suggest that traps are venting gas. Some of this gas could be biogenic, but the majority probably is not.

Additional considerations suggest that we see only a fraction of the evidence for gas leakage from crestal areas of traps. (1) Most traps have less closure at higher levels due to syn-tectonic sedimentation, differential compaction, or unconformities. At some level closure is eliminated and replaced with inclined strata. Leaking gas should migrate laterally off to the side of the structure when it reaches these inclined strata. This removes the evidence of gas leakage from directly above the trap. (2) The lateral extent of gas chimneys may be limited, and missed by a relatively coarse seismic grid. (3) At the other extreme, gas disturbance may be so diffuse and subtle over a structure that it is thought to be an inherent part of the quality of the seismic data. This is especially true of older data. (4) If gas were to leak directly up a fault plane into a higher carrier bed, it might either migrate to the surface or laterally off the structure without being detected. (5) Even though they performed as Class 2 and Class 3 Traps during charging, only traps in areas of active charging should show leakage. All inactive traps, which have been filled by previous charging, should show no leakage, although disruption of beds by previous leakage might still be detected.

In summary, there is probably more hydrocarbon leakage, especially gas, from traps than is generally recognized. In some of the best hydrocarbon provinces such as the Viking Graben, characterized by an over-abundance of hydrocarbons relative to aggregate volume of traps, strong present generation and good carrier beds, gas leakage does not necessarily indicate either a damaged seal or a gas accumulation. Indeed, gas leakage is likely to be a sign of a "healthy" trap with a strong probability for an oil charge.

The systematic way in which the three classes of trap distribute oil and gas has considerable implication for migration (Fig. 16). Cross-section 1 shows a classic "Gussow" migration chain of Class 1 Traps exposed to an excess of both oil and gass, but with all oil flushed to the seafloor by the excess gas. Cross-section 2 shows the same chain under the same conditions of exposure to excess oil and gass, but with the lowest trap in the chain being Class 2 rather than Class 1. Since a Class 2 Trap vents gas out of the chain and spills only oil updip, the chain above is saved for oil. Cross-section 3 shows the same chain but with a Class 3 Trap substituted as the lowest trap in the chain. Since a Class 3 Trap leaks both oil and gas the chain updip should be water-wet. Alternatively, tributary migration routes might bring oil generated at shallower levels into the upper part of the chain. Thus, in both the Class 2 and the Class 3 case the migration chain has an increased probability for oil, even though the lower end of the chain has had access to excess gas to flush oil from the entire chain. Note that there are also implications for traps stratigraphically above the one in question. Such a stratigraphically higher trap has an increased probability of being water-wet in cross-section 1, having just gas in cross-section 2 and having both gas and oil in cross-section 3.

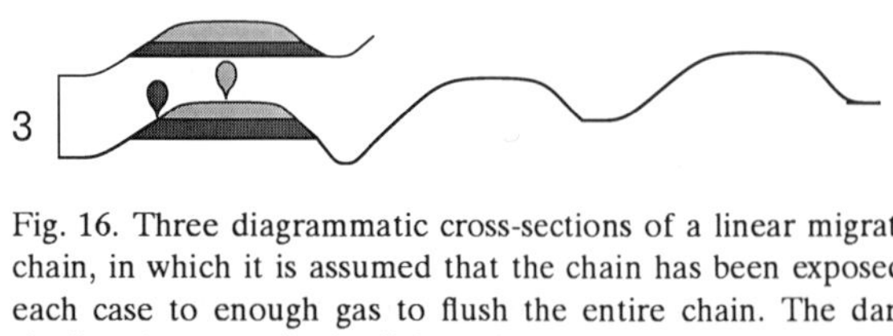

Fig. 16. Three diagrammatic cross-sections of a linear migration chain, in which it is assumed that the chain has been exposed in each case to enough gas to flush the entire chain. The darker shading denotes oil; the lighter denotes gas. The lowest trap in cross-section *1* is assumed to be Class 1; in cross-section *2*, Class 2; and in cross-section *3*, Class 3. Note that cross-sections 2 and 3 have an enhanced chance for oil even when exposed to an excess of gas. See text for additional discussion.

Conclusions

(1) The interplay between closure, sealing capacity, and buoyancy provides a fundamental control on the proportions of oil and gas in traps. If the trap has been exposed to an excess of both fluids, these relationships may provide the dominant control. Three trap classes are defined based on this relationship, using the height of the capacity gas column and the capacity two-phase column (the longest possible column) as thresholds. The position of the trap's spillpoint in relation to these thresholds determines the class of the trap. Each class has different trap-fill character and implications for traps updip and at shallower stratigraphic levels. Conversely, shallow traps give clues about traps downdip or at deeper stratigraphic levels.

(2) Class 1 Traps have excess sealing capacity and Class 3 Traps have a deficiency of sealing capacity relative to closure. Class 2 Traps have intermediate sealing capacity relative to closure. Class 1 Traps will contain gas, Class 2 Traps will contain both gas and oil, and Class 3 Traps will contain dominantly oil, even when the traps have been exposed to an excess of gas. Class 1 Traps spill gas and oil and leak neither; Class 2 Traps spill oil and leak gas; Class 3 Traps leak oil and gas and spill neither. Salient features of the trap classification are given in Table 1.

(3) Fluid contacts can be spill, charge, or pressure controlled. Pressure-controlled contacts equilibrate in mid-trap despite movement of oil and gas into and out of the trap.

(4) Because they have too much closure to allow gas flushing, Class 2 and Class 3 Traps have an enhanced probability for oil, even when exposed to excess gas, as long as they are not so deep that oil has been cracked to gas in the trap. Thus, large

closure and/or small sealing capacity are insurance against gas flushing and thereby increase the chance for oil. Indications of gas over a trap may enhance the probability for oil.

(5) The northern North Sea, with too much exposure to gas, and having generally good migration potential, still has some oil in the majority of traps. There appear to be enough Class 2 and Class 3 traps along the migration pathways to vent sufficient gas out of the main carrier bed and thus increase the proportion of oil relative to gas in the traps.

(6) In such an area, characterized by an overabundance of both oil and gas, strong present-day generation, and examples of long migration routes, trap performance may be the most important variable controlling oil and gas proportions within the traps. In areas of marginal generation or poor carrier beds, migration and/or the volumetrics of generation, in combination with Gussow-type reasoning may be equally or more important than trap performance.

(7) Troll-East is proposed as a Class 1 Trap, Oseberg as a Class 2 Trap, and Gullfaks, Ekofisk, and most of the chalk fields as Class 3 Traps.

(8) Inactive traps, those filled by a previous episode of charging, should have fill characteristics that were determined during the earlier charging. Inactive Class 2 and Class 3 Traps should not be presently leaking and therefore might be mistaken for active Class 1 Traps.

(9) Crestal leakage and Class 2 and Class 3 Traps are probably more common than is generally recognized. In some respects Class 2 Traps and Class 3 Traps are more important than Class 1 Traps. This is because of their enhanced probability for trapping oil and their favourable effect on multiple traps along a migration pathway, in venting gas and preserving oil. Class 2 Traps, being transitional, are expected to be less numerous than the other trap classes.

(10) Trap classes should respond differently to uplift. Class 1 Traps may be gas-flushed. Class 2 Traps should improve by gaining a higher oil column at the expense of the gas cap as leakage of the less dense gas accelerates. Class 3 Traps might lose both oil and gas if a preexisting free gas cap decreases. They might modestly improve if the only gas was in solution and it was forced out of solution by uplift.

(11) Preliminary statistical analysis of Norway's traps suggests: (a) a sub-equal distribution of nearly pure oil or pure gas traps, with few traps with mixed fills; (b) considerable difficulty in establishing spillpoint and closure; (c) considerable variation in seal capacity; (d) that most traps that appear to be filled-to-spill are gas traps and most that appear not to be filled-to-spill are oil traps; and (e) that there is little relationship between fill character and either depth of the trap or its distance from gas generation.

These distributions might be explained if overample generation in combination with good carrier beds have forced most of Norway's traps through their Gussow-evolution to equilibrium by closure versus sealing capacity principles. Although the relationship between closure and sealing capacity may be important in controlling distribution of fluids in Norway's traps, the data scatter will make precise prediction difficult. Nevertheless, traps with large closure, and traps with signs of gas leakage over them, should have an enhanced chance for oil.

Acknowledgements

Steve Buck, George Claypool, Jim Dixon, Pete Hansen, Jim Helwig, Doug Kirkland, Gene Leavitt, Pam Luttrell, John Snedden, Nick Theis and Joana Vizgirda read and commented on various versions of this manuscript. A.G. Doré and an anonymous reviewer improved it considerably during the editorial process. Evy Zenker helped with the typing and Kathryn Ward drafted the figures. I thank them all but retain responsibility for errors which remain. I thank Mobil Oil Corporation and Mobil Exploration Norway Inc. for permission to publish.

References

Berg, R.R., 1975. Capillary pressure in stratigraphic traps. Am. Assoc. Pet. Geol., Bull., 59: 939–956.

Campbell, C.J. and Ormaasen, E., 1987. The discovery of oil and gas in Norway: a historical synopsis. In: A.M. Spencer et al. (Editors), Geology of Norwegian Oil and Gas Fields. Graham and Trotman, London, pp. 1–37.

Cayley, G.T., 1987. Hydrocarbon migration in the central North Sea. In: J. Brooks and K. Glennie (Editors), Petroleum Geology of North West Europe. Graham and Trotman, London, pp. 549–555.

Dembicki, H. and Anderson, M.J., 1989. Secondary migration of oil: experiments supporting efficient movement of separate, buoyant oil phase along limited conduits. Am. Assoc. Pet. Geol., Bull., 73: 1018–1021.

D'Heur, M., 1984. Porosity and hydrocarbon distribution in the North Sea chalk reservoirs. Mar. Pet. Geol., 1: 211–238.

Durand, B., 1988. Understanding of HC migration in sedimentary basins (present state of knowledge). Org. Chem., 13: 445–459.

England, W.A., Mackenzie, A.S., Mann, D.M. and Quigley, T.M., 1987. The movement and entrapment of petroleum fluids in the subsurface. J. Geol. Soc. London, 144: 327–347.

Eriksen, T., Helle, M., Henden, J. and Rognebakke, A., 1987. Gullfaks. In: A.M. Spencer et al. (Editors), Geology of Norwegian Oil and Gas Fields. Graham and Trotman, London, pp. 273–286.

Foster, N.H. and Beaumont, E.A. (Compilers), 1988. Traps and Seals. Am. Assoc. Pet. Geol. Treatise of Petroleum Geology, Repr. Ser., 7, 2 Vols., 42 papers.

Gray, I., 1987. Troll. In: A.M. Spencer et al. (Editors), Geology of the Norwegian Oil and Gas Fields. Graham and Trotman, London, pp. 389–401.

Griggs, D.T. and Handin, J., 1960. Observations on fracture and a hypothesis of earthquakes. In: D.T. Griggs and J. Handin (Editors), Rock Deformation — A Symposium. Geol. Soc. Am. Mem., 42: 347–373.

Gussow, W.C., 1954. Differential entrapment of oil and gas: a fundamental principle. Am. Assoc. Pet. Geol., Bull., 38: 816–853.

Hovland, M. and Judd, A.G., 1988. Seabed Pockmarks and Seepages. Graham and Trotman, London, 293 pp.

Karlsson, W., 1986. The Snorre, Statfjord and Gullfaks oil fields and the habitat of hydrocarbons on the Tampen Spur, offshore Norway. In: A.M. Spencer (Editor), Habitat of Hydrocarbons on the Norwegian Continental Shelf. Norwegian Petroleum Society, Graham and Trotman, London, pp. 181–197.

Miles, J.A., 1990. Secondary migration routes in Brent sandstones of the Viking Graben and East Shetland Basin. Am. Assoc. Pet. Geol., Bull., 74: 1718–1735.

Nipen, O., 1987. Oseberg. In: A.M. Spencer et al. (Editors), Geology of Norwegian Oil and Gas Fields. Graham and Trotman, London, pp. 379–388.

Nyland, B., Jensen, L.N., Skagen, J.I. and Skarpnes, O., 1992. Tertiary uplift in the Barents Sea; magnitude, timing and consequences. In: R.M. Larsen, H. Brekke, B.T. Larsen and E. Talleraas (Editors), Structural Modelling and its Application to Petroleum Geology. Norwegian Petroleum Society (NPF), Special Publication 1. Elsevier, Amsterdam, pp. 153–162.

Pekot, L.J. and Gersib, G.A., 1987. Ekofisk. In: A.M. Spencer et al. (Editors), Geology of Norwegian Oil and Gas Fields. Graham and Trotman, London, pp. 73–87.

Schowalter, T.T., 1979. Mechanics of secondary hydrocarbon migration and entrapment. Am. Assoc. Pet. Geol., Bull., 63: 723–760.

Schowalter, T.T., 1991. Secondary petroleum migration in basin modelling. Int. Conf. Basin Modelling; Advances and Applications. Norwegian Petroleum Society, p. 23 (abstract)

Skagen, J., 1993. Effects on hydrocarbon potential caused by Tertiary uplift and erosion in the Barents Sea. In: T.O. Vorren, E. Bergsager, Ø.A. Dahl-Stamnes, E. Holter, B. Johansen, E. Lie and T.B. Lund (Editors), Arctic Geology and Petroleum Potential. Norwegian Petroleum Society (NPF), Special Publication 2. Elsevier, Amsterdam, pp. 711–719.

Spencer, A.M. and Larsen, V.B., 1990. Fault traps in the Northern North Sea. In: R.F.P. Hardman and J. Brooks (Editors), Tectonic Events Responsible for Britain's Oil and Gas Reserves. Geol. Soc. Spec. Publ., 55: 281–298.

Spencer, A.M. et al. (Editors), 1987. Geology of Norwegian Oil and Gas Fields. Graham and Trotman, London, 493 pp.

Thomas, B.M., Møller-Pedersen, P., Whitaker, M.F. and Shaw, N.D., 1985. Organic facies and hydrocarbon distributions in the Norwegian North Sea. In: B.N. Thomas et al. (Editors), Petroleum Geochemistry in Exploration of the Norwegian Shelf. Norwegian Petroleum Society, Graham and Trotman, London, pp. 3–26.

Tissot, B., 1987. Migration of hydrocarbons in sedimentary basins: a geological, geochemical and historical perspective. In: B. Doligez (Editor), Migration of Hydrocarbons in Sedimentary Basins. Technip, Paris, pp. 1–20.

Ungerer, P., Burrus, J., Doligez, B., Chenét, P.Y. and Bessis, F., 1990. Basin evaluation by integrated two-dimensional modeling of heat transfer, fluid flow, hydrocarbon generation, and migration. Am. Assoc. Pet. Geol., Bull., 74: 309–335.

Van den Bark, E. and Thomas, O.D., 1981. Ekofisk: first of the giant oil fields in western Europe. Am. Assoc. Pet. Geol., Bull., 65: 2341–2363.

Watts, N.L., 1983. Microfractures in chalks of Albuskjell Field, Norwegian Sector, North Sea: possible origin and distribution. Am. Assoc. Pet. Geol., Bull., 67: 201–234.

Watts, N.L., 1987. Theoretical aspects of cap-rock and fault seals for single- and two-phase columns. Mar. Pet. Geol., 4: 274–307.

JOHN K. SALES *Mobil Exploration Norway Inc., P.O. Box 510, N-4001 Stavanger, Norway*

Application of basin modelling on the Norwegian Shelf (extended abstract)

B. Nyland

After 26 years of exploration activity spread over 184 different exploration licenses on the Norwegian Shelf (Fig. 1), the size of discoveries made has decreased as the most promising prospects mapped have been drilled. The oil companies have therefore devoted an ever-increasing effort to the geological evaluation of areas in an attempt to discover additional resources. This is done by ascertaining the optimal conditions for the occurrence of reservoir and source rocks, and by understanding hydrocarbon generation and migration. To achieve this, basin modelling techniques have turned out to be a helpful tool in the evaluation process.

As a part of the regulations for the management of geological data from the Norwegian Shelf, the Norwegian Petroleum Directorate (NPD) has access to a large database of geological work performed by the oil companies. However, since these data are proprietary and strictly confidential, this paper will present only a general status of how the use of basin modelling techniques has developed within the companies in recent years as observed by the NPD. The main conclusions are summarized in Table 1, and are based on information collected from the oil companies' geological evaluation reports presented to the NPD as part of the application for new licenses from a selection of concession rounds.

The table shows the selected concession rounds and the number of companies involved in the application. The number of companies presenting one or more case studies using advanced integrated basin modelling programmes (1D and/or 2D) are tabulated. An indication of the number who have developed their own program, or who have used programs developed by consulting companies are also given. In addition, the number of companies *not* evaluating the source rock potential at all are given. The last column in the table refers to the number of companies modelling the migration and accumulation of hydrocarbons.

The part of the table referring to different maturity parameters is included to show how the companies prefer to present the level of maturation. The most commonly used parameters present maturity as a function of TTI/LOM, vitrinite reflectance (VR%), and transformation ratios based on kinetic modelling (TR). The last column shows the number of companies using temperature modelling in evaluating the hydrocarbon potential.

The value of more detailed analysis of source rock potential and migration has been widely accepted during the last 6–8 years of exploration, mainly as a result of the opening of new acreage north of 62°N in the 5th concession round.

Until 1979, exploration activity was restricted to the North Sea. The main problems here had been to establish play concepts that worked without too much analysis of the hydrocarbon potential. However, instead of finding an analogue to the prospective North Sea when moving northwards, the prospectivity turned out to be dependent on other factors not observed in the North Sea. Factors such as the source rock potential and/or the level of maturation became a problem in Trænabanken, Nordland II, and in parts of the western Barents Sea area, while high pressures in parts of the Haltenbanken area seemed to act locally as a migration barrier for hydrocarbons. In addition, the effects of Late Tertiary/Quaternary uplift and erosion in the Barents Sea, and along the Norwegian coast in general, seemed to be more important than earlier thought for the accumulation of hydrocarbons.

Starting with the 5th concession round, six of the twenty companies who delivered applications did not evaluate the source rock potential at all in the new area opened, while the remainder applied knowledge from the North Sea in their application. At this time, only two companies presented modelling studies using their own in-house basin modelling programs, modelling factors such as compaction and geothermal history.

Case studies using 2D basin modelling programs in which computer modelling is carried out along a geological profile rather than at a single point, were

Basin Modelling: Advances and Applications edited by A.G. Doré et al. NPF Special Publication 3, pp. 415–418, Elsevier, Amsterdam.

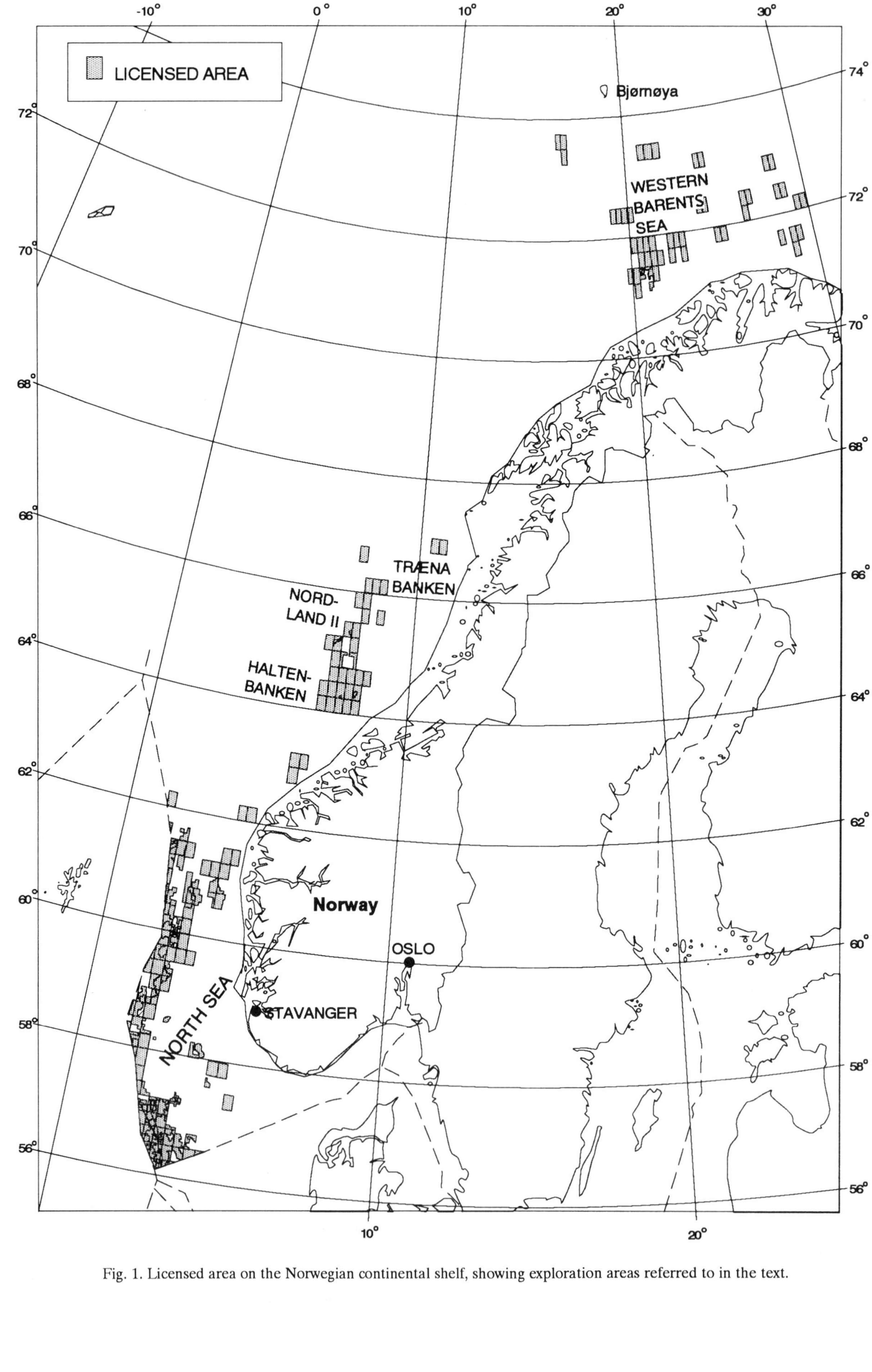

Fig. 1. Licensed area on the Norwegian continental shelf, showing exploration areas referred to in the text.

TABLE 1

The development of use of basin modelling on the Norwegian Shelf (the numbers are based on regional studies presented in a selection of concession rounds)

YEAR/ ROUND	NUMBER OF APPLICANTS	BASIN MOD. STUDIES		INHOUSE/ CONS.	NO SOURCE ROCK EVALUATION	MIGRATION MODELLING	MATURITY PARAMETERS USED			
		1D	2D				TTI/LOM	VR%	TR.	TEMP.
1982/5	20	2	-	2/-	6	-	2	4	-	-
1982/7	19	4	-	4/-	4	1	4	4	1	1
1984/8	19	4	-	4/-	2	2	6	4	-	1
1985/9	19	4	2	4/2	-	2	7	3	-	-
1987/11	13	10	3	9/2	-	4	10	9	1	1
1990/13	21	9	6	9/6	1	3	1	13	7	2

presented for the first time in the 9th round. This was also the first time that the reports referred to the use of commercial programs, and these were generally used in addition to the in-house versions.

The breakthrough in the use of basin modelling was during the 11th round, when ten companies presented 1D modelling studies. Of these, as many as nine companies used their own in-house developed program, one company used a modelling study done by a consultant, while one company used both. Three companies performed 2D basin modelling. In the 13th round, six companies presented 2D modelling studies. Nine companies refer to an in-house modelling program, while six of these, in addition to their own program, use a commercial program or a study performed by consultants. Unfortunately, the different programs are not used to compare case studies, but are used in separate analyses of different areas.

The importance of performing a detailed study on source rock potential is stressed in most of the geological reports given, and in the 13th round only one company did no evaluation of the source rock potential in its application.

Table 1 also indicates that studies of migration and calculation of volumes of hydrocarbons present to fill the traps are rare. The low number of these studies has been surprisingly constant during the last eight years.

One of the most important results from a basin modelling study is the maturation history of the source rocks in the area evaluated. In Table 1 the main parameters used to construct maturity maps are presented. No differentiation is here made between VR% and TTI calculated from advanced modelling programs and simple VR% vs depth plots, or Waples-Lopatin TTI calculations.

The main observation from the table is that a large proportion of the basin modelling programs developed up to the 11th round used the TTI equation to calculate source rock maturity. However, from 1987 to present, the use of TTI has decreased dramatically and has been rapidly replaced by the use of kinetic modelling. This is hardly surprising considering the increased attention focused on this topic in recent years, and in the 13th round seven companies presented transition ratio maps based on kinetic modelling.

The use of vitrinite reflectance as a measure of maturity was constant up to the 11th round, and has since then increased. However, the majority of the examples referred to in the table reflects simple vitrinite versus depth evaluation based on well data, and not results from basin modelling studies. One company has for some years used temperature maps from modelling studies as a measure of the level of maturation and it was joined by another company in the 13th round.

To conclude, and as indicated in the table, basin modelling studies have become common as an aid in hydrocarbon exploration since the 11th round. In addition to their use in concession rounds, they are also carried out as special studies under license work commitments, being used to model the causes for both discoveries and dry holes. Modelling studies have also been used to define new targets in existing licenses and to support requests for exemption from drilling.

However, to date, the correlation between basin modelling and discoveries is not impressive. The main observations are that in areas where the data quality is good, and where exploration activity is concentrated on finding reserves near existing infrastructures, such as in the North Sea, basin modelling has been a valuable aid. Even so, there are very few examples of systematic pre- and post-drilling assessments for specific areas, which would rigorously test the effectiveness of the tool. In new areas, the results from basin modelling sometimes appear to be in conflict with the desire to test new play concepts. The presence of large structures, which the companies argue they cannot afford not to drill, can result in the prognosis from basin modelling being overridden.

B. NYLAND *Norwegian Petroleum Directorate (NPD), P.O. Box 600, N-4001 Stavanger, Norway*

The influence of Tertiary and Quaternary sedimentation and erosion on hydrocarbon generation in Norwegian offshore basins

Birger Dahl and Jan H. Augustson

Hydrocarbon generation in a basin is mainly controlled by heatflow and subsidence history as well as by the kinetic properties of the source rock kerogens involved.

This study considers the onset of hydrocarbon generation from the Upper Jurassic–basal Cretaceous source rocks of three basins: the North Viking Graben, the Halten Terrace and the Hammerfest Basin using well data and pseudo-wells. Generation generally commences at three significantly different present-day depths of burial in these basins. To explain these differences three processes have been considered in detail, viz.:

(1) Heatflow histories have been determined for well sections using a combination of vitrinite reflectance and biological marker ratios. Regional heatflow variations through geological time have been reconstructed for the comparison of basins.

(2) Tertiary and Quaternary basin movements and associated erosional losses have been quantified. The key factors in this study were stratigraphic and seismic controls, and the application of selected thermal indicators featuring different, dynamic time and temperature ranges.

(3) Hydrocarbon generation kinetics for the source rocks have been determined and the result of kinetic modelling compared with hydrocarbon generation trends observed in exploration wells.

It is concluded that the differences in depth to oil generation between the basins arise mainly from the effect of Tertiary movement and differences in pre-Late Tertiary heatflow. Differences in source rock kerogen type have minimal effect.

Introduction

In the present paper, three Norwegian offshore basins have been selected for basin modelling to investigate the relationship between erosional episodes and hydrocarbon generation. The basins are: the northern Viking Graben (Færseth et al., 1986), the Halten Terrace (Hagevang and Rønnevik, 1986) and the Hammerfest Basin (Berglund et al., 1986). All these basins contain oils derived primarily from the Upper Jurassic–basal Cretaceous organic-rich shales. The location of these areas is shown in Fig. 1. The basins are highly prospective, and the observed depths to the onset of hydrocarbon generation are significantly different in the three basins. The $R_o = 0.5\%$ threshold seems to occur at a depth of 1800 m in the Hammerfest Basin, at about 3100 m in the Halten Terrace and at around 2600 m in the Viking Graben.

The Cretaceous and Early Tertiary geological histories of these basins do not differ sufficiently to explain the large variations observed in present-day generation states. We therefore propose that the inter basinal variations may be ascribed to one or to a combination of the three factors given below:

(1) Differences in subsidence history during the Cenozoic, and in particular, differences in the amount of erosion, represented by the major unconformities.

(2) Differences in the dominant kerogen type, which may lead to significant differences in the kinetic parameter values relating to hydrocarbon generation.

(3) Differences between the basins in palaeoheatflow histories.

One of the most difficult problems in the evaluation of sedimentary basins with respect to hydrocarbon potential is to obtain adequate control on subsidence history and, particularly, on the timing of erosional episodes. For these periods of basin history where no sedimentary record exists, advanced basin modelling techniques may provide a valuable tool for solving these problems in association with other techniques such as seismic velocity analysis, seismic truncation etc. in order to achieve a satisfactory result. The number of unknown parameters in the model used to characterize the basin must, however, be kept to a minimum. This means that basin

Basin Modelling: Advances and Applications edited by A.G. Doré et al. NPF Special Publication 3, pp. 419–431, Elsevier, Amsterdam.

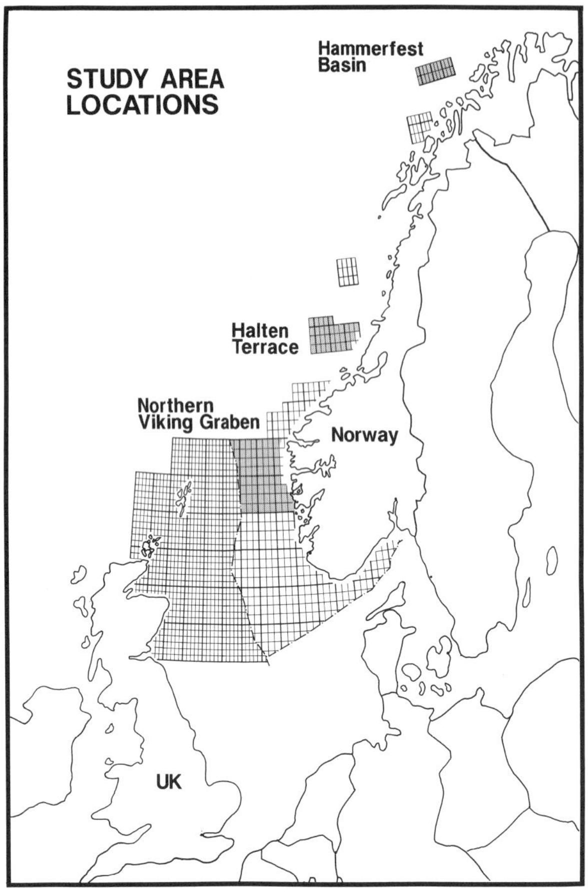

Fig. 1. Location map of the study areas, offshore Norway.

modelling techniques probably can only be successfully applied to the estimation of erosional losses in cases where an advanced understanding of the basin geology already exists. Ideally a basin should not be considered fully evaluated until the hydrocarbon generation histories are determined for the periods for which there is no sedimentary record.

In practice when attempting to estimate erosional losses by basin modelling, it is inevitable that most efforts are concerned with model optimization. One of the vital requirements in this situation is to use a wide selection of geochemical optimization parameters. This means that, in addition to the obligatory dense coverage of vitrinite reflectance measurements over the entire well modelled, biological marker ratios and pyrolysis data should also be employed.

For any basin the optimum parameters for kerogen conversion are the depletion curves of source rock versus depth since these represent the most direct indication of hydrocarbon generation from source rock layers. These curves are also ideal to use for the calibration of kinetic parameters used in basin simulation programs since they are derived from naturally matured source rocks.

A further objective of this study is therefore to establish geological, temperature history and kinetic input models in the three basins. By optimization of the models against a variety of maturity parameters it is intended to determine which factors are responsible for hydrocarbon generation thresholds occurring at significantly different depths of burial. Basin modelling is also used to quantify the Cenozoic erosional losses occurring in the three basins.

The basin simulation software; description and application

The evolution of sedimentary basins is herein quantified in terms of geologic, hydrodynamic, geothermal and hydrocarbon generation histories and the various sedimentary processes are characterized by mathematical equations. All processes have to be described by quantitative parameters and are contained in the conceptual model that provides the input to the simulation software. The conceptual model consists of chronostratigraphic units, lithological composition of units, palaeobathymetric estimates and palaeosediment–water interface temperatures. All of the geological processes taking place as a function of time are divided into events. These are characterized by a positive (deposition), zero (hiatus) or negative (erosion) thickness of lithology (or layer) that occurred during a specific time interval. The basin simulation software used in this study is an improved deterministic dynamic one-dimensional basin model based on earlier concepts (Yukler and Welte, 1980) further developed by A. Yükler. Application of this model in the northern North Sea has previously been reported by Dahl et al., 1987. The modelling software applies a modified Lopatin type subroutine for vitrinite reflectance calculations and a first-order kinetic model for sterane and hopane isomerization, similar to the model suggested by Mackenzie, 1984, for predictions of sterane isomerization.

In basin simulation work, model optimization is important. During this process the modelling input concepts are tested and refined by comparing and checking model output parameters with geological and geochemical well data. Running the basin simulation thereby provides feedback to the geoscientist concerning the geological interpretation and understanding that forms the basis for the model input.

Thermal maturity, hydrocarbon generation and source quality

It has been observed, as mentioned above, that the general depth versus maturity trends in the three basins studied are different. In Fig. 2, the

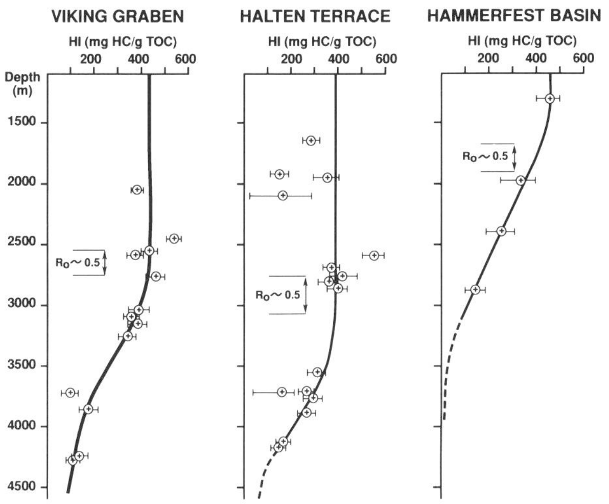

Fig. 2. The generation situation in the three basins at Present. This is shown by reduction in the average hydrogen index as a function of burial depth for Kimmeridge Clay Formation horizon equivalents. The bars across each point indicate the variation in the data suite used to obtain the average figure.

R_o = 0.5% thresholds are marked for comparative purposes. They are indicated as intervals since regional variations exist within each basin. (These R_o = 0.5% intervals are based on depth versus R_o trends in wells from the regions and not from discrete measurements in source rock horizons.) The depth differences between the 0.5% R_o boundary in the northern Viking Graben and the Halten Terrace are only moderate, and appear to be less than 500 m (Fig. 2). These R_o = 0.5% thresholds are comparable with other depth trends reported for these regions (e.g. Goff, 1983; Hagevang and Rønnevik, 1986). The R_o threshold of 0.5% occurs at 1800 m in the Hammerfest Basin, which is 800–1800 m or more shallower than in the other basins.

Figure 2 also contains generalized hydrocarbon generation curves for the Upper Jurassic and basal Cretaceous Kimmeridge Clay Formation equivalents based on hydrogen index, HI = S_2/TOC × 100 (mg HC/g TOC) versus depth of burial. The plotted HI data are average values calculated from source rock horizons of the same apparent source quality. The source rock quality selection criteria for the data used to establish these curves were visual kerogen typing and to some extent combined pyrolysis gas chromatography. In all three basins the Kimmeridge-equivalent source rocks are generally dominated by marine algal and amorphous organic matter whilst continental material represents only a small proportion of the kerogen (e.g. Goff, 1983; Cornford, 1984; Elvsborg et al., 1985; Berglund et al., 1986; Cohen and Dunn, 1987). However, some horizons had substantial amounts of terrestrially derived material. Some of these horizons, which were all characterised by lower HI values, were included in the data plots from the northern North Sea and the Halten Terrace. These points offset the general trend and are not taken into consideration when the generation curves are drawn (e.g. the point at 3700 m, on the Viking Graben plot and four of the five uppermost points in the Halten Terrace plot, Fig. 2). The plot from the Hammerfest Basin shows only data from four selected wells which are representative of a large database comprising most Upper Jurassic–basal Cretaceous horizons penetrated by wells in this region.

The curves drawn in Fig. 2 show the estimated average starting potential of the source rock systems in the three basins. These are: Viking Graben 420, Halten Terrace 390, and Hammerfest Basin 460 kg HC/tonne organic carbon, respectively. The differences in average source rock quality between the basins are therefore quite small. The best quality source material seems to be evident in the Hammerfest Basin and the poorest in the Halten Terrace region.

According to Fig. 2, significant hydrocarbon generation commences at 2600 m depth in the Viking Graben, at approx. 3100 m depth in the Halten Terrace and at between 1500 and 2000 m in the Hammerfest Basin. The onset of hydrocarbon generation is taken as the first inflection of the HI-curves against depth. Again, the differences between the basins are evident and follow closely those indicated by the R_o–maturity trends.

Results from other generation indicators (not included) such as normalized extracts (EOM/TOC) and production index PI = $S_1/(S_1+S_2)$ consistently support the data in Fig. 2. Likewise other maturity data, e.g. spore colouration and biomarkers, also support the R_o–maturity data in the three basins.

Model input and hypothesis testing

Present-day temperature gradient

Figure 3 shows a plot of some temperature data vs. depth for wells in the three basins. These data indicate that large differences in general thermal gradients do not exist between the basins. It should also be noted that the lithologies, which due to their different thermal conductivities can modify the thermal gradients, were fairly similar in the basins.

The general thermal gradient trends that can be drawn from this diagram are also consistent with reported data; e.g. Thomas et al. (1985), which showed that the temperature gradient in the northern Viking Graben varies between 30 and 40°C/km, and Cohen and Dunn (1987) which suggested the general temperature gradient in the Halten Terrace to be 39°C/km. Berglund et al. (1986), suggested temperature

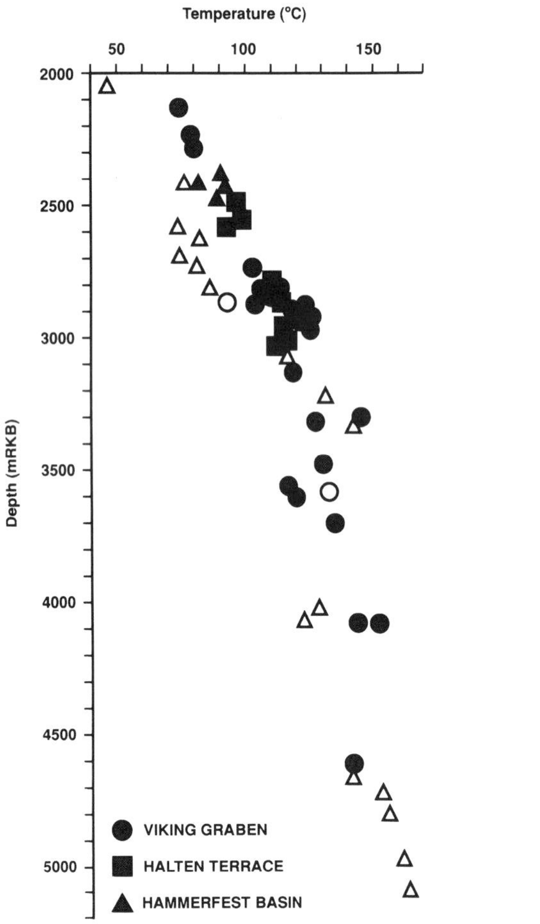

Fig. 3. Selected temperature data versus depth of burial from wells in the basins. Open symbols are BHT data; solid symbols are DST data.

gradients in excess of 35°C/km for the Hammerfest Basin.

Although the present-day heatflows in the basins are similar, palaeoheatflows may have been different and could have affected generation in the basins. The palaeoheatflow histories in the regions can be assessed by thermal history modelling, i.e. by studying different heatflow histories in the 1D basin simulations and calculating the evolution with respect to time and temperature of various thermally sensitive indicators such as vitrinite reflectance, isomerization of steranes and hopanes (C_{29} $\alpha\alpha\alpha$ 20R to C_{29} $\alpha\alpha\alpha$ 20S sterane conversion and C_{32} $\alpha\beta$ 22R to C_{32} $\alpha\beta$ 22S hopane conversion, respectively). When the calculated maturity matches the observed maturity then the suggested heatflow history is accepted providing that all other input parameters to the model such as lithologies, timing of the events, magnitudes of erosion etc. are correct.

It is possible, however, as demonstrated by Hermansen (this volume) and from the authors' experience, to derive reasonable fits with alternative heatflow histories. Therefore only heatflow histories that are geologically reasonable and in agreement with basin formation tectonics and subsidence should be accepted. This requires a close working relationship between the "modeller" and other geoscientists working in the same regions.

Tertiary and Quaternary stratigraphy

The most important periods with respect to hydrocarbon generation in the study areas were the Tertiary and Quaternary, since oil generation commenced in the Late Cretaceous or Early Tertiary in the Viking Graben, and in the Early Tertiary in the Halten Terrace and Hammerfest Basin.

Figure 4 depicts the generalized lithostratigraphy of the three basins with the major breaks in sedimentation indicated. Dating of the Late Tertiary and Quaternary in wells from the North Sea and the Halten Terrace has previously been poor. This is because of poor sample retention whilst drilling, or lack of interest in these sediments for their reservoir or source potential. For the purpose of the basin simulations, re-analysis and re-assessments of earlier work have been necessary. In the North Sea, these studies gave chronostratigraphies very close to those published by Rundberg and Smalley (1989), who based their results on Sr-isotope work. Most of the unconformities encountered represent erosional events. Those of longest duration are in the Late Tertiary (Fig. 4). The shortest sedimentation breaks are found in the Halten Terrace area. In the eastern Hammerfest Basin, Quaternary sediments rest unconformably on the Late Paleocene and Eocene. This is the largest break with no sedimentary record and represents a hiatus of more than 50 million years. The Pliocene and Pleistocene succession contains two short breaks in both the Viking Graben and the Halten Terrace. Sandy–shaly sediment, in

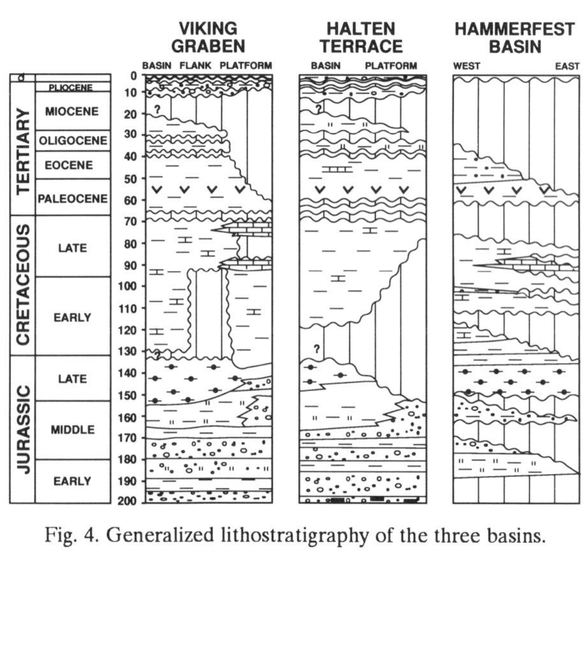

Fig. 4. Generalized lithostratigraphy of the three basins.

places more than 1000 m thick, was deposited between these breaks. This deposition occurred over a period of 2–3 million years. On seismic sections these Pliocene–Pleistocene sediments appear as a westward prograding system extending over the palaeoshelf away from Fennoscandia. The system was later peneplaned and the sediments deposited during the Late Quaternary (i.e. in the last 1 Ma) rest upon the unconformity.

The seismic expression and dating of these latter events are very similar in both the Viking Graben and the Halten Terrace.

Late Tertiary and Quaternary subsidence, uplift and erosion

In the process of model optimization and determination of heatflow histories in various wells, it is possible to assess the extent of sedimentation during the periods for which no sedimentary record survives. These estimations include timing of sedimentation (if sedimentation took place) and the timing and amounts of possible erosion.

The problem is solved by setting up various geologically plausible scenarios for deposition and erosion within these periods, based on estimates from regional geology and seismic stratigraphy. These scenarios for deposition and erosion within the sedimentation breaks are tested out on more than one well since maturity data or thermal indicators and well stratigraphy may be in error at some wells. It is equally important to use more than one type of thermal indicator, and preferably as many as possible.

As mentioned above, the most important unconformities in the Viking Graben and the Halten Terrace occur in the Miocene. In the Hammerfest Basin, the Oligocene to Pleistocene interval represents the most important unconformity. In this study the implications of these unconformities were considered carefully during the model optimisation process.

Published results suggest that erosion was taking place during the Late Tertiary (Miocene) in the northern Viking Graben (Dahl et al., 1987; Rundberg and Smalley, 1989). No magnitudes of this process or actual erosion rates are, however, suggested. No suggestions regarding Miocene or Pleistocene erosion in the Halten Terrace have so far been proposed.

In the Hammerfest Basin, Berglund et al. (1986) suggested a Late Tertiary uplift slightly in excess of 1000 m. In their model, Eocene sedimentation rates were extrapolated from those in the Paleocene. Erosion occurred during the Oligocene and Miocene and was very slow. The burial curve consequently takes a "V-shape" for subsidence, uplift and erosion as shown in Fig. 5A. Figure 5B suggests another scenario for Late Tertiary subsidence, uplift and erosion comprising rapid subsidence during the Oligocene. Subsidence was slow in the Miocene and Pliocene, but uplift and very rapid erosion occurred during the Pleistocene. These, and intermediate scenarios, were tested during heatflow optimization of wells in the Hammerfest Basin. Similar testing was carried out for the Miocene and Pliocene/Pleistocene unconformities in the northern Viking Graben and the Halten Terrace. Figure 6 shows the resulting subsidence curves during the last 50 million years from selected wells in the study areas. In the northern Viking Graben the basin is shown to have subsided quickly in Late Oligocene–Early Miocene. During the Mid Miocene, basin subsidence is shown to have decreased rapidly and deposition was slow or zero. This period was followed by rapid uplift and erosion of 100–200 m in the Oseberg area. Erosional magnitudes were larger in other parts of the basin, however, and particularly to the north. These move-

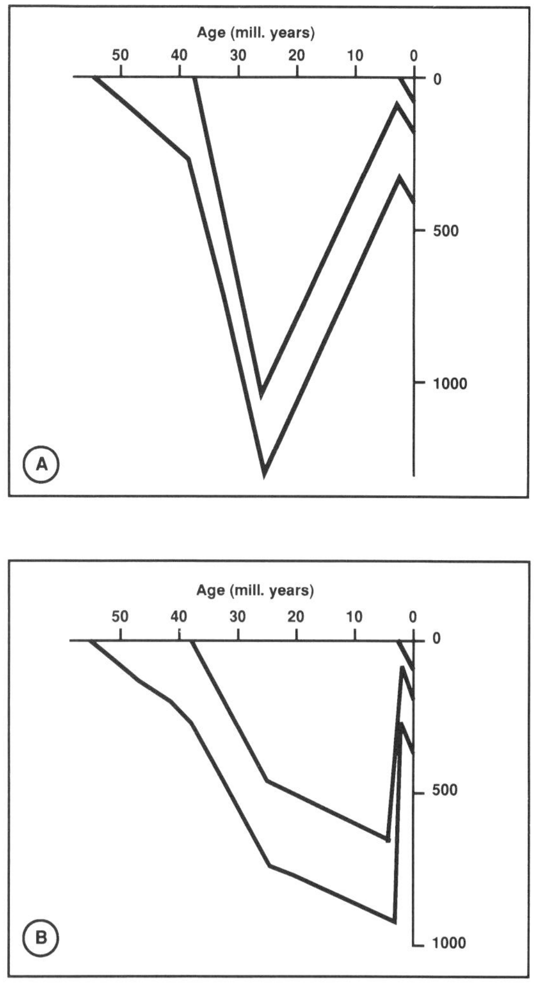

Fig. 5. Subsidence and uplift scenarios for Oligocene to Present, in the Hammerfest Basin. These scenarios and intermediates have been used to assess the most likely timing of subsidence, uplift and erosion in the area.

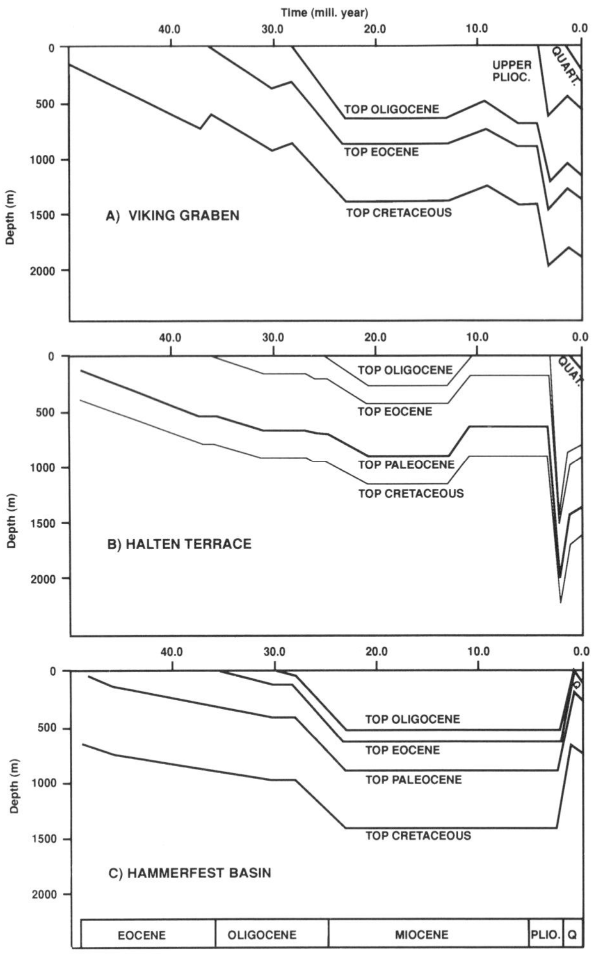

Fig. 6. Subsidence curves of selected horizons from three wells of each of the basins during the last 40 million years.

ments occurred simultaneously (or concomitantly) with an important phase of the Alpine Orogeny in Central Europe (Ziegler, 1982). After deposition of the sandy Utsira Foormation in the Late Miocene, rapid westward progradation occurred with deposition of up to 1000 m of sediments of Late Miocene age with one or two breaks in sedimentation. Between 1 and 2 million years ago (Late Pliocene–Early Pleistocene) this system was peneplaned either due to a sea-level change or uplift, and up to 150–200 m of sediment was eroded. The situation in the Halten Terrace is very similar to that of the northern Viking Graben, as shown in Fig. 6B, which is representative of a southern platform well in the region. It appears that the magnitude of Quaternary erosion was larger in this region relative to the northern North Sea. In addition, the Miocene depositional and erosional phases were probably slightly shifted in time and of shorter duration compared with the associated movements in the northern Viking Graben. Miocene erosional magnitudes were of the same order in the two basins.

When these areas are compared with the Hammerfest Basin (Fig. 6C) both similarities and differences exist. In the Hammerfest Basin, subsidence was rapid during the Late Oligocene–Early Miocene as in the other basins. In marked contrast to the Viking Graben and the Halten Terrace, however, there appears to have been little or no subsidence from the Middle Miocene to Late Pliocene. Rapid uplift and erosion (600 m) took place in the Pleistocene, possibly in association with some of the major glacial events. Subnormal pressures encountered in wells northwest of the Hammerfest Basin, where the erosion has been more substantial, suggest a fairly recent and very rapid loss of sediment in the region. This supports the scenario of Fig. 5B, and produces the calculated maturity trends that fit best with the observed data during the optimization procedure.

Figure 7 is an example of the match between the calculated and measured vitrinite reflectance and sterane and hopane isomerization data in the Viking Graben well referred to in Fig. 6. The use of three thermal indicators, with different dynamic temperature–time ranges, enhances the resolution with respect to quantitative aspects of erosional events.

Studies in the Hammerfest Basin using Apatite Fission Track Analysis have also been carried out, but the results were of little value in solving timing of subsidence, uplift and erosion.

Heatflow histories

The heatflow histories derived during the optimization process are compiled in Fig. 8. The three heatflow curves are for each area studied, using the wells shown in Fig. 6. The calculated heatflow, which increases and decreases during Jurassic and Cretaceous times, respectively, represents periods of crustal thinning. The number of wells penetrating pre-Jurassic sediments, for use in the assessment of heatflow history during the Jurassic and the Cretaceous, is limited. The subsidence curves associated with rifting and postrift cooling have therefore been used to determine the probable heatflow histories in combination with the few wells containing maturity parameters from pre-Jurassic sediments. These studies show differences between the basins. In the Halten Terrace, heatflow is interpreted to have increased earlier than in the Viking Graben, but to have decreased simultaneously with the Viking Graben. Heatflow increase in the Hammerfest Basin is interpreted to have been earlier than in the Halten Terrace area and to have decreased earlier than in either the Halten Terrace or Viking Graben.

A small increase in the heatflow was postulated for the Eocene (Dahl et al., 1987). This is to take

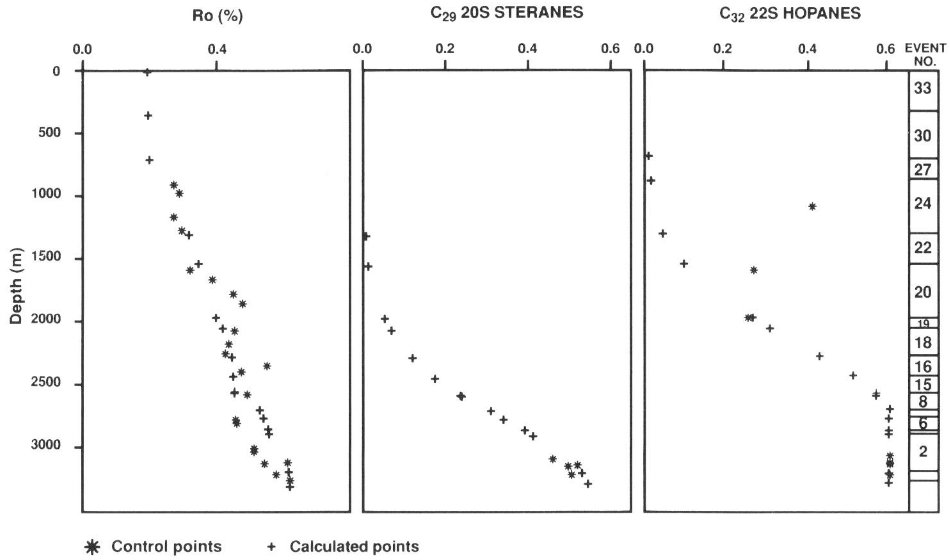

Fig. 7. Calculated and observed variations in vitrinite reflectance, C_{29}, 20S steranes and C_{32}, 22S hopanes from a North Sea well.

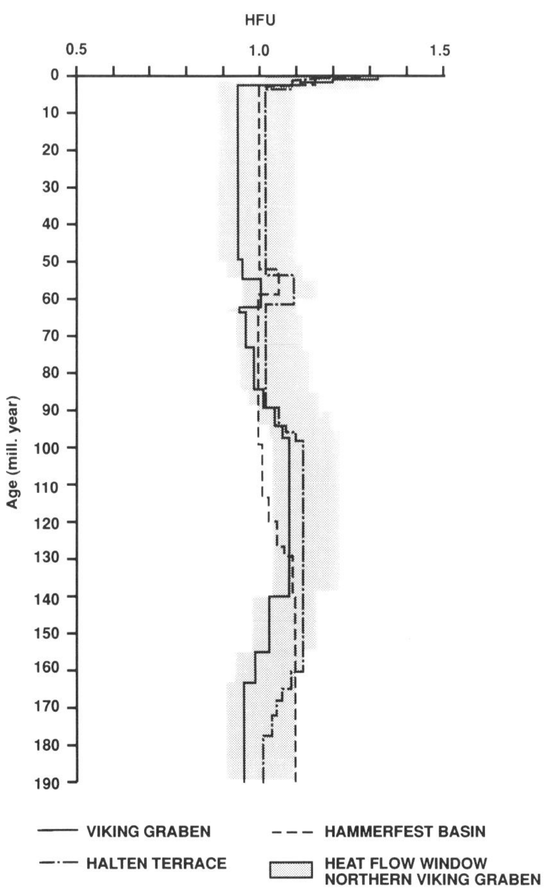

Fig. 8. Heatflow histories from selected wells in the study areas compared with heatflow variations in the northern Viking Graben. The Viking Graben heatflow variations are represented by the hatched area.

account of plate motions associated with widespread volcanism and opening of the North-East Atlantic Ocean (Doré and Gage, 1987).

In the Late Pliocene and Pleistocene, an apparent increase in heatflow is shown in all three basins, as suggested by Dahl et al., 1987. The phenomenon has been discussed by several authors (e.g. Jensen and Doré, Vik and Hermanrud, and Hermansen, this volume). This heatflow pulse is a necessary postulate using the basin modelling algorithm to match or predict the present-day temperatures in the simulated wells. Neither the maturity parameters nor the hydrocarbon generation are sensitive to this pulse if it is assumed to be of very short duration. It is possible that this increase is partly an artifact and may be a requirement to model present-day transient temperatures resulting from rapid ice unloading, associated destabilization of the geothermal gradient and movement of hot water from deeper parts of the basins. The relative magnitude of this pulse varies over the basins and is largest in flank positions where hydraulic continuity with the deepest part of the basins may be expected. With simulation at a time resolution of 10,000–20,000 years (in contrast to 1 Ma which is used in this study) and with ice loading and unloading during Quaternary, heatpulses of such magnitudes may not be required to match the present temperatures (A. Yükler, pers. commun., 1991).

The period of importance for hydrocarbon generation was the Tertiary and Quaternary. During this period, the basinal heatflow histories are interpreted to have been similar and to differ by 0.07 HFU, which is less than 10%. The Viking Graben well had a Tertiary heatflow of 0.95 HFU whilst the Hammerfest Basin and Halten Terrace wells had very similar values of 1.0 and 1.02 HFU, respectively. The hatched area in Fig. 8 represents the heatflow variations in the northern Viking Graben. Both wells from outside this region also fall within this heatflow "window". Regional variations in heatflow histories exist also in the Hammerfest Basin and the Halten Terrace. However, these regional variations overlap strongly with the Viking Graben heatflow variations shown in Fig. 8.

Consequently, it can be concluded that the general heatflow histories of the three basins were probably very similar and that heatflow cannot account for the observed discrepancies in the depth of burial for the various stages of HC generation.

Determination of hydrocarbon generation kinetics

It is well established that different kerogen types generate hydrocarbons over different temperature intervals (e.g. Tissot and Welte, 1978; Hunt, 1979). Mathematical models based on chemical kinetic principles to simulate the various generation processes have been established and integrated with basin models (Tissot and Espitalié, 1975; Yukler and Kokesh, 1984). Laboratory methods to calibrate activation energies and frequency factors on representative immature source rock samples have also been developed (Burnham et al., 1987; Ungerer and Pelet, 1987).

The laboratory methods used for the calibration of hydrocarbon generation models have inherent problems that may lead to severe errors in the hydrocarbon generation predictions (Dahl and Yukler, 1989; Nielsen and Dahl, 1991). It was therefore decided that, in this study, generation kinetic parameters would be calibrated against observed hydrocarbon generation rather than laboratory simulation data. An example of a hydrocarbon generation curve is shown in Fig. 2 and discussed in the associated text. These curves have been transferred to kerogen transformation plots as shown in Fig. 9. Since variations exist in the data points two alternative curves have been drawn representing the upper and lower boundaries for present-day hydrocarbon generation. The geohistory of a pseudo-well located in the Viking Graben has been simulated to give a present-day generation profile. The kinetic parameters in the simulation program have been varied over a number of runs until the present-day generation profile matches the "observed" transformation curve in Fig. 9. Comparison of the observed and predicted transformation is shown in Fig. 10. The kinetic parameters used to obtain the match shown in Fig. 10 are shown in Fig. 11. These parameters are within the same order of magnitude as parameters reported from laboratory kinetic measurements of type II kerogens (e.g. Burnham et al., 1987). It should be noted that regional heatflow variations have not been taken into account during this process, and the resulting kinetic parameters only approximately describe the hydrocarbon generation process. It should also be noted that this set of parameters is only one set of an infinite number of solutions, as change in energy distributions and fractions can be compensated for by changes in the frequency factor

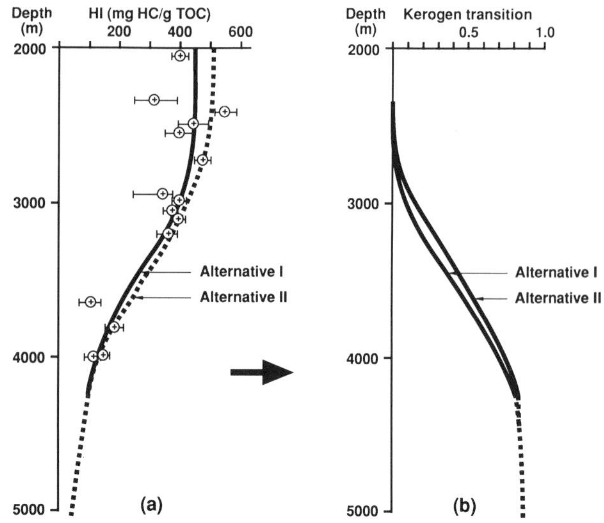

Fig. 9. Generalized hydrocarbon generation for the northern Viking Graben. (a) Average hydrogen indices from source rock horizons plotted versus depth as in Fig. 2. (b) Generalized kerogen conversion trends by transforming the curves in (a).

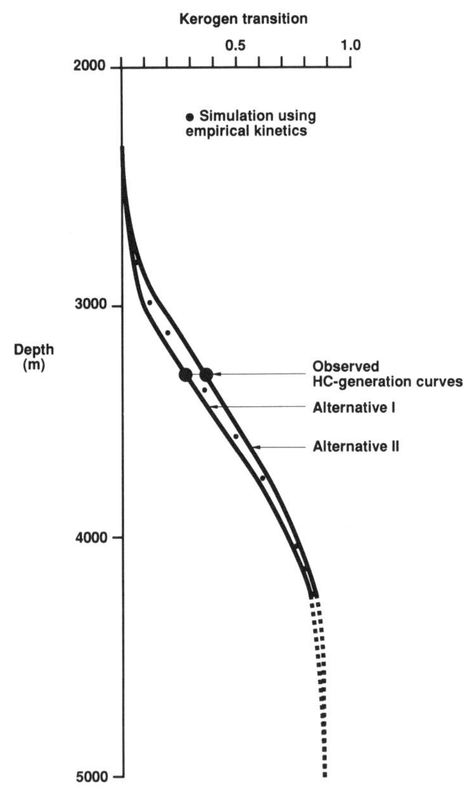

Fig. 10. Observed and predicted kerogen transformations, northern Viking Graben. The predicted conversions from the simulation of a pseudo-well are shown as solid points.

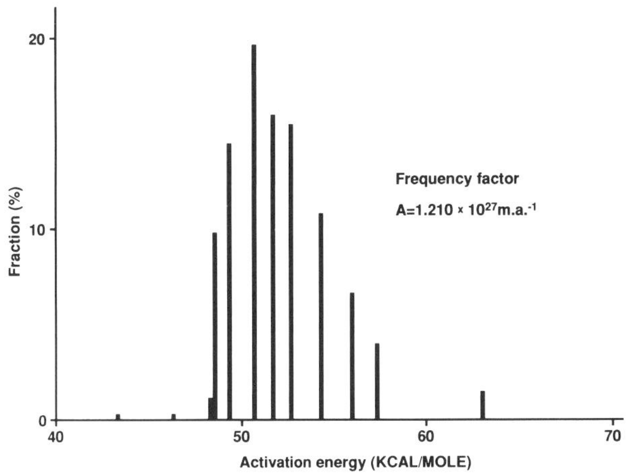

Fig. 11. Kinetic parameters used to obtain the predicted kerogen transformation in Fig. 10.

to give the same hydrocarbon generation prediction. The consequences multiple solutions may have with respect to errors in the prediction of generation histories have not yet been investigated.

The parameter set (Fig. 11) derived from the northern Viking Graben was tested out in pseudo-wells in the other two basins and the predictions compared with regional kerogen transformation curves. The average kerogen qualities are, as discussed above, fairly similar within the three basins. Consequently, it would be expected that the same set of kinetic parameters could be used to approximately predict the observed generation in the three basins. In the Halten Terrace, however, the generation predictions did not match the observed results. It was necessary, therefore, to shift the distribution of activation energies to higher values by 3 kcal/mol to obtain the satisfactory match shown in Fig. 12a. The frequency factor remained as for the results shown in Fig 11. In the Hammerfest Basin the same parameters predicted a kerogen conversion which matches the lower part of the observed curves, as shown in Fig 12b. This diagram also includes the predicted kerogen conversion for the Late Jurassic–Early Cretaceous Hekkingen Formation (Kimmeridge Clay Equivalent) in wells in the area. These points are close to the predicted conversion profile in the pseudo-well located in the western part of the Hammerfest Basin where the Upper Jurassic source rocks were more deeply buried. This exercise was carried out to compare the effect of basinal shifts in heatflow histories relative to the conversion predictions made by the pseudo-well used for kinetic calibration.

The observed differences in kinetic parameters used to match the observed data may be due to differences in source rock quality as indicated above (Fig. 2). The source rock in the Halten Terrace generally had the lowest hydrogen index because

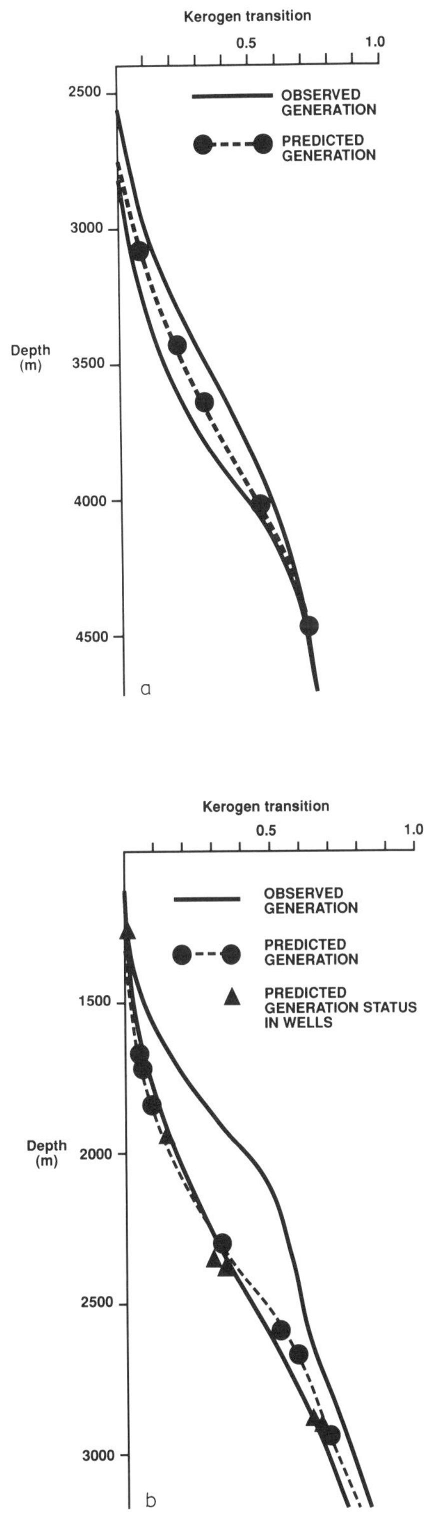

Fig. 12. Observed and predicted kerogen conversion in (a) the Halten Terrace and (b) the Hammerfest Basin. The observed conversion is obtained by transforming the curves in Fig. 2 as demonstrated in Fig. 9.

it contained more continentally derived material of Type III kerogen. This kerogen type has been shown to require higher activation energies for kerogen conversion (e.g. Tissot and Welte, 1978; Burnham et al., 1987). This may also explain the necessity to shift the distribution of activation energies of the North Sea kerogen (Fig. 11) to higher values to match observed generation at the Halten Terrace. Although the kinetic data obtained from the observed North Sea generation match the lower boundary of the observed generation in the Hammerfest Basin, a shift to lower activation energies would probably give a better match with the observed curve. However, as mentioned above, the derived kinetic parameters do not represent unique solutions. This means that other solutions with fewer discrepancies in the kinetic parameters might satisfy the observed generation in the three basins. A significant question is whether one set of parameters, within reasonable error limits, can match observed generation in all three basins. Another possibility is that different mixing of ideal kerogen end members (Dahl and Yukler, 1991) with associated kinetic parameters may match the observed generation trends in the three basins.

Observed discrepancies in the depth of onset of hydrocarbon generation between the Hammerfest Basin and the other two basins are not explained by differences in kinetic parameters. The differences in onset of generation between the North Sea and the Halten Terrace may be partly explained by such variations. However, a discrepancy in the R_0–maturity–depth trends, which appears to follow that of the hydrocarbon generation, also exists between the basins. The kinetics of R_0 evolution should be constant from basin to basin. The principal explanation for the differences must therefore lie in factors other than kinetics.

Implications of basin subsidence variations

The difference in subsidence between the Halten Terrace and the northern Viking Graben is not large. Miocene subsidence in the Viking Graben has been greater than in the Halten Terrace. In the Viking Graben the sediments have been retained at greater depths and higher temperature regimes for a longer period of time (compared with the Halten area) before the Mid–Late Miocene uplift and erosion occurred. The Pliocene–Pleistocene depositional episode may also have reached the Viking Graben earlier and held the sediments at depth longer than in the Halten Terrace. As a consequence of these minor differences in subsidence history, sediments at comparable present-day depths in the two areas will have experienced slightly different temperature histories. In the Viking Graben these sediments will have been retained within higher temperature regimes over a longer period of time and consequently reached higher thermal maturities and generated more hydrocarbons than their counterparts in the Halten Terrace area.

The difference between the Hammerfest Basin and the two other basins is obviously due to the lack of Pliocene–Pleistocene subsidence plus the recent rapid erosion. The Late Quaternary subsidence that occurred in this region was considerably less than the preceding erosion. The sediments older than Quaternary have been buried to greater depths and higher temperature regimes than at present, which explains why the observed hydrocarbon generation threshold is at such a shallow depth compared to the Viking Graben and Halten Terrace regions.

Implications for hydrocarbon generation

The differences (and similarities) in basin subsidence histories discussed above result in varying hydrocarbon generation histories, and these may have implications for the prospectivity of the three areas. The two basins in the south, which were rapidly subsiding basins in the Pliocene and Pleistocene, underwent a very effective hydrocarbon generation during the last 4–5 Ma. This is illustrated in Figs. 13 and 14 showing zones of kerogen transformation for the northern Viking Graben and the Halten Terrace, respectively, at 4 Ma and at Present. Both basins show a significant increase in hydrocarbon generation over this time interval. Although fairly rapid deposition also occurred in both basins during Late Oligocene–Early Miocene, the resulting generation efficiencies were quite different from those displayed in Figs. 13 and 14. The Mid–Late Miocene erosion was rapid and differential over the basins and probably led to tilting of oil–water contacts and partial loss of hydrocarbons in rotating structures. Structures with low relief configuration would have been particularly affected. These events gave rise to remnant oil–water contacts, which have been observed in some fields. The Miocene erosion may also have caused a lowering of cap rock efficiency in some traps, which may then have lost hydrocarbons. The Pliocene–Pleistocene subsidence, which resulted in a significant generation pulse, may have refilled the partially emptied structures. In the North Sea a large proportion of the hydrocarbons generated by this pulse are light oil, condensate and gas (Fig. 13) and this may explain why structures in some areas are dominated by gas and condensate. In the Halten Terrace the generation pulse in the Pliocene and Pleistocene is dominated by oil generation (Fig. 14) apart from the hinge zone to the west and northwest.

In the Hammerfest Basin, generation of oil and gas was effective during the Oligocene–Miocene subsidence. The generation rate slowed down in the

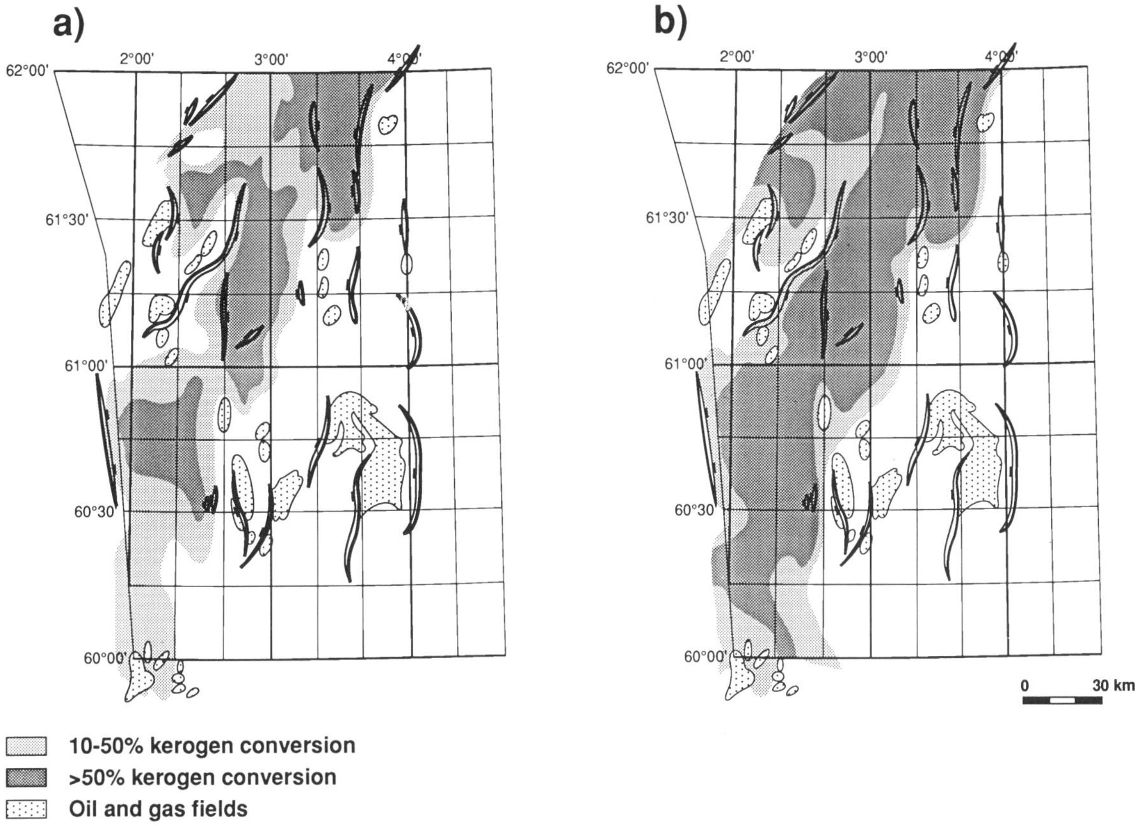

Fig. 13. Hydrocarbon generation in the northern Viking Graben, (a) Present and (b) 4 Ma.

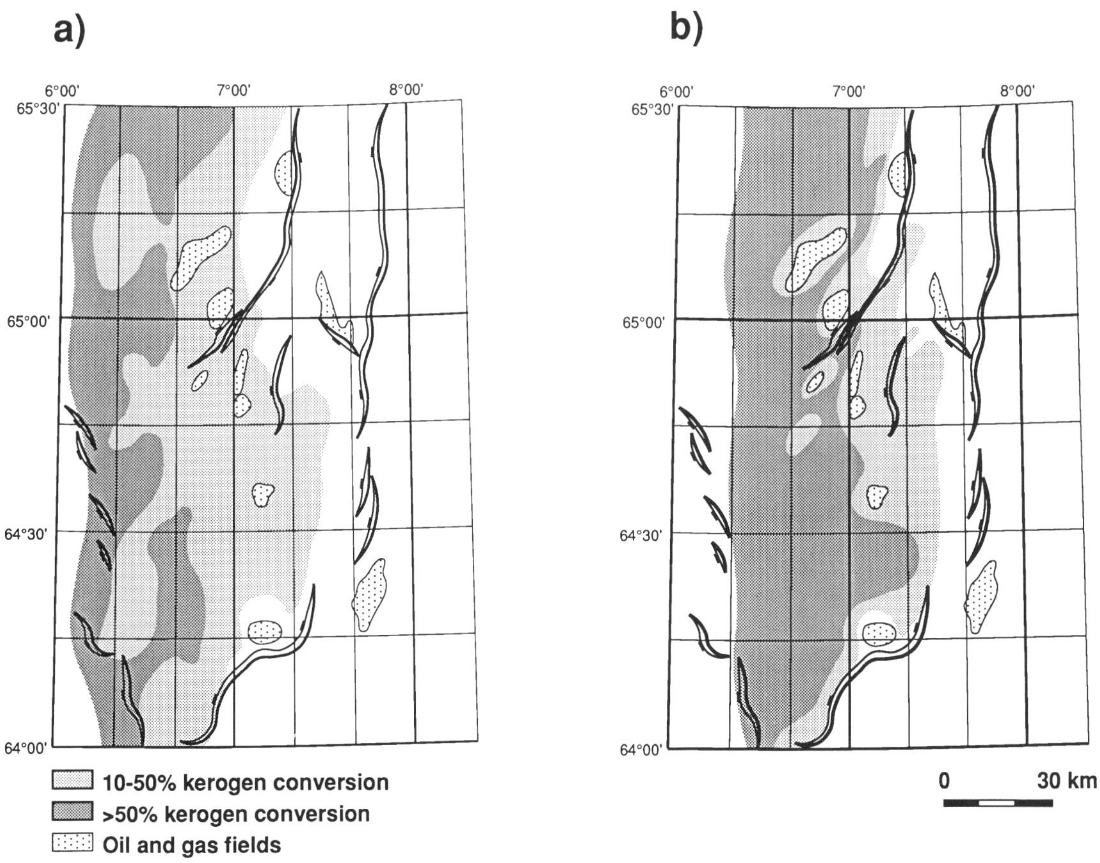

Fig. 14. Hydrocarbon generation in the Halten Terrace, (a) Present and (b) 4 Ma.

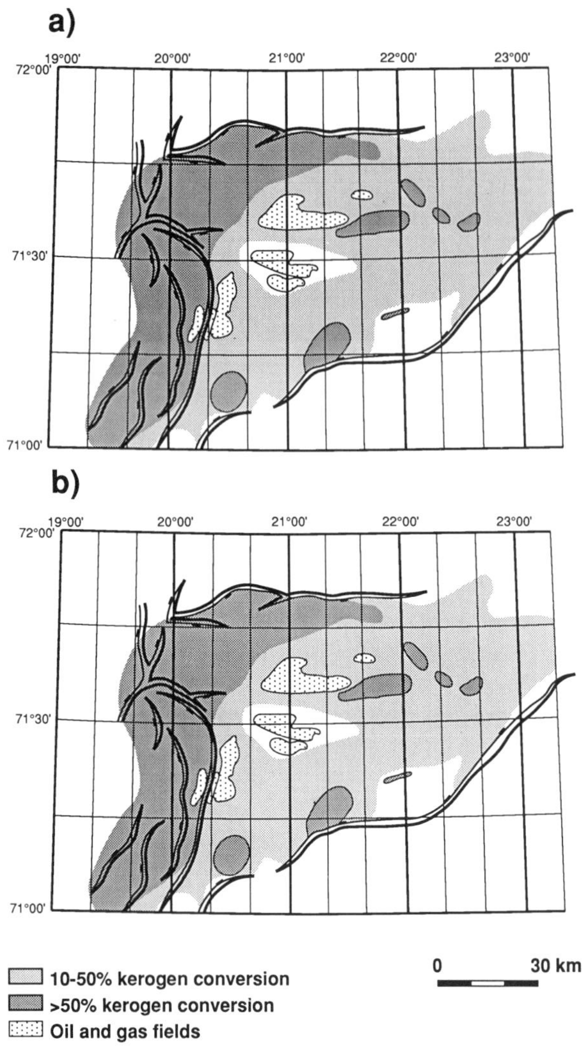

Fig. 15. Hydrocarbon generation situation in the Hammerfest Basin, (a) Present and (b) 4 Ma.

Middle Miocene and ceased during the Pleistocene uplift and erosion. During the last 4 Ma very little hydrocarbon generation has occurred as is shown in Fig. 15, which depicts the generation situation at 4 Ma and at Present in the Hammerfest Basin. During the erosional period the block rotation may have caused some fields to spill some of their hydrocarbon content. In areas to the north of the Hammerfest Basin the erosion has been strong enough to result in the subnormal pressures presently observed in some reservoirs. Subnormal pressures combined with possible tectonic disturbances may have caused seal failures, which in turn probably caused significant loss of hydrocarbons. The Late Quaternary subsidence was not large enough to cause generation of adequate amounts of hydrocarbons to refill the structures. The losses from, and partial refilling of structures in the Barents Sea area have been observed and documented by Augustson (1992). Since the basin subsidence and erosional losses during the Late Tertiary and Quaternary have been variable over the Hammerfest Basin and in the Barents Sea region in general, the amounts of hydrocarbons lost and the refilling of structures will vary from subregion to subregion.

Conclusions.

(1) The differences in thresholds of hydrocarbon generation and thermal maturities in the investigated basins were primarily caused by variations in the amounts of subsidence, uplift and erosion during the Late Tertiary and Quaternary.

(2) Kinetic parameters controlling the generation of hydrocarbons from the Upper Jurassic source rocks in the three basins have been determined. The parameters required to match observed generation in the Halten Terrace are different from those needed to match observed generation in the other two basins.

(3) Rapid sedimentation in the Pliocene and Pleistocene resulted in very efficient hydrocarbon generation in the northern Viking Graben and the Halten Terrace. In contrast, in the Hammerfest Basin, slow to nil sedimentation followed by rapid and extensive erosion during this period resulted in reduction and in some areas even cessation, of hydrocarbon generation. Losses from structures during this period may not have been effectively compensated for by freshly generated hydrocarbons.

Acknowledgements

The authors wish to thank Norsk Hydro A.S. for permission to publish this paper. Colleagues in Norsk Hydro Exploration and Norsk Hydro Research Center, Bergen are acknowledged for useful discussions and contributions. Special thanks are due to G. Butenko, S. Idil and I. Meisingset for discussions on the North Sea erosional episodes. We are also indebted to G.C. Speers and B.S. Cooper for kindly reading and commenting on this manuscript. C. Cornford, an anonymous referee and A.G. Doré are also acknowledged for advice which improved this manuscript.

References

Augustson, J.H., 1992. A method on classification of oil traps based on heavy oil content in cores with relevance to filling and drainage of Barents Sea oil bearing structures. In: T.O. Vorren, E. Bergsager, Ø.A. Dahl-Stamnes, E. Holter, B. Johansen, E. Lie and T.B. Lund (Editors), Arctic Geology and Petroleum Potential. Norwegian Petroleum Society (NPF), Special Publication 2, Elsevier, Amsterdam, pp. 691–702.

Berglund, L.T., Augustson, J., Færseth, R., Gjelberg, J. and Ramberg-Moe, H., 1986. The evolution of the Hammerfest

Basin. In: A.M. Spencer et al. (Editors), Habitat of Hydrocarbons on the Norwegian Continental Shelf. Graham and Trotman, London, pp. 319–338.

Burnham, A.K., Braun, R.L., Gregg, H.R. and Samoun, A.M., 1987. Comparison of methods for measuring kerogen pyrolysis rates and fitting kinetic parameters. Energy Fuel, 1: 452–458.

Cohen, M.J. and Dunn, M.E., 1987. The hydrocarbon habitat of the Haltenbank–Trænabank area offshore Mid Norway. In: J. Brooks and K. Glennie (Editors), Petroleum Geology of North West Europe. Graham and Trotman, London, pp. 1091–1104.

Cornford, C., 1984. Source rocks and hydrocarbons of the North Sea. In: K.W. Glennie (Editor) Introduction of the Petroleum Geology of the North Sea. Blackwell, London, pp. 171–204.

Cornford, C., 1990. Source rocks and hydrocarbons of the North Sea. In: K.W. Glennie (Editor), Introduction to the Petroleum Geology of the North Sea. Blackwell, London, 3rd ed.

Dahl, B. and Yukler, M.A., 1989. Calibration of hydrocarbon generation kinetics in the Viking Graben. Poster presentation at the 14th International Meeting on Organic Geochemistry, Paris.

Dahl, B. and Yukler, M.A., 1991. The role of petroleum geochemistry in basin modelling of the Oseberg area, North Sea. In: R.K. Merrill (Editor), AAPG Treatise of Petroleum Geology Handbook. Source and Migration Processes and Evaluation Techniques. American Association of Petroleum Geologists, Tulsa, Okla., pp. 65–85.

Dahl, B., Nysæther, E., Speers, G.C. and Yukler, M.A., 1987. Oseberg area — integrated basin modelling. In: J. Brooks and K. Glennie (Editors) Petroleum Geology of North West Europe. Graham and Trotman, London, pp. 1029–1038.

Doré, A.G. and Gage, M.S., 1987. Crustal alignments and sedimentary domains in the evolution of the North Sea, North East Atlantic Margin and Barents Shelf. In: J. Brooks and K. Glennie (Editors), Petroleum Geology of North West Europe. Graham and Trotman, London, pp. 1131–1148.

Elvsborg, A., Hagevang, T. and Throndsen, T., 1985. Origin of the gas condensate of the Midgard Field at Haltenbanken. In: B.M. Thomas et al. (Editors), Petroleum Geochemistry in Exploration of the Norwegian Shelf. Graham and Trotman, London, pp. 213–219.

Faerseth, R.B., Oppeboeen, K.A. and Sæboe, S., 1986. Trapping styles and associated hydrocarbon occurrence offshore Mid Norway. In: M.T. Halbouty (Editor), Future Petroleum Provinces of the World. Am. Assoc. Pet. Geol., Mem., 40: 585–598.

Goff, J.C., 1983. Hydrocarbon generation and migration from Jurassic source rocks in the East Shetland Basin and Viking Graben of the northern North Sea. J. Geol. Soc. London, 140: 445–474.

Hagevang, T. and Rønnevik, H., 1986. Basin development and hydrocarbon occurrence offshore Mid Norway. In: M.T. Halbouty (Editor), Future Petroleum Provinces of the Word. Am. Assoc. Pet. Geol., Mem., 40: 599–614.

Hermansen, D., 1993. Optimization of temperature history — aspects of vitrinite reflectance and sterane isomerization. In: A.G. Doré et al. (Editors), Basin Modelling: Advances and Applications. Norwegian Petroleum Society (NPF), Special Publication 3. Elsevier, Amsterdam, pp. 119–126 (this volume).

Hunt, J., 1979. Petroleum Geochemistry and Geology. Freeman and Company, San Francisco, Calif.

Jensen, R.P. and Doré, A.G., 1993. A recent Norwegian Shelf heating event–fact or fantasy? In: A.G. Doré et al. (Editors), Basin Modelling: Advances and Applications. Norwegian Petroleum Society (NPF), Special Publication 3. Elsevier, Amsterdam, pp. 85–106 (this volume).

Mackenzie, A.S., 1984. Applications of biological markers in petroleum geochemistry. In: J. Brooks and D.H. Welte (Editors), Advances in Petroleum Geochemistry, Vol. 1. Academic Press, London, pp. 115–214.

Nielsen, S.B. and Dahl, B., 1991. Confidence limits on kinetic models of primary cracking and implications for the modelling of hydrocarbon generation. Mar. Pet. Geol., 8: 483–492.

Rundberg, Y. and Smalley, P.C., 1989. High resolution dating of Cenozoic sediments from northern North Sea using $^{87}Sr/^{86}Sr$ stratigraphy. Am. Assoc. Pet. Geol., Bull., 73 (3): 298–308.

Thomas, B.M., Møller-Pedersen, O., Whitaker, M.F. and Shaw, N.D., 1985. Organic facies and hydrocarbon distributions in the Norwegian North Sea. In: B.M. Thomas et al. (Editors) Petroleum Geochemistry in Exploration of the Norwegian Shelf. Graham and Trotman, London, pp. 3–26.

Tissot, B. and Espitalié, J., 1975. L'évolution technique de la matière organique des sédiments: Application d'une simulation mathématique. Lev. Inst. Fr. Petr., 30: 743–777.

Tissot, B.P. and Welte, D.H., 1978. Petroleum Formation and Occurrence. Springer-Verlag, Berlin, 538 pp.

Ungerer, P. and Pelet, R., 1987. Exploration of kinetics of oil and gas formation from laboratory experiments to sedimentary basins. Nature, 327: 52–54.

Vik, E. and Hermanrud, C., 1993. Transient thermal effects of rapid subsidence in the Haltenbanken area. In: A.G. Doré et al. (Editors), Basin Modelling: Advances and Applications. Norwegian Petroleum Society (NPF), Special Publication 3. Elsevier, Amsterdam, pp. 107–117 (this volume).

Yukler, M.A. and Kokesh, F., 1984. A review of models used in petroleum resource estimation and organic geochemistry. In: J. Brooks and D.H. Welte (Editors), Advances in Petroleum Geochemistry, 1. Academic Press, London, pp. 69–113.

Yukler, M.A. and Welte, D.H., 1980. A three-dimensional deterministic dynamic model to determine geologic history and hydrocarbon generation and accumulation in a sedimentary basin. In: Fossil Fuels. 26th Int. Geol. Congr., Technip, Paris, pp. 267–285.

Ziegler, P.A., 1982. Geological Atlas of Western and Central Europe. Elsevier, Amsterdam, pp. 92–94.

BIGER DAHL *Norsk Hydro Research Center, N-5020 Bergen, Norway*

JAN H. AUGUSTSON *Norsk Hydro Research Exploration Office, Storaakeren 11, N-9400 Harstad, Norway*

Impact of data integration on basin modelling in the Barents Sea

N.J. Theis, H.H. Nielsen, J.K. Sales and G.J. Gail

Exploration wells in the Barents Sea show higher than expected levels of thermal maturity than is suggested by present-day depths and thermal gradients. Matching of predicted vitrinite trends for wells to actual vitrinite measurements requires appreciable amounts of regional post-Cretaceous uplift and erosion (750–1500 m in the Hammerfest Basin and 1900–3000 m for areas to the north and northwest). Reservoir rocks also show effects of deeper burial and have lower porosities and permeabilities than normal for their present-day depths.

Burial history profiles, as used in basin modelling, constrain timing of maturation and must also incorporate timing and amount of sediment removal normally associated with uplift. Integration of results from apatite fission track analyses, which suggest two stages of uplift, helps to refine burial history profiles. Tectonic analogy with other rifted-margin settings and changes in plate motion seen in the deep sea-floor magnetic anomaly record are also consistent with two stages of uplift.

Quantification of uplift and erosion outside of areas with well control was accomplished by incorporating results from seismic velocity analysis and calibrating these results to sediment removal determined from vitrinite reflectance modelling at control wells. Regional maps showing Tertiary sediment removal and effective temperature gradients were made and supply the input parameters for key prospects. "What if" scenarios, calculated using computerized basin modelling, determine the sensitivity of results on the variation of critical input parameters and help risk the prospects.

Integration of modelling techniques and data analyses is synergistic. Understanding the modelling process and developing an appreciation for the reliability of a prediction can be as important as getting the "right" answer.

Introduction

Drilling results from the widely spaced blocks awarded in the 11th and 12B Norwegian Licensing Rounds in the Norwegian Barents Sea were disappointing. However, some important geologic factors have become apparent. One factor that has widespread consequences is the extent and magnitude of regional post-Cretaceous uplift and erosion. Recent conferences hosted by the Norwegian Petroleum Society (1989 and 1990) and the Norsk Geologisk Forening (1990) have included many papers dealing with post-Cretaceous uplift and erosion. The magnitude, timing, and consequences of uplift have been discussed by Nyland et al. (1992). Much of the gas and large quantities of residual oil in the Barents Sea reservoirs may reflect oil displacement caused by gas expansion resulting from uplift and erosion (Skagen, 1990). In this paper, some methodology and techniques of data integration are directed toward understanding the uplift problem and assessing its implications for basin modelling in the Norwegian Barents Sea. The uplift problem and its consequences illustrate the impact of data integration on basin modelling. Integration of as many techniques as possible helps constrain the parameters used to solve or to understand basin modelling problems, and to prioritize areas for hydrocarbon exploration.

Basin modelling involves the bringing together of observations from different disciplines for the purpose of interpretation and prediction of underlying processes. Sensitivity analysis allows us to rank and prioritize the value of various observations. The goal is to assemble a complete, consistent, and well-constrained model from the data available even though some of the facts are seemingly unrelated. This paper is a status report, as exploration continues and new facts are discovered and integrated.

Ranges and sensitivities are stressed in basin modelling. The final answer (geologic reality) is not known, and under the circumstances, may not even be obtainable. However, basin modelling is a technique that permits us to place some constraints on geologic reality of parameters and predictions.

Setting

The Norwegian Barents Sea lies between the northern coast of Norway and Svalbard. Selected

Basin Modelling: Advances and Applications edited by A.G. Doré et al. NPF Special Publication 3, pp. 433–444, Elsevier, Amsterdam.

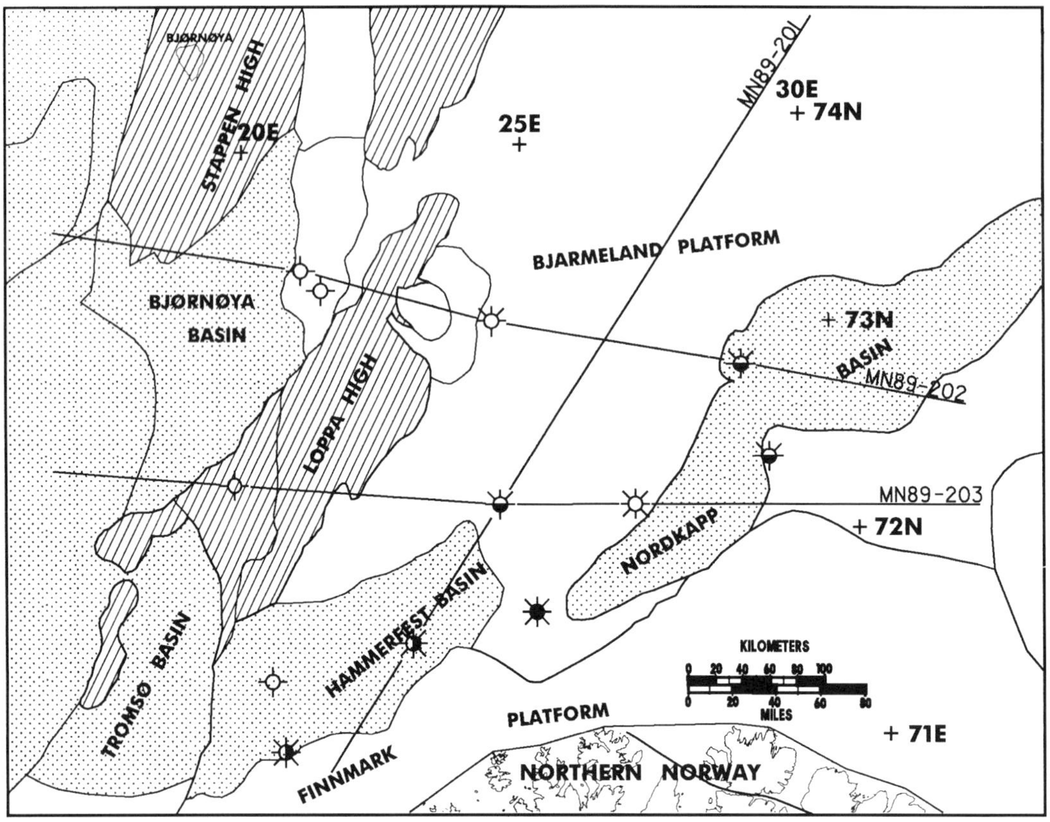

Fig. 1. Tectonic elements map of the Norwegian Barents Sea after Gabrielsen et al. (1990). Well symbols indicate locations of exploration wells included in this study. Locations of three, long regional seismic lines collected over the Barents Sea by the "Mobil Search" are indicated and were used to evaluate changes in interval velocities along the strike of the lines.

blocks within a portion of this area, that part between the coast of Norway and about 73°30′N, have been licensed for petroleum exploration. Several gas fields have been found in the Hammerfest Basin. Snøhvit, the largest and a major gas accumulation, was found in 1984 and has a 14-m-thick oil leg (Olsen and Linjordet, 1990). To date, after 46 exploration wells, no major oil fields have been found. This is a significant economic barrier because the entire Barents Sea area is north of the Arctic Circle and lacks important infrastructure.

The locations of the twelve exploration wells referred to in this paper are shown on the tectonic elements map (Gabrielsen et al., 1990) of the Norwegian Barents Sea (Fig. 1). These wells include initial commitment wells from the 11th and 12B Norwegian Licensing Rounds, and a few deeper wells from earlier drilling in the Hammerfest Basin. Also shown in Fig. 1 is the distribution of structural elements and the general tectonic grain of the Barents Sea area.

The Barents Sea: not analogous to the North Sea and Mid-Norway areas

The observations shown in Figs. 2 and 3 indicate different burial and thermal histories for the Barents Sea relative to Mid-Norway or the Norwegian North Sea. Such differences should be understood and incorporated into basin modelling.

Porosity degradation

Porosity values determined from analysis of sandstone core plugs are shown in Fig. 2. Plots of this

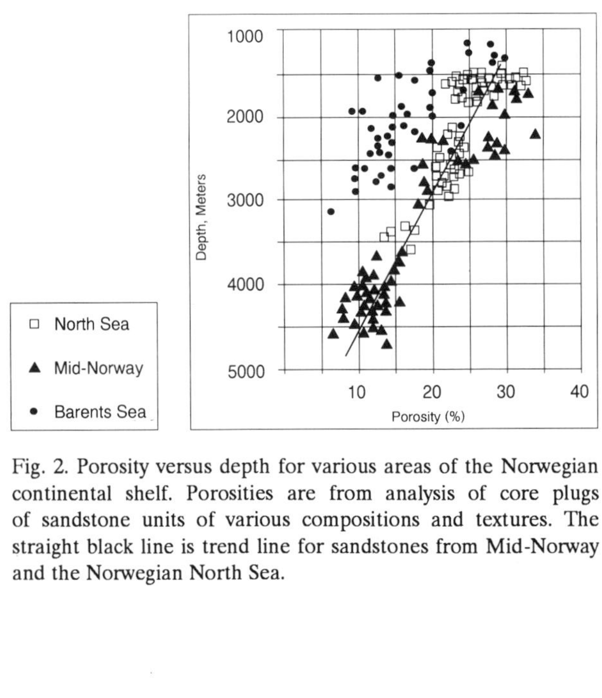

Fig. 2. Porosity versus depth for various areas of the Norwegian continental shelf. Porosities are from analysis of core plugs of sandstone units of various compositions and textures. The straight black line is trend line for sandstones from Mid-Norway and the Norwegian North Sea.

type are used to estimate and risk porosity values for reservoir units prior to drilling. Plotted values include a wide range of sandstone compositions and textures. Sandstones from the Mid-Norway area and those from the Norwegian North Sea generally show a dependable and predictable decrease in porosity with depth (compaction) as shown by the straight, projected black line in Fig. 2. The Barents Sea porosity values are lower than the Mid-Norway and North Sea values for comparable depths. These lower values are not due to cementation, but reflect differences in primary compaction. Barents Sea porosities could be produced by an additional burial depth of 500 to 2500 m, depending upon the point selected, and can be used to estimate uplift and erosion. Berglund et al. (1986) also found porosity reduction comparable to an additional burial depth of about 1000 m in the more limited area of the Hammerfest Basin in the Barents Sea. As exhibited in Fig. 2, considerably greater apparent amounts of removal due to uplift are indicated in other areas of the Barents Sea.

Maturation levels

One of the greatest impacts on exploration has been the recognition of the abnormally high thermal maturity of the sediments in the Barents Sea. Similar maturities are reached about 2000 m deeper in Mid-Norway and the North Sea. The depth to a vitrinite reflectance value of 1.35% is shown in Fig. 3 for several wells from the Barents Sea, Mid-Norway, and the North Sea. For the Barents Sea wells shown in Fig. 1, the posted values are from the basic geochemical reports for the respective wells and are simply the depths at which the measured R_o values are 1.35%. Each point in Fig. 3 represents one well. This high level of thermal maturity occurs at shallower depths than expected in the Barents Sea.

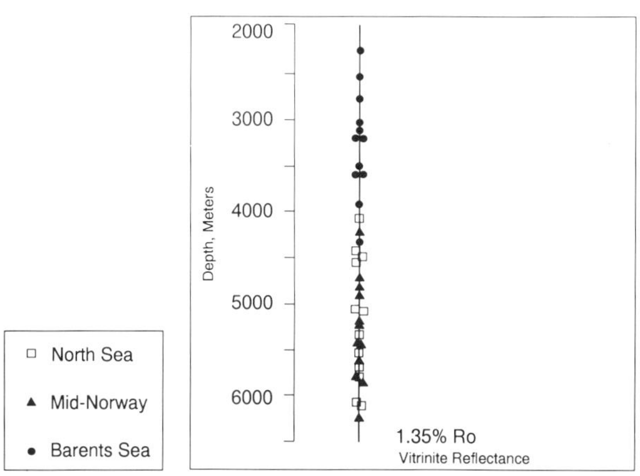

Fig. 3. Depth to a vitrinite reflectance of 1.35% showing a marked difference between the Barents Sea and other Norwegian offshore areas. Each point represents one well. Data are from the geochemistry report for the individual wells.

To contrast the maturity level in the Barents Sea with that in Mid-Norway and the Norwegian North Sea, twelve wells from each of those areas were also plotted in Fig. 3. The Mid-Norway and Norwegian North Sea wells were chosen to represent a broad geographical coverage of those areas. Few wells from either Mid-Norway or the Norwegian North Sea have sediments with measured vitrinite reflectance values as high as 1.35%. Therefore, the vitrinite trend for each well was plotted on semi-log paper, and projected with a straight edge along the line of best fit to a value of 1.35%.

Figure 3 has important implications for petroleum occurrence in the Barents Sea. If the depth at R_o = 1.35% is considered the base of the oil window, the cluster of points from the Barents Sea is about 2000 m shallower than the cluster of the points from Mid-Norway and the Norwegian North Sea. A significant observation is that the base of the oil window is shallower than 3600 m in ten out of the twelve Barents Sea wells used in this study. Both of the Barents Sea wells in which 1.35% vitrinite reflectance was encountered below 4000 m, had low effective temperature gradients, and one of the wells had minimal removal. Uplift and removal must be taken into account in reconstructing the geologic history of this area.

Geophysical indications

Relatively fast interval seismic velocities in the Barents Sea area were noted by geophysicists even before the regional extent of the uplift was recognized. Anomalously high seismic velocities are also associated with water-bottom sediments over some areas of the Barents Sea (J. Smith, Mobil, pers. commun., 1990) and give rise to water-bottom multiples on seismic sections.

High values seen in gravity modelling are also better explained by consideration of uplift of more compacted rocks, resulting in higher densities nearer the surface. Earlier, Mobil in-house gravity modelling, which used normal densities for the present-day depths, found poor fits in an area of substantial uplift along the northwest side of the Loppa High (Fig. 1). The stratigraphic section was more dense than could be explained by a normally compacted section. Higher densities attributed to additional compaction, caused by section now removed by uplift, results in better fits (J. Smith, Mobil, pers. commun., 1990).

To summarize various observations, the abnormal and rapid porosity degradation, the high maturation levels at shallow depths, the faster than normal seismic velocities, and the gravity models which require a more dense stratigraphic section, all suggest uplift and removal of stratigraphic section. In addition, seismic surveys demonstrate that sea-bottom muds which locally rest on almost flat-lying Cretaceous sediments, actually rest with regionally angular unconformity on rocks of various ages. Biostratigraphic data from wells indicate stratigraphic hiatuses of 95–100 Ma. The time gap depends upon the locality sampled, but the missing section is also consistent with considerable uplift and erosion.

Quantification of parameters

Any parameter that varies systematically with depth of burial can be calibrated and used to estimate uplift. Although many calibration techniques can be used, the resulting estimates still have uncertainties. Thus, confidence in estimates of uplift can be increased if several techniques are used and give reasonable agreement.

Vitrinite reflectance modelling

Vitrinite reflectance data ($\%R_o$) are readily available for most Barents Sea exploration wells and provide a useful constraint for basin modelling. Collection of $\%R_o$ data is standard practice for exploration wells, and the data are available by virtue of well trades or by pre-trade agreements which shared basic data between all Norwegian 11th Round and 12B Round participants. As with any measurement, the data need to be carefully evaluated for inconsistencies due to variations in kerogen/maceral type, sampling problems, or analytical procedure before any interpretation is undertaken.

The task is to get the maximum amount of information from the vitrinite data. At a given well location, the plot of $\%R_o$ against depth defines the vitrinite maturation gradient. Because the increase in reflectivity with depth is a temperature dependent and irreversible process, the slope of the vitrinite trend records the approximate maximum effective temperature gradient experienced by the sediments at the well location. The $\%R_o$ data can also be modelled to obtain information about the temperature history at the well location.

Two simple and quick methods for using R_o data to check for uplift are routinely used by operators and researchers working with Barents Sea data. One method involves plotting the vitrinite trend on semilog paper and projecting the trend to a value thought to represent a surface intercept and reading the amount of sediment removal from the graph (Dow, 1977). IKU (1990) used two different intercepts, a general and a Svalbard intercept, for predicting uplift from R_o data in the Barents Sea. The second method involves modelling the vitrinite reflectance by calculating a predicted vitrinite trend based on thermal exposure and matching the predicted trend to the measured vitrinite values. The approach of matching curves rather than projecting to a single hypothetical surface-intercept point appears to be inherently more reliable, and the modelling also allows for the deduction of temperature information from the slope of the vitrinite trend.

Our method of modelling vitrinite reflectance data provides an estimate of the amount of uplift and the effective temperature gradient. Both of these parameters can be further constrained by subsequent methods. We are not advocating the technique for sophisticated basin modelling, but to use the appropriate technique consistent with the accuracy of the result and level of effort required to complete the calculation.

Because not every vitrinite data point is reliable, vitrinite data should be evaluated before use. A critical examination of the histogram of vitrinite values for each individual sample can be made, in order to judge whether the data are statistically reliable, before use. Unreliable points should be eliminated.

Our preferred method of vitrinite modelling involves making reasonable assumptions that cause a predicted vitrinite reflectance curve to match the actual measured vitrinite trend from a given well. The predicted vitrinite reflectance curve is calculated as an option in our in-house basin modelling system and uses a single-component Arrhenius relationship empirically derived from field data (M.B. Ray, Mobil, pers. commun., 1991). Although not rigorously compared, Mobil's empirical relationship seems to give results comparable to the published kinetic model of Burnham and Sweeney (1989) or Sweeney and Burnham (1990) in this geologic setting. Calculations require the input of readily available data, namely a burial history curve and an initial estimate of effective geothermal gradient or temperature history. An initial temperature gradient could be derived from the present-day well temperatures, calculated from thermal conduction modelling and heat flow data, or obtained from regional maps. In any case, the final effective temperature gradient will be further constrained by the trial-and-error matching process described below.

The modelling of vitrinite reflectance involves estimation of the burial history curve for the section missing at an unconformity. A constraint on the estimate is that the combination of temperature (effective geothermal gradient) and burial history must produce the level of maturation indicated by the measured vitrinite reflectance. Two numbers must be determined: one is the effective temperature gradi-

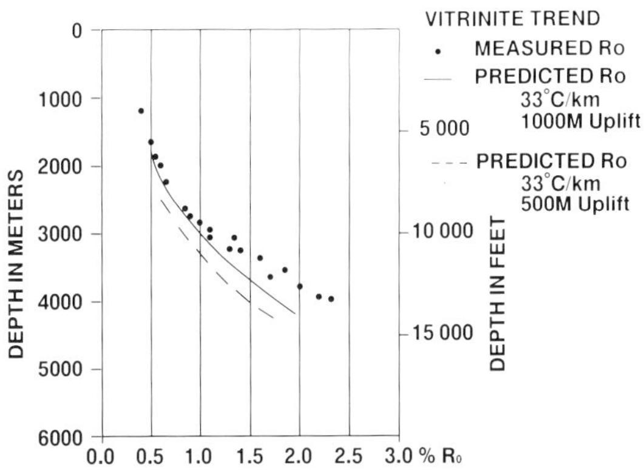

Fig. 4. Edited vitrinite trend of measured values (dots) from well "X" in the Barents Sea. Measured R_o% values are plotted against depth. Dashed line is a calculated vitrinite reflectance trend based on 33°C/km and 500 m of sediment removal. The measured values are more reflective than predicted by the calculation. Solid line is predicted vitrinite trend based on 33°C/km and 1000 m of sediment removal. The solid line almost matches the measured values at shallow depths, but is less reflective than measured values at deeper depths.

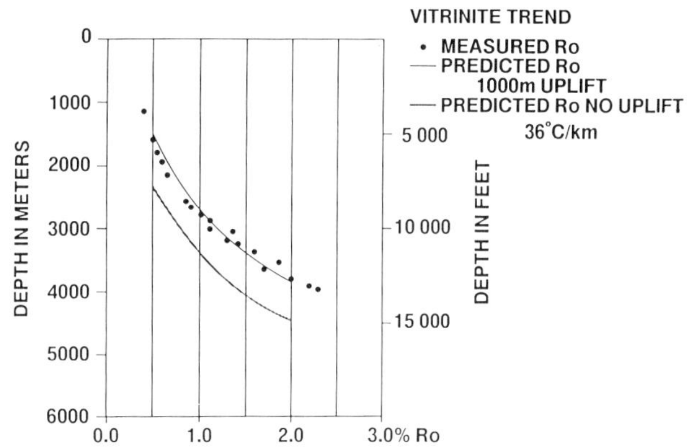

Fig. 5. A reasonable fit between the predicted vitrinite trend and the measured reflectance values was achieved with a geothermal gradient of 36°C/km and 1000 m of uplift. A predicted trend for no uplift and erosion is shown.

ent; the other is the amount of removal or uplift. Briefly, uplift is indicated by the amount of vertical shift required to obtain an optimum match. Final effective geothermal gradient is determined from the slope that gives optimum match. With both slope and magnitude of vitrinite reflectance as variables, a more constrained option exists on the combination of uplift and thermal gradient that can produce a match.

To model the vitrinite trend, it is necessary to start with a reasonably good vitrinite profile such as the one shown as discrete points in Fig. 4 for well "X" from the Barents Sea. The dashed line in Fig. 4 is a predicted vitrinite reflectance curve based on an assumed sediment removal of 500 m and an effective geothermal gradient of 33°C/km. The estimated burial history curve for the missing section beneath the sea-floor muds was assumed to have had a U-shape. That is, the first third of the missing time was spent subsiding during which 500 m of section was deposited. The next third of the time was stillstand. The last third of the time was spent in uplift, when 500 m of section was eroded. Figure 4 shows that the vitrinite is more reflective than predicted by the modelled burial history as shown by the fact that the measured R_o values (plotted as dots) are greater than the calculated R_o values (plotted as the dashed line). The solid line in Fig. 4 is the predicted vitrinite reflectance calculated for 1000 m of uplift and erosion at the same effective geothermal gradient as the example above. The fit with 1000 m of removal (solid line) is better than the fit with 500 m of removal (dashed line). Increasing the removal moved the predicted curve vertically. This produced a fit at shallow depths, but the vitrinite is more reflective than predicted at deeper depths.

The deeper part of the well will be more affected by an increase in temperature than the shallow part. The fit is further improved by using a higher temperature gradient. Figure 5 shows the increased maturation caused by an increase in effective geothermal gradient to 36°C/km. Note that the predicted curve bends more sharply to the right and matches the well data better. This is a reasonable fit for this data set. The above example shows how the vitrinite trend can be used to infer the amount of uplift and the effective geothermal gradient at a given well location.

A regional map of the uplift can be made by posting the uplift results and contouring the values. This preliminary contoured map (not shown) suggests that uplift is greater in the northwest, as much as 3 km in the area of the Stappen High (see also Wood et al., 1989). This preliminary uplift map will be integrated with the interval seismic velocity method as described below.

A regional map of the effective geothermal gradients was also constructed. On this map, spatially related wells were observed to have similar effective geothermal gradients, and thus could be used in a predictive way to evaluate prospects within the same tectonic elements.

%R_o of the Hekkingen Formation

The vitrinite-matching technique described above used the vitrinite trend from the entire well irrespective of age, depth, or formation of the unit which contained the vitrinite to estimate the amount of uplift. If the uplift estimates are valid and internally consistent, relationships should also be apparent for individual formations.

Plots of R_o versus present and inferred maximum depth for the Hekkingen Formation were

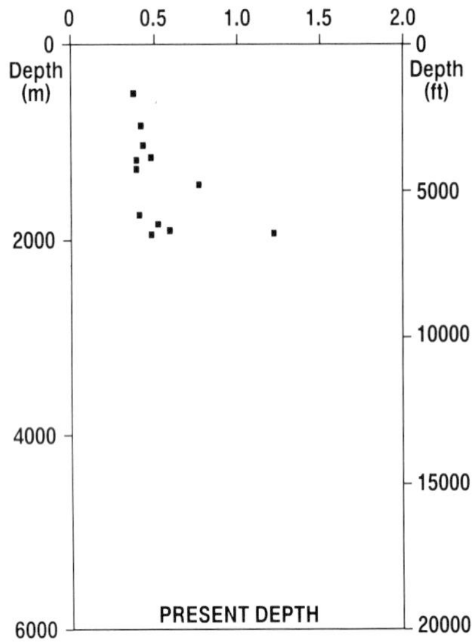

Fig. 6. Average vitrinite reflectance value for the Hekkingen Formation plotted against its present depth. Each point represents one well.

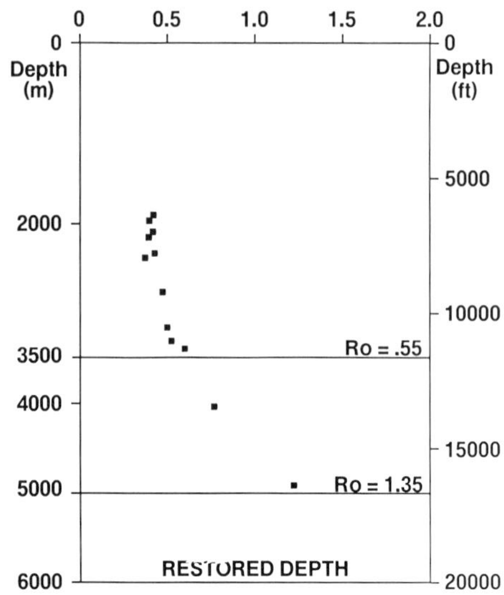

Fig. 7. Average vitrinite reflectance value for the Hekkingen Formation plotted against restored depth (present depth plus estimated uplift) for the same wells as in Fig. 6.

constructed to test the vitrinite-matching technique. These plots show an apparently random distribution (Fig. 6) until the data points are adjusted for uplift (Fig. 7). The Hekkingen Formation is a widespread Late Jurassic source rock in the Barents Sea. It is readily identified in well cuttings and has abundant vitrinite. Measurements from more than 20 particles are relatively easy to obtain from individual samples of this formation. The individual sample histograms are usually characterized by a sharp narrow peak.

Measured R_o values for the Hekkingen Formation are plotted against measured depth for the twelve Barents Sea wells in Fig. 6. Each point represents one well and is the average value for the formation in the respective well. The Hekkingen Formation is widespread, has a uniform age, thickness, and reliable vitrinite reflectance data. However, when plotted against present depth, the resulting raw data show no apparent relationship.

Figure 7 shows the same Hekkingen vitrinite data for the same wells but with calculated uplift added to the formation's present depth to arrive at the restored depth. The restored depth plot (Fig. 7) produces a smooth curve and resembles a vitrinite trend that curves slightly to the right with increasing depth. The plot resembles a typical vitrinite trend from an individual well where reflectivity increases with depth and thermal exposure. Thus, with the uplift taken into account, maturation is shown to be a function of depth (thermal exposure) as would be expected for all sample points in a unit of the same age. This suggests that reasonable predictions about maturation for this formation can be made if the maximum burial of this formation were known.

A simple maximum burial depth/maturation map for the Hekkingen Formation was constructed by adding the uplift map to a seismic depth map for this unit (Fig. 8). The map can be patterned according to the relationship between vitrinite reflectance and thermal maturity with respect to oil generation and destruction shown in Fig. 7. A value of 0.55 R_o% is considered as the beginning of oil generation in the Hekkingen Formation and corresponds to a maximum depth of 3500 m. Where the formation has experienced maximum depth (present depth plus amount of uplift) of less than 3500 m, the formation is thermally immature. If the combined present depth plus uplift is between 3500 and 5000 m, the formation has experienced oil window maturities. Further, if the present depth plus the amount of uplift exceeds 5000 m, the formation is post mature for oil generation. (This example expresses oil window maturity in terms of R_o, but other R_o limits could be chosen.) The accuracy of this map has been evaluated by comparing the prediction against the R_o of all Hekkingen Formation intersections from our Barents Sea well database. The rich, oil-prone Jurassic Hekkingen Formation, is immature in the entire eastern part of the Barents Sea, is within a window of oil maturation along the western edge of the Hammerfest Basin and along the western edge of the Loppa Ridge, and is gas-mature in the Tromsø and Bjørnøya Basins. This severely limits the distribution of viable oil prospects that can be Hekkingen-sourced. This observation has been further supported by oil–oil source correlation studies conducted independently (in-house studies).

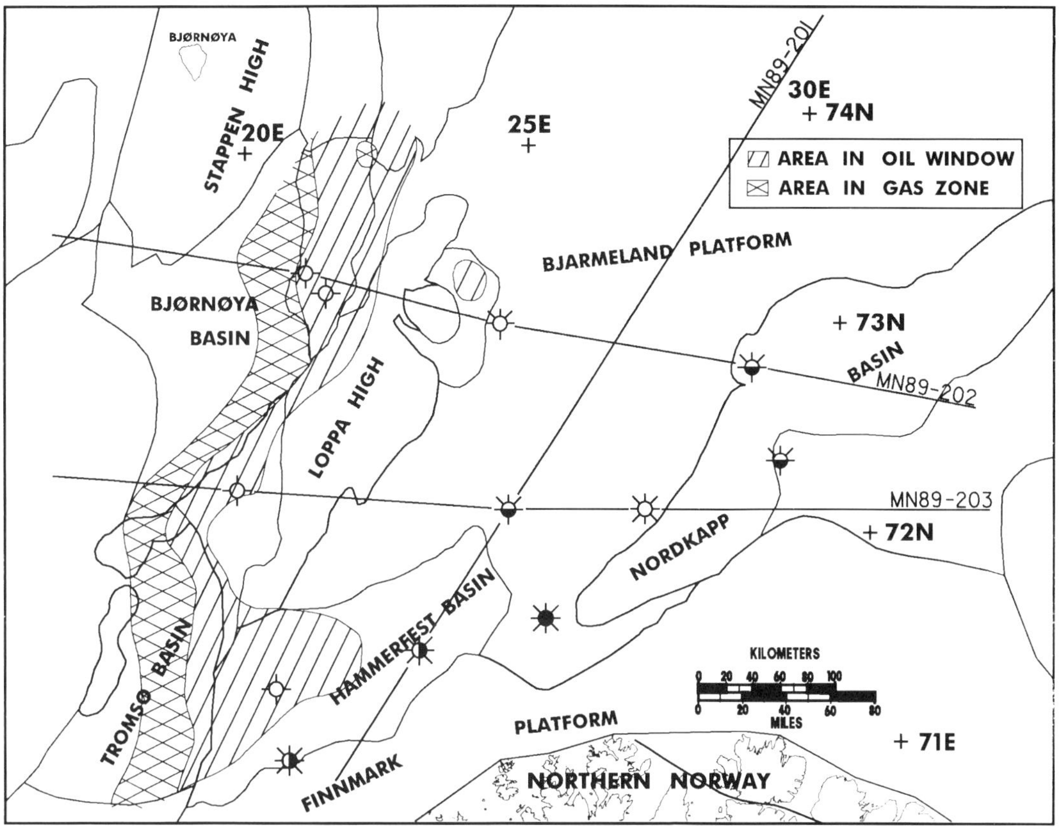

Fig. 8. Hekkingen Formation maturity map based on relationship in Fig. 7. The organic-rich oil-prone, Hekkingen Formation is immature in the entire eastern part of the Norwegian Barents Sea.

Apatite fission track analysis

Apatite fission track analysis (AFTA) is a technique which enables the determination of the magnitude of maximum palaeotemperatures and the time of cooling from maximum palaeotemperatures. AFTA thus provides timing information that is not available from vitrinite reflectance interpretation. AFTA is considered one of the best tools for separating and distinguishing multiple episodes of uplift that have been superimposed on a single unconformity surface. The basis of the technique and kinetic modelling methods have been described in a series of papers (Green et al., 1986, 1989a; Laslett et al., 1987; Duddy et al., 1988) with details of the use of the technique in Green et al. (1989b).

Maximum palaeotemperatures up to about 120°C (for average heating rates) can be determined using AFTA and thus are complementary with temperatures derived from vitrinite reflectance data up to this value. Variation of palaeotemperature with depth for a suite of well samples determined by either technique may be used to constrain the geothermal gradient at the time of maximum palaeotemperatures. In circumstances where a large degree of cooling has occurred, vitrinite reflectance will give better control on the geothermal gradient because vitrinite reflectance allows palaeotemperatures greater than 120°C to be determined. For example, Duddy et al. (1991) used the kinetic model of vitrinite reflectance published by Burnham and Sweeney (1989) (as implemented in the BasinMod software package of Platte River Associates) to determine geothermal gradients for a series of uplifted sequences with the time of cooling determined by AFTA. Thus, both the vitrinite and apatite-based techniques are more powerful when used together.

The exact burial history curve is unknown for missing section at unconformities. However, a constraint is that the burial history curve must produce the level of maturation as indicated by the vitrinite reflectance. In the vitrinite-matching models for well data, simplified assumptions for single-event burial and removal histories can be refined with integrated data.

Burial history curves predicted by apatite fission track analysis of material from Barents Sea wells suggest two stages of uplift (the first at roughly 65 Ma and the younger in the Neogene). Figure 9 shows the predicted vitrinite trends based on calculations for various burial history curves for well 7321/7-1. Both the single U-shaped unconformity burial history model, and the two-stage uplift model based on

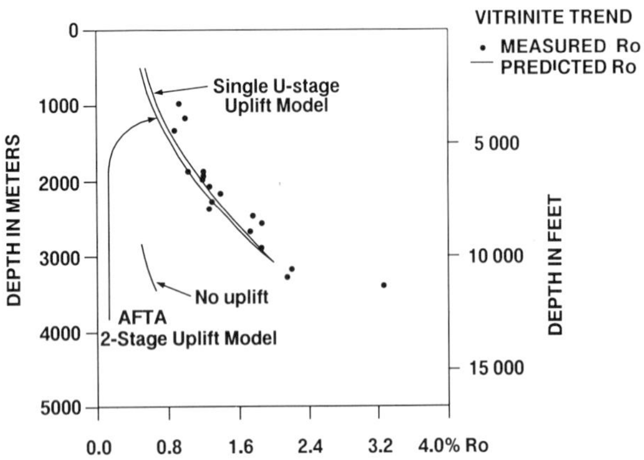

Fig. 9. Fit to vitrinite data using various uplift models. The two stages of uplift as indicated by apatite fission track analysis fits the vitrinite reflectance data equally well as the single U-shaped model described in the text. The predicted model with no uplift does not fit the measured vitrinite trend.

apatite fission track data fit the measured vitrinite data equally well. In contrast, a predicted vitrinite trend is also shown for a model with no uplift, which does not fit the measured vitrinite data. Tectonic analogy with other rifted-margin settings and changes in plate motion seen in the deep sea-floor magnetic anomaly record are also consistent with two stages of uplift.

Interval seismic velocity analysis

Mobil Exploration Norway used the Mobil Search seismic vessel to acquire three long, regional seismic lines averaging 500 km each (Fig. 1). One of their greatest values was simply to present a clearer regional picture. These lines were compressed 10 to 1 horizontally, and spectacularly enhance structures and basins. A second value of the regional seismic lines was that they provided a long line collected under the same parameters for seismic velocity analysis.

Porosity loss with depth of burial and the associated increase in interval velocity are commonly observed phenomena in sedimentary basins and have been known for a long time (Faust, 1951; Gassmann, 1951). Wyllie et al. (1956) demonstrated that interval velocities are inversely proportional to porosity. Therefore, a formation that has been uplifted and unloaded by erosion of the overlying sedimentary cover will have higher interval velocities due to the previously greater depths of burial. The seismic velocities constitute a dense grid of observations and are the only data available to connect estimations of uplift between the wells in the Barents Sea.

The regional estimation of uplift from seismic interval velocities is based on an empirical calibration of constants to uplift results from vitrinite studies in the wells. The assumptions for the calculations are: same lithology, pressure regime, and pore fluid composition throughout the area considered. In this analysis the gross geological interval from Base Jurassic to Base Smithian (Lower Triassic) has been used to estimate the amount of uplift from seismic interval velocities. The uplift estimated from this interval will only infer post-Jurassic maximum depth of burial. The maximum depth of burial estimated from seismic interval velocities is given by the expression (under the assumptions described above):

$$Z_{max} = (a \times V_{int})^p \tag{1}$$

where Z_{max} = maximum depth of burial of the Triassic megasequence; a and p are constants; V_{int} = interval velocity of Triassic megasequence.

$$\text{Apparent uplift} = Z_{max} - Z_{actual} \tag{2}$$

where Z_{actual} = actual depth of the Triassic megasequence.

This empirical formula was developed and used to estimate uplift in some areas of the Barents Sea. The equation applies to the Middle and Lower Triassic sequence and has been calibrated from the uplift at Barents Sea well locations where we were confident of the uplift value. With this formula (shown near the base of Fig. 10) uplift variations along the seismic line can be identified. Results for line MN89-201 (Fig. 10) are shown for the individual points as small +- signs and as the solid line which shows a three-point rolling mean.

Although the formula was calibrated at one well location, the uplift numbers calculated by the seismic interval velocity formula show general agreement at the other well locations in that part of the Barents Sea and provide support for use in other areas. The interval velocity method does not, however, agree

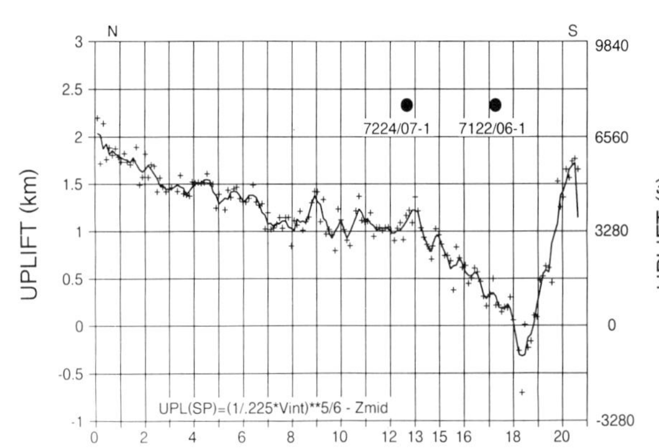

Fig. 10. Uplift profile plotted along the shot points (*SP*) of seismic line MN89-201. Individual points are shown as small +-signs. Variations in the data are smoothed by the solid line which represents a three-point rolling mean.

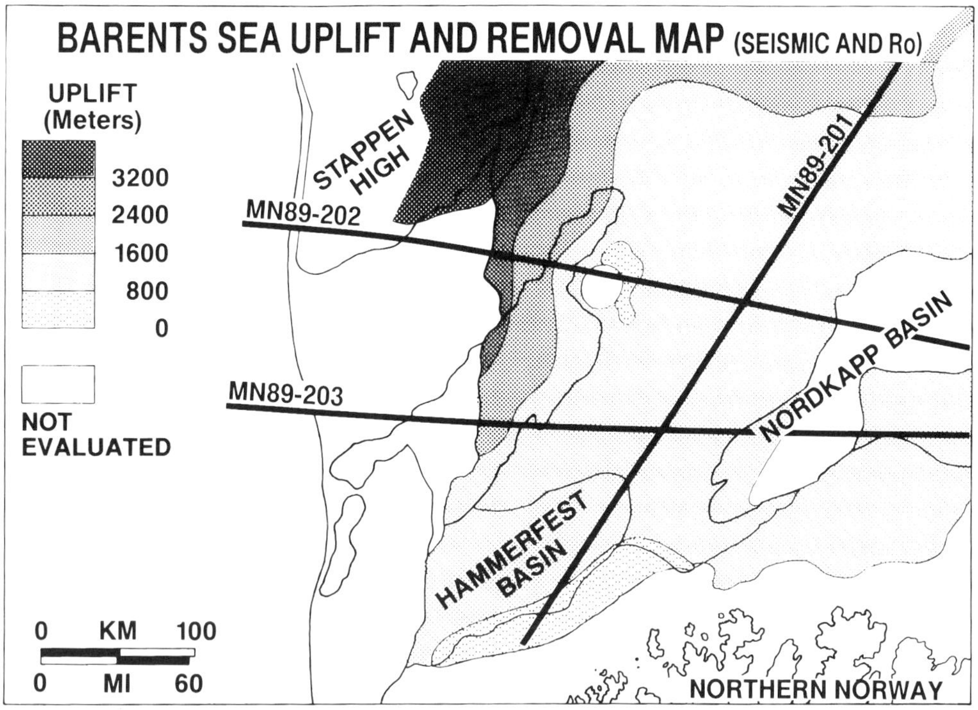

Fig. 11. Uplift map for the Barents Sea. Individual well locations were calculated by the vitrinite reflectance method. After calibrating at one location, the map was extended by the interval velocity method over a large grid of seismic lines. Points on the map have been compared. General agreement exists between the two approaches.

with the vitrinite-matching estimates in the Hammerfest Basin because of regional changes in properties such that they are no longer consistent with the assumptions. Nevertheless, the uplift map (Fig. 11) shows general agreement between uplift calculated by the interval velocity formula and vitrinite reflectance matching as discussed earlier.

Evaluation of prospects

All of the observations and quantifications reported in previous sections of this paper are used in basin modelling for prospect evaluation. The following example will discuss a 1D case and will emphasize the use of amount of uplift/removal and effective temperature gradient as input parameters for basin modelling directed toward prospect evaluation. First, the amount of uplift is estimated from the uplift and removal map (Fig. 11) or can be closely estimated from seismic velocities as in Fig. 10. Next, an effective geothermal gradient is estimated for use in the maturation modelling program. Knowing both the most likely uplift value and the most likely effective geothermal gradient allows prediction of the probable maturation of the prospect.

Basin modelling was applied to an example prospect interpreted as a Permian carbonate-reef complex. Depth to the reservoir is estimated to be 4300 m. An important question for this prospect, in view of the shallow depth of oil destruction in the Barents Sea, is whether the expected in-place hydrocarbons will be oil or gas. Various kinetic models were used to evaluate the expected source rock at the top of the Permian section and all have shown it to be post-mature for oil generation. The remaining question is whether the generated hydrocarbons at the reservoir level are cracked to gas within the prospect.

The oil deadline was calculated within our basin modelling system based on the input values of burial history (including uplift) and effective geothermal gradients. Other parameters are available from the seismic data and include interpretation, mapping, and depth conversion. For a model based on an effective geothermal gradient of 36°C/km and sediment removal (uplift) of 1000 m, the oil deadline is calculated to be 3700 m. Because 3700 m is shallower than the reservoir depth of 4300 m, only gas is expected in the prospect (Fig. 12).

An evaluation should also be made of the effect of input uncertainties on the modelling results. Sensitivity analysis indicates how the variation of input parameters changes the expected deadline result. If the effective geothermal gradient was 33°C/km, and the uplift was only 500 m, the calculated deadline for these assumptions is 4440 m. Because the reservoir is now above the deadline in this case (4300 m as opposed to a deadline of 4440 m) oil can now be preserved (Fig. 12).

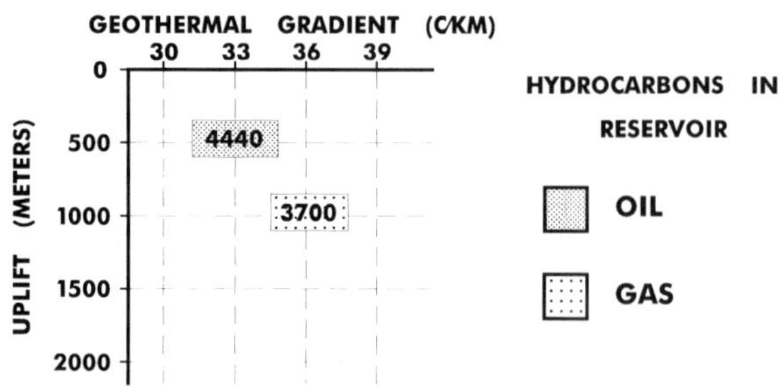

Fig. 12. Example of evaluating a Permian carbonate reservoir at a present depth of 4300 m. If the effective geothermal gradient is 36°C/km as indicated by regional maps and if the uplift is 1000 m, the oil deadline calculation indicates that oil would have been cracked below a present depth of 3700 m and gas would be expected in the reservoir. If the effective geothermal gradient is 33°C/km and the uplift is only 500 m, oil could exist above a present depth of 4440 m, and hence at the reservoir depth.

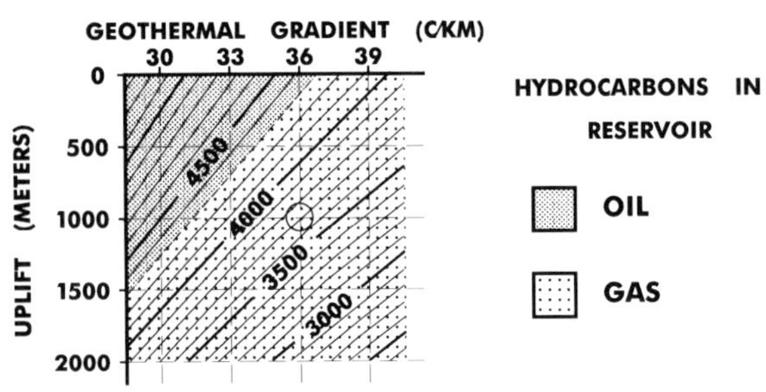

Fig. 13. An entire range of input parameters can be viewed in this figure and provides a feel for the interdependence of the input parameters and the sensitivities of different inputs. A lower temperature gradient or a smaller uplift would preserve oil to a greater depth. The judgment about oil versus gas was made relative to a reservoir at 4300 m. Therefore, the dashed line separating the patterns is the reservoir depth of 4300 m. The circle at the intersection of 36°C/km and 1000 m indicates the input parameters from regional maps and provides a "best guess" that gas will be the reservoired phase.

Figure 12 shows only the two examples cited above, whereas in Fig. 13 the entire figure is patterned either for oil or for gas based on the deadline calculation from basin modelling. The figure was made by calculating the node values and contouring the results. The judgment about oil versus gas was made relative to a reservoir at 4300 m. Therefore, the line separating the patterns is the reservoir depth of 4300 m. The circle at the intersection of 36°C/km and 1000 m uplift indicates the best estimate of input parameters from the regional maps. An entire range of input parameters can be viewed in this figure and provides a feel for the interdependence of the input parameters and sensitivities of different inputs. A lower temperature gradient or a smaller uplift number will preserve oil to a greater depth.

The above sensitivities only considered the variables of uplift and effective geothermal gradient. Other parameters that could be considered include seismic depth conversion (is the reservoir really at 4300 m?), age, and various kinetic models to name a few. The effects of variations in all of these parameters could also be risked by Monte Carlo techniques if warranted.

3D basin modelling

Recent advances in Mobil's in-house modelling capabilities allowed interactive sensitivity analyses based on multiple 3D maturity runs. Mapped structural, isopach, isochron, and temperature input were accommodated as arrays of either constant or variable values. The arrays were checked to ensure that surfaces stack appropriately in depth and time before maturity calculations were begun. Results were displayed for discrete areas and time intervals depending upon area and time frame of interest.

"Most likely" maturity results for the Hekkingen Formation in the Hammerfest Basin supported the general interpretation shown in Fig. 8. Burial histories for the most likely case assumed that maximum subsidence occurred prior to episodic uplift and erosion. Extensive input included arrays for bathymetry, structural surfaces for key shallow and deep horizons, removal and age of section now missing due to erosion, and temperature gradient estimates through time. Several standard seismic maps expressed in depth were included in the input process.

Other 3D modelling results were included but were not limited to arrays of palaeostructural and palaeotemperature data. Oil and gas volumetric estimations were calculated using specified inputs for thickness, organic richness, and migration threshold.

Summary

Vitrinite reflectance, sandstone porosity, and seismic velocities in the explored areas of the Barents Sea indicate greater depths of burial in the past, compared with similar data from oil fields of Mid-Norway and the North Sea. Once the removal (uplift) problem in the Barents Sea was identified, an effort was made to evaluate and quantify the potential effects through basin modelling. This was done by integrating several techniques as constraints on the burial history.

Apatite fission track analysis enables the separation of episodes of uplift in time and space, and provides an estimate of the amount of uplift. However, apatite fission track analysis requires the knowledge of temperature information or effective temperature gradient. Vitrinite reflectance data give information on total thermal exposure and on effective thermal gradient but require information on timing and burial history curves. Vitrinite reflectance gives a summary value for thermal exposure regardless of the number of events that contributed to that exposure.

Vitrinite reflectance, together with apatite fission track analysis, have the limitation that samples are derived from wells. Extrapolation between wells and evaluation of undrilled prospects is difficult. Seismic interval velocity analysis has the advantage that it can provide data and trends between wells, and can be used to extrapolate the other types of data.

All observed data and analytical methods which constrain basin simulations must be integrated and tested for internal consistency. Many other types of constraints (e.g., oil characterization and oil–source rock correlation) should also be evaluated as additional checks on the reliability of model predictions. Finally, our ability to determine "the answer" is inexact because our observations and understanding of underlying processes is incomplete. All "answers" must be viewed and risked in the context of a range of potentially correct solutions. Decisions based on model simulation reflect uncertainties associated with variable inputs and results.

Acknowledgements

This paper is the result of a true team effort by the Barents Sea Exploration Team supervised by O.K. Johansen to integrate data and to rank prospects in the Barents Sea. E.M. Leavitt, J.W. Snedden, G.E. Claypool, I.R. Duddy, and D.W. Waples have commented on various versions of this manuscript. Numerous colleagues commented and offered encouragement during the presentation and preparation of this paper. We thank them all but retain responsibility for errors that have remained.

Since the presentation of this paper, Mobil Exploration Norway Inc. has received permission from Statoil and Conoco to disclose the well name, 7321/7-1, for our apatite fission track results in Fig. 9. The drafting department at Mobil Exploration Norway Inc. prepared the illustrations.

We thank Mobil Exploration Norway Inc. for encouragement to participate in conferences and to publish. We also thank Mobil Oil Corporation for approval to publish.

References

Berglund, L.T., Augustson, J., Færseth, R., Gjellberg, J. and Ramberg-Moe, H., 1986. The evolution of the Hammerfest Basin. In: A.M. Spencer (Editor), Habitat of Hydrocarbons on the Norwegian Continental Shelf. Norwegian Petroleum Society, Graham and Trotman, London, pp. 319–354.

Burnham, A.K. and Sweeney, J.J., 1989. A chemical kinetic model of vitrinite reflectance maturation. Geochim. Cosmochim. Acta, 53: 2649–2657.

Dow, W.G., 1977. Kerogen studies and geological interpretations. J. Geochem. Explor., 7: 79–99.

Duddy, I.R., Green, P.F. and Laslett, G.M., 1988. Thermal annealing of fission tracks in apatite, 3. Variable temperature behaviour. Chem. Geol. (Isot. Geosci. Sect.), 73: 25–38.

Duddy, I.R., Green, P.F. and Bray, R.J., 1991. The application of apatite fission track analysis to thermal history reconstruction in basin modelling. Norwegian Petroleum Society, Stavanger, p. 59 (abstract).

Faust, L.Y., 1951. Seismic velocity as a function of depth and geologic time. Geophysics, 16: 192–206.

Gabrielsen, R.H., Færseth, R.B., Larsen, L.N., Kalheim, J.E. and Riis, F., 1990. Structural elements of the Norwegian continental shelf, Part I. The Barents Sea Region. Norw. Pet. Direct., Bull. 6, 33 pp.

Gassmann, F., 1951. Elastic waves through a packing of spheres. Geophysics, 16: 673–685.

Green, P.F., Duddy, I.R., Gleadow, A.J.W., Tingate, P.R. and Laslett, G.M., 1986. Thermal annealing of fission tracks in apatite, 1. A qualitative description. Chem. Geol. (Isot. Geosci. Sect.), 59: 237–253.

Green, P.F., Duddy, I.R., Laslett, G.M., Hegarty, K.A., Gleadow, A.J.W. and Lovering, J.F., 1989a. Thermal annealing of fission tracks in apatite, 4. Quantitative modelling techniques and extension to geological timescales. Chem. Geol. (Isot. Geosci. Sect.), 79: 155–182.

Green, P.F., Duddy, I.R., Gleadow, A.J.W. and Lovering, J.F., 1989b. Apatite fission track analysis as a paleotemperature indicator for hydrocarbon exploration. In: N.D. Naeser and T. McCulloh (Editors), Thermal History of Sedimentary Basins — Methods and Case Histories. Springer-Verlag, New York, N.Y., pp. 181–195.

IKU, 1990. Barents Sea uplift and erosion in Cenozoic time. IKU News, IKU Sintef Group, Trondheim, 4: 3.

Laslett, G.M., Green, P.F., Duddy, I.R. and Gleadow, A.J.W., 1987. Thermal annealing of fission tracks in apatite, 2. A quantitative analysis. Chem. Geol. (Isot. Geosci. Sect.), 65: 1–13.

Norsk Geologisk Forening, 1990. Post-Cretaceous uplift and sedimentation along the Western Fennoscandian Shield (abstract). NGF's Tectonic and Structural Geology Studies Group, TSGS 7th Annu. Meet., Stavanger (unpaginated program with abstracts).

Norwegian Petroleum Society, 1989. Abstracts, Workshop on Structural and Tectonic Modelling and its Application to Petroleum Geology, 18–20 October 1989, Stavanger.

Norwegian Petroleum Society, 1990. Abstracts, International Conference on Arctic Geology and Petroleum Potential, 15–17 August 1990, Tromsø. Norwegian Petroleum Society and University of Tromsø, 72 pp.

Nyland, B., Jensen, L.N., Skagen, J.I., Skarpnes, O. and Vorren, T., 1992. Tertiary uplift and erosion in the Barents Sea: magnitude, timing and consequences. In: R.M. Larsen, H. Brekke, B.T. Larsen and E. Talleraas (Editors), Structural and Tectonic Modelling and its Application to Petroleum Geology. Norwegian Petroleum Society (NPF), Special Publication 1. Elsevier, Amsterdam, pp. 153–162.

Olsen, R.G. and Linjordet, A., 1990. Snøhvit — a sleeping beauty, giant oil and gas fields of the decade 1978–1988. Norwegian Petroleum Society, Stavanger, p. 13 (abstract).

Skagen, J., 1990. Effects on hydrocarbon potential caused by Tertiary uplift and erosion in the Barents Sea. Abstr., Int. Conf. Arctic Geology and Petroleum Potential, 15–17 August

1990, Tromsø. Norwegian Petroleum Society, and University of Tromsø, p. 31a.

Sweeney, J.J. and Burnham, A.K., 1990. Evaluation of a simple model of vitrinite reflectance based on chemical kinetics. Am. Assoc. Pet. Geol. Bull., 74: 1559–1570.

Wood, R.J., Edrich, S.P. and Hutchinson, I., 1989. Influence of North Atlantic tectonics on the large-scale uplift of the Stappen High and Loppa High, western Barents Shelf. In: A.J. Tankard and H.R. Balkwill (Editors), Extensional Tectonics and Stratigraphy of North Atlantic Margins. Am. Assoc. Pet. Geol. Mem., 46: 559–575.

Wyllie, M.R.J., Gregory, A.R. and Gardner, G.H.F., 1956. Elastic wave velocities in heterogeneous and porous media. Geophysics, 21: 41–70.

N.J. THEIS *Mobil Exploration Norway Inc., P.O. Box 510, N-4001 Stavanger, Norway*
H.H. NIELSEN *Mobil Exploration Norway Inc., P.O. Box 510, N-4001 Stavanger, Norway*
J.K. SALES *Mobil Exploration Norway Inc., P.O. Box 510, N-4001 Stavanger, Norway*
G.J. GAIL *Mobil Exploration and Producing Services, Inc., P.O. 650232, Dallas, TX 75265, U.S.A.*

Mineralogical and diagenetic control on the thermal conductivity of the sedimentary sequences in the Bjørnøya Basin, Barents Sea

Willy Fjeldskaar, Edward Prestholm, Claudia Guargena and Martin Stephenson

The thermal conductivity of the different formations in the Bjørnøya Basin, Barents Sea has been estimated on the basis of a large database of measurements of vertical and horizontal thermal conductivity, mineralogy and porosity of standard core plugs. The thermal conductivities predicted by partial least square regression fall in the range 2.8–3.0 W/m K for the shales, with a significant anisotropy. The shale conductivity seems higher than is common in the North Sea. The thermal consequence of using North Sea shale conductivity values for the Bjørnøya Basin shales is dramatic: up to 40°C difference in the calculated temperature regime. The importance of knowing exactly the shale mineralogy when estimating the thermal conductivity is thus evident.

The most important diagenetic process with significant impact on the thermal conductivities in the Barents Sea is probably the transformation of Paleocene/Eocene opal to quartz. The corresponding increase in the thermal conductivities is approximately 100%. It is therefore also important to account for diagenetic factors in the study of palaeotemperatures.

Introduction

The palaeotemperature regime is considered important for many reasons. The temperature distribution of a sedimentary basin controls the generation, expulsion and migration of hydrocarbons, it governs the rate of diagenetic reactions, such as the transition from opal to quartz, and it controls the viscosity of the oil.

For a successful reconstruction of the palaeotemperature in a sedimentary basin it is imperative that the thermal properties of the rocks are known in detail. In some cases fluid flow will cause significant temperature perturbations (Domenico and Palciauskas, 1973). In offshore sedimentary basins located on the Norwegian Shelf, however, the effect of fluid flow on the temperature distribution is often regarded as insignificant (Hermanrud, 1986). One of the most important parameters for determining the temperature regime of an offshore sedimentary basin is the thermal conductivity of sedimentary rocks.

The main purpose of this paper is to present some aspects of the thermal conductivity of the sedimentary sequences in the Bjørnøya Basin, and their significance for the temperature distribution.

Geological setting

The Bjørnøya Basin, located at 70°30′–74°N and 18°–21°E, trends NE–SW and is separated from the Loppa High to the southeast by the Bjørnøyrenna Fault Complex, to the east by the Leirdjupet Fault Complex and is bounded to the northwest by the Stappen High (Fig. 1). To the west, the Sørvestnaget Basin is separated from the Bjørnøya Basin at a Jurassic level by a deeply buried high, with the Base Cretaceous estimated at more than 9 s TWT (Gabrielsen et al., 1990), and by a system of normal faults in the Tertiary section. At Cretaceous levels the southern Sørvestnaget Basin appears to be a continuation of the Bjørnøya Basin and both basins are characterized by large thicknesses of Cretaceous sediments, whereas in the northern part of the Sørvestnaget Basin, in the Vestbakken volcanic province, the age of the sediments is uncertain due to thick Paleocene lavas.

The Bjørnøyrenna Fault Complex forms the transition from the deep Bjørnøya Basin to the Loppa High where Triassic sediments subcrop the Quaternary. The fault complex consists of large rotated fault blocks of Triassic to Middle Jurassic sediments considered to represent an "early" syn-rift sequence overlain by "late" syn-rift wedges of Middle Jurassic to Early Cretaceous age and a post-rift sequence

Basin Modelling: Advances and Applications edited by A.G. Doré et al. NPF Special Publication 3, pp. 445–453, Elsevier, Amsterdam.

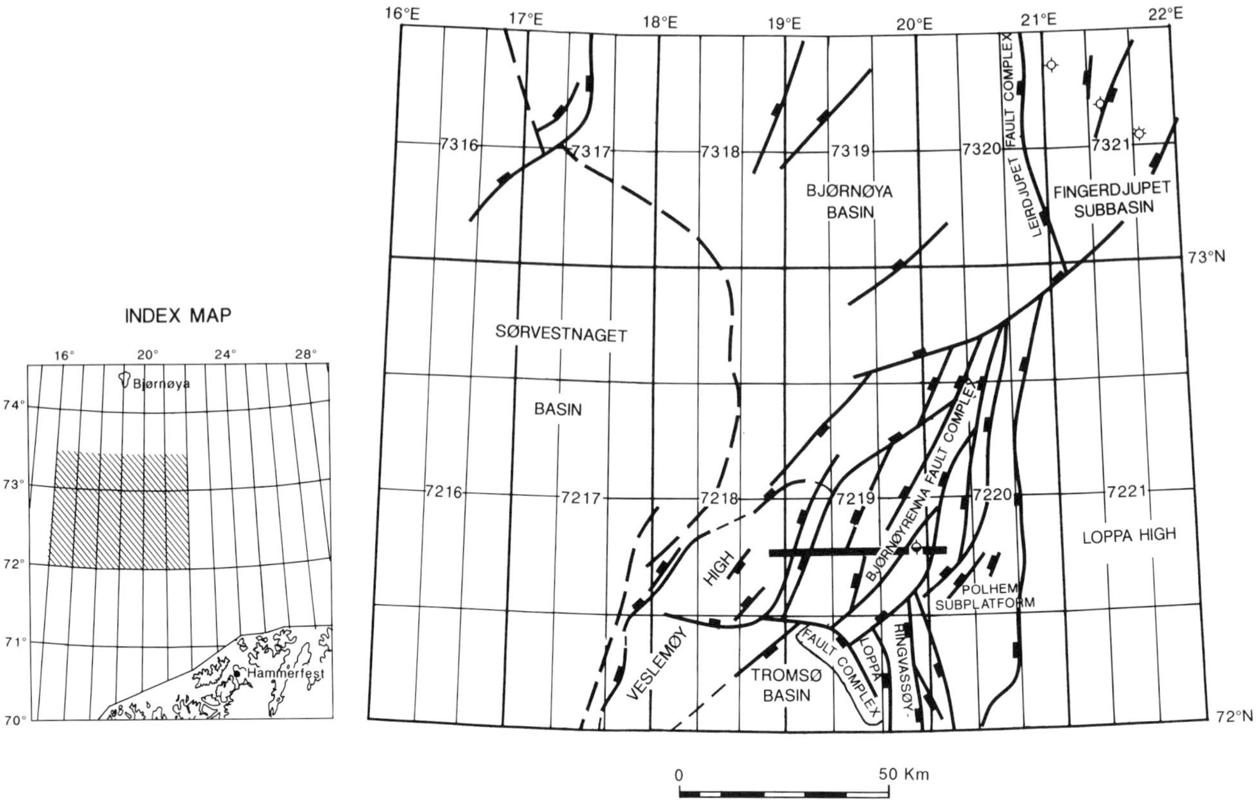

Fig. 1. Tectonic map of the Barents Sea. The seismic line selected for this study is shown by the E–W-oriented thick black line.

deposited from Early Cretaceous to present, representing the basin fill (Fig. 2). The seismic interpretation suggests that in the central parts of the Bjørnøya Basin, as in the Tromsø and Hammerfest basins (Prestholm, 1985), vast thicknesses of dominant shales accumulated during Early Cretaceous times, whereas the margins, over the Bjørnøyrenna Fault Complex (Fig. 2), received a rather thin sequence. However, this sequence also shows evidence of uplift and erosion.

Marginal marine reservoir sandstones are predicted at the Mid-Jurassic level, deposited during the "early" syn-rift tectonic phase and encountered in most explored areas of the western Barents Sea. Prograding clastic wedges can be identified in the seismic data at Late Jurassic, Cretaceous and Tertiary levels building out to the northwest from the Loppa High during the "late" syn-rift and post-rift tectonic phases. In the Bjørnøyrenna Fault Complex area these sediments occur at depths where reasonable reservoir qualities might be expected. The "late" syn-rift phase controlled the development of the primary oil-generating source rocks in the area.

Two-dimensional modelling

The modelling was carried out by using the two-dimensional basin modelling system developed at Rogaland Research (BMT: Basin Modelling Toolbox; Fjeldskaar et al., 1990). The depth-converted section of the Bjørnøya Basin selected for the modelling is shown in Fig. 2. The stratigraphy of the modelled profile is tied to well data.

There was severe erosion during the latest stage of the basin evolution, possibly up to 1.5 km, probably connected to the glaciations in the area during the last 2–3 m.y. (Riis and Fjeldskaar, 1992). The erosion and uplift of the area have strongly influenced the porosity distribution of the basin, and this was taken into account when the basin porosities were estimated (Fig. 3). The porosities have been calculated using the assumption of steady state porosity (according to the method described by Sclater and Christie, 1980). In this particular study the chosen values for the surface porosity of the lithologic units (shale 0.34, sandstones 0.38, carbonate 0.40) and their decay constant (shale 0.40, sandstones 0.30, carbonate 0.40) assume 1.5 km erosion.

The erosion has also influenced the temperature of the basin, causing a thermal non-equilibrium at present. This erosional effect has to be considered if a realistic reconstruction of the temperature regime is desired. In this paper, however, we do not aim at a precise temperature estimate in the basin, but rather the use of temperature calculations as a sensitivity test of the thermal conductivities. The present tem-

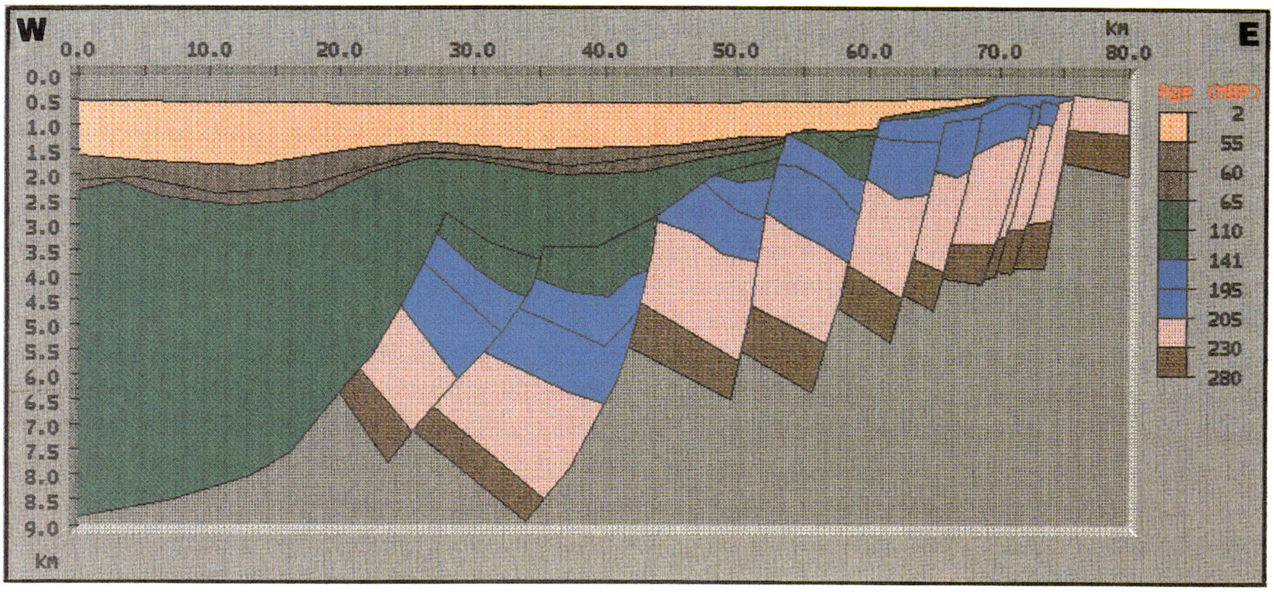

Fig. 2. Depth-converted seismic section through the Bjørnøya Basin.

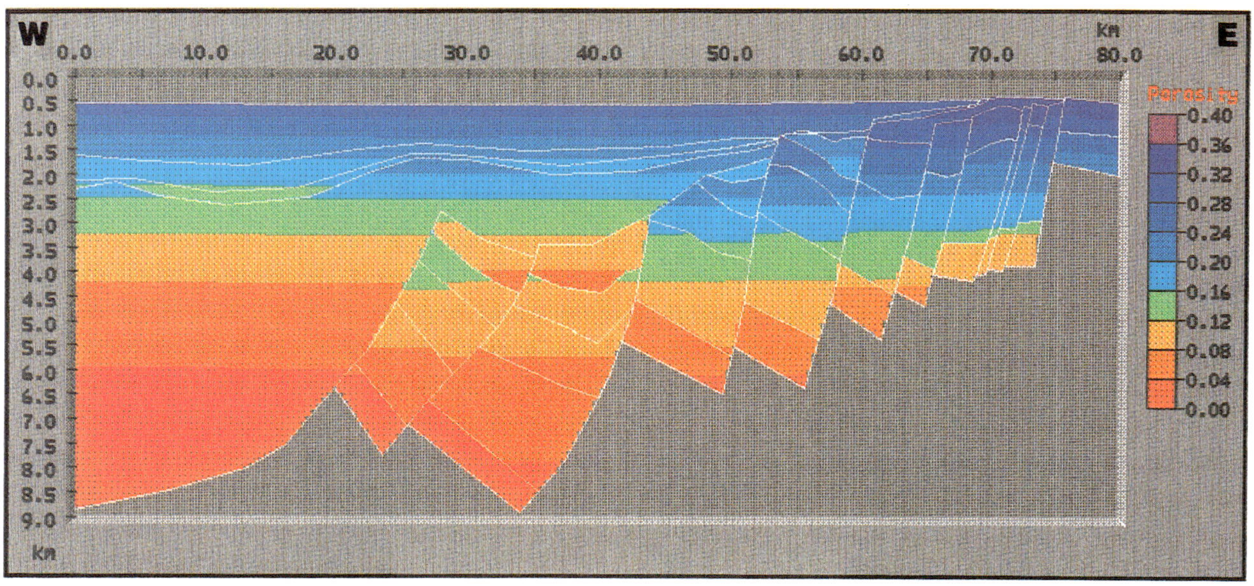

Fig. 3. Theoretical porosities in the Bjørnøya Basin, based on the given porosity–depth trends.

peratures in the Bjørnøya Basin will be higher than estimated by a steady state model, because of the late erosional event.

Thermal conductivities of sediments

The relation between mineralogical composition, porosity and thermal properties of rocks was studied using our in-house data base of measurements of the thermal conductivities of sedimentary rocks of different mineralogies, containing more than 270 rock samples from the Norwegian Continental Shelf (Prestholm et al., 1991). The thermal conductivities (in horizontal (K_h) and vertical (K_v) direction) were measured on standard core plugs by a needle-probe technique at the University of Aarhus (Kristiansen, 1982); the mineralogical composition was determined by XRD and thin section petrography, and porosity was measured by He-injection.

Quantification of the relation between the mineralogy/porosity of sedimentary rocks and the thermal conductivities was carried out by partial least-square regression. The measured content of each mineral, the TOC content, the porosity and the thermal conductivities (K_v and K_h) are variables in the regression (Table 1). The regression analysis establishes a quantified relationship (a calibration model) between the two sets of variables: (1) mineralogy/porosity, and (2) the thermal conductivities (K_v and K_h).

TABLE 1

The data base variables

1. Quartz
2. Alkali feldspar
3. Plagioclase
4. Mica
5. Chlorite
6. Illite/smectite
7. Kaolinite
8. Calcite
9. Dolomite/ankerite
10. Siderite
11. Pyrite
12. TOC
13. Porosity
14. Vertical conductivity, K_v
15. Horizontal conductivity, K_h

The calibration model is then used to predict unknown thermal conductivities once the mineralogical composition and porosity of a rock have been established (for more details, see Fjeldskaar et al., 1990).

In this study the predicted thermal conductivities were applied to study how mineralogical composition controls the present-day temperatures in the Bjørnøya Basin. The present-day temperatures were calculated applying a two-dimensional steady state finite difference conduction model (Fjeldskaar et al., 1990).

Mineralogical control on the thermal conductivities of the Bjørnøya Basin sediments

The mineralogical composition of the formations in the Bjørnøya Basin is mainly predicted from Barents Sea measured samples, well data and general information from the area (Table 2). Multivariate predictions based on this knowledge indicate that the thermal conductivities of the sandstones in the Bjørnøya Basin are 4.3–4.8 W/m K. These values are similar to standard values of thermal conductivities for quartz-rich sandstones (Brigaud et al., 1990).

The thermal conductivity predictions for the Barents Sea shales, however, reveal some important factors:

(1) The vertical thermal conductivities of the Barents Sea shales range from 2.8 to 3.0 W/m K (Fig. 4), which are significantly higher than for the North Sea shales (see Fjeldskaar et al., this volume). The mineralogical effect on the thermal conductivities is clearly seen in the Cretaceous sequence when using North Sea shale mineralogy with low quartz content (20%) and high content of kaolinite (40%) in the modelled profile (Fig. 5). The difference in vertical conductivity of up to 1.0 W/m K is mainly explained by the high content of thermally conductive quartz minerals in the Barents Sea shales.

The significance of the mineralogy effect is obvious when the steady state temperature regime is calculated. The calculated temperatures, keeping the boundary conditions constant (ocean bottom temperature of 5°C, basement heat flow of 78 mW/m^2, and basement conductivity of 3.0 W/m K) is significantly different for the two thermal conductivity models (Figs. 6 and 7): a difference of up to 40°C at 3 km depth.

(2) The predictions show that there are significant differences in the thermal conductivities of shales in the horizontal and the vertical directions, with an anisotropy (K_h/K_v) of up to 1.3 (Fig. 8). In clay minerals heat is transported more efficiently in the horizontal than the vertical direction, implying that the measured thermal conductivity anisotropy is mainly due to the predominant horizontal orientation of clay minerals in shales (Prestholm et al., 1991). Although significant, the thermal conductivity anisotropy is lower than for at least some of the North Sea shales (Fjeldskaar et al., this volume).

Diagenetic control on the thermal conductivity of the Barents Sea shales

For a palaeotemperature reconstruction of the Bjørnøya Basin it is also important to be aware of

TABLE 2

Mineralogy input data for the studied formations

Formation	Age (m.y.)	Lithology	Qtz	K-fsp	Plag	Mica	Mixed clay	Kaol	Carb	Pyr	TOC
1	2– 55	shale	57	10	10	2	3	10	5	2	1
2	55– 60	shale	57	10	10	2	3	10	5	2	1
3	60– 65	shale	57	10	10	2	3	10	5	2	1
4	65–110	shale	57	6	4	5	5	15	5	2	1
5	110–141	shale	54	6	4	11	7	15	0	2	1
6	141–195	shale/sand	80	2	1	2	3	1	3	1	7
7	195–205	shale/sand	62	10	8	5	5	1	3	1	5
8	205–230	shale/sand	53	10	10	5	5	1	10	1	5
9	230–280	carbonate	30	3	2	0	0	1	59	2	3

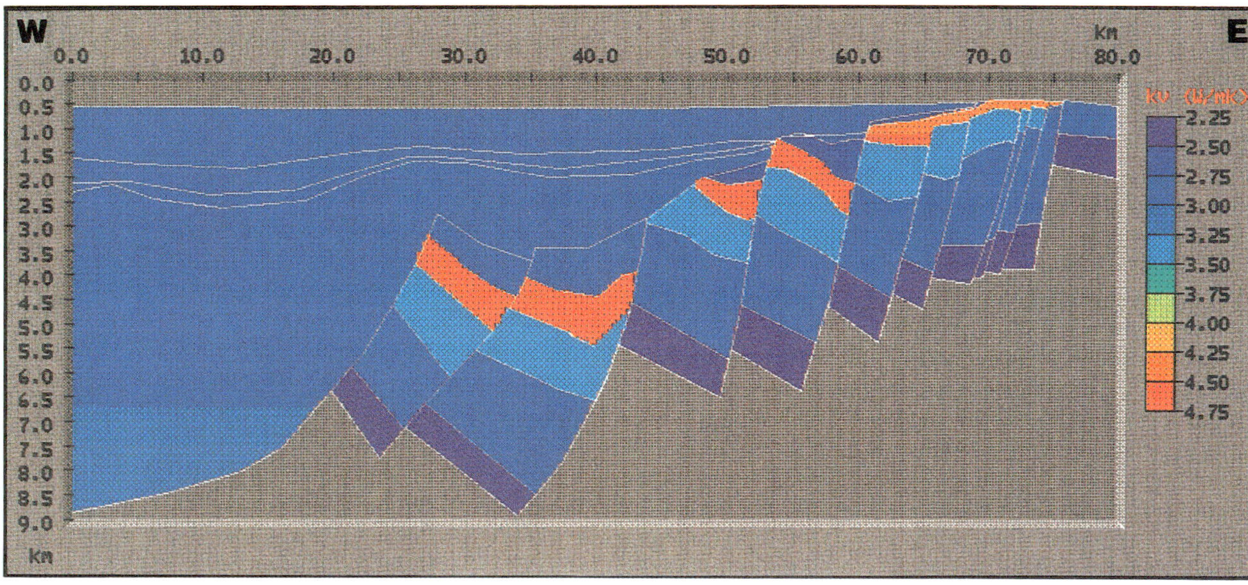

Fig. 4. Theoretical vertical thermal conductivities for the profile.

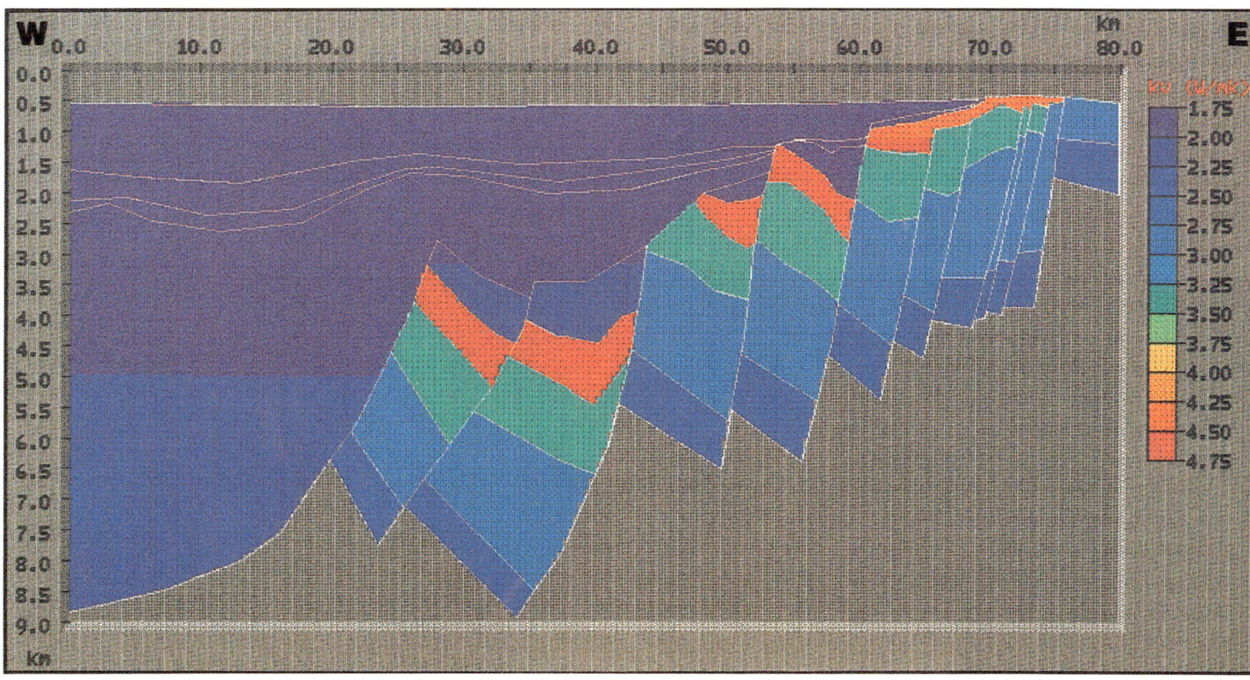

Fig. 5. Vertical thermal conductivities for the profile, assuming "standard" North Sea shale.

the diagenetic process transforming opal to quartz. This transformation has a significant influence on the thermal properties of the relevant formations.

During Paleocene/Eocene times thick sequences of opal-rich sediments accumulated. During heating of these sediments up to at least 60°C (e.g. by burial of approximately 1.5 km, as is the case with the Bjørnøya Basin), the opal will have been transformed to quartz (Murata et al., 1977; Pisciotto, 1981). In the conductivity database there are several examples of opal-CT-rich sediments (up to 95% opal-CT), which show that the thermal conductivity of a shale rich in opal-CT is much lower than for a quartz-rich shale: possibly 4–5 times lower. Thus the opal-rich sediments will give significantly higher temperatures in Late Tertiary time than would be estimated from present-day conductivity values, derived from the quartz-rich sediments.

The significance of the opal–quartz transformation for the temperature regime has been illustrated in Figs. 9 and 10, one calculated with the present conductivity values of quartz-rich shale, the other with opal-CT-rich shale in Paleocene/Eocene sediments. The difference in temperature at the base

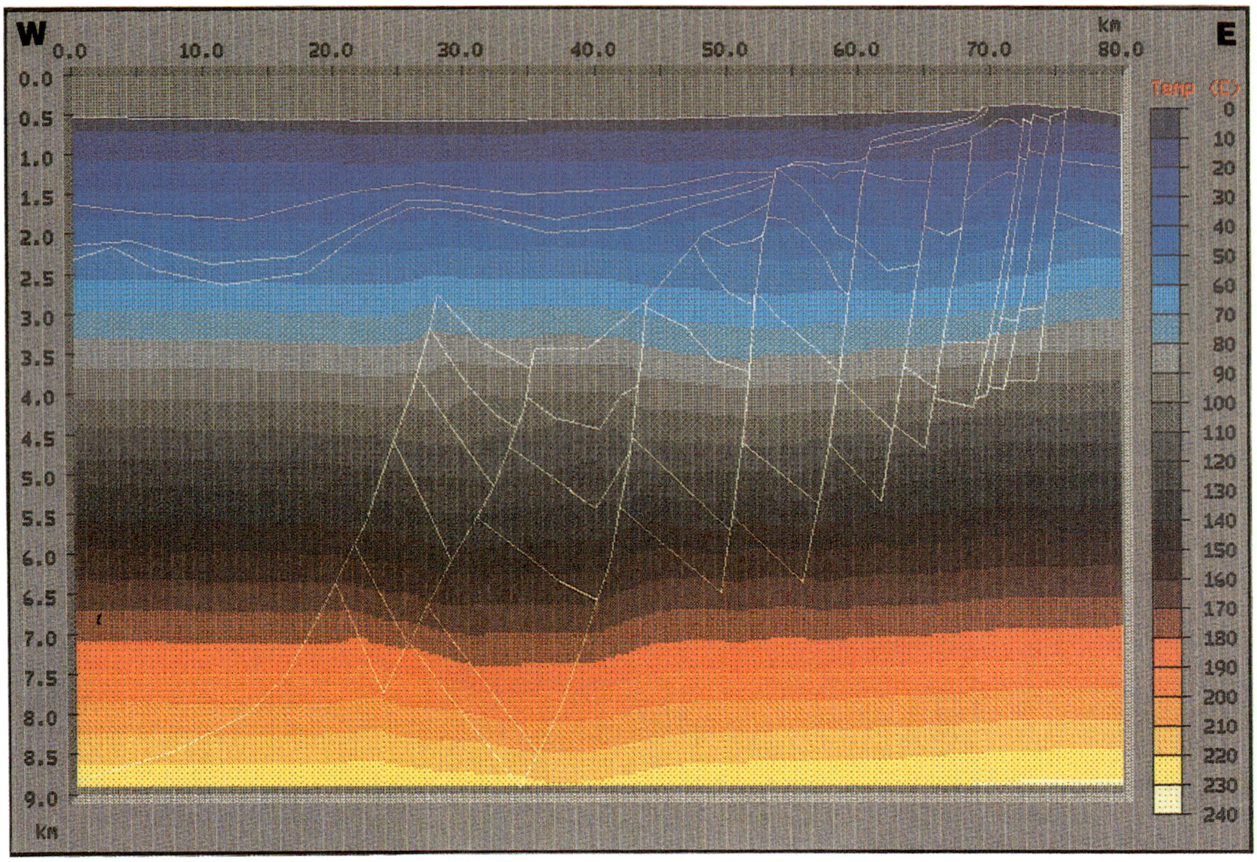

Fig. 6. Theoretical present-day temperatures in the Bjørnøya Basin, calculated by a heat flow of 78 mW/m^2 and sea floor temperature of 5°C.

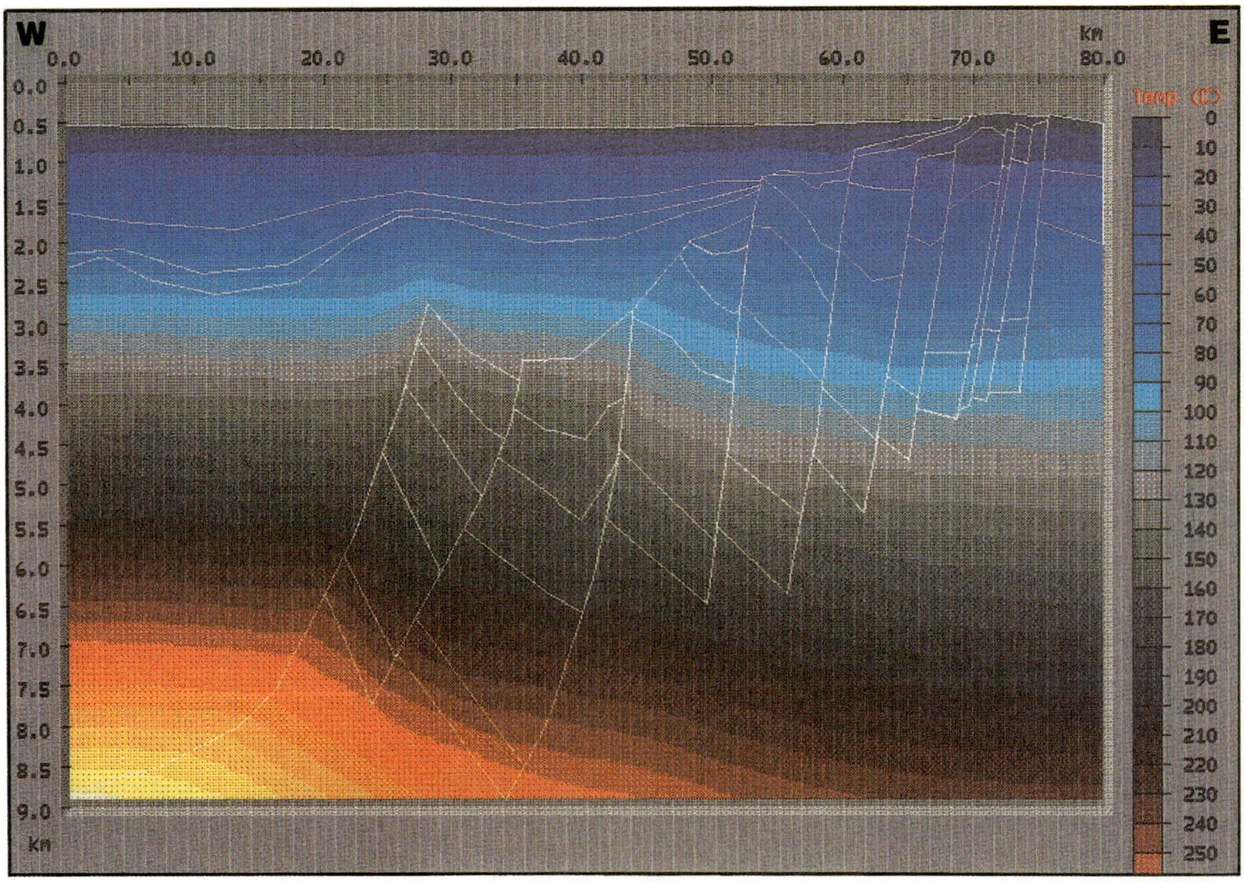

Fig. 7. Theoretical present-day temperatures in the Bjørnøya Basin, calculated by a heat flow of 78 mW/m^2 and sea floor temperature of 5°C, using thermal conductivities for "standard" North Sea shale.

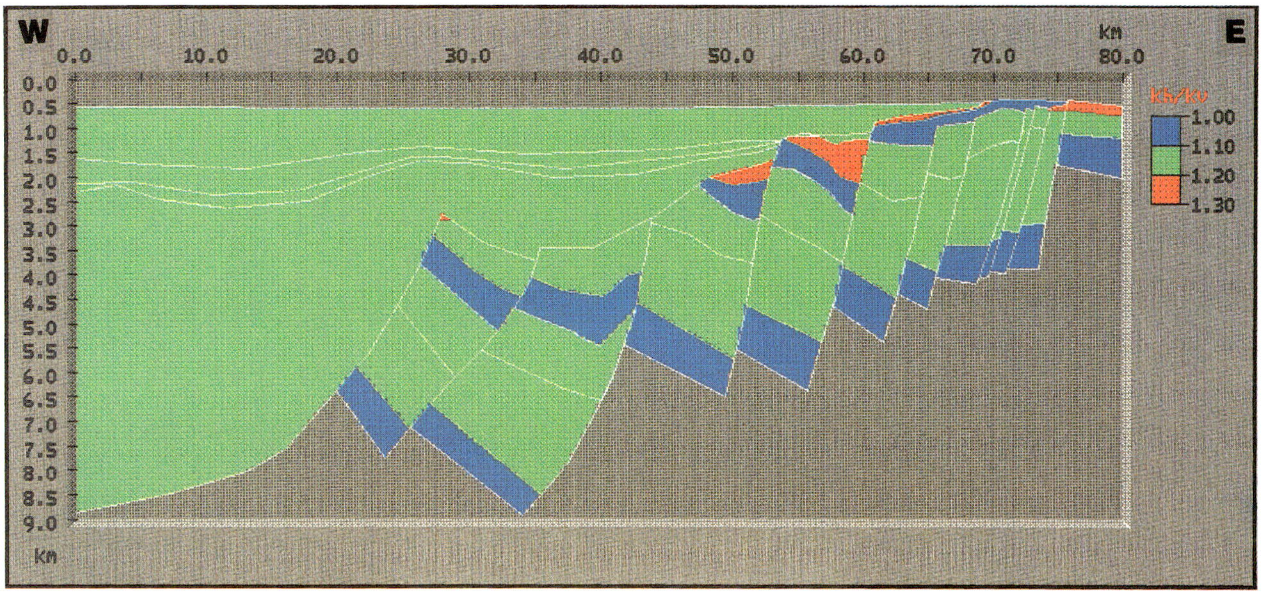

Fig. 8. Thermal conductivity anisotropy for the formations in the Bjørnøya Basin.

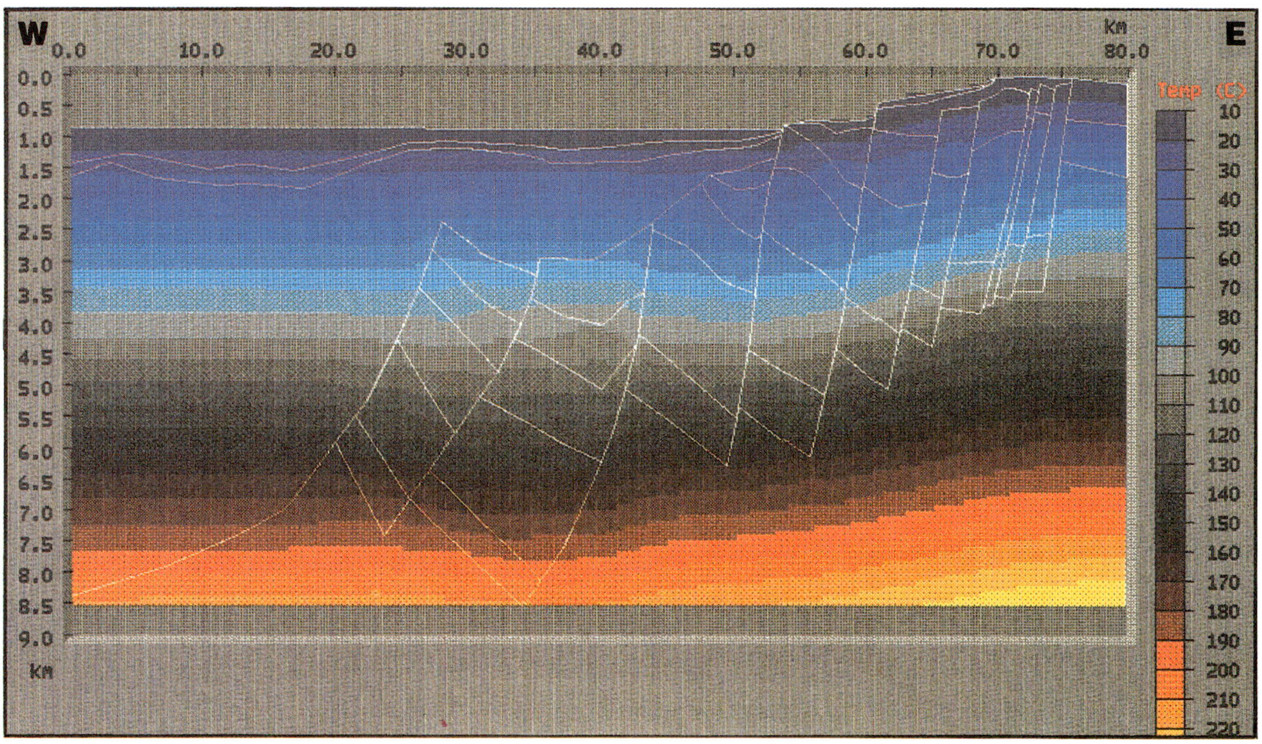

Fig. 9. Theoretical temperatures in the Bjørnøya Basin 55 m.y. ago, calculated by a heat flow of 78 mW/m^2 and a sea floor temperature of 10°C, using thermal conductivities for quartz-rich Paleocene/Eocene shale.

of the Paleocene–Eocene sediments is approximately 10°C.

Conclusions

In this study we have demonstrated the importance of quantifying the shale mineralogy for estimation of thermal conductivity of shale, because of the great variation in the mineralogy of shales.

The thermal conductivities of the sediments in the Bjørnøya Basin are calculated to be 2.8–3.0 W/m K for shales and 4.3–4.8 W/m K for sandstones. The thermal conductivities of shales are significantly anisotropic, with the horizontal conductivity 1.2–1.3 times higher than the vertical. The high quartz content in the Barents Sea shales gives relatively high conductivity, and relatively lower anisotropy, compared to shales in the North Sea.

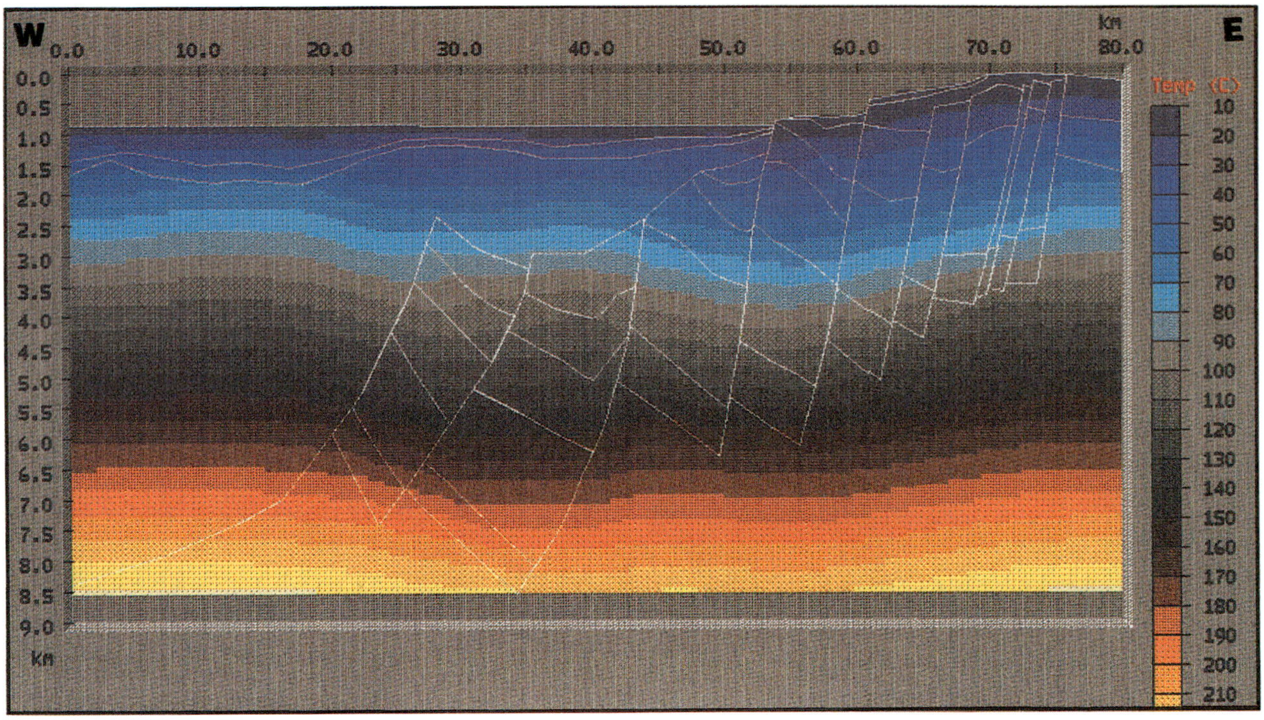

Fig. 10. Theoretical temperatures in the Bjørnøya Basin 55 m.y. ago, calculated by a heat flow of 78 mW/m^2 and a sea floor temperature of 10°C, using thermal conductivities for opal-CT-rich Paleocene/Eocene shale.

The important diagenetic transformation from opal to quartz in the Bjørnøya Basin has significant impact on the thermal conductivity of sediments, because the thermal conductivity of quartz is 4–5 times higher than for opal. The opal-rich sediments will thus give significantly higher temperatures (up to 10°C) in Late Tertiary time than would be estimated from conductivities for the present-day quartz-rich sediments.

The elevated temperatures and the period of occurrence are significant when considering the generation of hydrocarbons from source rocks. The transformation from kerogen to oil in a good source rock takes place over a depth range of approximately 1 km. The expulsion efficiency during the early phase of petroleum generation is mainly controlled by diffusion in kerogen matrix and is probably much lower than the expulsion efficiency encountered at the stages where a significant degree of oil saturation has occurred and the expulsion is a pressure-driven petroleum phase movement. Hence, the depth range of extensive petroleum expulsion is in the order of 500 m or a temperature range of 15°C. Erroneous temperature predictions by 10–15°C in late basin development may thus be crucial to the evaluation and prediction of the petroleum potential of a region with respect to timing of generation, entrapment and fluid phase.

Acknowledgements

The following institutions/oil companies are thanked for providing samples and funding the development of the database on thermal conductivities: Conoco Norway Inc., Esso Norge a.s., Norsk Agip A/S, Norske Fina A/S, Norwegian Petroleum Directorate, Statoil Norge AS and Saga Petroleum A/S.

References

Brigaud, F., Chapman, D.S. and Douran, S.L., 1990. Estimating thermal conductivity in sedimentary basins using lithologic data and geophysical well logs. Am. Assoc. Pet. Geol., Bull., 74: 1459–1477.

Domenico, P.A. and Palciauskas, V.V., 1973. Theoretical analysis of forced convective heat transfer in regional ground water flow. Bull. Geol. Soc. Am., 84: 3803–3814.

Fjeldskaar, W., Mykkeltveit, J., Christie, O.H.J., Johansen, H., Langfeldt, M., Tyvand, P., Skurve, O. and Bjørkum, P.A., 1990. Interactive 2D Basin Modelling on Workstations. Proc. SPE 20350 Petroleum Computer Conf., Denver, Colo., pp. 181–196.

Fjeldskaar, W., Prestholm, E., Guargena, C. and Gravdal, N., 1993. Isostatic and tectonic subsidence of the Egersund Sub-basin. In: A.G. Doré et al. (Editors), Basin Modelling: Advances and Applications. Norwegian Petroleum Society (NPF), Special Publication 3. Elsevier, Amsterdam, pp. 549–562 (this volume).

Gabrielsen, R.H., Færseth, R.B., Jensen, L.N., Kalheim, J.E. and Riis, F., 1990. Structural elements of the Norwegian continental shelf, Part I. The Barents Sea region. Norw. Pet. Direct., Bull., 6, 33 pp.

Hermanrud, C., 1986. On the importance to the petroleum generation of heating effects from compaction-derived water:

an example from the North Sea. In: J. Burrus (Editor), Thermal Modelling in Sedimentary Basins. Technip, Paris, pp. 247–269.

Kristiansen, J.I., 1982. The transient cylindrical probe method for determination of thermal parameters of earth material. Univ. Aarhus, Geoskr., 18, 154 pp.

Murata, K.J., Friedman, I. and Gleason, J.D., 1977. Oxygen isotope relations between diagenetic minerals in Monterey Shale, Temblor Range, California. U.S. Geol. Surv., J. Res., 3: 553–566.

Pisciotto, K.A., 1981. Notes on the Monterey rocks near Santa Maria, California. In: C.N. Isaacs (Editor), Guide to the Monterey Formation in the Californian Coastal Area, Ventura to San Luis Obisp. Am. Assoc. Pet. Geol., Pac. Sect., 52: 73–82.

Prestholm, E., 1985. A comparison of the Cretaceous sequences in the Barents Sea and on Svalbard. M.Sc. Thesis, Univ. Bergen, 243 pp. (unpublished).

Prestholm, E., Fjeldskaar, W., Balling, N. and Michelsen, J.K., 1991. Thermal conductivity anisotropy in sedimentary rocks, and its effect on the evaluation on petroleum potential. EAPG Meet., Florence (abstract).

Riis, F. and Fjeldskaar, W., 1992. On the magnitude of the Late Tertiary and Quaternary erosion and its significance for the uplift of Scandinavia and the Barents Sea. In: R.M. Larsen, H. Brekke, B.T. Larsen and E. Talleraas (Editors), Structural and Tectonic Modelling and its Application to Petroleum Geology. Norwegian Petroleum Society (NPF), Special Publication 1. Elsevier, Amsterdam, pp. 163–185.

Sclater, J.G. and Christie, P.A.F., 1980. Continental stretching: an explanation of the post-mid-Cretaceous subsidence of the Central North Sea basin. J. Geophys. Res., 85: 3711–3739.

WILLY FJELDSKAAR *Rogaland Research, P.O. Box 2503 Ullandhaug, N-4004 Stavanger, Norway*
EDWARD PRESTHOLM *Rogaland Research, P.O. Box 2503 Ullandhaug, N-4004 Stavanger, Norway*
CLAUDIA GUARGENA *Enterprise Oil Norge, P.O. Box 399, N-4001 Stavanger, Norway*
MARTIN STEPHENSON *Enterprise Oil Norge, P.O. Box 399, N-4001 Stavanger, Norway*

Issues in 3D sedimentary basin modelling and application to Haltenbanken, offshore Norway

D. Grigo, B. Maragna, M.T. Arienti, M. Fiorani, A. Parisi, M. Marrone, P. Sguazzero and A.S. Uberg

Sedimentary basin modelling is a significant technique in petroleum exploration. This approach permits the estimation of the timing and the amount of petroleum generation. Geology and organic geochemistry constitute the main theoretical framework of such basin modelling. To reconstruct the geological history of an area, several processes described by non-linear partial differential equations must be integrated in a systematic manner: the resulting large-scale modelling problem presents multiple computational issues.

A forward modelling code, SEBE-3, to simulate processes such as compaction, heat transfer, maturation of organic matter, petroleum generation, and migration has been subjected to a thorough analysis in order to obtain an integrated supercomputer/workstation implementation. The objectives of the work were: (1) the optimization of SEBE-3's numerical kernel with vectorized algorithms for the solution of the large, sparse, symmetric, indefinite, linear systems resulting from the implicit finite element discretization; and (2) the development of an advanced visualization environment, allowing simultaneously (a) the fast visualization of the relevant 3D scalar and vector fields on the supporting finite element grid, and (b) the navigation into SEBE-3's output data.

The validity of the approach is illustrated in a relevant test study on Haltenbanken, offshore Norway. The evolution of the sedimentary basin was studied using techniques of sequence stratigraphy which satisfy the needs of three-dimensional basin modelling. Sequence boundaries represent time surfaces and are used by the model in simulating sedimentation. Petrophysical characteristics of each depositional unit were determined through a multidisciplinary approach consisting of log analysis, sedimentology and seismic facies analysis. The 8 km grid used in the basic Haltenbanken modelling allows the computation only of the general pressure regime, but was sufficient to reconstruct the fluid flow evolution with time. The Permo-Triassic and Jurassic carrier beds show a general centrifugal flow trend, radiating from the depocentres. A basinwide strong increase of fluid flow rates has also been computed during the Pliocene and the Pleistocene. Such an increase is attributed to the higher sedimentation rates which characterize the Upper Pliocene depositional sequence. The fluid flow pattern estimated by the model exhibits some thermal anomalies, originating in the Jurassic and continuing up to the present, which match satisfactorily relevant well data over the study area. Such features in the space–time distribution of heat have strongly affected the source rock maturity evolution.

Introduction

Sedimentary basin modelling is a fundamental technique in petroleum exploration to estimate the timing and the amount of petroleum generation (Tissot and Welte, 1984; Allen and Allen, 1990). Geology and organic geochemistry constitute the main theoretical framework of basin modelling (Lopatin, 1971; Demaison and Moore, 1980; Magara, 1986; Mackenzie and Quigley, 1988). To reconstruct the geological history of an area, several processes described by non-linear partial differential equations must be integrated in a systematic manner (Doligez et al., 1986); the resulting large-scale modelling problem presents multiple computational issues (Menke and Murphy, 1989).

A new finite element modelling code, SEBE-3, three-dimensional in space and geologic-time-dependent, capable of simulating processes such as compaction, heat transfer, maturation of organic matter, petroleum generation and migration, has been subjected by the authors of this paper to a thorough analysis in order to obtain an efficient and integrated supercomputer–workstation implementation. This paper provides an account of this effort. The main objectives of the work were: (a) the reformulation of SEBE-3's numerical kernels with vectorized algorithms for the solution of the large, sparse, symmetric, indefinite, linear systems resulting from the implicit finite element discretization of the governing partial differential equations; (b) the development of an advanced visualization environment, allowing si-

Basin Modelling: Advances and Applications edited by A.G. Doré et al. NPF Special Publication 3, pp. 455–468, Elsevier, Amsterdam.

multaneously (1) the fast visualization of the relevant 3D scalar and vector fields on the supporting finite element grid, and (2) the navigation into SEBE-3's output data.

The paper is divided into two main sections, Methodology and Applications. The first section reviews the equations governing the simulation model, their numerical discretization, and the principal computational and visualization issues. The second section reports on a relevant study on Haltenbanken, offshore Norway, where the model and its methodology have been validated. On the base grid with size 8 km × 8 km utilized in the study, it has been possible to reconstruct successfully the pressure regime and the temporal variations of fluid flow. The computed three-dimensional thermal history has allowed a good understanding of the source rock maturity evolution.

Methodology

The modelling equations

The goal of the sedimentary basin model SEBE-3 is to integrate the physical and chemical laws for hydrocarbon generation, maturation and migration, with the geological and thermal basin dynamics in a three-dimensional, time-dependent simulation. The symbols used for the variables described below are given in the appendix.

Basin evolution. The sedimentary basin history is described by SEBE-3 as a temporal sequence of geological events: in each event sedimentation, erosion or stall can occur. If a sedimentation event occurs, the underlying layers are compacted due to the new accumulated sediments. In the case of erosion, the layers not eroded are not decompacted. A stall event is characterized by tectonic movements that vary the sea level with respect to the basin. Compaction is described as a kinematic phenomenon, without explicit computation of stresses. More precisely, compactive deformation is assumed to take place in the vertical z direction only, according to the law:

$$\nabla \cdot v_s = -\frac{\partial}{\partial z} s \frac{1-\phi_0}{1-\phi} \tag{1}$$

where s is the sedimentation rate, ϕ_0 is the sedimentation porosity, ϕ is the porosity, v_s is the velocity of the sedimentation matrix.

Thermal history. The basin thermal history is the key driving factor in the hydrocarbon generation. The main heat source is constituted by the flux from the underlying mantle, specified in the model at the sediment/basement interface. At the top of the basin the temperature is specified, while at all other boundaries zero flux is assumed. Within the basin the heat transfer is mainly conductive, as shown by the existence of vertical thermal gradients. Some deviations from such thermal gradients are observed, due to the existence of advective flows determined by (over)pressure gradients. The computation of the overpressure δp defined as deviation of pressure p from the hydrostatic approximation p_{H}, is therefore essential to obtain the fluid velocity driving the convective part in the temperature equation.

Using mass conservation and Darcy's law, the following overpressure equation is obtained, in a (Lagrangian) frame fixed with respect to the sediment matrix:

$$\phi\beta \frac{D(\delta p + p_{\mathrm{H}})}{Dt} + \nabla \cdot v_s = \nabla \left(\frac{\mathbf{K}}{\mu_w} \nabla \delta p \right) \tag{2}$$

where the (substantial) time derivative operator is represented by:

$$\frac{Df}{Dt} = \frac{\partial f}{\partial t} + v_s \cdot \nabla f \tag{3}$$

and the fluid velocity is represented by:

$$v_w = -\frac{\mathbf{K}}{\mu_w} \nabla \delta p \tag{4}$$

The permeability $\mathbf{K}$ is a diagonal tensor and it is a function of the porosity ϕ. It is expressed by the Kozeny-Carman law:

$$\mathbf{K} = \mathbf{K}^* \left(\frac{\phi}{\phi^*} \right)^3 \left(\frac{1-\phi^*}{1-\phi} \right)^2 \tag{5}$$

where $\mathbf{K}^*$, ϕ^* are, respectively, permeability and porosity at the present time.

If in the temperature equation the radioactive generation part is neglected, the total heat variation can be broken down into a convective part (first RHS term) and a conductive term (second RHS term):

$$(\rho C)\frac{DT}{Dt} = -\rho_w C_w v_w \cdot \nabla T + \nabla \cdot (\Lambda \nabla T) \tag{6}$$

The thermal conductivity varies with the porosity according to the law:

$$\Lambda = \Lambda_w^{\phi} \Lambda_s^{(1-\phi)} \tag{7}$$

Hydrocarbon generation. The oil and gas generation can be described using the Tissot-Espitalié kinetic model (Tissot and Espitalié, 1975). The following Arrhenius equations yield the quantity of type-i oil generated, w_{oi}, and type-i gas generated, w_{gi}, where the indices "oi" and "gi", respectively, denote oil or gas:

$$\frac{\mathrm{d}}{\mathrm{d}t}(w_{\mathrm{oi}}) = A_{\mathrm{i}}\,\mathrm{e}^{-E_{\mathrm{i}}/RT}(F_{\mathrm{oi}} - w_{\mathrm{oi}} - w_{\mathrm{gi}}) - A_{\mathrm{g}}\,\mathrm{e}^{-E_{\mathrm{g}}/RT}w_{\mathrm{oi}} \tag{8}$$

$$\frac{\mathrm{d}}{\mathrm{d}t}(w_{\mathrm{gi}}) = A_{\mathrm{g}}\,\mathrm{e}^{-E_{\mathrm{g}}/RT}w_{\mathrm{oi}} \tag{9}$$

where E_{i} is the activation energy for type-i oil, E_{g} for gas and the A's and F_{oi} are calibration constants.

The source terms for oil and gas are:

$$W_{\mathrm{o}} = \sum_i w_{\mathrm{oi}}, \quad W_{\mathrm{g}} = \sum_i w_{\mathrm{gi}} \tag{10}$$

Oil and gas saturation. Using the continuity of mass and Darcy's law, the following oil (i = o) and gas (i = g) saturation equations are obtained:

$$\phi\frac{D}{Dt}(\rho_{\mathrm{i}}S_{\mathrm{i}}) + \rho_{\mathrm{i}}S_{\mathrm{i}}\nabla\cdot v_{\mathrm{s}} = W_{\mathrm{i}} + \frac{\partial}{\partial x}\left(\rho_{\mathrm{i}}\frac{K_x K_{\mathrm{ri}}}{\mu_{\mathrm{i}}}\frac{\partial p}{\partial x}\right) + \frac{\partial}{\partial y}\left(\rho_{\mathrm{i}}\frac{K_y K_{\mathrm{ri}}}{\mu_{\mathrm{i}}}\frac{\partial p}{\partial y}\right) + \frac{\partial}{\partial z}\left(\rho_{\mathrm{i}}\frac{K_z K_{\mathrm{ri}}}{\mu_{\mathrm{i}}}\frac{\partial \delta p}{\partial z} + \rho_{\mathrm{i}}g\right) \tag{11}$$

$$S_{\mathrm{o}} + S_{\mathrm{w}} + S_{\mathrm{g}} = 1 \tag{12}$$

where K_{r} is relative permeability and the S's are saturations.

Discretization

The physical and chemical processes described by the non-linear partial differential eqs. 2, 6 and 11 are discretized by a finite element method (FEM) on a reference system fixed with respect to the sediment matrix. Time discretization is performed with a fully implicit backward difference scheme. The non-linear coupling between saturations, temperature and pressure is dealt with by a Newton-Raphson (outer) iterative method. Space discretization is performed with hexahedral 8-node and 20-node finite elements with constant and uniform horizontal projection $Dx \times Dy$. Convective terms in the temperature eq. 6 are handled with the so called *reverse particle tracking* technique. At each time step, the solution of two large, sparse *symmetric indefinite* linear systems, in the overpressure δp, and in the temperature T, corresponding to eqs. 2 and 6, is required. The discrete time-stepping of the ordinary differential eqs. 8 and 9 is accomplished by analytic integration. Saturations S_{i} are then explicitly advanced in time with eq. 11. The choice of the time step Dt is adaptive. For convenience, time steps of equal duration are grouped in macro time steps.

Numerical kernel

Two classes of methods are commonly used for the solution of sparse linear systems $Ax = b$ arising from the discretization of eqs. 2 and 6. The numerical literature has focused recently on the efficient reformulation of these methods on *vector computers*, especially those equipped with a hierarchical memory structure. The methods of the first class, known as *direct*, factorize the coefficient matrix A into the product of two matrices L and U, respectively, Lower triangular and Upper triangular. The computational cost and the memory occupation associated to these methods is high for general sparse matrices, because the two factors L and U do not keep the sparsity structure of the matrix A.

The methods of the second class, known as *iterative*, start from an initial guess of the solution, x_0, and generate a sequence of approximations x_k which converge under suitable conditions to the solution of the system. One of the most effective iterative methods for solving symmetric positive definite systems is a combination of the Conjugate Gradient method with some preconditioning techniques. Different generalizations have been proposed to deal with symmetric indefinite problems (Axelsson, 1985). The most promising of these for our purposes seem to be: Generalized Minimum RESidual (GMRES) (Saad and Schultz, 1986); Conjugate Gradient Stabilized (CGS) (Sonneveld, 1989); and Conjugate Gradient STABilized (CGSTAB) (Van der Vorst and Sonneveld, 1990).

As the applications for basin modelling may have various different levels of resolution and complexity, it is important to find solvers which are fast and stable for many different simulation problems. It turns out that preconditioning of iterative solvers is essential to fulfil the stability and the convergence requirements. In general, preconditioning can be viewed as finding a matrix P and solving the new preconditioned system $PA_x = Pb$ instead of $Ax = b$, where PA has better spectral properties than A. These properties guarantee a faster convergence and more stability. In this work we have experimented with multiple preconditioning strategies and in particular have considered incomplete factorization (Melhem, 1987; Meijerink et al., 1990). We have eventually selected the CGSTAB method with the Incomplete LU factorization (ILU) preconditioner, which has been shown to be the best choice for the symmetric indefinite problems arising from the simulation.

From an implementation point of view, the performance of the preconditioned CGSTAB, like all preconditioned iterative methods, depends strongly on the optimization of two computational kernels:

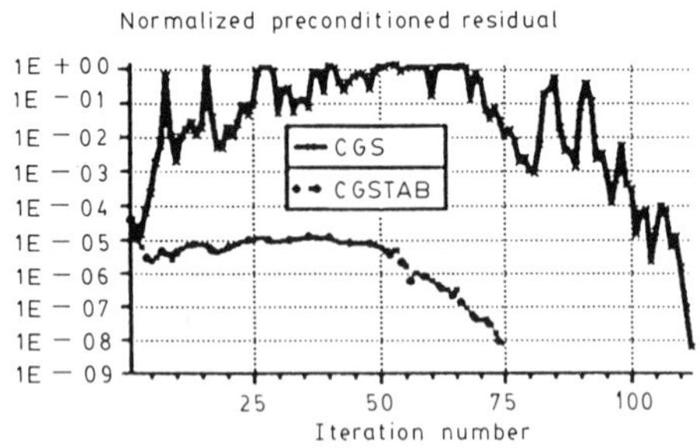

Fig. 1. Monotone convergence behaviour of the iterative method CGSTAB, preconditioned with ILU, when solving a pressure equation, contrasted with the oscillatory behaviour of CGS. Timings are reported in Table 4. Extensive comparisons of these two methods with more traditional and less effective methods, such as ORTHOMIN for example, are reported in the references.

(1) matrix–vector product; (2) solution of triangular systems.

Both kernels can be efficiently vectorized for problems with an irregular sparsity pattern, provided that a suitable data structure to store the coefficient matrix is employed. Effective vectorization of the solution of the sparse triangular systems, such as those arising when using ILU preconditioners, was achieved by a *wavefront* technique, ordering the unknowns in level sets of non-interconnected variables which can be solved concurrently (Anderson and Saad, 1988). To this purpose, we have introduced a suitable data structure, based on the layout of the non-zero coefficients and index vectors to vectorize both computational kernels (Paolini and Radicati, 1989). A reorder strategy for the vector of the unknowns was used to increase the data locality, which is crucial on machines with a hierarchical memory system. The convergent properties of the chosen algorithmical strategy are illustrated in Fig. 1.

Implementation issues

Basin simulation pipeline

The SEBE-3 code is the main component of a *basin simulation pipeline*. The first stage of the pipeline generates a file with the grid description and the data explicitly forcing the geological evolution of the basin. A second stage updates the grid file with the petrophysical parameter distribution. The *computational stage*, SEBE-3, is then invoked. A *post-processor* allows the visualization of SEBE-3's output files.

After the initialization part, the following processes are executed at each time step of SEBE-3.

(1) Update of the geometrical, physical and mechanical configuration of the basin:

- computation of the effects due to phenomena such as sedimentation, compaction, erosion or uplift; in this phase the finite element grid is rebuilt as the geometrical support for the successive computations.

(2) Update of the nodal values of overpressure and temperature:

- matrix assembly and solution of the overpressure linear system
- matrix assembly and solution of the temperature linear system; using the nodal overpressure values, the fluid velocity is computed; then the coefficients and the right-hand side of the temperature equation are computed; the matrix is assembled and the linear system is solved.

(3) Update of the hydrocarbon concentrations:

- using the temperature nodal values, the hydrocarbon concentrations and the source term for the migration equations are determined.

(4) Computation of oil and gas saturations:

- using the last values of overpressure, temperature and hydrocarbon concentrations, the coefficients and the right-hand term of the saturation equations are computed; in this way the complete physical and chemical information on the basin evolution is taken into account.

Tables 1–4 illustrate clearly the physical and computational dimensions of a simulation problem in a relevant case and the advantages of the approach taken. In particular, the last three tables indicate the large-scale nature of the computation and the effectiveness of the vectorized solvers.

TABLE 1

Dimensions of the sedimentary modelling problem for the Haltenbanken case on the 8 km × 8 km base grid

Basin x, y, z dimensions	80, 144, 9 km
Geologic time span	286 m.y.
Number of layers	14
Number of time steps	144
Grid horizontal size	8 × 8 km
Grid dimension (number of blocks)	2520
Matrix size	1403–11662

TABLE 2

CPU timings spent in the various sections of SEBE-3 for the unoptimized (time 1) CG-based original version and the optimized (time 2) CGSTAB-based code for the Haltenbanken case on the 8 km × 8 km base grid (timings are expressed in CPU min on an IBM 3090-E vector computer)

Component	Time 1	Time 2
Initialization	26.8	6.1
Coefficient computation + matrix assembly	142.0	99.2
Linear system solvers	130.3	80.8
Explicit computations	14.5	13.2
Other	2.7	2.7
Total	5 h, 15 min	3 h, 20 min

TABLE 3

Characteristics of a typical large matrix to be inverted, in the simulation of the Haltenbanken basin, discretized with a 2 km × 2 km grid

Matrix dimension	115587 × 115587
Non-zero element	3182286
Average number of elements/row	28

TABLE 4

Timings (in CPU min) on the IBM 3090-E vector computer for the solution of a typical large linear system occurring late in the simulation of the Haltenbanken basin discretized with a 2 km × 2 km grid

Solver	Iterations	Scal. time	Vect. time
CGS	112	20.3	11.6
CGSTAB	76	14.5	8.4

Two optimized iterative algorithms are compared: CGS and the more stable version of it, CGSTAB. The relative error-stopping criterion is 10^{-8}. An unoptimized scalar version requires an order of magnitude more CPU time.

It must be remarked that the choice of a non-optimized solver and/or the unavailability of vector hardware could imply an increase of computational cost by one order of magnitude, thus effectively ruling out large simulations such as those reported in Tables 3 and 4.

3D visualization

While the solution to the coupled fluid flow, thermal, chemical, structural and stratigraphic evolution equations with a finite element code such as SEBE-3 on fine simulation grids takes place naturally on a supercomputer, the visualization of the modelling output is accomplished typically in interactive fashion at the workstation level, allowing easy comparison of the simulation data with the geological observations. A menu-driven post-processor was built to solve two problems at the same time: to have a clear graphical representation of the data, and to have the possibility to navigate into the simulation output. Scalar or vector fields can be shown: scalar fields are colour-coded, and vector fields are represented using 3D arrows, superimposed on the finite element grid.

Since the FEM simulation code used irregular elements to reconstruct the basin geological layers, the visualization code must be able to reproduce such irregular elements without introducing any unwanted regridding or interpolation on regular grids before graphical rendering, as occurs in much general purpose software. To accomplish this goal the post-processor treats each block as a 3D object independent from the others: this way it is also possible to easily manage geological discontinuities. For high-quality display purposes, two different kinds of interpolation are available: the first is related to the assumed continuous variables, such as temperature or pressure, and the second to the variables that may be discontinuous through the layers, such as porosity or permeability. In the latter case the interpolation is allowed just within the layer. Multiple fields can also be shown, using contour surfaces and, on planes and surfaces, contour lines.

To understand the simulation it is essential to access the numerical value of the data and not just their colour representation. To accomplish the goal of navigating easily into the data, a control *volume* technique has been implemented. The user can select interactively on the grid a block that represents exactly one of the finite elements of the domain, and the values of the variable belonging to the selected block are displayed on the visualization screen. It is also possible to draw f–z plots that show a variable f as a function of depth z. To compare the simulation data with the same variables measured in the wells, the f–z plots show well parameters superimposed on the simulated ones. This is the easiest and most precise way to check if the simulation reproduces the physics of the problem.

Applications

The Haltenbanken study

The validity and physical response of the equations, the assumptions on which the model is grounded, and the stability of the numerical solutions were tested on synthetic examples before applying the model on a real case. The results prove the numerical model effective in simulating large-scale processes involved in the evolution of sedimentary basins, such as compaction, fluid flow, heat transfer by conduction and advection, petroleum formation and migration.

The validity of the approach is illustrated and discussed in a test study, on Haltenbanken (Fig. 2).

The area where the model was run has a surface extent of more than 100,000 km^2. The horizontal dimensions of the cells forming the 3D network are controlled using a 8 km regular, square grid.

The stratigraphic record is simplified on the basis of the resolution of the available regional seismic data, which allows the definition of fourteen isochronal events (Fig. 3).

Geological model

The sedimentary succession overlies a basement consisting of metamorphic rocks and intrusives of Caledonian age.

A sequence analysis of the Permo-Triassic succes-

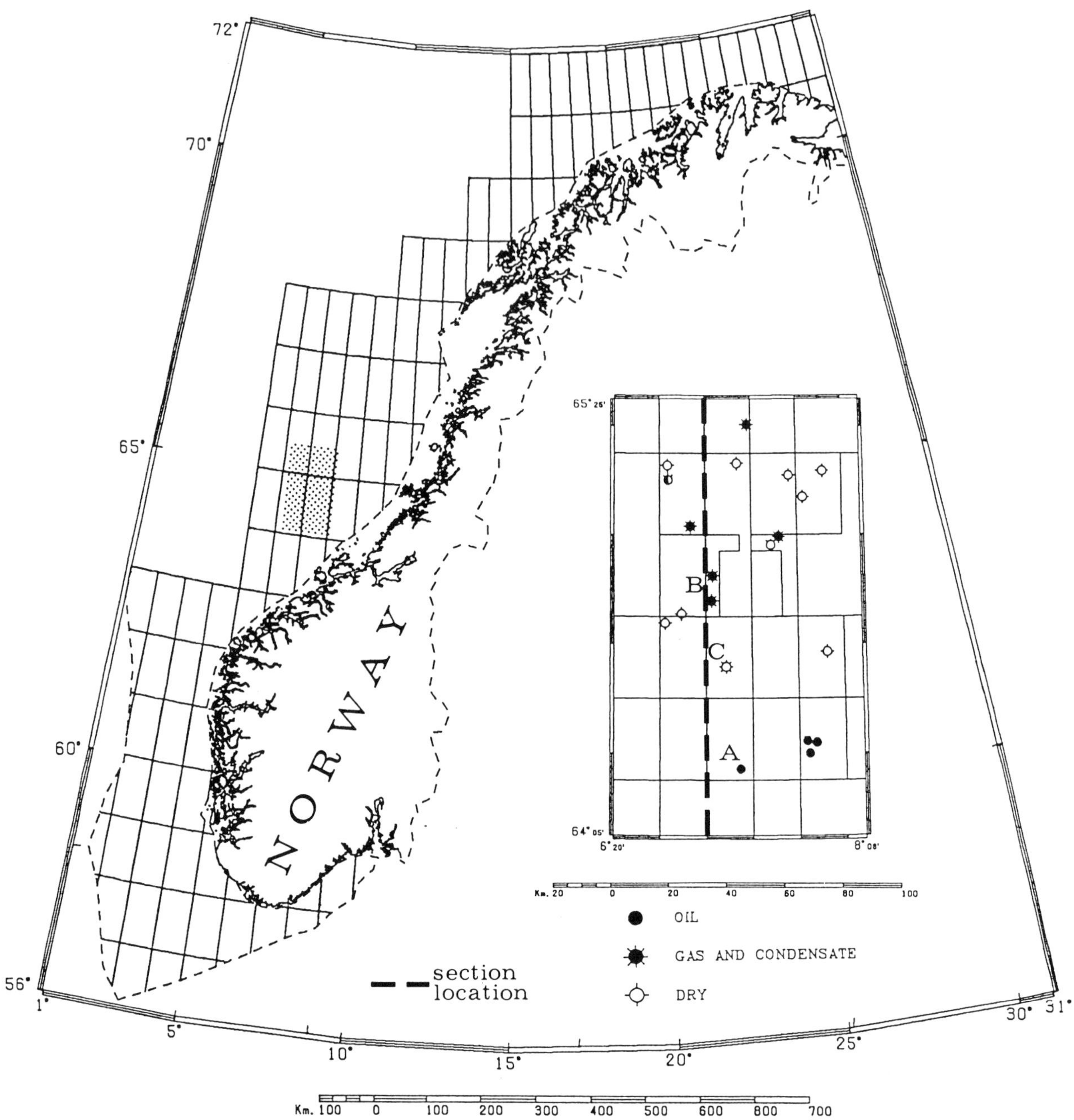

Fig. 2. Location map of the Haltenbanken area with wells and section mentioned in the text.

sion over the study area is difficult owing to lack of well data and the poor seismic resolution observed at deeper levels. The results of the Jurassic depositional sequence analysis are shown in Fig. 3; they are in agreement with previous studies concerning nearby areas (Vail and Todd, 1981).

Lower Cretaceous sediments are characterized by marly and condensed series which pass laterally and upward into claystone, locally silty toward the top. The Lower Cretaceous sedimentary succession suggests a distinct eustatic high-stand during the deposition of the Cromer Knoll Group, and therefore it could be correlated with the supercycles ranging from LZB-2 to UZA-2 (Haq et al., 1987).

During the Late Cretaceous and the Paleocene the deposition of silty shales with sandstone stringers (Shetland Group, Rogaland Group) took place. The main seismic surfaces of this age are represented by paraconformities. The accompanying low sedimentation rates indicate conditions of general still-stand and a correlation with the UZA-3 to TA-1 supercycles. The Paleocene succession shows many beds of intercalated volcaniclastics which tend to increase upward, and are related to the volcanic activity associated with the opening of the northern Atlantic much further to the east.

The Eocene to Early Oligocene sedimentation was predominantly argillaceous with minor sand-

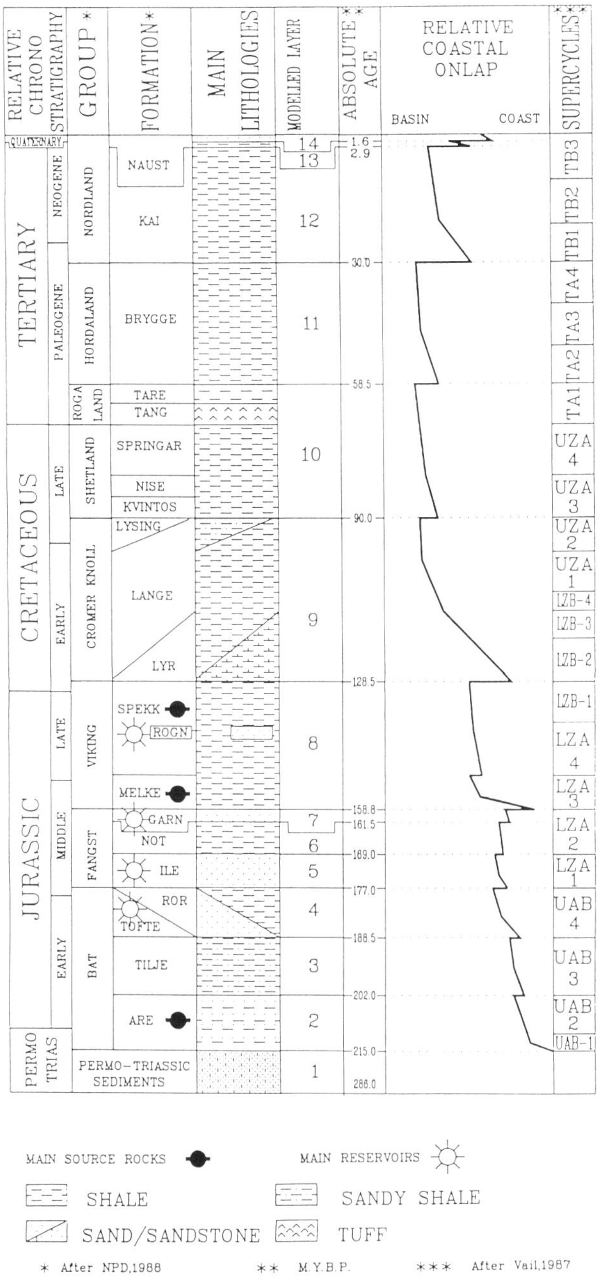

Fig. 3. Stratigraphic summary with subdivisions used during simulation (modelled layer) and main lithologies. See Table 6 for relevant conductivities.

stone and marls (Hordaland Group). This succession may be assigned to supercycles ranging from TA-2 to TA-4.

A moderate increase of coarser sediments is noted in the Upper Oligocene, Miocene and Lower Pliocene successions represented by shales with interbedded sandstones (Kai Formation) deposited in an open marine environment. Geometrical considerations coupled with the sedimentological characteristics of the Kai Formation demonstrate the correlation with the supercycle TB-1 and part of the supercycle TB-3.

The rate of sedimentation increased in the Late Pliocene with the deposition of a distinctive series of alternating fine sand and clay prograding towards the northwest (Naust Formation p.p.). The features of the Upper Pliocene succession permit the identification of a High Stand System Tract correlatable with a third-order cycle (3.7 of the supercycle TB-3).

Quaternary sediments, poorly sampled in wells, show a general increase in grain size with mainly sand and locally gravel (Naust Formation p.p.) related to a glaciomarine environment. The Quaternary section of the Naust Formation may be referred to third-order cycles (3.8 and 3.9 of the supercycle TB-3).

Geochemical model

The Middle and Late Jurassic shaly formations, the Spekk Formation and to a minor extent the Melke Formation, and the Late Triassic to Early Jurassic Åre Formation are considered potential source rocks in the area.

The kinetic parameters for kerogen degradation pertaining to each Late Jurassic unit are known as a result of optimization based on Rock-Eval pyrolysis runs on immature samples (Castelli et al., 1989); nevertheless, the use of experimentally determined parameters is made difficult owing to lack of seismic definition at the Melke–Spekk boundary, which prevents individual mapping of the two units. The problem is tentatively approached by choosing a set of kinetic parameters wide enough to account for the distribution encountered in both units, with modal values close to the experimental ones, and maintaining the same overall genetic potential. To this end, the kinetic parameters proposed by Tissot (1969) and Tissot and Espitalié (1975) for type II kerogen (Table 5) are considered to be reasonable approximations.

For the Åre Formation, no experimental values can be determined owing to the unavailability of immature samples. Type III kerogen kinetic parameters after Tissot and Espitalié are chosen to account for the peculiarity of the organic matter in this formation (Table 5). A set of six parallel independent reactions is therefore used for oil formation.

A single set of kinetic parameters is needed for gas formation; an activation energy of 57 kcal/mol is used.

Calibration

The lack of direct information about subsurface palaeotemperatures, simplifications assumed in restoring burial history, and extrapolation of local observations, all contribute to uncertainties that are introduced into the simulation.

These can be overcome by a trial and error ap-

TABLE 5

Kinetic parameters used in the simulation

Type II kerogen [a]			Type III kerogen [a]		
$E_{o,n}$ (kcal/mol)	$A_{o,n}$ (1/m.y.)	R_n (g/g)	$E_{o,n}$ (kcal/mol)	$A_{o,n}$ (1/m.y.)	R_n (g/g)
10	1.27×10^5	0.022	10	5.20×10^3	0.03
30	7.47×10^{16}	0.034	30	4.20×10^{16}	0.03
50	1.48×10^{27}	0.251	50	4.33×10^{25}	0.072
60	5.52×10^{29}	0.152	60	1.97×10^{32}	0.091
70	2.04×10^{35}	0.116	70	1.20×10^{33}	0.049
80	3.80×10^{35}	0.120	80	7.56×10^{31}	0.027

[a] $n = 1$ to 6.

proach. Three basic constraints are taken into account: present pressure regime, present temperatures, and present maturity trends known from the wells (Mackenzie and McKenzie, 1983; Yukler and Kokesh, 1984). At desired steps during the simulation the results in terms of pressure regime, temperature pattern or maturity can be checked against measured values, and the input parameters calibrated.

Present pressure regime. In the Haltenbanken simulation, the computed regional pore pressure pattern at the present time is compared with the gross regional pattern suggested by pressure regime in the wells. This provides a measure of the acceptability of the permeabilities and compaction trends assigned to the units in the basin.

The model yields a satisfactory reconstruction of the overpressure distribution within the Mesozoic units; the results for the Palaeozoic sequence might be questionable, but the subsurface data appear too scanty to allow a satisfactory comparison and discussion.

At the scale of the model, a precise comparison between the pore pressures computed in the Jurassic permeable units and the values measured in wells is not believed effective, mainly because the grid dimensions are only adequate to investigate large-scale phenomena. Moreover, the imposed geometry derived from a porosity vs depth approach for compaction (Baldwin and Butler, 1985), can affect the absolute value of the computed pressures.

Present temperatures and maturity trends. On the basis of well data, a 1D approach is used to approximate the present-day geothermal flow (Novelli et al., 1987). A trial and error approach based on the comparison between the measured and computed temperature trends allows the definition of the most likely present-day heat flow value, and simultaneous refining of the thermal parameters assigned to the sediments (Table 6).

TABLE 6

Thermal conductivities used in the simulation

Lithology	Matrix conductivity (cal cm^{-1} s^{-1} °C^{-1})
Clay	2.64×10^{-3}
Sand	5.0×10^{-3}
Sandstone	8.2×10^{-3}
Siltstone	7.6×10^{-3}
Limestone	8.7×10^{-3}
Marl	5.0×10^{-3}
Shale	4.5×10^{-3}
Sandy shale	5.0×10^{-3}
Tuff	8.0×10^{-3}

A similar method can be employed to restore the palaeogeothermal regime, i.e. the heat flow at the basement/sediment interface during the evolution of the basin. The evolution of thermal maturity under different hypotheses of thermal history is simulated; the constraint is provided by the trends of organic maturity measured in the wells.

The analysis of present temperature distribution, derived from BHT, RFT and DST measurements, shows an anomalous (high) temperature gradient within the Jurassic sequence (Fig. 4) in several wells in the area.

A striking feature in the 1D optimization is the need to raise the present-day heat flux, in order to generate temperature values that fit in with the measured trends, without any geological evidence of a regional heating event at the basement level. This is not enough in some cases to fit the temperature gradient of the Jurassic section (well B).

The same behaviour characterizes the organic matter maturity distribution, represented by the R_o trend in wells, where a strong increase at Jurassic level is evident (Fig. 5).

The 1D optimization for the wells of the Haltenbanken area suggests a scenario dominated by heat flow values around 1 HFU; the flux increases significantly only in the Late Jurassic. Cooling during

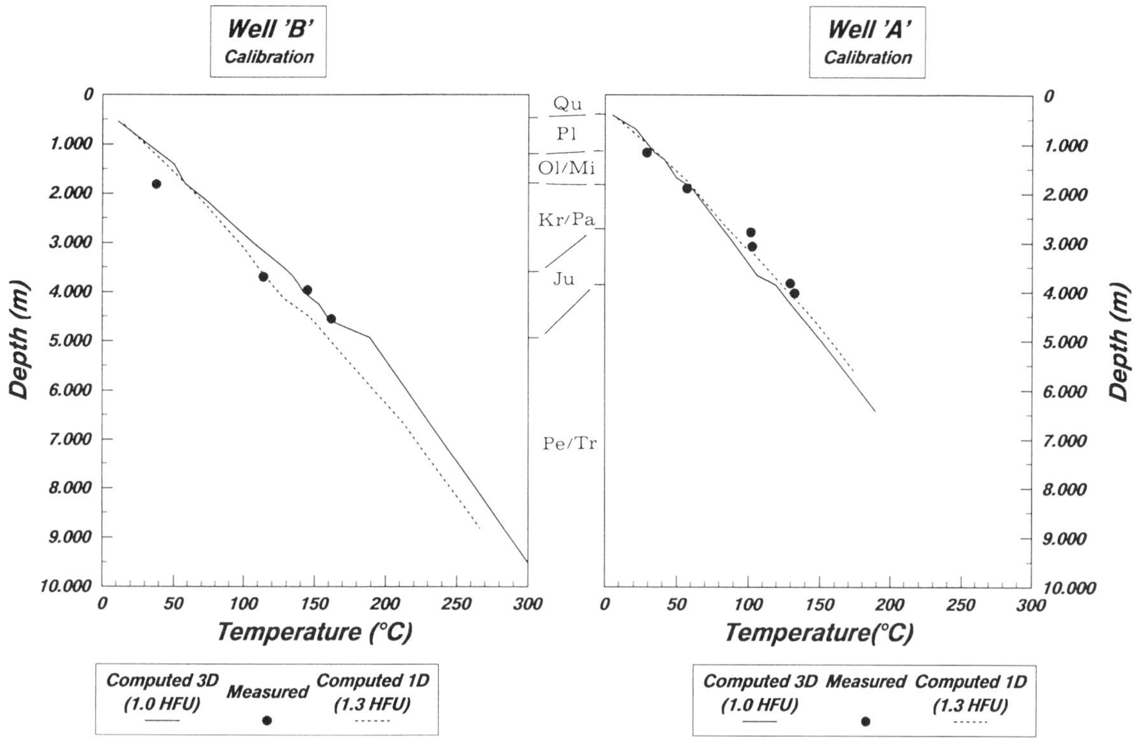

Fig. 4. Comparison between 1D and 3D present-temperature calibration on two wells located on the opposite side of the Jurassic depocentre (see map and section). Stratigraphy: *Qu* = Quaternary, *Pl* = Late Pliocene, *Ol/Mi* = Eocene–Oligocene–Miocene–Lower Pliocene, *Kr/Pa* = Cretaceous–Paleocene, *Ju* = Jurassic, *Pe/Tr* = Permian–Triassic.

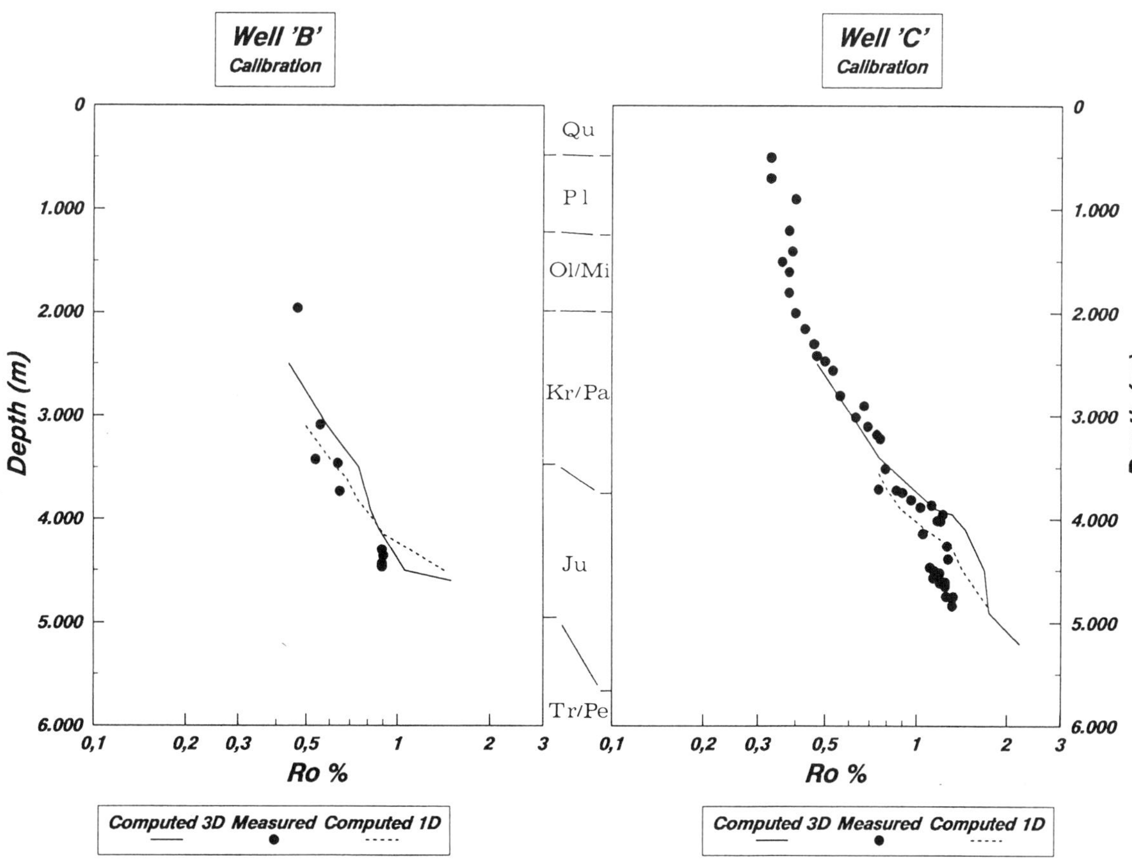

Fig. 5. Comparison between 1D and 3D present maturity trend calibration on two wells located on the northern side of the Jurassic depocentre (see map and sections). For stratigraphy see Fig. 4.

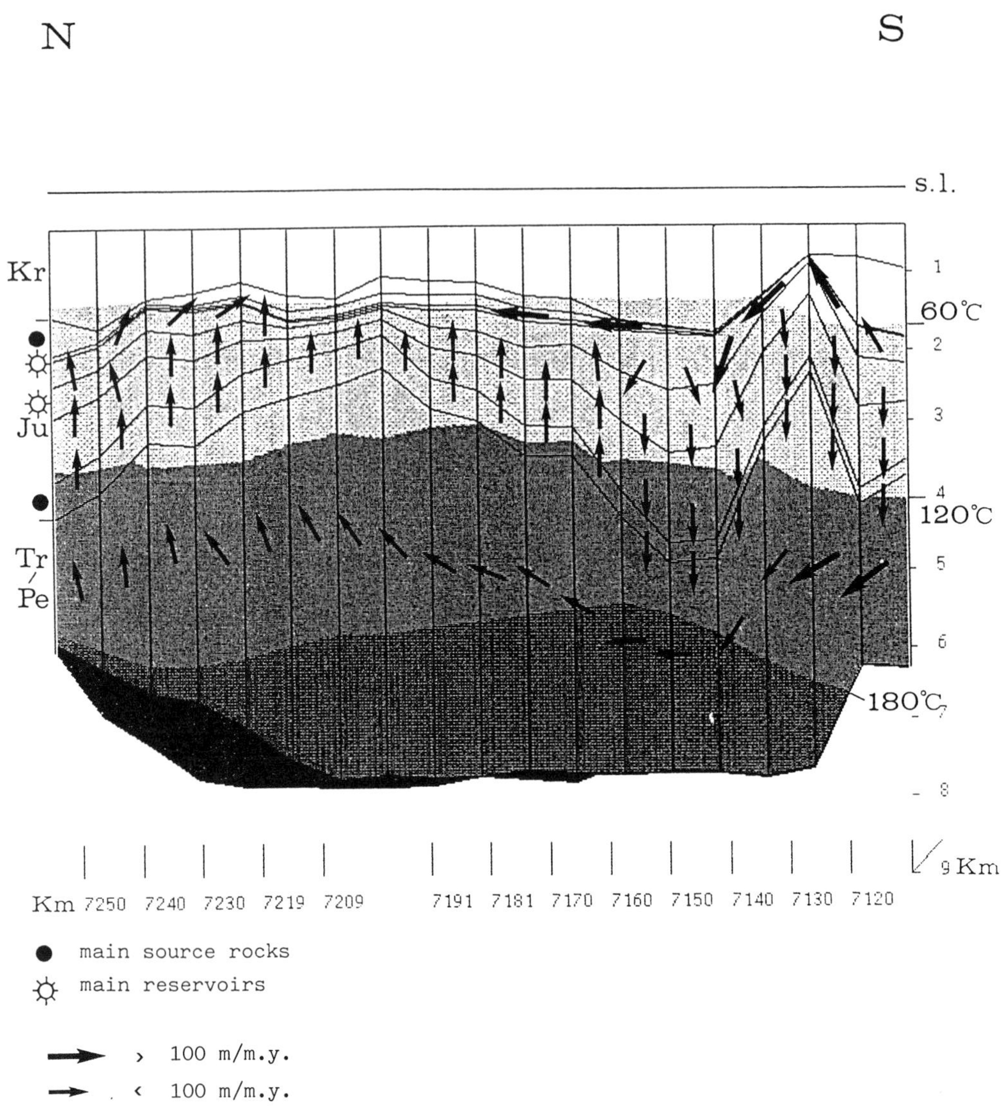

Fig. 6. North–south section (see map) showing computed fluid flow trend and temperature distribution at Lower Cretaceous time (90 Ma). Stratigraphy: *Kr* = Lower Cretaceous, *Ju* = Jurassic, *Tr/Pe* = Permian–Triassic.

the Early Cretaceous restores the average value of 1 HFU by the beginning of the Late Cretaceous. As for temperature this 1D simulation involves some difficulties in obtaining the best fit for the Jurassic maturity trend (well C).

Simulation

The grid dimensions utilized in the current simulation do not permit introduction of local permeability barriers resulting from small-scale faults. Otherwise, variable sediment loading causing differential compaction has established fluid flow gradients within the permeable layers. The hydrodynamic setting that results from simulation appears to have significant influence on the reconstruction of the thermal history of the area. The thermal evolution resulting from three-dimensional modelling allows a better reconstruction of petroleum generation in the study area.

Hydraulic regime. The Jurassic hydraulic regime resulting from simulation shows evidence of hydrostatic equilibrium. Low rates of sedimentation and fair permeabilities are considered to be factors responsible for this observation.

Hydrostatic conditions appear to have prevailed throughout the Cretaceous and a considerable part of the Tertiary. This is due to the fact that the low-permeability lithologies were deposited at slow rates of sedimentation, and thus the simulation shows only minor overpressuring.

The present-day hydrodynamic regime developed starting from the Late Pliocene with the occurrence of significant overpressuring within the Cretaceous–Tertiary succession. Such conditions prevailed up to present time and are found to characterize the northwestern sector of the area. This area is in fact the sedimentary depocentre of the Late Pliocene sequence. It is characterized by higher burial rates and consequent development of overpressuring, particularly within the poorly permeable lithologies. The pressure regime has not stabilized at present time, even though the Pleistocene records low sedimentation rates.

Well data in this area demonstrate anomalous high pressure gradients over the Cretaceous–Tertiary sequence. A maximum manifestation of these conditions is found over the northwestern sector of the area and it is matched by the computed pressure trend.

Fluid flow pattern. The higher compaction rates and resulting greater fluid expulsion are found to coincide with the sedimentary depocentres.

The initial modelled fluid flow within the basin was characterized by a centrifugal disposition with respect to the depocentre of the overlying Jurassic succession, located in the southern part of the basin.

A strong increase in the northward component of the flow, starting from the Early Cretaceous is observed (Fig. 6). This flow pattern persisted until the Early Pliocene (Fig. 7), after which a strong increase combined with a southward rotation of fluid flow can be identified (Fig. 8). This is related to higher compaction rates caused by the greater sediment loading of the Late Pliocene sequence, which causes a greater compaction rate in the northern sector.

At present time, permeable Permo-Triassic and Jurassic layers show, in general, a rise in fluid flow towards highs within the basin. Both differential compaction and permeability variations are found to be key elements responsible for the evolution of fluid flow within the Permo-Triassic and Jurassic section.

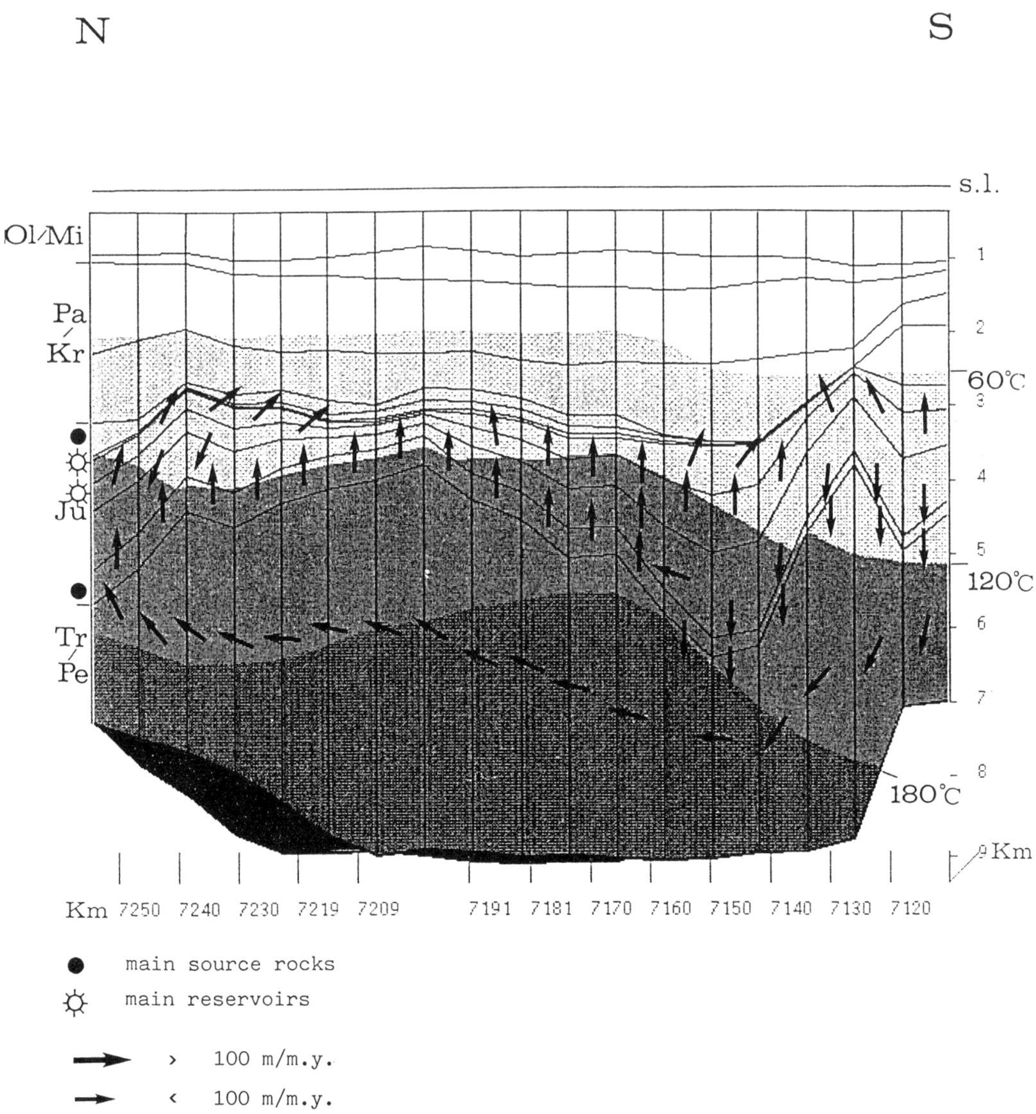

Fig. 7. North–south section (see map) showing computed fluid flow trend and temperature distribution at Lower Pliocene time (2.9 Ma). Stratigraphy: *Ol/Mi* = Eocene–Oligocene–Miocene–Lower Pliocene, *Pa/Kr* = Paleocene–Cretaceous, *Ju* = Jurassic, *Tr/Pe* = Permian–Triassic.

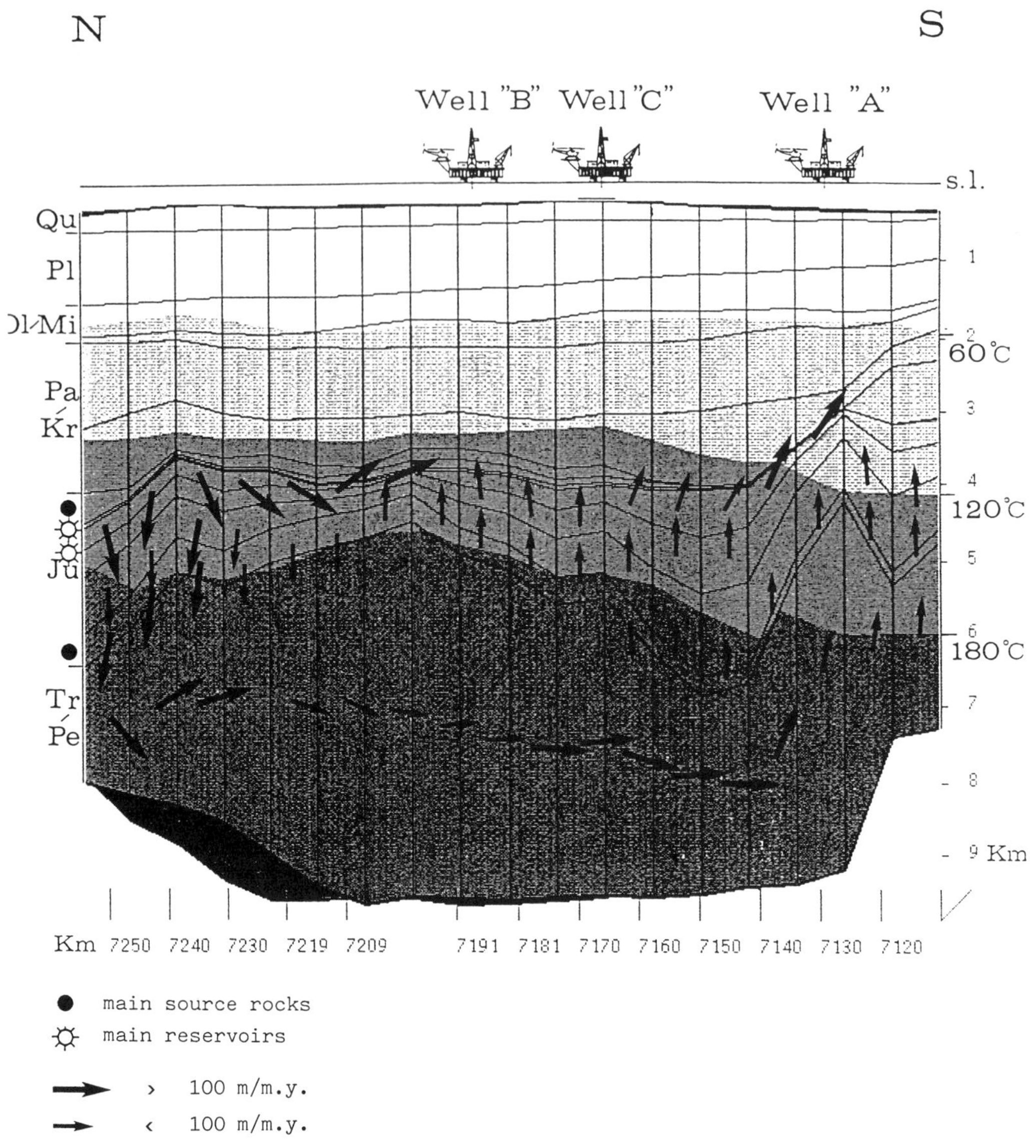

Fig. 8. North–south section (see map) showing computed fluid flow trend and temperature distribution at Present time. Stratigraphy: *Qu* = Quaternary, *Pl* = Late Pliocene, *Ol/Mi* = Eocene–Oligocene–Miocene–Lower Pliocene, *Pa/Kr* = Paleocene–Cretaceous, *Tr/Pe* = Permian–Triassic.

Temperature pattern. In the southern sector of the study area, a modelled thermal anomaly appears commencing from the Early Cretaceous in correspondence with the development of centrifugal fluid flow within the Permo-Triassic and Jurassic permeable units. This anomaly is located on the northern side of the Jurassic depocentre (Fig. 6).

During the Cretaceous the thermal anomaly spreads out, covering the northern sectors. A general heating of the central part of the basin can thus be identified. Conversely modelled downward fluid flow in the southern part seems to cool this part of the basin.

The thermal setting described above continues with little change well into the Tertiary. Temperature differences of up to 30°C are found to exist across the opposite sides of the Jurassic depocentre as indicated by strongly arched isotherms (Fig. 7).

During the Late Pliocene thermal anomalies are affected by rotation in fluid flow direction. The southward direction of fluid flow seems to smooth the thermal anomaly, increasing temperature along the southern side of the depocentre where the fluid flow has a strong upward component (Fig. 8).

The temperature trend of some wells (based on BHT and RFT measurements) of the central area and in particular the anomalous pattern of measured values within the Jurassic and Triassic sequences is easily reproduced by the computed temperature trend, showing the importance of the advective part of the 3D model (Fig. 4) with respect to the 1D simulation.

Hydrocarbon generation. The thermal regime described above was found to closely control the maturation of source rocks in the study area.

The case history was focused on the Upper Jurassic "oil prone" source rocks. The effects of the Upper Jurassic thermal event caused deeper parts of these units to enter the "oil window" during the Early Cretaceous. At this stage thermal anomalies resulting from fluid flow are found to exert a significant control on maturation. In fact, oil window areal distribution covers perfectly the extent of the thermal anomaly developed on the northern side of the Jurassic depocentre. This situation continued until today and resulted in the more mature area of source rocks being shallower to the north.

The reliability of this fluid flow influenced maturity distribution is proved by the fit with measured maturity parameters (Fig. 5). This confirms the effect of fluid flow on temperature evolution along the northern side of the Jurassic depocentre (well C).

Conclusions

Three-dimensional basin modelling carried out over the Haltenbanken area has produced some interesting outcomes.

This study, we believe, has demonstrated the superiority of a three-dimensional approach over a mono-dimensional one, particularly due to its ability to model fluid flow and its influence on thermal evolution with time. This is demonstrated by the modelled present-day temperature and maturity trend distribution within the Jurassic succession.

The presence of differing temperature at equivalent depths but areally separated locations during the evolution of the basin has caused variations in timing and extent of maturation. This contributes to a general northward displacement of more mature areas. This pattern fits in with the distribution of the hydrocarbon provinces in Haltenbanken (Fig. 2).

Acknowledgements

The authors thank Norsk Agip, and partners for giving permission to publish this paper and for technical contribution and assistance.

Appendix

List of the main variables

Overpressure equation

δp	overpressure
p	hydrostatic pressure
$\mathbf{K}$	permeability tensor
ϕ	porosity
μ_w	water viscosity
β	water compressibility factor

Temperature equation

T	temperature
$\Lambda_{w,s}$	thermal conductivity of rock and water
ρ_w	water density
C_w	water specific heat
v_w	fluid velocity
Λ, ρ, C	quantities related to wet rock

Hydrocarbon generation and saturation equations

K_x, K_y, K_z	diagonal components of the permeability tensor
μ_i	oil and gas viscosity
S_i	saturations
W_i	source terms

References

Allen, P.A. and Allen, J.R., 1990. Basin Analysis. Blackwell Scientific Publications, Oxford, pp. 263–308.

Anderson, E. and Saad, Y., 1988. Solving sparse triangular linear systems on parallel computers. Center of Supercomputing Research and Development, Austin, Texas, Rep. No. 794, June 1988.

Axelsson, O., 1985. A survey of preconditioned iterative methods for solving sparse systems of linear equations. BIT, 25: 166–187.

Baldwin, B. and Butler, C.O., 1985. Compaction Curves. Am. Assoc. Pet. Geol., Bull., 69 (4): 622–626.

Castelli, A., Chiaramonte, M.A., Beltrame, P.L., Carniti, P., Del Bianco, A. and Stroppa, F., 1989. Thermal degradation of kerogen by hydrous pyrolysis. A kinetic study. In: B. Durand and F. Behar (Editors), Advances in Organic Geochemistry, 1989. Pergamon Press, London, pp. 75–82.

Dalland, A., Worsley, D. and Ofstad, K. (Editors), 1988. A Lithostratigraphic Scheme for the Mesozoic and Cenozoic Succession Offshore Mid- and Northern Norway. Norw. Pet. Direct., Bull., 4.

Demaison, G.J. and Moore, G.T., 1980. Anoxic environments and oil source bed genesis. Am. Assoc. Pet. Geol., Bull., 64: 1178–1209.

Doligez, B., Bessis, F., Burrus, J., Ungerer, P. and Chénet, P.Y., 1986. Integrated numerical simulation of the sedimentary heat transfer, hydrocarbon formation and fluid migration in a sedimentary basin: the Themis model. In: J. Burrus (Editor), Thermal Modelling in Sedimentary Basins. Technip, Paris, pp. 174–195.

Haq, B.U., Hardenbol, J. and Vail, P.R., 1987. Chronology of fluctuating sea levels since the Triassic. Science, 235: 1156–1167.

Lopatin, N.V., 1971. Temperature and geologic time as a factor in coalification. Izv. Akad. Nauk SSSR, Ser. Geol., 3: 95–106.

Mackenzie, A.S. and McKenzie, D.P., 1983. Isomerization and aromatization of hydrocarbons in sedimentary basins formed by extension. Geol. Mag., 120: 417–470.

Mackenzie, A.S. and Quigley, T.M., 1988. Principles of geochemical prospect appraisal. Am. Assoc. Pet. Geol., Bull., 72: 399–415.

Magara, K., 1986. Geological Models of Petroleum Entrapment. Elsevier Applied Science Publishers, London, pp. 44–68.

Meijerink, J.A. and Van der Vorst, H., 1990. An iterative solution method for linear systems of which the coefficient is a symmetric M-matrix. Math. Comp., 31, No. 137: 148–162.

Melhem, R., 1987. Toward efficient implementation of preconditioned conjugate gradient methods on vector supercomputers. Int. J. Supercomp. Applic., 1 (1): 70–88.

Menke, W. and Murphy, B., 1989. The global basin research network as a testbed for empirical research on collaboration technology. Proposal to the National Science Foundation.

Novelli, L., Chiaramonte, M.A., Mattavelli, L., Pizzi, G., Sartori, L. and Scotti, P., 1987. Oil habitat in the Northwestern Po Basin. In: B. Doligez et al. (Editors), Migration of Hydrocarbons in Sedimentary Basins. Technip, Paris, pp. 27–58.

Paolini, G.V. and Radicati, G., 1989. Data structure to vectorize CG algorithms for general sparsity patterns. BIT, 29: 703–718.

Saad, Y. and Schultz, M.H., 1986. GMRES: a generalized minimal residual algorithm for solving nonsymmetric linear systems. Siam J. Sci. Stat. Comput., 7 (3): 856–869.

Sartori, L., Pizzi, G., Chiaramonte, M.A., Franzese, M. and Ghizzoni, E., 1990. Modello numerico tridimensionale ad elementi finiti per la simulazione dei processi di generazione, migrazione ed accumulo degli idrocarburi in un bacino sedimentario. X Convegno Nazionale Associazione Italiana di Meccanica Teorica ed Applicata, Pisa, October 2–5, 1990.

Sonneveld, P., 1989. CGS, A fast Lanczos type solver for nonsymmetric linear systems. Siam J. Sci. Stat. Comput., 10 (1): 36–52.

Tissot, B.P., 1969. Premières données sur les mécanismes et la cinétique de la formation du pétrole dans les sédiments: simulation d'un schéma réactionnel sur ordinateur. Rev. Inst. Fr. Pét., 24: 470–501.

Tissot, B.P. and Espitalié, J., 1975. L'évolution termique de la matière organique des sédiments: application d'une simulation mathématique. Rev. Inst. Fr. Pét., 30: 743–777.

Tissot, B.P. and Welte, D.H., 1984. Petroleum Formation and Occurrence. Springer-Verlag, Berlin, 2nd ed., pp. 583–608.

Vail, P.R. and Todd, R.G., 1981. Northern North Sea Jurassic unconformities, chronostratigraphy and sea-level changes from seismic stratigraphy. In: L.V. Illing and G.D. Hobson (Editors), Petroleum Geology of the Continental Shelf of North West Europe. Institute of Petroleum, Heyden, London, pp. 216–235.

Vail, P.R., Mitchum, R.M. and Thomson, S., 1977. Seismic stratigraphy and global changes of sea level, Part 3. Relative changes of sea level from coastal onlap. Am. Assoc. Pet. Geol., Mem., 26: 63–82.

Van der Vorst, H. and Sonneveld, P., 1990. CGSTAB: a more smoothly converging variant of CGS. Preprint of Delft University of Technology, May 21, 1990.

Yukler, M.A. and Kokesh, F., 1984. A review of models used in petroleum resource estimation and organic geochemistry. In: J. Brooks and D.H. Welte (Editors), Advances in Petroleum Geochemistry, Vol. I. Academic Press, London, pp. 344–348.

D. GRIGO *AGIP S.P.A., P.O. Box 12069–20120 Milan, Italy*
B. MARAGNA *AGIP S.P.A., P.O. Box 12069–20120 Milan, Italy*
M.T. ARIENTI *AGIP S.P.A., P.O. Box 12069–20120 Milan, Italy*
M. FIORANI *AGIP S.P.A., P.O. Box 12069–20120 Milan, Italy*
A. PARISI *IBM ECSEC, V.le Oceano Pacifico, Rome, Italy*
M. MARRONE *IBM ECSEC, V.le Oceano Pacifico, Rome, Italy*
P. SGUAZZERO *IBM ECSEC, V.le Oceano Pacifico, Rome, Italy*
A.S. UBERG *IBM Scientific Center, Thormøhlensgate 55, N-5008 Bergen, Norway*

Modelling the hydrocarbon system of the North Viking Graben: a case study

Fred W. Schroeder and Øyvind Sylta

A three-dimensional model of the hydrocarbon system within the North Viking Graben was developed. The objectives of this study were to gain quantitative insights into the local hydrocarbon system and to use the results to aid in the risking and ranking of prospects. The study area contains a number of fields and discoveries including Huldra, Veslefrikk, Oseberg, Brage, and Troll.

The model covers an area approximately 100 km by 250 km. Thirteen stratigraphic intervals from the seafloor down through the Middle Jurassic were mapped. Upper Jurassic organic-rich shales and Middle Jurassic coals were assumed as source intervals. The carrier was assumed to be Middle and Upper Jurassic sand-prone intervals. To model the hydrocarbon system, heat flow histories were calibrated at selected well locations and heat flow maps covering the entire study area were prepared. Then oil and gas generation histories were modelled for the two source intervals. Finally, secondary migration was modelled with consideration given to: trap capacities, expulsion efficiencies, oil lost during migration, oil cracked to gas, gas dissolved in oil, gas leaked through the top seal of traps, and oil and gas spillage.

Over forty modelling runs were made to fine-tune input parameters and check their sensitivities. Modelling results were compared with estimates of the capacities and in-place reserves of six fields and three discoveries. For the "optimum" model, major migration routes were determined and the fill histories of individual traps were modelled.

The model was used to assess the migration risk associated with several prospects. Relative rather than absolute results are important. One prospect was located on a major migration route and was modelled as containing oil in almost every run. A second prospect, also on a migration route, showed significant variations in terms of its ratio of oil to gas on different modelling runs. A third prospect was found to be consistently under-filled with gas. Based on these modelling results, these three prospects were assigned a risk and rank (priority) in terms of hydrocarbon generation and migration.

Introduction

Various geochemical techniques exist for assessing the potential of a source and modelling hydrocarbon generation. Techniques for defining a trap, modelling its development, and predicting its reservoir properties also are well advanced. Since most traps are stratigraphically and spatially removed from regions where hydrocarbon generation has taken place, secondary migration is necessary to complete the system. Unfortunately the processes controlling secondary migration are not well understood. The explorationist is forced to make qualitative estimates of how much oil and gas has migrated from the "kitchen" to the trap and along what pathways.

A few secondary migration modelling programs have been developed and applied to several basins (Welte and Yukler, 1981; Durand et al., 1983; Bethke, 1985; Ungerer et al., 1985, 1987, 1990; Nakayama, 1986; Doligez et al., 1986, 1987; Burrus and Audebert, 1990). Most programs model hydrocarbon generation, migration, and entrapment along a cross-section (i.e., they are two-dimensional models). One exception is the SEMI program (Sylta, 1987, 1991, and this volume). SEMI models secondary migration from a map perspective. A technical description of the SEMI modelling software is given in a separate paper (Sylta, this volume).

This paper describes a secondary migration study that used the SEMI program. Basic steps in the modelling procedure are outlined, components of the input model are summarized, examples of the generated output are shown as figures, and the application of the modelling results to aid in prospect evaluation is discussed. The theories, equations, and computational techniques used in SEMI have been published elsewhere (Sylta, 1987, 1990, and this volume); the focus of this paper in on the application of basin modelling techniques to a particular area.

Study area and objectives

The study area is located between 59°40′N and 62°00′N, and from 2°00′E to 4°00′E (Fig. 1). The migration model was constrained by the areal limits of the structure maps used as input. At the time

Basin Modelling: Advances and Applications edited by A.G. Doré et al. NPF Special Publication 3, pp. 469–484, Elsevier, Amsterdam.

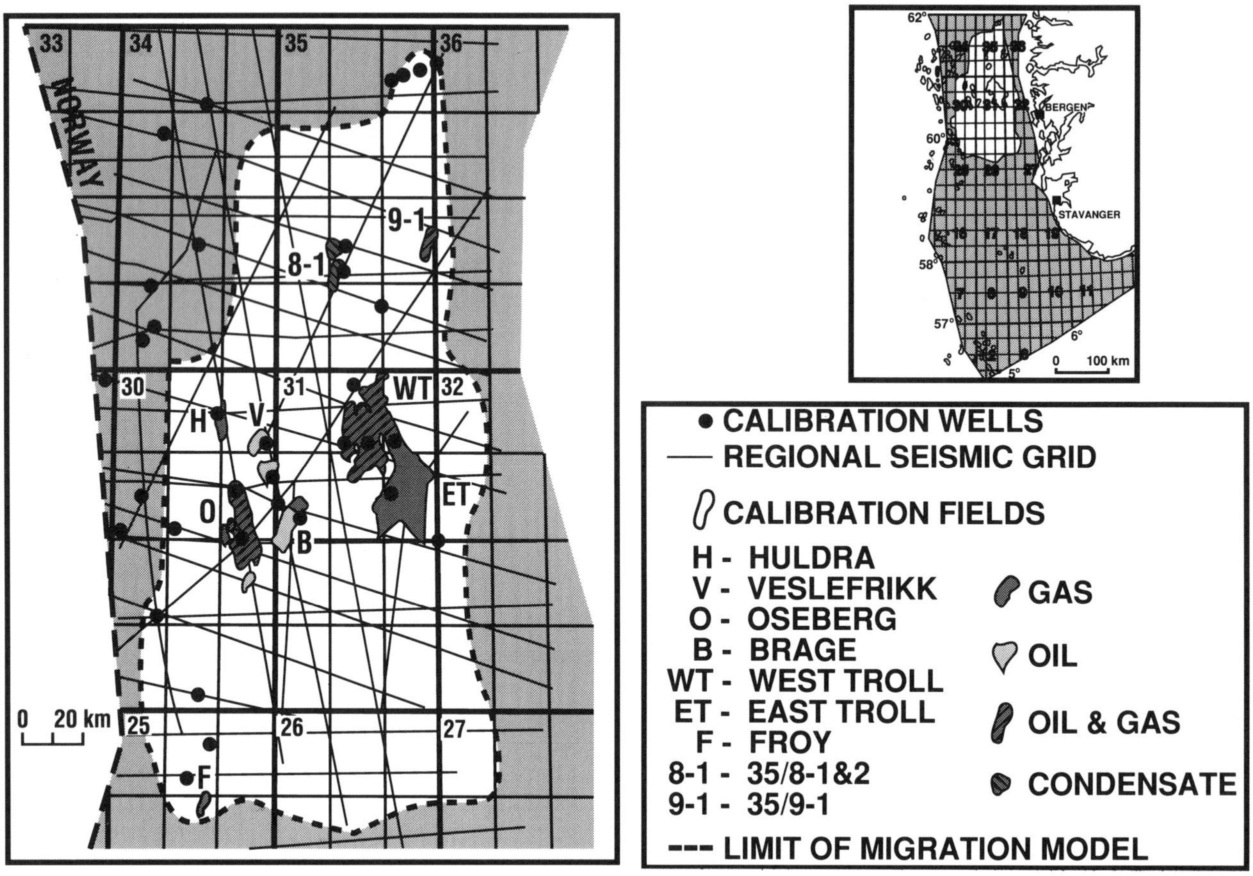

Fig. 1. Base map of the study area showing the location of the calibration wells, the grid of regional seismic lines, and the calibration fields and discoveries used in this study

of the study, Jurassic structure maps had not been finalized west of the Viking Graben axis. As a result the study area, bounded on the west by the graben axis, covers an area of approximately 100 km by 250 km.

The objectives of this study were to gain quantitative insights into the hydrocarbon system of this segment of the North Sea, and to use these insights to aid in the evaluation of several prospects. In particular, a number of prospects and leads lie within the study area. One way to evaluate these prospects and leads is to look at the results of the nearest exploration well. Would basin modelling provide information such that these prospects and leads might be evaluated differently?

Procedure

Figure 2 shows the four basic steps used in this study. The desired product, an assessment of prospects (step 4), requires the other three steps: compile the necessary data, model hydrocarbon generation, and model secondary migration. A basic description of each step follows. More detail as it relates to the North Viking Graben case study will be given in subsequent sections.

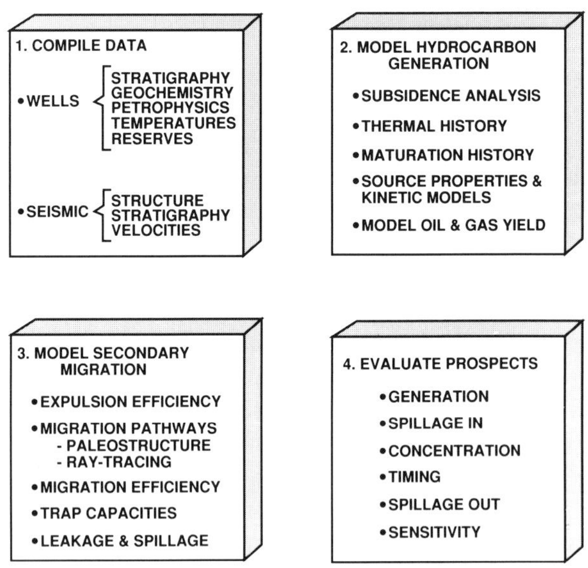

Fig. 2. Schematic showing the four basic steps used in this study.

Compile data

The most critical and time-consuming step is data compilation. Basic data come from wells (logs and samples) and from grids of seismic data. Wells provide information on the stratigraphy, geochemistry, petrophysics, current physical conditions, and in-place reserves in the vicinity of each well bore. Seismic data help to define the structure and stratigraphy beyond the limits of well control and provide seismic velocities for depth conversion of structure maps.

Model hydrocarbon generation

To model the history of hydrocarbon generation, one must have a model of the thermal history, a set of kinetic parameters that describe the transformation of organic matter into oil and gas, and geochemical properties (total organic carbon, hydrogen index, effective source thickness) for each source interval. Thermal histories should be calibrated against temperature and maturity (e.g., vitrinite reflectance) data from key wells. Well samples of the source intervals should be used to determine appropriate kinetic parameters and geochemical properties for the area being studied. The desired product is the amount of oil and gas generated within each grid of the migration model during each modelled time step.

Model secondary migration

The third step is to model secondary migration. Unfortunately this type of modelling is not as advanced as 1D thermal and hydrocarbon generation modelling. Migration must be modelled at least in 2D and ideally in 3D. A major limitation of 2D migration modelling is that conclusions can be misleading if there is significant migration into or out of the plane of the modelled 2D section. One conclusion of this study was that most traps filled from several directions making it dangerous to rely on 2D migration models.

To obtain at least a first approximation of secondary migration, equations and parameters to model expulsion, the location of migration pathways, migration efficiency, trap capacities, and the leakage and spillage of oil and gas through geologic time are necessary. Most of these processes are poorly understood and difficult to express mathematically. The parameters used in migration modelling software to approximate these processes are poorly constrained. Thus it is important to perform sensitivity analyses, vary input parameters within the range of perceived possibilities and monitor the change in the predicted output. Estimates of in-place reserves provide data against which the modelling results can be calibrated.

In Fig. 3 a schematic of a 3D carrier surface is used to illustrate the ray-tracing technique used by SEMI to model flow paths. SEMI would represent this surface as a series of grid cells in 3D space. Flowpaths out of each cell are computed by mapping the steepest upward gradient. This technique assumes that migration occurs near the top of a carrier unit

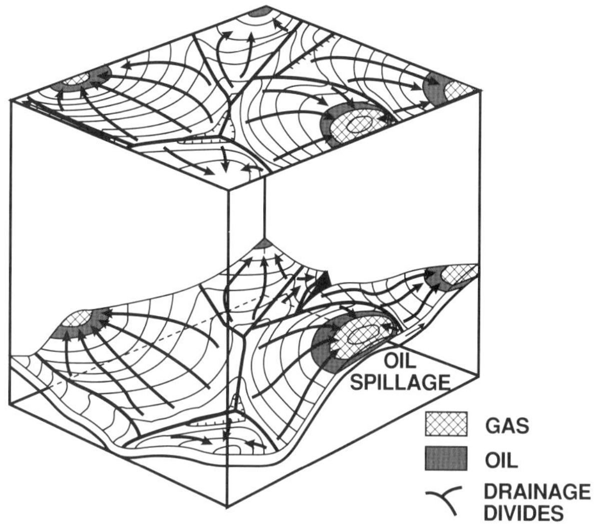

Fig. 3. Schematic of a surface assumed to be the top of a carrier horizon that illustrates SEMI's procedure of computing flow lines, drainage divides, and spill paths.

and that buoyancy is the primary factor in hydrocarbon migration (i.e., no significant hydrodynamic component of flow). Flow paths are extended until a point of flow convergence is reached at a structural culmination. Points of flow divergence mark drainage divides.

SEMI determines key parameters for the potential trap at each culmination, e.g., the position and depth of the crest and the synclinal or fault spill point. The capacity of each of these potential traps is obtained by computing the hydrocarbon pore volume between the crest and the spill point.

SEMI models secondary migration in two steps (Fig. 4). The first step, referred to as drainage, is the initial migration of hydrocarbons from the time they enter the carrier until they reach the nearest trap. The second step, referred to as spillage, involves those processes acting simultaneously once hydrocarbons reach a trap, e.g., leakage and spillage.

Drainage begins by summing the generated hydrocarbons throughout each drainage area during any given time step. The total available oil and gas must then be reduced to account for expulsion and migration efficiencies. More detail on how these efficiencies were defined will be provided in the next section.

Spillage begins with a determination of the volume of oil and gas that could be held by the trap. Oil was modelled to be incompressible in this model and thus does not change volume. Gas can leak out of the top of the trap, be dissolved in oil, or occur in a free gas state, i.e., as a gas cap. The density of free gas varies with pressure which is modelled as a function of depth. In terms of reservoir capacity, free

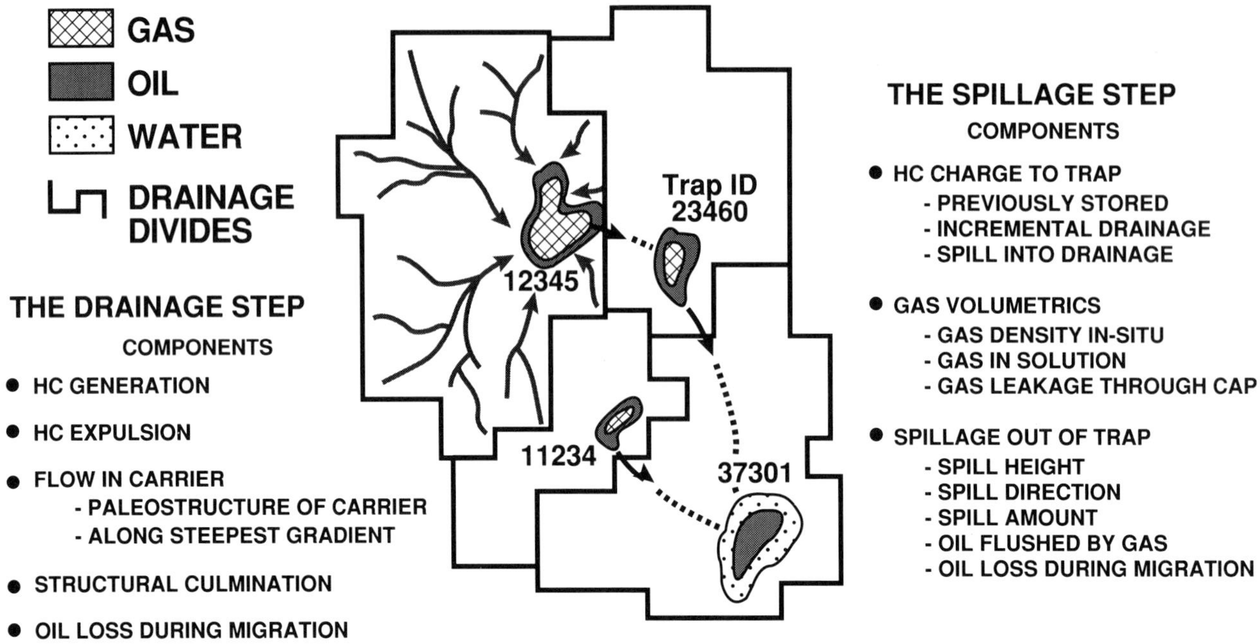

Fig. 4. Schematic of SEMI's drainage and spillage steps including the major processes that are modelled.

gas is considered to occupy space but gas dissolved in oil does not. If the volume of oil and free gas reaching a trap exceeds the trap capacity, first oil (with dissolved gas) and then free gas will spill. Spilt volumes are added to the hydrocarbon budget of the next updip drainage area.

Evaluate prospects

The final step is to use the model to evaluate the generation and migration risks of prospects and leads. Trap capacity in the model may be in error due to inaccuracies in the regional-scale maps used to build the model or as a result of the gridding procedures. Prospect-scale maps should be used to obtain a reliable trap capacity. The grids in the model can be adjusted so that the prospect has an appropriate capacity before oil and gas volumes are calculated.

The volumes of free gas and oil are monitored for several runs with slightly different input parameters. The smaller the difference in volumes on these sensitivity runs, the greater the confidence in the results.

Secondary migration modelling is not to the state that absolute numbers should be accepted as truth. A risk matrix helps to focus on the relative strengths and weaknesses of each prospect rather than the absolute numbers associated with a particular run.

North Viking Graben case study

The application of this procedure to the North Viking Graben study area will be summarized in the remainder of this paper. The discussion of the 1D hydrocarbon generation modelling will focus on an example well and the 3D migration modelling results will be shown for a 5 block area which includes the Huldra, Oseberg, and Brage fields.

TABLE 1

Time intervals used in the migration model

Interval number	Time step (Ma)	Chronostratigraphy
1	129 to 94	Lower Cretaceous
2	94 to 68	Upper Cretaceous
3	68 to 54	Paleocene
4	54 to 36	Eocene
5	36 to 29	Lower Oligocene
6	29 to 22	Upper Oligocene
7	22 to 6	Miocene
8	6 to 0	Plio-Pleistocene

In the final migration model, each cell has a dimension of 1 km by 1 km. In the time domain, eight geologic intervals were modelled (Table 1). Volumes of oil, free gas, and solution gas contained in each trap were modelled through time.

North Viking Graben data base

For this study, 35 wells were selected to build a data base for subsequent model input and calibration (Fig. 1). Primary criteria for well selection were the availability of temperature and organic maturation data. Geochemical data were compiled and analyzed to determine source rock properties. Petrophysical data were used to quantify properties of the reservoir and carrier units. Based on the well data and using

TABLE 2

Stratigraphic horizons mapped regionally and included in the migration model

Geologic Time (Ma)	Chronostratigraphy: Series	System	
0–2	Pleistocene	NEOGENE	TERTIARY
2–6	Pliocene	NEOGENE	TERTIARY
6–22	Miocene	NEOGENE	TERTIARY
22–29	U. Oligocene	PALEOGENE	TERTIARY
29–36	L. Oligocene	PALEOGENE	TERTIARY
36–54	Eocene	PALEOGENE	TERTIARY
54–68	Paleocene	PALEOGENE	TERTIARY
68–94	Upper	CRETACEOUS	
94–129	Lower	CRETACEOUS	
129–169	Middle to Upper	JURASSIC	

the basin architecture as a guide, maps of important properties (TOC, hydrogen index, etc.) were generated, digitized, and gridded at a grid-spacing of 4 × 4 km.

Stratigraphic data came from both well and seismic data. Regional seismic grids were used to map eleven major sequence boundaries throughout the study area (Table 2). All available well stratigraphy was tied to the seismic interpretation through synthetic seismograms. Time structure maps of each sequence boundary were digitized, converted to depth, and gridded at a grid-spacing of 1 × 1 km.

Estimates of in-place reserves were needed to calibrate the input parameters and functions used in the migration model. Published and in-house estimates of recoverable reserves for Huldra, Veslefrikk, Oseberg, Brage, West Troll, East Troll, Frøy, the 35/8-1 and 35/8-2 discoveries, and the 35/9-1 discovery (Fig. 1) were used to obtain estimates of in-place reserves.

North Viking Graben hydrocarbon generation

To obtain hydrocarbon generation histories, 1D models were used in two stages. In the first stage, 35 well locations were modelled and input parameters (e.g., palaeobathymetry, heat flow) were adjusted until predicted temperatures compared favorably with well data. In addition, predicted values of vitrinite reflectance, based on both an in-house kinetic model and one published by Larter (1988, 1989) were compared with observations. The input parameters used at each calibration site were mapped and gridded to ensure spatial consistency.

Two source intervals were considered in the modelling. The Upper Jurassic Draupne "hot" shales were assumed a source of both oil and gas. Data from 25 wells were used to develop maps of total organic carbon (TOC) and hydrogen index (HI). In-house procedures were used to convert present-day values to original values based on the temperature and burial histories. Coals within the Brent were the second modelled source. Since these coals contain terrestrial organic matter, they were assumed to generate only gas. A coal thickness map was prepared based on the interpretation of sonic logs from 33 wells. Due to limited data, the coals were assigned a TOC of 30% and a HI of 220 based on Ungerer et al. (1985). Other source intervals were thought to exist, but available data were not sufficient throughout the study area to prepare maps. Thus, one limitation of the final model is that only two source intervals were included.

In the second stage, 193 additional sites were "batch"-modelled by extracting stratigraphic data from the structure grids and the other required input parameters (e.g., palaeobathymetry, heat flow) from the grids that were built based on the 35 calibration sites. Hydrocarbon generation histories were obtained by combining the thermal histories, the source rock TOCs and HIs, and the kinetic models of Braun et al. (1990) for oil generation, MacKenzie and Quigley (1988) for primary gas generation, and Braun et al. (1988) for secondary gas generation. The generation histories were combined with grids of source thickness and richness (TOC) to obtain maps of oil and gas generation rates within each cell in the migration model for each time interval in units of kg m^{-2} Ma^{-1}.

Figure 5 illustrates the modelling technique used to develop a heat flow history. A backstripping approach was used to reconstruct total subsidence from present-day stratigraphy (van Hinte, 1978; Hardenbol et al., 1981). Temporal variations in eustasy and palaeobathymetry, along with the effects of compaction and erosion, were included. The component of subsidence due to crustal loading was computed and subtracted from the total subsidence to obtain an estimate of thermo-tectonic subsidence.

Thermo-tectonic subsidence can be used in divergent basins to interpret the magnitude of crustal thinning and infer the anomalous heat flow associated with rifting events (Royden et al., 1980; Sclater and Christie, 1980). A heat flow history can be developed and tested against available heat flow, well

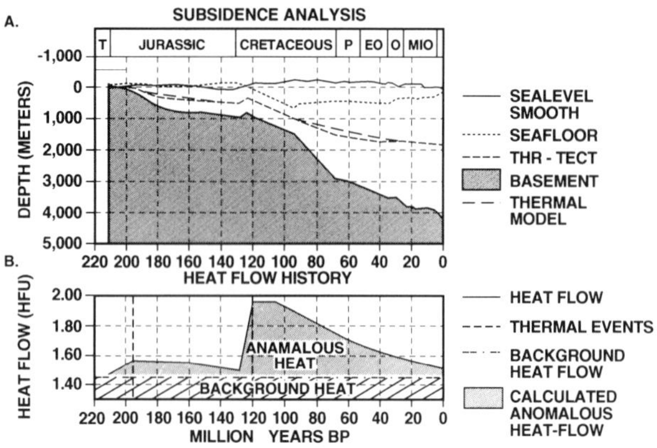

Fig. 5. An example of the subsidence and heat flow modelling done for each calibration well. An estimate of the component of subsidence due to crustal loading is subtracted from the total subsidence of the post-Triassic. The remainder, the thermo-tectonic subsidence, is used to infer the addition of heat due to crustal extension.

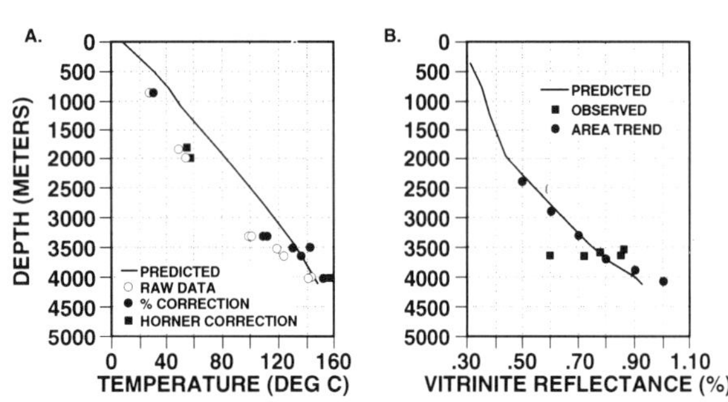

Fig. 6. An example of the heat flow calibration step in which predicted profiles of temperature and maturity (e.g., vitrinite reflectance) are compared with observations from the well and/or neighboring wells.

temperature data, and observed maturity indicators (e.g., vitrinite reflectance).

Figure 6 shows the present-day profiles of temperature and maturity resulting from the heat flow history shown in Fig. 5. The temperature match in the interval of interest (below 3000 m) is considered acceptable given the scatter in the observations. The maturity data from this and many wells have a limited depth range and show considerable variation. To compensate for these limitations, data from eight nearby wells that are in the same tectonic province were evaluated to develop a composite maturity profile. The circular points in Fig. 6, taken from the eight-well composite maturity profile, compare favorably with the predicted maturity profile.

Figure 7A shows the modelled temperature and maturity history for the Draupne source interval at the example well site. When this bed history is combined with TOC and HI data and the kinetic models (Braun et al., 1988, 1990; Mackenzie and Quigley, 1988), oil and gas yield histories can be obtained (Fig. 7B).

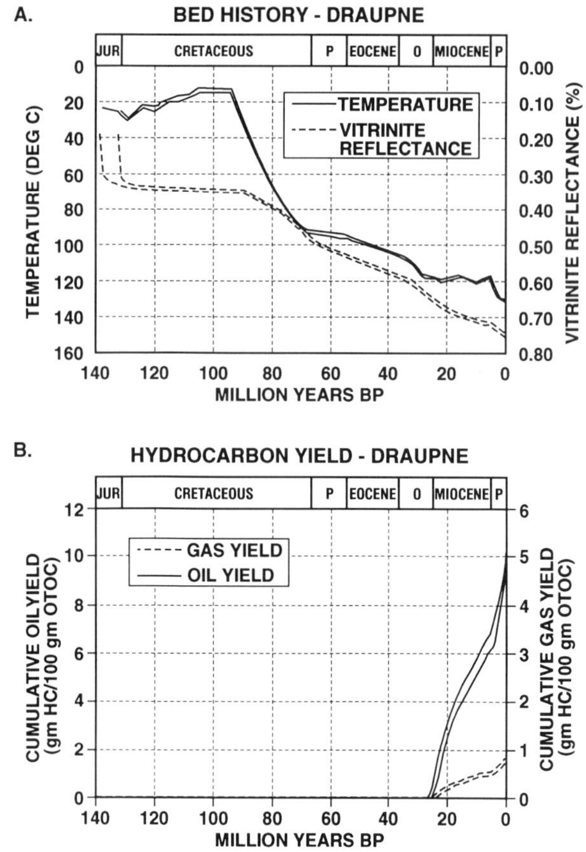

Fig. 7. The predicted temperature, maturation and generation history of the Upper Jurassic Draupne shales at the example well location. Double lines correspond to the top and base of the modeled interval.

North Viking Graben secondary migration

The secondary migration program SEMI (Sylta, 1987, 1991, and this volume) was used in this study to develop a pseudo-three-dimensional model of the North Viking Graben. It is not a full 3D model since only a single carrier/reservoir horizon is modelled at any one location through time. However, since we are interested in Jurassic prospects, this is not considered a major limitation for this study area.

As described by Sylta (1987, 1990, and this volume), the SEMI program uses gridded data and a set of user-defined parameters and functions to model the movement of hydrocarbons, first into a carrier unit, then updip to a structural trap, and, if there is any subsequent leakage or spillage, out of a trap. For this study, the subsurface area was represented by grid nodes with a 1-km spacing. Hydrocarbon generation, expulsion, and migration were modelled for 8 time intervals from earliest Cretaceous to the present (Table 1).

North Viking Graben model input

To formulate a model for computer input, the explorationist needs to consider how each component of the hydrocarbon system should be expressed. He must consider which intervals are effective sources, carriers, and reservoirs and how these elements interconnect. A conceptual model is developed to link the components together. The computer model provides a means to test this conceptual model against drill results and quantify its implications on undrilled structures. Some simplifications may be necessary due to the assumptions of the software being used or sparseness of data. Other simplifications may be adopted to optimize model run times. One strength of this type of modelling is that traditional or new concepts can be evaluated through sensitivity analyses.

The main elements of the conceptual model used to formulate input for the North Viking Graben are shown in Plate 1. The simplified Jurassic stratigraphy included four units: the Draupne hot shale, the Heather shale, the Sognefjord Formation, which is sand-prone around the Troll field, and the Brent Group, which is sand-prone throughout the study area.

Generation

The Draupne "hot shales" were considered to be a source for oil and gas while coals within the Brent were considered as a source of dry gas. Other potential sources, e.g., the Heather and formations older than the Brent, were not included as there were insufficient data to map geochemical parameters for these other potential sources.

Expulsion

For expulsion efficiency from the Draupne, an exponential function of the vertical distance between the base of the Draupne and the top of the carrier was used. Expulsion from the Brent coals was defined as a fixed percentage of the total generated gas.

The initial concept was that oil and gas generated in the Draupne would move down into the Brent and then migrate laterally to the traps. Gas generated from Middle Jurassic coals would also move laterally within the Brent to the traps. Draupne expulsion would only be significant when the vertical distance between the Draupne and the Brent was less than a few hundred meters.

The initial runs indicated major problems with these assumptions and simplifications. Very little oil was trapped in the known oil fields. The expulsion efficiency of the Draupne as a function of distance had to be increased while the expulsion efficiency of the Brent coals had to be decreased. These changes are largely an artifact of our simplifying assumptions. First, not all potential source intervals were included due to data limitations. Second, the model assumed hydrocarbon had to reach the carrier unit (Brent) before secondary migration commenced. It is more likely that thin silt and sand layers much closer to the source intervals can serve as initial paths for secondary migration. The expulsion efficiencies required to match known reserves had to be abnormally high so as to counteract these limitations.

Carrier

In the study area, most hydrocarbons are reservoired in Middle Jurassic Brent Group and older sandstones. The main exception is the Troll field in which the Upper Jurassic Sognefjord Formation is the primary reservoir. In the model, the top of the Brent was defined as the carrier except where the overlying Sognefjord Formation contains more than 5% sand. Thus, in the model, the carrier jumps stratigraphic level in the vicinity of Troll (Plate 1).

Hydrocarbon loss

To model the oil retained in the carrier, the equations and constants developed by Sylta (1987, and this volume) were used. The most important parameter in this equation is permeability. Unfortunately, we had limited data on porosity or permeability within the study area. Simple porosity versus depth and permeability versus porosity functions were used in the model. Overpressure was not modelled and no corrections in the porosity or permeability estimate were made in regions of known overpressure.

Since mapping was performed at a regional scale, many small structures were not properly mapped. As a result, a significant amount of oil has to be "lost" in the model in unmapped traps and dead ends. An initial estimate of 0.25 m^3 m^{-2} was used for the amount of "lost" oil in unmapped traps and dead ends.

Gas leakage through the top of the traps varies with the height of the gas column and a user-defined gas leakage factor (Sylta, 1991). The gas leakage factor derived by Sylta (this volume) was used as an initial estimate.

The parameters affecting secondary migration efficiency (or hydrocarbon loss) were varied in a series of sensitivity runs. In particular, the porosity versus permeability function, the volume assumed to be lost in "dead ends", and the gas leakage factor were varied systematically; the predicted hydrocarbon volumes and compositions from each run were compared to known reserves to determine a set of "optimum" parameters.

Hydrocarbon spill

An empirical approach was adopted to evaluate whether faults are open or closed to migrating hydrocarbons (Fig. 8). When major faults were defined

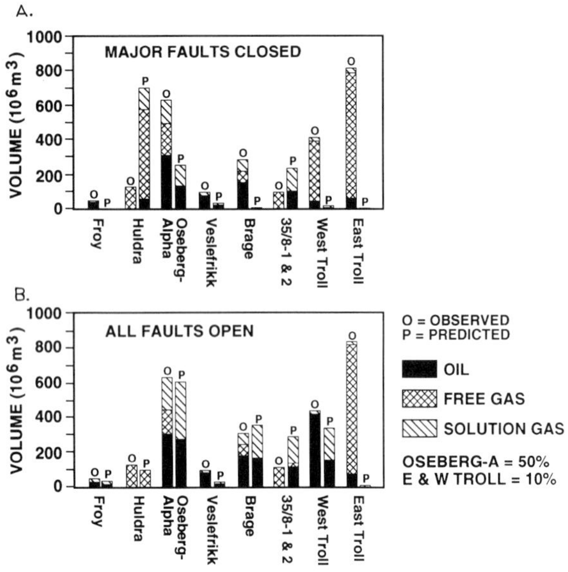

Fig. 8. Comparison of observed and predicted trapped hydrocarbons for two runs in which all parameters remain constant except fault leak. (A) All faults are modelled as open. (B) Major faults are modelled as closed.

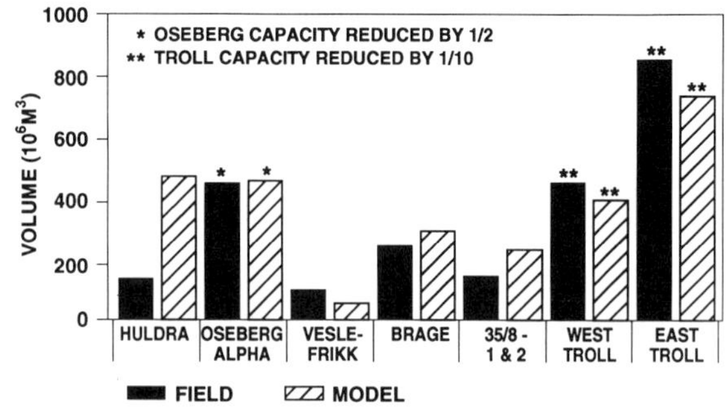

Fig. 9. Comparison of estimates of trap capacities derived from reserve estimates (solid bars) and the trap capacities in the model (hatchured bars).

to be closed to migrating hydrocarbons, too much hydrocarbon was trapped on the low-side of the closed faults and too little was trapped on the high-side as compared to known fields. When all faults were defined to be open, a much better match to known reserves was obtained. Based on these results, subsequent runs assumed the faults to be open at the time scale associated with secondary migration.

The spill point for each trap was based on the shallowest fault-related leak point or synclinal spill point. If hydrocarbon charge exceeded trap capacity, there would be spillage first of oil with dissolved gas and then of free gas.

North Viking Graben model calibration

Reservoir calibration consists of two phases. First, trap capacities for the calibration fields and discoveries must be compared to predicted capacities. Next the reported hydrocarbon phases in each trap must be compared to predicted phases.

Trap capacities

There was an acceptable fit to the capacities computed from the reported reserves when a hydrocarbon saturation of 70% was assumed for the reservoirs (Fig. 9). Mismatches in trap capacities are primarily due to mapping and gridding inaccuracies. Some editing of the structure grids was done to improve the match. The only significant mismatch is in Huldra where several small traps within the model coalesced with the main structure at Huldra.

Troll merits a few additional comments. The present-day capacity modelled for Troll agrees with the field estimates. In the model, however, trap capacity decreased and then increased through the Tertiary. This loss of capacity is believed to be an artifact of the simplifications used to handle the significant Neogene erosion at the crest of Troll. Part, but not all, of the mismatch in gas at Troll is attributed to an artificial tilting of the trap in the model which resulted in gas spillage.

Hydrocarbon distribution

The second phase of the reservoir calibration focused on the parameters controlling the volumes of hydrocarbons expelled, migrated, leaked, and lost. Over 40 modelling runs were made to evaluate the effects of individual parameters. Modelled in-place reserves were compared with estimated in-place reserves at the calibration fields and discoveries.

Figure 10 compares selected modelling runs for three major fields. Note that runs 30 to 35 have too little oil in Oseberg-alpha and in Brage but are the only runs that have an acceptable match to the gas in Troll. The model could be calibrated to get an acceptable match at Oseberg-alpha and Brage or at Troll; a single set of parameters to match all three was not found.

There are several possible explanations for this dichotomy. Oseberg and Brage are nearer the graben than Troll and so differences may relate to short- versus long-distance migration. The relative importance of the two sources for Oseberg–Brage may be different from that for Troll. Another explanation, one we favor, is that the loss of capacity during the Tertiary at Troll due to the way in which the erosion was handled in the model resulted in unrealistic amounts of trapped gas being spilt out of the Troll structure.

North Viking Graben model output

Several maps focusing on five blocks in and around Oseberg are included here. These illustrate how the

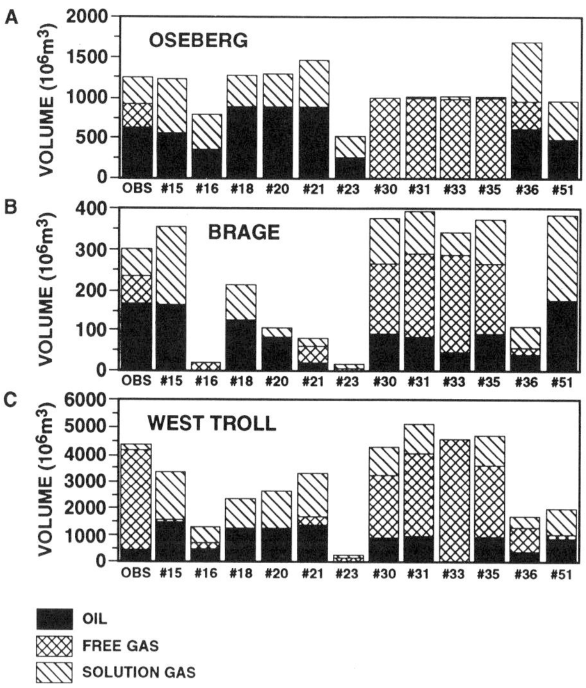

Fig. 10. Comparison of observed and predicted trapped hydrocarbons in (A) Oseberg-alpha, (B) Brage, and (C) West Troll for selected runs.

modelling results can be applied to solve exploration issues.

Plate 2 shows the present-day structure map of the top of the carrier with superimposed flow lines, drainage divides, and traps. Figure 11 is a corresponding trap statistics map. This map lists the capacity, percent hydrocarbon fill, and the volumes of oil, gas in solution, and free gas for each trap exceeding a user-defined minimum capacity. Similar maps were generated corresponding to the end of each modelled interval (Table 1).

A color-coded map indicating the rate of oil flow through each cell of the model during the Upper Cretaceous is shown in Plate 3 (warm colors indicate low flow rates, cool colors indicate high flow rates). Plate 4 is an oil flow rate map for the Pleistocene. It is apparent that the location of primary oil generation (the swath of warm colors) has moved through time from the graben axis up towards the platform. Active spill paths are marked by "rivers" of cool colors.

The predicted fill history of individual fields and prospects were displayed. For example, Fig. 12 shows for Oseberg-alpha the subsidence of its crest and

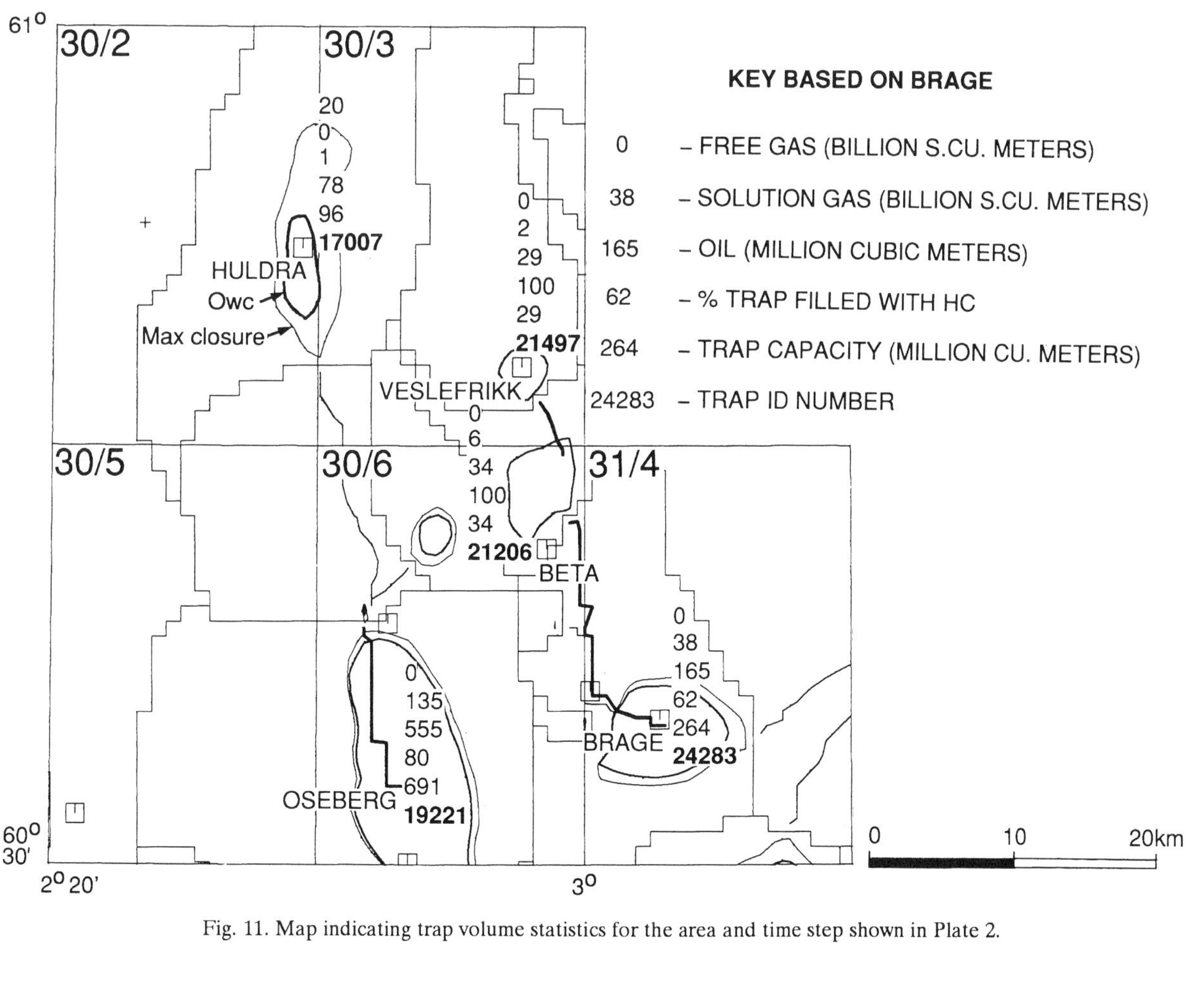

Fig. 11. Map indicating trap volume statistics for the area and time step shown in Plate 2.

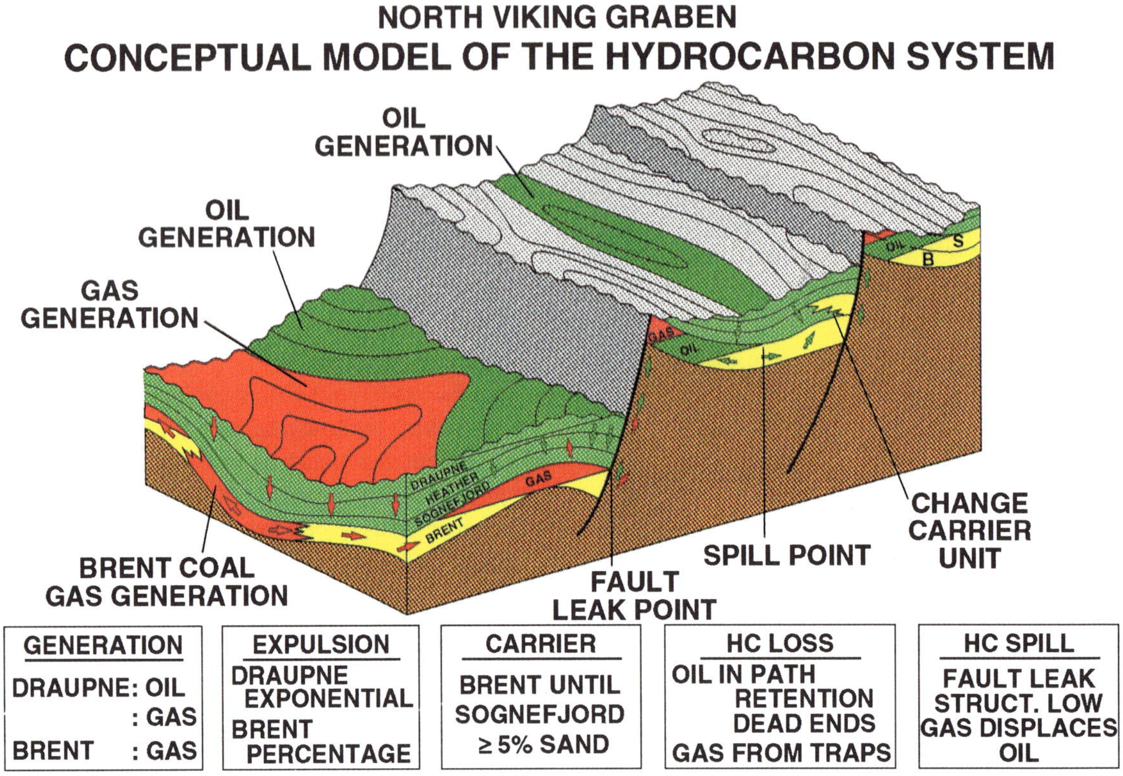

Plate 1. Schematic illustrating the main elements of the concepts used in defining the North Viking Graben hydrocarbon system.

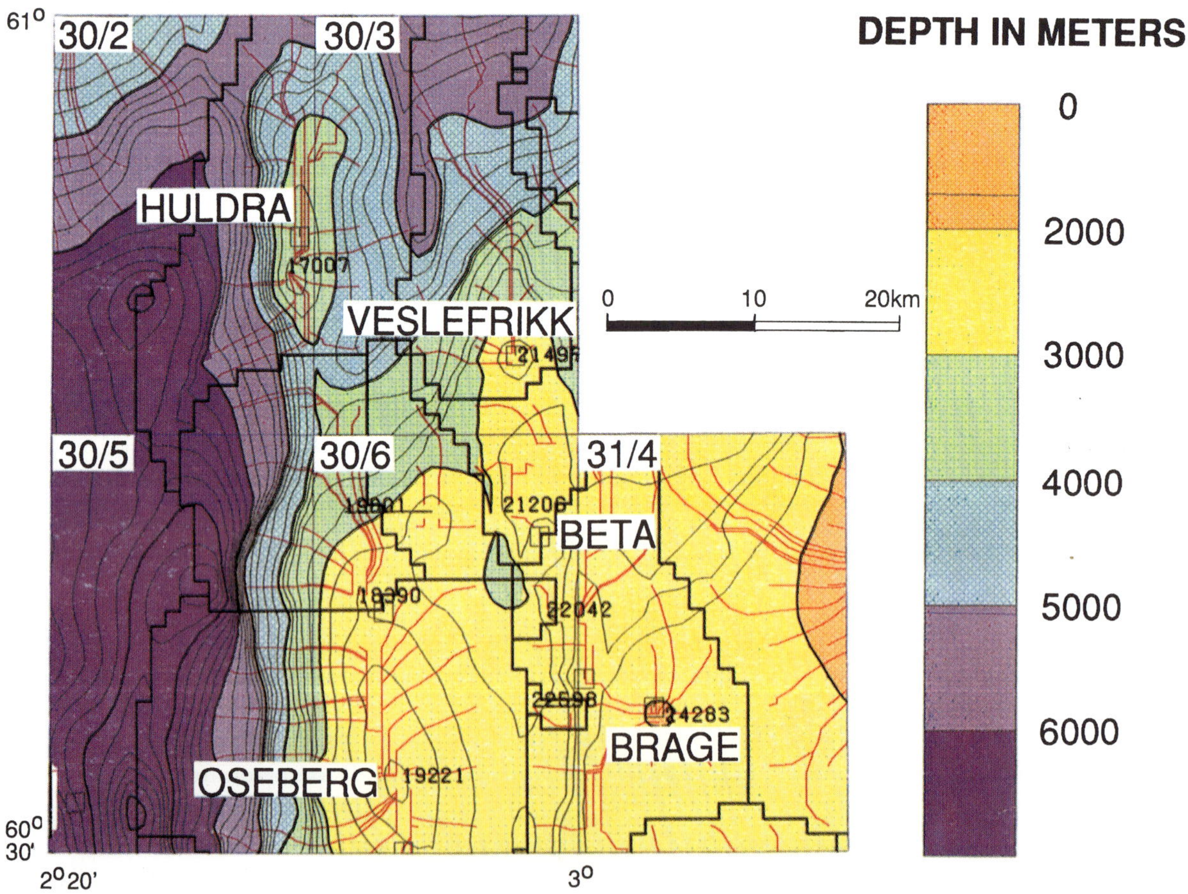

Plate 2. Present-day structure map of the carrier with superimposed flow lines, drainage divides, and structural closures (i.e., traps) for a selected portion of the total model.

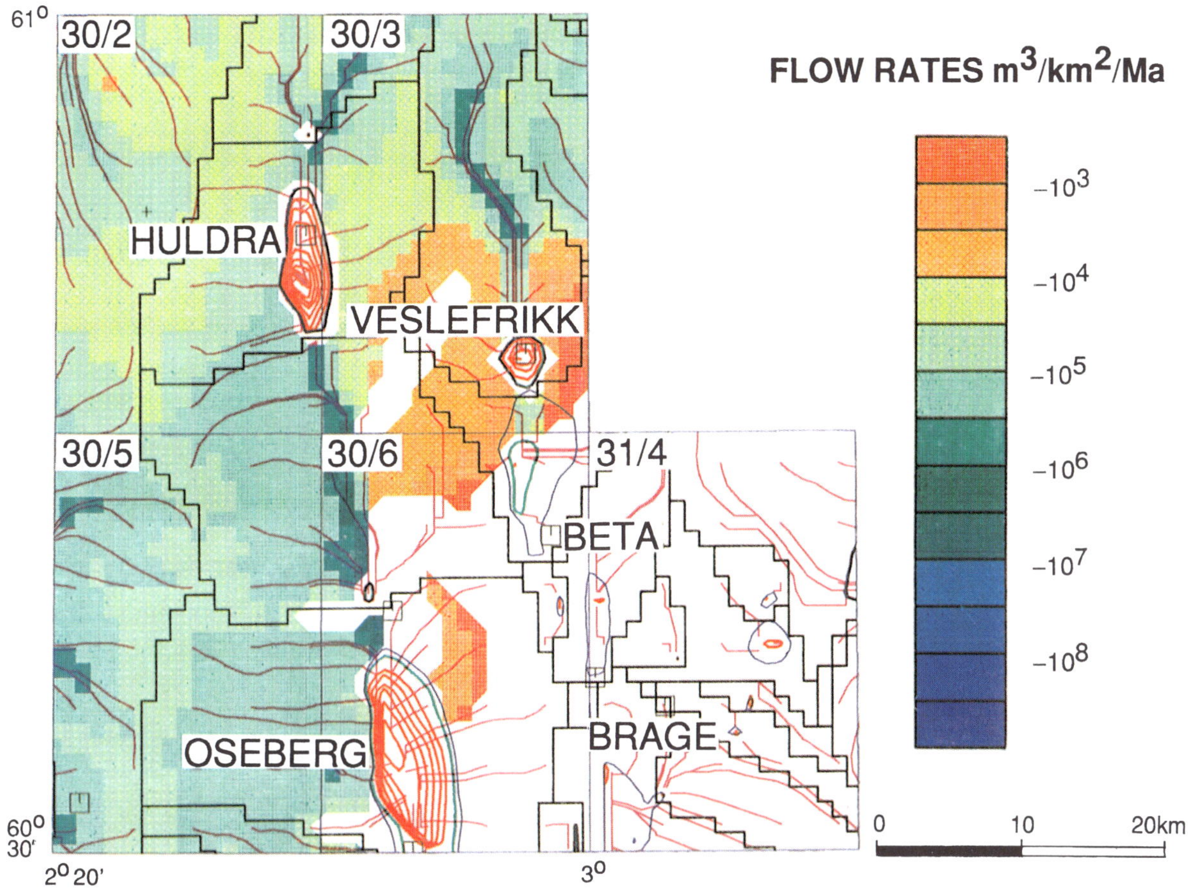

Plate 3. Map showing color-coded oil flow rates through each cell of a selected portion of the total model for the interval from 94 Ma to 68 Ma (Upper Cretaceous). The area of primary migration is near the axis of the graben and a few spill paths are apparent by "fingers" of high flow rates.

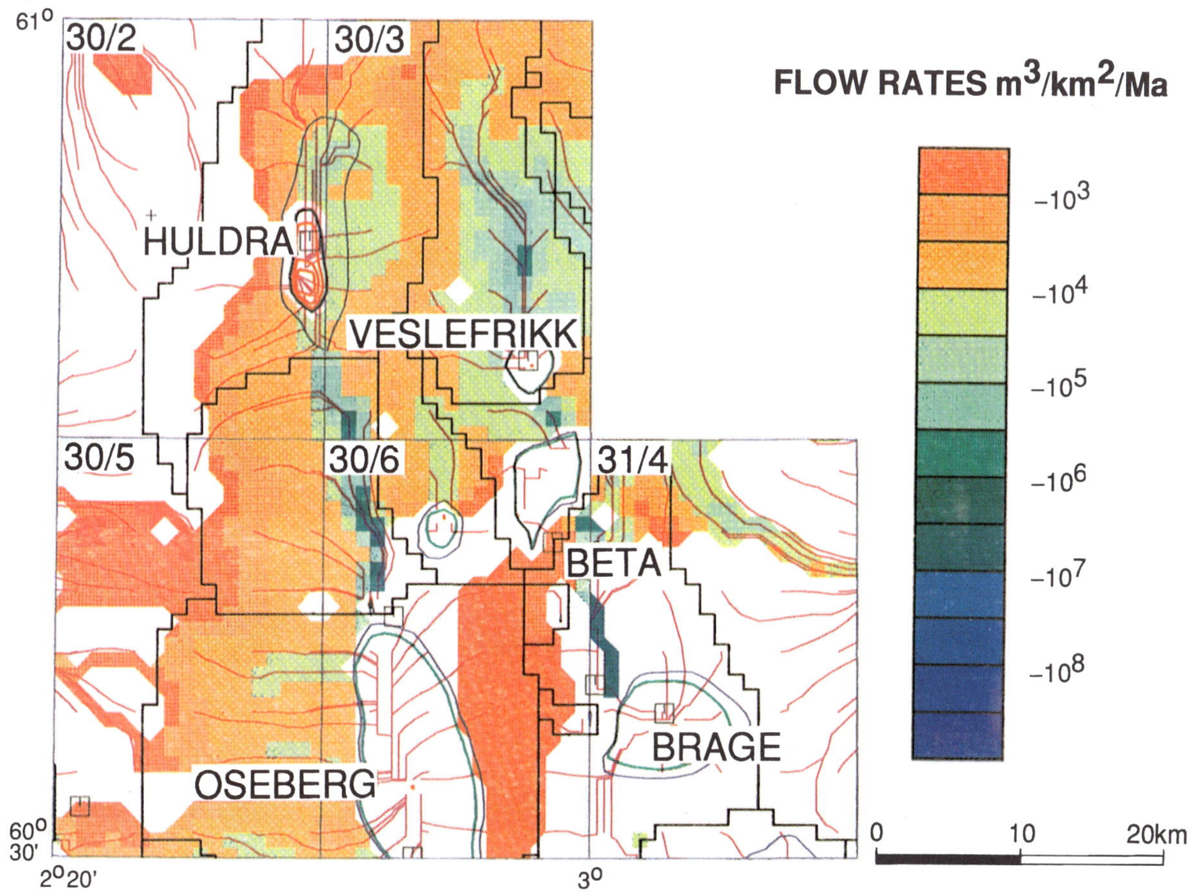

Plate 4. Map showing color-coded oil flow rates through each cell of a select portion of the total model for the interval from 6 Ma to 0 Ma (Plio-Pleistocene). The area of primary migration has moved towards the platform and major spill paths are apparent.

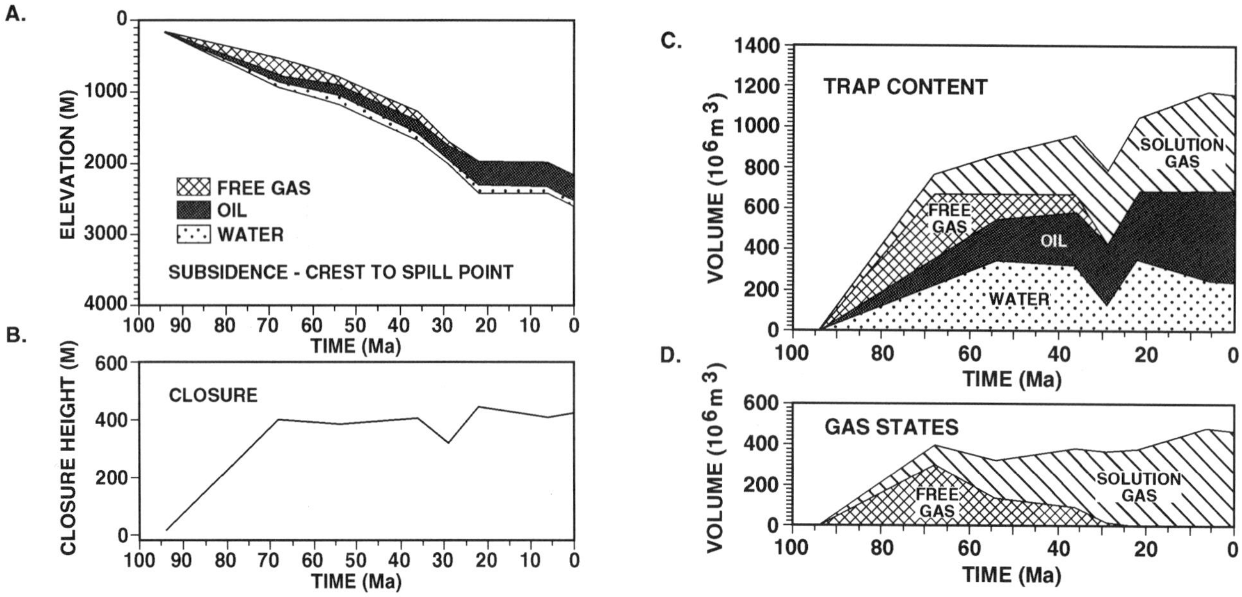

Fig. 12. Modelled trap fill history for Oseberg-alpha. (A) The subsidence of the modelled crest and spill point. (B) The amount of closure, vertical distance between crest and spill point. (C) The volumes contained within the trap. (D) The volumes of free and solution gas.

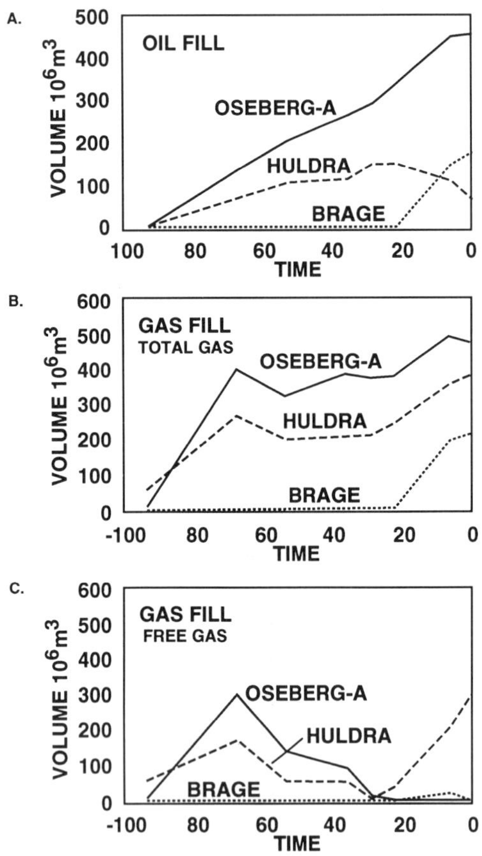

Fig. 13. Comparison of the trap fill history of (A) oil, (B) total gas, and (C) free gas for Oseberg-alpha, Huldra, and Brage.

spill point, the fluid volumes within the trap, the gas states, and the amount of closure.

Predicted reservoired fluid histories for three traps are shown in Fig. 13. Huldra contained oil, but excess free gas subsequently flushed the oil out of the trap. Brage began to fill much after Huldra and Oseberg. These fill histories were anticipated; however, qualitative supposition has been replaced with a quantitative model that can be interrogated (i.e., different assumptions can be tested).

Prospect evaluation

The final step is to use the model to evaluate the generation and migration risks of several prospects. Often qualitative risks have been assigned prior to basin modelling. The model developed here allows one to test these ideas and either add support to the assigned risk or suggest the need to reevaluate the risk.

In the case of the North Viking Graben, four types of traps were envisioned. Three of the four are illustrated in Fig. 13. The "Huldra-type" trap has gone past its maximum oil content. Due to high maturation levels within the drainage area, an excess of gas arrived at the trap and has flushed oil out of the trap. The "Oseberg-type" trap is at or near maximum oil content due to a combination of generation within its local drainage area and spillage from traps further down the migration path. The "Brage-type" trap has little oil due to maturation within its immediate drainage area but has a significant amount of oil that was flushed from deeper traps. The fourth trap type has little or no oil due to local generation or spillage from deeper traps. This might be called the "dry-hole" type.

Three prospects within the study area were being evaluated at the time of this study. Two prospects were assigned to the "Huldra-type" based on the nearest drill results. The third was mapped with a large capacity but was thought to be shielded from mature source rock by an intervening horst, a "dry-hole"-type trap.

Trap fill histories were generated for the three prospects based on the "optimum" parameters. For the first two prospects, drainage divides played a key role. One was modelled to be at or near maximum oil fill (Oseberg-type rather than Huldra-type). The second was modelled to be at or just past maximum oil fill (transitional between Oseberg-type and Huldra-type). For the third prospect, the suspicion that it was under-filled was supported. The modelling indicated that the qualitative risk assigned to one prospect was correct but the risk assigned the other two merited reconsideration.

One method to summarize the results of this evaluation is to use a risk matrix. Risk matrices are constructed by raising a series of critical questions and then using the modelling results to obtain an understanding of the range of possible answers to these questions. Questions that might be raised include: How much of the drainage area for a given prospect is mature? Could significant volumes of oil or gas have spilled into the prospect's drainage area from further down the migration path? Do flow paths converge towards the prospect? How does the evolution of trap capacity compare with the timing of migration? Does the hydrocarbon content of the prospect vary significantly for runs with slightly different input parameters?

Table 3 shows a risk matrix in which the five questions posed above are summarized by five words: generation, spillage, concentration, timing, and sensitivity, respectively. The evaluation of these components of risk based on the modelling is shown for these three prospects. A "+" indicates an increased confidence level that the prospect will be an economic discovery. A "–" indicates a negative impact on the prospect. A "?" indicates that the modelling results were not conclusive. Those components thought not to apply to the prospect are left blank.

Figure 14 shows the modelled trap capacities for the three prospects for selected modelling runs. Run 51 is the "optimum" run, but the histograms indicate the variability of the results with changes in the input parameters.

Prospect A

Prospect A is at a focusing point for flow from a mature area and trap capacity existed when migration began. Spillage from down-dip traps is not a significant factor. A negative component is that there was a significant variation in the oil to gas ratio on different modelling runs (Fig. 14A). The

TABLE 3

Risk matrix for generation and secondary migration

	Prospect A	Prospect B	Prospect C
Generation: within drainage area	+	+	–
Spillage in: from down-dip traps	N/A	N/A	N/A
Concentration: trap vs. migration	+	+	+
Spillage out: if excess HC charge	?	?	
Sensitivity: variability of results on different runs	?	+	+

+ = element with positive impact on the prospect; – = element with negative impact on the prospect; ? = element considered important but inconclusive; N/A = element not considered to be important.

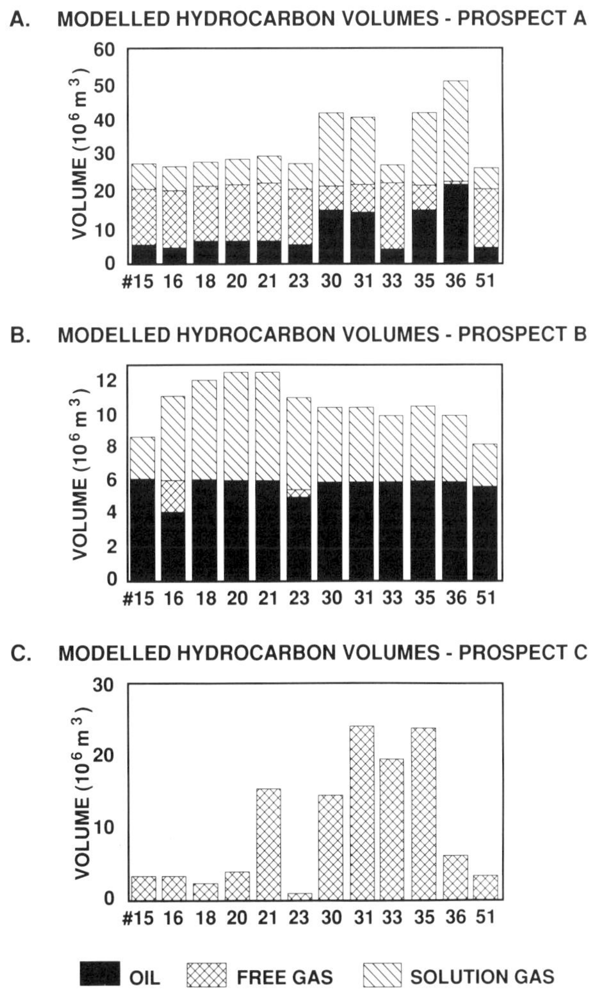

Fig. 14. Comparison of predicted trapped hydrocarbons in prospects A, B, and C for selected runs. This indicates how sensitive the results are to slight changes in the input.

trap was modelled as having 20% of its capacity filled with oil for the "optimum" model. This varied between 15% and 95%. Thus the chance that hydrocarbons reached the prospect is relatively high; but the ratio of oil to free gas can not be predicted with confidence. The drainage area for this prospect was distinct from that of the nearest drilled structure. This was one of the prospects for which the modelling suggested a more optimistic evaluation.

Prospect B

The evaluation of Prospect B is quite similar to Prospect A with one major exception, the sensitivity of hydrocarbon content to changes in model parameters (Fig. 14B). The oil to gas ratio is consistently modelled to be high for Prospect B. This trap was modelled as having 100% of its capacity filled with oil for the "optimum" model. The minimum case was 65% oil fill. Thus the chance that hydrocarbons reached this prospect is high and a high oil to gas ratio can be predicted with greater confidence. The drainage area for this prospect was also distinct from that of the nearest drilled structure. The modelling indicated that a more optimistic evaluation was in order.

Prospect C

Although Prospect C is at a focusing point for flow, its drainage area is predominantly immature. There is no significant spillage from down-dip traps so the hydrocarbon charge is quite limited. Trap timing is early but peak generation has probably not occurred. All the modelling runs indicated only a small amount of gas in the prospect, probably sourced from the Brent coals (Fig. 14C). Thus the risk that enough migration has occurred is very high. The initial suspicion that the structure was shielded from oil generated in the Draupne was supported by the modelling.

We hasten to point out that we have been discussing only the risks associated with hydrocarbon generation and migration. Other risks, such as reservoir quality and seal, must be considered before a final prospect seriatim can be made.

Conclusions

This study demonstrates that even though many of the migration parameters are not well constrained (e.g., expulsion efficiency, oil loss during migration), it is possible to calibrate some parameters and test the sensitivity of trap fill for prospects and leads. A good set of reservoir data from a number of fields and/or discoveries in the area is required. In this study, the available reservoir data allowed us to investigate oil expulsion efficiency. Data or time constraints prevented us from experimenting with other parameters, e.g., the volume of oil lost in small traps during migration. The results from the various modelling runs suggested a more optimistic evaluation than first thought for two prospects and supported the initial prognosis for a third. Through sensitivity analyses, a degree of confidence in the predictions was obtained.

One way to demonstrate the importance of migration modelling is to summarize the contribution this study made on our view of the North Viking Graben hydrocarbon system. Four main points can be made.

(1) A common way to qualitatively assign migration risk to a prospect was to look at the nearest drilled structure. Two of the three prospects were modelled to have significantly different drainage areas than their nearest drilled structure. In both cases, the modelling suggested a better chance for oil than initially assumed.

(2) It was thought that any hydrocarbons generated within the Draupne would not enter the Brent carrier if the distance between the Draupne and the Brent exceeded a critical value, on the order of a hundred meters. This would exclude the Draupne as a source throughout most of the center of the graben. It was further thought that if the Draupne near the graben axis was an effective source, so much gas would have entered the system that all the oil in Oseberg would have been flushed out. The modelling showed that oil generation was less voluminous than expected. To match the known oil reserves, expulsion efficiency had to exceed 75%. Gas from the Draupne did not flush the oil in the known fields but flushing did occur in the model if the contribution from the Brent coals exceeded 30% of its generated gas.

(3) There was a basic understanding of how traps had filled and spilled based on present-day structure maps of the tops of reservoir units. This was updated with a model of palaeodrainage and palaeospill directions. Trap fill history diagrams helped quantify the timing of hydrocarbon movement. Drainage areas around Brage have shifted considerably through time. The modelling suggests that most of the hydrocarbons in Troll have come from the northwest.

(4) The pressure gradients across several major faults suggest that these fault segments are at least partially sealing at present. Does it follow that these faults have been totally closed to hydrocarbon flow during the past? The modelling was able to address this question and the results indicate that the faults may inhibit hydrocarbon flow but do not totally prevent it on a geological time scale.

Based on this study, some general statements can be made on the ways in which migration modelling will impact exploration. First, this type of modelling leads to a thorough integration of structural, stratigraphic, geochemical, petrophysical, and geophysical data into a consistent framework. This alone can have a major positive impact. During the model calibration phase, elements of the explorationist's conceptual model can be tested quantitatively. When using state-of-the-art modelling software, many of the important processes in the formation of an oil or gas field are modelled. The resulting model can be tested by comparing predictions with observations at known accumulations and dry holes. This will link more closely exploration and production departments. Finally, the model can be used to make sure that the conditions required for a successful discovery are understood and that there is an acceptable degree of certainty that these conditions exist. If more than one opportunity exists, prospects can be ranked against each other with more objectivity.

Acknowledgements

The authors wish to express their thanks to the geologists and geophysicists at Esso Norge in Stavanger who made this study possible. In particular, we acknowledge the work and cooperation of B.D. Wiggins, T.R. Nardin, T. Valheim, H. Bolås, S.A. Kristoffersen, M.A. Moore, T.S. Ringstad, and J. Seekatz. They thank Esso Norge, Exxon Production Research, and IKU for allowing this work to be published. We are grateful for the comments and suggestions of the reviewers that resulted in an improved paper.

References

Bethke, C.M., 1985. A numerical model of compaction-driven groundwater flow and heat transfer and its application to the paleohydrology of intracratonic sedimentary basins. J. Geophys. Res., 90: 6817–6828.

Braun, R.L., Burnham, A.K. and Sweeney, J.J., 1988. Petroleum Geochemistry Project, Annual Report, October 1987 to September 1988. LLNL Internal Rep., UCID-21-303-88-4.

Braun, R.L., Burnham, A.K., Reynolds, J.G. and Clarkson, J.E., 1990. Pyrolysis kinetics for lacustrine and marine source rocks by programmed micropyrolysis. Energy Fuels, 5: 192–204.

Burrus, J. and Audebert, F., 1990. Thermal and compaction processes in a young rifted basin containing evaporites: Gulf of Lions, France. Am. Assoc. Pet. Geol., Bull., 74: 1420–1440.

Doligez, B., Bessis, F., Burrus, J., Ungerer, P. and Chénet, P.Y., 1986. Integrated numerical modelling of sedimentation, heat transfer and fluid migration in a sedimentary basin: the THEMIS model. In: J. Burrus (Editor), Thermal Modelling in Sedimentary Basins. Technip, Paris, pp. 173–195.

Doligez, B., Ungerer, P., Chénet, P.Y., Burrus, J., Bessis, F. and Bessereau, G., 1987. Numerical modelling of sedimentation heat transfer, hydrocarbon formation and fluid migration in the Viking Graben, North Sea. In: J. Brooks and K. Glennie (Editors), Proceedings of the 3rd Conference on Petroleum Geology of Northwest Europe. Graham and Trotman, London, pp. 1039–1048.

Durand, B., Ungerer, P., Chiarelli, A. and Oudin, J.L., 1983. Modélisation de la migration de l'huile; application à deux exemples de bassins sédimentaires. In: Proceedings of the 11th World Petroleum Congress. Wiley and Sons, New York, N.Y., PD 1(1): 3–15.

Hardenbol, J., Vail, P.R. and Ferrer, J., 1981. Interpreting paleoenvironments, subsidence history and sea level changes of passive margins from seismic and biostratigraphy. Oceanol. Acta. Proc. 26th Int. Geol. Congr., Geology of Continental Margins, pp. 33–44.

Larter, S., 1988. Some pragmatic perspectives in source rock geochemistry. Mar. Pet. Geol., 5: 194–204.

Larter, S., 1989. Chemical models of vitrinite reflectance evolution. Geol. Rundsch., 78 (1): 349–359.

MacKenzie, A.S. and Quigley, T.M., 1988. Principles of geochemical prospect appraisal. Am. Assoc. Pet. Geol., Bull., 72: 399–415.

Nakayama, K., 1986. Two-Dimensional Basin Analysis for Petroleum Exploration. Ph.D. Dissertation, University of South Carolina, Columbia, S.C., 214 pp.

Royden, L., Sclater, J.G. and von Herzen, R.P., 1980. Continental margin subsidence and heat flow: important parameters in formation of petroleum hydrocarbons. Am. Assoc. Pet. Geol., Bull., 64: 173–187.

Sclater, J.G. and Christie, P.A.F., 1980. Continental stretching: an explanation of the post-mid-Cretaceous subsidence of the Central North Sea Basin. J. Geophys. Res., 85: 3711–3739.

Sylta, Ø., 1987. SEMI — a program for the modelling of buoyance driven, secondary migration of oil and gas by means of a ray-tracing technique. IKU Rep. 25.2403.00/01/87, 51 pp. (open).

Sylta, Ø., 1991. Modelling of secondary migration and entrapment of a multicomponent hydrocarbon mixture using equation-of-state modelling techniques. Proc. Petroleum Migration Meet., Burlington House, London, September 1989.

Sylta, Ø., 1993. New techniques and their application in the analysis of secondary migration. In: A.G. Doré et al. (Editors), Basin Modelling: Advances and Applications. Norwegian Petroleum Society (NPF), Special Publication 3. Elsevier, Amsterdam, pp. 385–398 (this volume).

Sylta, Ø., Christiansen, J.B., Leith, L.T., Lippard, S.J., Liu, G., Raa, A.M., Ritter, U. and Smelror, M., 1990. South Viking Graben Basin Modeling Study, 1990. Institute for Continental Shelf and Petroleum Technology Research (IKU), Rep. 21.3472.00/01/90 (restricted).

Ungerer, P., Chiarelli, A. and Oudin, J.L., 1985. Modelling of petroleum genesis and migration with a bidimensional computer model in the Frigg sector, Viking Graben. In: B.M. Thomas et al. (Editors), Petroleum Geochemistry in Exploration of the Norwegian Shelf. Norwegian Petroleum Society (NPF), Graham and Trotman, London, pp. 121–129.

Ungerer, P., Doligez, B., Chénet, P.Y., Burrus, J., Bessis, F., Lafargue, E., Giroir, G., Heum, O. and Eggen, S., 1987. A 2D model of basin scale migration by two-phase fluid flow; application to some case studies. In: B. Doligez (Editor), Migration of Hydrocarbons in Sedimentary Basins. Technip, Paris, pp. 415–456.

Ungerer, P., Burrus, J., Doligez, B., Chénet, P.Y. and Bessis, F., 1990. Basin evaluation by integrated two-dimensional modelling of heat transfer, fluid flow, hydrocarbon generation, and migration. Am. Assoc. Pet. Geol., Bull., 47: 309–335.

van Hinte, J.E., 1978. Geohistory analysis — application of micropaleontology in exploration. Am. Assoc. Pet. Geol., Bull., 62: 201–220.

Welte, D.H. and Yukler, M.A., 1981. Petroleum origin and accumulation in basin evolution: a quantitative model. Am. Assoc. Pet. Geol., Bull., 65: 1387–1396.

FRED W. SCHROEDER *Exxon Production Research Company, P.O. Box 2189, Houston TX 772252-2189, U.S.A.*
ØYVIND SYLTA *IKU Petroleum Research, N-7034 Trondheim, Norway*

History of hydrocarbon emplacement in the Oseberg Field determined by fluid inclusion microthermometry and temperature modelling

Olav Walderhaug and Willy Fjeldskaar

Petrographic examination of sandstone samples from the Brent Group in well 30/6-10 in the Oseberg Field reveals that hydrocarbon inclusions in diagenetic quartz cement are common in the Tarbert, Etive and Oseberg Formations. 300 measurements of homogenization temperatures indicate that hydrocarbon inclusions in the Tarbert Formation were trapped at temperatures from 80 to 85°C and up to 94°C whereas hydrocarbon inclusions in the Etive and Oseberg Formations were trapped at temperatures between 90°C and 106°C. These temperatures of entrapment suggest that oil had started entering the Tarbert Formation at this well location at temperatures of 80–85°C. In the Etive and Oseberg Formations the measured homogenization temperatures indicate that oil was present when temperatures reached around 90°C. Probable minor pressure corrections for homogenization temperatures below 85°C recorded in the Tarbert Formation suggest that a gas cap did not form in the Oseberg reservoir before temperatures exceeded 85°C in the Tarbert Formation in the studied well. This is so because after a gas cap had formed, the fluids in the reservoir would probably be saturated with gas, and inclusions containing fluids that were gas-saturated at the time of entrapment should not have pressure corrections.

In order to translate the above temperatures into dates, temperature modelling has been performed with the basin modelling program BMT (Basin Modelling Toolbox). Results from the temperature modelling suggest that filling of the Tarbert Formation with oil at the 30/6-10 location had started approximately 32–27 m.y. ago, and that oil has probably been present for at least 10 m.y. in the Etive and Oseberg Formations at this well location. If there really is a pressure correction for homogenization temperatures lower than 85°C in the Tarbert Formation, then the gas cap in the Oseberg Field has probably formed after 27 m.y. B.P.

Introduction

Diagenetic minerals within sandstone reservoirs such as quartz overgrowths, albitized feldspar and carbonate cements may contain hydrocarbon fluid inclusions (Burruss et al., 1983; Horsfield and McLimans, 1984; Pagel et al., 1986; Walderhaug, 1990). Entrapment temperatures for the hydrocarbon inclusions can be determined by microthermometric techniques (e.g. Burruss, 1981), and the entrapment temperatures indicate the temperatures at which hydrocarbons entered a reservoir. If the temperature history of the studied reservoir can be determined, then temperatures of entrapment for hydrocarbon inclusions can be translated to times of entrapment. Constraints can thus be placed on the hydrocarbon emplacement history of a reservoir, and this may in turn aid the understanding of hydrocarbon generation and migration in the area.

In the present study, entrapment temperatures for hydrocarbon inclusions in the Brent Group in the Oseberg Field (Fig. 1) have been determined, and a time versus temperature history for the studied Brent Group sandstones has been generated with the BMT basin modelling program (Fjeldskaar et al., 1990). The combined results from the fluid inclusion study and the temperature history modelling have allowed time of oil emplacement in the different sandstones of the Brent Group to be determined, and constraints have also been placed on the probable time of formation of the gas cap in the Oseberg Field.

Sample material

Samples for the fluid inclusion study were taken from well 30/6-10 where almost the whole Brent Group has been cored between 2461 m RKB (metres below kelly bushing) and 2575 m RKB. Depositional environments and stratigraphy of the Brent Group in well 30/6-10 are shown in Fig. 2. The gas–oil contact is located within the Ness Formation at 2520 m RKB, and the oil–water contact is located at 2723–2742 m RKB (Nipen, 1987), i.e. below the base of the Brent Group in well 30/6-10. Examination of 29 thin-sections from the cored interval revealed that hydrocarbon inclusions are abundant within quartz

Basin Modelling: Advances and Applications edited by A.G. Doré et al. NPF Special Publication 3, pp. 485–497, Elsevier, Amsterdam.

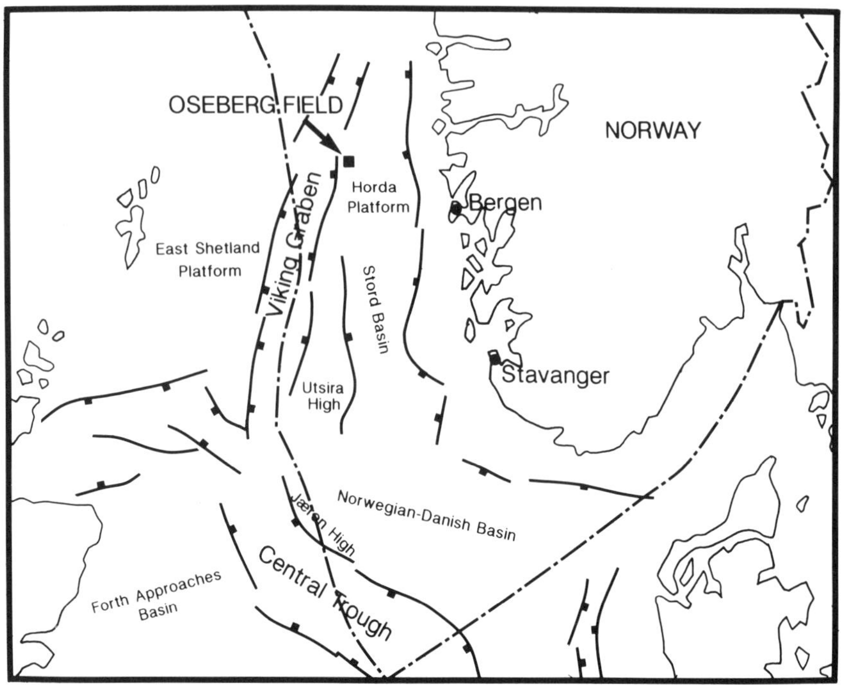

Fig. 1. Location of the Oseberg Field.

overgrowths in the lower half of the Tarbert Formation, common in the Oseberg Formation and quite common in the Etive Formation. Hydrocarbon inclusions were only found in one of eight samples from the Ness Formation and only one inclusion was present in the one sample containing inclusions. Aqueous inclusions also occur in the Tarbert, Etive and Oseberg Formations, but are much rarer than hydrocarbon inclusions.

Table 1 gives the petrographic composition of the eight fine- to coarse-grained sandstone samples used for fluid inclusion studies. In addition to detrital quartz and K-feldspar, clastic constituents in the samples include minor amounts of clay clasts and muscovite plus traces of plagioclase, biotite and heavy minerals. Quartz overgrowths account for 2.3–6.3% of sample volumes, and authigenic kaolinite and pyrite are present in amounts of up to 1.7% and 8.7%, respectively. The dominant source for quartz cement is thought to be dissolution of detrital quartz at stylolites evolved from clay-rich and mica-rich laminae plus dissolution of quartz at grain contacts.

Methods

All thin-sections were examined with a standard polarization microscope with fluorescence accessories in order to locate inclusions and to determine if they consist of pore water or hydrocarbons. Hydrocarbon inclusions are easily identified by their fluorescence. During fluorescence microscopy exciting radiation with wave lengths of 330–380 nm was used together with a 420 nm barrier filter. The inclusions were photographed and numbered in order to enable relocation of inclusions for cathodoluminescence study or other analyses after homogenization temperature had been determined.

The samples used for homogenization temperature measurements consisted of polished thin-sections with thicknesses of 50 μm and of approximately 100 μm thick rock slices impregnated with green epoxy and polished on both sides. The measurements were performed with a Linkam THM 600 heating stage and a Linkam TMS 90 control unit. The heating stage was calibrated with Merck standards with melting points of 70°C and 135°C. Homogenization of inclusions was observed with a Nikon 40× extra long working distance objective and 15× eyepieces. The optical path also contained a 1.25× accessory lens. Total magnification during observation of homogenization was thus 750×.

During measurements of homogenization temperatures heating rates of 6°C/min were used until homogenization was imminent. The temperature was then raised by steps of 0.3–0.5°C and kept constant for 30–60 s between each temperature increase. Homogenization temperature was taken as the mean value of the highest temperature where a vapour bubble was observed and the lowest temperature where a vapour bubble was not observed. When homogenization was considered to have taken place, temperature was lowered a few degrees in order

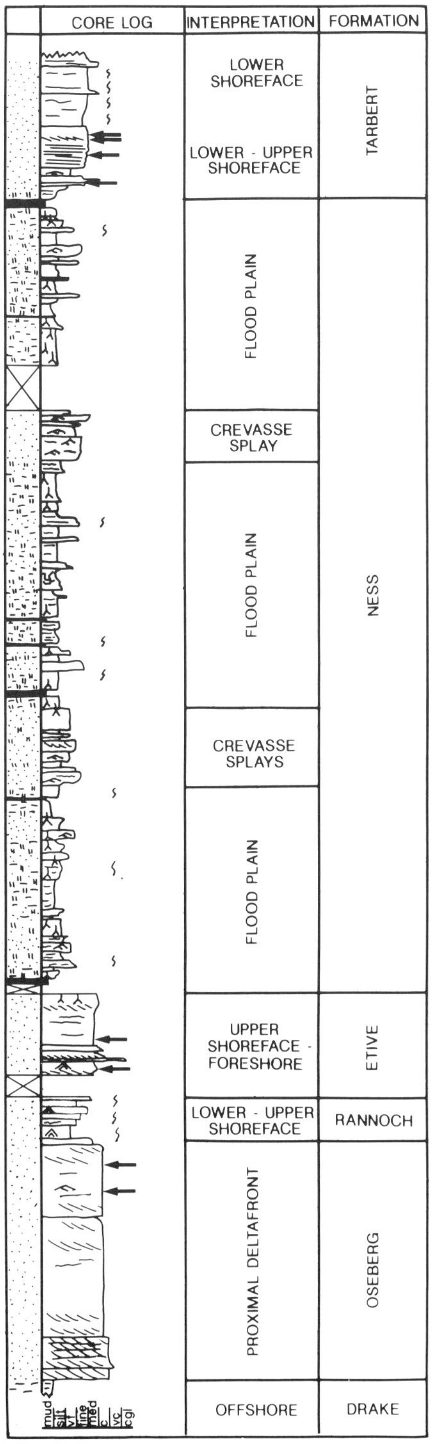

Fig. 2. Sedimentological log of the Brent Group in well 30/6-10. Arrows mark the location of samples used for fluid inclusion studies. Top of cored interval: 2461 m RKB, bottom: 2575 m RKB.

to determine whether a vapour bubble reappeared. If a vapour bubble did reappear, this showed that homogenization had not taken place, since the inclusions required cooling to at least 10°C below their homogenization temperatures before vapour bubbles reformed. By applying this procedure (Pecher, 1981), precise measurements of homogenization temperatures for the smallest inclusions (3–4 μm) could be obtained. Reproducibility of measurements is approximately $\pm$1°C, i.e. the experimental errors in determinations of homogenization temperatures are far smaller than the spread of the homogenization temperatures.

In order to verify that inclusions really were located in quartz cement and not within quartz clasts, cathodoluminescence photography was performed. Cathodoluminescence photography was carried out with a Technosyn Cold Cathode luminescence unit mounted on a Nikon Optiphot polarization microscope with a Nikon Microflex UFX-II photo unit. Kodacolor Gold 400 ASA and Kodacolor VR 1000 ASA films were used, accelerating voltages were in the range 8–10 kV, beam currents were between 250 μA and 345 μA and vacuums between 0.055 and 0.08 Torr. 4× and 10× objectives were used. Cathodoluminescence photography was always performed after homogenization temperatures were determined in case the cathodoluminescence photography had an effect on homogenization temperatures.

Description of inclusions

All studied hydrocarbon and aqueous inclusions are located within quartz overgrowths. The hydrocarbon inclusions are up to 60 μm in diameter, but most inclusions are smaller than 15 μm (Fig. 3). The largest inclusions appear to be sheet-like. The outline of the inclusions varies from roughly spherical and elliptical to complex forms with irregular protrusions (Fig. 3). None of the hydrocarbon inclusions have the shape of negative crystals. Aqueous inclusions are similar in outline to the hydrocarbon inclusions, but no aqueous inclusions larger than 25 μm were found.

Both hydrocarbon and aqueous inclusions in quartz overgrowths are concentrated close to the boundaries between quartz clasts and quartz overgrowths. Cathodoluminescence photography confirms that inclusions are located within quartz cement and not within quartz clasts. Commonly, inclusions are so abundant at this location that they form a dust rim around the whole, or around parts, of the detrital quartz grains. Hydrocarbon inclusions commonly occur in groups, i.e. single hydrocarbon inclusions are rare compared to groups of around 5–50 inclusions in a single overgrowth. In several samples from the lower part of the Tarbert Formation most quartz grains have overgrowths containing hydrocarbon inclusions (Fig. 4).

Apparent thicknesses of quartz overgrowths in the Brent Group in 30/6-10 are typically in the range 10–60 μm, many of the larger apparent thicknesses are probably greater than real thicknesses outside the

TABLE 1

Petrographic composition of samples used for microthermometry

	Depth (m RKB):							
	2467.60	2468.40	2469.95	2471.95	2547.85	2550.85	2558.98	2561.15
Clasts:								
quartz	44.7	56.0	50.7	49.0	60.0	54.3	57.3	55.3
K-feldspar	2.0	2.7	4.7	1.7	5.7	4.3	7.3	6.3
plagioclase		trace				0.3	trace	0.3
muscovite	0.3	0.3	1.7	5.3	trace	trace	0.7	0.3
biotite			trace			trace		
heavy minerals	trace	trace	trace	trace	trace		0.3	
clay clasts	0.7	0.3	trace	1.0	1.0	2.7	1.7	0.7
clay matrix	1.0	0.3		0.7	3.0	6.3	0.7	4.7
Cement:								
quartz	4.3	4.0	5.7	6.3	4.3	2.3	6.0	3.3
kaolinite	1.7	1.0	0.7	1.0	0.3	0.7	trace	
pyrite	8.7	2.7	trace	5.3	trace	0.7	1.3	1.3
Porosity:								
primary	36.0	31.3	35.7	29.3	23.3	27.7	24.3	26.7
dissolution	0.7	1.3	1.0	0.3	2.3	0.7	0.3	1.0
Grain size (mm):	0.25	0.3	0.2	0.2	0.5	0.3	0.8	0.7

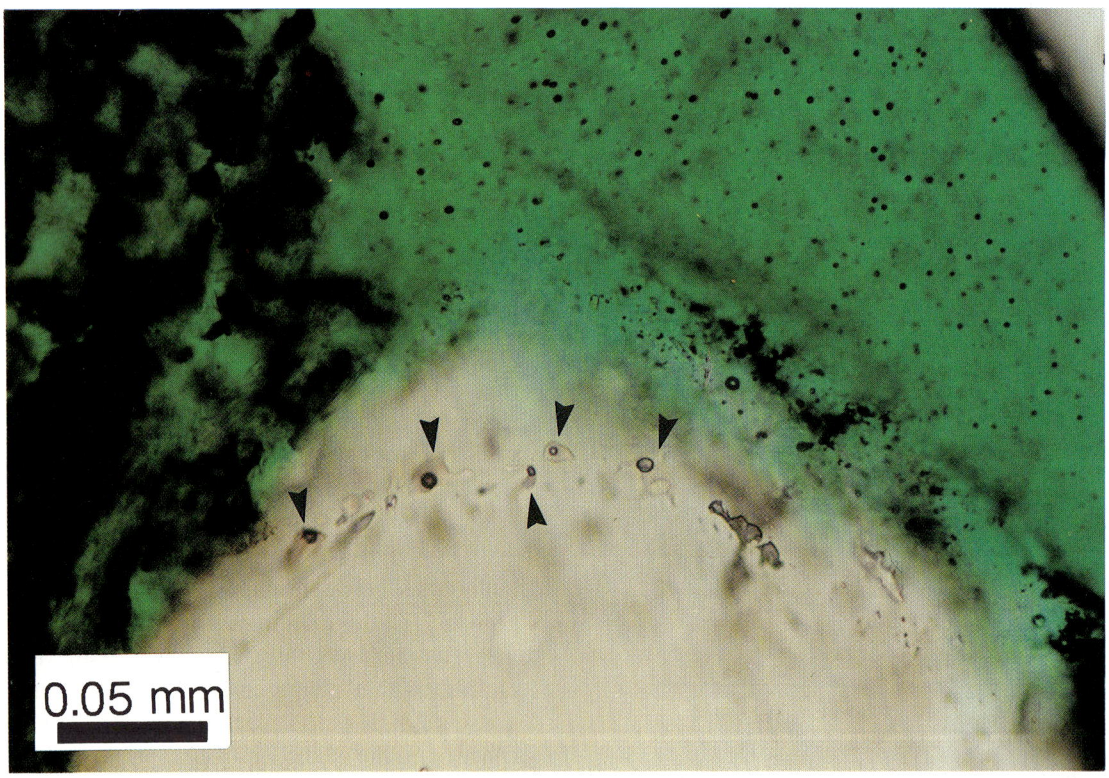

Fig. 3. Hydrocarbon inclusions in quartz overgrowth marked by arrows. Inclusions contain a yellowish brown liquid and a gas bubble at room temperature. Sample from 2561.15 m RKB in the Oseberg Formation; green area is porosity. Photograph taken at 22°C.

plane of the thin-section/polished plate. Up to 50 μm of quartz cement seems to have been precipitated on top of hydrocarbon inclusions in some cases, but usually only 5–20 μm of quartz cement appears to have been precipitated after inclusions were trapped.

At 20°C the hydrocarbon inclusions contain a pale yellowish brown liquid and a vapour bubble. The aqueous inclusions contain a colourless liquid and a vapour bubble. The vapour bubbles in aqueous inclusions make up less than 10% of the cross-sectional area of the inclusions. In hydrocarbon inclusions vapour bubbles are larger and may constitute 10–20% of inclusion areas. Vapour bubbles are usually mobile at 20°C.

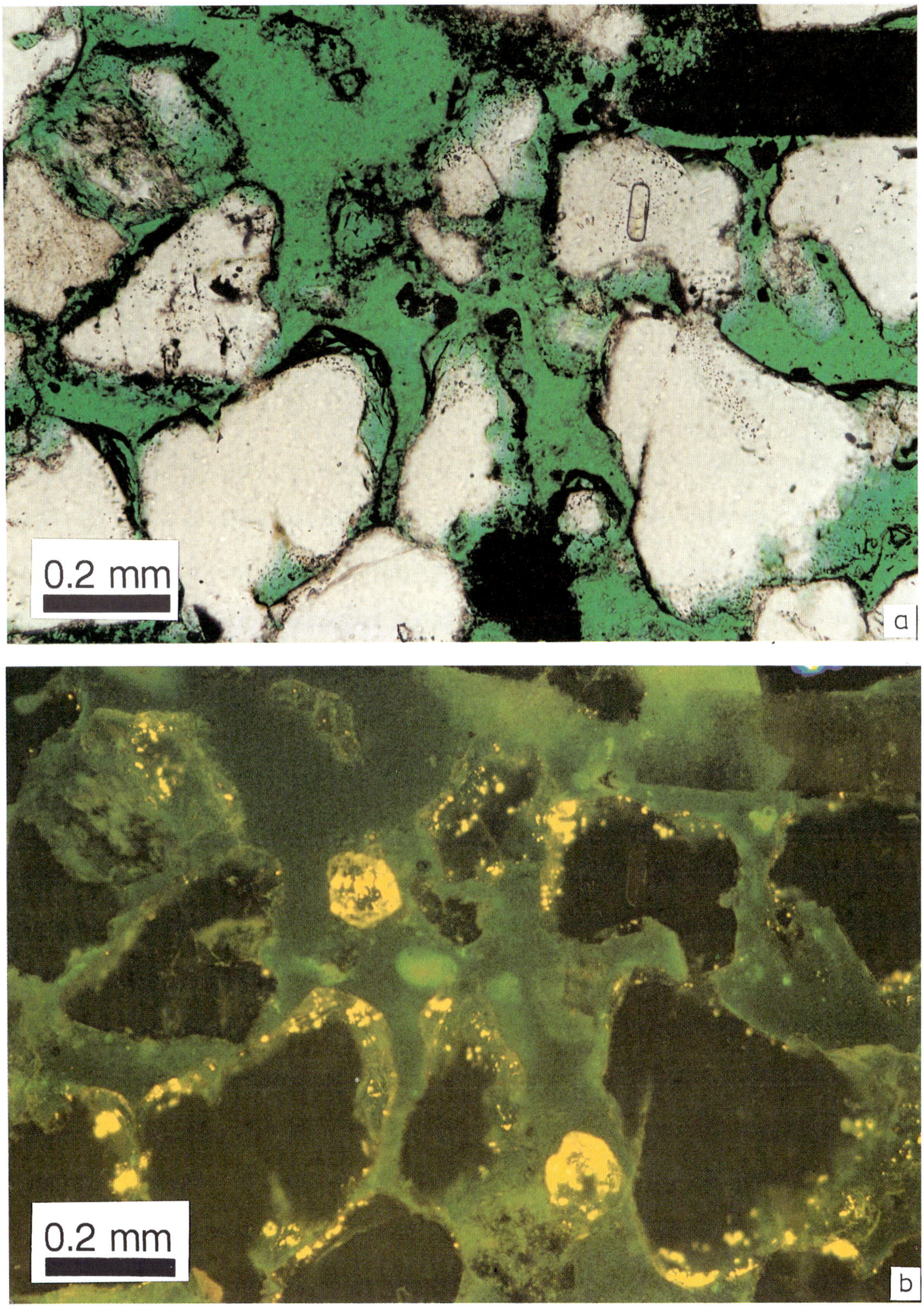

Fig. 4. (a) A sample from 2468.40 m RKB in the Tarbert Formation in plane polarized light. (b) The same area as (a) in blue light. In blue light, hydrocarbon inclusions have a bright yellow fluorescence and are easily located. White grains are quartz clasts with overgrowths; green areas are porosity.

Most of the inclusions large enough to permit determination of homogenization temperature have a pale yellow fluorescence colour in ultraviolet (330–380 nm) light and a deep yellow fluorescence colour in blue (400–440 nm) light (Fig. 4). The fluorescence shows that the inclusions contain hydrocarbons and not brine. Non-fluorescent inclusions are probably filled with brine.

Homogenization temperatures

Unless formed very close to the surface in the vadose zone, fluid inclusions in sedimentary rocks are normally completely filled with pore water or oil at the time of entrapment. When inclusions are cooled to room temperature, a vapour/gas bubble forms in the inclusions due to thermal contraction of the trapped liquid and release of dissolved gas as pressure drops. When the inclusion is heated in the laboratory, the process is reversed, and the gas bubble shrinks and eventually disappears. The temperature where the gas bubble disappears is the homogenization temperature of the inclusion.

If the liquid in an inclusion was saturated with gas at the time of entrapment, a gas bubble will be present at all temperatures below the temperature of entrapment during heating of the inclusion in the laboratory. For gas-saturated inclusions, homogenization temperatures are therefore equal to temperatures of entrapment. However, if the trapped liquid was not saturated with gas at the time of entrapment, the gas bubble in the inclusion may disappear several tens of degrees centigrade below the inclusion's entrapment temperature. The difference between the homogenization temperature and the true entrapment temperature of the inclusion is termed the pressure correction. Pressure corrections can be estimated by various methods, but are usually difficult to determine exactly. However, pressure corrections are commonly small or negligible for inclusions in hydrocarbon reservoirs on the Norwegian shelf due to the common presence of large amounts of dissolved gases, such as methane, in the pore fluids.

The homogenization temperatures measured in this study are illustrated as histograms in Fig. 5. All inclusions homogenized to a liquid phase upon heating.

Homogenization temperatures for hydrocarbon and aqueous inclusions in the Tarbert Formation fall in the intervals 79.5–93.3°C and 80.2–92.3°C, respectively. There seems to be a tendency for aqueous inclusions to have slightly higher homogenization temperatures than hydrocarbon inclusions (Fig. 5). Fifteen of the measurements were performed on aqueous inclusions and 88 on hydrocarbon inclusions.

The 98 homogenization temperatures measured for hydrocarbon inclusions in the Etive Formation fall in the range 89.2–105.1°C except for one measurement of 83.3°C. Only one homogenization temperature has been measured for aqueous inclusions in the Etive Formation (104.2°C).

The 96 homogenization temperatures measured for hydrocarbon inclusions in the Oseberg Formation are between 90.5°C and 107.4°C except for one spurious measurement of 120.5°C. Aqueous inclusions (5 measurements) homogenized between 94.5°C and 106.7°C.

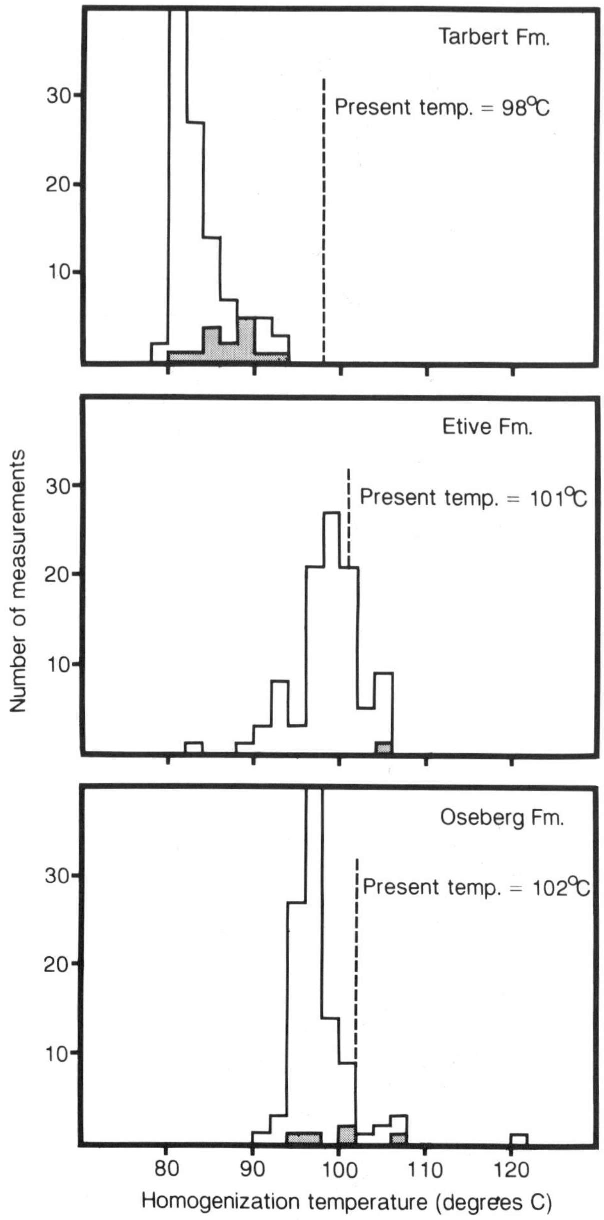

Fig. 5. Measured homogenization temperatures for hydrocarbon (white) and aqueous (grey) inclusions in quartz overgrowths in the Tarbert, Etive and Oseberg Formations.

All measurements performed in the Tarbert Formation are below present formation temperature of 98°C. In the Etive and Oseberg Formations 10% and 5% of measurements are 1–5°C above present formation temperatures of 101°C and 102°C, respectively. Formation temperatures have been calculated from the temperature at 2497 m below mean sea level given by Nipen (1987) by assuming a present sea floor temperature of 5°C and by applying the resulting mean geothermal gradient of 0.04°C/m. The homogenization temperatures recorded in the Etive and Oseberg Formations suggest that present formation temperatures may possibly be a few degrees higher than the values calculated from the data given by Nipen (1987).

Pressure corrections

Since the Oseberg reservoir has a gas cap, it is reasonable to assume that the oil trapped in the inclusions has a high gas content and that pressure corrections may be small or zero. Furthermore, since the measured homogenization temperatures are close to present formation temperature, addition of a considerable pressure correction to most of the homogenization temperatures would imply that inclusions formed at temperatures considerably above present formation temperature. This seems unlikely due to the simple burial history of the studied sandstones and due to the lack of any hydrothermal mineralization or other evidence for previous thermal events that could have raised formation temperature above present values. Dahl et al. (1987) have previously concluded that the Oseberg reservoir is at present at the maximum temperature that it has been subjected to (see their fig. 8).

Although there is little doubt that pressure corrections for the recorded homogenization temperatures cannot be large, some aspects of the recorded data from the Tarbert Formation suggest that pressure corrections of up to around 5°C should be added to most of the homogenization temperatures for hydrocarbon inclusions from this formation. Figure 5 illustrates that aqueous inclusions from the Tarbert Formation have homogenization temperatures with a mean value around 5°C higher than the modal value for the hydrocarbon homogenization temperatures in the Tarbert Formation. There is no obvious reason for assuming that aqueous inclusions should form at higher temperatures than the hydrocarbon inclusions, especially since both types of inclusions occur within the same overgrowths and thus appear to have formed at the same time. On the other hand, if aqueous and hydrocarbon inclusions form simultaneously and at a time when pore fluids are not saturated with gas, pressure corrections for the hydrocarbon inclusions will typically be larger than for the aqueous inclusions due to the different PVT properties of brine and hydrocarbons (e.g. Narr and Burruss, 1984; Pagel et al., 1986; Jourdan et al., 1987). The homogenization temperatures recorded in the Tarbert Formation therefore suggest that homogenization temperatures from this formation, in the range 80–85°C, may have pressure corrections of 1–5°C. In the Etive and Oseberg Formations very few aqueous inclusions were found, and it is difficult to decide whether the same difference between homogenization temperatures for aqueous and hydrocarbon inclusions is present in these two formations.

Lower homogenization temperatures for both aqueous and hydrocarbon inclusions in the Tarbert Formation compared to homogenization temperatures recorded in the Etive and Oseberg Formations could in principle be a result of larger pressure corrections for the Tarbert data, i.e. lower contents of dissolved gas in the Tarbert Formation pore fluids at the time of inclusion entrapment. However, all of the difference between homogenization temperatures in the Tarbert Formation versus the Oseberg plus Etive Formations probably cannot be a result of differing pressure corrections; if the inclusions in all three formations were trapped at the same time, it seems unreasonable to suppose that the fluids in the Tarbert Formation contained less dissolved gas than the fluids in the underlying Etive and Oseberg Formations.

Temperatures of hydrocarbon emplacement

Homogenization temperatures for hydrocarbon inclusions in the Tarbert Formation fall between 79.5°C and 93.3°C. Since pressure corrections are probably small for homogenization temperatures below 85°C and absent for higher homogenization temperatures, oil was probably present in the Tarbert Formation when temperatures reached 80–85°C, although it is difficult to say what the oil saturation was. The lack of homogenization temperatures higher than 94°C in the Tarbert Formation may indicate that hydrocarbon saturations had increased to a value that stopped quartz cementation when temperatures reached around 94°C. However, it could also be that the lack of higher homogenization temperatures is due to inclusions preferentially forming at the boundaries between clasts and overgrowths. Since formation of inclusions and quartz cementation seems to have continued up to the present formation temperature in the Etive and Oseberg Formations, the lack of homogenization temperatures above 94°C is not considered to be a strong indication for a termination of quartz precipitation at this temperature.

The homogenization temperatures recorded in the Etive and Oseberg Formations are systematically higher than the homogenization temperatures from the Tarbert Formation. In the Etive and Oseberg Formations, 99% of oil inclusions homogenize between 90°C and 107°C. This indicates that oil was present when temperatures of 90°C were reached, although some oil may possibly also have been present at lower temperatures since quartz cementation may have started after the first oil entered the Etive and Oseberg Formations. Few homogenization temperatures are higher than 100°C. This might point to reduced quartz cementation and high oil saturations after temperatures reached 100°C, although the effect of selective inclusion-formation at clast/overgrowth boundaries is probably a better explanation of the form of the homogenization temperature distribution.

When free gas is present in the top of a reservoir, the oil and irreducible water below the gas cap is probably saturated with gas. If hydrocarbon or aqueous inclusions form from a gas-saturated liquid, no pressure correction of their homogenization temperatures is required. Therefore, if fluid inclusions from a reservoir with a gas cap can be shown to have pressure corrections, it probably follows that the gas cap was not present at the time when the inclusions formed. In the case of the Oseberg reservoir, the difference in homogenization temperatures for aqueous and hydrocarbon inclusions in the Tarbert Formation (Fig. 5) suggests that a pressure correction of up to approximately 5°C should be applied to the lowest (80–85°C) homogenization temperatures recorded in this formation. If this is the case, then the gas cap in the Oseberg reservoir was probably not established before temperatures in the Tarbert Formation reached around 85°C (implicitly assuming that gas saturations in the reservoir have increased through time). However, more measurements of homogenization temperatures of aqueous inclusions would be required to substantiate that there really is a systematic difference between homogenization temperatures for aqueous and hydrocarbon inclusions in the Tarbert Formation, i.e. that there is a pressure correction for some of the Tarbert data.

Time of hydrocarbon emplacement

A 2D temperature history simulation has been performed with the basin modelling program BMT (Basin Modelling Toolbox) in order to translate the temperatures for the various hydrocarbon filling events into dates. A general description of the BMT program is given in Fjeldskaar et al. (1990).

Figure 6 shows the profile used for the temperature history simulation. The profile is a slightly modified version of the profile OSB 121 published by Dahl et al. (1987). Ages for the various formations on the profile have been taken from the Norwegian Petroleum Directorate well data summary sheets and from completion logs.

Measurements of thermal conductivities of sedimentary rocks from the Norwegian continental shelf have been performed on a large number of core samples, covering a wide variety of mineralogical compositions (Prestholm et al., 1991; Fjeldskaar et al., this volume). Analysis of this data base has shown that the vertical and horizontal thermal conductivities of sediments may be predicted by partial least squares regression from the mineralogies and porosities of the sediments (Fjeldskaar et al., 1990). The horizontal and vertical conductivities of the modelled formations (Fig. 7) were predicted from the mineralogies, surface porosities and exponential decay constants given in Table 2. The thermal conductivity of the basement (i.e. pre-Triassic) is assumed to be isotropic and having a value of 3.0 W/m K.

Based on present thermal conductivities (Fig. 7) and present formation and sea floor temperatures in well 30/6-10 (98°C at the top of the Brent Group and 5°C at the sea floor), present heat flow across the profile was calculated as 69 mW/m^2. The calculated present temperature distribution across the profile is shown in Fig. 8.

Palaeo-heat flow values were calculated from isostatic and tectonic modelling. The observed tectonic subsidence is quantified indirectly by a backstripping technique, which includes isostatic correction. By comparison of the tectonic subsidence with theoretical subsidence curves, a stretching factor may be found (McKenzie, 1978; Sclater and Christie, 1980). For the Oseberg area, backstripping calculations have been performed separately for the western and eastern parts of the profile assuming Airy compensation. The theoretical tectonic subsidence was calculated by a simple McKenzie (1978) approximation for the same locations. The resulting palaeo-heat flow (Table 3) shows a heat flow high of 77 mW/m^2 caused by the Jurassic/Cretaceous tectonic event followed by decreasing heat flow to a present value of 69 mW/m^2.

Palaeo-sea floor temperatures have been based on the temperature curve of Nilsson (1972) and were set to 20°C at 57 m.y. B.P., 13°C at 23 m.y. B.P., 9°C at 10 m.y. B.P., 6°C at 2 m.y. B.P., and 5°C at present. Between these points, sea floor temperatures were estimated by assuming a linear rate of change. Palaeo-water depths have been set to 100 m for shales and marls and to 20 m for shallow-marine sandstones. The generated temperature versus time history for the top of the Tarbert Formation and for the Oseberg Formation in well 30/6-10 is shown in Fig. 9.

Figure 9 illustrates that temperatures equal to the lowest entrapment temperatures for oil inclusions in the Tarbert Formation (80–85°C) were reached approximately 32–27 m.y. ago. This suggests that at the well location studied, filling of the Tarbert Formation

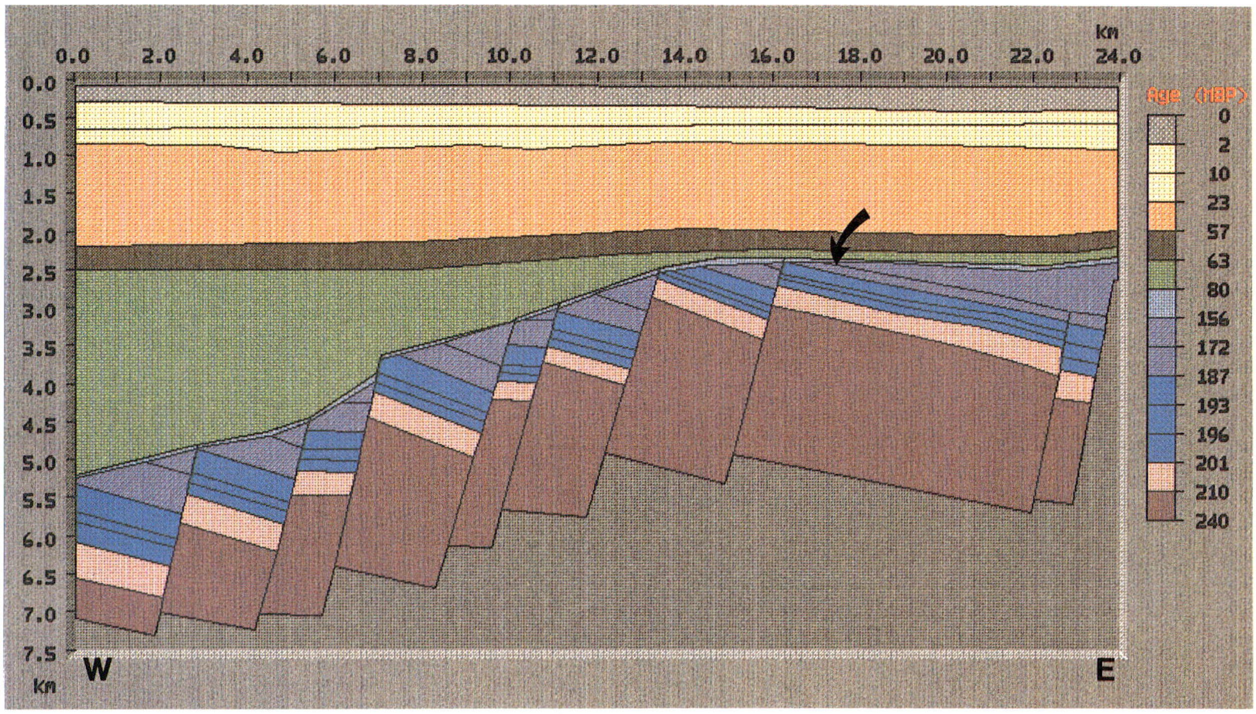

Fig. 6. Modelled profile across the Oseberg Field. The location of the top of the Tarbert Formation in the studied well corresponds to a position at the tip of the arrow. Ages for the boundaries of the modelled horizons are given in m.y. B.P. at the right-hand side of the profile. Formation and Group names for the modelled horizons are given in Table 2.

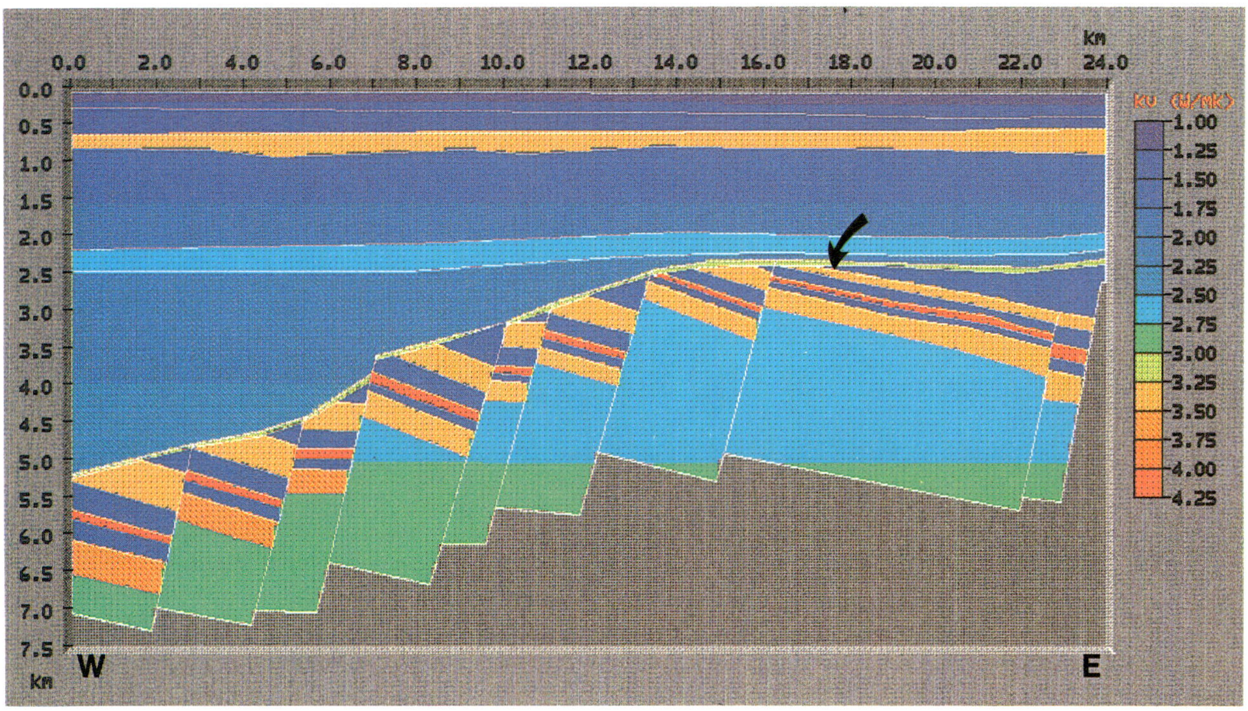

Fig. 7. Calculated present vertical conductivities for the modelled profile. The colour code for the vertical conductivities is given at the right-hand side of the profile. The formation boundaries (white lines) are the same as in Fig. 6. The top of the Tarbert Formation in the studied well is located at the tip of the arrow.

with oil had started at 32–27 m.y. B.P. The lowest entrapment temperatures for oil inclusions in the Etive and Oseberg Formations are probably around 90°C. This temperature was reached at 26 m.y. B.P. in these formations (Fig. 9), but, according to the performed modelling, temperatures then increased to 94°C at 23 m.y. B.P., dropped to 90°C at 10 m.y. B.P., and again reached 94°C at 2 m.y. B.P.

TABLE 2

Mineralogies and porosity versus depth gradients

Horizon:	1	2	3	4	4	5	5	6	6	7
Age (m.y. B.P.):	0–2	2–5	10–23	23–57	23–57	57–63	57–63	63–85	63–85	156–163
Age (stratigraphic):	Pleistocene	Pliocene	Miocene	Oligocene–Eocene	Oligocene–Eocene	Paleocene	Paleocene	Paleocene–Upper Cretaceous	Paleocene–Upper Cretaceous	Upper Jurrasic
Group/Formation:	Nordland	Nordland	Utsira	Hordaland	Hordaland	Rogaland	Rogaland	Shetland	Shetland	Heather
Lithology:	shale	shale	sandstone	shale	sandstone	tuff	shale	shale	chalk	shale
Lithology (%):	100	100	100	90	10	30	70	60	40	100
Quartz	25	25	63	22	63	1	68	22	10	20
K-feldspar	3	3	11	3	11	0	0	3	0	6
Plagioclase	5	5	5	5	5	26	3	5	0	5
Mica	20	20	10	22	10	0	3	22	5	15
Chlorite	8	8	4	8	4	50	5	6	0	2
Mixed layer clay	8	8	3	10	3	8	6	8	5	5
Kaolinite	29	29	2	28	2	0	11	34	0	40
Calcite	2	2	0	2	0	5	0	0	80	1
Dolomite	0	0	1	0	1	0	0	0	0	0
Siderite	0	0	1	0	1	1	0	0	0	0
Hematite	0	0	0	0	0	0	0	0	0	0
Pyrite	0	0	0	0	0	9	4	0	0	6
TOC	0	0	0	1	0	0	0	1	0	1
Surface porosity	0.63	0.63	0.49	0.63	0.49	0.60	0.63	0.63	0.70	0.63
Exp. constant	0.51	0.51	0.27	0.51	0.27	0.50	0.51	0.51	0.71	0.51

Horizon:	8	8	9	10	11	12	12	13	13
Age (m.y. B.P.):	172–187	172–187	187–193	193–196	196–201	201–210	201–210	210–240	210–240
Age (stratigraphic):	Middle Jurrasic	Middle Jurrasic	Toarcian	Pliensbackian	Pliensbackian–Sinemurian	Sinemurian–Rhaetian	Sinemurian–Rhaetian	Triassic	Triassic
Group/Formation:	Brent	Brent	Drake	Cook	Bu./Am.	Statfjord	Statfjord		
Lithology:	shale	sandstone	shale	sandstone	shale	shale	sandstone	shale	sandstone
Lithology (%):	20	80	100	100	100	20	80	35	65
Quartz	20	76	20	75	20	32	76	11	44
K-feldspar	2	11	6	10	6	3	10	4	15
Plagioclase	1	3	5	3	5	10	5	14	20
Mica	17	2	15	3	15	9	2	51	4
Chlorite	1	0	2	0	2	5	0	17	1
Mix. layer clay	11	0	5	0	5	5	0	1	1
Kaolinite	43	6	40	7	40	35	7	0	9
Calcite	0	0	1	0	1	0	0	0	0
Dolomite	0	0	0	0	0	0	0	0	6
Siderite	2	0	0	0	0	1	0	0	0
Hematite	0	0	0	0	0	0	0	2	0
Pyrite	3	2	6	2	6	0	0	0	0
TOC	4	0	0	0	0	1	0	0	0
Surface porosity	0.63	0.49	0.63	0.49	0.63	0.63	0.49	0.63	0.49
Exp. constant	0.51	0.27	0.51	0.27	0.51	0.51	0.27	0.51	0.27

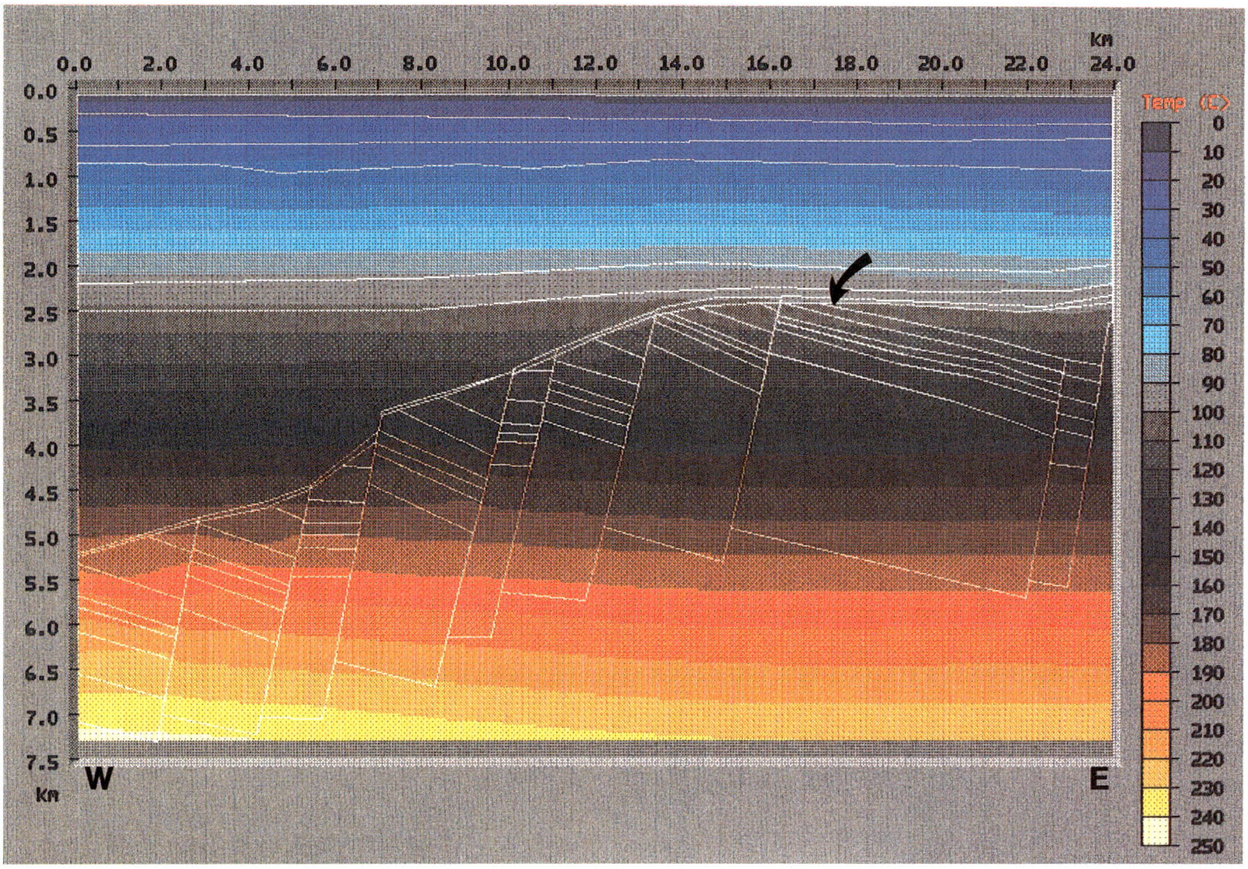

Fig. 8. Present temperature distribution for the modelled profile based on calculated thermal conductivities and heat flows. The temperature distribution is calibrated against present temperature at the top of the Tarbert Formation (tip of the arrow) in the studied well. Formation boundaries (white lines) are the same as in Fig. 6.

TABLE 3

Calculated heat flow values

Time (m.y. B.P.)	Heat flow (mW/m^2) west	Heat flow (mW/m^2) east
0	69	69
2	70	69
10	72	70
23	73	71
57	76	72
63	77	73

During the last 2 m.y., temperature increased from 94°C to 102°C. This implies that oil inclusions in the Etive and Oseberg Formations may have started to form as early as 26 m.y. B.P., although it cannot be excluded that they only started forming at 10 m.y. B.P. The modelled temperature history indicates that inclusions with homogenization temperatures above 94°C (i.e. 92% of the measurements) formed during the last 2 m.y. This last observation might suggest that inclusions in the Oseberg and Etive Formations have formed during the last 10 m.y. and not as far back as 26 m.y. B.P. In summary, the data seem to indicate that oil has been present in the Oseberg and Etive Formations at this well location for at least 10 m.y., and possibly for 26 m.y.

Inclusions with homogenization temperatures above 94°C are absent in the Tarbert Formation. According to Fig. 9, this means that inclusions stopped forming in the Tarbert Formation around 1–2 m.y. ago. Whether this implies that hydrocarbon saturations increased to values that inhibited quartz cementation in the Tarbert Formation at this time or whether it simply means that inclusion-free quartz was being precipitated is difficult to say; however, since inclusions seem to have continued to form up to the present day in the oil-filled Etive and Oseberg Formations, it seems quite likely that the lack of homogenization temperatures above 94°C is just a result of precipitation of inclusion-free quartz at higher temperatures. However, if lack of inclusions formed at temperatures above 94°C is due to termination of quartz cementation, then this implies that the gas–oil contact moved below the Tarbert Formation in well 30/6-10 after 2 m.y. B.P., since termination of cementation would have been due to either high oil saturations or to gas saturation of

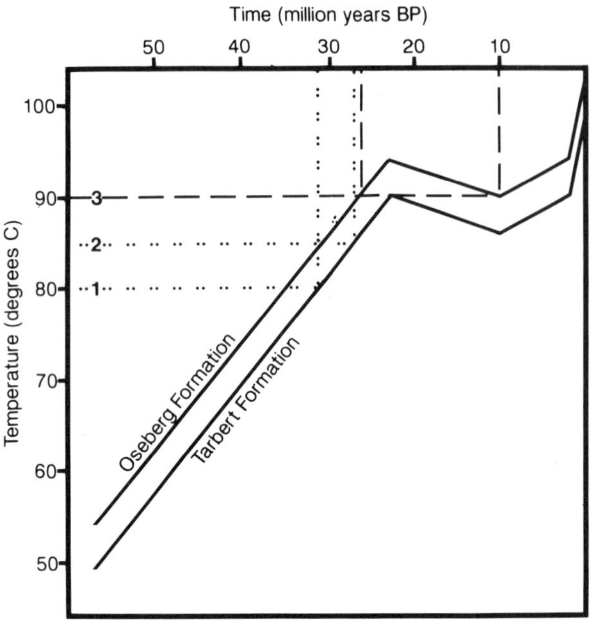

Fig. 9. Temperature versus time curve for the Tarbert Formation and the Etive and upper Oseberg Formations in well 30/6-10. The dotted lines illustrate that temperatures of 80°C (*1*) and 85°C (*2*) in the Tarbert Formation correspond to ages of approximately 32 m.y. B.P. and 27 m.y. B.P., respectively. The dashed lines illustrate that temperatures of 90°C in the Etive Formation and upper part of the Oseberg Formation correspond to an age of at least 10 m.y. B.P. and possibly as much as 26 m.y. B.P. according to the performed modelling.

the sandstones (assuming that gas saturations have increased with time).

If there really is a small pressure correction for the lowest (<85°C) homogenization temperatures in the Tarbert Formation, and if this implies the absence of a gas cap at this time, then a gas cap was not present in the Oseberg reservoir before 27 m.y. B.P. since a temperature of 85°C was reached at 27 m.y. B.P. in the Tarbert Formation according to the thermal modelling.

The times of oil emplacement calculated above are of course only valid for 30/6-10 and other wells in the same structural positions, i.e. located approximately at the present gas–oil contact. Oil would probably have been present for some time at the crest of the structure before the Brent Group sandstones were filled with oil at the location of well 30/6-10, and oil emplacement will have taken place later at lower structural positions. Furthermore, there are uncertainties in the various input data for the temperature model, in addition to the fact that a steady state temperature model has been used. The error introduced by assuming thermal equilibrium is, however, not thought to be significant as there have probably not been significant erosional events nor very rapid sedimentation at the location of the Oseberg Field during the Tertiary. The uncertainties in the temperature history of the Oseberg reservoir are greatest during the period 23–2 m.y. B.P. when the rate of heating or cooling was low, and relatively small errors in the temperature history curve could produce errors of 5 m.y. in the dating of events recorded by the fluid inclusion data. Before 23 m.y. B.P. and after 2 m.y. B.P., heating rates were relatively high, and minor errors in the temperature history curve would not have a major effect on the dating of the hydrocarbon filling events that took place in these periods.

Conclusions

Measured homogenization temperatures for hydrocarbon inclusions indicate that oil had entered the Tarbert Formation in well 30/6-10 when temperatures reached 80–85°C, and that oil was present in the Etive and Oseberg Formations when temperatures reached 90°C. The temperature history modelling indicates that temperatures of 80–85°C in the Tarbert Formation were reached 32–27 m.y. ago, and oil is therefore thought to have entered the Tarbert Formation in well 30/6-10 by the Late Oligocene. In the Etive and Oseberg Formations temperatures of 90°C were reached 26 m.y. ago. Temperatures then increased to approximately 94°C at 23 m.y. B.P., dropped to 90°C around 10 m.y. B.P., and subsequently increased up to present values. Oil has therefore probably been present in the Etive and Oseberg Formations in the studied well for at least 10 m.y., i.e. from the Late Miocene, but may have been present as early as the Late Oligocene. The studied well is located at the present gas–oil contact in the Oseberg Field, and filling of the studied formations with oil will have taken place earlier in wells closer to the crest of the structure and later in wells located further down-flank than 30/6-10.

Probable presence of small pressure corrections for inclusions in the Tarbert Formation with homogenization temperatures below 85°C, suggests that the reservoir fluids were not saturated with gas before temperatures exceeded 85°C in the Tarbert Formation at the location of well 30/6-10. It is therefore suggested that the gas cap in the Oseberg Field was not present before 27 m.y. B.P., i.e. not before latest Oligocene times.

Acknowledgements

This paper is based on results from a project funded by Amoco Norway. Their permission to publish results from this project is gratefully acknowledged. The Norwegian Petroleum Directorate is thanked for providing samples from well 30/6-10.

References

Burruss, R.C., 1981. Hydrocarbon fluid inclusions in studies of sedimentary diagenesis. In: L.S. Hollister and M.L. Crawford (Editors), Short Course in Fluid Inclusions: Applications to Petrology. Mineralogical Association of Canada, Calgary, Alta., pp. 138–156.

Burruss, R.C., Cercone, K.R. and Harris, P.M., 1983. Fluid inclusion petrography and tectonic burial history of the Al Ali No. 2 well: evidence for the timing of diagenesis and oil migration, northern Oman Foredeep. Geology, 11: 567–570.

Dahl, B., Nysæther, E., Speers, G.C. and Yükler, A., 1987. Oseberg area — integrated basin modelling. In: J. Brooks and K. Glennie (Editors), Petroleum Geology of North West Europe. Graham and Trotman, London, pp. 1029–1038.

Fjeldskaar, W., Mykkeltveit, J., Johansen, H., Langfeldt, J.M., Christie, O.H.J., Tyvand, P., Skurve, O. and Bjørkum, P.A., 1990. Interactive 2D Basin Modelling on Workstations. Proc. SPE 20350 Petroleum Computer Conf., Denver, Colo., pp. 181–196.

Fjeldskaar, W., Prestholm, E., Guargena, C. and Stephenson, M., 1993. Mineralogical and diagenetic control on the thermal conductivity of the sedimentary sequences in the Bjørnøya Basin, Barents Sea. In: A.G. Doré et al. (Editors), Basin Modelling: Advances and Applications. Norwegian Petroleum Society (NPF), Special Publication 3. Elsevier, Amsterdam, pp. 445–453 (this volume).

Horsfield, B. and McLimans, R.K., 1984. Geothermometry and geochemistry of aqueous and oil-bearing fluid inclusions from Fateh Field, Dubai. Org. Geochem., 6: 733–740.

Jourdan, A., Thomas, M., Brevart, O., Robson, P., Sommer, F. and Sullivan, M., 1987. Diagenesis as the control of the Brent sandstone reservoir properties in the Greater Alwyn area (East Shetland Basin). In: J. Brooks and K. Glennie (Editors), Petroleum Geology of North West Europe. Graham and Trotman, London, pp. 951–961.

McKenzie, D., 1978. Some remarks on the development of sedimentary basins. Earth Planet. Sci. Lett., 40: 25–32.

Narr, W. and Burruss, R.C., 1984. Origin of reservoir fractures in Little Knife Field, North Dakota. Am. Assoc. Pet. Geol., Bull., 68: 1087–1100.

Nilsson, T., 1972. Pleistocen. Berlingska Boktryckeriet, Lund, 508 pp. (in Swedish).

Nipen, O., 1987. Oseberg. In: A.M. Spencer et al. (Editors), Geology of the Norwegian Oil and Gas Fields. Graham and Trotman, London, pp. 379–387.

Pagel, M., Walgenwitz, F. and Dubessy, J., 1986. Fluid inclusions in oil and gas-bearing sedimentary formations. In: J. Burrus (Editor), Thermal Modelling in Sedimentary Basins. Technip, Paris, pp. 565–583.

Pecher, A., 1981. Experimental decrepitation and re-equilibration of fluid inclusions in synthetic quartz. Tectonophysics, 78: 567–583.

Prestholm, E., Fjeldskaar, W., Balling, N., Michelsen, J.K. and Walderhaug, O., 1991. Thermal conductivity anisotropy in sedimentary rocks and its effect on the evaluation of petroleum potential. In: Proc. 3rd EAPG Conf., Florence, 26–30 May, 1991 (abstract).

Sclater, J.G. and Christie, P.A.F., 1980. Continental stretching: an explanation of the post-mid-Cretaceous subsidence of the central North Sea basin. J. Geophys. Res., 85: 3711–3739.

Walderhaug, O., 1990. A fluid inclusion study of quartz cemented sandstones from offshore mid-Norway — possible evidence for continued quartz cementation during oil emplacement. J. Sediment. Petrol., 60: 203–210.

OLAV WALDERHAUG *Rogaland Research, P.O. Box 2503 Ullandhaug, N-4004 Stavanger, Norway*

WILLY FJELDSKAAR *Rogaland Research, P.O. Box 2503 Ullandhaug, N-4004 Stavanger, Norway*

Modelling of expulsion and secondary migration along the southwestern margin of the Horda Platform

Arild Skjervøy and Øyvind Sylta

An integrated basin modelling study was carried out to support the evaluation of three prospects in Block 30/12, with mapped structural closures at Top Brent level. The probability for entrapment of hydrocarbons and the prediction of relative volumes of trapped oil and gas were largely undetermined prior to the study due to the long-distance secondary migration required for hydrocarbons to accumulate and the low-relief structuring. The modelling was aimed at determining the kitchen and drainage areas for hydrocarbons migrating into the area of Block 30/12 through time. The quantities, types and the timing of hydrocarbon generation and expulsion from each source rock interval were investigated. The integration of the expulsion histories with secondary migration and entrapment modelling enabled a prediction of the types and amounts of hydrocarbons that migrated into the block and accumulated in the 30/12 prospects. The modelling results were calibrated to the volumes of oil and gas in place in the J-structure in the adjacent Block 30/9.

Oil expulsion is assumed to be saturation-dependent; expulsion is initiated when hydrocarbon yields from labile kerogen exceed magnitudes equivalent to a critical pore saturation. The Viking Group source rocks contribute from 83% to 100% of the expelled oil over the modelled range of critical saturations from 16% to 40%. The contribution from the Draupne Formation alone is 52% to 100% of the expelled oil. The oils that were generated in the Heather Formation, and the Brent and Dunlin Groups were mainly retained in the source rocks due to inferior expulsion capabilities and were later subjected to secondary cracking. Significant oil expulsion started in the Paleocene and continued through to the present. The modelled expulsion efficiencies are higher along the Viking Graben flanks than along the graben axis.

Arguments can be made for both upwards and downwards oil expulsion from the Viking Group, and both were modelled as different migration scenarios. All modelling cases indicate that oil is the main phase in the 30/12 prospects, although different expulsion cases (saturation thresholds) and migration scenarios produce different amounts of trapped free gas and associated gas.

The secondary migration and hydrocarbon entrapment modelling highlighted previously unidentified risks associated with the prospects. The modelling resulted in mutual exclusivity of the two smaller prospects and reduced trapping capacity of the largest prospect, based on the evaluation and modelling of sealing versus non-sealing faults. The most optimistic fault sealing scenario implies a capacity for 423×10^6 Sm^3 of oil trapped in the largest prospect, 233×10^6 Sm^3 of oil in the medium-sized prospect and no trapping in the smallest. In contrast, the most pessimistic fault sealing scenario would lead to insignificant volumes of gas and no oil accumulating in the medium-sized prospect, with no hydrocarbons in the other two prospects.

Background

Numerous basin modelling studies, addressing many aspects of the petroleum prospectivity of areas within the Norwegian Continental Shelf, have been published over the last decade. Many of these have been oriented towards the Viking Graben, Lomre Terrace and Horda Platform areas. Academic and industrial interest in basin modelling along with continued exploration drilling have resulted in a growing understanding of the location of mature kitchen areas, classification of potential source rocks, the habitat of oil and gas accumulations and the integrated effect of time and temperature on maturation of organic matter and the process of hydrocarbon generation.

This paper is a result of an integrated basin modelling study which was carried out to support the evaluation of prospects in Block 30/12 on the Norwegian Continental Shelf. The block was offered in the 13th round of license applications in 1990. It is located on the southwestern margin of the Horda Platform, to the east of the Viking Graben (Fig. 1).

Several hydrocarbon discoveries have been made in Brent sandstones in the vicinity of Block 30/12. These include discoveries of oil and free gas in Block 30/9 to the north. The short distance to the Oseberg and Brage oil fields to the north and accumulations in the Frigg area to the south and southwest should also be noted.

Three potential prospects with structural closure at the Top Brent horizon were identified in the initial seismic mapping of the block; these are the A, B and C prospects (Fig. 1). In spite of the proven

Basin Modelling: Advances and Applications edited by A.G. Doré et al. NPF Special Publication 3, pp. 499–537, Elsevier, Amsterdam.

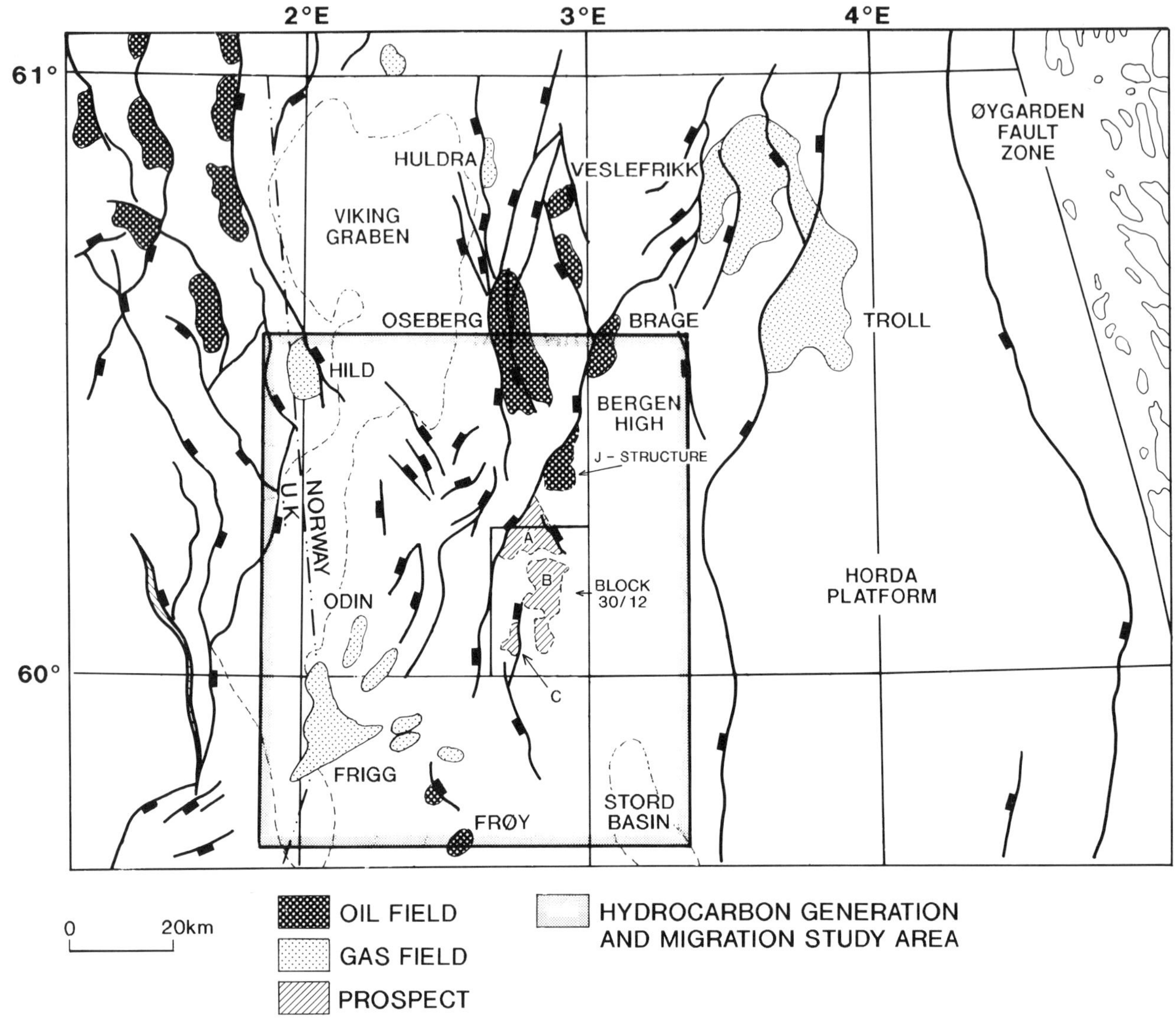

Fig. 1. Location map.

prospectivity of nearby blocks, we found Block 30/12 to be quite a challenge in a basin modelling context. The prospects within the block require long-distance migration since they are offset from mature kitchen areas. The block is characterized by low-relief structuring which implies that migration pathways may have changed several times throughout the geological history. The proven accumulations in adjacent Block 30/9 include oils which, from geochemical evidence, may originate from different kitchen areas. Finally, although there is a tendency for predominance of oil accumulations in this part of the Horda Platform Margin, significant quantities of free gas are also trapped. Thus, the likelihood for accumulation of oil versus that of gas was largely unresolved prior to the study.

An integrated basin modelling study was therefore performed with the objectives of addressing the following questions.

Which were the kitchen- and drainage areas for the 30/12-prospects through geological time?

To what degree do different source rocks contribute to the oil expulsion in the area?

What was the timing of hydrocarbon generation and expulsion?

Were sufficient volumes of hydrocarbons generated to migrate into, and fill, the 30/12 prospects?

What conduits for long-distance migration may be supported and what are the implications of different carrier bed scenarios?

What are the most likely reservoired hydrocarbon phases in the prospects (oil, oil and free gas or gas only)?

Does the migration modelling support the existence of each prospect or are there limiting factors that reduce their trapping potential and/or probability for coexistence?

Introduction

The integrated basin modelling study area encompassed about 7500 km^2 and included portions of the Viking Graben, the Horda Platform and the northern part of the Stord Basin (Fig. 1). The study included source rock evaluation, burial history evaluation and modelling, thermal history modelling, maturation and kerogen transformation modelling, expulsion modelling and modelling of secondary migration. The study also included special projects in seismic interpretation, regional petrophysical evaluation of the Brent Group, GC-Time Slice Pyrolysis for source rock kinetics and tectonic modelling.

An overview of the project is given in Fig. 2. Schematically we can subdivide the project into three consecutive phases.

In the first phase a calibrated heatflow and burial history model is defined by tectonic modelling and maturation modelling. The model is calibrated to maturity (R_o, T_{max} and hydrogen index) and temperature (BHT) data from wells. Prior to proceeding to the next phase, the interpretation of the organic geochemical data base and preparation of a source rock model is required. The source rock model includes yield parameters, types of kerogen, weight and volume fractions of kerogen, isopachs and kinetic parameters for each source rock unit defined. The seismic data is interpreted and depth converted in order to provide input to the modelling of pseudo-wells within the study area. Thus, the first phase defines all the needed models and parameters for the modelling of hydrocarbon generation through interpretation, modelling and calibration to control data.

The second phase involves modelling of hydrocarbon generation and oil expulsion. Expulsion is modelled as a function of hydrocarbon saturation, which follows from the burial and compaction history together with the hydrocarbon generation history. The modelling provides expulsion histories for each source rock unit at each well and pseudo-well location. This output is mapped as a series of time slices, resulting in cumulative expulsion maps expressed in kg/m^2. The modelling of oil expulsion also allows for the modelling of secondary cracking of retained oil, and resulting generated amounts of gas are also mapped. Backstripping of the depth-converted maps is carried out in order to obtain palaeostructure maps for secondary migration modelling. Reservoir petrophysical, thickness and pressure data for the secondary migration modelling are defined. In short, the second phase provides additional data necessary to model secondary migration.

The third and final phase is secondary migration modelling. The modelling routines are described in Sylta (1991, and this volume). The output from the secondary migration modelling includes flow patterns and filling and spilling histories of the prospects through time. The model is calibrated by matching the modelled reserves and known amounts of oil and gas in place in the 30/9-J structure. Based on the achievement of a satisfactory match between modelled and known reserves in the 30/9-J structure, a prediction of the amounts of oil and gas in the A, B and C prospects (Fig. 1) can be inferred.

In order to keep this paper within a reasonable size, we wish to focus primarily on the hydrocarbon generation, expulsion and secondary migration. The text is subdivided into four main sections: source rock potential; petroleum generation modelling concepts; expulsion; secondary migration.

These are followed by a discussion of gas volume uncertainties and conclusions. The Appendix describes maturity correction procedures for source rock yield parameters, the numerical technique for expulsion modelling and includes definitions of all parameters used. Source rock input and modelling parameters are summarized in Tables A1 and A2 in the Appendix.

Source rock potential

Well data in the study area restrict potential petroleum source rocks to the Jurassic section. The source potential for the entire section encompassing the Dunlin through the Viking Group was addressed (Fig. 3). Rock-Eval data (642 analyses available) and visual kerogen data (total of 89 analyses available) from cuttings, sidewall cores and core chips in twelve wells were used to assess the richness, quality and likely yield products of the four intervals that were differentiated, i.e. the Draupne Formation, the Heather Formation, the Brent Group and the Dunlin Group.

Non-random sample coverage and large vertical variations of properties within each source rock interval precluded the mapping and evaluation of lateral variations in source rock parameters. However, compilation of all available data allowed for an overall differentiation between four major source rock units, both with respect to richness and type of organic matter (Figs. 3, 4 and 5). The large standard deviations are due to vertical variations, insufficient or non-random sample coverage and possible lateral variations.

Dunlin Group (Figs. 3, 4a and 5a)

The average TOC in the Dunlin Group argillites is 2.88% (±2.26). Some samples exhibit relatively high hydrogen indices. The spread in the data is extensive,

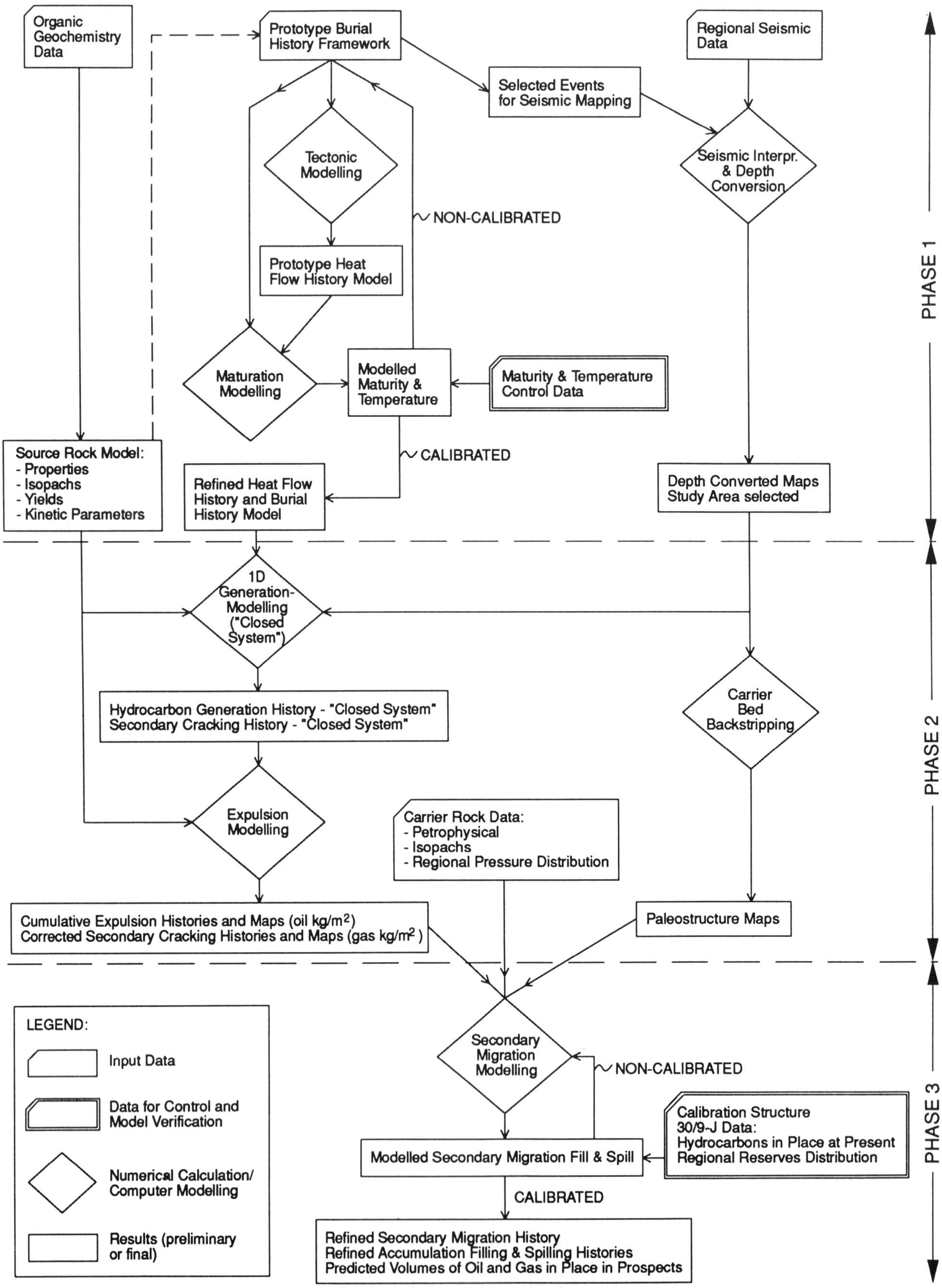

Fig. 2. Schematic organization of the integrated basin modelling study.

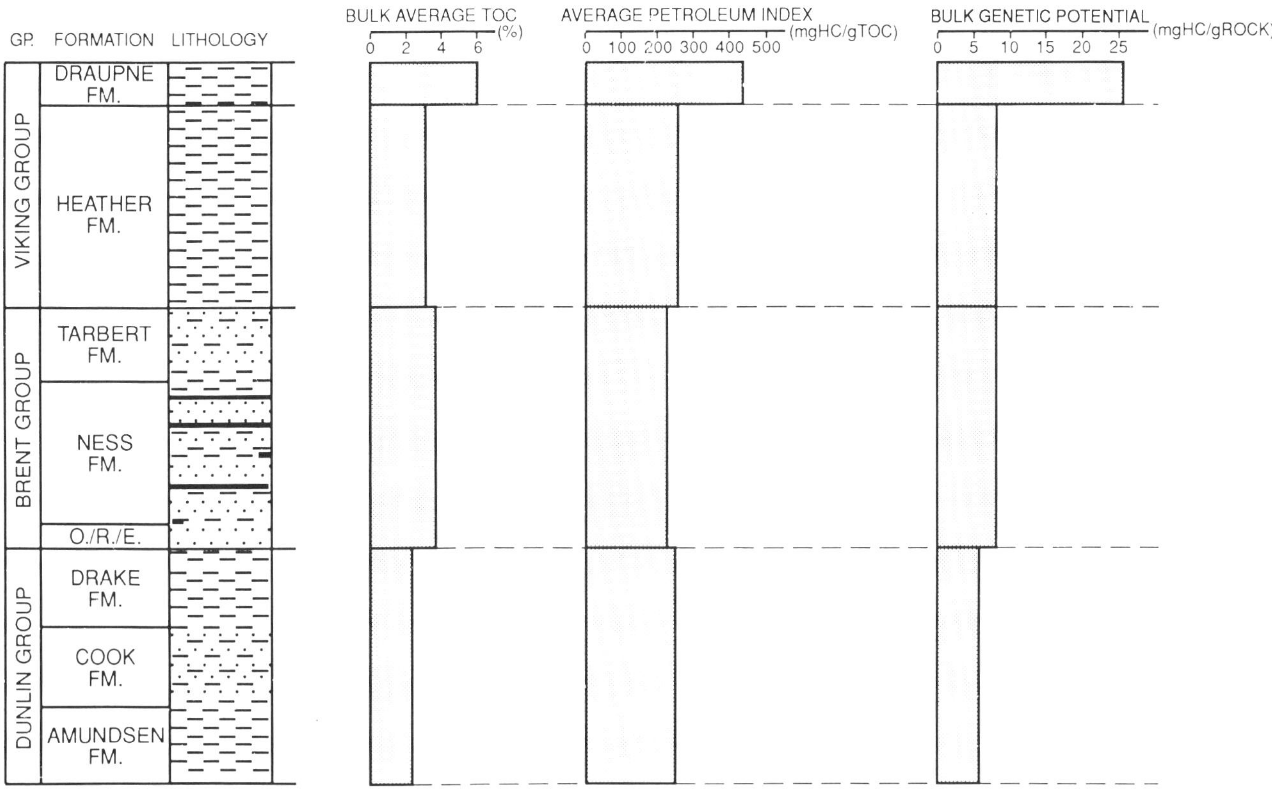

Fig. 3. Source rock summary. Average values for each formation are shown. The TOC and genetic potential for Brent and Dunlin are diluted with sandstone; the richness of the shales and coals are given in Fig. A1 and Table A2.

ranging from 0 to 415 mg HC/g TOC, after correction for maturity and free hydrocarbons (herein referred to as "initial petroleum index" (HI_o), Appendix). The petroleum index averages 246 mg HC/g TOC ($\pm$105) and an average initial genetic potential of 7.08 mg HC/g ROCK is inferred.

The visual kerogen data indicate a predominance of vitrinite and generally low percentages of liptinite. The liptinite fraction ranges from 2 to 54%, constituting on average about 30% of the kerogen. Since liptinite has a generation potential 5 to 7 times larger than that of vitrinitic macerals, the generative contribution from liptinite will dominate.

The best source rock intervals are generally observed in the Drake Formation; however, there are significant contributions also from the Burton and Amundsen Formations. A marked increase in organic richness with depth is observed in the Amundsen in a number of wells.

The Dunlin Group gross thicknesses range from 100 m to more than 600 m within the area. Sandstones, mainly within the Cook Formation, constitute from 1 to 38% of the Dunlin Group thickness, averaging 19%.

Brent Group (Figs. 3, 4b and 5b)

Significant source rock potential is observed in both coal beds and shales. The initial petroleum index (HI_o) is distinctly higher in the coals (309 mg HC/g TOC $\pm$ 86) than in the shales (217 mg HC/g TOC $\pm$ 99). It is not clear whether this is due to stronger mineral matrix effects in the shales (Hartman-Stroup, 1987) or to systematic variations in the kerogen quality. There is no indication of the latter in the visual kerogen data. The TOC averages 56.83% ($\pm$18.30) and 5.31% ($\pm$5.83) implying average genetic potentials of 170.74 and 11.52 mg HC/g ROCK for coals and shales, respectively.

The structured component in the liptinite fraction is usually prominent. As structured liptinite is generally more resistant to thermal stress than its amorphous counterpart (Horsfield, 1989), the remaining generation potential is expected to be higher in the Brent Group kerogens at high levels of maturity than in the Dunlin, Heather and Draupne kerogens. The evolution path envelopes (Fig. 4b) support this, since a flatter, and hence, more refractory envelope is produced from the Brent Group Rock-Eval data (see later discussions). The maceral composition varies significantly, with the liptinite fraction ranging from 10 to 82%, averaging 45% (Fig. 5b).

The coals and the richest shales are mostly located in the Ness Formation. The Brent Group gross thickness ranges from 0 m (eroded) to about 600 m. The percentages of net source rock thicknesses (shales and coals) are in the 22 to 49% range, averaging 38%.

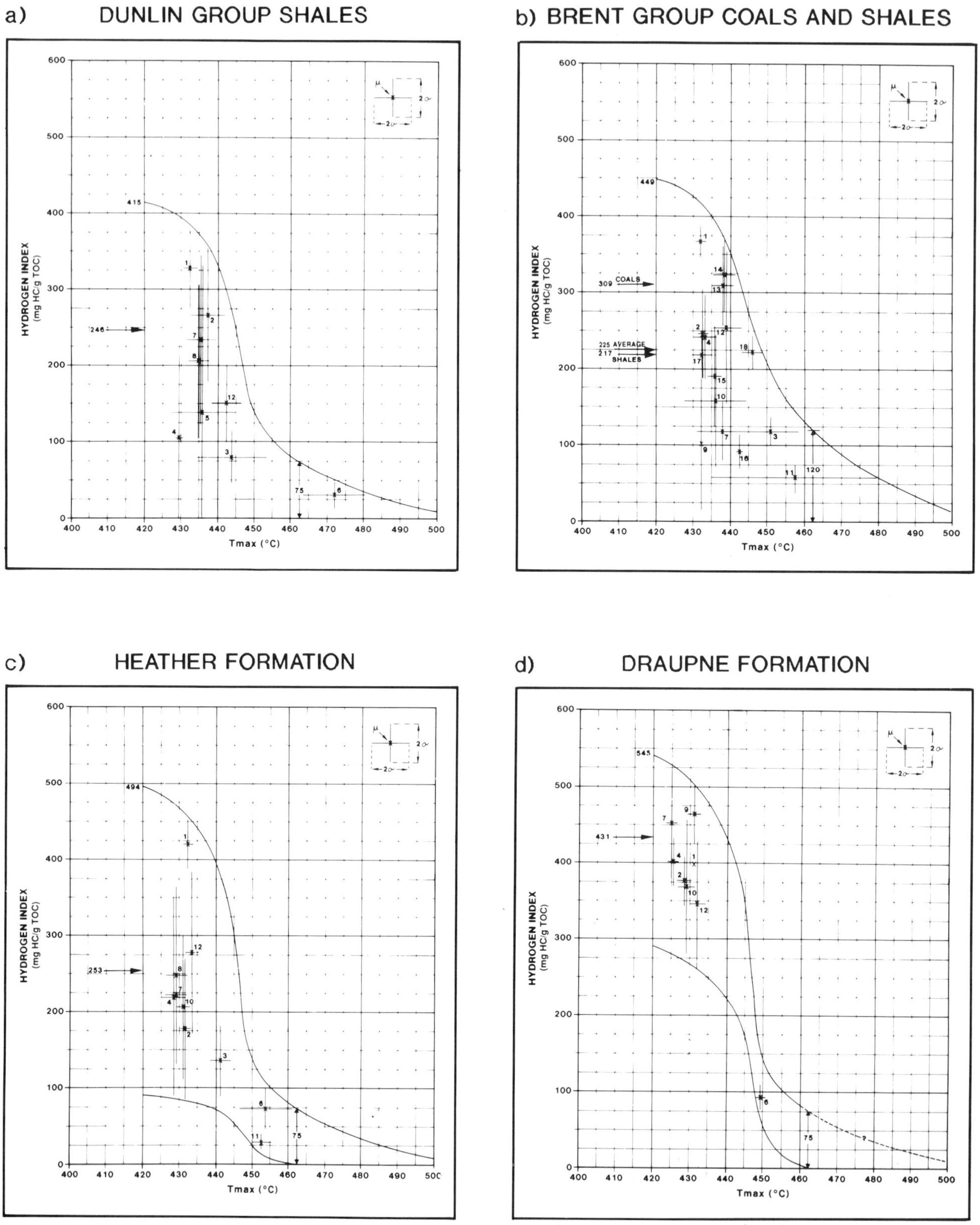

Fig. 4. Source rock evolution path envelopes for the Dunlin Group, Brent Group, Heather Formation and Draupne Formation. Posted numbers are well codes. The average ± the standard deviation is plotted for the data population from each well.

Heather Formation (Figs. 3, 4c and 5c)

The Heather Formation exhibits more regionally consistent data than the two potential source rock sections previously described. However, variance at a particular location is significant (Fig. 4c), implying strong vertical variation in potential; source quality and richness generally increase upwards. The average initial petroleum index equals 253 mg HC/g TOC

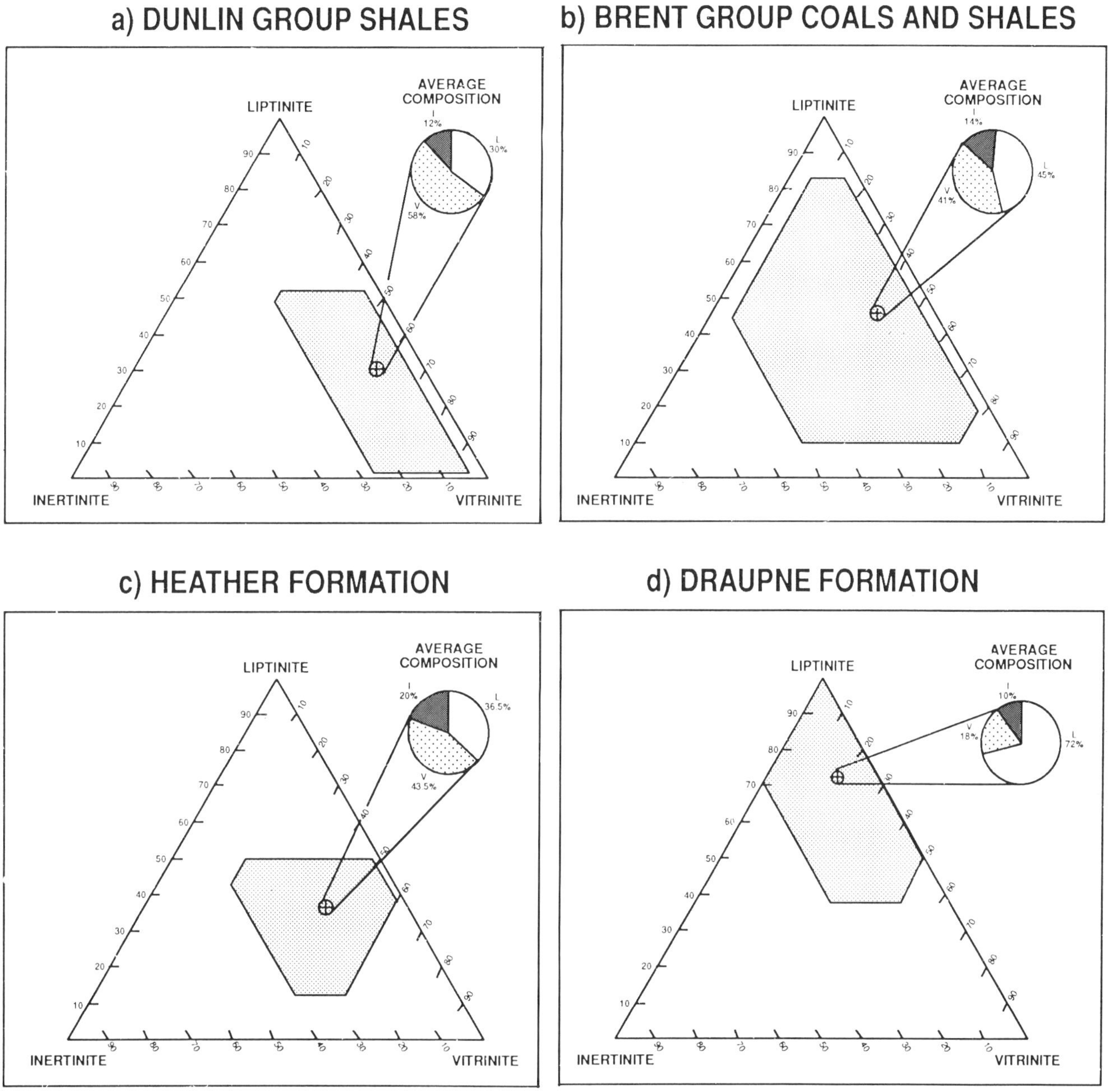

Fig. 5. Source rock visual kerogen compositions.

(±115) and the TOC is 3.14% (±1.64). An average genetic potential of 7.94 mg HC/g ROCK is thus implied.

The visual kerogen data show a less variable composition than those for the Brent and Dunlin kerogens, and the spread is less than 40% with respect to any of the components. The average composition has 36.5% liptinite, mostly amorphous.

The thickness ranges from 0 m (depositional thinning and local footwall erosion of rotated fault blocks) to more than 800 m locally, but thicknesses in the range between 100 and 500 m are generally more representative for the mature kitchen areas.

Draupne Formation (Figs. 3, 4d and 5d)

The Draupne Formation contains the richest argillites in the evaluated Jurassic section. High hydrogen indices and little variance both in a vertical and regional sense are depicted in Fig. 4d. The initial petroleum index averages 431 mg HC/g TOC (±56) which in combination with the average TOC of 5.93% (±1.58) implies an average genetic potential of 25.56 mg HC/g ROCK.

The Visual Kerogen data (Fig. 5d) exhibit a strong predominance of liptinitic material (40 to 100%, 72% on average), in agreement with the Rock-Eval data. The liptinitic fraction is dominated by kerogen of amorphous character.

The Draupne Formation ranges in thickness from a few tens of metres along the graben flanks (locally zero) to more than 150 m in the deep graben. Although this may appear modest in comparison to the volumes of potential source rock described for the Dunlin and Brent Groups and the Heather Formation, the superior concentration of organic matter in the Draupne Formation, as discussed later, outweighs the impact of inferior rock volume.

Petroleum generation modelling

Since the original publication on Rock-Eval pyrolysis by Espitalié et al. (1977) and the earliest kinetic schemes for the "original" three main kerogen types (I, II and III; Tissot and Welte, 1978; Tissot and Espitalié, 1975), classification schemes based on pyrolysis data have grown in complexity and sophistication, as have published mathematical models for kerogen transformation to petroleum. However, the main principles remain unchanged. Kerogen hydrogen richness remains one of the most widely used criteria for the identification of oil-proneness, and various usages of the Arrhenius equation are still regarded as empirically adequate for the modelling of kerogen transformation in response to thermal stress through geological time. More recently, kerogen has been classified according to resistance to thermal stress, resulting in the two categories "labile" and "refractory" (Cooles et al., 1986; Mackenzie and Quigley, 1988). Labile kerogens are in general oil-prone, especially when hydrogen-rich. The oil proneness of hydrogen-rich refractory kerogens (certain exinitic macerals) is debated and it can be speculated that the end of their oil generation window is likely to be defined by oil stability, i.e. secondary cracking (Horsfield, 1989), rather than by their initial potential as indicated by pyrolysates. Labile types are generally exhausted prior to reaching a Rock-Eval T_{max} maturity of 460–465°C, while the refractory types (both hydrogen-rich and hydrogen-poor) retain potential beyond 500°C (Espitalié, 1986).

Source rock kinetic data were obtained for this study through time-slice pyrolysis-GC analyses on wet cuttings from the Viking Group in the North Viking Graben (60°15′N to 61°30′N) at immature locations. This method, developed by Conoco Exploration Research Services, is a modified method of programmed temperature pyrolysis based on the work of Weitkamp and Gutberlet (1968). Similar principles were also applied later by Sekhar and Ternan (1979). It is noteworthy that, in a kinetic sense, the Draupne Formation and Heather Formation kerogens are very similar, not only with respect to kinetic parameters but also with respect

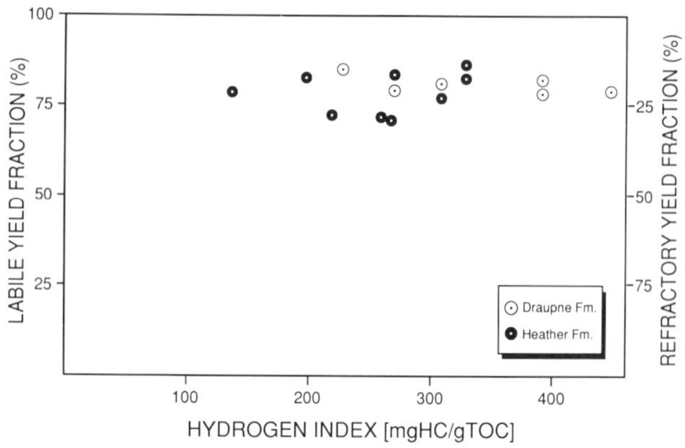

Fig. 6. Yield fractions versus hydrogen indices based on time-slice pyrolysis–GC data from the Viking Group (T.T.Y. Ho, pers. commun., 1990).

to the yield fractions from labile and refractory kerogen. The relative magnitudes of labile versus refractory yields are independent of the hydrogen indices (Fig. 6). Consequently, the hydrogen index cannot be used to predict or estimate the relative amounts of labile and refractory kerogen yields.

The refractory yields are realized at a thermal stress that is well beyond that of secondary cracking. The potential for oil generation must therefore be attributed to the labile fraction. In the following sections, the labile yield is assumed to represent the generation of oils that are undersaturated with respect to gas during catagenesis (Horsfield et al., 1990) and may thus be expelled as a liquid phase.

The time-slice pyrolysis results from the Viking Group samples were used to create a composite set of reactions representative of labile and refractory kerogens (Appendix, Table A1). Due to lack of immature samples for time-slice pyrolysis of Brent and Dunlin samples, a kinetic model was created on the basis of their evolution path envelopes in combination with the Viking Group kinetic parameters. The evolution path envelope for the Dunlin (Fig. 4a) supports a degradation rate with increasing maturity similar to that of the Heather (compare the maximum enveloping curves for each source rock unit in Figs. 4a and 4c). As pointed out earlier, the Brent Group kerogens exhibit higher resistance to catagenesis than the three other units as reflected in the slower degradational rate of the evolution path envelope (Fig. 4b). Any remaining yield potential at maturities of about 460–465°C T_{max} should be refractory. On this basis it was observed that the refractory yield fraction in the Brent needed to be increased by 60% compared to that of the Dunlin and Heather kerogens (Appendix, Table A1). This agrees with time-slice pyrolysis results from coals of Early to Middle Jurassic age on Haltenbanken (T.T.Y. Ho, pers. commun., 1990)

The secondary cracking reaction was based on a relationship between E_a and A for oil destruction, proposed by Conoco research workers based on observations of laboratory and refinery data (Appendix, Table A1). The primary oil and gas potentials of labile kerogens of the respective source rocks are based on the evolution path envelopes together with the visual kerogen data. The primary oil value for the Brent Group has been corrected for the adsorptive capacity of the coals (Ejadawe, 1986). The retained oil (non-expelled) is cracked to gas, resulting in 50% gas and 50% residue (char). Gas dissolved in the retained oil is not cracked. The reactive kerogen classification and modelling concept, and the fate of the yield products in the process of catagenesis and secondary cracking are shown in Fig. 7.

Figure 8 demonstrates the transformation ratio curve for the composite kinetic data (Draupne, Heather and Dunlin) and the petroleum products generated from the Draupne at increasing maturity. Secondary cracking is shown acting upon the entire generated oil volume (i.e. closed system). The correlation to vitrinite reflectance was carried out using the parameters of Burnham and Sweeney (1989).

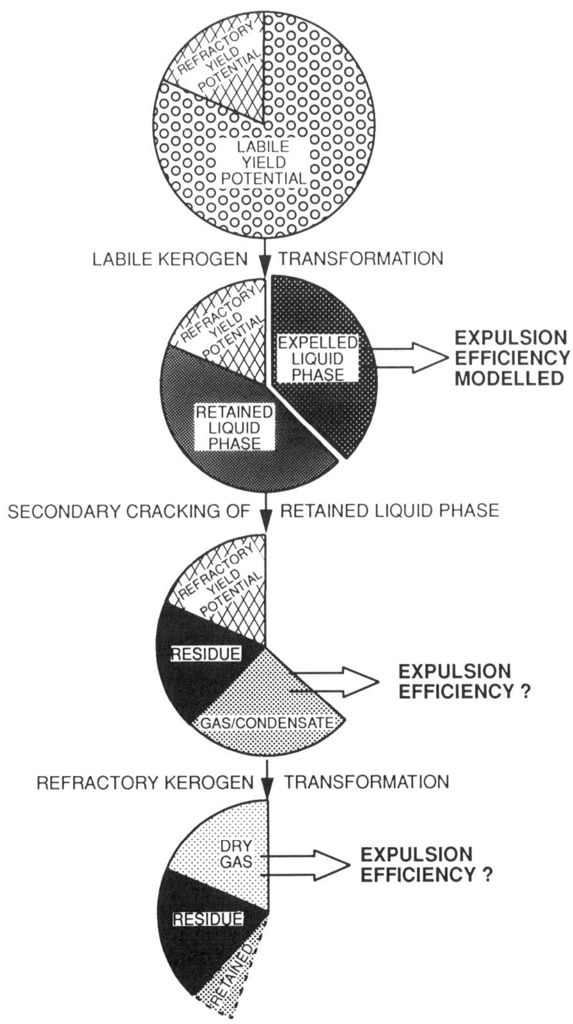

Fig. 7. Reactive kerogen classification, showing yields in relation to kerogen transformation, expulsion, retention and secondary cracking.

The window for oil existence occurs within the range between 0.50 and 1.38% R_o (Fig. 8). It embraces an upper zone (0.50 to 0.92% R_o, primary cracking) defined by oil generation and a lower zone (0.92 to 1.38% R_o, secondary cracking) defined by oil destruction. Electrical resistivity measurements on mature Draupne equivalent source rocks (Goff, 1983) are highest at about 0.8% R_o, and suggests that at this maturity a maximum oil saturation has been reached. In addition, Thomas et al. (1985) observed that the Draupne Formation kerogen quality declines to a "pseudo-vitrinitic" composition and post-maturity for oil generation prior to reaching 1.0% R_o. These observations indicate, firstly, that a significant portion of the labile kerogen must have been transformed into petroleum at 0.8% R_o and, secondly, that all of the labile kerogen is exhausted prior to reaching a maturity of 1.0% R_o. In both cases, there is agreement with the kinetic model shown in Fig. 8.

The maturity at which expulsion is shown to begin, varies throughout the study area due to the use of a saturation-dependent expulsion model (see following sections). The expulsion window indicated in Fig. 8 encompasses the maturity range within which expulsion was possible from the Draupne Formation. The expulsion windows of the Heather, Brent and Dunlin were much narrower due to their lower genetic potential.

Expulsion

Background

The source rock evaluations amply document a huge generation potential from labile kerogen contained in the Jurassic section. The liptinite content and often high hydrogen indices (Figs. 4 and 5) suggest kerogen oil-proneness within all four source rock intervals. A superior richness and quality is observed in the Draupne Formation, but its rock volume is relatively small. Estimates of the total generation potential for each of the source rock units within the respective areas subjected to catagenesis (see areas in Figs. 14 through 17) indicate that in total about 64 billion tonnes of products (oil, gas and residue from cracking) resulted from the transformation of labile kerogen. This potential can be attributed to the source rocks as follows: Draupne 17%, Heather 29%, Brent 30% and Dunlin 24%.

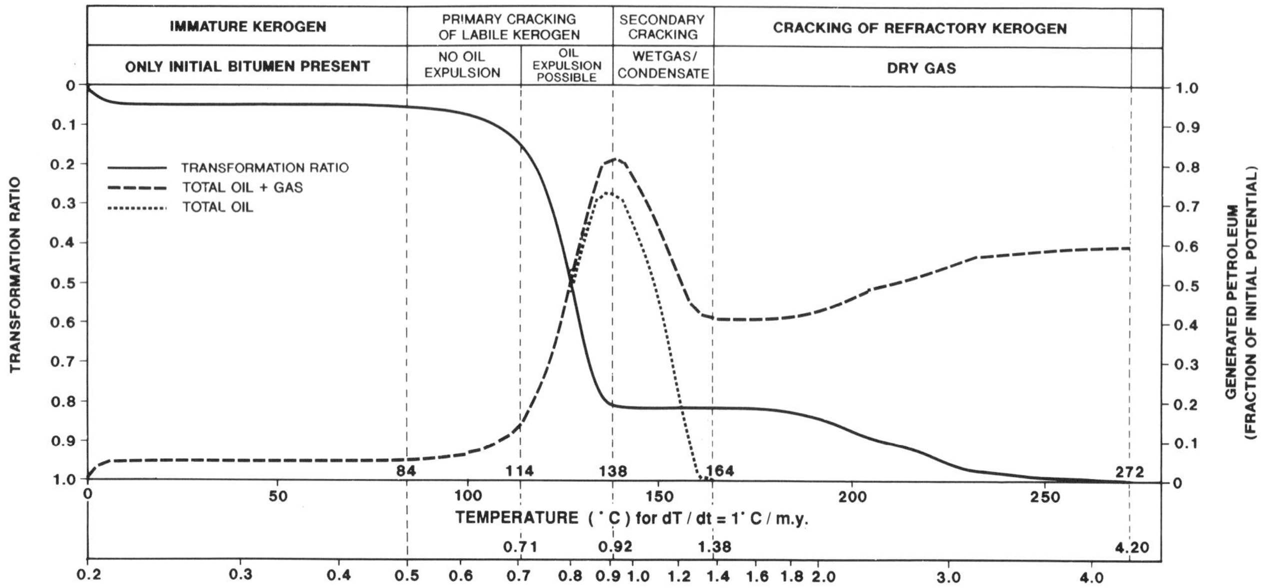

Fig. 8. Labile and refractory kerogen transformation in relation to petroleum generation rates, secondary cracking rates, temperature and vitrinite reflectance. The effect of expulsion is not shown (i.e. closed system). The width of the expulsion window refers to the Draupne Formation. Vitrinite reflectance is derived using parameters from Burnham and Sweeney (1989).

Published studies on oil–source rock correlations identify the Viking Group in general, and often specifically the Draupne Formation, as the source for reservoired oils (Dahl and Speers, 1985; Schou et al., 1985). Hence, it is apparent that selective processes promote the availability of Viking Group sourced oils for entrapment. This is largely attributed to the role of expulsion in our modelling.

Two modes of expulsion frequently cited are the matrix-controlled and the microfracture-dominated modes. The conditions under which either of the two is dominant is still a point of contention. Primary rock porosity is held to be sufficient for the flow of the petroleum-rich phase in the matrix-controlled mode. The microfracturing mode requires that the internal fluid pressure must exceed the mechanical strength of the rock before fluids can escape. A third model, proposed by Stainforth and Reinders (1990) involves petroleum diffusion through organic matter networks. The model requires the organic matter to be distributed in a continuous 3D network. This is argued to be achieved for source rocks where the TOC typically exceeds 1–2%. We discard this hypothesis for use in this study because the source rocks included herein (with perhaps the exception of the Draupne Formation) exhibit significant inhomogeneity on the vertical scale, and a continuous 3D network of organic matter seems implausible.

Published studies (Leythaeuser et al., 1987, 1988; Mackenzie et al., 1987) argue that expulsion from the Kimmeridge Clay in the Brae Field area follows the matrix-controlled mode, and furthermore, that the main driving force in the thick shale sequences in the Brae area is compaction. These conclusions supported the use of a saturation-dependent technique to address the quantitative modelling of expulsion in this study.

Methodology

It has often been suggested that a minimum critical saturation of oil is necessary for expulsion to initiate; however, the absolute values are debated. The numerical values of critical saturation quoted in the literature are generally around 20% (Durand, 1988). The efficiency of expulsion at and beyond this point is unclear.

We address the saturation range from 16% to 40%. The limits for the saturation range are defined by S_1 and EOM data, respectively, from a mature well within the study area. 107 S_1 readings and 20 EOM readings in the Draupne and Heather Formations were used to constrain the saturation range.

Oil expulsion was modelled for three different cases that were defined by saturation criteria. The minimum expulsion case (40% saturation required for expulsion to initiate) and the maximum case (16% saturation required for expulsion to initiate) define the range of saturations encompassed in the volumetric output. For simplicity we refer to these cases as $S_{o,irr} = 0.40$ and $S_{o,irr} = 0.16$, respectively. Both cases are characterized by a zero expulsion rate until the critical saturation has been reached and an expulsion rate of 1 for excess volumes (i.e. expulsion of excess volumes is 100% efficient, see Fig. 9a).

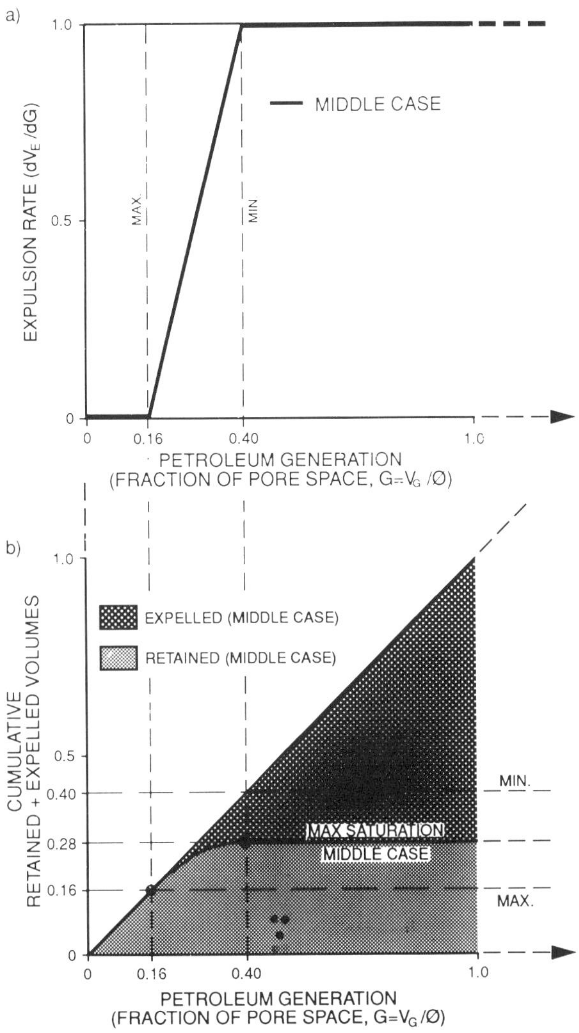

Fig. 9. The expulsion model: (a) boundary conditions, (b) impact on generated volumes. Note: G can become larger than 1.

The middle case follows an alternative numerical scheme in an attempt to account for the vertical inhomogeneity reflected by the source rock parameters in the geochemical data base. For convenience we introduce the parameter G which is the volume fraction of generated petroleum (V_G) to the available pore space (Φ), $G = V_G/\Phi$ (see Appendix). Hence, prior to the initiation of expulsion G equals the average petroleum saturation in the source rock. When considering source rock units of complex inhomogeneous character and variable properties in the vertical scale, it is likely that the rock unit experiences a buildup phase even after the initiation of expulsion before 100% efficiency is achieved. It follows then that parts of the volumes locally exceeding the maximum saturation in richer zones of the source rock contribute to saturate leaner zones until the maximum achievable saturation ($S_{o,irr}$) has been reached throughout. Having reached this point, any further retention of later generated petroleum becomes impossible. This scenario is highlighted in Fig. 9 (bold line) and is described as follows. Expulsion initiates when $G = 0.16$ but is 100% efficient ($dV_E/dG = 1$) only when $G > 0.40$. The latter threshold corresponds to a maximum pore saturation of 28% because we allow the expulsion rate to increase linearly in the interval $0.16 \leq G \leq 0.40$. Within this interval, oil retention accompanies expulsion until saturation has reached its maximum value of 28% (Fig. 9b). The minimum and maximum lines in Fig. 9 represent the $S_{o,irr} = 0.40$ and $S_{o,irr} = 0.16$ cases, respectively.

Volumetric modelling requires quantification of the expulsion efficiency (E). E is the ratio of the expelled volume (V_E) to the generated volume (G). V_E can be determined from the boundary conditions described above through the mathematical integration of the three line segments in Fig. 9a, with respect to G (see Appendix for formulae and definitions). The integrated curve is shown in Fig. 10. Based on the preceding discussions of source rock properties, the kinetic modelling and modelling of expulsion, we observe that four major factors influence liquid phase expulsion efficiency. These ares: labile genetic potential, thermal maturity, degree of compaction (i.e. porosity) and hydrocarbon density. The sensitivity of G to hydrocarbon density and porosity is shown for each of the source rock units in Fig. 11. The ranges obtained for G at the end of labile kerogen transformation are represented with bars. The high end of the porosity ranges represents the deepest areas in the Viking Graben where catagenesis took place early and at shallow depths (Fig. 12). The low end represents areas along the basin flanks where catagenesis took place recently and therefore the end of the labile kerogen transformation window was located at depths in excess of 4000 m below sea floor (e.g. Fig. 13).

Source rock expulsion properties

Volumetric criteria are used in the expulsion modelling, and the source rock input parameters must therefore include volume fractions of kerogen and bulk densities of the rocks (Appendix, Table A2). In contrast to Draupne and Heather, which were modelled to consist of shale and kerogen, the Brent and Dunlin Groups also include significant amounts of sandstone. When considering expulsion in the Brent coals and shales and the Dunlin shales we were faced with the conceptual choice of whether or not to account for losses of oil in the sandstones. In the secondary migration modelling (Sylta, 1991, and this volume) the oil migrates along the top of the carrier bed, which in this study is the Tarbert Formation. The shales and coals are spread out over most of

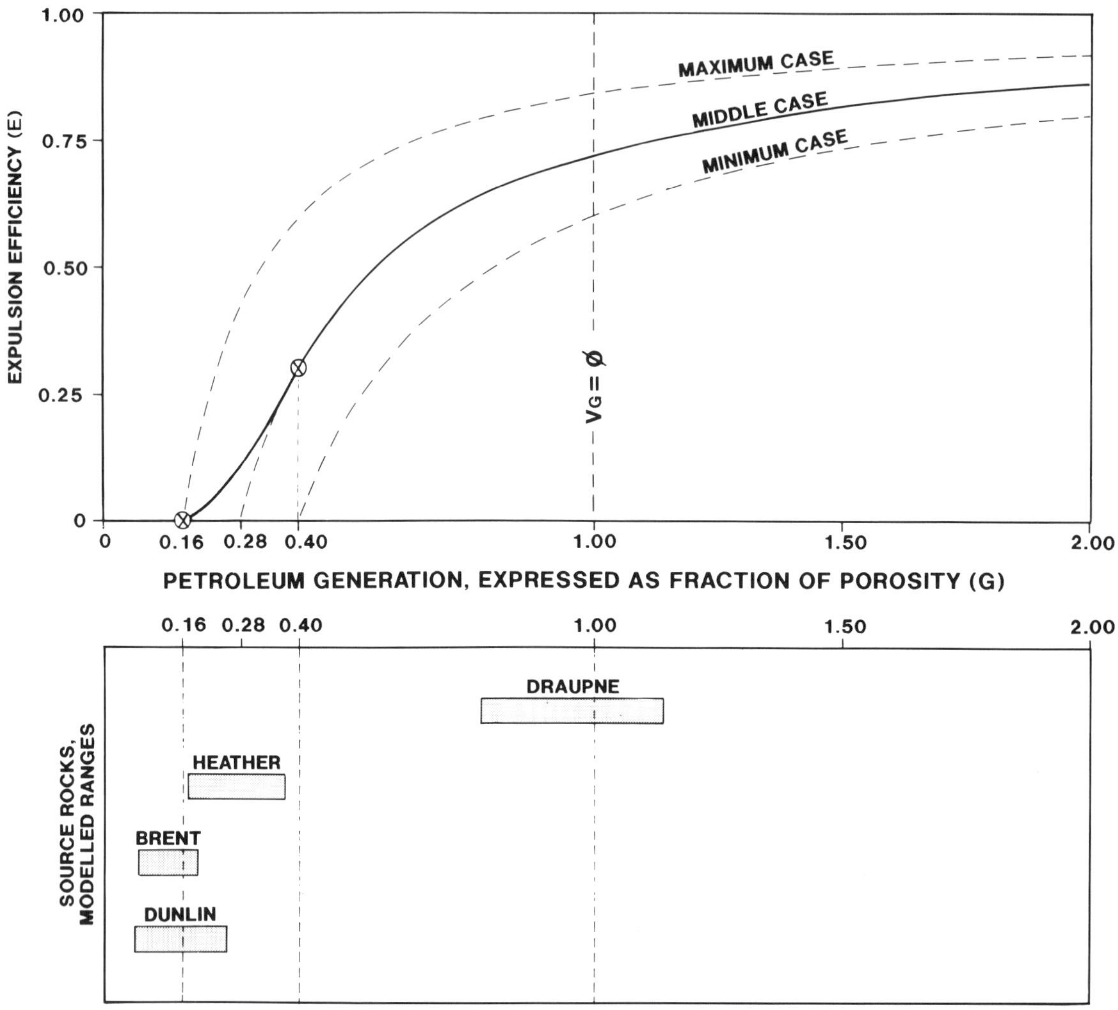

Fig. 10. Expulsion efficiency (E) as a function of petroleum generation (G) expressed as a fraction of porosity. Modelled ranges of G for each source rock unit are shown.

the vertical section and are intermixed with numerous sandstone beds of variable thickness. It would not be correct to assume that the hydrocarbons that were primarily expelled from the coals and shales would move to the top of the Tarbert Formation without incurring losses. We therefore modelled the Brent and Dunlin Groups as bulk source units; i.e. the concentration of organic matter was modelled to be dispersed throughout the whole formation. This choice causes a reduction in genetic potential due to the "dilution" by the sandstones (see Appendix, Table A2) and the sandstone porosities must be saturated to the same degree as the shales before a supply of oil to the top of the Tarbert is established. Hence, expulsion is in this context not regarded as the release of petroleum from a shale or coal into a sandstone, but rather as the net effect of numerous such processes on the vertical scale, ultimately resulting in an excess volume of oil available for lateral migration at the top of the Brent sandstone sequence.

The labile kerogen fraction is generally thought to yield a liquid petroleum which, at the pressures and temperatures experienced during catagenesis, is undersaturated with respect to gas. Cited values for "GOR" from source rocks range from 0.05 to 0.30 kg/kg (235 to 1500 scf/bbl) and these vary with organic matter type and maturity (England and Mackenzie, 1989; Horsfield et al., 1990). In this study, a constant fixed subsurface petroleum density of 0.7 g/cm^3 has been used in the modelling of liquid phase expulsion. No attempts were made to vary the density because the molecular composition of both oil and gas within the source rock, together with variations in GOR in the course of catagenesis, are largely unknown. The sensitivity of G to hydrocarbon densities within the range between 0.5 to 0.9 g/cm^3 is shown in Fig. 11.

The source rock pore space reaches its maximum oil saturation prior to or at the end of the labile kerogen transformation, i.e. prior to the onset of oil destruction by secondary cracking. Hence the total

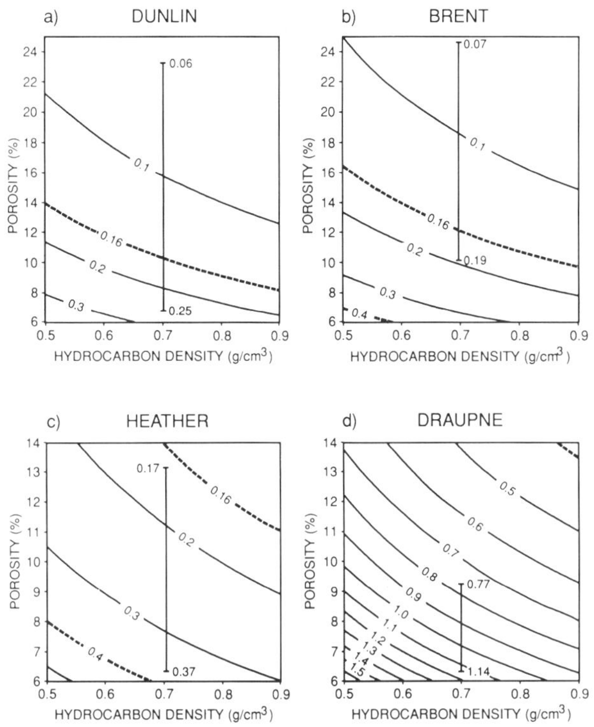

Fig. 11. Sensitivity of G (contours) to porosity and hydrocarbon density for the Dunlin Group, Brent Group, Heather Formation and Draupne Formation. The modelled ranges of G obtained in the study are represented by the bars. The labile generation potential and lithological characteristics for each unit are inherent in the plots.

potential for oil expulsion is controlled by the labile genetic potential and the degree of compaction reached at this particular level of maturity. Expulsion is enhanced when catagenesis occurs at greater depths since less pore space needs to be saturated. The Heather, Brent and Dunlin source rock units are particularly sensitive to the degree of compaction as their genetic potential is only moderate to marginal. In consequence, based on our modelling, the Brent and Dunlin source units are incapable of expelling any liquids at shallower depth than 3950 m and 3700 m, respectively (below sea floor). Heather is capable of expelling oil already at 2900 m (below sea floor). The elevated thresholds for expulsion in the Dunlin and Brent Groups are due to the presence of sandstones, which compact more slowly than shale. The Brent source rock unit, which has a higher labile genetic potential than the Dunlin Group (5.75 versus 4.67 mg HC/g ROCK), is a poorer source rock for liquid expulsion due to the higher proportion of sandstones. In all cases the Draupne Formation is the dominant source rock for oil expulsion, exhibiting expulsion efficiencies from 0.64 to 0.76 (middle case). In comparison, Heather, Brent and Dunlin exhibit ranges between 0.005 and 0.25, 0 and 0.01 and 0 and 0.08, respectively (see Fig. 10). Large volumes of retained liquids in the Heather

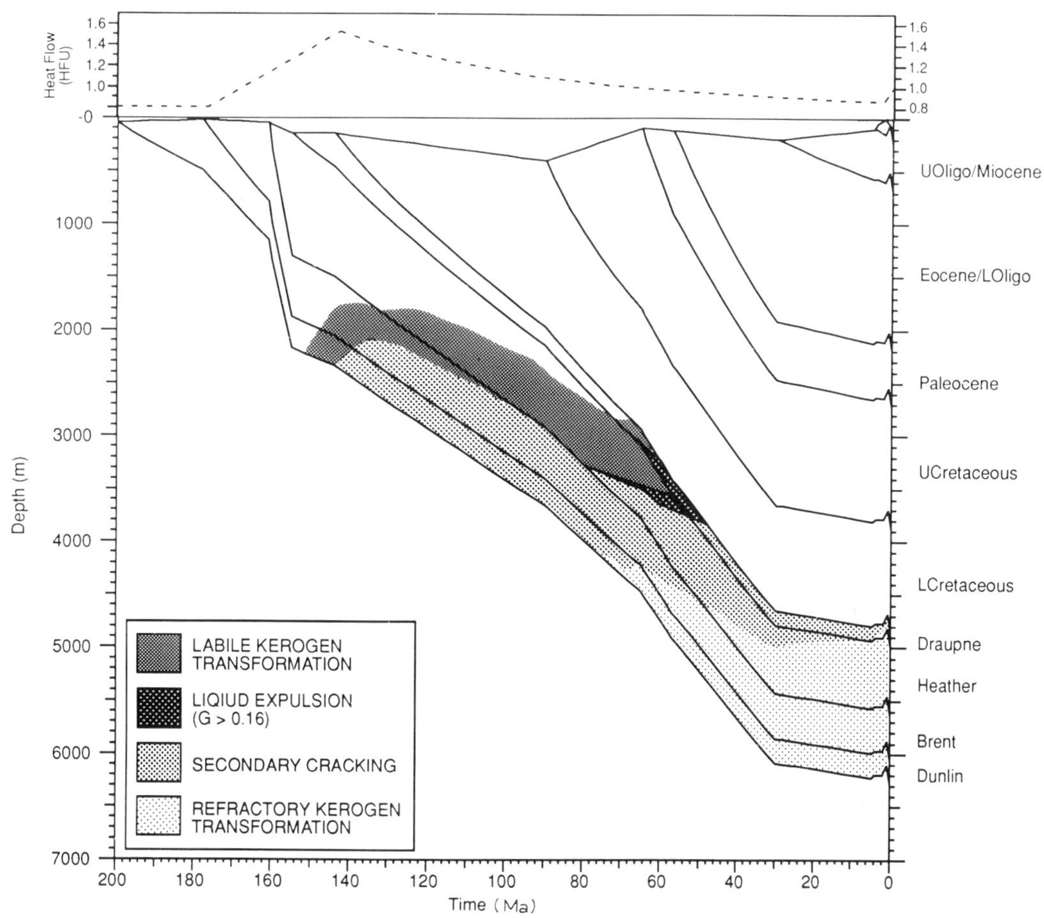

Fig. 12. Burial and maturation histories, graben axis.

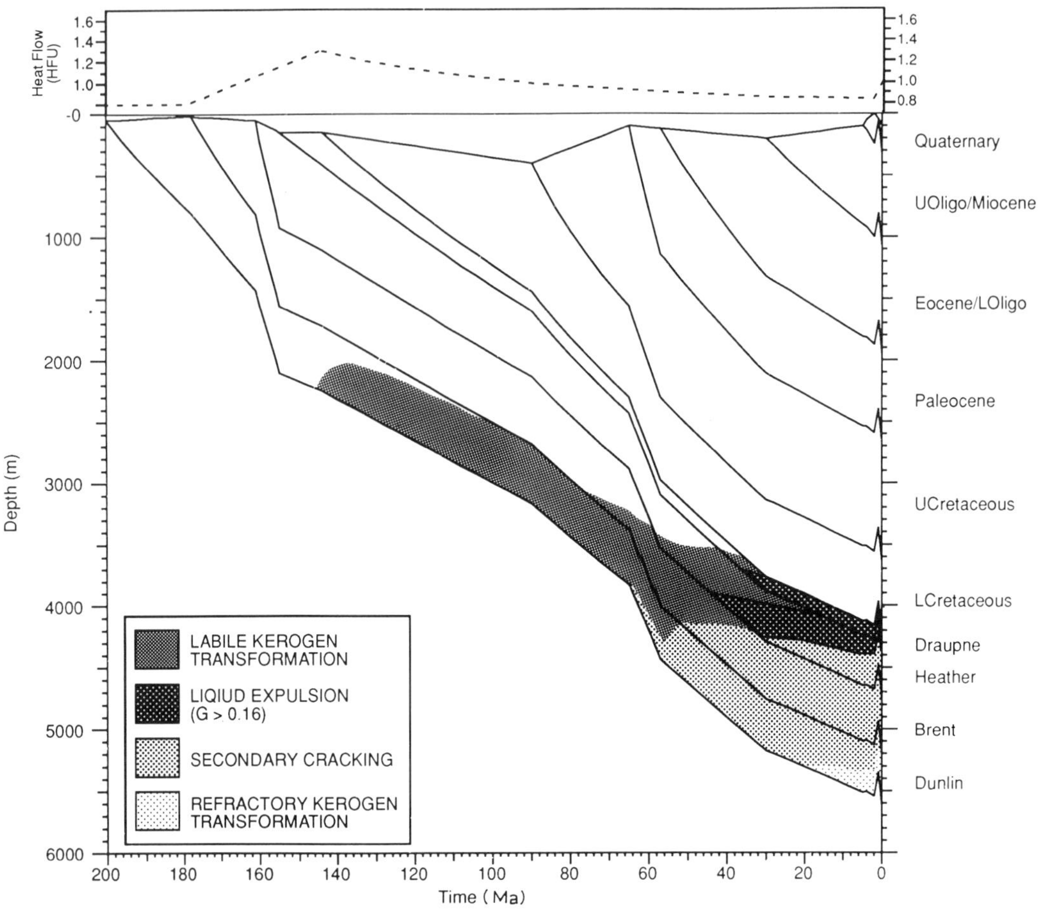

Fig. 13. Burial and maturation histories, basin flank south.

Formation, Brent and Dunlin Groups result in a tremendous potential for generation of gas by secondary cracking.

Maturation and expulsion histories

The commercially available Platte River Associates BASINMOD-software, version 2.45, was used to model burial history and maturation. A "finite rate extension" (Jarvis and McKenzie, 1980) type of heatflow history model was calibrated to computed tectonic subsidence histories (I.M. Walker, pers. commun., 1990). We used the resulting heatflow, which decays exponentially from the end of rifting (Figs. 12 and 13).

The present day bottom hole temperatures (BHT) in the wells are too high to explain the present low level of maturity (vitrinite, T_{max}, hydrogen index) and indicate that cooler thermal conditions may have prevailed in most of the Tertiary. This observation may indicate a high transient thermal regime occurring in recent geological time. The best match in a regional context is obtained by using 2 Ma as initiation for a linear increase of "apparent" heatflow through to the present. Further discussions on the nature and credibility of this late heating event at Haltenbanken are given by Jensen and Doré (this volume).

Figures 12 and 13 summarize the burial, generation and expulsion histories at a deep basinal and a shallow basinal location. The maturity windows displayed in the burial history diagrams are based on transformation ratios in accordance with the definitions in Fig. 8. For the Heather Formation, Brent Group and Dunlin Group liquid expulsion occurs close to the base of the oil window (end of labile kerogen transformation). The time and depth range for expulsion from the entire section is therefore narrow, particularly in the deep basinal areas where the Dunlin and Brent source rock units reach catagenesis at depths too shallow for expulsion to occur.

Cumulative volumetric calculations and mapping of liquid phase expulsion were carried out at 65, 57, 30, 2 and 0 Ma. The time slices correspond to the seismic events that were mapped and used in the backstripping process in order to create palaeorepresentations of the migration surfaces. Calculations of gas generated by secondary cracking of retained petroleum and primary cracking of refractory kerogen were carried out and mapped for the same time slices.

The Dunlin and Brent Groups were subjected to strong catagenesis as early as the Late Jurassic in the deepest parts of the Viking Graben due to high syn- and post-rifting heatflows (Fig. 12). However, the rapid decay in heatflow (cooling) during the Early Cretaceous, in combination with relatively low sedimentation rates, produced net heating rates which were close to zero. The net result was a reduction in catagenic intensity.

By the Late Cretaceous the cooling rate had become significantly reduced and the basin entered into a more intense phase of catagenesis in response to increased sedimentation rates, which continued through the Early Tertiary. The Heather and Draupne Formations experienced sufficient catagenesis in the deepest parts of the basin, from about 80 and 68 Ma, respectively, for expulsion of oil to initiate. At the beginning of the Tertiary only small volumes in local parts of the basin had been expelled (Figs. 16 and 17). At 65 Ma, having failed to reach the required saturation for liquid expulsion, Brent and Dunlin were producing only gas by

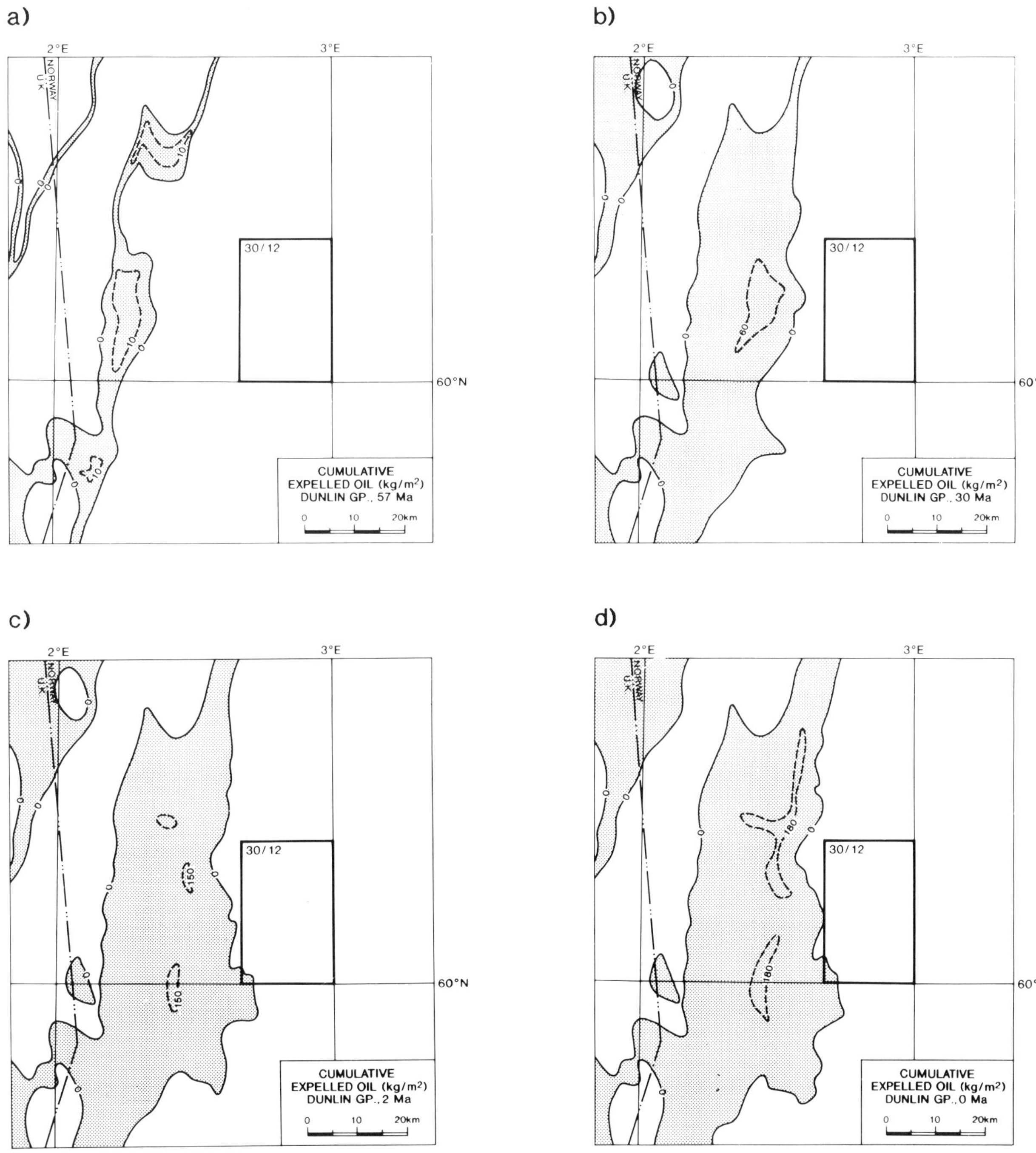

Fig. 14. Cumulative quantities of expelled oil from the Dunlin Group. Middle case at (a) 57 Ma, (b) 30 Ma, (c) 2 Ma and (d) 0 Ma. No oil expelled prior to 60 Ma.

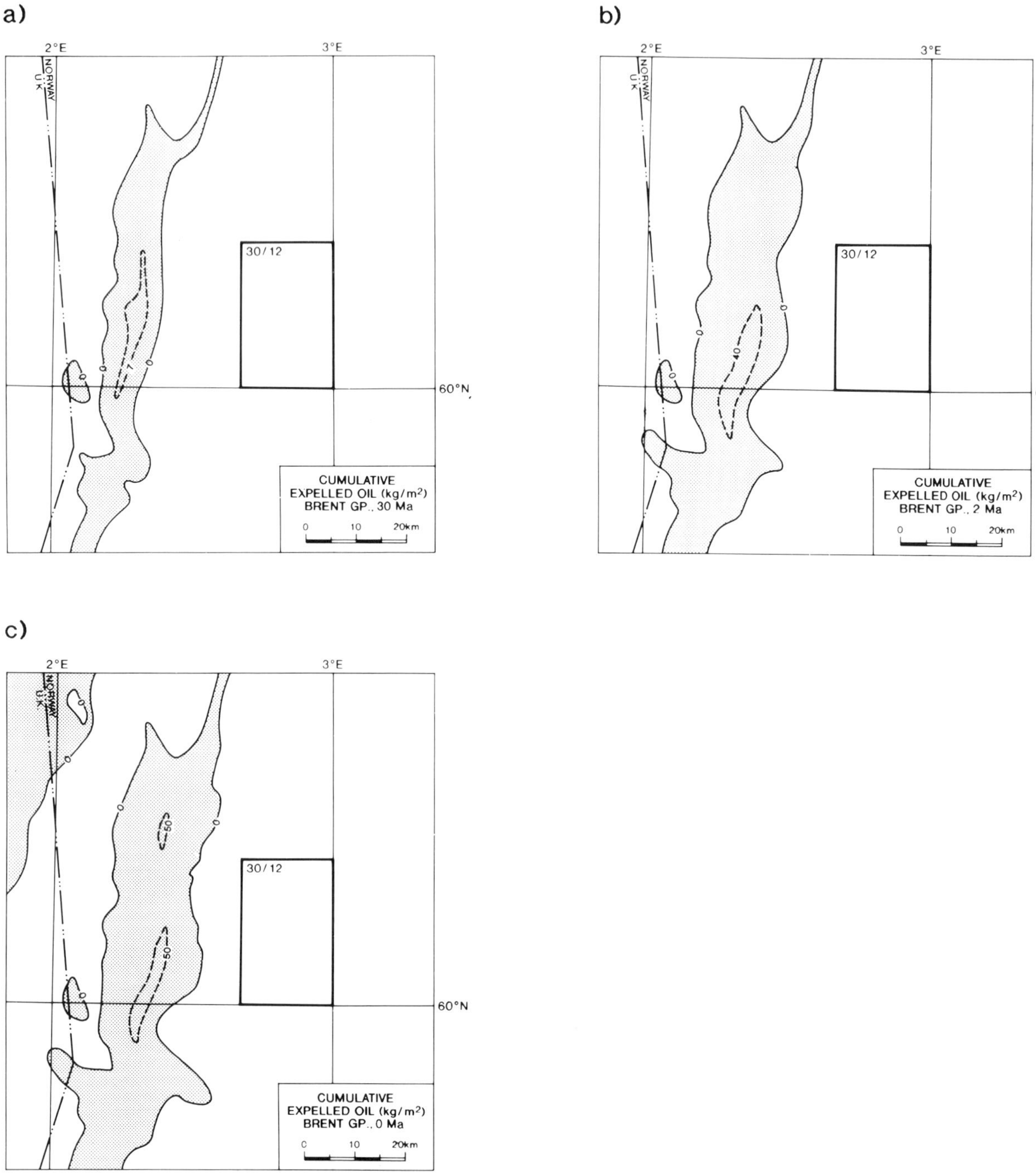

Fig. 15. Cumulative quantities of expelled oil from the Brent Group. Middle case at (a) 30 Ma, (b) 2 Ma and (c) 0 Ma. No oil expelled prior to 54 Ma.

secondary cracking throughout the mature area. Locally, intense secondary cracking also occurred in the Heather Formation. The pre-Tertiary was, as a consequence, characterized by gas generation (see Fig. 18).

In Early Tertiary times, large areas were subjected to catagenesis and by 30 Ma significant volumes of oil had been expelled from the Draupne and Heather Formations (Figs. 16 and 17). By this time the basin had cooled sufficiently for labile kerogen transformation to take place in a depth window of about 3600–4400 m (Fig 13). Hence, due to the lesser pore space available for retention, liquid phase expulsion had become possible also from the Brent and Dunlin Groups (Figs. 14 and 15), starting earlier in the Dunlin Group due to its slightly more favourable lithological composition. From 57 Ma secondary cracking was intense in the deeper parts of the basin (Fig. 18).

Between 30 Ma and 2 Ma labile kerogen catagenesis continued to occur at great depths, and favourable conditions for liquid expulsion from all source units

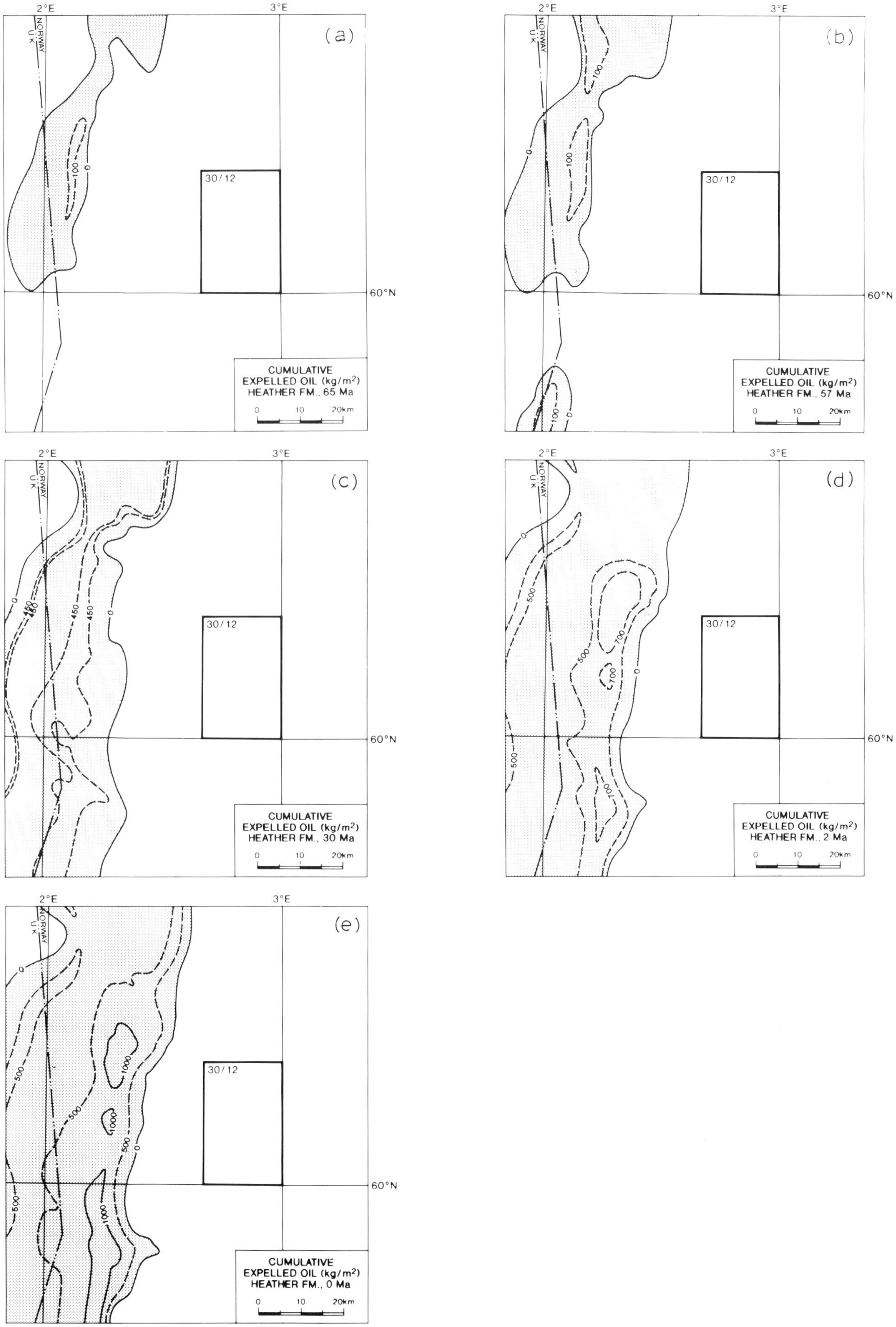

Fig. 16. Cumulative quantities of expelled oil from the Heather Formation. Middle case at (a) 65 Ma, (b) 57 Ma, (c) 30 Ma, (d) 2 Ma and (e) 0 Ma. No oil expelled prior to 80 Ma.

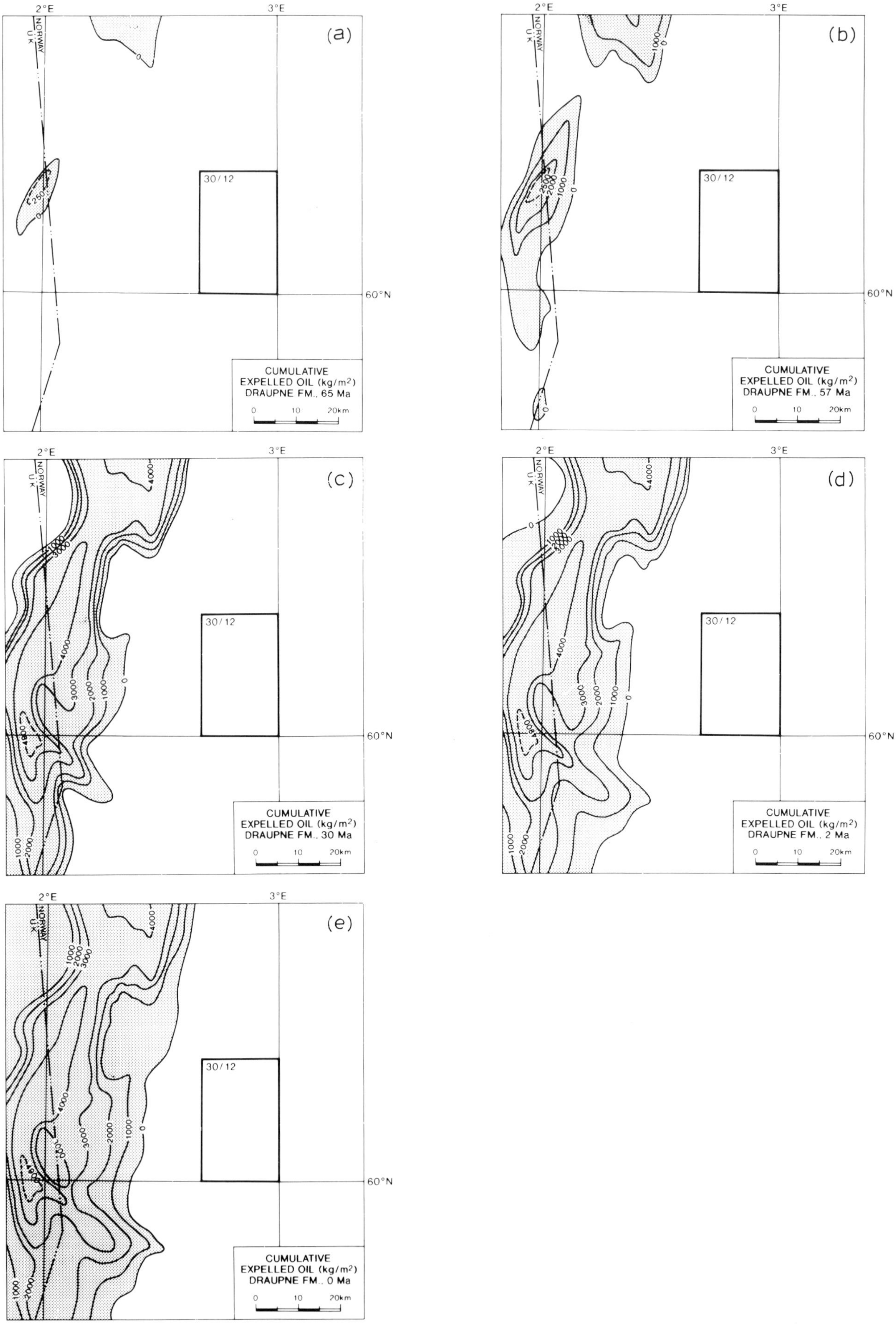

Fig. 17. Cumulative quantities of expelled oil from the Draupne Formation. Middle case at (a) 65 Ma, (b) 57 Ma, (c) 30 Ma, (d) 2 Ma and (e) 0 Ma. No oil expelled prior to 68 Ma.

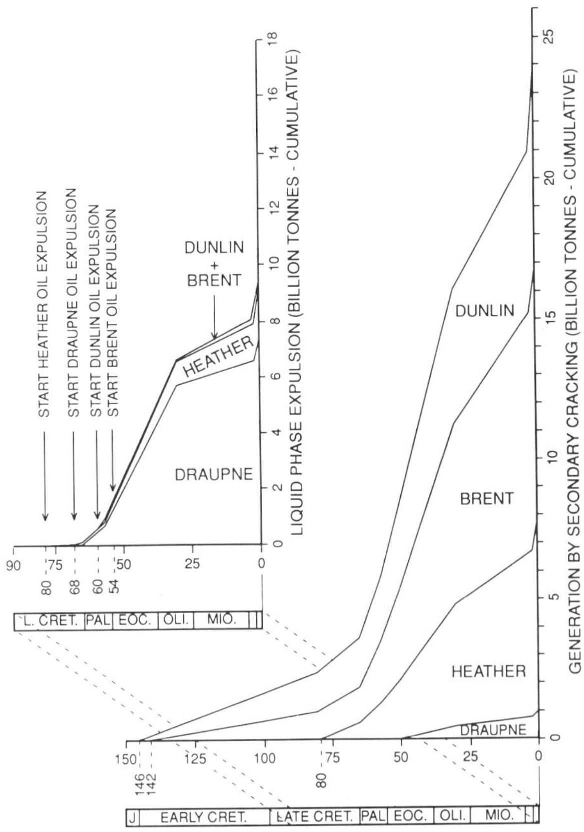

Fig. 18. Oil expulsion and gas generation histories of the study area. Middle case for expulsion.

along the graben flanks resulted. A slight decline in reaction rates is predicted during this period due to slower sedimentation, but still a significant amount of oil was expelled. High sedimentation rates still prevailed in the southeast (Fig. 13).

From 2 Ma to present, the increasing heatflow accelerated the generation rates throughout the area. Despite the short time span, significant volumes of hydrocarbons were generated during this period (Fig. 18).

The northern part of the Stord Basin, at the southeast corner of the study area, is not mature enough for expulsion from any of the Jurassic source rocks. Hermanrud et al. (1990) concluded likewise. The base of the Dunlin Group commenced to mature in the deepest part during the Quaternary; however, the Dunlin Group richness is far too modest to allow expulsion. Unless the quality and/or richness of the Dunlin Group improves dramatically towards the Stord Basin, the Viking Graben can be regarded as the only kitchen area for hydrocarbons expelled from Jurassic source rocks, within the study area (Fig. 1).

Figures 14 through 17 are simplified versions of the maps which resulted from the generation and liquid phase expulsion modelling. These were used as input to the secondary migration modelling. The posted values represent the middle expulsion case. Similar map series were made for secondary cracking and generation from refractory kerogen. Generation of gas from refractory kerogen was not found to be volumetrically significant (see Fig. 20), and only occurred in the deeper parts of the basin. The cumulative gas yields for each source unit are shown at 0 Ma in Fig. 19. The hydrocarbon products from secondary cracking are included in the gas in the migration modelling (see discussions below). Gas expulsion efficiency is treated as one of the unknowns in the secondary migration computations.

The cumulative oil expulsion and gas generation histories for the middle expulsion case are illustrated in Fig. 18. A calculated total of 65.1 billion tonnes of products resulted from the transformation of kerogen (including 0.9 billion tonnes from refractory kerogen) in the different source rock units as shown in Fig. 20. The total quantities of expelled oil, retained oil, gas and residue from secondary cracking are also shown for the three expulsion cases. We note that the expelled amount of oil increases from 6.1 to 17.2 billion tonnes over the saturation range encompassed between the maximum ($S_{o,irr}$ = 16) and minimum ($S_{o,irr}$ = 40) expulsion cases. The contribution of expelled oil from non-Draupne source rocks increases accordingly such that the relative contribution from Draupne is reduced from 100% in the minimum case to only 52% in the maximum case. Since the Draupne Formation is held to be the dominant source rock based on independent evidence (see preceding discussions on oil–source rock correlations), the consideration of saturation thresholds lower than 16% (maximum expulsion case) was needless.

Secondary migration

Input data

Modelling of secondary migration was carried out with the computer program described by Sylta (1991, and this volume). Backstripped palaeorepresentations of the Base Cretaceous and the Top Brent Group were computed from mapped seismic horizons, representing Base Tertiary (65 Ma), Top Paleocene (57 Ma), Mid-Oligocene (30 Ma), Base Quaternary (1 Ma) and Sea Floor (2 Ma and 0 Ma). The horizons correlate to Base MU1, Base MU2, Intra MU5, Base MU8 and Top MU8, respectively, from Rundberg (1989). Data from Rundberg and Smalley (1989) contributed to the absolute dating of the Mid-Oligocene and Base Quaternary events.

The secondary migration oil loss function (Sylta, 1991) used in the simulations is dependent on a num-

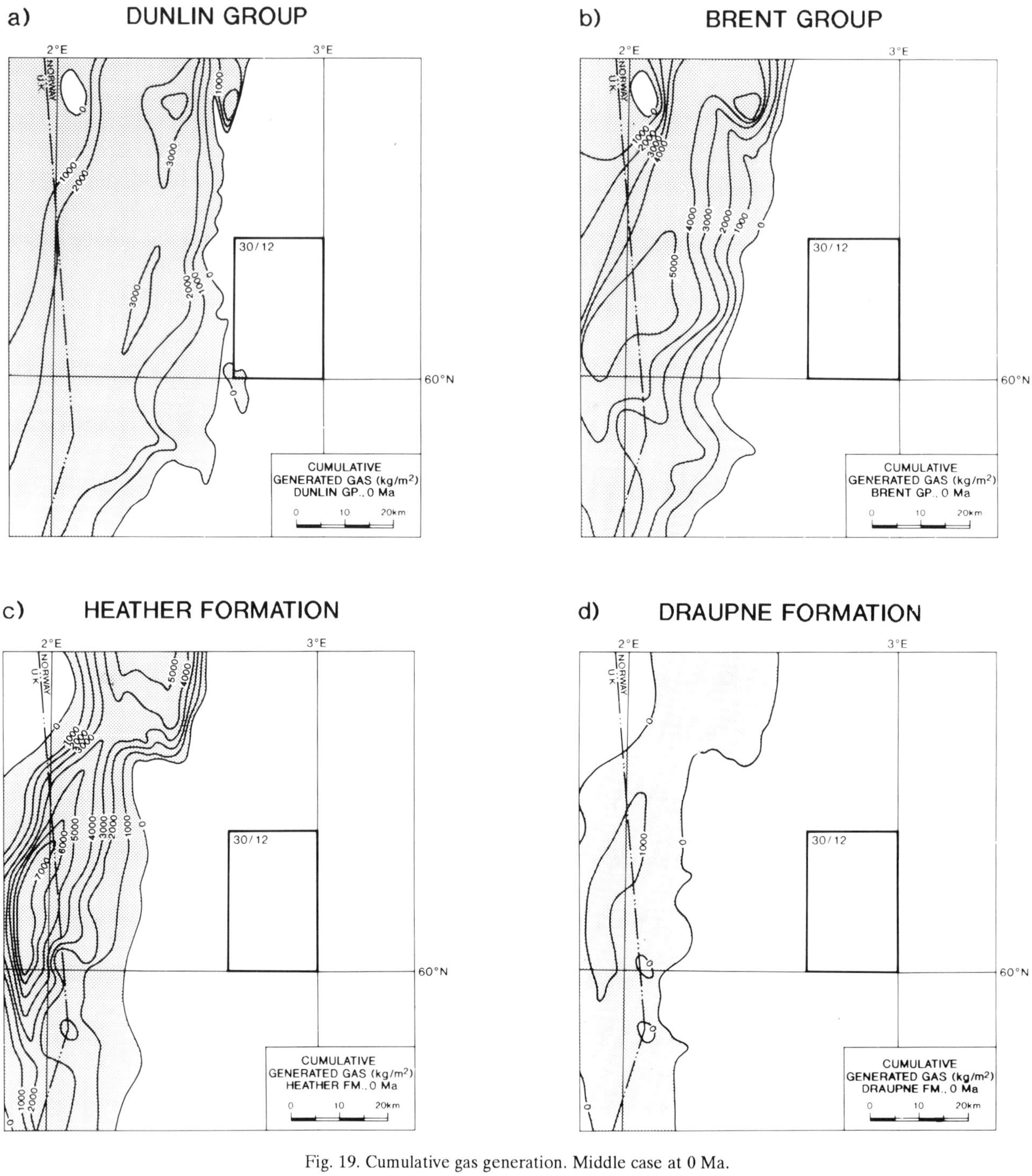

Fig. 19. Cumulative gas generation. Middle case at 0 Ma.

ber of petrophysical parameters (e.g. porosity, saturations, permeability and tortuosity). A comprehensive petrophysical study on Brent Group porosities and permeabilities in the Viking Graben area (59°45′ to 62°N) was carried out before the migration study. All available data from cored Brent sandstones on the Norwegian Sector were used. A significant number of UK wells was also included. Porosity and permeability versus depth relationships were derived for the Tarbert, Ness, Etive, Rannoch, Oseberg and Broom sands. Buoyancy-driven secondary migration implies that migration occurs in a zone at the top of the carrier bed, and the relationships derived for the Tarbert Formation are therefore held to be the most realistic estimates of the petrophysical properties of the Brent carrier beds. Briefly, the study concluded that porosity decreases linearly with depth. A reasonable approximation for the porosity vs. depth is:

$$\Phi = 0.57\,e^{-0.42Z} \qquad (1)$$

where Z = depth in km, and Φ = porosity (fraction)

Permeability of Brent sandstones is independent of pore pressure and its reduction with depth is

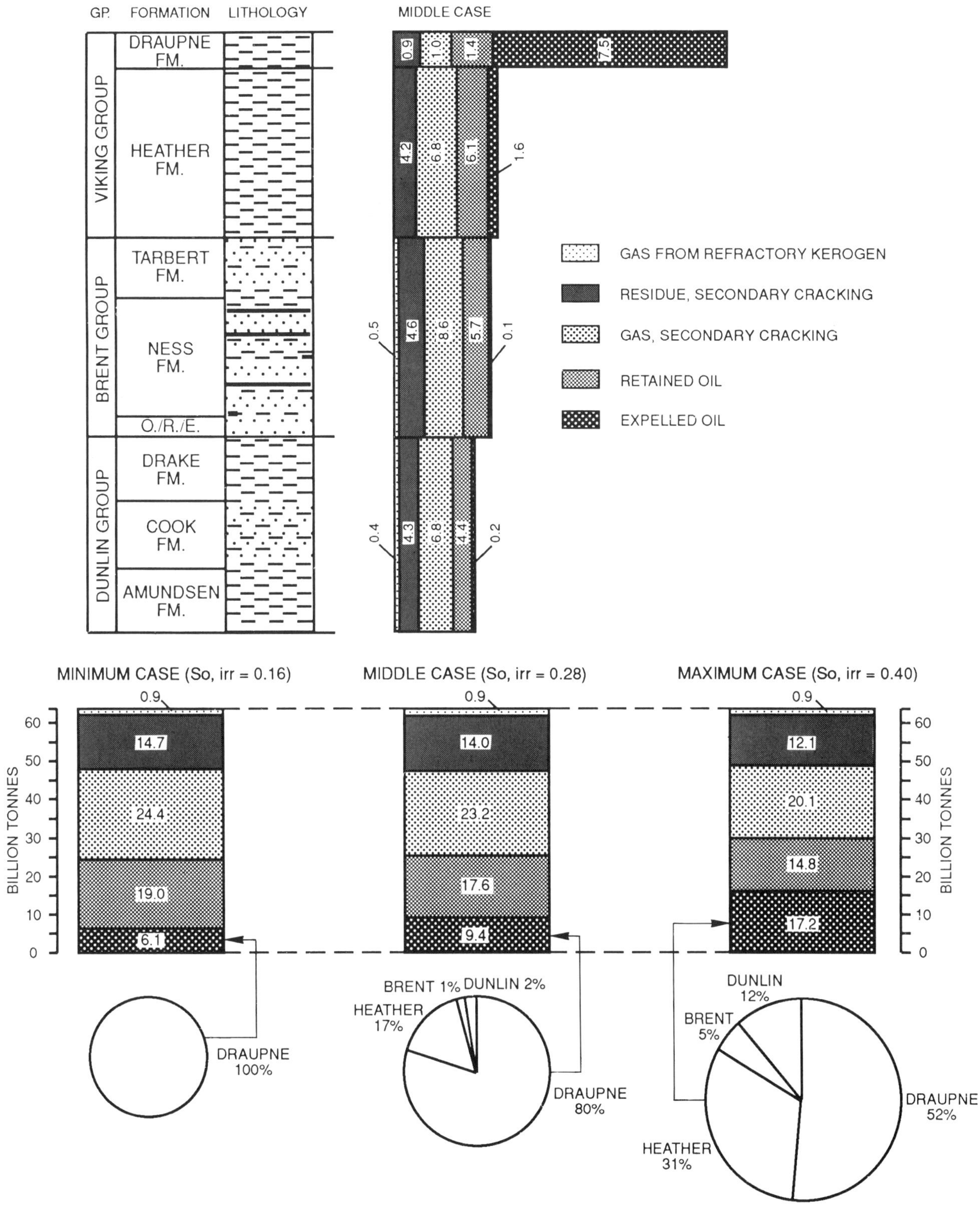

Fig. 20. Cumulative yields within the study area in relation to source rock units, generation products and expulsion cases. Total yield from labile kerogen: 64.2×10^9 tonnes, total yield from refractory kerogen: 0.9×10^9 tonnes.

exponential. The average permeability for Tarbert is represented by:

$$\log K = 7.323 - 1.774Z \qquad (2)$$

where K = permeability in mD.

The initial oil saturation threshold for secondary migration in the Brent sands is modelled to be 50%, the irreducible hydrocarbon saturation 50%, relative permeability 0.3, tortuosity 2, and microtrap losses were taken to be 0.25 km^3/km^2 (see Sylta, this volume). Gas expulsion efficiency and gas loss rates through cap rocks were treated as unknowns, and could be varied as part of the calibration to know accumulations.

Expulsion and migration scenarios

Differing cases for oil expulsion have been described in the previous section. The middle expulsion case was used as the primary case in the modelling of secondary migration. The maximum expulsion case was also modelled for comparison. The minimum expulsion case shows less variation from the middle case in terms of relative volumes of generated oil and gas. An inter-relationship between the different expulsion cases exists and a high oil expulsion efficiency reduces the potential for gas generation (through secondary cracking in the source rock) and vice versa. The sensitivity to expulsion was therefore considered adequately covered in a qualitative sense by two cases. For this reason, and because of time constraints, the minimum case was not pursued.

A more critical issue is the direction of movement of hydrocarbons expelled from the Viking Group. In most published studies, expulsion is dominantly held to be directed downwards into the Brent Group (e.g. Cornford et al., 1986; Mackenzie and Quigley, 1988), aided by interdigitation or juxtaposition of source and carrier beds (Field, 1985).

It is difficult to visualize oil generated in the Draupne Formation moving downwards, opposite to the forces of buoyancy, through several hundreds of metres of Heather shale. In basinal settings, as the Draupne becomes mature and sufficiently saturated for oil expulsion, the basal parts of the Heather Formation have already entered into a stage of severe secondary cracking (see Figs. 12 and 13).

Downwards expulsion may likewise be inhibited by overpressuring in the Brent sandstones. Buhrig (1989) shows that the Triassic and Jurassic reservoir systems exhibit pore pressures which clearly indicate that pressure communication is poor from the basin centre to the flanks in the North Viking Graben and East Shetland Basin areas. This lateral pressure differential may have initiated in the Late Cretaceous, starting in the deepest depocentres and subsequently expanding laterally through time. Buhrig (1989) identifies the main mechanisms of overpressuring as rapid loading and generation of gas from deeply buried kitchen areas. Faults along the graben margins may cause inefficient lateral pressure dissipation.

We suggest that other pathways for the movement of oil from Upper Jurassic source rocks in the basin centre to the flanks, and subsequently down into the Brent carrier system, may exist. Since the movement of fluids can only occur in a direction of decreasing fluid potential (Hubbert, 1953), the zone where oil enters into the Brent sandstones is constrained to where the fluid potential in the Brent is lower than in the stratigraphically shallower conduits for oil discharge from the basin to the flanks. At such places oil dissipates through favourable pathways (faults, points of juxtaposition and truncation) and into the Brent Group sandstones. This view is in agreement with that of Doligez et al. (1987) who proposed that lateral migration away from the graben occurs within Cretaceous strata.

In the Oseberg area, pore pressures are much higher in the Cretaceous strata than in the underlying Brent sandstones (Buhrig, 1989 and Doligez et al., 1987). At this structural position, in contrast to the basin in the west, a potential for downward movement of fluids exists. The Oseberg Field is located in a pressure gradient regime termed the "restricted system" by Buhrig (1989). This zone, along the east flank of the Viking Graben, contains moderately geopressured reservoirs which are considered to be transitional between a highly overpressured ("closed") system to the west and a hydrostatically pressured ("open") system to the east. The "restricted system" is therefore the zone where pressure dissipation occurs. Pressure gradient data from wells were used in order to better constrain the transitional zone defined by Buhrig (1989) in the study area (Fig. 22). We propose that the entrance of migrating oils from the Viking Group into the Brent system may take place within this zone.

Following the above arguments, primary and secondary migration were modelled for two main scenarios described below.

Brent sandstones as exclusive carrier bed. The Brent scenario concurs with Sylta (this volume). Secondary migration is confined to the Brent sandstones and faults, and is driven by buoyancy. Expulsion is directed towards the Brent Group *in situ* and the yields from all four source rock units are included. This model is referred to in subsequent sections as the "Brent migration scenario" (see Fig. 21).

Base Cretaceous as carrier bed. A second scenario models migration along the Base Cretaceous Unconformity within the overpressured region ("closed system") in order to simulate upwards expulsion and shallower lateral migration. The hydrocarbons are modelled to pass into the Brent sandstones along the boundary of the closed system. Thus, migration from the unconformity and into the

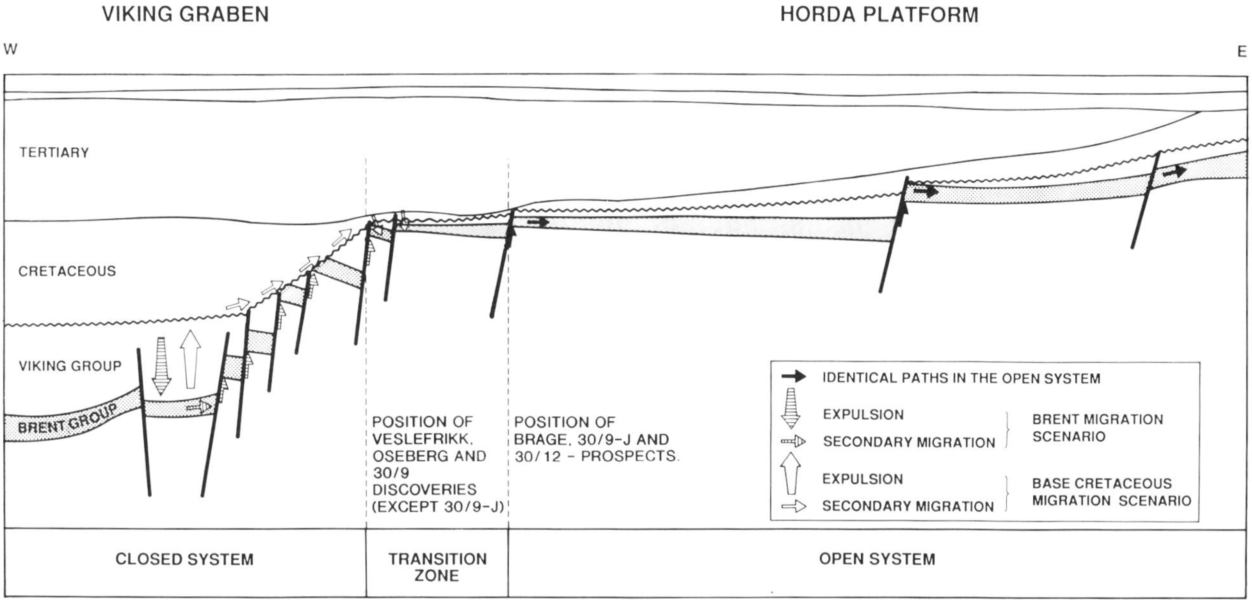

Fig. 21. Concepts of the combined expulsion and migration scenarios.

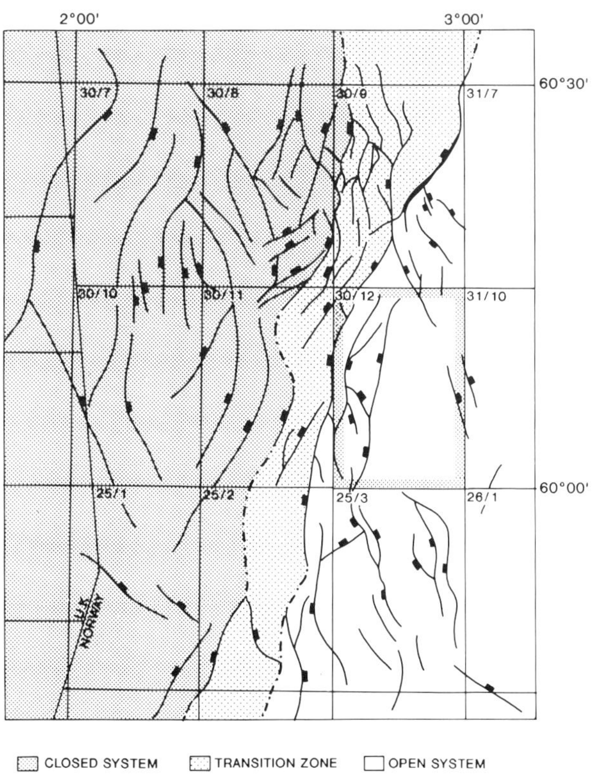

Fig. 22. Pressure domains of the Brent Group sandstones (adapted from Buhrig, 1989) in the environs of the prospect area. Closed system: pressure gradient to Top Brent >2.00 psi/m. Open system: pressure gradient to Top Brent <1.55 psi/m.

Brent is simulated as migration towards lower pressure (fluid potential). The concept is illustrated in Fig. 21. The Top Brent surface was used as carrier in the transition zone. In the open system zone the conditions for secondary migration are modelled in the same way as in the Brent migration scenario.

The gas yields from the Brent and Dunlin are encased in the overpressured system and the escape of gas is impeded. Thus, only yields from the Draupne and Heather Formations can enter into this migration mode. However, the Dunlin has locally reached the oil expulsion window outside the overpressured zone and contributes some hydrocarbons (Fig. 14).

Migration along the Base Cretaceous Unconformity surface is simulated as occurring along a 25 cm thick conduit. This is a completely different situation from that of secondary migration in sandstones, and the secondary migration loss function (Sylta, 1991) does not apply. For this reason only losses in microtraps and dead ends were modelled, and the zone must be fully saturated before migration starts. Although this migration scenario is a hybrid of migration first along the Base Cretaceous Unconformity and later in the Brent sands, it is referred to in later sections as the "Base Cretaceous migration scenario".

Results

The discussion below is essentially geared to identifying those conditions that would make the main prospects addressed (A, B and C of Fig. 23) viable, and to establishing the factors necessary for an optimal reserves case.

The J-structure (Fig. 23) is ideal for calibration. Petroleum migrating into the structure flows through prospect A (Fig. 23). Furthermore, several indirect relationships which have implications for J, exist between the A-, B- and C-prospects. The presence of a free gas cap in the J-structure allows us to

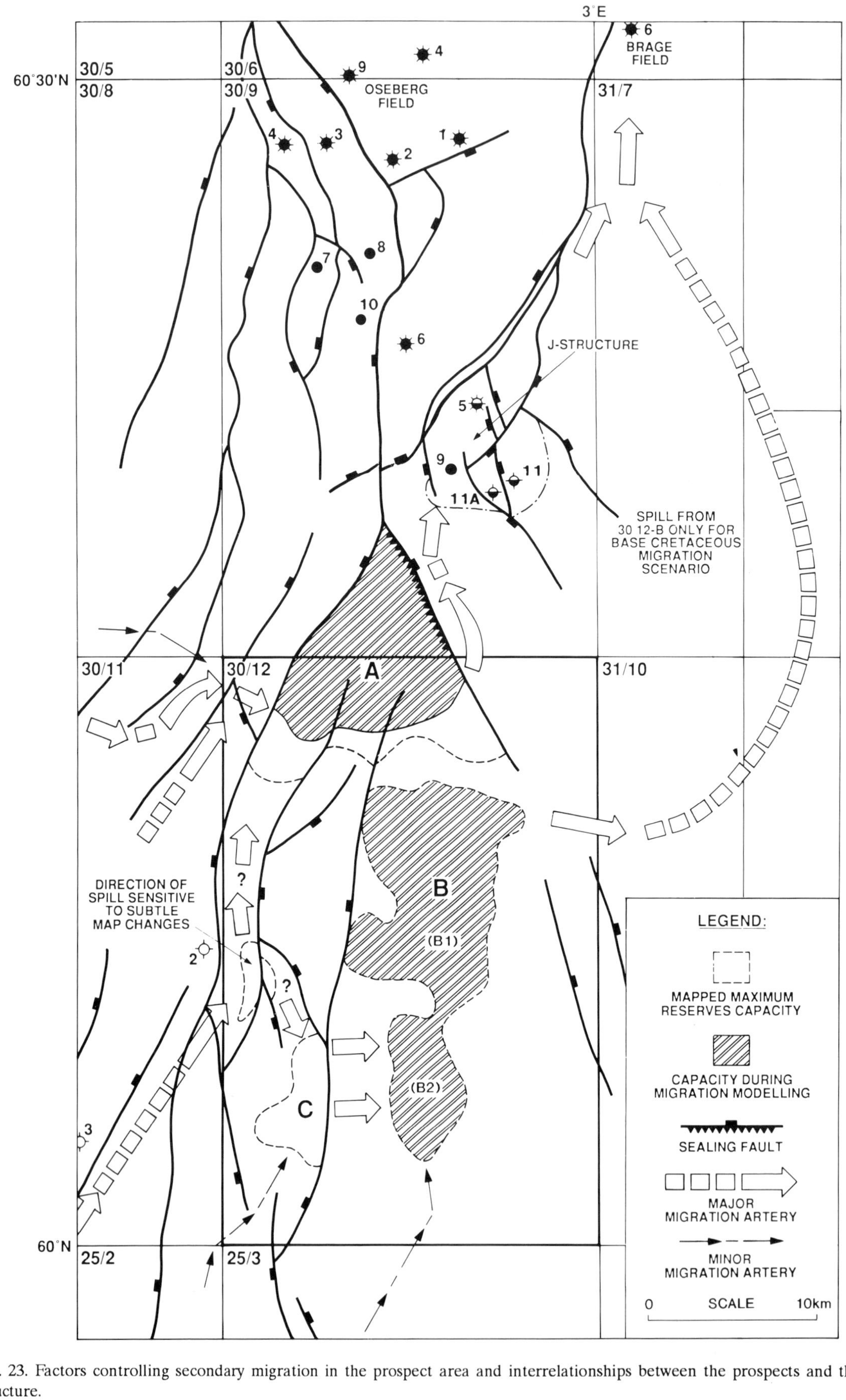

Fig. 23. Factors controlling secondary migration in the prospect area and interrelationships between the prospects and the calibration structure.

constrain the probability of oil (versus gas) in the prospects to the south. The amount of free gas in J was estimated to be 10×10^9 Sm^3, which is used in the calibrations. However, sensitivity tests were carried out on the modelling results for possible variations in the amount of gas. The J-structure spills towards the Brage Field to the north.

Prospect A reaches its maximum potential when its eastern fault is non-permeable. It is very important to note that if the whole fault seals, the A-prospect spills into the B-prospect which again spills east of J, resulting in total shielding of J in the Late Tertiary (after 30 Ma) and an absence of accumulated hydrocarbons in the Brent scenario. Hence, the presence of hydrocarbons in J constrains the maximum possible potential in A. The modelling was therefore carried out with a shortened version of the fault seal to the east of A. This modelling version was pursued consistently in both scenarios in order to form the basis for meaningful comparison. Note, however, that the Base Cretaceous scenario will allow A to be modelled with maximum capacity, since both A, B and J are filled prior to 30 Ma and thus before the spill path from B bypasses the J structure (Figs. 24 and 25).

The C-prospect requires a sealing fault in order to form a viable prospect. However, such a sealing fault completely shields the B-prospect from sourcing from the west. This results in only minor amounts of hydrocarbons reaching the B-prospect. Hence, the existence of a viable C-prospect tends to severely reduce, or even exclude, the prospectivity of the B-prospect. In order to examine possible volumetrics and phase relationships in the B-prospect, the existence of a possible C-prospect was discarded. This allows investigation of the largest potential for accumulation (Fig. 23). The mutual exclusivity of the B- and C-prospects should, however, be noted.

Examination of the maps in the area to the west of the C-prospect illustrates what is probably the most important risk pertaining to the B-prospect. Subtle changes in the mapping in this area could shield the B- and C-prospects by deviating migration routes northwards, particularly during the Late Tertiary (Fig. 23). However, the preferred interpretation of the area around the prospects provides a scenario whereby hydrocarbons have been supplied to the B-prospect throughout most of the Tertiary. Table 1 summarizes optimized volumetric results, based on this scenario. These are considered to be the most realistic cases for the Brent and Base Cretaceous migration scenarios.

In all modelled cases, sufficient amounts of oil and gas migrated into the area to fill the A-prospect. The fill of the B-prospect is far more sensitive to the migration scenarios and variability in expulsion, because it is located in a more shielded position with respect to secondary migration.

The Brent scenario generally leads to oil with a free gas cap in prospects A and B. However, in the maximum expulsion case ($S_{o,irr} = 0.16$), all the gas in B is dissolved in the oil. It is interesting to note that the final migration pulse (2–0 Ma) is responsible for re-establishing a gas cap in A, which only contains oil and dissolved gas at 2 Ma (Figs. 26d and 26e). In contrast, a gas cap is present in B from 30 Ma until the present. The Brent carrier scenario leads to insufficient migration into the B-prospect, which is not filled to spill point (Table 1). Gas expulsion efficiencies modelled with the Brent carrier scenario need to be very low (0.07–0.165) in order to match the hydrocarbon phases and volumes in the J-structure.

The Late Tertiary represents a time predominated by gas migration in the Brent carrier scenario, and gas partly replaces the oil. In contrast, this is a time of intense oil migration in the Base Cretaceous scenario and free reservoir gas (Figs. 27d and 27e) is dissolved.

The Base Cretaceous scenario results in oil without a gas cap in the A- and B-prospects, and all traps are 100% filled. When the maximum expulsion case ($S_{o,irr} = 0.16$) is used, insufficient gas volumes are available to honour the hydrocarbon composition in the J-structure (gas expulsion efficiency above 100% required), whereas the middle expulsion case ($S_{o,irr} = 0.28$) yields a more realistic gas expulsion efficiency (about 23%, see Table 1).

Sensitivity

We address the interrelationship between the free gas volume in the J-structure and the volumes of gas (dissolved + free gas) in the A- and B-prospects. Note that when the B-prospect in some modelling cases is not filled to spill, the prospect is divided into two parts: the large northern lobe B1 (Table 1), and the smaller southern lobe B2. The parameters which have been varied in the simulations are: (1) carrier bed scenario (Brent, Base Cretaceous); (2) expulsion model (middle case, $S_{o,irr} = 0.28$; maximum case, $S_{o,irr} = 0.16$); (3) gas expulsion factor (0.06–1.30); (4) leakage of gas through cap rock (0 to 10×10^{-6} $m^2\ kg^{-1}\ Ma^{-1}$)

We use the simulations that yield hydrocarbon volumes close to the optimum cases in order to quantify the uncertainty of the prospect estimates with respect to variations in the gas expulsion factor. In the sensitivity evaluation, the gas volumes for the A-, B1- and B2-prospects are compared to the amount of free gas in place in the J-structure. We conclude that

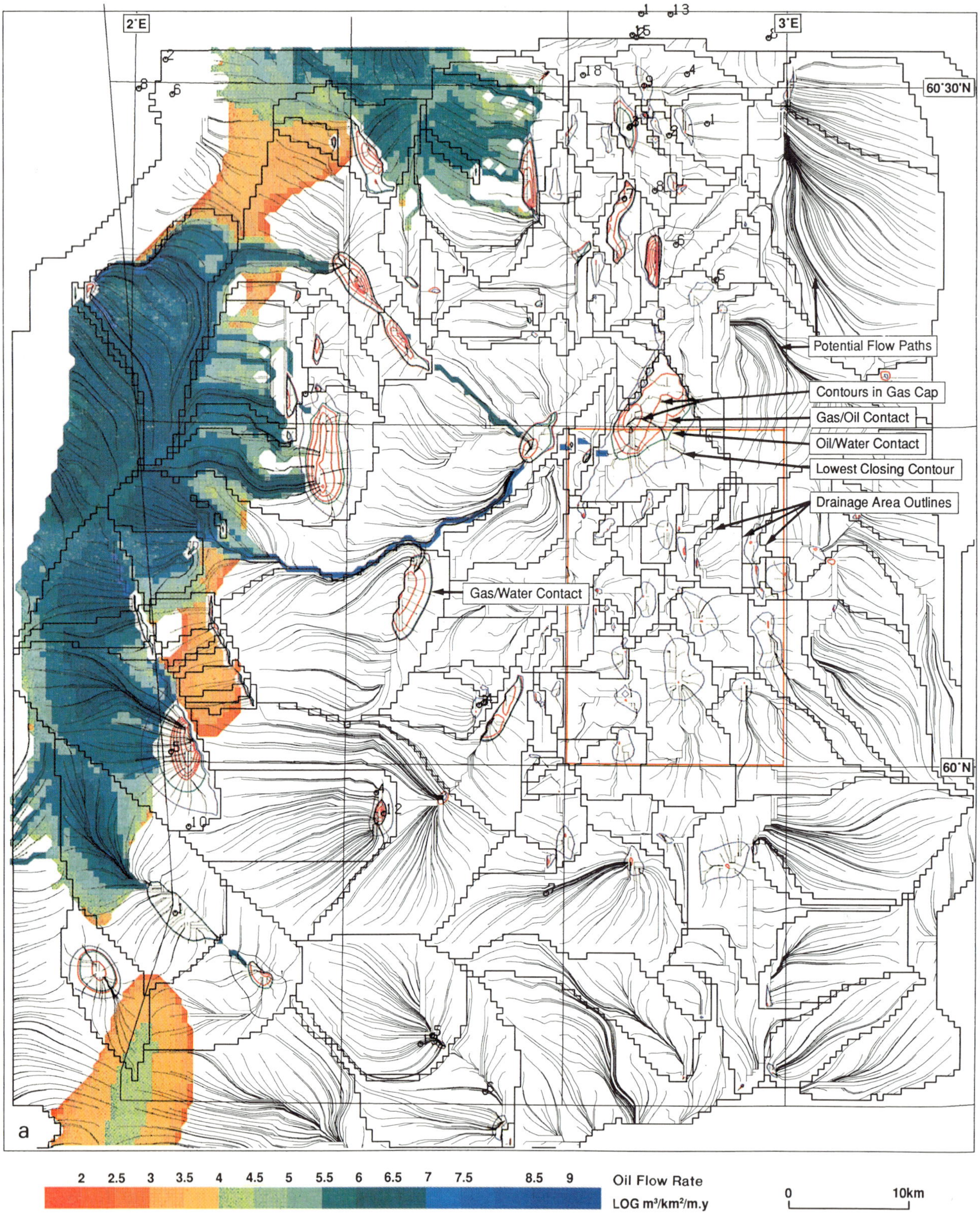

Fig. 24. Brent scenario. (a) Oil flow rates from 65 to 57 Ma, accumulations at 57 Ma. Block 30/12 is highlighted in red. Location of prospects is shown in Figs. 26 and 27. Zig-zag lines outline distinct drainage areas.

in many cases small variations in the gas expulsion factor result in large differences in the amount of free gas in the J-structure and much smaller changes in the amounts of gas in the prospects.

Only the Brent migration scenario produces free gas in the A-prospect. Free gas is only modelled in the B1-prospect for some of the Brent scenarios. We concluded in the previous section that the most likely Brent scenario yields gas and oil in the B1-prospect, but we now add that the prospect may also be an oil

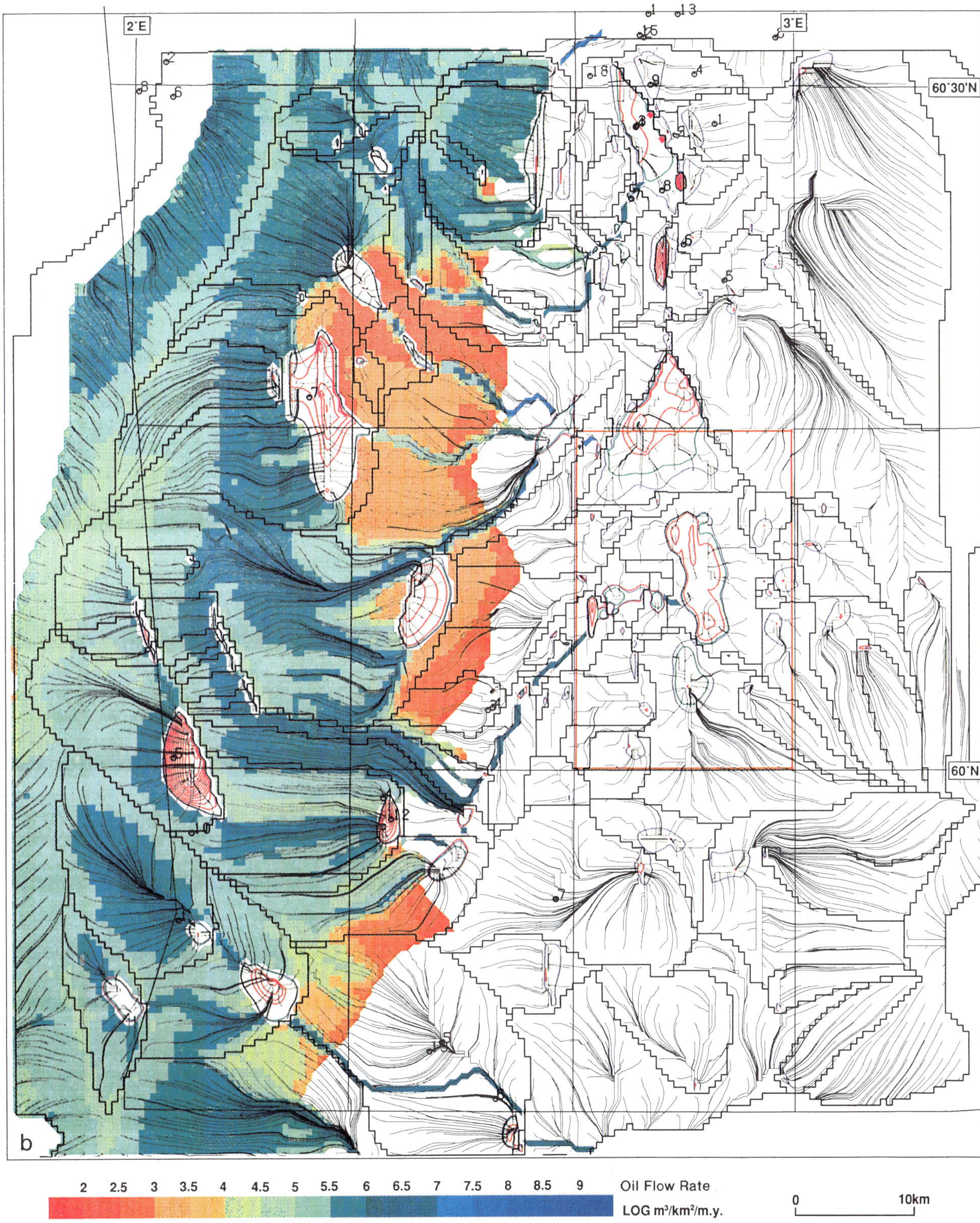

Fig. 24 (continued). (b) Oil flow rates from 57 to 30 Ma, accumulations at 30 Ma.

prospect with all the gas in solution if the maximum case for oil expulsion ($S_{o,irr}$ = 0.16) applies. The B2-prospect (B south) contains free gas in the Brent migration scenarios. The prospect is less sensitive to uncertainties in the free gas estimates in J than prospects A and B1.

Gas volume uncertainties

There are large factors of uncertainty associated with the volumes of gas that can be expelled and subjected to secondary migration. Gas expulsion efficiencies, as referred to herein, should account for

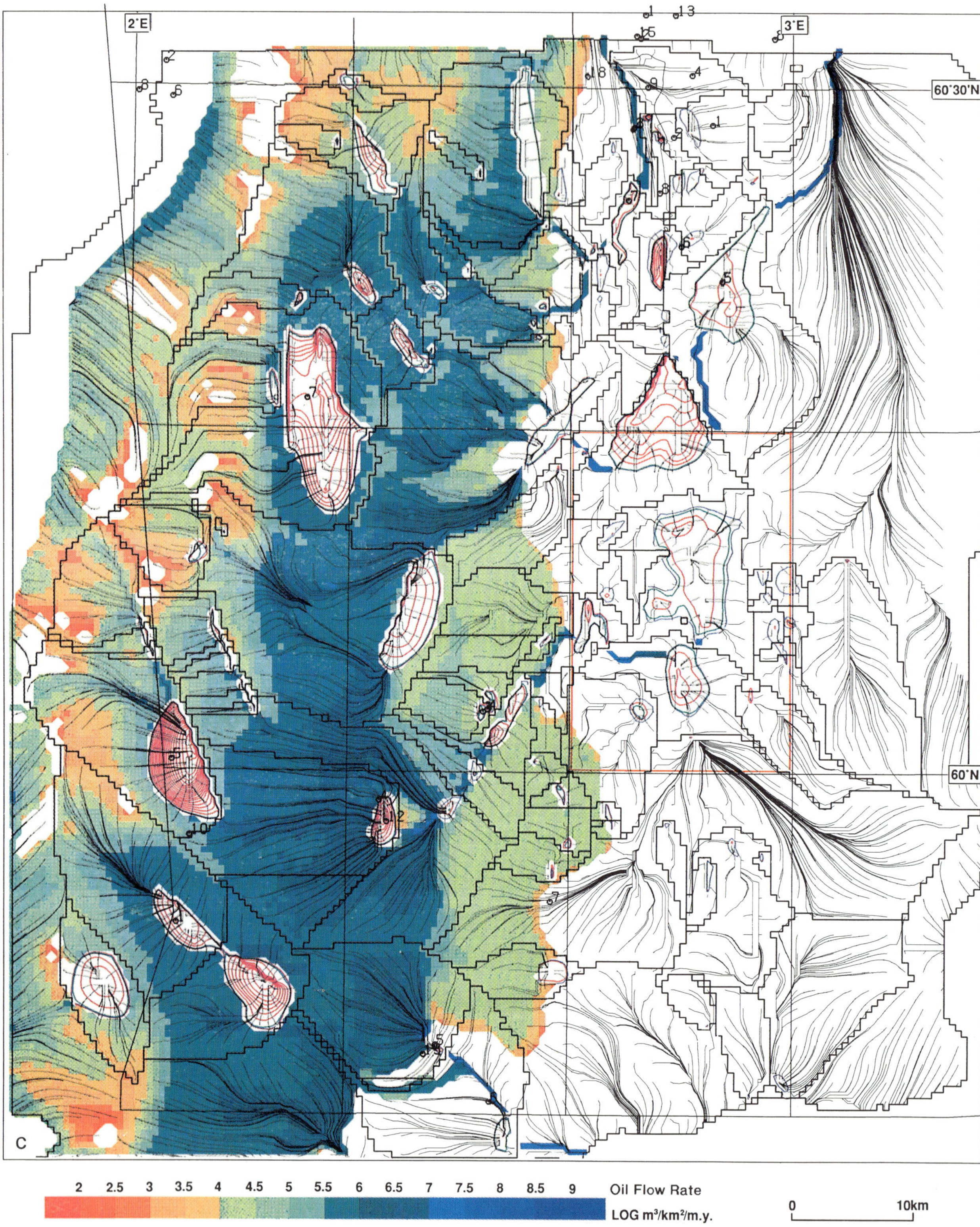

Fig. 24 (continued). (c) Oil flow rates from 2 to 0 Ma, accumulations at 0 Ma.

volumes retained in the source rock, volumes lost by diffusion and volumes lost by dissolution in formation waters at great depths.

Mackenzie and Quigley (1988) proposed that secondary cracking of unexpelled oil invariably leads to high expulsion efficiencies of gas or gas condensate, based on earlier observations (Cooles et al., 1986). They argue that their expulsion efficiency ("PEE") approaches 1 as the transformation ratio ("PGI") approaches 1. However, this relationship is a direct consequence of the nature of the algebraic scheme of Cooles et al. (1986) which relies upon accurate

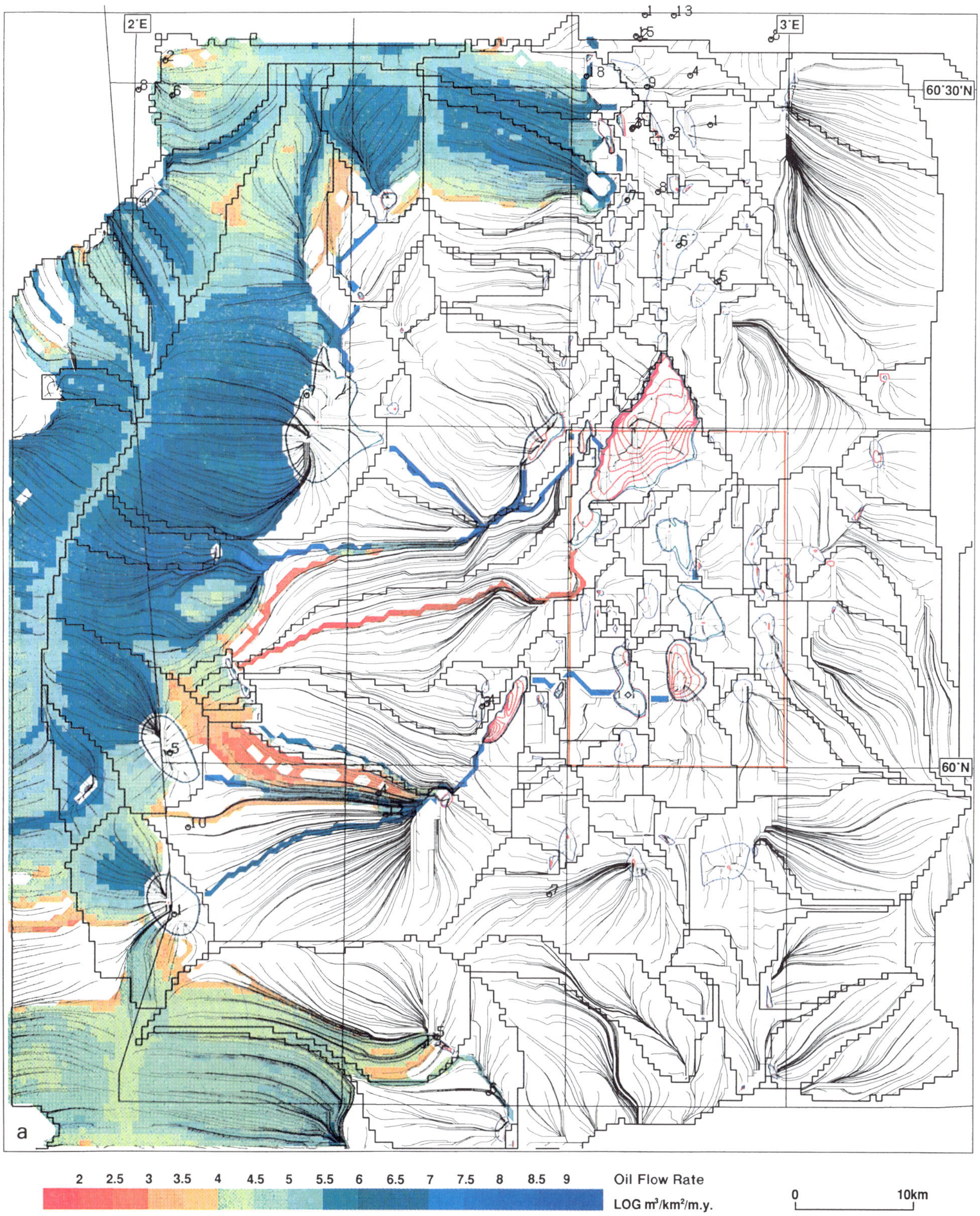

Fig. 25. Base Cretaceous scenario. (a) Oil flow rates from 65 to 57 Ma, accumulations at 57 Ma. Accumulations in the closed system area do not represent true traps (see Figs. 21 and 22 and text). Block 30/12 is highlighted in red. Location of prospects is shown in Figs. 26 and 27. Zig-zag lines outline distinct drainage areas.

assessment of the source rock oil and gas concentration at the full range of maturities. Particularly at high maturities, when the volatile component is high, their algebraic scheme becomes vulnerable to degassing during sampling and sample handling. Hence, at high maturities the hydrocarbon concentration in the source rock tends to be underestimated and too high expulsion efficiencies may be estimated.

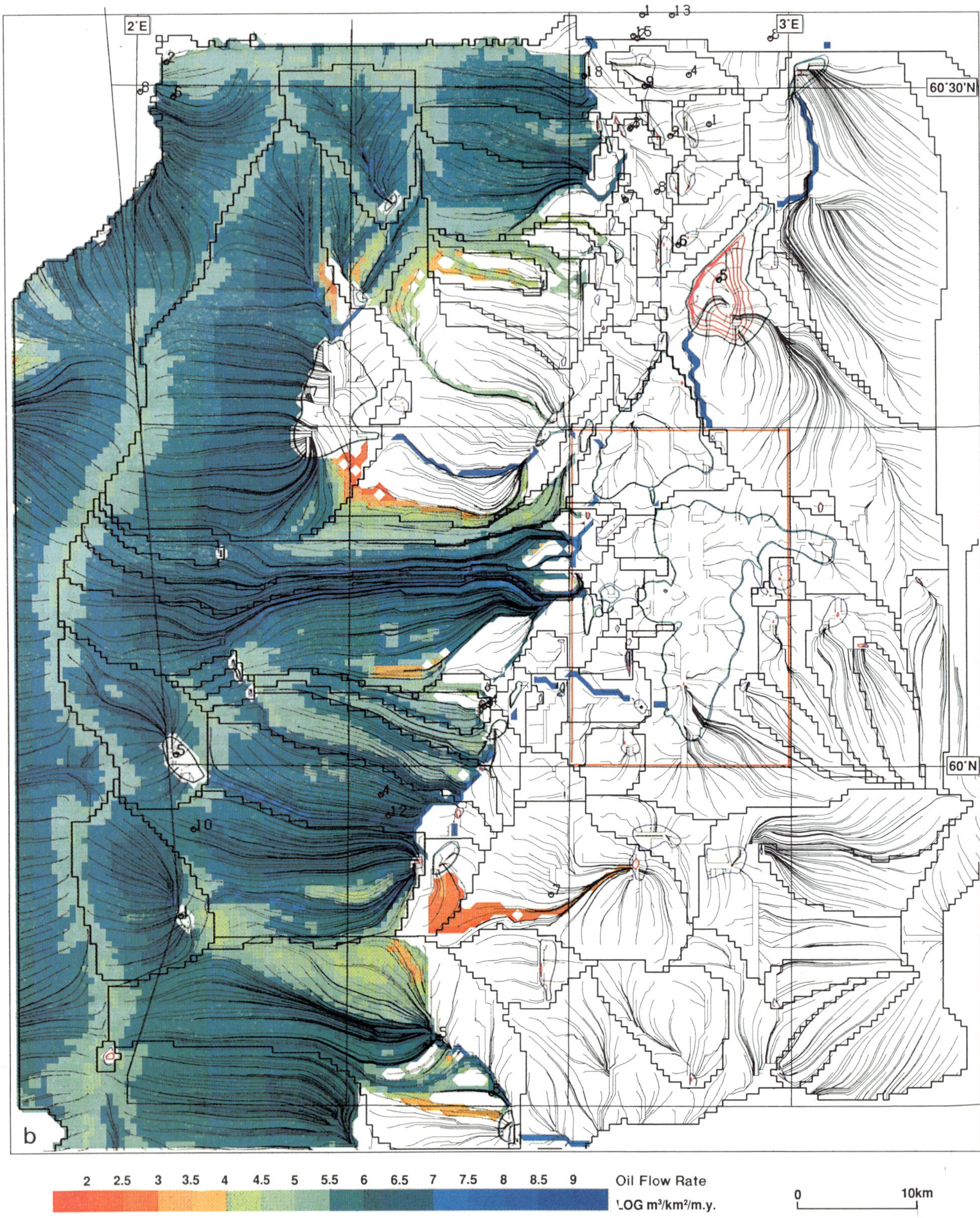

Fig. 25 (continued). (b) Oil flow rates from 57 to 30 Ma, accumulations at 30 Ma.

Furthermore, they assume that the inert carbon mass is constant, and hence do not take into account the production of char (residue) during secondary cracking. This assumption results in an overestimate of the initial generation potential for highly mature samples, since they hold inert carbon and initial generation potential to be proportional. Their calculated expulsion efficiencies for highly mature samples will therefore be too high.

It follows from our numerical scheme (expulsion, secondary cracking) that the generated products from secondary cracking would mostly be retained

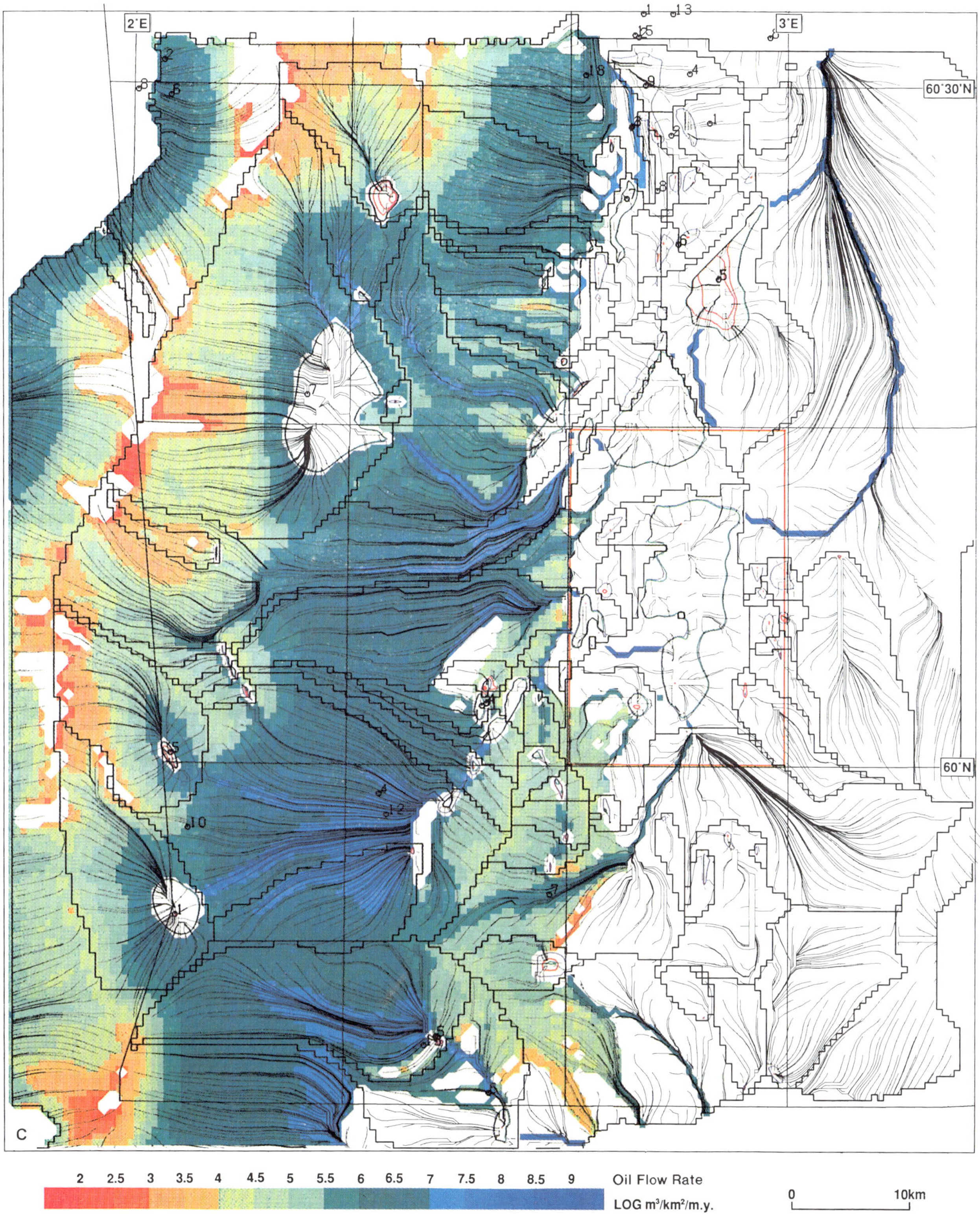

Fig. 25 (continued). (c) Oil flow rates from 2 to 0 Ma, accumulations at 0 Ma.

within the source rock, since expulsion is modelled to be a consequence of excess volumes resulting from hydrocarbon generation and compaction. The estimates of excess volume during secondary cracking become uncertain because the hydrocarbon fluid destruction results in two end products, namely char and a fluid which is less dense than the parent fluid. The volumetric outcome and expulsion efficiency during this process depend on the properties of the retained oil, the relative proportions of char and gas resulting from cracking, the subsurface pressure and the porosity reduction in response to continued

TABLE 1

Volumes of oil and gas in place (standard conditions) and % fill for favoured cases (Figs. 26 and 27). Significance of migration scenarios. Gas expulsion efficiencies required to honour the hydrocarbon composition in the 30/9-J structure

Migration scenario	Expulsion case	Gas expulsion efficiency required for calibration to 30/9-J	A-prospect				B-prospect			
			oil (10^6 Sm^3)	free gas (10^9 Sm^3)	total gas (10^9 Sm^3)	fill (%)	oil (10^6 Sm^3)	free gas (10^9 Sm^3)	total gas (10^9 Sm^3)	fill (%)
Brent Carrier	$S_{o,irr} = 0.28$	0.070	125.0	74.3	111.8	100	105.0	9.2	42.5	80 [a]
Base Cretaceous	$S_{o,irr} = 0.28$	0.230	423.0	0.0	93.4	100	233.0	0.0	21.8	100

[a] B is not filled to spill and consists of two separate accumulations, B1 and B2 (see Fig. 26). The 80% refers to fill in the larger northern portion of the 30/12-B (B1). B2 is filled to spill (see Fig. 24c).

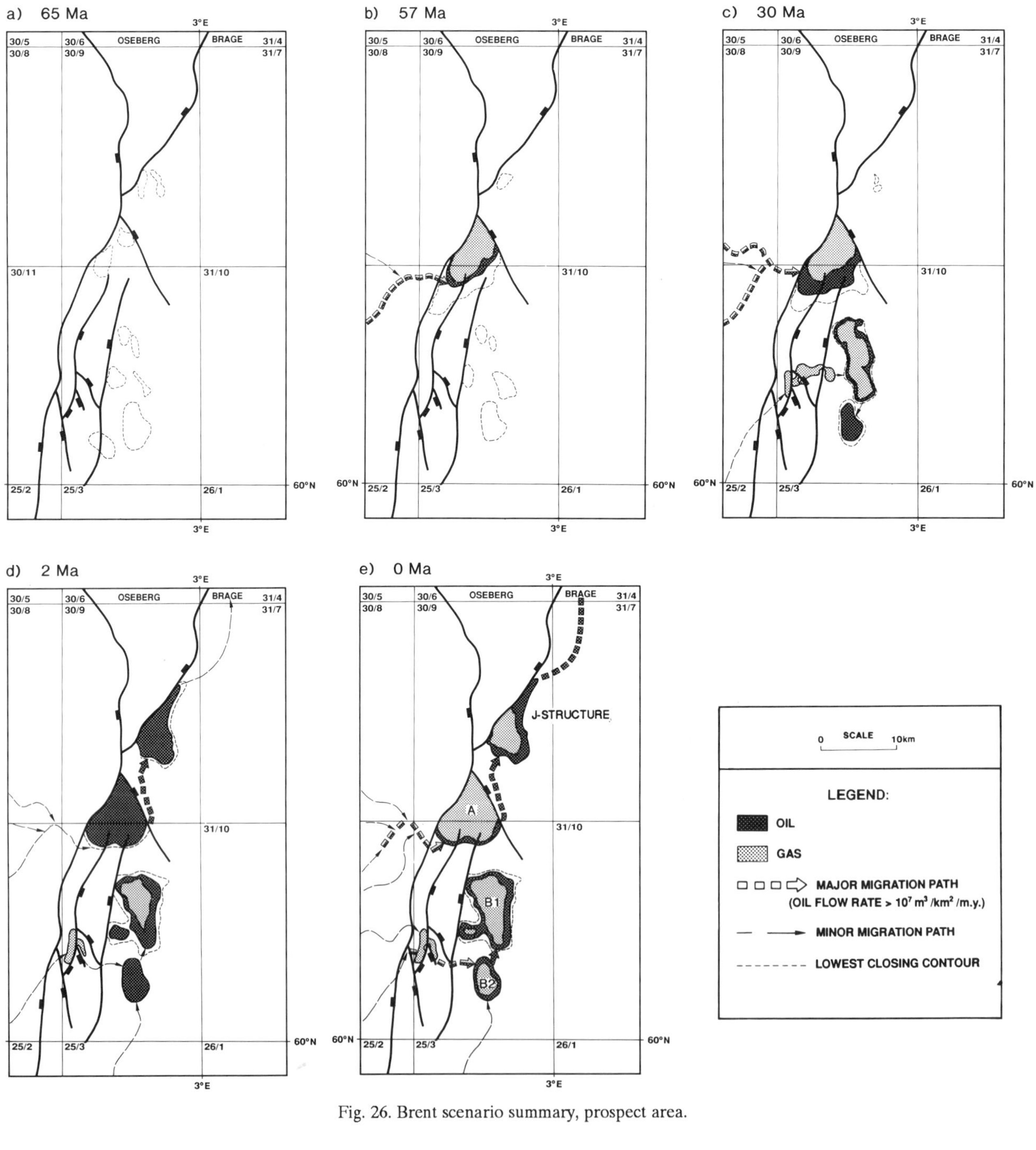

Fig. 26. Brent scenario summary, prospect area.

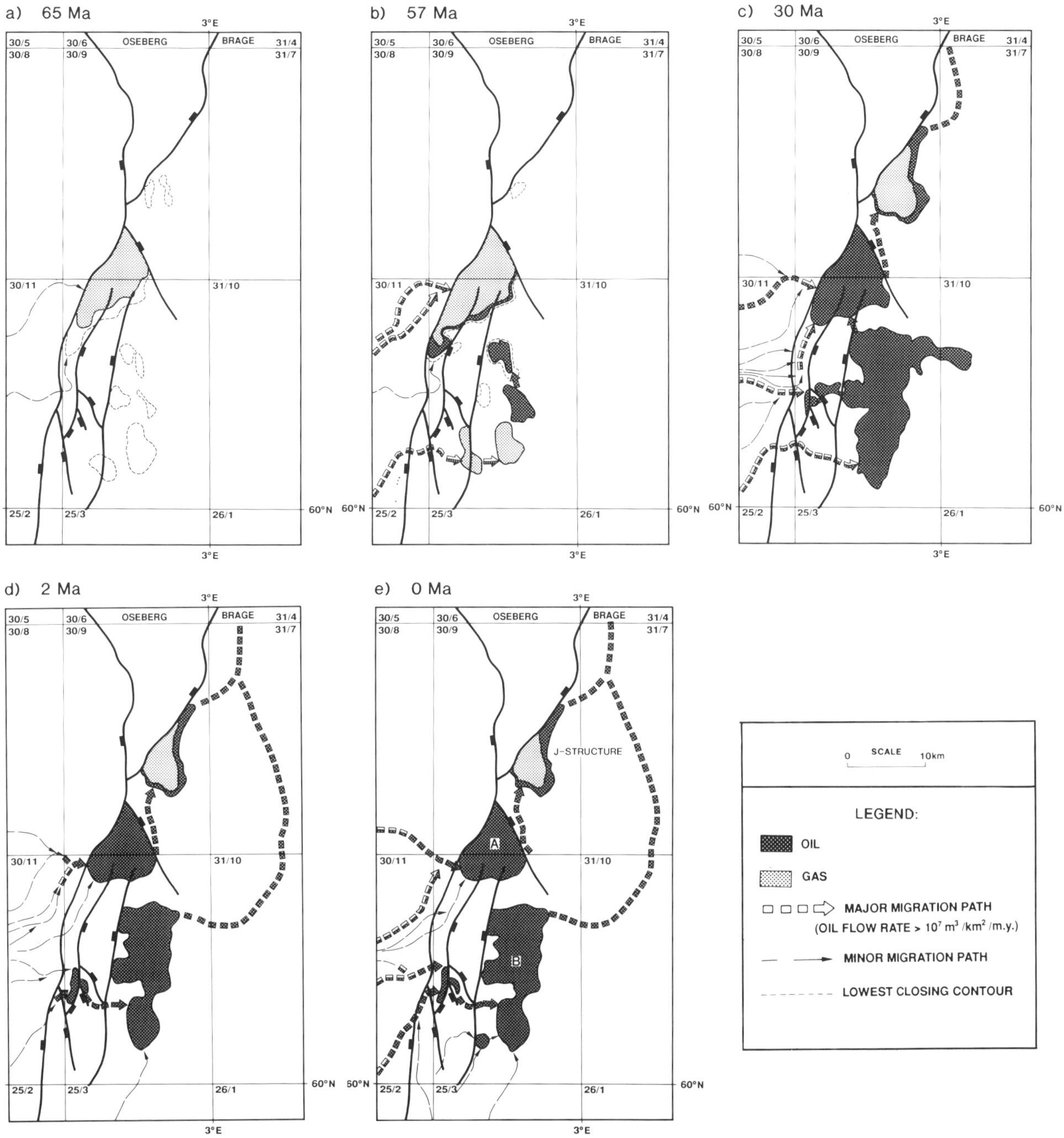

Fig. 27. Base Cretaceous scenario summary, prospect area.

compaction. Condensate-saturated gas at subsurface pressures of 50 MPa or more is generally denser than 0.4 g/cm^3 (England et al., 1987). One can calculate a hydrocarbon volume reduction by a factor of 0.875 from the top to the base of the secondary cracking window (in the study area at present about 4100 to 5000 m below sea floor), by assuming that the hydrocarbon density is reduced from 0.7 to 0.4 g/cm^3 and that the oil destruction results in equal proportions of hydrocarbons and char. The porosity is reduced by a factor of 0.6 over the same depth range (assuming that compaction is not inhibited by overpressuring). The fraction of hydrocarbon volume to the pore space has increased by a factor of $0.875/0.600 = 1.46$. Hence, for source rocks that are fully saturated at the onset of secondary cracking, an excess volume of about 30%, $(1.46 - 1)/1.46$, that may be expelled during secondary cracking, is inferred.

According to the model outcomes, only 0.9 billion tonnes of gas were generated from refractory kerogen within the study area throughout the modelled geological history. This suggests that the study area has only locally entered into a stage of significant transformation of refractory kerogen. Not even in the deepest parts along the graben axis has the base of the Dunlin Group realized more than about half of its refractory kerogen generation potential. This implies that the gas contribution from refractory kerogen is only of secondary importance to the total gas yield in the study area.

Gas dissolution in subsurface formation waters influences the volume of gas available to secondary migration, as does diffusion. About 8 to 11 Sm^3 of methane may be dissolved per m^3 of formation waters at a depth of about 5000 m (Hunt, 1979) depending on temperature and pressure. Maximov et al. (1984) reported gas concentrations as high as 17.5 Sm^3/m^3 at 63 MPa pressure. Increasing salinities tend to reduce these numbers. In the study area, subsurface formation waters may therefore dissolve relatively large quantities of methane. The central parts of the graben were subjected to catagenesis during relatively shallow burial depths (2–3 km, Fig. 12). The expulsion modelling suggests that these parts of the basin are more gas prone than the flanks. However, due to high porosities (high end porosities in Fig. 11) relatively more formation water was available to dissolve gas and later, upon continued compaction, carry it away in solution.

The modelled gas expulsion needed to honour the hydrocarbon composition of the J-structure is highly dependent on the migration scenarios. The "Base Cretaceous migration scenario" is fed only by the Draupne and Heather Formation in the "closed system" area. The mid-case ($S_{o,irr}$ = 0.28) for liquid phase expulsion implies that about 23–35% of the subsequently generated gas in the Heather and Draupne Formations needs to be expelled for successful calibration to the hydrocarbon volumes in the J-structure. The maximum case ($S_{o,irr}$ = 0.16) for liquid phase expulsion results in too small quantities of retained oil in the source rock for gas generation, since modelled gas expulsion efficiencies in the order of 120%, i.e. more than what is available, are required to honour the hydrocarbon composition in the J-structure.

The "Brent migration scenario" is fed by oil and gas yields from all source rocks throughout the kitchen area (Figs. 14–18). Due to the huge gas potential from the Brent and the Dunlin Groups, average basin-wide gas expulsion efficiencies as low as 7% (middle case, $S_{o,irr}$ = 0.28) and 17% (maximum case, $S_{o,irr}$ = 0.16) are required to match the hydrocarbon composition in the J-structure. The contrasting pressures in the Brent sandstones in the graben and at the graben flanks may reflect inefficient escape of fluids from the deep basin to the basin flanks along Brent sandstones, adding to the uncertainty of the scenario.

Conclusions

(1) The volumetric modelling of liquid phase expulsion demonstrates that the oils trapped in the study area originated chiefly from the Viking Group. Our present knowledge is limited about several factors that impact gas expulsion; gas expulsion efficiency was therefore used as a variable. The amounts of oil and gas in place in the calibration structure were honoured in the different migration scenarios, and different gas expulsion efficiencies were obtained. The gas expulsion efficiency can thus not be uniquely determined in this study. Better control on secondary cracking products would be helpful since most of the generated gas originates from secondary cracking in our modelling. The modelling generally indicates that much of the gas that was generated by secondary cracking in the basin never reached the flanks mainly due to retention in the source rocks, but losses caused by dissolution in formation waters and vertical diffusion may also have been significant. Inhibited lateral escape of fluids through the Brent Group sandstones, reflected by basin-wide pressure compartmentalization, also reduces the amount of gas available for lateral migration to the Viking Graben flanks.

(2) Modelling shows that the conditions for oil expulsion improve towards the graben flanks where catagenesis occurred at greater depths of burial, and where less pore space was available for oil retention. Hydrocarbon generation and expulsion mainly occurred during Tertiary times.

(3) We question whether it is likely that expulsion from the Draupne Formation was directed downwards, through hundreds of metres of Heather shale. Where Heather is thicker than 200–300 m, its maturity is too high to allow hydrocarbons to pass without being affected by secondary cracking. Hence, it is suggested that expulsion may be directed upwards and discharged from the basin to the flanks through shallower conduits as modelled in the Base Cretaceous scenario.

(4) Oil is modelled to be the main hydrocarbon phase in the A- and B-prospects. The investigated expulsion directions and secondary migration scenarios point to different conclusions with respect to the amount of free gas in the prospects.

(5) The secondary migration modelling identified potential risks associated with the prospects. The modelling indicated that all three prospects cannot exist simultaneously since B and C are mutually exclusive. A similar relationship was indicated between the A-prospect and the J-structure, indicating that the trapping capacity of A must be reduced in order to prevent the J-structure from being shielded from

migrating hydrocarbons. The modelling was carried out according to our most optimistic trapping scenario under the given conditions, and this indicates a capacity for 423 × 10^6 Sm^3 of oil in A, 233 × 10^6 Sm^3 of oil in B and no capacity in C. The most pessimistic trapping scenario within the prospect area would suggest shielding of the B- and C-prospects from migrating hydrocarbons due to sealing faults to the west of the prospects and a non-viable A-prospect, resulting in only insignificant volumes of gas migrating into the B-prospect from the south.

(6) The fill and spill histories within the prospective area are quite sensitive to differences between the Brent scenario (expulsion from Viking Group source rocks directed downwards) and the Base Cretaceous scenario (expulsion from Viking Group source rocks directed upwards). The Base Cretaceous scenario allows the prospects to fill earlier (filled to spill point prior to 30 Ma) and suggests a reduction of the amount of free gas within the J-structure throughout the Late Tertiary. In the Brent scenario, hydrocarbons do not reach J until in the Late Tertiary (between 30 and 2 Ma) and the Late Tertiary is characterized by dissolution of the free gas caps in the oil (30 to 2 Ma) followed by a final pulse dominated by gas migration (2 to 0 Ma). This reinstated the free gas caps in A and J, and a growth of the free gas cap in B took place.

Acknowledgements

We would like to acknowledge Anne Mette Raa (IKU) for her contribution to the secondary migration modelling, Ian M. Walker (Conoco Norway Inc.) for his contributions in tectonic modelling and isopaching of source rocks and reservoir rocks, Adrian J. Machell (Conoco Norway Inc.) for the regional petrophysical data for Brent sandstones and Thomas T.Y. Ho and his colleagues at Conoco Exploration Research Services at Ponca City for the time-slice pyrolysis results. Last, but not least, we express our acknowledgements to Donald Dorn-Lopez (Conoco Norway Inc.) and Robert M. White (Consultant) for the interpretation of the seismic data and the generation of the maps that were used in this study.

Appendix

A listing of all parameters used is provided at the end of the Appendix.

Corrections for maturity effects in the source rock input data

Initial source rock yield parameters, reflecting the initial genetic potential of the kerogen prior to hydrocarbon generation and expulsion, are required in volumetric modelling. We correct the average hydrogen indices and TOC values obtained for each source rock unit (Dunlin Shales, Brent Shales, Brent Coals, Heather Formation and Draupne Formation) in every well prior to computing final average values for the study area.

We define the "petroleum index" (H_o) as the initial ratio of the genetic potential ($S_1 + S_2$) to the TOC (C). For immature samples, where it is believed that expulsion not yet has initiated, the petroleum index is given by the sum of the hydrogen index and the ratio of S_1 to the TOC (A1). The petroleum indices obtained from immature samples were used to constrain the evolution path envelope at a reference maturity of 420°C T_{max}.

For mature samples, the hydrogen indices were projected back to the reference maturity of 420°C T_{max} by using the evolution path envelopes as charts. Since typical S_1 values for immature samples are inherent in the envelope at 420°C T_{max}, the corrected hydrogen index (H') accounts for free hydrocarbons (assuming identical kerogen types). Thus we define H' as the petroleum index for mature samples (A2). In summary:

$$H_o = H + S_1/C, \quad \text{immature samples} \quad \text{(A1)}$$

$$H_o = H', \quad \text{mature samples} \quad \text{(A2)}$$

The correction for total organic carbon follows from the difference between the observed hydrogen index (H) and H_o, assuming that the difference, after accounting for S_1, reflects the amounts of hydrocarbons that have been generated from the kerogen and expelled from the source rock. Hence, the fraction by which the amount of carbon has been reduced can be approximated. We remove the effect of free hydrocarbons (S_1) in the observed data by substracting 0.825 S_1 from the TOC (0.825 is used as an approximate weight fraction of carbon to the total weight of hydrocarbons):

$$C_o = \frac{C - 0.825\,S_1}{1 - 0.825\left(H_o - H - \frac{S_1}{C}\right)} + 0.825\,S_1 \quad \text{(A3)}$$

All parameters in eq. A3 are weight fractions (g/g).

The initial source rock yield parameters computed for the study area are shown in Fig. A1.

Generated hydrocarbon volumes

The product of C_o and H_o defines the initial genetic potential of the source rock. The transformation ratio (r) is a function of temperature and time and results from the computer modelling using the kinetic parameters in Table A1 (Arrhenius'

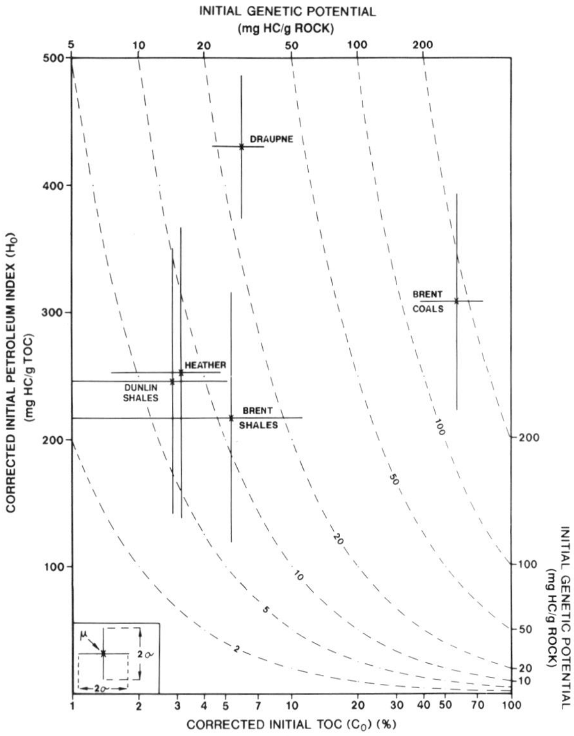

Fig. A1. Petroleum potential of Jurassic shales and coals. The regional average within the study area ± the standard deviation is shown. Curves refer to genetic potential.

equation). The volume of hydrocarbon produced is defined by the equation:

$$V_G = C_o H_o r \frac{\rho_R}{\rho_{HC}} \quad \text{(A4)}$$

where ρ_R is the bulk density of the source rock, ρ_{HC} is the hydrocarbon density and V_G is the ratio of hydrocarbon volume to the rock volume.

Expulsion modelling formulae

A listing of input data for modelling expulsion efficiency is given in Table A2. For convenience we introduce the parameter G which denotes the ratio of the generated hydrocarbon volume to the porosity.

$$G = \frac{V_G}{\Phi} \quad \text{(A5)}$$

Porosity is modelled as a function of compaction during burial (Sclater and Christie, 1980). We model liquid phase expulsion for critical oil saturations from 16% ($S_{o,irr} = 0.16$, maximum case) to 40% ($S_{o,irr} = 0.40$, minimum case). The maximum and minimum cases are defined by zero expulsion rate ($dV_E/dG = 0$) for generated hydrocarbon volumes less than $S_{o,irr}$. Hence, G must exceed $S_{o,irr}$ for expulsion to initiate. Expulsion is 100% efficient for excess volumes ($dV_E/dG = 1$). The expelled volume (V_E) can thus be defined as:

$$\text{If } G > S_{o,irr}: \; V_E = G - S_{o,irr}, \text{ else: } V_E = 0 \quad \text{(A6)}$$

TABLE A1

Kinetic modelling parameters and yield fractions

Reaction No.	E_a (kcal/mol)	A (1/m.y.)	Yield fraction Draupne, Heather and Dunlin	Yield fraction Brent[a]	Classification
1	23.2	0.840×10^{18}	0.050	0.043	initial bitumen
2	36.6	0.625×10^{21}	0.016	0.014	labile fraction
3	41.7	0.743×10^{23}	0.021	0.018	labile fraction
4	47.6	0.156×10^{27}	0.019	0.016	labile fraction
5	52.3	0.744×10^{28}	0.288	0.249	labile fraction
6	55.3	0.154×10^{30}	0.421	0.364	labile fraction
Secondary cracking	55.4	0.340×10^{28}	–	–	secondary cracking
7	58.0	0.101×10^{27}	0.077	0.123	refractory fraction
8	64.4	0.328×10^{28}	0.080	0.128	refractory fraction
9	71.9	0.275×10^{30}	0.022	0.035	refractory fraction
10	78.4	0.939×10^{31}	0.006	0.010	refractory fraction

Source rock	Generative potential in % of total labile genetic potential		Initial petroleum index (mg/g TOC)	Labile fraction	Refractory fraction	% Gas in secondary cracking products
	primary oil	primary gas				
Draupne	94	6	431	0.815	0.185	50
Heather	76	24	253	0.815	0.185	50
Brent[a]	70	30	225	0.704	0.296	50
Dunlin	78	22	246	0.815	0.185	50

[a] The data for the Brent Group refer to the composite potential of both coals and shales. The Brent Group was also corrected for the adsorptive capacity of the coals.

TABLE A2

Input parameters for the modelling of expulsion efficiency

Lithostratigraphic unit	Mass-related parameters				Volumetric parameters			
	average TOC (%)	average petroleum index (mg HC/g TOC)	genetic potential (mg HC/g ROCK)	labile potential (mg HC/g ROCK)	kerogen (%)	shale (%)	sand (%)	bulk density (g/cm^3)
Draupne Formation	5.93	431	25.56	20.83	12.26	87.74	–	2.40
Heather Formation	3.14	253	7.94	6.47	6.50	93.50	–	2.48
Brent shales [a]	(5.31)	(217)	(11.52)	(8.11)	(10.69)	(89.31)	(–)	(2.42)
Brent coals [a]	(56.83)	(309)	(175.60)	(123.62)	(75.40)	(24.60)	(–)	(1.57)
Brent Group bulk	3.63	225	8.17	5.75	7.30	30.70	62.00	2.42
Dunlin shales [a]	(2.88)	(246)	(7.08)	(5.77)	(5.97)	(94.03)	(–)	(2.49)
Dunlin Group bulk	2.33	246	5.73	4.67	4.83	76.17	19.00	2.49

Values used in the calculations: kerogen density = 1.25 g/cm^3 (Ungerer et al., 1983); shale and sandstone matrix densities = 2.70 and 2.65 g/cm^3, respectively (Sclater and Christie, 1980); pure shale bulk density = 2.564 g/cm^3 ($\Phi = 8\%$); pure sand bulk density = 2.485 g/cm^3 ($\Phi = 10\%$). The latter values are believed to be representative for a depth of burial of about 4000 m below sea floor.
a Bracketed values are included for information and comparison only (not used).

Note that the expulsion rate (dV_E/dG) can only assume the values 1 or 0 in the maximum and minimum cases.

The expulsion efficiency (E) is defined as the ratio of the expelled volume to the generated volume, hence it follows that:

$$\text{If } G > S_{o,irr}: \quad E = 1 - \frac{S_{o,irr}}{G}, \quad \text{else: } E = 0 \tag{A7}$$

The curves for the maximum case ($S_{o,irr} = 0.16$) and the minimum case ($S_{o,irr} = 0.40$) are given in Figs. 9 and 10.

The middle case introduces a concept which allows the expulsion rate (dV_E/dG) to increase linearly from zero to 1 with increasing hydrocarbon generation over a window defined by $S_i \leq G \leq G_e$, where S_i is the critical saturation required for expulsion to initiate and G_e corresponds to the generation volume where $dV_E/dG = 1$. The expelled volume is then defined as:

$$\text{If } G < S_i: \quad V_E = 0 \tag{A8}$$

$$\text{If } S_i \leq G \leq G_e: \quad V_E = \int_{S_i}^{G} \frac{(G - S_i)}{(G_e - S_i)} dG \tag{A9}$$

$$\text{If } G_e < G: \quad V_E = G - \left(S_i + \frac{G_e - S_i}{2}\right) \tag{A10}$$

where $(G - S_i)/(G_e - S_i)$ is the linear expression for dV_E/dG when $S_1 < G < G_e$ (see Fig. 9a) and $(G_e - S_i)/2$ defines the area of the curve for dV_E/dG between S_i and G_e. The maximum achievable saturation of the source rock ($S_{o,irr}$) thus equals $S_i + (G_e - S_i)/2$. The generalised expulsion formulae for the middle case concept become:

$$\text{If } G < S_i: \quad E = 0 \tag{A11}$$

$$\text{If } S_i \leq G \leq G_e: \quad E = \left[\frac{\int_{S_i}^{G} \frac{(G - S_i)}{(G_e - S_i)} dG}{G}\right] \tag{A12}$$

$$\text{If } G_e < G: \quad E = 1 - \frac{\left(S_i + \frac{G_e - S_i}{2}\right)}{G} \tag{A13}$$

We calculate the volumes in the middle case by choosing $S_i = 0.16$ and $G_e = 0.40$, which is equivalent to $S_{o,irr} = 0.28$.

$$\text{If } G < 0.16: \quad E = 0 \tag{A14}$$

$$\text{If } 0.16 \leq G \leq 0.40: \quad E = \frac{G}{0.48} - \frac{2}{3} + \frac{0.16}{3G} \tag{A15}$$

$$\text{If } 0.40 < G: \quad E = 1 - \frac{0.28}{G} \tag{A16}$$

The curves for the middle case are highlighted in Figs. 9 and 10.

Parameters

H	hydrogen index (g HC/g TOC)
H'	corrected hydrogen index (g HC/g TOC)
H_o	petroleum index (g HC/g TOC)
C	total organic carbon (g C/g ROCK)
C_o	initial total organic carbon (g C/g ROCK)
V_G	volume of hydrocarbons generated per unit rock volume (cm^3 HC/cm^3 ROCK)
r	modelled transformation ratio
ρ_R	rock density (g/cm^3)
ρ_{HC}	hydrocarbon density (g/cm^3)
Φ	porosity (volume fraction)
G	generated hydrocarbon volume expressed as fraction of pore space (cm^3 HC/cm^3 Φ)
$S_{o,irr}$	maximum achievable oil saturation
dV_E/dG	expulsion rate

V_E	volume expelled expressed as fraction of porosity (cm^3/cm^3)
E	expulsion efficiency (cm^3 expelled/cm^3 generated)
S_i	saturation threshold for initiation of expulsion (middle case)
G_e	generation volume required for 100% efficient expulsion rate (middle case)

Discussion

Dr. P. Ungerer (IFP, France):

Question 1: *Did you apply secondary reactions to the unexpelled fraction only, or to the unexpelled and expelled petroleum?*

Authors' reply: Modelling of secondary cracking in the source rocks was handled in the modelling routines for generation and expulsion of oil and gas. The expelled oil was subjected to secondary cracking in traps during modelling of secondary migration. Cracking of oil while migrating is not modelled because the secondary migration is considered to be fast (less than 1 Ma duration from source to trap).

Question 2: *Did you use S_1 or bitumen content to constrain the amount of unexpelled petroleum?*

Authors' reply: The amount of unexpelled petroleum was constrained by both the S_1 and bitumen content. One well in our data base encounters both the Draupne and Heather Formations at significant maturity (labile kerogen transformation nearing completion, well 6 in Figs. 3c and 3d). The S_1-record (64 analyses) was investigated down to a depth within the Heather Formation where secondary cracking is believed to initiate according to the kinetic modelling parameters. The S_1-record over the subject interval shows no significant depth-related variation in the saturations. Furthermore, the S_1 and TOC data appear to be independent. We propose that these observations may indicate that a subsurface maximum saturation has been reached. The S_1-values consistently fall in the 1.8 to 2.7 mg HC/g ROCK range. The mean of 2.2 mg HC/g ROCK was calculated to be equivalent to 16% of the pore volume using parameters that are consistent with the modelling (see also Appendix, Table A2) and correcting for the proportion of oil lost during the laboratory measurement procedure (C_{15+} fraction, 35%, Cooles et al., 1986). Unfortunately, these samples are all cuttings and the losses of free hydrocarbons induced by drilling and circulating cuttings up to the surface, sample handling and preparation (including washing) are not accounted for. We therefore believe that the true subsurface saturation may be significantly higher and chose the value of 16% to represent the lower end of the modelled saturation range.

The EOM data that were investigated in the subject well (20 analyses) are widely scattered between 15,000 ppm to less than 5000 ppm. The overall mean is 7539 ppm. The EOM mean corresponds to a volume equivalent to a saturation of about 36%. Correcting for losses incurred by laboratory procedures (C_{15-} fraction, 35%, Cooles et al., 1986) we obtained a volume equivalent to 55% of the pore space. The EOM data are significantly skewed. In fact, while the mean is reflected by 7539 ppm, the median is only 5851 ppm, which corresponds to a saturation of about 42% which we feel is more representative for the data. The EOM has high proportions of NSO-compounds and asphaltenes (42–90%, averaging 61%). All of these heavier compounds can hardly be regarded as mobile. Hence, we adopt 40% to constrain the upper end of the modelled saturation range.

References

Buhrig, C., 1989. Geopressured Jurassic reservoirs in the Viking Graben: modelling and geological significance. Mar. Pet. Geol., 6: 31–48.

Burnham, A.K. and Sweeney, J.J., 1989. A chemical model of vitrinite maturation and reflectance. Geochim. Cosmochim. Acta, 53: 2649–2657.

Cooles, G.P., Mackenzie, A.S. and Quigley, T.M., 1986. Calculation of masses of petroleum generated and expelled from source rocks: In: D. Leythaeuser and J. Rullkötter (Editors), Advances in Organic Geochemistry 1985. Pergamon Press, Oxford, pp. 235–246.

Cornford, C., Needham, C.E.J. and de Walque, L., 1986. Geochemical habitat of North Sea oils. In: A.M. Spencer et al. (Editors), Habitat of Hydrocarbons on the Norwegian Continental Shelf. Norwegian Petroleum Society. Graham and Trotman, London, pp. 39–54.

Dahl, B. and Speers, G.C., 1985. Organic geochemistry of the Oseberg Field, I. In: B.M. Thomas et al. (Editors), Petroleum Geochemistry in Exploration of the Norwegian Shelf. Norwegian Petroleum Society. Graham and Trotman, London, pp. 185–197.

Doligez, B., Ungerer, P., Chénet, P.Y., Burrus, J., Bessis, F. and Bessereau, G., 1987. Numerical modelling of sedimentation, heat transfer, hydrocarbon formation and fluid migration in the Viking Graben, North Sea. In: K. Brooks and K. Glennie (Editors), Petroleum Geology of North West Europe. Graham and Trotman, London, pp. 1039–1048.

Durand, B., 1988. Understanding of HC migration in sedimentary basins (present state of knowledge). Org. Geochem., 13 (1–3): 445–459.

Ejadawe, J.E., 1986. The expulsion criterion in the evaluation of the petroleum source beds of the Tertiary Niger Delta. J. Pet. Geol., 9 (4): 439–450.

England, W.A. and Mackenzie, A.S., 1989. Some aspects of the organic geochemistry of petroleum fluids. Geol. Rundsch., 78 (1): 291–303.

England, W.A., Mackenzie, A.S., Mann, D.M. and Quigley, T.M., 1987. The movement and entrapment of petroleum fluids in the subsurface. J. Geol. Soc. London, 144: 327–347.

Espitalié, J., 1986. Use of Tmax as maturation index for different types of organic matter. Comparison with vitrinite reflectance.

In: J. Burrus (Editor), Thermal Modelling in Sedimentary Basins. Technip, Paris, pp. 475–496.

Espitalié, J., Laporte, J.L., Madec, M., Marquis, F., Leplat, P., Paulet, J. and Boutefeu, A., 1977. Méthode rapide de caractérisation de roches mères de leur potentiel pétrolier et de leur degré d'évolution. Rev. Inst. Fr. Pét., 32: 23–42.

Field, J.D., 1985. Organic geochemistry in exploration of the northern North Sea. In: B.M. Thomas et al. (Editors), Petroleum Geochemistry in Exploration of the Norwegian Shelf. Norwegian Petroleum Society. Graham and Trotman, London, pp. 39–57.

Goff, J.C., 1983. Hydrocarbon generation and migration from Jurassic rocks in the East Shetland Basin and Viking Graben of the northern North Sea. J. Geol. Soc. London, 140: 445–474.

Hartman-Stroup, C., 1987. The effect of organic matter type and organic carbon content on Rock-Eval hydrogen index in oil shales and source rocks. Org. Geochem., 11: 351–369.

Hermanrud, C., Eggen, S. and Larsen, R.M., 1990. Investigation of the thermal regime of the Horda Platform by basin modelling. Implications for the hydrocarbon potential of the Stord Basin, northern North Sea. In: A.M. Spencer (Editor), Generation, Accumulation and Production of Europe's Hydrocarbons. Spec. Publ. European Association of Petroleum Geoscientists, 1, Oxford University Press, London, pp. 65–73.

Horsfield, B., 1989. Practical criteria for classifying kerogens: some observations from pyrolysis–gas chromatography. Geochim. Cosmochim. Acta, 53: 891–901.

Horsfield, B., Larter, S. and Düppenbecker, K.F.A., 1990. Gas/oil ratios in source rocks. AAPG Annu. Convention 1990, Am. Assoc. Pet. Geol., Bull., 74 (5): 677 (abstr.).

Hubbert, M.K., 1953. Entrapment of petroleum under hydrodynamic conditions. Am. Assoc. Pet. Geol., Bull., 37: 1954–2026.

Hunt, J.M., 1979. Petroleum Geochemistry and Geology. Freeman, San Francisco, Calif., 617 pp.

Jarvis, G.T. and McKenzie, D.P., 1980. Sedimentary basin formation with finite extension rates. Earth Planet. Sci. Lett., 48: 42–52.

Jensen, R.P. and Doré, A.G., 1993. A recent Norwegian Shelf heating event — fact or fantasy? In: A.G. Doré et al. (Editors), Basin Modelling: Advances and Applications. Norwegian Petroleum Society (NPF), Special Publication 3. Elsevier, Amsterdam, pp. 85–106 (this volume).

Leythaeuser, D., Schaefer, R.G. and Radke, M., 1987. On the primary migration of petroleum. Proc. of the 12th World Petroleum Congress, Spec. Pap., 2. John Wiley and Sons, Chichester, pp. 227–236.

Leythaeuser, D., Schaefer, R.G. and Radke, M., 1988. Geochemical effects of primary migration of petroleum in Kimmeridge source rocks from Brae field area, North Sea, I. Gross composition of C_{15+}-soluble organic matter and molecular composition of C_{15+}-saturated hydrocarbons. Geochim. Cosmochim. Acta, 52: 701–713.

Mackenzie, A.S. and Quigley, T.M., 1988. Principal of geochemical prospect appraisal. Am. Assoc. Pet. Geol., Bull., 72 (4): 399–415.

Mackenzie, A.S., Price, I., Leythaeuser, D., Mueller, P., Radke, M. and Schaefer, R.G., 1987. The expulsion of petroleum from Kimmeridge clay source-rocks in the area of the Brae oilfield, UK continental shelf. In: J. Brooks and K.W. Glennie (Editors), Proc. 3rd Conf. Petroleum Geology of North West Europe. Graham and Trotman, London, pp. 865–878.

Maximov, S.P., Zolotov, A.N. and Lodzhevskaya, M.I., 1984. Tectonic conditions for oil and gas generation and distribution on ancient platforms. J. Pet. Geol., 7 (3): 329–340.

Rundberg, Y., 1989. Tertiary Sedimentary History and Basin Evolution of the Norwegian North Sea between 60° and 62°N — An Integrated Approach. Dr. Ing. Thesis, University of Trondheim-NTH (NTH-Trykk, 1989), 292 pp.

Rundberg, Y. and Smalley, P.C., 1989. High-resolution dating of Cenozoic sediments from northern North Sea using $^{87}Sr/^{86}Sr$ stratigraphy. Am. Assoc. Pet. Geol., Bull., 73 (3): 298–308.

Schou, L., Eggen, S. and Schoell, M., 1985. Oil–oil and oil–source rock correlation, northern North Sea. In: B.M. Thomas et al. (Editors), Petroleum Geochemistry in Exploration of the Norwegian Shelf. Norwegian Petroleum Society. Graham and Trotman, London, pp. 101–121.

Sclater, J.G. and Christie, P.A.F., 1980. Continental stretching: an exploration of the post mid Cretaceous subsidence of the Central North Sea Basin. J. Geophys. Res., 85: 3711–3739.

Sekhar, M.V.C. and Ternan, M., 1979. Pyrolysis of pitch derived from hydrocracked Athabasca bitumen. Fuel, 58: 94–98.

Stainforth, J.G. and Reinders, J.E.A., 1990. Primary migration of hydrocarbons by diffusion through organic matter networks, and its effect on oil and gas generation. Advances in Organic Geochemistry 1989. Org. Geochem., 16 (1–3): 61–74.

Sylta, Ø., 1991. Modelling of secondary migration and entrapment of a multicomponent hydrocarbon mixture using equation of state and ray-tracing modelling techniques. In: W.A. England and A.J. Fleet (Editors), Petroleum Migration. Geol. Soc. London, Spec. Publ., 59: 111–122.

Sylta, Ø., 1993. New techniques and their applications in the analysis of secondary migration. In: A.G. Doré et al. (Editors), Basin Modelling: Advances and Applications. Norwegian Petroleum Society (NPF), Special Publication 3. Elsevier, Amsterdam, pp. 385–398 (this volume).

Thomas, B.M., Møller-Pedersen, P., Whitaker, M.F. and Shaw, N.D., 1985. Organic facies and hydrocarbon distributions in the Norwegian North Sea. In: B.M. Thomas et al. (Editors), Petroleum Geochemistry in Exploration of the Norwegian Shelf. Norwegian Petroleum Society. Graham and Trotman, London, pp. 3–26.

Tissot, B. and Espitalié, J., 1975. L'évolution thermique de la matière organique des sédiments. Applications d'une simulation mathématique. Rev. Inst. Fr. Pét., 30: 743–777.

Tissot, B. and Welte, D.H., 1978. Petroleum Formation and Occurrence. Springer-Verlag, Berlin, 538 pp.

Ungerer, P., Behar, F. and Deschamps, D., 1983. Tentative calculation of the overall volume of organic matter; implications for primary migration. In: M. Bjørøy et al. (Editors), Advances in Organic Geochemistry 1981. Wiley, Chichester, pp. 129–135.

Weitkamp, A.W. and Gutberlet, L.C., 1968. Application of a micro-retort to problems in shale pyrolysis. Am. Chem. Soc. Div. Pet. Chem., Prepr., 13 (2).

ARILD SKJERVØY *Conoco Norway Inc., P.O. Box 488, N-4001 Stavanger, Norway*

ØYVIND SYLTA *IKU Petroleum Research,, N-7034 Trondheim, Norway*

Basin modelling of hydrocarbon charge in the Egersund Basin, Norwegian North Sea: pre- and post-drilling assessments

Hilary Irwin, Christian Hermanrud, Ely Mydland Carlsen, Jan Vollset and Ingemar Nordvall

The Egersund Basin is a marginally mature basin of restricted size. Small, local kitchens of varying depth and size within the area source nearby structures. Some of these structures contain minor oil accumulations, and some are dry.

An integrated basin modelling study was performed prior to drilling in block 9/2. This study included stochastic volumetric calculations of hydrocarbon charge using Latin Hypercube sampling based on a probability distribution of input parameters.

The model has been adapted to discard outcomes from simulations if they are not in accordance with observations made from tested prospects. The model was applied to block 9/2 during progressive exploration drilling of three structures, Gamma, Alpha and Beta East.

Prior to the first well the model indicated an expected charge to the Gamma structure of 44×10^6 Sm^3 with 40% probability of a dry hole. When information from already drilled prospects in the area was used to discard outcomes, the expected charge increased by only 10%, but the probability of a dry hole decreased to 20%.

Results from new wells in block 9/2 produced new information about basin modelling parameters such as temperature, erosion and source rock potential which led to revised maturity maps. Most importantly, information about actual in-place hydrocarbon volumes was used to restrict the modelled outcomes of volumetric calculations. As a result, the uncertainty of modelling hydrocarbon charge to undrilled structures was significantly decreased. The probability of charging the Beta East structure increased from 50 to 80% although the expected volume was reduced by half (16×10^6 Sm^3 modelled before 9/2-1 and 9/2-1; 8×10^6 Sm^3 after).

Examination of the frequency distributions of accepted input data reveals a shift in the loss on secondary migration (saturation of carrier bed) to higher values to accommodate observed volumes of hydrocarbons in place. Other input parameters showed no effect.

As exploration continues and further constraints are made on the model it will become an increasingly important tool for assessing undrilled structures in this marginally mature area where accurate estimates of the hydrocarbon charge are critical.

Introduction

Numeric simulation of the processes leading to petroleum formation and occurrence (commonly termed basin modelling) has become widely used in hydrocarbon exploration. In recent years sensitivity analysis has been applied to basin modelling addressing the effect of variations in input parameters on the results of modelling heat transport and hydrocarbon generation (Ritter et al., 1987; Ritter, 1988; Cao and Lerche, 1989, 1990). The uncertainty associated with all input parameters to basin modelling exercises and volumetric calculations is also thoroughly discussed by Hermanrud et al. (1990). The latter study revealed that parameters most critical in calculations of hydrocarbon generation are also most critical to the calculation of recoverable hydrocarbons, whilst additional uncertainties introduced when hydrocarbon migration is modelled are secondary in comparison. In particular, uncertainty in the initial source rock potential contributes hugely to the uncertainty inherent in volumetric calculations of generated and accumulated hydrocarbons.

Hermanrud et al. (1990) concluded that a probability distribution of outcomes from volume calculations which accommodate uncertainties gives a better basis for evaluating prospects. Furthermore, the uncertainties of the modelling results can only be reduced by constraining the parameters to which the modelling results are most sensitive. The present study takes this work further and presents a new approach to hydrocarbon charge volume calculations which constrains input parameters by taking into account discovered hydrocarbon volumes in structures sourced by kitchens of slightly varying depths and magnitudes. The method is applied to the Egersund Basin.

Basin Modelling: Advances and Applications edited by A.G. Doré et al. NPF Special Publication 3, pp. 539–548, Elsevier, Amsterdam.

Geological setting

The Egersund Basin is situated approximately 100 km west of the southern part of Norway and forms a small extensional basin between the Stavanger Platform and the Sele and Flekkefjord Highs (Fig. 1). It lies to the east of the main extensional zone forming the Central and Viking Grabens.

Exploration history

The first drilling in the general area took place as early as 1966 when well 8/3-1 was found to be dry. Subsequent drilling six years later led to the discovery of the Brisling and then the Bream Fields which contain oil but are not filled to spill point. Between 1970 and 1977 wells were drilled in blocks 9/4, 8/1, 17/11 and 18/11. All were found to be dry. In 1979 a minor amount of oil was discovered in well 18/10-1 on a structure also not filled to spill. In 1980, the dry well 17/12-3, a neighbour to Brisling and Bream, was drilled. Further dry wells include 8/3-2 (1982) and 9/3-1 (1987). In 1988 drilling began in block 9/2 where well 9/2-1 discovered oil in the Gamma structure. 9/2-2 on the neighbouring Alpha structure was dry, whilst in 1989 oil was discovered in well 9/2-3, drilled on the Beta East structure. This latter discovery is also not believed to be filled to spill point.

It thus appears that limited amounts of oil have been generated and accumulated in the Egersund Basin.

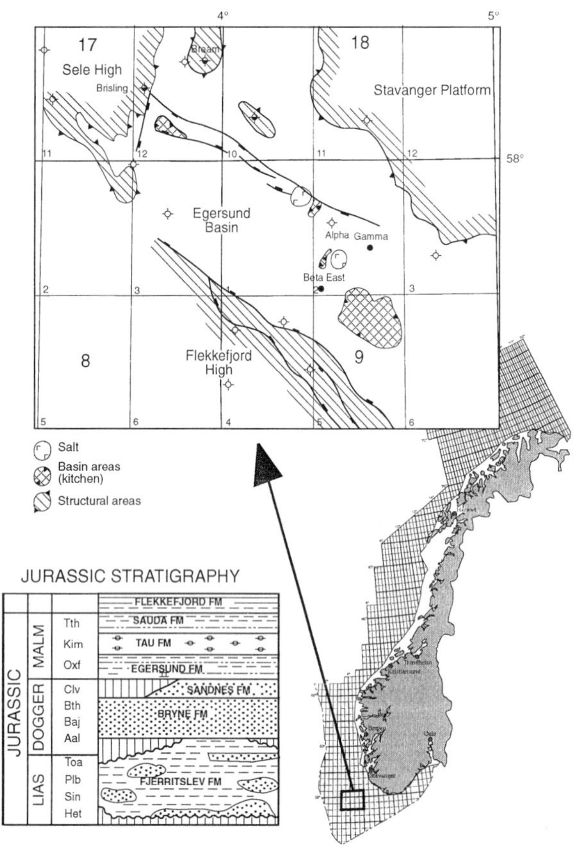

Fig. 1. Generalised map of the Egersund Basin showing the location of block 9/2 and other oil discoveries.

Source rock

Several different formations have been identified as possible source rocks (Ritter et al., 1987). In this study the Upper Jurassic Tau Formation (Fig. 1) is assumed to be the only significant source rock supplying oil to structures in block 9/2. This is substantiated by oil–source correlation studies which have been performed for Statoil.

Source rock potential in the basinal area (9/4-4) can reach 29 mg HC/g rock from type II kerogen with a hydrogen index up to 740 mg HC/g org. C (Ritter et al., 1987). Subsequent drilling in block 9/2 revealed source potentials reaching over 70 mg HC/g rock in some samples. Primary migration from the Tau Formation was probably relatively efficient considering its organic richness and may therefore have been effective even at early maturity levels.

The source rocks are, today, relatively shallowly buried (about 3500 m to top Tau Formation in the kitchen area to 9/2) and therefore only marginally mature or immature in most of the basin. This is reflected in the discovery record for the basin.

The source rocks were previously believed to be generally at their maximum depth of burial even though seismic evidence suggested that parts of the Egersund Basin were moderately uplifted in Tertiary times. The present study takes into account the results of recent work which suggests up to about 400 m Tertiary uplift and erosion has affected block 9/2 (Jensen and Schmidt, 1992). The source rocks are modelled to be presently somewhat shallower than their maximum burial depth. This implies that source rock temperatures are not at their highest at present. Reduction of surface temperatures in Late Cenozoic times will also have had a cooling effect (Hermanrud et al., 1990).

Reservoir rocks and carrier beds

The Tau Formation is sandwiched between two relatively thick packs of leaner shales and this has implications for the concept of primary migration used. Applying the principles outlined by Cooles et al. (1986), a high percentage of primary expulsion will be expected to occur in a downward direction towards the porous and permeable Sandnes Formation of the Middle Jurassic which comprises the carrier beds and reservoir rocks in the block (Fig. 1). Pressure profiles in the form of mud weight equivalents can be interpreted as supporting this, but are of dubious reliability. The separation of the source rock interval from the carrier beds by the somewhat leaner Egersund Formation forms an obstacle to hydrocarbon migration. Thus, saturation of the pores by migrating hydrocarbons is required before significant migration to the effective carrier bed develops. The occurrence of north–south faulting relieves this situation in parts of block 9/2, particularly along the drainage route up to the Gamma structure (9/2-1). In addition, loss on migration may be alleviated by *in situ* generation of limited amounts of hydrocarbons from the Egersund Formation (S_2 around 10 mg HC/g rock, type II/III kerogen), thus relieving the saturation problem. On the whole, long-distance migration is not necessary in this area and loss on migration is expected to be moderate.

Modelling strategy

The in-house Statoil basin modelling program, GEOSIM, which has been described by Hermanrud et al. (1991) was used for modelling the kitchen areas sourcing the structures in block 9/2 according to the method described in Hermanrud et al. (1990). The modelling procedure was as follows:

(1) The 1D version of GEOSIM was used and the thermal histories modelled at different locations turned out to be almost identical. The model was calibrated against available temperature data from local wells.

(2) The modelled thermal histories were used as input to a module which simulates hydrocarbon generation by a series of first-order kinetic reactions to give the present-day transformation ratios.

(3) As thermal histories were so uniform, transformation ratios were extrapolated throughout the kitchen areas and were subsequently applied in volumetric calculations for generated and migrated hydrocarbons following the principles of Cooles et al. (1986).

The volumetric calculations involve specification of several input parameters (indicated in Table 1) which are used to calculate the volume of generated hydrocarbons, the volume of hydrocarbons reaching the carrier beds, and the hydrocarbon charge to structures. The method requires estimation of a value for S_1 which is assumed to be equivalent to the saturation threshold which must be reached before expulsion occurs (Hermanrud et al., 1990) as well as of the percentage of generated hydrocarbons which migrate in the direction of the carrier bed. The transformation ratios were grouped into five different depth intervals for the purpose of being used as input parameters to the volumetric calculations (together with the other input parameters required for the volumetric calculations).

Stochastic modelling was used to describe the uncertainty of the volume calculations by assigning a probability distribution to each input parameter. The probability distributions were previously determined and explained by Hermanrud et al. (1990). The vol-

TABLE 1

Input parameters with uncertainties as applied in this study

Input data	Probability distributions [a]	
	pre-9/2-1	post-9/2-1
Initial potential (g HC/g rock)	T (0.02, 0.031, 0.051)	TN (0.03, 0.003, 0.1)
Source rock thickness (m)	N (40, 7)	N (50, 5)
Source rock saturation before migration (g HC/g rock)	N (0.006, 0.001)	N (0.007, 0.001)
Migration to carrier bed (a)	T (0.5, 0.8, 1)	T (0.5, 0.75, 0.8)
(b)	T (0.3, 0.5, 0.7)	T (0.3, 0.5, 0.7)
Thickness of carrier bed (m)	N (150, 25)	N (180, 5)
Porosity of carrier bed	T (0.05, 0.15, 0.2)	N (0.18, 0.05)
Saturation of carrier bed	T (0.01, 0.025, 0.1)	(0.004 + 0.02) × LN (1.5, 0.8)
Hydrocarbon density (g cm^{-3})	N (0.55, 0.1)	N (0.75, 0.1)
Source rock density (g cm^{-3})	N (2.4, 0.1)	N (2.4, 0.1)
Areas of maturity intervals in kitchen	dependent N	dependent N revised
Areas of drainage segments	dependent N	dependent N revised

[a] N = normal (mean, σ); T = triangular (minimum, most likely, maximum); TN = truncated normal (mean, σ, minimum, maximum); LN = lognormal (mean, σ).
(a) = Gamma.
(b) = others.

umetric calculations were then performed by Latin Hypercube sampling (McWilliams, 1987) from the input distributions for over 500 iterations to produce probability distributions for volumes of generated and primary migrated hydrocarbons as well as hydrocarbon charge to the structures. The results can be presented either in the form of probability distributions of calculated volumes or as cumulative probability curves of the volume distributions. The expected volume marked on the cumulative probability curves is the mean of the volume distribution.

Selection of input parameters can be simply constrained by implementing a yes/no test of the input selection against observations; for example, to test maturity the selected transformation ratio can be compared with observations. If the selection is acceptable the volumes are calculated, if not the program begins again by selecting new input parameters. An example of a flow diagram for the stochastic model is given in Fig. 2 with a schematic illustration of output.

The main problem is usually that we do not know with certainty what the probability distribution around the input parameters should be, but we can compare the outcome of drilling with the results of the model and then implement our yes/no test. The size of a new discovery can be used as a test of the selected input parameters for every iteration. If the input parameters selected produce a result which matches observations, then the program continues and carries out volume calculations for the undrilled structures. If not, the calculation is aborted and a new selection of input parameters made (see below).

The model can then be modified at different stages of exploration and undrilled prospects remodelled using new information. Subsequent tests can be added after each drilling result and the data constrained even further. Cumulative probability curves for the hydrocarbon charge obtained for the undrilled structures can be compared with results obtained without constraints. In addition frequency distributions of the "accepted" input parameters can be inspected and compared to "unacceptable" parameters to gain insight into the sensitivity of the model, to constrain input parameters and to derive further information about processes affecting the volume calculations.

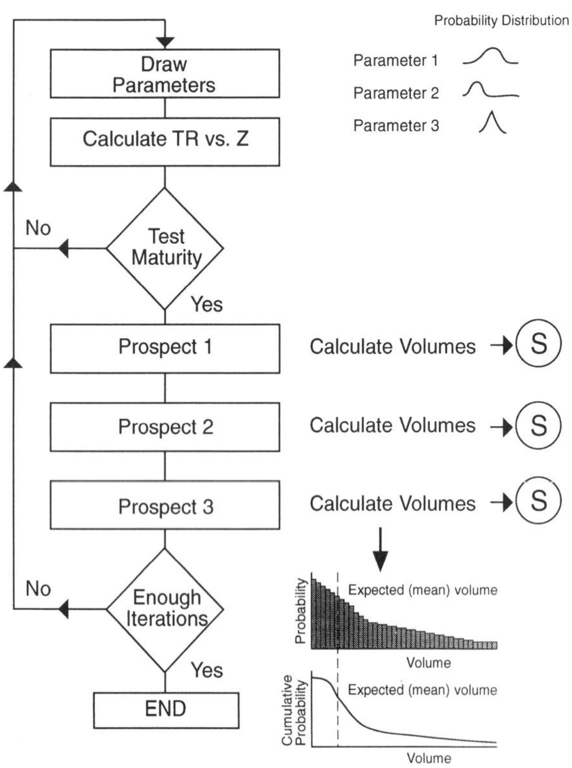

Fig. 2. Flow diagram for the stochastic model including a test for maturity and schematic illustration of the output volumes.

Application to the Egersund Basin

Basin modelling and volumetric calculations of hydrocarbon charge to the structures in block 9/2 were carried out prior to drilling well 9/2-1 on the Gamma structure. The input data (specified in Table 1) were based on inspection and evaluation of geological and geochemical data then available in the area. Results from this modelling are shown in Fig. 3.

Results of drilling at that time indicated that oil had been generated and migrated in the kitchen area to Brisling and Bream. It was desirable to use these external data in some way to constrain the volumetric modelling. It was decided that the maturity/depth relationship (in terms of transformation ratio) could be constrained. A maximum depth for the Brisling and Bream kitchen area had been derived and based on the fact that oil had been generated and migrated in the basin it was possible to assign a minimum level for the transformation ratio experienced by the deepest part of this basin. This could be used to constrain the transformation ratio when modelling hydrocarbon charge in block 9/2. A transformation ratio of 0.3 at 3500 m was chosen. A flow diagram for the stochastic modelling is shown in Fig. 4. Results are shown in Fig. 5.

Well 9/2-1 was drilled on the Gamma structure. It discovered oil and provided new geological and geochemical data. The present best estimate of oil in place is 32 $\times$ 10^6 Sm^3. Initial source rock potential was revised. New virgin rock temperatures were available for calibrating the basin modelling. In this case they were approximately 10°C lower than previously used (Fig. 6). In addition to this it was becoming increasingly clear that Tertiary uplift and

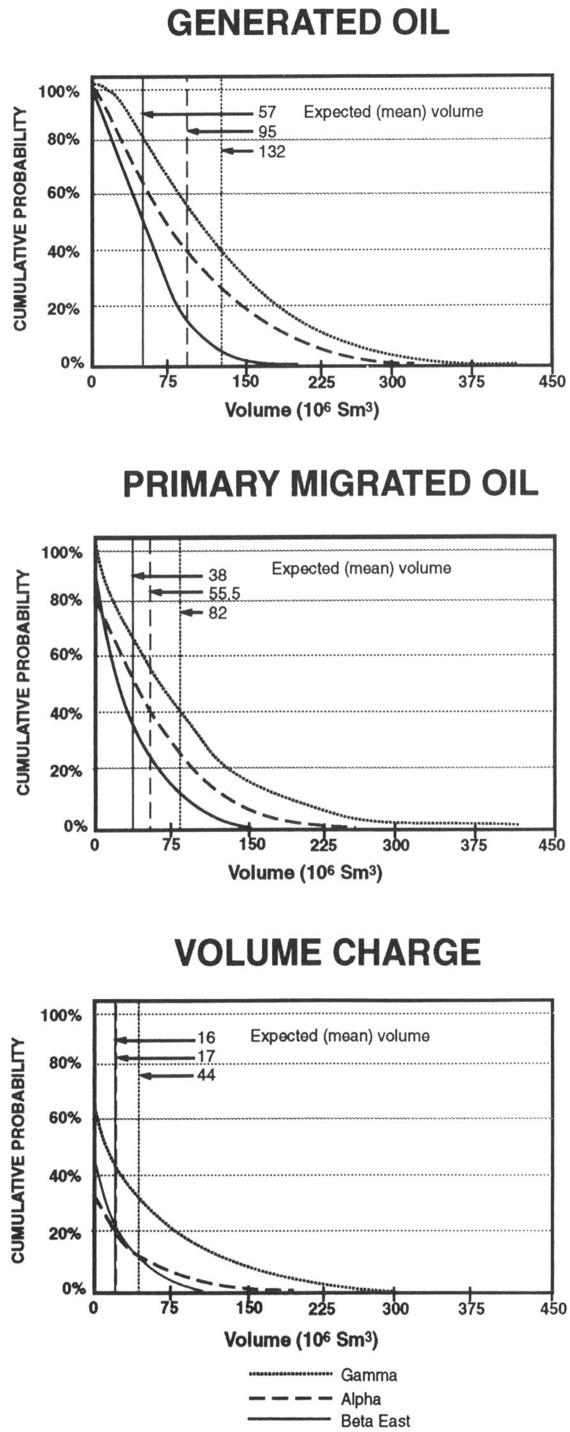

Fig. 3. Cumulative probability of volumes of generated, primary migrated and charged hydrocarbons to three structures in block 9/2: pre-9/2-1 without the Brisling and Bream maturity check.

erosion had significantly affected block 9/2. Maturity of 0.8 to 0.9 on the vitrinite scale was found at around 3500 m in well 9/2-1. Maximum burial to about 4000 m would not be unusual for this maturity in the Central Graben. The transformation ratio was remodelled and mapped using the new input data and calibration points, and taking into

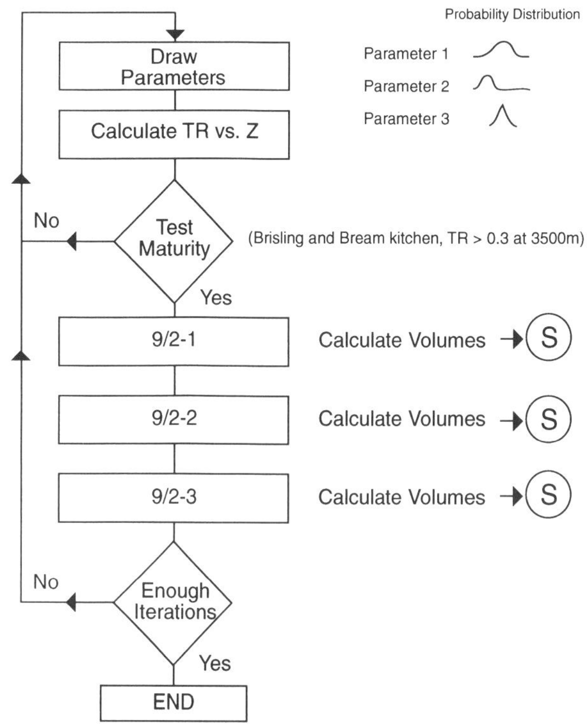

Fig. 4. Flow diagram for the stochastic model prior to drilling 9/2-1, with the Brisling and Bream maturity check. *S* = statistics; *TR* = transformation ratio; *Z* = depth.

account 450 m of Tertiary uplift and erosion.

Stochastic modelling was carried out with input parameters modified according to new information and results from the revised basin model (Table 1). A new maturity test was implemented based on local data from 9/2-1. A volume test was carried out after charge calculations for 9/2-1 (Fig. 7). If the result was in accordance with observations (e.g. over 30×10^6 Sm^3) then volume calculations for the Alpha (9/2-2) and Beta East (9/2-3) structures were carried out using the same basic input. If the Gamma structure is assumed not be filled to spill point, this gives us control not only on the minimum charge (around 30×10^6 Sm^3), but also an estimate of the maximum charge (around 50×10^6 Sm^3). Using this information an alternative test was constructed (see flow diagram in Fig. 7). The acceptable results must lie between 30 to 50×10^6 Sm^3 for the calculation of charges to Alpha and Beta East to be carried out. Results are shown in Fig. 8.

After 9/2-2 was found to be dry, the volume charge for Beta East was calculated using an additional test as demonstrated in the flow diagram in Fig. 9. It was estimated that there could be updip reserves in the structure with a maximum volume of 1×10^6 Sm^3 and this volume was used to further constrain input parameters for the volume calculations for Beta East. Results of this modelling are shown in Fig. 10.

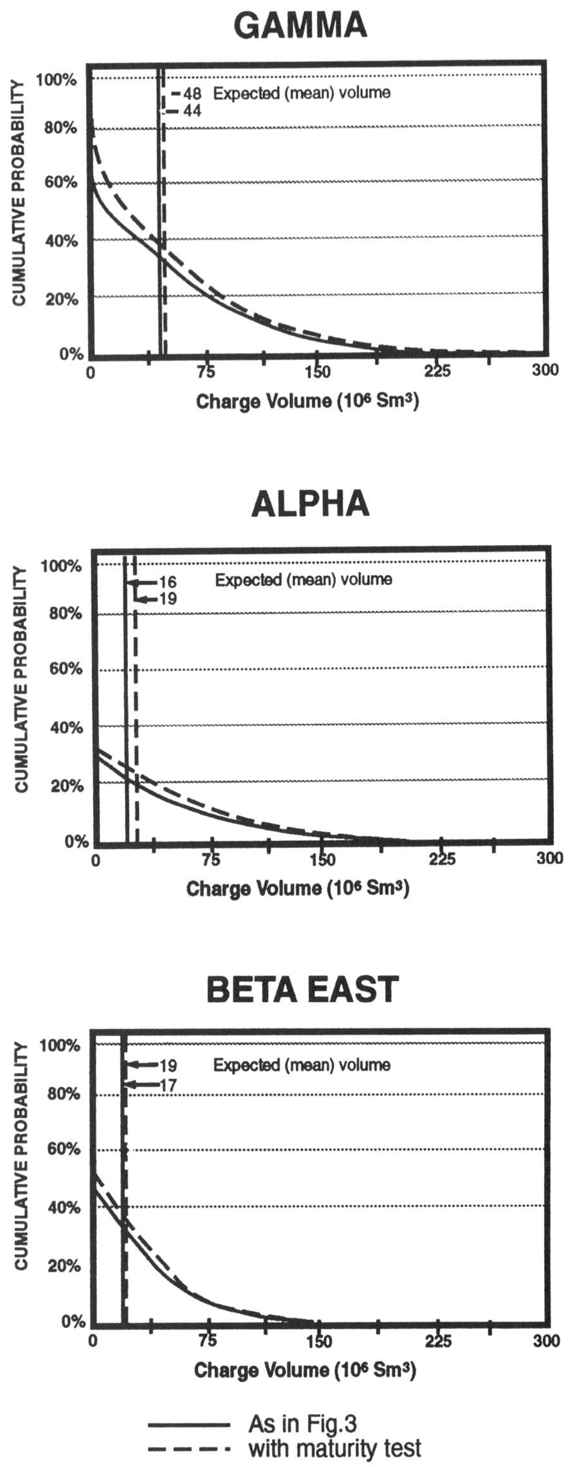

Fig. 5. Cumulative probability of volume charge to three structures in block 9/2: pre-9/2-1 including the Brisling and Bream maturity test.

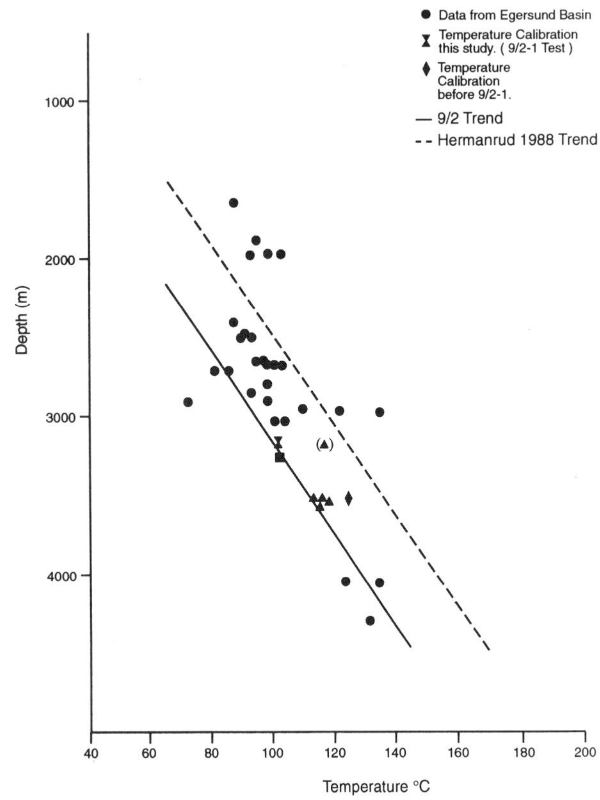

Fig. 6. Egersund Basin present-day temperature–depth relationship.

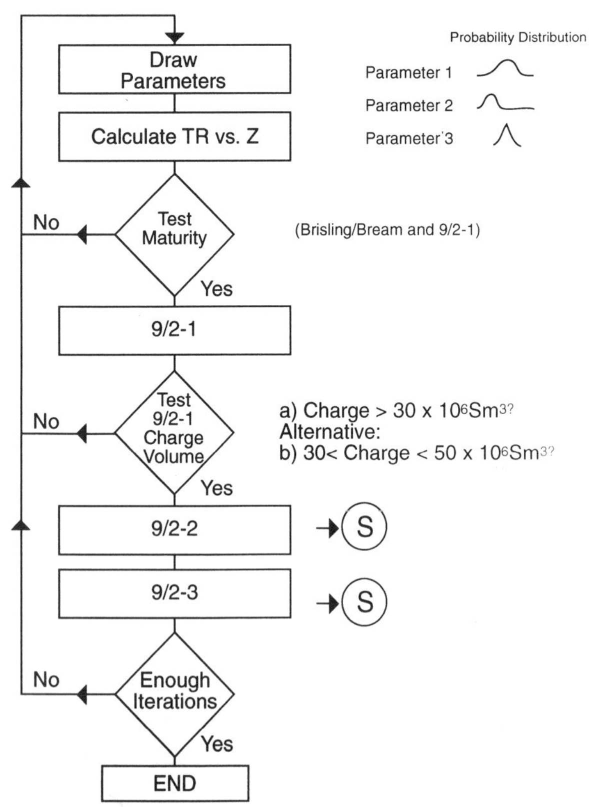

Fig. 7. Flow diagram for the stochastic model: post-9/2-1 results. S = statistics; TR = transformation ratio; Z = depth.

Results

Pre-9/2-1

Volumetric calculations of charge to the three structures in block 9/2 prior to drilling gave results presented in Fig. 5, which clearly show that the Gamma prospect was the most favourable with a

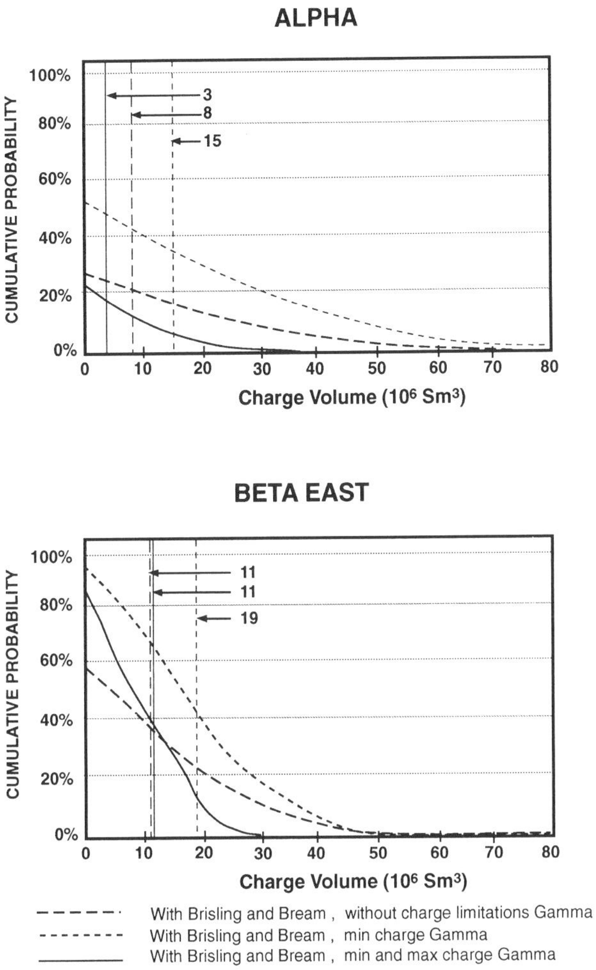

Fig. 8. Cumulative probability of volumetric charge calculations for the Alpha and Beta East structures: post-9/2-1, pre-9/2-2.

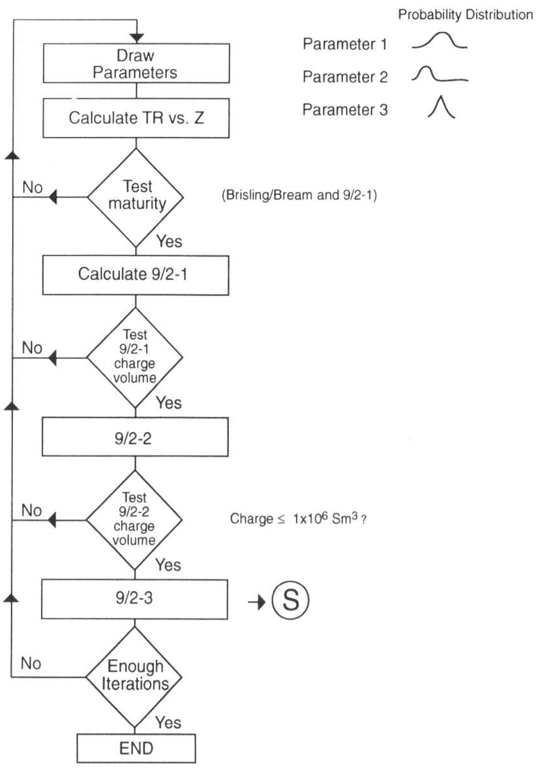

Fig. 9. Flow diagram for the stochastic model, post-9/2-2.

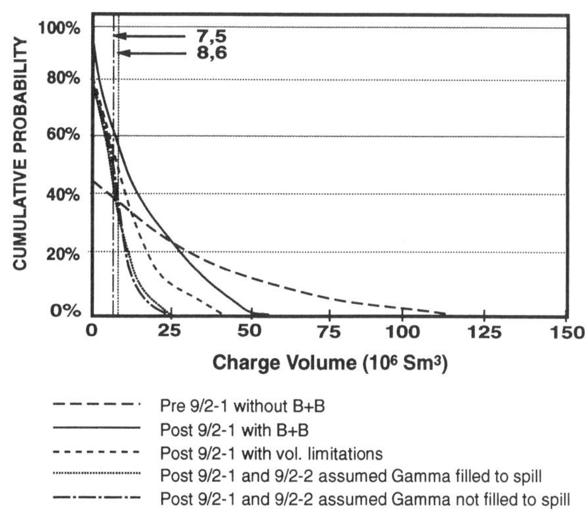

Fig. 10. Cumulative probability of volumetric charge calculations to the Beta East structure: pre- and post-9/2-1 and 9/2-2.

most probable charge of 44 × 10^6 Sm^3 oil. The most probable charges to the Alpha and Beta East structures were considerably smaller volumes (ca. 16 and 17 × 10^6 Sm^3, respectively). The cumulative probability curve also shows around 40% chance of a dry hole for Gamma, and around 60% to 70% for the Beta East and Alpha prospects. All three structures have significant high side charge volumes greater than the trapping capacities, implying a significant chance of spill from the structures.

Implementation of the test for maturity using external data from the Brisling and Bream kitchen caused a 9 to 18% increase in the expected charge. Calculations for Gamma increased to 48 × 10^6 Sm^3 and for Alpha and Beta East to 19 × 10^6 Sm^3. There was a decrease in the probability of a dry hole, particularly for the Gamma structure (40–20%). The reason for the small difference in volume is probably the similarity in maturity between the 9/2 kitchens and the Brisling and Bream kitchen.

Post-9/2-1

Cumulative probability curves for the calculated charge, taking into account the results from 9/2-1 well (such as temperature, erosion etc.) are shown in Fig. 8. The expected charge to Alpha is reduced from 16 to 8 × 10^6 Sm^3 with a 75% chance of a dry hole using the new basin modelling results. The effect of introducing a minimum charge to Gamma increased the expected volume in Alpha to 15 × 10^6 Sm^3 and reduced the chance of a dry hole to 45%.

TABLE 2

Volumetric calculations; the conditions and results of the models

	Pre-9/2-1		Post-9/2-1			Post-9/2-2	
Model number:	1	2	3	4	5	6	7
Conditions:							
Maturity (a)	–	×	×	×	×	×	×
(b)	–	–	×	×	×	×	×
Minimum volume Gamma	–	–	–	30	30	30	30
Maximum volume Gamma	–	–	–	–	50	–	50
Maximum volume Alpha	–	–	–	–	–	1	1
Results:							
Gamma	44	48	36	36	36	36	36
Alpha	16	19	8	15	3	15	2
Beta East	17	19	11	19	11	9	8

(a) = Brisling and Bream maturity test.
(b) = 9/2-1 maturity test.

The effect of introducing the additional constraint of a maximum possible charge to Gamma, assuming it is not filled to spill, was to considerably reduce the high side potential calculated for the other two prospects. The maximum possible volume to the Alpha structure was halved to 40×10^6 Sm^3 and the most probable charge to Alpha was reduced to 3×10^6 Sm^3. The results now showed 80% probability of no hydrocarbons reaching the structure. 9/2-2 was dry with possible reserves updip being less than 1×10^6 Sm^3 oil in place. 9/2-2 was drilled on the Alpha structure although with maximum constraints it almost inevitably had to be dry. However, the probability of a dry hole was significantly less if the Gamma structure was assumed to be filled to spill.

Post-9/2-2

The expected charge to Beta East was reduced to 11×10^6 Sm^3 using the new basin modelling results after drilling 9/2-1 (Fig. 8). Constraining the volume calculation results using the minimum charge to Gamma increased the most probable volume charge to 19×10^6 Sm^3 whilst maximum charge limitations reduced the expected volume again to 11×10^6 Sm^3. The latter modelling resulted in only 20% probability of a dry structure and considerable reduction of the high side potential to 30×10^6 Sm^3. Further constraint using the results from 9/2-2 resulted in an expected volume charge of 9×10^6 Sm^3 to Beta East assuming Gamma is filled to spill and 8×10^6 Sm^3 if not (Fig. 10). The model now indicated 80% probability of hydrocarbons reaching the structure and a maximum volume of 25×10^6 Sm^3. 9/2-3 was drilled on the Beta East structure and discovered hydrocarbons estimated at around 7×10^6 Sm^3.

The results of the volumetric calculations, using different constraints on the model, are summarised in Table 2.

Discussion and conclusions

The Brisling and Bream maturity test eliminated selected maturity input; those that did not conform to the assumption that for hydrocarbon migration a transformation ratio of >0.3 must have been experienced by the parts of the basin deeper than 3500 m. The 9/2-1 minimum volume test eliminated the combinations of input which gave rise to small calculated volumes and therefore "guaranteed" maturity, expulsion and secondary migration. The addition of the 9/2-1 maximum volume constraint eliminated the input which was responsible for the extreme high side calculations. Implementation of the 9/2-2 test further restricted the high side results. Input data not constrained were: the areas of maturity zones, areas of drainage segments, and the value for the expulsion efficiency. As mentioned earlier, the latter was assessed to be different and less advantageous for the Alpha and Beta East structures (0.5) than for the Gamma structure (0.7) (Table 1).

A record of the frequency distribution of accepted and rejected input data was made and studied for information about the constrained input data. Imposing the Brisling and Bream maturity test caused rejection of only 13% of the input data selections. Addition of the 9/2-1 test, i.e. minimum charge $\geq 30 \times 10^6$ Sm^3 caused rejection of 72% of the input data selections. Further implementation of an upper limit for the charge to Gamma (50×10^6 Sm^3) caused rejection of a further 10% of all iterations. Implementation of the 9/2-2 test increased rejection to 85% of all iterations and necessitated over 10,000

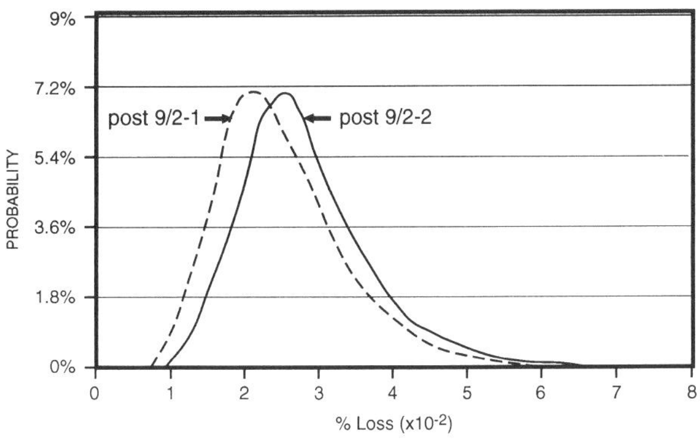

Fig. 11. Frequency distribution for accepted values of secondary loss on migration: pre- and post-9/2-2.

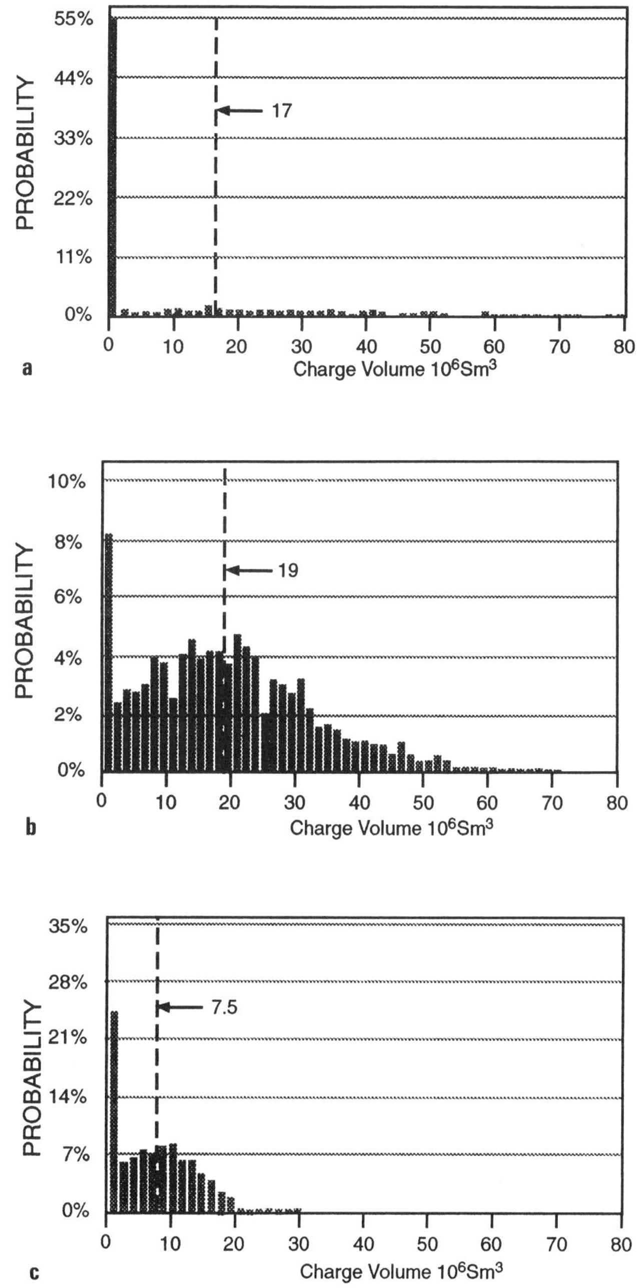

Fig. 12. Probability distribution of volume charge to the Beta East structure. (a) Pre-9/2-1. (b) Post-9/2-1: Maturity tests; Gamma test minimum volume. (c) Post-9/2-2: Maturity tests; Gamma test minimum and maximum volume; Alpha test maximum volume.

iterations to achieve a representative distribution of "acceptable" results for the undrilled Beta East structure.

The record of the "correct" input should give concrete limits for the quantification of initial source potential, loss on secondary migration etc. However, comparison of the frequency distribution of rejected input with that of the accepted input for a given simulation indicated very little difference. Small differences were observed; for example, the higher values for loss on secondary migration did not occur in the distribution for accepted input but were observed for rejected input. This may be explained by the smaller population describing the accepted input. If more iterations were included, these high values could conceivably be observed. The conclusion was that there was no significant difference, overall, between the accepted and rejected input.

Comparison of "accepted" input data between simulations with different constraints can provide important information. Figure 11 shows clearly a shift in the frequency distribution for selected values of loss on secondary migration towards higher values when the results from 9/2-2, drilled on the Alpha structure, were taken into account. This is interesting from the point of view that spill from the Alpha structure may have occurred before Late Tertiary burial gave rise to the present structural picture (i.e. loss on migration may have been unusually high). Although this was not taken into account when defining input it seems to be accounted for by the model.

To conclude, the procedure described above has the aim of defining the combination of input parameters that give volume charge calculations in accordance with observations, and in this way constraining these parameters for use in volume calculations for undrilled structures. The improvement in accuracy is impressive and is illustrated in Fig. 12. To reiterate: 8 to 9 $\times$ 10^6 Sm^3 oil was calculated as the most probable charge to Beta East with only 20% probability of a dry hole. This is a considerable improvement on the pre-9/2-1 calculations which gave a best estimate of 17 $\times$ 10^6 Sm^3 charge with 60% probability of a dry hole. To begin with, it was uncertain whether oil would be discovered and what the quantity was likely to be (over 150 $\times$ 10^6 Sm^3 was possible). Afterwards it was fairly certain that we would discover oil but in small amounts (maximum ca. 25 $\times$ 10^6 Sm^3). An estimated volume of 7 $\times$ 10^6 Sm^3 oil in place was discovered. The accuracy is even more impressive taking into consideration the simplifications that were made when constructing the model. For example, the possibility of changes in drainage direction and size and geometry of the structures since hydrocarbon generation and migration was not included in the model, even though such variations are not unreasonable considering that Tertiary uplift, erosion and subsequent burial have taken place. We conclude therefore, that attempts to constrain risk and charge volumes of the types described in this paper are definitely worthwhile, despite the fact that every possible variable cannot be accounted for.

Acknowledgements

The authors thank Statoil and partners in Pl 114 (Saga, Petrobras and Deminex) for permission to publish data from block 9/2 and the results of this study. Thanks also go to Britt Sissel Todnem, Venche Mellomstrand and Birte Garvik for preparing the figures, and Mette Kristensen for help with preparation of the manuscript. We are grateful for the helpful comments of the referees who reviewed the manuscript.

References

Cao, S. and Lerche, I., 1989. Sensitivity analysis of basin modeling with applications. In: F. Agterberg (Editor), Applications of Statistics in the Earth Sciences. Can. Geol. Soc. Pap., 89-9: 489–504.

Cao, S. and Lerche, I., 1990. Basin modeling: Applications of Cao, sensitivity analysis. J. Pet. Sci. Eng., 4: 83–104.

Cooles, G.P., Mackenzie, A.S. and Quigley, T.M., 1986. Calculation of petroleum generated and expelled from source rocks. In: D. Leythauser and J. Rullkötter (Editors), Advances in Organic Geochemistry 1985. Org. Geochem., 10: 234–245.

Hermanrud, C., 1988. Determination of Formation Temperature from Downhole Measurements. Ph.D. Thesis, Univ. of South Carolina, Columbia, S.C., 203 pp.

Hermanrud, C., Eggen, S., Jacobsen, T., Carlsen, E.M. and Pallesen, S., 1990. On the accuracy of modelling hydrocarbon generation and migration: the Egersund Basin oil find, Norway. In: Advances in Organic Geochemistry 1989. Org. Geochem., 16: 389–399.

Hermanrud, C., Eggen, S. and Larsen, R.M., 1991. Investigation of the thermal regime of the Horda platform by basin modelling; implications for the hydrocarbon potential of the Stord basin, northern North Sea. In: A.M. Spencer (Editor), Generation, Accumulation and Production of Europe's Hydrocarbons. Special Publication of the European Association of Petroleum Geoscientists, 1, Oxford University Press, London, pp. 65–73.

Jensen, L.N. and Schmidt, B.J., 1992. Late Tertiary uplift and erosion in the Skagerrak area; magnitude and consequences. In: L.N. Jensen, F. Riis and R. Boyd (Editors), Post-Cretaceous Uplift and Sedimentation along the Western Fennoscandian Shelf. N.G.T., 72: 275–279.

McWilliams, T.P., 1987. Sensitivity analysis of geologic computer models: a formal procedure based on Latin Hypercube sampling. Math. Geol., 19: 81–90.

Ritter, U., Leith, T.L., Griffiths, C.M. and Schou, L., 1987. Hydrocarbon generation and thermal evolution in parts of the Egersund Basin, northern North Sea. In: C. Beaumont and A.J. Tankard (Editors), Sedimentary Basins and Basin Forming Mechanisms. Can. Soc. Pet. Geol. Mem., 12: 75–85.

Ritter, U., 1988. Modelling of hydrocarbon generation patterns in the Egersund Sub-Basin, North Sea. In: L. Mattavelli and L. Novelli (Editors), Advances in Organic Geochemistry 1987. Org. Geochem., 13: 165–174.

HILARY IRWIN *Statoil A/S, P.O. 300, N-4001 Stavanger, Norway*
CHRISTIAN HERMANRUD *Statoil A/S, P.O. 300, N-4001 Stavanger, Norway*
ELY MYDLAND CARLSEN *Statoil A/S, P.O. 300, N-4001 Stavanger, Norway*
JAN VOLLSET *Statoil A/S, P.O. 300, N-4001 Stavanger, Norway*
INGEMAR NORDVALL *Statoil A/S, P.O. 300, N-4001 Stavanger, Norway*

Isostatic and tectonic subsidence of the Egersund Basin

Willy Fjeldskaar, Edward Prestholm, Claudia Guargena and Nina Gravdal

The palaeo-heatflow is, together with the thermal conductivities of the sedimentary rocks, the most important parameter for the determination of the temperature history of a sedimentary basin. The parameters are essential for an accurate estimation of the hydrocarbon generation and timing of expulsion. These critical parameters are, however, often a weak point in basin modelling activities.

The thermal conductivities of the sedimentary rocks in the Egersund Basin have been predicted by partial least-square regression based on measurements of the thermal conductivity and mineralogy/porosity. The predictions suggest vertical thermal conductivities of shale-rich formations in the range 1.3 to 2.0 W/m K, with significant anisotropy, that increases with burial depth, up to 1.4. The present-day heatflow is estimated to be 52 mW/m^2 by calibrating the theoretical temperature regime with well data.

The development of sedimentary basins of the extensional type is usually explained by two processes: tectonic subsidence due to thinning of the lithosphere, and isostatic subsidence due to sediment loading. Modelling of the tectonic subsidence is carried out to estimate the amount of lithosphere thinning and the related palaeo-heatflow of the area. To determine the effect of these two processes in the Egersund Basin, North Sea, a two-dimensional modelling tool working on depth-converted seismic sections is used.

The observed tectonic subsidence of the area indicates two periods with significant increase of subsidence rate, probably caused by tectonic events; one of Triassic age, the other of Late Jurassic age. The estimated amount of relative crustal thinning depends indirectly on the flexural rigidity of the lithosphere. The present crustal thickness of the area puts restrictions to the flexural rigidity; an upper limit is estimated to be 10^{22} N m (elastic thickness $t_e \leq 10$ km). This is much lower than has previously been estimated for Scandinavia from post-glacial uplift, possibly indicating a significant decrease in the flexural rigidity towards the west, into the North Sea.

Introduction

Traditionally, basin modelling studies focus on the modelling of generation, expulsion and migration of hydrocarbons, aspects which are covered by strong research activity in the scientific community as well as in the oil companies. One critical, but often weak point, however, is the determination of the input parameters for temperature history estimation. These parameters, such as palaeo-heatflow, thermal conductivity of sedimentary rocks and palaeo-surface temperature, are essential for a more accurate estimation of the hydrocarbon generation and timing of expulsion. The uncertainty of the modelling results can be reduced significantly by constraining these parameters.

The importance of the palaeo-heatflow has also been pointed out in sensitivity studies undertaken for the Egersund Basin (Ritter et al., 1987; Ritter, 1988; Hermanrud et al., 1990). It was found that the transformation of hydrocarbons in the area was significantly affected by the palaeo-heatflow. Hermanrud et al. (1990) also found that the modelled transformation ratio is practically insensitive to changes in the thermal conductivity of the sediments. However, this conclusion is highly governed by the fact that the calculated temperatures in the Egersund Basin were calibrated by observed temperatures of the source rock itself. In general, the thermal conductivity of sedimentary rocks is one of the most important parameters for determination of the temperature regime in a basin.

The aim of this study has been to constrain the thermal conductivities of the sedimentary rocks and the palaeo-heatflow, in order to reduce the uncertainty of the predicted palaeo-temperature regime in the Egersund Basin, southern North Sea. The thermal conductivity of sedimentary rocks is modelled by partial least-square regression on the basis of a large database of measurements on mineralogy and thermal conductivity of core samples, and the palaeo-heatflow is estimated by a two-dimensional isostatic and tectonic model.

Geological setting

The Egersund Basin is a relatively unexplored Mesozoic narrow marginal basin trending NE–SW

Basin Modelling: Advances and Applications edited by A.G. Doré et al. NPF Special Publication 3, pp. 549–562, Elsevier, Amsterdam.

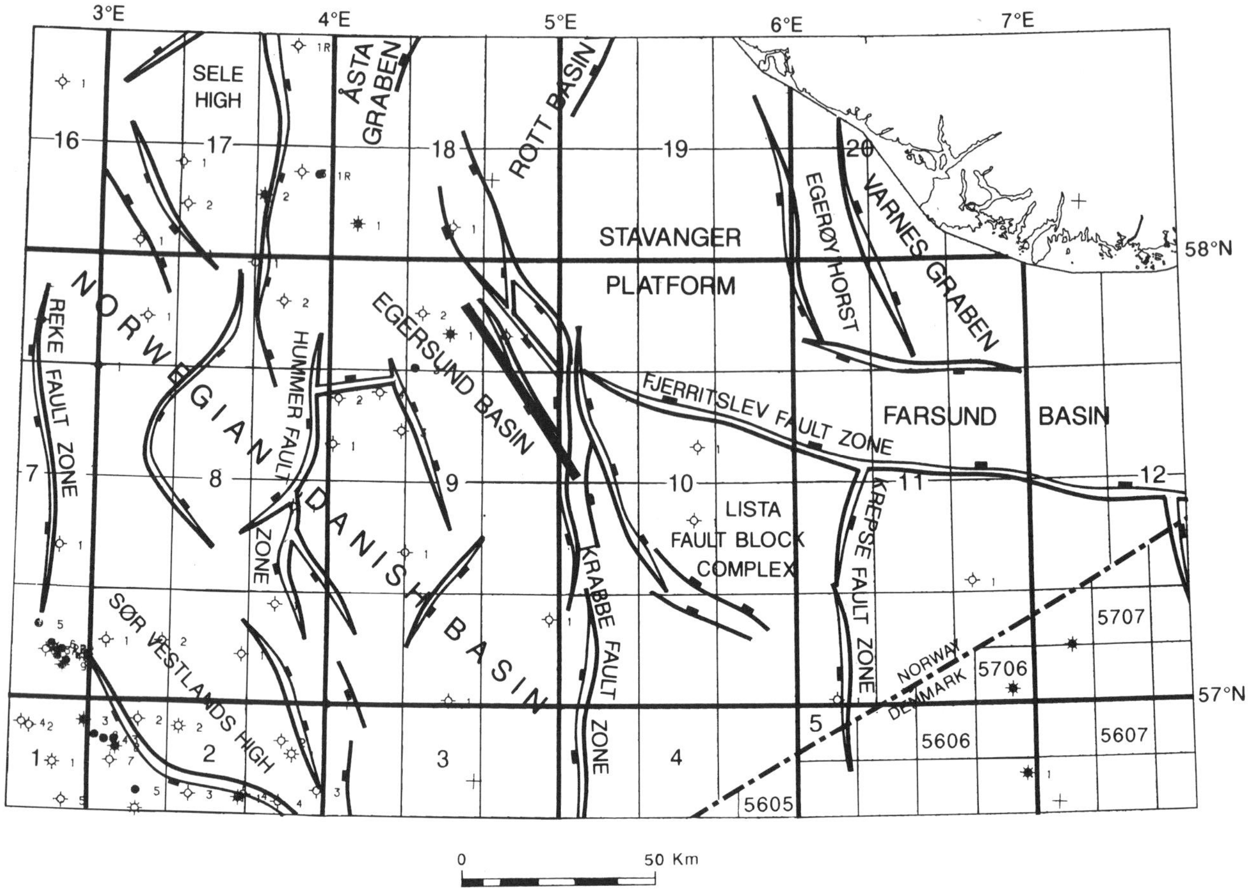

Fig. 1. Tectonic map of Norwegian Danish Basin. The seismic line selected for the study is shown by the solid black line running NW–SE.

and lying to the east of the Central Graben. The basin is bordered to the northeast by the Precambrian Stavanger Platform. The northwestern and southeastern basin boundaries are defined by the Sele High and Lista Fault Complex, respectively, whereas the southern basin extension is characterized by a gentle dip into the Norwegian–Danish Basin (Fig. 1).

The expression of Late Palaeozoic tectonic activity is seen by the Sele High, Lista Fault Complex and the extension of the Fjerritslev Fault Zone into the southern part of the Stavanger Platform (Skjerven et al., 1983). Large portions of the area are underlain by a layer of Late Permian (Zechstein) salt. The main tectonic features played an important role in the Early Mesozoic basin development, notably the facies associated with salt movement and the thickening of Triassic sequences to the east and southeast of the basin (Lervik et al., 1989). Fault activity during the Triassic was minor and restricted to old zones of weakness. Regional subsidence accommodated vast amounts of continental deposits and the increase in overburden suggests the initiation of salt movement.

During the Early Jurassic most of the area was probably inundated but was subjected to erosion in Middle Jurassic times, evident in the northeastern part of the basin. Except for continued salt movements, no particular rifting pulse is observed during this period. Although Lervik et al. (1989) suggest that "Late Jurassic rifting did not reactivate Triassic faulting" it is apparent that, at the onset of the Late Jurassic, tectonic activity accentuated preexisting relief. Transtension along pre-established zones of weakness trending NW–SE, axial to the basin, and NE–SW at the southeastern basin margin, led to the development of the Egersund Basin (Pegrum, 1984). Zechstein evaporites act as a zone of decollement for these basin margin faults. The NW–SE-trending fault system running through the basin delineates the southwestern boundary of the main Upper Jurassic/Lower Cretaceous basin development. This is a highly complicated zone where a salt swell triggered by old topography is coupled with salt withdrawal and Late Jurassic/Early Cretaceous growth faulting.

During the Middle and Late Cretaceous the basin was structurally inverted along the same NW–SE-trending lineaments that created the basin. Due to salt growth, inversion along the southwesternmost lineament is only observed north of the salt piercement. Structural inversion may be a result of transpressional movements of basement faults within the Fjerritslev Fault Zone (Pegrum, 1984). However, if the Egersund Basin is considered to have formed strictly as a result of a NE–SW-trending extensional vector, as proposed by Roberts et al. (1990), the structural inversion would be due to reversed dip slip reactivation.

The inversion tectonics seem to have come to an end during the Early Tertiary and, with the exception of salt growth and the Pliocene/Pleistocene erosional event, the Tertiary is characterized by almost continuous subsidence and accumulation of clastic sequences. Possible minor pulses of uplift and erosion in the Tertiary cannot be attributed to any thermal pulse and are considered to be mainly due to isostatic adjustments.

Modelling

The modelling has been carried out by using the basin modelling system developed at Rogaland Research (BMT, Basin Modelling Toolbox). A detailed description of the system is given in Fjeldskaar et al. (1990). BMT is a modular system for geological, geophysical and geochemical basin modelling, working on depth-converted seismic sections. The depth-converted section across the Egersund Basin selected for the modelling is shown in Fig. 2. The modelled profile is tied to well 9/2-1, which, based on biostratigraphy, gives the ages of the different seismic sequences. This well, together with the interpreted seismic section, indicates a very thin salt layer at the base of the Mesozoic sequence. However, the amount of salt was probably much more significant prior to salt movements (Taylor, 1984). Minor faulting related to post-Late Jurassic movements has been ignored as these faults appear to have little influence on the temperature history.

The present-day temperature regime in the basin was calculated by a two-dimensional steady-state finite-difference conduction model. The assumption that the heat is transported primarily by conduction is a reasonable assumption, as the calculated contribution from compaction-derived water to heat transport in offshore sedimentary basins generally is less than 1/1000 relative to the contribution from conduction (Hermanrud, 1986).

In addition to the burial history, the input data required to run the thermal history model are the ocean bottom temperature history, the heatflow from the earth's crust into the basin and the thermal conductivity of the sediments.

Thermal conductivities of rocks

For a successful reconstruction of the temperature regime in a sedimentary basin it is imperative that the thermal conductivities of the rocks are known in detail. The relation between the mineral composition, porosity and the thermal conductivity of sedimentary rocks has been studied as part of our in-house database of measurements. By partial least-square regression analysis the correlation between the thermal conductivities of rocks and their mineralogies/porosities has been quantified, enabling the predic-

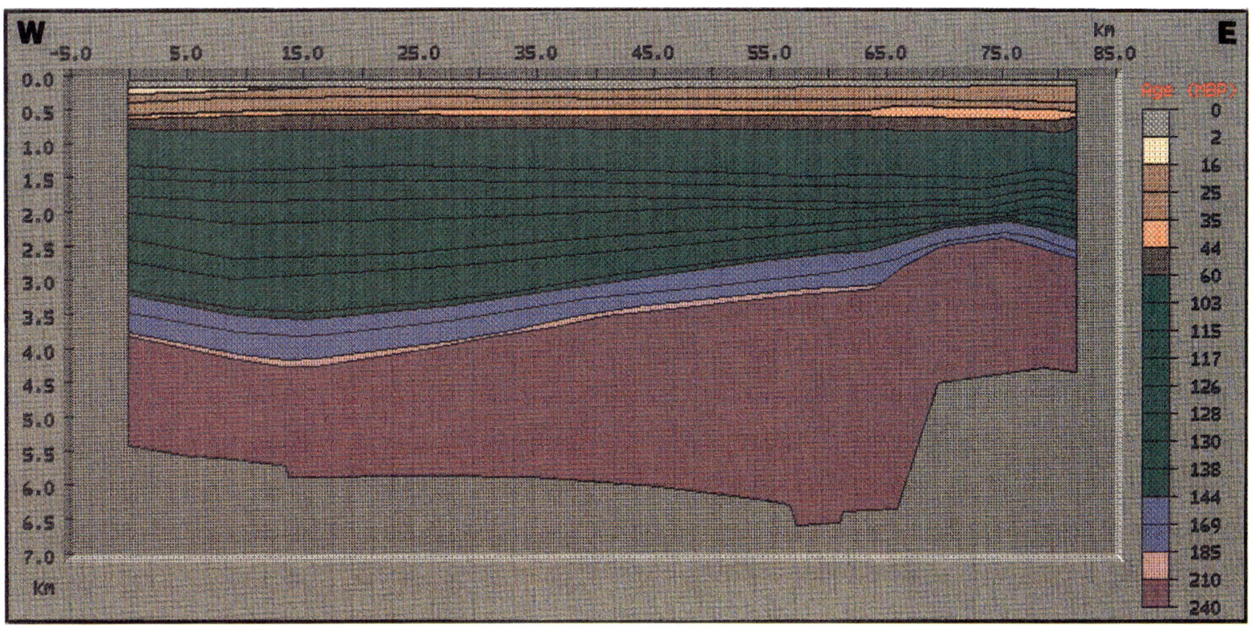

Fig. 2. Depth-converted seismic section of the Egersund Basin.

TABLE 1

Input data for reconstruction of thermal regime in the Egersund Basin.

1. Lithologies, water depths, surface porosities and exponential decay constants for the modelled formations

Formation	Age (m.y.)	Lithology	Water depth	Surface porosity	Exponential constant
1	0– 2	sandstone	100	0.49	0.27
2	2– 16	shale	100	0.50	0.40
3	16– 25	shale	120	0.63	0.51
4	26– 35	shale	150	0.50	0.40
5	35– 44	shale	150	0.63	0.51
6	44– 60	shale	200	0.50	0.40
7	60–103	shale	200	0.71	0.70
8	103–115	shale	120	0.63	0.51
9	115–117	shale	120	0.63	0.51
10	117–126	shale	110	0.63	0.51
11	126–128	shale	110	0.63	0.51
12	128–130	shale	100	0.63	0.51
13	130–138	shale	120	0.63	0.51
14	138–144	shale	120	0.63	0.51
15	144–169	shale/sandstone	120	0.50	0.40
16	169–185	shale/sandstone	20	0.50	0.40
17	185–210	shale	60	0.63	0.51
18	210–240	shale/sandstone	0	0.56	0.39

2. Inferred mineralogy of the formations given as percentages

Formation	Qtz	K-fsp	Plag	Mica	Mixed clay	Kaol	Calc	Dol	Sid	Pyr	TOC
1	50	10	10	10	10	10	0	0	0	0	0
2	30	9	9	8	10	30	1	0	0	2	2
3	20	8	8	7	10	40	2	0	0	2	3
4	30	8	9	10	8	30	2	0	0	1	2
5	20	10	10	10	9	35	2	0	0	1	3
6	30	8	8	10	9	30	2	0	0	1	2
7	15	1	2	1	2	2	75	1	1	0	0
8	20	8	8	9	10	35	4	1	0	2	3
9	20	8	8	9	10	35	4	1	0	2	3
10	20	7	7	8	8	35	10	1	0	1	3
11	20	8	8	8	8	35	10	0	0	1	2
12	20	10	10	10	10	35	2	0	0	1	2
13	20	10	10	10	10	35	2	0	0	1	2
14	20	9	8	9	10	35	2	0	0	1	6
15	40	9	9	10	8	20	2	0	0	1	1
16	40	10	10	7	10	20	1	0	0	1	1
17	20	9	10	10	10	35	2	0	0	1	3
18	50	12	12	5	5	13	2	0	0	0	1

tion of unknown thermal conductivities once the mineralogical composition and porosity of the rocks are given (Fjeldskaar et al., 1990; Prestholm et al., 1991; Fjeldskaar et al., this volume).

The present-day porosities are calculated under the assumption of an exponential decay of porosity with depth, which means that the porosity is derived from the surface porosity and a lithology-dependent exponential decay constant (Table 1; Fig. 3). The inferred mineralogical composition of the formations in the Egersund Basin is mainly based on data from well 9/2-1 (Table 1). The predictions show that the vertical thermal conductivities of the shale-rich formations lie between 1.3 and 2.0 W/m K, whereas the sand-rich formations have vertical conductivities up to 2.75 W/m K (Fig. 4). There are significant differences in the vertical (K_v) and horizontal (K_h) thermal conductivities, especially in shales, with an anisotropy (K_h/K_v) up to 1.4 (Fig. 5). There is a clear tendency towards increased thermal conductivity anisotropy as a function of depth, probably due to increased alignment of the sheet minerals as porosity decreases.

Present-day heatflow

When the thermal conductivities have been estimated, and the present-day ocean bottom temperature is known (here set to 5°C), the modelling of the present-day temperature regime depends on one unknown: the present-day heatflow. The present-day

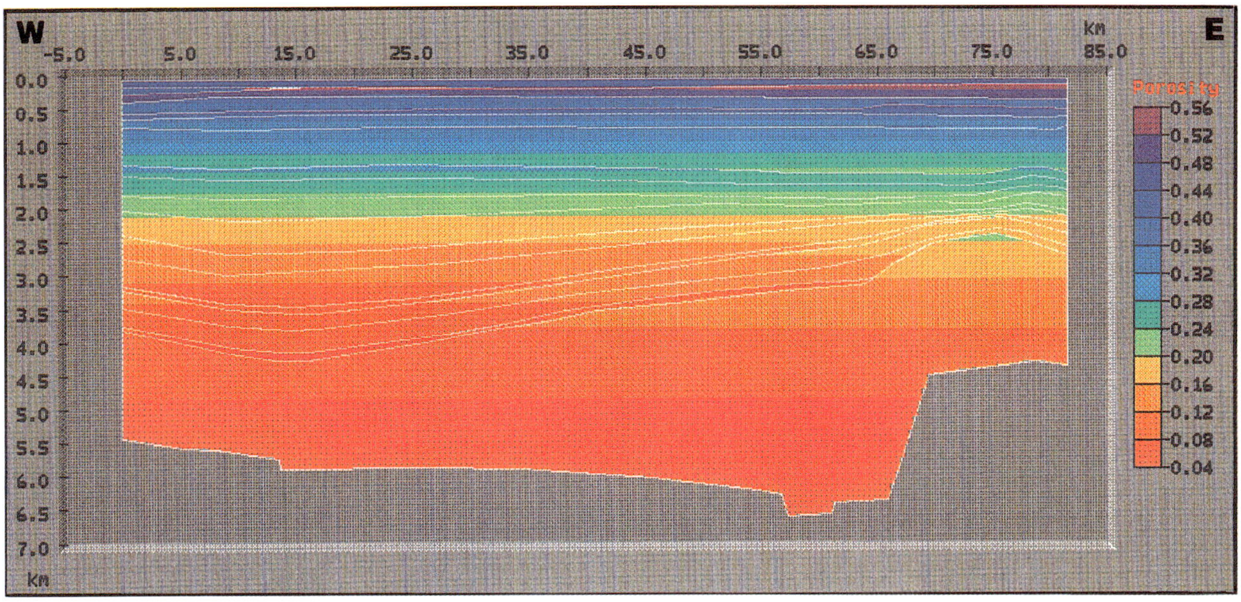

Fig. 3. Theoretical porosities in the Egersund Basin.

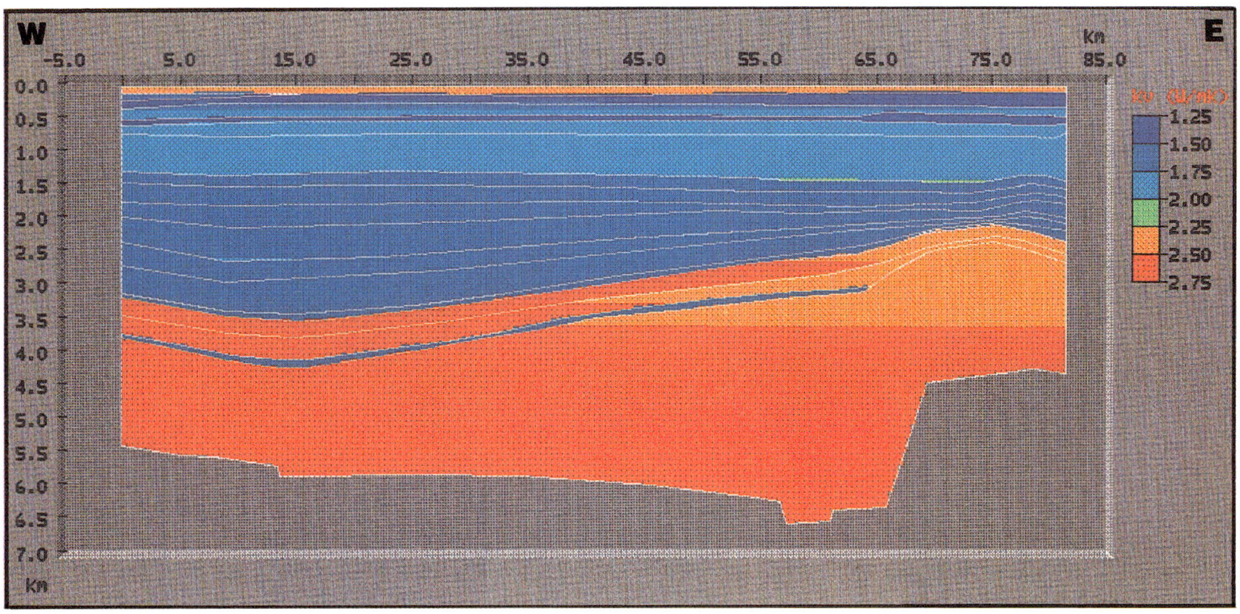

Fig. 4. Theoretical vertical thermal conductivities for the Egersund Basin predicted from the mineralogy and porosity of the sedimentary rocks.

heatflow is found by calibration of the theoretical temperature regime with observed temperatures from well 9/2-1. The calibration temperature is 100°C at 3000 m depth, which is in the lower range of the values given by Hermanrud et al. (1990). By tuning the present-day heatflow until there is a firm fit to the calibration temperature, we found a present-day heatflow of 52 mW/m^2 (Fig. 6).

Geohistory

The evolution of the sedimentary basin was reconstructed, taking into account the progressive compaction of the sediments, under the assumption of: (1) steady-state porosity/depth trends, i.e. the porosity of the sediments depending on their burial depth and lithology only; and (2) preservation of the volume of the solid matrix during compaction. Assuming an exponential decay of porosity with depth, the porosity is as mentioned above derived from the depth, the surface porosity and the exponential decay constant. The palaeo-water depth, which is an important parameter for the reconstruction of the basin evolution, is inferred from lithological and palynological evidence, and is assumed to be constant over the profile (Table 1).

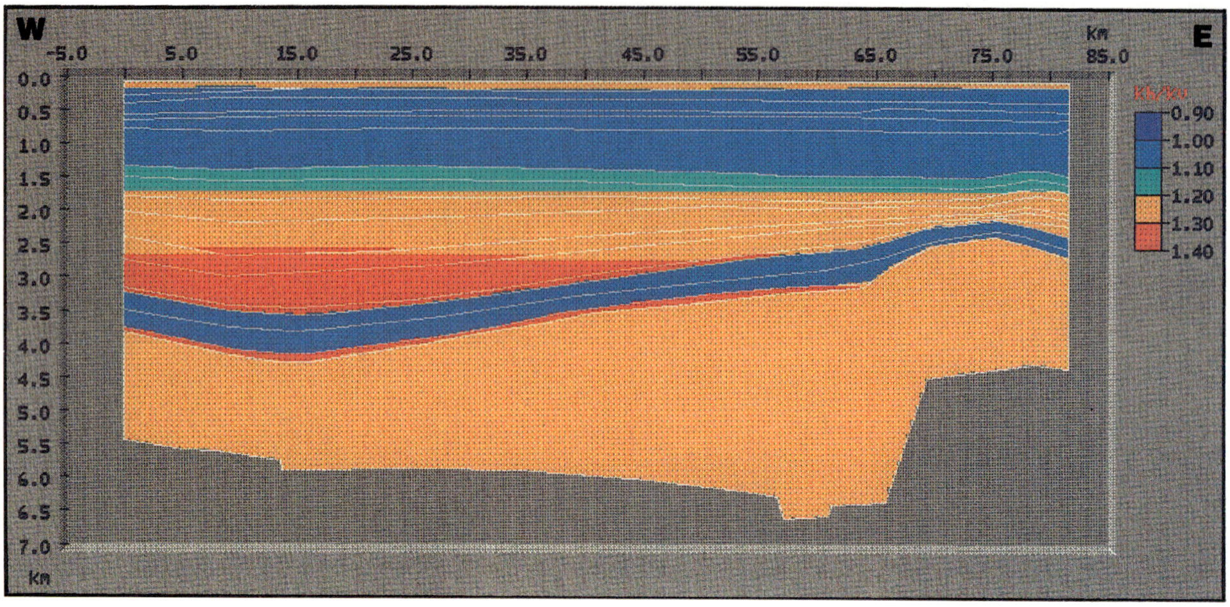

Fig. 5. Thermal conductivity anisotropy for the sedimentary rocks.

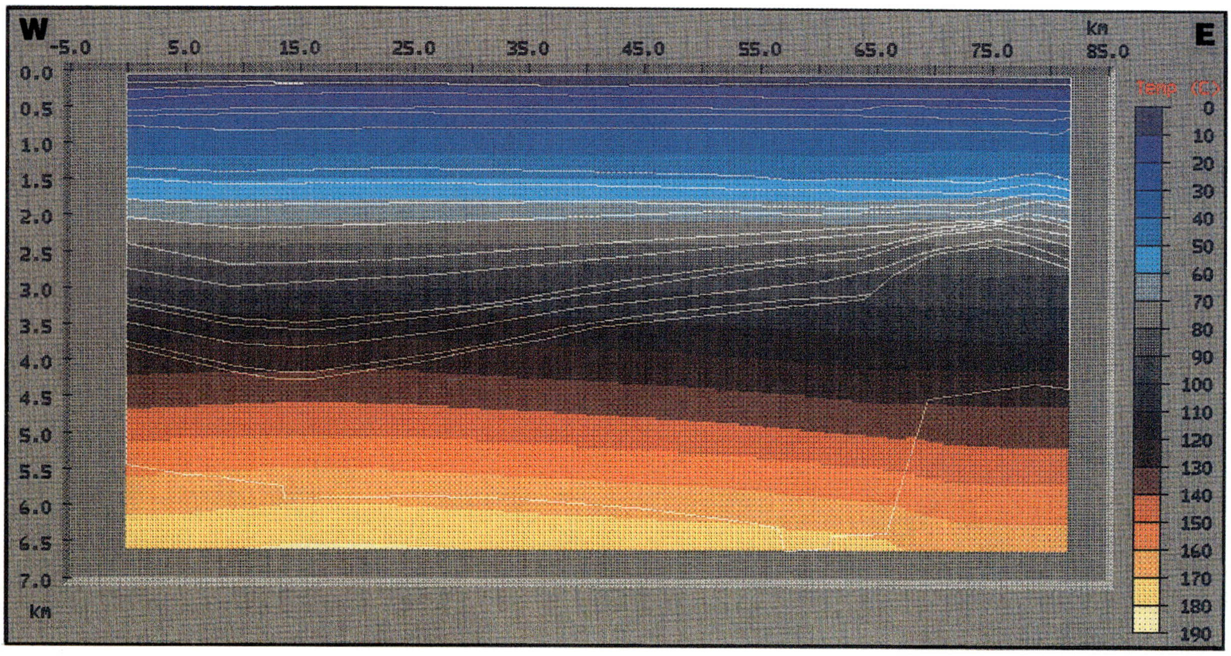

Fig. 6. Theoretical present temperatures in the Egersund Basin. The present-day ocean bottom temperature is assumed to be 5°C.

Palaeo-heatflow

The development of extensional sedimentary basins, like the Egersund Basin, is explained by two processes: (1) tectonic subsidence due to thinning of the lithosphere, and (2) isostatic subsidence due to sediment loading.

The tectonic subsidence, i.e. the subsidence a basin would undergo if it was not loaded with sediments, is quantified indirectly by the backstripping technique, which includes correction for the earth's response to sedimentary loads (Steckler and Watts, 1978). By comparison of the calculated tectonic subsidence with theoretical subsidence curves, a stretching factor may be found (McKenzie, 1978; Sclater and Christie, 1980). This is in turn used to estimate palaeo-heatflow.

Isostatic subsidence

The isostatic subsidence is calculated under the assumption that over short time periods the sedimentary loads will be partly balanced by the elastic strength of the lithosphere and partly by buoyant forces in the mantle. From data on post-glacial uplift it is clear that the asthenospheric viscosity is approximately 10^{19} Pa s, and the relaxation time

TABLE 2

Parameter values used in the calculations

Poisson's ratio	ν	$= 0.25$
Young's modulus	E	$= 10.3 \times 10^{10}$ N/m^2
Lamé's parameter	μ	$= 4.1 \times 10^{10}$ N/m^2
Asthenospheric viscosity	η_0	$= 10^{19}$ Pa s
Asthenospheric density	ρ	$= 3300$ kg/m^3
Asthenospheric temperature	T_1	$= 1380$°C
Lithospheric thickness	a	$= 90$ km
Lithospheric conductivity	K	$= 3.4$ W/m K
Thermal diffusivity	κ	$= 0.804$ mm^2/s
Thermal expansion coeff.	γ	$= 3.28 \times 10^{-5}$ °C^{-1}
Matrix density shale	ρ_{sh}	$= 2720$ kg/m^3
Matrix density sand	ρ_{sa}	$= 2650$ kg/m^3
Water density	ρ_w	$= 1000$ kg/m^3
Surface temperature	T_0	$= 0$°C

for the flexural response is less than ten thousand years (Fjeldskaar and Cathles, 1991). With respect to development of sedimentary basins the lithosphere flexure is assumed to be an instantaneous process. Over long time periods the viscous properties of the lithosphere will significantly affect the isostatic subsidence. In the limit ($t \rightarrow \infty$), the response approaches that of the Airy model of isostasy. The isostatic model is described in detail in Appendix A.

The isostatic subsidence history for the Egersund Basin was calculated under the assumption of an instantaneous deposition of each formation and standard matrix density values (Table 2). The theoretical isostatic subsidence history will be different for different models of the rheology of the lithosphere. The elastic thickness of the lithosphere will significantly affect the isostatic response, because loads of short wavelength will be supported by the lithosphere. The greater the elastic thickness, the more significant part of the load will be balanced by the lithosphere. We have first examined some cases for a pure elastic lithospheric (extremely high lithosphere viscosity, relative viscosity $\eta = 10^6$; other parameters given in Table 2). In the central part of the Egersund Basin the isostatic subsidence was up to 4.4 km, assuming a thin elastic lithosphere (Fig. 7). For elastic thickness varying between 0 and 10 km the resulting isostatic subsidence was not significantly different. The largest subsidence was for an elastic thickness of 5 km. An elastic thickness of more than 15 km gave considerably lower isostatic response; an elastic thickness of 40 km gave less than 3 km of isostatic subsidence in this part of the basin.

It has been shown in laboratory experiments that the viscosity of the lithosphere varies strongly with depth, the mean value of viscosity lies between 10^{23} Pa s and 10^{24} Pa s (Meissner and Kusznir, 1987). Isostatic calculations for the Egersund Basin were also run using a viscosity of 5×10^{23} Pa s ($\eta = 5 \times 10^4$); however, this does not change the general isostatic response considerably, except for the case of a thicker elastic lithosphere (Fig. 8).

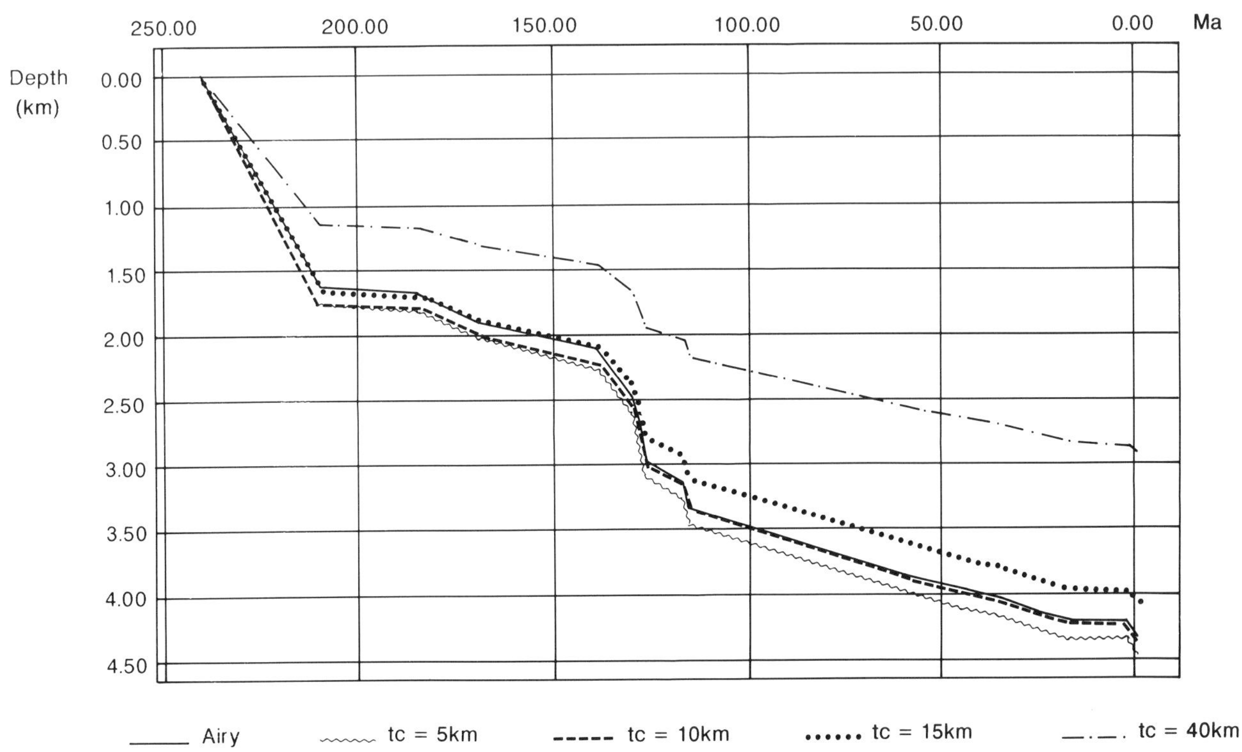

Fig. 7. Isostatic subsidence history of the central parts of the Egersund Basin, assuming an elastic lithosphere of various thicknesses.

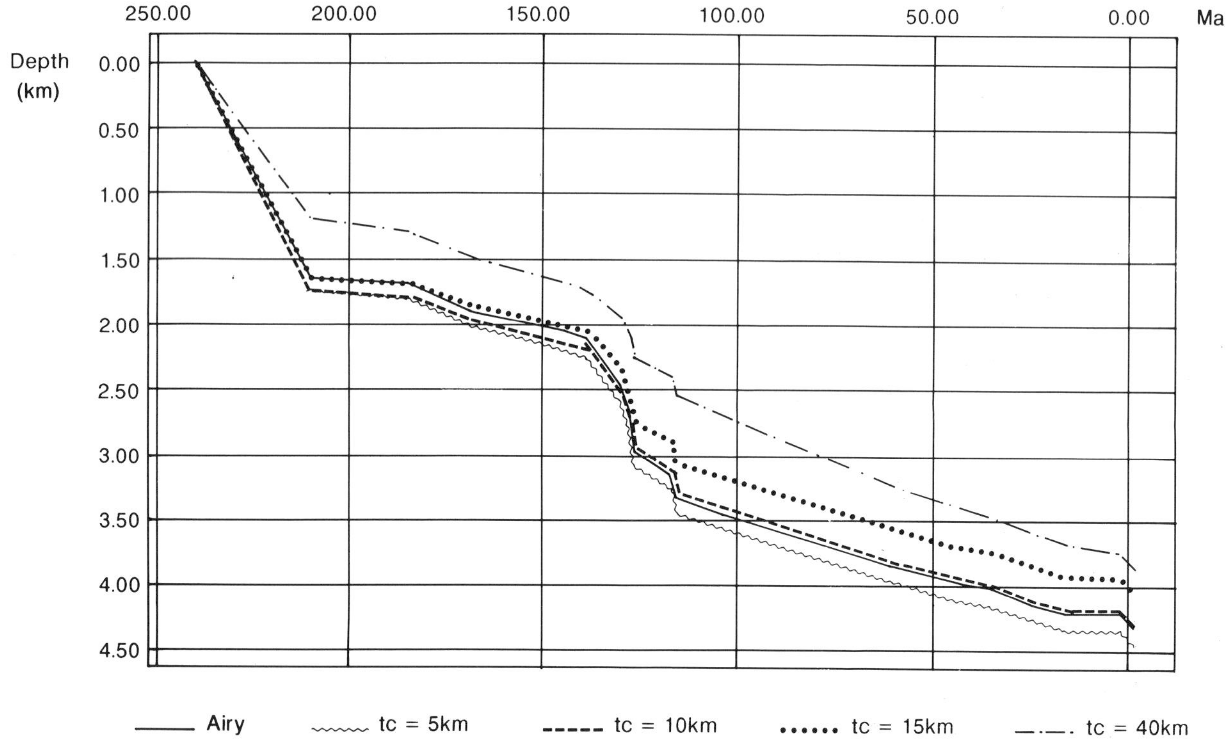

Fig. 8. Isostatic subsidence history of the central parts of the Egersund Basin, assuming a visco-elastic lithosphere, of various thicknesses, but all with viscosity 5.0 × 10^{23} Pa s.

Observed tectonic subsidence

The observed tectonic subsidence, which is the total subsidence minus the isostatic subsidence, was estimated by the backstripping technique described in Appendix B. Having obtained the observed tectonic subsidence curves for different locations within the basin, they are compared with theoretically deduced tectonic subsidence curves (described below), giving estimates of the stretching factors of the area, and the palaeo-heatflow.

Theoretical tectonic subsidence

Theoretical tectonic subsidence curves are commonly deduced using the "McKenzie model". McKenzie (1978) showed that sedimentary basins could form when the lithosphere is stretched, resulting in reduced crustal thickness and upwelling of hot mantle material. After the stretching event the surface will subside due to thermal contraction of the lithosphere.

The model for the theoretical tectonic subsidence used here is a two-dimensional non-uniform extensional model, implying that the crustal thinning and the lithosphere heating do not necessarily have a one-to-one relationship. Non-uniform (or modified) extension was first proposed by Royden and Keen (1980) and Sclater et al. (1980). The possible mechanisms leading to non-uniform extension are discussed in Royden (1986).

Extension of sedimentary basins can be divided into crustal stretching and sub-crustal stretching. The crustal stretching causes basin subsidence by isostasy, which is simulated by the isostatic model described above (Appendix C for details).

It seems to be widely accepted that the North Sea rift system has undergone two significant periods of lithospheric extension, one of Permo–Triassic and one of Late Jurassic age (Roberts et al., 1990). The observed basement subsidence of the Egersund Basin, in accordance with the general picture, shows two periods with significant increase in subsidence rates, one of Triassic age, the other of Jurassic age (Fig. 9). These are interpreted as effects of tectonic events of the area. The crustal stretching factors are, for both events, assumed to change over the profile to give a best fit with the observed tectonic subsidence. Both stretching events seem to involve the most significant crustal thinning in central parts of the Egersund Basin. Estimated stretching factor differences across a basin may, however, be somewhat artificial in areas of significant faulting. The reason is that faulting and tilting of blocks result in different tectonic subsidence profiles for closely adjacent points where the driving forces are nearly the same. To minimize this effect the centre of the basin (away from major faults) is chosen as the main modelling location. Over the Egersund Basin the mean values of the estimated crustal stretching

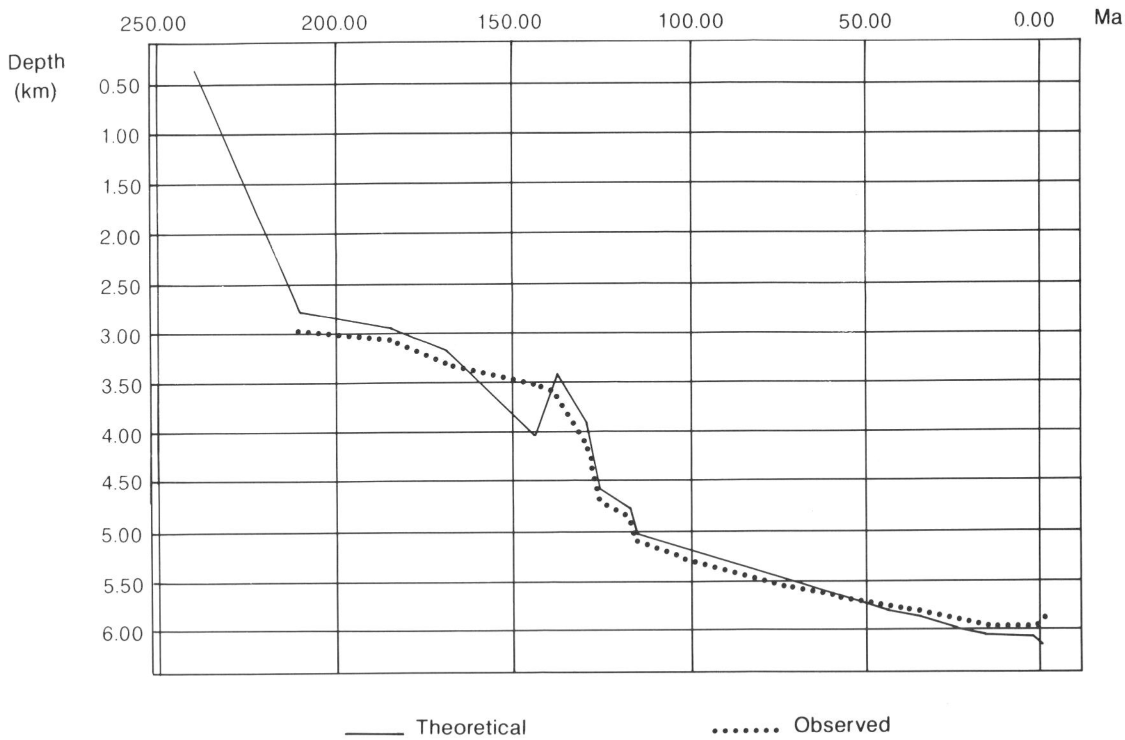

Fig. 9. Total subsidence versus total theoretical basement subsidence (isostatic plus tectonic subsidence) in central parts of the basin, assuming an elastic thickness of 10 km, and a lithosphere viscosity of 5.0×10^{23} Pa s.

factors are $\delta_1 = 1.30$ (Triassic age) and $\delta_2 = 1.25$ (Jurassic age).

The theoretical total cumulative thinning of an initial crustal thickness of approximately 35 km predicts a present crustal thickness in the area (22 km), which is slightly less than the present crustal thickness according to Barton and Wood (1984) (approximately 24 km), and the 28 km suggested by Ritter et al. (1987). The stretching factors given above are thus in the upper range of what could be justified by the observations. The observed present crustal thickness of the area puts a restriction on the modelled flexural rigidity, because the observed tectonic subsidence is greater for a high-rigidity lithosphere (or a large elastic thickness). For the Egersund Basin a large elastic thickness gives a higher observed tectonic subsidence than could be justified by the present crustal thickness. Calibration with the observed present crustal thickness in the area showed that the best-fitting rheological model has a lithosphere of elastic thickness less than 10 km (flexural rigidity less than 10^{22} N m).

The sub-crustal stretching, causing time-dependent movements due to cooling of the lithosphere towards thermal equilibrium, is treated by a transient analytical solution of the heat equation in two dimensions (Appendix C). The subsidence is caused by thermal contraction of the sub-crustal lithosphere. In the Egersund Basin the sub-crustal stretching factors are assumed to be $\beta_1 = 1.4$ and $\beta_2 = 1.3$, with a mean value over the entire basin roughly similar to the crustal stretching factors above (parameters given in Table 2).

Total theoretical basement subsidence (isostatic plus tectonic subsidence), assuming an elastic thickness of 10 km and the above stretching factors, gives a very satisfactory fit with the total observed subsidence in the area (Fig. 9). The calculated palaeoheatflow (Appendix C), based on the stretching factors above, shows two significant heating events, the first (240 Ma) up to more than 70 mW/m^2 and the second (140 Ma) a heatflow of more than 65 mW/m^2 (Fig. 10). As mentioned above, the crustal stretching factors are in the upper range of what can be justified by the present crustal thickness in the area. Accordingly, the estimated amount of heating is also assumed to be in the upper range of what is reasonable for this area.

Discussion

Factors that have not been considered are the uplift and erosion during the Tertiary, as suggested by Jensen and Schmidt (1990). The dating of the younger sequences in well 9/2-1 is rather uncertain. However, accepting the palaeontological dating that is available, several episodes of minor erosion have occurred in the Tertiary and one in Late Cretaceous. Shale compaction curves from well 9/2-1 suggest

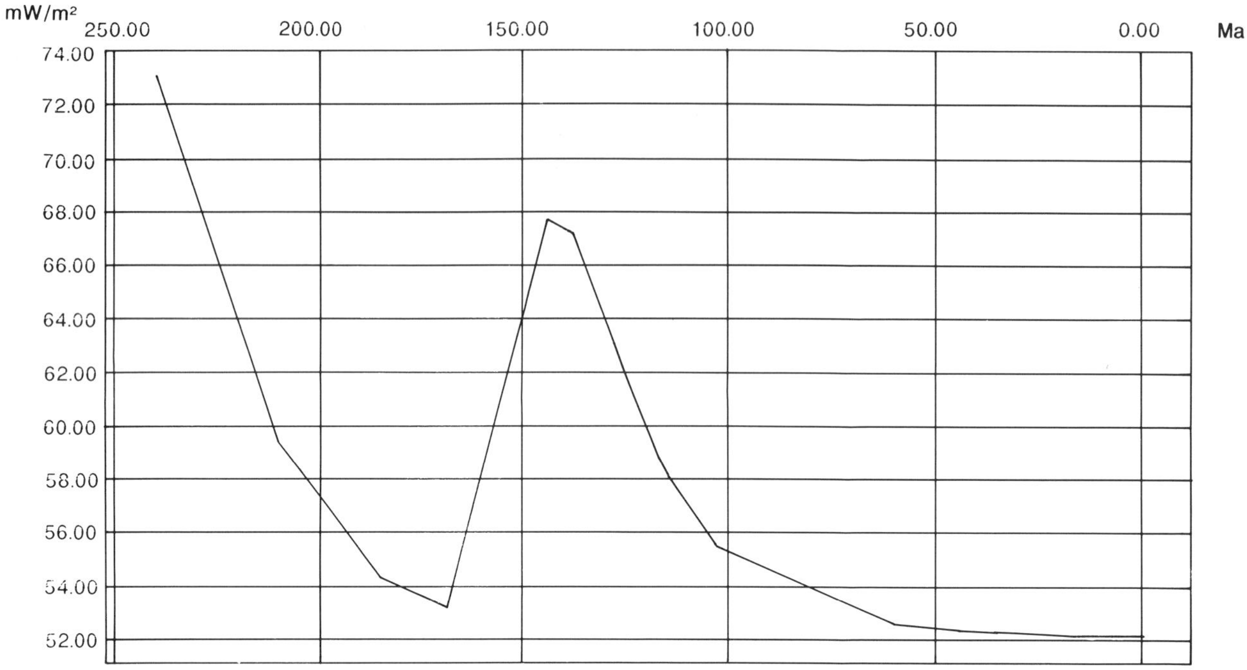

Fig. 10. Palaeoheatflow curve for central parts of the Egersund Basin.

that a total of approximately 600 m of erosion has taken place since the Early Cretaceous. Jensen and Schmidt (1992) have found the Late Tertiary erosion in this area to be 200–400 m, with iso-erosion contours parallel to the present coastline. This erosion may be related to glaciations in Pleistocene/Pliocene times (Riis and Fjeldskaar, 1992). The well stratigraphy and calculated sedimentation rates indicate that there might have been several erosional episodes during the Miocene/Oligocene. Some of these possible pulses of erosion may be related to local salt movements; however, they have not been considered in the modelling, as they have practically no impact on the more regional modelled tectonic subsidence.

The most important factor that has not been considered in the modelling is the impact of salt withdrawal. Salt withdrawal at the depocentre of the basin obviously affected the accommodation/sedimentation rates after deposition of the Late Jurassic Tau Formation. Salt withdrawal in the basin may have caused additional subsidence, which in the modelling is incorrectly ascribed to crustal thinning, therefore over-estimating the true value of crustal thinning.

Isostatic subsidence depends to a certain degree upon the assumed rheological model of the lithosphere. The approach used here to include the lithosphere in isostatic adjustment calculations was pioneered by Walcott (1970). Based on long-term isostatic adjustment of small-scale loads at various locations over the globe, Walcott concluded that the flexural rigidity of the lithosphere ranged from 5×10^{22} to 4×10^{23} N m, but the flexural rigidity over short time scales could be larger (6–9 $\times 10^{24}$ N m). In the present study the flexural rigidity of the lithosphere in the Egersund Basin seems to be well constrained by the present crustal thickness of the area. It is suggested that the flexural rigidity of the lithosphere is less than 10^{22} N m ($t_e \leq 10$ km). This is in accordance with the suggested values of Thorne and Watts (1989), which are in the range 10^{28}–10^{29} dyne cm^{-1} (10^{21}–10^{22} N m; $t_e \approx$ 5–10 km) for this area, and also with Barton and Wood (1984) who suggested a 5 km elastic lithosphere thickness for the North Sea. However, this is much lower than suggested from a rough analysis of the gravity anomalies in Fennoscandia; in the range of 1 to 5×10^{24} N m ($t_e \approx$ 50–90 km; Cathles, 1975).

One of the best sources of information on the physical properties of the earth's lithosphere and mantle is provided by data on post-glacial uplift. Based on a three-dimensional visco-elastic model, data on the last glaciation in northern Europe and data on the post-glacial uplift in Fennoscandia, Fjeldskaar and Cathles (1991) concluded the flexural rigidity to be equal to or less than 10^{24} N m ($t_e \leq 50$ km). They found some indications for a decreasing flexural rigidity towards the west; but, in any case, their value is close to two orders of magnitude larger than the suggested value for the Egersund Basin. The difference could be due to the following facts: (1) the post-glacial study gives an over-estimate of the flexural rigidity, possibly caused by an over-estimated ice thickness; (2) the flexural rigidity is decreasing drastically towards the west, from onshore into the North Sea; and (3) the flexural rigidity is time-dependent (as suggested by Walcott, 1970), i.e. the flexural rigidity is higher over a short time scale (a glacial cycle) than over a longer time-span (basin evolution). At the present we are not able to favour any one of these hypotheses. However, the crust and seismic lithosphere seem to decrease in thickness towards the west (Husebye et al., 1986), so it is reasonable to assume a certain decrease also in the elastic thickness.

Conclusion

The observed tectonic subsidence of the area shows two periods with significant increase of subsidence rate, probably caused by tectonic events; one of Triassic age, the other of Jurassic age. The first event gave a maximum heatflow of 73 mW/m^2, the other a maximum heatflow of 68 mW/m^2. The heatflow values are assumed to be maximum values, partly because salt withdrawal is not taken into account in the modelling. The present heatflow of the area is 52 mW/m^2.

The estimated amount of thinning depends indirectly on the flexural rigidity of the lithosphere. The present crustal thickness of the area puts restrictions on the flexural rigidity; an upper limit is estimated to be 10^{22} N m (elastic thickness $\leq$ 10 km). This is much lower than has previously been estimated for Scandinavia from post-glacial uplift, possibly indicating a significant decrease in the flexural rigidity towards the west, into the North Sea.

Acknowledgements

We thank one anonymous referee for constructive reviews of an earlier version of this paper. Thanks also to Åse Brit Tunheim who drew some of the figures.

Appendix A — Isostatic subsidence

Subsidence by flexure

If a load is applied to the elastic lithosphere covering the asthenosphere, part of the applied load will be supported by the lithosphere, part by the buoyant forces of the asthenosphere beneath acting through the lithosphere. Loads of short wavelength are supported by the lithosphere so that the lithosphere acts as a lowpass filter. The characteristics of this filter depend on the elastic strength of the lithosphere. The elastic strength of the lithosphere can be expressed by a parameter called the flexural rigidity (D). The relationship between the elastic thickness t_e of the lithosphere and the flexural rigidity D is given by the formula:

$$D = \frac{E\,t_e^3}{12\,(1-\nu^2)} \tag{A1}$$

where ν is Poisson's ratio and E is Young's modulus.

For a flat Earth approximation (which is appropriate for calculations concerning sedimentary basins) the regional isostatic compensation by flexure $s(k)$ due to a harmonic load $f(k)$ is achieved by subsidence (Cathles, 1975):

$$s(k) = \frac{f(k)}{\rho g\,\alpha(k)} \tag{A2}$$

where ρ is the density of the asthenosphere, g is the gravity, k is the wavenumber and $\alpha(k)$ is the "lithosphere filter":

$$\alpha(k) = \frac{\frac{2\mu k}{\rho g}(Sh^2 - k^2 t_e{}^2) + (Ch\,Sh + k t_e^2)}{Sh + k t_e Ch} \tag{A3}$$

where μ is Lame's parameter, $Sh = \sinh k t_e$, $Ch = \cosh k t_e$.

Subsidence by viscous flow

The rate of viscous subsidence into the lithosphere is different for different wavelengths of load. The subsidence rate may be expressed by the relaxation time, which is the time for an exponentially decaying function to reduce to $1/e$ of its initial amplitude.

If the viscous flow is assumed to be restricted to a lithosphere of uniform viscosity and of thickness a (in this case $a = t_e$), the relaxation time $\tau(k)$ for the subsidence is (Cathles, 1975):

$$\tau(k) = \tau_0(k) z \tag{A4}$$

where

$$\zeta = \frac{\eta\,(Ch^2 + k^2 a^2)}{Ch\,Sh - ka} \tag{A5}$$

η is the ratio of the lithospheric viscosity to the asthenospheric viscosity, $Sh = \sinh ka$, $Ch = \cosh ka$, and:

$$\tau_0(k) = \frac{2\eta_0 k}{\rho g} \tag{A6}$$

where η_0 is the asthenospheric viscosity.

Total isostatic subsidence

The modelled isostatic subsidence is the sum of the immediate flexural response followed by the time-dependent viscous subsidence. In the frequency domain the response is:

$$s(k,t) = \frac{p(k)\rho_s}{\rho\alpha(k)}\left\{1 + [\alpha(k) - 1]\,e^{-t/\tau(k)}\right\} \qquad (A7)$$

where ρ_s is the sediment density and $p(k)$ is the Fourier transformed sediment thickness.

The isostatic deflection is the inverse Fourier transform (F^{-1}) of $s(k, t)$:

$$S(x,t) = F^{-1}[s(k, t)] \qquad (A8)$$

Appendix B — Backstripping model

Backstripping involves the removal of sediment and water loads. The result is the calculation of the depth to basement in the absence of sediment loads. This is the so-called tectonic subsidence Y, which is the difference between the total decompacted sediment thickness (P) and the Earth's response to this load (S). Under the assumption of no change of palaeo-water depth the observed tectonic subsidence is given by:

$$Y(x,t) = P(x,t) - S(x,t) \qquad (B1)$$

The observed tectonic subsidence is then (Fjeldskaar and Pallesen, 1989):

$$Y(x,t) = F^{-1}\left[p(k)\left(1 - \frac{\rho_s}{\rho}\left\{\alpha(k)^{-1} + [1 - \alpha(k)^{-1}]\,e^{-t/\tau(k)}\right\}\right)\right] \qquad (B2)$$

Appendix C — Crustal and sub-crustal stretching

Crustal stretching

The crustal thinning generates an isostatic load, which (assuming Airy compensation) is:

$$S_i{}^0(x) = t_c\,\frac{\rho - \rho_c}{\rho} - \left(1 - \frac{1}{\delta(x)}\right) \qquad (C1)$$

where t_c is the crustal thickness, $\delta(x)$ is the crustal stretching factor and ρ_c is the crustal density. The regional isostatic response is calculated from $S_i^0(x)$ by applying the lithospheric filter given in eq. A1.

Sub-crustal stretching

The thermal movements, caused by the sub-crustal stretching, is approximated by a transient analytical solution of the heat equation. The same solution is used to obtain the surface heat flux over time $Q(x,t)$. The method is a generalization to two dimensions (x and z) of the method given in one dimension (z) by McKenzie (1978). The model used here is described in Fjeldskaar et al. (1990), which we follow closely here.

Assuming a lithosphere (of thickness a) with constant diffusivity, no heat production in the computational region, and K (thermal conductivity of the lithosphere) independent of depth z, the heat equation has the form:

$$T(x,z,t) = T_0 + \Delta T\left[1 - \frac{z}{a} + \frac{4\beta}{\pi^3} \times \left(\frac{\pi W}{2L} + \sum_{m=1}^{\infty}\frac{\sin\frac{m\pi W}{L}}{m}\cos\frac{2m\pi x}{L}\,e^{-4m^2\pi^2\kappa t/L^2}\right) \times \left(\sum_{n=1}^{\infty}\frac{\sin\frac{n\pi}{1-1/\beta}}{n^2}\sin\frac{n\pi z}{a}\,e^{-n^2\pi^2\kappa t/a^2}\right)\right] \qquad (C2)$$

where β is the sub-crustal stretching factor, W is the extensional width of the sub-crustal lithosphere, L is the width of the basin and κ is the diffusivity.

The corresponding heat flux is given by:

$$Q(x,t) = \frac{K\Delta T}{a}\left[1 + \frac{4\beta}{\pi^2} \times \left(\frac{\pi W}{2L} + \sum_{m=1}^{\infty}\frac{\sin\frac{m\pi W}{L}}{m}\cos\frac{2m\pi x}{L}\,e^{-4m^2\pi^2\kappa t/L^2}\right) \times \left(\sum_{n=1}^{\infty}\frac{\sin\frac{n\pi}{\beta}}{n}\,e^{-n^2\pi^2\kappa t/a^2}\right)\right] \qquad (C3)$$

Following McKenzie (1978) and Jarvis and McKenzie (1980) the subsidence due to thermal contraction in terms of $e(x,t)$ is determined. $e(x,t)$ is defined as the elevation of the upper surface of the lithosphere relative to that of $t \to \infty$.

The expression for $e(x,t)$ is obtained by equating the pressure at points on the x-axis at all times $t \geq 0$. Under the assumption that the temperature in the crust is so close to T_0 that t_c (the thickness of the crust) may be considered constant for $t \geq 0$, we obtain the following expression for $e(x,t)$:

$$e(x,t) = \frac{8a \cdot \gamma \cdot \beta \cdot \Delta T}{\pi^4}$$

$$\times \left(\frac{\pi W}{2L} + \sum_{m=1}^{\infty} \frac{\sin \frac{m\pi W}{L}}{m} \cos \frac{2m\pi x}{L} \, \mathrm{e}^{-4m^2\pi^2\kappa t/L^2} \right)$$

$$\times \left(\sum_{n=1}^{\infty} \frac{\sin \frac{(2n+1)\pi}{\beta}}{(2n+1)^3} \, \mathrm{e}^{-(2n+1)^2\pi^2\kappa t/a^2} \right) \quad \text{(C4)}$$

where γ is the thermal expansion coefficient.

The subsidence due to conduction, $S_q(x,t)$ is given by:

$$S_q(x,t) = e(x,0) - e(x,t) \quad \text{(C5)}$$

In the case of two stretching phases there will be two periods of subsidence due to conduction, S_{q1} and S_{q2}. The total subsidence due to conduction is then $S_{q1}(x,t) + S_{q2}(x,t)$.

References

Barton, P. and Wood, R., 1984. Tectonic evolution of the North Sea basin: crustal stretching and subsidence. Geophys. J. R. Astron. Soc., 79: 987–1022.

Cathles, L., 1975. The Viscosity of the Earth's Mantle. Princeton University Press, Princeton, N.J., 386 pp.

Fjeldskaar, W. and Cathles, L., 1991. Rheology of mantle and lithosphere inferred from post-glacial uplift in Fennoscandia. In: R. Sabadini, K. Lambeck and E. Boschi (Editors), Glacial Isostasy, Sea-Level and Mantle Rheology. Kluwer, Dordrecht, pp. 1–19.

Fjeldskaar, W. and Pallesen, S., 1989. The application of a viscoelastic lithosphere model to isostatic subsidence in back-stripping In: J. Collinson (Editor), Correlation in Hydrocarbon Exploration. Norwegian Petroleum Society, Graham and Trotman, London, pp. 53–59.

Fjeldskaar, W., Mykkeltveit, J., Christie, O.H.J., Johansen, H., Langfeldt, J.M., Tyvand, P., Skurve, O. and Bjørkum, P.A., 1990. Interactive 2D Basin Modelling on Workstations. Proc. SPE 20350 Petroleum Computer Conference, Denver, Colo., pp. 181–196.

Fjeldskaar, W., Prestholm, E., Guargena, C. and Stephenson, M., 1993. Mineralogical and diagenetic control on the thermal conductivity of the sedimentary sequences in the Bjørnøya Basin, Barents Sea. In: A.G. Doré et al. (Editors), Basin Modelling: Advances and Applications. Norwegian Petroleum Society (NPF), Special Publication 3. Elsevier, Amsterdam, pp. 445–453 (this volume).

Hermanrud, C., 1986. On the importance to the petroleum generation of heating effects from compaction-derived water: an example from the North Sea. In: J. Burrus (Editor), Thermal Modelling of Sedimentary Basins. Technip, Paris, pp. 247–269.

Hermanrud, C., Eggen, S., Jacobsen, T., Carlsen, E.M. and Pallesen, S., 1990. On the accuracy of modelling hydrocarbon generation and migration: the Egersund Basin oil find, Norway. Advances in Organic Geochemistry 1990. Org. Geochem., 16: 389–399.

Husebye, E.S., Hovland, J., Christoffersen, A., Åstrøm, K., Slunga, R. and Lund, C.-E., 1986. Tomographical mapping of the lithosphere and asthenosphere beneath southern Scandinavia and adjacent areas. In: D.A. Galson and St. Muller (Editors), The European Geotravers, Part 2. Tectonophysics, 128: 229–250.

Jarvis, G.T. and McKenzie, D.P., 1980. Sedimentary basin formation with finite extension rates. Earth Planet. Sci. Lett., 48: 42–52.

Jensen, L.N. and Schmidt, B.J., 1992. Late Tertiary uplift and erosion in the Skagerrak area; magnitude and consequences. Nor. Geol. Tidsskr., 72: 275–279.

Lervik, K.S., Spencer, A.M. and Warrington, G., 1989. Outline of Triassic stratigraphy and structure in the central and northern North Sea. In: J. Collinson (Editor), Correlation in Hydrocarbon Exploration. Norwegian Petroleum Society, Graham and Trotman, London, pp. 173–189.

McKenzie, D., 1978. Some remarks on the development of sedimentary basins. Earth Planet. Sci. Lett., 40: 25–32.

Meissner, R. and Kusznir, N.J., 1987. Crustal viscosity and the reflectivity of the lower crust. Ann. Geophys., 5B, 4: 365–374.

Pegrum, R.M., 1984. The extension of the Tornquist Zone in the Norwegian North Sea. Nor. Geol. Tidsskr., 64: 39–68.

Prestholm, E., Fjeldskaar, W., Balling, N. and Michelsen, J.K., 1991. Thermal conductivity anisotropy in sedimentary rocks, and its effect on the evaluation on petroleum potential. EAPG Meet., Florence (abstract).

Riis, F. and Fjeldskaar, W., 1992. On the magnitude of the Late Tertiary and Quaternary erosion and its significance for the uplift of Scandinavia and the Barents Sea. In: R.M. Larsen, H. Brekke, B.T. Larsen and E. Talleraas (Editors), Structural and Tectonic Modelling and its Application to Petroleum Geology. Norwegian Petroleum Society (NPF), Special Publication 1. Elsevier, Amsterdam, pp. 163–185.

Ritter, U., 1988. Modelling of hydrocarbon generation patterns in the Egersund Sub-Basin, North Sea. Advances in Organic Geochemistry 1987. Org. Geochem., 13: 165–174.

Ritter, U., Leith, T.L., Griffiths, C.M. and Schou, L., 1987. Hydrocarbon generation and thermal evolution in parts of the Egersund Basin, Northern North Sea. In: C. Beaumont and A.J. Tankards (Editors), Sedimentary Basins and Basin-Forming Mechanisms. Can. Soc. Pet. Geol., Mem., 12: 75–85.

Roberts, A.M., Yielding, G. and Badley, M.E., 1990. A kinetic model for the orthogonal opening of the Late Jurassic North Sea rift system, Denmark–Mid Norway. In: D.J. Blundell and A.D. Gibbs (Editors), Tectonic Evolution of the North Sea Rifts. Oxford University Press, London, pp. 180–199.

Royden, L., 1986. A simple method for analyzing subsidence and heat flow in extensional basins. In: J. Burrus (Editor), Thermal Modelling of Sedimentary Basins. Technip, Paris, pp. 49–73.

Royden, L. and Keen, C.E., 1980. Rifting process and thermal evolution of the continental margin of eastern Canada determined from subsidence curves. Earth Planet. Sci. Lett., 51: 343–361.

Sclater, J.G. and Christie, P.A.F., 1980. Continental stretching: an explanation of the post-mid-Cretaceous subsidence of the Central North Sea basin. J. Geophys. Res., 85: 3711–3739.

Sclater, J.G., Royden, L., Horvath, F., Burchfield, B.C., Semken, S. and Stegena, L., 1980. The formation of the intra-Carpathian basins as determined from subsidence data. Earth Planet. Sci. Lett., 51: 139–162.

Skjerven, J., Riis, F. and Kalheim, J.E., 1983. Late Paleozoic to Early Cenozoic structural development of the south-southeastern Norwegian North Sea. In: J.P.H. Kaasschieter and T.J.A. Reijers (Editors), Petroleum Geology of the Southeastern North Sea and the Adjacent Onshore Areas. Geol. Mijnbouw, 62: 35–45.

Steckler M.S. and Watts, A.B., 1978. Subsidence of the Atlantic-type continental margin off New York. Earth Planet. Sci. Lett., 41: 1–13.

Taylor, J.C.M., 1984. Late Permian–Zechstein. In: K.W. Glennie (Editor), Introduction to the Petroleum Geology of the North Sea. Blackwell, Oxford, pp. 87–111.

Thorne, J.A. and Watts, A.B., 1989. Quantitative analysis of North Sea subsidence. Am. Assoc. Pet. Geol., Bull., 73: 88–116.

Walcott, R.I., 1970. Flexural rigidity, thickness, and viscosity of the lithosphere. J. Geophys. Res., 75: 3941–3954.

WILLY FJELDSKAAR *Rogaland Research, P.O. Box 2503, Ullandhaug, N-4004 Stavanger, Norway*
EDWARD PRESTHOLM *Rogaland Research, P.O. Box 2503, Ullandhaug, N-4004 Stavanger, Norway*
CLAUDIA GUARGENA *Enterprise Oil Norge, P.O. Box 399, N-4001 Stavanger, Norway*
NINA GRAVDAL *Enterprise Oil Norge, P.O. Box 399, N-4001 Stavanger, Norway*

The importance of kinetic parameters in modelling generation by cracking of oil to gas — a case study in 1D from well 2/4-14

Pål Andresen, Nigel Mills, H.-J. Schenk and Brian Horsfield

An important reason for conducting basin modelling studies is to use the temperature history to estimate the generation of hydrocarbons in space and time. This is most commonly done using the kinetic approach of Tissot and Espitalié (1975). This approach consists of using the Arrhenius equation for a first-order reaction for several parallel reactions (chosen to represent the formation of "oil" from primary cracking of kerogen) with different activation energies and frequency factors. A limitation to this approach, certainly in many commercially available modelling software packages, has been the modelling/prediction of gas generation. The general assumption is that the oil remains in the source rock and is subsequently cracked to gas at higher temperatures and pressures. Cracking to gas has historically been approached using one reaction with one set of kinetic parameters.

One purpose of this paper is to show the great importance of the kinetic constants in the Arrhenius equation (A and E) when trying to model the cracking of oil to gas. To illustrate this, Norwegian well 2/4-14 was chosen as an example. This well, which is situated in the Norwegian Central Graben just north of the Ekofisk field, is well suited to such a study as it was drilled to approximately 4700 m depth where it encountered hydrocarbons under high pressure and temperature conditions. The modelling of this well demonstrates that small variations in the constants used, particularly the activation energy, result in large variations in the predicted hydrocarbon generation histories (specifically the stability of oil/generation of gas).

The second purpose of this paper is to present new and hopefully more realistic values for the kinetic parameters controlling the cracking of oil to gas. To generate a new and more representative set of kinetic parameters, an experimental laboratory study involving micro-scale sealed-vessel (mssv) pyrolysis of oil was undertaken. The results from this study suggest that cracking of oil to gas may be treated as a series of parallel first-order reactions, in the same manner as primary generation of petroleum.

Introduction

The Central Graben is a proven hydrocarbon province situated in the southern part of the Norwegian continental shelf, bordering the offshore exploration waters of Denmark in the southeast, and the UK in the west/southwest. An outline of the structural geology of this area can be found in Gowers and Sæbøe (1985). The majority of accumulations in the area are reservoired in the Cretaceous chalk sequence and several of these fields are described in D'Heur (1986) and Sørensen et al. (1986). Other (Jurassic) accumulations are described in Spencer et al. (1986). Figure 1 shows the general location of the study area, and gives an overview of the producing fields, including Ekofisk, the second largest field in Norwegian waters. The northern part of block 2/4 was relinquished by the Phillips Petroleum licence group, and subsequently awarded to licence PL 146 with Saga Petroleum as operator in concession round 12A in 1988. The first well drilled by Saga in this licence, 2/4-14, forms the basis of the input data for this study.

The model

To perform computerized basin modelling, we used a programme developed by Gesellschaft für Integrierte Explorations Systeme (IES) in Jülich, Germany. The methodology behind the programme has been described by Welte and Yukler (1981), Welte and Yalçin (1986) and Yalçin (1991). Examples of case studies applying this software package are given by Mo et al. (1989) and Wygrala (1988).

The programme requires an input model consisting of a number of layers or events with a defined temporal and spatial extent, which are then assigned values for the seven types of input data required. These input data comprise lithology, porosity, thickness and a deposition/non-deposition/erosion flag for each layer, plus estimates of waterdepth, sediment/water interface temperature and heatflow through time. The conceptual model comprising all these data is then subjected to an optimizing procedure to obtain a calibrated temperature history. The optimizing procedure involves matching observed and modelled thicknesses by adjusting present-day porosities,

Basin Modelling: Advances and Applications edited by A.G. Doré et al. NPF Special Publication 3, pp. 563–571, Elsevier, Amsterdam.

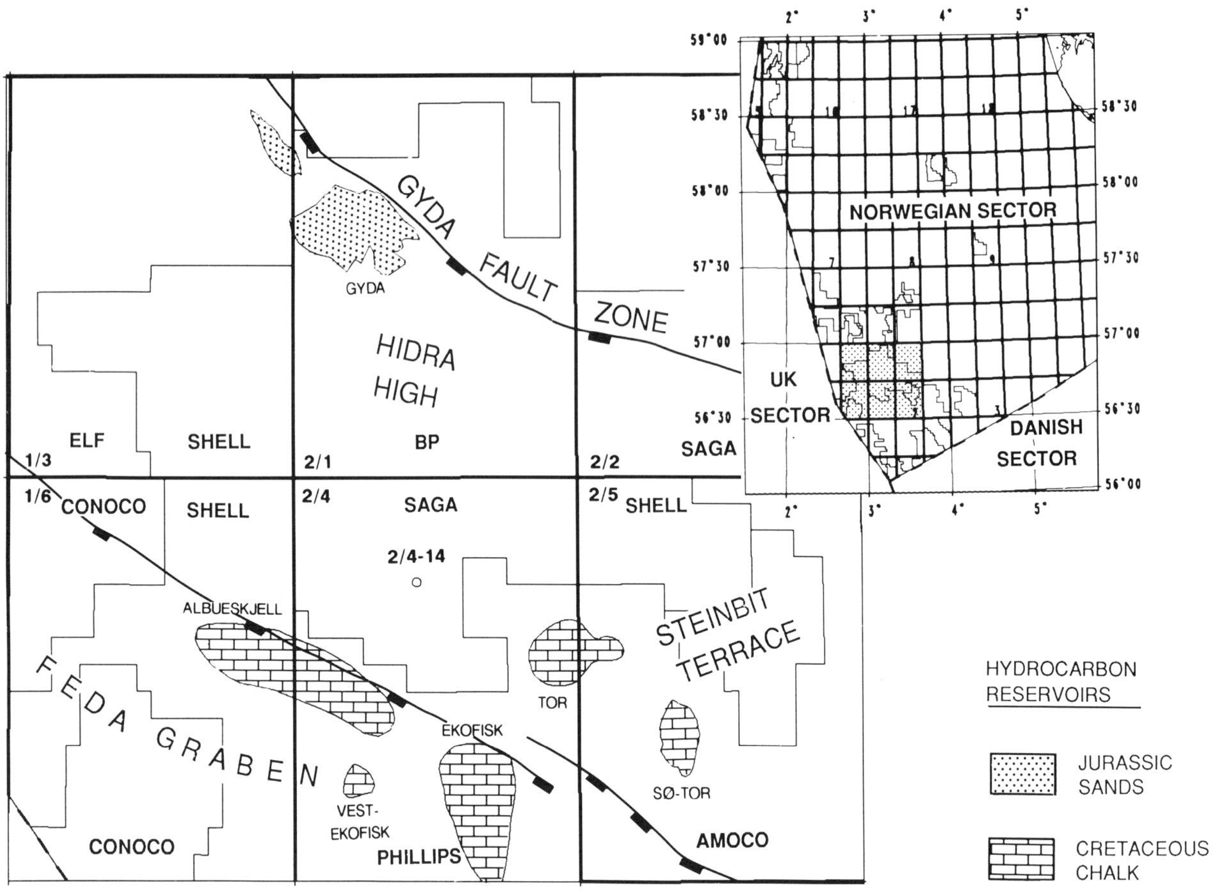

Fig. 1. Map of study area, showing an outline of structural elements, blocks, fields, licences and operators in the vicinity of the studied well 2/4-14.

and constructing a heatflow history by calibrating to various thermal indicators, such as vitrinite reflectance, hopane/sterane isomerization and sterane aromatization. The programme utilizes a forward method for modelling, and no regression/inversion techniques are involved. For well 2/4-14 the initial work was performed prior to drilling in order to predict the hydrocarbon phase that could be expected in the reservoir. Thus, thermal indicator data were not available, and a general heatflow model based on observations of rifting phases in the area, observed present-day temperatures and thermal indicator data in other nearby wells, was applied. An example of the calibrated model results for temperature and vitrinite reflectance, together with observed data points for a well near 2/4-14, is shown in Fig. 2. The calibrated temperature history constructed for a well can be used as input in other calculations, such as that for hydrocarbon generation based on a kinetic approach.

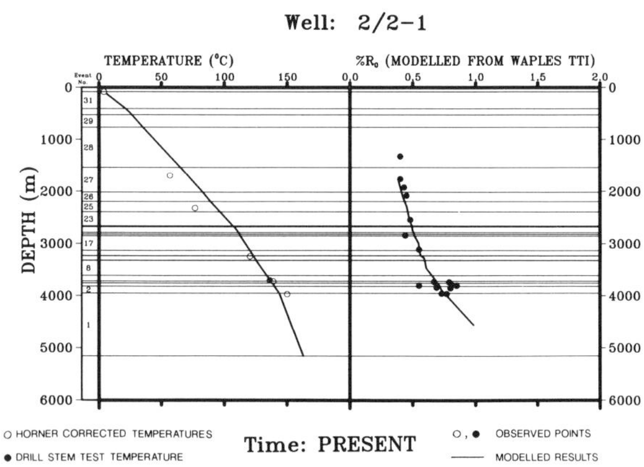

Fig. 2. Comparison plot of modelled results and observed values for the two parameters temperature and vitrinite reflectance, for a well near 2/4-14. (Constituting one of the wells on which the heatflow history was based.)

The conceptual model for well 2/4-14, is summarized in Fig. 3. The well has been subdivided into 38 events extending from the Permian (248 Ma) to the present day (0 Ma), with absolute ages based on the timescale of Harland et al. (1982). This timescale has been revised, Harland et al. (1989), but the new revised version was not available at the time of the study. Lithologically the well contains mostly sands, shales and mixtures of the two, except for the Late Permian Zechstein salt and the approximately one kilometre thick Chalk Group. A long non-depositional event in Late Triassic/Early Juras-

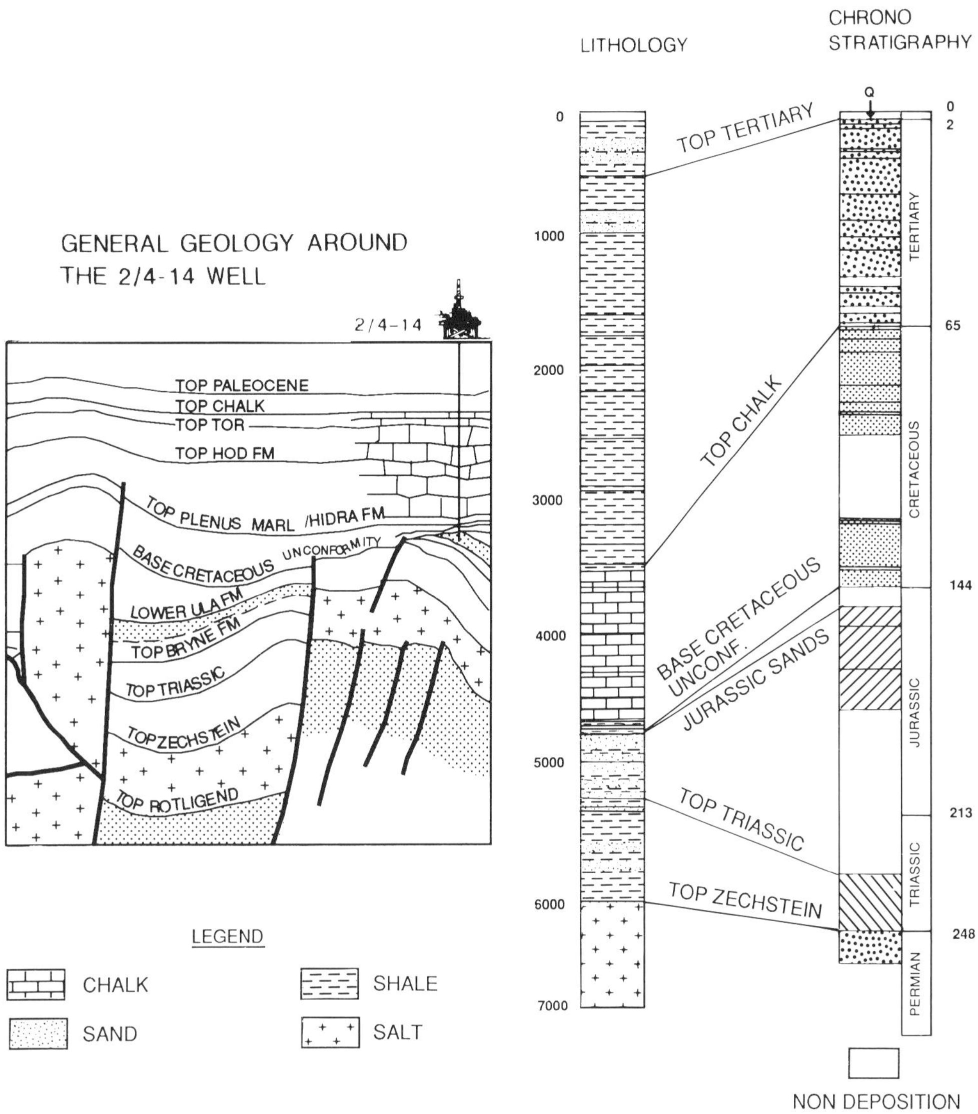

Fig. 3. Geology of the area around 2/4-14, and an outline of the conceptual model used in the study.

sic time, and a few shorter events of non-deposition/slight erosion at a later time, are modelled for this well. No major erosion (in the order of hundreds of metres) has been observed in this area. The heatflow history, which was obtained by forward modelling techniques based on measurements from nearby wells, shows a peak of 1.45 HFU in the Jurassic, decreasing to a constant value of 1.15 HFU in the Cretaceous and Tertiary, with an increase to 1.30 HFU in the Quaternary to match measured present-day temperatures.

Results of the simulation

Based on the input data of the conceptual model described above, several simulation runs were performed, optimizing the modelled thicknesses to match the observed values in the well. The calibrated and optimized results of the simulation were then used as a basic temperature history to evaluate the sensitivity of the cracking kinetic parameters.

The temperature history is shown in Fig. 4, versus depth for the entire well, and versus time for the basal layer, Zechstein. The figure shows that the temperature gradient is higher in the upper part of the well than in the lower part, with an average of 32°C/km for the entire well, and as much as 37°C/km for the upper three kilometres. This is due to the varying thermal properties of different lithologies at different burial depths. The plot versus time, covering the last 50 m.y., shows an increase from 100°C to over 200°C during this interval, for the Zechstein layer. This profile suggests that had this layer been a source rock, temperatures high enough for significant hydrocarbon generation would only have been reached in the last 50 m.y. in this area.

Also included in Fig. 4 is a plot versus depth and time for a temperature history based on the

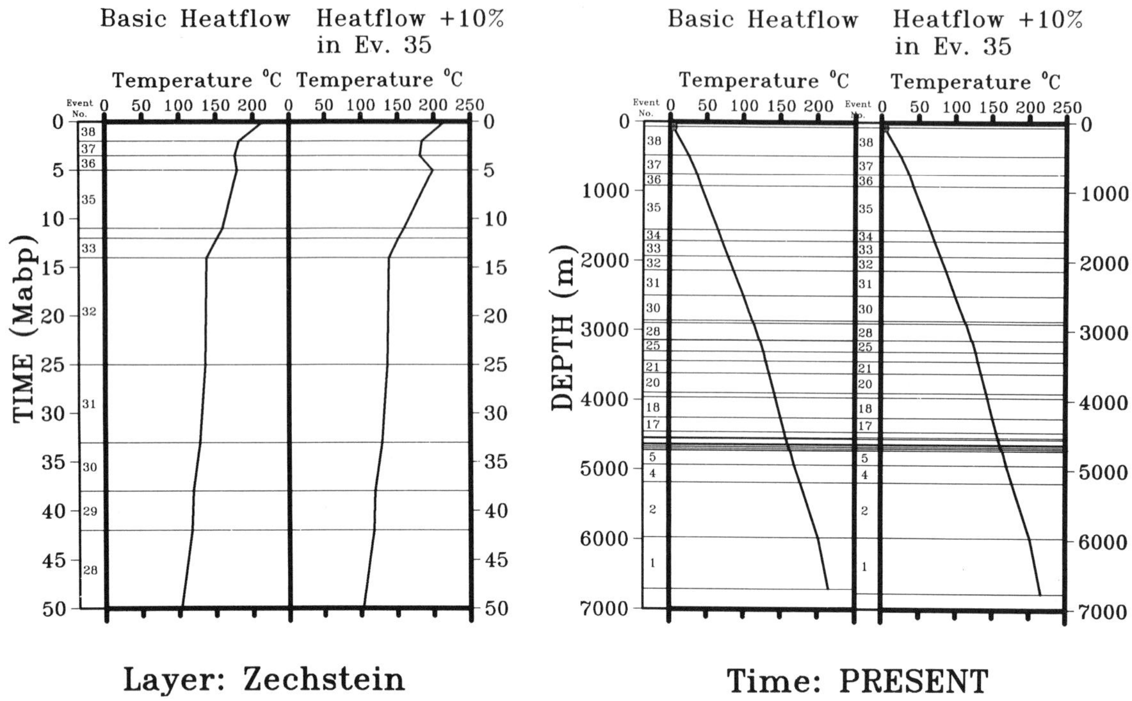

Fig. 4. Plot of the temperature history for the last 50 m.y., and temperature versus depth at the present for 2/4-14. Also included are the same plots for an increased heatflow (by 10%) in event 35.

same model and input data, except for one change, a 10% increase in heatflow during event 35. Both of these heatflow histories are used in the kinetic calculations as an example of parameter change in the same order of magnitude as the differences in the kinetic parameters, to illustrate the relative importance of changing various parameters. Figure 4 shows that versus depth there is no difference between the two heatflow histories with respect to present-day temperature. However, versus time there is a significant difference for event 35 where the temperature increases more since the heatflow is 10% higher for this event. In event 36 where the heatflow is "set back to normal" and there is little net deposition, the temperature falls back to the level of the basic heatflow history. This implies that the time required for equilibration with a change in heatflow is very short in geological terms, given that the model calculations apply.

Hydrocarbon generation — the kinetic approach

The use of a kinetic first-order Arrhenius equation to calculate the generation of oil and gas is described by Tissot and Espitalié (1975) and Tissot and Welte (1984). The model they describe involves a series of first-order reactions, all obeying the Arrhenius equation $k = A\,e^{-(E/RT)}$, with different values for the constants A (frequency or pre-exponential factor) and E (activation energy). For the subsequent reaction of cracking of oil to gas, they give one set of values for the kinetic parameters, implying that one reaction covers this process. The values given by Tissot and Welte (1984) were accepted as a reasonable estimate by several companies, and were used for a long time, even though it is loosely stated that "an activation energy of 50 kcal/mol seems convenient to account for gas generation in the deep parts of the sedimentary basins" (Tissot and Welte, 1984, p. 589).

Later it seemed clear from oil discoveries at depths greater than 4000 m that the models based on Tissot and Welte's (1984) cracking kinetics would underestimate the "oil floor", i.e. the greatest depth in a particular basin at which one could expect to discover oil. Several papers were published at that time suggesting kinetic data that might be more applicable for generation of oil and gas, e.g. Burnham and Braun (1985), Sweeney et al. (1986), Ungerer et al. (1986). The paper by Quigley et al. (1987) was considered by the authors to contain kinetic data that could be applied to the temperature models used in the North Sea basins. The kinetic parameters for cracking of oil to gas, given by Quigley et al. (1987) were derived by closed-vessel pyrolysis of Bathonian source rock samples from NE Scotland, and activation energies were about 10% higher than those of Tissot and Welte (1984) (Table 1).

Several authors have recently studied the problem of modelling cracking of oil to gas, coming up with

TABLE 1

Kinetic parameters for cracking of oil to gas, taken from literature and used in this and other studies performed at Saga Petroleum a.s.

Tissot and Welte (1984)	Quigley et al. (1987)
$A = 8.18 \times 10^{25}$ Ma^{-1}	$A = 3.15 \times 10^{26}$ Ma^{-1}
E = 50 kcal/mol	E = 55 kcal/mol

various values for the kinetic parameters. For the modelling described in this paper we have applied the data in Table 1. For further information the reader is referred to the reviews by Ungerer et al. (1988), and Ungerer (1990).

Simulation results

The results of the hydrocarbon generation modelling with respect to gas are shown in Fig. 5 as a plot of generation potential versus depth for the kinetic parameters of Tissot and Welte (1984). The generation potential is the proportion of gas generated from an oil precursor. The programme calculates a generation potential for every layer, regardless of lithology, since only the temperature history is needed as input in the kinetic calculations. The generation potential of gas is relative to the generation potential of oil from kerogen, which explains why total exhaustion of gas generation from oil occurs at a transformation ratio of about 0.3 g/g TOC. However, note that this case is used for illustration purposes only; no quantitative calculation of hydrocarbons generated based on TOC content was carried out. This model suggests that there is no gas generation above 3700 m, and that there is a rapid increase in realization of the gas potential between 4000 and 4500 m. The total gas generation potential is almost exhausted at 5000 m depth.

Included for comparison in Fig. 5 is the same plot for the case with an increase in heatflow of 10% at event 35. This case is chosen as a representative example of the response of the results from the kinetic models when one particular input parameter, other than the kinetic parameters, is varied by 10%. Several other sensitivity test cases were tried but none showed any more significant influence on the results of the kinetic models than the one included in this paper. As is clearly evident from Fig. 5, the only difference is in a slight increase in the generation potential at the same depth in the zone of rapid gas generation (between 4000 and 4500 m). The overall shape of the curve (the starting point, rapid gas generation and total realization of the gas potential) is the same for the two temperature histories.

This demonstrates the need for new and better kinetic parameters for the cracking of oil to gas, since it shows that on the basis of the optimized conceptual model in this area, most of the oil should have been cracked to gas at depths of around 4500 m. However, oil was found reservoired at depths of about 4600 m in 1987 in nearby well 2/12-1 (Söderström et al., 1991), thus giving a clear indication that the kinetic models need refining.

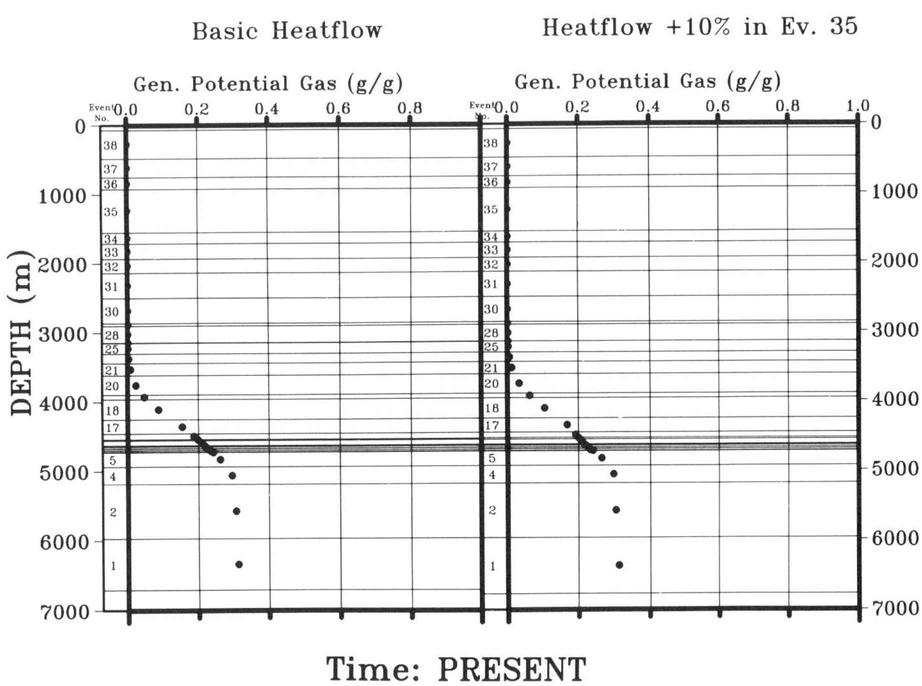

Fig. 5. Plot of the generation potential of gas from cracking of oil for the two temperature cases shown in Fig. 3. Kinetic parameters after Tissot and Welte (1984).

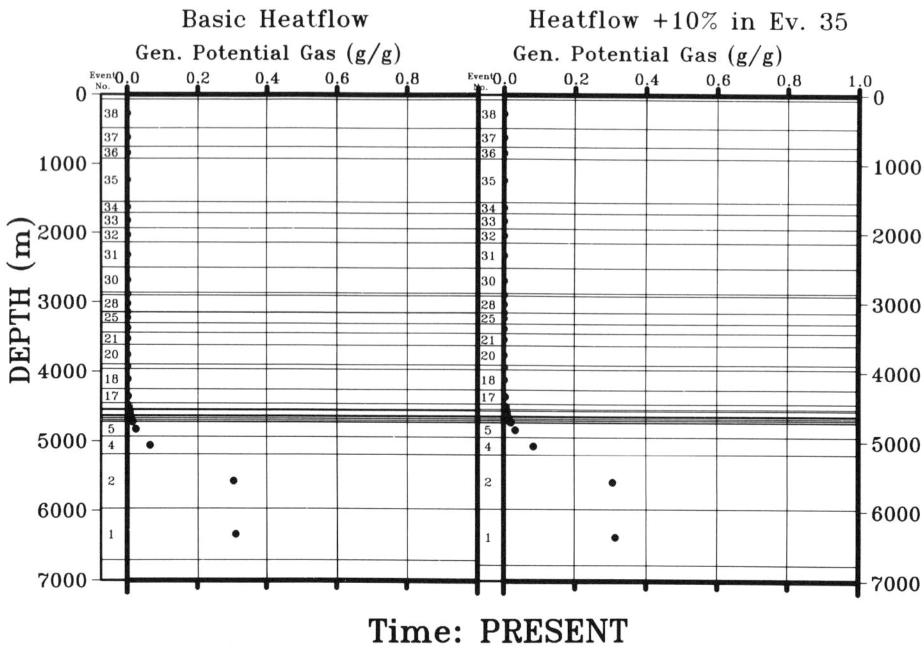

Fig. 6. Plot of the generation potential of gas from cracking of oil for the two temperature cases shown in Fig. 3. Kinetic parameters after Quigley et al. (1987).

As previously mentioned, the data presented by Quigley et al. (1987) were found by the authors to be applicable to the North Sea basins. However, Quigley et al. (1987) also treated the cracking of oil to gas as one single reaction with one set of kinetic parameters governing this reaction. The results of using these data in modelling 2/4-14 are shown in Fig. 6. This plot shows that there is no gas generation by cracking of oil above 4700 m, and that the rapid realization of the potential occurs between 5000 and 5300 m. Nearly total exhaustion occurs at 5600 m. This result is vastly different from the case based

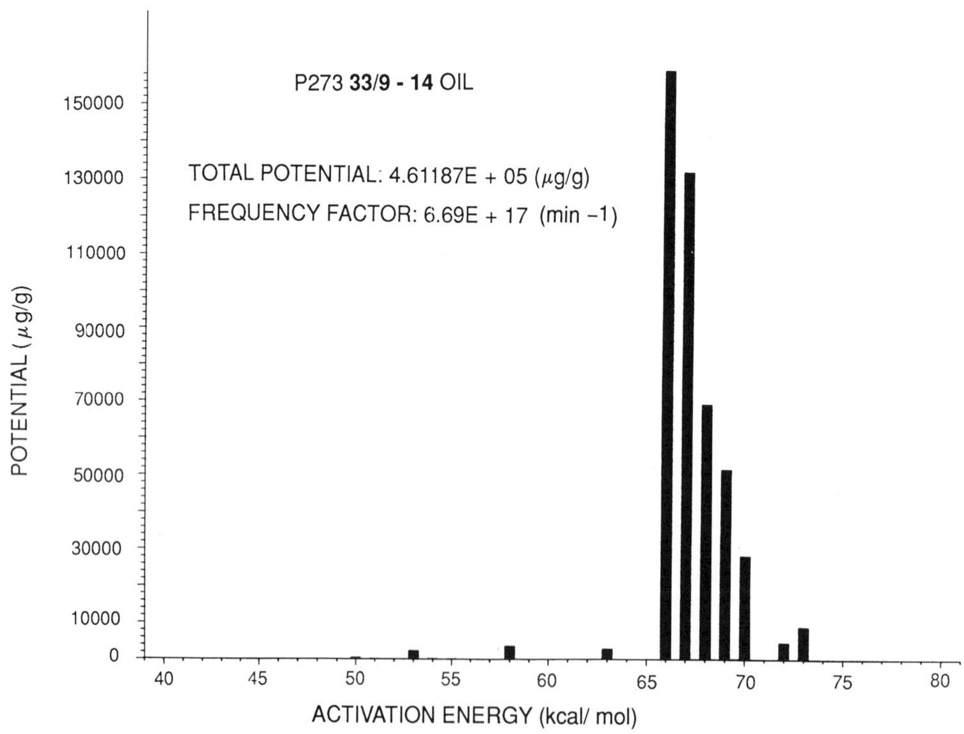

Fig. 7. Distribution of activation energies for the cracking of a North Sea oil sample to gas, as derived from the experiments carried out by the KFA.

on Tissot and Welte's (1984) kinetic data (Fig. 5), and this observation is of extreme importance when appraising prospects at depths greater than 4000 m. Again, the second half of the figure shows the same plot with the heatflow increased by 10% in event 35, and the results are similar to Fig. 5, with a slight increase in the generation potential at the same depth in the zone of rapid generation (between 5000 and 5300 m).

Laboratory work

On finding this vast difference in modelling results dependent upon which kinetic parameters one chose to use, we decided to perform laboratory simulations to try to determine realistic parameters for cracking of oil to gas in the North Sea basin. The method chosen was to decide upon a representative relatively immature ("uncracked") oil from the North Sea, and run a series of experiments on this to determine the kinetic parameters for the cracking of this oil. The product of the cracking experiments could then be compared to the oil found in well 2/4-14. Unfortunately, well 2/4-14 ran into serious operational problems, and only a small amount of liquid was collected from the shale shaker, prior to the onset of the operational problems. No further samples could be collected, and up to the time of writing, a proper sample (RFT, DST) has not been obtained. However, the part of the experiments to be done on the immature oil was carried out. It was performed at the laboratory of the ICH5 at the Kernforschungsanlage (KFA) in Jülich, Germany. The method used was micro-scale sealed-vessel pyrolysis (mssv), in which small samples of oil are sealed in glass tubes and subjected to controlled heating, and the products are analysed in a gas-chromatograph for each temperature step. The method and results are given in detail in Horsfield et al. (1993).

The experimental approach was based on the realization that one set of kinetic data for the cracking of oil to gas is insufficient, and that this process can be better modelled as a series of parallel first-order kinetic reactions with separate kinetic constants for each, as for the generation of oil from kerogen. Figure 7 shows the results of this study, i.e. the distribution of activation energies for the cracking of the oil sample (from Norwegian well 33/9-14) to gas, with a peak at 66–68 kcal/mol. These are substantially higher values than other work has indicated. Obviously, more work on different oil samples is necessary to resolve the question of whether these values are representative of oils from all over the North Sea area.

Comparison of results from different kinetic models

The results from the three different kinetic parameter-sets are shown in Fig. 8, and the difference in depth of onset and peak gas generation from cracking of oil to gas is clearly visible. To further visualize the effect of these results, Table 2 lists the percent increase in activation energy for the three

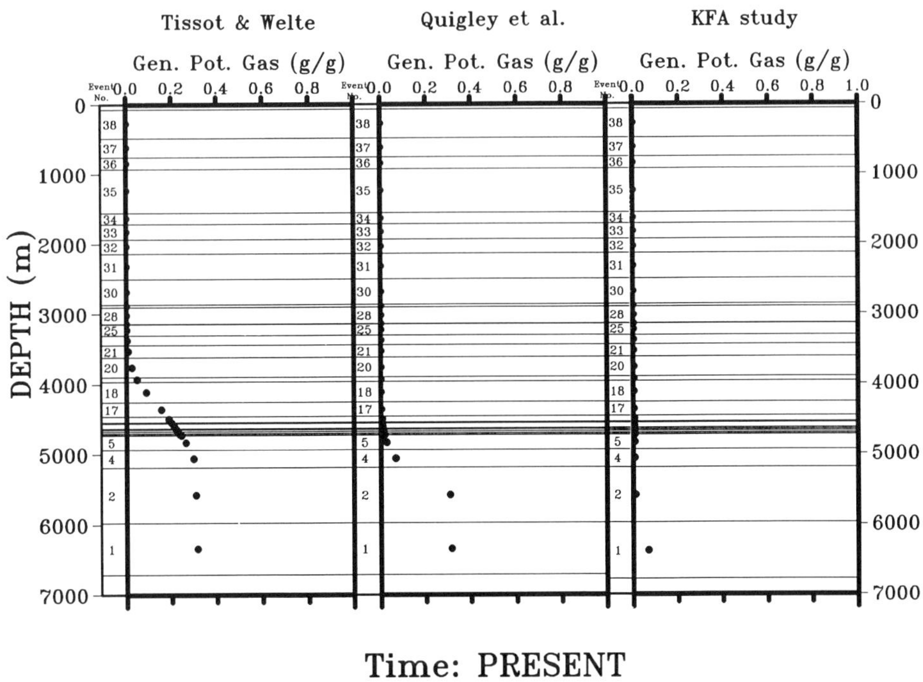

Fig. 8. Comparison plot of the generation potential of gas from cracking of oil, for the three different sets of kinetic parameters.

TABLE 2

Comparison between the different kinetic parameters for cracking of oil to gas, and the impact they had on the depth to onset of gas generation given the thermal history of this study

Data source	*E* (kcal/mol)	Δ	Depth to onset of gas generation (m)	Δ
Tissot and Welte	50		4130	
Quigley et al.	55	10%	5110	23.7%
KFA study	~67	34%	~6500	~57%

The percent change (Δ) is relative to the Tissot and Welte (1984) data.

kinetic models, together with the percent increase in depth to onset of gas generation (characterized, somewhat arbitrarily, by a generation potential of 0.1). The data from Quigley et al. (1987) represent a 10% increase in activation energy with respect to Tissot and Welte (1984), but give a 23.7% increase in the depth to onset of gas generation, whereas our new data represent a ~34% increase in activation energy, resulting in a ~57% increase in the depth to onset of gas generation. These results clearly demonstrate the importance of obtaining the most accurate possible kinetic parameters for oil-cracking. One of the most important aims of exploration basin modelling is to predict the hydrocarbon phases present in any given prospect today. Explorationists risk arriving at vastly different conclusions depending upon which set of kinetic parameters are chosen.

Conclusion

This study shows that kinetic parameters are by far the most important (sensitive) input parameters in modelling generation of oil and gas. If there is any uncertainty in these parameters, then it will make other uncertainties insignificant in comparison. This implies that some effort could be wasted in trying to determine any other input parameters with great certainty, if the kinetic parameters are poorly determined. However, if the kinetic parameters are well known the relative importance of uncertainties in other parameters is greater, and work on the determination of optimum values should be continued.

Acknowledgements

The authors wish to extend their thanks to Saga Petroleum a.s. for permission to publish this paper, to numerous Saga staff members for input data, and to Jan Slettebø who drafted the figures. The manuscript benefitted from constructive criticism by an anonymous referee, and the editor.

Discussion

M. Bagge, Intera–ECL Petroleum Technologies, U.K.: *Do the authors think that the kinetic parameters determined in this study are equally valid for oil cracking to gas in a source rock and in the reservoir? Were similar analytical measurements performed on oil in the presence of clays to assess if clay minerals have any effect on the kinetic parameters?*

Authors' reply. The kinetic parameters were determined from oil contained in glass tubes, and so they will probably be more valid for reservoir conditions than source rock conditions. However, our study was specifically aimed at reservoir conditions. No experiments were undertaken with clays present during this particular study, but these should certainly be carried out to test the effect they might have on the kinetic parameters.

References

Burnham, A.K. and Braun, R.L., 1985. General kinetic model of oil and shale pyrolysis. In Situ, 9 (1): 1–23.

D'Heur, M., 1986. The Norwegian chalk fields. In: A.M. Spencer et al. (Editors), Habitat of Hydrocarbons on the Norwegian Continental Shelf. Norwegian Petroleum Society, Graham and Trotman, London, pp. 77–90.

Gowers, M.B. and Sæbøe, A., 1985. On the structural evolution of the Central Through in the Norwegian and Danish sectors of the North Sea. Mar. Pet. Geol., 2: 298–318.

Harland, W.B., Cox, A.V., Llewellyn, P.G., Pickton, C.A.G., Smith, A.G. and Walters, R., 1982. A Geologic Time-Scale. Cambridge Earth Sciences Series, Cambridge University Press, Cambridge, 131 pp.

Harland, W.B., Armstrong, R.L., Cox, A.V., Craig, L.E., Smith, A.G. and Smith, D.G., 1989. A Geologic Time-Scale. Cambridge University Press, Cambridge, 263 pp.

Horsfield, B., Schenk, H.-J., Mills, N. and Welte, D.H., 1993. Closed-system programmed-temperature pyrolysis for simulating the conversion of oil to gas in a deep petroleum reservoir: compositional and kinetic findings. Org. Geochem. (in press).

Mo, E.S., Throndsen, T., Andresen, P., Bäckström, S.A., Forsberg, A., Haug, S. and Tørudbakken, B., 1989. A dynamic deterministic model of hydrocarbon generation in the Midgard Field drainage area offshore Mid-Norway. Geol. Rundsch., 78 (1): 305–317.

Quigley, T.M., Mackenzie, A.S. and Gray, J.R., 1987. Kinetic theory of petroleum generation. In: B. Doligez (Editor),

Migration of Hydrocarbons in Sedimentary Basins. Technip, Paris, pp. 649–666.

Söderström, B., Forsberg, A., Holtar, E. and Rasmussen, B.A., 1991. The Mjølner Field: a deep Upper Jurassic oil field in the Central North Sea. First Break, 9 (4): 156–171.

Sørensen, S., Jones, M., Hardman, R.F.P., Leutz, W.K. and Schwarz, P.H., 1986. Reservoir characteristics of high- and low-productivity chalks from the Central North Sea. In: A.M. Spencer et al. (Editors), Habitat of Hydrocarbons on the Norwegian Continental Shelf. Norwegian Petroleum Society, Graham and Trotman, London, pp. 91–110.

Spencer, A.M., Home, P.C. and Wiik, V., 1986. Habitat of hydrocarbons in the Jurassic Ula trend, Central Graben, Norway. In: A.M. Spencer et al. (Editors), Habitat of Hydrocarbons on the Norwegian Continental Shelf. Norwegian Petroleum Society, Graham and Trotman, London, pp. 111–128.

Sweeney, J.J., Burnham, A.K. and Braun, R.L., 1986. A model of hydrocarbon maturation in the Uinta Basin, Utah, USA. In: J. Burrus (Editor), Thermal Modelling in Sedimentary Basins. Technip, Paris, pp. 547–564.

Tissot, B.P. and Espitalié, J., 1975. L'évolution de la matière organique des sédiments: application d'une simulation mathématique. Rev. Inst. Fr. Pét., 30: 743–777.

Tissot, B.P. and Welte, D.H., 1984. Petroleum Formation and Occurrence. Springer-Verlag, Berlin, 2nd ed., 699 pp.

Ungerer, P., 1990. State of the art of research in kinetic modelling of oil formation and expulsion. Org. Geochem., 16 (1–3): 1–25.

Ungerer, P., Espitalié, J., Marquis, F. and Durand, B., 1986. Use of kinetic models of organic matter evolution for the reconstruction of paleotemperatures. Application to the Gironville well (France). In: J. Burrus (Editor), Thermal Modelling in Sedimentary Basins. Technip, Paris, pp. 531–546.

Ungerer, P., Behar, F., Villalba, M., Heum, O.R. and Audibert, A., 1988. Kinetic modelling of oil cracking. Org. Geochem., 13 (4–6): 857–868.

Welte, D.H. and Yalçin, M.N., 1986. Formation and occurrence of petroleum in sedimentary basins as deduced from computer-aided basin modelling. In: S.P. Kumar et al. (Editors), Int. Conf. Petroleum Geochemistry and Exploration in the Afro–Asian Region. ONGC, Dehra Dun, pp. 1–21.

Welte, D.H. and Yukler, M.A., 1981. Petroleum origin and accumulation in basin evolution — a quantitative model. Am. Assoc. Pet. Geol., Bull., 65: 1387–1396.

Wygrala, B.P., 1988. Integrated computer-aided basin modelling applied to analysis of hydrocarbon generation history in a Northern Italian Oil Field. Org. Geochem., 13 (1–3): 187–198.

Yalçin, M.N., 1991. Basin modelling and hydrocarbon exploration. J. Pet. Sci. Eng., 5: 379–398.

PÅL ANDRESEN *Saga Petroleum A/S, P.O. Box 490, N-1301 Sandvika, Norway*

NIGEL MILLS *Saga Petroleum A/S, P.O. Box 490, N-1301 Sandvika, Norway*

H.-J. SCHENK *Kernforschunganlage Jülich (KFA), P.O. Box 1913, D-5170 Jülich, Germany*

BRIAN HORSFIELD *Kernforschunganlage Jülich (KFA), P.O. Box 1913, D-5170 Jülich, Germany*

Basin modelling in an overthrust area of Austria

Franz Schmidt and L. Tufan Erdoğan

A basin modelling study has been performed on part of the overthrust area in the northwestern part of eastern Austria. Starting with the deposition of Malm (Upper Jurassic), the area, in general, was a stable platform. The main tectonic event was the emplacement of a thick nappe system during the Neogene when the area suffered major subsidence.

The emplacement of the nappes has produced large excess pressures. This was followed by a long erosional period during which the source rocks generated their maximum oil and were probably subjected to a meteoric water invasion. Lateral excess head distributions for the source horizons through time indicate a generally north-northeasterly movement of formation fluids.

The geothermal gradient shows a general straight-line character becoming higher below the nappes.

The vitrinite reflectance data show that the source interval, the Early Oligocene, is middle to late mature in the southern and central parts of the area, and immature in the north.

The source rock started generating hydrocarbons in the central and southern sections of the area during the erosional period following the emplacement of nappes. The main oil generation started later and at greater depths as one moved northwards, because of the fact that the thickness of the thrust decreases towards the north. The base of the gas window has not been penetrated in the area and it is believed that PW4, the southernmost pseudowell, has an oil window between 4100 and 5950 m.

The overall heating rate in this part of the basin is low. Therefore, the study area has a better chance of containing gas accumulations rather than oil as the potential for secondary cracking is high.

Introduction

Two important hydrocarbon provinces exist in Austria. One is the Vienna Basin, a post-Alpine Neogene downwarp, superimposed on the nappes of the Flysch and Calcareous Alps. The other area of interest is the Molasse Basin. It is an east–west-trending trough of sediments at the northern front of the Alpine overthrust. The Austrian part of the Molasse Basin is divided into two parts, the first being to the east of the Bohemian Massif, the other to the west. This particular study refers to the western part of the Molasse (Fig. 1).

The first commercial use of hydrocarbons in this area was at the end of the last century when gas was detected in water wells. After the Second World

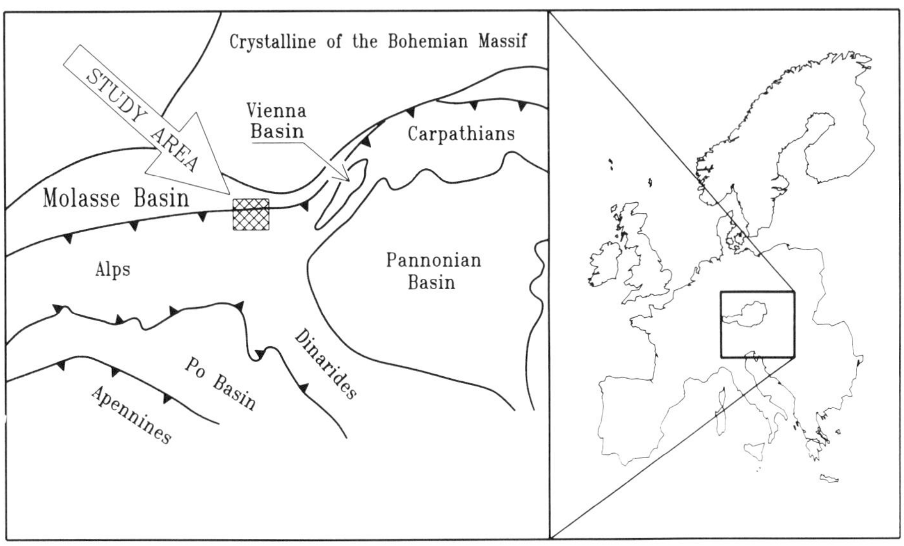

Fig. 1. The study area within the regional geological framework.

Basin Modelling: Advances and Applications edited by A.G. Doré et al. NPF Special Publication 3, pp. 573–581, Elsevier, Amsterdam.

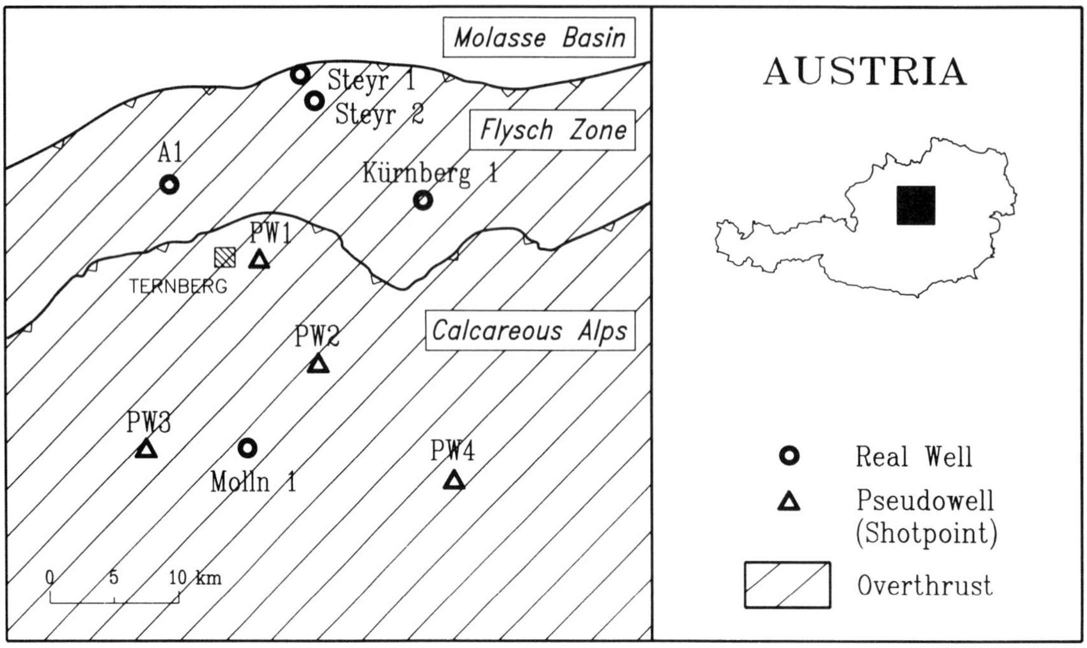

Fig. 2. The study area, showing location of wells and pseudowells.

War, exploration activity increased with the first oil discovery in 1956 (Polesny, 1983). Currently, some 70 oil and gas fields are known. The initial discoveries were in the Eocene sediments and in the Late Oligocene sands of the Molasse Basin. The later ones were made in the underlying autochthonous sediments within the Molasse Basin. Wells which penetrated the Alpine nappe system (the first of these was drilled in 1965) proved the existence of autochthonous sediments underneath these nappes. This knowledge and the positive results of earlier wells encouraged further intra-Alpine exploration.

In this study the hydrocarbon potential of an overthrusted area located in the Flysch and Calcareous Alps in the northwestern part of eastern Austria is evaluated (Fig. 2). The crystalline basement in the area, which is referred to as the Bohemian Massif, is covered with autochthonous sediments of Mesozoic age, starting with the Dogger sandstones and limestones and Cretaceous sandstones and marls. These underlie Tertiary deposits of the Molasse Basin composed mainly of clastics which are finally overridden by the Alpine nappe system, made up of Flysch and Calcareous Alps (Fig. 3).

An attempt to quantitatively model and analyze a sedimentary basin in the petroleum industry should finally concentrate on the quantification of the evolution of its geological framework in terms of its hydrocarbon potential. The ÖMV Basin Modelling approach, based on the "Yükler Model" as it is commercially called, has been applied in the quantification of hydrocarbon potential in the study area. This multi-dimensional, deterministic model uses all available data in geology, geophysics, hydrodynamics, geothermics, rock mechanics and geochemistry in a synergistic manner (Yükler and Welte, 1980; Yükler and Kokesh, 1984; Yükler, 1987).

Geological evolution

The crystalline Bohemian Massif is discordantly overlain by the sediments of the Dogger and Malm. These sediments were deposited in the southwestern part of the area during an overall transgression in water depths from 2 to 50 m, while a hiatus period prevailed in the north. The basin shallowed during the deposition of the late Malm limestone and an erosional period followed during the Early Cretaceous. The succeeding Cenomanian sandstones were deposited in water depths of 10 to 30 m. During the Turonian, marls and sandstones were sedimented in water depths of 40 to 50 m. Turonian, Coniacian and Santonian sandstones and marls followed in similar water depths. Emergence during Late Cretaceous and early Tertiary caused an erosional period which was followed by deposition of the Eocene limestone at depths varying between 10 and 30 m. This was the beginning of the Molasse sedimentation which lasted until the area was overridden by the Alpine nappes at the beginning of the Neogene (Wessely, 1987).

The Molasse Basin is located to the north of the Alps and extends from Switzerland to the eastern border of Austria. It is an asymmetrical foreland basin which deepens to the south and is filled with predominantly clastic sediments. The Molasse Basin is the foredeep of the Alpine nappes moving from

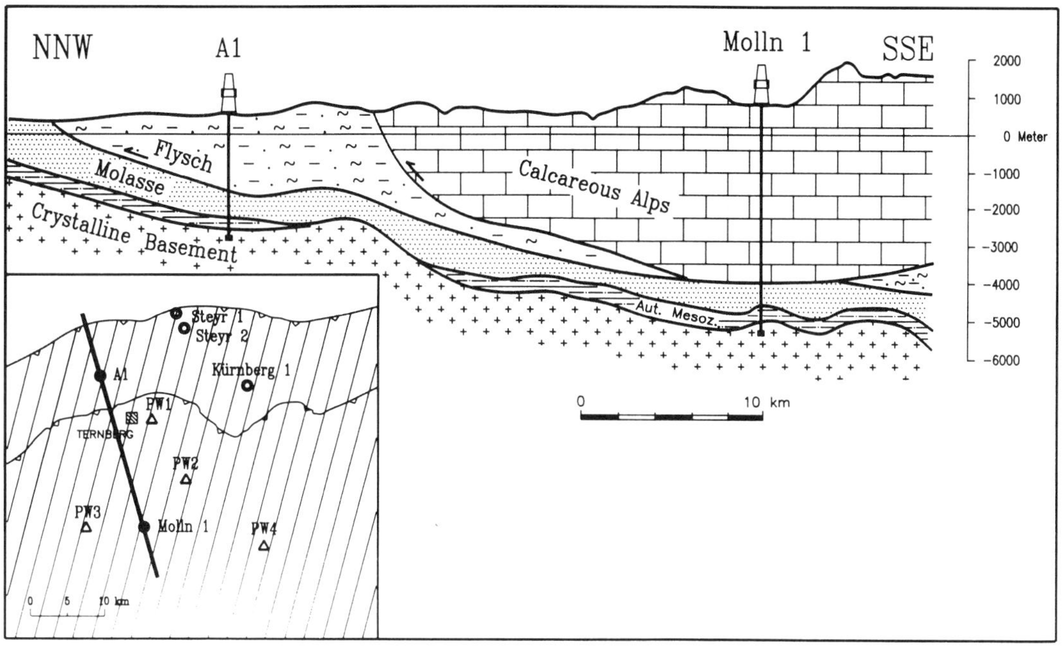

Fig. 3. North–south geological cross-section.

south to north. The sediment input into the basin comes both from the north, from the crystalline basement, and from the south but predominantly from the south, derived from the advancing nappe system.

The study area started subsiding at high rates, initially being 100 m deep during deposition of the Early Oligocene source rock, 200 m during the Middle, and later 400 m deep during the Late Oligocene series. The first nappe, consisting of Flysch sediments, was emplaced at a depth of 200 m and the second, Calcareous Alps, at a shallower depth of 100 m. The whole area was later subjected to an extensive erosional period which was followed by the localized deposition of fluvial and fluvio-glacial Quaternary sediments which in turn were subjected to later (recent) erosion (Fig. 4).

In the mathematical model the heat flows are initially assigned based on the commonly accepted tectonic history of the basin. They are later optimized to their correct time distribution by means of matching the calculated maturities and temperatures with the measured ones. The distribution of the heat flow over the area through geologic time thus obtained shows no significant reason to indicate tectonic zonings or subdivisions within the basin. They decrease slightly, in the order of 0.05 heat flow units (HFU; 1 HFU = 41.7 mW/m^2) northwards, towards the shallower parts of the basin.

The optimized heat flow distribution within the stratigraphic sequence against time (Fig. 4) implies that the formation of the basin started with tensional forces involved during the deposition of the Dogger (Middle Jurassic) which continued to a decreasing extent until the deposition of the Malm about 150 Ma. The area was a stable platform after the short compressional phase resulting in erosion following the deposition of the Malm. This continued until the deposition of the Turonian–Coniacian–Santonian up to 88 Ma. A short-lived and weak tensional phase which was later followed by stable platform conditions coincided with the deposition of these units. Starting with the deposition of the Early Oligocene and reaching its peak during the deposition of the Middle to Late Oligocene series, the area suffered a major east–west extension. After the emplacement of the nappe system during the Early Miocene, a ma-

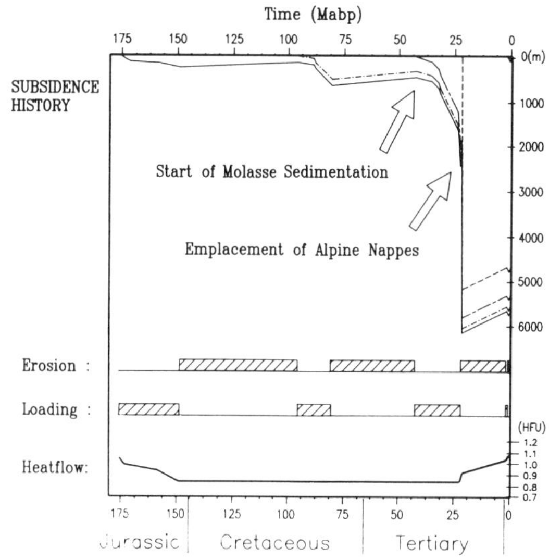

Fig. 4. The calculated heat flow history and geological evolution of the area.

jor uplift and erosional phase followed. Starting with the Quaternary, the area entered a tensional regime which is reflected in the present-day heat flow.

Source rocks and hydrocarbons

The major source rocks in the area are found in the Early Oligocene section (Figs. 5 and 7). The Lattorfian (lowermost Oligocene) Fischschiefer and the Rupelian Bändermergel (Fig. 8) are the two main sources encountered. The Fischschiefer is a dark-brown finely laminated marl with bituminous odour when crushed. The data indicate a good-quality source rock. The mean TOC content is 2.4% with a maximum of 10%. The organic matter is mainly of Type II Kerogen. The other source rock, the Rupelian Bändermergel, has an average TOC of 1.6% and the organic matter is of an intermediate Type II–III Kerogen, although closer to Type III.

The laboratory results show that the Fischschiefer is of a source rock quality that allows effective generation and expulsion when subjected to moderate to high heating rates. The lower quality interval, i.e. Bändermergel, can only expel liquid hydrocarbons under high heating rate conditions. Otherwise gas is the main type of hydrocarbon to be expected from this source rock. The measured maturities from rock samples are generally around a vitrinite reflectance of 2% R_o, but the distributions of the R_o-readings are not uniform and suggest that the samples include reworked material.

The origin of the gas encountered in shallow stratigraphic units in wells located in the Molasse is mostly biogenic. According to the isotopic studies no biogenic signature of gas is recorded for wells within the overthrust such as in Molln 1 and Grünau 1 (west of the study area). The gas isotopes of samples

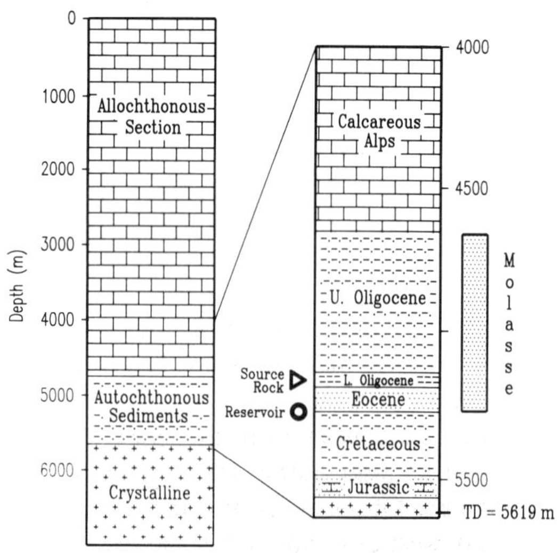

Fig. 5. Generalized stratigraphy in Molln 1 well.

obtained from tests of the well Molln 1 allow an estimation of the state of maturity at which the gas was generated. According to Faber (1987), this maturity is around 1.5% R_o.

Significant oil production from mainly the Eocene and Cenomanian units can be observed 10 km to the northwest of the area in the Molasse Basin. Oil was recovered from tests in various wells within the overthrust, such as A1, Grünau 1, Kirchdorf 1, Steinfelden 1 and Steyr 2. Oil-impregnated cores are recorded in well Kürnberg 1.

Oil is encountered in places more than 50 km north of the Flysch front. This requires a very effective migration mechanism. Parallel to that a continuous change from south to north in oil properties would be expected, and this is in fact observed. The oils show a decreasing API-gravity trend towards the north. The character of both the oils and source rock extracts changes with the distance from the nappe front. The most mature oil samples are produced from wells within the thrusted area itself, whereas the least mature samples are the northernmost ones.

The source rocks encountered in various wells in the foreland of the nappes are immature for hydrocarbon generation, so all hydrocarbons of the upper Austrian Molasse are expected to come from places of higher maturity, i.e. below the nappe systems in the south.

Input data

The area under consideration is poorly explored with scarce well data. The available data consist of five wells and four pseudowells. The wells west of the western limit of the study area have also been checked and their geochemical data incorporated.

For computational reasons, the stratigraphy has been subdivided into time slices, each with uniform rheological properties. The chronostratigraphic interval of 180 m.y., covering a thickness of up to 6000 m has been subdivided into a total of nineteen layers in all wells and pseudowells.

The Calcareous Alps, in reality, consist of not one, but several layers with different lithological and hence petrophysical properties. It had to be input as a one layer nappe, because of the time-stratigraphic nature of the model data structure. The higher nappe sheets were emplaced earlier than the lower ones. Therefore, a weighted-average technique was applied to estimate a single lithology describing the whole mass, based on the thickness and lithology of individual sheets. This, naturally, inherited the error derived from the fact that rheological properties, rather than thicknesses, played a more important role in the process of compaction.

Subsidence history

Sedimentation rates in the area were generally low, in the order of 0.7–15 m/m.y. Malm and Turonian to Santonian sandstones and marls have normal sedimentation rates changing in between 13 and 50 m/m.y., and Oligocene sediments have higher rates reaching 100–600 m/m.y. The subsidence curves (Fig. 4) indicate a generally stable platform with slow to normal subsidence rates during the deposition of Dogger, Malm and Cretaceous. The basin started subsiding rapidly with the deposition of the Late Oligocene sediments and large subsidence rates were encountered with the emplacement of the nappes. For the last 21.5 m.y., the basin has stabilized except for a short interval of break in the Quaternary fluvial sedimentation in places. The erosional phases cannot always be clearly identified on the subsidence curves because of the generally very low erosional rates, in the order of 1.3 to 4 m/m.y.

The thickness distributions of the interval from the Malm to the thrust slice of the Calcareous Alps have been plotted to cross-check the subsidence history and derive a two-dimensional view of subsidence and compaction all over the area.

The Eocene and Late Cretaceous sandstones constitute the reservoir horizons. These will be referred to as the "Eocene reservoir" hereafter. The reservoir section (Fig. 6) has an overall trend of thinning northeast and southwards with a maximum thickness of 300 m around Molln 1 and PW4. The orientation of minimum and maximum thickness axes parallels the thrust front.

The depths to the top of the reservoir horizon have been plotted, starting with the time of first nappe emplacement till the Recent to observe major tectonic disturbances and the change of structural configuration through time. These maps show a smooth and general trend of shallowing northwards through geologic time. The basin deepened along an axis passing through wells PW4, PW2, PW1 and PW3.

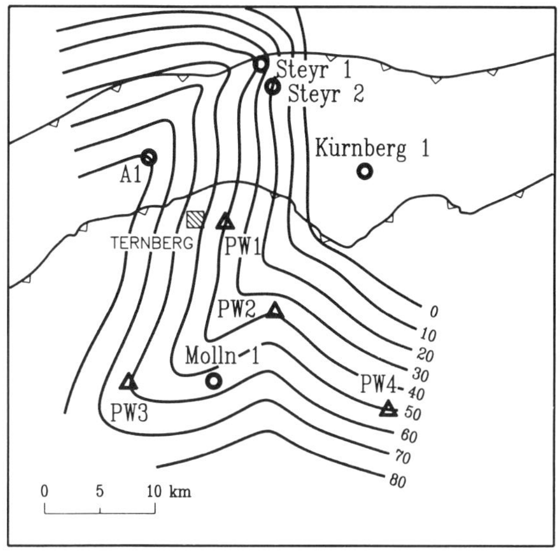

Fig. 7. Present-day thickness distribution of the Oligocene source rocks, Fischschiefer and Bändermergel.

As stated earlier, the source rocks in the area are found in the Early Oligocene section (Figs. 7 and 8). The source rocks of the Early Oligocene have a collective SW–NE-trending minimum thickness axis, where all the other formations have their maximum.

The "Calcareous Alps" nappe thins northwards due to both its original tectonic geometry and subsequent erosion. The amounts eroded decrease northwards, with the exception of another maximum erosion trend around well PW1.

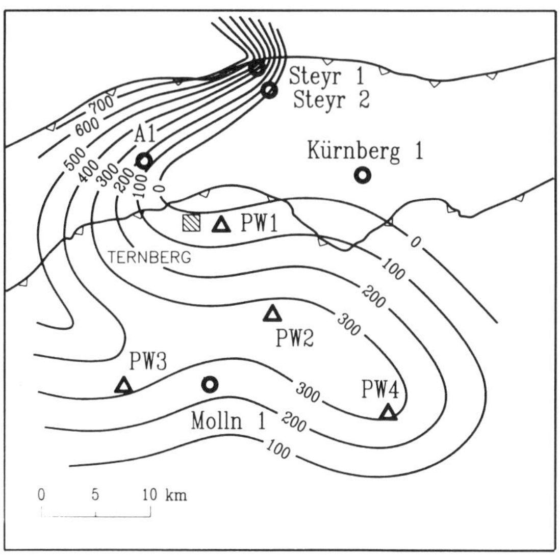

Fig. 6. Present-day thickness distribution of the Eocene reservoir (contours in metres).

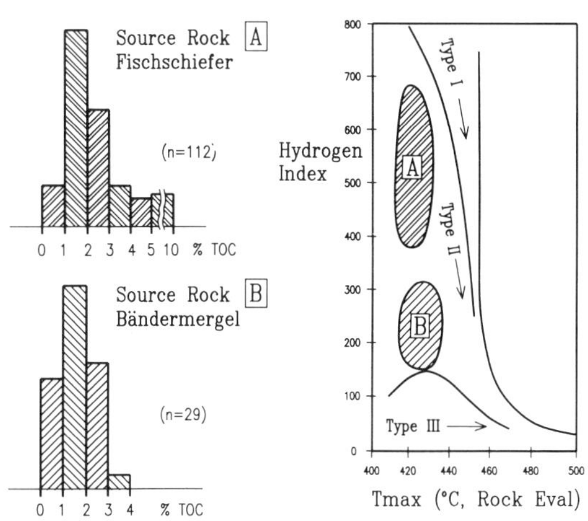

Fig. 8. The Oligocene source rocks; quantity and quality assessment (see stratigraphy, Fig. 5).

Porosity and excess pressure history

When the nappes were emplaced, 21.5 Ma, excess pressures in Molln 1, PW4, PW3 and PW2 are modelled to have reached values of 60 to 900 bar at depths of about 5000 m. The formation fluids at those depths (which coincided with the Early Oligocene source rock) were charged mainly downwards in Molln 1 and upwards in the rest of the wells, because of the existing low potential zones above and below that interval. During the erosional phase following the emplacement of the nappes, the whole sequence was hydrostatic and hence became susceptible to meteoric water invasion. It is to be noted here that this was the period of maximum oil generation within the source rock in the area. With the localized deposition of the Quaternary sediments, more fluids were charged upwards and downwards into the lower potential zones. Presently, the system is hydrostatic. In wells PW1 and A1, the system became hydrostatic after the deposition of the Late Oligocene Puchkirchen Formation during which the calculated excess pressures were around 5 bar.

Total hydraulic heads as the direct indicator of fluid potentials over the study area have been plotted through time for the source rock and the reservoir. At present, the fluid head distribution inside the reservoir indicates a northerly fluid movement.

The calculated total porosities for the Early and Middle Oligocene series are lower than their measured ones, indicating a possibility of dissolution during diagenesis.

Thermal history

Temperature data from wells Molln 1 and A1 were utilized to optimize the calculated temperature values. The main source of information came from the bottom hole temperatures as indicated on the well-logs. Because of lack of sufficient data in the "time-since-circulation" information, only a few of these dynamic temperatures could be converted into static. Therefore, the dynamic quality of the rest had to be considered and the calculated temperatures were higher than the measured data.

Another source of information was the DST bottom hole temperature data which were limited in number and confined to a very narrow interval. These too were dynamic in nature, but more realistic than those from the log operations, because the additional cooling effect of the mud circulation before logging did not need to be taken into account.

The geothermal gradient shows a general straight-line character becoming higher below the nappes. This is a typical case in such areas where a huge mass buries the basin to great depths in very short intervals of time. The fact that the maximum oil generation was tied to the emplacement of nappes in the area also supports this gradient increase.

Thermal maturation history

The vitrinite measurements in the Molln 1 well covered the interval from 5137 to 5559 m. The measured values show that the Late Oligocene is late mature, the source horizon Early Oligocene middle to late mature, and the Eocene and Turonian overmature. The vitrinite reflectance measurements in the A1 well showed the source rock to be immature.

The most consistent data came from the well Grünau 1, west of the study area, and show that the autochthonous section is early to middle-late mature (0.51% R_o at 3650 m to 0.80% R_o at 5166 m). An excellent match exists between these values and the ones calculated for the same depth range in Molln 1.

Hydrocarbon generation and migration

Just after the emplacement of the nappes, the source rocks within the study area were in an immature state. Most of the oil generated inside the source rocks following the sudden emplacement of the nappes was cracked before it could leave the source rocks. At the end of the erosion that followed, the source rock started generating condensate and gas in Molln 1 and PW4, peak oil in PW3 and PW2, and early oil in PW1. The source rock in well A1 during that time was still in an immature phase. At the end of the localized Quaternary deposition, the source rock in PW1 passed into the main oil generation phase, while the rest of the area remained unaffected. Presently, the source rock in PW3 is very close to maximum oil generation (Fig. 9).

The main oil generation phase started later and at a greater depth as one goes northwards. PW4 has an economical oil window between 4100 and 5950 m, whereas its base is more than 5200 m both in Molln 1 and PW2. None of the wells currently drilled have reached the base of the economical gas window.

The overall heating rate in this part of the basin is low, i.e. less than 1°C/m.y. Under low heating rate conditions, very little, if any, oil can be expelled out of the source rock and most of the liquid hydrocarbons stay inside it. These then crack into gas as the area is buried suddenly by the emplacement of the nappes. Therefore, the study area seemingly has a better chance of containing prolific gas accumulations rather than those of oil.

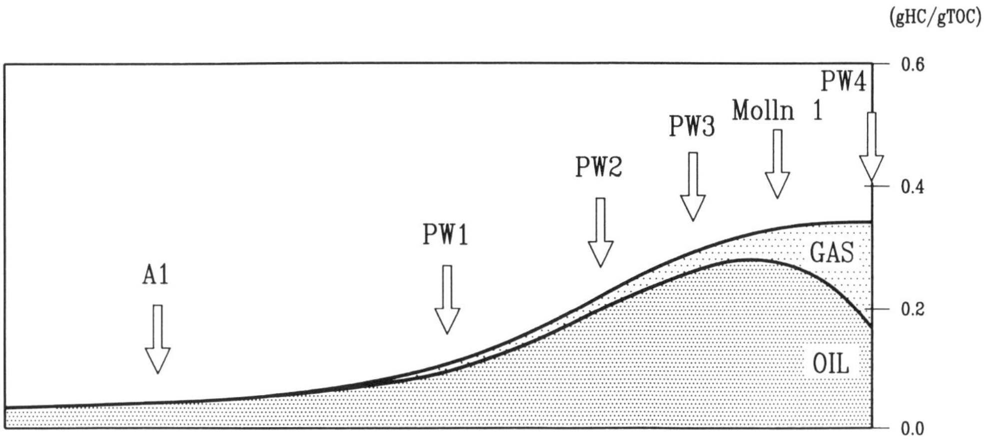

Fig. 9. Calculated present-day hydrocarbon generation in the wells within the study area.

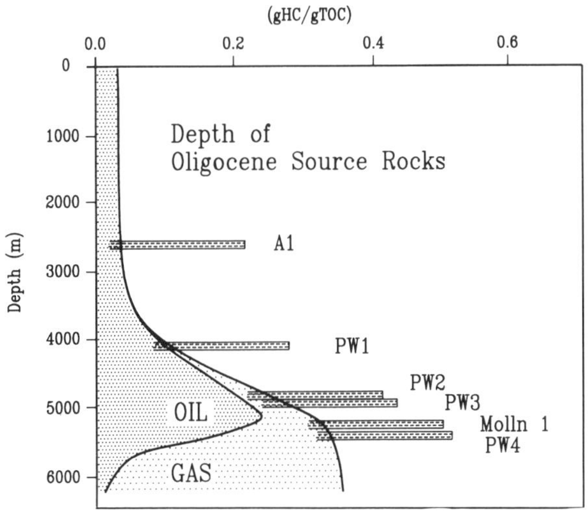

Fig. 10. Oil and gas windows relative to the encountered depths of the Oligocene source rocks.

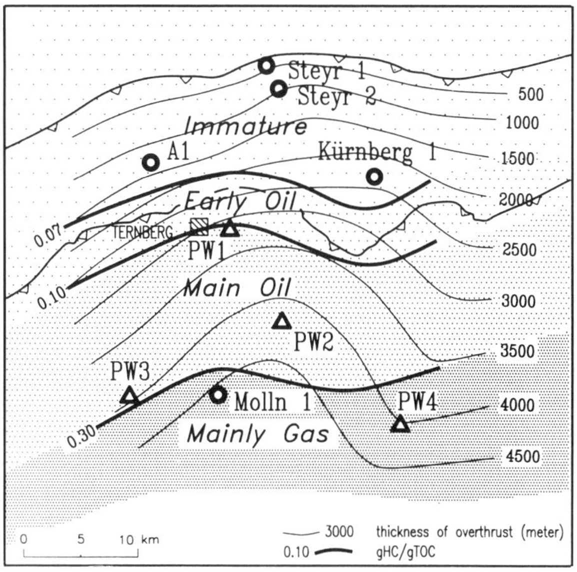

Fig. 11. Hydrocarbon generation zones with respect to thickness of overthrust, present day.

The source rock intervals (Fig. 10) in Molln 1 and PW4 are currently in the state of gas and condensate generation, whereas in PW3, PW2 and PW1 they are in the main oil generation phase (in decreasing order), and in A1 in immature state. It should be noted, however, that only the northern part of the Molasse Basin has been studied. The most important kitchen areas are to the south under thick allochthonous cover where seismic data acquisition is problematic (Fig. 11). Because of the active hydrodynamic forces involved in the basin, a roughly northwestward migration of the hydrocarbons should be expected.

These facts, together with the fact that there is a long-distance migration in the area, could have made it possible for the reservoir horizons in wells other than Molln 1 to contain hydrocarbon accumulations.

The idea of long-distance migration of hydrocarbons from the kitchen areas south of the study area into the far north Molasse Basin oil-and-gas-fields, are supported by the active hydrodynamic forces involved inside the source and reservoir horizons (Fig. 12). The equipotential maps plotted for the reservoir section show a northward movement of formation fluids and the API-gravity distribution of the oils inside the Molasse Basin fields clearly define a northwards decrease which is an outstanding example of the differential entrapment principle (Gussow, 1955; Erdoğan and Akgül, 1980, 1981). In one single oil field, the vertical distribution of the API-gravities of the oils within the oil column shows a northerly tilt within the structural limits of the oil field (Fig. 13). All these observations show that long-distance migration is mainly a product of the active hydrodynamic conditions inside the reservoir horizon, which seems to communicate laterally over large areas.

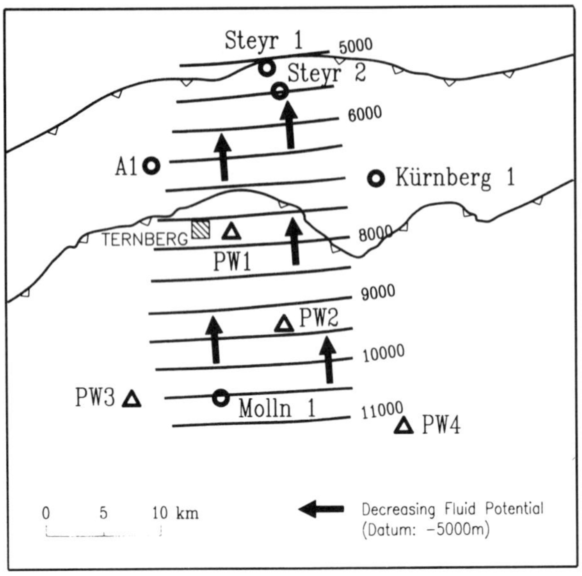

Fig. 12. Present-day total hydraulic head distribution (metres).

Conclusions

The study was primarily aimed at the evaluation of the hydrocarbon potential of the area. There is a general lack of reliable data and the data points available are considerable distances apart. Because the studied area covers only a part of a basin, detailed quantitative geochemical (volumetric) conclusions could not be derived. The chronostratigraphic sequence, starting from Dogger until present has been subdivided into nineteen layers and primary geological data (such as ages, durations, lithologies, thicknesses, sedimentary environments, etc.) have been input for each of them. Initially assumed parameters such as heat flow and present total porosities have been corrected as a result of temperature, maturity and compaction history optimization processes.

The distribution of optimized heat flow over the study area through geologic time shows no indication of major tectonic zoning. Tensile stresses were mainly active during the initial periods of basin formation and the deposition of the Late Oligocene. Starting with the deposition of the Malm, the area, in general, was a stable platform. The main tectonic event has been the emplacement of the thick nappe system, during which the area suffered major subsidence.

The emplacement of the nappes produced large excess pressures at depths of about 5000 m, causing the fluids in that interval to be ejected upwards and downwards. This was followed by a long erosional period during which the source rocks generated their maximum oil and were probably subjected to a meteoric water invasion. Lateral excess head distributions for the source horizons through time indicate a generally north-northeasterly movement of formation fluids.

The difficulty in choosing a single lithologic type for the complete nappe system has introduced an error range in the order of $\pm 4\%$ for the top of the oil window.

The temperature data as obtained from well logs and temperature logs inherited a dynamic character and drill-stem-test bottom hole temperatures were again lower than the static. The vertical temperature

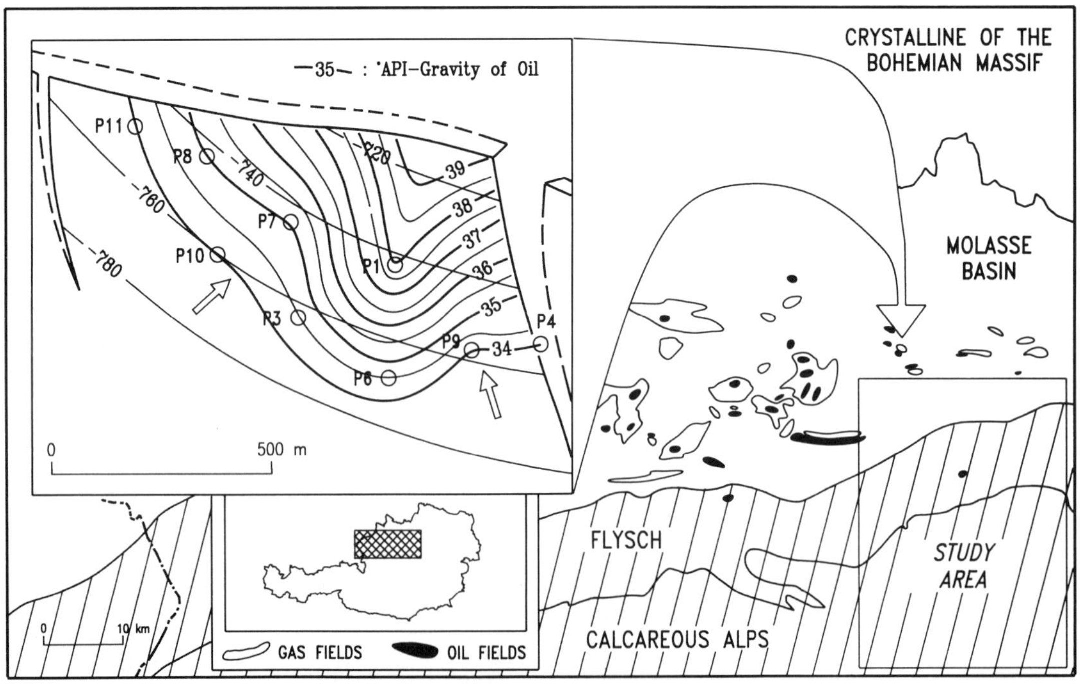

Fig. 13. Oil API-gravity distribution in the Piberbach oil field north of the study area.

distribution as calculated and checked against the static temperature data shows a general straight-line character down to the nappes and a general increase below them.

The vitrinite reflectance data show that the source interval, the Early Oligocene, is middle to late mature in the southern and central parts of the area, and immature in the north. The source rock started generating hydrocarbons in the central and southern sections of the area during the erosional period following the emplacement of nappes. The main oil generation started later and at greater depths moving northwards, because the controlling factor, i.e. the thickness of the thrust, decreases towards the north (Fig. 11). The base of the economical gas window has not been penetrated in the area and it is believed that well PW4 has an oil window between 4100 and 5950 m.

The calculated maturities in Molln 1 are lower than those derived from carbon isotopes (according to Faber, 1987) of test gas. Therefore, it can be stated that the gas produced in Molln 1 was not (completely) generated locally, but migrated from locations of higher maturity.

The source rocks within the Molasse Basin are immature. So the oil and gas produced in the Molasse Basin must have migrated to the north from underneath the nappes. The geographical location of the onset of the oil window coincides more or less with the front of the Calcareous Alps.

The existence of active hydrodynamics as the main cause for the migration of oil and gas is suggested by the API-gravity distribution both within the Molasse and within a particular field. The geochemical zoning (least mature oil in the north, most mature oil within autochthonous sediments in the overthrusted area) further supports this conclusion.

The overall heating rate in this part of the basin is low. Therefore, the study area has a better chance of containing gas accumulations than oil.

References

Erdoğan, L.T. and Akgül, A., 1980. Oil migration and re-entrapment model. Proc. 5th Biennual Petroleum Congress of Turkey, pp. 41–50.

Erdoğan, L.T. and Akgül, A., 1981. An oil migration and re-entrapment model for the Mardin Group reservoirs of southeast Anatolia. J. Pet. Geol., 4: 57–75.

Faber, E., 1987. Isotope geochemistry of gaseous hydrocarbons. Erdöl Erdgas Z., 103: 210–218.

Gussow, W.C., 1955. Time of migration of oil and gas. Am. Assoc. Pet. Geol., Bull., 39: 548–574.

Polesny, H., 1983. Distribution of the oil and gas fields of the Upper Austrian Molasse Basin. Erdöl Erdgas Z., 99: 90–102.

Wessely, G., 1987. Mesozoic and Tertiary evolution of the Alpine–Carpathian foreland in eastern Austria. Tectonophysics, 137: 45–59.

Yükler, A.M., 1987. How essential is quantitative basin modelling in petroleum exploration. Proc. 7th Biannual Petroleum Congress of Turkey, pp. 392–404.

Yükler, A.M. and Kokesh, F., 1984. A review of models used in petroleum resource estimation and organic geochemistry. In: J. Brooks and D. Welte (Editors), Advances in Petroleum Geochemistry, Vol. 1. Academic Press, London, pp. 69–113.

Yükler, A.M. and Welte, D., 1980. A three-dimensional deterministic dynamic model to determine geologic history and hydrocarbon generation, migration and accumulation in a sedimentary basin. In: Fossil Fuels. Technip, Paris, pp. 267–285.

FRANZ SCHMIDT *ÖMV Aktiengesellschaft Basin Modelling Group, Gerasdorferstraße 151, A-1210 Vienna, Austria*

L. TUFAN ERDOĞAN *ÖMV Aktiengesellschaft Basin Modelling Group, Gerasdorferstraße 151, A-1210 Vienna, Austria*

Geological–geochemical model of hydrocarbon systems generation in the deepest formations of the western Siberian Basin

N.N. Nemchenko, A.S. Rovenskaya and L.Sh. Girshorn

The West Siberian Depression is one of the world's largest petroleum provinces. The further increase of commercial hydrocarbon resources in this area will be tied to the development of ultra-deep horizons. The prediction of different phase states in such accumulations has been carried out using a variety of geologic–geochemical indicators. A new type of hydrocarbon system — high-temperature undersaturated gas condensates with high condensate content and recovery factors for liquid hydrocarbons approaching 100% — has been established. Further occurrences of this system are predicted in deep Jurassic levels.

Introduction

The further increase of commercial oil and gas reserves in western Siberia depends upon deeper geological horizons than hitherto exploited. These are confined to the northern region of the West Siberian oil and gas province (Fig. 1), where the maximum thickness of the Mesozoic–Cenozoic cover is 6–18 km. Jurassic, Triassic and Palaeozoic sediments have been penetrated in this area by isolated wells drilled to depths of over 4.5 km (Nadymskaya well 7; Yamburgskaya well 700; Urengoyskaya wells 264, 266, 282).

Commercial deep oil and gas discoveries have been made in Jurassic sediments of the Middle Ob region and commercial oil and gas is produced from Palaeozoic sediments, for example on the Yamal Peninsula and in the Nadym–Purovsk oil and gas region. Thus there is ample justification for exploration of oil and gas prospects in deeper parts of the western Siberian Basin. Jurassic sediments form the main prospective interval. In existing wells these sediments are penetrated to depths of 4–4.5 km and attain a thickness of up to 2 km.

At present it is possible to make only a qualitative estimate of the oil and gas potential of Palaeozoic sediments because of the low level of investigation. More quantitative estimations depend on improved efficiency in exploratory drilling. The prediction of possible pool distributions and their phase state is also of great importance.

The distribution of hydrocarbon reservoirs in the deep subsurface depends on primary genetic zoning. The selective generation of liquid or gaseous hydrocarbons depends on the maturity and composition (humic, sapropelic) of the organic source matter (the source rock facies). It also depends on the organic richness of the source rocks. Vertical migration processes that can affect genetic zoning within lower horizons of the sedimentary cover are rather limited.

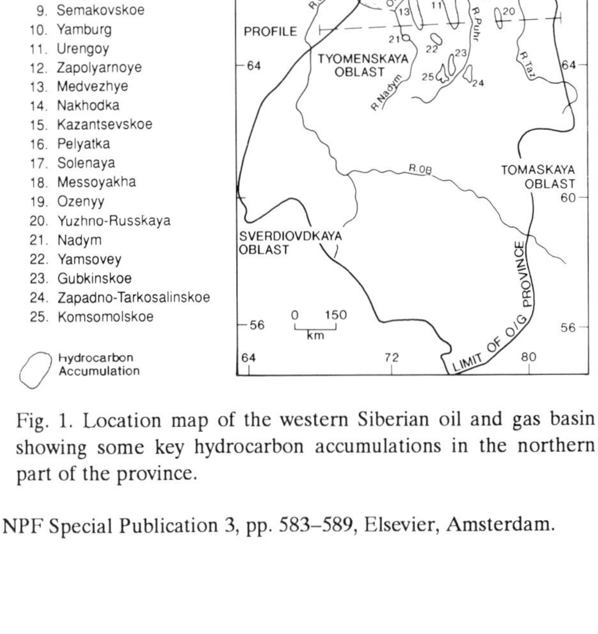

Fig. 1. Location map of the western Siberian oil and gas basin showing some key hydrocarbon accumulations in the northern part of the province.

Basin Modelling: Advances and Applications edited by A.G. Doré et al. NPF Special Publication 3, pp. 583–589, Elsevier, Amsterdam.

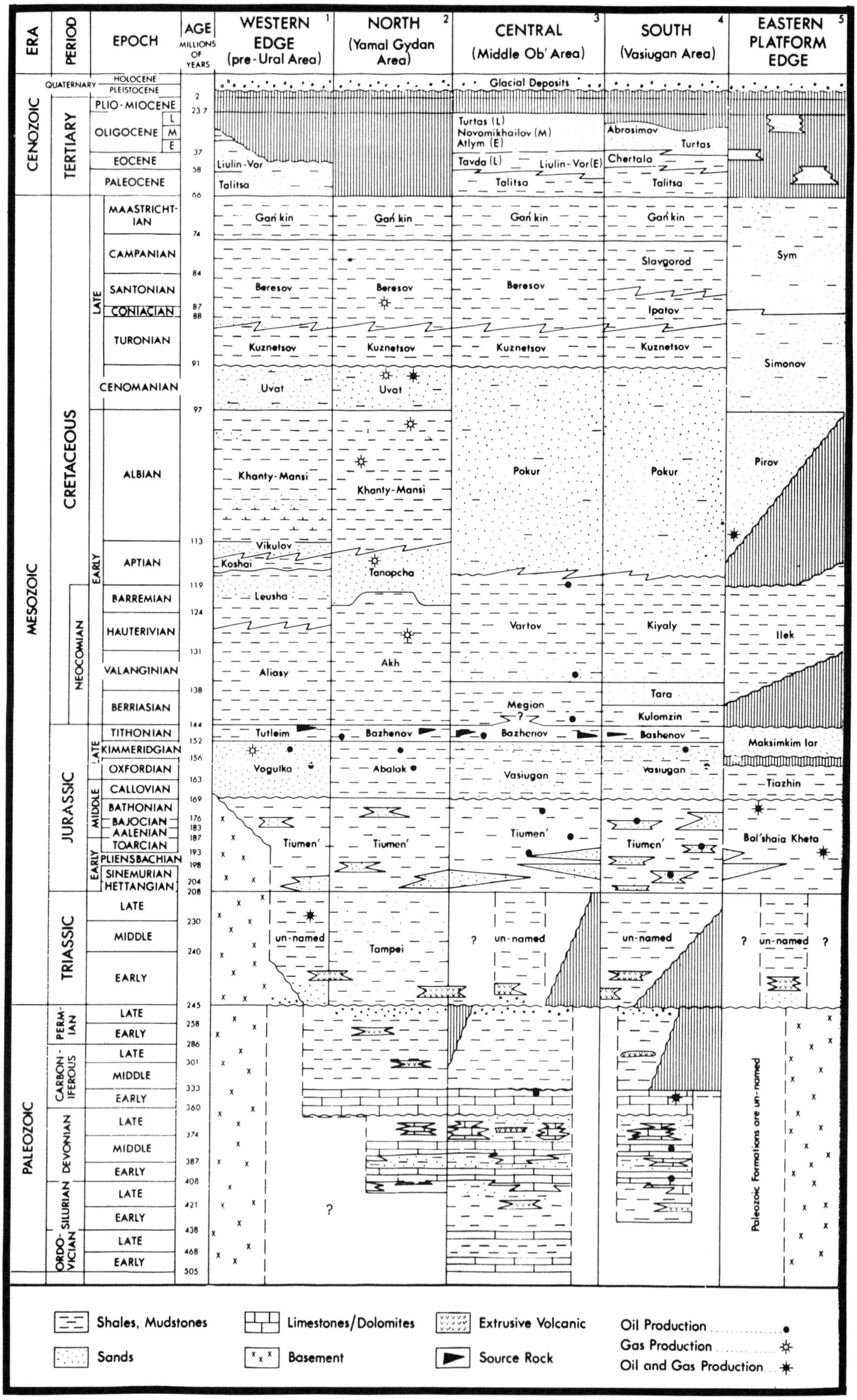

Fig. 2. Stratigraphy of the western Siberian Basin.

Geology

The generalized stratigraphy and basin configuration for western Siberia are shown in Figs. 2 and 3, respectively.

Western Siberian geophysical studies provide data to analyze the stratigraphy of the deeper stratigraphic zones. Seismic reflection data are the most informative, and indicate that at a depth of 5000 m (Fig. 4) the deep zones in most of the area studied

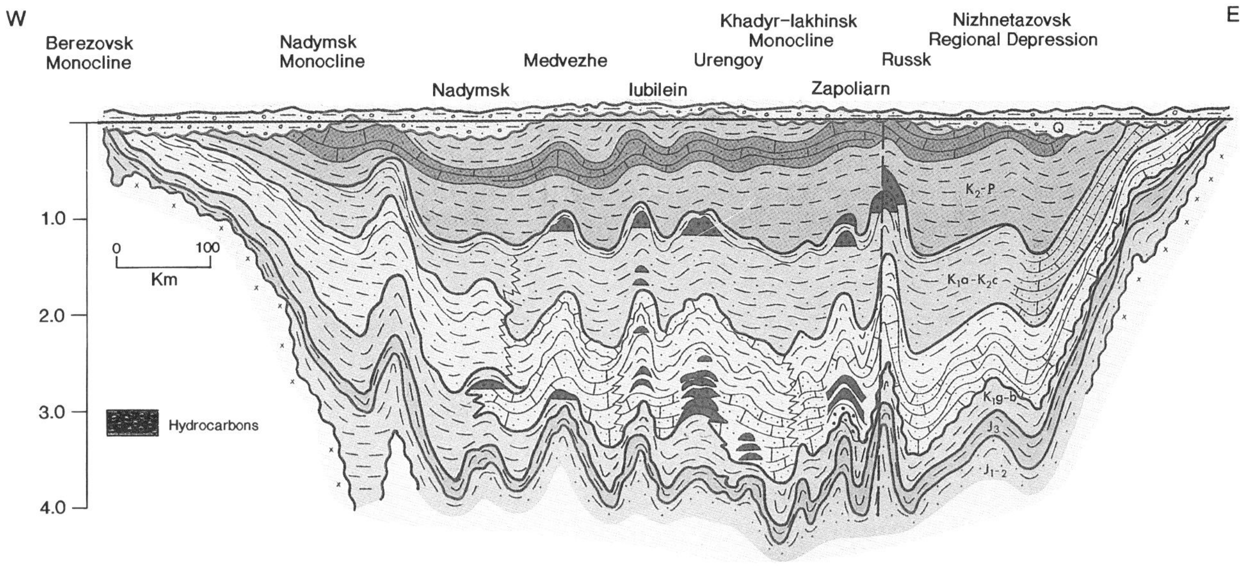

Fig. 3. East–west geological cross-section over the western Siberian Basin. For location see Fig. 1. J_{1-2} = Lower–Middle Jurassic; J_3 = Upper Jurassic; K_1g–b = Lower Cretaceous Hauterivian–Barremian; K_1a–K_2c = uppermost part of the Lower Cretaceous to Cenomanian; K_2–P = uppermost Cretaceous to Palaeogene; Q = Quaternary.

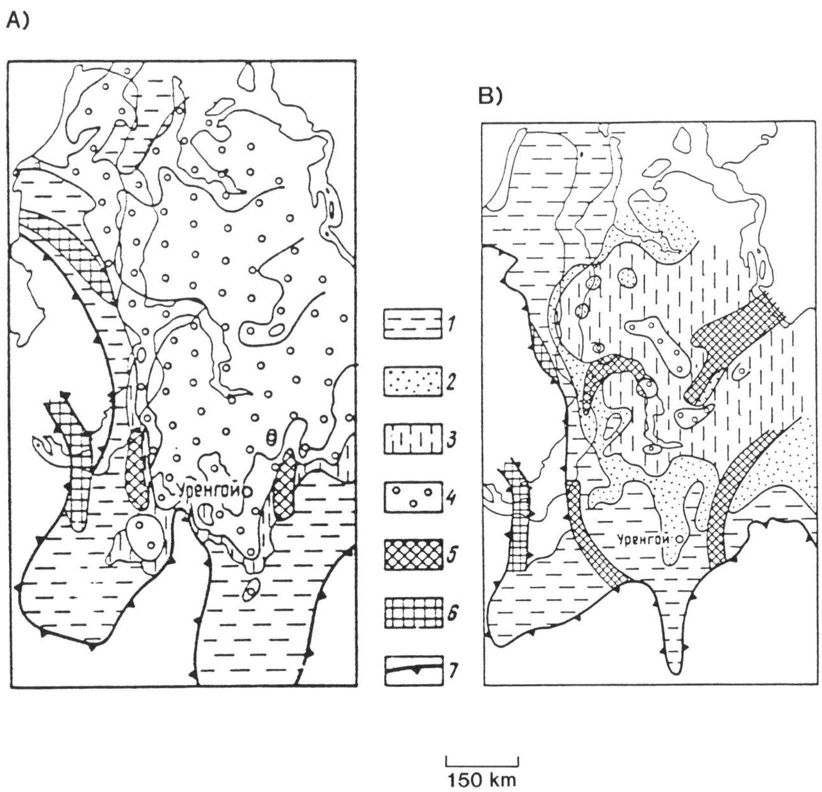

Fig. 4. Formation distribution at deep levels in western Siberia: (A) at 5000 m depth; (B) at 6000 m depth. Key to formations: *1* = Lower–Middle Palaeozoic carbonate–terrigenous; *2* = Upper Palaeozoic terrigenous–coaly–volcanic; *3* = Triassic continental argillaceous sandstone; *4* = Lower Jurassic marine and coastal-marine arenaceous–argillaceous; *5* = Palaeozoic terrigenous–carbonate–volcanic; *6* = Lower Mesozoic terrigenous–carbonate–volcanic; *7* = outline of basement complexes. The town shown in Russian script is Urengoy.

are represented by the Lower Jurassic, marine to coastal-marine Hetang Formation of Sinemurian to Toarcian age. Within the Nadym–Puhr oil and gas province this formation is underlain by a thin interval of the Triassic Tampenian series. The Lower–Middle Palaeozoic carbonate/clastic succession is developed in the northwest and in the north (in the Yamal and Gydan oil and gas province).

Deeper in the section (6000 m depth) the thickness of the Jurassic sediments significantly decreases (Fig. 4). It remains only in the central part of the northern area and disappears in the Yamal and Gydan peninsulas at a depth of approximately 6000 m, but continental Triassic deposits are well developed and form isolated troughs. In some areas the Upper Palaeozoic Tungusian series, a succession of coal-bearing clastics and volcanics, occurs. Deeper in the basin at a depth of approximately 7000 m, the Lower Jurassic formation practically disappears, but the Tampenian series and the Lower–Middle Palaeozoic carbonates/clastics remain.

The prediction of formation types is verified by drilling results and core data from some deep wells. For example, the Nadymskaya well (5000 m depth) terminated in Lower Palaeozoic deposits and the Yamburgskaya well 700 (6000 m depth) did not penetrate the Lower Jurassic. In the Urengoy area, however, deep wells penetrated Lower–Middle Jurassic deposits of the Tiumen and Yamburg series at 3.8–5.5 km depth.

Lower Jurassic marine to coastal-marine clastic formations contain sapropelic–humic organic matter of mixed origin. Arctic region sediments were deposited in normal marine basins, but sediments of the central-north area (Urengoy area) were deposited in near-shore subcontinental conditions that again caused the accumulation of sapropelic–humic organic matter. The Triassic Tampenian series was deposited in continental conditions that resulted in the accumulation of humic organic matter.

The upper Palaeozoic is represented by clastic strata of the Tungusian series with volcanogenic–coaly intercalations, and the lower Palaeozoic by carbonate/clastic strata deposited in a variable continental/marine depositional environment which resulted in a mixed sapropelic/humic organic facies development.

Hydrocarbon phase types and maturation

The study of source rock catagenesis (maturity) to depths of 5000 m has shown that the deep-seated zones in the Nadym–Puhr oil and gas province and the northwestern part of the Yamal Province are characterized by maximum vitrinite reflectance values corresponding to the AK_1 stage of catagenesis, while deep-seated zones within the Puhr–Tazovsk oil and gas province are characterized by minimum values, corresponding to the MK_3 catagenesis stage. Figure 5 shows these catagenetic stages and gives their equivalent vitrinite reflectance values. Table 1 compares vitrinite reflectance values and formation temperatures for some key reservoirs.

There is a relationship between the degree of catagenesis (maturity), thickness of sediments below 5000 m level (from 5000 m depth to the basement), geothermal gradient and gradient of vitrinite reflectance change. "Reduced" catagenetic zonation (relatively high gradients) is characteristic of the western part of the Nadym–Puhr and the northern part of the Yamal oil and gas provinces (AK_1 catagenesis phase). Here there are rather high geothermal gradients and minimum thicknesses of the sedimentary cover (1–3 km). The vast territory within the Taz and Gydan peninsulas is characterized by an "extended" zonation of catagenesis and also by rather low geothermal gradients (MK_3 phase) with maximum thicknesses of sedimentary cover (over 10 km). Most of the Nadym–Taz syncline and the western part of the Yamal–Gydan syncline are characterized by the MK_4 catagenesis stage and 6–8 km sedimentary cover thickness (Fig. 5).

Studies of catagenesis and stratigraphic development in the deeper parts of the basin allow us to predict the hydrocarbon phase and to define promising zones for the discovery of specific hydrocarbon phase types. The most likely zone for oil discoveries is the MK_3 stage of catagenesis where the section is represented by Lower Jurassic marine to coastal-marine deposits (with sapropelic and sapropelic–humic organic matter). The MK_3 stage of catagenesis within

TABLE 1

Depth, vitrinite reflectance and reservoir temperature relations for some key accumulations in western Siberia

Area	Depth (m)	R_o (%)	T_x (°C)
Yamalskaya oil/gas area (northern part):			
Kharasveyskaya	2000	0.62	96
Bovenenkovskaya	3000	0.94	112
Kharasveyskaya	4000	1.60	160
Nadym–Purskaya oil/gas area:			
Gubkinskaya	3000	1.00	100
Nadymskaya	4000	1.27	121
Nadymskaya	5000	2.40	160
Urengoyskaya	5000	2.0–2.4	140
Puhr–Tazovskaya oil/gas area:			
Russkaya	2000	0.48	58
Russkaya	3000	0.5	61
Tazovskaya	4000	0.62	75
Tazovskaya	5000	0.73	92

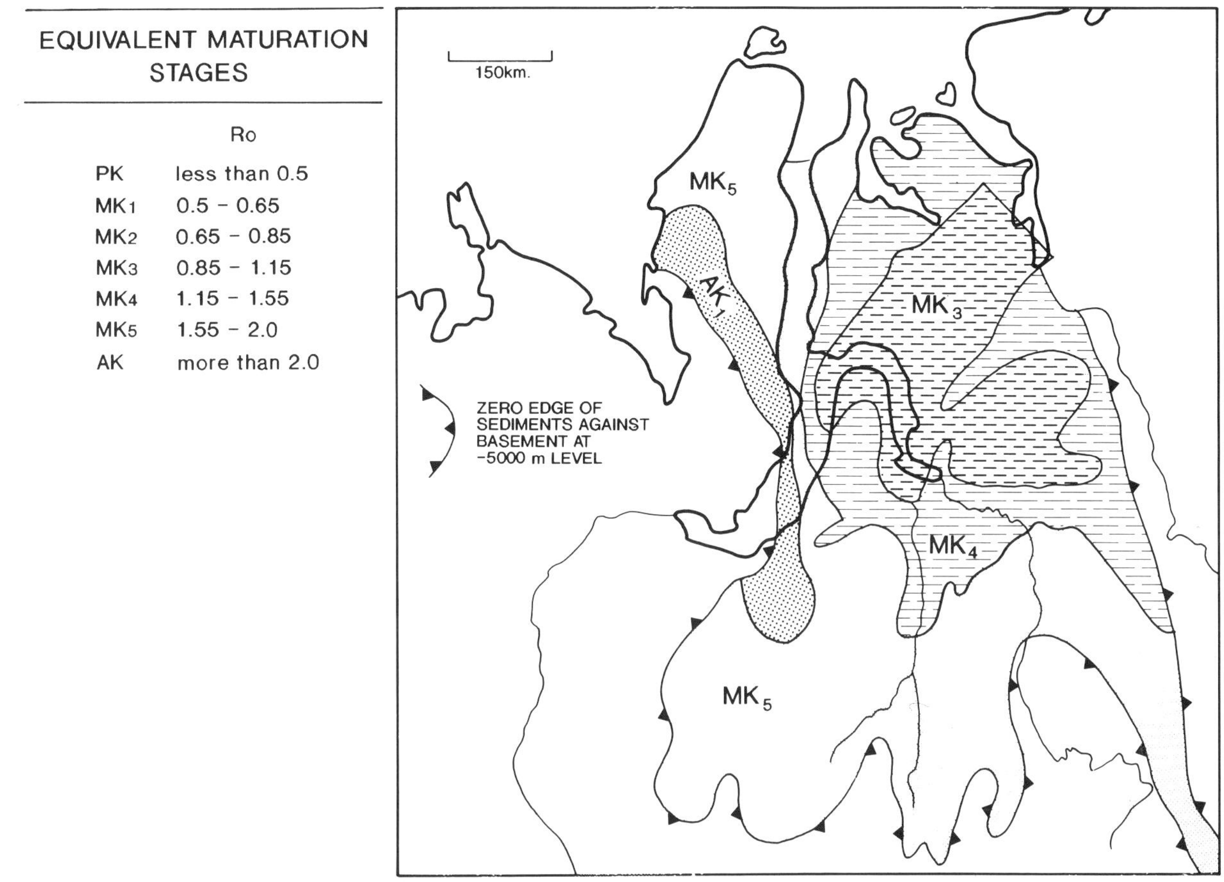

Fig. 5. Level of maturation at a depth of 5000 m in western Siberia. The relationship between the maturation classification used in western Siberia and vitrinite reflectance is shown.

Lower Jurassic marine to coastal-marine formations holds promise for finding oil- and gas-condensate, gas-condensate and gas pools. In these zones at high catagenesis the type of formation does not define the phase state of hydrocarbons. Zones with predominant development of oil–gas-condensate pools are predicted in the Puhr–Taz, Ust–Yenisey and Gydan provinces. Oil pools of high gas–oil ratio are possible in the Puhr–Taz province and the development of mainly gas-condensate pools is likely in the Yamal and Nadym–Puhr provinces. (Fig. 5)

Geochemistry

Extensive studies of geochemical factors to determine the detailed hydrocarbon composition of the gasoline fraction, and higher molecular weight alkane fraction, of oil and condensate in producing areas of western Siberia has allowed us to predict the hydrocarbon phase state in deep zones of the northern part of the West Siberian oil and gas province.

Established relations of hydrocarbon compounds of 2-Mh and 3-Mh (2-methylheptane/3-methylheptane) paraffin, of the napthenic and aromatic series (cyclopentane/cyclohexane, meta + para-xylol/ethylbenzene), and of i-C_{19}/i-C_{20} rather precisely indicate zones of different hydrocarbon phase state. Minimum values (<1) of meta + para-xylol/ethylbenzene ratios indicate an oil zone. Maximum values (>10) are indicative of the gas-condensate zones. Intermediate values (1–5) are characteristic of oil–gas-condensate zones. The 1.6–1.7 values of the 2-Mh/3-Mh ratio specify the oil zone, 1.1–1.3 the gas-condensate zone, and 1.3–1.7 values the oil–gas-condensate zone. Minimum values of the i-C_{19}/i-C_{20} ratio (<1) indicate an oil zone. Maximum values (>5) indicate the zone of gas-condensate, and intermediate values (3–5) the zone of oil–gas-condensate.

Table 2 presents data on the composition of the gasoline and the higher molecular weight alkane fraction of oils and condensates produced from Lower–Middle Jurassic deposits in deep wells in the Urengoy Field. From hydrocarbon composition and phase-genetic indices, oils and condensates can be referred to the lower "intermediate" oil–gas-condensate zone and to the zone of gas-condensate pool development. The accumulation of oil reservoirs with high "methane" content (see Table 2) and high gas–oil ratio is predicted within the Puhr–Taz province. The established zonation is consistent

TABLE 2

Geochemical characteristics of hydrocarbon zones in Lower–Middle Jurassic strata in western Siberia

Region	Depth (m)	Benzene fractions						Higher mol. wt. alkane fractions (i-C_{19}/i-C_{20})	T (°C)	Organic matter catagenic stage
		HC composition (%)			HC correlations					
		n- and iso-paraffins	naphthenes 5- and 6-member	arenes	2-Mh/3-Mh [a]	cyclo-pentane/cyclo-hexane	meta + para-xylol/ethylben-zene			
Oil zone (methane):										
Pokachevskaya	2725–2763	73.04	24.06	2.90	2.16	0.70	0.79	0.88	98	MK_2
Shirokovskaya	2886–2892	79.58	25.84	3.58	1.76	0.676	0.01	0.72	97	MK_2
Kogolymskaya	2803–2807	62.03	34.10	3.87	1.65	0.60	0.89	0.88	93	MK_3
Oil/gas condensate zone (methane–napthene):										
Talinskaya	2666–2692	65.19	28.14	6.67	1.40	1.72	1.70	1.11	108	MK_3
EM-Egovskaya	2619–2675	69.21	29.06	1.73	1.92	1.61	3.54	1.01	107	MK_4
	2369–2500	59.62	31.83	8.55	1.60	1.39	2.96	0.93	102	MK_4
Gas condensate zone (naphtene–methane and napthene):										
Kharasaveyskaya	3170–3202	53.74	40.46	5.80	1.16	2.06	8.35	3.83	123	MK_4
Urengoyskaya	3710–3730	54.63	39.68	15.69	–	4.98	11.56	4.00	113	MK_5
Samburgskaya	4320–4380	52.89	45.92	1.18	1.14	2.78	17.51	3.97	130	MK_5
Urengoyskaya [b]	4753–5034	65.72	32.58	1.69	1.012	0.811	12.93	3.39	140	MK_5
Urengoyskaya	4504–4520	41.69	49.34	8.99	0.930	0.768	12.44	–	140	MK_5

[a] 2-Methylheptane/3-Methylheptane.
[b] Deeply buried zone (more than 4.5 km).

with the change of organic matter catagenesis. The oil zone is characterized by MK_1–MK_3 catagenesis stages, the oil–gas-condensate zone by MK_4, MK_5–AK_1 stages.

The combined analysis of seismic and geological/geochemical data (formation types, study of organic matter catagenesis, individual hydrocarbon composition of gasoline fractions and high alkane fractions of oil and condensates) allows us to predict source rock maturity in deeply buried northern zones of the Lower–Middle Jurassic in western Siberia for possible development of oil–gas-condensate and gas-condensate systems. The area of mainly oil–gas-condensate generation is predicted in the Puhr–Taz and Gydan province, and the area of mainly gas-condensate generation is characteristic of the Yamal (north part) and Nadym–Puhr provinces.

The ratio of liquid and gaseous hydrocarbons is of great importance in obtaining precise values of potential hydrocarbon resources in deep basins. The phase state of multicomponent reservoir hydrocarbon systems is defined by their composition, physical properties and temperature/pressure conditions. Studies of phase diagrams for local deep hydrocarbon accumulations indicate the existence of systems in which the reservoir temperature does not allow for a pressure at which condensation can begin. The distinctive feature of these systems is that at reservoir temperature they cannot be condensed at any pressure down to hydrostatic. The condensate content in these systems is up to 1000 g/Sm^3 and above. In western Siberia such a reservoir was discovered at deep levels in the Urengoy Field (5000 m depth, 99 MPa formation pressure, 140°C formation temperature). The condensate content is 760 g/Sm^3.

A characteristic feature of the above systems is their undersaturation at high contents of liquid hydrocarbons. This allows production processes to be devised such that the recovery factor is close to 100%. Studies of the hydrocarbon system phase state in northern areas of western Siberia suggest that this type of accumulation may be common where reservoir pressure exceeds 40 MPa and temperature exceeds 120°C.

Conclusions

Studies of the probability of pool occurrence and hydrocarbon phase state have made it possible to predict hydrocarbon types in the deeper horizons of western Siberia. The areas thought most likely for liquid hydrocarbon accumulations are the South Gydan, Taz, Ust–Yenisey, Chasel and Tolkin regions in the central and northern parts of the province. Within these regions oil, condensate and gas pools are expected. They are associated with Jurassic and Triassic clastic deposits of 1500–4000 m thickness.

The Urengoy, Gubkin, Yamal, north Gydan (Nadym, east Taz) and Russko–Chasel regions are mainly promising for gas-condensate, oil–gas-condensate and gas pools. The thickness of prospective Jurassic and Triassic clastic deposits in these regions is 1000–1600 m. In the Nurmin (southern part), Nadym (western part) and Gubkin (southern part) regions, the potential for the occurrence of oil and gas is considered to be low. Dry gas pools are possible in Palaeozoic terrigenous-carbonate rocks of various ages, up to 800–1000 m in thickness.

Areas outlined by the 120°C isotherm, for example the northwestern part of the Yamal oil and gas region, are prospective areas for new types of pool (undersaturated gas-condensate accumulations of high condensate content) at the Jurassic level. Taking into account the Jurassic oil and gas interval thickness (1500–2500 m) and high reservoir temperatures at its base (up to 200°C) one should expect the more extensive occurrence of undersaturated gas-condensate in deeper horizons of the northern part of the western Siberian Basin. Zones of reservoir pressures and temperatures over 40 MPa and 120°C, respectively, are consistent with areas with a possible condensate content of up to 1000 g/m^3 and more. Generally, possible zones featuring gas-condensate systems with a high liquid hydrocarbon content are delineated by zones of abnormally high reservoir pressures. The recognition of these new types of accumulation will give rise to increased liquid hydrocarbon recovery, both due to their high liquid content and the prevention of condensate losses during exploitation.

Discussion

Philippe Ungerer, IFP, France: *What is the role of permafrost in contributing to the sealing efficiency of caprocks in the Urengoy area?*

Authors' reply: Permafrost does not serve as a caprock for deep-seated fields, but it can influence the thermal regime. It is not important as a caprock in Urengoy, which is sealed by the Turonian. Other, shallower fields are sealed by permafrost. An example is Messoyshkoe, where the reservoir is in the Cenomanian at a depth of 600–700 m.

N.N. NEMCHENKO *State Commision for Natural Resources, B. Poljanka 54, Moscow, Russia*
A.S. ROVENSKAYA *Institute of Geology and Exploitation of Combustible Fuels, Moscow, Russia*
L.Sh. GIRSHORN *Imalneftegas Geophysicq, Salekhard, Russia*

Main factors controlling regional oil and gas potential in the West Arctic, former USSR

B.K. Ostisty and S.I. Cheredeev

Periods of fundamental structural rearrangement — of the West Arctic as a whole or of separate large areas — were accompanied by changes in boundaries of sedimentary basins and in sedimentation regimes. These conditions resulted in the present-day scattered distribution (both areally and vertically) of oil and gas occurrences.

The main factors controlling the present regional oil and gas potential in the area are associated with the geodynamic evolution of the region, especially with sedimentation and geothermal conditions. Analysis of these factors has resulted in an estimation of speculative resources in the West Arctic. We have taken into account all the data concerning discovered resources in the region. Proven and potential resources are illustrated in map form for the principal structural units of the region.

Introduction

The oil and gas potential of the Pechora–Barents–Kara metaplatform area is a consequence of the long-term geological evolution, stepwise formation and reformation of separate petrogeological elements — from large zones of oil accumulations to local hydrocarbon accumulations.

Major periods of structural deformation of the whole or parts of the region influenced the development of sedimentary basins and sedimentation regimes, resulting in the present-day scattered distribution (areal and vertical) of oil and gas occurrences.

The structural framework of the Barents Sea Shelf was presented in detail as a joint presentation with Statoil at the Arctic Geology Conference in Tromsø, August 1990 (Ostisty and Fedorovsky, 1992). In this paper we will give a brief overview of these data.

The raw data on which maps and concepts presented in this paper are based (Fig. 1) comprise the following: (1) marine seismic in the Barents Sea (both reflection and refraction); (2) well data; (3) geochemical studies in wells and along profiles; and (4) geological data from the surrounding land and islands.

Sediment thickness in the centre of the basin reaches 20 km (Fig. 2). The structure of the cover is illustrated in Fig. 3 which comprises contour maps of two regional reflectors. Figure 3A shows depth to near Top Jurassic. Figure 3B indicates depth to the top of the Permo–Carboniferous carbonates.

The main structural elements of the region are shown in Fig. 4. They are as follows: Admiralty High, Central Barents High, North Barents and South Barents Basins separated by the Ludlov Saddle.

The structure of the Jurassic sediments is fairly disturbed except along the Ludlov Saddle. Little structural contrast in the Central Barents High and absence of Jurassic sediments on the larger part of Admiralty High are typical.

The main factors controlling the present-day regional oil and gas potential in the area under consideration are associated with the geodynamic evolution of the region, especially sedimentation and geothermal conditions.

Geodynamic evolution

When assessing the factors attributed to the regional geodynamic evolution of this area it should be kept in mind that the boundary between the South Barents Basin and adjacent Pechora and Northern Kara area is arbitrary. These two areas probably interacted during the development of major zones of oil and gas accumulations, especially in the Triassic–Jurassic interval when hydrodynamic communication was possible between them. This idea is illustrated by the backstripped geoseismic sections shown in Fig. 5.

It is shown in Fig. 5 that in the pre-Upper Permian period the Kola–Kanin zone of uplift completely separated the Pechora and the Barents Basins. In the post-Permian period the Pechora block was involved in the large-scale subsidence of the Barents Basin and as a result the Pechora–Kanin zone of palaeo-uplift was broken up. Thus, in the post-Permian period the Pechora and the Barents synclines were

Basin Modelling: Advances and Applications edited by A.G. Doré et al. NPF Special Publication 3, pp. 591–597, Elsevier, Amsterdam.

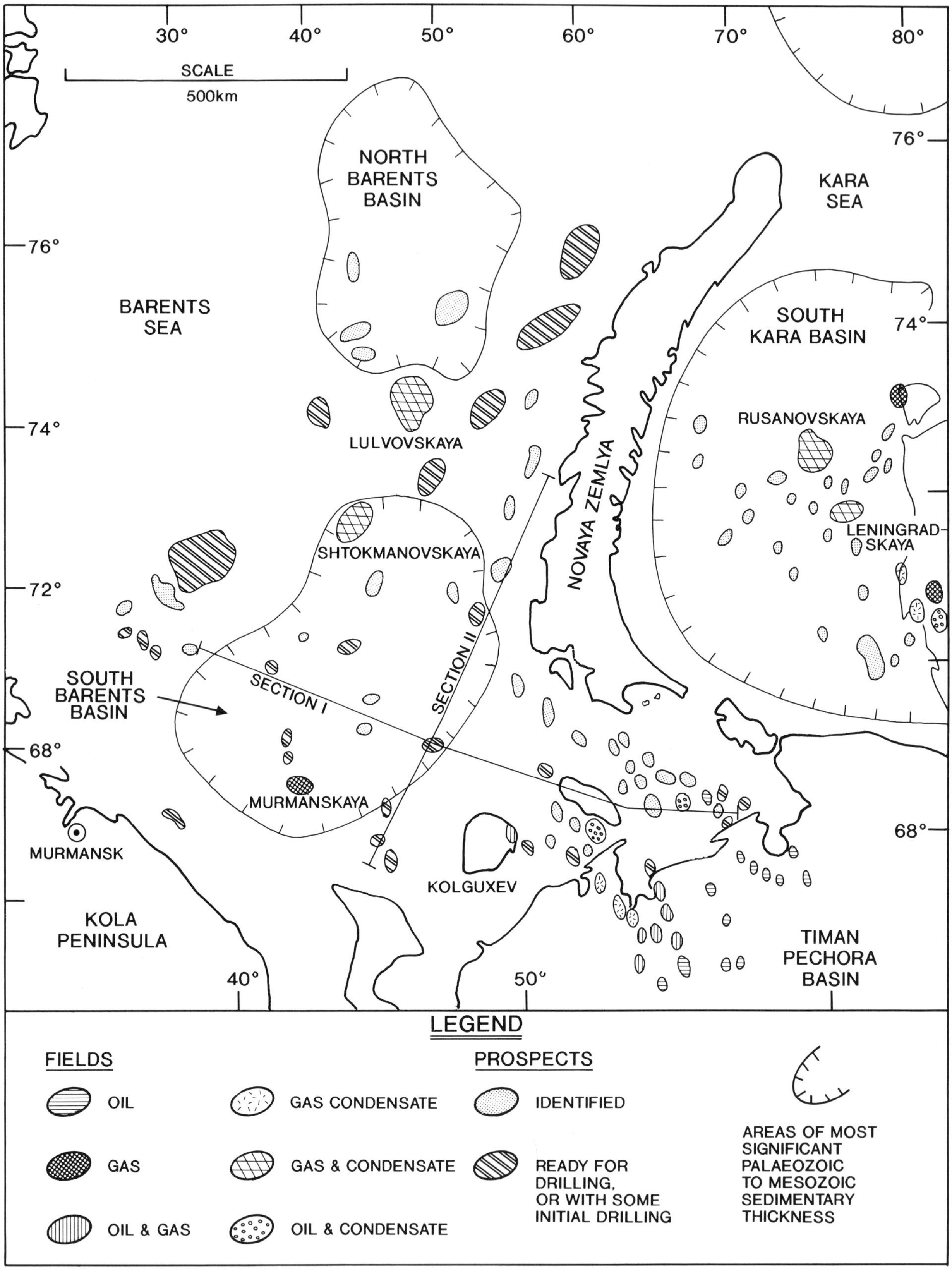

Fig. 1. Location map of the Barents–Kara region, showing oil, gas and condensate discoveries. Undrilled prospects and their state of readiness for drilling are also indicated. The density of seismic coverage in the area is shown in Ostisty and Fedorovsky (1992, fig. 1). Cross-sections *I* and *II* refer to Figs. 5 and 6.

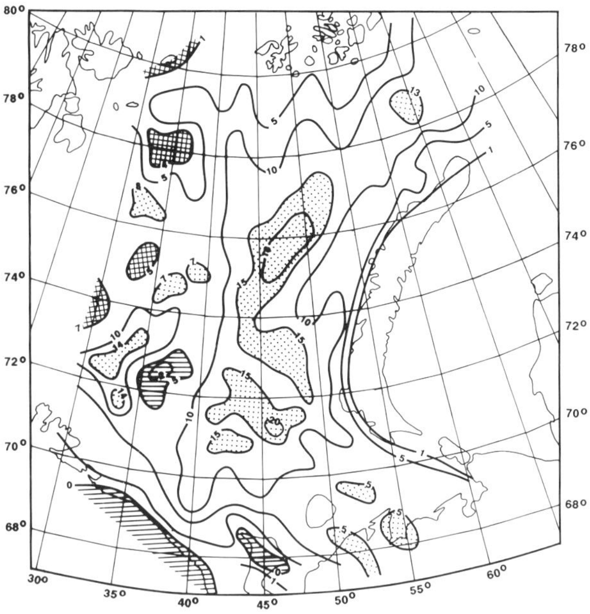

Fig. 2. Isopachytes of the post-basement sedimentary cover. Contours in kilometres.

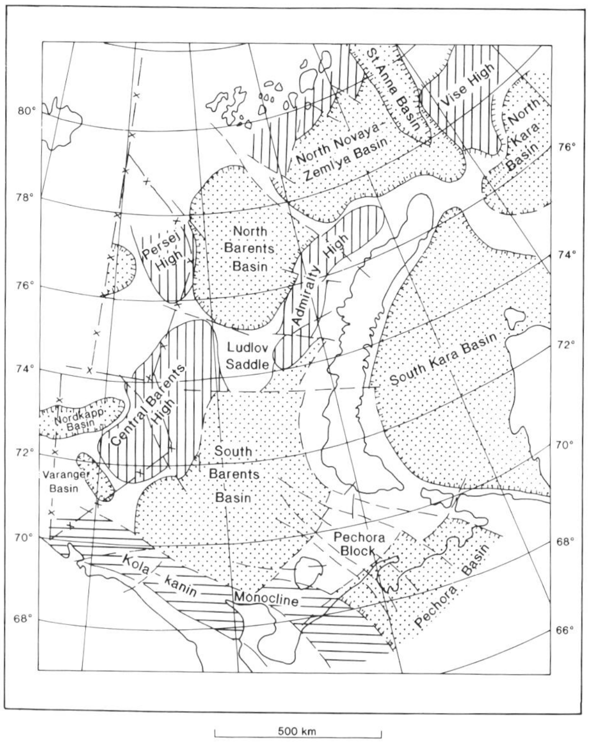

Fig. 4. Main structural elements of the Barents and Kara Seas.

united into a common petroleum megabasin comprising three basins that earlier were separate.

The Barents syncline was a single basin in the pre-Upper Permian period, but during the Triassic period the Ludlov Saddle developed in its central part, dividing it into the North Barents and South Barents Basins.

The processes of unification and separation of petroleum basins and the overall patterns of regional subsidence have resulted in the following favourable geodynamical factors:

(1) A favourable temperature–pressure regime and development of hydrocarbon migration paths,

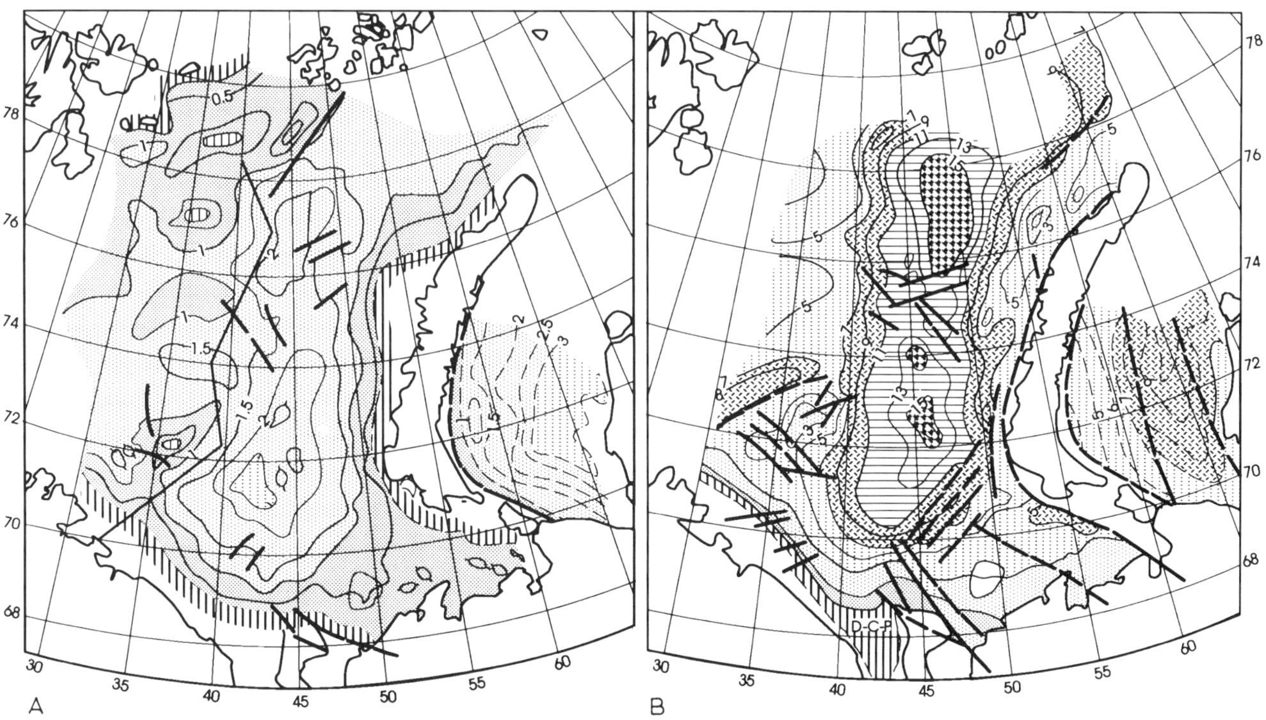

Fig. 3. Depth (in km) to some key reflectors in the Barents–Kara region. (A) Near Top Jurassic. (B) Top of Permo-Carboniferous carbonates.

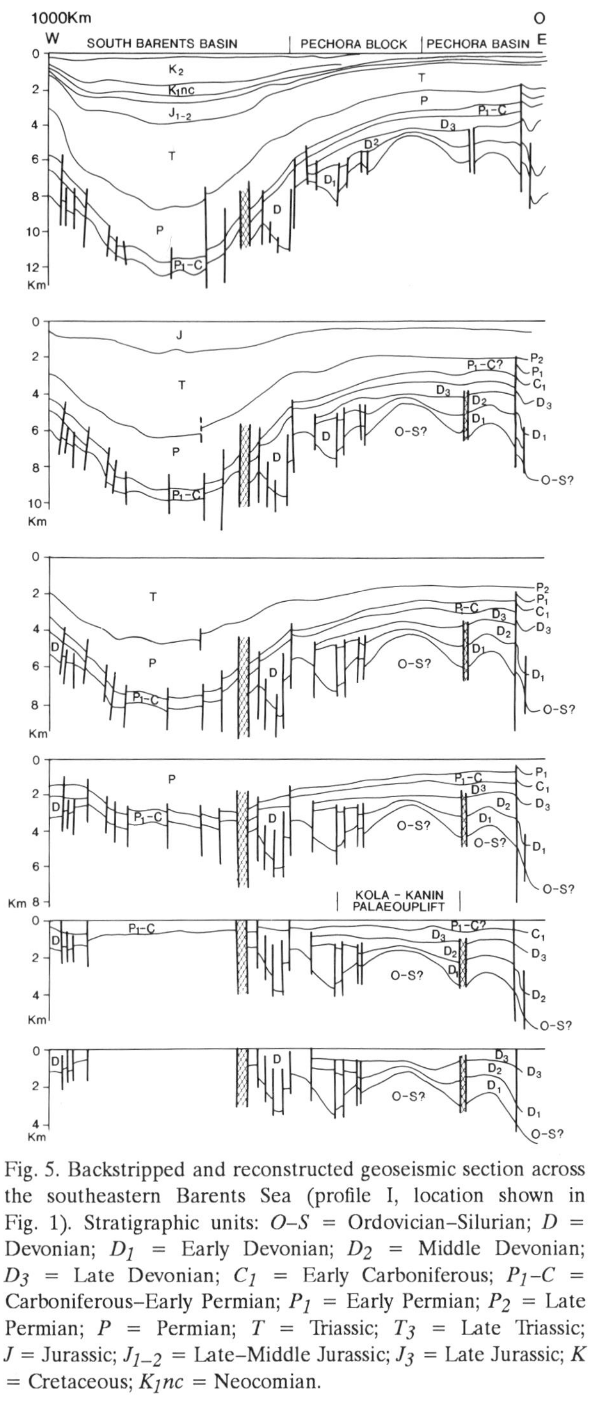

Fig. 5. Backstripped and reconstructed geoseismic section across the southeastern Barents Sea (profile I, location shown in Fig. 1). Stratigraphic units: *O–S* = Ordovician–Silurian; *D* = Devonian; D_1 = Early Devonian; D_2 = Middle Devonian; D_3 = Late Devonian; C_1 = Early Carboniferous; P_1–*C* = Carboniferous–Early Permian; P_1 = Early Permian; P_2 = Late Permian; *P* = Permian; *T* = Triassic; T_3 = Late Triassic; *J* = Jurassic; J_{1-2} = Late–Middle Jurassic; J_3 = Late Jurassic; *K* = Cretaceous; K_1nc = Neocomian.

especially on the edges of rift troughs resulting from thinning of old continental crust in the Palaeozoic and Mesozoic.

(2) Favourable conditions for lateral migration of hydrocarbons in basins from the Permian through to the Cretaceous.

(3) Predominance of conditions suitable for the generation of hydrocarbons in the central parts of the basins for very long periods of time.

(4) Alternation of "primary" accumulative hydrocarbon zoning due to differential vertical movements of synclines during the Mesozoic–Cenozoic.

Sedimentation characteristics

The essential factors are as follows:

(1) Accumulation of a sedimentary sequence 15 km thick in the main areas of hydrocarbon generation due to high sedimentation rates in the Palaeozoic–Mesozoic.

(2) High rates of sediment burial with undercompaction in the Triassic period, which resulted in the vertical thickening of the diagenetic and catagenetic zones and the migration of dissolved hydrocarbons.

Sedimentation is illustrated on the graphs of palaeo-tectonic movements for two profiles shown in Fig. 6. The two left-hand graphs show sedimentation rates in the Permo–Carboniferous, Upper Permian, Triassic and Jurassic periods (in mm/1000 years). The central graphs show relative subsidence rates for the same complexes. The vertical axes of the graphs show the ratio between thickness of the complex and total thickness of sedimentary cover. On the right, absolute subsidence amplitude along the line of the profile is shown for each of the units.

The graphs show that the tectonic regime was most stable in the Carboniferous–Early Permian interval. Sedimentation rates were not high (up to 15 mm/1000 y.) and constant within the region. Subsidence was more active in the Late Permian, and sedimentation rates exceeded 150 mm/1000 y. in the larger part of the South Barents Basin. In the Triassic period, the picture was the same; rates decreased a little, but still remained high. A new tectonic–sedimentary cycle started in the Early Jurassic. Subsidence was moderate. Sedimentation rates were not more than 35 mm/1000 y. and relatively constant.

(3) High sedimentation rates prohibited the full realization of the oil-generating potential in the thermal range 60–150°C, owing to subsidence to great depths and into zones of temperatures above 150°C, where gas would be generated.

(4) Favourable accumulation–seal relationships in sediments of a wide stratigraphic range, the great variety of trap types and predominance of favourable hydrodynamic conditions are attributed to the instability of sedimentation setting.

Geothermal conditions

The geothermal regime in the region controlled the position of source rock maturation zones, generation zones and distribution of hydrocarbon type. Key observations are:

(1) Elevated heat flow in the deepest parts of the South and North Barents synclines and in the Prednovozemelskaya transition zone.

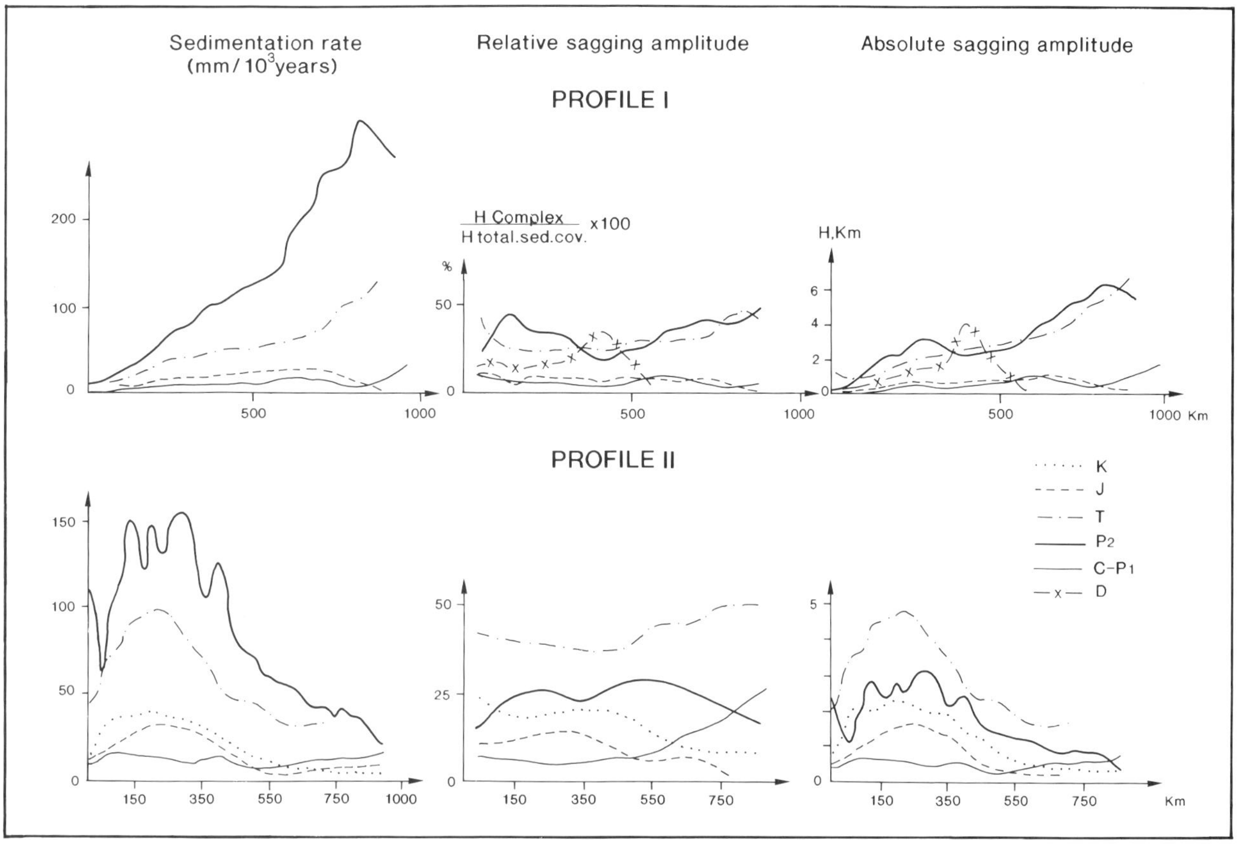

Fig. 6. Graphs illustrating palaeotectonic movements and sedimentation along profiles I and II (location shown on Fig. 1). The left-hand graphs show sedimentation rates for the main stratigraphic units. The central graphs indicate relative subsidence (i.e. the ratio between the thickness of a stratigraphic unit and that of the total sedimentary cover). The right-hand graphs show total subsidence amplitude for each unit. Symbols for stratigraphic units are as shown in Fig. 5.

(2) The supposed decrease of rock temperature at the depth of 4–5 km of 30–50°C during the Cenozoic, characteristic of sedimentary basins where the main subsidence was completed in the Mesozoic period. This could lead to the liberation into the free phase of great quantities of gas from oil and interstitial waters, and the reorganization of oil deposits into oil–gas–condensate and oil-and-gas, as well as the extraction by free gas of additional volumes of oil.

Figure 7 shows the depth of palaeo-isothermal lines for the interval immediately preceding the Cenozoic period. The maps have been constructed for temperatures of 60°C, 130°C and 150°C. These are important control points in the process of organic matter maturation. The comparison of these maps testifies that there are favourable conditions for petroleum generation within the interval 2 to 8 km.

Petroleum potential of the region

Analysis of the above-mentioned factors has resulted in the estimation of speculative resources in the Barents Sea region. We took into account all the data on discovered resources in this region.

The results of our forecast are shown on a map (Fig. 8). Five regions (A–E) were evaluated. The map contains the following information: mainly oil-bearing areas (oil proportion > 60%); mainly gas-bearing areas (oil proportion < 40%); gas–oil-bearing areas.

Figure 8 describes the relative importance of the five areas, and the type of hydrocarbon phase expected (see figure caption for explanation).

The largest hydrocarbon resources (approximately 50%) are concentrated in the South Barents Basin (B) and Central Barents High (E). Both of these areas are mainly gas-bearing. Hydrocarbons are mainly to be found in the Jurassic and Triassic strata.

The main oil resources (> 80%) are also concentrated in these areas and in addition the Prinovozemelskaya area (D). In all of these areas oil prospects are mainly in Triassic sediments (> 60%). The remainder of the resources in areas B and E are concentrated in Jurassic–Cretaceous strata and in the Permo–Carboniferous for the D area.

The Pechora block (A) contains approximately 10% of the hydrocarbon resources of the region. This

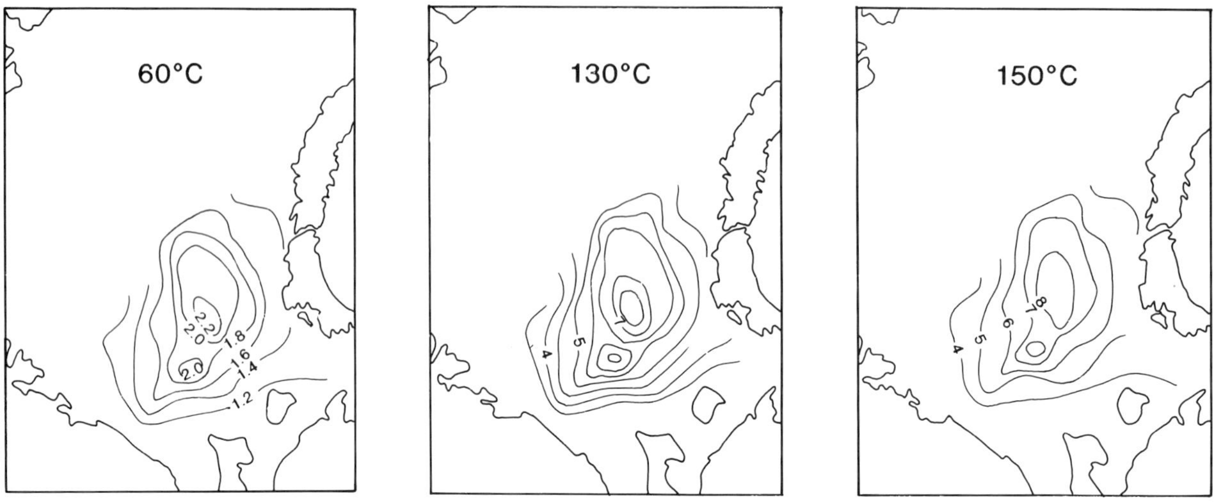

Fig. 7. Palaeo-isothermal lines as modelled for the eastern Barents Sea in immediate pre-Cenozoic time. Palaeo-depths to 60°C, 130°C and 150°C geotherms are indicated.

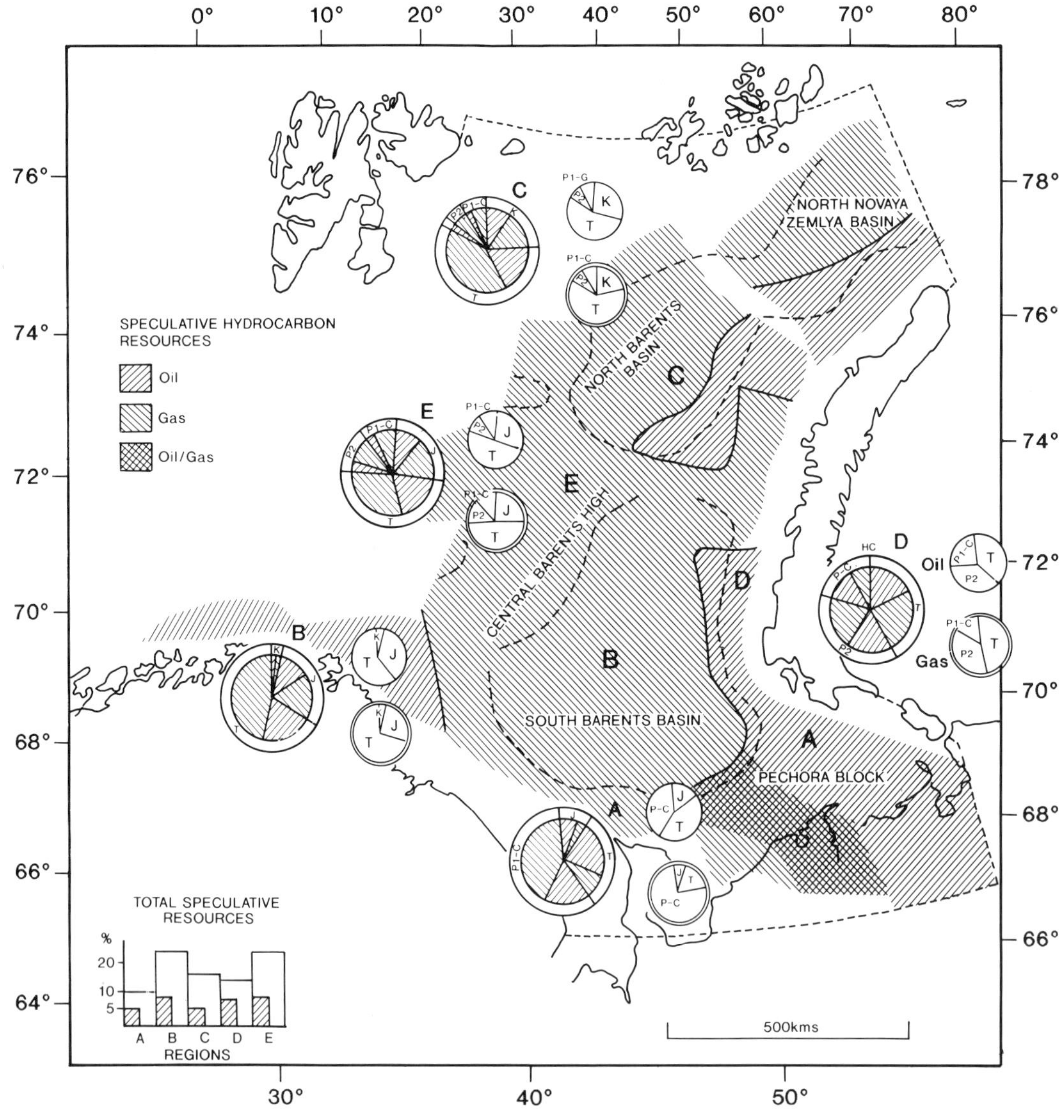

Fig. 8. Speculative hydrocarbon resources of the Barents Sea. Resources are shown for structural divisions *A* to *E*. Large pi-diagrams show the relative importance of the main formations in containing hydrocarbons, and the predominant hydrocarbon type expected in each formation. Small pi-diagrams treat oil and gas separately, and show the expected distribution of each phase between the formations. Bar charts show the expected proportion of total hydrocarbon resources within each area.

is mainly an oil-bearing region. The North Barents Basin (C) is the least explored area. According to our estimation it contains up to 15% of the resources of the region. The area is thought to be mainly gas-bearing, with most of the resources in Triassic strata.

Reference

Ostisty, B.K. and Fedorovsky, Y.F., 1992. Main results of oil and gas prospecting in the Barents and Kara Seas inspire optimism. In: T.O. Vorren et al. (Editors), Arctic Geology and Petroleum Potential. Norwegian Petroleum Society (NPF), Special Publication 2. Elsevier, Amsterdam, pp. 243–252.

B.K. OSTISTY *Arctic Centre, Kola Science Centre, Institute of Economy, Murmansk -23, Russia*
S.I. CHEREDEEV *Arctic Centre, Kola Science Centre, Institute of Economy, Murmansk -23, Russia*

The petroleum systems of the South Caspian Basin

A.A. Narimanov

The South Caspian hydrocarbon-bearing basin is an intermontane depression situated between the Greater and Lesser Caucasus and Talysh Mountains in Azerbaijan, and the Elburz and Kopet Dag Mountains in Iran and Turkmenia, respectively.

Over 620 oil and gas accumulations have been discovered in Quaternary to Miocene sediments. Of these only ten rather small accumulations occur in Quaternary and Miocene rocks. The main volume of hydrocarbons is located in Pliocene reservoirs. The Pliocene is the main target for future exploration.

Despite the production of 1300 million tonnes of oil, the genesis and source rock systems for the hydrocarbons are still unknown.

The model for hydrocarbon accumulations, outlined in this paper, is based on: a palaeotectonic analysis of the basin structure; a study of the nature and distribution of the sedimentary fill to the basin; thermal analysis and research into the geochemical characteristics of the oils, gases and condensates produced. The model explains many of the peculiarities of oil and gas distribution, and leads to the prediction that the basin is prospective to a depth of up to 9 km.

Introduction

The geological history of the South Caspian intermontane basin has been studied for many years, allowing us to give a relatively complete picture of its origin and subsequent development. Here the various distinguishable systems of sedimentary rock volumes are identified and ascribed to distinct and separate oil- and gas-bearing formations. These formations are genetically connected with each other in time and space by palaeotectonic and palaeogeographical conditions, which were suitable for their subsequent development as hydrocarbon source rocks and reservoirs.

The Middle Pliocene contains the main regional oil- and gas-bearing sediments of the South Caspian Basin (Figs. 1, 2 and 3). Within the basin certain regularities have been identified in the distribution of the hydrocarbon accumulations in the Middle Pliocene sediments. Until now many questions have remained unanswered, in spite of the discovery and development of a great number of oil and gas-condensate fields, and the production of more than 1300 million tonnes of oil. These questions include the ages of the traps and hydrocarbon accumulations, the timing, mechanism and scales of hydrocarbon migration and the location of the generating hydrocarbon kitchens. Only with answers to these questions can clear accounts of the origins of the oil and gas deposits be given.

For the first time this paper presents the formation and location of the hydrocarbon accumulations within the basic framework of palaeotectonic and formation analysis. Different methods have been used to assess the timing of the formation of traps and of the accumulation of hydrocarbons within them. Data are used from organic matter pyrolysis, trace element and organic chemistry of oils and bituminoids in the rocks, as well as analysis of the distribution patterns of oil and gas accumulations. A new model for the formation of the hydrocarbon deposits of the South Caspian Basin is presented based on these results.

History of trap formation

Figures 4–8 illustrate the structural styles of several of the major hydrocarbon accumulations in the South Caspian Basin. Palaeotectonic analysis of the South Caspian Basin shows a four-phase history:

(1) The *Mesozoic* stage of tectogenesis played the main role in originating the basin, and its further development as a geosynclinal area. Deep faults, developed in Mesozoic time, exerted a considerable influence on the deposition and structural development of the Cenozoic formations, and in particular on the Pliocene and Quaternary.

(2) The Pliocene structural traps originated at the end of *Middle Pliocene* time, when their dislocation became appreciable and dips along the flanks of the structures reached about 5–6°. Tectonic activity and the movement of large blocks along the pre-Pliocene

Basin Modelling: Advances and Applications edited by A.G. Doré et al. NPF Special Publication 3, pp. 599–608, Elsevier, Amsterdam.

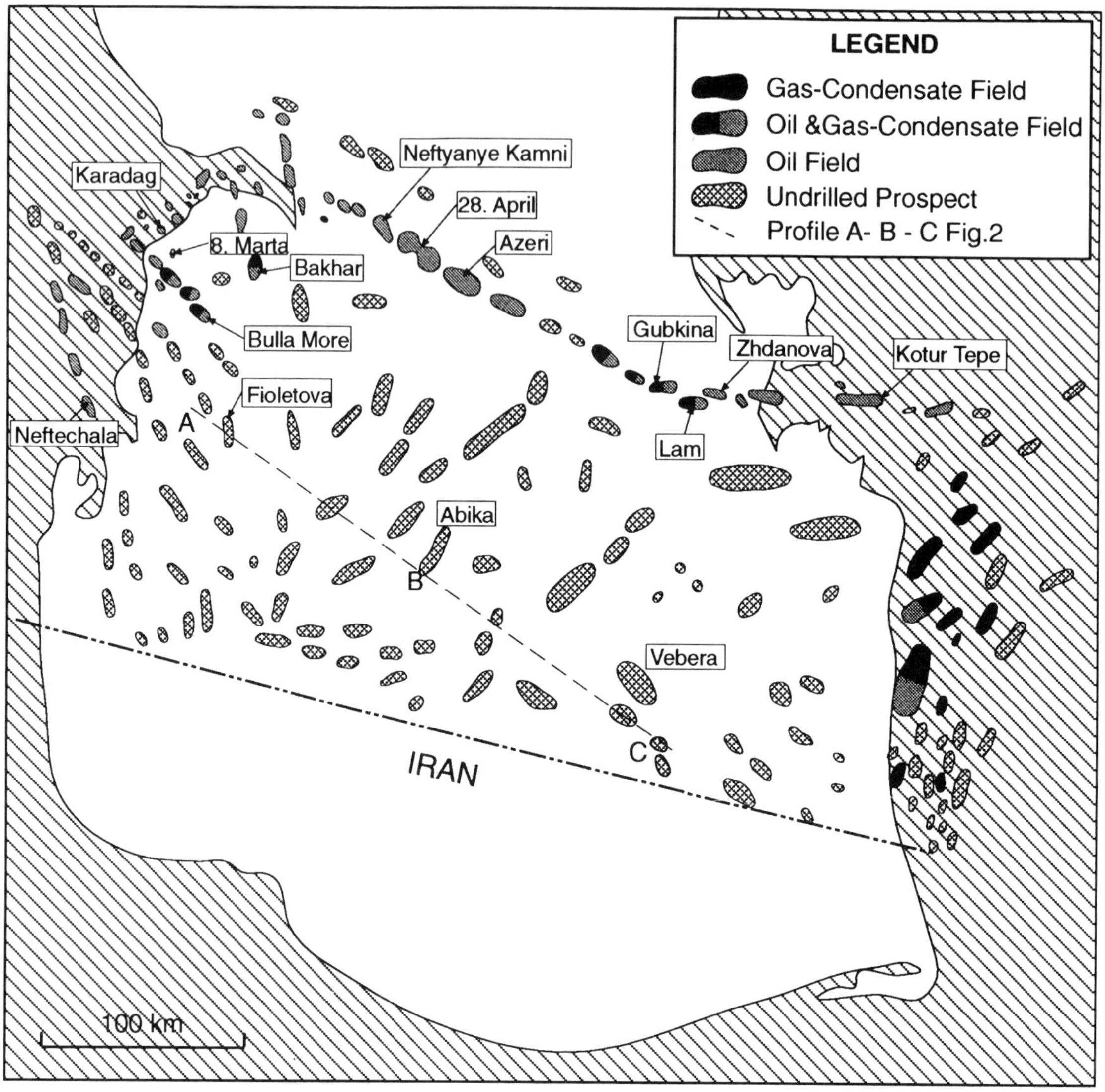

Fig. 1. The hydrocarbon fields and prospects of the South Caspian Basin.

faults took place from the beginning of Pliocene time until the deposition of the Kirmakinskaya group (early Middle Pliocene). These movements did not form large structural traps.

(3) At the end of *Apsheron (Late Pliocene)* time, a sharp increase in folding processes resulted in structures with flank dips of 8–16°, and led to sharp elevations of the crests of many of the anticlinal folds, and their deep erosion. By this time the Pliocene–Quaternary structural features had essentially been formed.

(4) Tectonic activity in the basin reached its peak in *Quaternary* time. The range of vertical tectonic movements increased up to 1000–1500 m. The most dislocated structures were those near the basin rim, where dip inclinations frequently reached up to 60–70dgr, and fold cores were complicated by diapirs composed of clayey Oligocene–Miocene strata. Intensive subsidence of the middle part of the basin resulted in submergence of the structures near the edge of the present-day shelf, and consequently these structures avoided denudation and breaching. Within them the traps are relatively tight folds, but are less complex and have reliable impermeable cap rocks.

As a result of the active tectonic processes the Pliocene–Quaternary sequence of the South Caspian Basin shows a wide range of different fold structures such as diapirs, monoclines, reverse faults, and general brecciation. Most structures were derived from tectonic features inherited from the pre-Pliocene foundation. Rootless and submerged folds are less common. All these tectonic structures are grouped in separate zones. The distribution of these zones has been used in ranking the various parts of the South Caspian Basin.

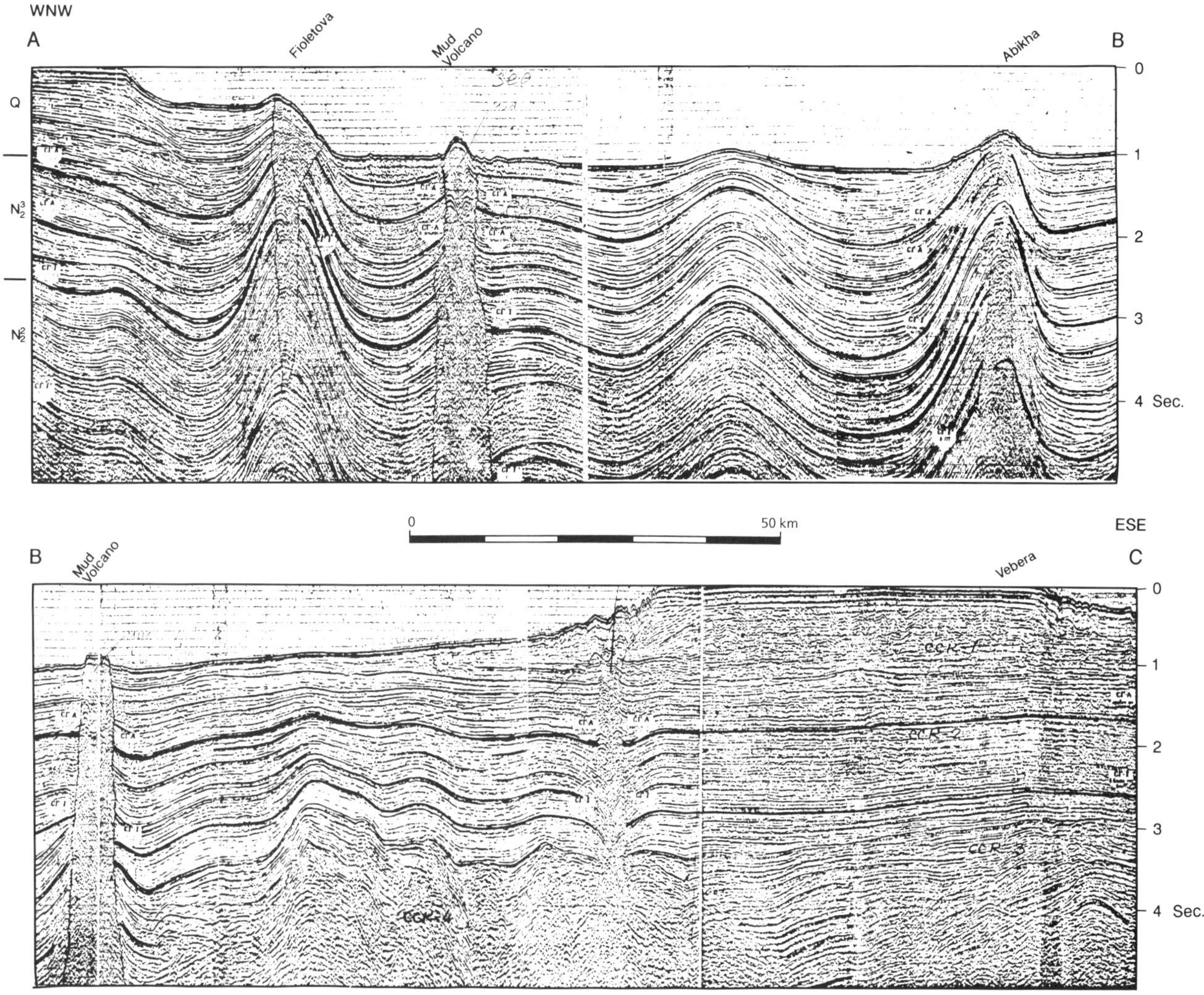

Fig. 2. Regional seismic line across the South Caspian Basin. See Fig. 1 for location.

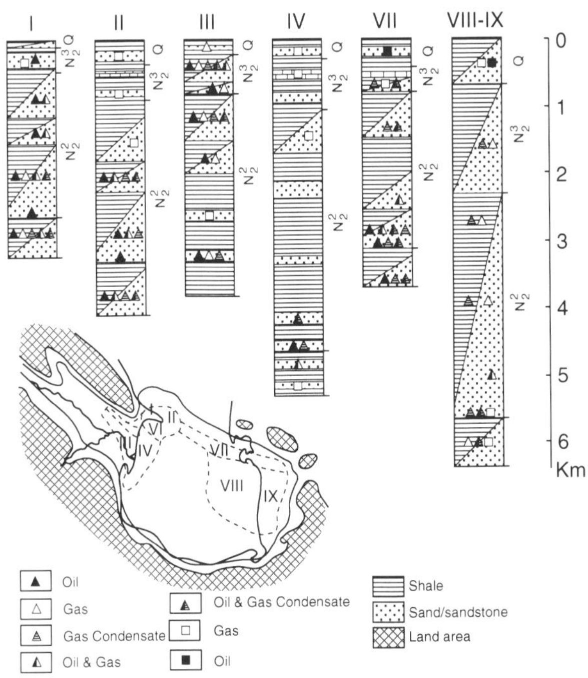

Fig. 3. Stratigraphic scheme of the Pliocene and Quaternary sequences of the South Caspian Basin. From Bagir-Zade et al. (1988, fig. 19).

Timing of hydrocarbon generation and entrapment

There are several viewpoints concerning the origin and trapping of the hydrocarbons in the South Caspian Basin. Many of these are not supported by the latest research reported in this paper. Namely:

(1) It has been generally thought that the hydrocarbons in the Middle Pliocene sediments are syngenetic: i.e. that the main volume of the oil and gas accumulated as a result of lateral migration. In this model the absence of oil and gas in well developed traps is explained by unfavourable geochemical conditions during burial of the organic matter. Some researchers hold the opinion that the Pliocene hydrocarbon accumulations are of an epigenetic character.

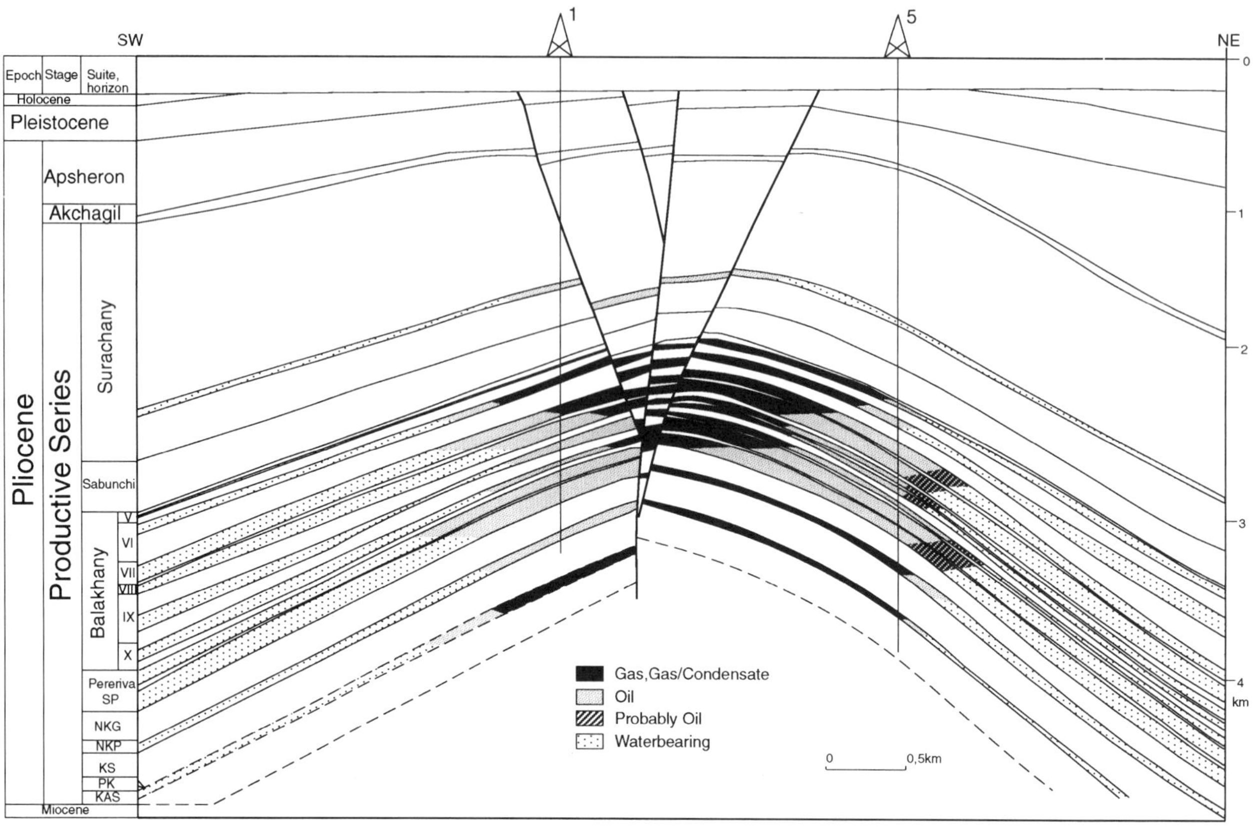

Fig. 4. Geological profile of the Azeri field.

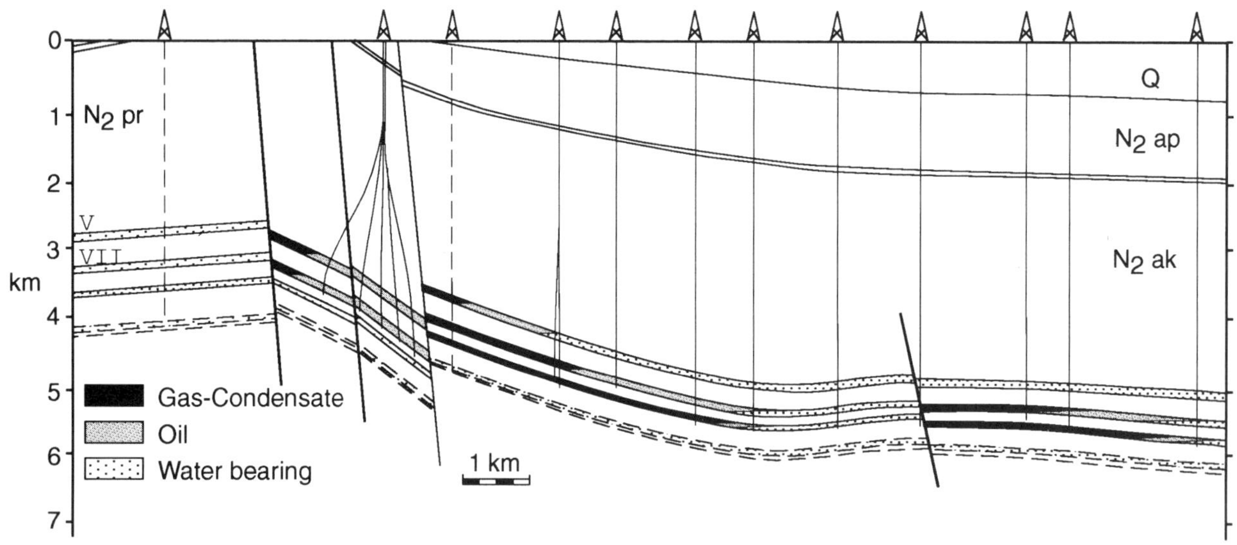

Fig. 5. Geological profile of the Duvanny More field.

(2) It has also been suggested that the known deposits were formed in the Middle Pliocene period. In this view, younger (Pliocene–Quaternary age) traps are considered to be non-prospective. The basin has also been zoned according to the hydrocarbon phase: based on hypsometric evidence of uplift of the oil and gas accumulations, and geographical evidence.

The conclusion of this paper is that the main period of hydrocarbon migration into the traps was in Pliocene–Quaternary time. This conclusion is based on the analysis of palaeotectonic activity, which has shown that most traps formed relatively late (Late Pliocene–Quaternary time), and the results of research on the paroxysmal activity of mud volcanoes, assessed from their observed activity in the sedimentary section.

The timing of trap formation has been analysed using assessments of trap volume changes with time

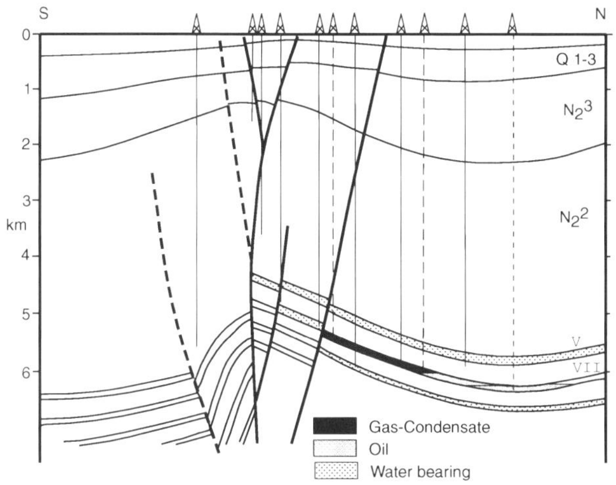

Fig. 6. Geological profile of the Bulla More field. From Bagir-Zade et al. (1988, fig. 28).

for the Bulla More field as an example (Figs. 1, 6 and 7). This field is optimally located on the shelf area, where fold formation took place against a background of steady submergence, which prevented severe erosion of any stratigraphic interval. The severe dislocation of structures located at the basin margins resulted in the disruption, uplift and deep erosion of the strata there. Consequently these margin structures do not preserve an adequate stratigraphic record to assess the palaeotectonic development.

Calculations using the volume method were carried out for the well prospected northeast flank of the Bulla More field. These calculations show that a completely filled Horizon VII reservoir trap would have contained little more than 37 milliard m^3 of gas at the end of the Middle Pliocene. This compares with the known present-day volume of 148 milliard m^3 gas.

In Late Pliocene time, deposition of the Apsheron suite and considerable tectonic activity led to the development of a monoclinal structure, and to a marked increase in potential trap volume. The trap volume of the Horizon VII reservoir increased so much during Late Pliocene time that it could contain a volume of gas nearly three times that at the end of Middle Pliocene time: i.e. 144 milliard m^3 gas.

Reverse calculations demonstrate that, under present-day thermobaric conditions, the volume of a possible gas accumulation, accumulated before the Middle Pliocene, would occupy an area of only 21 km^2, compared with an actual field area of about 80 km^2.

Similar calculations for the liquid phase have been made for some other oil and gas accumulations in the basin. These calculations show that more than half of the oil and gas volume under consideration must have charged the traps no earlier than Late Pliocene–Quaternary time.

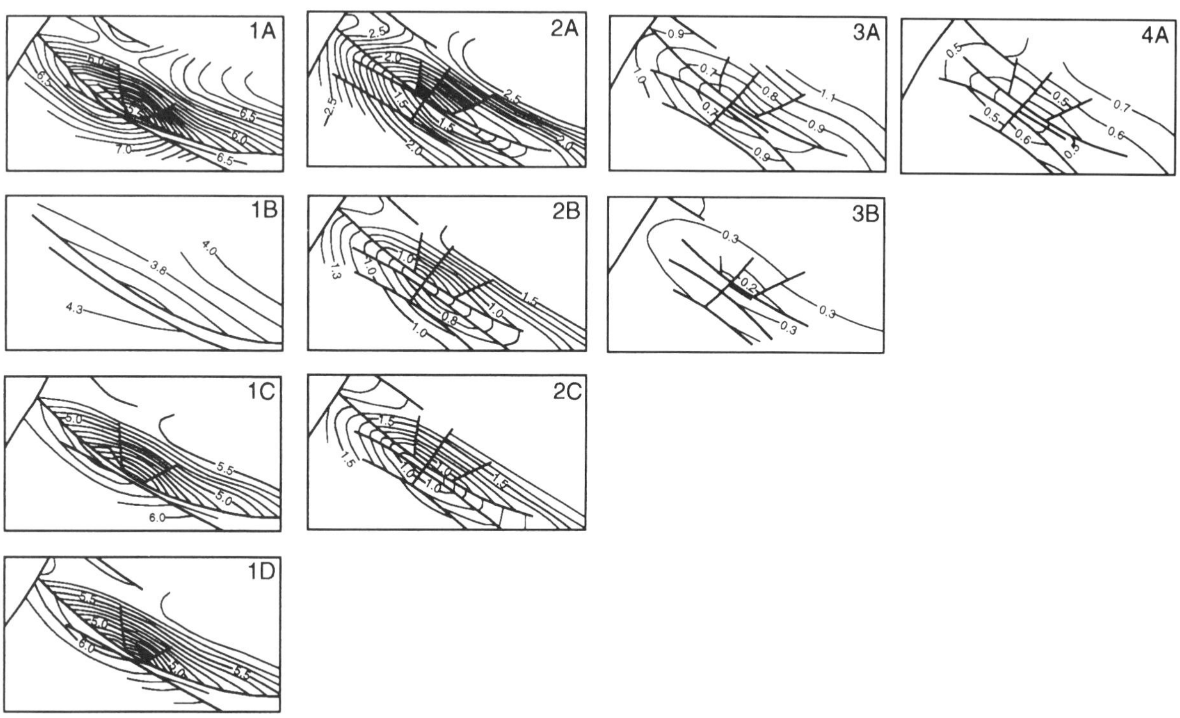

Fig. 7. Structural evolution of the Bulla More field. Structure contours in kilometres on the following horizons: *1* = top productive series, *2* = top Upper Pliocene, *3* = post-Pliocene, *4* = top Middle Pleistocene. The horizons are shown at the following times: *A* = present, *B* = Early Pleistocene, *C* = Late Pliocene, *D* = Middle Pliocene. From Bagir-Zade et al. (1988, fig. 23).

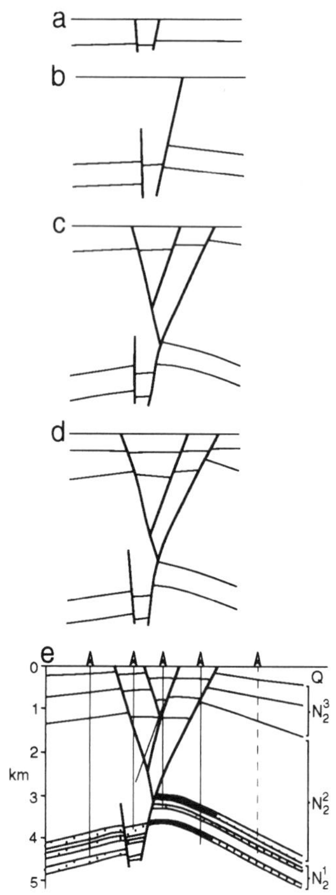

Fig. 8. Structural evolution of the Zhdanov Bank field. (a) End of early Middle Pliocene. (b) End of Middle Pliocene. (c) Middle Late Pliocene. (d) End Pliocene. (e) Recent. From Bagir-Zade et al. (1988, fig. 24).

The conclusion from the above analysis is that in the South Caspian Basin the Late Pliocene–Quaternary stage was the main time of formation of the traps in the Middle Pliocene reservoirs. Therefore the main volume of hydrocarbons must have migrated into these traps during the period from Late Pliocene up to Holocene time. There is evidence indicating a hydrocarbon migration phase during early Middle Pliocene time. However, extensive traps for oil and gas were not present in this period, and this phase of migration could not have played an important role in the formation of hydrocarbon accumulations in the Pliocene rocks.

Source of the hydrocarbons

The main hydrocarbon-bearing Middle Pliocene section of the basin is poor in organic matter. According to the results of pyrochromatographic analyses, the hydrocarbon generation potential of this sequence is considerably less than that of the underlying Palaeogene sediments (i.e. 0.3–0.6 kg/tonne compared with 2.4 kg/tonne). An increased content of organic matter is observed only in the lower layers of the Middle Pliocene (0.18–1.1 kg/tonne). These low values do not provide much support for a syngenetic character of the oil in the Middle Pliocene reservoirs. Analytical data for the trace element content and the isotope chromatography of oils within the multilayer deposits, confirm the identical content and/or consistency in compositional trends in the vertical section. This suggests the prevailing role of vertical hydrocarbon migration in the formation of these deposits.

Based on the identity of the oils, it is possible to divide the majority of discovered deposits into three geographical groups: Nizhnekurinskaya depression; Bakinsky archipelago–Apsheron peninsula–Apsheron archipelago; Pribalkhansky region.

Sedimentary and petroleum sub-provinces

The Middle Pliocene sequence is often regarded as a single widespread formation. However, more detailed research has identified several regions with distinct characteristics of their oil and gas properties. The boundaries to these regions of different lithofacies are not tectonic in nature.

The known oil deposits of the Apsheron hydrocarbon-bearing area (Surakhany, Balakhany–Sabunchy–Ramany, Neftyanye Kamni, Gunashli fields etc.) are related to a characteristic type of sediment lithofacies. The accumulations of the Nizhnekurinskaya depression (Kyurovdag, Kyursanga, Neftechala fields etc.) are associated with a different type of lithofacies. Yet a third type occurs in the fields of Turkmenistan (Nebit Gad, Kotur Tepe, Cheleken, Banka Lam fields etc.).

These areas of different lithofacies have been distinguished on several factors. In the Nizhnekurinskaya depression the sand reservoirs occur predominantly in the upper part of the Middle Pliocene interval and are characterised by average trap fullness. In the Apsheron area the reservoirs occur in the middle and lower parts of the Middle Pliocene sequence. Here the Pereriva Suite and Horizon X of the Balakhanskaya group often contain the largest volume of oil and gas reserves, and are filled to spill point. The reservoirs of the Turkmenian part of the Apsheron–Pribalkhanskaya zone occur in the upper part of the Middle Pliocene, and have an average amount of trap fill, often being restricted to the crestal zones. Figures 9 and 10 show the distribution of these lithofacies zones in the South Caspian Basin.

Such areas of uniform sedimentary facies can serve as the essential basis for high-grading the basin. Distinguishing them is possible by re-establishing the original conditions of sediment accumulation. Simplification is unavoidable when determining the areal

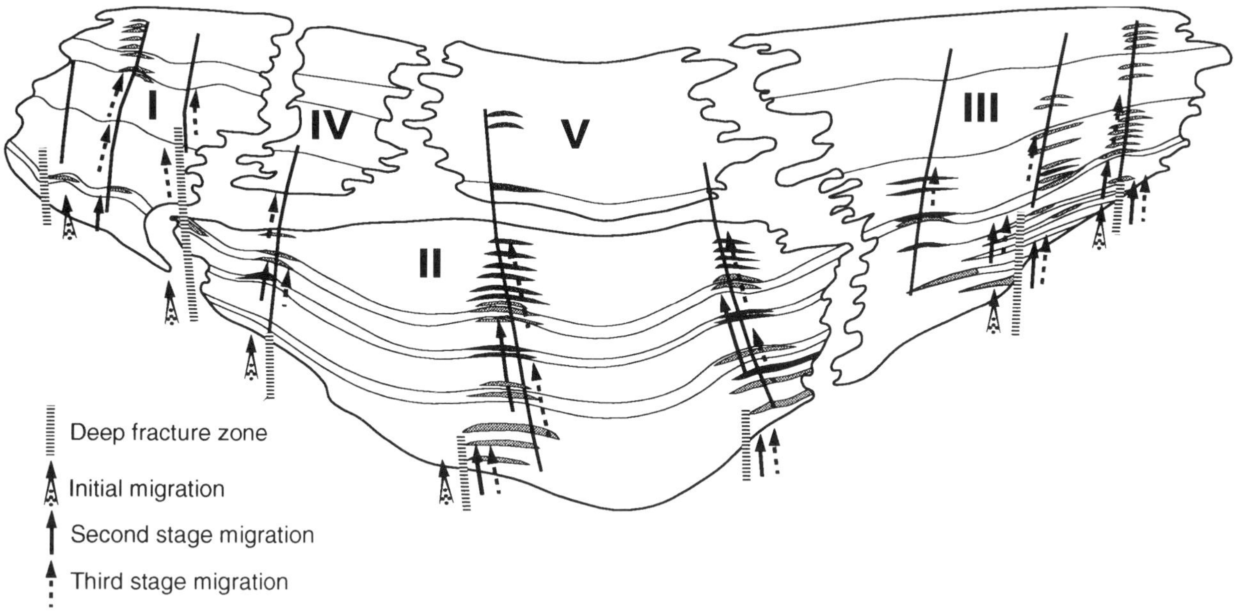

Fig. 9. Schematic profile of the separate lithofacies zones of the South Caspian Basin.

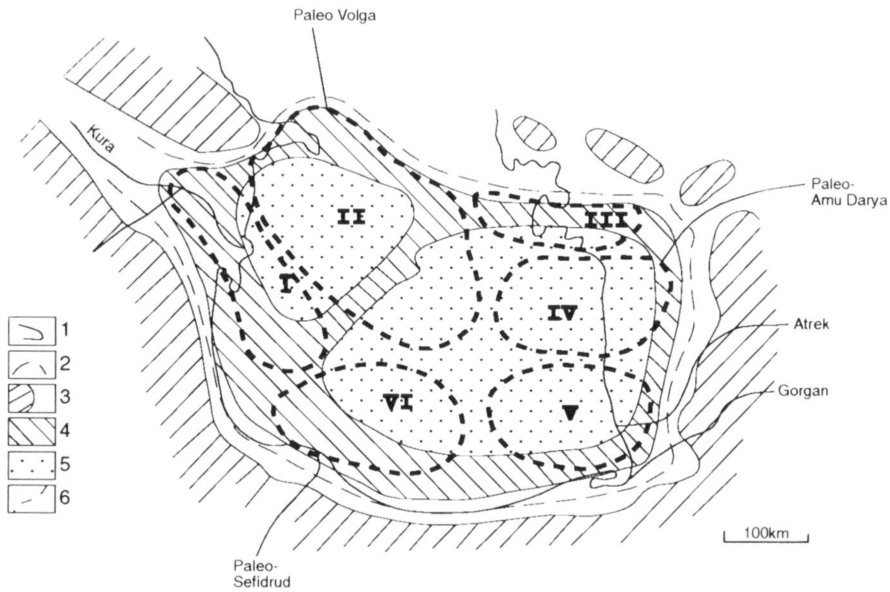

Fig. 10. Map of the separate lithofacies zones of the South Caspian Basin.

distribution of sediments of such a large stratigraphic interval, because of the dynamics of sedimentation in space and time. It is also obvious that the boundary zones between the different areas will have mixed sediment types, often with a sharp deterioration in reservoir properties. The Middle Pliocene sediments were deposited in an intermontane depression by the erosion of the surrounding mountain frame, and the introduction of clastic material by several large rivers: the Palaeo-Volga, the Palaeo-Kura, the Palaeo-Amu Darya and others. These separate sediment supplies generated distinct isolated types of sediments.

A detailed analysis of the sediments indicates that there are at least six independent lithofacies systems present within the Middle Pliocene sediments of the South Caspian Basin. A number of significant conclusions follow from the identification of these separate lithofacies systems.

(1) The presence of isolated sediment systems with their own sediment supply sources renders difficult, if not impossible, attempts to create a unified correlation system for the entire Middle Pliocene. Subdivision of the sequence must be carried out separately for each of the sediment systems. However, this does not preclude the identification of one of

more bench marks, indicated by a regional event such as a sharp change in geotectonic regime. Such a change occurred at the end of the deposition of the lower part of the Middle Pliocene, after which the sandstone of the Pereriva Suite was deposited in the Apsheron area. This event corresponds to the deposition of Horizon VII in the Bakinsky archipelago, Horizon VIII in the red-coloured section of the West Turkmenian depression, and Horizon XX in the Nizhnekurinskaya depression. This implies that Horizon VII of the Bakinsky archipelago correlates with the Pereriva Suite of the Apsheronskaya formation. The sand deposition of the Apsheronskaya lithofacies zone reached its maximum development at this time (deposition of the Pereriva Suite sands), and extended over the northern part of the Bakinsky archipelago and the Pribalkhanskaya zone as far east as the Barinov field.

(2) Separate sediment systems and the distinct lithological zones which border them are not consistent with widespread large-scale (hundreds of kilometres) lateral migration of hydrocarbons along the same reservoir unit.

(3) The presence of local sediment systems explains the existence of neighbouring zones which have distinct and differing hydrocarbon phases. Moreover, the variation in hydrocarbon distribution within a tectonic zone is also explained, e.g. within the Apsheron–Pribalkhanskaya zone.

(4) The existence of distinct sediment systems requires an appropriate exploration approach, such as the drilling of at least one parametric well before commencing exploration.

In conclusion, the existence of isolated lithofacies sedimentary systems within one large hydrocarbon-bearing sequence must be taken into account in analyzing the oil–gas geological zonation of the basin. The zonal distribution of the South Caspian Basin on this basis has allowed the distinguishing of regions with differing prospectivity.

Synthesis — petroleum geological history

A new concept for the development of the hydrocarbons in the South Caspian is presented in this paper, based on the analysis of the geotectonic development of the basin, and an analysis of its Pliocene–Quaternary sedimentary section (formation analysis, hydrocarbon generation, migration and accumulation conditions and preservation history within the traps) (Fig. 11).

The first stage of the diagenesis of the clastic sediments of the lower part of the Middle Pliocene (pre-Kirmakinskaya group) was accompanied by active tectonic movements. These involved broad-scale

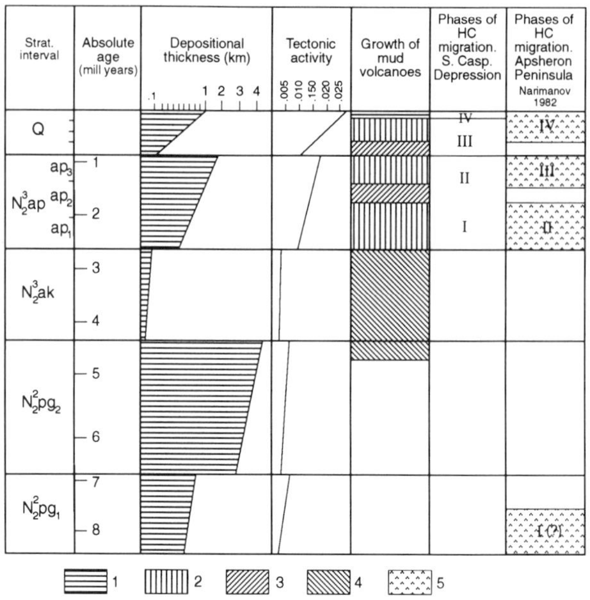

Fig. 11. Timing of the main elements of petroleum geology relevant to the entrapment of hydrocarbons. Key to mud volcano activity: *1* = intense, *2* = strong, *3* = weak, *4* = insignificant. From Bagir-Zade et al. (1988, fig. 22).

subsidence during the final stage of geosynclinal development of the South Caspian Basin. In addition, movement of the major tectonic blocks along deep-seated faults, developed in the Mesozoic–Palaeogene–Miocene strata, generated minor faults and weak folds in the overlying Pliocene strata. The first phase of enhanced hydrocarbon migration took place along the fissure zones of these deep faults during the early Middle Pliocene, and the weakly deformed sediments became locally charged. This process was enhanced by numerous small faults. Considerable quantities of hydrocarbons generated in this phase were lost or underwent oxidation and degassing.

A further migration of hydrocarbons into the Middle Pliocene sediments took place during Apsheron time (Late Pliocene). Active, vertical hydrocarbon migration occurred for the first time. However, traps were only weakly developed at this time and most of the hydrocarbons escaped to the surface and were lost.

The third stage of migration started in late Apsheron time and continued till the end of Pliocene time. Hydrocarbon migration was accompanied by extensive folding deformation. Vertical migration and localised lateral migration within the separate lithofacies systems led to the accumulation of hydrocarbons in structural traps and regional stratigraphic traps. By the end of this stage, all deposits had essentially been formed, and the majority is thought to have been gaseous.

The fourth stage began in Middle Pleistocene time and continues to the present day. This stage brought major changes in the structural appearance of the basin, and led to a high degree of tectonic fracturing and complexity, along with deep erosion of sediments in the crestal parts of the folds (in places down to the level of the lower part of the Middle Pliocene). This high level of tectonic activity was accompanied by the enhanced migration of additional hydrocarbons into the Middle Pliocene reservoirs, and the replenishment of the large traps in the deeper parts of the basin. In the elevated, severely deformed, marginal zones of the basin, extensive redistribution of the hydrocarbons occurred together with considerable degassing. Several hydrocarbon accumulations were destroyed by erosion and exposure. This fourth stage of oil and gas accumulation played a major role in forming the hydrocarbon accumulations in the deeply buried structures.

This interpretation of the hydrocarbon accumulation process provides answers to many questions, including those concerned with the evaluation of oil and gas prospects in the little explored regions of the basin. This model also precludes the simplistic notion that the deeper central part of the South Caspian Basin is exclusively gas-prone.

Conclusions

The following conclusions can be made.

(1) The greater part of the hydrocarbons in the South Caspian Basin is contained in Middle Pliocene clastic reservoirs. These reservoirs are the main objectives of ongoing and future exploration.

(2) Structural traps predominate, although many have a stratigraphic component as sands pinch-out up the flanks of several of the structures. A variety of structural styles occur, ranging from anticlinal folds to monoclines, with various degrees of reverse faulting and brecciation. Many structures are penetrated by mud diapirs and associated with mud volcanoes. The structures in the Middle Pliocene occur along clearly identifiable trends associated with underlying deep-seated fault zones.

(3) Most structural traps at Middle Pliocene levels were developed in Late Pliocene–Quaternary

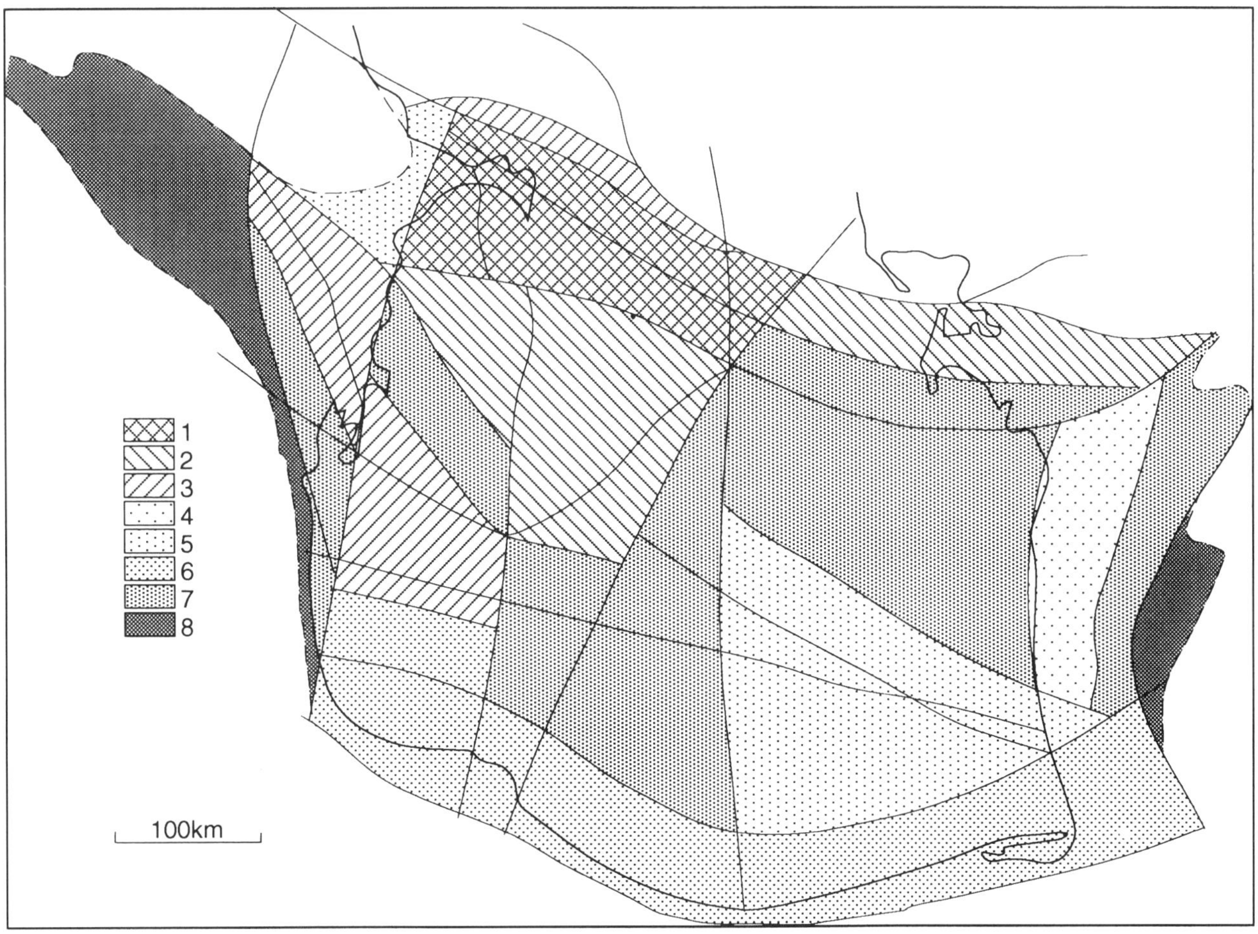

Fig. 12. Prospectivity of the regions of the South Caspian Basin from best (*1*) to worst (*8*).

time. Thus the main volume of hydrocarbons must have migrated into the traps during the time span from Late Pliocene to the present day. Hydrocarbon generation and migration appears to have been polyphase, from early Middle Pliocene time to the present day. The traps were mainly charged by vertical migration of hydrocarbons.

(4) The Middle Pliocene section is poor in organic matter and unlikely to be the main source rock for the extensive hydrocarbons contained in the Middle Pliocene reservoirs. The source rock(s) have yet to be clearly identified.

(5) Analysis of the distribution patterns of the Middle Pliocene clastic sediments and the hydrocarbons demonstrates the existence of at least six independent lithofacies systems/zones. Many of these were derived from the major palaeo-rivers that flowed into the basin, i.e. Palaeo-Volga, Palaeo-Kura, Palaeo-Amu Darya.

(6) The recognition of distinct lithofacies zones with different hydrocarbon distribution patterns provides the basis for high-grading the basin (Fig. 12). Furthermore the model of hydrocarbon generation and migration precludes the simplistic concept that the deeper central parts of the basin are exclusively gas-prone.

Acknowledgements

The preparation of this article for publication was aided by P. Wells (BP, London) and O. Skarpnes and A.M. Spencer (Statoil, Stavanger).

References

Bagir-Zade, F.M., Narimanov, A.A. and Babayev, F.R., 1988. Geological and Geochemical Features of the Caspian Sea Fields. Nedra Press, Moscow.

Ismailov, K.A., Idrisov, V.G. and Tagiev, E.A., 1972. Productive Deposits of the Lower Kura Valley and the Baku Archipelago. Academy of Sciences, Azerbaijan.

Narimanov, A.A., 1982. The time of formation of the Bulla-More gas-condensate deposit. Azerbaijanskoye neftyanoye khozyaystvo, 9: 12–17.

Narimanov, A.A., 1983. The main phases of oil and gas migration in the Pliocene–Quaternary complex of deposits within the western margin of the Southern Caspian Basin. EI VNEE Gazproma, Geologiya, bureniye razvedkha gazovykh mestorozhdeniy, Moscow, 15: 1–4.

A.A. NARIMANOV *Kaspmorneftegas, Republic of Azerbaijan*

Inverse elastic flexural modelling of sedimentary basins: a case study from the Colville Trough, National Petroleum Reserve, Alaska

Jeff Cogan, Ian Lerche and Branch Russell

An inverse elastic flexural model is applied to the Colville Trough, National Petroleum Reserve, Alaska (NPRA). Five parameters dictate the bending of basement strata: flexural rigidity (D), bending moment (M), load (V), dip angle (θ), and length (L), as well as the choice of a reference (hinge) point. The optimum values for these parameters are determined by an inversion technique that adjusts the parameter values so that the difference between predicted and observed subsidence is minimized. This procedure avoids the trial and error approach inherent in forward models. The sensitivity of the flexural system to variations of the parameter values is also discussed.

A north–south seismic dip section from the NPRA is modeled and the average difference between predicted and observed values is 560 m, about 6% of the maximum depth of the basin. The close fit indicates that the subsidence of the Colville Trough is due to elastic flexure of the Franklinian basement in response to thrusting of the Brooks Range. The results also suggest that the peripheral bulge, which forms the present-day Barrow Arch, has migrated roughly 200 km from the south. In addition, the model results support geological data which indicate that the hinge point (termination of Franklinian strata) lies about 100 km to the north of the section.

Introduction

The first models of lithospheric compensation were proposed by Pratt (1855) and Airy (1855) and were based on the principle of isostasy — the idea that crustal loads are in hydrostatic equilibrium with the upper mantle. These models assume that the lithosphere has no strength; in effect, crustal blocks of various thicknesses and/or densities float on fluid mantle material. On the other hand, models of compensation by flexure are based on the assumption that the lithosphere has some finite strength. Loads are supported by the mechanical strength, or rigidity, of the lithosphere as well as by buoyant forces in the mantle (Turcotte, 1979).

Three broad categories of flexural models have been proposed: elastic, viscoelastic and thermally controlled flexure. Linear elastic models are based on Hooke's law, such that a given strain is directly proportional to the stresses producing the strain (Turcotte and Schubert, 1982). Under this condition deformation occurs instantaneously when a given stress is applied and there is complete recovery when the stress is removed (no permanent deformation). Deformation is not dependent on time. Viscoelastic models express strain as a function of both stress and time. Following Maxwell's law, strain is taken to decrease exponentially with time (Turcotte and Schubert, 1982). In this case there is a relaxation of stress with time, resulting in permanent deformation of the material. Thermal flexure models are based on elastic or viscoelastic theory with the additional assumption that the rigidity is related to heat transfer within the material (Haxby et al., 1976; McMullen et al., 1981; Quinlan and Beaumont, 1984; McMullen and Mohraz, 1987). The strength of the material decreases with increasing temperature.

Previous works have suggested that pure elastic flexure cannot account for the observed deflection of the lithosphere. Beaumont (1981) produced poor results initially in the Alberta Basin using an elastic model. However, his results were greatly improved by employing a viscoelastic model fitted to the same data. Similarly, Quinlan and Beaumont (1984) had little success modelling the Appalachian Basin using elastic or viscoelastic theory; acceptable results were achieved only after the incorporation of thermal effects. Likewise, Patton and O'Connor (1988), using an elastic model, determined an "anomalously low" flexural rigidity for the northern Oman Mountain Foredeep (D = 10^{22} N m) and attributed this finding to compressional stresses and/or mechani-

Basin Modelling: Advances and Applications edited by A.G. Doré et al. NPF Special Publication 3, pp. 609–621, Elsevier, Amsterdam.

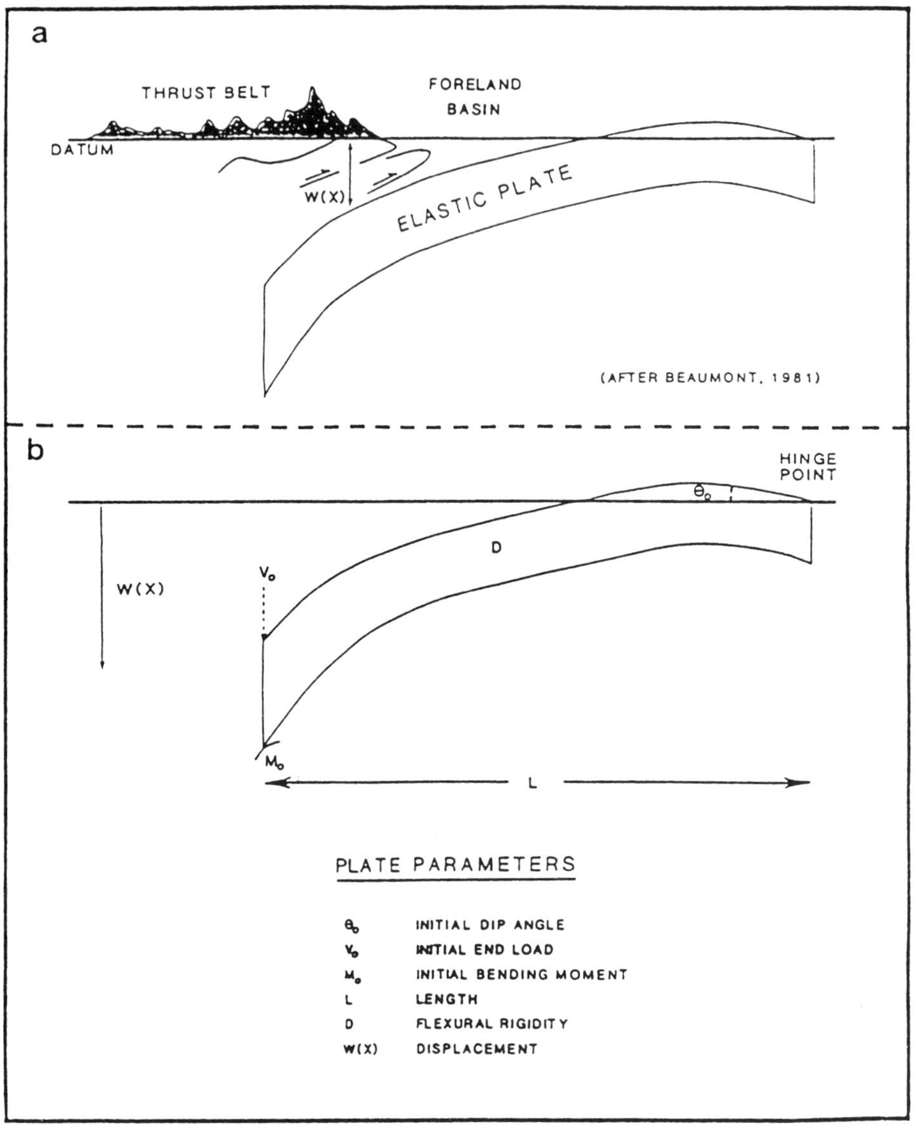

Fig. 1. (a) Flexural model in its geological setting. (b) Illustration of flexural parameters.

cal weakness of the lithosphere. In every case, the basic elastic model was modified by incorporating secondary effects.

It should be noted, however, that the elastic models employed by the above investigators and others (Jacobi, 1981; Jordan, 1981; Karner and Watts, 1983; Royden and Karner, 1984) assume that the lithosphere is an infinite plate. If the lithosphere is indeed infinite in lateral extent it must have a horizontal planar configuration prior to loading, so that the plate cannot have an initial dip angle or bending moment. Hence, only two parameters control the bending of an infinite elastic plate: the flexural rigidity (D), and the load (V), a situation which limits the number of plate geometries that can be produced.

Physically, it is more appropriate to deal with a lithospheric plate of finite length. In this case, three more parameters enter the elastic model: plate length (L), initial dip angle (θ), and initial bending moment (m) (Fig. 1). Displacements must be calculated relative to some reference point, so a fixed hinge point is required at one end. The physical significance of the hinge point is that it represents some geological boundary, either structural or stratigraphic. As a result of these modifications, there is no intrinsic necessity for the plate to be flat prior to loading (although a planar pre-loaded plate is permitted). The unloaded plate configuration, whether flat or not, is an indication of the basement geometry that existed prior to any sediment or thrust loading. The allowed variation in pre-loaded basement shapes can produce greater variability in the final (loaded) basement geometry than an initially flat (infinite) plate.

If a finite elastic beam model can produce acceptable fits to observed data, there is no need to invoke thermal or viscoelastic effects. Certainly, heating and stress relaxation play a role in flexure, but one proposal of this study is to isolate the elastic component of flexure in the Colville Trough. It is necessary to understand the influence of elastic forces on flexure in isolation before attempting to apply more complicated models.

Model description

Fig. 1a is an illustration of the model in its geological setting. The basement is assumed to act as an elastic plate of finite length and constant thickness which is anchored at one end in cantilever fashion. The anchored end is taken to remain stationary throughout the development of the basin. Thrusting and sediment loading cause the remainder of the plate to bend according to the applied stresses. Immediately in front of the thrust front a depression, or foredeep, develops in which sediments are deposited. Cratonward of the foredeep the basement is flexed upward forming a peripheral bulge. This configuration is not static because, as the thrust front moves, the peripheral bulge migrates ahead of the front in the direction of thrusting (Patton and O'Connor, 1988). The shape and position of the bulge are a function of both the loading history and the properties of the elastic plate. Sedimentation in the basin can be strongly affected by the development of the peripheral bulge (Cogan et al., 1993).

Flexure of a finite elastic beam of length L which is anchored at one end due to the force of a single applied end load is dictated by five parameters: flexural rigidity, D; bending movement, M; dip angle of the anchored end, θ; end load, V; and length, L (Appendix; Fig. 1b). Application of this model is limited by the assumption that the total load of the basin can be approximated by a point load applied at the end of the plate. Lateral variations in load are not accounted for, a situation certainly encountered in sedimentary basins. Hence, the single load model has to be modified to account for differential lateral loading (Cogan et al., 1993).

The model runs in a forward sense by specifying values for the five flexural parameters and the observed load on the basement. USGS seismic line 6, a dip section, is used to estimate the present-day load. The basement horizon is picked and the hinge point is taken as the highest point of the basement at the northern edge of the section. It is possible that the true hinge point is located beyond the northern limit of the section; this case is discussed in a later section. The sediment and thrust load overlying the basement are divided into a series of blocks of equal width (Fig. 2). The lines between the blocks are drawn from the seismic datum to the top of the basement. We assume that the porosity decreases exponentially with depth:

$$\phi(z) = \phi_o \exp\left(\frac{-z}{l}\right) \tag{1}$$

where $\phi(z)$ is the porosity at a given depth z, ϕ_o is the surface and l is a scaling length (Sclater and Christie, 1980). The values of the surface porosity and scaling length are not known in this study, so we use Sclater and Christie's suggested values of $\phi_o = 0.56$ and $l = 2300$ m, for sandy shales. The total load within each block is then calculated as follows:

$$V = gA \int_o^z \{\rho_w\,\phi(z) + \rho_s\,[1 - \phi(z)]\}\,dz \tag{2}$$

where V is the total load within the block, g is the acceleration due to gravity, A is the cross-sectional area of the block, ρ_s is the density of the solid material, and ρ_w is the density of water. In this

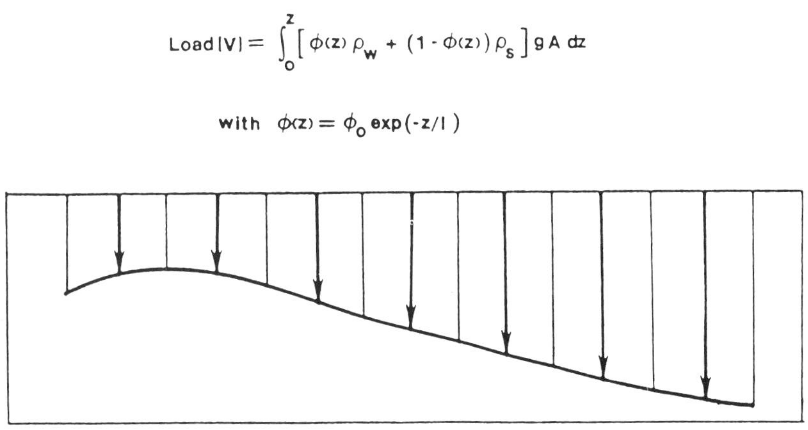

Fig. 2. Determination of point loads from observed data. The present-day load on the basement is divided into a discrete number of blocks of equal width. The load within each block is calculated and applied as a point load in the center of the block. Refer to text (eqs. 1 and 2) for an explanation of the mathematics on the figure.

fashion the total load on the system is broken up into a series of equally spaced point loads.

Although we can estimate the present-day load on the basement, we cannot measure the values of the flexural parameters D, M, V, L, and θ that existed in the past prior to loading. Hence, in a forward model we are forced to determine the optimum parameter values by a series of trial and error adjustments. Estimates are made for each parameter, the forward model is run, and the predicted and observed curves are compared. In the event that the curves do not match, one or more parameters are varied and the model is run again. An inverse approach has been developed which automatically determines the optimum values for the parameters on the basis of available data. A detailed description of the inverse model is given in Cogan et al. (1993). A theoretical basement profile is calculated for the observed loading scheme with the parameter values set at some initial values. The theoretical profile is compared to the observed profile using a standard least-squares fit. The average degree of mismatch is represented by the mean square residual (MSR). The lower the MSR, the better the fit. (The unit of MSR is length squared, but we take the logarithm of MSR in order to simplify the calculations. Henceforth, all MSR values refer to the logarithm of a number whose unit is meters squared.) After an MSR is calculated for the initial parameter guesses a tomographic technique is employed to adjust the parameter values in the appropriate direction (Fig. 3). This technique is a discrete inverse method in the manner of Menke (1984). Each parameter is raised and lowered by a small amount and the MSR is calculated for both of these values. These MSRs are compared to the initial MSR and whichever one is lowest indicates the direction that the parameter must be adjusted. All five parameters are adjusted in this manner for a number of iterations specified by the user. It should be noted that the length, L, of the plate is treated differently in the inverse approach than in the forward mode. It is possible that the plate extends beyond the limits of the data, but cannot be shorter than the observed end of the plate (Fig. 4). Therefore, length is renamed added length, referring to the segment of plate added on the end of the data point furthest from the hinge.

When the best values of the parameters have been determined the model is run with these values for a final time. The theoretical basement profile is determined and the final MSR is calculated. In addition, the pre-loading basement profile is computed. The unloaded profile is determined by running the forward model with the best parameters without applying the present-day load observed on the seismic section. The pre-loading basement geometry is important because it represents the basement configuration prior to any sediment or thrust loading in the section. The position of the peripheral bulge prior to observed thrust/sediment loading may have a significant effect on the subsequent development of the basin (Cogan et al., 1993).

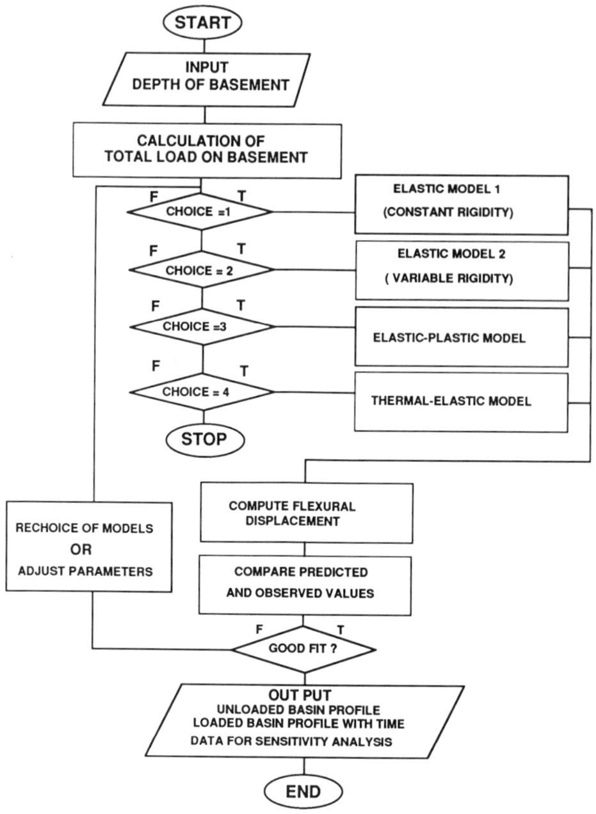

Fig. 3. Flow chart illustrating the inversion technique.

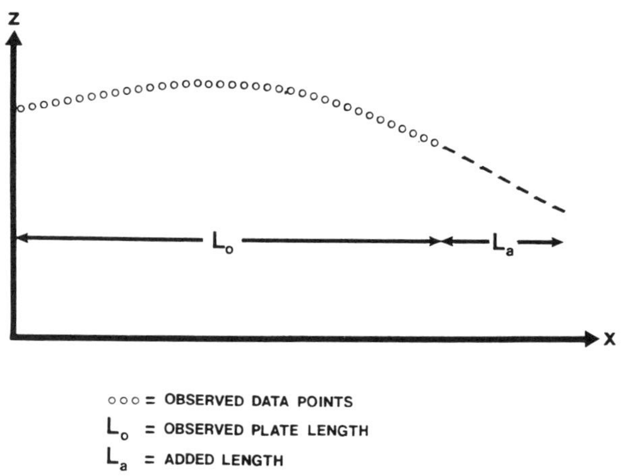

Fig. 4. Relationship between the observed length of the plate and the added length. It is possible that the plate extends beyond the observed data points, but the plate cannot be shorter than the length observed in the data.

Although all five flexural parameters have a bearing on the deformation of the basement they do not

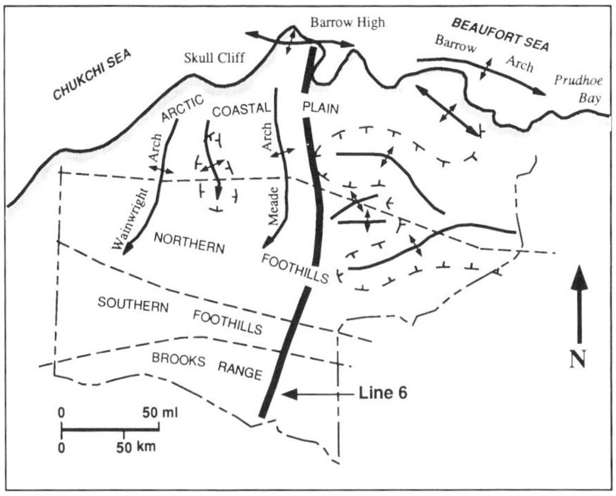

Fig. 5. Location map, NPRA.

make equal contributions to the final result. The flexural system is more sensitive to variations of some parameters than to variations of others. This effect is illustrated by linear search diagrams. After the best parameter values have been determined one parameter is allowed to vary in a linear fashion around its best value while the others are held fixed at their best values. At each increment the MSR is calculated and a plot of parameter value versus MSR is generated. The shape of the curve is an indication of how well the parameter can be determined. If the curve is relatively flat small departures from the minimum do not affect the fit to any great extent; the value of the parameter can lie within a broad range and produce essentially the same final result. If the curve is steep, and there is a sharply defined minimum, small departures from that point will cause a significant change in MSR so the value of the parameter is more tightly constrained.

Regional geology

Located on the North Slope of Alaska between the Brooks Range and the Beaufort Sea, the National Petroleum Reserve, Alaska (NPRA), has been one of the most intensely explored hydrocarbon tracts in the United States (Fig. 5). It covers nearly 36,000 square miles (1.08×10^5 km^2) of Arctic Alaska and contains a thick sequence of pre-Devonian through Tertiary sediments with proven petroleum reserves (Kirschner et al., 1983). Detailed discussions of the geological development have been published by Carter et al. (1977), Grantz et al. (1979), Bird (1981), Carmen and Hardwick (1983), and others. A brief summary of geological data relevant to the present study is presented below. Although the structural and stratigraphic aspects of the NPRA will be treated separately, there is some overlap in the following discussion since structure and stratigraphy are genetically related.

Structure

In a broad sense, the NPRA can be divided into four structural provinces which appear as E–W-trending belts in plan view (Fig. 6).

Brooks Range. Situated along the southern border of the NPRA, the Brooks Range is an imbricate thrust belt that was transported from the south in Jurassic to Cretaceous time (Churkin et al., 1979). The deformation was probably the product of a colli-

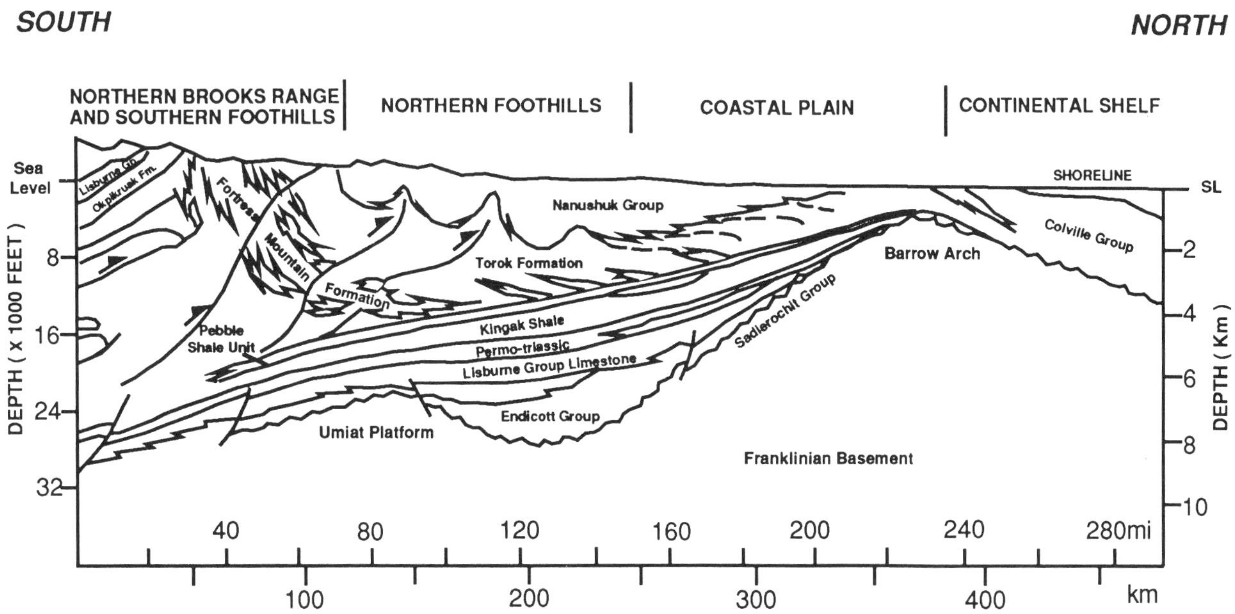

Fig. 6. Generalized north–south cross-section of the NPRA.

sion of a microcontinental block or volcanic arc, now represented by part of the southern Brooks Range. Most of the displaced strata belong to the Ellesmerian sequence and have been subjected to up to 320 miles (580 km) of crustal shortening (Armagnac, 1985).

Foothills Province. This unit exhibits a fold-thrust structural style similar to that of the Brooks Range, but the deformation here is less intense (Armagnac, 1985). Unlike the Brooks Range, the Foothills Province contains mainly Brookian sequence rocks (Carter et al., 1977). In general, the deformation becomes less severe towards the north although narrow fault-related anticlines are commonly found near the northern border of the province (Kirschner et al., 1983).

Arctic Coastal Province. In general, the Arctic Coastal Province is made up of two units: the Arctic coastal plain and the Arctic platform. The coastal plain contains mostly flat-lying shallow-marine sediments of the Brookian sequence (Armagnac, 1985). Dipping south beneath the coastal plain and the Foothills Province, the Arctic platform features undeformed to mildly folded strata of both the Ellesmerian and Brookian sequences (Kirschner et al., 1983).

Barrow Arch. The Barrow Arch is a major southeast-plunging basement high extending nearly 350 miles (640 km) along the Beaufort Sea coast. The arch was uplifted and eroded in the Lower Cretaceous in response to Brooks Range thrusting to the south (Bushnell, 1981). Hence, the Barrow Arch is a peripheral bulge that formed by flexure of the basement strata.

In addition to the major structural provinces there are several basement-involved structures that have played important roles in both the structural and sedimentological development of the NPRA. Beneath the coastal and foothill provinces, the Franklinian basement has been disrupted by a series of horsts and grabens that formed in Devonian times (Armagnac, 1985). The Meade, Ikpikpuk, and Umiat basins, and the Meade and Wainwright arches are all associated with this extensional event. In the Mesozoic, compression became the dominant tectonic force as the Brooks Range was thrust up from the south. A foredeep, the Colville Trough, formed in front of the Brooks Range in response to the thrust load and was filled with sediments derived mainly from the Brooks Range (Churkin et al., 1979).

Stratigraphy

It is common practice to group the rocks of the NPRA into three major sequences: the Franklinian, Ellesmerian, and Brookian sequences. They are differentiated on the basis of source-area direction and are roughly correlated with major tectonic developments in the region (Carter et al., 1977). The sequences and their various sub-units are summarized in Fig. 7; a short description of each major sequence is given below.

Franklinian sequence (Cambrian–Devonian). According to Jamison et al. (1980), the Franklinian sequence constitutes economic basement for petroleum potential in the NPRA. For the most part, the Franklinian contains deformed, mildly metamorphosed greywackes and argellites which are intruded in places by granitic plutons (Armagnac, 1985). The sediments were shed from what is now northern Alaska as the area was uplifted in the Late Devonian Ellesmerian (Antler) Orogeny (Grantz et al., 1979). A period of erosion and subsidence followed, resulting in regional truncation of Franklinian strata by the end of the Devonian (Carter et al., 1977).

Ellesmerian sequence (Mississippian–Jurassic). The Late Devonian regional unconformity separates the Ellesmerian sequence from the Franklinian rocks below. Generally speaking, the Ellesmerian sequence is made up of platform carbonates and clastics (marine and non-marine) which were derived from a northern source, presumably from a location near the present Barrow Arch (Jamison et al., 1980). The Ellesmerian attains a maximum thickness of about 20,000 feet (7000 m) in the southern NPRA; to the north the beds thin and onlap onto the Barrow Arch (Armagnac, 1985).

The Endicott Group, a series of Mississippian non-marine clastics with interbedded carbonates, represents the base of the Ellesmerian sequence (Grantz et al., 1979). Endicott Group rocks rest unconformably on Franklinian strata, largely filling in the Franklinian graben basins (Kirschner et al., 1983). The Lisburne Group (Pennsylvanian), a marine carbonate sequence with interspersed clastics, was deposited on top of the Endicott forming a broad carbonate platform (Kirschner et al., 1983).

Following an Early Permian regression, the Sadlerochit Group and the Shublik Formation, dominantly clastic units of Late Permian to Triassic age, were deposited unconformably on the pre-Permian section. Both units tend to fine downdip, grading into siltstone and shale to the south (Carter et al., 1977).

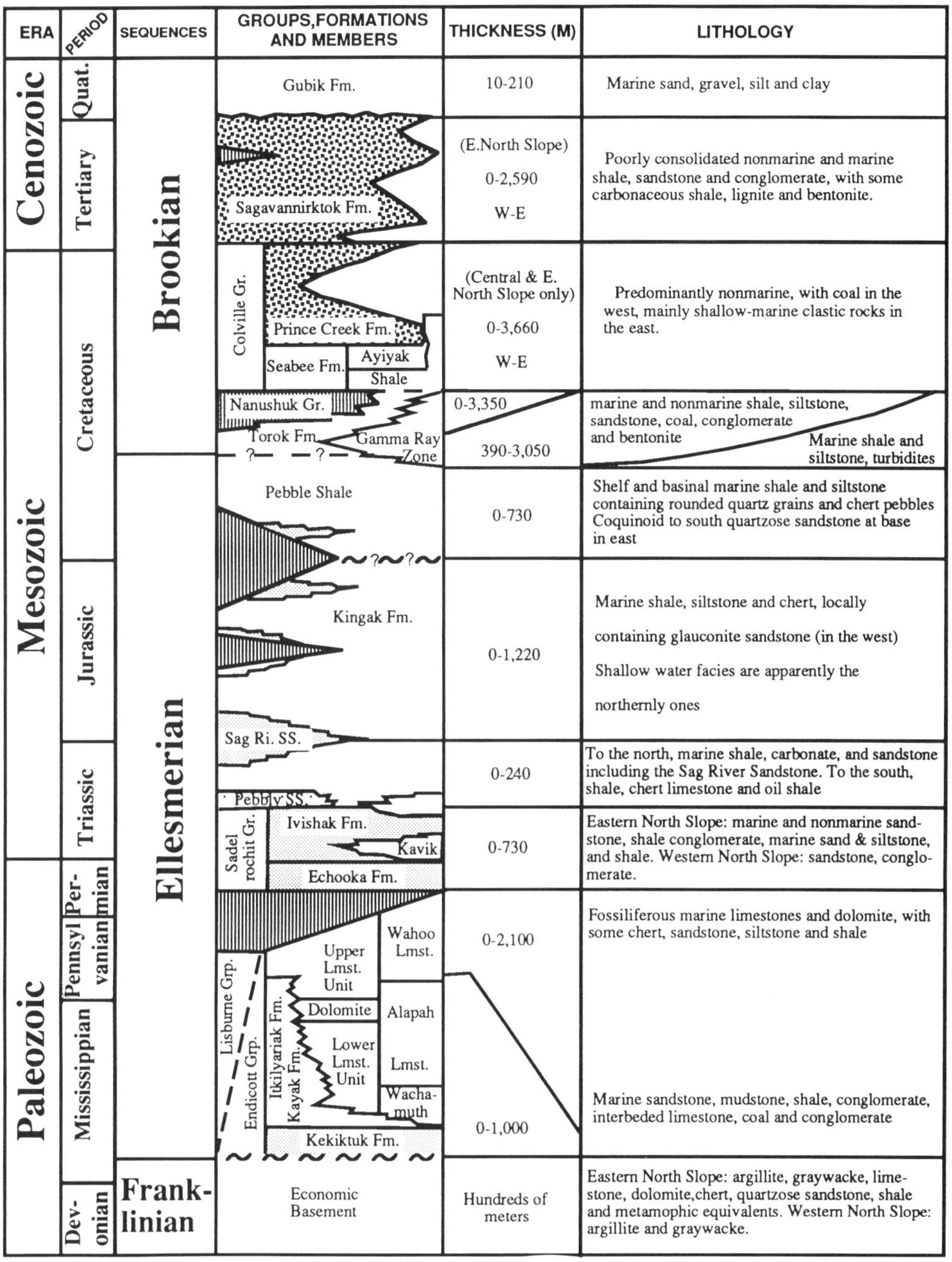

Fig. 7. Generalized stratigraphic chart, NPRA.

A thick series of largely shale units (Kingak Shale and Pebble Shale units) was deposited from Jurassic to Early Cretaceous times. The Kingak Shale contains minor amounts of sandstone and siltstone and is unconformably overlain by the Pebble Shale Formation, an important marker unit noted for its basal conglomerate (Kirschner et al., 1983).

Brookian sequence (Cretaceous–Recent). The deposition of the Brookian sequence marks an abrupt shift in source-area direction. As thrusting progressed from the south, northern sources were replaced by southern sources. Large volumes of flysch and molasse were shed from the encroaching Brooks orogen into the migrating Colville foredeep (Grantz et al., 1979). Continued sedimentation filled the Colville Trough, eventually overtopping the Barrow Arch (Kirschner et al., 1983).

Depositionally, the Brookian sequence is characterized by a series of prograding clastic wedges separated by thick shale units. The Fortress Mountain Formation (Early Cretaceous), Nanushuk Group (Middle Cretaceous), Colville Group (Late Cretaceous), and the Sagavanirktok Formation (Early Tertiary) represent the major clastic wedges in the sequence.

Runs and results

USGS line 6 traverses the NPRA in a N–S direction from the Beaufort Sea to the northern margin of the Brooks Range (Fig. 8). Each of the four structural provinces and all three major stratigraphic sequences are represented in the section. To the north, sediments of the Ellesmerian and Brookian sequences onlap the southward-dipping Franklinian basement, thinning northward onto the Barrow Arch. The Late Devonian unconformity between the Ellesmerian and the Franklinian is clearly visible, separating high-amplitude continuous reflections (Ellesmerian) from the more weakly continuous faulted basement reflectors. The highly reflective Pebble Shale unit marks the boundary between the Ellesmerian and Brookian sequences. As noted earlier, deformation increases southward in the section. Several fault-related anticlines are evident in the Brookian rocks of the Foothills Province. Further south, the Brooks Range Province is represented by an extremely chaotic zone of reflectors as would be expected in such a severely deformed orogen. Only one fault, a steeply dipping thrust, can be clearly distinguished in the Brooks Range part of the section.

The hinge point is located on the right margin of the section at a depth of about 1500 m. From this point the supra-Franklinian load has been divided into 73 blocks of 3 km width, giving a total length of 215 km. The loads and observed depths of each block were calculated and input to the model which was then run to determine the best flexural parameters. The surface porosity is assumed to be 0.56 and the scaling length is taken to be 2300 m (Sclater and Christie, 1980). Table 1 shows the estimates and ranges as well as the final values for the five parameters. After 40 iterations the final MSR was 5.5, an improvement of 40% over the initial MSR of 9.24; further runs did not significantly improve this value. An MSR of 5.5 represents an average mismatch of about 560 m between the predicted and observed basement depth values. The maximum depth of the Colville Trough is on the order of 10,000 m, so the average mismatch is roughly 6% of the total depth of the basin. This close fit suggests that the Colville Trough has developed dominantly as a flexural response to the loading of the basement.

As mentioned before, the location of the hinge point may be out of the section. If the Barrow Arch is indeed a peripheral bulge, the true hinge point should be located some distance north of the end of the section. In order to test this conjecture, the position of the hinge is allowed to vary in a linear fashion (with the other parameters held fixed at their best values) northwards from its initial position at the edge of the section (Fig. 9). After the best hinge position has been determined, the five flexural parameters are once again inverted to determine their best values for the new hinge position. The optimum hinge position is 105 km north of the edge of the section, giving a final MSR of 5.48 (average mismatch of 550 m). The best values of the flexural rigidity and bending moment were changed slightly by the new hinge position; the other parameters were unaffected (Table 2). Although the fit is better with the new hinge position, the improvement over the

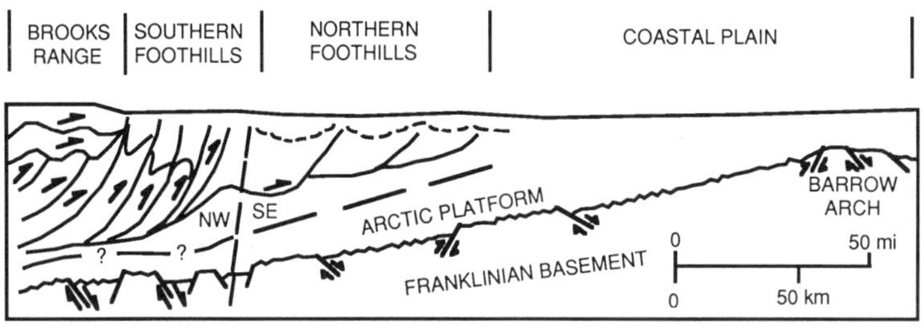

Fig. 8. Simplified interpretation of USGS seismic line 6 (from Kirschner et al., 1983).

TABLE 1

Initial estimates and ranges of input parameters

Parameter	Initial guess	Minimum	Maximum	Best value
D (N m)	$10 \times 10^{22.0}$	$10 \times 10^{21.0}$	$10 \times 10^{24.0}$	$10 \times 10^{23.38}$
M (N)	$10 \times 10^{17.0}$	$10 \times 10^{16.0}$	$10 \times 10^{18.0}$	$10 \times 10^{16.98}$
V (N/m)	$10 \times 10^{10.0}$	$10 \times 10^{8.0}$	$10 \times 10^{15.0}$	$10 \times 10^{13.40}$
$\tan \theta$	0.0	−0.1	0.1	0.002
L (km)	0.0	0.0	270.0	0.0

Initial MSR = 9.24; final MSR = 5.50.

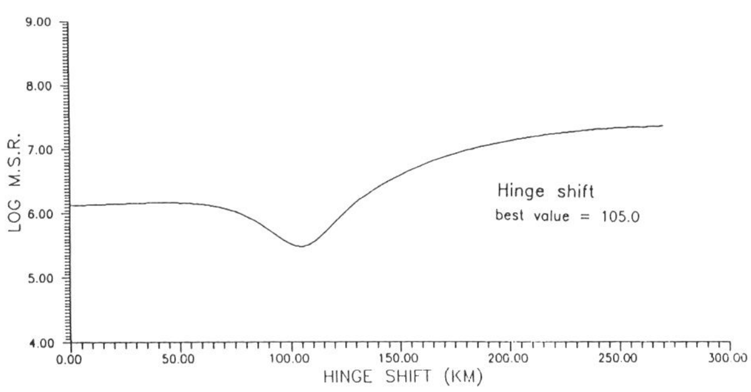

Fig. 9. Diagram illustrating the sensitivity of basement subsidence to variation of the hinge point position.

TABLE 2

Variation of parameters with hinge point shifted 105 km

Parameter	Best value	
	with original hinge	with shifted hinge
D (N m)	$10 \times 10^{23.38}$	$10 \times 10^{23.31}$
M (N)	$10 \times 10^{16.98}$	$10 \times 10^{17.04}$
V (N/m)	$10 \times 10^{13.40}$	$10 \times 10^{13.40}$
$\tan\theta$	0.002	0.002
L (km)	0.0	0.0

Initial MSR = 8.0; final MSR = 3.6.

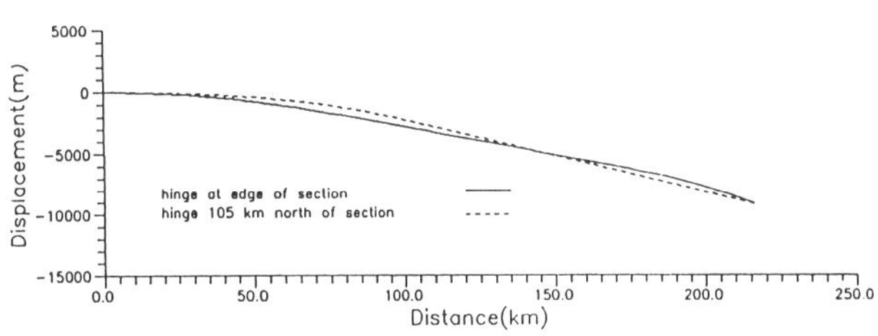

Fig. 10. Predicted basement profiles with hinge point chosen at two possible positions.

first run is negligible (less than 1% change in MSR) (Fig. 10). Hence, the validity of either situation cannot be determined by the model alone.

Figures 11–13 show the predicted, observed and unloaded basement profiles with the hinge point located on the edge of the section. Northward extension of the hinge point does not significantly change the shape of the curves. The unloaded profile represents the basement configuration prior to any

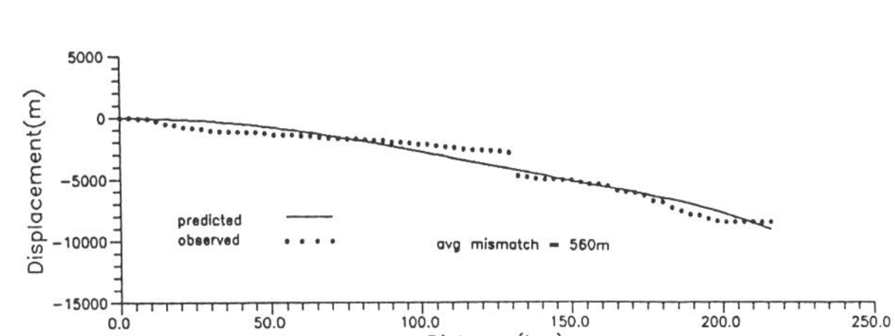

Fig. 11. Predicted vs. observed basement profiles.

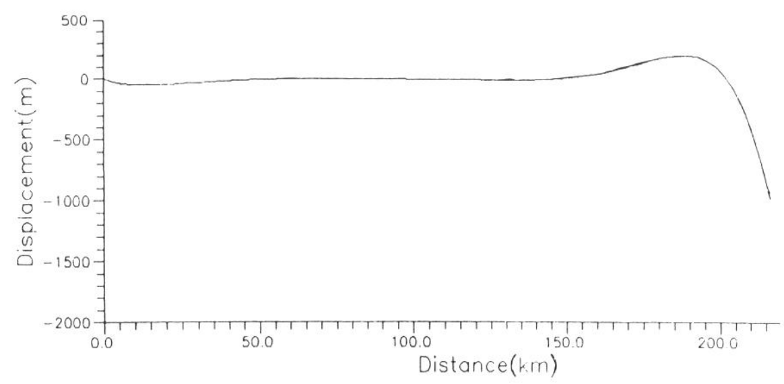

Fig. 12. Unloaded basement profile. This is the configuration of the basement prior to sediment/thrust loading.

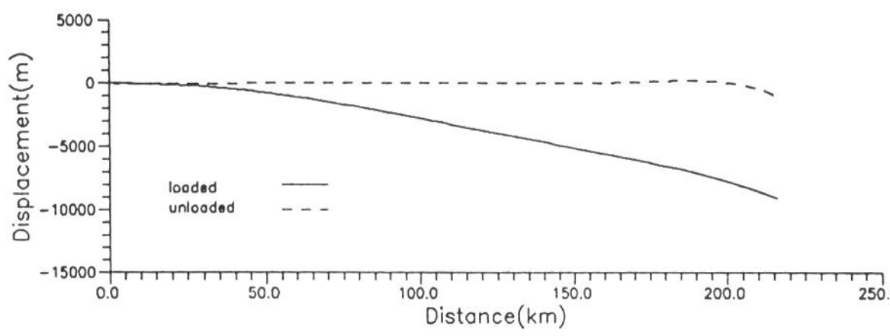

Fig. 13. Theoretical and unloaded basement profiles.

sediment or thrust loading, about 350 m.y. B.P. At this time a bulge roughly 50 km wide with a maximum height of nearly 200 m existed about 190 km from the hinge point. This bulge is not observed on the present-day section. Presumably, it has migrated northward, perhaps contributing to the formation of the Barrow Arch, in response to the thrusting of the Brooks Range from the south.

Parameter sensitivity

The system is most sensitive to variations of the flexural rigidity, bending moment and dip angle, and all three are well determined by the model (Figs. 14–16). Small departures from the best values cause significant changes in fit; if any one of the three parameters is raised or lowered by one half of an order of magnitude the average mismatch increases by more than 2000 m. The MSR is constant for log initial load values less than 12.0 and increases

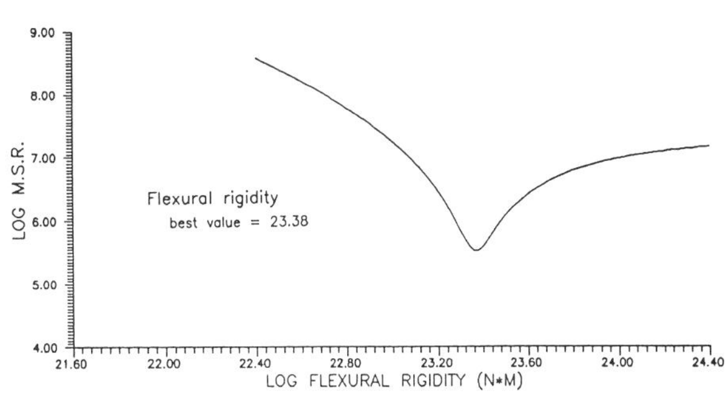

Fig. 14. Diagram illustrating the sensitivity of basement subsidence to variation of the flexural rigidity.

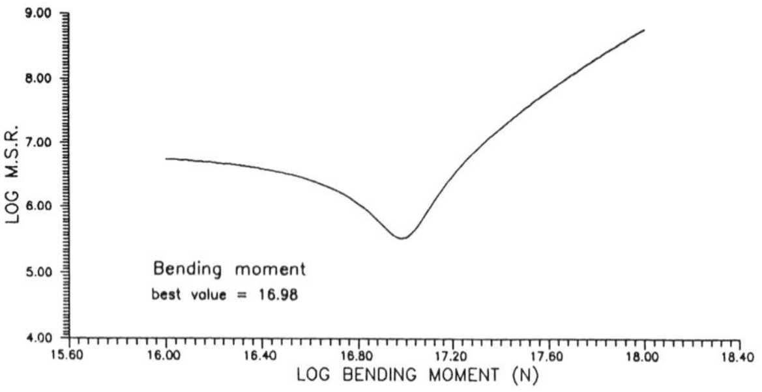

Fig. 15. Diagram illustrating the sensitivity of basement subsidence to variation of the bending moment.

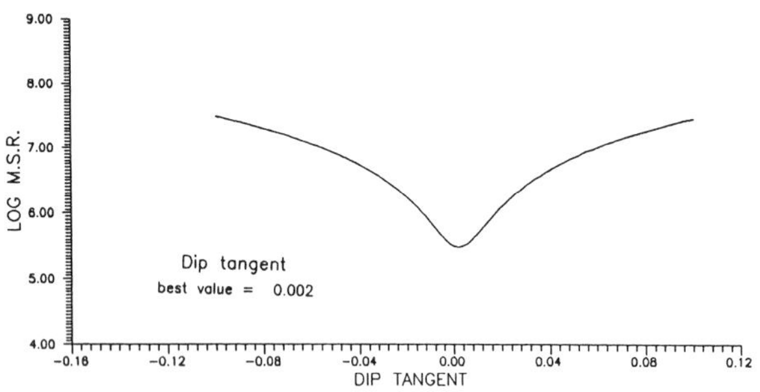

Fig. 16. Diagram illustrating the sensitivity of basement subsidence to variation of the dip tangent.

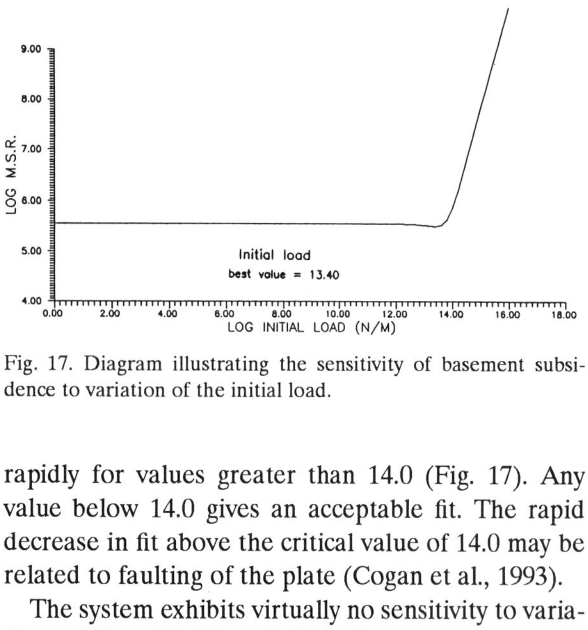

Fig. 17. Diagram illustrating the sensitivity of basement subsidence to variation of the initial load.

rapidly for values greater than 14.0 (Fig. 17). Any value below 14.0 gives an acceptable fit. The rapid decrease in fit above the critical value of 14.0 may be related to faulting of the plate (Cogan et al., 1993).

The system exhibits virtually no sensitivity to variations of the added length (Fig. 18). Although the best value is 0, higher values produce nearly the same fit to the observed data. In this application the length of the plate has little influence on the flexure of the basement.

One of the assumptions of the model is that the hinge point remains stationary as the plate is loaded. In essence, it is a reference point relative to which all displacement occurs. As mentioned above, there is no obvious termination of the Franklinian basement

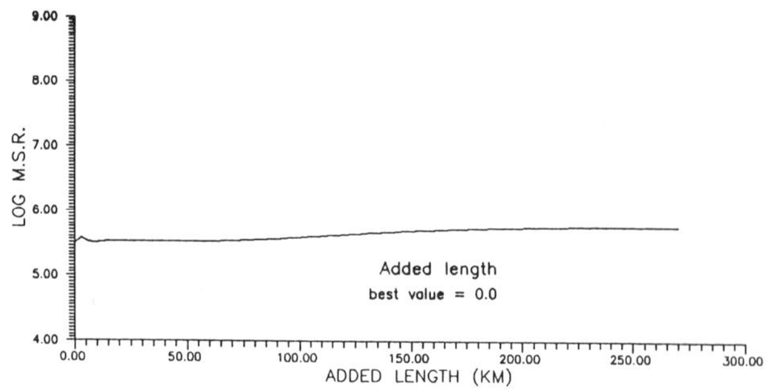

Fig. 18. Diagram illustrating the sensitivity of basement subsidence to variation of the added length.

to the north on the section, so the hinge point was originally taken at the northern margin of the line and the model was run. Next, the model was modified to allow for variations in hinge position and was run a second time. The optimum hinge position thus determined is 105 km north of the margin of the line. This value is in good agreement with the observed distance of roughly 100 km from the end of line 6 to the Hinge Line Fault zone, a major break in the Franklinian strata (Thurston and Theiss, 1987). Although the degree of fit between theoretical predictions and observed data is nearly the same for either hinge point position, it is likely that the hinge lies on the Hinge Line Fault Zone since this zone represents the termination of the Franklinian basement.

Conclusions

Three conclusions can be reached from this study.

(1) The average fit of 560 m between the theoretical and observed basement depths is only 6% of the maximum depth of the Colville Trough. This result suggests that the subsidence of the Franklinian basement in the NPRA is largely due to elastic flexure. It has been shown elsewhere (Cogan et al., 1993) that flexure occurs on a horizontal scale of roughly 10 km; the present study indicates that elastic flexure operates on a scale of hundreds of kilometers as well. Apparently flexure is a scale-independent mechanism. Perhaps some structural highs that have been identified as fault-induced anticlines are actually flexural anticlines.

(2) The model results support the contention that peripheral bulges migrate in front of thrust belts in the direction of thrusting. Prior to any sediment or thrust loading the basement exhibits a bulge roughly 200 km south of the northern margin of the section. This bulge is not observed today and has probably migrated northward, forming the Barrow Arch.

(3) The model results are relatively insensitive to the hinge position, although an optimum hinge

position near or on the Hinge Line Fault Zone is preferred. The optimum parameter values are: $\log D = 23.4$ N m, $\log M = 17.0$ N, $\log V = 13.4$ N/m, $\tan\theta = 0.002$, $L = 0.0$ km. The flexural rigidity value of 3.10^{23} N m is in accordance with other published values: 10^{24}–10^{25} N m (Beaumont, 1981), 10^{22} N m (Patton and O'Connor, 1988), 10^{25} N m (Quinlan and Beaumont, 1984), and 10^{23} (Jordan, 1981).

Acknowledgements

This study was supported, in part, by the Industrial Associates of the Basin Modelling Group at the University of South Carolina.

Appendix A — Elastic model

Flexure of a finite elastic beam of length L which is anchored at one end due to the force of a single applied end load is governed by the equation:

$$D\frac{\partial^4 w(x)}{\partial x^4} + (\rho_m - \rho_w)gw(x) = 0 \qquad \text{(A1)}$$

where D is the flexural rigidity of the beam, $w(x)$ is the vertical displacement of the beam, x is the horizontal position along the beam, ρ_m is the density of the mantle, ρ_w is the density of the overlying material, g is the acceleration due to gravity (Turcotte, 1979). Four boundary conditions are necessary for solution of the fourth-order eq. A1. Referring to Fig. 1b on the anchored end ($x = 0$) no displacement occurs and the dip angle of the beam of that point is θ. On the free end ($x = L$) the bending moment is M and the end load is V. If the densities of the mantle and overlying material are known, five parameters dictate the bending of the beam: D, M, V, θ, and L.

The flexural rigidity, D, of the beam is a measure of its resistance to bending. With all other parameters held constant, a higher flexural rigidity will cause the plate to bend less. Flexural rigidity is independent of the other parameters but is a function of the elastic properties of the plate (Turcotte, 1979):

$$D \equiv \frac{Eh^3}{12\,(l - \nu^2)} \qquad \text{(A2)}$$

where E is Young's modulus, ν is Poisson's ratio and h is the thickness of the plate.

The bending movement M is related to both the rigidity D and curvature at the end of the plate:

$$M = -D\frac{\partial^2 w(x)}{\partial x^2} = \frac{D}{R}, \qquad \text{on } x = L \qquad \text{(A3)}$$

where R is the radius of curvature.

The dip angle and end load are related to the first and third derivatives of eq. A1, respectively:

$$\begin{aligned} \tan\theta &= \frac{\partial w(x)}{\partial x}, && \text{on } x = 0; \\ V &= D\frac{\partial^3 w(x)}{\partial x^3}, && \text{on } x = L \end{aligned} \qquad \text{(A4)}$$

If all five parameters are known, vertical displacement as a function of horizontal distance from the hinge can be calculated and a theoretical basement profile plotted.

Appendix B — Numerical inversion procedure

The solution of eq. A1 in Appendix A is of the form:

$$\begin{aligned} w(x) = {} & \exp(x/\alpha)[c_1\cos(x/\alpha) + c_2\sin(x/\alpha)] \\ & + \exp(-x/\alpha)[c_3\cos(x/\alpha) + c_4\sin(x/\alpha)] \end{aligned} \qquad \text{(B1)}$$

where

$$\alpha = \left[\frac{4D}{g(\rho_m - \rho_s)}\right]^{1/4}$$

is a scaling length, x is horizontal distance, and c_1, c_2, c_3 and c_4 are integration constants. The values of c_1, c_2, c_3 and c_4 depend on the five flexural parameters; M, V, and $\tan\theta$ enter eq. B1 linearly, while D and L enter nonlinearly. A least-squares fit of predicted versus measured vertical displacement ($w(x)$) is used to estimate the best values of M, V, θ, D, and L:

$$X^2 = -\frac{1}{N}\sum_{i=1}^{N}[w_i\,(\text{measured}) - w(M,\nu,\theta,D,L,x_i)]^2 \qquad \text{(B2)}$$

where N is the total number of loads on the system. The flexural parameters are varied until X^2 is at a minimum. Commonly, a minimum is obtained by differentiating X^2 with respect to unknown parameters and requiring zero values for the derivatives, i.e.:

$$\begin{aligned} &F_1 = \frac{\partial X^2}{\partial D}; \quad F_2 = \frac{\partial X^2}{\partial M}; \quad F_3 = \frac{\partial X^2}{\partial V}; \\ &F_4 = \frac{\partial X^2}{\partial \theta}; \quad F_5 = \frac{\partial X^2}{\partial L} \end{aligned} \qquad \text{(B3)}$$

should all be set to zero. We can calculate the derivatives F_1–F_5 analytically, but there are two problems which make it undesirable to use only the zeros of the derivatives F_1–F_5 to compute the minimum of X^2. First, the accuracy with which we can calculate the values of D, M, V, θ, and L is limited by the interpolation scheme used for determining the zeros of F_1–F_5. The accuracy may well be less than the degree to which X^2 is minimal. Second, physically we require D and L to be positive, but have not imposed that *a priori* constraint on the solution.

To avoid these problems simultaneously, we proceed as follows. Suppose we have chosen the initial trial values $D(0)$, $M(0)$, $L(0)$, $V(0)$, and $\theta(0)$. For each iteration we find new variables $D(n)$, $M(n)$, $L(n)$, $V(n)$ and $\theta(n)$ by:

$$\begin{aligned} D(n) &= D(n-1)\ \exp\left[-A\frac{\partial y}{\partial D(n-1)}\right] \\ M(n) &= M(n-1)\ \exp\left[-B\frac{\partial y}{\partial M(n-1)}\right] \\ L(n) &= L(n-1)\ \exp\left[-C\frac{\partial y}{\partial L(n-1)}\right] \\ V(n) &= V(n-1)\ \exp\left[-D\frac{\partial y}{\partial V(n-1)}\right] \\ \theta(n) &= \theta(n-1)\ \exp\left[-E\frac{\partial y}{\partial \theta(n-1)}\right] \end{aligned} \tag{B4}$$

where $y = X^2/X_0^2$ and A, B, C, D and E are constants determined to enable the program user to specify the size of the first step. X_0^2 represents the initial trial of D, M, L, V, and θ. This procedure ensures that at each iteration D and L remain positive, and that each iteration will systematically reduce y, so that all derivatives and X^2 are reduced at each iteration. The precise number of iterations required to produce a minimum depends upon the initial values chosen for the flexural parameters, but convergence to a fixed minimum always occurs.

References

Airy, G.B., 1855. On the computation of the effect of the attraction of mountain masses as disturbing the apparent astronomical latitude of stations of geodetic surveys. Philos. Trans. R. Soc. London, 145: 101–104.

Armagnac, C.M., 1985. The Determination of Paleoheat Flow in the National Petroleum Reserve, Alaskam through the Inversion of Vitrinite Reflectance. Master's Thesis. University of South Carolina, Columbia, S.C., 161 pp.

Beaumont, C., 1981. Foreland basins. Geophys. J. R. Astron. Soc., 65: 291–329.

Bird, K.J., 1981. Petroleum exploration of the North Slope, Alaska: In: J.F. Mason (Editor), Petroleum Geology in China, Principal Lectures Presented to the United Nations International Meeting on Petroleum Geology, Beijing, China, 18–25 March, 1980. Penn Well Books, Tulsa, Okla., pp. 233–248.

Bushnell, H., 1981. Unconformities — key to North Slope oil. Oil Gas J., 79: 114–118.

Carmen, G.J. and Hardwick, P., 1983. Geology and regional setting of the Kuparuk oil field, Alaska. Am. Assoc. Pet. Geol., Bull., 67: 1014–1031.

Carter, A.D., Mull, C.G., Bird, K.J. and Powers, R.B., 1977. The petroleum geology and hydrocarbon potential of Naval Petroleum Reserve No. 4, North Slope, Alaska. U.S. Geol. Surv., Open File Rep., 77-475, 61 pp.

Churkin, M., Jr., Nokleberg, W.J. and Huie, C., 1979. Collision-deformed Paleozoic continental margin, western Brooks Range, Alaska. Geology, 7: 379–383.

Cogan, J., Lerche, I., Dorman, J.T. and Kanes, W., 1993. Flexural plate inversion: application to the Carpathian Foredeep, Czechoslovakia. Modern Geol. (in press).

Grantz, A., Eittreim, S. and Dintar, D.A., 1979. Geology and tectonic development of the continental margin north of Alaska. Tectonophysics, 59: 263–291.

Haxby, W.F., Turcotte, D.L. and Bird, J.M., 1976. Thermal and mechanical evolution of the Michigan Basin. Tectonophysics, 36: 57–75.

Jacobi, R.D., 1981. Peripheral bulge — a causal mechanism for the Lower/Middle Ordovician unconformity along the western margin of the Northern Appalachians. Earth Planet. Sci. Lett., 56: 245–251.

Jamison, H.C., Brockett, L.D. and McIntosh, R.A., 1980. Prudhoe Bay — a 10 year perspective. Am. Assoc. Pet. Geol., Bull., 64: 289–314.

Jordan, T.E., 1981. Trough loads and foreland basin evolution, Cretaceous, western United States. Am. Assoc. Pet. Geol., Bull., 65: 2506–2520.

Karner, G.D. and Watts, A.B., 1983. Gravity anomalies and flexure of the lithosphere at mountain ranges. J. Geophys. Res., 88: 10449–10477.

Kirschner, C.E., Gryc, C., and Molenaar, C., 1983. Regional seismic lines in the National Petroleum Reserve of Alaska. In: A. W. Bally (Editor), Seismic Expression of Structural Styles. Am. Assoc. Pet. Geol., Stud. Geol. Ser., 15 (1): Sect. 1.2, pp. 5–14.

McMullen, R.J. and Mohraz, B., 1987. Axisymmetric finite element analysis of an elastic lithosphere subjected to thermal loading. Geophys. J. R. Astron. Soc., 88: 325–343.

McMullen, R.J., Hodge, D.S. and Cozzarelli, F.A., 1981. A technique for incorporating geothermal gradients and nonlinear creep into lithospheric flexure models. J. Geophys. Res., 86: 1745–1753.

Menke, W., 1984. Geophysical Data Analysis: Discrete Inverse Methods. Academic Press, New York, N.Y., 260 pp.

Patton, T.L. and O'Connor, S.J., 1988. Cretaceous flexural history of Northern Oman Mountain Foredeep, United Arab Emirates. Am. Assoc. Pet. Geol., Bull., 72: 797–809.

Pratt, J.H., 1855. On the attraction of the Himalayan Mountains, and of the elevated regions beyond them, upon the plumb line in India. Philos. Trans. R. Soc. London, 145: 53–100.

Quinlan, G., 1984. Postglacial rebound and the focal mechanisms of eastern Canadian earthquakes. Can. J. Earth Sci., 21: 1018–1023.

Quinlan, G.M. and Beaumont, C., 1984. Appalachian thrusting, lithospheric flexure, and the Paleozoic stratigraphy of the Eastern Interior of North America. Can. J. Earth Sci., 21: 973–996.

Royden, L. and Karner, G.D., 1984. Flexure of lithosphere beneath Apennine and Carpathian foredeep basins: evidence for an insufficient topographic load. Am. Assoc. Pet. Geol., Bull., 68: 704–712.

Sclater, J.G. and Christie, P.A.F., 1980. Continental stretching: an explanation of the post-mid-Cretaceous subsidence of the central North Sea basin. J. Geophys. Res., 85: 3711–3739.

Thurston, D.K. and Theiss, L.A., 1987. Geologic report for the Chukchi Sea planning area, Alaska; regional geology, petroleum geology, and environmental geology. U.S. Department of the Interior, Anchorage, Alaska, OCS Rep. MMS 87-0046, 193 pp.

Turcotte, D.L., 1979. Flexure. Adv. Geophys., 21: 51–87.

Turcotte, D.L. and Schubert, G., 1982. Geodynamics: Applications of Continuum Physics to Geological Problems. Wiley, New York, N.Y., 450 pp.

JEFF COGAN *1718 35th Street, N.W. Washington, DC 20007, U.S.A.*

IAN LERCHE *Department of Geological Sciences, University of South Carolina, Columbia, SC 29208, U.S.A.*

BRANCH RUSSELL *Marathon Oil Co., P.O. Box 629, Littleton, CO 80160, U.S.A.*

Analysis of the Beaufort–MacKenzie Basin, Canada: burial, thermal and hydrocarbon histories

J. Tang and I. Lerche

The burial, thermal and hydrocarbon histories of the Jurassic–Tertiary sedimentary succession were investigated for three structural zones of the Beaufort–Mackenzie Basin using a one-dimensional comprehensive burial history model. The reconstructed geohistories of seven wells indicate two stages of tectonics in the study area: (1) rifting from the Jurassic to Early Cretaceous in structural zone I and subsequent extension in structural zone II; and (2) post-rifting uplift and erosion as well as thermal subsidence. The maximum burial of Jurassic/Cretaceous and Tertiary sediments corresponds to the rifting episode. The Late Eocene to Late Miocene uplift caused at least 1000 to 2000 m of sediment to be removed in the study area. The Late Miocene to Pleistocene thermal subsidence resulted in an accumulation of thick fluvial-delta sediments (the Iperk Sequence) from the Mackenzie Delta to offshore of the Beaufort Sea.

Heating episodes from around 150 to 20 Ma in the Tuktoyaktuk Peninsula province and from around 60 to 10 Ma in the central and western part of the basin are indicated by inversion of vitrinite reflectance data from seven wells.

Comparison of the thermal maturation pattern with oil and gas occurrences in seven of the wells indicates that hydrocarbons in Cretaceous reservoirs may have migrated vertically from the Husky Formation, while the hydrocarbons in the Tertiary reservoirs may have undergone extensive lateral migration from offshore, where the Richards Sequence source rock has a deeper burial history than onshore.

Introduction

Two fundamental approaches to oil and gas exploration have been developed: (1) methods designed to locate hydrocarbons *in situ* from direct measurements ahead of the drill; (2) methods focusing on the progressive evolution of a sedimentary basin considering the degree of thermal alteration of organic material, and the timing of hydrocarbon generation, migration, and accumulation, which provide an assessment of hydrocarbon prospectivity. Hydrocarbon occurrences depend on three major factors: (1) the sediment burial history of the basin, which is governed by many factors, including sediment supply, mechanical and chemical compaction, tectonic forces, erosional events, intrusive events, sea-level changes, and so on; (2) the structural evolution of the basin; and (3) the thermal evolution of sediments in the basin, which controls the thermal maturation of organic materials, and hydrocarbon generation, migration and accumulation.

In this paper, a one-dimensional thermal indicator tomography model (He and Lerche, 1989) has been used to simulate the evolutionary processes of the Beaufort–Mackenzie Basin, including burial, thermal and hydrocarbon generation histories of the Jurassic–Tertiary succession. The specific objectives of this study are: (1) to reconstruct the burial paths of Jurassic–Tertiary strata, the amounts of erosion, and basement subsidence, from which we interpret the basin evolution; (2) to determine the thermal maturation levels of Jurassic–Tertiary source rocks; and (3) to investigate possible relationships between the maturity of source rocks and present-day oil and gas occurrences. Seven wells have been chosen for this study (Fig. 1).

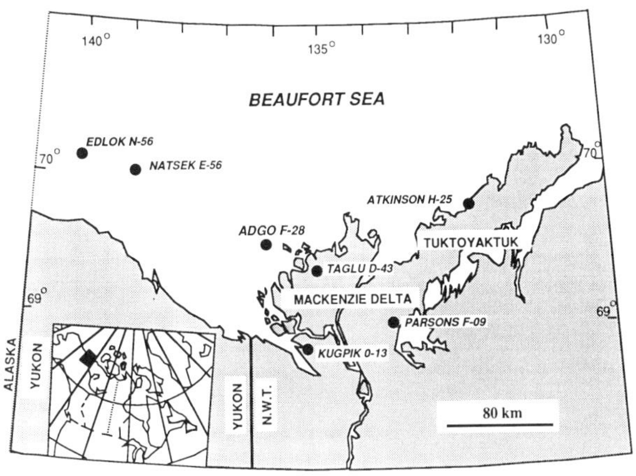

Fig. 1. Location map of the Beaufort–Mackenzie Basin showing the wells to be modelled (modified from McNeil, 1990).

Basin Modelling: Advances and Applications edited by A.G. Doré et al. NPF Special Publication 3, pp. 623–640, Elsevier, Amsterdam.

Regional setting

Figure 2 shows the geological setting of the Beaufort–Mackenzie Basin which lies at the junction of three crustal elements (Jones et al., 1980). The basin is bounded to the east by the Eskimo Lakes Fault Zone and to the southwest by the Brooks Range; to the north is the Canada Basin. It is suggested that the development of the Beaufort–Mackenzie Basin is related to Arctic plate tectonics (Young et al., 1976). For details on strata distribution and tectonic evolution of the study area the reader is referred to the comprehensive volumes of Young et al. (1976), Dixon et al. (1985), Dixon and Dietrich (1988), and Dietrich et al. (1989a).

It is emphasized here that there are two opposing viewpoints on tectonic evolution of the study area. One is that the tectonic evolution of the region began with extension from Early Jurassic to Early Cretaceous, and that a compressional history was dominant from later Cretaceous to the present day (Dietrich et al., 1989b). The other suggests that the extensional history, lasting to the Middle Miocene, and a later thermal subsidence, resulted in a 14 to 16 km thick sedimentary deposition from later Cretaceous to the present day (Enachescu, 1990).

The stratigraphic framework and nomenclature in the Beaufort–Mackenzie Basin are shown in Fig. 3. In terms of the characteristics of lithology, facies distributions and thickness variations of Jurassic to Tertiary strata, Dixon et al. (1985) divided them into five tectonostratigraphic assemblages, similar to Young's division of three tectonic–sedimentary

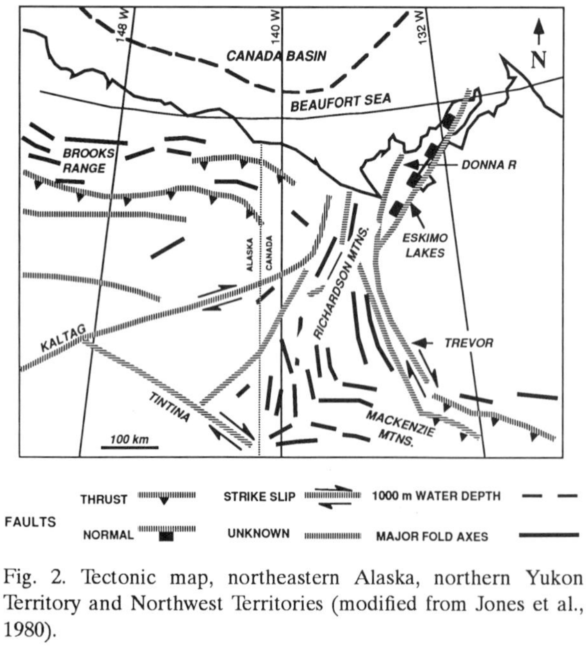

Fig. 2. Tectonic map, northeastern Alaska, northern Yukon Territory and Northwest Territories (modified from Jones et al., 1980).

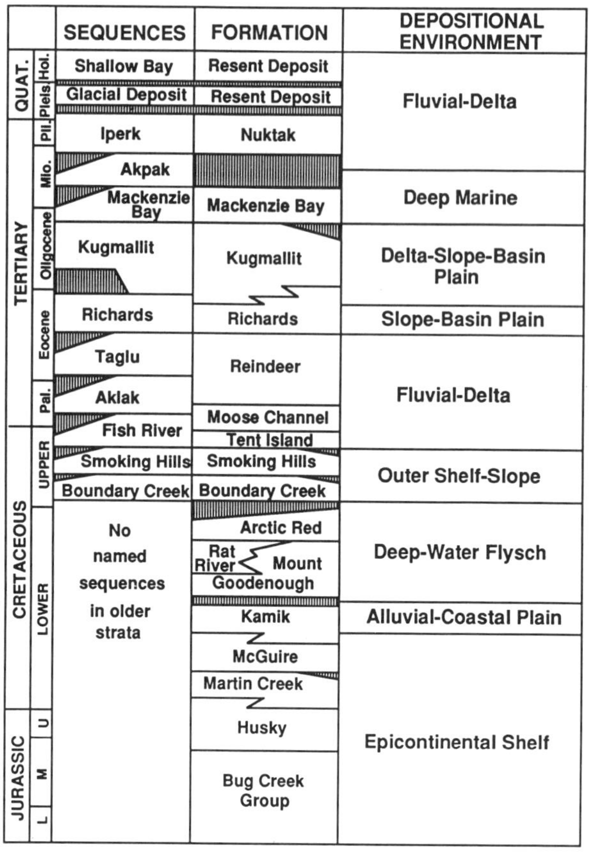

Fig. 3. Stratigraphic framework in the Beaufort–Mackenzie Basin (modified from Dixon et al., 1988).

phases for Triassic to Tertiary strata in the study area (Young et al., 1976, figs. 4–16).

In the Jurassic to middle Hauterivian assemblage, sediments were deposited on an epicontinental shelf (Dixon et al., 1985). In the eastern part of the basin, the Husky Formation is considered to be source unit, and the overlying Parsons sandstone is the reservoir unit (Dixon et al., 1985; Cote et al., 1975). In the period from late Hauterivian to Albian (the second assemblage), the sediments represent the deposition of low-energy shelf and deep-water flysch. A major regional unconformity separates Jurassic to middle Hauterivian strata from late Hauterivian to Albian strata. The depositional history is superimposed on an extensional phase of the tectonic development of the Beaufort–Mackenzie Basin.

The Late Cretaceous to Pleistocene strata are divided into three other tectonostratigraphic assemblages, which are separated by the Late Eocene and Late Miocene unconformities (Dixon et al., 1985). The third assemblage contains six sequences (Fig. 3). The Boundary Creek and Smoking Hills sequences of the later Cretaceous strata consist of dark grey to black, organic-rich shales. Organic carbon content is in a range of 4–12%, which represents deposition in a low-energy anoxic marine environment, probably in an outer shelf to slope setting (Dixon et al., 1985). The Fish River sequence is considered to be the first major fluvio-deltaic complex deposited in the Beaufort–Mackenzie Basin (Dixon et al., 1985). The Reindeer and Richards sequences of Early/Middle Paleocene to Late Eocene strata have been interpreted to be prodelta to delta-front to delta-plain deposits and a mudstone-dominant succession of prodelta to slope deposits, respectively (Dixon et al., 1985). In this assemblage the source rock appears to be the Richards sequence (Snowdon, 1987, 1988; Issler and Snowdon, 1990). The Reindeer and Fish River sequences are the reservoirs in this assemblage (Dixon et al., 1985).

The fourth assemblage consists of the Oligocene Kopanoar, Kugmallit sequences, the Late Oligocene to Late Miocene Mackenzie Bay and Akpak sequences. The former two sequences represent delta-plain to slope-distal basin plain deposits in which hydrocarbons have been found in the Beaufort Sea (Dixon et al., 1985). Both Mackenzie and Akpak sequences represent mud-dominant deep-water deposits beneath the outer Beaufort Sea shelf and are considered as seals of the Kugmallit reservoir (Dixon et al., 1985; Enachescu, 1990).

The final assemblage consists of the Late Miocene to Pleistocene Iperk sequence, and represents a fluvio-delta depositional complex which rests unconformably on the older strata (Dixon et al., 1985).

Structural framework

Based on the distribution, geometry and type of faults in the area and surrounding area, the Beaufort–Mackenzie Basin can be divided into three structural zones (Jones et al., 1980; Enachescu, 1990).

Zone I lies in the southeast, overlying the faulted northwest flank of the Aklavik Arch (Fig. 4; Jones et al., 1980). Proterozoic and Palaeozoic sedimentary rocks, and overlying basal Jurassic and Lower Cretaceous clastics, are downfaulted toward, and occasionally away from, the basin. The overlying Upper Cretaceous and Palaeogene sequences are progressively less affected by faulting. Faults include the Eskimo Lakes Fault (Fig. 2; Jones et al., 1980) and the Parsons Lake Fault. The Eskimo Lakes Fault terminates against, and is interpreted to be offset by, the Donna River Fault (Fig. 2; Jones et al., 1980). Major Columbian tectonic elements, such as the Aklavik Arch and Cache Creek Uplift, do not continue across the Donna River Fault and appear to be offset by it. The Parsons Lake gas field and the Atkinson oil discovery lie onshore within zone I (Jones et al., 1980).

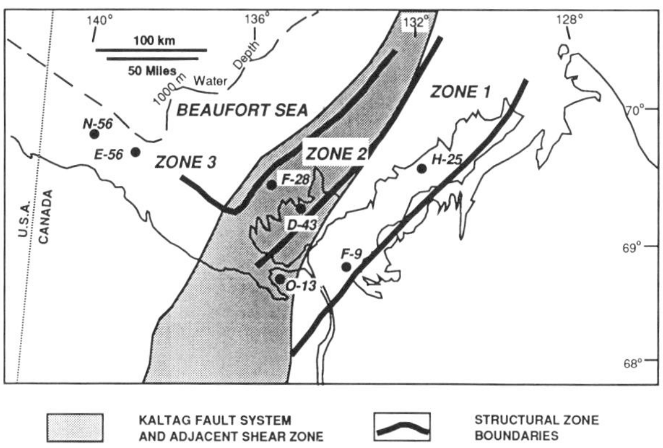

Fig. 4. Structural subdivisions, Beaufort–Mackenzie Basin (modified from Jones et al., 1980).

In *Zone II*, north of the seaward extension of the Donna River Fault, Tertiary and possibly Upper Cretaceous clastic sediments are cut by listric growth faults, most of which flatten basinward at depth in thick shale sections. The zone overlies the inferred seaward continuation of the Kaltag Fault and associated shear zone. The listric faults are clearly syn-sedimentary and have no direct connection with deep underlying structures. The Taglu gas field in the Mackenzie Delta lies within zone II (Jones et al., 1980).

Zone III, to the northwest, including zone IV of Jones et al. (1980), is characterized by shale-cored anticlines which are east–west and southeast–northwest trending, in which faults are few and relatively minor (Fig. 4; Jones et al., 1980).

Source rocks

The Richards sequence of Tertiary age, and the Smoking Hills and Boundary Creek Formations of Upper Cretaceous age, as well as the Arctic Red, Mount Goodenough, McGuire and Husky Formations of Jurassic–Lower Cretaceous age, have been considered to be potential source rocks in the study area (Dixon et al., 1985; Issler and Snowdon, 1990).

TOC histograms for the Arctic Red, Mount Goodenough, Husky and McGuire Formations (Fig. 5c) indicate that all contain sufficient organic carbon for hydrocarbon generation. The organic matter is dominated by terrestrial (type III) material (Dixon et al., 1985).

For the Upper Cretaceous source rocks, the Smoking Hills and Boundary Creek Formations contain significant amounts of organic carbon (Fig. 5b) with most of the samples containing more than 3% TOC. The organic matter is composed of a mixture of type III (terrestrial) and type II (marine planktonic) material (Dixon et al., 1985). In structural zones II and III the source rock is the Tertiary Richards sequence in which the TOC histograms (Fig. 5a)

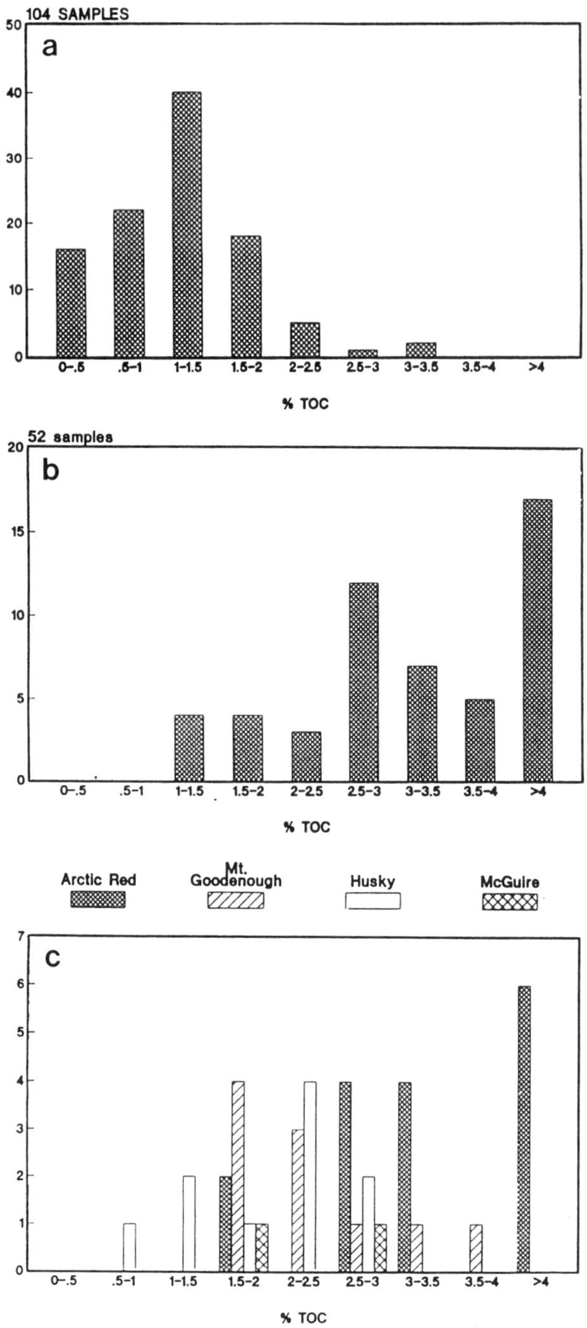

Fig. 5. Total organic carbon content: (a) Richards Formation; (b) Smoking Hills and Boundary Creek Formations; (c) Arctic Red, Mount Goodenough, Husky and McGuire Formations. (After Snowdon and Brooks, 1985.)

indicate a moderate, ubiquitous organic content with modal values in the 1–2% range. It is possible that the source quality improves in the offshore (to the north) where it has not been penetrated, although the Richards sequence has been classified as a poor to marginal source rock where sampled (Dixon and Dietrich, 1988; Issler and Snowdon, 1990). The organic matter in all of the Tertiary delta sediments is totally dominated by higher land plant debris and thus type III kerogen (Dixon et al., 1985).

Basin modelling

So far more than 170 wells have been drilled and 42 oil and gas fields have been discovered in Jurassic and Lower Cretaceous sandstone reservoirs along a northeastern trend of the Tuktoyktuk Peninsula, and in Tertiary strata which are distributed in the area of the Mackenzie River delta and extend seaward to the north (Meneley, 1986; Enachescu, 1990). The oil price crisis of the mid-80's negatively affected the exploration activity in the Canadian north. With the Amauligak J-44 and Kigark J-54 discoveries, however, oil and gas exploration activity is picking up again in the Beaufort–Mackenzie Basin. In order to possibly reduce the geological risk in exploration for oil and gas in this area, a method focusing on the dynamic evolution of a sedimentary basin is used here to understand the evolutionary processes of the basin. The method, called one-dimensional thermal indicator tomography model (He and Lerche, 1989; Lerche, 1990), is applied to seven wells in the Beaufort–Mackenzie Basin. The model is based on a 1D flow/compaction model (Cao, 1985; Cao et al., 1986; Cao and Lerche, 1987). The model consists of three separate components including dynamic models for geohistory, thermal history and hydrocarbon generation history.

From the geohistory model the following results can be obtained: basement subsidence, structural evolution, determination of erosion thickness and age of an unconformity, change of porosity, permeability, fluid flow rate and pore pressure with time and depth. The thermal history model evaluates the heat flow change through geological time and gives temperature evolution with time and depth. It is also possible to predict the behaviour of a series of thermal maturity indicators with any given burial history. The hydrocarbon generation model simulates the hydrocarbon generation history from Jurassic to Recent formations of the Beaufort–Mackenzie Basin based on the temperature history, burial history, and the kerogen content and type.

Data for basin modelling

Data from seven publicly released wells comprise the foundation of this paper. In structural zone I, three wells have been chosen (Atkinson H-25, Parsons F-09, and Kugpik 0-13); two wells are located in structural zone II (Adgo F-28 and Taglu D-43); two wells are in structural zone III (Edlok N-56 and Natsek E-56) (Fig. 4). The input data for the geohistory model are present-day formation thicknesses, ages of each formation, lithology, water depth at deposition for each formation, and present-day porosity with depth, if available. The input data used in the thermal history model are vitrinite reflectance and bottom hole temperature or temperature gradient. The surface temperature for each well is set at the average 2°C for the area. The data for temperature gradient come from the work of Majorowicz and Dietrich (1989). To run the hydrocarbon generation model, the geochemical character of Jurassic to Tertiary shales (Dixon et al., 1985; Snowdon, 1987; Issler and Snowdon, 1990) indicates that type III kerogen should be chosen for the Tertiary shales and types II and III kerogens for the Jurassic and Cretaceous shales.

Modelling procedure and result interpretation

The basic procedure to run the model is: first obtain the calculated thicknesses, which best fit with input thicknesses, in the geohistory model. As pointed out by Cao and Lerche (1987) the simulation of sediment compaction is crucial before running the other models, because the precision of simulation of sediment compaction affects either directly or indirectly the modelling of thermal and hydrocarbon generation histories. Second, the vitrinite reflectance data from the seven wells are input as thermal maturation indicators. Then the profiles of measured and calculated vitrinite reflectance are brought into minimum discord by adjusting heat flow parameters, and also by changing the burial paths by adjusting the thicknesses of eroded sediments (He and Lerche, 1989). The point is that if vitrinite reflectance is traced across an unconformity in a measured section, the trend of the vitrinite reflectance profile will show a broad change at the unconformity, where vitrinite reflectance values above the unconformity are markedly less than those below the unconformity. The vitrinite reflectance gradient above the unconformity may be different from the vitrinite reflectance gradient below the unconformity due to a varying thermal history. We determine the heat flow during the unconformity period by using the vitrinite reflectance data above and below. Then we vary the amount of sediment removed until the calculated vitrinite reflectance profile matches the measured vitrinite reflectance data on either side of the unconformity (Fig. 6). Finally, after each parameter is obtained by this minimization scheme (Table 1, eq. 3),

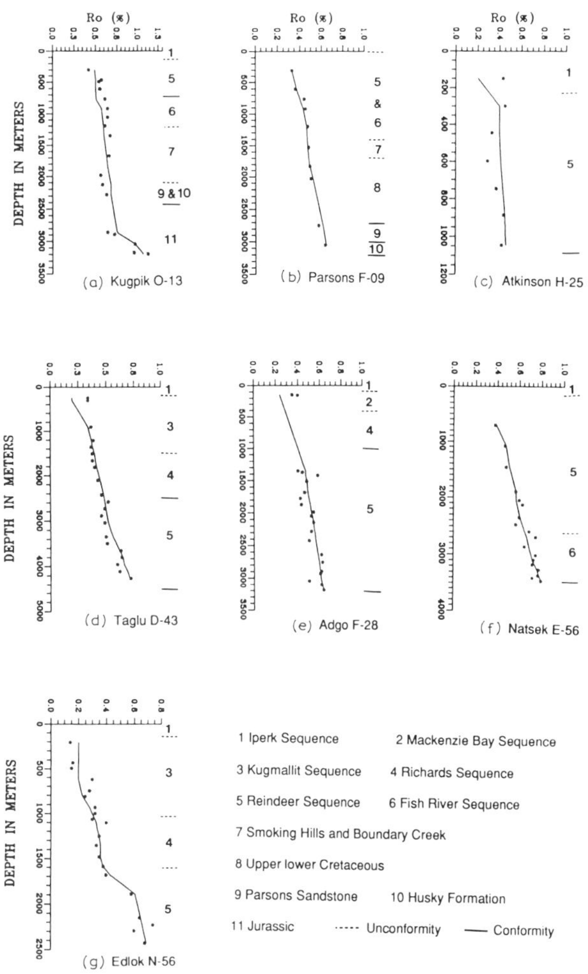

Fig. 6. Reflectance vs. depth profiles for seven wells in the Beaufort–Mackenzie Basin. Asterisks are measured vitrinite values; solid lines are predicted vitrinite reflectance profiles.

numerical reconstruction of the burial, thermal and hydrocarbon generation histories are then obtained. The errors are estimated by the MSR (mean square residual):

$$\mathrm{MSR} = \frac{1}{N}\sum_{i=1}^{N}[\mathrm{VR_c}(z_i) - \mathrm{VR_m}(z_i)]^2 \tag{1}$$

where N is the number of measurements of vitrinite reflectance, and $\mathrm{VR_c}$ and $\mathrm{VR_m}$ are the calculated and measured vitrinite reflectance at depth z_i, respectively. For example, after the best values of parameters were determined in the burial history

TABLE 1

Errors between measured and predicted vitrinite reflectance for seven wells

Well:	H-25	F-09	0-13	D-43	F-28	E-56	N-56
MSR[a]	0.17×10^{-1}	0.55×10^{-3}	0.54×10^{-2}	0.34×10^{-2}	0.45×10^{-2}	0.15×10^{-2}	0.24×10^{-2}

[a] Mean square residual, representing the average degree of mismatch between a series of predicted and measured vitrinite reflectance values. The lower the MSR, the better the fit.

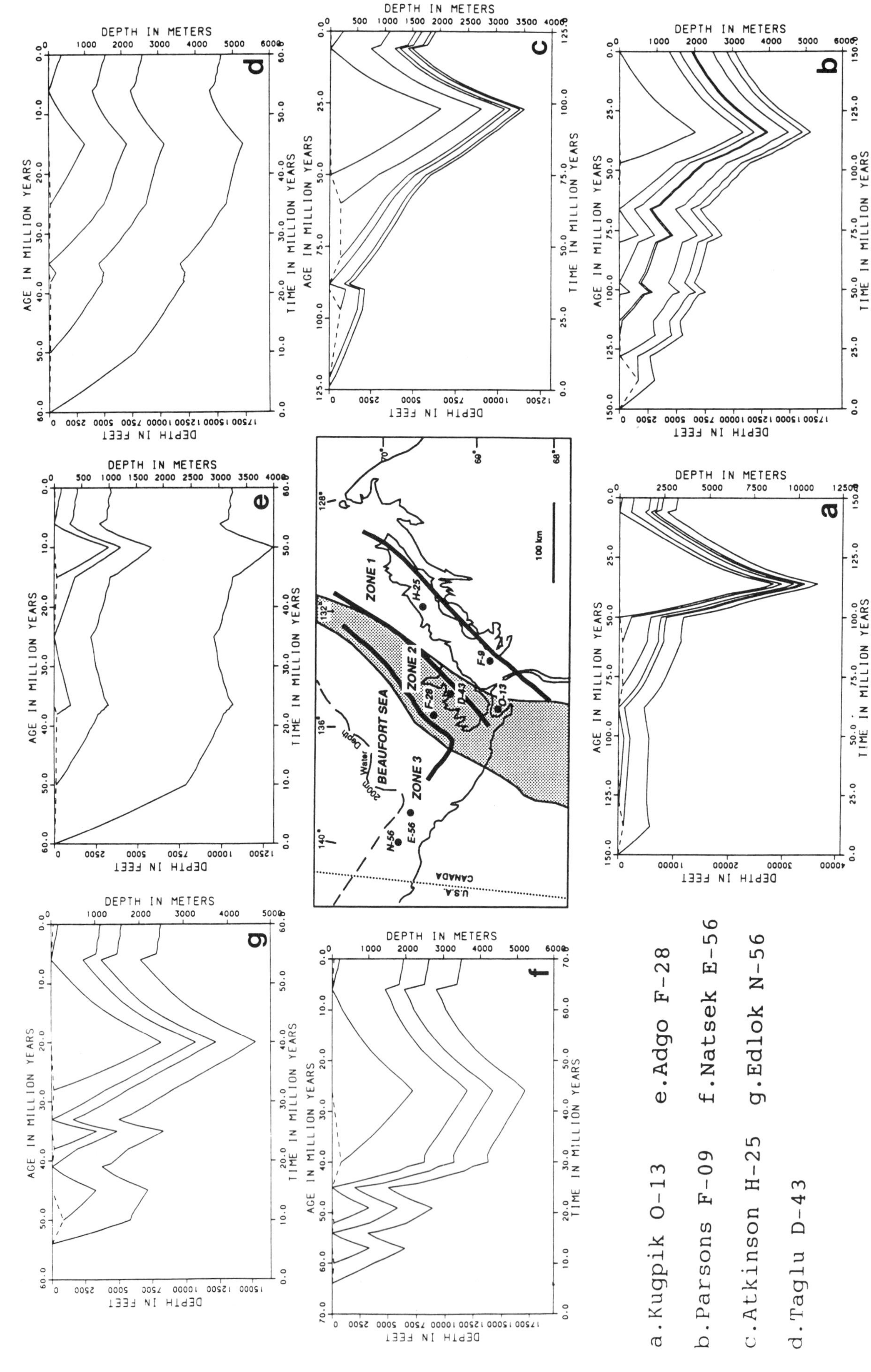

Fig. 7. Geohistory plots of the seven wells.

modelling, the smallest MSR was 0.5×10^{-3}, which corresponds to the best fit between observed and predicted vitrinite reflectance values (Fig. 6b), with the best estimate of erosion thickness at 2000 m. If the MSR is doubled relative to its minimum value, the range in thickness of eroded section is between 2000 ± 1000 m at this level of uncertainty. Note that although there are several major unconformities in the Cretaceous/Tertiary strata (Fig. 6) the vitrinite reflectance (R_o) recorded (0.33 to 0.43% for the Tertiary section of the three structural zones near, or at, the present surface) indicates that there are two deep erosional events: in the Late Eocene (structural zone I) and in the Late Miocene (structural zones II and III). The vertical distributions of vitrinite reflectance profiles for the seven wells show that the thermal gradient is higher in the upper section than in the lower section. For example, the vitrinite reflectance–depth profile of the Parsons F-09 well (Fig. 6b) is characteristically of order 0.25%/km in the Tertiary rocks and of order 0.09%/km in the Cretaceous rocks. We reconstructed the palaeoheat flow variation with time by inversion (Lerche, 1990) which best solves this variation, including the erosional events.

The following sections provide descriptions of the preliminary modelling results for the Beaufort–Mackenzie Basin.

Geohistory of the Beaufort–Mackenzie Basin. From the inverse method (He and Lerche, 1989), the depositional and tectonic history of the Beaufort–Mackenzie Basin is reconstructed by plotting curves of depth of burial versus geological age for each stratigraphic unit (Fig. 7). In structural zone I, the curves were constructed from three single wells (Atkinson H-25, Parsons F-09 and Kugpik 0-13, to the base of the Jurassic). In structural zone II, the curves were constructed from two single wells (Taglu D-43 and Adgo F-28, to the base of the Paleocene). In structural zone III, the curves were constructed from two single wells (Natsek E-56 and Edlok N-56, to the base of the Paleocene). The burial curve patterns in structural zone I (Fig. 7) show that maximum burial of the Jurassic, Cretaceous and Early Paleocene strata throughout most of the structural zone occurred during Eocene–Oligocene time. For example, at well 0-13, the Jurassic Husky Formation was buried about 11 km deep. However, Eocene uplift and erosion elevated it to its present-day depth of 3 km. This uplift and erosion event must have been responsible for the removal of 6000–8000 m of sediments. To the northeast (structural zone I), the eroded thickness decreases to 2000 m, and the erosion event lasted to Late Miocene. The Iperk sequence then began to be deposited.

As shown in Fig. 7, the maximum burial of the Tertiary strata in structural zone II occurred during the Late Miocene, and then, because of the Late Miocene uplift and erosional event, about 1000–2000 m of sediments was eroded. Unlike the burial curves in structural zone II, the maximum depth of burial of Tertiary strata in structural zone III may not have been reached until about Late Oligocene to Early Miocene time. The amount of erosion is determined to be about 2000 thick using vitrinite reflectance following the method outlined in He and Lerche (1989).

Figure 8 shows the basement evolution of the Beaufort–Mackenzie Basin with time. The total basement subsidence is divided into two parts, one caused by the weight of sediment and water on the basement and the other caused by tectonic forces, calculated from the isostatic rebound equation (Steckler and Watts, 1978):

$$Y = S\frac{\rho_m - \rho_s}{\rho_m - \rho_w} + Wd - \Delta S_L\frac{\rho_m}{\rho_m - \rho_w} \tag{2}$$

where Y is the tectonically caused basement subsidence, S is the sediment thickness, Wd is the water depth at the time of sediment deposition, ΔS_L is the change in elevation of mean sea level, ρ_m is the mantle density, ρ_w is the water density and ρ_s is the average sediment density. From subsidence patterns in structural zone I, the tectonic subsidence did not undergo any sudden change before the early-Late Cretaceous, while during the post-Late Cretaceous the tectonic subsidence obviously increased. For example, the tectonic basement subsided about 500 m up to the early-Late Cretaceous and subsided about 1500 m at the end of the Late Oligocene in well Atkinson H-25 (Fig. 8). Thereafter there was Late Miocene uplift, followed by thermal subsidence during deposition of the Iperk sequence. In structural zone II a sudden change of tectonic subsidence occurred from Paleocene to Oligocene, whereas the tectonic subsidence in structural zone III underwent a complex phase from Paleocene to the present day.

The regional implication is that the tectonics in the study area underwent two stages: the first was a rifting stage, caused by crustal extension, which continued into the Early Miocene, and which had commenced prior to the Cretaceous; the second stage was one of post-rifting uplift and erosion which may be caused by increased regional palaeoheat flow (as described in the next section). The results of the isostatic rebound model agree with results obtained using a flexural model (Tang and Lerche, 1991). The evidence for a major component of extensional history and later thermal subsidence is supported by the data, but a late compressional component cannot be ruled out by the available evidence.

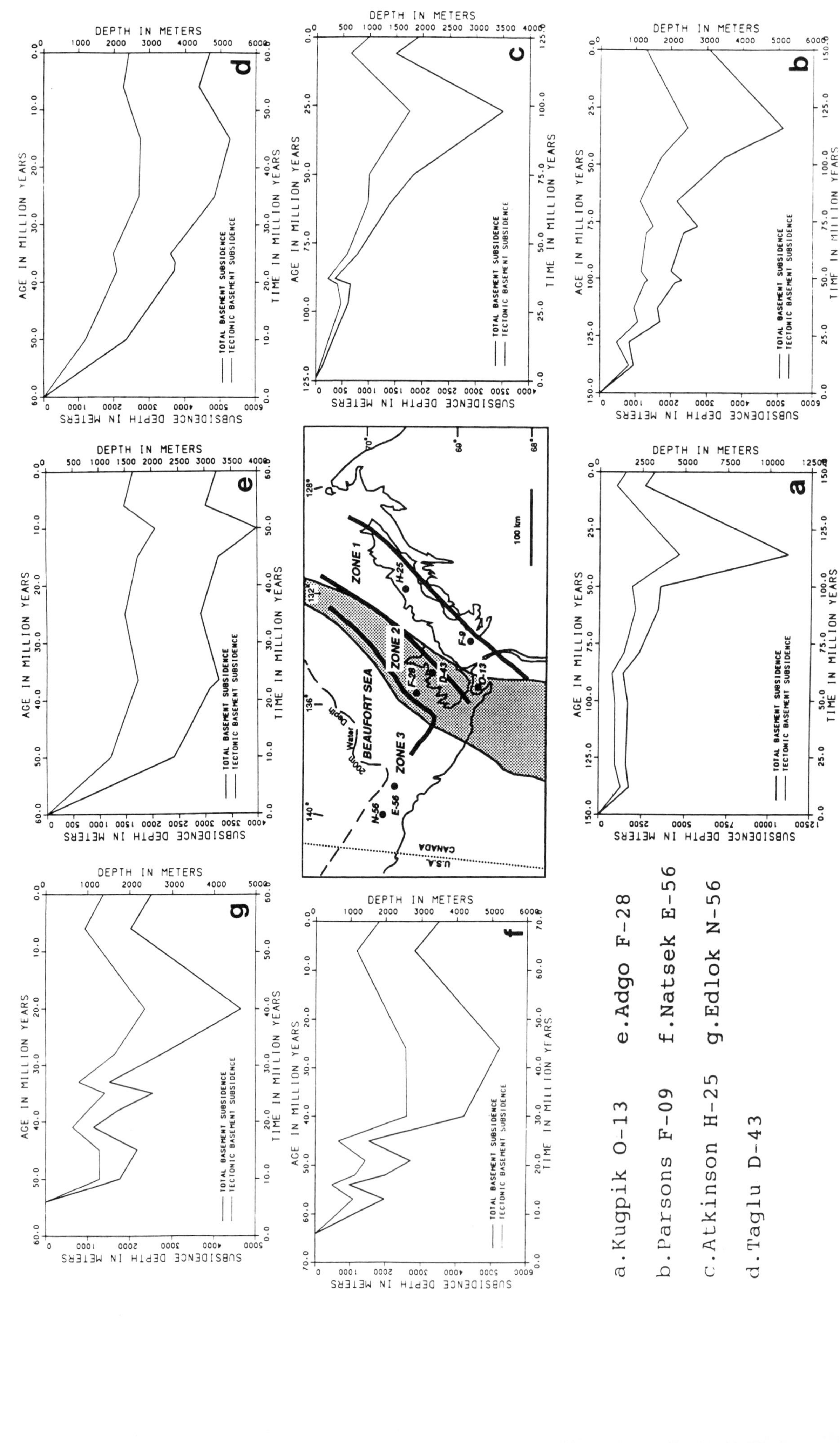

Fig. 8. Basement subsidences with time for the seven wells.

Thermal history of the Beaufort–Mackenzie Basin — palaeoheat flow and palaeotemperature. For reconstruction of the thermal history of the study area, an inverse method called thermal indicator tomography (He and Lerche, 1989) was used to determine the variations in heat flow and temperature with time.

The basement heat flow is described by:

$$Q(t) = Q_0 \exp\left[\beta t + a_1 \sin\left(\frac{\pi t}{t_{\max}}\right) + a_2 \sin\left(\frac{2\pi t}{t_{\max}}\right) + a_3 \sin\left(\frac{3\pi t}{t_{\max}}\right)\right] \quad (3)$$

where $Q(t)$ is the basement heat flow, Q_0 is the present-day heat flow, b, a_1, a_2 and a_3 are parameters to be determined by minimizing the difference between predicted and measured vitrinite reflectance with depth, and $t_{\max}$ is the age of the deepest sediment for which thermal indicator information is available.

The temperature of each layer at time t and depth $z(t)$ is given by the conventional conduction formula:

$$T[t, z(t)] = T_s + Q(t) \int_0^{z(t)} dz/k(z) \quad (4)$$

where T_s is the surface temperature and $k(z)$ is the sedimentary thermal conductivity at depth $z(t)$ described by:

$$k(z) = (k_{f\,l})^{\phi(z)} (k_s)^{[1-\phi(z)]} \quad (5)$$

where $k_{f\,l}$ is the thermal conductivity of the pore fluid, k_s is the thermal conductivity of the solid rock, and $\phi(z)$ is the porosity at depth z (Lewis and Rose, 1970; Andrews-Speed et al., 1984; Cao and Lerche, 1987).

In our study vitrinite reflectance data from the seven wells were chosen as the thermal indicator. Plots of measured and predicted vitrinite reflectance (Fig. 6) show very good fits except for well Atkinson H-25 (Table 1). The variation of vitrinite reflectance values with depth in well H-25 is abnormal (R_o = 0.43 at 149 m and 0.42 at 1045 m, Fig. 6c). Some of the vitrinite reflectance values are probably from reworked kerogen. Therefore, the error between measured and predicted values with depth is relatively larger for this well (Table 1) than for the other wells. Figures 9 and 10 give the heat flow and temperature patterns with time for the seven wells.

In structural zone I, the reconstructed heat flow histories for the three wells show a warming trend from 150 Ma to 20 Ma and then a cooling trend from 20 Ma to the present day. In structural zones II and III the heat flow patterns show a warming trend from 60 Ma to 10 Ma and then a cooling trend to the present day except for well N-56 for

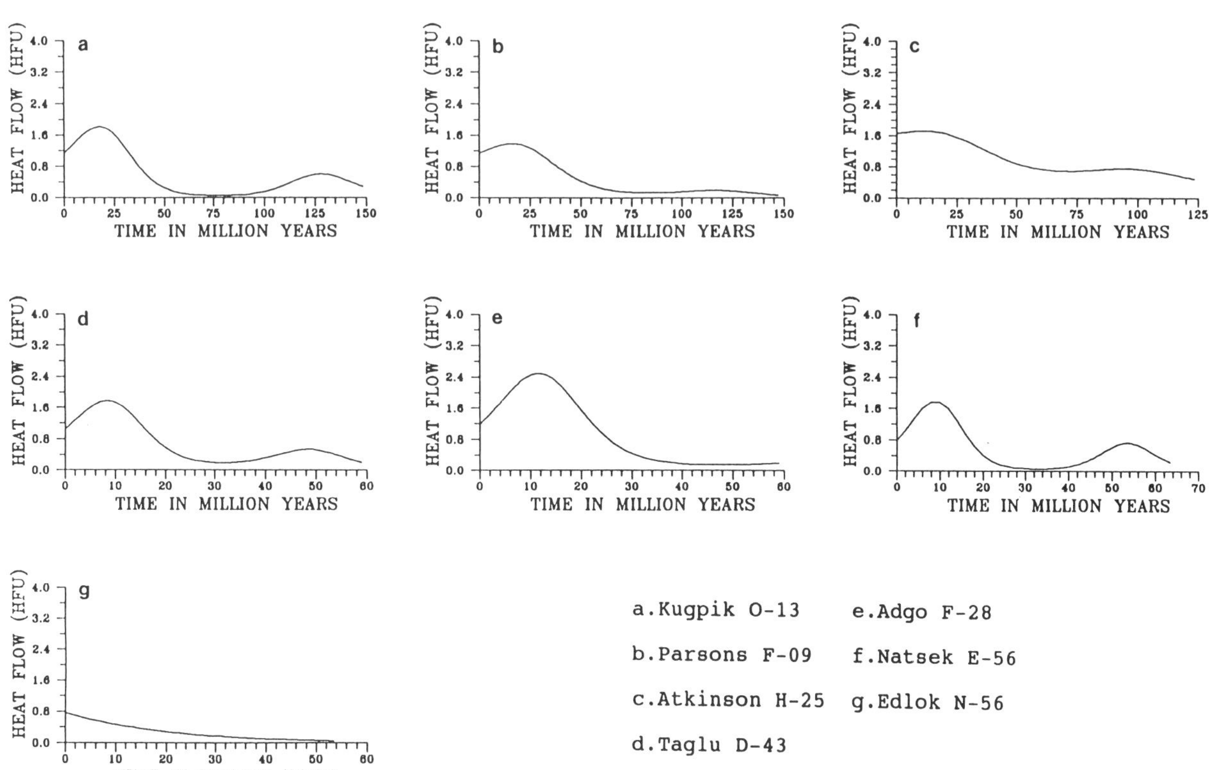

Fig. 9. Heat flow evolution with time for the seven wells.

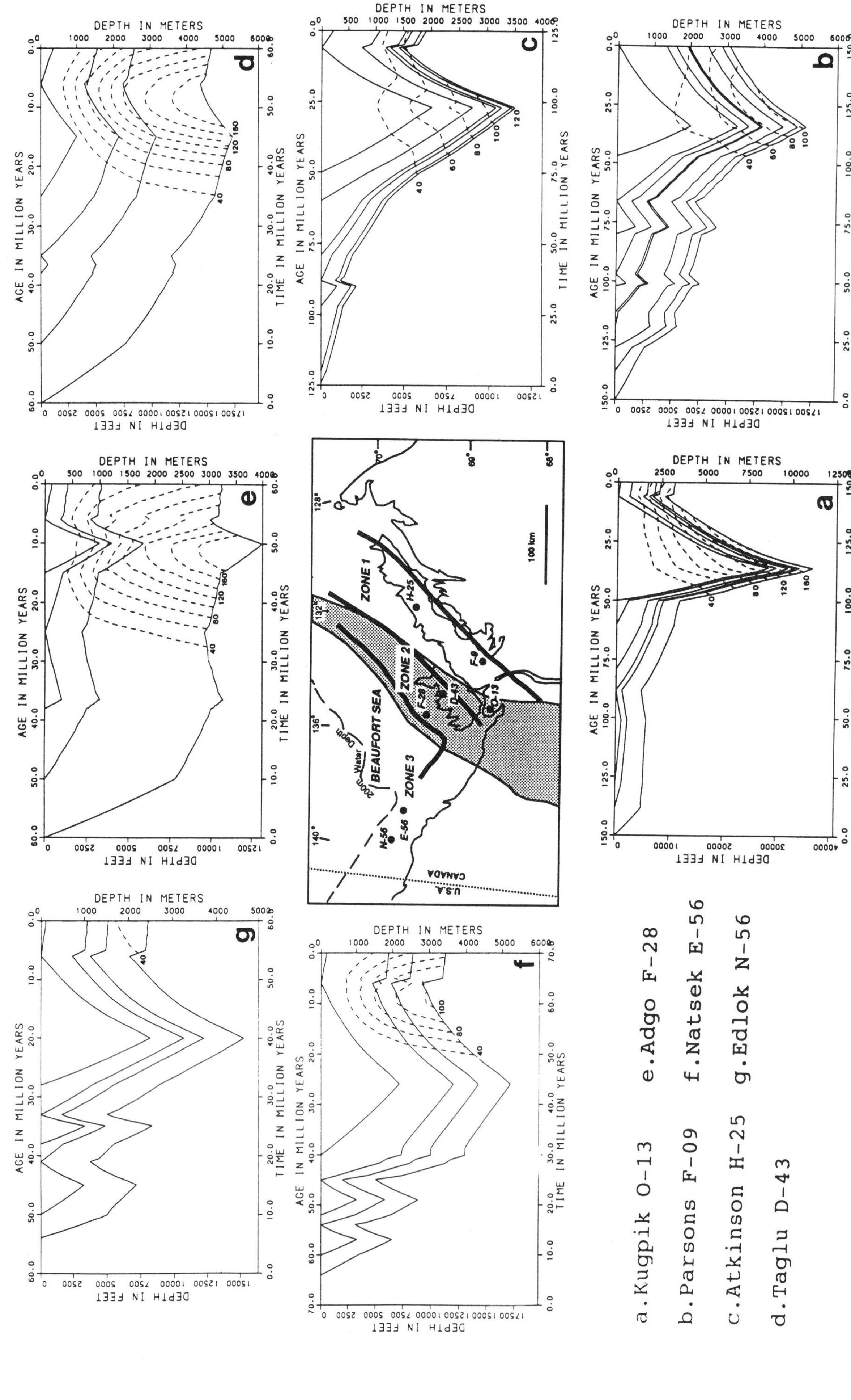

Fig. 10. Temperature changes with time and depth for the seven wells.

which the heat flow pattern shows only a warming trend from past to present (Fig. 9), probably because the vitrinite data do not permit fine enough scale resolution to determine the peak of any heat pulse at around 10 Ma. The temperature variation pattern with time and depth (Fig. 10) shows a trend similar to that of the heat flow. The temperature anomaly at 20 Ma in structural zone I, and at 10 Ma in structural zones II and III, result from high heat flow during the above periods. The problem is to determine the cause of the heat flow anomaly. Comparisons between the subsidence patterns (Fig. 8), thermal histories (Figs. 9 and 10), and burial histories (Fig. 7) suggest that the increase of heat flow from 150 Ma to 20 Ma in structural zone I may be related to Late Jurassic–Early Cretaceous extension along this region, representing a continuation of the Early Cretaceous rift phase. The geological evidence for the extension of this region is that large marginal faults and a series of subparallel normal faults are developed. The increase of heat flow from 60 Ma to 10 Ma in structural zones I and II may represent a final rifting episode. Associated with the final rifting event is the development of a series of gravity structures (shale diapirism and normal growth faults) in the structural zones. After these two heating stages, the study area began to uplift and erode until 6 Ma. Finally, the thermal subsidence of Late Miocene time permitted deposition of the Iperk sequence. Therefore, the anomaly patterns of heat flow at this area may reflect the migration of a heat source from the basin margin to the basin center. Also, the heating episode resulted in temperature anomalies which play an important role in hydrocarbon generation and occurrence in the study area.

Thermal maturity of source rocks. The maturity of source rocks in a sedimentary basin is a function of temperature variations with time and the burial depth of source rocks. Here we used vitrinite reflectance data to simulate the maturation history of the source rocks in seven wells.

Vitrinite reflectance variations with time and depth can be modelled using the equation (Lerche, 1988a, b):

$$R[z(t)] = \left[R_*^{1/2} + \alpha\,\mathrm{TTI}(t)\right]^2 \qquad (6)$$

where α is determined from:

$$\alpha = \frac{\sum_i [R(z_i)^{1/2} - R_*^{1/2}]}{\sum_i \mathrm{TTI}(t_i)} \qquad (7)$$

where $R(z_i)$ is measured vitrinite reflectance, R_* is vitrinite reflectance value at the depositional surface (in our case, $R_* = 0.2\%$) and where:

$$\mathrm{TTI}(t_i) = \int_0^t \exp[(T - T_c)/T_d]\,\mathrm{d}t, \quad T > T_c \qquad (8)$$

with the integrand zero in $T < T_c$, where T is the temperature along the burial depth, and T_c and T_d are fixed at 295 K and 200 K, respectively. The values of T_c and T_d are determined from 20 wells from the North Sea, 21 wells from the Australian Northwest Shelf, 6 wells from Indonesia, and 21 wells from the National Petroleum Reserve of Alaska (Lerche et al., 1984).

A vitrinite reflectance of 0.6 and a TTI of 15 (Waples, 1980) have been chosen as marking the onset of the "oil window". Figure 11 shows the evolutionary history of vitrinite reflectance for the seven wells, from which the time when a given source rock enters the "oil window" is readily obtained. For example, in structural zone I, the model predicts that the Upper Jurassic Husky Formation began to enter the oil window during the Late Oligocene–Early Miocene at the locations of Atkinson H-25 and Kugpik 0-13, whereas at the location of Parsons F-09, the Husky Formation began to enter the oil window in the Late Miocene. In structural zones II and III the Richards sequence never entered the oil window except at the location of the well Natsek E-56. Figure 12 shows the evolutionary history of the Waples (1980) TTI for the seven wells. The Jurassic Husky Formation entered the oil window during the Early Oligocene in structural zone I. The Richards sequence in structural zone II entered the oil window during the Early/Middle Miocene, whereas the Richards sequence in structural zone III has not yet entered the oil window.

Hydrocarbon generation history. To simulate the history of kerogen degradation and hydrocarbon generation in the Beaufort–Mackenzie Basin, the model of Tissot (see e.g. Tissot and Welte, 1978) has been used. Figures 13 and 14 show the hydrocarbon generation histories for the seven wells. From Fig. 13 oil began to be generated during the Late Eocene/Oligocene in structural zone 1, at a rate of about 50 mg/g kerogen per Ma. In structural zones II and III the base of the Richards sequence did not begin to generate oil until the Middle Miocene, except for well N-56, with a small generation rate of 1 mg/g kerogen per Ma. Figure 14 shows the gas generation histories in the study area, indicating that in structural zone I the Jurassic/Cretaceous source rocks at the location of well 0-13 entered the gas window during the Early Oligocene. At the location of well H-25, the Jurassic and Lower Cretaceous source

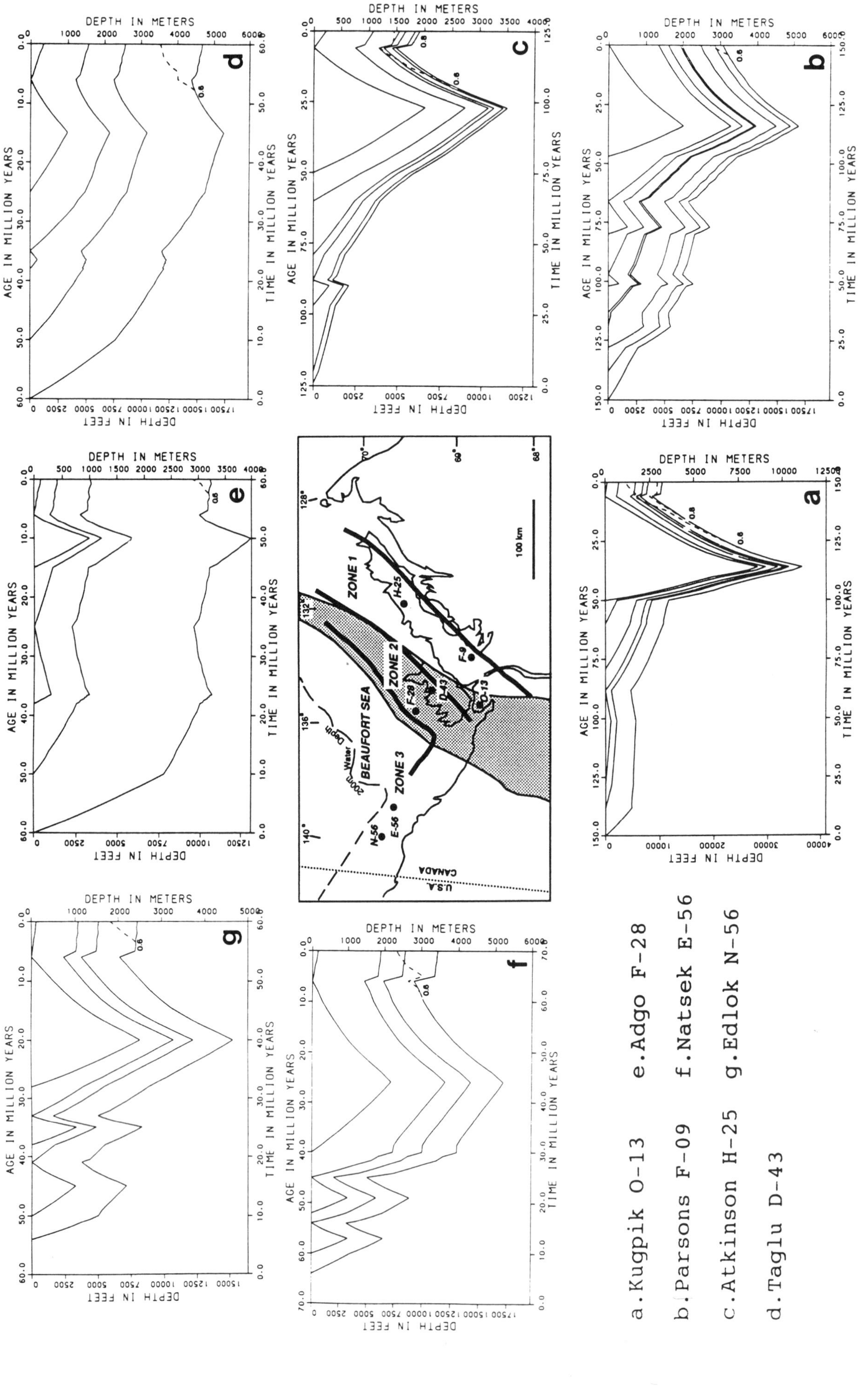

Fig. 11. Predicted vitrinite reflectance histories for the seven wells.

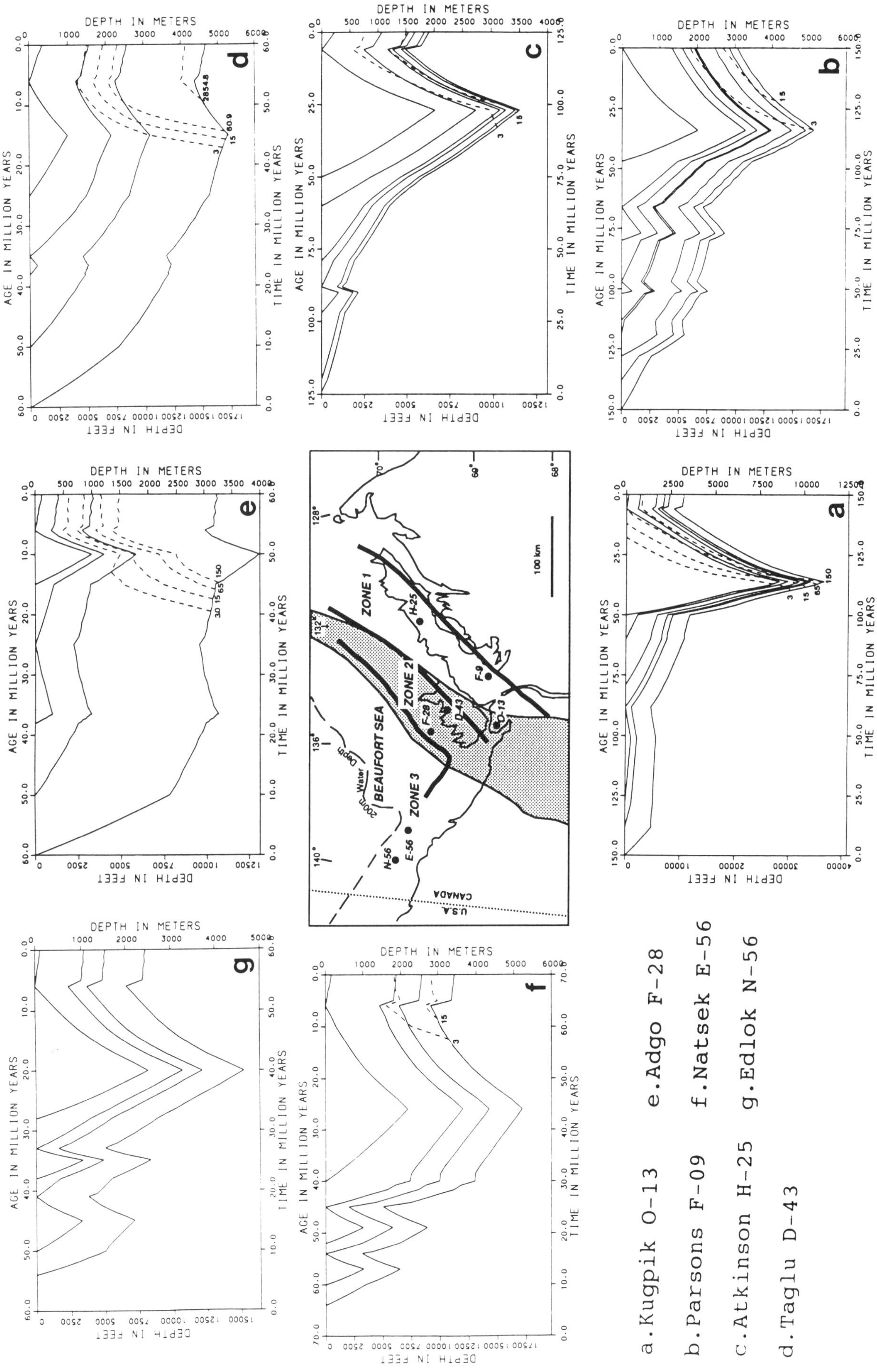

Fig. 12. Time–temperature index (TTI) histories for the seven wells.

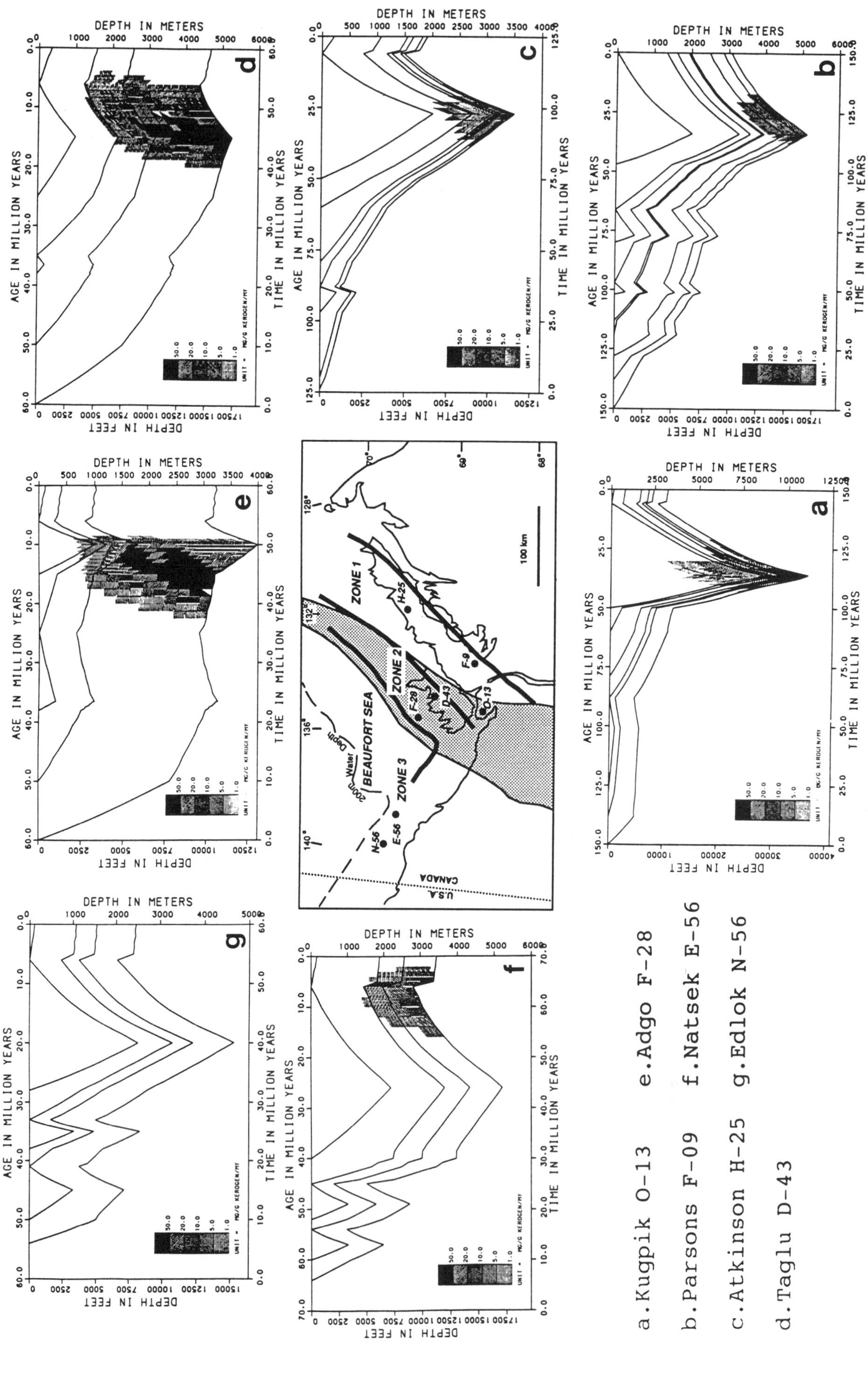

Fig. 13. Oil generation histories for the seven wells.

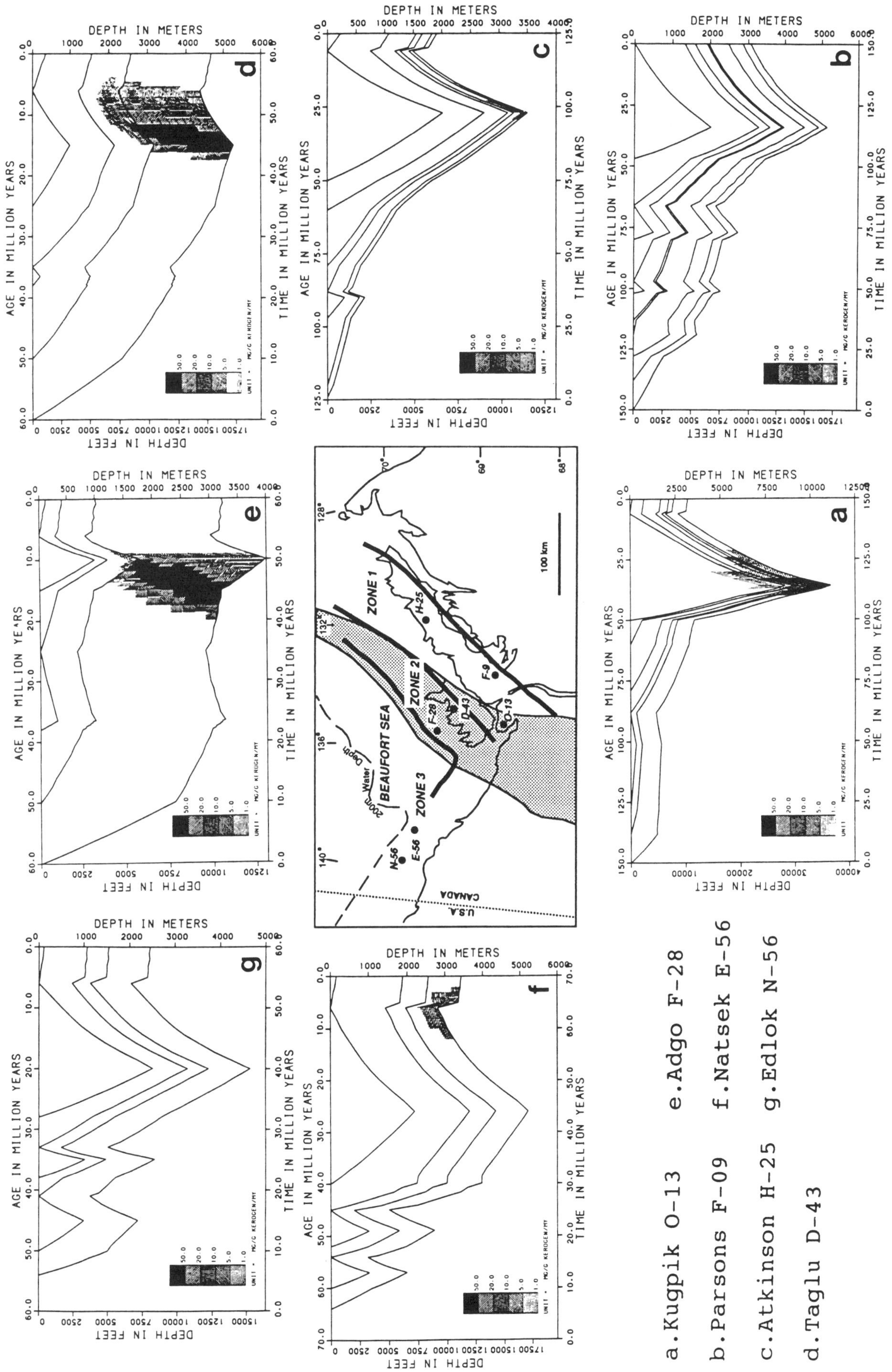

Fig. 14. Gas generation histories for the seven wells.

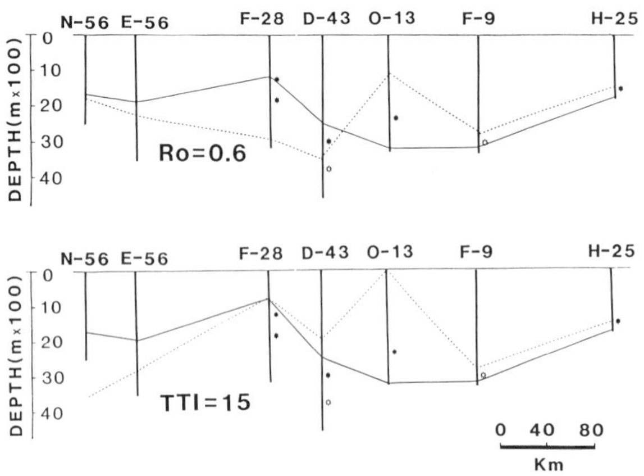

Fig. 15. Correlation of present-day maturity depth from predicted vitrinite reflectance and TTI. The solid lines represent the bottom boundaries of source rocks (see text), the dashed lines represent vitrinite reflectance of 0.6 and a Waples' time–temperature index of 15, respectively.

rocks just entered the gas window during the Early Oligocene, whereas the Jurassic Husky Formation has not yet entered the gas window at the location of well F-09. In structural zone II, the Richards sequence began to enter the gas window about 15 Ma (Middle Miocene). However, the sequence has not entered the gas window at the present day in structural zone III.

Figure 15 shows the correlation of present-day maturity depth of the seven wells from vitrinite reflectance and Waples TTI. The solid lines represent the bottom boundaries of source rocks with different geological ages; the dashed lines stand for vitrinite reflectance of 0.6, and a Waples TTI of 15. From Fig. 15 the oil window extends from 1100 to 2800 m in structural zone I, implying that the Jurassic and Cretaceous source rocks (Smoking Hills, Boundary Creek, Arctic Red, Mount Goodenough, Husky and McGuire Formations) have entered the oil window at the locations of Atkinson H-25 and Kugpik 0-13. At the location of Parsons F-09, the Jurassic and Lower Cretaceous source rocks have entered the oil window. In structural zones II and III the oil window extends from 1800 to 3500 m, indicating that the Tertiary source rocks (the Richards sequence) have not entered the oil window based on the predicted vitrinite reflectance (R_o = 0.6). However, in terms of the maturity level of TTI = 15, the bottom and middle of the Richards sequence have entered the oil window at the locations of Adgo F-28 and Taglu-43, respectively. Combining the maturity level (R_o = 0.6, TTI = 15) (Fig. 15), the TOC of the Jurassic/Cretaceous/Tertiary source rocks (Fig. 5), and the hydrocarbon generation history (Figs. 13 and 14), in relation to the observed present-day hydrocarbon occurrences, the oil of the Tuktoyaktuk Peninsula area (wells 0-13, F-09 and H-25) has a local Jurassic and Cretaceous source, while the gas in the reservoirs of the Parsons Lake area results from extensive migration. As well as the above factors, considerations of the relation between reservoir and source rocks suggest that the Tertiary oil and gas reservoirs (Fish River, Aklak and Taglu sequences, Fig. 3) are more likely to have been sourced from the Jurassic and Cretaceous strata onshore in structural zone II. Toward the north (offshore), the deeper burial of the Richards sequence may provide a hydrocarbon source for the overlying reservoir rocks (Kopanoar and Kugmallit sequences). This basic conclusion is supported by observations of hydrocarbon occurrences, and by the work of Issler and Snowdon (1990) who point out that the base of the Richards sequence entered the oil window during the Late Oligocene and that peak hydrocarbon generation is presently occurring in the middle of the sequence at both the Arluk E-90 and Amauligak J-44 well locations. These observations are consistent with biomarker studies that fingerprint the Richards sequence as the source for offshore Tertiary oils, and the observation that these oils are more mature than their host rocks and therefore have migrated from deeper formations into the shallower reservoirs. Because of the lower maturity level of the Tertiary source rocks in structural zone III compared to structural zones I and II, exploration for oil and gas should probably be concentrated on both the central part, and on the Tuktoyaktuk Peninsula, of the study area.

Conclusions

A general picture of the subsidence, thermal, and hydrocarbon generation histories of the Beaufort–Mackenzie Basin has been developed using downhole well data and a comprehensive burial history model. The results of this study are now summarized.

(1) The subsidence history model simulates the burial paths of Jurassic to Pleistocene strata in the three structural zones, including erosion thickness and basement motion behaviour. The maximum burial of Jurassic/Cretaceous and Tertiary sediments was during the Eocene to Oligocene, and Late Miocene, respectively. After the deep burial the Late Eocene uplift and erosion (which lasted until Late Miocene) caused at least 2000 m of sediment to be removed in structural zone I. The Late Miocene uplift and erosion event in the central and western parts of the study area resulted in 1000–2000 m of sediment being removed.

Based on interpreted basement behaviour and reconstructed palaeoheat flow patterns, two stages of tectonics have been recognized in the study area.

The first, a rifting stage, caused the development of Eskimo Lakes Fault and Parsons Lake Fault in structural zone I during the Late Jurassic–Early Cretaceous period. The subsequent extension resulted in the development of a series of gravity structures in structural zone II. The second, post-rifting uplift and erosion and thermal subsidence, may have been caused by palaeoheat flow variations from Late Miocene to the present day.

(2) Based on interpretations of vitrinite reflectance data, heating episodes occurred from 150 to 20 Ma in structural zone I and 60 to 10 Ma in structural zones II and III. The temperature anomaly is likely to have resulted from the higher heat flow during the Late Oligocene–Late Miocene.

(3) The Richards sequence of Tertiary age and the Husky Formation of Jurassic age have been considered to be major potential source rocks in the study area. Peak hydrocarbon generation for the Husky Formation and the Richards sequence in the study area occurred during the Late Eocene to Late Oligocene, and Middle to Late Miocene, respectively.

(4) The hydrocarbon generation model results suggest that hydrocarbon potential is higher in both the central part of the study area (onshore and offshore) and in the Tuktoyaktak Peninsula than in the west part of the study area (structural zone III). The one-dimensional model assumes that fluid movement was dominantly vertical; lateral migration of hydrocarbons is not considered. Due to tectonic activity during Late Miocene time lateral migration of oil and gas may have significantly affected accumulation or removal of hydrocarbons in the study area.

Acknowledgements

The work reported here was supported by the Industrial Associates of the Basin Analysis Group at the University of South Carolina. We wish to thank the Petroleum Geology and Basin Analysis Section, Alberta Geological Survey, Alberta Research Council for their very generous supply of data which enabled this study to be completed. We also wish to thank Brian Hitchon and Stefan Bachu for their critical comments on a prior draft of this paper.

References

Andrews-Speed, C.P., Oxburgh, E.R. and Cooper, B.A., 1984. Temperature and depth-dependent heat flow in West North Sea. Am. Assoc. Pet. Geol., Bull., 68: 1764–1781.

Armagnac, C., Bucci, J., Kendall, C.G.St.C. and Lerche, I., 1989. Estimating the thickness of sediment removed at an unconformity using vitrinite reflectance data. In: N.D. Naeser and T.H. McCulloh (Editors), Thermal History of Sedimentary Basins, Methods and Case Histories. Springer-Verlag, Berlin, 319 pp.

Cao, S., 1985. Quantitative Dynamic Model for Basin Analysis and Its Application to the Northern North Sea Basin. M.S. Thesis, University of South Carolina, Columbia, S.C., 94 pp.

Cao, S. and Lerche, I., 1987. Geohistory, thermal history and hydrocarbon generation of the northern North Sea Basin. Energy Explor. Exploit., 5: 315–355.

Cao, S., Glezen, W.H. and Lerche, I., 1986. Fluid flow, hydrocarbon generation and migration: a quantitative model of dynamical evolution in sedimentary basins. Proc. Offshore Technology Conf., 2: 267–276.

Cote, R.P., Lerand, M.M. and Rector, R.J., 1975. Geology of the Lower Cretaceous Parsons Lake Gas Field, Mackenzie Delta, Northwest Territories. In: C.J. Yorath, E.R. Parker and D.J. Glass (Editors), Canada's Continental Margins. Can. Soc. Pet. Geol., Mem., 4: 613–632.

Dietrich, J.R., Dixon, J., McNeil, D.H. and Snowdon, L.R., 1989a. Geology, biostratigraphy and organic geochemistry. Brief supplementary notes to accompany a one-day short course sponsored by the Canadian Society of Petroleum Geologists, Calgary, Alta., 15 June 1989, 18 pp.

Dietrich, J.R., Dixon, J., McNeil, D.H., McIntyre, D.J., Snowdon, L.R. and Cameron, A.R., 1989b. The geology, biostratigraphy and organic geochemistry of the Natsek E-56 and Edlok N-56 wells, western Beaufort Sea. Geol. Surv. Can., Pap., 89-1G: 133–157.

Dixon, J., 1986. Cretaceous to Pleistocene stratigraphy and paleogeography, northern Yukon and northwestern District of Mackenzie. Bull. Can. Pet. Geol., 34: 49–70.

Dixon, J. and Dietrich, J.R., 1988. The nature of depositional and seismic boundaries in Cretaceous–Tertiary strata of the Beaufort–Mackenzie Basin. In: D.P. James and D.A. Leckie (Editors), Sequences, Stratigraphy, Sedimentology: Surface and Subsurface. Can. Soc. Pet. Geol., Mem., 15: 63–72.

Dixon, J., Dietrich, J.R., McNeil, D.H., McIntyre, D.J., Snowdon, L.R. and Brooks, P., 1985. Geology, biostratigraphy and organic geochemistry of Jurassic to Pleistocene strata, Beaufort–Mackenzie area, northwest Canada. Course Notes, Can. Soc. Pet. Geol., 65 pps.

Enachescu, M.E., 1990. Structural setting and validation of direct hydrocarbon indicators for Amauligak Oil Field, Canadian Beaufort Sea. Am. Assoc. Pet. Geol., Bull., 74: 41–59.

Gunther, P.R., 1976a. A study employing optical methods to evaluate organic metamorphism and oil-generating potential of sediments in the Mackenzie Delta area, District of Mackenzie Project 740032, Report of Activities, Part C. Geol. Surv. Can., Pap., 76-1C: 143–152.

Gunther, P.R., 1976b. Palynomorph color and dispersed coal particle reflectance from three Mackenzie Delta boreholes. Geosci. Man, 15: 35–39.

He, Z. and Lerche, I., 1989. Inversion of multiple thermal indicators: quantitative methods of determining palaeoheat flux and geological parameters, IV. Case histories using thermal indicator tomography. Math. Geol., 21: 523–541.

Issler, D.R. and Snowdon, L.R., 1990. Hydrocarbon generation kinetics and thermal modelling, Beaufort–Mackenzie Basin. Bull. Can. Pet. Geol., 38: 1–16.

Jones, P.B., Brache, J. and Lentin, J.K., 1980. The geology of the 1977 offshore hydrocarbon discoveries in the Beaufort–Mackenzie Basin, N.W.T. Bull. Can. Pet. Geol., 28: 81–102.

Lerand, M.M., 1973. Beaufort Sea. In: R.S. McCrossan (Editor), The Future Petroleum Provinces of Canada. Can. Soc. Pet. Geol., Mem., 1: 315–386.

Lerche, I., 1988a. Inversion of multiple thermal indicators: quantitative methods of determining palaeoheat flux and geological parameters, I. Theoretical development for palaeoheat flux. Math. Geol., 20: 1–36.

Lerche, I., 1988b. Inversion of multiple thermal indicators: quantitative methods of determining palaeoheat flux and geological parameters, II. Theoretical development for chemical, physical and geological parameters. Math. Geol., 20: 73–96.

Lerche, I., 1990. Basin Analysis: Quantitative Methods, Vol. 1. Academic Press, San Diego, Calif., 559 pp.

Lerche, I., Yarzab, R.F. and Kendall, C.G.St., 1984. Determination of palaeoheat flux from vitrinite reflectance data. Am. Assoc. Pet. Geol., Bull., 68: 1704–1717.

Lewis, C.R. and Rose, S.C., 1970. A theory relating high temperatures and overpressures. J. Pet. Technol., 22: 11–16.

Majorowicz, J.A. and Dietrich, J.R., 1989. Comparison of the geothermal and organic maturation gradients of the central and southwestern Beaufort–Mackenzie Basin, Yukon and Northwest Territories. Geol. Surv. Can., Pap., 89-1G: 63–67.

McNeil, D.H., 1990. Stratigraphy and paleoecology of the Eocene Stellarina Assemblage Zone (pyrite diatom steinkerns) in the Beaufort–Mackenzie Basin, Arctic Canada. Bull. Can. Pet. Geol., 38: 17–27.

McWhae, J.R., 1986. Tectonic history of northern Alaska, Canadian Arctic, and Spitsbergen regions since Early Cretaceous. Am. Assoc. Pet. Geol., Bull., 70: 430–450.

Meneley, R.A., 1986. Oil and gas fields in the east coast Arctic Basins of Canada In: H.T. Halbouty (Editor), Future Petroleum Provinces of the World. Am. Assoc. Pet. Geol., Mem., 40.

Snowdon, L.R., 1987. Organic properties and source rock potential of two early Tertiary shales, Beaufort–Mackenzie Basin. Bull. Can. Pet. Geol., 35: 212–232.

Snowdon, L.R., 1988. Hydrocarbon migration in Mackenzie Delta sediments. Bull. Can. Pet. Geol., 36: 407–412.

Snowdon, L.R. and Brooks, P., 1985. Part C, Organic Geochemistry. In: J. Dixon, J.R. Dietrich, D.H. McNeil and L.R. Snowdon (Editors), Geology, Biostratigraphy and Organic Geochemistry of Jurassic to Pleistocene Strata, Beaufort–Mackenzie Area, Northwest Canada. Course Notes, Can. Soc. Pet. Geol., 65 pp.

Steckler, M.S. and Watts, A.B., 1978. Subsidence of the Atlantic-type continental margin off New York. Earth Planet. Sci. Lett., 41: 1–13.

Tang, J. and Lerche, I., 1991. Tertiary flexural evolution of the Beaufort–Mackenzie Basin, Canada. Mar. Pet. Geol., 9: 245–255.

Tissot, B. and Welte, D.H., 1978. Petroleum Formation and Occurrence. Springer-Verlag, Berlin, 635 pp.

Waples, D.W., 1980. Time and temperature in petroleum formation: application of Lopatin's method to petroleum exploration. Am. Assoc. Pet. Geol., Bull., 64: 916–926.

Willumsen, P.S. and Cote, R.P., 1982. Tertiary sedimentation in the southern Beaufort Sea, Canada. In: A.F. Embry and H.R. Balkwill (Editors), Arctic Geology and Geophysics. Can. Soc. Pet. Geol., Mem., 8: 43–54.

Young, F.G., Myhr, D.W. and Yorath, C.J., 1976. Geology of the Beaufort–Mackenzie Basin. Geol. Surv. Can., Pap., 76-11, 63 pp.

J. TANG *Department of Geological Sciences, University of South Carolina, Columbia, SC 29208, U.S.A.*

I. LERCHE *Department od Geological Sciences, University of South Carolina, Columbia, SC 29208, U.S.A.*

Palaeohydrology of the Gulf of Mexico Basin: development of compactional overpressure and timing of hydrocarbon migration relative to cementation

Lori L. Summa, Robert J. Pottorf, Theresa F. Schwarzer and Wendy J. Harrison

In an ongoing effort to improve our understanding of hydrocarbon accumulations in the United States Gulf of Mexico Basin, we are evaluating the extent to which computer models of basin palaeohydrology and equilibrium mineral stabilities can be used to help predict hydrocarbon migration, and fluid/rock interaction. As part of the evaluation, we have developed an integrated model for evolution of compactional overpressure and relative timing of hydrocarbon migration along three regional cross sections in East and South Texas and South Louisiana. For each cross section, we constructed a 2D fluid flow model using the Basin2 computer program of C.M. Bethke. In addition, we collected pressure, temperature, and fluid chemistry data to help constrain the model, and assessed the geochemistry and thermal maturity of potential source rocks and oils. We are continuing to analyze authigenic cements, and evaluate mineral stabilities in order to help constrain the fluid flow models and establish the timing of cementation relative to hydrocarbon migration.

The modelling results are partially controlled by choice of input parameters: increasing shale permeability by 10× at a given porosity can increase the modelled depth to top overpressure by 2 km, supporting the need for continued efforts to acquire good measured data. However, we conclude that in the Gulf of Mexico, overpressures can be explained largely by compaction disequilibrium; contributions from aquathermal expansion, smectite dehydration and hydrocarbon generation are not required to model overpressure development in this setting. We also suggest that extensive overpressures developed at different times along the three cross sections: during the Early Oligocene in South Texas, and the Early Miocene in Louisiana and East Texas. Assuming that hydrocarbon migration occurred along major growth faults, variations in overpressure timing may have influenced the complex trap-fill histories of major Oligocene reservoirs in the Gulf: at the time of peak migration, potential traps in South Texas were probably overpressured, whereas traps in South Louisiana may have been normally pressured.

Introduction

As part of an integrated effort to improve predictions of hydrocarbon distribution, we have modelled the evolution of subsurface pressures and the potential interactions between pore waters and rocks for three areas within the Tertiary stratigraphic section of the United States Gulf Basin (Fig. 1). The distribution of overpressure is an important control on hydrocarbon migration, maturation, and entrapment, as well as a variety of authigenic cementation processes involving fluid flow. As such, developing a better understanding of the past and present distribution of overpressure is important, particularly in the Gulf Coast Basin where kilometer-scale vertical migration of hydrocarbons through overpressure is required to produce the observed distribution of oil and gas (Sassen, 1990; Land, 1991).

Our palaeohydrologic models were constructed using the Basin2 computer program (Bethke, 1985; Bethke et al., 1988). Basin2 is a finite difference program for modelling groundwater flow in two dimensions. The program accounts for basin subsidence and fill, compaction of sediments during burial, and subsurface movement of pore fluids. Fluid pressure

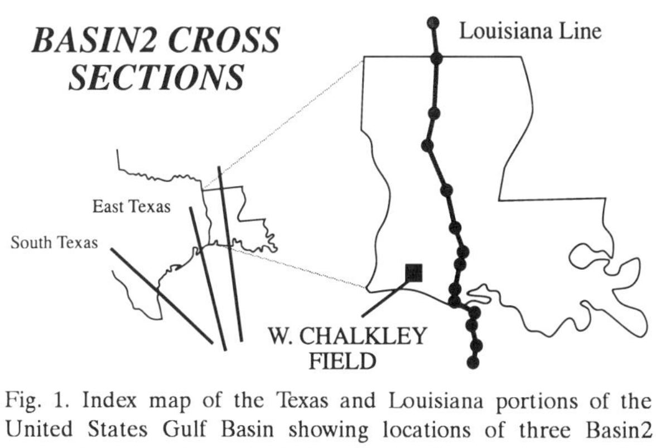

Fig. 1. Index map of the Texas and Louisiana portions of the United States Gulf Basin showing locations of three Basin2 regional cross-sections. Inset shows in detail the onshore portion of the most recently constructed model in Louisiana. Shaded dots mark the locations of well control. The location of Exxon's West Chalkley Field is also shown in the shaded box.

Basin Modelling: Advances and Applications edited by A.G. Doré et al. NPF Special Publication 3, pp. 641–656, Elsevier, Amsterdam.

distribution is calculated as a function of depth, time, and lateral position by solving an equation of flow through a deforming medium (Bethke et al., 1988).

Detailed discussion of our modelling approach and results of initial modelling efforts in East Texas can be found in Harrison and Summa (1991). This report summarizes the results of recent modelling efforts in South Louisiana as an example of our approach, and compares the evolution of overpressures in South Louisiana with results from previous Gulf Basin models. We conclude the discussion with a preliminary evaluation of the extent to which two-dimensional groundwater flow models can be used to predict fluid–rock interactions, and cement precipitation.

South Louisiana geologic framework

The Tertiary of South Louisiana was generally characterized by periods of rapid clastic influx, separated by periods of relative quiescence. The resulting sedimentary section consists of a series of prograding wedges made up of sand-rich non-marine and shallow-marine sediments, separated by thick packages of shale-dominated marine sediments (Fig. 2). Major periods of progradation, characterized by thick sand wedges, occurred during the Late Paleocene to Early Eocene (Wilcox Formation), Oligocene (Frio Formation), Miocene, and Plio-Pleistocene.

The principal depositional wedges contain a complex assemblage of fluvial, coastal plain, nearshore marine, and offshore lithologies subdivided into three principal lithofacies by Sharp et al. (1988) for the purpose of regional analysis of hydraulic properties (Fig. 3). The first is a sand-dominated lithofacies that contains >30% sand and sandstone. It occurs in the updip non-marine and shallow-marine portions of the section and displays the highest relative hydraulic conductivities. The second major lithofacies consists of mixed sand and shale and contains approximately 5–30% sand and sandstone. This lithofacies represents the transition between nearshore and offshore depositional systems, and displays variable hydrologic properties, depending on the lateral and vertical continuity of individual sand bodies. The third lithofacies consists of >95% mud and shale and forms the most volumetrically significant portion of the depositional system. These fine-grained sediments are characterized by consistently low hydraulic conductivities. Rocks of the sandy and mixed sand and shale lithofacies that contain >10% sandstone, e.g. Wilcox and Frio Formations, are the principal aquifers, whereas marine mudstone and shale lithofacies that contain <5–10% sandstone generally form confining units, or aquitards.

The Tertiary depositional history of South Louisiana is complicated by contemporaneous growth faulting and development of salt structures. Growth faults and salt domes cause additional complexi-

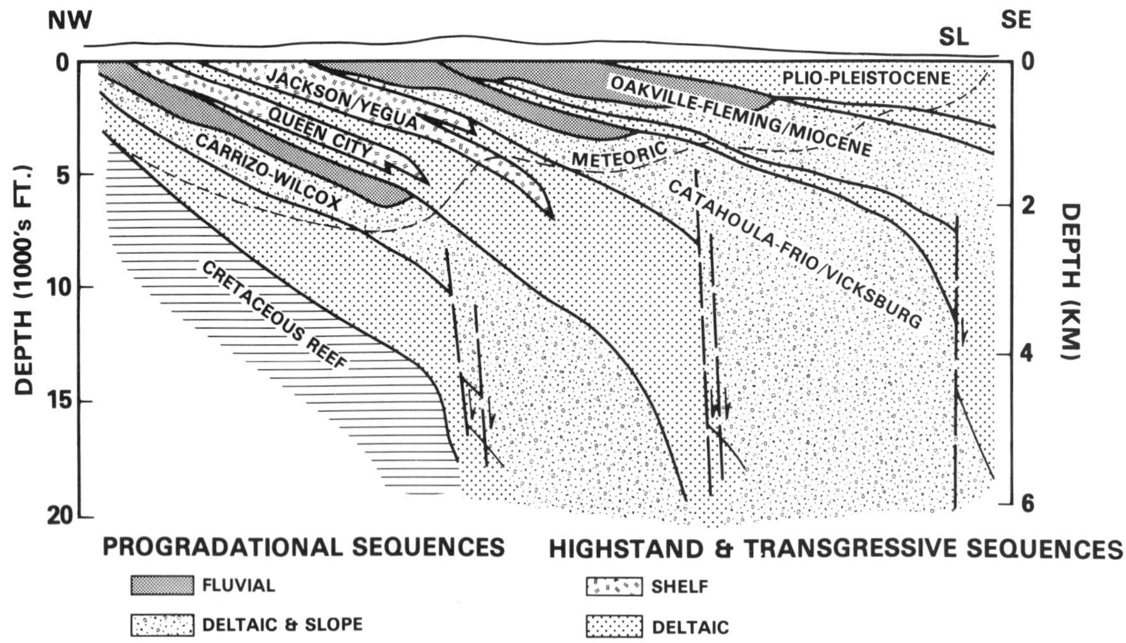

Fig. 2. Generalized stratigraphic cross-section for the Texas and Louisiana Gulf Coast, showing the distribution of the principal aquifers and confining units. The principal aquifers are the Eocene Carrizo–Wilcox Formations, the Oligocene Catahoula–Frio Formations, and the Miocene Oakville–Fleming Formations. The limits of meteoric incursion are approximately 1–2 km (after Galloway et al., 1982; Harrison and Summa, 1991).

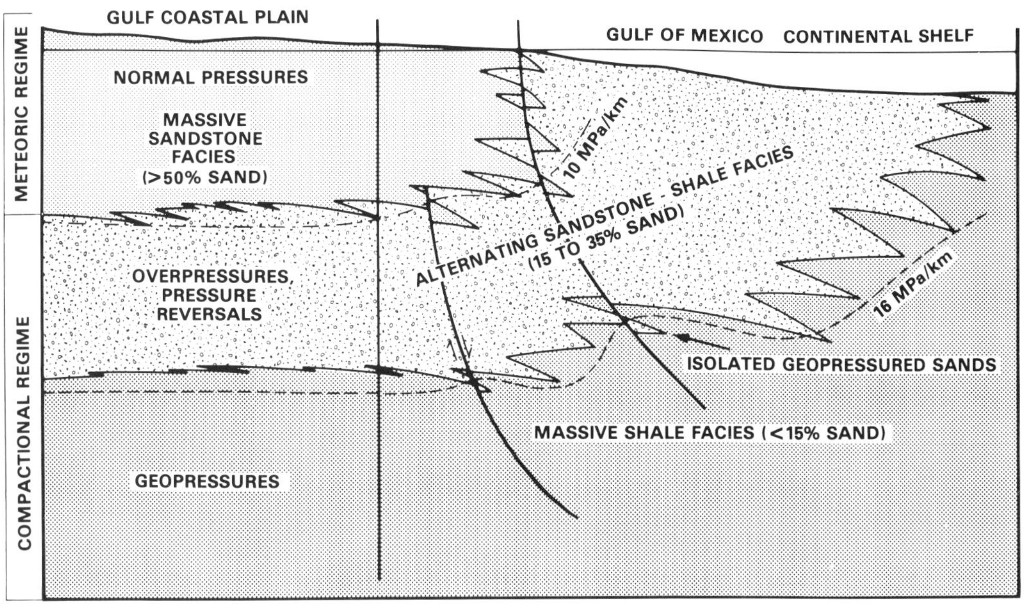

Fig. 3. Generalized lithofacies distribution for Gulf Coast sediments. Massive sandstones are generally normally pressured, whereas massive shales are predominantly geopressured. The alternating sand and shale lithofacies is commonly characterized by normally pressured sands and geopressured shales. The top to geopressure rises in a basinward direction as a consequence of basinward progradation through time (after Norwood and Holland, 1974; Wallace et al., 1979; Harrison and Summa, 1991).

ties in the distribution of aquifers and aquitards by resulting in the isolation of deltaic and near-shore marine sands against impermeable marine shales and salt domes. During some periods in their development, both growth faults and salt-related faults may also have provided conduits for the escape of compaction-driven fluids, with marked local effects on the fluid pressure distribution.

Present-day hydrologic characteristics

The characteristics of present-day hydrologic regimes are used to constrain the fluid flow models. In South Louisiana, as in most of the Gulf Basin, two principal hydrologic regimes are recognized: (1) a compactional regime characterized predominantly by upward moving fluids derived from expulsion of water trapped during sediment compaction, and (2) a meteoric regime characterized by basinward movement of fluids derived from surface water infiltration. Sediments in the compactional regime are commonly overpressured. In this report, sediments with average pressure gradients of 0.4–0.7 psi/ft (9.0–13.5 ppg mud weight equivalent) are described as being in the "transition zone" between hydrostatic and hard overpressures. Sediments with average gradients in excess of 0.7 psi/ft (13.5 ppg) are considered to be "hard overpressured" or "geopressured".

The following characteristics of the overpressured compactional regime are used to constrain the modelled distribution of overpressures. First, a typical column of rocks in the Gulf Basin is normally pressured at burial depths of less than 2–3 km, and undergoes a relatively rapid transition to hard overpressure over a depth range of a km or less (Fig. 4). Although several variables may contribute to the observed distribution of overpressures, pressure data suggest that depositional facies may be a primary control. Laterally continuous, fluvial sands that allow rapid fluid movement are generally hydrostatically pressured, and marine mudstones are generally overpressured (Fig. 3). Interbedded sands and shales

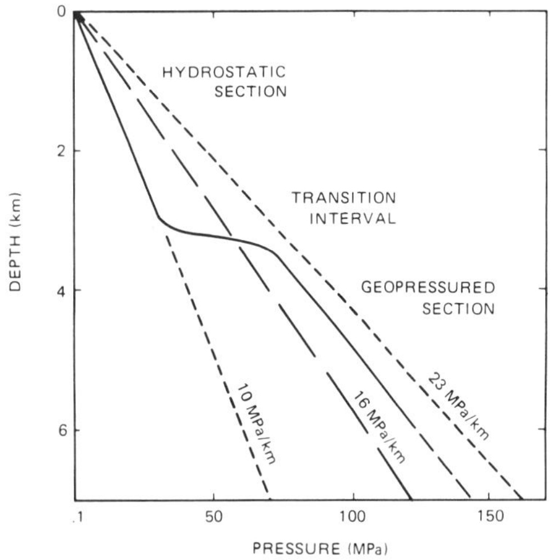

Fig. 4. Generalized nearshore trend of pressure vs. depth. The thickness of the transition interval varies depending on location. Onshore, the transition from normal pressure to geopressure can occur within a few hundred meters. Offshore, the transition interval is generally thicker, and less well defined (after Bethke, 1986; Harrison and Summa, 1991).

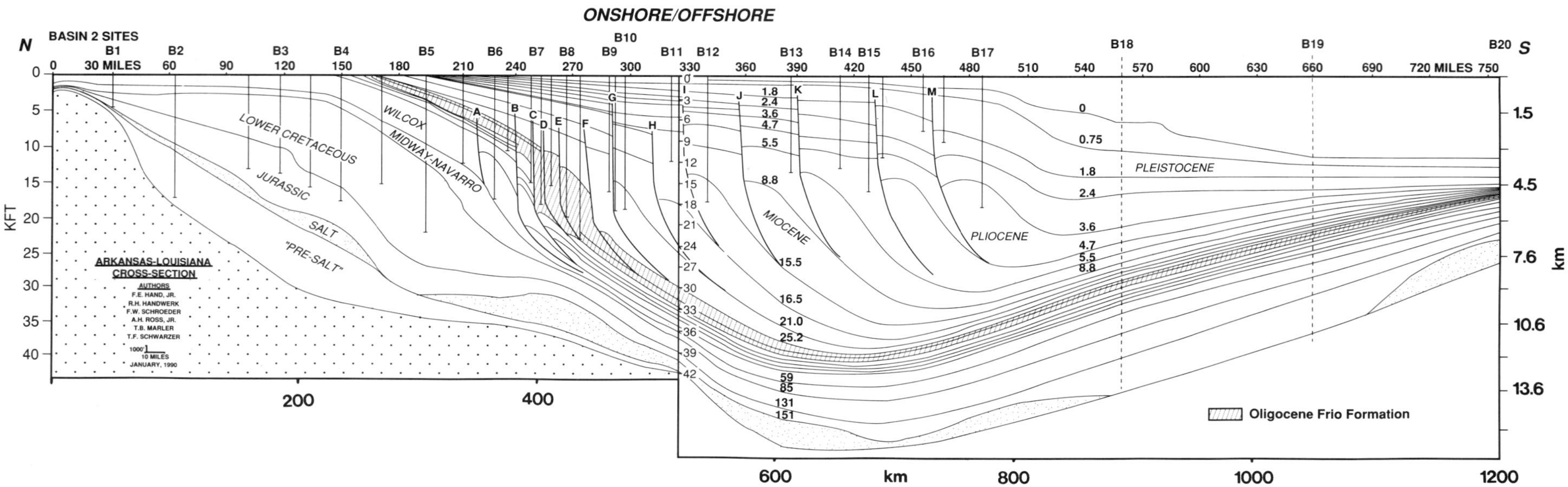

Fig. 5. Input stratigraphic cross-section for the Louisiana Basin2 model (location shown in Fig. 1). The stratigraphic section has been subdivided into 27 major horizons to account for the principal hydrologic characteristics. The cross-section is approximately 1200 km (750 mi) in lateral extent, and 17 km (50,000 ft) at its maximum depth. The locations of Basin2 well control points are shown across the top (*B1–B20*), as are the well penetration depths. Principal growth faults are also shown (labeled *A–M*), although they are not modelled by Basin2. The cross-hatched area marks the Oligocene Frio Formation.

typical of nearshore marine, deltaic and coastal plain environments often contain the transition zone between hydropressure and geopressure, and may be characterized by normally pressured sands, and overpressured shales.

The depth to the first occurrence of geopressured sediments generally rises through stratigraphically younger sediments in a basinward direction. Onshore sediments may not become geopressured above 3–4 km, but offshore sediments commonly become geopressured as shallow as 1–2 km, as shown schematically in Fig. 3. Drilling records also indicate that the thickness of the transition zone between hydrostatically pressured sediments and geopressured sediments increases in a basinward direction. Onshore, the transition from hydropressure to geopressure can occur within several hundred feet, whereas offshore, the transition zone is less well defined and extends over a wider depth interval.

The above pattern of normally pressured sands and geopressured shales can be complicated by movement along growth faults and intrusion of salt domes. Isolated fluvial and shallow-marine sands on the downdip sides of growth faults may become geopressured, presumably because they are juxtaposed against impermeable shales (e.g. Fowler, 1970). In contrast, some high permeability growth faults may enhance pressure dissipation in the sedimentary section (e.g. Berg and Habeck, 1982). Changes in fluid density associated with salt dome emplacement may also influence subsurface pressures, although these effects are not well documented (e.g. Hanor and Sassen, 1990).

The following characteristics of the present-day Gulf Coast meteoric regime are used to constrain the modelled extent of downward flowing meteoric fluids. In onshore sediments of South Louisiana, meteoric water, as defined by salinities of $<35,000$ ppm total dissolved solids, reaches depths of approximately 1–1.5 km in the Frio, Wilcox, and Miocene (Hanor and Sassen, 1990). Little evidence of freshwater incursion is found in offshore Miocene and Plio-Pleistocene sediments, even those relatively close to potential recharge areas onshore (Land et al., 1987; Land and Macpherson, 1989). However, several authors have cited evidence that waters of meteoric origin may have reached depths of 10,000 ft. (3 km) during past sea level lowstands (Lundegard, 1985; Fisher and Land, 1986; Land et al., 1987).

Model input

The fluid flow models described in this report have been constructed using Basin2 (Bethke, 1985). Input to the Basin2 models consists of standard stratigraphic and rock property data. Stratigraphic input data include formation thicknesses, lithologies, ages of formation tops, and palaeobathymetry. Rock property data include thermal conductivities, porosity versus effective depth, and porosity versus permeability, for each lithologic type.

Figure 5 shows the input stratigraphic cross section, which was constructed from well log and seismic data by the Onshore Exploration and Offshore Alaska Divisions of Exxon USA. The line runs from Dallas County, Arkansas offshore to the Sigsbee Escarpment, through the center of the Nodosaria blanpiedi Embayment, an area of active exploration interest containing fewer salt structures than most of South Louisiana. Primary stratigraphic data came from 24 deep wells on file with Exxon USA. Regional maps of depth to magnetic basement and thickness of the Mesozoic Louann salt were used to constrain total sediment fill (H. Yorston, unpubl. data). Below 7.5 km the stratigraphic interpretation is poorly constrained and based primarily on the concept that each depositional sequence thins toward the basin. Palaeobathymetric estimates are based on Exxon USA palaeontologic reports for all of the intervals penetrated by wells. Extrapolation of water depths to intervals deeper than the available well control was accomplished through geohistory analysis (F.W. Schroeder, unpubl. data).

The stratigraphic units are grouped according to their hydrologic characteristics. However, each unit actually includes a range of lithologies and formation thicknesses in order to accommodate lateral facies changes and formation pinchouts. Tertiary formations are described in the greatest detail. The Wilcox, Frio and Pliocene have each been subdivided into two major time slices, and the Miocene has been subdivided into four major time slices in order to address details of the palaeohydrology during these periods of rapid sedimentation.

Basin2 cannot model relative movement along faults or intrusion of salt domes. Therefore, neither feature is represented quantitatively in the input stratigraphy. Displacement of stratigraphic units along major growth faults is represented by changes in sediment thickness. Salt domes are not modelled. It is assumed that the domes are not a regionally continuous regional barrier to fluid flow (Woodbury et al., 1973) and that in a three-dimensional model, fluids would flow around the domes. Because salt domes are not included, the potential for regional fluid transport by kilometer-scale thermohaline convection observed adjacent to several domes (Hanor, 1987; Hanor and Sassen, 1990) has not been assessed.

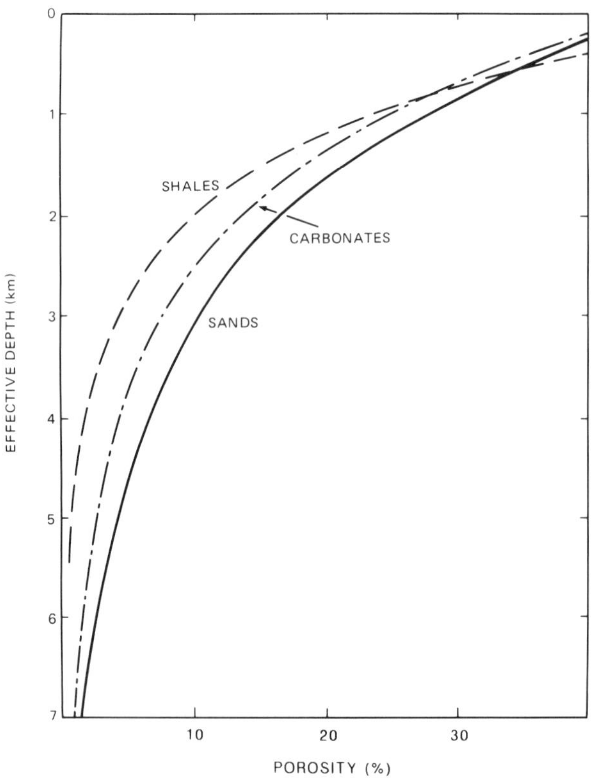

Fig. 6. Porosity vs. effective depth relationship used in the South Louisiana and previous Basin2 models (after Harrison and Summa, 1991).

Porosity versus effective stress relationships used in all three Gulf Basin models are shown in Fig. 6. The porosity data are based on empirical porosity decline curves constructed by Perrier and Quiblier (1974), and are consistent with Gulf Coast sandstones. Note that these porosities are reversible for changes in effective depth. In other words, if the effective depth decreases due to a decrease in effective stress associated with the onset of overpressuring, the porosity decline relationship is reversible. For a detailed discussion of the porosity versus effective depth concept, see Bethke (1985).

Porosity versus permeability relationships used in South Louisiana are also similar to those used in the previous models. Figure 7a shows average core plug porosity–permeability relationships for Gulf Coast sandstones. The choice of the relationship used in South Louisiana is based on: (1) the argument that regional permeabilities of the scale being modelled in Basin2 should be higher than average core plug permeabilities due to large-scale lithologic heterogeneities and regional fracture systems (Garven and Freeze, 1984; Bethke, 1989), and (2) the requirement that present-day modelled porosities, permeabilities and pressures match observed porosities, permeabilities, and pressures. Vertical permeabilities in sandstones were calculated assuming a permeability anisotropy of $k_x/k_z = 2.5$ (Bethke, 1985), where x is strata-parallel and z is vertical. Figure 7b shows estimates from several authors of porosity–permeability relationships in shales. The relationship used in this study is intermediate between data obtained by several authors and represents the shale permeability that best reproduces the present-day observed fluid pressures. Sensitivity analyses showing the effects of changing the input shale permeabilities are discussed below. To account for the control exerted by less permeable shale layers on cross-formational flow, vertical permeabilities have been calculated using an anisotropy of $k_x/k_z = 100$ (Byerlee and Morrow, 1983).

Permeability versus porosity relationships for mixed sand–shale lithologies are derived empirically. For mixed lithologies with $\geq$30% sand, lateral permeabilities were calculated as an arithmetic mean of the permeabilities of sand and shale end-members; a geometric mean was used for lithologies with $<$30% sand. Vertical permeabilities were calculated using a harmonic mean for all lithologic combinations of sand and shale. For a detailed discussion of these choices, see Harrison and Summa (1991). Because of the regional extent of the input cross section, lateral variations in permeability due to the presence of growth faults are not accounted for. However, accurate reconstruction of the present-day pressure distribution proved to be possible without accounting for fault-related permeability changes.

Sources of additional rock property data such as thermal conductivities, heat capacities, and enthalpies, which are needed to solve the heat transfer equations, were taken from Bethke (1985). A geothermal gradient of 25°C/km and a basal heat flow of 1.3 HFU were used as input data for the thermal calculations.

Model results

The results reported for South Louisiana are of regional scale. The nodal blocks over which pore pressure evolution and meteoric incursion were calculated are approximately 40 km in the strata-parallel direction and 200 m in a vertical direction. Small-scale variations in pressures or flow rates associated with local heterogeneities in lithology, cemented zones, etc. are not considered.

Overpressure generation mechanisms considered in the calculations include compactional disequilibrium and thermal expansion of pore fluids. However, sensitivity analyses (Harrison and Summa, 1991) suggest that aquathermal expansion has only a minor effect on the total pressure generated in a Gulf Coast-type model. Terms for additional sources of

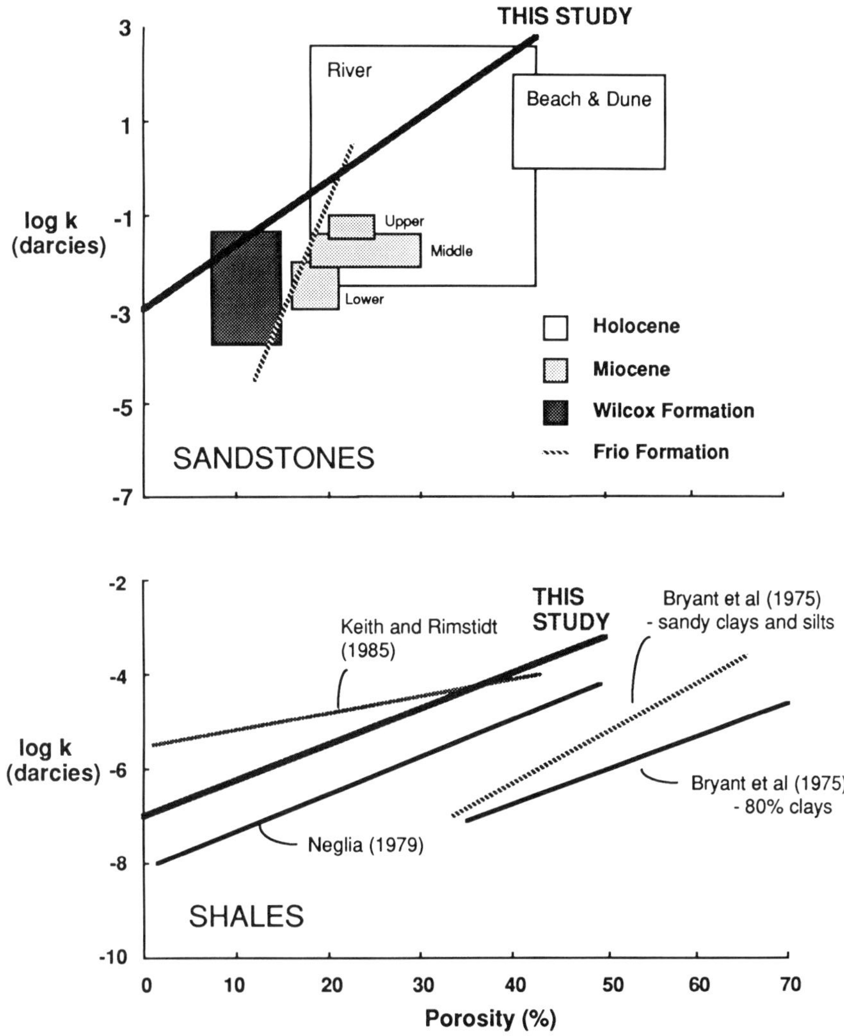

Fig. 7. Porosity vs. permeability relationships used in the South Louisiana study. The top diagram shows sandstone data compiled for Gulf Basin sandstones, and is identical to the relationship used in previous Basin2 models. The bottom diagram shows data compiled for Cenozoic shales by a number of different authors. The porosity vs. permeability relationship used in this study allows higher permeabilities than previous Gulf Coast Basin2 models, in order to get the best match between modelled and observed present-day geopressures.

fluid pressure generated from smectite dehydration and overmaturation of hydrocarbons are not included. For a discussion of the possible effects of these variables, see Harrison and Summa (1991), and Bethke (1986).

Model predictions for pressure distribution, extent of meteoric incursion, thermal maturity, and porosity have been compared with present-day observed data as a means of constraining the model calculations. Comparisons of present-day observed pressures and fluid salinities serve to illustrate this method. Output pressures are shown in Fig. 8. Figure 9 illustrates the comparison between observed and modelled pressure distribution. The observed pressures consist of mud weight values for wells along the cross section, and additional infill wells within the cross section area. With the exception of the updip area, the present-day pressure distribution is correctly reproduced within the limits of the model resolution. There is an extensive zone of overpressures, predominantly offshore. Predicted pressure gradients vary from hydrostatic to approximately 0.85 psi/ft, and true lithostatic pressures are not observed. The first occurrence of abnormally pressured sediments occurs where the rocks become shale-dominated, at burial depths of approximately 2.0–2.5 km in the onshore Tertiary section, and rises to approximately 1.0–1.5 km in offshore marine shales. These results suggest that the observed overpressures can be explained largely by compaction disequilibrium and that contributions from smectite dehydration and hydrocarbon generation are not required to model overpressure development in this setting.

The failure of the model to predict the updip pressure distribution is not well understood, although it has been observed in the previous two Gulf Coast

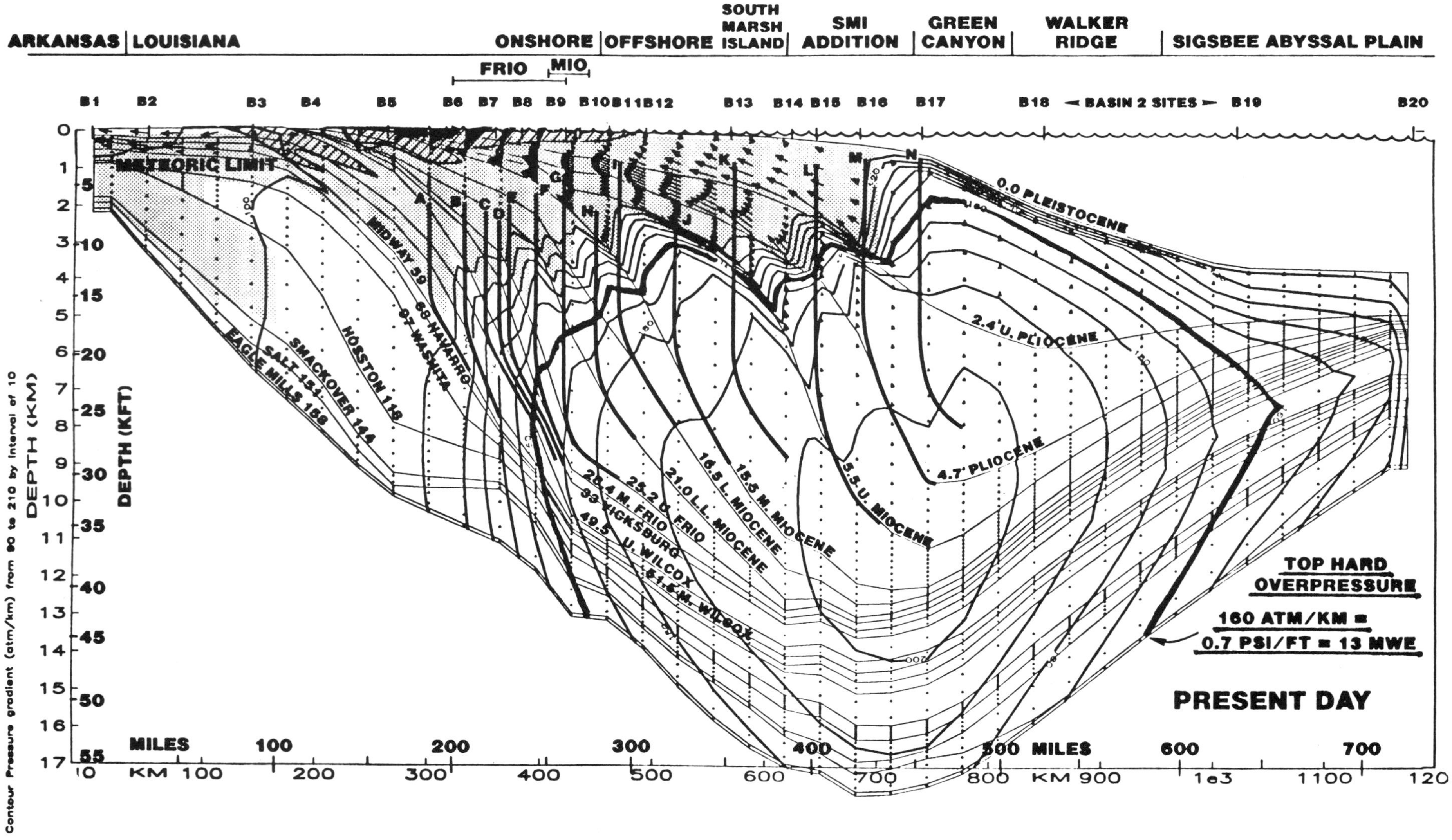

Fig. 8. Present-day Basin2 output model for South Louisiana. Principal stratigraphic units and their ages are labeled. Contours are pressure gradients in atm/km. The top of hard overpressure is marked by the bold line. Fluid flow velocities and directions relative to the rock framework are shown by the arrows. Each arrow is placed at the center of a nodal block over which the hydrologic properties of the section have been averaged. The stippled area includes rocks with greater than 30% sand. Notice that most of these sediments are normally pressured. The hachured area at the top of the section shows the basinward limits of meteoric incursion. Input control points are shown across the top, and the major faults have been sketched in and labeled *A–K*.

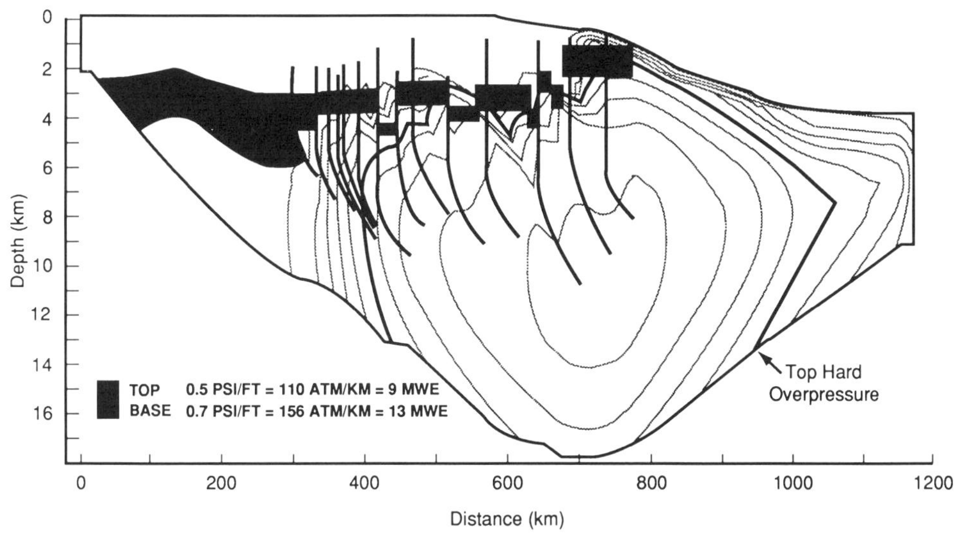

Fig. 9. Present-day observed pressure gradient compared to Basin2 predicted gradient. The darkened area shows the observed thickness of the pressure transition zone from 110 to 156 atm/km (0.5–0.7 psi/ft). Note the close match between observed and modelled values except in the updip portion of the section, where the modelled pressure gradient is lower than the observed gradient.

Basin2 models. In part, it may be an artifact of the no-flow boundary at the base of the section. Also, the rock properties data were defined mainly for Tertiary units, and may not correctly reproduce the properties of older, more compacted and presumably more brittle rocks. Sensitivity analyses suggest that if the updip strata are given model permeabilities equivalent to shales, the anomaly can be eliminated — however, there is no evidence to suggest that the observed permeabilities in this region should all be low. Finally, the anomaly may result from an inadequate description of the compaction state in the updip part of the section that cannot be described by a simple porosity versus effective stress function. Efforts are currently underway to address this question.

Figure 10 illustrates the comparison between the modelled meteoric regime, defined by basinward

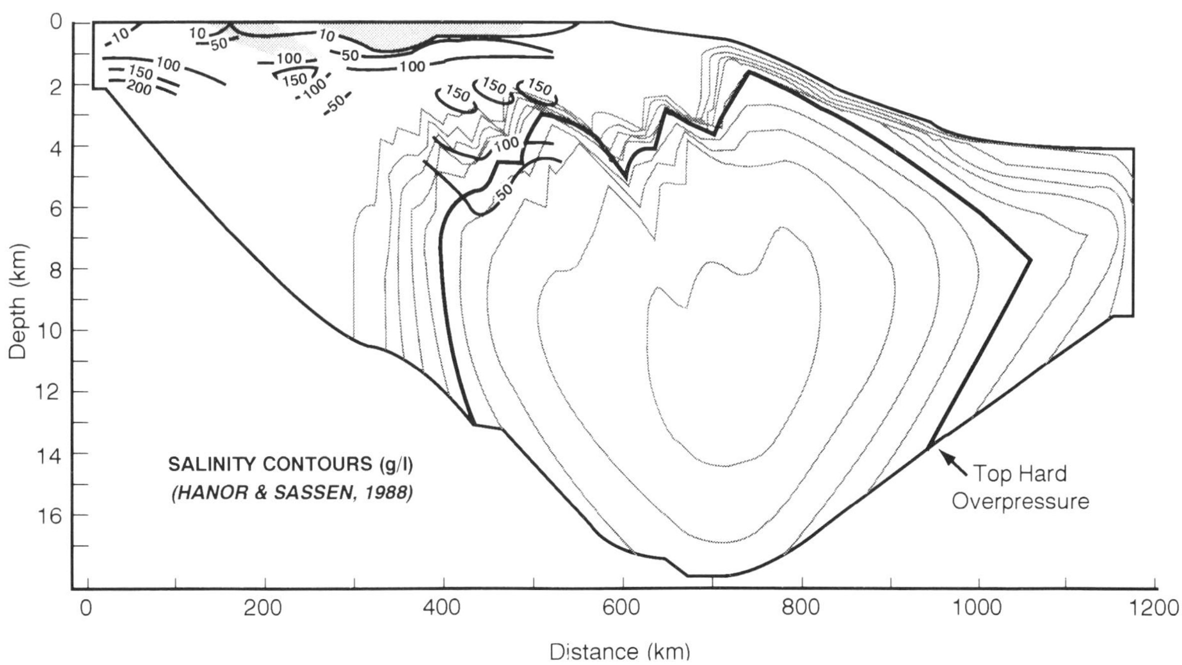

Fig. 10. Present-day observed salinities (Hanor and Sassen, 1990) compared to the predicted basinward extent of downward flowing meteoric waters (stippled). Seawater salinity is 35 g/l. Note that there is generally a close match between the modelled extent of meteoric incursion and salinities less than seawater, although modelled meteoric incursion in the updip portion of the cross-section is slightly deeper than observed.

flow, and the observed meteoric regimes, defined by salinities of <35 g/l, i.e. seawater. Measured salinity data were collected by Hanor and Sassen (1990) along a cross-section in central Louisiana just east of the Basin2 line. The modelled meteoric regime is found in the upper kilometer of the basin, and the correct Tertiary aquifers are reproduced. The modelled salinities are also in good agreement with present-day salinities for the shallowest portion of the section. Note that the measured salinities first increase rapidly with depth, and then decrease. The increase is thought to be due to the influence of salt dissolution, which was not modelled with Basin2 (Hanor and Sassen, 1990). The subsequent decrease is thought to reflect the influence of lower-salinity waters derived from shale compaction.

Basinward flow predominates in the meteoric regime, as shown by the flow velocity arrows (Fig. 8). The maximum meteoric flow velocities average 10–60 cm/yr, comparable to observed flow rates in similar topographic settings (Fogg, 1986). In the compactional regime, flow vectors are dominantly vertical in the overpressured, shale-rich portion of the section where fluids move mainly across the isobars. In the shallower, sandier portions of the basin, flow is predominantly strata-parallel in the updip direction, and flow rates increase with decreasing depth and increasing sand percentage. The maximum compactional flow velocities average only a few centimeters per year, also comparable to observed flow rates in similar sections (e.g. Bredehoeft et al., 1988). Note that flow rates are calculated relative to the rock framework, and not relative to some fixed elevation such as sea level.

Sensitivity analyses have been used to test how the match between predicted and observed fluid pressures is affected by modification of the assumed porosities, permeabilities, and permeability anisotropy. Changing the sandstone permeability has relatively little effect on the modelled overpressures, since sand abundance is minor compared to shale abundance. However, shale permeability assumptions are critical for accurate reconstruction of overpressures. Figure 11 shows the effects on depth to geopressure that result from order of magnitude changes in the shale permeability. The relationship used in this model (Fig. 7b) produces the middle horizon. Increasing the shale permeability by an order of magnitude results in an essentially normally pressured basin. Decreasing the shale permeability by an order of magnitude results in hard overpressure right at the sediment/water interface. Additional analyses indicate that decreasing shale permeability anisotropy from $k_x/k_z = 100$ to $k_x/k_z = 10$ for the input porosity–permeability relationship shown in

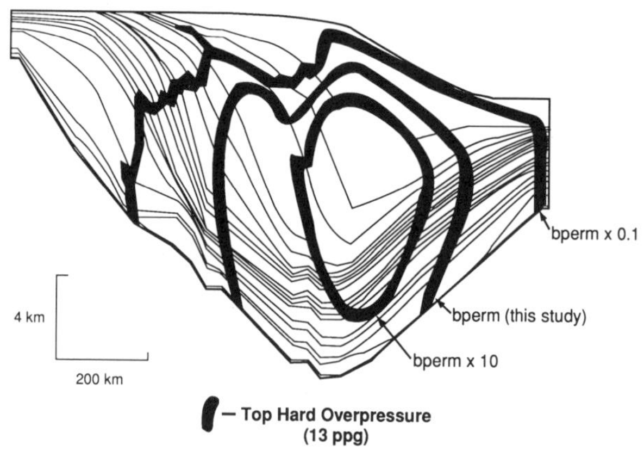

Fig. 11. Sensitivity analysis showing the effect on top overpressure of an order of magnitude change in the input shale permeability, denoted by *bperm*. In the Basin2 program, bperm is the permeability at a porosity of zero. Increasing the permeability an order of magnitude results in an essentially normally pressured basin, whereas decreasing the permeability an order of magnitude results in hard overpressures essentially at the sediment–water interface.

Fig. 7b allows more fluid to escape in a vertical direction and results in a normally pressured basin. The significant pressure gradient change that results from using either higher or lower permeabilities, along with the close agreement between the modelled and observed pressures for a shale porosity–permeability relationship of $\log k = 8\phi - 6.9$ and $k_x/k_z = 100$, supports the choice of these permeability values for the South Louisiana model. The extreme sensitivity of the model to the choice of input shale properties also emphasizes the need for development of better techniques to measure shale permeability.

Variations in Gulf Coast geopressure evolution and implications for hydrocarbon migration

If we assume that the observed geopressures in the Gulf Basin resulted largely from compactional disequilibrium, then regional variations in sedimentation rates could have caused corresponding variations in overpressure history in different parts of the basin. As shown in Fig. 12, there has been a clockwise migration of major Tertiary depocenters from South Texas in the Early Tertiary toward East Texas and South Louisiana in the Miocene and Plio-Pleistocene. Reconstructions of maximum excess pressures within the Oligocene Frio Formation (Fig. 13) suggest that in South Louisiana, extensive overpressures began to develop in the Miocene and expanded rapidly during the Plio-Pleistocene. In contrast, high sedimentation rates during the Early Tertiary of South Texas apparently caused the onset of geopressures much earlier during the Eocene, while more northern areas of the basin still had pre-

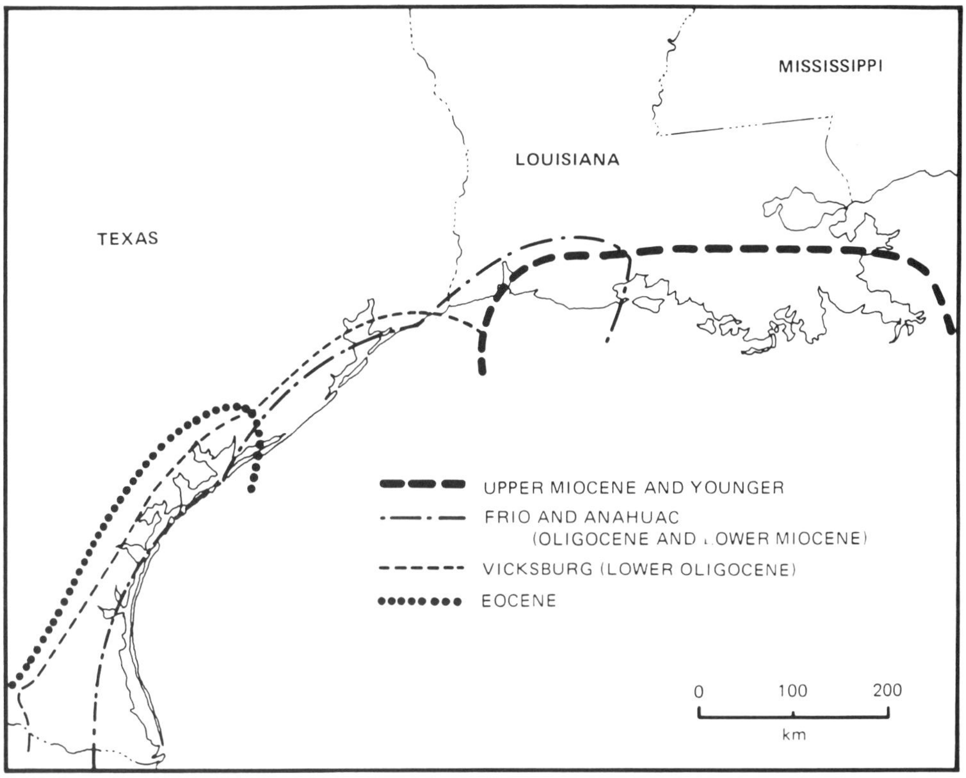

Fig. 12. Variations in principal depocenters. South Texas was the principal focus of sediment deposition in the Early Tertiary, whereas East Texas and South Lousiana are characterized by thick sections of Miocene and younger rocks (after Hardin, 1962; Harrison and Summa, 1991).

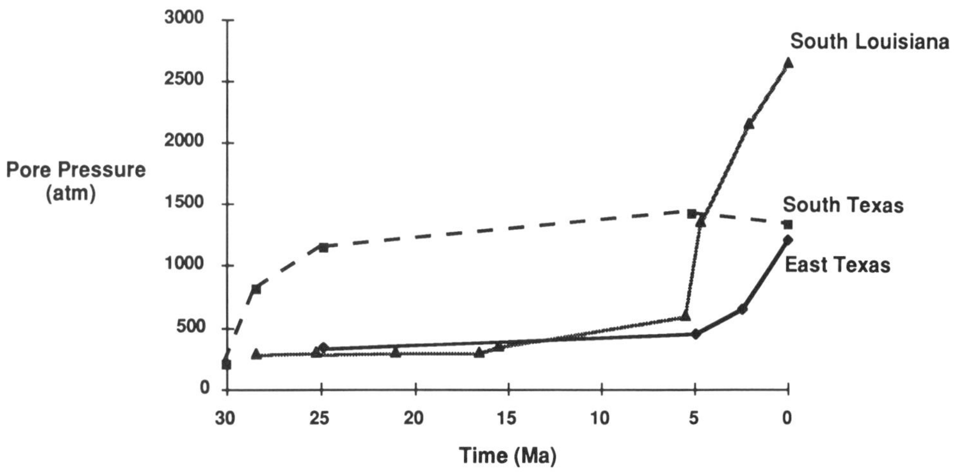

Fig. 13. Maximum modelled excess pressures for the Frio Formation in South Texas, East Texas and Louisiana. Note that sediments apparently became overpressured in South Texas during the Oligocene, much earlier than in either East Texas or Louisiana, where significant overpressures did not develop until late in the Miocene.

dominantly normally pressured sediments (Fig. 13). Pressures in South Texas also appear to be declining at the present in response to current low sedimentation rates, whereas pressures in the northern Gulf Basin are apparently continuing to increase.

We argue that under certain circumstances, overpressured traps may be less likely to be filled with hydrocarbons than normally pressured traps. This argument is based on the assumption that the migration conduit can have lower pressures than the trap, which is the case for at least some of the potential migration pathways in the Gulf Coast (see, for example, Berg and Habeck, 1982). If one makes this assumption, then evaluating the timing of overpressure generation relative to hydrocarbon migration is critical to risk assessment for hydrocarbon migration. Existing hypotheses for hydrocarbon migration timing and mechanisms in the South Louisiana area suggest that hydrocarbons may have migrated from the Eocene Wilcox Formation into major Frio and Miocene reservoirs sometime during the Miocene (Nunn and Sassen, 1986; Sassen et al., 1988). During most of the proposed migration period, many of the prospective sands within the Frio and Miocene intervals should have been in transitional overpressure (Fig. 13), suggesting that there may have been only minor risk of traps being by-passed. In South Texas, at least some hydrocarbon migration is thought to have occurred during the Oligocene (Galloway et al., 1982; T.F. Schwarzer, unpubl. data). As suggested in Fig. 13, many prospective traps may have been overpressured during the Oligocene, and therefore at greater risk of not containing hydrocarbons.

Application of fluid flow models to cement prediction

Diagenetic cements precipitate in the pore systems of sandstones as a result of interaction between grains and pore fluids. Therefore, it may be possible to use fluid flow models such as Basin2 to predict areas that are more susceptible to fluid flow, and therefore more susceptible to volumetrically significant cementation. Conversely, the observed chemistry, distribution and timing of cements can theoretically be used to better constrain model predictions of fluid flow in the past, by providing independent measurements of palaeotemperature, fluid volume, fluid chemistry, and in some cases, palaeopressure.

Application of these techniques is currently underway in South Louisiana. In this discussion, we compare predicted fluid flow velocities for the Frio Formation (Fig. 14) with observed cement timing and volume in Frio sands from the West Chalkley Field (Fig. 1). Basin2 fluid flow predictions for the Frio show a brief period of meteoric incursion followed by minor compactional flow for the remainder of the Frio burial history (Fig. 15). Predicted flow velocities suggest that compactional flow rates averaged 0–0.2 cm/yr relative to the rock framework

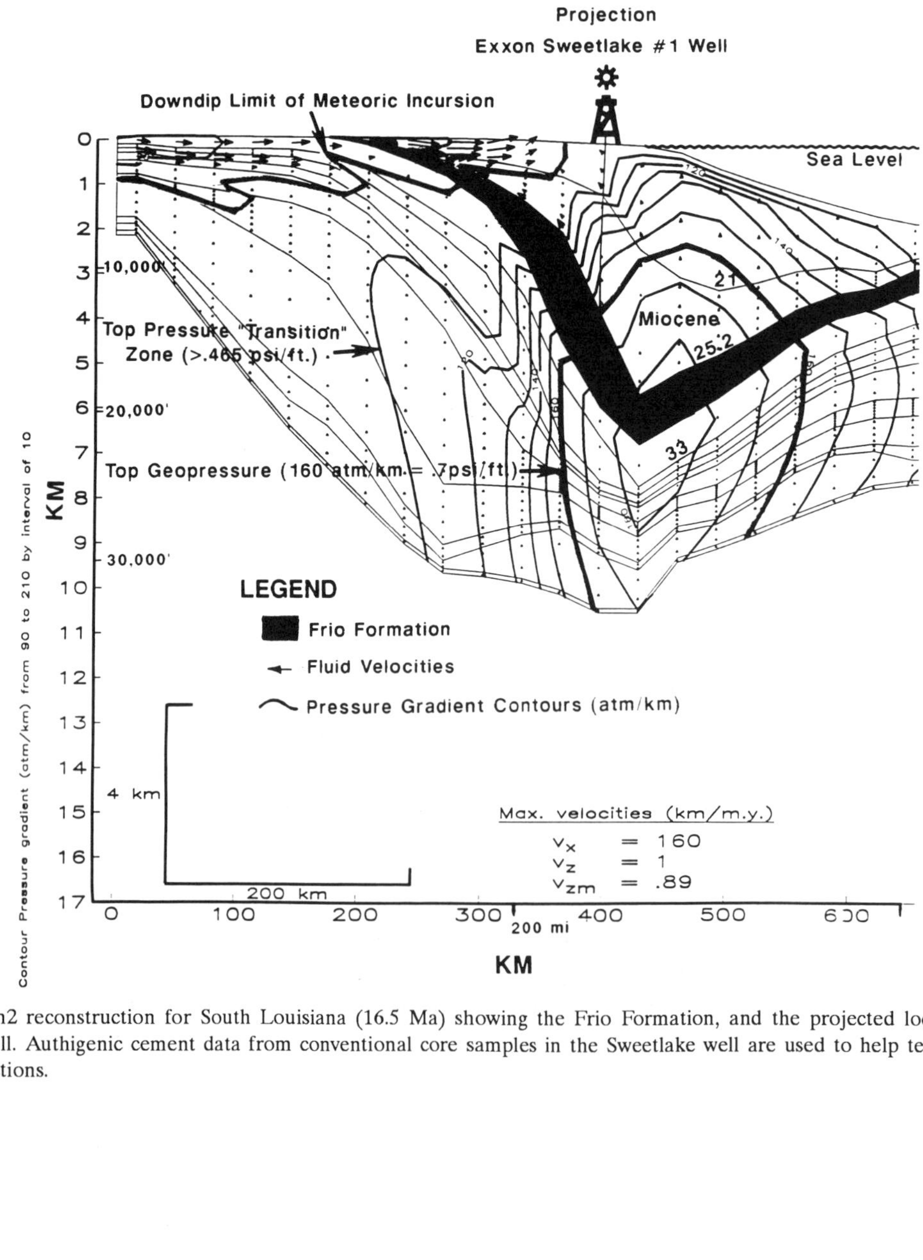

Fig. 14. Basin2 reconstruction for South Louisiana (16.5 Ma) showing the Frio Formation, and the projected location of the Exxon Sweetlake well. Authigenic cement data from conventional core samples in the Sweetlake well are used to help test and constrain the Basin2 predictions.

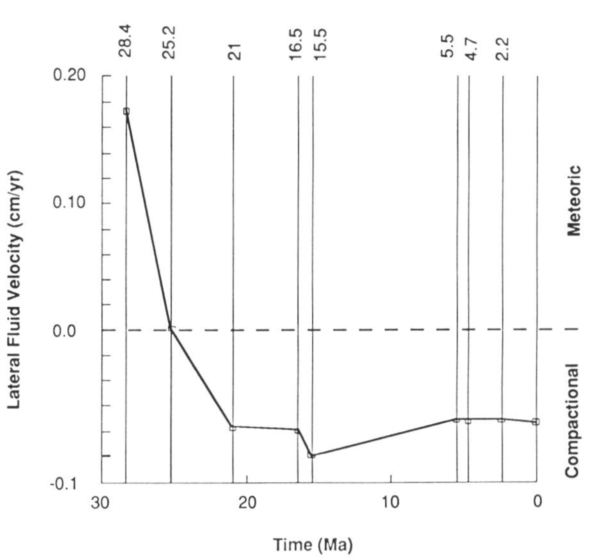

Fig. 15. Basin2 predicted flow velocities, relative to the rock framework, through time for the Frio at the location of the Exxon Sweetlake projection. Positive velocities are downward moving waters. Negative velocities are upward moving waters. As shown, a brief period of meteoric incursion was followed by approximately 25 m.y. of relatively slow compactional flow.

(Fig. 15). If the observed cementation in the West Chalkley Field occurred as a result of the predicted fluid flow, then one might expect the authigenic cement assemblage to have the following characteristics. There should be minor early cement associated with meteoric incursion, but porosity should not be significantly degraded, since freshwater flushing occurred over a brief period of time, and only at shallow depths. Sediment burial should have produced only minor additional cementation, based on the assumption that the predicted flow rates for compactional waters would have been too slow to result in significant cement precipitation. As such, the remainder of the cementation would have occurred in what was essentially a closed system, with small volumes of burial cements whose compositions reflect increasing temperature and pore water salinity due solely to increased burial.

As shown in Fig. 16, the Frio in the West Chalkley Field contains a broad assemblage of cements, consisting predominantly of quartz, albite, illite, kaolinite, and analcime. Average cement volumes are relatively low. Quartz is the most common cement, ranging from 2 to 15% of the rock volume, and averaging approximately 8%. The remainder of the cements make up less than a few percent of the rock volume. Present-day porewaters are extremely saline, with total dissolved solids of 150,000–165,000 ppm, based on formation water analyses from a nearby geothermal well (Karkalits and Hankins, 1981).

The average intergranular volume (i.e. porosity plus cement, or the space between the grains) is approximately 26%. This value corresponds to expected values for Cenozoic Gulf Coast sandstones buried to 1.5–2.0 km in the absence of cement (Paxton et al., 1990), and suggests that cementation occurred largely at burial depths greater than 1.5–2.5 km. This is consistent with the Basin2 prediction that South Louisiana rocks should contain only minor early cement associated with early meteoric influx.

Kaolinite was apparently the first cement to precipitate, consistent with predicted early meteoric infiltration. The "ragged" nature of kaolinite grains may indicate that it is partially dissolved (Fig. 16). Kaolinite emplacement was followed by precipitation of illite, albite, quartz, and analcime (Fig. 16) and the concurrent dissolution of detrital K-feldspar, not shown. The complete dissolution of K-feldspar and the ragged texture of the kaolinite suggest that the fluids were buffered by the silicate assemblage K-feldspar + kaolinite + quartz reacting to form illite. With the exception of analcime, this assemblage is consistent with simple burial to increased depth and temperature (see, for example, Smith and Ehrenburg, 1989; Hutcheon and Abercrombie, 1990), as predicted by Basin2.

The observed volume of quartz cement is not explained. According to Basin2, less than a few hundredths of a percent of quartz cement would have precipitated from rising, cooling compactional waters. However, an average of 8% quartz cement is observed. Either the calculated compactional flow rates are incorrect, or quartz cement precipitated through a mechanism other than rising compactional waters. Precipitation of the observed volume of quartz cement would have required orders of magnitude higher compactional flow rates than predicted, leading to the conclusion that quartz precipitated by another mechanism.

Investigation of the quartz precipitation mechanism is currently underway. However, the presence of analcime, and the saline nature of the present-day pore fluids suggests that these fluids may have interacted with deeply buried evaporites, and been introduced into the Frio section along nearby growth faults. As shown in Fig. 16, analcime cement was precipitating at the same time as quartz cement, suggesting that fluid flow along the growth fault also occurred at the same time as some quartz cement was precipitating.

These analyses suggest that two-dimensional flow models such as Basin2 can be used to infer the presence or absence of cements likely to be precipitated in the meteoric or compactional regimes, e.g. kaolinite. However, such a model cannot presently be used

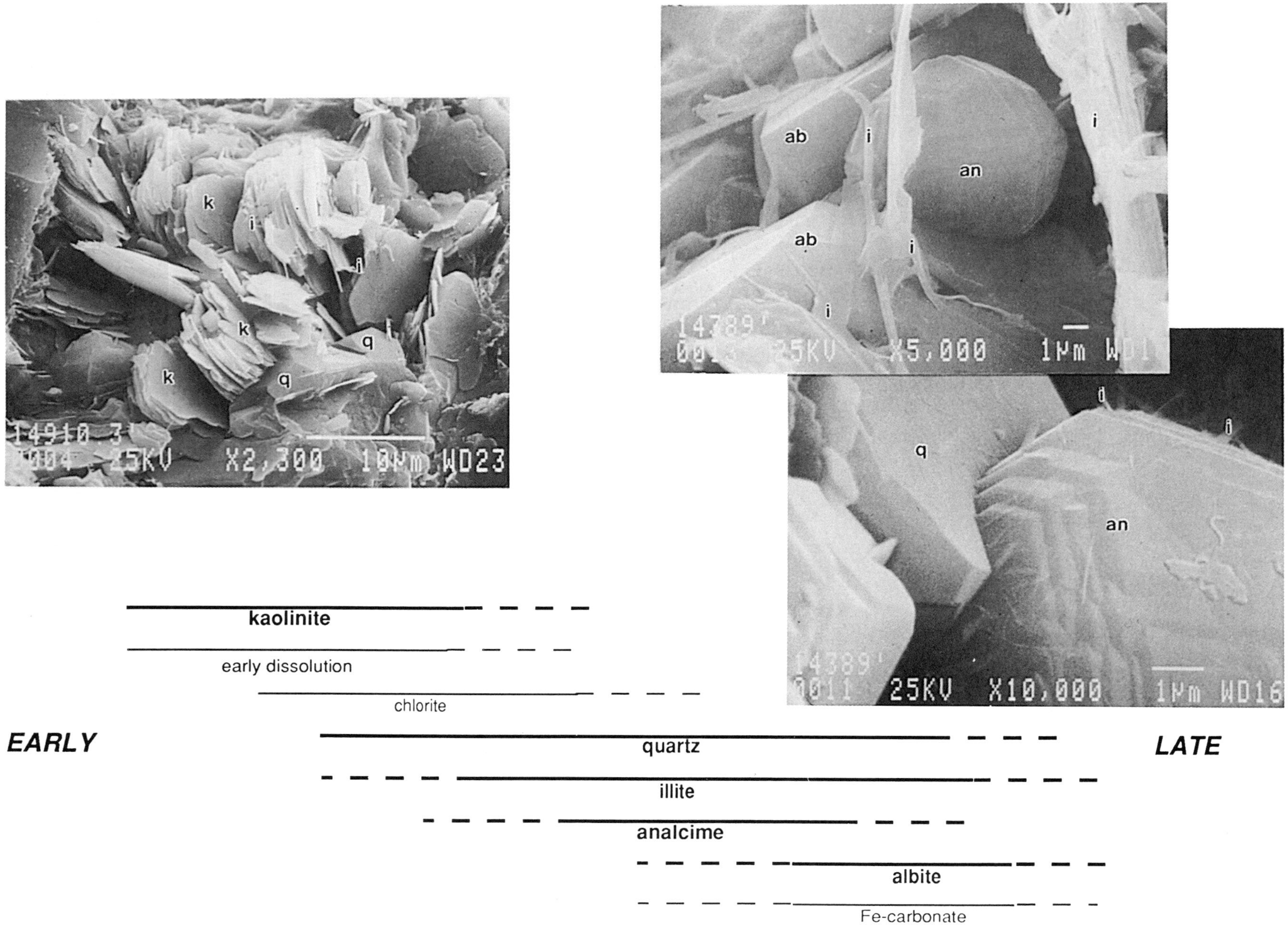

Fig. 16. Generalized paragenetic sequence for the Exxon Sweetlake well. Principal diagenetic events include precipitation of early kaolinite, followed by precipitation of quartz, illite, albite, and analcime, and dissolution of K-feldspar. Note that kaolinite grains appear to be "ragged", indicating partial dissolution. Also note precipitation of illite on top of kaolinite. On the right-hand side, note relative timing of analcime and quartz precipitation. Quartz has continued to precipitate after analcime precipitation was complete.

to explain cement volume. It may be necessary to introduce the capability to model flow along faults and fractures in order to explain the cement volumes observed in some areas.

Conclusions

This study, and preceding Basin2 modelling efforts, have demonstrated that two-dimensional numerical modelling of regional palaeohydrology can be used to understand the evolution of overpressuring and the interactions between different hydrologic regimes in the Gulf of Mexico. The models further suggest that the Gulf Basin was characterized by different overpressure histories and may therefore have been characterized by different trap-fill histories. In South Louisiana, a core of overpressured sediments developed during the Miocene, and continued to expand basinward during the Plio-Pleistocene. However, Frio and Miocene sands remained largely in transitional overpressure throughout much of their burial history, suggesting that there may be relatively minor risk of traps being by-passed. Overpressures developed during the Eocene in South Texas, expanded in the Oligocene in response to rapid deposition of the Frio Formation, and then began to decline. As such, many potential traps in South Texas were overpressured throughout much of their burial history, and may have been at a greater risk of being by-passed. Finally, preliminary analysis of petrographic and geochemical data suggests that two-dimensional flow models such as Basin2 can be used to assess cement types present in the Gulf Coast Tertiary, but the capability to model flow along faults and fractures may be needed to explain cement volumes in some areas.

Acknowledgements

We gratefully acknowledge the Onshore Division of Exxon U.S.A. for providing much of the input data needed to construct the Basin2 models, and for giving us permission to publish the results. Input stratigraphic cross sections were constructed by F.E. Hand, A.H. Ross, and F.W. Schroeder.

References

Berg, R.R. and Habeck, M.F., 1982. Abnormal pressures in the lower Vicksburg, McAllen Ranch Field, South Texas. Trans. Gulf Coast Assoc. Geol. Soc., 32: 247–253.

Bethke, C.M., 1985. A numerical model of compaction driven groundwater flow and heat transfer and its application to the palaeohydrology of intracratonic sedimentary basins. J. Geophys. Res., 90: 6817–6828.

Bethke, C.M., 1986. Inverse hydrologic analysis of the distribution and origin of Gulf Coast-type geopressured zones. J. Geophys. Res., 91: 6535–6545.

Bethke, C.M., 1989. Modelling subsurface flow in sedimentary basins. Geol. Rundsch., 78: 124–154.

Bethke, C.M., Harrison, W.J., Upson, C. and Altaner, S.P., 1988. Supercomputer analysis of sedimentary basins. Science, 239: 261–267.

Bredehoeft, J.D., Djevanshir, R.D. and Beltiz, K.R., 1988. Lateral fluid flow in a compacting sand–shale sequence: south Caspian basin. Am. Assoc. Pet. Geol., Bull., 72: 416–424.

Bryant, W.R., Hottman, W. and Trabant, P., 1975. Permeability of unconsolidated and consolidated marine sediments, Gulf of Mexico. Mar. Geotechnol., 1: 1–14.

Byerlee, J.D. and Morrow, C.A., 1983. Final report on the permeability of geopressured Gulf Coast shale cores. U.S. Geol. Surv., Open-File Rep., 83-0180, 9 pp.

Fisher, R.S. and Land, L.S., 1986. Diagenetic history of Eocene Wilcox sandstones, South-Central Texas. Geochim. Cosmochim. Acta, 50: 551–561.

Fogg, G.E., 1986. Groundwater flow and sand body interconnectedness in a thick, multiple-aquifer system. Water Resour. Res., 22: 679–694.

Fowler, W.A., 1970. Pressures, hydrocarbon accumulation, and salinities — Chocolate Bayou Field, Brazoria County, Texas. J. Pet. Technol., 22: 411–423.

Galloway, W.E., Hobday, D.K. and Magara, K., 1982. Frio Formation of the Gulf Coast basin: depositional systems, structural framework, and hydrocarbon origin, migration, distribution and exploration potential. The University of Texas at Austin, Bur. Econ. Geol., Rep. Invest., 122, 78 pp.

Garven, G. and Freeze, R.A., 1984. Theoretical analysis of the role of groundwater flow in the genesis of stratabound ore deposits. Am. J. Sci., 284: 1085–1124.

Hanor, J.S., 1987. Kilometer-scale thermohaline overturn of pore waters in the Louisiana Gulf Coast. Nature, 327: 501–503.

Hanor, J.S. and Sassen, R., 1990. Evidence for large-scale vertical and lateral migration of formation waters, dissolved salt, and crude oil in the Louisiana Gulf Coast. In: D. Schumacher and B.F. Perkins (Editors), Gulf Coast Oils and Gases: Their Characteristics, Origin, Distribution, and Exploration and Production Significance. Proc. 9th Annu. Res. Conf., Soc. Econ. Paleontol. Mineral. Found., pp. 283–296.

Hardin, G.C., 1962. Notes on Cenozoic sedimentation in the Gulf Coast geosyncline, USA. In: E.H. Rainwater and R.P. Zingula (Editors), Geology of the Gulf Coast and Central Texas and Guidebook of Excursions. Houston Geological Society, 1962 Annual Meeting of the Geological Socity of America and Associated Societies, pp. 1–15.

Harrison, W.J. and Summa, L.L., 1991. Paleohydrology of the Gulf of Mexico Basin. Am. J. Sci., 291: 109–176.

Hutcheon, I. and Abercrombie, H., 1990. Carbon dioxide in clastic rocks and silicate hydrolysis. Geology, 18: 541–544.

Karkalits, O.C. and Hankins, B.E., 1981. Chemical analysis of gas and brine from MGT-DOE Sweet Lake design well. Geopressured-Geothermal Energy, U.S. Gulf Coast, Conf. Proc., DOE Rep., 811026: 331–334.

Keith, L.A. and Rimstidt, J.D., 1985. A numerical compaction model of overpressuring in shales. J. Int. Assoc. Math. Geol., 17: 115–135.

Land, L.S., 1991. Evidence for vertical movement of fluids, Gulf Coast sedimentary basin. Geophys. Res. Lett., 18: 919–922.

Land, L.S. and Macpherson, G.L., 1989. Geochemistry of formation water, Plio-Pleistocene reservoirs, offshore, Louisiana. Trans. Gulf Coast Assoc. Geol. Soc., 39: 421–430.

Land, L.S., Milliken, K.L. and McBride, E.F., 1987. Diagenetic evolution of Cenozoic sandstones, Gulf of Mexico sedimentary basin. Sediment. Geol., 50: 195–225.

Lundegard, P.D., 1885. Carbon Dioxide and Organic Acids: Origin and Role in Burial Diagenesis (Texas Gulf Coast Tertiary). Ph.D. Dissertation, University of Texas, Austin, Texas, 145 pp.

Neglia, S., 1979. Migration of fluids in sedimentary basins. Am. Assoc. Pet. Geol., Bull., 63: 573–597.

Norwood, E.M. and Holland, D.S., 1974. Lithofacies mapping, a descriptive tool for ancient delta systems of Louisiana outer continental shelf. Trans. Gulf Coast Assoc. Geol. Soc., 24: 175–188.

Nunn, J.A. and Sassen, R., 1986. The framework of hydrocarbon generation and migration, Gulf of Mexico continental slope. Trans. Gulf Coast Assoc. Geol. Soc., 36: 257–262.

Paxton, S.T., Szabo, J.O., Calvert, C.S. and Ajdukiewicz, J.M., 1990. Preservation of primary porosity in deeply buried sandstones: a new play concept from the Cretaceous Tuscaloosa sandstone of Louisiana. Am. Assoc. Pet. Geol., Bull., 74: 737 (abstract).

Perrier, R. and Quiblier, J., 1974. Thickness changes in sedimentary layers during compaction history: methods of quantitative evaluation. Am. Assoc. Pet. Geol., Bull., 58: 507–520.

Sassen, R., 1990. Lower Tertiary and Upper Cretaceous source rocks in Louisiana and Mississippi: implications to Gulf of Mexico crude oil. Am. Assoc. Pet. Geol., Bull., 74: 857–878.

Sassen, R., Tye, R.S., Chinn, E.W. and Lemoin, R.C., 1988. Origin of crude oil in the Wilcox Trend of Louisiana and Mississippi: evidence of long-range migration. Trans. Gulf Coast Assoc. Geol. Soc., 38: 27–34.

Sharp, J.M., Jr., Galloway, W.E., Land, L.S., McBride, E.F., Blanchard, P.E., Bodnar, D.P., Dutton, S.P., Farr, M.R., Gold, P.B., Jackson, T.J., Lundegard, P.D., Macpherson, G.L. and Milliken, K.L., 1988. Diagenetic processes in northwestern Gulf of Mexico Sediments. In: G.V. Chilingarian and K.H. Wolf (Editors), Diagenesis, II. Developments in Sedimentology, Vol. 43. Elsevier, Amsterdam, pp. 43–113.

Smith, J.T. and Ehrenberg, S.N., 1989. Correlation of carbon dioxide abundance with temperature in clastic hydrocarbon reservoirs: relationship to inorganic chemical equilibrium. Mar. Pet. Geol., 6: 129–135.

Wallace, R.H., Jr., Kraemer, T.F., Taylor, R.E. and Wesselman, J.B., 1979. Assessment of geopressured–geothermal resources in the northern Gulf of Mexico basin. U.S. Geol. Surv. Circ., 790: 132–155.

Woodbury, H.O., Murray, I.B., Pickford, P.J. and Akers, W.H., 1973. Pliocene and Pleistocene depocenters, outer continental shelf, Louisiana and Texas. Am. Assoc. Pet. Geol., Bull., 57: 2428–2439.

LORI L. SUMMA — *Exxon Production Research Company, P.O. Box 2189, Houston, TX 77252-2189, U.S.A.*
ROBERT J. POTTORF — *Exxon Production Research Company, P.O. Box 2189, Houston, TX 77252-2189, U.S.A.*
THERESA F. SCHWARZER — *Exxon Company U.S.A., P.O. Box 4778, Houston, TX 77060-2544, U.S.A.*
WENDY J. HARRISON — *Department of Geology and Geological Engineering, Colorado School of Mines, Golden, CO 80401, U.S.A.*

References index

Subject index